# Encyclopedia of Metalloproteins

Robert H. Kretsinger • Vladimir N. Uversky
Eugene A. Permyakov
Editors

# Encyclopedia of Metalloproteins

Volume 4

S–Z

With 1109 Figures and 256 Tables

*Editors*
Robert H. Kretsinger
Department of Biology
University of Virginia
Charlottesville, VA, USA

Vladimir N. Uversky
Department of Molecular Medicine
College of Medicine
University of South Florida
Tampa, FL, USA

Eugene A. Permyakov
Institute for Biological Instrumentation
Russian Academy of Sciences
Pushchino, Moscow Region, Russia

ISBN 978-1-4614-1532-9    ISBN 978-1-4614-1533-6 (eBook)
ISBN 978-1-4614-1534-3 (print and electronic bundle)
DOI 10.1007/ 978-1-4614-1533-6
Springer New York Heidelberg Dordrecht London

Library of Congress Control Number: 2013931183

© Springer Science+Business Media New York 2013
This work is subject to copyright. All rights are reserved by the Publisher, whether the whole or part of the material is concerned, specifically the rights of translation, reprinting, reuse of illustrations, recitation, broadcasting, reproduction on microfilms or in any other physical way, and transmission or information storage and retrieval, electronic adaptation, computer software, or by similar or dissimilar methodology now known or hereafter developed. Exempted from this legal reservation are brief excerpts in connection with reviews or scholarly analysis or material supplied specifically for the purpose of being entered and executed on a computer system, for exclusive use by the purchaser of the work. Duplication of this publication or parts thereof is permitted only under the provisions of the Copyright Law of the Publisher's location, in its current version, and permission for use must always be obtained from Springer. Permissions for use may be obtained through RightsLink at the Copyright Clearance Center. Violations are liable to prosecution under the respective Copyright Law.
The use of general descriptive names, registered names, trademarks, service marks, etc. in this publication does not imply, even in the absence of a specific statement, that such names are exempt from the relevant protective laws and regulations and therefore free for general use.
While the advice and information in this book are believed to be true and accurate at the date of publication, neither the authors nor the editors nor the publisher can accept any legal responsibility for any errors or omissions that may be made. The publisher makes no warranty, express or implied, with respect to the material contained herein.

Printed on acid-free paper

Springer is part of Springer Science+Business Media (www.springer.com)

# About the Editors

**Robert H. Kretsinger** Department of Biology, University of Virginia 395, Charlottesville, VA 22904, USA

**Robert H. Kretsinger** is Commonwealth Professor of Biology at the University of Virginia in Charlottesville, Virginia, USA. His research has addressed structure, function, and evolution of several different protein families. His group determined the crystal structure of parvalbumin in 1970. The analysis of this calcium binding protein provided the initial characterization of the helix, loop, helix conformation of the EF-hand domain and of the pair of EF-hands that form an EF-lobe. Over seventy distinct subfamilies of EF-hand proteins have been identified, making this domain one of the most widely distributed in eukaryotes.

Dr. Kretsinger has taught courses in protein crystallography, biochemistry, macromolecular structure, and history and philosophy of biology, and has served as chair of his department and of the University Faculty Senate.

His "other" career as a sculptor began, before the advent of computer graphics, with space filling models of the EF-hand (www.virginiastonecarvers.com). He has also been an avid cyclist for many decades.

# Preface

Metal ions play an essential role in the functioning of all biological systems. All biological processes occur in a milieu of high concentrations of metal ions, and many of these processes depend on direct participation of metal ions. Metal ions interact with charged and polar groups of all biopolymers; those interactions with proteins play an especially important role.

The study of structural and functional properties of metal binding proteins is an important and ongoing activity area of modern physical and chemical biology. Thirteen metal ions – sodium, potassium, magnesium, calcium, manganese, iron, cobalt, zinc, copper, nickel, vanadium, tungsten, and molybdenum – are known to be essential for at least some organisms. Metallo-proteomics deals with all aspects of the intracellular and extracellular interactions of metals and proteins. Metal cations and metal binding proteins are involved in all crucial cellular activities. Many pathological conditions are correlated with abnormal metal metabolism. Research in metallo-proteomics is rapidly growing and is progressively entering curricula at universities, research institutions, and technical high schools.

*Encyclopedia of Metalloproteins* is a key resource that provides basic, accessible, and comprehensible information about this expanding field. It covers exhaustively all thirteen essential metal ions, discusses other metals that might compete or interfere with them, and also presents information on proteins interacting with other metal ions. *Encyclopedia of Metalloproteins* is an ideal reference for students, teachers, and researchers, as well as the informed public.

**Vladimir N. Uversky** Department of Molecular Medicine, College of Medicine, University of South Florida, Tampa, FL 33612 USA

**Vladimir N. Uversky** is an Associate Professor at the Department of Molecular Medicine at the University of South Florida (USF). He obtained his academic degrees from Moscow Institute of Physics and Technology (PhD in 1991) and from the Institute of Experimental and Theoretical Biophysics, Russian Academy of Sciences (DSc in 1998). He spent his early career working mostly on protein folding at the Institute of Protein Research and Institute for Biological Instrumentation, Russia. In 1998, he moved to the University of California, Santa Cruz, where for six years he studied protein folding, misfolding, protein conformation diseases, and protein intrinsic disorder phenomenon. In 2004, he was invited to join the Indiana University School of Medicine as a Senior Research Professor to work on intrinsically disordered proteins. Since 2010, Professor Uversky has been with USF, where he continues to study intrinsically disordered proteins and protein folding and misfolding processes. He has authored over 450 scientific publications and edited several books and book series on protein structure, function, folding and misfolding.

# Acknowledgements

We extend our sincerest thanks to all of the contributors who have shared their insights into metalloproteins with the broader community of researchers, students, and the informed public.

**Eugene A. Permyakov**, Institute for Biological Instrumentation, Russian Academy of Sciences, Pushchino, Moscow Region, Russia

**Eugene A. Permyakov** received his PhD in physics and mathematics at the Moscow Institute of Physics and Technology in 1976, and defended his Doctor of Sciences dissertation in biology at Moscow State University in 1989. From 1970 to 1994 he worked at the Institute of Theoretical and Experimental Biophysics of the Russian Academy of Sciences. From 1990 to 1991 and in 1993, Dr. Permyakov worked at the Ohio State University, Columbus, Ohio, USA. Since 1994 he has been the Director of the Institute for Biological Instrumentation of the Russian Academy of Sciences. He is a Professor of Biophysics and is known for his work on metal binding proteins and intrinsic luminescence method. He is a member of the Russian Biochemical Society. Dr. Permyakov's primary research focus is the study of physico-chemical and functional properties of metal binding proteins. He is the author of more than 150 articles and 10 books, including *Luminescent Spectroscopy of Proteins* (CRC Press, 1993), *Metalloproteomics* (John Wiley & Sons, 2009), and *Calcium Binding Proteins* (John Wiley & Sons, 2011). He is an Academic Editor of the journals *PLoS ONE* and *PeerJ*, and Editor of the book *Methods in Protein Structure and Stability Analysis* (Nova, 2007).

In his spare time, Dr. Permyakov is an avid jogger, cyclist, and cross country skier.

# Section Editors

**Sections: Physiological Metals: Ca; Non-Physiological Metals: Pd, Ag**

**Robert H. Kretsinger**  Department of Biology, University of Virginia, Charlottesville, VA, USA

**Section: Physiological Metals: Ca**

**Eugene A. Permyakov**  Institute for Biological Instrumentation, Russian Academy of Sciences, Pushchino, Moscow Region, Russia

**Sections: Metalloids; Non-Physiological Metals: Ag, Au, Pt, Be, Sr, Ba, Ra**

**Vladimir N. Uversky** Department of Molecular Medicine, College of Medicine, University of South Florida, Tampa, FL, USA

**Sections: Physiological Metals: Co, Ni, Cu**

**Stefano Ciurli** Laboratory of Bioinorganic Chemistry, Department of Pharmacy and Biotechnology, University of Bologna, Italy

## Sections: Physiological Metals: Cd, Cr

**John B. Vincent** Department of Chemistry, The University of Alabama, Tuscaloosa, AL, USA

## Section: Physiological Metals: Fe

**Elizabeth C. Theil** Children's Hospital Oakland Research Institute, Oakland, CA, USA
Department of Molecular and Structural Biochemistry, North Carolina State University, Raleigh, NC, USA

**Sections: Physiological Metals: Mo, W, V**

**Biswajit Mukherjee** Department of Pharmaceutical Technology, Jadavpur University, Kolkata, India

**Sections: Physiological Metals: Mg, Mn**

**Andrea Romani** Department of Physiology and Biophysics, Case Western Reserve University, Cleveland, OH, USA

### Sections: Physiological Metals: Na, K

**Sergei Yu. Noskov** Institute for BioComplexity and Informatics and Department for Biological Sciences, University of Calgary, Calgary, AB, Canada

### Section: Physiological Metals: Zn

**David S. Auld** Harvard Medical School, Boston, Massachusetts, USA

**Sections: Non-Physiological Metals: Li, Rb, Cs, Fr**

**Sergei E. Permyakov**  Protein Research Group, Institute for Biological Instrumentation, Russian Academy of Sciences, Pushchino, Moscow Region, Russia

**Sections: Non-Physiological Metals: Lanthanides, Actinides**

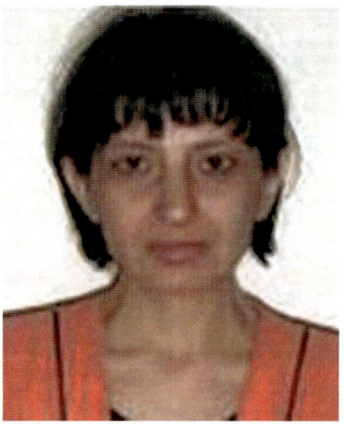

**Irena Kostova**  Department of Chemistry, Faculty of Pharmacy, Medical University, Sofia, Bulgaria

**Sections: Non-Physiological Metals: Sc, Y, Ti, Zr, Hf, Rf, Nb, Ta, Tc, Re, Ru, Os, Rh, Ir**

**Chunying Chen** Key Laboratory for Biological Effects of Nanomaterials and Nanosafety of CAS, National Center for Nanoscience and Technology, Beijing, China

**Sections: Non-Physiological Metals: Hg, Pb**

**K. Michael Pollard** Department of Molecular and Experimental Medicine, The Scripps Research Institute, La Jolla, CA, USA

**Sections: Non-Physiological Metals: Al, Ga, In, Tl, Ge, Sn, Sb, Bi, Po**

**Sandra V. Verstraeten** Department of Biological Chemistry, University of Buenos Aires, Buenos Aires, Argentina

# List of Contributors

**Satoshi Abe** Department of Biomolecular Engineering, Graduate School of Bioscience and Biotechnology, Tokyo Institute of Technology, Yokohama, Japan

**Vojtech Adam** Department of Chemistry and Biochemistry, Faculty of Agronomy, Mendel University in Brno, Brno, Czech Republic

Central European Institute of Technology, Brno University of Technology, Brno, Czech Republic

**Olayiwola A. Adekoya** Pharmacology Research group, Department of Pharmacy, Institute of Pharmacy, University of Tromsø, Tromsø, Norway

**Paul A. Adlard** The Mental Health Research Institute, The University of Melbourne, Parkville, VIC, Australia

**Magnus S. Ågren** Department of Surgery and Copenhagen Wound Healing Center, Bispebjerg University Hospital, Copenhagen, Denmark

**Karin Åkerfeldt** Department of Chemistry, Haverford College, Haverford, PA, USA

**Takashiro Akitsu** Department of Chemistry, Tokyo University of Science, Shinjuku-ku, Tokyo, Japan

**Lorenzo Alessio** Department of Experimental and Applied Medicine, Section of Occupational Health and Industrial Hygiene, University of Brescia, Brescia, Italy

**Mamdouh M. Ali** Biochemistry Department, Genetic Engineering and Biotechnology Division, National Research Centre, El Dokki, Cairo, Egypt

**James B. Ames** Department of Chemistry, University of California, Davis, CA, USA

**Olaf S. Andersen** Department of Physiology and Biophysics, Weill Cornell Medical College, New York, NY, USA

**Gregory J. Anderson** Iron Metabolism Laboratory, Queensland Institute of Medical Research, PO Royal Brisbane Hospital, Brisbane, QLD, Australia

**Janet S. Anderson** Department of Chemistry, Union College, Schenectady, NY, USA

**João Paulo André** Centro de Química, Universidade do Minho, Braga, Portugal

**Claudia Andreini** Magnetic Resonance Center (CERM) – University of Florence, Sesto Fiorentino, Italy

Department of Chemistry, University of Florence, Sesto Fiorentino, Italy

**Alexey N. Antipov** A.N. Bach Institute of Biochemistry Russian Academy of Sciences, Moscow, Russia

**Tayze T. Antunes** Kidney Research Center, Ottawa Hospital Research Institute, University of Ottawa, Ottawa, ON, Canada

**Varun Appanna** Department of Chemistry and Biochemistry, Laurentian University, Sudbury, ON, Canada

**Vasu D. Appanna** Department of Chemistry and Biochemistry, Laurentian University, Sudbury, ON, Canada

**Cristina Ariño** Department of Analytical Chemistry, University of Barcelona, Barcelona, Spain

**Vladimir B. Arion** Institute of Inorganic Chemistry, University of Vienna, Vienna, Austria

**Farukh Arjmand** Department of Chemistry, Aligarh Muslim University, Aligarh, UP, India

**Fabio Arnesano** Department of Chemistry, University of Bari "Aldo Moro", Bari, Italy

**Joan L. Arolas** Proteolysis Lab, Department of Structural Biology, Molecular Biology Institute of Barcelona, CSIC Barcelona Science Park, Barcelona, Spain

**Afolake T. Arowolo** Department of Biochemistry, Microbiology & Biotechnology, Rhodes University, Grahamstown, Eastern Cape, South Africa

**Nebojša Arsenijević** Faculty of Medical Sciences, University of Kragujevac, Centre for Molecular Medicine, Kragujevac, Serbia

**Samuel Ogheneovo Asagba** Department of Biochemistry, Delta State University, Abraka, Delta State, Nigeria

**Michael Aschner** Department of Pediatrics, Division of Pediatric Clinical Pharmacology and Toxicology, Vanderbilt University Medical Center, Nashville, TN, USA

Center in Molecular Toxicology, Vanderbilt University Medical Center, Nashville, TN, USA

Center for Molecular Neuroscience, Vanderbilt University Medical Center, Nashville, TN, USA

The Kennedy Center for Research on Human Development, Vanderbilt University Medical Center, Nashville, TN, USA

**Michael Assfalg** Department of Biotechnology, University of Verona, Verona, Italy

**William D. Atchison** Department of Pharmacology and Toxicology, Michigan State University, East Lansing, MI, USA

# List of Contributors

**Bishara S. Atiyeh** Division of Plastic Surgery, Department of Surgery, American University of Beirut Medical Center, Beirut, Lebanon

**Sílvia Atrian** Departament de Genètica, Facultat de Biologia, Universitat de Barcelona, Barcelona, Spain

**Christopher Auger** Department of Chemistry and Biochemistry, Laurentian University, Sudbury, ON, Canada

**David S. Auld** Harvard Medical School, Boston, MA, USA

**Scott Ayton** The Mental Health Research Institute, The University of Melbourne, Parkville, VIC, Australia

**Eduard B. Babiychuk** Department of Cell Biology, Institute of Anatomy, University of Bern, Bern, Switzerland

**Petr Babula** Department of Natural Drugs, Faculty of Pharmacy, University of Veterinary and Pharmaceutical Sciences Brno, Brno, Czech Republic

**Damjan Balabanič** Ecology Department, Pulp and Paper Institute, Ljubljana, Slovenia

**Wojciech Bal** Institute of Biochemistry and Biophysics, Polish Academy of Sciences, Warsaw, Poland

**Graham S. Baldwin** Department of Surgery, Austin Health, The University of Melbourne, Heidelberg, VIC, Australia

**Cynthia Bamdad** Minerva Biotechnologies Corporation, Waltham, MA, USA

**Mario Barbagallo** Geriatric Unit, Department of Internal Medicine and Medical Specialties (DIMIS), University of Palermo, Palermo, Italy

**Juan Barceló** Lab. Fisiología Vegetal, Facultat de Biociencias, Universidad Autónoma de Barcelona, Bellaterra, Spain

**Khurram Bashir** Graduate School of Agricultural and Life Sciences, The University of Tokyo, Tokyo, Japan

**Partha Basu** Department of Chemistry and Biochemistry, Duquesne University, Pittsburgh, PA, USA

**Andrea Battistoni** Dipartimento di Biologia, Università di Roma Tor Vergata, Rome, Italy

**Mikael Bauer** Department of Biochemistry and Structural Biology, Lund University, Chemical Centre, Lund, Sweden

**Lukmaan Bawazer** School of Chemistry, University of Leeds, Leeds, UK

**Carine Bebrone** Centre for Protein Engineering, University of Liège, Sart–Tilman, Liège, Belgium

Institute of Molecular Biotechnology, RWTH–Aachen University, c/o Fraunhofer IME, Aachen, Germany

**Konstantinos Beis** Division of Molecular Biosciences, Imperial College London, London, South Kensington, UK

Membrane Protein Lab, Diamond Light Source, Harwell Science and Innovation Campus, Chilton, Oxfordshire, UK

Research Complex at Harwell, Harwell Oxford, Didcot, Oxforsdhire, UK

**Catherine Belle** Département de Chimie Moléculaire, UMR-CNRS 5250, Université Joseph Fourier, ICMG FR-2607, Grenoble, France

**Andrea Bellelli** Department of Biochemical Sciences, Sapienza University of Rome, Rome, Italy

**Gunes Bender** Department of Biological Chemistry, University of Michigan Medical School, Ann Arbor, MI, USA

**Stefano Benini** Faculty of Science and Technology, Free University of Bolzano, Bolzano, Italy

**Stéphane L. Benoit** Department of Microbiology, The University of Georgia, Athens, GA, USA

**Tomas Bergman** Department of Medical Biochemistry and Biophysics, Karolinska Institutet, Stockholm, Sweden

**Lawrence R. Bernstein** Terrametrix, Menlo Park, CA, USA

**Marla J. Berry** Department of Cell & Molecular Biology, John A. Burns School of Medicine, University of Hawaii at Manoa, Honolulu, HI, USA

**Ivano Bertini** Magnetic Resonance Center (CERM) – University of Florence, Sesto Fiorentino, Italy

Department of Chemistry, University of Florence, Sesto Fiorentino, Italy

**Gerd Patrick Bienert** Institut des Sciences de la Vie, Universite catholique de Louvain, Louvain-la-Neuve, Belgium

**Andrew N. Bigley** Department of Chemistry, Texas A&M University, College Station, TX, USA

**Luis M. Bimbo** Division of Pharmaceutical Technology, University of Helsinki, Helsinki, Finland

**Ohad S. Birk** Head, Genetics Institute, Soroka Medical Center Head, Morris Kahn Center for Human Genetics, NIBN and Faculty of Health Sciences, Ben Gurion University, Beer Sheva, Israel

**Ruth Birner-Gruenberger** Institute of Pathology and Center of Medical Research, Medical University of Graz, Graz, Austria

**Cristina Bischin** Department of Chemistry and Chemical Engineering, Babes-Bolyai University, Cluj-Napoca, Romania

**Florian Bittner** Department of Plant Biology, Braunschweig University of Technology, Braunschweig, Germany

**Jodi L. Boer** Department of Biochemistry and Molecular Biology, Michigan State University, East Lansing, MI, USA

**Judith S. Bond** Department of Biochemistry and Molecular Biology, Pennsylvania State University College of Medicine, Hershey, PA, USA

**Martin D. Bootman** Life, Health and Chemical Sciences, The Open University Walton Hall, Milton Keynes, UK

**Bhargavi M. Boruah** CAS Key Laboratory of Pathogenic Microbiology and Immunology, Institute of Microbiology, Chinese Academy of Sciences, Beijing, China

Graduate University of Chinese Academy of Science, Beijing, China

**Sheryl R. Bowley** Division of Hemostasis and Thrombosis, Beth Israel Deaconess Medical Center, Harvard Medical School, Boston, MA, USA

**Doreen Braun** Institute for Experimental Endocrinology, Charité-Universitätsmedizin Berlin, Berlin, Germany

**Davorka Breljak** Unit of Molecular Toxicology, Institute for Medical Research and Occupational Health, Zagreb, Croatia

**Leonid Breydo** Department of Molecular Medicine, Morsani College of Medicine, University of South Florida, Tampa, FL, USA

**Mickael Briens** UPR ARN du CNRS, Université de Strasbourg, Institut de Biologie Moléculaire et Cellulaire, Strasbourg, France

**Joan B. Broderick** Department of Chemistry and Biochemistry, Montana State University, Bozeman, MT, USA

**James E. Bruce** Department of Genome Sciences, University of Washington, Seattle, WA, USA

**Ernesto Brunet** Dept. Química Orgánica, Facultad de Ciencias, Universidad Autónoma de Madrid, Madrid, Spain

**Maurizio Brunori** Department of Biochemical Sciences, Sapienza University of Rome, Rome, Italy

**Susan K. Buchanan** Laboratory of Molecular Biology, National Institute of Diabetes and Digestive and Kidney Diseases, US National Institutes of Health, Bethesda, MD, USA

**Gabriel E. Büchel** Institute of Inorganic Chemistry, University of Vienna, Vienna, Austria

**Živadin D. Bugarčić** Faculty of Science, Department of Chemistry, University of Kragujevac, Kragujevac, Serbia

**Melisa Bunderson-Schelvan** Department of Biomedical and Pharmaceutical Sciences, Center for Environmental Health, The University of Montana, Missoula, MT, USA

**Jean-Claude G. Bünzli** Center for Next Generation Photovoltaic Systems, Korea University, Sejong Campus, Jochiwon–eup, Yeongi–gun, ChungNam–do, Republic of Korea

École Polytechnique Fédérale de Lausanne, Institute of Chemical Sciences and Engineering, Lausanne, Switzerland

**John E. Burke** Medical Research Council, Laboratory of Molecular Biology, Cambridge, UK

**Torsten Burkholz** Division of Bioorganic Chemistry, School of Pharmacy, Saarland State University, Saarbruecken, Germany

**Bruce S. Burnham** Department of Chemistry, Biochemistry, and Physics, Rider University, Lawrenceville, NJ, USA

**Ashley I. Bush** The Mental Health Research Institute, The University of Melbourne, Parkville, VIC, Australia

**Kunzheng Cai** Institute of Tropical and Subtropical Ecology, South China Agricultural University, Guangzhou, China

**Iván L. Calderón** Laboratorio de Microbiología Molecular, Universidad Andrés Bello, Santiago, Chile

**Glaucia Callera** Kidney Research Center, Ottawa Hospital Research Institute, University of Ottawa, Ottawa, ON, Canada

**Marcello Campagna** Department of Public Health, Clinical and Molecular Medicine, University of Cagliari, Cagliari, Italy

**Mercè Capdevila** Departament de Química, Facultat de Ciències, Universitat Autònoma de Barcelona, Cerdanyola del Vallés (Barcelona), Spain

**Fernando Cardozo-Pelaez** Department of Pharmaceutical Sciences, Center for Environmental Health Sciences, University of Montana, Missoula, MT, USA

**Bradley A. Carlson** Molecular Biology of Selenium Section, Laboratory of Cancer Prevention, National Cancer Institute, National Institutes of Health, Bethesda, MD, USA

**Silvia Castelli** Department of Biology, University of Rome Tor Vergata, Rome, Italy

**Tommy Cedervall** Department of Biochemistry and Structural Biology, Lund University, Chemical Centre, Lund, Sweden

**Sudipta Chakraborty** Department of Pediatrics and Department of Pharmacology, and the Kennedy Center for Research on Human Development, Vanderbilt University Medical Center, Nashville, TN, USA

**Henry Chan** Department of Molecular Biology, Division of Biological Sciences, University of California at San Diego, La Jolla, CA, USA

**N. Chandrasekaran** Centre for Nanobiotechnology, VIT University, Vellore, Tamil Nadu, India

**Loïc J. Charbonnière** Laboratoire d'Ingénierie Moléculaire Appliquée à l'Analyse, IPHC, UMR 7178 CNRS/UdS ECPM, Strasbourg, France

**Malay Chatterjee** Division of Biochemistry, Department of Pharmaceutical Technology, Jadavpur University, Kolkata, West Bengal, India

**Mary Chatterjee** Division of Biochemistry, Department of Pharmaceutical Technology, Jadavpur University, Kolkata, West Bengal, India

**François Chaumont** Institut des Sciences de la Vie, Universite catholique de Louvain, Louvain-la-Neuve, Belgium

**Juan D. Chavez** Department of Genome Sciences, University of Washington, Seattle, WA, USA

**Chi-Ming Che** Department of Chemistry, State Key Laboratory of Synthetic Chemistry and Open Laboratory of Chemical Biology of the Institute of Molecular Technology for Drug Discovery and Synthesis, The University of Hong Kong, Hong Kong, China

**Elena Chekmeneva** Department of Chemistry, University of Sheffield, Sheffield, UK

**Di Chen** The Developmental Therapeutics Program, Barbara Ann Karmanos Cancer Institute, and Departments of Oncology, Pharmacology and Pathology, School of Medicine, Wayne State University, Detroit, MI, USA

**Hong-Yuan Chen** National Key Laboratory of Analytical Chemistry for Life Science, School of Chemistry and Chemical Engineering, Nanjing University, Nanjing, China

**Jiugeng Chen** Laboratory of Plant Physiology and Molecular Genetics, Université Libre de Bruxelles, Brussels, Belgium

**Sai-Juan Chen** State Key Laboratory of Medical Genomics, Shanghai Institute of Hematology, Rui Jin Hospital Affiliated to Shanghai Jiao Tong University School of Medicine, Shanghai, China

**Zhu Chen** State Key Laboratory of Medical Genomics, Shanghai Institute of Hematology, Rui Jin Hospital Affiliated to Shanghai Jiao Tong University School of Medicine, Shanghai, China

**Robert A. Cherny** The Mental Health Research Institute, The University of Melbourne, Parkville, VIC, Australia

**Yana Chervona** Department of Environmental Medicine, New York University Medical School, New York, NY, USA

**Christopher R. Chitambar** Division of Hematology and Oncology, Medical College of Wisconsin, Froedtert and Medical College of Wisconsin Clinical Cancer Center, Milwaukee, WI, USA

**Hassanul Ghani Choudhury** Division of Molecular Biosciences, Imperial College London, London, South Kensington, UK

Membrane Protein Lab, Diamond Light Source, Harwell Science and Innovation Campus, Chilton, Oxfordshire, UK

Research Complex at Harwell, Harwell Oxford, Didcot, Oxforsdhire, UK

**Samrat Roy Chowdhury** Department of Pharmaceutical Technology, Jadavpur University, Kolkata, West Bengal, India

**Stefano Ciurli** Department of Agro-Environmental Science and Technology, University of Bologna, Bologna, Italy

**Stephan Clemens** Department of Plant Physiology, University of Bayreuth, Bayreuth, Germany

**Nansi Jo Colley** Department of Ophthalmology and Visual Sciences, UW Eye Research Institute, University of Wisconsin, Madison, WI, USA

**Gianni Colotti** Institute of Molecular Biology and Pathology, Consiglio Nazionale delle Ricerche, Rome, Italy

**Giovanni Corsetti** Division of Human Anatomy, Department of Biomedical Sciences and Biotechnologies, Brescia University, Brescia, Italy

**Max Costa** Department of Environmental Medicine, New York University Medical School, New York, NY, USA

**Jos A. Cox** Department of Biochemistry, University of Geneva, Geneva, Switzerland

**Adam V. Crain** Department of Chemistry and Biochemistry, Montana State University, Bozeman, MT, USA

**Ann Cuypers** Centre for Environmental Sciences, Hasselt University, Diepenbeek, Belgium

**Martha S. Cyert** Department of Biology, Stanford University, Stanford, USA

**Sabato D'Auria** National Research Council (CNR), Laboratory for Molecular Sensing, Institute of Protein Biochemistry, Naples, Italy

**Verónica Daier** Departamento de Química Física/IQUIR-CONICET, Facultad de Ciencias Bioquímicas y Farmacéuticas, Universidad Nacional de Rosario, Rosario, Argentina

**Charles T. Dameron** Chemistry Department, Saint Francis University, Loretto, PA, USA

**Subhadeep Das** Division of Biochemistry, Department of Pharmaceutical Technology, Jadavpur University, Kolkata, West Bengal, India

**Nilay Kanti Das** Department of Dermatology, Medical College, Kolkata, West Bengal, India

**Rupali Datta** Department of Biological Sciences, Michigan Technological University, Houghton, MI, USA

**Benjamin G. Davis** Chemistry Research Laboratory, Department of Chemistry, University of Oxford, Oxford, UK

**Dennis R. Dean** Department of Biochemistry, Virginia Tech University, Blacksburg, VA, USA

**Kannan Deepa** Biochemical Engineering Laboratory, Department of Chemical Engineering, Indian Institute of Technology Madras, Chennai, Tamil Nadu, India

**Claudia Della Corte** Unit of Liver Research of Bambino Gesù Children's Hospital, IRCCS, Rome, Italy

**Simone Dell'Acqua** REQUIMTE/CQFB, Departamento de Química, Faculdade de Ciências e Tecnologia, Universidade Nova de Lisboa, Caparica, Portugal

Dipartimento di Chimica, Università di Pavia, Pavia, Italy

**Hakan Demir** Department of Nuclear Medicine, School of Medicine, Kocaeli University, Umuttepe, Kocaeli, Turkey

**Sumukh Deshpande** Institute for Biocomplexity and Informatics, Department of Biological Sciences, University of Calgary, Calgary, AB, Canada

**Alessandro Desideri** Department of Biology, University of Rome Tor Vergata, Rome, Italy

Interuniversity Consortium, National Institute Biostructure and Biosystem (INBB), Rome, Italy

**Patrick C. D'Haese** Laboratory of Pathophysiology, University of Antwerp, Wilrijk, Belgium

**José Manuel Díaz-Cruz** Department of Analytical Chemistry, University of Barcelona, Barcelona, Spain

**Saad A. Dibo** Division Plastic and Reconstructive Surgery, American University of Beirut Medical Center, Beirut, Lebanon

**Pavel Dibrov** Department of Microbiology, University of Manitoba, Winnipeg, MB, Canada

**Adeleh Divsalar** Department of Biological Sciences, Tarbiat Moallem University, Tehran, Iran

**Ligia J. Dominguez** Geriatric Unit, Department of Internal Medicine and Medical Specialties (DIMIS), University of Palermo, Palermo, Italy

**Delfina C. Domínguez** College of Health Sciences, The University of Texas at El Paso, El Paso, TX, USA

**Rosario Donato** Department of Experimental Medicine and Biochemical Sciences, University of Perugia, Perugia, Italy

**Elke Dopp** Institute of Hygiene and Occupational Medicine, University of Duisburg-Essen, Essen, Germany

**Melania D'Orazio** Dipartimento di Biologia, Università di Roma Tor Vergata, Rome, Italy

**Q. Ping Dou** The Developmental Therapeutics Program, Barbara Ann Karmanos Cancer Institute, and Departments of Oncology, Pharmacology and Pathology, School of Medicine, Wayne State University, Detroit, MI, USA

**Ross G. Douglas** Zinc Metalloprotease Research Group, Division of Medical Biochemistry, Institute of Infectious Disease and Molecular Medicine, University of Cape Town, Cape Town, South Africa

**Annette Draeger** Department of Cell Biology, Institute of Anatomy, University of Bern, Bern, Switzerland

**Gabi Drochioiu** Alexandru Ioan Cuza University of Iasi, Iasi, Romania

**Elzbieta Dudek** Department of Biochemistry, University of Alberta, Edmonton, AB, Canada

**Todor Dudev** Institute of Biomedical Sciences, Academia Sinica, Taipei, Taiwan

**Henry J. Duff** Libin Cardiovascular Institute of Alberta, Calgary, AB, Canada

**Evert C. Duin** Department of Chemistry and Biochemistry, Auburn University, Auburn, AL, USA

**R. Scott Duncan** Vision Research Center and Departments of Basic Medical Science and Ophthalmology, School of Medicine, University of Missouri, Kansas City, MO, USA

**Michael F. Dunn** Department of Biochemistry, University of California at Riverside, Riverside, CA, USA

**Serdar Durdagi** Institute for Biocomplexity and Informatics, Department of Biological Sciences, University of Calgary, Calgary, AB, Canada

**Kaitlin S. Duschene** Department of Chemistry and Biochemistry, Montana State University, Bozeman, MT, USA

**Ankit K. Dutta** School of Molecular and Biomedical Science, University of Adelaide, Adelaide, South Australia, Australia

**Naba K. Dutta** Ian Wark Research Institute, University of South Australia, Mawson Lakes, South Australia, Australia

**Paul J. Dyson** Institut des Sciences et Ingénierie Chimiques, Ecole Polytechnique Fédérale de Lausanne (EPFL) SB ISIC-Direction, Lausanne, Switzerland

**Brian E. Eckenroth** Department of Microbiology and Molecular Genetics, University of Vermont, Burlington, VT, USA

**Niels Eckstein** Federal Institute for Drugs and Medical Devices (BfArM), Bonn, Germany

**David J. Eide** Department of Nutritional Sciences, University of Wisconsin-Madison, Madison, WI, USA

**Thomas Eitinger** Institut für Biologie/Mikrobiologie, Humboldt-Universität zu Berlin, Berlin, Germany

**Annette Ekblond** Cardiology Stem Cell Laboratory, Rigshospitalet University Hospital, Copenhagen, Denmark

**Jean-Michel El Hage Chahine** ITODYS, Université Paris-Diderot Sorbonne Paris Cité, CNRS UMR 7086, Paris, France

List of Contributors

**Alex Elías** Laboratorio de Microbiología Molecular, Departamento de Biología, Universidad de Santiago de Chile, Santiago, Chile

**Jeffrey S. Elmendorf** Department of Cellular and Integrative Physiology and Department of Biochemistry and Molecular Biology, and Centers for Diabetes Research, Membrane Biosciences, and Vascular Biology and Medicine, Indiana University School of Medicine, Indianapolis, IN, USA

**Sanaz Emami** Department of Biophysics, Institute of Biochemistry and Biophysics (IBB), University of Tehran, Tehran, Iran

**Vinita Ernest** Centre for Nanobiotechnology, VIT University, Vellore, Tamil Nadu, India

**Miquel Esteban** Department of Analytical Chemistry, University of Barcelona, Barcelona, Spain

**Christopher Exley** The Birchall Centre, Lennard-Jones Laboratories, Keele University, Staffordshire, UK

**Chunhai Fan** Laboratory of Physical Biology, Shanghai Institute of Applied Physics, Shanghai, China

**Marcelo Farina** Departamento de Bioquímica, Centro de Ciências Biológicas, Universidade Federal de Santa Catarina, Florianópolis, SC, Brazil

**Nicholas P. Farrell** Department of Chemistry, Virginia Commonwealth University, Richmond, VA, USA

**Caroline Fauquant** iRTSV/LCBM UMR 5249 CEA-CNRS-UJF, CEA/Grenoble, Bât K, Université Grenoble, Grenoble, France

**James G. Ferry** Department of Biochemistry and Molecular Biology, Eberly College of Science, The Pennsylvania State University, University Park, PA, USA

**Ana Maria Figueiredo** Instituto de Pesquisas Energeticas e Nucleares, IPEN-CNEN/SP, Sao Paulo, Brazil

**David I. Finkelstein** The Mental Health Research Institute, The University of Melbourne, Parkville, VIC, Australia

**Larry Fliegel** Department of Biochemistry, University of Alberta, Edmonton, AB, Canada

**Swaran J. S. Flora** Division of Pharmacology and Toxicology, Defence Research and Development Establishment, Gwalior, India

**Juan C. Fontecilla-Camps** Metalloproteins; Institut de Biologie Structurale J.P. Ebel; CEA; CNRS; Université J. Fourier, Grenoble, France

**Sara M. Fox** Department of Pharmacology and Toxicology, Michigan State University, East Lansing, MI, USA

**Ricardo Franco** REQUIMTE FCT/UNL, Departamento de Química, Faculdade de Ciências e Tecnologia, Universidade Nova de Lisboa, Caparica, Portugal

**Stefan Fränzle** Department of Biological and Environmental Sciences, Research Group of Environmental Chemistry, International Graduate School Zittau, Zittau, Germany

**Christopher J. Frederickson** NeuroBioTex, Inc, Galveston Island, TX, USA

**Michael Frezza** The Developmental Therapeutics Program, Barbara Ann Karmanos Cancer Institute, and Departments of Oncology, Pharmacology and Pathology, School of Medicine, Wayne State University, Detroit, MI, USA

**Barbara C. Furie** Division of Hemostasis and Thrombosis, Beth Israel Deaconess Medical Center, Harvard Medical School, Boston, MA, USA

**Bruce Furie** Division of Hemostasis and Thrombosis, Beth Israel Deaconess Medical Center, Harvard Medical School, Boston, MA, USA

**Roland Gaertner** Department of Endocrinology, University Hospital, Ludwig-Maximilians University Munich, Munich, Germany

**Sonia Galván-Arzate** Departamento de Neuroquímica, Instituto Nacional de Neurología y Neurocirugía Manuel Velasco Suárez, Mexico City, DF, Mexico

**Livia Garavelli** Struttura Semplice Dipartimentale di Genetica Clinica, Dipartimento di Ostetrico-Ginecologico e Pediatrico, Istituto di Ricovero e Cura a Carattere Scientifico, Arcispedale S. Maria Nuova, Reggio Emilia, Italy

**Carolyn L. Geczy** Inflammation and Infection Research Centre, School of Medical Sciences, University of New South Wales, Sydney, NSW, Australia

**Emily Geiger** Department of Biological Sciences, Michigan Technological University, Houghton, MI, USA

**Alayna M. George Thompson** Department of Chemistry and Biochemistry, University of Arizona, Tucson, AZ, USA

**Charles P. Gerba** Department of Soil, Water and Environmental Science, University of Arizona, Tucson, AZ, USA

**Miltu Kumar Ghosh** Department of Pharmaceutical Technology, Jadavpur University, Kolkata, West Bengal, India

**Pramit Ghosh** Department of Community Medicine, Medical College, Kolkata, West Bengal, India

**Saikat Ghosh** Department of Pharmaceutical Technology, Jadavpur University, Kolkata, West Bengal, India

**Hedayatollah Ghourchian** Department of Biophysics, Institute of Biochemistry and Biophysics (IBB), University of Tehran, Tehran, Iran

**Jessica L. Gifford** Department of Biological Sciences, Biochemistry Research Group, University of Calgary, Calgary, AB, Canada

**Danuta M. Gillner** Department of Chemistry, Silesian University of Technology, Gliwice, Poland

**Mario Di Gioacchino** Occupational Medicine and Allergy, Head of Allergy and Immunotoxicology Unit (Ce.S.I.), G. d'Annunzio University, Via dei Vestini, Chieti, Italy

**Denis Girard** Laboratoire de recherche en inflammation et physiologie des granulocytes, Université du Québec, INRS-Institut Armand-Frappier, Laval, QC, Canada

**F. Xavier Gomis-Rüth** Proteolysis Lab, Department of Structural Biology, Molecular Biology Institute of Barcelona, CSIC Barcelona Science Park, Barcelona, Spain

**Harry B. Gray** Beckman Institute, California Institute of Technology, Pasadena, CA, USA

**Claudia Großkopf** Department Chemicals Safety, Federal Institute for Risk Assessment, Berlin, Germany

**Thomas E. Gunter** Department of Biochemistry and Biophysics, University of Rochester School of Medicine and Dentistry, Rochester, NY, USA

**Dharmendra K. Gupta** Departamento de Bioquímica, Biología Celular y Molecular de Plantas, Estación Experimental del Zaidín, CSIC, Granada, Spain

**Nikolai B. Gusev** Department of Biochemistry, School of Biology, Moscow State University, Moscow, Russian Federation

**Mandana Haack-Sørensen** Cardiology Stem Cell Laboratory, Rigshospitalet University Hospital, Copenhagen, Denmark

**Bodo Haas** Federal Institute for Drugs and Medical Devices (BfArM), Bonn, Germany

**Hajo Haase** Institute of Immunology, Medical Faculty, RWTH Aachen University, Aachen, Germany

**Fathi Habashi** Department of Mining, Metallurgical, and Materials Engineering, Laval University, Quebec City, Canada

**Alice Haddy** Department of Chemistry and Biochemistry, University of North Carolina, Greensboro, NC, USA

**Nguyêt-Thanh Ha-Duong** ITODYS, Université Paris-Diderot Sorbonne Paris Cité, CNRS UMR 7086, Paris, France

**Jesper Z. Haeggström** Department of Medical Biochemistry and Biophysics (MBB), Karolinska Institute, Stockholm, Sweden

**James F. Hainfeld** Nanoprobes, Incorporated, Yaphank, NY, USA

**Sefali Halder** Department of Pharmaceutical Technology, Jadavpur University, Kolkata, West Bengal, India

**Boyd E. Haley** Department of Chemistry, University of Kentucky, Lexington, KY, USA

**Raymond F. Hamilton Jr.** Department of Biomedical and Pharmaceutical Sciences, Center for Environmental Health, The University of Montana, Missoula, MT, USA

**Heidi E. Hannon** Department of Pharmacology and Toxicology, Michigan State University, East Lansing, MI, USA

**Timothy P. Hanusa** Department of Chemistry, Vanderbilt University, Nashville, TN, USA

**Edward D. Harris** Department of Nutrition and Food Science, Texas A&M University, College Station, TX, USA

**Todd C. Harrop** Department of Chemistry, University of Georgia, Athens, GA, USA

**Andrea Hartwig** Department Food Chemistry and Toxicology, Karlsruhe Institute of Technology, Karlsruhe, Germany

**Robert P. Hausinger** Department of Biochemistry and Molecular Biology, Michigan State University, East Lansing, MI, USA

Department of Microbiology and Molecular Genetics, 6193 Biomedical and Physical Sciences, Michigan State University, East Lansing, MI, USA

**Hiroaki Hayashi** Department of Dermatology, Kawasaki Medical School, Kurashiki, Japan

**Xiao He** CAS Key Laboratory for Biomedical Effects of Nanomaterials and Nanosafety & CAS Key Laboratory of Nuclear Analytical Techniques, Institute of High Energy Physics, Chinese Academy of Sciences, Beijing, China

**Yao He** Institute of Functional Nano & Soft Materials, Soochow University, Jiangsu, China

**Kim L. Hein** Centre for Molecular Medicine Norway (NCMM), University of Oslo Nordic EMBL Partnership, Oslo, Norway

**Claus W. Heizmann** Department of Pediatrics, Division of Clinical Chemistry, University of Zurich, Zurich, Switzerland

**Michael T. Henzl** Department of Biochemistry, University of Missouri, Columbia, MO, USA

**Carol M. Herak-Kramberger** Unit of Molecular Toxicology, Institute for Medical Research and Occupational Health, Zagreb, Croatia

**Christian Hermans** Laboratory of Plant Physiology and Molecular Genetics, Université Libre de Bruxelles, Brussels, Belgium

**Griselda Hernández** New York State Department of Health, Wadsworth Center, Albany, NY, USA

**Akon Higuchi** Department of Chemical and Materials Engineering, National Central University, Jhongli, Taoyuan, Taiwan

Department of Reproduction, National Research Institute for Child Health and Development, Setagaya-ku, Tokyo, Japan

Cathay Medical Research Institute, Cathay General Hospital, Hsi–Chi City, Taipei, Taiwan

**Russ Hille** Department of Biochemistry, University of California, Riverside, CA, USA

**Alia V. H. Hinz** Department of Chemistry, Western Michigan University, Kalamazoo, MI, USA

**John Andrew Hitron** Graduate Center for Toxicology, University of Kentucky, Lexington, KY, USA

**Miryana Hémadi** ITODYS, Université Paris-Diderot Sorbonne Paris Cité, CNRS UMR 7086, Paris, France

**Christer Hogstrand** Metal Metabolism Group, Diabetes and Nutritional Sciences Division, School of Medicine, King's College London, London, UK

**Erhard Hohenester** Department of Life Sciences, Imperial College London, London, UK

**Andrij Holian** Department of Biomedical and Pharmaceutical Sciences, Center for Environmental Health, The University of Montana, Missoula, MT, USA

**Richard C. Holz** Department of Chemistry and Biochemistry, Loyola University Chicago, Chicago, IL, USA

**Charles G. Hoogstraten** Department of Biochemistry and Molecular Biology, Michigan State University, East Lansing, MI, USA

**Ying Hou** Key Laboratory for Biomechanics and Mechanobiology of the Ministry of Education, School of Biological Science and Medical Engineering, Beihang University, Beijing, China

**Mingdong Huang** Division of Hemostasis and Thrombosis, Beth Israel Deaconess Medical Center, Harvard Medical School, Boston, MA, USA

**David L. Huffman** Department of Chemistry, Western Michigan University, Kalamazoo, MI, USA

**Paco Hulpiau** Department for Molecular Biomedical Research, VIB, Ghent, Belgium

**Amir Ibrahim** Plastic and Reconstructive SurgeryBurn Fellow, Massachusetts General Hospital / Harvard Medical School & Shriners Burn Hospital, Boston, USA

**Mitsu Ikura** Ontario Cancer Institute and Department of Medical Biophysics, University of Toronto, Toronto, Ontario, Canada

**Andrea Ilari** Institute of Molecular Biology and Pathology, Consiglio Nazionale delle Ricerche, Rome, Italy

**Giuseppe Inesi** California Pacific Medical Center Research Institute, San Francisco, CA, USA

**Hiroaki Ishida** Department of Biological Sciences, Biochemistry Research Group, University of Calgary, Calgary, AB, Canada

**Vangronsveld Jaco** Centre for Environmental Sciences, Hasselt University, Diepenbeek, Belgium

**Claus Jacob** Division of Bioorganic Chemistry, School of Pharmacy, Saarland State University, Saarbruecken, Germany

**Sushil K. Jain** Department of Pediatrics, Louisiana State University Health Sciences Center, Shreveport, LA, USA

**Peter Jensen** Department of Dermato-Allergology, Copenhagen University Hospital Gentofte, Hellerup, Denmark

**Klaudia Jomova** Department of Chemistry, Faculty of Natural Sciences, Constantine The Philosopher University, Nitra, Slovakia

**Raghava Rao Jonnalagadda** Chemical Laboratory, Central Leather Research Institute (Council of Scientific and Industrial Research), Chennai, Tamil Nadu, India

**Hans Jörnvall** Department of Medical Biochemistry and Biophysics, Karolinska Institutet, Stockholm, Sweden

**Olga Juanes** Dept. Química Orgánica, Facultad de Ciencias, Universidad Autónoma de Madrid, Madrid, Spain

**Sreeram Kalarical Janardhanan** Chemical Laboratory, Central Leather Research Institute (Council of Scientific and Industrial Research), Chennai, Tamil Nadu, India

**Paul C. J. Kamer** School of Chemistry, University of St Andrews, St Andrews, UK

**Tina Kamčeva** Laboratory of Physical Chemistry, Vinča Institute of Nuclear Sciences, University of Belgrade, Belgrade, Serbia

Laboratory of Clinical Biochemistry, Section of Clinical Pharmacology, Haukeland University Hospital, Bergen, Norway

**ChulHee Kang** Washington State University, Pullman, WA, USA

**Kazimierz S. Kasprzak** Chemical Biology Laboratory, Frederick National Laboratory for Cancer Research, Frederick, MD, USA

**Jane Kasten-Jolly** New York State Department of Health, Wadsworth Center, Albany, NY, USA

**Jens Kastrup** Cardiology Stem Cell Laboratory, Rigshospitalet University Hospital, Copenhagen, Denmark

The Heart Centre, Cardiac Catheterization Laboratory, Rigshospitalet University Hospital, Copenhagen, Denmark

**Prafulla Katkar** Department of Biology, University of Rome Tor Vergata, Rome, Italy

**Fusako Kawai** Center for Nanomaterials and Devices, Kyoto Institute of Technology, Kyoto, Japan

**Jason D. Kenealey** Department of Biomolecular Chemistry, University of Wisconsin, Madison, WI, USA

**Bernhard K. Keppler** Institute of Inorganic Chemistry, University of Vienna, Vienna, Austria

**E. Van Kerkhove** Department of Physiology, Centre for Environmental Sciences, Hasselt University, Diepenbeek, Belgium

**Kazuya Kikuchi** Division of Advanced Science and Biotechnology, Graduate School of Engineering, Osaka University, Suita, Osaka, Japan

Immunology Frontier Research Center, Osaka University, Suita, Osaka, Japan

**Michael Kirberger** Department of Chemistry, Georgia State University, Atlanta, GA, USA

**Masanori Kitamura** Department of Molecular Signaling, Interdisciplinary Graduate School of Medicine and Engineering, University of Yamanashi, Chuo, Yamanashi, Japan

**Rene Kizek** Department of Chemistry and Biochemistry, Faculty of Agronomy, Mendel University in Brno, Brno, Czech Republic

Central European Institute of Technology, Brno University of Technology, Brno, Czech Republic

**Nanne Kleefstra** Diabetes Centre, Isala clinics, Zwolle, The Netherlands

Department of Internal Medicine, University Medical Center Groningen, Groningen, The Netherlands

Langerhans Medical Research Group, Zwolle, The Netherlands

**Judith Klinman** Departments of Chemistry and Molecular and Cell Biology, California Institute for Quantitative Biosciences, University of California, Berkeley, Berkeley, CA, USA

**Michihiko Kobayashi** Graduate School of Life and Environmental Sciences, Institute of Applied Biochemistry, The University of Tsukuba, Tsukuba, Ibaraki, Japan

**Ahmet Koc** Department of Molecular Biology and Genetics, Izmir Institute of Technology, Urla, İzmir, Turkey

**Sergey M. Korotkov** Sechenov Institute of Evolutionary Physiology and Biochemistry, The Russian Academy of Sciences, St. Petersburg, Russia

**Peter Koulen** Vision Research Center and Departments of Basic Medical Science and Ophthalmology, School of Medicine, University of Missouri, Kansas City, MO, USA

**Nancy F. Krebs** Department of Pediatrics, Section of Nutrition, University of Colorado, School of Medicine, Aurora, CO, USA

**Zbigniew Krejpcio** Division of Food Toxicology and Hygiene, Department of Human Nutrition and Hygiene, The Poznan University of Life Sciences, Poznan, Poland

The College of Health, Beauty and Education in Poznan, Poznan, Poland

**Robert H. Kretsinger** Department of Biology, University of Virginia, Charlottesville, VA, USA

**Artur Krężel** Department of Protein Engineering, Faculty of Biotechnology, University of Wrocław, Wrocław, Poland

**Aleksandra Krivograd Klemenčič** Faculty of Health Sciences, University of Ljubljana, Ljubljana, Slovenia

**Peter M. H. Kroneck** Department of Biology, University of Konstanz, Konstanz, Germany

**Eugene Kryachko** Bogolyubov Institute for Theoretical Physics, Kiev, Ukraine

**Naoko Kumagai-Takei** Department of Hygiene, Kawasaki Medical School, Okayama, Japan

**Anil Kumar** CAS Key Laboratory for Biomedical Effects of Nanoparticles and Nanosafety, National Center for Nanoscience and Nanotechnology, Chinese Academy of Sciences, Beijing, China

Graduate University of Chinese Academy of Science, Beijing, China

**Thirumananseri Kumarevel** RIKEN SPring-8 Center, Harima Institute, Hyogo, Japan

**Valery V. Kupriyanov** Institute for Biodiagnostics, National Research Council, Winnipeg, MB, Canada

**Wouter Laan** School of Chemistry, University of St Andrews, St Andrews, UK

**James C. K. Lai** Department of Biomedical & Pharmaceutical Sciences, College of Pharmacy and Biomedical Research Institute, Idaho State University, Pocatello, ID, USA

**Maria José Laires** CIPER – Interdisciplinary Centre for the Study of Human Performance, Faculty of Human Kinetics, Technical University of Lisbon, Cruz Quebrada, Portugal

**Kyle M. Lancaster** Department of Chemistry and Chemical Biology, Cornell University, Ithaca, NY, USA

**Daniel Landau** Department of Pediatrics, Soroka University Medical Centre, Ben-Gurion University of the Negev, Beer Sheva, Israel

**Albert Lang** Department of Molecular and Cell Biology, California Institute for Quantitative Biosciences, University of California, Berkeley, Berkeley, CA, USA

**Alan B. G. Lansdown** Faculty of Medicine, Imperial College, London, UK

**Jean-Yves Lapointe** Groupe d'étude des protéines membranaires (GÉPROM) and Département de Physique, Université de Montréal, Montréal, QC, Canada

**Agnete Larsen** Department of Biomedicine/Pharmacology Health, Aarhus University, Aarhus, Denmark

**Lawrence H. Lash** Department of Pharmacology, Wayne State University School of Medicine, Detroit, MI, USA

**David A. Lawrence** Department of Biomedical Sciences, School of Public Health, State University of New York, Albany, NY, USA

Laboratory of Clinical and Experimental Endocrinology and Immunology, Wadsworth Center, Albany, NY, USA

**Peter A. Lay** School of Chemistry, University of Sydney, Sydney, NSW, Australia

**Gabriela Ledesma** Departamento de Química Física/IQUIR-CONICET, Facultad de Ciencias Bioquímicas y Farmacéuticas, Universidad Nacional de Rosario, Rosario, Argentina

**John Lee** Department of Biochemistry and Molecular Biology, University of Georgia, Athens, GA, USA

**Suni Lee** Department of Hygiene, Kawasaki Medical School, Okayama, Japan

**Silke Leimkühler** From the Institute of Biochemistry and Biology, Department of Molecular Enzymology, University of Potsdam, Potsdam, Germany

**Herman Louis Lelie** Department of Chemistry and Biochemistry, University of California, Los Angeles, CA, USA

**David M. LeMaster** New York State Department of Health, Wadsworth Center, Albany, NY, USA

**Joseph Lemire** Department of Chemistry and Biochemistry, Laurentian University, Sudbury, ON, Canada

**Thomas A. Leonard** Max F. Perutz Laboratories, Vienna, Austria

**Alain Lescure** UPR ARN du CNRS, Université de Strasbourg, Institut de Biologie Moléculaire et Cellulaire, Strasbourg, France

**Solomon W. Leung** Department of Civil & Environmental Engineering, School of Engineering, College of Science and Engineering and Biomedical Research Institute, Idaho State University, Pocatello, ID, USA

**Bogdan Lev** Institute for Biocomplexity and Informatics, Department of Biological Sciences, University of Calgary, Calgary, AB, Canada

**Aviva Levina** School of Chemistry, University of Sydney, Sydney, NSW, Australia

**Huihui Li** School of Chemistry and Material Science, Nanjing Normal University, Nanjing, China

**Yang V. Li** Department of Biomedical Sciences, Heritage College of Osteopathic Medicine, Ohio University, Athens, OH, USA

**Xing-Jie Liang** CAS Key Laboratory for Biomedical Effects of Nanoparticles and Nanosafety, National Center for Nanoscience and Nanotechnology, Chinese Academy of Sciences, Beijing, China

**Patrycja Libako** Faculty of Veterinary Medicine, Wrocław University of Environmental and Life Sciences, Wrocław, Poland

**Carmay Lim** Institute of Biomedical Sciences, Academia Sinica, Taipei, Taiwan

Department of Chemistry, National Tsing Hua University, Hsinchu, Taiwan

**Sara Linse** Department of Biochemistry and Structural Biology, Lund University, Chemical Centre, Lund, Sweden

**John D. Lipscomb** Department of Biochemistry, Molecular Biology, and Biophysics, University of Minnesota, Minneapolis, MN, USA

**Junqiu Liu** State Key Laboratory of Supramolecular Structure and Materials, College of Chemistry, Jilin University, Changchun, China

**Qiong Liu** College of Life Sciences, Shenzhen University, Shenzhen, P. R. China

**Zijuan Liu** Department of Biological Sciences, Oakland University, Rochester, MI, USA

**Marija Ljubojević** Unit of Molecular Toxicology, Institute for Medical Research and Occupational Health, Zagreb, Croatia

**Mario Lo Bello** Department of Biology, University of Rome "Tor Vergata", Rome, Italy

**Yan-Chung Lo** The Genomics Research Center, Academia Sinica, Taipei, Taiwan

Institute of Biological Chemistry, Academia Sinica, Taipei, Taiwan

**Lingli Lu** MOE Key Laboratory of Environment Remediation and Ecological Health, College of Environmental & Resource Science, Zhejiang University, Hangzhou, China

**Roberto G. Lucchini** Department of Experimental and Applied Medicine, Section of Occupational Health and Industrial Hygiene, University of Brescia, Brescia, Italy

Department of Preventive Medicine, Mount Sinai School of Medicine, New York, USA

**Bernd Ludwig** Institute of Biochemistry, Goethe University, Frankfurt, Germany

**Quan Luo** State Key Laboratory of Supramolecular Structure and Materials, College of Chemistry, Jilin University, Changchun, China

**Jennene A. Lyda** Department of Pharmaceutical Sciences, Center for Environmental Health Sciences, University of Montana, Missoula, MT, USA

**Charilaos Lygidakis** Regional Health Service of Emilia Romagna, AUSL of Bologna, Bologna, Italy

**Jiawei Ma** Key Laboratory for Biomechanics and Mechanobiology of the Ministry of Education, School of Biological Science and Medical Engineering, Beihang University, Beijing, China

**Jian Feng Ma** Plant Stress Physiology Group, Institute of Plant Science and Resources, Okayama University, Kurashiki, Japan

**Megumi Maeda** Department of Biofunctional Chemistry, Division of Bioscience, Okayama University Graduate School of Natural Science and Technology, Okayama, Japan

**Axel Magalon** Laboratoire de Chimie Bactérienne (UPR9043), Institut de Microbiologie de la Méditerranée, CNRS & Aix-Marseille Université, Marseille, France

**Jeanette A. Maier** Department of Biomedical and Clinical Sciences L. Sacco, Università di Milano, Medical School, Milano, Italy

**Robert J. Maier** Department of Microbiology, The University of Georgia, Athens, GA, USA

**Masatoshi Maki** Department of Applied Molecular Biosciences, Graduate School of Bioagricultural Sciences, Nagoya University, Nagoya, Japan

**R. Manasadeepa** Department of Pharmaceutical Technology, Jadavpur University, Kolkata, West Bengal, India

**David J. Mann** Division of Molecular Biosciences, Department of Life Sciences, Imperial College London, South Kensington, London, UK

**G. Marangi** Istituto di Genetica Medica, Università Cattolica Sacro Cuore, Policlinico A. Gemelli, Rome, Italy

**Wolfgang Maret** Metal Metabolism Group, Diabetes and Nutritional Sciences Division, School of Medicine, King's College London, London, UK

**Bernd Markert** Environmental Institute of Scientific Networks, in Constitution, Haren/Erika, Germany

**Michael J. Maroney** Department of Chemistry, Lederle Graduate Research Center, University of Massachusetts at Amherst, Amherst, MA, USA

**Brenda Marrero-Rosado** Department of Pharmacology and Toxicology, Michigan State University, East Lansing, MI, USA

**Christopher B. Marshall** Ontario Cancer Institute and Department of Medical Biophysics, University of Toronto, Toronto, Ontario, Canada

**Dwight W. Martin** Department of Medicine and the Proteomics Center, Stony Brook University, Stony Brook, NY, USA

**Ebany J. Martinez-Finley** Department of Pediatrics, Division of Pediatric Clinical Pharmacology and Toxicology, Vanderbilt University Medical Center, Nashville, TN, USA

Center in Molecular Toxicology, Vanderbilt University Medical Center, Nashville, TN, USA

**Jacqueline van Marwijk** Department of Biochemistry, Microbiology & Biotechnology, Rhodes University, Grahamstown, Eastern Cape, South Africa

**Pradip K. Mascharak** Department of Chemistry and Biochemistry, University of California, Santa Cruz, CA, USA

**Anne B. Mason** Department of Biochemistry, University of Vermont, Burlington, VT, USA

**Hidenori Matsuzaki** Department of Hygiene, Kawasaki Medical School, Okayama, Japan

**Jacqueline M. Matthews** School of Molecular Bioscience, The University of Sydney, Sydney, Australia

**Andrzej Mazur** INRA, UMR 1019, UNH, CRNH Auvergne, Clermont Université, Université d'Auvergne, Unité de Nutrition Humaine, Clermont-Ferrand, France

**Paulo Mazzafera** Departamento de Biologia Vegetal, Universidade Estadual de Campinas/Instituto de Biologia, Cidade Universitária, Campinas, SP, Brazil

**Michael M. Mbughuni** Department of Biochemistry, Molecular Biology, and Biophysics, University of Minnesota, Minneapolis, MN, USA

**Joseph R. McDermott** Department of Biological Sciences, Oakland University, Rochester, MI, USA

**Megan M. McEvoy** Department of Chemistry and Biochemistry, University of Arizona, Tucson, AZ, USA

**Astrid van der Meer** Interfaculty Reactor Institute, Delft University of Technology, Delft, The Netherlands

**Petr Melnikov** Department of Clinical Surgery, School of Medicine, Federal University of Mato Grosso do Sul, Campo Grande, MS, Brazil

**Gabriele Meloni** Division of Chemistry and Chemical Engineering and Howard Hughes Medical Institute, California Institute of Technology, Pasadena, CA, USA

**Ralf R. Mendel** Department of Plant Biology, Braunschweig University of Technology, Braunschweig, Germany

**Mohamed Larbi Merroun** Departamento de Microbiología, Facultad de Ciencias, Universidad de Granada, Granada, Spain

**Albrecht Messerschmidt** Department of Proteomics and Signal Transduction, Max-Planck-Institute of Biochemistry, Martinsried, Germany

**Marek Michalak** Department of Biochemistry, University of Alberta, Edmonton, AB, Canada

Faculty of Medicine and Dentistry, University of Alberta, Edmonton, AB, Canada

**Isabelle Michaud-Soret** iRTSV/LCBM UMR 5249 CEA-CNRS-UJF, CEA/Grenoble, Bât K, Université Grenoble, Grenoble, France

**Radmila Milačič** Department of Environmental Sciences, Jožef Stefan Institute, Ljubljana, Slovenia

**Glenn L. Millhauser** Department of Chemistry and Biochemistry, University of California, Santa Cruz, Santa Cruz, CA, USA

**Marija Milovanovic** Faculty of Medical Sciences, University of Kragujevac, Centre for Molecular Medicine, Kragujevac, Serbia

**Shin Mizukami** Division of Advanced Science and Biotechnology, Graduate School of Engineering, Osaka University, Suita, Osaka, Japan

Immunology Frontier Research Center, Osaka University, Suita, Osaka, Japan

**Cristina Paula Monteiro** Physiology and Biochemistry Laboratory, Faculty of Human Kinetics, Technical University of Lisbon, Cruz Quebrada, Portugal

**Augusto C. Montezano** Kidney Research Center, Ottawa Hospital Research Institute, University of Ottawa, Ottawa, ON, Canada

**Pablo Morales-Rico** Department of Toxicology, Cinvestav-IPN, Mexico city, Mexico

**J. Preben Morth** Centre for Molecular Medicine Norway (NCMM), Nordic EMBL Partnership, University of Oslo, Oslo, Norway

**Jean-Marc Moulis** Institut de Recherches en Sciences et Technologies du Vivant, Laboratoire Chimie et Biologie des Métaux (IRTSV/LCBM), CEA–Grenoble, Grenoble, France

CNRS, UMR5249, Grenoble, France

Université Joseph Fourier–Grenoble 1, UMR5249, Grenoble, France

**Isabel Moura** REQUIMTE/CQFB, Departamento de Química, Faculdade de Ciências e Tecnologia, Universidade Nova de Lisboa, Caparica, Portugal

**José J. G. Moura** REQUIMTE/CQFB, Departamento de Química, Faculdade de Ciências e Tecnologia, Universidade Nova de Lisboa, Caparica, Portugal

**Mohamed E. Moustafa** Department of Biochemistry, Faculty of Science, Alexandria University, Alexandria, Egypt

**Amitava Mukherjee** Centre for Nanobiotechnology, VIT University, Vellore, Tamil Nadu, India

**Biswajit Mukherjee** Department of Pharmaceutical Technology, Jadavpur University, Kolkata, West Bengal, India

**Balam Muñoz** Department of Toxicology, Cinvestav-IPN, Mexico city, Mexico

**Francesco Musiani** Department of Agro-Environmental Science and Technology, University of Bologna, Bologna, Italy

**Joachim Mutter** Naturheilkunde, Umweltmedizin Integrative and Environmental Medicine, Belegarzt Tagesklinik, Constance, Germany

**Bonex W. Mwakikunga** Council for Scientific and Industrial Research, National Centre for Nano–Structured, Pretoria, South Africa

Department of Physics and Biochemical Sciences, University of Malawi, The Malawi Polytechnic, Chichiri, Blantyre, Malawi

**Chandra Shekar Nagar Venkataraman** Condensed Matter Physics Division, Materials Science Group, Indira Gandhi Centre for Atomic Research, Kalpakkam, Tamil Nadu, India

**Hideaki Nagase** Kennedy Institute of Rheumatology, Nuffield Department of Orthopaedics, Rheumatology and Musculoskeletal Sciences, University of Oxford, London, United Kingdom

**Sreejayan Nair** University of Wyoming, School of Pharmacy, College of Health Sciences and the Center for Cardiovascular Research and Alternative Medicine, Laramie, WY, USA

**Manuel F. Navedo** Department of Physiology and Biophysics, University of Washington, Seattle, WA, USA

**Tim S. Nawrot** Centre for Environmental Sciences, Hasselt University, Diepenbeek, Belgium

**Karel Nesmerak** Department of Analytical Chemistry, Faculty of Science, Charles University in Prague, Prague, Czech Republic

**Gerd Ulrich Nienhaus** Institute of Applied Physics and Center for Functional Nanostructures (CFN), Karlsruhe Institute of Technology (KIT), Karlsruhe, Germany

Department of Physics, University of Illinois at Urbana–Champaign, Urbana, IL, USA

**Crina M. Nimigean** Department of Anesthesiology, Weill Cornell Medical College, New York, NY, USA

Department of Physiology and Biophysics, Weill Cornell Medical College, New York, NY, USA

Department of Biochemistry, Weill Cornell Medical College, New York, NY, USA

**Yasumitsu Nishimura** Department of Hygiene, Kawasaki Medical School, Okayama, Japan

**Naoko K. Nishizawa** Graduate School of Agricultural and Life Sciences, The University of Tokyo, Tokyo, Japan

Research Institute for Bioresources and Biotechnology, Ishikawa Prefectural University, Ishikawa, Japan

**Valerio Nobili** Unit of Liver Research of Bambino Gesù Children's Hospital, IRCCS, Rome, Italy

**Nicholas Noinaj** Laboratory of Molecular Biology, National Institute of Diabetes and Digestive and Kidney Diseases, US National Institutes of Health, Bethesda, MD, USA

**Aline M. Nonat** Laboratoire d'Ingénierie Moléculaire Appliquée à l'Analyse, IPHC, UMR 7178 CNRS/UdS ECPM, Strasbourg, France

**Sergei Yu. Noskov** Institute for Biocomplexity and Informatics, Department of Biological Sciences, University of Calgary, Calgary, AB, Canada

**Wojciech Nowacki** Faculty of Veterinary Medicine, Wroclaw University of Environmental and Life Sciences, Wrocław, Poland

**David O'Connell** University College Dublin, Conway Institute, Dublin, Ireland

**Masafumi Odaka** Department of Biotechnology and Life Science, Graduate School of Technology, Tokyo University of Agriculture and Technology, Koganei, Tokyo, Japan

**Akira Ono** Department of Material & Life Chemistry, Faculty of Engineering, Kanagawa University, Kanagawa-ku, Yokohama, Japan

**Laura Osorio-Rico** Departamento de Neuroquímica, Instituto Nacional de Neurología y Neurocirugía Manuel Velasco Suárez, Mexico City, DF, Mexico

**Patricia Isabel Oteiza** Departments of Nutrition and Environmental Toxicology, University of California, Davis, Davis, CA, USA

**Takemi Otsuki** Department of Hygiene, Kawasaki Medical School, Okayama, Japan

**Rabbab Oun** Strathclyde Institute of Pharmacy and Biomedical Sciences, University of Strathclyde, Glasgow, UK

**Vidhu Pachauri** Division of Pharmacology and Toxicology, Defence Research and Development Establishment, Gwalior, India

**Òscar Palacios** Departament de Química, Facultat de Ciències, Universitat Autònoma de Barcelona, Cerdanyola del Vallès (Barcelona), Spain

**Maria E. Palm-Espling** Department of Chemistry, Chemical Biological Center, Umeå University, Umeå, Sweden

**Claudia Palopoli** Departamento de Química Física/IQUIR-CONICET, Facultad de Ciencias Bioquímicas y Farmacéuticas, Universidad Nacional de Rosario, Rosario, Argentina

**Tapobrata Panda** Biochemical Engineering Laboratory, Department of Chemical Engineering, Indian Institute of Technology Madras, Chennai, Tamil Nadu, India

**Lorien J. Parker** Biota Structural Biology Laboratory, St. Vincent's Institute of Medical Research, Fitzroy, VIC, Australia

Department of Biochemistry and Molecular Biology, Bio21 Molecular Science and Biotechnology Institute, The University of Melbourne, Parkville, VIC, Australia

**Michael W. Parker** Biota Structural Biology Laboratory, St. Vincent's Institute of Medical Research, Fitzroy, VIC, Australia

Department of Biochemistry and Molecular Biology, Bio21 Molecular Science and Biotechnology Institute, The University of Melbourne, Parkville, VIC, Australia

**Marianna Patrauchan** Department of Microbiology and Molecular Genetics, College of Arts and Sciences, Oklahoma State University, Stillwater, OK, USA

**Sofia R. Pauleta** REQUIMTE/CQFB, Departamento de Química, Faculdade de Ciências e Tecnologia, Universidade Nova de Lisboa, Caparica, Portugal

**Evgeny Pavlov** Department of Physiology & Biophysics, Faculty of Medicine, Dalhousie University, Halifax, NS, Canada

**V. Pennemans** Biomedical Institute, Hasselt University, Diepenbeek, Belgium

**Harmonie Perdreau** Centre for Molecular Medicine Norway (NCMM), Nordic EMBL Partnership, University of Oslo, Oslo, Norway

**Alice S. Pereira** Departamento de Química, Faculdade de Ciências e Tecnologia, Requimte, Centro de Química Fina e Biotecnologia, Universidade Nova de Lisboa, Caparica, Portugal

**Eulália Pereira** REQUIMTE, Departamento de Química e Bioquímica, Faculdade de Ciências da Universidade do Porto, Porto, Portugal

**Eugene A. Permyakov** Institute for Biological Instrumentation, Russian Academy of Sciences, Pushchino, Moscow Region, Russia

**Sergei E. Permyakov** Protein Research Group, Institute for Biological Instrumentation of the Russian Academy of Sciences, Pushchino, Moscow Region, Russia

**Bertil R. R. Persson** Department of Medical Radiation Physics, Lund University, Lund, Sweden

**John W. Peters** Department of Chemistry and Biochemistry, Montana State University, Bozeman, MT, USA

**Marijana Petković** Laboratory of Physical Chemistry, Vinča Institute of Nuclear Sciences, University of Belgrade, Belgrade, Serbia

**Le T. Phung** Department of Microbiology and Immunology, University of Illinois, Chicago, IL, USA

**Roberta Pierattelli** CERM and Department of Chemistry "Ugo Schiff", University of Florence, Sesto Fiorentino, Italy

**Elizabeth Pierce** Department of Biological Chemistry, University of Michigan Medical School, Ann Arbor, MI, USA

**Andrea Pietrobattista** Unit of Liver Research of Bambino Gesù Children's Hospital, IRCCS, Rome, Italy

**Thomas C. Pochapsky** Department of Chemistry, Rosenstiel Basic Medical Sciences Research Center, Brandeis University, Waltham, MA, USA

**Ehmke Pohl** Biophysical Sciences Institute, Department of Chemistry, School of Biological and Biomedical Sciences, Durham University, Durham, UK

**Joe C. Polacco** Department of Biochemistry/Interdisciplinary Plant Group, University of Missouri, Columbia, MO, USA

**Arthur S. Polans** Department of Ophthalmology and Visual Sciences, UW Eye Research Institute, University of Wisconsin, Madison, WI, USA

**K. Michael Pollard** Department of Molecular and Experimental Medicine, The Scripps Research Institute, La Jolla, CA, USA

**Charlotte Poschenrieder** Lab. Fisiología Vegetal, Facultad de Biociencias, Universidad Autónoma de Barcelona, Bellaterra, Spain

**Thomas L. Poulos** Department of Biochemistry & Molecular Biology, Pharmaceutical Science, and Chemistry, University of California, Irivine, Irvine, CA, USA

**Richard D. Powell** Nanoprobes, Incorporated, Yaphank, NY, USA

**Ananda S. Prasad** Department of Oncology, Karmanos Cancer Center, Wayne State University, School of Medicine, Detroit, MI, USA

**Walter C. Prozialeck** Department of Pharmacology, Midwestern University, Downers Grove, IL, USA

**Qin Qin** Wise Laboratory of Environmental and Genetic Toxicology, Maine Center for Toxicology and Environmental Health, Department of Applied Medical Sciences, University of Southern Maine, Portland, ME, USA

**Thierry Rabilloud** CNRS, UMR 5249 Laboratory of Chemistry and Biology of Metals, Grenoble, France

CEA, DSV, iRTSV/LCBM, Chemistry and Biology of Metals, Grenoble Cedex 9, France

Université Joseph Fourier, Grenoble, France

**Stephen W. Ragsdale** Department of Biological Chemistry, University of Michigan Medical School, Ann Arbor, MI, USA

**Frank M. Raushel** Department of Chemistry, Texas A&M University, College Station, TX, USA

**Frank Reith** School of Earth and Environmental Sciences, The University of Adelaide, Centre of Tectonics, Resources and Exploration (TRaX) Adelaide, Urrbrae, South Australia, Australia

CSIRO Land and Water, Environmental Biogeochemistry, PMB2, Glen Osmond, Urrbrae, South Australia, Australia

**Tony Remans** Centre for Environmental Sciences, Hasselt University, Diepenbeek, Belgium

**Albert W. Rettenmeier** Institute of Hygiene and Occupational Medicine, University of Duisburg-Essen, Essen, Germany

**Rita Rezzani** Division of Human Anatomy, Department of Biomedical Sciences and Biotechnologies, Brescia University, Brescia, Italy

**Marius Réglier** Faculté des Sciences et Techniques, ISM2/BiosCiences UMR CNRS 7313, Aix-Marseille Université Campus Scientifique de Saint Jérôme, Marseille, France

**Oliver-M. H. Richter** Institute of Biochemistry, Goethe University, Frankfurt, Germany

**Agnes Rinaldo-Matthis** Department of Medical Biochemistry and Biophysics (MBB), Karolinska Institute, Stockholm, Sweden

**Lothar Rink** Institute of Immunology, Medical Faculty, RWTH Aachen University, Aachen, Germany

**Alfonso Rios-Perez** Department of Toxicology, Cinvestav-IPN, Mexico city, Mexico

**Rasmus Sejersten Ripa** Cardiology Stem Cell Laboratory, Rigshospitalet University Hospital, Copenhagen, Denmark

Cluster for Molecular Imaging and Department of Clinical Physiology, Nuclear Medicine and PET, Rigshospitalet University Hospital, Copenhagen, Denmark

**Marwan S. Rizk** Deptartment of Anesthesiology, American University of Beirut Medical Center, Beirut, Lebanon

**Nigel J. Robinson** Biophysical Sciences Institute, Department of Chemistry, School of Biological and Biomedical Sciences, Durham University, Durham, UK

**João B. T. Rocha** Departamento de Química, Centro de Ciências Naturais e Exatas, Universidade Federal de Santa Maria, Santa Maria, RS, Brazil

**Juan C. Rodriguez-Ubis** Dept. Química Orgánica, Facultad de Ciencias, Universidad Autónoma de Madrid, Madrid, Spain

**Harry A. Roels** Louvain Centre for Toxicology and Applied Pharmacology, Université catholique de Louvain, Brussels, Belgium

**Andrea M. P. Romani** Department of Physiology and Biophysics, School of Medicine, Case Western Reserve University, Cleveland, OH, USA

**S. Rosato** Struttura Semplice Dipartimentale di Genetica Clinica, Dipartimento di Ostetrico-Ginecologico e Pediatrico, Istituto di Ricovero e Cura a Carattere Scientifico, Arcispedale S. Maria Nuova, Reggio Emilia, Italy

**Barry P. Rosen** Department of Cellular Biology and Pharmacology, Florida International University, Herbert Wertheim College of Medicine, Miami, FL, USA

**Erwin Rosenberg** Institute of Chemical Technologies and Analytics, Vienna University of Technology, Vienna, Austria

**Amy C. Rosenzweig** Departments of Molecular Biosciences and of Chemistry, Northwestern University, Evanston, IL, USA

**Michael Rother** Institut für Mikrobiologie, Technische Universität Dresden, Dresden, Germany

**Benoît Roux** Department of Pediatrics, Biochemistry and Molecular Biology, The University of Chicago, Chicago, IL, USA

**Namita Roy Choudhury** Ian Wark Research Institute, University of South Australia, Mawson Lakes, South Australia, Australia

**Jagoree Roy** Department of Biology, Stanford University, Stanford, USA

**Kaushik Roy** Division of Biochemistry, Department of Pharmaceutical Technology, Jadavpur University, Kolkata, West Bengal, India

**Marian Rucki** Centre of Occupational Health, Laboratory of Predictive Toxicology, National Institute of Public Health, Praha 10, Czech Republic

**Anandamoy Rudra** Department of Pharmaceutical Technology, Jadavpur University, Kolkata, West Bengal, India

**Giuseppe Ruggiero** National Research Council (CNR), Laboratory for Molecular Sensing, Institute of Protein Biochemistry, Naples, Italy

**Kelly C. Ryan** Department of Chemistry, Lederle Graduate Research Center, University of Massachusetts at Amherst, Amherst, MA, USA

**Lisa K. Ryan** New Jersey Medical School, The Public Health Research Institute, University of Medicine and Dentistry of New Jersey, Newark, NJ, USA

**Janusz K. Rybakowski** Department of Adult Psychiatry, Poznan University of Medical Sciences, Poznan, Poland

**Ivan Sabolić** Unit of Molecular Toxicology, Institute for Medical Research and Occupational Health, Zagreb, Croatia

**Kalyan K. Sadhu** Division of Advanced Science and Biotechnology, Graduate School of Engineering, Osaka University, Suita, Osaka, Japan

**Anita Sahu** Institute of Pathology and Center of Medical Research, Medical University of Graz, Graz, Austria

**P. Ch. Sahu** Condensed Matter Physics Division, Materials Science Group, Indira Gandhi Centre for Atomic Research, Kalpakkam, Tamil Nadu, India

**Milton H. Saier Jr.** Department of Molecular Biology, Division of Biological Sciences, University of California at San Diego, La Jolla, CA, USA

**Jarno Salonen** Laboratory of Industrial Physics, Department of Physics, University of Turku, Turku, Finland

**Abel Santamaría** Laboratorio de Aminoácidos Excitadores, Instituto Nacional de Neurología y Neurocirugía Manuel Velasco Suárez, Mexico City, DF, Mexico

**Luis F. Santana** Department of Physiology and Biophysics, University of Washington, Seattle, WA, USA

**Hélder A. Santos** Division of Pharmaceutical Technology, University of Helsinki, Helsinki, Finland

**Dibyendu Sarkar** Earth and Environmental Studies Department, Montclair State University, Montclair, NJ, USA

**Louis J. Sasseville** Groupe d'étude des protéines membranaires (GÉPROM) and Département de Physique, Université de Montréal, Montréal, QC, Canada

**R. Gary Sawers** Institute for Microbiology, Martin-Luther University Halle-Wittenberg, Halle (Saale), Germany

**Janez Ščančar** Department of Environmental Sciences, Jožef Stefan Institute, Ljubljana, Slovenia

**Marcus C. Schaub** Institute of Pharmacology and Toxicology, University of Zurich, Zurich, Switzerland

**Sara Schmitt** The Developmental Therapeutics Program, Barbara Ann Karmanos Cancer Institute, and Departments of Oncology, Pharmacology and Pathology, School of Medicine, Wayne State University, Detroit, MI, USA

**Paul P. M. Schnetkamp** Department of Physiology & Pharmacology, Hotchkiss Brain Institute, University of Calgary, Calgary, AB, Canada

**Lutz Schomburg** Institute for Experimental Endocrinology, Charité – University Medicine Berlin, Berlin, Germany

**Gerhard N. Schrauzer** Department of Chemistry and Biochemistry, University of California, San Diego, La Jolla, CA, USA

**Ruth Schreiber** Department of Pediatrics, Soroka University Medical Centre, Ben-Gurion University of the Negev, Beer Sheva, Israel

**Ulrich Schweizer** Institute for Experimental Endocrinology, Charité-Universitätsmedizin Berlin, Berlin, Germany

**Ion Romulus Scorei** Department of Biochemistry, University of Craiova, Craiova, DJ, Romania

**Lucia A. Seale** Department of Cell & Molecular Biology, John A. Burns School of Medicine, University of Hawaii at Manoa, Honolulu, HI, USA

**Lance C. Seefeldt** Department of Chemistry and Biochemistry, Utah State University, Logan, UT, USA

**William Self** Molecular Biology & Microbiology, Burnett School of Biomedical Sciences, University of Central Florida, Orlando, FL, USA

**Takashi Sera** Department of Applied Chemistry and Biotechnology, Graduate School of Natural Science and Technology, Okayama University, Okayama, Japan

**Aruna Sharma** Laboratory of Cerebrovascular Research, Department of Surgical Sciences, Anesthesiology & Intensive Care medicine, University Hospital, Uppsala University, Uppsala, Sweden

**Hari Shanker Sharma** Laboratory of Cerebrovascular Research, Department of Surgical Sciences, Anesthesiology & Intensive Care medicine, University Hospital, Uppsala University, Uppsala, Sweden

**Honglian Shi** Department of Pharmacology and Toxicology, University of Kansas, Lawrence, KS, USA

**Xianglin Shi** Graduate Center for Toxicology, University of Kentucky, Lexington, KY, USA

**Satoshi Shinoda** JST, CREST, and Department of Chemistry, Graduate School of Science, Osaka City University, Sumiyoshi-ku, Osaka, Japan

**Maksim A. Shlykov** Department of Molecular Biology, Division of Biological Sciences, University of California at San Diego, La Jolla, CA, USA

**Siddhartha Shrivastava** Rensselaer Nanotechnology Center, Rensselaer Polytechnic Institute, Troy, NY, USA

Center for Biotechnology and Interdisciplinary Studies, Rensselaer Polytechnic Institute, Troy, NY, USA

**Sandra Signorella** Departamento de Química Física/IQUIR-CONICET, Facultad de Ciencias Bioquímicas y Farmacéuticas, Universidad Nacional de Rosario, Rosario, Argentina

**Amrita Sil** Department of Pharmacology, Burdwan Medical College, Burdwan, West Bengal, India

**Radu Silaghi-Dumitrescu** Department of Chemistry and Chemical Engineering, Babes-Bolyai University, Cluj-Napoca, Romania

**Simon Silver** Department of Microbiology and Immunology, University of Illinois, Chicago, IL, USA

**Britt-Marie Sjöberg** Department of Biochemistry and Biophysics, Stockholm University, Stockholm, SE, Sweden

**Karen Smeets** Centre for Environmental Sciences, Hasselt University, Diepenbeek, Belgium

**Stephen M. Smith** Departments of Molecular Biosciences and of Chemistry, Northwestern University, Evanston, IL, USA

**Małgorzata Sobieszczańska** Department of Pathophysiology, Wroclaw Medical University, Wroclaw, Poland

**Young-Ok Son** Graduate Center for Toxicology, University of Kentucky, Lexington, KY, USA

**Martha E. Sosa Torres** Facultad de Quimica, Universidad Nacional Autonoma de Mexico, Ciudad Universitaria, Coyoacan, Mexico DF, Mexico

**Jerry W. Spears** Department of Animal Science, North Carolina State University, Raleigh, NC, USA

**Sarah R. Spell** Department of Chemistry, Virginia Commonwealth University, Richmond, VA, USA

**Christopher D. Spicer** Chemistry Research Laboratory, Department of Chemistry, University of Oxford, Oxford, UK

St. Hilda's College, University of Oxford, Oxford, UK

**Alessandra Stacchiotti** Division of Human Anatomy, Department of Biomedical Sciences and Biotechnologies, Brescia University, Brescia, Italy

**Jan A. Staessen** Study Coordinating Centre, Department of Cardiovascular Diseases, KU Leuven, Leuven, Belgium

Unit of Epidemiology, Maastricht University, Maastricht, The Netherlands

**Maria Staiano** National Research Council (CNR), Laboratory for Molecular Sensing, Institute of Protein Biochemistry, Naples, Italy

**Anna Starus** Department of Chemistry and Biochemistry, Loyola University Chicago, Chicago, IL, USA

**Alexander Stein** Hubertus Wald Tumor Center, University Cancer Center Hamburg (UCCH), University Hospital Hamburg-Eppendorf (UKE), Hamburg, Germany

**Iryna N. Stepanenko** Institute of Inorganic Chemistry, University of Vienna, Vienna, Austria

**Martin J. Stillman** Department of Biology, The University of Western Ontario, London, ON, Canada

Department of Chemistry, The University of Western Ontario, London, ON, Canada

**Walter Stöcker** Johannes Gutenberg University Mainz, Institute of Zoology, Cell and Matrix Biology, Mainz, Germany

**Barbara J. Stoecker** Department of Nutritional Sciences, Oklahoma State University, Stillwater, OK, USA

**Edward D. Sturrock** Zinc Metalloprotease Research Group, Division of Medical Biochemistry, Institute of Infectious Disease and Molecular Medicine, University of Cape Town, Cape Town, South Africa

**Minako Sumita** Department of Biochemistry and Molecular Biology, Michigan State University, East Lansing, MI, USA

**Kelly L. Summers** Department of Biology, The University of Western Ontario, London, ON, Canada

**Raymond Wai-Yin Sun** Department of Chemistry, State Key Laboratory of Synthetic Chemistry and Open Laboratory of Chemical Biology of the Institute of Molecular Technology for Drug Discovery and Synthesis, The University of Hong Kong, Hong Kong, China

**Claudiu T. Supuran** Department of Chemistry, University of Florence, Sesto Fiorentino (Florence), Italy

**Hiroshi Suzuki** Department of Biochemistry, Asahikawa Medical University, Asahikawa, Hokkaido, Japan

**Q. Swennen** Biomedical Institute, Hasselt University, Diepenbeek, Belgium

**Ingebrigt Sylte** Medical Pharmacology and Toxicology, Department of Medical Biology, University of Tromsø, Tromsø, Norway

**Yoshiyuki Tanaka** Graduate School of Pharmaceutical Sciences Tohoku University, Sendai, Miyagi, Japan

**Shen Tang** Department of Chemistry, Georgia State University, Atlanta, GA, USA

**Akio Tani** Research Institute of Plant Science and Resources, Okayama University, Kurashiki, Okayama, Japan

**Pedro Tavares** Departamento de Química, Faculdade de Ciências e Tecnologia, Requimte, Centro de Química Fina e Biotecnologia, Universidade Nova de Lisboa, Caparica, Portugal

**Jan Willem Cohen Tervaert** Clinical and Experimental Immunology, Maastricht University, Maastricht, The Netherlands

**Tiago Tezotto** Departamento de Produção Vegetal, Universidade de São Paulo/Escola Superior de Agricultura Luiz de Queiroz, Piracicaba, SP, Brazil

**Elizabeth C. Theil** Children's Hospital Oakland Research Institute, Oakland, CA, USA

Department of Molecular and Structural Biochemistry, North Carolina State University, Raleigh, NC, USA

**Frank Thévenod** Faculty of Health, School of Medicine, Centre for Biomedical Training and Research (ZBAF), Institute of Physiology & Pathophysiology, University of Witten/Herdecke, Witten, Germany

**David J. Thomas** Pharmacokinetics Branch – Integrated Systems Toxicology Division, National Health and Environmental Research Laboratory, U.S. Environmental Protection Agency, Research Triangle Park, NC, USA

**Ameer N. Thompson** Department of Anesthesiology, Weill Cornell Medical College, New York, NY, USA

Department of Physiology and Biophysics, Weill Cornell Medical College, New York, NY, USA

**Rüdiger Thul** School of Mathematical Sciences, University of Nottingham, Nottingham, UK

**Jacob P. Thyssen** National Allergy Research Centre, Department of Dermato-Allergology, Copenhagen University Hospital Gentofte, Hellerup, Denmark

**Milon Tichy** Centre of Occupational Health, Laboratory of Predictive Toxicology, National Institute of Public Health, Praha 10, Czech Republic

**Dajena Tomco** Department of Chemistry, Wayne State University, Detroit, MI, USA

**Hidetaka Torigoe** Department of Applied Chemistry, Faculty of Science, Tokyo University of Science, Tokyo, Japan

**Rhian M. Touyz** Kidney Research Center, Ottawa Hospital Research Institute, University of Ottawa, Ottawa, ON, Canada

Institute of Cardiovascular & Medical Sciences, BHF Glasgow Cardiovascular Research Centre, University of Glasgow, Glasgow, UK

**Chikashi Toyoshima** Institute of Molecular and Cellular Biosciences, The University of Tokyo, Tokyo, Japan

**Lennart Treuel** Institute of Applied Physics and Center for Functional Nanostructures (CFN), Karlsruhe Institute of Technology (KIT), Karlsruhe, Germany

Institute of Physical Chemistry, University of Duisburg–Essen, Essen, Germany

**Shweta Trivedi** Department of Animal Science, North Carolina State University, Raleigh, NC, USA

**Thierry Tron** iSm2/BiosCiences UMR CNRS 7313, Case 342, Aix-Marseille Université, Marseille, France

**Chin-Hsiao Tseng** Department of Internal Medicine, National Taiwan University College of Medicine, Taipei, Taiwan

Division of Endocrinology and Metabolism, Department of Internal Medicine, National Taiwan University Hospital, Taipei, Taiwan

**Tsai-Tien Tseng**  Center for Cancer Research and Therapeutic Development, Clark Atlanta University, Atlanta, GA, USA

**Samantha D. Tsotsoros**  Department of Chemistry, Virginia Commonwealth University, Richmond, VA, USA

**Petra A. Tsuji**  Department of Biological Sciences, Towson University, Towson, MD, USA

**Hiroshi Tsukube**  JST, CREST, and Department of Chemistry, Graduate School of Science, Osaka City University, Sumiyoshi-ku, Osaka, Japan

**Sławomir Tubek**  Institute of Technology, Opole, Poland

**Raymond J. Turner**  Department of Biological Sciences, University of Calgary, Calgary, AB, Canada

**Toshiki Uchihara**  Laboratory of Structural Neuropathology, Tokyo Metropolitan Institute of Medical Science, Tokyo, Japan

**Takafumi Ueno**  Department of Biomolecular Engineering, Graduate School of Bioscience and Biotechnology, Tokyo Institute of Technology, Yokohama, Japan

**Christoph Ufer**  Institute of Biochemistry, Charité – Universitätsmedizin Berlin, Berlin, Germany

**İrem Uluisik**  Department of Molecular Biology and Genetics, Izmir Institute of Technology, Urla, İzmir, Turkey

**Balachandran Unni Nair**  Chemical Laboratory, Central Leather Research Institute (Council of Scientific and Industrial Research), Chennai, Tamil Nadu, India

**Vladimir N. Uversky**  Department of Molecular Medicine, University of South Florida, College of Medicine, Tampa, FL, USA

**Joan Selverstone Valentine**  Department of Chemistry and Biochemistry, University of California, Los Angeles, CA, USA

**Marian Valko**  Department of Chemistry, Faculty of Natural Sciences, Constantine The Philosopher University, Nitra, Slovakia

Faculty of Chemical and Food Technology, Slovak Technical University, Bratislava, Slovakia

**J. David van Horn**  Department of Chemistry, University of Missouri-Kansas City, Kansas City, MO, USA

**Frans van Roy**  Department for Molecular Biomedical Research, VIB, Ghent, Belgium

Department of Biomedical Molecular Biology, Ghent University, Ghent, Belgium

**Marie Vancová** Institute of Parasitology, Biology Centre of the Academy of Sciences of the Czech Republic and University of South Bohemia, České Budějovice, Czech Republic

**Jaco Vangronsveld** Centre for Environmental Sciences, Hasselt University, Diepenbeek, Belgium

**Antonio Varriale** National Research Council (CNR), Laboratory for Molecular Sensing, Institute of Protein Biochemistry, Naples, Italy

**Milan Vašák** Department of Inorganic Chemistry, University of Zürich, Zürich, Switzerland

**Claudio C. Vásquez** Laboratorio de Microbiología Molecular, Departamento de Biología, Universidad de Santiago de Chile, Santiago, Chile

**Oscar Vassallo** Department of Biology, University of Rome Tor Vergata, Rome, Italy

**Claudio N. Verani** Department of Chemistry, Wayne State University, Detroit, MI, USA

**Nathalie Verbruggen** Laboratory of Plant Physiology and Molecular Genetics, Université Libre de Bruxelles, Brussels, Belgium

**Sandra Viviana Verstraeten** Department of Biological Chemistry, IQUIFIB (UBA-CONICET), School of Pharmacy and Biochemistry, University of Buenos Aires, Argentina, Buenos Aires, Argentina

**Ramon Vilar** Department of Chemistry, Imperial College London, South Kensington, London, UK

**John B. Vincent** Department of Chemistry, The University of Alabama, Tuscaloosa, AL, USA

**Hans J. Vogel** Department of Biological Sciences, Biochemistry Research Group, University of Calgary, Calgary, AB, Canada

**Vladislav Volarevic** Faculty of Medical Sciences, University of Kragujevac, Centre for Molecular Medicine, Kragujevac, Serbia

**Anne Volbeda** Metalloproteins; Institut de Biologie Structurale J.P. Ebel; CEA; CNRS; Université J. Fourier, Grenoble, France

**Eugene S. Vysotski** Photobiology Laboratory, Institute of Biophysics Russian Academy of Sciences, Siberian Branch, Krasnoyarsk, Russia

**Anne Walburger** Laboratoire de Chimie Bactérienne (UPR9043), Institut de Microbiologie de la Méditerranée, CNRS & Aix-Marseille Université, Marseille, France

**Andrew H.-J. Wang** Institute of Biological Chemistry, Academia Sinica, Taipei, Taiwan

**Jiangxue Wang** Key Laboratory for Biomechanics and Mechanobiology of the Ministry of Education, School of Biological Science and Medical Engineering, Beihang University, Beijing, China

**Xudong Wang** Department of Pathology, St Vincent Hospital, Worcester, MA, USA

**John Wataha** Department of Restorative Dentistry, University of Washington HSC D779A, School of Dentistry, Seattle, WA, USA

**David J. Weber** Department of Biochemistry and Molecular Biology, University of Maryland School of Medicine, Baltimore, MD, USA

**Nial J. Wheate** Faculty of Pharmacy, The University of Sydney, Sydney, NSW, Australia

**Chris G. Whiteley** Graduate Institute of Applied Science and Technology, National Taiwan University of Science and Technology, Taipei, Taiwan

**Roger L. Williams** Medical Research Council, Laboratory of Molecular Biology, Cambridge, UK

**Judith Winogrodzki** Department of Microbiology, University of Manitoba, Winnipeg, MB, Canada

**John Pierce Wise Sr.** Wise Laboratory of Environmental and Genetic Toxicology, Maine Center for Toxicology and Environmental Health, Department of Applied Medical Sciences, University of Southern Maine, Portland, ME, USA

**Pernilla Wittung-Stafshede** Department of Chemistry, Chemical Biological Center, Umeå University, Umeå, Sweden

**Bert Wolterbeek** Interfaculty Reactor Institute, Delft University of Technology, Delft, The Netherlands

**Simone Wünschmann** Environmental Institute of Scientific Networks in Constitution, Haren/Erika, Germany

**Robert Wysocki** Institute of Experimental Biology, University of Wroclaw, Wroclaw, Poland

**Shenghui Xue** Department of Biology, Georgia State University, Atlanta, GA, USA

**Xiao-Jing Yan** Department of Hematology, The First Hospital of China Medical University, Shenyang, China

**Xiaodi Yang** School of Chemistry and Material Science, Nanjing Normal University, Nanjing, China

**Jenny J. Yang** Department of Chemistry, Georgia State University, Atlanta, GA, USA

Natural Science Center, Atlanta, GA, USA

**Vladimir Yarov-Yarovoy** Department of Physiology and Membrane Biology, Department of Biochemistry and Molecular Medicine, School of Medicine, University of California, Davis, CA, USA

**Katsuhiko Yokoi** Department of Human Nutrition, Seitoku University Graduate School, Matsudo, Chiba, Japan

**Vincenzo Zagà** Department of Territorial Pneumotisiology, Italian Society of Tobaccology (SITAB), AUSL of Bologna, Bologna, Italy

**Carla M. Zammit** School of Earth and Environmental Sciences, The University of Adelaide, Centre of Tectonics, Resources and Exploration (TRaX) Adelaide, Urrbrae, South Australia, Australia

CSIRO Land and Water, Environmental Biogeochemistry, PMB2, Glen Osmond, Urrbrae, South Australia, Australia

**Lourdes Zélia Zanoni** Department of Pediatrics, School of Medicine, Federal University of Mato Grosso do Sul, Campo Grande, MS, Brazil

**Huawei Zeng** United States Department of Agriculture, Agricultural Research Service, Grand Forks Human Nutrition Research Center, Grand Forks, ND, USA

**Cunxian Zhang** Department of Pathology, Women & Infants Hospital of Rhode Island, Kent Memorial Hospital, Warren Alpert Medical School of Brown University, Providence, RI, USA

**Chunfeng Zhao** Institute for Biocomplexity and Informatics and Department of Biological Sciences, University of Calgary, Calgary, AB, Canada

**Anatoly Zhitkovich** Department of Pathology and Laboratory Medicine, Brown University, Providence, RI, USA

**Boris S. Zhorov** Department of Biochemistry and Biomedical Sciences, McMaster University, Hamilton, ON, Canada

Sechenov Institute of Evolutionary Physiology and Biochemistry, Russian Academy of Sciences, St. Petersburg, Russia

**Yubin Zhou** Department of Chemistry, Georgia State University, Atlanta, GA, USA

Division of Signaling and Gene Expression, La Jolla Institute for Allergy and Immunology, La Jolla, CA, USA

**Michael X. Zhu** Department of Integrative Biology and Pharmacology, The University of Texas Health Science Center at Houston, Houston, TX, USA

**Marcella Zollino** Istituto di Genetica Medica, Università Cattolica Sacro Cuore, Policlinico A. Gemelli, Rome, Italy

# S100 Proteins

Rosario Donato[1], Carolyn L. Geczy[2] and David J. Weber[3]
[1]Department of Experimental Medicine and Biochemical Sciences, University of Perugia, Perugia, Italy
[2]Inflammation and Infection Research Centre, School of Medical Sciences, University of New South Wales, Sydney, NSW, Australia
[3]Department of Biochemistry and Molecular Biology, University of Maryland School of Medicine, Baltimore, MD, USA

## Definition

Most S100 proteins are $Ca^{2+}$-binding proteins involved in $Ca^{2+}$-signal transduction. They have a well-conserved EF-hand $Ca^{2+}$-binding motif and a second atypical EF-hand. S100s form stable symmetric homodimers and in the presence of appropriate binding targets, dissociation constants for $Ca^{2+}$-binding reach physiological levels. S100 proteins are generally constitutively expressed in a cell-specific manner. Several are induced by growth factors, cytokines, or Toll-like receptor (TLR) ligands, in processes associated with stress responses, an activated innate immune system, tumorigenesis, and/or tissue repair. In addition to functions as intracellular regulators, many S100 proteins act extracellularly and particular posttranslational modifications can promote changes in extracellular function. Receptors have been elusive, but include the receptor for advanced glycation end products (RAGE), $N$-glycans and TLRs.

## Background

Several classes of $Ca^{2+}$-binding proteins have evolved from a common ancestor to handle intracellular $Ca^{2+}$ (Carafoli and Klee 1999; Permyakov and Kretsinger 2009). Of these, some have evolved to extrude $Ca^{2+}$ outside the cell or toward intracellular $Ca^{2+}$ stores. These proteins (e.g., $Ca^{2+}$-ATPases) cooperate to maintain cytosolic free $Ca^{2+}$ concentrations low under resting conditions (~100 nM), thereby avoiding $Ca^{2+}$ precipitation and/or excess $Ca^{2+}$ signal activity, and the $Ca^{2+}$-ATPase associated with the endoplasmic reticulum acts to constitute a $Ca^{2+}$ reserve pool to be released into the cytoplasm according to the cell's needs. Some other $Ca^{2+}$-binding proteins (e.g., calsequestrin and calretinin) characterized by a low $Ca^{2+}$-binding affinity and a high $Ca^{2+}$-binding capacity, and located within $Ca^{2+}$ stores, entrap the ion and make it available for release when required. Still other $Ca^{2+}$-binding proteins, mostly found in the cytoplasm, either buffer $Ca^{2+}$ during the course of $Ca^{2+}$ transients as a result of their high $Ca^{2+}$-binding affinity (e.g., parvalbumin and S100G) or act as transducers of the so-called $Ca^{2+}$ signal. This latter group represents a large fraction of $Ca^{2+}$-binding proteins, which once activated by $Ca^{2+}$, can interact with other proteins, thereby regulating a large variety of functions. Calmodulin, troponin-C, and most S100 proteins belong to the class of $Ca^{2+}$-binding proteins involved in $Ca^{2+}$-signal transduction, and they all have a well-conserved calcium-binding motif termed the EF-hand.

Since the E- and F-helices in the helix-loop-helix calcium-binding domain of parvalbumin were characterized by Kretsinger and colleagues (Moews and Kretsinger 1975), there have been over 400 crystal

**S100 Proteins, Fig. 1** Coordinating residues for the canonical (site 2) and S100 EF-hand (site 1) for the S100 protein, S100B. In the table below the residues typically at each position of the EF-hand are illustrated. It should be noted that the S100 EF-hand has 14 rather than 12 residues

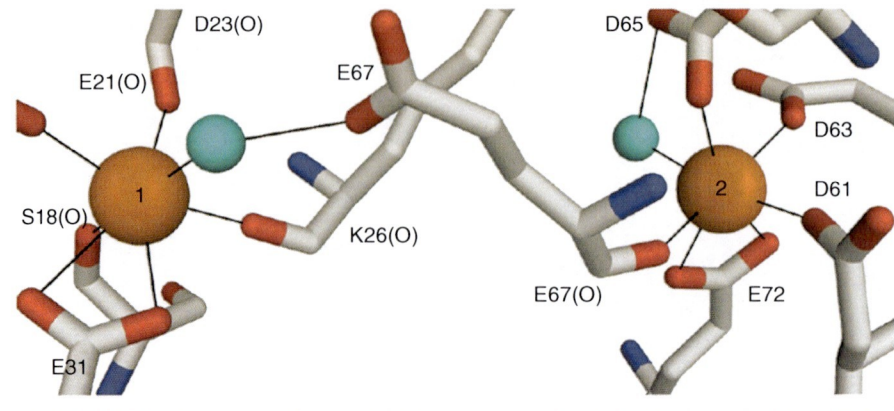

| Position | 1 | 2 | 3 | 4 | 5 | 6 | 7 | 8 | 9 | 10 | 11 | 12 |
|---|---|---|---|---|---|---|---|---|---|---|---|---|
| Ligand Coordinate[a] | X | | Y | | Z | | $-Y$ | | $-X$ | | | $-Z$ |
| Canonical EF-hand[b] | D | Basic | D/N | G | D/N | G | Any | Hyd | O | Hyd | Acidic | E |
| S100 Canonical EF-hand | D | | N | | D | | E(O) | | $H_2O$ | | | E |
| S100 Pseudo EF-hand | S | | E | | D | | K(O) | | $H_2O$ | | | E |
| S100B (EF1)[c] Pseudo EF-hand | S18(O) | G19 | E21(O) | G22 | D23(O) | K24 | K26(O) | L27 | $H_2O$ | K28 K29 | S30 | E31 |
| S100B (EF2) Canonical EF-hand | D61 | S62 | D63 | G64 | D65 | G66 | E67(O) | C68 | $H_2O$ | F69 | Q70 | E72 |

and NMR structures of EF-hand $Ca^{2+}$-binding proteins deposited into the protein data bank (PDB). Typically, EF-hand $Ca^{2+}$-binding domains are arranged in pairs and held together by a very short antiparallel β-strand. The canonical EF-hand has 12-residues with six or seven backbone or sidechain oxygen ligands coming from residues in positions 1, 3, 5, 7, and 12 (bidentate) of the helix-loop-helix domain (Fig. 1) (Strynadka and James 1989). However, the S100 family provides a very unique set of EF-hand proteins since one of the EF-hand domains in the pair (termed the pseudo- or S100-hand) has 14 rather than 12 residues, and many ligands for calcium to $Ca^{2+}$ are backbone carbonyl oxygen atoms rather than from sidechain Asn, Asp, Gln, or Glu residues as is typically found in the canonical EF-hand (Kligman and Hilt 1988). Likewise, the dimeric nature of S100 proteins does not allow for the movement of the exiting helix upon $Ca^{2+}$-binding as is found for other EF-hand proteins (i.e., CaM, TnC, etc) (Yap et al. 1999). Instead it is the entering helix (helix 3), which rotates upon binding $Ca^{2+}$ and exposes a hydrophobic patch in the core of this $Ca^{2+}$-loaded form that is necessary for interacting with its specific protein targets (Fig. 2) (Zimmer et al. 2003).

A question that has arisen is "how can a cell have a large number of intracellular $Ca^{2+}$-binding proteins available, which need to work in the nM to very low micromolar range (i.e., 100 nM to 2 μM), without sequestering all the free $Ca^{2+}$ ions necessary for signaling biological events?" The mechanistic details are still being characterized but in most cases, EF-hand-binding proteins, including S100s, typically do not bind $Ca^{2+}$ very tightly in the absence of their biological target (Fig. 2) (Zimmer and Weber 2010). Thus, only when the EF-hand protein, appropriate $Ca^{2+}$ levels, and the biological target are available

**S100 Proteins, Fig. 2** Ribbon diagram illustrating the rotation of helix 3 upon the addition of calcium and a table listing the degree of movement upon calcium and target protein and/or drug binding. The dissociation constants are from Zimmer and Weber (2010)

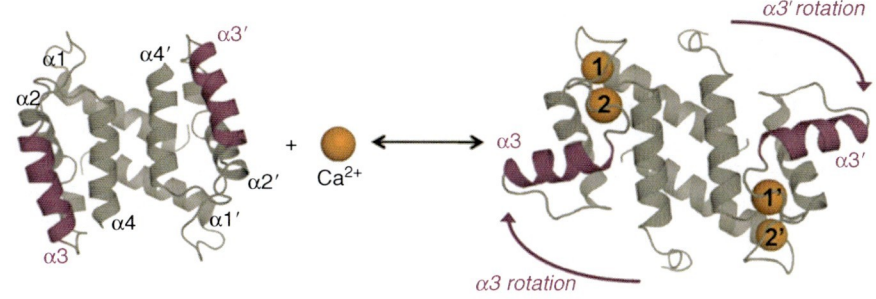

| Protein | Kd, µM (EF-1, EF-2)[a] | α3 rotation[b] (Ca$^{2+}$, apo) | Target (Kd, µM) | Target Reference | α3 rotation[c] (Target, Ca$^{2+}$) |
|---|---|---|---|---|---|
| S100A1 | –, 27 | 27° | TRTK$^{Pep}$ 24 µM | 2 | 4° |
|  |  |  | RyR$^{Pep}$ 8 µM | 3 | –11° |
| S100A2 | –, 470 | 33° |  |  |  |
| S100A3 | –, 4000 |  |  |  |  |
| S100A4 (MTSI) | –, 2.6 | 52° | TFP 55 µM[d] | 4 | 1° |
|  |  |  | PCP 56 µM[d] | 4 | 3° |
| S100A5 | 160, 0.2 | 46° |  |  |  |
| S100A6 (Calcyclin) | –, 3 | 38° | Siah1$^{Pep}$ 5 µM | 5 | –5° |
| S100A7 (Psoriasin) | –, 1 |  |  |  |  |
| S100A10 (p11) | No Ca$^{2+}$-binding |  |  |  |  |
| S100A11 (Calgizzarin) | –, 500 | 25° |  |  |  |
| S100A12 | –, 50 | 35° |  |  |  |
| S100A13 | 400, 8 | 30° | IL1A 3 µM | 6 | –16° |
|  |  |  | FGF1 2 µM | 7 | 0° |
|  |  |  | Amlexanox 20 nM | 8 | –22° |
| S100A16 | X, 430 | 6° |  |  |  |
| S100B | >350, 56 | 61° | p53$^{Pep}$ <24 µM | 9 | –7° |
|  |  |  | TRTK$^{Pep}$ 3 µM | 10 | –3° |
|  |  |  | Pentamidine 53 µM | 11 | 1° |
|  |  |  | Ndr$^{Pep}$ 20 µM | 12 | –1° |
|  |  |  | S44 20 µM | 13 | –3° |
|  |  |  | SBi132 80 µM | 14 | 1° |
|  |  |  | SBi279 2 mM | 14 | 1° |
|  |  |  | SBi523 120 µM | 14 | 3° |
|  |  |  | SC0067 700 µM | 15 | 4° |
|  |  |  | SC0322 55 µM | 15 | –3° |
| S100G (Calbindin) | 0.004, 0.004[1] | –4° |  |  |  |
| S100P | 800, 2 | 16° |  |  |  |

simultaneously is the protein complex sufficiently folded as necessary for sequestering Ca$^{2+}$ tightly (i.e., at the nM concentrations). In summary, a "functional binding and folding event occurs," which involves interactions between the target protein and loops/helices of the EF-hand proteins. It is the protein-protein interaction itself that is necessary to complete the folding process for the complex, so that the EF-hand-binding protein is then able to appreciably sequester free $Ca^{2+}$ at physiologically relevant free $Ca^{2+}$ ion concentrations (Charpentier et al. 2010; Zimmer and Weber 2010). In the absence of a biological target, the EF-hand signaling protein has little or no effect on free $Ca^{2+}$ levels since it is in effect "not fully folded" and cannot bind $Ca^{2+}$ tightly until the target protein complex forms (Rustandi et al. 1999; Zimmer and Weber 2010). The "almost folded" form of the S100-$Ca^{2+}$ protein complex exhibits significant amounts of mobility on both fast and slow timescales by NMR in the absence of bound target, which is likely the reason why the EF-hand $Ca^{2+}$-binding protein does not have a high affinity for $Ca^{2+}$ (Charpentier et al. 2010; Rustandi et al. 1999; Zimmer and Weber 2010).

There are numerous examples of functional binding and folding (FBF) for S100s, CaM, TnC, and other canonical EF-hand proteins involved in protein-protein interactions; however, the allosteric effects on $Ca^{2+}$-binding after a protein target interaction are quite different for S100s (versus other EF-hands) because S100s are very stable symmetric homodimers, and because they have a very different conformational change than those observed for canonical EF-hands (Yap et al. 1999). For example, definitive $Ca^{2+}$-binding data in the presence and absence of target is reported for several S100s including S100B, S100A1, S100A4, S100A6, and others. In all cases, the presence of target significantly reduced the dissociation constant for $Ca^{2+}$-binding and/or $Ca^{2+}$ off-rates for the S100 (Zimmer and Weber 2010). S100A10 does not conform to other family members since it lacks a functional EF-hand $Ca^{2+}$-binding domain, so its target protein interactions are all independent of $Ca^{2+}$ (Donato 2001; Heizmann 2002). Likewise, it is also clear that while peptides targets do show this trend, they are often not sufficiently intact to induce the same "functional binding/folding" effect as found for full-length proteins. For example, target peptides derived from the ryanodine receptor were shown to lower the dissociation constant of S100A1 by about a factor of 10, whereas full-length RyR enabled S100A1 to interact with $Ca^{2+}$ at 100 nM free $Ca^{2+}$ concentration. This represents an over 100-fold lowering of the dissociation constant for $Ca^{2+}$-binding to S100A1 when compared to binding in the absence of target (Prosser et al. 2008; Wright et al. 2008), and makes this S100 interaction physiologically relevant within the cytoplasm (Berridge et al. 2003).

**S100 Proteins, Table 1** S100 protein synonyms

| S100 protein | Synonyms |
|---|---|
| S100A1 | S100α |
| S100A2 | CAN19; S100L; MGC111539 |
| S100A3 | S100E |
| S100A4 | 42A; 18A2; CAPL; FSP1; MTS1; p9KA; pEL98, metastasin, calvasculin |
| S100A5 | S100D |
| S100A6 | 2A9; PRA; 5B10; CABP; CACY; calcyclin |
| S100A7 | Psoriasin; CAAF2; BDA11; PSOR1; S100A7c |
| S100A8 | P8; CFAG; calgranulin A (CAGA); MRP8; CP-10; CFAg; 60B8AG |
| S100A9 | P14; CFAG; calgranulin B (CAGB); LIAG; MRP14; 60B8AG |
| S100A8/S100A9 | L1Ag; MAC387; Calprotectin |
| S100A10 | 42C; P11; p10; GP11; ANX2L; calpactin 1 light chain; CLP11; Ca[1]; ANX2LG; MGC111133 |
| S100A11 | MLN70; S100C; calgizzarin |
| S100A12 | p6; calgranulin C (CAGC); CGRP; MRP6; CAAF1; EN-RAGE |
| S100A13 | None |
| S100A14 | BCMP84; S100A15 |
| S100A15 | BCMP84; koebnerisin |
| S100A16 | AAG13; S100F; DT1P1A7; MGC17528 |
| S100B | NEF; S100; S100β |
| S100G | CABP; CABP1; CALB3; CABP9K; MGC138379 |
| S100P | MIG9 |
| S100Z | Gm625; S100ζ |

Contrary to calmodulin and troponin-C whose activities are restricted to the intracellular milieu, a large number of S100 proteins also exert extracellular regulatory activities. Indeed, certain S100 proteins are secreted and/or released to regulate activities of target cells in a paracrine and autocrine manner. Thus, several S100 proteins (Table 1) act as intracellular regulators (Table 2) and extracellular signals (Table 3). Importantly, S100 proteins are only expressed in vertebrates, and exhibit somewhat cell-specific distribution (Donato 2001). Of the 24 human S100 genes, 19, i.e., S100A1 to S100A16 plus a number of S100A pseudogenes, are located within chromosome 1q21, S100A11P maps to chromosome 7q22–q3, S100B to chromosome 21q22, S100G to chromosome Xp22, S100P to chromosome 4p16, and S100Z to chromosome 5q13 (Marenholz et al. 2004).

**S100 Proteins, Table 2** Functions regulated by intracellular S100 proteins

| S100 protein | Functions |
| --- | --- |
| S100A1 | S100A1 interacts with the sarcoplasmic reticulum ATPase (SERCA2a) and the ryanodine receptor 2 (RyR2) in the heart, resulting in improved $Ca^{2+}$ handling and contractile performance (Rohde et al. 2010). S100A1 also targets the cardiac sarcomere and mitochondria, leading to reduced pre-contractile passive tension and enhanced oxidative energy generation (Rohde et al. 2010). S100A1 deficiency results in abnormal SR $Ca^{2+}$ content and fluxes, accelerated deterioration of cardiac performance, and transition to heart failure (Rohde et al. 2010). In skeletal myofibers, S100A1 binds to RyR1 and potentiates its open probability and plays role in skeletal muscle excitation-contraction coupling (Rohde et al. 2010). S100A1 also interacts with the giant sarcomeric kinase, titin, with potential improvement of sarcomeric compliance |
| S100A2 | S100A2 expression is downregulated in many cancers and the loss in nuclear expression has been associated with poor prognosis. Binding of S100A2 to p53 transactivation domain and p53 activation represent a potential mechanism of action of S100A2 (van Dieck et al. 2009). However, S100A2 is upregulated in some cancers |
| S100A3 | S100A3 is highly expressed in hair root cells and some astrocytomas. S100A3 has been proposed to have a role in epithelial cell differentiation and $Ca^{2+}$-dependent hair cuticular barrier formation. It may protect hair from oxidative damage due to high Cys content |
| S100A4 | S100A4 expression is associated with patient outcome in a number of tumor types via stimulation of cell survival, motility, and invasion (Boye and Maelandsmo 2010) |
| S100A5 | Although S100A5 is upregulated in bladder cancers and recurrent grade I meningiomas, its biological function is unknown |
| S100A6 | S100A6 has been implicated in cell proliferation, cytoskeletal dynamics, and tumorigenesis. S100A6 interacts with calcyclin-binding protein/Siah-1-interacting protein, a component of ubiquitin ligase involved in the ubiquitination of β-catenin. S100A6 also inhibits the interaction between the heat shock proteins (Hsp70 and Hsp90) and Sgt1 or Hop, which suggests a potential role for S100A6 in the cellular response to different stress factors. In this respect, the presence of S100A6 favors apoptosis in some cells, but limits it in others. S100A6 interacts with caldesmon, calponin, tropomyosin, and kinesin light chain |
| S100A7 | S100A7 promotes aggressive features in breast cancer by binding to c-Jun activation domain-binding protein 1, thereby stimulating Akt and NF-κB (West and Watson 2010). Proinflammatory cytokines upregulate S100A7 expression in human breast cancer (West and Watson 2010) |
| S100A8 | S100A8 (Ghavami et al. 2009; Goyette and Geczy 2010) inhibits differentiation-dependent telomerase activity in a keratinocyte cell line in a $Ca^{2+}$-dependent manner and scavenges intracellular reactive oxygen species (ROS) generated by phagocytic neutrophils; it may stabilize nitric oxide in neutrophils. S100A8 reduces p38 MAPK phosphorylation of S100A9 in neutrophils |
| S100A9 | S100A9 (Ghavami et al. 2009; Goyette and Geczy 2010) abrogates S100A8-induced reduction in telomerase activity, inhibits myeloid (dendritic cell and macrophage) differentiation and accumulation of myeloid-derived suppressor cells in pathological responses via intracellular ROS generation, is a p38 MAPK target, phosphorylated after phagocyte activation, reduces microtubule polymerization and F-actin cross-linking by the S100A8/S100A9 complex, mediates $Ca^{2+}$ signaling associated with inflammatory agonist-induced IP3-mediated $Ca^{2+}$ release in neutrophils, is induced by oncostatin M in MCF-7 breast tumor cells and is essential for oncostatin M-induced growth repression, mediates transformation and proliferation of human aortic smooth muscle cells, is a p53 transcriptional target and mediates the p53 apoptosis pathway in esophageal squamous cell carcinoma, and is involved in glutathione metabolism in activated neutrophils |
| S100A8/ S100A9 | S100A8/S100A9 (Ghavami et al. 2009; Goyette and Geczy 2010) inhibits casein kinase I and II, a role in myeloid cell differentiation has been suggested, interacts with nuclear factors, transports unsaturated fatty acids and arachidonic acid, and promotes NADPH oxidase activation in phagocytes by interaction with p67*phox* and Rac-2. Phagosomal ROS production requires extracellular $Ca^{2+}$ entry mediated by S100A8/A9 as $Ca^{2+}$ sensor. S100A8 and S100A9 overexpression in HaCaT keratinocytes increases NADPH oxidase activity and enhances ROS levels; in hepatocellular carcinoma cells, co-expression promotes malignant progression by induction of ROS, downregulation of p38 MAPK signaling, and cell survival and resistance to TNF-induced apoptosis. S100A8/A9 modulates $Ca^{2+}$-dependent interactions with cytoskeletal components during migration, degranulation, phagocytosis of activated monocytes and neutrophils; the tetramer promotes microtubule polymerization and F-actin cross-linking. S100A8/S100A9 expression associated with a macrophage subtype associated with low antimycobacterial activity; S100A8-/S100A9-expressing epithelial cells resist invasion by *Porphyromonas gingivalis, Listeria monocytogenes*, and *Salmonella typhimurium*. S100A8/S100A9 co-expression in ductal carcinomas of breast associated with poor tumor differentiation, vessel invasion, node metastasis; annexin 6 is involved in the $Ca^{2+}$-dependent expression of S100A8/S100A9 on surface of tumor cells; this complex may mediate cell membrane–regulated events. S100A8/S100A9 may mediate pathological differentiation of psoriatic keratinocytes; interacts with keratin intermediate filaments, and may modulate wound healing |

(*continued*)

**S100 Proteins, Table 2** (continued)

| S100 protein | Functions |
|---|---|
| S100A10 | S100A10 tethers certain transmembrane proteins to annexin A2, thereby assisting their traffic to the plasma membrane and/or their firm anchorage at certain membrane sites (Cdc42, tetrodotoxin-resistant sodium channel Nav 1.8, background two-pore domain potassium channel TWIK-related acid-sensitive K, acid-sensing ion channel ASIC1a, actin-binding protein AHNAK, tissue-type plasminogen activator, serotonin 1B receptor) (Rescher and Gerke 2008). S100A10 is induced by neurotrophins, is downregulated in human and rodent depressive-like states, and is implicated in the mechanism of action of antidepressant drugs and electroconvulsive seizures, in part due to its interaction with specific serotonin receptors (Warner-Schmidt et al. 2010) |
| S100A11 | When phosphorylated by protein kinase C-α, $Ca^{2+}$-bound S100A11 inhibits cell growth via binding to nucleolin, translocation to the nucleus, and activation of the cell cycle modulator $p21^{WAF1/CIP1}$, and stimulates cell growth by enhancing the level of EGF family proteins. S100A11 also binds Rad54B, a DNA-dependent ATPase involved in recombinational repair of DNA damage |
| S100A12 | S100A12 (Goyette and Geczy 2010) may modulate interactions between cytoskeletal elements and membranes. S100A12 inhibits aggregation of aldolase and GAPDH and may have $Ca^{2+}$-dependent chaperone/anti-chaperone-like functions. S100A12 may play a role in vascular remodeling; expressed in human aortic aneurysms. Overexpression causes several vascular smooth muscle cell (VSMC) dysfunctions such as increased pro-MMP-2 generation, increased phosphorylation and nuclear translocation of Smad2, and modulation of mitochondrial function (Hofmann Bowman et al. 2010). VSMC from mice overexpressing S100A12 in these cells have increased NADPH oxidase-mediated generation of peroxide, possibly via interaction with Nox-1 |
| S100A13 | S100A13 plays an important role in the stress-induced release of FGF-1 and IL-1α |
| S100A14 | S100A14 might function as a cancer suppressor working in the p53 pathway |
| S100A15 | No intracellular role for S100A15 has been reported, the protein mainly acting as an extracellular factor |
| S100A16 | S100A16 is upregulated in several tumors. However, no intracellular role for S100A16 has been reported |
| S100B | S100B (Donato et al. 2009) acts as a stimulator of cell proliferation and migration and an inhibitor of apoptosis and differentiation, which might have important implications during brain, cartilage, and skeletal muscle development and repair, activation of astrocytes in the course of brain damage and neurodegenerative processes, and of cardiomyocyte remodeling after infarction, as well as in melanomagenesis and gliomagenesis. S100B's binding partners within cells are tubulin and the microtubule-associated τ protein, the actin-binding protein caldesmon, type III intermediate filament subunits, membrane-bound guanylate cyclase, the small GTPase Rac1 and Cdc42 effector IQGAP1, Src kinase, the serine/threonine protein kinase Ndr, the tumor suppressor p53, intermediates upstream of IKKβ/NF-κB, the giant phosphoprotein AHNAK/desmoyokin, the E3 ligase hdm2, dopamine D2 receptor, the mitochondrial AAA ATPase ATAD3A. S100B is highly expressed in astrocytes (Donato et al. 2009), and elevation of S100B serum levels positively correlates with mood disorders and schizophrenia (Rothermundt et al. 2010). Serum levels of S100B are of prognostic value in patients with cutaneous melanoma (Mocellin et al. 2008) and breast cancer. Serum levels of S100B are an outcome predictor in severe traumatic brain injury |
| S100G | S100G acts as cytosolic $Ca^{2+}$ buffers in many tissues, resulting in modulation of $Ca^{2+}$ adsorption |
| S100P | S100P interacts with ezrin/radixin/moesin, thereby activating ezrin and promoting the transendothelial migration of tumor cells. S100P also interacts with the scaffolding protein IQGAP, thereby reducing IQGAP signaling |
| S100Z | S100Z is downregulated in several tumors, but no functional roles have been reported for it |

## S100 Proteins in the Extracellular Fluids

Various S100 proteins are found in body fluids, including serum, urine, seminal plasma, saliva, sputum, cerebrospinal fluid, and in feces and abscess fluid and most are associated with active disease states. Some, such as the S100A8/S100A9 complex and S100B, are considered biomarkers for particular disease processes (Dassan et al. 2009; Mocellin et al. 2008). Secretion of some S100 proteins is stimulated by particular cell activators. For example, 5-$HT_{1A}$ receptor agonists, glutamate, adenosine, and lysophosphatidic changes in extracellular $Ca^{2+}$ and $K^+$ levels trigger release of S100B from astrocytes (Donato et al. 2009), and release of S100A4 from human pulmonary artery smooth muscle cells (Donato 2007). Metabolic/oxidative stress induces release from several cells and activation of some cell types, or cell adhesion, can also promote secretion.

S100 proteins lack a leader sequence and are not secreted via the classical Golgi pathway and mechanisms remain somewhat unclear. S100A8/S100A9 may be passively released from necrotic myeloid cells, or actively secreted following translocation to

**S100 Proteins, Table 3** Functions regulated by extracellular S100 proteins

| S100 protein | Functions |
|---|---|
| S100A1 | Extracellular translocation seen after heart ischemia. S100A1 enhanced $Ca^{2+}$ influx in cultured ventricular cardiomyocytes. S100A1 enhanced Cav1 channel currents in a PKA-dependent manner, prolonged action potentials, and amplified action potential–induced $Ca^{2+}$ transients in neurons |
| S100A2 | S100A2 is chemotactic for eosinophils. S100A2 may be involved in calcification of cartilage/bone |
| S100A3 | No reported extracellular function |
| S100A4 | S100A4 has effects on numerous cell types (Donato 2007). S100A4 released by tumor or stroma cells triggers prometastatic cascades. S100A4 modifies the cytoskeleton and focal adhesions of tumor cells, downregulates the pro-apoptotic Bax and the angiogenesis inhibitor thrombospondin-1 genes; increases production of matrix-metalloproteinases (MMP) by endothelial and tumor cells; induces tube formation in endothelial cells in a RAGE-independent pathway, possibly through interaction with annexin II and accelerated plasmin formation; stimulates production of cytokines, particularly granulocyte colony-stimulating factor and eotaxin-2 from T lymphocytes. S100A4 may stimulate infiltration of T cells into primary tumors. S100A4 may promote TCRγΔ T-cell mediated lysis and negatively regulate matrix mineralization/calcification. S100A4/RAGE interactions increase MMP-13 production from articular chondrocytes (Donato 2007). S100A4 may be cardioprotective: promotes smooth muscle cell motility and proliferation, possibly mediates neointimal hyperplasia and arterial muscularization; promotes cardiac myocyte growth, survival, and differentiation. Oligomeric S100A4 forms (tetrameric or more) promote growth and survival of cultured neurons, possibly protective after injury; regulates astrocyte motility. ERK1/2 is activated in many of the S100A4-responsive cells and may modulate growth and survival responses of S100A4 |
| S100A5 | No reported extracellular function |
| S100A6 | S100A6 is expressed by many cells and tumors. It may regulate secretory processes: stimulates secretion of lactogen II by trophoblasts and insulin release from pancreatic islet cells. S100A6 inhibited histamine release by mast cells. S100A6 modulates RAGE-dependent survival of neuroblastoma cells by triggering apoptosis and generation of ROS through JNK activation |
| S100A7 | S100A7 (psoriasin) is overexpressed in inflammatory skin diseases and is induced in keratinocytes by IL-17 and IL-22. It has roles in antimicrobial responses and innate immunity. S100A7 adheres to, and reduces *Escherichia coli* survival; the hinge region (amino acids 35–80) is sufficient for full activity. S100A7-RAGE binding and zinc-dependent chemotactic activity for lymphocytes, monocytes, and granulocytes; acts synergistically with S100A15. Mice expressing elevated amounts of doxycycline-regulated mS100A7/A15 in skin keratinocytes have an exaggerated inflammatory response characterized by leukocyte infiltration and elevated levels of T helper 1 and T helper 17 proinflammatory cytokines, linked to the pathogenesis of psoriasis. S100A7 stimulates ROS generation from neutrophils. S100A7 promotes α-secretase activity (via promotion of ADAM (a disintegrin and metalloproteinase)-10) from primary cortico-hippocampal neuron cultures and in brain; may prevent generation of amyloidogenic peptides in Alzheimer's disease |
| S100A8 | Murine S100A8 (Ehrchen et al. 2009; Goyette and Geczy 2010) and its hinge domain are chemotactic for leukocytes at picomolar concentrations; it causes actin polymerization; a G-protein-coupled receptor is implicated. Human S100A8 is not chemotactic for monocytes and is chemotactic for neutrophils and for periodontal ligament cells S100A8 suppresses expression of high affinity β-2 integrin epitope on neutrophils induced by S100A9, thereby influencing adhesion. S100A8 has an anti-inflammatory effect by triggering oxidation-sensitive repulsion of neutrophils. S-nitrosylated S100A8 reduces mast cell activation, leukocyte transmigration in the microcirculation, and shuttles NO. Human S100A8 scavenges oxidants and a protective function is proposed; chemotactic properties of murine S100A8 are modified by oxidation. S100A8 suppresses mast cell activation by allergen by reducing intracellular ROS required for signaling and suppressed eosinophil migration and symptoms of allergic inflammation in an acute asthma model. S100A8 upregulates and activates MMPs and aggrecanase enzymes from chondrocytes, suggesting a role in pericellular matrix degradation (van Lent et al. 2008). S100A8 is a TLR-4 ligand that activates cytokine production by murine bone marrow cells (Vogl et al. 2007). S100A8 activates FcγRI and FcγRIV on macrophages through the activation of TLR-4 |
| S100A9 | S100A9 in the presence of $Zn^{2+}$ and $Ca^{2+}$ is a RAGE ligand and a TLR-4 ligand (Vogl et al. 2007); other studies indicate that S100A9 does not activate phagocytes via TLR-4 (Goyette and Geczy 2010). S100A9 affects leukocyte tissue invasion. Human S100A9 is chemotactic for neutrophils (Goyette and Geczy 2010) and activates expression of high affinity β-2 integrin epitope on neutrophils. S100A9 activates adhesion to fibronectin (Goyette and Geczy 2010) that is inhibited by S-glutathionylated S100A9. S100A9 induces degranulation of secretory and specific/gelatinase granules from neutrophils. S100A9 (Goyette and Geczy 2010) decreases PMA-triggered peroxide production by BCG-activated macrophages, is mitogenic for fibroblasts; S100A9 stimulates IL-8 release from epithelial cells, and S100A9 may mediate dystrophic calcification, and is incorporated into urinary calcium oxalate crystals. S100A9 C-terminal peptide reduces spreading and phagocytosis of adherent macrophages induced by |

(*continued*)

**S100 Proteins, Table 3** (continued)

| S100 protein | Functions |
|---|---|
| | proteinase-activated receptor-1 agonists and this peptide, or S100A9, suppresses macrophage activation upon ingestion of apoptotic neutrophils. S100A9 C-terminal peptide also modulates primary afferent nociceptive signals by inhibiting activation of N-type voltage–operated calcium channels and may reduce pain responses in inflammation |
| S100A8/ S100A9 | The S100A8/S100A9 complex is also known as calprotectin. It has antimicrobial properties, chiefly mediated by chelation of $Zn^{2+}$ and $Mn^{2+}$. However, S100A8/S100A9 increased *Mycobacterium tuberculosis* growth in vitro, suggesting a role in immunopathogenesis of tuberculosis (Goyette and Geczy 2010). Human S100A9/S100A9 is chemotactic for neutrophils (Goyette and Geczy 2010); S100A8/A9 dissociates pancreatic acinar cell-cell contacts which is $Ca^{2+}$-dependent. S100A8/S100A9 influences migration of other cell types, including myeloid-derived suppressor cells and some tumor cells, and may facilitate tumor cell invasion. S100A8/S100A9 (Goyette and Geczy 2010) may activate proinflammatory cytokine production by human monocytes and macrophages via the NF-κB and p38 MAPK pathways and may induce NO production by macrophages. S100A8/S100A9 potentiates the TLR-4 response to LPS but does not directly activate phagocytes via TLR-4 (Vogl et al. 2007). On the other hand, S100A8/S100A9 may suppress acute inflammation by binding and modulating activities of proinflammatory cytokines (Goyette and Geczy 2010). S100A8/S100A9 stimulation of $CD8^+$ T lymphocytes from individuals with lupus erythematosus upregulated IL-17, suggesting a role in development of autoreactive lymphocytes (Loser et al. 2010). S100A8/S100A9 (Goyette and Geczy 2010) promotes HIV-1 transcriptional activity and viral replication in infected $CD4^+$ T lymphocytes and inhibits immunoglobulin synthesis of lymphocytes. S100A8/S100A9 is elevated in psoriatic lesions; the complex induces production of a number of cytokines and chemokines in normal human keratinocytes; low concentrations stimulate keratinocyte growth (Nukui et al. 2008). S100A8/S100A9 stimulates proinflammatory properties of endothelial cells, possibly mediated by RAGE and potentiates activation by advanced glycation end products. S100A8/S100A9 delivers arachidonic acid to endothelium via CD-36-mediated uptake and stabilizes and protects leukotriene A(4) from nonenzymatic hydrolysis, possibly increasing availability of bioactive leukotrienes. S100A8/S100A9 (Ghavami et al. 2009) inhibits proliferation and differentiation of C2C12 myoblasts and induces caspase-3-dependent apoptosis. Growth inhibitory activities of relatively high S100A8/S100A9 concentrations against a variety of normal cell types (such as macrophages, bone marrow cells, lymphocytes, fibroblasts) are reported, and the complex has apoptosis-inducing activity to numerous tumor cell lines. S100A8/S100A9 can reduce mitochondrial membrane potential, causing Smac/Diablo and Omi/HtrA2 (without cytochrome C) release and reduced Drp1 expression which inhibits mitochondrial fission machinery and induces cell death by altering the balance between pro- and anti-apoptotic proteins in some cells. S100A8/S100A9 may also promote autophagy-like death; apoptosis and autophagy may involve translocation of BNIP3, a BH3 only pro-apoptotic Bcl2 family member, to mitochondria, and cross talk between mitochondria and lysosomes and generation of reactive oxygen species (Ghavami et al. 2009). $Zn^{2+}$ chelation may be another mechanism (Ghavami et al. 2009). S100A8/S100A9 at low concentrations promotes growth of some tumor cells through RAGE signaling and NF-κB activation (Ghavami et al. 2009). S100A8/S100A9 binding to carboxylated glycans on RAGE activates NF-κB and proliferation of colon cancer cells (Turovskaya et al. 2009). Binding to RAGE on prostate cancer cells activates the MAPK pathway (Ghavami et al. 2009). S100A8 and S100A9 influence cardiomyocyte contractility by causing RAGE-dependent decreases in $Ca^{2+}$ flux (Boyd et al. 2008). S100A8/S100A9-activated melanoma cells overexpress MMPs -2, -9, and -14. On the other hand, S100A8/S100A9 inhibits MMPs by sequestering $Zn^{2+}$ from their active sites (Goyette and Geczy 2010). S100A8 and S100A9 inhibit the spontaneous and stimulated oxidative burst of neutrophils, possibly mediated by P1 adenosine receptors |
| S100A10 | Heterotetrameric complexes of S100A10 with annexin A2 serve as an extracellular binding partner for pathogens and host proteins. S100A10 binds tissue plasminogen activator (tPA) and plasminogen via C-terminal lysine residues, promoting tPA-dependent plasmin production. Plasmin binds S100A10 at a distinct site; the S100A10-plasmin complex stimulates plasmin autoproteolysis, thereby providing a highly localized transient pulse of plasmin activity at the cell surface. S100A10 mediates macrophage recruitment in response to inflammatory stimuli by activating pro-MMP-9 that in turn promotes plasmin-dependent invasion in vivo |
| S100A11 | S100A11 localizes in the cytosol of luteal cells in the mouse ovary, and oviductal epithelial cells; suppresses fertilization through its action on cumulus cells. S100A11 is induced/released by chondrocytes cultured with IL-1β, TNF-α, and CXCL8. S100A11 promotes hypertrophic chondrocyte differentiation, stimulates RAGE-dependent type X collagen and IL-8 production by reticular chondrocytes. Transamidation generates covalently bonded S100A11 homodimer that can signal through RAGE and the p38 MAPK pathway to accelerate chondrocyte hypertrophy and matrix catabolism and this may promote osteoarthritis progression |
| S100A12 | There is no S100I2 in rodent genomes. S100A12 is constitutively expressed in neutrophils; TNF, IL-6, and endotoxin induce the gene in monocytes/macrophages, LPS on smooth muscle cells (Pietzsch and Hoppmann 2009). S100A12 inhibits growth and motility of filarial parasites by binding paramyosin. The C-terminal peptide is |

*(continued)*

**S100 Proteins, Table 3** (continued)

| S100 protein | Functions |
|---|---|
|  | antimicrobial and antifungal; $Zn^{2+}$ enhances activity. Low concentrations of S100A12 and its hinge domain are chemotactic for monocytes and mast cells; a G-protein-coupled receptor is implicated. Higher levels of S100A12 activate mast cells and potentiate IgE-mediated activation in a RAGE-independent manner. Hexameric S100A12 forms multimeric RAGE complexes. Bovine S100A12 stimulates RAGE-dependent TNF-α and IL-1β production from murine BV-2 microglial cells, IL-2 from lymphocytes and ICAM-1 and VCAM expression on endothelial cells (Hofmann et al. 1999). S100A12 enhances Mac-1 integrin affinity and L-selectin shedding from neutrophils and modulates neutrophil release from bone marrow (Pietzsch and Hoppmann 2009; Goyette and Geczy 2010). S100A12 stimulates neurite outgrowth of rat hippocampal neurons via RAGE ligation and activation of PLC, PLK, CAM-kinase II, and MAPK pathways (Donato 2007). S100A12 inhibits MMP-3 and -9 by chelating $Zn^{2+}$ from their active sites (Goyette and Geczy 2010). S100A12 overexpression in smooth muscle cells in mice leads to aortic aneurysms; linked to leukocyte influx, increased IL-6 in response to LPS, and increased latent MMP-2 levels. Overexpression of S100A12 in vascular smooth muscle cells in mice promotes atherosclerotic plaque remodeling and nodular calcification, possibly by influencing osteoblastic genes in a feedback mechanism involving RAGE (Hofmann Bowman et al. 2011). $Cu^{2+}$ sequestration by S100A12 may modulate redox (Pietzsch and Hoppmann 2009) |
| S100A13 | S100A13 promotes intracellular translocation of S100A13, possibly via RAGE binding on endothelial cells. S100A13 is involved in the nonclassical secretion of fibroblast growth factor (FGF-1); the complex may contribute to angiogenesis |
| S100A14 | No reported extracellular function |
| S100A15 | S100A15 is expressed in keratinocytes in inflamed skin; induced by LPS, IL-1β, and Th-1 cytokines. It is chemotactic for monocytes and granulocytes, possibly via A G-protein-coupled receptor; acts synergistically with S100A7 in leukocyte recruitment in vitro and in vivo. Human S100A15 has antimicrobial activity against *E. coli* |
| S100A16 | No reported extracellular function |
| S100B | S100B secreted or released from astrocytes has different (trophic and toxic) effects on neurons, astrocytes, and microglia depending on the concentration (Donato et al. 2009). Up to a few nanomolar amounts, S100B protects neuronal cells against neurotoxic stimuli whereas at micromolar doses, it kills neurons via RAGE engagement in both cases. S100B stimulates astrocyte proliferation at low doses and supports inflammatory activities in astrocytes at high doses. S100B attenuates microglia activation and activates microglia in a RAGE-dependent manner at low and high doses, respectively (Donato et al. 2009). After permanent middle cerebral artery occlusion in S100B transgenic (TG) mice, infarct volumes are significantly increased during the first post-infarct days and astrogliosis is enhanced compared with controls (Donato et al. 2009). Moreover, S100B TG mice show increased susceptibility to perinatal hypoxia-ischemia (Donato et al. 2009), and overexpression of S100B accelerates Alzheimer disease-like pathology with enhanced astrogliosis and microgliosis (Mori et al. 2010). Outside the nervous system, S100B released from injured skeletal muscle tissue is a potent stimulator of myoblast proliferation and inhibitor of myoblast differentiation enhancing bFGF/FGFR1 signaling and blocking RAGE's promyogenic signaling (Donato et al. 2009). Upon forming complexes with TLR-2 ligands, S100B inhibits TLR-2 via RAGE, through a paracrine epithelial cells/neutrophil circuit that restrains pathogen-induced inflammation; however, upon binding to nucleic acids, S100B activates intracellular TLR-3/9 eventually resolving danger-induced inflammation via transcriptional inhibition of S100B. S100B becomes expressed in the cardiomyocytes surviving an infarct under the action of catecholamines, thereby modulating left ventricular remodeling (Tsoporis et al. 2010). At relatively high doses, S100B causes cardiomyocyte apoptosis RAGE dependently (Tsoporis et al. 2010). S100B-RAGE interactions have an important role in vascular smooth muscle cell proliferation in diabetes and neovascular macular disease (Donato et al. 2009). In bronchial epithelial cells, S100B is upregulated at early stages of fungal infection via MyD88-dependent activation of canonical NF-κB and downregulated via TLR-3/9-dependent activation of noncanonical NF-κB at late stages. S100B null mice exhibit enhanced spatial and fear memories as well as enhanced long-term potentiation (LTP) in the hippocampal CA1 region, and perfusion of hippocampal slices with S100B reverses the levels of LTP to those of the wild-type slices (Donato et al. 2009). This suggests that extracellular S100B might play a role as a regulator of synaptic plasticity, although the molecular mechanism underlying this activity remains to be elucidated |
| S100G | No reported extracellular function |
| S100P | S100P can mediate tumor growth, drug resistance, and metastasis through RAGE binding on cancer cells (Arumugam and Logsdon 2010) |
| S100Z | No reported extracellular function |

the membrane, in a process requiring an intact microtubule network and protein kinase C activation. S100 proteins may themselves have roles in nonclassical secretion. These proteins have diverse affinities to lipid structures that allow translocation across the plasma membrane following cell stress or activation. For example, S100A13-lipid interactions and formation of a multiprotein complex with fibroblast growth factor 1 (which also lacks a classical secretion sequence) and synaptotagmin peptides allows release, possibly mediated by N-type $Ca^{2+}$ channel activity, or flip-flop to the extracellular compartment via annexin binding. The heterotetrameric S100A10 and annexin II complex is associated with von Willebrand factor secretion from endothelial cells. Other mechanisms include vesicular S100A8/S100A9 release in neutrophil extracellular traps (NETs), triggered by production of reactive oxygen species (ROS) from dying neutrophils. NETs are chromatin fibers (histones and DNA) bound with antimicrobial proteins that deliver high local concentrations to pathogens.

## Inducers of S100 Proteins

Numerous S100 genes are induced in a somewhat cell-specific manner, by appropriate growth factors, cytokines, and Toll-like receptors (TLR) ligands. In these circumstances, they are generally secreted, and may function as extracellular alarmins or damage-associated molecular pattern factors that principally mediate functions of the innate immune system, stimulate cancer cell locomotion, and/or participate in tissue repair (Donato 2007; Donato et al. 2009). The extracellular functions of S100 proteins are summarized in Table 3.

## S100 Receptors

The oligomeric forms of some S100 proteins and their putative binding partners can determine function. For example, S100A8 and S100A9 have functions that can be dependent on or independent of the heterocomplex (see Table 3). In some circumstances, more highly oligomerized S100 proteins may be more functionally efficient and in some cases high concentrations are required for activation whereas other functions depend on very low amounts (Donato 2007), indicating different receptor affinities. Divalent cation binding can also determine functional outcomes. In addition, some S100 proteins are structurally altered and particular posttranslational modifications can promote changes in function. These include oxidation products of S100A8 and or S100A9 generated by nitric oxide, oxygen-free radicals, and hypohalous acids.

Receptors mediating the extracellular functions of S100 proteins have been elusive and remain a matter of debate. Both non-receptor- and receptor-mediated endocytosis are implicated. For example, exogenous S100A1 is internalized into neurons via multiple endocytic pathways and delivered to early/recycling endosomes, Golgi apparatus, late endosomes, and lysosomes. Although many effects may be mediated by the receptor for advanced glycation end products (RAGE), others are not. Structural studies of some S100 proteins indicate at least three recognition sites within two distinct surfaces that may accommodate multiple binding partners that result in complex interactions, or binding to specific ligands on different target cells. This is supported by reports that some S100 functions reside within the divergent hinge domains between the $Ca^{2+}$-binding regions and can be mimicked by relevant peptides whereas others require homo- or hetero-S100 complexes. Structural and the binding data suggest that octameric S100B triggers RAGE by receptor dimerization (Ostendorp et al. 2007). Furthermore, some S100s, such as S100A6, preferentially bind the C2 domain of RAGE rather than the V and/or C1 domains that bind S100B, indicating another layer of receptor complexity. Some outcomes of RAGE signaling are summarized in Donato (2007).

S100A12 was the first S100 protein for which RAGE was the designated receptor on myeloid cells (Hofmann et al. 1999) although RAGE participation is controversial and other receptors including $N$-glycans (including glycosylated RAGE) a G-protein-coupled receptor and scavenger receptors are implicated (Donato 2007; Ghavami et al. 2009; Pietzsch and Hoppmann 2009; Goyette and Geczy 2010). Receptors for S100A8 and/or S100A9 include RAGE (particularly in responses of tumor cells), heparan sulfate proteoglycans and $N$-glycans, TLR4, and CD36 on endothelial cells (see Table 3). Other examples include effects of S100A4 on neurite outgrowth that depend on binding to heparan sulfate proteoglycans and a putative Gαq-coupled receptor.

## Cross-References

- Bacterial Calcium Binding Proteins
- Biological Copper Transport
- Calcium ATPase
- Calcium in Health and Disease
- Calcium in Heart Function and Diseases
- Calmodulin
- Calsequestrin
- EF-Hand Proteins
- Troponin
- Zinc-Binding Sites in Proteins

## References

Arumugam T, Logsdon CD (2010) S100P: a novel therapeutic target for cancer. *Amino Acids 41*:893–899.

Berridge MJ, Bootman MD, Roderick HL (2003) Calcium signaling: dynamics, homeostasis and remodeling. *Nature Rev Mol Cell Biol 4*:517–529.

Boye K, Maelandsmo GM (2010) S100A4 and metastasis: a small actor playing many roles. *Am J Pathol 176*: 528–535.

Boyd JH, Kan B, Roberts H et al (2008) S100A8 and S100A9 mediate endotoxin-induced cardiomyocyte dysfunction via the receptor for advanced glycation end products. *Circ Res 102*:1239–1246.

Charpentier TH, Thompson LE, Liriano MA et al (2010) The effects of CapZ peptide (TRTK-12) binding to S100B-$Ca^{2+}$ as examined by NMR and X-ray crystallography. *J Mol Biol 396*:1227–1243.

Carafoli E, Klee C (Eds) (1999) *Calcium as a cellular regulator*. New York: Oxford University Press.

Corbin BD, Seeley EH, Raab A et al (2008) Metal chelation and inhibition of bacterial growth in tissue abscesses. *Science 319*:962–965.

Dassan P, Keir G, Brown MM (2009) Criteria for a clinically informative serum biomarker in acute ischaemic stroke: a review of S100B. *Cerebrovasc Dis, 27*:295–302.

Donato R (2001) S100: a multigenic family of calcium-modulated proteins of the EF-hand type with intracellular and extracellular functional roles. *Int J Biochem Cell Biol 33*:637–668.

Donato R (2007) RAGE: a single receptor for several ligands and different cellular responses: the case of certain S100 proteins. *Curr Mol Med 7*:711–724.

Donato R, Sorci G, Riuzzi F et al (2009) S100B's double life: intracellular regulator and extracellular signal. *Biochim Biophys Acta 1793*:1008–1022.

Ehrchen JM, Sunderkötter C, Foell D, Vogl T, Roth J (2009) The endogenous toll-like receptor 4 agonist S100A8/S100A9 (calprotectin) as innate amplifier of infection, autoimmunity, and cancer. *J Leukoc Biol 86*:557–566.

Ghavami S, Chitayat S, Hashemi M, et al (2009) S100A8/A9: a Janus-faced molecule in cancer therapy and tumorgenesis. *Eur J Pharmacol 625*:73–83.

Goyette J, Geczy CL. (2010) Inflammation-associated S100 proteins: new mechanisms that regulate function. *Amino Acids 41*:821–842.

Heizmann CW (2002) The multifunctional S100 protein family. *Meth Mol Biol 172*:69–80.

Hiratsuka S, Watanabe A, Aburatani H, Maru Y (2006) Tumour-mediated upregulation of chemoattractants and recruitment of myeloid cells predetermines lung metastasis. *Nat Cell Biol 8*:1369–1375.

Hofmann MA, Drury S, Fu C, et al (1999) RAGE mediates a novel proinflammatory axis: a central cell surface receptor for S100/calgranulin polypeptides. *Cell 97*:889–901.

Hofmann Bowman MA, Wilk J, Heydemann A et al (2010) S100A12 mediates aortic wall remodeling and aortic aneurysm. *Circ Res 106*:145–154.

Hofmann Bowman MA, Gawdzik J, Bukhari U et al (2011) S100A12 in vascular smooth muscle accelerates vascular calcification in apolipoprotein e-null mice by activating an osteogenic gene regulatory program. *Arterioscler Thromb Vasc Biol 31*:337–344.

Kligman D, Hilt DC (1988) The S100 protein family. *Trends Biochem Sci 13*:437–443.

Lesniak W, Filipek A, Donato R (2009) S100a6. *UCSD-Nat Molecule Page.* doi:10.1038/mp.a002122.01.

Lin J, Yang Q, Wilder PT et al (2010) The calcium-binding protein S100B down-regulates p53 and apoptosis in malignant melanoma. *J Biol Chem 285*:27487–27498.

Loser K, Voskort M, Lueken A, et al (2010) The Toll-like receptor 4 ligands Mrp8 and Mrp14 are crucial in the development of autoreactive $CD8^+$ T cells. *Nat Med 16*:713–717.

Marenholz I, Heizmann CW, Fritz G (2004) S100 proteins in mouse and man: from evolution to function and pathology (including an update of the nomenclature). *Biochem Biophys Res Commun 322*:1111–1122.

Mocellin S, Zavagno G, Nitti D (2008) The prognostic value of serum S100B in patients with cutaneous melanoma: a meta-analysis. *Int J Cancer 123*:2370–2376.

Moews PC, Kretsinger RH (1975) Refinement of the structure of carp muscle calcium-binding parvalbumin by model building and difference Fourier analysis. *J Mol Biol 91*:201–225.

Mori T, Koyama N, Arendash GW et al (2010) Overexpression of human S100B exacerbates cerebral amyloidosis and gliosis in the Tg2576 mouse model of Alzheimer's disease. *Glia 58*:300–314.

Nukui T, Ehama R, Sakaguchi M et al (2008) S100A8/A9, a key mediator for positive feedback growth stimulation of normal human keratinocytes. *J Cell Biochem 104*:453–464.

Ostendorp T, Leclerc E, Galichet A et al (2007) Structural and functional insights into RAGE activation by multimeric S100B. *EMBO J26*:3868–3878.

Permyakov EA, Kretsinger RH (2009) Cell signaling, beyond cytosolic calcium in eukaryotes. *J Inorg Biochem 103*:77–86.

Pietzsch JS, Hoppmann S (2009) Human S100A12: a novel key player in inflammation? *Amino Acids 36*:381–389.

Prosser BL, Wright NT, Hernandez-Ochoa EO et al (2008) S100A1 binds to the calmodulin-binding site of ryanodine receptor and modulates skeletal muscle excitation-contraction coupling. *J Biol Chem 283*:5046–5057.

Rescher U, Gerke V (2008) S100A10/p11: family, friends and functions. *Pflugers Arch 455*:575–582.

Rohde D, Ritterhoff J, Voelkers M et al (2010) S100A1: a multifaceted therapeutic target in cardiovascular disease. *J Cardiovasc Transl Res* 3:525–537.

Rothermundt M, Ahn JN, Jörgens S (2010) S100B in schizophrenia: an update. *Gen Physiol Biophys* 28:F76–F81.

Rustandi RR, Baldisseri DM, Drohat AC, Weber DJ (1999) Structural changes in the C-terminus of $Ca^{2+}$-bound rat S100B (ββ) upon binding to a peptide derived from the C-terminal regulatory domain of p53. *Protein Sci* 8:1743–1751.

Sparvero LJ, Asafu-Adjei D, Kang R et al (2009) RAGE (Receptor for advanced glycation endproducts), RAGE ligands, and their role in cancer and inflammation. *J Transl Med* 7:17.

Strynadka NC, James MN (1989) Crystal structures of the helix-loop-helix calcium-binding proteins. *Annu Rev Biochem,* 58:951–998.

Tsoporis JN, Mohammadzadeh F, Parker TG (2010) Intracellular and extracellular effects of S100B in the cardiovascular response to disease. *Cardiovasc Psychiatry Neurol* 2010, 206073. doi:10.1155/2010/206073.

Turovskaya O, Foell D, Sinha P et al (2009) RAGE, carboxylated glycans and S100A8/A9 play essential roles in colitis-associated carcinogenesis. *Carcinogenesis* 29:2035–2043.

van Dieck J, Teufel DP, Jaulent AM et al (2009) Posttranslational modifications affect the interaction of S100 proteins with tumor suppressor p53. *J Mol Biol* 394:922–930.

van Lent PL, Grevers LC, Blom AB et al (2008) Stimulation of chondrocyte-mediated cartilage destruction by S100A8 in experimental murine arthritis. *Arthritis Rheum* 58:3776–3787.

Vogl T, Tenbrock K, Ludwig S et al (2007) Mrp8 and Mrp14 are endogenous activators of Toll-like receptor 4, promoting lethal, endotoxin-induced shock. *Nat Med* 13:1042–1049.

Warner-Schmidt JL, Chen EY, Zhang X et al (2010) A role for p11 in the antidepressant action of brain-derived neurotrophic factor. *Biol Psychiatry* 68:528–535.

West NR, Watson PH (2010) S100A7 (psoriasin) is induced by the proinflammatory cytokines oncostatin-M and interleukin-6 in human breast cancer. *Oncogene* 29:2083–2092.

Wright NT, Prosser BL, Varney KM et al (2008) S100A1 and calmodulin compete for the same binding site on ryanodine receptor. *J Biol Chem* 283:26676–26683.

Yap KL, Ames JB, Swindells MB, Ikura M (1999) Diversity of conformational states and changes within the EF-hand protein superfamily. *Proteins* 37:499–507.

Zimmer DB, Wright SP, Weber DJ (2003) Molecular mechanisms of S100-target protein interactions. *Microsc Res Tech* 60:552–559.

Zimmer DB, Weber DJ (2010) The calcium-dependent interaction of S100B with its protein targets. *Cardiovasc Psychiatry Neurol.* doi:10.1155/2010/728052.

# S-Adenosyl-L-methionine:Arsenic(III) Methyltransferase

▶ Arsenic Methyltransferases

# Samarium

Takashiro Akitsu
Department of Chemistry, Tokyo University of Science, Shinjuku-ku, Tokyo, Japan

## Definition

A lanthanoid element, the fifth element (cerium group) of the f-elements block, with the symbol Sm, atomic number 62, and atomic weight 150.36. Electron configuration [Xe] $4f^6 6s^2$. Samarium is composed of stable ($^{144}$Sm, 3.07%; $^{150}$Sm, 7.38%; $^{152}$Sm, 26.75%; $^{154}$Sm, 22.75%) and four radioactive ($^{146}$Sm; $^{147}$Sm, 14.99%; $^{148}$Sm, 11.24%; $^{149}$Sm, 13.82% ) isotopes. Discovered by L. de Boisbaudran in 1879. Samarium exhibits oxidation states III and II; atomic radii 180 pm, covalent radii 199 pm; redox potential (acidic solution) $Sm^{3+}/Sm$ −2.414 V; $Sm^{3+}/Sm^{2+}$ −1.000 V; electronegativity (Pauling) 1.17. Ground electronic state of $Sm^{3+}$ is $^6H_{6/2}$ with S = 5/2, L = 5, J = 5/2 with λ = 240 $cm^{-1}$. Most stable technogenic radionuclide $^{148}$Sm (half-life 7 × $10^{15}$ years). The most common compounds: $Sm_2O_3$, $SmF_3$ and $SmCl_3$. Biologically, samarium is of high toxicity, and $LD_{50}$ is less than 50 mg of $Sm_2O_3$ for animals to cause respiratory stoppage immediately (Atkins et al. 2006; Cotton et al. 1999; Huheey et al. 1997; Oki et al. 1998; Rayner-Canham and Overton 2006).

## Cross-References

▶ Lanthanide Ions as Luminescent Probes
▶ Lanthanide Metalloproteins
▶ Lanthanides and Cancer
▶ Lanthanides in Biological Labeling, Imaging, and Therapy
▶ Lanthanides in Nucleic Acid Analysis
▶ Lanthanides, Physical and Chemical Characteristics

## References

Atkins P, Overton T, Rourke J, Weller M, Armstrong F (2006) Shriver and Atkins inorganic chemistry, 4th edn. Oxford University Press, Oxford/New York

Cotton FA, Wilkinson G, Murillo CA, Bochmann M (1999) Advanced inorganic chemistry, 6th edn. Wiley-Interscience, New York

Huheey JE, Keiter EA, Keiter RL (1997) Inorganic chemistry: principles of structure and reactivity, 4th edn. Prentice Hall, New York

Oki M, Osawa T, Tanaka M, Chihara H (1998) Encyclopedic dictionary of chemistry. Tokyo Kagaku Dojin, Tokyo

Rayner-Canham G, Overton T (2006) Descriptive inorganic chemistry, 4th edn. W. H. Freeman, New York

## SAP, Aminopeptidase from *Streptomyces Griseus*

▶ Zinc Aminopeptidases, Aminopeptidase from Vibrio Proteolyticus (Aeromonas proteolytica) as Prototypical Enzyme

## Sarco(Endo)plasmic Reticulum $Ca^{2+}$-ATPase, SERCA

▶ Calcium ATPases

## Sarcoplasmic Calcium-Binding Protein

▶ Sarcoplasmic Calcium-Binding Protein Family: SCP, Calerythrin, Aequorin, and Calexcitin

## Sarcoplasmic Calcium-Binding Protein Family: SCP, Calerythrin, Aequorin, and Calexcitin

Jos A. Cox
Department of Biochemistry, University of Geneva, Geneva, Switzerland

## Synonyms

Aequorin; Calerythrin; Calexcitin; EF-hand proteins; Sarcoplasmic calcium-binding protein

## Definition

A family of intracellular calcium-binding proteins, spread over all forms of life, from bacteria to vertebrate neuronal tissues, have a very similar three-dimensional structure, characteristic, and unique for this family, and involved in a variety of functions including intracellular calcium buffering, bioluminescence, enzymatic activity and associative learning.

## Introduction

Calcium ions ($Ca^{2+}$) are essential for the normal function of the cell. A transient rise of the intracellular $Ca^{2+}$ concentration is achieved by its entering the cell through various plasma membrane $Ca^{2+}$ channels and/or by being liberated from various internal stores after an appropriate stimulus. The rise and fall of intracellular $Ca^{2+}$ is very versatile and different in different cell types, with very local ion gradients, hot spots, waves, and oscillations. ▶ Calcium-binding proteins (CaBP) have been selected evolutionarily with a range of affinities, kinetics, and capacities to intervene in cell-specific $Ca^{2+}$ signals. Most of the CaBPs involved in cell signaling (activation of membrane conductances, secretion, contraction, regulation of enzymes and genes) are present in the cytosol. The cytosol contains not only activator CaBPs (e.g., calmodulin, troponin C) for signaling but usually also a potent $Ca^{2+}$-buffering system composed of specialized CaBPs (e.g., parvalbumin, calbindin), which shape the $Ca^{2+}$ signal.

To date, four superfamilies of cytosolic CaBPs have been identified: (1) ▶ EF-hand CaBPs, (2) annexins, (3) C2 motif-containing proteins, and (4) gelsolin and villin. Among these, the EF-hand CaBPs are the most diversified and numerous. They contain two to six helix-loop-helix motifs with the loop acting directly in the coordination of $Ca^{2+}$ (and of $Mg^{2+}$). These motifs form pairs via short antiparallel β-sheets in the $Ca^{2+}$-binding loops; in the functional pairs, the two $Ca^{2+}$ ions are at a distance of 1 nm. The great majority of EF-hand CaBPs contain two such pairs, e.g., ▶ calmodulin, troponin C, myosin light chain, centrin, neuronal calcium sensor (NCS), parvalbumin, and sarcoplasmic $Ca^{2+}$-binding protein (SCP), but in several of these proteins, one or more EF hands are not functional anymore, i.e., they do not bind

$Ca^{2+}$ due to evolutionary mutation of one or more of the seven $Ca^{2+}$-coordinating, oxygen-containing amino acid residues, or due to truncating mutations as in parvalbumin. SCPs are members of the four EF-hand-containing CaBPs, with two pairs of pairs. They will be described in some detail in this entry in terms of their biochemical, biophysical, structural, and functional properties. The mini review published in 1995 (Hermann and Cox 1995) will be complemented with information on the SCP-related proteins calerythrin, aequorin, and calexcitin.

SCPs were discovered in the 1970s in the search for parvalbumins in crayfish muscle (Cox et al. 1976). The protein found was sufficiently different from parvalbumin to warrant its classification as a new EF-hand subfamily. It was isolated from the sarcoplasm, i.e., from the supernatant after centrifugation of a muscle homogenate extracted in physiologic ion salt concentrations. A quantitative analysis of the extractability of total amphioxus SCP indicates that it is a completely soluble protein. Their inability to form stable complexes with amphiphilic peptides or to bind to proteins when mixed with muscle extracts suggests that SCPs are not involved in protein-protein interaction. During further investigations in many phyla of the invertebrate kingdom, it became clear that parvalbumins are not expressed in invertebrates and that SCPs may be functional analogs of parvalbumin. Soluble SCPs were also discovered in other crustacea, in mollusks, protochordates, and insects, but comparison of invertebrate SCP sequences shows an unusually high degree of variability among these proteins, with only nine residues common to all species.

SCPs have polypeptide molecular weights of 20–22 kDa; they are monomeric with the exception of crayfish SCP which forms a very stable homodimer of 44 kDa; SCPs are acidic with isoelectric points of 4.7–5.3. They contain about 57% α-helical structure. SCPs possess four EF-hand domains per 22 kDa of which two to three are functional. Contrary to well-known CaBPs as calmodulin, troponin C, myosin light chain, centrin, NCS, calcineurin B, calpains, in the whole sarcoplasmic calcium-binding protein family, the four EF hands are not evenly distributed over the polypeptide chain: the length of the linkers between the EF hands decreases from EF-I/EF-II (20 aa), EF-II/EF-III (12 aa) to EF-III/EF-IV (5–6 aa); correspondingly, the α-helices surrounding the different $Ca^{2+}$-binding loops differ much more in length than in the above cited proteins. Similar to parvalbumin, in the absence of $Mg^{2+}$, the affinity of SCPs for $Ca^{2+}$ is quite high with a $K_D$ of about $10^{-7}$ M, and at physiological $Mg^{2+}$ levels of ca. 1 mM, the affinity for $Ca^{2+}$ decreases by more than one order of magnitude. In contrast to parvalbumin, SCPs show both positive and negative cooperativity in $Ca^{2+}$ binding, which is reenforced in the presence of $Mg^{2+}$.

## Crustacean SCPs

Crayfish (*Astacus pontastacus*) SCP (CSCP) was fairly well characterized in 1975–1982. It is a homodimer with four $Ca^{2+}$-$Mg^{2+}$-mixed sites and two $Ca^{2+}$-specific sites. Sequence analysis of CSCP indicated that the fourth EF-hand motif in each polypeptide does not bind $Ca^{2+}$ due to mutations in the $Ca^{2+}$-coordinating ligands. Comparison of the SCP concentrations in different muscles of crayfish shows a ratio of 1:10:40 for heart, claw, and tail muscle, with a $Ca^{2+}$-binding capacity of ca. 400 µM per kg wet weight for tail, which is considered to be a fast muscle. This is similar to parvalbumin in fish fast muscle. During ion exchange chromatography of sarcoplasm of different crustacea, CSCPs emerge in three peaks in the ratio of 14:2:1 for crayfish, 7:2:1 for lobster, and 3:2:1 for shrimp. However, there are only two different polypeptide chains α and β, which appear in the form of three dimers: $α_2$, αβ, and $β_2$. The tendency to dimerize is very strong since the monomer concentration is below the limit of detection, and the dimer is still stable in 6 M urea but not at 8 M. All three isotypes from both crayfish and lobster display the same metal-binding properties. The sequence of the crustacean β CSCP is presently not elucidated, but the differences between crayfish α and β seem minor from the HPLC profile after tryptic digestion. In crayfish α SCP, the conformational changes are concomitant with binding of the first $Ca^{2+}$ (per monomer), likely to the $Ca^{2+}$-specific sites. An SCP of the red swamp crayfish *Procambarus clarkii* was recently cloned and characterized in 2006. It shows 92.8% identity with α CSCP of *A. pontastacus*, with the greatest difference occurring in the region between EF-II and EF-III. In the shrimp *Penaeus orientalis*, the isotypes α and β were sequenced and found to show ca. 18% of sequence differences distributed all over the polypeptide chain.

In a search on allergen proteins in the Pacific white shrimp *Litopenaeus vannamei*, its SCP was cloned, which is a potent allergen specially for children (▶ Calcium in Health and Disease), and shows 93.8% sequence identity with *Penaeus* SCP α.

## Protochordate SCPs

Amphioxus SCP (ASCP) was first described in 1978. In contrast to crayfish SCP, the protein is monomeric, but sequence analysis revealed that the same three EF-hand motifs are functional, i.e., EF-I, EF-II, and EF-III. In the absence of $Mg^{2+}$, the affinity for $Ca^{2+}$ is very high with a $K_D$ of about $10^{-8}$ M; one $Mg^{2+}$ ion binds to amphioxus SCP with a $K_D$ of about $10^{-5}$ M. This $Mg^{2+}$ competes with $Ca^{2+}$ for only one of the ion-binding sites. In the protochordate amphioxus (*Branchiostoma lanceolatum*), there are several isotypes which can be separated by ion exchange chromatography. A total of seven isoforms was recorded with ASCP I and II, making up each 40–45% of the total amount and the more acid minor isoforms together 10–20%. Sequence analysis revealed that all the differences are restricted to nine positions in a 17-residue-long segment (the stretch 20–36), thus strongly suggesting that the isoforms are generated by alternative splicing of the primary RNA transcript with a mutually exclusive pattern. Interestingly, in this respect, two lineages of ASCP isoforms can be distinguished: an α lineage with ASCP I, III, and IV and a β lineage with ASCP II, V, VI, and VII. The $Ca^{2+}$-binding properties of the four more abundant SCPs were identical. Polyclonal antibodies raised against each ASCP I and II reacted equally with all the isoforms.

The crystal structure of ASCP I at 2.4 Å resolution was reported in 1993 (Cook et al. 1993). Despite the paired pair organization of its four EF hands, it showed a number of differences with previously described CaBPs, differences related to (1) the differences in the length of the linkers, (2) the absence of an extended α-helix between the two halves, and (3) the pairing of a functional to a nonfunctional EF-hand site. ASCP I is a compact molecule of approximately $25 \times 40 \times 40$ Å. The linker between EF-II and EF-III contains a tight turn which brings the two halves closely together in a mouth-to-mouth alignment, thus hiding nearly all the 23 aromatic residues in a dense hydrophobic core. The nonfunctional EF-IV has an overall conformation similar to a functional site, but important details in liganding are lacking, and no short β-pleated sheet is present in the C-terminal half. The long C-terminal loop contains a short region of $3_{10}$ helix, and three hydrophobic residues close to the C-terminal end are tucked into the hydrophobic core. As almost all members of the sarcoplasmic calcium-binding protein family, EF-I has an unusual Asp in the −Z position instead of the usual Glu, which provides a bidentate ligand: the first loop is indeed smaller than the second and third.

## Annelid SCPs

Sandworm (*Nereis diversicolor*) SCP (NSCP) was described in 1981 (Cox and Stein 1981). Sequence analysis showed that EF-I, EF-III, and EF-IV are canonical; however, EF-II is very different from a canonical, functional $Ca^{2+}$-binding site. Interestingly, this SCP binds three $Ca^{2+}$ in a noncooperative way with high affinity ($K_D = 6 \times 10^{-9}$ M) and binds also three $Mg^{2+}$ ions (▶ Magnesium Binding Sites in Proteins), at the same sites as $Ca^{2+}$. In the presence of 1 mM $Mg^{2+}$, the affinity for $Ca^{2+}$ is decreased to a $K_D$ of $10^{-7}$ M, but high positive cooperativity ($n_H$ equals 2.4) is induced. Microcalorimetric studies showed that binding of the first cation, either $Ca^{2+}$ or $Mg^{2+}$, induces enthalpy and conformational changes that are much larger than those of the subsequent ion-binding steps. This first $Ca^{2+}$ ion likely binds to EF-I. A study on the isolated N- and C-terminal domains (halves) of NSCP indicated that the N-terminal half with its single active EF-I is the stable structural nucleus of the protein, that the isolated domains form homodimers ($N_2$ and $C_2$) and that in the presence of $Ca^{2+}$, the equimolar mixture of the N- and C-terminal halves forms a heterodimeric molecule (N–C), with several properties identical to those of the native NSCP, except high affinity $Ca^{2+}$ binding. NMR experiments revealed a highly compact and rigid structure with the two halves in close contact through their hydrophobic surfaces.

The crystal structure of NSCP at 2-Å resolution was reported in 1992 (Vijay-Kumar and Cook 1992). NSCP is compact with dimensions of $25 \times 35 \times 40$ Å (Fig. 1). The nonfunctional EF-II has conserved several structural characteristics of a canonical EF hand, including a short β-strand paired to that of

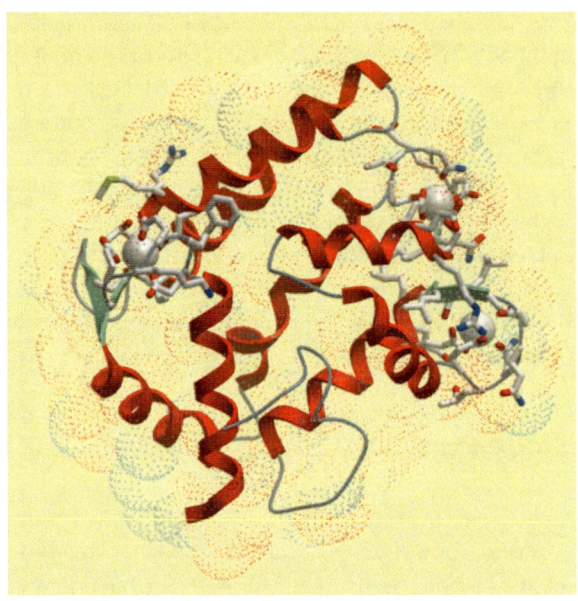

**Sarcoplasmic Calcium-Binding Protein Family: SCP, Calerythrin, Aequorin, and Calexcitin, Fig. 1** Structure of NSCP (PDB file 2SCP). The overall configuration is a C-terminal half (two $Ca^{2+}$) as an acorn in its cup (N-terminal half with one $Ca^{2+}$). In each half, there is a short β-pleated sheet (*green*). The N-terminus is at *bottom left* Drawn with assistance of Dr. E. Permyakov

EF-I, but at the C-terminal end, this loop is distinctly larger than a usual $Ca^{2+}$-binding loop. The first EF-hand site possesses an unusual Asp in the $-Z$ position (a bidentate ligand) instead of the usual Glu; as a consequence, the angle between the α-helices A and B is unusually small, and the first loop is smaller than the third and fourth, as in amphioxus SCP. There is a strong hydrophobic core with 20 aromatic residues. Similar to ASCP, the C-terminal residues 160–167 are hydrophilic and form a large loop on the surface of the molecule followed by hydrophobic residues which are tucked into the central hydrophobic core. The similarities between the three-dimensional structure of *Nereis* and amphioxus SCPs are striking considering the low degree of sequence similarity between the two proteins and the fact that the location of the nonfunctional EF hand is different in the two molecules (site IV in amphioxus and site II in *Nereis* SCP).

Removal of $Ca^{2+}$ from the NSCP crystals leads to a collapse of the crystal structure. NMR experiments enabled a better insight into the apo state and the $Ca^{2+}$-induced conformational changes. Apo-NSCP shows no elements of regular secondary or tertiary structure. Addition of one $Ca^{2+}$ to EF-I determines large spectral changes bringing the molecule into a conformation that is very close to the native Ca state. The molten globule structure of the apo state was confirmed by studies using thermal and chemical denaturation and the interaction with fluorescent hydrophobic probes. The highly fluctuating apo-NSCP is reminiscent of an equilibrium intermediate between the random coil and molten globule. Among CaBPs, this highly fluctuating state is quite unusual as NMR studies on metal-free calmodulin and troponin C showed that metal removal maintains the secondary and tertiary structure.

## Mollusk SCPs

The presence of SCP in the scallop striated muscle (*Patinopecten*) SCP was reported in 1983 (Collins et al. 1983). The monomeric protein possesses both $Ca^{2+}$-specific and $Ca^{2+}$-$Mg^{2+}$-mixed sites with a $K_D$ less than $10^{-8}$ M for $Ca^{2+}$ and $3 \times 10^{-6}$ M for $Mg^{2+}$. SCP of another scallop (*Patinopecten yessoensis*) binds only two $Ca^{2+}$ ions, and its amino acid sequence revealed that only EF-I and EF-III have canonical sequences to bind $Ca^{2+}$. These proteins are presently less well characterized.

## Insect SCPs

Using the same procedures as for crayfish, sandworm, and amphioxus, no SCP could be isolated or even detected in locust (*Schistocerca gregaria*) flight and leg muscle. But studies in 1992 and 1997 revealed the presence of a 23-kDa CaBP in *Drosophila melanogaster*, either in neuronal or muscular tissues. Later work revealed that, in fact, from the fruit fly two distinct cDNA for this protein were isolated: dSCP1 is abundantly present in thorax muscle, and dSCP2 is present exclusively in neuronal tissue. The former protein, present in different copies in the genome of the fly, has 55% sequence identity to shrimp SCP α and has canonical EF-I, EF-II, and EF-III $Ca^{2+}$-binding loops; in contrast, dSCP2, a single copy in the genome, possesses 27% of sequence identity to amphioxus SCP I and likely is a member of the calexcitin subfamily (see later).

## Calerythrin

Calerythrin, a 20-kDa $Ca^{2+}$-binding protein isolated from the gram-positive bacterium *Saccharopolyspora* (formerly called *Streptomyces*) *erythraea* (▶ Bacterial Calcium Binding Proteins), was the first prokaryotic protein shown to possess canonical EF-hand motifs. Among the four EF-hands, only EF-II does not possess a canonical $Ca^{2+}$-binding loop. Amino acid sequence analysis (Cox and Bairoch 1988), $Cd^{2+}$ NMR experiments, and global fold determination revealed that despite the rather low sequence identity (33.3% with NSCP), the protein structurally belongs to the family of sarcoplasmic calcium-binding proteins. The spacing of the EF hands displays the same irregularity between the EF hands as in NSCP with linkers of 20, 10, and 2 residues, respectively. The protein possesses three high-affinity $Ca^{2+}$-binding sites, two of which are strongly cooperative. Its NMR solution structure was reported in 2003 (Tossavainen et al. 2003). It confirmed that as in NSCP, EF-II is not active. As in NSCP, the two pairs of EF-hand loops are on opposite sides of the molecule. EF-I, which has an Asp instead of Glu in the −Z position, shows a smaller loop and a smaller A/B angle. The superposed structures of calerythrin and NSCP show an rms difference of 4.5 Å, due to a slightly different mutual orientation of the two halves. Contrary to NSCP, the short β-pleated sheet in the N-terminal half is not clearly visible in calerythrin, as in the case of the C-terminal half of amphioxus ASCP.

It is noteworthy that a protein highly homologous (66% sequence identity) to calerythrin is present in the prokaryote *Streptomyces ambofaciens*. Disruption of this protein or overexpression in *S. Ambofaciens* does not cause detectable change in phenotype. Similarly, we have no clues as to the role of calerythrin in *S. erythraea*. The homology to SCPs seems to speak for a $Ca^{2+}$-buffering role rather than a role as activator.

## Aequorin

A marked sequence homology exists also between NSCP and aequorin, the bioluminescent calcium-binding protein from the coelenterate *Aequorea aequorea* (Cox and Bairoch 1988). The four EF hands are asymmetrically distributed on the polypeptide chain with a pattern as in NSCP. EF-II is unable to bind $Ca^{2+}$. Aequorin consists of a 22-kDa apoprotein, the hydrophobic prosthetic coelenterazine moiety, and an oxygen molecule. Similar to NSCP, aequorin also binds three $Ca^{2+}$ ions with strong positive cooperativity and subsequently undergoes a series of reactions leading to the emission of photons. Thus, a well-defined biochemical function is attributed to this protein, in contrast to the $Ca^{2+}$-buffering role of genuine SCPs. Because it is very sensitive to $Ca^{2+}$ ions and does not interact with other proteins, aequorin is widely used as a probe to monitor intracellular levels of free $Ca^{2+}$ in living cells. Aequorin was purified in 1962, and the crystal structure of metal-free aequorin with bound coelenterazine was reported in 2000 (Head et al. 2000). The protein scaffold consists of four EF-hand domains arranged in pairs to form the compact globular molecule. A hydrophobic surface is present in each half of the molecule, situated within a cup formed by each pair of EF hands. These two cups come together in a mouth-to-mouth alignment, rather than in the offset arrangement of calmodulin gripping its target. The overall conformation is most similar to the homologous NSCP, but contrary to NSCP, there is in the center of aequorin a coelenterazine-binding cavity with a volume of 600 Å$^3$. The cavity is entirely filled with the substrate 2-hydroperoxycoelenterazine, which is oxidatively decarboxylated to coelenteramide. The crystal structures of the metal-free photoproteins, aequorin, and obelin, a very similar protein from *Obelia longissima*, with bound coelenteramide, and of their $Ca^{2+}$-bound form without any prosthetic coelenterazine group were reported. The $Ca^{2+}$-loaded structures retain the same compact scaffold and characteristic two-domain fold as the metal-free proteins with a very similar hydrophobic cavity. Without $Ca^{2+}$, EF-I has a configuration similar to that of the $Ca^{2+}$-loaded site, whereas EF-III and especially EF-IV have to move their coordinating residues a lot in order to become active. The absence of $Ca^{2+}$ binding to EF-II was confirmed in all the photoproteins. Thus, even if the general scaffold of the photoproteins is very similar to that of NSCP, the dynamic structure is very different. It is intriguing how a protein that buffers the $Ca^{2+}$ concentration in muscle cells has evolved to produce bioluminescence in photocytes.

## Calexcitin

The most recent CaBP, reported to be a member of the SCP family, is calexcitin, a protein which in squid plays a role in associative learning through inhibition of K$^+$ channels and activation of the ryanodine receptor, both in a Ca$^{2+}$-dependent manner (Nelson et al. 1999). Calexcitin has GTPase activity, and the recombinant protein binds Ca$^{2+}$ and is a high-affinity substrate for protein kinase C. The protein occurs in two isoforms: A and B, with the major difference residing in the last ten C-terminal residues, where the GTP seems to be anchored in calexcitin A. The sequence identity between A and B is 93%; however, calexcitin B was reported to have no GTP-binding site nor GTPases activity. Calexcitin from the optic lobe of squid (*Loligo pealei*) has four EF hands with the EF loops I, II, and III having canonical sequences, whereas EF loop IV is clearly inactive; the distribution of the EF hands over the polypeptide chain is asymmetric as in the genuine SCPs. The sequence identity of calexcitin A with that amphioxus muscle ASCP I is 31%; the sequence of calexcitin B displays 45% sequence identity with the neuronal form dSCP2 and 18% with the muscular form dSCP1 of *Drosophila melanogaster*. This likely explains why studies using antibodies raised against amphioxus SCP I reacted both in the neuronal and muscular tissues in *Drosophila* and in *Aplysia californica*.

The neuronal form of *Drosophila* SCP has 61% sequence identity with the juvenile hormone-selective diol kinase (JHDK) of the insect *Manduca sexta*. The latter kinase contains three canonical EF hands, EF-I (with Asp in the bidentate −Z position, as in genuine SCPs), EF-II, and EF-III. Its three-dimensional structure was modeled using NSCP as a template. The rms difference of 4.1 Å is reasonably good since a hydrophobic cavity was predicted in JHDK, although no such cavity is present in NSCP. The cavity was postulated to harbor ATP and the substrate. The role of Ca$^{2+}$ in JHDK is enigmatic since its enzymatic activity is Ca$^{2+}$ independent.

Calexcitin B is monomeric and binds three Ca$^{2+}$ ions with a K$_D$ of $8 \times 10^{-7}$ M and positive cooperativity (n$_H$ = 1.6) and one Mg$^{2+}$ ion with a K$_D$ of $10^{-5}$ M. As in the case of SCPs, the relation between the Ca$^{2+}$ and Mg$^{2+}$ interaction is complex: the data can be best interpreted as follows: calexcitin B possesses one Ca$^{2+}$-specific site and two Ca$^{2+}$/Mg$^{2+}$-mixed sites. Of the mixed sites, one is very sensitive to Mg$^{2+}$, whereas the other displays a weak Mg$^{2+}$ effect (Gombos et al. 2003). The thiol reactivity, circular dichroism, Trp fluorescence, thermal stability, and urea denaturation all indicate that the protein displays three distinct structures based on the type of ion that is bound, as do also the genuine SCPs. From ANS binding experiments, it can be inferred that the metal-free protein has molten globule characteristics, the Ca$^{2+}$ form displays a hydrophobic groove, likely involved in protein-protein interaction, and the Mg$^{2+}$ form is very compact without a hydrophobic patch.

The three-dimensional structure at 1.8 Å resolution of neuronal calexcitin from *Loligo pealei* was reported in 2006 (Erskine et al. 2006). It is a compact molecule of $28 \times 42 \times 42$ Å with a very pronounced hydrophobic core and a scaffold similar to that of ASCP. The eight α-helices, which are parts of the four EF hands, are of different length, and the long C-terminal loop contains a short $3_{10}$ helix followed by four hydrophobic residues that tuck in between the α-helix 7 and 8 of the inactive EF-hand IV. This region is strongly conserved in all calexcitins and may be directly implicated in their GTPase activities. The protein shows a sharp turn in the linker between EF-II and EF-III. EF-IV is paired to EF-III although there is no evidence of a β-pleated sheet between these two sites, as in ASCP.

## Conclusion

It is striking that all the members of the sarcoplasmic calcium-binding protein family are present only in invertebrates and in prokaryotes. It is also striking that the "genuine" SCPs are present only in muscle and given its abundance serve to buffer the intracellular Ca$^{2+}$ concentration, as do parvalbumins in vertebrate muscle. They apparently have no other well-defined function. The prokaryotic CaBPs very likely have the same physiological role. In contrast, aequorin in the photocyte is a Ca$^{2+}$-dependent enzyme with oxidative decarboxylase activity. It serves a precise role of communication in a very restricted group of invertebrates. More recently, a new regulatory CaBP of enzymes and channels, calexcitin, was discovered. It seems to be restricted to neuronal tissue in mollusks and insects. There, the Ca$^{2+}$-loaded calexcitin inhibits the activity of K$^+$ channels and activates the ryanodine receptor, a calcium channel in the endoplasmic reticulum; both actions are implicated in associative learning. Another member of the SCP family, more close to calexcitin, has a diol kinase activity in insects, but this enzymatic activity seems not to be regulated by $Ca^{2+}$. The reason why the SCP family is so successful from prokaryotes to amphioxus, a species very close to the vertebrates, and why this family was not retained in the vertebrates, is not understood. Despite their completely different functions, despite the low sequence homology, all these proteins have the same unique three-dimensional structure. If progress in SCP biology was rather slow in the 1970s, much has been learned in the last 10 years including unexpected twists and extensions, and much progress is expected in the next decade.

## Cross-References

▶ Bacterial Calcium Binding Proteins
▶ Calcium in Health and Disease
▶ Calcium-Binding Proteins
▶ Calmodulin
▶ EF-Hand Calcium-Binding Proteins
▶ Magnesium Binding Sites in Proteins

## References

Collins JH, Johnson DJ, Szent-Gyorgyi AG (1983) Purification and characterization of a scallop sarcoplasmic calcium-binding protein. Biochemistry 22:341–345

Cook WJ, Jeffrey LC, Cox JA, Vijay-Kumar S (1993) Structure of a sarcoplasmic calcium-binding protein from amphioxus refined at 2.4 Å resolution. J Mol Biol 229:461–471

Cox JA, Bairoch A (1988) Sequence similarities in calcium-binding proteins. Nature 331:491

Cox JA, Stein EA (1981) Characterisation of a new sarcoplasmic calcium-binding protein with magnesium-induced cooperativity in the binding of calcium. Biochemistry 20:5430–5436

Cox JA, Wnuk W, Stein EA (1976) Isolation and properties of a sarcoplasmic calcium-binding protein from crayfish. Biochemistry 15:2613–2618

Erskine PT, Beaven GDE, Hagan R, Findlow IS, Wermer JM, Wood SP, Vernon J, Giese KP, Fox G, Cooper JB (2006) Structure of the neuronal protein calexcitin suggests a mode of interaction in signalling pathways of learning and memory. J Mol Biol 357:1536–1547

Gombos Z, Durussel I, Ikura M, Rose DR, Cox JA, Chakrabartty A (2003) Conformational coupling of $Mg^{2+}$ and $Ca^{2+}$ on the three-state folding of calexcitin B. Biochemistry 42:5531–5539

Head JF, Inouye S, Teranishi K, Shimomura O (2000) The crystal structure of the photoprotein aequorin at 2.3 Å resolution. Nature 405:372–376

Hermann A, Cox JA (1995) Sarcoplasmic calcium-binding proteins. Comp Biochem Physiol 111B:337–345

Nelson TJ, Zhao WQ, Yuan S, Favi A, Pozzo-Miller L, Alkon DL (1999) Calexcitin interacts with neuronal ryanodine receptors. Biochem J 341:423–433

Tossavainen H, Permi P, Annila I, Kilpelainen I, Drakenberg T (2003) NMR solution structure of calerythrin, an EF-hand calcium-binding protein of *Saccharopolyspora erythraea*. Eur J Biochem 270:2505–2512

Vijay-Kumar S, Cook WJ (1992) Structure of a sarcoplasmic calcium-binding protein from *Nereis diversicolor* refined at 2.0 Å resolution. J Mol Biol 224:413–426

## Sarcoplasmic Reticulum (SR)

▶ Calcium Sparklets and Waves

## Scandium and Albumin

Xiao He
CAS Key Laboratory for Biomedical Effects of Nanomaterials and Nanosafety & CAS Key Laboratory of Nuclear Analytical Techniques, Institute of High Energy Physics, Chinese Academy of Sciences, Beijing, China

## Definition

This contribution refers to the features of albumin and its interactions with scandium. Albumin is commonly found in blood plasma, and is unique to other plasma proteins in that it is not glycosylated. Albumin is the most abundant plasma protein in mammals and is produced in the liver and forms a large proportion of all plasma proteins. The human version is human serum albumin, and it normally constitutes about 60% of human plasma proteins. Human serum albumin is a single peptide chain of 585 amino acids, held in three homologous domains by 17 disulfide bonds. Within each domain are two long loops plus one shorter loop. The S–S bonds provide stability while the intervening peptide strands allow for flexibility. The configuration includes 67% alpha helix and 10% beta turn.

The main function of albumin is to regulate the colloidal osmotic pressure needed for proper distribution of body fluids between intravascular compartments and body tissues. It also binds water, cations (such as $Ca^{2+}$, $Na^+$, and $K^+$), fatty acids, hormones, bilirubin, and drugs. The metal-transport site of serum albumin is one of the most extensively studied metal-binding sites of proteins, and albumin has been shown to bind a variety of essential and toxic metal ions, including $Ca^{2+}$, $Co^{2+}$, $Ni^{2+}$, $Cu^{2+}$, $Zn^{2+}$, and $Cd^{2+}$ (Bal et al. 1998; Banerjee and Lahiri 2009).

The binding of ScEDDHA to albumin is five times higher than that of the $ScCl_3$ while there is no binding of ScNTA, ScEDTA, ScDTPA, and ScCDTA to this protein. The ultrafiltration assays showed that the Sc chelated of high stability constant, such as ScEDTA, ScCDTA, and ScDTPA are completely filterable from human serum and do not bind to crystalline albumin. It also showed that ScEDDHA binds very effectively to albumin in equilibrium dialysis studies. Since EDDHA has phenolic groups, Rosoff and Spencer (1979) proposed that EDDHA could bind to the side chain of albumin forming a bridge between Sc and the protein. The concept is further supported by spectrophotometric studies, which showed that concentration of the EDDHA at equilibrium parallels that of the Sc.

## Cross-References

- ▶ α-Lactalbumin
- ▶ Iron Proteins, Transferrins and Iron Transport
- ▶ Parvalbumin
- ▶ Penta-EF-Hand Calcium-Binding Proteins
- ▶ Scandium, Biological Effects
- ▶ Scandium, Interactions with Actin
- ▶ Scandium, Interactions with Globulin
- ▶ Scandium, Interactions with Myosin Subfragment 1
- ▶ Scandium, Interactions with Nucleotide and Nucleic Acids
- ▶ Scandium, Interactions with Transferrin
- ▶ Scandium, Physical and Chemical Properties
- ▶ Zinc Homeostasis, Whole Body

## References

Bal W, Christodoulou J, Sadler P, Tucker A (1998) Multi-metal binding site of serum albumin. J Inorg Biochem 70:33–39

Banerjee A, Lahiri S (2009) Albumin metal interaction: a multielemental radiotracer study. J Radioanal Nucl Chem 279:733–741

Rosoff B, Spencer H (1979) Binding of rare earths to serum proteins and DNA. Clin Chim Acta 93:311–319

# Scandium, Biological Effects

Xiao He
CAS Key Laboratory for Biomedical Effects of Nanomaterials and Nanosafety & CAS Key Laboratory of Nuclear Analytical Techniques, Institute of High Energy Physics, Chinese Academy of Sciences, Beijing, China

## Definition

The biological effects of scandium refer to the bioactivity, behavior, and toxicity of Scandium element or compounds in cells, tissues, organs, and organisms. In general, scandium is present in low abundance in soil and water bodies, and it is not an essential element for organisms. The bioavailability of scandium is very low, so it is considered as a metal of low toxicity. Nevertheless, the contents of scandium are detectable in a wide variety of tissues in many species including human beings. Several biochemical effects of Sc have been recognized for decades and widely utilized for the replacement of toxic heavy metals in technological applications. However, Sc is not entirely free of toxicity. Mechanisms of Sc toxicity may include substitution for essential metals, enzyme inhibition, elemental dyshomeostasis, and other indirect effects. Therefore, it is necessary to study the biological effects of scandium.

Previous studies on the metabolism of scandium were initiated through possible hazards arising from its presence in nuclear explosion sites, and $^{46}Sc$ was always used as radioactive tracer. Lachine et al. (1976) studied the acute toxicity, differential distribution in various tissues, and elimination of $^{46}ScCl_3$ and $^{46}Sc$-EDTA in mice. The $LD_{50}$ doses for $ScCl_3$ at 24 h postexposure are 440 and 24 mg/kg, respectively, via intraperitoneal and intravenous injection, and 720 and 108 mg/kg, respectively, for Sc-EDTA. $ScCl_3$ is extensively deposited in the liver and spleen, while Sc-EDTA is rapidly taken up by the kidney with subsequent elimination via the urine. Whole-body desaturation kinetics for Sc-EDTA are found to fit a three compartmental model. When human is exposed to the weak chelated scandium via intravenous injection, the plasma level of Sc decreases very slowly; it is excreted principally via the intestine. While the strong chelated Sc is administered, Sc disappears rapidly from the vascular space and urinary excretion is high, with 82% of the given dose being excreted in 24 h (Rosoff and Spencer 1965).

In male wistar rats exposed to scandium chloride via intraperitoneal injection, only 0.0063% of administered Sc is dose-dependently excreted via urine within 24 h (Tanida et al. 2009). The low Sc elimination via kidney might reflect the overflow of administered Sc from the storage capacity of the retention in the reticuloendothelial phagocytic clearance system. The administration of Sc would induce a significant decrease of urine volume and creatinine, and a significant increase of β-2-microglobulin and N-acetyl-beta-D-glucosaminidase. These results suggest that the urinary excretion of Sc can be a useful tool for monitoring Sc exposure in occupational and environmental health screenings. The formation of Sc colloidal conjugates that deposit in glomeruli might be the cause of a reduction of the glomerular filtration rate.

The toxic effects of scandium on liver, kidney, lung, eyes, skin, and other organs were described in detail by Horovitz (2000). Histopathological examination of the heart, lung, liver, kidney, spleen, pancreas, adrenals, and small intestine of rats revealed no substantial changes that could be related to the ingestion of $ScCl_3$ at dietary levels. $ScCl_3$ could dose-dependently depress the intestinal tonus and contractility, terminating in complete ileal paralysis in rabbits, and block the circular and longitudinal muscle contractions in guinea pigs. Application of $ScCl_3$ to rabbit eyes could induce immediate or delayed effects on the cornea, conjunctiva, and iris. Direct application of $ScCl_3$ produces no reaction on intact rabbit skin whereas a maximum irritation index of 8 on abraded skin. Intravenous injection of $ScCl_3$ would not cause skin calcification at the site of injection, but might cause a sensitization with extensive thrombosis of the ophthalmic and intracranial veins.

Although it is quite different from the typical view about scandium's biological effects in the literature, a few reports suggested that scandium is not cytotoxic and does not have a deleterious effect on cell metabolic activity (Herath et al. 2005). Irrespective of the surface topography promoted HOS cell attachment on scandium oxide at a level comparable to the control. Quantitatively, a significant increase in cell proliferation was observed on $Sc_2O_3$ compared to the control. Furthermore, $Sc_2O_3$ was shown to be nontoxic, be able to maintain cell viability and support cell growth and proliferation. However, the above results are only preliminary findings and further study is needed.

Scandium has similar phytotoxic effects to those caused by aluminum. It has adverse effects on mitosis and root elongation (Blamey et al. 2011; Kopittke et al. 2009). Another report showed that Sc is tenfold more toxic to wheat (*Triticum aestivum*) roots than mononuclear aluminum on a concentration basis (Kinraide 1991). The speciation of Sc in this representative experiment is unknown, but if polynuclear hydroxy-Sc species are present, they are about as toxic as the exceedingly toxic triskaidekaaluminium $(AlO_4Al_{12}(H2O)_{12})^{7+}$. Chlorosis and substantially reduced chlorophyll concentration were observed in young leaves of honey locust (Gleditsia triacanthos L.) and loblolly pine (Pinus taeda L.) exposed to 10–100 μM and 10–50 μM Sc, respectively (Yang et al. 1989). The mechanism for the phytotoxicity of scandium remains obscure but may be linked to changes in water uptake, cell turgor, and cell wall extensibility (Kopittke et al. 2009; Mckenna et al. 2010; Reid et al. 1996).

The microorganisms are susceptible to scandium at high concentrations (at levels of mM), although the mechanism underlying the inhibitory effect is not fully understood. *Streptomyces* spp. (*S. coelicolor* and *S. lividans*), typical soil bacteria, are most susceptible (minimum inhibitory concentration (MIC) = 1 mM), while *Mycobacterium smegmatis* (MIC = 1.5 mM) and *Escherichia coli, Staphylococcus aureus,* and *Bacillus subtilis* (MICs = 3 mM) are less susceptible (Kawai et al. 2007). Rogers (1987) found that $Sc^{3+}$, as a free ion, lacks antibacterial activity in pathogenic serotypes of *Escherichia coli*. However, when $Sc^{3+}$ is introduced as complex of enterochelin, *Escherichia coli* is immediately killed. The $Sc^{3+}$-enterochelin complex could also inhibit growth of *Pseudomonas aeruginosa* in serum and exert a therapeutic effect on *Pseudomonas aeruginosa* infections in mice. The bacteriostasis could be reversed upon addition of deferri-pseudobactin (Rogers 1987). As an analog of $Fe^{3+}$, Sc could inhibit growth of the *Pseudomonas* in iron-limited media, and the trivalent metals are listed in order of decreasing toxicity as follows: Ga > In > Sc > Cr > Y > Al (Fekete and Barton 1991). In contrast to bacteria, the eukaryotic microorganisms *Saccharomyces cerevisiae* (MIC = 5 mM) and Aspergillus oryzae (MIC = 15 mM) are relatively resistant to scandium.

At a low concentration (10–100 μM), culture with scandium causes antibiotic overproduction by 2–25-fold in *Streptomyces coelicolor* A3(2), *Streptomyces antibioticus*, and *Streptomyces griseus* (Kawai et al. 2007). Scandium is also effective in activating the dormant ability to produce actinorhodin in *Streptomyces lividans*. The effects of scandium are exerted at the level of transcription of pathway-specific positive regulatory genes, as demonstrated by marked upregulation of *act*II-ORF4 in *Streptomyces lividans* cells exposed to scandium. The bacterial alarmone, guanosine 5′-diphosphate 3′-diphosphate, is essential for actinorhodin overproduction provoked by scandium. The compelling effect of low levels of scandium on antibiotic production implies that scandium functions in situ as a factor that induces or stimulates the production of secondary metabolites, which can include pigments, mycotoxins, phytotoxins, and antibiotics. Although the mechanism of action remains to be clarified, it is possible that scandium acts on the ribosome, eventually leading to modulation of ribosomal function, because addition of scandium immediately reduced the level of ppGpp, which is synthesized on the ribosome.

Scandium is a potential antiviral agent. Sulfonated scandium diphthalocyanine ($ScPc_2(SO_3)_4^{4-}$), with less toxic than remantadine, has a pronounced antiviral activity (in respect of the flu virus and Rous sarcoma) (Kulvelis et al. 2007).

## Cross-References

▶ Lanthanide Metalloproteins
▶ Lanthanides, Toxicity
▶ Scandium, Interactions with Nucleotide and Nucleic Acids
▶ Scandium, Physical and Chemical Properties

## References

Blamey F, Kopittke P, Wehr J et al (2011) Recovery of cowpea seedling roots from exposure to toxic concentrations of trace metals. Plant Soil. doi:10.1007/s11104-010-0655-0:1-14
Fekete F, Barton L (1991) Effects of iron(III) analogs on growth and pseudobactin synthesis in a chromiumtolerant *Pseudomonas* isolate. Biometals 4:211–216
Herath H, Silvio L, Evans J (2005) Scandia-A potential biomaterial? J Mater Sci Mater Med 16:1061–1065
Horovitz C (2000) Biochemistry of scandium and yttrium. Kluwer/Plenum, New York
Kawai K, Wang G, Okamoto S et al (2007) The rare earth, scandium, causes antibiotic overproduction in *Streptomyces* spp. FEMS Microbiol Lett 274:311–315
Kinraide T (1991) Identity of the rhizotoxic aluminium species. Plant Soil 134:167–178
Kopittke P, Mckenna B, Blamey F et al (2009) Metal-induced cell rupture in elongating roots is associated with metal ion binding strengths. Plant Soil 322:303–315
Kulvelis Y, Lebedev V, Torok D et al (2007) Structure of the water salt solutions of DNA with sulfonated scandium diphthalocyanine. J Struct Chem 48:740–746
Lachine E, Noujaim A, Ediss C et al (1976) Toxicity, tissue distribution and excretion of $^{46}ScCl_3$ and $^{46}Sc$-EDTA in mice. Int J Appl Radiat Isot 27:373–377
Mckenna B, Kopittke P, Wehr J et al (2010) Metal ion effects on hydraulic conductivity of bacterial cellulose–pectin composites used as plant cell wall analogs. Physiol Plant 138:205–214
Reid R, Rengel Z, Smith F (1996) Membrane fluxes and comparative toxicities of aluminium, scandium and gallium. J Exp Bot 47:1881–1888
Rogers H (1987) Bacterial iron transport as a target for antibacterial agents. In: Winkelmann G, Helm D, Neilands J (eds) Iron transport in microbes, plants and animals. VCH Verlagsgesellschaft, Weinheim
Rosoff B, Spencer S (1965) Metabolism of scandium-46 in man. Int J Appl Radiat Isot 16:479–485
Tanida E, Usuda K, Kono K et al (2009) Urinary scandium as predictor of exposure: effects of scandium chloride hexahydrate on renal function in rats. Biol Trace Elem Res 130:273–282
Yang C, Schaedle M, Tepper H (1989) Phytotoxicity of scandium in solution culture of loblolly pine (*Pinus taeda* L.) and honey locust (*Gleditsia triacanthos* L.). Environ Exp Bot 29:155–164

# Scandium, Interactions with Actin

Xiao He
CAS Key Laboratory for Biomedical Effects of Nanomaterials and Nanosafety & CAS Key Laboratory of Nuclear Analytical Techniques, Institute of High Energy Physics, Chinese Academy of Sciences, Beijing, China

## Definition

This entry refers to the features of actin and its interactions with scandium. Actin, a globular, roughly 42-kDa protein, is the component of the cytoskeletal system that allows movement of cells and cellular processes. It is the most abundant protein in the typical eukaryotic cells where it may be present at concentrations of over 100 μM. It is also one of the most highly conserved proteins, differing by no more than 20% in species as diverse as algae and humans. In vertebrates, three main groups of actin isoforms, alpha, beta, and gamma have been identified.

The elements from $Ce^{3+}$ to $Ho^{3+}$ have the ability to induce the formation of actin tubes and microcrystals, whereas $La^{3+}$ and the heaviest ions, $Er^{3+}$ to $Lu^{3+}$, do not. The previous work suggested that actin tubes are induced by trivalent cations, principally on the basis of their binding stoichiometry, which in turn is determined by ionic radius. $Sc^{3+}$ is not able to induce crystalline actin tubes but form amorphous aggregates, because the ionic radius of $Sc^{3+}$ (87 pm when the coordination number is 8) is smaller than the smallest lanthanide cations ($Ln^{3+}$). The high resolution $^1$H-NMR spectroscopy showed that the first $Sc^{3+}$ binds to a site on actin that is inaccessible to $Mg^{2+}$, $Y^{3+}$, and $Ln^{3+}$. However, the second $Sc^{3+}$ to bind to actin behaves exactly like $Y^{3+}$ and $Ln^{3+}$. When $Sc^{3+}$ saturates its binding sites on actin and when the ionic strength is raised to 0.1 M with KCl at pH 6.9, $Sc^{3+}$ binds with a ratio of 8:1 and induced amorphous actin aggregates (Barden et al. 1981).

## Cross-References

▶ Scandium, Biological Effects
▶ Scandium and Albumin
▶ Scandium, Interactions with Globulin
▶ Scandium, Interactions with Myosin Subfragment 1
▶ Scandium, Interactions with Nucleotide and Nucleic Acids
▶ Scandium, Interactions with Transferrin
▶ Scandium, Physical and Chemical Properties

## References

Barden J, Curmi P, Dos Remedios C (1981) Crystalline actin tubes: III. The interaction of scandium and yttrium with skeletal muscle actin. Biochim Biophys Acta 671:25–32

# Scandium, Interactions with Globulin

Xiao He
CAS Key Laboratory for Biomedical Effects of Nanomaterials and Nanosafety & CAS Key Laboratory of Nuclear Analytical Techniques, Institute of High Energy Physics, Chinese Academy of Sciences, Beijing, China

## Definition

This entry refers to the features of globulin and its interactions with scandium. Globulin is one of the three types of serum proteins, the others being albumin and fibrinogen. Globulins can be divided into three fractions based on their electrophoretic mobility. Most of the α- and β-globulins are synthesized by the liver, whereas γ-globulins are produced by lymphocytes and plasma cells in lymphoid tissue. α- and β-globulins are transport proteins, serve as substrates upon which other substances are formed, and perform other diverse functions. γ-globulins have a vital role in natural and acquired immunity to infection.

Scandium could bind to α-globulin. Rosoff and Spencer (1979) studied the binding of two Sc chelates to α-globulin and albumin in comparison to its chloride salt. The binding of ScEDDHA to α-globulin is about one-third that of the $ScCl_3$ and the binding of the ScNTA is 1/20 that of $ScCl_3$. $ScCl_3$ and ScNTA are proved to bind more strongly to α-globulin than to albumin in equilibrium dialysis studies. The β-globulin in its pure form and its fraction in plasma are both shown to bind $Sc^{3+}$ (Perkins 1966).

## Cross-References

▶ Iron Proteins, Transferrins and Iron Transport
▶ Scandium, Interactions with Globulin
▶ Scandium, Biological Effects
▶ Scandium, Interactions with Actin
▶ Scandium and Albumin
▶ Scandium, Interactions with Myosin Subfragment 1
▶ Scandium, Interactions with Nucleotide and Nucleic Acids

- Scandium, Interactions with Transferrin
- Scandium, Physical and Chemical Properties

## References

Perkins D (1966) The interaction of trivalent metal ions with human transferrin. In: Peeters J (ed) Protides of the biological fluids, vol 14. Elsevier, Amsterdam

Rosoff B, Spencer H (1979) Binding of rare earths to serum proteins and DNA. Clin Chim Acta 93:311–319

# Scandium, Interactions with Myosin Subfragment 1

Xiao He
CAS Key Laboratory for Biomedical Effects of Nanomaterials and Nanosafety & CAS Key Laboratory of Nuclear Analytical Techniques, Institute of High Energy Physics, Chinese Academy of Sciences, Beijing, China

## Definition

This entry refers to the features of myosin subfragment 1 (S1) and its interactions with scandium. S1 is the globular head region of the myosin produced by proteolytic cleavage of the intact myosin molecule. It contains one heavy chain compactly folded into a motor domain and an extended regulatory or light chain domain wrapped by two light chains. S1 contains the ATPase and actin binding sites. The energy for muscle contraction is derived from the cyclic attachment and detachment of S1 to the actin filament with the concomitant hydrolysis of MgATP. The series of events occurred during the contractile cycle is: MgATP rapidly releases myosin from actin by binding to the ATPase site of myosin; then, myosin hydrolyzes ATP to form a relatively stable intermediate myosin-ADP·Pi; and finally, actin recombines with this complex and dissociates the products, thereby forming the original actin-myosin complex.

Gopal and Burke (1995) found that the existence of scandium (presumably $ScF_x$) could lead to effective trapping of ADP in S1 in both the presence and absence of $Mg^{2+}$ with concomitant inactivation of the ATPase properties. Analysis for the amount of scandium present in these complexes indicated that 2 mol of scandium are bound to per mol of S1. No high affinity binding site exists for Sc when ADP is absent. It implies that the ATPase-inhibitory complexes with myosin S1 are ternary complexes designated Myosin-MgADP·$ScF_x$ and Myosin-ADP·$ScF_x$. At 4°C and 25°C in the presence of actin, the complex formed with ADP appears to have a higher stability than the one formed with MgADP.

## Cross-References

- Iron Proteins, Transferrins and Iron Transport
- Scandium and Albumin
- Scandium, Biological Effects
- Scandium, Interactions with Actin
- Scandium, Interactions with Globulin
- Scandium, Interactions with Nucleotide and Nucleic Acids
- Scandium, Interactions with Transferrin
- Scandium, Physical and Chemical Properties

## References

Gopal D, Burke M (1995) Formation of stable inhibitory complexes of myosin subfragment 1 using fluoroscandium anions. J Biol Chem 270:19282–19286

# Scandium, Interactions with Nucleotide and Nucleic Acids

Xiao He
CAS Key Laboratory for Biomedical Effects of Nanomaterials and Nanosafety & CAS Key Laboratory of Nuclear Analytical Techniques, Institute of High Energy Physics, Chinese Academy of Sciences, Beijing, China

## Definition

This entry refers to the interactions between scandium element/compounds and nucleic acids. Metal ion-nucleotide complexes have significant chemical and biological properties. Their structures have been studied extensively by various physical techniques such as NMR, IR, UV, and others.

Shyy et al. (1985) studied the interaction of scandium with adenosine 5′-triphosphate (ATP) by $^{17}$O NMR, $^{31}$P NMR, and $^1$H NMR, and confirmed the formation of 1:2 Sc(III)/ATP complex. The Sc(III)/ATP complexes are mixtures of rapidly exchanging diastereomers. The $^{17}$O NMR results showed that binding of Sc induced a small chemical shift effect and a large broadening effect for the line widths of [α-$^{17}$O$_2$]ATP, [β-$^{17}$O$_2$]ATP, and [γ-$^{17}$O$_3$] ATP. Comparison of the relative magnitudes of the line-broadening effect for all of the three phosphates of ATP suggested that the predominant macroscopic structure of Sc$^{3+}$(ATP)$_2$ is the α, β, γ-tridentate. Such a conclusion is further supported by $^1$H NMR. The strength of the interaction between ATP and Sc$^{3+}$ was much stronger than with Mg$^{2+}$ or Mn$^{2+}$, due to the decreased ionic radius. The ATP complex of scandium also functions as a slow-binding inhibitor of the yeast hexokinase reaction at pH 8.0 (Horovitz 2000).

In the pH range of 6.5–8.8, scandium(III)-8-hydroxyquinoline (8-HQL) can form ternary fluorescence systems with nucleic acids including native and thermally denatured calf thymus DNA, fish sperm DNA, and yeast RNA (Huang et al. 1997). All the ternary systems can emit fluorescence of 490 ~ 496 nm depending on the kind of nucleic acids. The ternary complexes of S(III)-8-HQL have larger fluorescence quantum yield and longer fluorescence lifetime than the binary complex. The analytical results suggested that the Sc(III)-8-HQL method is sensitive to the determination for DNA, but not for RNA.

Phthalocyanine is a synthetic analog of porphyrins. It forms complex compounds metallophthalocyanines by reactions with metal salts. Kulvelis et al. (2007) found that sulfonated scandium diphthalocyanine could interact with DNA and packaging of DNA molecules occurs in water-salt solution after addition of the stated dyes. The complex of ScPc$_2$(SO$_3$)$_4^{4-}$-DNA is stable over the temperature range 20–40°C. Complexation of the negatively charged DNA molecules with ScPc$_2$(SO$_3$)$_4^{4-}$ may be due to the formation of a double layer of ions around DNA, with the first positively charged layer of counterions attracting the negatively charged diphthalocyanine molecules. One can also assume that the conformation of DNA changes under the action of diphthalocyanine because of changes in its secondary structure.

## Cross-References

▶ Lanthanide Metalloproteins
▶ Lanthanides, Toxicity
▶ Scandium, Biological Effects
▶ Scandium, Physical and Chemical Properties

## References

Horovitz C (2000) Biochemistry of scandium and yttrium. Kluwer/Plenum, New York

Huang C, Li K, Tong S (1997) Fluorescence features of scandium (III)-8-hydroxyquinoline-nucleic acid systems and their analytical applications. Chin J Anal Chem 25:759–764

Kulvelis Y, Lebedev V, Torok D et al (2007) Structure of the water salt solutions of DNA with sulfonated scandium diphthalocyanine. J Struct Chem 48:740–746

Shyy Y, Tsai T, Tsai M (1985) Metal-nucleotide interactions. 3. $^{17}$O, $^{31}$P, and $^1$H NMR studies on the interaction of scandium (III), lanthanum (III), and lutetium (III) with adenosine 5′-triphosphate. J Am Chem Soc 107:3478–3484

# Scandium, Interactions with Transferrin

Xiao He
CAS Key Laboratory for Biomedical Effects of Nanomaterials and Nanosafety & CAS Key Laboratory of Nuclear Analytical Techniques, Institute of High Energy Physics, Chinese Academy of Sciences, Beijing, China

## Definition

This entry refers to the features of transferrin and its interactions with scandium. Transferrin is a glycoprotein found in blood plasma and milk that binds iron very tightly but reversibly to control the level of free iron in biological fluids. It has a molecular weight of around 80 kDa and contains two specific high-affinity Fe(III) binding sites. The liver is the main source of manufacturing transferrin, but other sources, e.g., the brain, also produce this molecule. The main role of transferrin is to transport iron through the blood to the liver, spleen, and bone marrow.

The transferrin was shown to bind a number of trivalent metal ions, including scandium (Perkins 1966). In vitro experiments showed that Sc$^{3+}$ is bound specifically onto the iron-binding sites of transferrin, conferring on the protein similar stability to denaturants and chemical reagents as iron (Ford-Hutchinson and Perkins 1971). The binding of each $Sc^{3+}$ involves the ionization of two phenolic tyrosyl residues. Further tests with human serum showed that all the protein-bound $Sc^{3+}$ was bound to transferrin with concentrations of $Sc^{3+}$ less than the free iron-binding capacity of the serum. General binding of $Sc^{3+}$ to all protein fractions was observed when the concentrations of $Sc^{3+}$ were greater than the free iron-binding capacity. In vivo results showed that 75% of the $Sc^{3+}$ in the plasma was associated with the small molecular weight band and 25% with the transferrin at 10 min after the injection of radioactive $^{46}Sc$. At 1 h post-injection, these figures were reversed and 24 h post-injection 95% of the metal was bound to the transferrin. After 48 h all detectable radioactivity was associated with the transferrin band. This work shows that transferrin plays a major role in the transport of scandium in the plasma and is the only scandium-serum protein complex demonstrable.

In a later report, the binding constants for $Sc^{3+}$ binding to human serum transferrin were determined and the bicarbonate-independent binding constants were $14.6 \pm 0.2$ ($\log K_1^*$, C-lobe) and $13.3 \pm 0.3$ ($\log K_2^*$, N-lobe), respectively (Li et al. 1996). It suggests that even though $Sc^{3+}$ has only slightly larger ionic radius than $Fe^{3+}$ (0.075 nm vs. 0.065 nm), it binds to the C-lobe and N-lobe sites much more weakly.

In another work, high-field quadrupolar NMR spectroscopy was employed to monitor the binding of $Sc^{3+}$ to the metal ion binding sites in chicken ovotransferrin (Aramini and Vogel 1994). In the presence of carbonate, two $^{45}Sc$ and $^{13}C$ signals were observed, which could be assigned using the proteolytic half-molecules of ovotransferrin to bound $Sc^{3+}$ and $^{13}CO_3^{2-}$ onto both metal ion binding sites of the protein. No site preference for $Sc^{3+}$ binding to ovotransferrin was found when either carbonate or oxalate served as the synergistic anion. The competition experiments showed that ovotransferrin binds $Sc^{3+}$ with a higher affinity than $Al^{3+}$ and even $Ga^{3+}$.

## Cross-References

▶ Iron Proteins, Transferrins and Iron Transport
▶ Scandium and Albumin
▶ Scandium, Biological Effects
▶ Scandium, Interactions with Actin
▶ Scandium, Interactions with Globulin
▶ Scandium, Interactions with Myosin Subfragment 1
▶ Scandium, Interactions with Nucleotide and Nucleic Acids
▶ Scandium, Physical and Chemical Properties

## References

Aramini J, Vogel H (1994) A scandium-45 NMR study of ovotransferrin and its half-molecules. J Am Chem Soc 116:1988–1993

Ford-Hutchinson A, Perkins D (1971) The binding of scandium ions to transferrin in vivo and in vitro. Eur J Biochem 21:55–59

Li H, Sadler P, Sun H (1996) Rationalization of the strength of metal binding to human serum transferrin. Eur J Biochem 242:387–393

Perkins D (1966) The interaction of trivalent metal ions with human transferrin. In: Peeters J (ed) Protides of the biological fluids, vol 14. Elsevier, Amsterdam

# Scandium, Physical and Chemical Properties

Fathi Habashi
Department of Mining, Metallurgical, and Materials Engineering, Laval University, Quebec City, Canada

Although scandium is the first member of the group of metals that include the lanthanides it is not recovered from rare earth minerals such as monazite or bastnasite but is recovered mainly as a by-product from uranium ores and ilmenite processing. It has many things in common with aluminum, for example, it is as light as aluminum (density of aluminum 2.70 density of scandium 2.99), it is trivalent, and all its compounds are also colorless. However, it is not a reactive metal like aluminum but it is the first member of the transition metals hence there is difference in the electronic structure of both metals (Fig. 1). It is located immediately below aluminum in the Periodic Table.

## Physical Properties

| Atomic number | 21 |
|---|---|
| Atomic weight | 44.96 |

(continued)

| | |
|---|---|
| Relative abundance in Earth's crust, % | $5 \times 10^{-4}$ |
| Density, g/cm$^3$ | 2.989 |
| Melting point, °C | 1,540 |
| Boiling point, °C | 2,832 |
| Crystal structure | Hexagonal |
| Atomic radius, pm | 162 |
| Heat of fusion, kJ·mol$^{-1}$ | 14.1 |
| Heat of vaporization, kJ·mol$^{-1}$ | 332.7 |
| Molar heat capacity, J·mol$^{-1}$·K$^{-1}$ | 25.52 |
| Thermal expansion, μm/(m·K) | 10.2 |

| Li | Be | | | |
|---|---|---|---|---|
| 2 | 2 | | | |
| 1 | 2 | | | |
| Na | Mg | Al | | |
| 2 | 2 | 2 | | |
| 8 | 8 | 8 | | |
| 1 | 2 | 3 | | |
| K | Ca | Sc | Ti | V |
| 2 | 2 | 2 | 2 | 2 |
| 8 | 8 | 8 | 8 | 8 |
| 8 | 8 | 9 | 10 | 11 |
| 1 | 2 | 2 | 2 | 2 |
| Rb | Sr | Y | Zr | Nb |
| 2 | 2 | 2 | 2 | 2 |
| 8 | 8 | 8 | 8 | 8 |
| 18 | 18 | 18 | 18 | 18 |
| 8 | 8 | 9 | 10 | 11 |
| 1 | 2 | 2 | 2 | 2 |
| Cs | Ba | La† | Hf | Ta |

**Scandium, Physical and Chemical Properties, Fig. 1** A portion of the Periodic Table showing the position of scandium and its neighbors: the typical metals (*red*) and the transition metals (*yellow*)

## Chemical Properties

Although scandium has two electrons in the outermost shell and is therefore expected to be divalent it is however, like aluminum, is trivalent in all its compounds. Like aluminum, the oxide $Sc_2O_3$ and the hydroxide $Sc(OH)_3$ are amphoteric:

$$Sc(OH)_3 + 3OH^- \rightarrow Sc(OH)_6^{3-}$$
$$Sc(OH)_3 + 3H^+ + 3H_2O \rightarrow [Sc(H_2O)_6]^{3+}$$

Scandium is produced by reducing scandium fluoride with calcium:

$$2ScF_3 + 3Ca \rightarrow 2Sc + 3CaF_2$$

The metal has little commercial importance. The major application is expected to be as an alloying element with aluminum since it is a light metal. The combination of high strength and light weight makes Al-Sc alloys suitable for a number of applications. However, titanium being much more common, and similar in lightness and strength, is much more widely used. The Russian military aircrafts Mig 21 and Mig 29 used aluminum scandium alloys. In Europe and the USA, scandium-containing alloys have been evaluated for use in structural parts in airplanes.

## References

Cotton S (2006) Lanthanide and actinide chemistry. Wiley, New York

Favorskaya L (1969) Chemical technology of scandium. Kazakh Research Institute of Mineral Resources, Alma-Ata, Kazakhstan

Habashi F (2008), Researches on rare earths. History and technology. Métallurgie Extractive Québec, Québec City. Distributed by Laval University Bookstore.www.zone.ul.ca

Horovitz CT, Gschneidner KA Jr (1975) Scandium, its occurrence, chemistry, physics, metallurgy, biology, and technology. Academic, New York

McGill I (1997) Rare earth metals. In: Habashi F (ed) Handbook of extractive metallurgy. Wiley-VCH, Weinheim, pp 1693–1741

## Scheelite

▶ Tungsten Cofactors, Binding Proteins, and Transporters in Biological Systems

## SCN

▶ Sodium Channels, Voltage-Gated

## SDVs – Silicon Deposition Vehicles

▶ Silicon, Biologically Active Compounds

## Sec

▶ Selenium, Biologically Active Compounds

## SECIS

▶ Selenium, Biologically Active Compounds

## Second Messenger

▶ Calcium in Nervous System

## Second Messenger Signaling

▶ C2 Domain Proteins

## Second Zinc Ion

▶ Zinc Structural Site in Alcohol Dehydrogenases

## Secondary Active Transporter

▶ Sodium-Coupled Secondary Transporters, Structure and Function

## Secondary Transporter

▶ Sodium-Coupled Secondary Transporters, Structure and Function

## Secondary-Active Transporters

▶ Sodium/Glucose Co-transporters, Structure and Function

## Secretase

▶ Zinc Adamalysins

## Secretory Phospholipase $A_2$ Group IB

▶ Platinum (IV) Complexes, Inhibition of Porcine Pancreatic Phospholipase A2

## Sel K

▶ Selenoprotein K

## SelD

▶ Selenophosphate Synthetase

## Selectivity

▶ Potassium Channel Diversity, Regulation of Potassium Flux across Pores
▶ Potassium Channels, Structure and Function

## Selenide

▶ Selenophosphate Synthetase

## Selenite Channel

▶ Selenium and Aquaporins

## Selenium

▶ Selenium in Human Health and Disease
▶ Tellurite-Detoxifying Protein TehB from *Escherichia coli*

# Selenium and Aquaporins

Gerd Patrick Bienert and François Chaumont
Institut des Sciences de la Vie, Universite catholique de Louvain, Louvain-la-Neuve, Belgium

## Synonyms

Aquaporin-mediated selenite transport; Major intrinsic proteins and selenite transport; Selenite channel

## Definition

Aquaporins or major intrinsic proteins (MIPs) are transmembrane channel proteins, which facilitate the passive and bidirectional diffusion of water and/or small and noncharged compounds across biological membranes. Aquaporins are found in organisms of all kingdoms of life and are present in all main subcellular membrane systems. The substrate specificity/spectra of aquaporins are highly isoform-dependent. Selenite is one of the major bioaccessible selenium compounds for plants. A member of the nodulin26-like intrinsic protein (NIP) aquaporin subfamily from rice (*Oryza sativa*) was shown to facilitate the diffusion of undissociated selenite molecules across the plasma membrane. To date, aquaporins represent the only known selenite transport proteins in plants.

## Selenium in Plants

Selenium (Se) is an essential microelement in most eukaryotes, including mammals and humans (Rayman 2000). The essentiality of Se is mostly based on the formation of selenoproteins (e.g., glutathione peroxidases and thioredoxin reductases), which are playing essential roles as antioxidant enzymes protecting cells from oxidative damages (Lobanov et al. 2009). Plant-based food products represent the main source of dietary Se, but interestingly, Se seems not to be essential for higher plants, which might explain why Se concentrations in staple food plants are rather low (Zhu et al. 2009).

One promising strategy to counteract Se deficiency in humans is Se biofortification of staple food plants. This implicates the breeding of cultivars, which take up more Se and translocate it to and accumulate it in edible plant parts. This strategy requires a detailed molecular and physiological understanding of Se uptake and allocation mechanisms, especially because elevated concentrations of Se are toxic for plants (Zhu et al. 2009). As rice represents one of the most important staple foods worldwide, a lot of attention has been attributed to enhance biofortification of Se in this species.

Selenate and selenite represent the two major inorganic Se species occurring in soils which are bioavailable for plants.

## Transmembrane Transport Pathways for Selenate in Plants

In most aerated soils, selenate is exclusively occurring in its ionic forms ($HSeO_4^-$ and $SeO_4^{2-}$) but not as a noncharged $H_2SeO_4$ molecule due to its very low pKa values. Selenate is chemically very similar to sulfate. Sulfate transporters from different plant species were shown to efficiently transport selenate (Sors et al. 2005). Selenate uptake is therefore higher in sulfur-starved plants, in which the expression of sulfate transporters is upregulated. The crosstalk between selenate and sulfur uptake makes this transport system a less favored target for biofortification approaches. A potentially disadvantageous modulation of transport routes of an essential macronutrient might entail detrimental consequences on important traits outweighing the advantageous effect of a higher Se accumulation.

## Transmembrane Transport Pathways for Selenite

The second major inorganic Se compound, which represents the dominant species of Se in paddy soil rice fields, is selenite. Selenite is a weak acid with a $pKa_1$ and $pKa_2$ value of 2.57 and 6.6, respectively. Depending on the pH of the solution, selenite exists as $H_2SeO_3$, $HSeO_3^-$, and $SeO_3^{2-}$. Therefore, in most soils and at physiological pH ranges, the equilibrium is greatly shifted towards the ionic forms ($HSeO_3^-$ and $SeO_3^{2-}$), while the undissociated form ($H_2SeO_3$) only becomes prevalent at acidic pH. Until recently, no selenite transporter had been identified in any organism, and selenite was thought to cross biological membranes by passive diffusion. However, in 2010, two types of transport proteins were identified to be involved in selenite transport, aquaporins in plants (Zhao et al. 2010) and proton-coupled monocarboxylate transporters in yeast (McDermott et al. 2010). It is assumed that selenite is adventitiously transported by these proteins due to a high similarity in chemical features determining the substrate specificity of the respective transport protein. The selenite-transporting proton-coupled monocarboxylate transporter Jen1p of *Saccharomyces cerevisiae* is usually involved in the transport of carboxylic acids, such as formate, acetate, pyruvate, and lactate (McDermott et al. 2010). Yet, it is not known whether monocarboxylate transporters from other organisms than yeast are able to transport selenite too. Future studies will reveal the impact of these transport proteins on biologically relevant selenite transport processes.

## Aquaporins Are Involved in the Uptake of Selenite by Plants

Aquaporins are membrane proteins facilitating the diffusion of water and small uncharged solutes (see ▶ Aquaporins and Transport of Metalloids) (reviewed in Hachez and Chaumont 2010). A physiological study demonstrated that the application of typical aquaporin inhibitors, like mercury chloride ($HgCl_2$) and silver nitrate ($AgNO_3$), to rice roots inhibited significantly the uptake of selenite (Zhang et al. 2006).

Another study investigating selenite uptake kinetics of maize roots came also to the conclusion that aquaporins are involved in selenite transport (Zhang et al. 2010). When the growth solution was adjusted to pH 3, at which most of the selenite occurs as undissociated $H_2SeO_3$ molecules, selenite uptake followed a linear kinetics suggesting a passive channel-mediated transport mechanism. At this pH, the selenite uptake was greatly inhibited by aquaporin inhibitors ($HgCl_2$ and $AgNO_3$). In contrast, at pH 5, at which most of the selenite molecules are deprotonated, an active transport mechanism was indicated by a saturated uptake kinetic behavior.

## Aquaporins Belonging to the NIP Subfamily are Important for Selenite Uptake in Rice

Using elegant molecular and physiological approaches, selenite uptake by rice roots in paddy soils was demonstrated to be mediated by the silicon influx transporter OsNIP2;1 (synonymous with OsLsi1; standing for "low silicon 1"), which had been identified to be essential for the physiological uptake of silicon (Ma 2010) (see ▶ Silicon and Aquaporins). When rice plants expressing the nonfunctional OsNIP2;1 isoform (the so-called *lsi1* loss-of-function mutants) were grown in medium supplied with selenite, they contained significantly less Se in their shoots and xylem sap compared to the wild-type plants (Zhao et al. 2010). However, when selenate was used instead, similar Se concentrations were measured in both plant types indicating that selenite but not selenate is transported by OsNIP2;1. The fully protonated form of selenite ($H_2SeO_3$) is the most likely substrate of OsNIP2;1 as, in short-term uptake assays, the selenite uptake rate greatly increased under more acidic conditions. When the external growth solution was adjusted to pH 3.5, selenite uptake was reduced to 5% in *lsi1* mutant compared to wild-type plants, suggesting that, in these conditions, OsNIP2;1 represents the major pathway for selenite transport into the roots (Zhao et al. 2010). The permeability of OsNIP2;1 proteins to selenite was independently confirmed in uptake experiments using the yeast system (Zhao et al. 2010).

OsNIP2;1 belongs to the NIPIII subgroup which forms a distinct ar/R selectivity filter among aquaporins consisting of glycine, serine, glycine, and arginine (see ▶ Silicon and Aquaporins) (Danielson and Johanson 2010; Ma 2010). The relative small sizes of these residues potentially form a larger constriction region compared to other NIP subgroups or microbial and mammalian aquaglyceroporins. That selenite shares chemical features with other hydroxylated metalloid species does go along with the observation that these compounds share a common NIP-channel-mediated transmembrane pathway (see ▶ Arsenic and Aquaporins, ▶ Boron and Aquaporins, ▶ Silicon and Aquaporins) (Bienert et al. 2008).

## Concluding Remarks

Based on the few existent data about aquaporin-mediated selenite transport, it is probably too early to speculate whether selenite is only adventitiously transported by aquaporins due to molecular substrate similarity or whether these metalloid channels are also actively involved in the physiological regulation of selenium homeostasis. However, the fact that OsNIP2;1 is responsible for 95% of the uptake of selenite into rice roots strongly suggests that aquaporins might have a large impact on selenite accumulation and/or distribution in plants. Selenite transport capacity of plant aquaporins will be of high interest for future biofortification strategies with the aim to breed cultivars enriched in Se and counteract Se deficiency in humans. Furthermore, it has to be resolved whether MIP isoforms in organisms other than plants are permeable for selenite and whether these isoforms are involved in the regulation of transmembrane selenite transport processes. It will be especially interesting to focus on organisms for which selenium represents an essential microelement.

## Funding

This work was supported by grants from the Belgian National Fund for Scientific Research (FNRS), the Interuniversity Attraction Poles Programme–Belgian Science Policy, and the "Communauté française de Belgique–Actions de Recherches Concertées". GPB was supported by a grant from the FNRS.

## Cross-References

▶ Aquaporins and Transport of Metalloids
▶ Arsenic and Aquaporins
▶ Boron and Aquaporins
▶ Silicon and Aquaporins

## References

Bienert GP, Schüssler MD, Jahn TP (2008) Metalloids: essential, beneficial or toxic? Major intrinsic proteins sort it out. Trends Biochem Sci 33:20–26

Danielson JAH, Johanson U (2010) Phylogeny of major intrinsic proteins. Landes Bioscience-Springer, New York

Hachez C, Chaumont F (2010) Aquaporins: a family of highly regulated multifunctional channels. Landes Bioscience-Springer, New York

Lobanov AV, Hatfield DL, Gladyshev VN (2009) Eukaryotic selenoproteins and selenoproteomes. Biochim Biophys Acta 1790:1424–1428

Ma JF (2010) Silicon transporters in higher plants. Landes Bioscience-Springer, New York

McDermott JR, Rosen BP, Liu Z (2010) Jen1p: a high affinity selenite transporter in yeast. Mol Biol Cell 21:3934–3941

Rayman MP (2000) The importance of selenium to human health. Lancet 356:233–241

Sors TG, Ellis DR, Salt DE (2005) Selenium uptake, translocation, assimilation and metabolic fate in plants. Photosynth Res 86:373–389

Zhang LH, Shi WM, Wang XC et al (2006) Genotypic difference in selenium accumulation in rice seedlings and correlation with selenium content in brown rice. J Plant Nutrition 29:1601–1618

Zhang L, Yu F, Shi W et al (2010) Physiological characteristics of selenite uptake by maize roots in response to different pH levels. J Plant Nutr Soil Sci 173:417–422

Zhao XQ, Mitani N, Yamaji N et al (2010) Involvement of silicon influx transporter OsNIP2;1 in selenite uptake in rice. Plant Physiol 153:1871–1877

Zhu YG, Pilon-Smits EA, Zhao FJ et al (2009) Selenium in higher plants: understanding mechanisms for biofortification and phytoremediation. Trends Plant Sci 14:436–442

# Selenium and Glutathione Peroxidases

Christoph Ufer
Institute of Biochemistry, Charité – Universitätsmedizin Berlin, Berlin, Germany

## Synonyms

Apoptotic cell death; Cancer; Inflammation; Reactive oxygen species; Spermatogenesis

## Definition

Selenium-containing Glutathione Peroxidases constitute a family of enzymes that carry a selenocysteine residue in their polypeptide chain and that catalyze the reduction of hydroperoxides to the corresponding alcohols at the expense of reduced glutathione.

## Introduction

About 30 years ago in 1982, Ursini and his colleagues identified a protein factor that prevented lipid peroxidation in the presence of glutathione. In the following years, it was established that this protein termed phospholipid hydroperoxide glutathione peroxidase (phGPx)

contains selenium in equimolar amounts (Ufer and Wang 2011). By today, a whole family of Glutathione Peroxidases (GPx; EC 1.11.1.9) has been characterized comprising of at least eight known isoforms in mammals. Some of these GPx family members are selenium-dependent, whereas other enzymes of the GPx family do not contain selenium (Herbette et al. 2007). Moreover, Glutathione Peroxidases have been identified in virtually all branches of life either as selenoproteins or as selenium-independent GPx (Margis et al. 2008). The non-selenium isoforms carry a cysteine residue instead of the selenocysteine residue in their polypeptide chain.

Since its discovery great progress has been made in defining the molecular principles of GPx activity and in elucidating the biological functions in the cell and whole organisms. Since most of our current understanding of selenium-dependent GPxs has been obtained in mammals, this entry will mainly focus on mammalian organisms.

## Selenium-Dependent Glutathione Peroxidases

Glutathione peroxidases catalyze the reduction of hydroperoxide substrates (R-OOH) to the corresponding alcohols (R-OH) at the expense of a reducing equivalent (Toppo et al. 2009). The hydroperoxide substrates can be either hydrogen peroxide ($H_2O_2$), which is reduced to water ($H_2O$), or complex lipophilic hydroperoxides (R-OOH) such as phospholipid or cholesterol hydroperoxides. The reducing equivalent, which is employed in most cases, is glutathione (GSH). However, in addition to glutathione, several GPx isoenzymes accept other thiol-containing substrates such as thioredoxins. The catalytic cycle is described as a ping-pong mechanism and starts with the oxidation of the active site selenolate by the hydroperoxide substrate. The resulting selenenic acid derivative is then re-reduced by two molecules of reduced glutathione in two consecutive steps to regenerate the enzyme in its ground state ready for the next catalytic cycle. Thus, the selenocysteine moiety constitutes the vital component of GPx enzymatic activities. If the selenocysteine is replaced by a cysteine residue, drastically reduced enzymatic activities have been observed, and supporting observations have been made non-selenium-type GPx enzymes.

Most organisms that have been screened for GPx expression were found to contain either selenium-dependent GPxs or selenium-independent GPxs or both. The human and the murine genome codes for eight GPx isoforms (designated GPx1-8), out of which five are selenium dependent in humans (GPx1-4, GPx6) and four in mice (GPx1-4). Mammalian selenium-dependent GPx enzymes are either homotetrameric or monomeric enzymes with a subunit size of about 20 kDa. Selenium-dependent GPxs are not only found in mammals but also in other vertebrates such as chicken and zebra fish. Outside the vertebrate branch selenium-dependent GPxs have been identified in a parasitic helminth, nematodes, arthropods, algae, and even in viral proteomes (Margis et al. 2008).

Selenium-dependent GPxs are characterized by the presence of an unusual amino acid in their peptide chain – selenocysteine (Sec). This so-called 21st amino acid is incorporated co-translationally into the growing peptide chain by an unusual mechanism (Papp et al. 2007). There is no canonical coding triplet specifically assigned for Sec. Instead, an opal codon (TGA), which normally initiates translational termination, is used as a place holder. A stem-loop structure (SECIS, *Sec insertion sequence*) in the 3′-untranslated region of a eukaryotic selenoprotein mRNA facilitates the re-coding of the stop codon and the subsequent insertion of Sec into the growing polypeptide chain. This mechanism is worth mentioning since it entails two major consequences: (a) The presence of a stop codon within the coding sequence makes selenoprotein enzymes a classical target for the nonsense-mediated RNA decay pathway. This pathway normally subjects RNAs bearing inappropriate stop codons to degradation and prevents the formation of abnormally truncated proteins. Somehow this pathway appears to be masked in mRNAs coding for selenoproteins to facilitate selenoprotein biosynthesis. (b) The so-called SECIS element of the various GPx isoenzymes exhibits individual features that potentially allow for a finely tuned regulation of each individual selenoprotein. In fact, such finely tuned regulation of GPx isoenzyme expression is observed, when selenium in the organism becomes scarce (Brigelius-Flohe 1999). Whereas the mRNAs of some GPx isoforms quickly disappear following selenium deficiency, others are retained in the organism in a tissue-specific manner. This phenomenon has been termed selenoprotein hierarchy and the reader is referred back to this in the following sections describing the individual features of GPx isoforms. However, despite the knowledge on the Sec incorporation machinery, the molecular mechanisms underlying the fascinating phenomenon of the selenoprotein hierarchy have not yet been fully elucidated.

Their ability to detoxify hydrogen peroxide has implicated GPx enzymes as crucial components of the cellular antioxidative defense system. Hydrogen peroxide, together with the superoxide anion and the hydroxyl radical, are termed reactive oxygen species (ROS) (Ufer et al. 2010). These highly reactive molecules are able to modify virtually all macromolecules in the cell including membranes, proteins, and nucleic acids and thereby impair the function of these molecules, which subsequently may result in abnormal cell behavior or cell death. ROS are constantly attacking the living organism from exogenous sources but are also actively produced in significant amounts within the cell (Ufer et al. 2010). However, the concepts on the biological role of ROS have changed over the recent years (Ufer et al. 2010). Whereas ROS have originally been considered as largely deleterious to the cell, it has now become rather clear that ROS constitute vital signaling molecules in physiological processes. Similarly, the biological roles of the different GPx enzymes have been shown to be far more complex than the simple concept of antioxidative defense suggests. Nonetheless, the antioxidative capacities are vital characteristics of GPx enzymes. The following sections are aimed at summarizing the biological implications of the individual selenium-dependent GPx enzymes with a focus on the mammalian isoenzymes.

*Glutathione Peroxidase 1 (GPx1)* is a homotetrameric enzyme, which is expressed in most tissues and cell types, including erythrocytes, liver, lung, and kidney (Brigelius-Flohe 1999; Ufer and Wang 2011). Due to its ubiquitous expression GPx1 is considered the major antioxidative enzyme within the GPx protein family. It is thought to reduce hydrophilic peroxide species such as hydrogen peroxide but also organic hydroperoxides (Toppo et al. 2009). GPx1 is a mainly cytoplasmic enzyme, but nuclear or even mitochondrial localization of the GPx1 enzyme has been reported, where it is believed to protect the nuclear or mitochondrial DNA (Ufer and Wang 2011). However, the molecular mechanisms that confer the insertion of GPx1 into these organelles have not yet been convincingly addressed.

Targeted abrogation of GPx1 expression in mice has not induced a major phenotype (Ufer and Wang 2011). Instead, GPx1$^{-/-}$ mice develop normally and are able to compensate mild oxidative stress. This finding is in line with the low ranking of GPx1 in the selenoprotein hierarchy, which dictates that GPx1 expression drops quickly following selenium depletion (Brigelius-Flohe 1999). However, the finding that an apparently important antioxidant enzyme appears to be dispensable is not surprising. Vertebrate organisms appear to have developed multilevel firewalls against oxidative stress to ensure the toxic molecules are kept in control. Thus, genetic knockouts of many antioxidative enzymes failed to induce major phenotypes, possibly, because the lack of one antioxidative enzyme is at least partially compensated for by another (Ufer et al. 2010).

*Glutathione Peroxidase 2 (GPx2)* is a homotetrameric enzyme, which is dominantly expressed in the gastrointestinal (GI) tract and the liver, where this isoenzyme is believed to protect the GI tract from hydrogen peroxide produced in the gut (Ufer and Wang 2011). Interestingly, the GI tract and the liver also express high amounts of monoamine oxidases (MAOs), which degrade ingested and potentially harmful amines (Ufer et al. 2010). The enzymatic reaction of MAOs generates hydrogen peroxide as a by-product. Concluding from these findings, it may be speculated that GPx2 protects these tissues not only from ingested hydroperoxides but also from hydrogen peroxide produced by the MAO reaction or by other reactions of oxidative metabolism. Selenium depletion studies have revealed that GPx2 mRNA expression is kept constant if not increased following selenium scarcity, which implies a vital role of this enzyme in GI epithelia (Brigelius-Flohe 1999). In contrast, abrogation of GPx2 expression in GPx2$^{-/-}$ did not produce a majorly aberrant phenotype (Ufer and Wang 2011). However, GPx2 deficiency coincides with an elevated incidence of UV-induced squamous cell carcinomas. This finding is further aggravated in GPx1$^{-/-}$/GPx2$^{-/-}$ double knockout mice, which are characterized by a high incidence of intestinal inflammation and increase tumor development (Ufer and Wang 2011). The reduced antioxidative capacity of cells of gastrointestinal epithelia results in an accumulation of genetic and epigenetic errors, which is believed to promote cancer development and progression in these tissues. These data imply that GPx1 can compensate to a certain extent the lack of GPx2 in GPx2$^{-/-}$ animals.

When this compensatory mechanism is switched off in a double knockout, the antioxidative firewall is significantly weakened, which results in abnormal cellular behavior.

*Glutathione Peroxidase 3 (GPx3)* is a secreted plasma protein and acts as a homotetrameric enzyme. Its dominant source of expression are cells of the proximal tubuli of the kidney, but also other tissues such as the placenta, the lung, and the mammary gland have been shown to secrete GPx3 enzymes (Brigelius-Flohe 1999). Very little is known about the biological implications of this isoform. The first puzzle was that glutathione levels in the plasma are too low to be a good substrate for the completion of the GPx2 catalytic cycle. Thus, it has been speculated that this enzyme also accepts glutaredoxin and thioredoxins as electron donors (Toppo et al. 2009). The biological role of this enzyme is still quite opaque. Selenium depletion causes the GPx3 messenger to disappear quickly compared to the other GPx enzymes (Brigelius-Flohe 1999). Targeted GPx3 knockouts produced viable offspring with no apparent phenotype (Ufer and Wang 2011). These data imply that GPx3 expression is either dispensable, which is less likely, or its expression exerts its biological relevance only under certain circumstances. However, such conditions that rely on GPx3 function have not yet satisfyingly been defined.

*Glutathione Peroxidase 4 (GPx4)* is the most dazzling enzyme of the vertebrate GPx protein family and recent research has given many new insights into GPx functioning (Ufer and Wang 2011). Expression of GPx4 has been detected in virtually all tissues, but the GPx4 protein is massively enriched in spermatogenic cells of the postpubertal testis (Ursini et al. 1997). The mechanisms for the dramatic increase of GPx4 expression in these cells have not yet been sufficiently addressed.

By today three different GPx4 isoforms have been shown to be generated from one joint GPx4 gene (Conrad et al. 2007). These GPx4 isoenzymes are identical with respect to the catalytic domains, but differ in their N-terminal sequences, which are believed to direct the GPx4 isoenzymes to different subcellular compartments. In fact, GPx4 protein has been detected in the cytoplasm, in the nucleus, and in mitochondria. However, more recent experimental data have suggested that GPx4 can be inserted into these organelles independently of their N-terminal targeting signals (Ufer and Wang 2011). Nonetheless, the GPx4 gene produces various mRNA species that are generated by alternative transcriptional initiation and/or alternative splicing. This becomes even more complicated by the presence of alternative in-frame translational start sites in some of these mRNAs. The reasons for the coding multiplicity of the GPx4 gene have not yet been satisfyingly addressed. However, it was shown recently that the sequences, which are specific to the alternative transcripts, serve as means to allow for isoform-specific expression regulation on a posttranscriptional level.

In contrast to the other GPx enzymes, which are tetrameric, GPx4 exists as a small monomeric enzyme. This characteristic and the lack of an exposed surface loop on its protein surface, which frames the reactive center and potentially limits access of substrates to the active site in other GPx enzyme, are believed to be the structural basis for the broad substrate specificity of the GPx4 enzyme (Toppo et al. 2009; Ufer and Wang 2011). In fact, GPx4 reduces not only hydrogen peroxide but also complex lipophilic hydroperoxides such as phospholipid and cholesterol hydroperoxides, even when these are incorporated into membranes or lipoproteins (Toppo et al. 2009). Moreover, the GPx4 enzyme also exhibits a broad substrate specificity also toward its reducing substrate. Although glutathione is a good and available substrate in most cells, it becomes limited in certain stages of sperm maturation. In developing sperm cells, which have massively elevated levels of GPx4 protein, GPx4 becomes less discriminating toward its reducing substrate and starts oxidizing protein thiols such as chromatin and eventually GPx4 itself (Ursini et al. 1997). Thus, in maturing sperm cells the GPx4 enzyme transforms itself into an enzymatically inactive structural component, which has been shown to be vital for the integrity of the mature sperm cell and thereby for male fertility. However, the source of hydroperoxides that initiates the dramatic increase of GPx4 activity during this process has not yet been convincingly identified.

In addition to its involvement in sperm development, the unique properties of the GPx4 enzyme have been implicated GPx4 activity in a plethora of processes that by far exceed a simple antioxidative function. Following selenium depletion GPx4 expression is comparably stable in particular in the testis and the brain, which ranks GPx4 high in the selenoprotein hierarchy and suggests a strong dependence of vertebrate organisms on this enzyme (Brigelius-Flohe 1999). This was further confirmed by various targeted genetic knockout approaches. GPx4 knockout mice fail to develop viable offspring and developing GPx4$^{-/-}$ embryos undergo intrauterine death by gestational day E7.5 (Ufer and Wang 2011). Thus, the biological roles of GPx4 cannot be easily compensated for by other antioxidative enzymes. Mice that are heterozygous for GPx4 in contrast are viable, but show an enhanced sensitivity toward oxidative stress. One of the underlying molecular mechanisms that cause embryonic lethality in GPx4 deficient mice was suggested to be a dysregulation of apoptotic cell death (Ufer and Wang 2011) and this type of cell death is a vital physiological process during normal embryo development. Indeed, GPx4 deficiency is accompanied by an induction of apoptosis and this is in line with the concept of GPx4 to act as an anti-apoptotic enzyme (Ufer and Wang 2011). Different mechanisms of action have been shown to be involved in the anti-apoptotic properties of the GPx4 enzyme. GPx4 activity has been associated with the prevention of oxidative modifications of the mitochondria-specific lipid cardiolipin as well as of the mitochondrial protein AIF (apoptosis-inducing factor) (Ufer and Wang 2011). Oxidative modification of cardiolipin and AIF has been shown to trigger certain apoptotic signaling cascades, and this can be ameliorated by GPx4 action.

In addition, cellular behavior such as cell death signaling can be altered by changes in gene expression. Signaling cascades that lead to altered gene expression depend on regulatory protein factors including transcription factors, and these may be sensitive toward the cellular redox state by means of redox-sensitive protein thiol groups (Ufer et al. 2010). Indeed, GPx4 has been shown to modulate the activity of transcription factors such as nuclear factor kappa B (NFκB). NFκB is a key regulator of many biological events in particular inflammation. By dampening NFκB activity GPx4 has thus been reported to be an anti-inflammatory enzyme. In addition, GPx4 has been depicted as a natural antagonist of lipoxygenases (LOX) and cyclooxygenases (COX) (Kuhn and Borchert 2002). These enzymes catalyze the oxidative modification of lipids such as arachidonic acid, which is the key step in the generation of leukotrienes and prostaglandins. These lipid mediators are vital regulators of inflammatory processes. LOX as well COX enzymes need a certain hydroperoxide tonus for their enzymatic activity. Thus, by lowering the cellular hydroperoxide tonus GPx4 is able inhibit LOX and COX activities and thereby to suppress leukotriene and prostaglandin biosynthesis. However, the precise role of GPx4 in inflammatory processes is not yet clearly defined.

*Glutathione Peroxidase 6 (GPx6)* is found as a fifth selenium-dependent GPx in humans and is present in rodents as a selenium-independent enzyme (Toppo et al. 2009). GPx6 is a cytoplasmic protein exclusively expression in the olfactory epithelium. Very little is known about this selenoprotein and the biological function of this enzyme is not yet understood.

## Conclusion and Outlook

The progress of GPx research has taken some surprising turns. Little of the original idea of GPx as merely antioxidative enzymes has survived into the presence. This is also largely due to the changed conception of reactive oxygen species from being mostly deleterious to being vital signaling molecules. Yet, the production of reactive oxygen species must be regulated and contained, and GPx1 certainly is a major antioxidative defense mechanism of the cell. On the other hand, GPx activity is now considered to be important for the regulation of cellular signaling events that affect differentiation, proliferation, and cell death. It may thus be speculated that the detoxification of potentially hazardous reactive oxygen species is to a certain extent just a side reaction of the various regulatory function of GPx enzymes. In other word, reactive oxygen species drive the enzymatic activities of GPx enzymes and thereby allow for the formation of the mitochondrial capsule in developing sperm cells in the case of GPx4, for instance. Similar observations have been made in pro-oxidative enzymes such as amine oxidases, which generate hydrogen peroxide as a by-product of their enzymatic activities breaking down amines. And in the case of amine oxidases, hydrogen peroxide has been shown to be vital to explain the physiological functions of this group of enzymes. Thus, GPx activities are maybe just driven by the presence of hydrogen peroxides simply because in our aerobic environment the chances are low that hydrogen peroxides become limiting. However, GPx enzymes may serve as sensors for critical levels of peroxides and their regulatory activities may subsequently trigger the appropriate cellular response. Another surprise of the past few years of research was that "Glutathione" Peroxidases may be a bit of a misnomer, since many GPx enzymes accept thioredoxins or other protein thiols rather than glutathione. It is therefore vital to avoid stereotypical thinking, when trying to walk new paths in GPx research. New targeted genetic approaches may give new insights into the highly specialized tasks of the individual GPx enzymes and these experimental setups may provide useful data for understanding the progression of pathophysiological conditions such as atherosclerosis and cancer.

## Cross-References

▶ Iron Proteins, Mononuclear (non-heme) Iron Oxygenases
▶ Peroxidases
▶ Selenium and the Regulation of Cell Cycle and Apoptosis
▶ Selenium in Human Health and Disease
▶ Selenium, Biologically Active Compounds

## References

Brigelius-Flohe R (1999) Tissue-specific functions of individual glutathione peroxidases. Free Radic Biol Med 27(9–10):951–965

Conrad M, Schneider M et al (2007) Physiological role of phospholipid hydroperoxide glutathione peroxidase in mammals. Biol Chem 388(10):1019–1025

Herbette S, Roeckel-Drevet P et al (2007) Seleno-independent glutathione peroxidases. More than simple antioxidant scavengers. FEBS J 274(9):2163–2180

Kuhn H, Borchert A (2002) Regulation of enzymatic lipid peroxidation: the interplay of peroxidizing and peroxide reducing enzymes. Free Radic Biol Med 33(2):154–172

Margis R, Dunand C et al (2008) Glutathione peroxidase family - an evolutionary overview. FEBS J 275(15):3959–3970

Papp LV, Lu J et al (2007) From selenium to selenoproteins: synthesis, identity, and their role in human health. Antioxid Redox Signal 9(7):775–806

Toppo S, Flohe L et al (2009) Catalytic mechanisms and specificities of glutathione peroxidases: variations of a basic scheme. Biochim Biophys Acta 1790(11):1486–1500

Ufer C, Wang CC (2011) The roles of glutathione peroxidases during embryo development. Front Mol Neurosci 4:12

Ufer C, Wang CC et al (2010) Redox control in mammalian embryo development. Antioxid Redox Signal 13(6):833–875

Ursini F, Maiorino M et al (1997) Phospholipid hydroperoxide glutathione peroxidase (PHGPx): more than an antioxidant enzyme? Biomed Environ Sci 10(2–3):327–332

# Selenium and Iodothyronine Deiodinases

Ulrich Schweizer and Doreen Braun
Institute for Experimental Endocrinology, Charité-Universitätsmedizin Berlin, Berlin, Germany

## Synonyms

5′-DI (type I-5′-deiodinase), 5′-DII (type II-5′-deiodinase), 5-DIII (type III-5-deiodinase); D1-3 (widely used, but may be confused with dopamine D1 (D2, D3)-receptors); Deiodinase 1-3 (Dio1-3, official symbols); *Dio1* (official gene symbol); Thyroxine deiodinase (TXD); Type I (II, III)-iodothyronine deiodinase

## Definition

Deiodinases are enzymes capable of iodine elimination from iodothyronines. Iodothyronines are amino acid derivatives that are generated exclusively in the thyroid gland, where tyrosine residues in thyroglobulin are first iodinated at ring positions 3 and 5 and subsequently condensed via an ether bond. Iodothyronine molecules are proteolytically liberated from thyroglobulin and secreted. The major product of the thyroid gland, thyroxine, is a prohormone and needs conversion by 5′-deiodination to yield the nuclear receptor-activating triiodothyronine. Termination of iodothyronine action is achieved by 5-deiodination. In their catalytic centers, deiodinases carry selenocysteine and thus belong to the family of selenoproteins. They are different from iodotyrosine dehalogenase, a non-selenium-containing flavoenzyme which deiodinates mono- and diiodotyrosine, byproducts of thyroid hormone biosynthesis.

## Introduction

Tissues can adjust their local thyroid hormone status by expression of deiodinases. Thyroid hormones play a pivotal role for vertebrate development. They influence energy and intermediate metabolism, e.g., gluconeogenesis, lipogenesis, and glycogenolysis and are critical regulators of energy expenditure. Metamorphosis of tadpoles to frogs is regulated by thyroid hormones and involves degradation/involution of tail and gills, the maturation of lung and intestine, and the reshaping of the nervous system. Also human development depends on thyroid hormone action as obvious in pediatric patients suffering from untreated congenital hypothyroidism, a severe neurodevelopmental condition formerly called cretinism.

The thyroid gland produces and releases the inactive prohormone $T_4$ (thyroxine) and to a lesser extent the biologically active $T_3$. After transporter-mediated import into target cells, thyroid hormones mediate their biological function via the $T_3$ receptor (TR), a ligand-activated transcription factor located in the nucleus (Schweizer et al. 2008b). Activation of the inactive prohormone $T_4$ is induced by the cleavage of the 5′-iodine atom from the phenolic outer ring of iodothyronines, a process called outer ring deiodination (ORD). Removal of a 5-iodine atom from the tyrosyl ring is called inner ring deiodination (IRD). Historically, these activities are called 5′- or 5-deiodination, but it is now known that the enzymes remove the second, 3′- or 3-iodine atom as well. The more general ORD abbreviation is adhered to here (Fig. 1). Enzymes responsible for deiodination reactions are called iodothyronine deiodinases or just deiodinases.

## Deiodinases are Selenium-Dependent Enzymes

The family of iodothyronine deiodinases consists of three selenium-dependent enzymes (Dio1, Dio2, and Dio3). A plethora of different abbreviations has been used for deiodinase activities. Since all enzymes have been cloned, common gene names will be adhered to. Around 1990 several laboratories independently showed that hepatic Dio1 (deiodinase 1) activity was selenium-dependent (Köhrle et al. 2005). Expression cloning in *Xenopus* oocytes of Dio1 from rat liver revealed a cDNA containing an *in-frame* UGA codon, similar as in the two previously cloned selenoproteins, glutathione peroxidase 1 and glycine reductase (Berry et al. 1991b). Failure to express functional Dio1 from a shortened cDNA clone lacking the 3′-untranslated region of the mRNA revealed the *SelenoCysteine Insertion Sequence* (SECIS) element (Berry et al. 1991a). Mutation of the TGA codon to TGT for cysteine resulted in more than tenfold lower activity compared to wild-type Dio1. These and other findings demonstrated that the *in-frame* TGA codon encodes selenocysteine which is involved in catalysis. The Dio1 cDNA was used to clone other deiodinases by homology (for a comparison of these enzymes see Table 1). All three deiodinases are integral membrane proteins with a single N-terminal transmembrane helix, and several conserved amino acids, including the catalytic selenocysteine, a histidine, and a more distant cysteine (Fig. 2). Available evidence suggests that all deiodinases are functional as homodimers and require membrane association for activity (Bianco et al. 2002).

## Catalytic Mechanism

The catalytic mechanism of deiodinases is not completely understood. While there is agreement that selenocysteine, possibly as selenolate, plays a critical role in catalysis, the presumed selenenyl-iodide has not yet been demonstrated. Mutational analyses have pointed to a catalytic role of a conserved cysteine in Dio1. This amino acid may serve as a resolving cysteine liberating iodide through formation of a selenenyl-sulfide. This would be consistent with the observed *ping-pong* kinetics in vitro using DTT (Dithiothreitol) as reducing cofactor. Thioredoxin and glutathione have been suggested as physiological reducing cofactors, but there is no consensus on the issue. The invariant histidine is required for catalysis and may serve as the base donating the proton during deiodination (Kuiper et al. 2005). A proposed catalytic mechanism is depicted in Fig. 3.

## Deiodinase 1

Dio1 was first identified as a 5′-deiodinase catalyzing ORD. However, it also carries IRD activity. Its activity is highest in liver, kidney, and thyroid. In kinetic experiments, the enzyme follows *ping-pong* kinetics suggesting an enzyme-substrate complex or activated enzyme intermediate. Type I deiodinase is sensitive to propyl-thio-uracil (PTU), iopanoic acid, and auro-thio-glucose.

The human *DIO1* gene is located on chromosome 1 and leads to an mRNA size of about 1.8 kb. The mRNA encodes a protein of 249 amino acids and 28.9 kD.

**Selenium and Iodothyronine Deiodinases, Fig. 1** Deiodination pathways of iodothyronines. Thyroxine contains four iodine atoms. 5′-deiodination, or more generally outer ring deiodination (ORD), of $T_4$ leads to formation of $T_3$ the thyroid hormone receptor-activating ligand. 5-deiodination, i.e., inner ring deiodination (IRD), of T4 leads to formation of reverse $T_3$, $rT_3$, an inactive iodothyronine metabolite. Further IRD or ORD proceeds over one of three isomers of diiodothyronines to monoiodothyronines and eventually, thyronine

Dio1 is found at the plasma membrane with its C-terminus located in the cytosol.

Some experiments suggested that hepatic Dio1 activates thyroxine, but several lines of evidence suggest that it is involved in iodothyronine clearance and iodine recycling:
- The preference of Dio1 for $rT_3$ and sulfated iodothyronine metabolites ($T_4S$, $T_3S$, $T_2S$) over $T_3$ and $T_4$.
- Hepatic Dio1 mRNA expression is positively regulated by $T_3$. If $T_4$ ORD activity was predominant, hyperthyroidism would lead to a counterintuitive *feed-forward* loop. Induction by $T_3$, conversely, would be more compatible with a homeostatic predominance of IRD.
- Dio1 can mediate ORD and IRD and it is unclear how the enzyme should choose the appropriate activity under different physiological conditions.
- The phenotype of *Dio1*-KO (knock out) mice does not reveal a role of Dio1 in $T_3$ homeostasis. In contrast, fecal excretion of iodothyronine is increased due to reduced iodine recycling.

Dio1 activity and selenium content are high in the thyroid. It has been speculated that thyroidal Dio1 is involved in $T_3$ release from the gland or selenium-dependent peroxidases are required to protect the organ from peroxides (Schweizer et al. 2008a). However, targeted inactivation of selenoprotein biosynthesis in thyrocytes has not revealed an essential role of selenoproteins in thyrocytes under normal or iodine-deficient conditions, despite increased oxidative damage (Chiu-Ugalde et al. 2012). Dio1 is capable of thyronamine (TAM) deiodination. TAM, in particular 3-$T_1$AM and $T_0$AM are decarboxylated iodothyronine derivatives. While pharmacological effects have been observed and TAM can be measured

**Selenium and Iodothyronine Deiodinases, Table 1** The human iodothyronine deiodinases

|  | Deiodinase 1 | Deiodinase 2 | Deiodinase 3 |
|---|---|---|---|
| Function | ORD; IRD | ORD | IRD |
| Substrate preference | $rT_3 > T_2S >> T_4$ (ORD) $T_4S > T_3S > T_3, T_4$ (IRD) | $T_4 > rT_3$ | $T_3 > T_4$ |
| Chromosome | 1p32-p33 | 14q21.3 | 14q32 |
| mRNA | 1,876 bp (NM_000792.5) | 6,379 bp (NM_000793.5) | 2,120 bp (NM_001362.3) |
| Protein | 249aa; 28.9kD (NP_000783.2) | 273aa; 30.5kD (NP_000784.2) | 304aa; 31.5kD (NP_001353.4) |
| Half-life | Several hours | >1 h | Several hours |
| Subcellular localization | Plasma membrane; catalytic site is located in the cytosol | ER membrane; catalytic site is located in the cytosol | Plasma membrane; catalytic site is located in the cytosol |
| Tissue distribution | Liver; kidney; thyroid gland; pituitary | Brain; pituitary; hypothalamus; cochlea; skeletal muscle; brown adipose tissue | Brain; skin; retina; placenta; pregnant uterus; fetal tissues |
| Cofactor | Endogenous: unknown, possibly GSH, Trx | Endogenous: unknown | Endogenous: unknown |
|  | Artificial: DTT | Artificial: DTT | Artificial: DTT |
| Inhibitors | PTU; iopanoic acid; gold | Iopanoic acid | Iopanoic acid; gold |
| Mechanism | Ping-pong | Sequential | Sequential |
| Regulation | ↑T3 (2 TREs) | ↑Hypothyroidism | ↑Thyroid hormones |
|  | ↑TSH | ↑Cold exposure | ↑EGF; FGF; Shh |
|  | ↑Retinoic acid | ↑cAMP | ↑MERK/ERK pathway |
|  | ↑cAMP | ↑Insulin; glucagon | ↓Hypothyroidism |
|  | ↑GH | ↑Catecholamines | ↓GH |
|  | ↓Se deficiency | ↓$T_4$;$T_3$;$rT_3$ | ↓Glucocorticoids |

in vivo, their physiological roles are still uncertain (Piehl et al. 2011).

Inherited *DIO1* deficiencies have not been described so far.

## Deiodinase 2

Type II-deiodinase activity is defined as a PTU-insensitive 5′-deiodinase activity. Dio2 has a 700-fold greater catalytic efficiency for 5′-ORD than Dio1 and thus is likely responsible for most of the peripheral $T_4$ to $T_3$ conversion. In humans, skeletal muscle is a significant source of extrathyroidally generated plasma $T_3$. Expression of Dio2 in brain, cochlea, or brown adipose tissue is believed to control cellular $T_3$ content.

The human *DIO2* gene is located on chromosome 14. Its mRNA contains more than 6.3 kb with its UGA codon separated from the SECIS element by approximately 5 kb, the longest separation found in eukaryotic selenoproteins. The 273 amino acids long protein is located at the endoplasmic reticulum membrane with its active center and C-terminus located in the cytosol. Dio2 mRNA expression is induced by hypothyroidism and thus homeostatically increases local $T_3$ levels.

*Dio2* gene expression is positively regulated via cAMP and CREB, e.g., by adrenergic stimulation in brown adipose tissue, where Dio2 produces increased $T_3$ levels and $T_3$-dependent induction of uncoupling protein 1 (UCP1). A negative posttranslational regulator is the substrate, $T_4$ itself. Catalytic activity stimulates ubiquitin-dependent degradation at the proteasome.

Similar to Dio1, the endogenous cofactor for 5′-ORD of iodothyronines is unknown, albeit DTT serves in vitro as reducing cofactor. Dio2 displays sequential reaction kinetics, which correlated with the identity of the second amino acid after the selenocysteine: deiodinases with proline at this position show sequential kinetics, while serine confers *ping-pong* kinetics. Exchange of serine for proline in Dio2 or proline for serine in Dio1 changes kinetics accordingly, suggesting that IRD and ORD mechanisms are not fundamentally different.

The *Dio2*-KO mouse displays partial hypothyroidism in tissues depending on Dio2-dependent $T_4$ to $T_3$ conversion. The animals have elevated serum TSH (thyroid stimulating hormone) and $T_4$ levels while the $T_3$ levels remain unaltered. TSH (thyrotropin) is

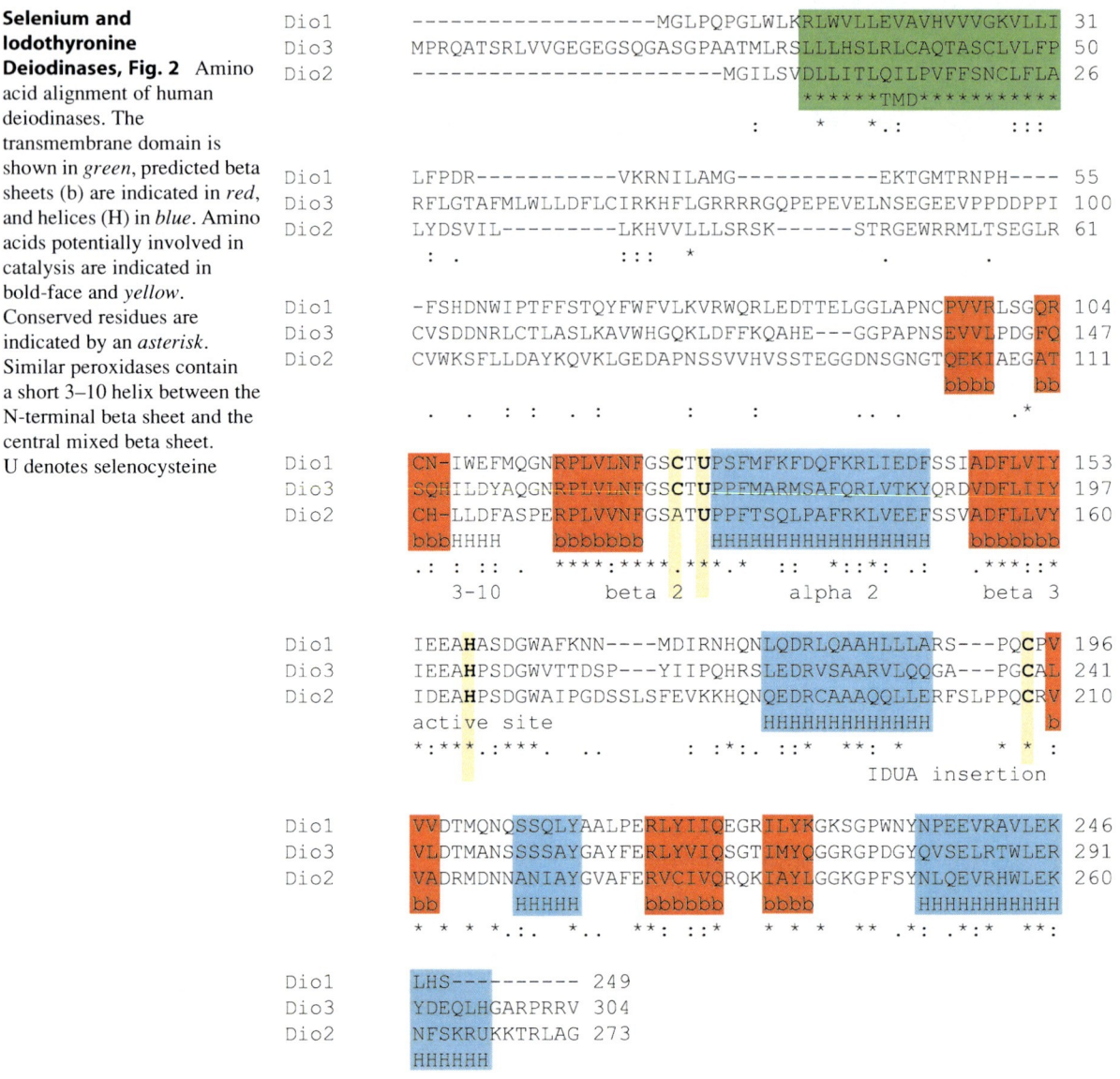

**Selenium and Iodothyronine Deiodinases, Fig. 2** Amino acid alignment of human deiodinases. The transmembrane domain is shown in *green*, predicted beta sheets (b) are indicated in *red*, and helices (H) in *blue*. Amino acids potentially involved in catalysis are indicated in bold-face and *yellow*. Conserved residues are indicated by an *asterisk*. Similar peroxidases contain a short 3–10 helix between the N-terminal beta sheet and the central mixed beta sheet. U denotes selenocysteine

secreted by the pituitary after stimulation with hypothalamic TRH (thyrotropin releasing hormone) and activates the thyroid gland to produce thyroid hormones. In turn, increasing concentrations of thyroid hormones inhibit the release of TSH and TRH in the pituitary or hypothalamus, respectively. Dio2 deficiency in the pituitary leads to $T_4$ resistance in *Dio2*-KO mice explaining the elevated TSH levels that are suppressed after $T_3$, but not $T_4$ treatment. Elevated plasma $T_4$ results from increased thyroidal secretion, an adjustment made to maintain normal serum $T_3$ levels. *Dio2*-KO mice are unable to maintain their body temperature after cold exposure due to impaired energy expenditure in brown adipose tissue. *Dio2*-KO mice display hearing impairment due to a developmental defect of the cochlea.

A SNP (single nucleotide polymorphism) has been identified in the human genome that affects circulating thyroid hormone levels.

## Deiodinase 3

Dio3 is an obligate IRD deiodinase. It terminates $T_4$ and $T_3$ action by removal of the 5-iodine atom from the tyrosyl ring. Thus, Dio3 is a key player in protecting

**Selenium and Iodothyronine Deiodinases, Fig. 3** Proposed catalytic mechanism of deiodinases. In analogy to the mechanism of selenium-dependent glutathione peroxidases, a selenolate is believed to abstract an iodonium from the *ring* while a proton is donated by a general base (B). The selenenyl-iodide intermediate may be reduced by a conserved cysteine liberating the iodide ion and forming a selenenyl-sulfide intermediate. A reducing cofactor like DTT or GSH regenerates the enzyme

tissues from excess thyroid hormones, e.g., during hyperthyroid conditions or locally at specific developmental stages (Huang and Bianco 2008).

The human *DIO3* gene is located on chromosome 14 and codes for an mRNA of more than 2,100 bp resulting in translation of a protein of 304 amino acids. The gene contains no introns, but an antisense transcript. Dio3 protein is located at the plasma membrane in a large number of fetal tissues, e.g., liver, brain, lung, heart, intestine, and skin, while expression of Dio3 in postnatal tissues is more limited, e.g., to brain, skin, and retina. Dio3 activity is also found in the human placenta and the pregnant uterus where it plays a role in limiting fetal tissue exposure to thyroid hormones.

As reported for Dio2, Dio3 reaction kinetics follow a sequential mechanism with DTT as reducing cofactor. Again, the physiologic cofactor has not been identified. Dio3 is positively regulated by thyroid hormones. Thus, the enzyme homeostatically maintains adequate thyroid hormone levels during hyperthyroidism or thyrotoxic states. Conversely, Dio3 is downregulated during hypothyroidism to preserve active $T_3$ and its precursor $T_4$. The positive regulation of Dio3 by growth factors like EGF and FGF and morphogens like *sonic hedgehog* (Shh) is compatible with the observed involvement of Dio3 in developmental and regenerative processes (Dentice et al. 2005). In fact, Dio3 is elevated in wound healing or other disease states thus protecting tissues from the differentiating activity of $T_3$ or reducing energy expenditure. For example, after hepatectomy or heart ischemia, serum and liver $T_3$ and $T_4$ levels drop due to the reactivation of Dio3 in "stem-like" liver cells.

*Dio3*-KO mice exhibit significant perinatal mortality, impaired growth, and reduced fertility. In the retina, cone development is diminished by 80% due to apoptotic cell death caused by excessive thyroid hormone exposure. In addition, *Dio3*-KO mice showed increased $T_4$ levels, decreased $T_3$ levels, and a normal TSH content in serum – compatible with central hyperthyroidism during the critical period of setpoint definition in the hypothalamus.

## Deiodinases in Disease

The non-thyroidal illness syndrome (NTI or low $T_3$ syndrome) is associated with a fall in circulating free $T_3$ level and an increase in total $rT_3$ concentration. This observation is compatible with induction of an IRD enzyme, Dio1 and/or Dio3. During chronic illness, Dio3 activity is elevated while Dio2 and Dio1 activities are decreased. Inflammatory cells like neutrophilic polymorphonuclear cells (PMN) exhibit elevated Dio3 activity. It is believed that PMNs require iodide for bacterial phagocytosis and use 5-deiodination of iodothyronines as source for the halogen.

A striking example of pathogenic Dio3 overexpression is observed in hemangiomas that elicit a syndrome called consumptive hypothyroidism, where circulating thyroid hormones are quantitatively consumed by IRD (Huang et al. 2000).

The first syndrome associated with congenital impairment of deiodinase expression in humans is associated with mutations in SECIS-binding protein 2 (SBP2). In this syndrome, selenoprotein expression is impaired and, remarkably, patients are identified because of their impaired thyroid hormone metabolism and growth retardation. Patients are characterized by hearing impairment (as observed in *Dio2*-KO mice) and high $rT_3$ (as observed in *Dio1*-KO mice). DIO3-deficiency is not apparent in the patients (Dumitrescu et al. 2010).

## Summary

Deiodinases are selenoenzymes capable of iodide elimination from iodothyronines. The three deiodinase enzymes are clearly important as mediators of physiological adaption to changing demands of local and systemic thyroid hormone levels. Surprisingly, the phenotype of mice deficient in both ORD deiodinases (*Dio1/Dio2*-deficient mice) is rather mild challenging the view that deiodinase-mediated activation of thyroxine to $T_3$ is essential(Galton et al. 2009; Schweizer et al. 2008b). Mutations in the selenoprotein biosynthetic factor SECIS-binding protein 2 in humans, however, lead to a syndrome of thyroid hormone insufficiency associated with growth retardation and hearing impairment.

## Cross-References

▶ Selenoproteins and the Biosynthesis and Activity of Thyroid Hormones
▶ Selenoproteins and Thyroid Gland

## References

Berry MJ, Banu L, Chen YY, Mandel SJ, Kieffer JD, Harney JW, Larsen PR (1991a) Recognition of UGA as a selenocysteine codon in type I deiodinase requires sequences in the 3′ untranslated region. Nature 353:273–276

Berry MJ, Banu L, Larsen PR (1991b) Type I iodothyronine deiodinase is a selenocysteine-containing enzyme. Nature 349:438–440

Bianco AC, Salvatore D, Gereben B, Berry MJ, Larsen PR (2002) Biochemistry, cellular and molecular biology, and physiological roles of the iodothyronine selenodeiodinases. Endocr Rev 23:38–89

Chiu-Ugalde J, Wirth EK, Klein MO, Sapin R, Fradejas N, Renko K, Schomburg L, Köhrle J, Schweizer U (2012) Thyroid function is maintained despite increased oxidative stress in mice lacking selenoprotein biosynthesis in thyroid epithelial cells. Antioxid Redox Signal, in press

Dentice M, Bandyopadhyay A, Gereben B, Callebaut I, Christoffolete MA, Kim BW, Nissim S, Mornon JP, Zavacki AM, Zeold A, Capelo LP, Curcio-Morelli C, Ribeiro R, Harney JW, Tabin CJ, Bianco AC (2005) The hedgehog-inducible ubiquitin ligase subunit WSB-1 modulates thyroid hormone activation and PTHrP secretion in the developing growth plate. Nat Cell Biol 7:698–705

Dumitrescu AM, Di Cosmo C, Liao XH, Weiss RE, Refetoff S (2010) The syndrome of inherited partial SBP2 deficiency in humans. Antioxid Redox Signal 12:905–920

Galton VA, Schneider MJ, Clark AS, St Germain DL (2009) Life without thyroxine to 3,5,3′-triiodothyronine conversion: studies in mice devoid of the 5′-deiodinases. Endocrinology 150:2957–2963

Huang SA, Bianco AC (2008) Reawakened interest in type III iodothyronine deiodinase in critical illness and injury. Nat Clin Pract Endocrinol Metab 4:148–155

Huang SA, Tu HM, Harney JW, Venihaki M, Butte AJ, Kozakewich HP, Fishman SJ, Larsen PR (2000) Severe hypothyroidism caused by type 3 iodothyronine deiodinase in infantile hemangiomas. N Engl J Med 343:185–189

Köhrle J, Jakob F, Contempre B, Dumont JE (2005) Selenium, the thyroid, and the endocrine system. Endocr Rev 26:944–984

Kuiper GG, Kester MH, Peeters RP, Visser TJ (2005) Biochemical mechanisms of thyroid hormone deiodination. Thyroid 15:787–798

Piehl S, Hoefig CS, Scanlan TS, Köhrle J (2011) Thyronamines–past, present, and future. Endocr Rev 32:64–80

Schweizer U, Chiu J, Köhrle J (2008a) Peroxides and peroxide-degrading enzymes in the thyroid. Antioxid Redox Signal 10:1577–1592

Schweizer U, Weitzel JM, Schomburg L (2008b) Think globally: act locally. New insights into the local regulation of thyroid hormone availability challenge long accepted dogmas. Mol Cell Endocrinol 289:1–9

# Selenium and Muscle Function

Alain Lescure and Mickael Briens
UPR ARN du CNRS, Université de Strasbourg, Institut de Biologie Moléculaire et Cellulaire, Strasbourg, France

## Synonyms

Cardiovascular disorders; Cardiomyopathy; Chagas disease; Exudative diathesis; Keshan disease; Mallory Body-like Desmin-Related Myopathy (MB-DRM);

Multiminicore disease (MmD); Myopathy with Congenital Fiber Type Disproportion (CFTD); Myotonic dystrophies; Nutritional muscular dystrophy; Rigid lamb syndrome; Rigid Spine Muscular Dystrophy (RSMD1); Selenoprotein 15 kDa (Sep15); Selenoprotein K (SelK); Selenoprotein N (SelN, SepN); Selenoprotein S (SelS); Selenoprotein T (SelT); SEPN1 gene; SEPN1-related myopathies (*SEPN1*-RM); White muscle disease

## Definition

Selenium deficiencies have been linked to different forms of cardiac and skeletal muscle diseases in human and livestock, and selenium supplementation has been proved to be protective against the emergence of these disorders. Selenium is incorporated as selenocysteine into several enzymes that play a critical role in oxidative-reduction homeostasis. Several of these enzymes are supposed to play important function for muscle physiology, including oxidative stress defense and calcium handling. Notably, mutations in the gene coding for the selenoprotein N have been shown to cause different forms of inherited muscular disorders in human.

## Introduction

Shortly after selenium has acquired the status of essential trace element, a number of publications pointed to its important role in muscle development and maintenance. Selenium deficiency was linked to muscular disorders affecting both cardiac and skeletal muscles in humans and livestock (Rederstorff et al. 2006). This syndrome called nutritional muscular dystrophy is characterized by weakness and muscle pain. Development of the diseases is strongly influenced by diet and selenium supplementation was proved to be protective. Identification of the complete selenoproteome in humans and animals paved the way to a better understanding of selenium contribution to muscle physiology. However, targeted therapeutic benefits of selenium supplementation still await a mechanistic characterization of specific selenoprotein's role in these tissues.

## Selenium Function in Human Cardiovascular Disorders

Vascular endothelial cells are sensitive to oxidative stress damages, which are the causes for cardiovascular diseases such as artherosclerosis, hypertension, and congestive heart failure. Since selenoproteins are important contributors to oxidative stress defense, the use of selenium in prevention and treatment of cardiovascular disease has been considered for many years (de Lorgeril and Salen 2006; Bellinger et al. 2009). Selenoprotein's function in cardiovascular diseases has been investigated in several experimental models demonstrating an important role for the glutathione peroxidases, GPx1 GPx3 GPx4 in the protection of platelet, vascular endothelial, and smooth muscle cells against oxidative stress, cell damage, and apoptosis. In addition, increased GPxs expression has been shown to reduce ischemia-/reperfusion-induced oxidative damage and apoptosis of cardiac cells. Furthermore, another selenoprotein, the thioredoxin reductase or TrxR, being an important contributor to the intracellular oxidative regulation has been shown to participate to the regulation of cardiomyocytes' signaling and differentiation. SelK is another selenoprotein abundantly expressed in the heart and has antioxidant function in cardiomyocytes. Despite these promising results in experimental models, clinical and epidemiology studies examining selenium benefits to cardiovascular disease mortality have provided contradictory results (Bellinger et al. 2009). Further understanding of the mechanistic contribution of selenoproteins to cardiac cell physiology and regulation will help to solve the confusing clinical observations.

Keshan disease is a cardiomyopathy occurring mainly in the north-eastern area of China, in regions with low selenium soils level. Experimental studies proved a dual etiology for this disease: low selenium intake and infection by the enterovirus *Coxsakie*. Actually, it is the low GPx expression associated to the host selenium status, which influences the host response to the pathogen and induces the appearance of pathogenic mutations in the virus genome, raising its virulence and causing heart damages (Beck 2007). Another cardiac disease occasioned by selenium status and parasite infection is the Chagas disease. The initial invasion by the intracellular parasite *Trypanosoma cruzi* is followed by progressive inflammatory destruction of heart, muscles, nerves, and gastrointestinal tract tissue. Patients with the Chagas disease and low selenium level have a higher risk to develop heart dysfunction (Jelicks et al. 2011).

## Selenium Role in Muscular Disease in Humans

Decreasing selenium concentration during the evolution of the disease has been observed in blood of patients with myotonic dystrophies. However, beneficial effect of selenium treatment provided no conclusive evidences. Similarly, different responses to selenium supplementation in Duchenne or Becker muscular dystrophies suggest different etiologies (Rederstorff et al 2006). Low selenium level associated to muscle pains and congestive cardiomyopathy have been reported in HIV-infected patients or subjects on prolonged parental nutrition (Chariot and Bignani 2003). Interestingly, several reports suggest a possible connection between selenium and vitamin E deficiencies in muscular disorders, but the mechanism remains largely elusive (Rederstorff et al. 2006). All together these reports stress the important role for selenium in the evolution of muscular diseases. However, the situation is confusing and elucidation of the role played by selenoproteins in muscle should help to get a better understanding of selenium contribution and define the conditions for adequate treatments.

A proof for a direct link between selenium and muscle disorders was finally provided: Mutations in the gene coding for one selenium containing protein, the selenoprotein N or SelN, account for a group of inherited muscular disorder in human (Moghadaszadeh et al. 2001). SelN was initially discovered by a bioinformatic screen based on the identification of its SECIS element (Lescure et al. 1999), next SelN orthologs have been identified in all vertebrates and in some invertebrates, including echinoderms, annelids, or arthropods. The *SEPN1* gene codes for a protein of 590 amino acids, harboring a selenocysteine residue as part of a SCUG motif. This motif is reminiscent of the GCUG catalytic site found in the thioredoxin reductase. SelN has been found to be an *N*-glycosylated transmembrane protein specifically associated with the endoplasmic reticulum, ER. SelN enzymatic activity is presently unknown (Castets et al. 2012).

## Selenoprotein N Deficiency Causes a Group of Muscular Disease in Human

The SEPN1-related myopathies (*SEPN1*-RM) were initially characterized as four independent clinical entities: Rigid Spine Muscular Dystrophy (RSMD1; OMIM #602770), the classical form of Multiminicore disease (MmD; OMIM #255320), Mallory Body-like Desmin-Related Myopathy (MB-DRM; OMIM #602771), and myopathy with Congenital Fiber Type Disproportion (CFTD; OMIM #255310). These disorders are characterized by an early-onset clinical phenotype including predominant affection of axial muscles and generalized muscle atrophy, evolving to severe scoliosis and respiratory insufficiency. Mutations in the *SEPN1* gene have been successively identified in RSMD1, MmD, MB-DRM, and CFTD families. All mutations identified are autosomal recessive, consistent with a loss of function (Arbogast and Ferreiro 2010). It is noteworthy that identical mutations can lead to variable degrees of severity or different clinical descriptions, suggesting a progressive manifestation of the disease and the contribution of modifying factors that may modulate the evolution of the disease.

Mutations in the gene coding for the SECIS binding protein SBP2, a specific translation factor for selenocysteine insertion, have recently been linked to *SEPN1*-RM. Two nonsense mutations were identified in patients with multisystemic disease characterized by thyroid dysfunction, muscle weakness, scoliosis, and mild respiratory insufficiency (Castets et al. 2012). Partial loss of SBP2 function was shown to alter the expression of many selenoproteins, including SelN. However, previous reports on *SBP2* mutations had been described in patients with thyroid defects but presenting no muscle disorders, suggesting that the severity of the pathology, related to the set of selenoproteins affected, depends on the type of mutation.

## Animal Models for Selenoprotein N Deficiency

SelN has been found to be expressed in all tissues, contrasting with the muscle-specific affection associated to its absence in human. However, it is expressed at a higher level in fetal tissues, in agreement with the early onset of the *SEPN1*-RM diseases. During development, SelN has been shown to be mainly expressed in the zebrafish somites or in the mouse myotome, two embryonic structures that contain the muscle precursors. These observations led to the hypothesis that

SelN could play an important role during embryogenesis, in a critical step of muscle differentiation and organization. Accordingly, *Sepn1*-depleted zebrafish obtained by injection of antisense RNA analogs displayed somites' disorganization and alteration of muscles architecture, as well as defect in slow fibers' differentiation. In contrast, SelN has been shown to be dispensable for muscle development in mouse: *Sepn1* knockout mice are indistinguishable from wild-type and exhibit no particular muscular defects, neither in terms of histology nor in their performance (Castets et al. 2012). On the other hand, the *Sepn1*-deficient mice submitted to a force swimming test, a physical challenge associated to a general stress context, developed muscle dysfunction, partly recapitulating the patients' symptoms. The mutant mice displayed a severe curvature of the spine and major alterations of the paravertebral muscles, revealing a sensitivity of exercised SelN-deficient mouse muscle to recurrent stress. These studies clearly show that SelN is involved in a pathway mostly limiting in normal muscles' function, in particular in the muscles of trunk.

Another study on the *Sepn1* knockout mice demonstrates the importance for SelN in the maintenance of the satellite cells, quiescent muscle progenitors involved in adult muscle repair. SelN expression is strongly upregulated in mononucleated muscle precursors during muscle regeneration and SelN-deficient muscles failed to regenerate after recurrent injuries, due to the depletion of satellite cell pool. Most importantly, a reduced satellite cell pool has also been observed in muscle biopsies from *SEPN1*-RM patients and worsening with age (Castets et al. 2012).

## Molecular Pathways Affected by SelN Dysfunction

SelN has been shown to play an important role in muscle development and maintenance, yet our understanding of SelN function at the molecular and cellular levels is coarse and incomplete. Based on the presence of selenocysteine residue in its predicted catalytic site, SelN is supposed to be an enzyme involved in an oxidation-reduction reaction. *In silico* predictions also revealed the presence in the promoter of the *SEPN1* gene of putative binding sites for the transcription factors NF-KB, an ER stress response element, and AP-1, an oxidation-sensitive transcription factor. Recent data indicate that SelN is involved in oxidative stress control. Studies on cultured cells reported an increased level of oxidized proteins, representative of an increased basal oxidative activity, in SelN-deficient myotubes from SEPN1-RM patients or SelN knockdown C2C12 cells. In addition, SelN-depleted cells exhibited an increased susceptibility to $H_2O_2$ treatment. This oxidative stress sensitivity was abrogated by a pretreatment of the SelN-deficient cells with the antioxidant N-acetylcysteine (NAC) (Arbogast and Ferreiro 2010). Therefore, it has been postulated that SelN may participate in the cell oxidative defense in muscles, by buffering reactive oxygen species (ROS) and participating in the repair of damaged macromolecules, such as proteins. SelN might also contribute to ER stress oxidative defense by regulating secreted protein's maturation, folding, trafficking, and stability. A similar function has already been described for a group of other selenoproteins of the ER, such as SelS, SelK, and Sep15. In agreement with this hypothesis, accumulation of tubular aggregates has been described in muscles of *Sepn1* knockout mice submitted to a force swimming test, consistent with an increased ER stress. On the other hand, the antioxidant function of SelN could also be limited to specific targets with critical function in muscle.

The ryanodine receptor 1, RyR1, is a calcium transport channel involved in muscle contraction, whose activity is regulated by the oxidative status of exposed cysteines. It has been demonstrated that SelN and RyR1 colocalize in the terminal cisternae of ER rabbit muscle and that RyR1 activity or its ryanodine affinity was modified in the absence of SelN in both human and zebra fish extracts. Complementation of the protein extracts with unmodified SelN protein restored normal ryanodine binding and response to the oxidative status. In agreement with a deregulation of RyR activity, SelN-deficient zebra fish embryos displayed an abnormal calcium handling in the Kupffer's vesicle, and myotubes from *SEPN1*-RM patients exhibited a higher cytoplasmic calcium concentration. Importantly, other selenoproteins, such as the selenoprotein T, have been also shown to contribute to intracellular calcium concentration control. SelN may either act as a chaperon controlling RyR folding or be involved in the repair of RyR under oxidative stress conditions (Castets et al. 2012).

## Muscle Diseases Linked to Selenium Deficiency in Livestock

Before being recognized as an essential element, selenium was considered for a long time as a poisonous compound. Several examples reported livestock intoxication due to consumption of selenium accumulator plants. On the contrary, animals fed with selenium-deficient diets show typical disorders associated with muscle syndromes. White muscle disease affecting skeletal and cardiac muscle has been described in pigs, cattle, sheep, poultry, horses, and fish (Oldfield 1987; Rederstorff et al. 2006). Fiber alterations and abnormal calcium deposition in tissues are characteristics of this disease, and have been shown to be cured by selenium supplementation. The white muscle disease has been linked to an altered expression of selenoprotein W. Other forms of nutritional muscular dystrophy have been associated to a lack of selenium: in poultry, exudative diathesis, a muscle disorder characterized by fluid accumulation in connective tissue; in sheep, the rigid lamb syndrome, an affection with spinal muscle rigidity resembling the symptoms described in *SEPN1*-RM patients. Notably, there are several indications to show that a combination of selenium deficiency and low vitamin E supply exacerbates the symptoms previously described (Rederstorff et al. 2006; Surai 2002).

Selenium is provided to animals mainly as selenomethionine through diet grains and plants or as feed supplement (sodium selenite or selenomethionine enriched yeasts). After absorption and metabolism, selenium present in muscle is almost exclusively converted into the two amino acids, selenomethionine and selenocysteine. Selenomethionine is nonspecifically incorporated into proteins in place of methionine, whereas selenocysteine is specifically incorporated into selenoproteins. The selenomethionine and selenocysteine ratio in muscle has been demonstrated to be linked to its metabolic activity. Hence, oxidative muscles showed higher proportions of selenocysteine than glycolytic muscles, associated to higher redox activity and selenoprotein requirement (Juniper et al. 2011).

## Cross-References

▶ Selenium and Glutathione Peroxidases
▶ Selenium in Human Health and Disease
▶ Selenium, Biologically Active Compounds
▶ Selenoprotein K
▶ Selenoprotein Sep15

## References

Arbogast S, Ferreiro A (2010) Selenoproteins and protection against oxidative stress: selenoprotein N as a novel player at the crossroads of redox signaling and calcium homeostasis. Antioxid Redox Signal 12:893–904

Beck MA (2007) Selenium and vitamin E status: impact on viral pathogenicity. J Nutr 137:1338–1340

Bellinger FP, Raman AV, Reeves MA, Berry MJ (2009) Regulation and function of selenoproteins in human disease. Biochem J 422:11–22

Castets P, Lescure A, Guicheney P, Allamand V (2012) Selenoprotein N in skeletal muscle: from diseases to function. J Mol Med (Berl) [Epub ahead of print]

Chariot P, Bignani O (2003) Skeletal muscle disorders associated with selenium deficiency in humans. Muscle Nerve 27:662–668

de Lorgeril M, Salen P (2006) Selenium and antioxidant defenses as major mediators in the development of chronic heart failure. Heart Fail Rev 11:13–17

Jelicks LA, de Souza AP, Araújo-Jorge TC, Tanowitz HB (2011) Would selenium supplementation aid in therapy for chagas disease? Trends Parasitol 27:102–105

Juniper DT, Phipps RH, Bertin G (2011) Effect of dietary supplementation with selenium-enriched yeast or sodium selenite on selenium tissue distribution and meat quality in commercial-line turkeys. Animal 5:1751–1760

Lescure A, Gautheret D, Carbon P, Krol A (1999) Novel selenoproteins identified in silico and in vivo by using a conserved RNA structural motif. J Biol Chem 274:38147–38154

Moghadaszadeh B, Petit N, Jaillard C et al (2001) Mutations in SEPN1 cause congenital muscular dystrophy with spinal rigidity and restrictive respiratory syndrome. Nat Genet 29:17–18

Oldfield J (1987) The two faces of selenium. J Nutr 117:2002–2008

Rederstorff M, Krol A, Lescure A (2006) Understanding the importance of selenium and selenoproteins in muscle function. Cell Mol Life Sci 63:52–59

Surai PF (2002) Selenium in poultry nutrition 1. Antioxidant properties, deficiency and toxicity. Worlds Poult Sci J 58:333–347

# Selenium and the Control of Cell Growth

▶ Selenium and the Regulation of Cell Cycle and Apoptosis

# Selenium and the Regulation of Cell Cycle and Apoptosis

Huawei Zeng
United States Department of Agriculture, Agricultural Research Service, Grand Forks Human Nutrition Research Center, Grand Forks, ND, USA

## Synonyms

Selenium and the control of cell growth; The effect of selenium on cell proliferation

## Definition

Selenium (Se) is an essential trace mineral nutrient for humans and animals. At nutritional doses, Se functions through selenoproteins, several of which are oxidant-defense enzymes. At supranutritional doses, small Se metabolite molecules (e.g., $H_2Se$ and $CH_3SeH$) are the major functional pools in vivo (Allan et al. 1999). The regulation of cell cycle and apoptosis is a key mechanism by which both selenoproteins and Se metabolites exert their biological function. Health effects of Se include cardiovascular disease (CVD), cancer, diabetes, inflammatory disorders, male fertility, and others (Brigelius-Flohe 2008; Fairweather-Tait et al. 2011).

## Introduction

Selenium was discovered by Berzelius in 1817, and the essentiality of Se was suggested by the early experiment that trace amounts of Se protected against liver necrosis in vitamin E deficient rats in the mid-1950s. The benefits of Se include protection against cancer, heart diseases, muscle disorders, immunity, and age-related diseases. The health-promoting properties of Se are due to vital functions of selenoproteins, and Se compounds have been considered to act as Se donors for selenoprotein biosynthesis which is required for normal physiology. However, chemoprevention, in most cases, is obtained with small Se-metabolite molecules such as selenide and methylselenol through supranutritional intake (Brozmanova et al. 2010). Investigations into the protective role of Se have been undertaken for many years both in animal and case-control, but not all nutritional intervention studies show that high Se intakes effectively reduce cancer risk. This is because there is a relatively narrow margin between Se intakes that result in deficiency or toxicity. Various underlying mechanisms have been presented which, however, may differ depending on the chemical form of Se and the time point of intervention (Jackson and Combs 2008; Rayman et al. 2009).

The regulation of cell cycle and apoptosis is the most fundamental biological process related to Se's health-promoting properties. It is well recognized that Se plays a critical role in cell proliferation but the mechanism remains elusive. In humans and animals, cell proliferation must be regulated to maintain tissue homeostasis. The eukaryotic cell cycle is divided into four major phases as follows: the G1 phase before DNA replication, the periods of DNA synthesis (S phase), the G2 phase before cell division, and cell division (M phase). The cell cycle is a conserved mechanism by which eukaryotic cells replicate themselves. Apoptosis, a programmed cell death, enables an organism to eliminate unwanted and defective cells during normal development, turnover, and pathological conditions. There are several putative mechanisms by which Se modulates cell cycle and apoptosis, and these mechanistic aspects may reveal the relationship between Se and health outcomes, including cancer and inflammatory disorders.

## Selenium Metabolism

The chemical and physical properties of Se are very similar to sulfur. Selenium is biologically active in a variety of covalent compounds including inorganic salts, amino acids, and methylated compounds. The compounds available for use as Se supplements include the inorganic forms (sodium selenite/selenate)

---

Disclaimer The U.S. Department of Agriculture, Agricultural Research Service, Northern Plains Area, is an equal opportunity/affirmative action employer and all agency services are available without discrimination.Mention of a trademark or proprietary product does not constitute a guarantee or warranty of the product by the U.S. Department of Agriculture and does not imply its approval to the exclusion of other products that may also be suitable. This work was supported by the US Department of Agriculture.

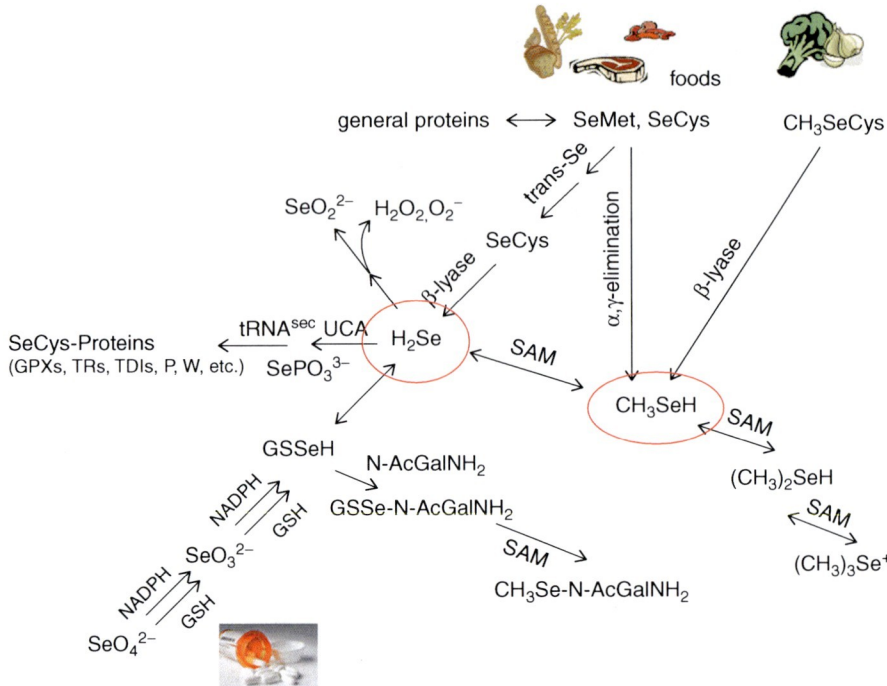

**Selenium and the Regulation of Cell Cycle and Apoptosis, Fig. 1** Proposed pathways for the metabolism of biologically important selenomolecules (Adapted from Combs 2004 and Rayman 2005)

and organic forms (SeMet and SeCys). Although not all of these forms are metabolized alike, the metabolism of both organic and inorganic Se forms shows certain similarities (Fig. 1). There are two major Se metabolic pathways, and selenide is the branch point molecule. The selenoprotein synthesis is the first pathway which includes the co-translational biosynthesis of SeCys and its incorporation into specific selenoproteins. This process is tightly regulated, and requires at least five unique gene products (four proteins/enzymes and a unique tRNA). In addition, each selenoprotein mRNA must contain two specific mRNA elements (a novel use of the UGA codon plus a unique SECIS element) for Se incorporation to take place. Selenoproteins are found in all three kingdoms (prokaryotic, eukaryotic, and archaea), and the mechanism of SeCys synthesis and incorporation is generally the same in all three kingdoms. Because selenoprotein expression is tightly regulated, the second pathway is that Se in excess of these needs enters an excretory pathway, methylation and sugar-derivation of selenides from the major excretory products. Selenide is converted in the liver to a selenosugar, which is the major urinary form. Selenide can also be methylated using S-adenosylmethionine (SAM) by either methyl transferases to form methyl selenol, trimethyl selenonium ion, and dimethylselenide. The excretion of Se occurs by the methylation or sugar derivation of selenides. The pattern of these excretory forms as well as the total amount of Se excreted is influenced by both the level of Se intake and the physiological Se status (Sunde 2006).

## The Effect of Se on Cell Proliferation at Nutritional Doses

Selenium intake plays a critical role in human health including protection against cancer, heart diseases, and other metabolic disorders. Selenoproteins are the key factors in Se essentiality, and about 25 different SeCys-containing selenoproteins have so far been observed in human tissues. Many are involved in oxidative stress protection or in maintaining cellular redox balance. The sulfur amino acids, cysteine and methionine, are the main targets of reactive oxygen species in proteins. Selenoproteins are differentially expressed in specific tissues and developmental stages, and in response to various environmental stimuli. Because of its chemical similarity to sulfur and stronger nucleophilicity and acidity, selenium is an extremely efficient catalyst of reactions between sulfur and oxygen. For example, it has been shown that the expression levels of the selenoproteins GPx1, GPx4, SEPS1, Sep15, SEPP1, and TXNRD1 are related to cardiovascular disease, pre-eclampsia, and cancer. Thus, Se would guarantee normal physiology. The extreme situations of deficiency and nutritional doses are relatively straightforward. The recommended daily allowance for Se is 55 μg/day for healthy Americans. At nutritional levels, selenoproteins exhibit a wide range of cellular function such as cellular antioxidative protection and redox regulation, and the best examples are that the most abundant selenoproteins in mammals are glutathione peroxidases (GPXs, [GPX1, 2, 3, 4, 6]). These GPX enzymes are redox regulators, eliminate hydrogen peroxide and damaging lipid and phospholipid hydroperoxides generated in vivo by free radicals and other oxygen-derived species at the expense of glutathione (Hawkes and Alkan 2010). At concentrations of nmol/L, Se is essential for certain mammalian cell growth cells in culture. For example, Se-deficient lymphocytes are less able to proliferate in response to a mitogen, but the response can be improved by Se supplementation (Fig. 2). The observation is consistent with the reports that Se is essential and directly affects both the innate and acquired immune systems. Recent studies demonstrate that Se (nmol/L) upregulates multiple key cell cycle–related gene mRNA levels, and enhances total cellular phosphorylated proteins. In addition, Se leads to the promotion of cell cycle progression, particularly the G2/M transition, and/or a reduction of apoptosis primarily in G1 cells, which is essential to cell growth and proliferation (Zeng 2002). Certain mechanisms have been suggested to explain the anticarcinogenic action of selenoproteins in reducing DNA damage, oxidative stress reduction, reduced inflammation, detoxification, improved immune response, increased tumor suppressor protein p53, inactivation of protein kinase C, alteration in DNA methylation, cell cycle arrest, induction of apoptosis of cancer cells, and inhibition of angiogenesis. However, reduced activity of selenoproteins might result in a compensatory increase of non-Se-dependent antioxidants to counteract the damaging effect of oxidative stress. These observations suggest that selenoproteins participate in multiple molecular pathways and that their expression may be tightly associated with complex regulatory networks and signaling.

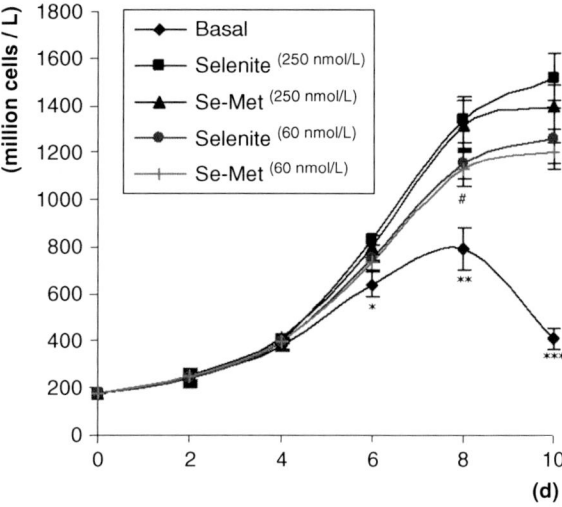

**Selenium and the Regulation of Cell Cycle and Apoptosis, Fig. 2** Effect of selenium on the growth of HL-60 cells in the absence of selenium (basal) or in the presence of selenite or selenomethionine (Se-Met), respectively. *Panel A*: HL-60 cell growth curve. Values are means ± SD, $n = 4$. Cell density was greater in HL-60 cells supplemented with selenite or Se-Met than in selenium-deficient cells (*$P < 0.03$ at 0.25 Se μmol/L); (**$P < 0.01$ at 0.25 Se μmol/L; $P < 0.05$ at 0.06 Se μmol/L); (***$P < 0.002$ at 0.25 Se μmol/L; $P < 0.01$ at 0.06 Se μmol/L). (Adapted from Zeng 2002)

Although the deprivation of Se can reduce the protection against oxidative stress and impair immunocompetence, certain cancer cells appear to have acquired a selective survival advantage that is apparent under conditions of Se deficiency and oxidative stress. A recent report showed that most hepatocellular carcinoma cell lines (10 of 13), breast cancer cell lines (11 of 14), colon cancer cell lines (8 of 10), and all melanoma cell lines tested were resistant to se deficiency-induced cell death (Imak et al. 2003). Interestingly, the more recent data also indicated that Se deficiency has little effect on the growth of certain colon cancer cell lines such as Caco-2 cells even though GPX activity of Se-deficient cell was 11 mU/mg protein compared to 140 mU/mg protein in cell supplemented with Se. However, Se may exert its anticancer property by increasing the expression of humoral defense gene and tumor suppressor–related genes while decreasing the expression of pro-inflammatory genes in the same Caco-2 cells.

Therefore, the proliferation of certain cell types such as immune cells is extremely sensitive to Se-deprivation while others are not. However, Se may still modulate the gene expression in those cells which tolerate Se deficiency. Thus, an adequate Se status with optimal selenoprotein expression might contribute to the prevention of cancer initiation (Zeng 2009).

## The Effect of Se on Cell Proliferation at Supranutritional Doses

Although the recommended daily allowance for Se is 55 μg/day for healthy Americans, doses of 100–200 μg Se/day have been reported to inhibit genetic damage and cancer development in human subjects, and about 400 μg Se/day is considered an upper safe limit. Thus, doses of 100–200 μg Se/day are considered supranutritional doses which is above the nutritional requirement that support the maximal expression of selenoproteins but lower than the toxic level. Similarly, Se is an essential micronutrient at levels of ~0.1–0.2 μg/day in the diets of experimental animal but it becomes a toxic at levels exceeding 5 μg Se/day. In addition, Se at ~0.05–0.2 μmol/L, but not at >1 μmol/L, is necessary for growth of many human cells in culture. Data from various preclinical and epidemiological studies and from some clinical trials justify consideration of Se as a potential chemopreventive agent for certain cancers. Both inorganic and organic Se compounds can be antitumorigenic at supranutritional doses. The inhibition of carcinogen activation, reduction of angiogenesis, and regulation of cell cycle and apoptosis are key proposed Se's anticancer mechanisms. It has been documented that Se compounds can interact with thiols, and hydrogen selenide ($H_2Se$) and methylselenol ($CH_3SeH$) are major pools of Se metabolites which are greatly increased because of supranutritional doses. The active anticancer Se metabolite is likely a monomethylated Se species, possibly methylselenol, and that the chemopreventing efficacy of a given Se compound might depend on rate of its metabolic conversion to that active Se form. Methylselenol displays greater preventing efficacy than selenide in rodent carcinogenesis model. This is because selenide and methylselenol induce distinct types of biochemical and cellular responses. Selenide precursors such as selenite and SeCys generate superoxide, induce DNA single breaks (genotoxicity), and cause S-phase cell arrest/ROS-induced apoptosis. In contrast, metabolic precursors of methylselenol (methylselenocyanate, SeMSC) cause G1-phase cell arrest as there is a decrease in the cdk2 kinase activity accompanied by a decrease in cyclin E-cdk2 content. Earlier studies had also found that metabolic precursors of methylselenol reduced cell proliferation, as determined by $^3$H-thymidine incorporation into DNA of the cells at the G1 phase

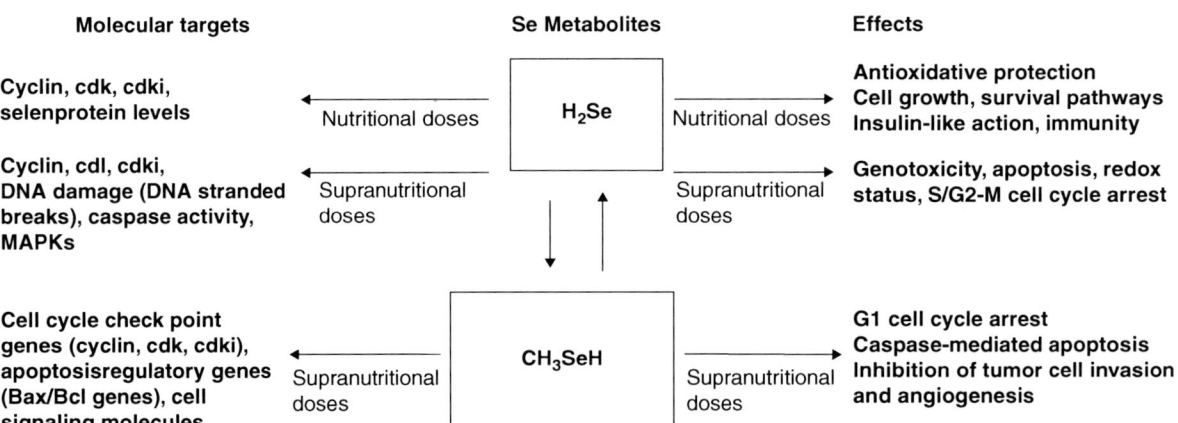

**Selenium and the Regulation of Cell Cycle and Apoptosis, Fig. 3** Molecular targets and cellular effects of hydrogen selenide and methylselenol (Adapted from Lu and Jiang 2001)

of the cell cycle, whereas methylselenol precursor rapidly blocked DNA synthesis and arrested cells in the S phase. Although methylselenol also induces the same G1 cell cycle arrest, its molecular targets are likely cell- or tissue-type dependent, and it involves different cyclin/cdk/cdki and upstream mitogenic signaling pathways without inducing DNA single-strand breaks. Microarray profiling analyses showed that Se-treated cells altered the expression of several genes related to cell cycle/apoptosis in a manner related to cancer prevention. The Se treatment upregulated genes involved in phase II detoxication enzymes, Se-binding proteins, and apoptosis. In addition, Se also resulted in the downregulation of genes related to phase I–activating enzymes and cell proliferation (EI-Bayoumy and Sinha 2005). The proposed mechanism of Se's anticancer property is that methylselenol caused cell cycle arrest, and led to an increase in the G1 and G2 fractions with a concomitant drop in the S-phase. In addition, it also activates the caspase cascade and apoptosis in the tumor cells. Furthermore, these observations were extended to a parallel comparison study in which methylselenol showed stronger potential of inhibiting cell proliferation/survival signals in certain cancerous colon cells when compared with that in noncancerous colon cells. Overall, these data consistently support that methylselenol, produced by supranutritional intake, is a critical anticancer Se metabolite.

## Summary

The dominant forms of Se found in foods are SeMet and SeCys, with smaller amounts of methylated selenides. Dietary supplements may include inorganic Se-salts such as selenite and selenate. As an important micronutrient, Se plays a critical role in Se's biological function, and its biological efficacy depends on both dose and chemical form. At nutritional doses, Se is an essential component of SeCys in selenoproteins, and it promotes cell cycle progression and prevents cell death. In contrast, at supranutritional doses that are greater than the nutritional requirement but not toxic, Se induces cell cycle arrest and apoptosis. The Se-regulation of selenoprotein expression, cell cycle, and apoptosis is the major mechanism by which Se exerts its cellular functions (Fig. 3).

## Cross-References

▶ Selenium in Human Health and Disease
▶ Selenium-Binding Protein 1 and Cancer

## References

Allan BC, Lacourciere GM, Stadman TC (1999) Responsiveness of selenoproteins to dietary selenium. Annu Rev Nutr 19:1–6

Brigelius-Flohe R (2008) Selenium compounds and selenoproteins in cancer. Chem Biodivers 5:389–395

Brozmanova J, Manikova D, Vlckova V, Chovanec M (2010) Selenium: a double-edged sword for defense and offence in cancer. Arch Toxicol 84:919–938

Combs GF Jr (2004) Status of selenium in prostate cancer prevention. Br J Cancer 91:195–199

El-Bayoumy K, Sinha R (2005) Molecular chemoprevention by selenium: a genomic approach. Mutat Res 591:224–236

Fairweather-Tait SJ, Bao Y, Broadley MR, Collings R, Ford D, Hesketh JE, Hurst R (2011) Selenium in human health and disease. Antioxid Redox Signal 14:1337–1383

Hawkes WC, Alkan Z (2010) Regulation of redox signaling by selenoproteins. Biol Trace Elem Res 134:235–251

Irmak MB, Ince G, Ozturk M, Cetin-Atalay R (2003) Acquired tolerance of hepatocellular carcinoma cells to selenium deficiency: a selective survival mechanism? Cancer Res 63:6707–6715

Jackson MI, Combs GF Jr (2008) Selenium and anticarcinogenesis: underlying mechanisms. Curr Opin Clin Nutr Metab Care 11:718–726

Lu J, Jiang C (2001) Antiangiogenic activity of selenium in cancer chemoprevention: metabolite-specific effects. Nutr Cancer 40:64–73

Rayman MP (2005) Selenium in cancer prevention: a review of the evidence and mechanism of action. Proc Nutr Soc 64:527–542

Rayman MP, Combs GF Jr, Waters DJ (2009) Selenium and vitamin E supplementation for cancer prevention. JAMA 301:1876

Sunde RA (2006) Selenium. In: Bowman BA, Russell RM (eds) Present knowledge in nutrition, 9th edn. ILSI Press, Washington, DC

Zeng H (2002) Selenite and selenomethionine promote HL-60 cell cycle progression. J Nutr 132:674–679

Zeng H (2009) Selenium as an essential micronutrient: roles in cell cycle and apoptosis. Molecules 14:1263–1278

## Selenium Donor Protein

▶ Selenophosphate Synthetase

# Selenium in Human Health and Disease

Lucia A. Seale and Marla J. Berry
Department of Cell & Molecular Biology, John A. Burns School of Medicine, University of Hawaii at Manoa, Honolulu, HI, USA

## Synonyms

Selenium; Selenocysteine; Selenomethionine; Selenoprotein; Selenoproteome

## Definition

Selenium is a nonmetal trace element with atomic number 34 and atomic mass of 78.96. It is mostly found in nature as inorganic selenite, or organic amino acids selenomethionine and selenocysteine. Selenocysteine is present in the catalytic sites of selenoproteins. Humans have 25 selenoproteins. Most of the selenoproteins with identified function work as antioxidant enzymes, with the exception of the deiodinases, selenoprotein P, and selenophosphate synthethase 2, which regulate thyroid hormone levels, transport selenium through the bloodstream, and autoregulate the synthesis of selenoproteins themselves, respectively. Selenium and selenoproteins are implicated in several aspects of human health and disease, such as male fertility; neurological, endocrine, and immune function; and cancer. These implications have been widely studied and are discussed in this entry.

## Introduction

Selenium (Se) was discovered in 1817 by Swedish chemist John Berzelius, who isolated the new rare element from reddish mud and named it after the Greek mythological Moon goddess Selene. For decades Se was mainly recognized as a toxic and carcinogenic element for livestock animals and humans, respectively. Se toxicity, or selenosis, is defined by the chronic intake of more than 400 µg daily. Selenosis is characterized in humans by garlic breath, nail and hair loss, disorders of the nervous system, and paralysis (Rayman 2012). This perspective did not change until 1957, when it was reported that rats deficient in Se and vitamin E developed liver necrosis.

Severe Se deficiency is found in northeastern areas of China, leading to Keshan disease, a reversible congestive cardiomyopathy that may be caused by a Coxsackie virus infection on a background of Se deficiency (Schweizer et al. 2011). Keshan disease was virtually eradicated by supplementation of Se to the affected populations. Se deficiency along with iodine deficiency contributes to Kashin-Beck disease, a degenerative osteoarthropathy found in northeast China, Mongolia, Siberia, and North Korea. Both Keshan and Kashin-Beck diseases are associated with disturbances of selenoprotein expression and function.

Nevertheless, Se toxicity and severe Se deficiency are considered to be rare pathologies. More commonly, slight changes in Se availability are found in the human population, and these moderate changes are the ones implicated in far more disease states. On the other hand, the role of Se deficiency and/or supplementation in the pathogenesis of several diseases has been increasingly recognized.

## Se Intake

Se intake varies widely throughout the world, ranging from 7 to 4,990 µg per day, with mean values of 40 µg in Europe and from 93 (women) to 134 µg (men) in the USA (Rayman 2012). The world's most Se-deficient areas are found in China and New Zealand, although the latter has overcome the deficiency by importing Se-rich wheat from Australia. Nevertheless, the local values still correlate strongly with the existing amounts in the soil or livestock in the surrounding area. It is worth noting that the higher mean intake reported for the USA is closely related to the fact that almost 50% of the American population takes dietary supplements that are rich in Se.

The World Health Organization recommends a daily intake of 34 µg for adult men and 26 µg for adult women. These values are slightly lower for children (6–21 µg) and higher for pregnant or lactating women (28–42 µg). The calculation for such recommendations is based on values of Se deficiency that are able to epidemiologically cause a significant impairment in health, like in areas of China where Keshan disease is prevalent due to low Se in the soil (Yang et al. 1984). However, for the proper maintenance of Se metabolism, researchers still struggle to achieve a consensus on a common recommended value.

On a study in Keshan disease areas of China, it was found that providing 41 μg of daily Se intake to a very Se-deficient person was sufficient to bring the plasma Se levels to a satisfactory level. Nevertheless, other researchers will define an adequate intake as the amount to guarantee unrestricted selenoprotein expression; this ranges from 40 to 55 μg daily.

The 2005 United States Department of Agriculture National Nutrient Database for Standard Reference listing of foods with high concentrations of Se includes Brazil nuts, most seafood, barley, wheat, and certain meat products (turkey, chicken breast, duck). However, one should remember that the Se content of crops varies greatly depending on the surrounding environment in which it grows.

Dietary Se is found in various chemical forms, and its efficient uptake and metabolism will depend on which form it was ingested. Inorganic Se consists predominantly of selenite and selenate, both water soluble and strong oxidizers. Organic forms of Se include predominantly the amino acids selenomethionine (SeMet) and selenocysteine (Sec), although rare forms such as selenoneine, Se-methylselenocysteine, and g-glutamyl-Se-methylselenocysteine may have important biological roles that remain to be unveiled (Rayman 2012). A comparison between the uptake of inorganic and organic forms of Se demonstrated that approximately 57% of selenite is absorbed by the gastrointestinal tract, while ~97% of SeMet is uptaken by the same route. Once these forms are absorbed, their metabolic fate is similar; they are converted to selenide in order to act in selenoprotein biosynthesis. Neither Sec nor SeMet accumulate in biological systems, being promptly converted into selenide via recycling of Sec or the transsulfuration pathway led by the action of Sec lyase or Cystathione γ-lyase, respectively.

The molecular structures of the amino acids cysteine (Cys) and Sec are also analogs, but Sec is exclusively inserted into genetically unique selenoprotein families, requiring a recoding of the UGA stop codon, and specific factors that function in its co-translational incorporation into selenoproteins.

The variability of Se pools in different tissues of the body may be associated with the variations in hierarchy of selenoprotein production on several levels, including transcription, translation, and turnover. When Se is limiting, prioritization of the synthesis of some selenoproteins occurs over others, as well as protection of the Se pool in some organs, such as brain, testis, and thyroid. Tissues that uptake more Se, such as liver and kidneys, are likely to possess tighter regulation of selenoprotein biosynthesis.

## Selenium, Selenoproteins, and Human Health

Humans have 25 selenoproteins (Kryukov et al. 2003) and mice have 24 (summarized in Table 1), all containing the highly reactive Sec in the active site. Sec is usually deprotonated at physiological pH, which renders its highly reactive state, and increases the efficiency of selenoprotein enzyme activity. Most selenoproteins participate in reactions involving redox state changes, thus their function is largely to promote antioxidant defense or detoxification. There are three main families of selenoproteins in humans: glutathione peroxidases (GPx), thioredoxin reductases (TrxR), and iodothyronine deiodinases (Dio). SPS2 is a factor involved in the process of Sec incorporation, thus displaying an auto-regulatory function. Besides these, an alphabet soup of selenoproteins including Selenoproteins H, K, M, N, O, P, R, S, T, V, and W was identified primarily through bioinformatics approaches. A brief description of these selenoproteins is given in Table 1.

Se is transported through the circulatory system bound to plasma proteins such as selenoalbumin, and as Sec residues incorporated into the primary sequence of glutathione peroxidase 3 (GPx3) and Selenoprotein P (Sepp1). GPx3 plasma levels depend on Se status and individual hypoxic state, and its pool encompasses approximately 20% of the circulating Se. In addition, GPx3 regulates the bioavailability of nitric oxide produced by the platelets and vascular cells, and is one of the highest expressed selenoproteins in the thyroid gland, an organ that maintains high Se content even during Se deficiency. Sepp1 is a unique vertebrate selenoprotein containing 10 Sec residues per protein molecule in mice and humans. The Sepp1 primary sequence contains one Sec in the amino terminal region of the protein and a cluster of Secs in the middle to the carboxyterminal region. The liver is the primary site of Sepp1 production. In hepatocytes, Sepp1 can potentially bypass the Golgi apparatus and be secreted as an incompletely processed form. This secretory process is dependent on extracellular $Ca^{2+}$ as well as $Ca^{2+}$ from the endoplasmic reticulum (ER). After Sepp1 is released into the bloodstream, it functions as the main Se carrier for other tissues in the body. Sepp 1 is responsible for homeostatic redistribution of tissue

**Selenium in Human Health and Disease, Table 1** Summary of Selenoproteins and their functions and health-related characteristics

| Selenoprotein | Function and health-related characteristics |
|---|---|
| Glutathione peroxidase 1 (GPx1) | Very sensitive to Se status. Reduces hydroperoxides, thus avoiding propagation of free radicals and reactive oxygen species. Overexpression of GPx1 in mice leads to hyperinsulinemia and insulin resistance. Deficiency causes cardiomyopathies. |
| Glutathione peroxidase 2 (GPx2) | Mostly found in the gastrointestinal tract. Helps to maintain intestinal mucosal integrity and anti-apoptosic action in colon crypts. |
| Glutathione peroxidase 3 (GPx3) | Produced in the kidneys and released to bloodstream, where it is a Se biomarker. Important for cardiovascular protection, perhaps through modulation of nitric oxide levels; antioxidant function in the thyroid gland. |
| Glutathione peroxidase 4 (GPx4) | Expression is relatively resistant to dietary Se changes. GPx4 acts on phospholipids as antioxidant and a sensor of oxidative stress and pro-apotic signals in cells. It is a structural protein in sperm. |
| Glutathione peroxidase 6 (GPx6) | Importance is unknown. Expression restricted to the olfactory epithelium of adults. |
| Thioredoxin reductase type I (TrxR1) | Genetic deletion in mice is embryonic lethal. Controls the reduced state of thioredoxin, a class of small redox proteins. Modulates TNFα signaling and NFκ B activation. |
| Thioredoxin reductase type II (TrxR2) | Genetic deletion in mice is embryonic lethal. Controls and regulates redox state of mitochondria. Highly expressed in reproductive tissues. May play a role in redo-regulated cell signaling. |
| Thioredoxin reductase type III (TrxR3) | Testes-specific expression. Reduces glutathione disulfide. May play a role in sperm maturation. Specific physiologic function is undefined. |
| Type 1 Iodothyronine deiodinase (Dio1) | Important for systemic activation of thyroid hormone. |
| Type 2 Iodothyronine deiodinase (Dio2) | Important for peripheral activation of thyroid hormone. Involved in body thermal estability, in auditory function and in the pituitary feedback regulation of thyroid hormone levels. |
| Type 3 Iodothyronine deiodinase (Dio3) | Inactivates thyroid hormone. Highly expressed in placenta and fetal tissues, prevents fetal overexposure to T4. |
| 15-kDa selenoprotein (Sep 15) | ER resident. Gene polymorphism linked to gastric cancer. Involved in colon cancer tumor growth. |
| Selenoprotein H (SelH) | Nucleolar transcription factor. Overexpression in neuronal cells protects against UV-induced oxidative damage. |
| Selenoprotein I (SelI) | Possibly involved in phospholipid biosynthesis. |
| Selenoprotein K (SelK) | Involved in immune response and in ER-related protein misfolding mechanisms. |
| Selenoprotein M (SelM) | Regulates $Ca^{2+}$ homeostasis in neurons. |
| Selenoprotein N (SelN) | Potential role in early muscle formation; involved in calcium mobilization from ryanodine-receptors of the ER; mutations lead to multiminicore disease, rigid spine muscular dystrophy and other myopathies. |
| Selenoprotein O (SelO) | Contains a Cys-X-X-Sec motif suggestive of redox function, but physiological role remains unknown. |
| Selenoprotein P (Sepp1) | Serves as a biomarker for Se status. Se transport to brain and testes. Increased serum levels correlate with insulin resistance. Additional function as intracellular antioxidant in phagocytes. Brain levels linked to Alzheimer's and Parkinson's Disease. |
| Selenoprotein R or Methionine-R-sulfoxide reductase B1 (SelR or MsrB1) | Catalyzes the Zn-depent reduction of methionine-R-sulfoxides. |
| Selenoprotein S (SelS) | May be involved in ER stress. Anti-inflammatory function. Expression is regulated by glucose levels. Gene polymorphisms linked to colorectal and gastric cancers, and to risk of pre-eclampsia and ischaemic stroke. |
| Selenoprotein T (SelT) | ER protein involved in calcium mobilization for neuropeptide PACAP release from the pituitary. May be involved in keratinocyte function and release. |
| Selenoprotein V (SelV) | Testes-specific expression. Function is unknown. |
| Selenoprotein W (SelW) | Highly dependent on adequate dietary Se levels as well as levels of Sepp1. Putative antioxidant role, perhaps important in muscle growth and differentiation. |
| Selenophosphate synthetase 2 (SPS2) | Involved in synthesis of all selenoproteins, including itself. |

Se reserves to vital tissues, such as brain and testis. The specific mechanism of Se transport into the cells is unknown, but there is evidence of interaction between Sepp1 and Apolipoprotein-E receptor 2 (ApoER2) in the brain and testis, and interaction between Sepp1 and megalin in the kidneys. Whole-body disruption of Sepp1 expression in a mouse model revealed more details of the hierarchical regulation of selenoprotein transcription. Brain and testis, as well as a subset of selenoproteins, appear to be more affected by the absence of Se transport than others.

For human health, most nutrients are regarded to act in a U-shaped curve profile, and Se is not an exception to this rule. For instance, there is a narrow optimal concentration value where one can weigh the risks and benefits from Se intake. Deviations from this small range will affect several aspects of human health and lead to the problems discussed below. Moreover, due to the action of selenoproteins in most tissues of the body and their crucial participation in anti-inflammatory mechanisms and cell detoxification, this class of proteins is involved in the pathogenesis of several human diseases.

The most researched role of Se in human health is related to cancer prevention (Davis et al. 2012). Despite the failure of the Selenium and Vitamin E Cancer Prevention Trial (SELECT) to prove such benefit for prostate cancer (Klein et al. 2011), rigorous analysis of other clinical trials in comparison to SELECT have demonstrated that the resulting benefits of Se supplementation on cancer prevention depend ultimately on baseline Se levels in participating subjects. Se reduced total cancer incidence when baseline Se levels were lower than 106 μg/L. Furthermore, Se supplementation has been shown to reduce the incidence of liver, esophageal, pancreatic, prostatic, colon, and mammary carcinogenesis.

Due to the antioxidant function of several selenoproteins, one could safely predict they would be involved in various aspects of cancer cell physiology. In fact, gene polymorphisms in GPx2, GPx4, and Sepp1 have been implicated in colorectal cancer, while SelS and Sep15 polymorphisms were implicated in lung and gastric cancer, respectively (Davis et al. 2012). Using animal models, researchers have shown that GPx1 increases lead to reduction of tumorigenicity following injection of cancer cells; whereas if both GPx1 and GPx2 genes are disrupted, the animal will spontaneously develop colon tumors. On the other hand, certain selenoproteins may help promote cancer (Yoo et al. 2012 For instance, Sep15 knockdown in colon carcinoma cells from mice and humans inhibited cell and tumor growth, and increased TrxR1 expression occurs in thyroid cancer as well as in oral squamous cell carcinoma.

Se supplementation has also been associated with benefits in sepsis response during critical illness. Low Se levels and low plasma GPx and Sepp1 have been correlated with systemic inflammatory response syndrome and with increased mortality in intensive care (Fairweather-Tait et al. 2011).

The neurological dysfunctions and male sterility present in Sepp1 KO mice highlighted the crucial importance of maintaining Se homeostasis in the brain and testes, which appear to be tissues that are hierarchically protected against low circulating Se levels. Dietary Se deficiency affects the Se concentrations in these tissues the least; however, when extreme experimental conditions such as selective sequestration or genetic knockouts are employed to induce unusually severe Se deficiencies, the brain and neuroendocrine tissues are the most affected.

Increases in reactive oxygen species have been linked to the pathogenesis of several neurodegenerative diseases, including Alzheimer's and Parkinson's Diseases. Not surprisingly, the levels of selenoproteins GPx4 and Sepp1 as well as their gene expression patterns in different brain cell types are altered in individuals with both pathologies.

The fact that an entire group of selenoproteins, the deiodinases, is the main regulator of thyroid function exposes the deep relationship between this trace element and endocrine function. Low Se levels have been associated with thyroid disorders such as goiter (Berry and Larsen 1992). In addition, mutations in the SBP2 gene in humans were shown to lead to decreases in Dio2 levels and consequent abnormal thyroid metabolism. Mice lacking tRNA$^{[Ser]Sec}$ in the thyroid epithelial cells, thus lacking the ability to synthesize selenoproteins in this tissue, still maintain normal thyroid function despite the increases in oxidative stress due to a lack of GPxs and TrxRs.

Furthermore, mice with targeted disruption of Dio2 lack proper adaptive thermoregulation and are more prone to high-fat-induced obesity. Interestingly, a lack of the Dio3 gene in mice disrupts proper production of insulin in pancreatic β-cells, causing an insulin resistant phenotype as well. Overexpression of GPx1 also has been shown to induce hyperinsulinemia and insulin resistance in mice. Moreover, liver-specific disruption of the whole selenoproteome in mice elevates plasma cholesterol levels, suggesting a role of selenoproteins in lipid metabolism and metabolic syndrome.

Despite the evidence in mice studies, the effects of Se on glucose metabolism in humans are an unresolved issue among researchers. Se has been shown to adversely affect glucose homeostasis, despite being considered a safe supplement with potential to prevent several human diseases, including cancer (Davis et al. 2012). The Third National Health and Nutrition Examination Survey (NHANES III) found a positive association between serum Se and prevalence of type 2 diabetes in American adults. However, three additional case–control studies found reduced incidence of type 2 diabetes upon high Se status (Rayman 2012). In the SELECT and Nutritional Prevention of Cancer (NPC) trials (Stranges et al. 2007), subjects were administered Se supplements for assessment of effects on the incidence of prostate and skin cancers, respectively. After 7 years, both trials were prematurely terminated. In the SELECT, it was revealed that those subjects who received high doses of supplemental Se presented a trend, although not significant, toward increased risk of glucose intolerance. However, a 10-year follow-up analysis of SELECT participants confirmed the lack of influence of Se supplementation in the outcome of type 2 diabetes prevalence and no effect on prostate cancer protection. In contrast, a post hoc analysis of the NPC trial showed a significant increase in risk of type 2 diabetes in those supplemented with 200 μg of Se (Rayman 2012).

One possible reason for such discrepancy is the value of baseline serum Se among trial participants. The baseline Se status of men on the SELECT trial was 136 μg/L, while in the NPC trial, the baseline Se was 114 μg/L. The higher baseline of SELECT may have caused selenoprotein expression to plateau or surpass a threshold of risk. Trial participants in the SELECT study probably could not benefit from Se supplementation because their Se system was already saturated.

The randomized United Kingdom Prevention of Cancer by Intervention with Selenium (UK PRECISE) pilot study unveiled the effects of Se on lipid profile. Early studies have suggested that low serum Se could increase cardiovascular morbidity and mortality, and the association of those effects with lipid content was investigated by the UK PRECISE study. The pilot study revealed that Se supplementation up to 300 mg/day was only modestly beneficial to plasma cholesterol. The baseline serum Se of this study, however, was 88.8 μg/L, considerably lower than in the SELECT or NPC trials. In terms of benefits of Se supplementation for glucose and lipid profile, it reinforces the importance of the U-shaped association, where most benefits fall in a midrange concentration, while harm is prevalent in the extreme sides of the curve (Rayman 2012). Combined, these clinical controversies underline our limited understanding of the effects of Se supplementation, specifically regarding its influence on glucose and lipid metabolism and on the metabolic syndrome and type 2 diabetes outcomes.

In addition to the roles played by selenoproteins in energy and thyroid metabolism, Se is essential for testosterone biosynthesis and normal development of spermatozoa. In testes, Se is present as part of GPx4, a protein that is crucial for the architecture of the spermatozoan midpiece and a major protector against oxidative damage of spermatozoan DNA (Ursini et al. 1999). Subfertile/infertile men displayed improvements in sperm quality upon Se supplementation. Such improvement, however, was dependent on baseline Se levels of the analyzed population.

The data are scarce on the relationship between Se and female fertility. However, Se, which is transported through the placenta, is required for proper fetal development. Selenoproteins Sepp1 and Dio3 are highly expressed in the placenta, thus suggesting that Se levels are relevant to fetal development at least at a certain stage of pregnancy.

Lastly, immune function also appears to be affected by Se levels. Se-deficient mice are more susceptible to viral infection. Se supplementation suppresses HIV replication induced by tumor necrosis factor alpha and protects against Coxsackie virus infection and virulence. Deletion of the GPx1 gene rendered mice vulnerable to Coxsackie virus-mediated myocarditis as a consequence of increased mutations in the viral genome. Moreover, selenoprotein deficiency caused by T-cell-specific deletion of tRNA$^{[Ser]Sec}$ in mice leads to a dysfunctional adaptive immune response.

## Cross-References

▶ Arsenic, Mechanisms of Cellular Detoxification
▶ Artificial Selenoproteins

- ► Mercury Neurotoxicity
- ► Mercury Toxicity
- ► Parvalbumin
- ► Peroxidases
- ► Scandium, Physical and Chemical Properties
- ► Selenium and Aquaporins
- ► Selenium and Iodothyronine Deiodinases
- ► Selenium and Glutathione Peroxidases
- ► Selenium and Muscle Function
- ► Selenium and the Regulation of Cell Cycle and Apoptosis
- ► Selenium, Biologically Active Compounds
- ► Selenium-Binding Protein 1 and Cancer
- ► Selenocysteinopathies
- ► Selenophosphate Synthetase
- ► Selenoprotein K
- ► Selenoprotein P
- ► Selenoprotein Sep15
- ► Selenoproteins and the Biosynthesis and Activity of Thyroid Hormones
- ► Selenoproteins and Thyroid Gland

## References

Berry MJ, Larsen PR (1992) The role of selenium in thyroid hormone action. Endocr Rev 13(2):207–219

Davis CD, Tsuji PA et al (2012) Selenoproteins and cancer prevention. Annu Rev Nutr. [epub ahead of print]

Fairweather-Tait SJ, Bao Y et al (2011) Selenium in human health and disease. Antioxid Redox Signal 14(7):1337–1383

Klein EA, Thompson IM Jr et al (2011) Vitamin E and the risk of prostate cancer: the selenium and vitamin E cancer prevention trial (SELECT). JAMA 306(14):1549–1556

Kryukov GV, Castellano S et al (2003) Characterization of mammalian selenoproteomes. Science 300(5624):1439–1443

Rayman MP (2012) Selenium and human health. Lancet 379(9822):1256–1268

Schweizer U, Dehina N et al (2011) Disorders of selenium metabolism and selenoprotein function. Curr Opin Pediatr 23(4):429–435

Stranges S, Marshall JR et al (2007) Effects of long-term selenium supplementation on the incidence of type 2 diabetes: a randomized trial. Ann Intern Med 147(4):217–223

Ursini F, Heim S et al (1999) Dual function of the selenoprotein PHGPx during sperm maturation. Science 285(5432):1393–1396

Yang GQ, Zhu LZ et al (1984) Human selenium requirements in China. In: Combs GF, Spallholz JE, Levander OA, Oldfield JE (eds) Selenium in biology and medicine. AVI Van Nostrand, New York, pp 589–607

Yoo MH, Carlson BA et al (2012) Selenoproteins harboring a split personality in both preventing and promoting cancer. In: Hatfield DL, Berry MJ, Gladyshev VN (eds) Selenium: its molecular biology and role in human health. Springer, New York

# Selenium, Biologically Active Compounds

Leonid Breydo
Department of Molecular Medicine, Morsani College of Medicine, University of South Florida, Tampa, FL, USA

## Synonyms

Sec – Selenocysteine; SECIS – Selenocysteine insertion element

## Definition

Selenium is a heavy analog of sulfur and its derivatives have similar chemical properties to those of sulfur. However, selenium compounds are more nucleophilic and are more stable in lower oxidation states. These differences lead to selenorganic compounds playing an important role in maintenance of redox balance in the physiological environment.

## Introduction

Selenium is a heavier analog of sulfur. Chemical properties of selenium and sulfur atoms (bond energies, ionization potentials, atom sizes) are nearly identical. Thus, most organoselenium compounds are the analogs of organosulfur ones.

However, there are a few substantial differences in chemical properties between sulfur and selenium. Due to the presence of an additional filled electron shell, selenium accommodates additional electrons better than sulfur. It has a stronger preference for lower oxidation states compared to sulfur ($Se^{+6}$ is much less stable compared to $S^{+6}$). Higher tolerance of selenium for the negative charge also results in lower $pK_a$ values of selenols compared to thiols. For example, selenocysteine ($pK_a$ 5.25) is ionized at physiological conditions while cysteine ($pK_a$ 8.25) is not (Rezanka and Sigler 2008). Existence of selenocysteine as a highly nucleophilic anion at neutral pH leads to its utilization by a variety of enzymes. In addition, high reactivity of selenocysteine and other selenides allow them to form adducts with cysteine residues on proteins resulting in both beneficial and toxic effects.

## Selenorganic Compounds In Vivo

Selenium is present in the environment either in elemental form or in the form of selenide ($Se^{2-}$), selenite ($Se^{4+}$), or selenate ($Se^{6+}$). Higher oxidation states are favored at higher pH and in more oxidizing conditions. Selenate and selenite are absorbed via distinct pathways by both animals and plants: Selenate is actively absorbed utilizing the sulfate absorption mechanism, while selenite utilizes the phosphate transport mechanism in plants and is passively absorbed via intestinal lining in animals. While two types of sulfate transporters are present in plants (high affinity and low affinity), only high affinity transporters were shown to transport selenate (Terry et al. 2000).

Selenium metabolism starts with reduction of selenate by either ATP sulfurylase or dedicated selenate reductase (in some species of bacteria). ATP sulfurylase converts sulfate into adenosine phosphosulfate, and selenate can substitute for sulfate in this reaction (Fig. 1). The resulting adenosine phosphoselenate is either reduced by glutathione to selenodiglutathione or reduced by APS reductase to selenite (Terry et al. 2000).

Further reduction of selenite involves nonenzymatic reaction with glutathione or other thiols (Fig. 2). Briefly, the reaction involves addition of two equivalents of thiol to selenite and subsequent reduction of the resulting intermediate to selenodiglutathione (GSSeSG) (Combs and Combs 1984). Selenodiglutathione is structurally similar to glutathione trisulfide but, unlike the trisulfide, is relatively stable to nonenzymatic reduction. At low glutathione concentrations GSSeSG can react with cysteine residues of proteins resulting in their inactivation. This process may play a role in selenium toxicity. Selenodiglutathione is further reduced by glutathione reductase utilizing NADPH as a cofactor. Reduction is a two-step process with the first equivalent of NADPH reducing GSSeSG to selenopersulfide, GSSeH. This unstable

**Selenium, Biologically Active Compounds, Fig. 1** Reduction of selenate to selenite

**Selenium, Biologically Active Compounds, Fig. 2** Glutathione-dependent reduction of selenite

**Selenium, Biologically Active Compounds, Fig. 3** Methylation of hydrogen selenide

intermediate is reduced by the enzyme to hydrogen selenide ($H_2Se$). Selenite can also be reduced directly to selenide with sulfite reductase.

Hydrogen selenide is one of the most toxic selenium compounds active at levels well below 0.1 mg/l (Tarze et al. 2007). Hydrogen selenide exists as a monoanion at physiological pH and is highly nucleophilic and highly susceptible to oxidation. The mechanism of its toxicity is not well understood but is likely to involve reduction of cysteine disulfide bonds (Fig. 2) (Woods and Klayman 1974). In addition, selenide was shown to directly reduce heme iron thus inactivating heme-dependent enzymes (Bjornstedt et al. 1996).

Detoxification of hydrogen selenide is a crucial process in the biological systems and it is usually achieved by its methylation. As it is often the case in selenium metabolism, existing pathways of sulfide methylation are utilized in selenide methylation (Fig. 3). S-adenosyl methionine (SAM) is used as a methyl donor and at least some of the methyltransferases that methylate aromatic thiols have been shown to catalyze the methylation of selenide as well (Ranjard et al. 2002). The initial product is methyl selenol ($CH_3SeH$) that can be further methylated to dimethylselenide ($Me_2Se$, DMSe) or oxidized to dimethyldiselenide ($CH_3SeSeCH_3$, DMDSe). The ratio of these compounds depends on selenium concentration with DMSe predominantly produced at low selenium levels and DMDSe at high selenium levels. Methylation of selenide to methyl selenol is reversible but subsequent methylation steps are essentially irreversible. DMSe can get methylated further to produce trimethylselenium cation although this process is relatively slow. Alternatively, DMSe may be oxidized to the respective selenoxide and selenone although these reactions have been poorly characterized (Terry et al. 2000).

Selenide (or its glutathione adduct, GSSeH) can replace thiol as a nucleophile in the biosynthesis of cysteine to yield selenocysteine (Sec). Sec can be converted to selenomethionine utilizing the methionine biosynthetic pathway (Fig. 4). Both selenocysteine and selenomethionine can be further methylated to the respective Se-methyl derivatives. Methylation of selenocysteine (catalyzed by a specific selenocysteine methyltransferase) is important for certain plants that accumulate significant quantities of selenium as it prevents selenocysteine from being incorporated into proteins instead of cysteine. Se-methylselenocysteine serves as a reservoir of other, more biologically active selenium-containing compounds (Ganther 1999). In a reaction catalyzed by one of several lyases, low concentrations of methyl selenol are released. This compound is believed to have anticancer effects by promoting apoptosis and slowing down cell growth (Ganther 1999). Similar reaction can occur for Se-methyl selenomethionine which can be enzymatically cleaved to yield dimethylselenium.

## Selenium-Containing Proteins

About 25 human proteins contain selenium in the form of selenocysteine (Sec). Most of these proteins are enzymes involved in redox reaction with a selenocysteine residue at the active site. Functions of several of these proteins, such as selenoproteins N, P, and S, are still not fully understood (Rayman 2012). Selenomethionine is also found in many proteins at low levels but its incorporation is nonspecific and occurs due to its competition with methionine for the respective aminoacyl tRNA synthetase (Rayman 2008).

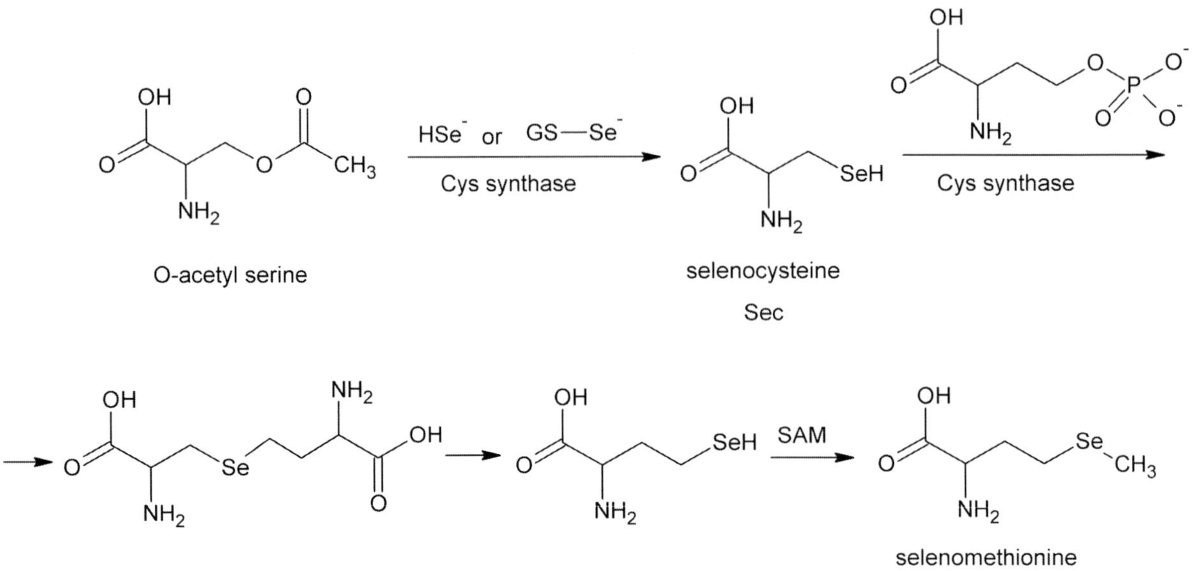

**Selenium, Biologically Active Compounds, Fig. 4** Synthesis of selenium-containing amino acids

Selenocysteine is incorporated into the proteins through a unique co-translation mechanism. Selenocysteine is encoded in mRNA by UGA that usually functions as a stop codon. However, if SECIS element (selenocysteine insertion element) is present nearby, in the presence of selenium it directs the cell to translate the UGA codon as selenocysteine. SECIS element is an RNA sequence 60 nucleotides in length that adopts a hairpin structure. In the presence of selenium it binds to SBP2 (SECIS binding protein 2) that forms a complex with a Sec-specific elongation factor $EF_{Sec}$. These proteins direct incorporation of Sec into proteins in response to the UGA codon.

Since concentration of free Sec in the cell is quite low, it is synthesized directly on tRNA (Fig. 5). Sec tRNA is initially loaded with serine. Serine is then O-phosphorylated by a specific kinase. Selenide then becomes phosphorylated as well via reaction with ATP to yield selenophosphate in a reaction catalyzed by selenophosphate synthetase. Sec synthase then uses the selenophosphate as a nucleophile to displace the phosphate on O-phosphoserine bound to Sec tRNA to form active Sec tRNA (Turanov et al. 2011).

Lower pKa and higher reduction potential of selenocysteine makes this amino acid highly suitable for proteins involved in antioxidant activity, such as glutathione peroxidase (described in more detail elsewhere in this volume). Glutathione peroxidase contains a selenocysteine residue in its active site that is involved in metal-independent redox cycling. Organic or inorganic peroxides oxidize the selenocysteine to the respective selenenic acid (RSeOH) and glutathione reduces it back to selenocysteine. The enzyme thus decreases the levels of peroxides and protects the cell from oxidative stress.

Selenocysteine plays a similar role in thioredoxin reductase in higher eukaryotes. This enzyme is structurally similar to glutathione reductase and contains one selenocysteine and three cysteines in the active site (Zhong et al. 2000). Selenocysteine is required for reduction of physiological substrates of the enzyme. Compared to bacterial enzymes that contain only cysteines, eukaryotic thioredoxin reductase is sensitive to the redox environment of the cell and can reduce a broader range of substrates.

Another group of selenocysteine-containing enzymes are iodothyronine deiodinases. They are important in production and regulation of thyroid hormones. These enzymes remove the iodine atoms from the thyroid hormone precursor thyroxine to form

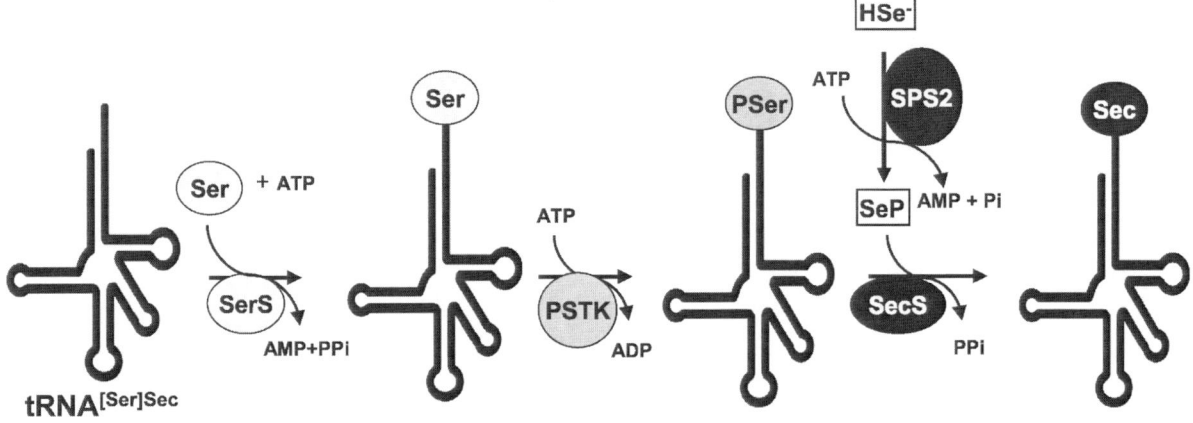

**Selenium, Biologically Active Compounds, Fig. 5** Biosynthesis of selenocysteine-tRNA in eukaryotes. SerS is the seryl-tRNA synthetase and PSTK is the $O$-phosphoseryl-tRNA$^{[Ser]Sec}$ kinase (Adopted from Xu et al. 2007)

the active hormone triiodothyronine. Removal of another iodine atom from the active hormone inactivates it. Selenocysteine is located in the active site of the enzyme and is directly involved in iodine removal. Replacement of selenocysteine with cysteine significantly increases the Michaelis constant for the enzyme and decreases its turnover rate (Kuiper et al. 2003).

## Biological Activity of Selenium Derivatives

Selenium is a required micronutrient for both animals and plants. Selenium deficiency affects the synthesis of selenoproteins that play important roles in many physiological processes. Certain selenoproteins (e.g., glutathione peroxidases) are especially important as their synthesis is prioritized at low levels of selenium ($< 80$ µg/l). In humans, optimum concentration of selenium in plasma is around 100–150 µg/l (Rayman 2012). Optimum selenium levels are associated with improved function of immune system (higher T-cell count, improved survival of AIDS patients), improved cognition, lower incidence of infertility and cardiovascular disease. Many of these benefits are believed to be due to higher levels of selenium-containing enzymes (especially glutathione peroxidases) that reduce oxidative stress. Higher selenium levels are also protective against a variety of cancers (Ganther 1999). Several redox-active selenoproteins (Sep15, glutathione peroxidases, thioredoxin reductases) are believed to play an important role in chemoprotection against cancer. In addition, direct reactions of selenols and other selenium derivatives, such as GSSeSG, with cysteine residues in proteins may contribute to protection from cancer by altering the disulfide bonds in transcription factors or other regulatory proteins (Ganther 1999).

At high concentrations ($> 150$ µg/l in human plasma), selenium becomes toxic. As discussed above, this toxicity is primarily due to glutathione depletion and direct modification of cysteine residues on proteins by selenide and various selenols. In addition, selenocysteine can be incorporated into the proteins instead of cysteine resulting in alterations to their tertiary structure and catalytic activity.

## Conclusions

Selenium is a heavier analog of sulfur and their chemical properties are fairly similar. However, selenium compounds are more nucleophilic and are more stable in lower oxidation states. These differences lead to selenorganic compounds playing an important role in maintenance of redox balance in the physiological environment. The most important physiological role of selenium is as a source of selenocysteine to a variety of selenoproteins. In addition, selenium-containing compounds are potent reducing agents that may be protective at low concentrations and toxic at higher concentrations.

## Cross-References

▶ Selenium and Glutathione Peroxidases
▶ Selenium and Iodothyronine Deiodinases
▶ Selenium in Human Health and Disease
▶ Selenocysteinopathies
▶ Selenoproteins in Prokaryotes

## References

Bjornstedt M, Odlander B, Kuprin S, Claesson HE, Holmgren A (1996) Selenite incubated with NADPH and mammalian thioredoxin reductase yields selenide, which inhibits lipoxygenase and changes the electron spin resonance spectrum of the active site iron. Biochemistry 35:8511–8516

Combs GF Jr, Combs SB (1984) The nutritional biochemistry of selenium. Annu Rev Nutr 4:257–280

Ganther HE (1999) Selenium metabolism, selenoproteins and mechanisms of cancer prevention: complexities with thioredoxin reductase. Carcinogenesis 20:1657–1666

Kuiper GG, Klootwijk W, Visser TJ (2003) Substitution of cysteine for selenocysteine in the catalytic center of type III iodothyronine deiodinase reduces catalytic efficiency and alters substrate preference. Endocrinology 144: 2505–2513

Ranjard L, Prigent-Combaret C, Nazaret S, Cournoyer B (2002) Methylation of inorganic and organic selenium by the bacterial thiopurine methyltransferase. J Bacteriol 184: 3146–3149

Rayman MP (2008) Food-chain selenium and human health: emphasis on intake. Br J Nutr 100:254–268

Rayman MP (2012) Selenium and human health. Lancet 379:1256–1268

Rezanka T, Sigler K (2008) Biologically active compounds of semi-metals. Phytochemistry 69:585–606

Tarze A, Dauplais M, Grigoras I, Lazard M, Ha-Duong NT, Barbier F, Blanquet S, Plateau P (2007) Extracellular production of hydrogen selenide accounts for thiol-assisted toxicity of selenite against *Saccharomyces cerevisiae*. J Biol Chem 282:8759–8767

Terry N, Zayed AM, De Souza MP, Tarun AS (2000) Selenium in higher plants. Annu Rev Plant Physiol Plant Mol Biol 51:401–432

Turanov AA, Xu XM, Carlson BA, Yoo MH, Gladyshev VN, Hatfield DL (2011) Biosynthesis of selenocysteine, the 21st amino acid in the genetic code, and a novel pathway for cysteine biosynthesis. Adv Nutr 2:122–128

Woods TS, Klayman DL (1974) Cleavage of sulfur-sulfur bonds with sodium hydrogen selenide. J Org Chem 39: 3716–3720

Xu XM, Carlson BA, Mix H, Zhang Y, Saira K, Glass RS, Berry MJ, Gladyshev VN, Hatfield DL (2007) Biosynthesis of selenocysteine on its tRNA in eukaryotes. PLoS Biol 5:e4

Zhong L, Arner ES, Holmgren A (2000) Structure and mechanism of mammalian thioredoxin reductase: the active site is a redox-active selenolthiol/selenenylsulfide formed from the conserved cysteine-selenocysteine sequence. Proc Natl Acad Sci USA 97:5854–5859

# Selenium, Physical and Chemical Properties

Fathi Habashi
Department of Mining, Metallurgical, and Materials Engineering, Laval University, Quebec City, Canada

Selenium is a by-product of the extraction of copper and nickel from their sulfide ores. It is concentrated in the anodic slimes of electrorefining of these metals. Red amorphous selenium is produced by reducing selenous acid with sulfur dioxide, by crystallization from solution, or by quenching of selenium vapor or molten selenium. It consists of $Se_8$ rings similar to elemental sulfur, soluble in carbon disulfide giving red solutions from which selenium can be crystallized. Red selenium is an electrical nonconductor. On heating to >100°C, is transformed into the stable, gray, hexagonal metal-like form in an exothermic reaction. In the vapor phase, red selenium exists mainly as $Se_8$ rings (Fig. 1a). At higher temperature these decompose into smaller units, eventually forming $Se_2$ molecules.

The gray, hexagonal crystals have metallic appearance and are composed of helical chains which melt at 220°C (Fig. 1b). It is a semiconductor whose electrical conductivity increases by a factor >1,000 under the influence of light (photoconductivity). The electrical conductivity is strongly dependent on purity. Thus, the conductivity of selenium, including the poorly conducting amorphous form, can be increased considerably by traces (a few parts per million) of halide ions or by alloying elements such as tellurium or arsenic. Because the electrical properties are sensitive to the presence of crystal defects, the purity of selenium, which can be deliberately controlled by doping or alloying with other

## Selenium, Physical and Chemical Properties,

**Fig. 1** (a) Structure of red selenium. (b) Structure of gray selenium

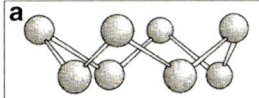
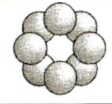
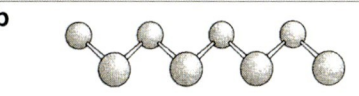

elements, plays an important role in its use in electrical and electronic applications.

## Physical Properties

| | |
|---|---|
| Atomic number | 34 |
| Atomic weight | 78.96 |
| Relative abundance in Earth's crust, % | $9 \times 10^{-6}$ |
| Density of gray Se at 25°C, g/cm³ | 4.189 |
| Melting point of gray Se, °C | 220 |
| Boiling point (101.3 kPa; 760 mm Hg), °C | $684.7 \pm 0.03$ |
| Atomic radius, nm | 0.140 |
| Ionic radii, nm | |
| $Se^{2-}$ | 0.191 |
| $Se^{4+}$ | 0.050 |
| $Se^{6+}$ | 0.042 |
| Ionization energy, kJ/mol | 940 |

## Chemical Properties

In its compounds, selenium exhibits the oxidation states $-2$, $+4$, and $+6$. Selenium burns in air to form selenium dioxide, which has the smell of rotten radishes. Selenium is oxidized by nitric acid to selenous acid. Hot, concentrated sulfuric acid dissolves selenium, giving a green color and forming polymeric selenium cations, for example:

$$Se_8 + 3H_2SO_4 \rightarrow Se_8^{2+} + 2HSO_4^- + SO_2 + 2H_2O$$

With chlorine, vigorous oxidation to selenium tetrachloride occurs. Selenium reacts with electropositive elements (e.g., many metals) to form selenides and is therefore strongly corrosive, especially at high temperature. With hydrogen, the toxic gas hydrogen selenide, $H_2Se$, is formed. Selenium combines with sulfur to form nonstoichiometric alloys. It dissolves in solutions of strong alkali, disproportionating into selenide and selenite. It dissolves in alkali-metal sulfides, sulfites, and cyanides, with formation of addition compounds.

## References

Langner BE (1997) In: Habashi F (ed) Handbook of extractive metallurgy. Wiley-VCH, Weinheim, pp 1557–1570

## Selenium-Binding Protein 1 and Cancer

Cunxian Zhang[1] and Xudong Wang[2]
[1]Department of Pathology, Women & Infants Hospital of Rhode Island, Kent Memorial Hospital, Warren Alpert Medical School of Brown University, Providence, RI, USA
[2]Department of Pathology, St Vincent Hospital, Worcester, MA, USA

## Synonyms

Cancer: carcinoma; Selenium-binding protein 1: SELENBP1

## Definitions

Selenium-binding protein 1 is an intracellular protein that binds selenium covalently and is thought to mediate the actions of selenium. Cancer is any malignant growth or tumor characterized by abnormal and uncontrolled proliferation of neoplastic cells that disrupt adjacent tissue and have the capacity of spreading to other parts of the body through lymphatic vessels or blood stream.

## Selenium and Selenium-Containing Proteins

Selenium is an essential trace mineral with a potent anticancer effect. Epidemiological studies have shown a correlation between high cancer rates and low dietary selenium intake (Klein 2004). The relationship of selenium with cancer was further demonstrated by clinical trials revealing an effect of dietary selenium supplementation on the reduction of total cancer mortality, total cancer incidence, and incidences of prostate, lung, colorectal, and skin cancers. The proposed mechanisms for the anticancer actions of selenium include antioxidant activity, altered carcinogen metabolism, inhibition of tumor proliferation and invasion, and induction of apoptosis. Indeed, experimental studies have shown that selenium inhibits proliferation and promotes apoptosis in cancer cells.

The actions of selenium are thought to be mediated by selenium-containing proteins. Selenium-containing proteins in humans fall into two distinct categories. One class refers to selenoproteins that are formed by combination of selenocysteine and other amino acids (Gladyshev 2006). Selenocysteine is an amino acid similar to cysteine except that sulfur is replaced by selenium. In humans, there are about 25 different selenoproteins, one example of which is glutathione peroxidase that has been shown to possess an antioxidant role and is able to protect cell membranes from damage by free radicals released from hydrogen peroxide formed during normal cellular metabolism.

The other class refers to selenium-binding proteins including selenoprotein P, liver fatty acid binding protein, 58 kDa selenium-binding protein, and mouse 56 kDa selenium-binding protein. The homologue of the mouse 56 kDa selenium-binding protein in humans is selenium-binding protein 1 (SBP1), a 56 kDa protein with selenium-binding properties. The human SBP1 gene is located at q21-22 on chromosome 1 and encodes a 472 amino acid protein present in both the nucleus and the cytoplasm (Chang et al. 1997). Unlike selenoproteins in which selenium is incorporated into amino acids sequence via the TGA codon, SBP1 binds selenium covalently. SBP1 is an alpha-beta protein with loop regions characterized by the presence of intrinsically unordered segments. SBP1 consists of four cysteine residues, of which Cys57 is the only cysteine capable of binding selenium.

Identified initially in the liver, SBP1 is present in a variety of other tissue types, including the heart, brain, prostate, lung, kidney, ovaries, thyroid, and gastrointestinal tract. The high degree conservation of SBP1 in different organisms suggests that this protein may play a role in fundamental biological processes. In this regard, SBP1 has been shown to facilitate intra-Golgi protein transport, protrusive cell motility, and ubiquitination/deubiquitination-mediated protein degradation pathways. SBP1 has also been implicated in cancer development and progression.

## SBP1 in Cancer

Carcinogenesis involves the various roles of oncogenes and tumor suppressor genes that encode proteins engaged in the control of cell growth and differentiation. Because structural and functional alterations of these genes have been recognized as etiologic factors of cancer and are frequently used as biomarkers of disease progression, identification of these genes and their changes is critical to the understanding of cancer biology and behavior. As a newly recognized tumor-related gene, SBP1 has increasingly gained attention in cancer research.

Although the exact functions of SBP1 are unclear, altered SBP1 expression in various cancers indicates an important role of SBP1 in the development of cancer. Compared to that in normal tissues, SBP1 expression is markedly decreased in cancers of multiple organs including the lung, colorectum, stomach, esophagus, liver, thyroid, and ovaries (Huang et al. 2006; Stammer et al. 2008). A summary concerning altered expression of SBP1 in various tumors is presented in Table 1. Hepatocellular carcinoma displays a decreased SBP1 expression compared to adjacent benign liver. In the stomach, there is a progressive decrease in the expression of SBP1 from low-grade dysplasia (93%), to high-grade dysplasia (69%), and finally to adenocarcinomas (37%). Progressive loss of SBP1 expression from Barrett's esophagus to adenocarcinoma also supports the role of SBP1 in malignant transformation and cancer progression. In ovarian

**Selenium-Binding Protein 1 and Cancer, Table 1** Decreased expression of SBP1 in various human tumors

| Tumor site | Tumor type | Techniques used | References |
|---|---|---|---|
| Colon | Colorectal carcinoma | cDNA microarray, Western blot, IHC | Li et al. (2008) |
| Liver | Hepatocellular carcinoma | IHC | Raucci et al. (2011) |
| Stomach | Gastric carcinoma | RT-PCR, Western blot, IHC, tissue microarray, | Xia et al. (2011); Zhang et al. (2011) |
| Esophagus | Adenocarcinoma | Microarray, RT-PCR, Western blot | Silvers et al. (2010) |
| Thyroid | Papillary carcinoma | Western blot, IHC | Brown et al. (2006) |
| Lung | Adenocarcinoma | IHC, Western blot, oligonucleotide microarray | Chen et al. (2004) |
| Prostate | Prostate cancer cells | Northern blot | Yang and Sytkowski (1998) |
| Ovary | Serous borderline tumor and invasive serous carcinoma | IHC | Huang et al. (2006); Zhang et al. (2010a) |
| Uterus | Leiomyoma | Western blot, IHC | Zhang et al. (2010b) |
| Pleura | Mesothelioma | RT-PCR, IHC | Pass et al. (2004) |

Abbreviations: *RT-PCR* reverse transcriptase polymerase chain reaction, *IHC* immunohistochemistry

lesions ranging from serous cystadenoma, serous borderline tumor, to low-grade serous adenocarcinoma, the strongest immunoreactivity of SBP1 was seen in the epithelial cells of flat cyst wall and primary papillae, followed by secondary papillae of the hierarchical structures (Zhang et al. 2010a). Micropapillae and invasive carcinoma exhibited a near complete loss of SBP1 expression. Figure 1 illustrates the progressive loss of SBP1 from simple layer epithelium to papillary proliferations and finally to low-grade serous adenocarcinoma. In these ovarian lesions, progressive loss of SBP1 expression correlated with increasing epithelial proliferation and papillary complexity, supporting a role of SBP1 in the development and progression of neoplasia. In addition to cancers, uterine leiomyoma also shows a diminished expression of SBP1 compared to adjacent normal uterine smooth muscle, suggesting a role of SBP1 in tumorigenesis of uterine leiomyoma and a possibility of using selenium in the prevention and treatment of this common neoplasm of the female genital tract. Loss or reduced expression of SBP1 in cancers not only indicates an active role of this protein in tumorigenesis but also suggests its tumor suppressor functions.

In clinical practice, the assessment of tumor grade and stage is a critical part of workup. Tumor grade is closely related to cell differentiation; usually, well-differentiated tumors are of low grade, and poorly differentiated tumors are of high grade. Stage is determined by primary tumor size, depth of invasion, and the presence or absence of local and distant metastasis. In general, patients with high-grade and high-stage cancers do more poorly than those with low-grade and low-stage tumors. Several studies have shown that SBP1 may be used as a biomarker of tumor phenotype, grade, and stage. For example, SBP1 is expressed in a relatively slow-growing, androgen-sensitive human prostate cancer cell line (equivalent to well-differentiated tumor) but not in a rapidly growing, androgen-insensitive prostate cancer cell line (equivalent to poorly differentiated tumor) (Yang and Sytkowski 1998), indicating a relationship between SBP1 expression and tumor cell phenotype. An association between a gradual loss of SBP1 expression and increased malignant grades is seen in hepatocellular carcinoma. In gastric cancer, lower level of SBP1 expression is associated with poor differentiation, high tumor stage, and lymph node metastasis. Loss of SBP1 expression in lung adenocarcinoma also correlates with tumor high grade and stage (Chen et al. 2004). These findings suggest that SBP1 expression may be used to supplement conventional staging.

Several studies have indicated a role of SBP1 expression in predicting patient's prognosis. In patients with stage III colon cancer, lower level of SBP1 expression is associated with a shorter disease-free interval and a shorter overall survival. Other striking evidence comes from cancers of the stomach,

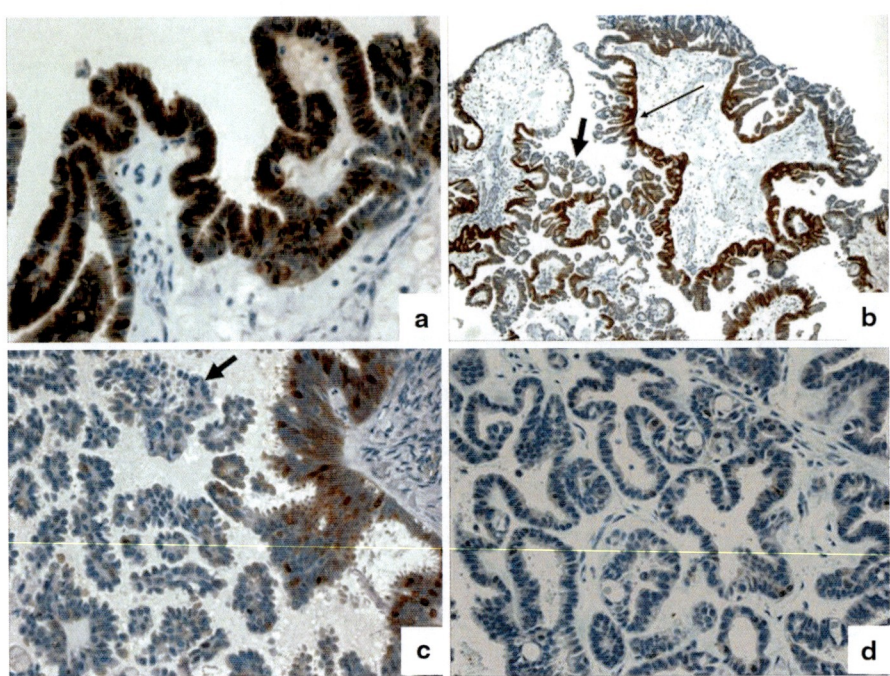

**Selenium-Binding Protein 1 and Cancer, Fig. 1** Immunohistochemical studies on SBP1 expression in various ovarian lesions. (**a**) Strong immunoreactivity is seen in the single layer epithelial lining cells of serous cystadenoma. (**b**) The stain is also strong in the lining epithelial cells of the primary papillae (*long thin arrow*), but is significantly reduced in the secondary papillae (*short thick arrow*) of serous borderline tumor. (**c**) Micropapillae of serous borderline tumor show almost complete loss of immunostain (*short thick arrow*). (**d**) In low-grade serous carcinoma, the immunostain is negative. Magnifications: ×400 in (**a**), (**c**), and (**d**); ×100 in (**b**)

ovary, pleural mesothelioma, and lung where low level of SBP1 expression is associated with poor patient's prognosis (Chen et al. 2004). Furthermore, SBP1 expression seems to be an independent prognostic factor in these cancers. These studies imply that assessment of SBP1 expression can contribute to risk stratification in cancers and may further define prognostic subgroups.

Treatment decisions are based on the likelihood of tumor response, but the efficiency of systemic therapy in an individual cancer is difficult to foresee. Interestingly, several lines of evidence indicate a promising possibility of using SBP1 expression to predict treatment response. Loss of SBP1 has a significant negative impact on the ability of selenium to control tumor cell growth and chemotherapeutic agents to kill cancer cells. With restoration of SBP1 expression, colon cancer cells become more responsive to $H_2O_2$ with increased apoptosis and decreased migration and growth. Also, when SBP1 is overexpressed, esophageal cancer cells respond to selenium treatment with increased apoptosis, enhanced cellular senescence, and enhanced cisplatin cytotoxicity (Silvers et al. 2010). Studies using real time quantitative polymerase chain reaction have demonstrated a correlation between a decreased expression of the chromosome 1q21 gene cluster and a resistance of esophageal adenocarcinoma to chemoradiotherapy, implying a correlation between reduced SBP1 expression and treatment resistance, as SBP1 gene is included within the chromosome 1q21 gene cluster.

The effect of SBP1 in cancer development may be partly attributed to its role in cellular differentiation. In normal colon, the highest level of SBP1 expression is seen in the terminally differentiated epithelial cells on the luminal surface of crypts, indicating a relationship of SBP1 with cell differentiation. In cultured colon cancer cells, a reduced expression of SBP1 by small interfering RNA was coupled with a decrease in the expression of carcinoembryonic antigen, an epithelial differentiation marker located on chromosome 19. The paralleled reduction of SBP1 and carcinoembryonic antigen by small RNA treatment supports an effect of SBP1 on cell differentiation.

The proposed mechanisms of SBP1 in cell differentiation seem to involve the expression of another important group of cellular proteins within the cell differentiation complex. Many of the proteins within this cell differentiation complex are involved in cellular maturation and differentiation. The locus of chromosome where SBP1 gene is located also contains genes that encode proteins within the cell differentiation complex. The physical proximity between the genes encoding SBP1 and proteins within the cell differentiation complex result in the frequent co-expression of these genes, explaining at least in part the relationship between the expression of SBP1 and cellular differentiation.

## Regulation of SBP1 Expression

Gene regulation can occur at various points in the process from DNA to protein, including DNA sequence alteration (gene mutation), epigenetic modification, and regulations of transcription and translation. Epigenetic modification refers to functionally relevant alterations to the genome without changes in the nucleotide sequence. It causes gene silencing through two common mechanisms: histone covalent modulation and DNA methylation. DNA methylation in humans occurs almost exclusively at the cytosine residue in the cytosine-guanine dinucleotides (CpG islands) and is mediated by DNA methyltransferases. Hypermethylation of CpG islands at the promoter of tumor suppressor genes switches these genes off, serving as a common mechanism of gene silencing in cancer and an early event in cancer development.

The SBP1 gene contains CpG islands near the promoter region, providing a structural basis for gene methylation. Indeed, methylation of SBP1 promoter has been identified as a mechanism for the reduced expression of SBP1 in several cancers. For example, hypermethylation of SBP1 promoter is seen in colon cancer (Pohl et al. 2009). This mechanism is further supported by an increased promoter activity and restored SBP1 expression through demethylation of the SBP1 promoter by treatment with the demethylating agent 5'-Aza-2'-Deoxycytidine in human colon cancer cells. DNA methylation has also been proposed as a mechanism for diminished SBP1 expression in esophageal adenocarcinoma (Silvers et al. 2010). However, the reduced expression of SBP1 in lung adenocarcinoma is not attributed to DNA methylation, indicating the existence of other mechanisms. In fact, alternative splicing of mRNA serves as another mechanism for the reduced expression of SBP1 in esophageal adenocarcinoma.

Given the role of SBP1 in carcinogenesis, identification of agents that can alter the expression of SBP1 would seem important in cancer prevention and treatment. In this regard, androgen has been shown to decrease the expression of SBP1 in a slow-growing, androgen-sensitive prostate cancer cell line by an unknown mechanism (Yang and Sytkowski 1998). In ovarian cancer cells, selenium treatment has been shown to increase SBP1 expression (Huang et al. 2006). Since DNA hypermethylation is a proposed mechanism for reduced expression of SBP1 in several cancers and selenium is a DNA demethylating agent, it is possible that selenium increases SBP1 expression through demethylation of SBP1 gene promoter.

## Future Directions

Despite significant discoveries concerning the role of SBP1 in cancer development and progression, the exact mechanisms of SBP1 in carcinogenesis remain unclear. Major efforts should be made to characterize the SBP1 pathway, using cultured cells and animal models. Animal models are particularly helpful as they allow us to analyze the cascade of events in the SBP1 pathway leading to cancer. The information obtained by these analyses will contribute to the creation of new targeted chemoprevention and chemotherapy. Since high level of SBP1 expression predicts a favorable treatment response, identifying agents that regulate SBP1 expression would be of paramount importance in cancer therapy. With its function to increase SBP1 expression, selenium appears to be a promising chemopreventive and chemotherapeutic agent that can be combined with other drugs to maximize chemotherapeutic effect and overcome the limitation of any single agent. Combination of drugs can also minimize the side effects of each individual drug because synergistic effects of combined therapy allow the dose of individual agent to be reduced. Further investigations are required to identify other agents that can modulate SBP1 expression to optimize the new generation of molecular target-directed therapies.

## References

Brown LM, Helmke SM, Hunsucker SW, Netea-Maier RT, Chiang SA, Heinz DE, Shroyer KR, Duncan MW, Haugen BR (2006) Quantitative and qualitative differences in protein expression between papillary thyroid carcinoma and normal thyroid tissue. Mol Carcinog 45:613–626

Chang PWG, Tsui SKW, Liew CC, Lee CY, Waye MMY, Fung KP (1997) Isolation, characterization, and chromosomal mapping of a novel cDNA clone encoding human selenium binding protein. J Cell Biochem 64:217–224

Chen G, Wang H, Miller CT, Thomas DG, Gharib TG, Misek DE, Giordano TJ, Orringer MB, Hanash SM, Beer DG (2004) Reduced selenium-binding protein 1 expression is associated with poor outcome in lung adenocarcinomas. J Pathol 202:321–329

Gladyshev VN (2006) Selenoproteins and selenoproteomes. In: Hatfield DL, Berry MJ, Gladyshev VN (eds) Selenium: its molecular biology and role in human health, 2nd edn. Springer Science + Business Media LLC, Philadelphia, pp 99–114

Huang KC, Park DC, Ng SK, Lee JY, Ne X, Ng WC, Bandera CA, Welch WR, Berkowitz RS, Mok SC, Ng SW (2006) Selenium binding protein 1 in ovarian cancer. Int J Cancer 118:2433–2440

Klein EA (2004) Selenium: epidemiology and basic science. J Urol 171:S50–S53

Li T, Yang W, Li M, Byun D, Tong C, Nasser S, Zhuang M, Arango D, Mariadason JM, Augenlicht LH (2008) Expression of selenium-binding protein 1 characterizes intestinal cell maturation and predicts survival for patients with colorectal cancer. Mol Nutr Food Res 52:1289–1299

Pass HI, Liu Z, Wali A, Bueno R, Land S, Lott D, Siddiq F, Lonardo F, Carbone M, Draghici S (2004) Gene expression profiles predict survival and progression of pleural mesothelioma. Clin Cancer Res 10:849–859

Pohl NM, Tong C, Fang W, Bi X, Li T, Yang W (2009) Transcriptional regulation and biological functions of selenium-binding protein 1 in colorectal cancer in vitro and in nude mouse xenografts. PLoS One 4:e7774

Raucci R, Colonna G, Guerriero E, Capone F, Accardo M, Castello G, Costantini S (2011) Structural and functional studies of the human selenium binding protein-1 and its involvement in hepatocellular carcinoma. Biochim Biophys Acta 1814:513–522

Silvers AL, Lin L, Bass AJ, Chen G, Wang Z, Thomas DG, Lin J, Giordano TJ, Orringer MB, Beer DG, Chang AC (2010) Decreased selenium-binding protein 1 in esophageal adenocarcinoma results from posttranscriptional and epigenetic regulation and affects chemosensitivity. Clin Cancer Res 16:2009–2021

Stammer K, Edassery SL, Barua A, Bitterman P, Bahr JM, Hales DB, Luborsky JL (2008) Selenium-binding protein 1 expression in ovaries and ovarian tumors in the laying hen, a spontaneous model of human ovarian cancer. Gynecol Oncol 109:115–121

Xia YJ, Ma YY, He XJ, Wang HJ, Ye ZY, Tao HQ (2011) Suppression of selenium-binding protein 1 in gastric cancer is associated with poor survival. Hum Pathol 42:1620–1628

Yang M, Sytkowski AJ (1998) Differential expression and androgen regulation of the human selenium-binding protein gene HSP56 in prostate cancer cells. Cancer Res 58:3150–3153

Zhang C, Wang YE, Zhang P, Liu F, Sung CJ, Steinhoff MM, Quddus MR, Lawrence WD (2010a) Progressive loss of selenium-binding protein 1 expression correlates with increasing epithelial proliferation and papillary complexity in ovarian serous borderline tumor and low-grade serous carcinoma. Hum Pathol 41:255–261

Zhang P, Zhang C, Wang X, Liu F, Sung CJ, Quddus MR, Lawrence WD (2010b) The expression of selenium-binding protein 1 is decreased in uterine leiomyoma. Diag Pathol 5:80

Zhang J, Zhan N, Dong WG (2011) Altered expression of selenium-binding protein 1 in gastric carcinoma and precursor lesions. Med Oncol 28:951–957

## Selenium-Binding Protein 1: SELENBP1

▶ Selenium-Binding Protein 1 and Cancer

## Selenium-Containing Protein Mimic

▶ Artificial Selenoproteins

## Selenocysteine

▶ Selenium in Human Health and Disease

## Selenocysteine-Containing Proteins in Bacteria and Archaea

▶ Selenoproteins in Prokaryotes

## Selenocysteine-Containing Proteins, Selenoproteins

▶ Selenoproteins and Thyroid Gland

# Selenocysteine-Related Diseases

▶ Selenocysteinopathies

# Selenocysteinopathies

Ohad S. Birk
Head, Genetics Institute, Soroka Medical Center Head, Morris Kahn Center for Human Genetics, NIBN and Faculty of Health Sciences, Ben Gurion University, Beer Sheva, Israel

## Synonyms

Diseases of selenoprotein deficiency; Genetic/hereditary and nonhereditary diseases of selenoprotein deficiency; Selenocysteine-related diseases

## Definition

Selenocysteinopathies are human diseases caused by defects in formation and function of selenium-containing proteins (selenoproteins).

## Disorders of Selenium Deficiency

Selenoproteins (selenium-containing proteins) have a wide range of pleiotropic effects, ranging from antioxidant and anti-inflammatory effects to the production of active thyroid hormone (Rayman 2012). Selenium deficiency has been implicated in neuromuscular disorders, male infertility, decreased immune and thyroid function, as well as increased risk for cancer, infection, and HIV progression (Rayman 2012; Papp et al. 2007; Huang et al. 2012). Notably, myxedematous cretinism (mental and growth retardation) has been attributed to combined selenium and iodine deficiency (Rayman 2012), while a fatal form of Coxsackie B–induced cardiomyopathy (Keshan disease) is found in selenium-poor areas in China and can be reversed by selenium supplementation (Rayman 2012). Higher selenium status or selenium supplementation has antiviral effects, is essential for successful male and female reproduction, and reduces the risk of autoimmune thyroid disease (Rayman 2012; Papp et al. 2007; Huang et al. 2012). Prospective studies have shown some benefit of higher selenium status on the risk of prostate, lung, colorectal, and bladder cancers, but findings from trials have been mixed. It is noteworthy that supplementation of people who already have adequate intake with additional selenium might increase their risk of type 2 diabetes (Rayman 2012).

Selenoproteins are essential in the maintenance of optimal brain functions via redox regulation, and therapies targeting specific selenoproteins influencing redox regulation are suggested to delay the onset of neurological disorders and improve quality of life of patients already affected (Rayman 2012; Papp et al. 2007). Dietary deficiency of the trace element selenium has been long associated with neurological diseases: decreased expression of several selenoproteins has been found in pathological specimens of age-associated neurodisorders, including Parkinson's disease, Alzheimer's disease, and epilepsy (Rayman 2012).

## Hereditary Selenocysteinopathies

Aside from dietary effects of selenium, there are hereditary selenium-related neurological syndromes that are **not** due to defects in selenium dietary intake (Agamy et al. 2010; Schoenmakers et al. 2010). The essential micronutrient selenium is found in the human body within proteins as the 21st amino acid, selenocysteine (Sec). It is Sec that acts in most of the functional roles of selenium. Thus, even if sufficient amounts of selenium are available, if Sec is not formed, de facto selenium deficiency emerges. While all other amino acids are generated independent of their tRNA and are only later transferred by their appropriate tRNAs, Sec is the only genetically encoded amino acid in humans whose biosynthesis occurs on its cognate tRNA. Also unique is the genetic code for Sec: while all other 20 amino acids in the human have their own genetic codes, Sec is encoded by the stop codon UGA, which normally dictates translation termination. Twenty-five human genes have in the 3′-untranslated region (UTR) of their mRNA a unique Sec insertion sequence (SECIS) that in concert with several protein factors (including a specific elongation factor EFSec and the SECIS binding protein 2, SBP2) dictates Sec incorporation by UGA, rather than translation termination. The 25 selenoproteins encoded by these genes are the only human proteins that contain the amino acid Sec (Papp et al. 2007; Bellinger et al. 2009).

Mutations in the essential enzyme SepSecS that catalyzes the final step of Sec formation cause a severe neurological phenotype of autosomal recessive progressive cerebellocerebral atrophy (PCCA) (Agamy et al. 2010). Two different SepSecS founder mutations, common in Jews of Iraqi and Moroccan ancestry, disrupt the sole route to Sec biosynthesis in humans. The PCCA phenotype is of nondysmorphic profound mental retardation: beginning with normal head circumference and brain MRI at birth, progressive microcephaly, psychomotor retardation, and severe spasticity ensue by less than 1 year of age (Agamy et al. 2010). Myoclonic or generalized tonic-clonic seizures are often observed as well. Repeat magnetic resonance imaging shows progressive cerebellar atrophy followed by cerebral atrophy involving both white and gray matter, with thinning of corpus callosum.

Sec incorporation into selenoproteins is mediated by a multi-protein complex that includes SBP2. Following our elucidation of SepSecS mutations underlying PCCA, Schoenmakers et al. (2010) described two unrelated patients – each with compound heterozygosity for two different mutations in SBP2 leading to reduced synthesis of most of the 25 known human selenoproteins. Both patients had mild global psychomotor and speech delay, accompanied with varying degrees of other abnormalities: azoospermia with failure of the latter stages of spermatogenesis, axial muscular dystrophy, photosensitivity likely due to cutaneous deficiencies of antioxidant selenoenzymes, increased cellular reactive oxidative species (ROS), and susceptibility to ultraviolet radiation–induced oxidative damage, reduced levels of selenoproteins in peripheral blood cells associated with impaired T lymphocyte proliferation, abnormal mononuclear cell cytokine secretion, and telomere shortening. Paradoxically, raised ROS in affected subjects was associated with enhanced systemic and cellular insulin sensitivity (Schoenmakers et al. 2010).

Both PCCA and this latter phenotype are expected to stem from depletion of all or most selenoproteins. The human selenoproteome is comprised of 25 members, whose biological functions have been implicated in diverse human diseases ranging from cardiovascular and endocrine disorders to abnormalities in immune responses and cancer (Papp et al. 2007; Bellinger et al. 2009). Five human glutathione peroxidases (GPxs) and three thioredoxin reductases (TrxRs) are selenoenzymes that orchestrate the human antioxidant defense mechanisms (Lu and Holmgren 2009). They protect the cell from reactive oxygen species, such as hydrogen peroxide and lipid hydroperoxides, and are also responsible for recycling nutritional antioxidants, such as vitamin C, vitamin E, and coenzyme Q. Sec is also the catalytic residue in three iodothyronine deiodinases, the enzymes that potentiate circulating thyroxine (T4) by converting it to its active cellular form, triiodothyronine (T3), in peripheral tissues (Lu and Holmgren 2009). Many of the selenoproteins have homologues in which Sec is replaced by cysteine; however, this replacement is detrimental for the catalytic activity of selenoenzymes (Lobanov et al. 2009).

While the SepSecS mutations seen in PCCA fully abrogate formation of all selenoproteins, and the SBP2 mutations lead to reduced synthesis of most of the 25 known human selenoproteins, there are "selenoproteinopathies" that are due to mutations in specific members of the 25 selenoprotein family. One such example is selenoprotein N (*SEPN1*): its deficiency causes several inherited neuromuscular disorders collectively termed SEPN1-related myopathies, characterized by early onset, generalized muscle atrophy, and muscle weakness affecting especially axial muscles and leading to spine rigidity, severe scoliosis, and respiratory insufficiency (Castets et al. 2012). SelN is ubiquitously expressed, is located in the membrane of the endoplasmic reticulum, and is essential for muscle regeneration and satellite cell maintenance in mice and humans; however, its function remains elusive (Castets et al. 2012). Different mutations of *SEPN1* can cause either congenital myopathy with fiber-type disproportion (similar to that seen in SBP2 deficiency) and insulin resistance or rigid spine syndrome (known also as rigid spine muscular dystrophy 1/severe classic form of minicore myopathy/Desmin-related myopathy with Mallory bodies). Rigid spine syndrome (Arbogast et al. 2009) is an autosomal recessive disease of diffuse muscle weakness and hypotonia characterized by joint contractures with a flat thorax and limited neck flexion, high arched palate, and nasal voice, combined with short stature, failure to thrive, and delayed motor development. Histologically, there is merosin-positive muscular atrophy with variation in muscle fiber size and hyaline plaques. Poorly defined "minicore" regions are found within type 1 and type 2 muscle fibers, in which there is sarcomeric disorganization and absent mitochondria with no oxidative activity. This severe disease that ensues in infancy leads to nocturnal hypoventilation and usually culminates in death due to respiratory failure before adulthood. Ex vivo cultures of myoblasts derived from patients with *SEPN1* mutations show oxidative and nitrosative stress with increased intracellular ROS and nitric oxide. This phenotype is ameliorated by treatment with the antioxidant *N*-acetylcysteine (Arbogast et al. 2009).

## Mechanisms of the Neurological Phenotypes of Selenocysteinopathies

The neurological phenotypes found in the selenocysteinopathies, as well as demonstrated in the progressive brain atrophy in PCCA, are in line with the known crucial role of selenium and selenoproteins in brain development and function. Many of the selenoproteins are expressed in the human brain and are essential for normal neuronal development and activity (Zhang et al. 2008; Wirth et al. 2010). In mice, all known selenoprotein mRNAs are expressed in the brain. Neurons in olfactory bulb, hippocampus, cerebral cortex, and cerebellar cortex are exceptionally rich in selenoprotein gene expression, in particular GPx4, SelK, SelM, SelW, Sep15, and SelP. The expression patterns of the Sec machinery genes SBP2, PSTK, and SecP43 (a tRNA$^{Sec}$-binding protein) correlate with selenoprotein gene-enriched structures (Zhang et al. 2008). Selenoproteins have been implicated in removing reactive oxidative species, protein folding, degradation of misfolded membrane proteins, and control of cellular calcium homeostasis, all processes known to be deregulated in neurodegenerative diseases (Papp 2007). Compared to other organs, the selenium content of the brain is not particularly high (Zachara et al. 2001). Nevertheless, under dietary selenium deficiency, circulating selenoprotein P (SelP) ensures the preferential transport of selenium reserves to the brain at the expense of other organs (Nakayama et al. 2007). It is, thus, crucial that selenium brain content remains stable (Nakayama et al. 2007). Indeed, knockout of SelP (Hill et al. 2003) or its brain receptor ApoER2 (Burk et al. 2007) in mice leads to cerebral selenium deficiency and neurodegeneration phenotypes, such as ataxia, seizures, and motor abnormalities with severe spasticity (Valentine et al. 2008). It should be noted that SelP is itself a selenoprotein, and thus, in the absence of Sec (as in SepSecS mutations in PCCA) or in the absence of the tRNA for Sec (tRNA$^{Sec}$), one would expect a neurological phenotype that is at least as severe as that in the case of isolated deletion of SelP. Indeed, similar to the PCCA human phenotype, mouse experiments demonstrated that neuron-specific deletion of tRNA$^{Sec}$ ablating brain expression of all 24 selenoproteins resulted in a neurodevelopmental and degenerative phenotype of astrogliosis in the cerebral cortex, cerebellum, and hippocampus (Wirth et al. 2010). Of the various selenoproteins, Gpx4 has a particularly important role in the ROS-scavenging network, protecting the brain from oxidative insult by protecting against lipid hydroperoxide accumulation, a characteristic of neurodegenerative diseases (Yoo et al. 2010). In mice, neuron-specific GPx4 depletion causes neurodegeneration in vivo and ex vivo (Yoo et al. 2010). Another selenoprotein, SelM, was recently suggested to have an important role in protecting against oxidative damage in the brain and may potentially function in calcium regulation (Reeves et al. 2010).

Treatment modalities based on our understanding of the molecular basis of selenocysteinopathies are not trivial and are yet to emerge. It can be well assumed that there are further selenocysteinopathies yet to be unraveled. With the growing availability of genome-wide association studies and next-generation sequencing, one can expect that polymorphisms in selenoprotein pathways will be found to be associated with common neurodegenerative diseases, while detrimental mutations in genes of this pathway will likely be demonstrated to cause severe monogenic neurological diseases.

## Cross-References

- ▶ Selenium and Glutathione Peroxidases
- ▶ Selenium and Iodothyronine Deiodinases
- ▶ Selenium and Muscle Function
- ▶ Selenium and the Regulation of Cell Cycle and Apoptosis
- ▶ Selenium in Human Health and Disease
- ▶ Selenium, Biologically Active Compounds

- Selenium-Binding Protein 1 and Cancer
- Selenophosphate Synthetase
- Selenoprotein K
- Selenoprotein P
- Selenoprotein Sep15
- Selenoproteins and the Biosynthesis and Activity of Thyroid Hormones
- Selenoproteins in Prokaryotes
- Selenoproteins and Thyroid Gland

## References

Agamy O, Ben Zeev B, Lev D et al (2010) Mutations disrupting selenocysteine formation cause progressive cerebello-cerebral atrophy. Am J Hum Genet 87(4):538–544

Arbogast S, Beuvin M, Fraysse B et al (2009) Oxidative stress in SEPN1-related myopathy: from pathophysiology to treatment. Ann Neurol 65:677–686

Bellinger FP, Raman AV, Reeves MA et al (2009) Regulation and function of selenoproteins in human disease. Biochem J 422:11–22

Burk RF, Hill KE, Olson GE et al (2007) Deletion of apolipoprotein E receptor-2 in mice lowers brain selenium and causes severe neurological dysfunction and death when a low-selenium diet is fed. J Neurosci 27:6207–6211

Castets P, Lecsure A, Guicheney P, Allamand V (2012) Selenoprotein N in skeletal muscle: from disease to function. J Mol Med. (Pubmed ahead of print)

Hill KE, Zhou J, McMahan WJ et al (2003) Deletion of selenoprotein P alters distribution of selenium in the mouse. J Biol Chem 278:13640–13646

Huang Z, Rose AH, Hoffmann PR (2012) The role of selenium in inflammation and immunity: from molecular mechanisms to therapeutic opportunities. Antioxid Redox Signal 16(7):705–743

Lobanov AV, Hatfield DL, Gladyshev VN (2009) Eukaryotic selenoproteins and selenoproteomes. Biochim Biophys Acta 1790:1424–1428

Lu J, Holmgren A (2009) Selenoproteins. J Biol Chem 284:723–727

Nakayama A, Hill KE, Austin LM et al (2007) All regions of mouse brain are dependent on selenoprotein P for maintenance of selenium. J Nutr 137:690–693

Papp LV, Lu J, Holmgren A, Khanna KK (2007) From selenium to selenoproteins: synthesis, identity and their role in human health. Antioxid Redox Signal 9(7):775–806

Rayman MP (2012) Selenium and human health. Lancet 379:1256–1268

Reeves MA, Bellinger FP, Berry MJ (2010) The neuroprotective functions of selenoprotein M and its role in cytosolic calcium regulation. Antioxid Redox Signal 12(7):809–818

Schoenmakers E, Agostini M, Mitchell C (2010) Mutations in the selenocysteine insertion sequence–binding protein 2 gene lead to a multisystem selenoprotein deficiency disorder in humans. J Clin Invest 120(12):4220–4235

Valentine WM, Abel TW, Hill KE et al (2008) Neurodegeneration in mice resulting from loss of functional selenoprotein P or its receptor apolipoprotein E receptor 2. J Neuropathol Exp Neurol 67:68–77

Wirth EK, Conrad M, Winterer J et al (2010) Neuronal selenoprotein expression is required for interneuron development and prevents seizures and neurodegeneration. FASEB J 24(3):844–852

Yoo MH, Gu X, Xu XM et al (2010) Delineating the role of glutathione peroxidase 4 in protecting cells against lipid hydroperoxide damage and in Alzheimer's disease. Antioxid Redox Signal 12(7):819–827

Zachara BA, Pawluk H, Bloch-Boguslawska E et al (2001) Tissue level, distribution, and total body selenium content in healthy and diseased humans in Poland. Arch Environ Health 56:461–466

Zhang Y, Zhou Y, Schweizer U et al (2008) Comparative analysis of selenocysteine machinery and selenoproteome gene expression in mouse brain identifies neurons as key functional sites of selenium in mammals. J Biol Chem 283:2427–2438

# Selenomethionine

- Selenium in Human Health and Disease

# Selenophosphate Synthase

- Selenophosphate Synthetase

# Selenophosphate Synthetase

Nicholas Noinaj and Susan K. Buchanan
Laboratory of Molecular Biology, National Institute of Diabetes and Digestive and Kidney Diseases,
US National Institutes of Health, Bethesda, MD, USA

## Synonyms

SelD; Selenide; Selenium donor protein; Selenophosphate synthase; Water dikinase

## Definition

Selenophosphate synthetase is an enzyme that catalyzes the formation of selenophosphate, the selenium donor in the biosynthesis of the amino acid selenocysteine.

Selenophosphate synthetase (SPS) is a metalloprotein responsible for catalyzing the formation of selenophosphate, a selenium-containing precursor utilized in the biosynthesis of selenocysteine, commonly referred to as the 21st amino acid.

Newly synthesized proteins that carry a seleno-modified amino acid such as selenocysteine or selenomethionine are often termed selenoproteins (Rother et al. 2001). Examples include the enzymes thioredoxin reductase and glutathione peroxidase, both of which act as antioxidants for the clearance of potentially harmful reactive oxygen species. Selenoproteins can be found in many organisms from bacteria to humans and are expressed in many tissues within the human body. Mutations in genes for selenoproteins can lead to a number of disorders including abnormal hormone metabolism and rigid spine muscular dystrophy (Brown and Arthur 2001), bringing to the surface the importance of selenium intake in maintaining proper health and wellness.

Selenium is an essential nutrient that has been linked to several human diseases related to hormone regulation, immune response, protection from reactive oxygen species, and brain function (Chen and Berry 2003). Deficiencies in selenium intake have been implicated in several health disorders over the past two decades including cancer, cardiovascular disease, male infertility, asthma, and viral infections. Like other biologically important elements, selenium can be toxic at high levels and is typically transported in an inorganic state rather than in its free elemental form.

Selenium has been shown to be incorporated into enzymes and other proteins either as a selenocysteine (Turanov et al. 2011), which is inserted at specific positions in the protein's sequence where it usually participates in the protein's catalytic function, or as a selenomethionine, which results from random substitution of selenium in place of sulfur in methionine but typically does not affect on an enzyme's activity. In the bacteria *Escherichia coli*, incorporation of selenocysteine is mediated by a UGA codon-directed co-translation, involving four genes, *selA*, *selB*, *selC*, and *selD*, with SelD producing the gene product SPS, a 37 kDa protein that functions as a biological homodimer (Leinfelder et al. 1988). Many of the studies on SPS have been performed in bacteria; however, homologous systems for the biosynthesis of selenophosphate are present in other organisms including humans.

SPS is a metalloprotein that is dependent on both magnesium and potassium for activity and uses ATP, selenide, and water to catalyze the formation of AMP, orthophosphate, and selenophosphate in a two-step reaction (Lacourciere 1999) (Fig. 1). The first step in the reaction is the hydrolysis of the γ-phosphate which is transferred to the selenide to form selenophosphate. Next, the β-phosphate of the resulting ADP is then hydrolyzed to yield the additional products

**Selenophosphate Synthetase, Fig. 1** *Biosynthesis of selenophosphate (SeP) and its role in the formation of selenoproteins.* (**a**) Reaction pathway for selenophosphate biosynthesis. Overall reaction is shown in *black* while Rxn 1 and Rxn 2 are shown in *gray*. (**b**) Hydrolysis of ATP for the biosynthesis of selenophosphate and the other products orthophosphate and AMP. (**c**) Utilization of selenophosphate for the production of selenoproteins orthophosphate and AMP. Until recently, it was unclear why SPS hydrolyzes both high energy bonds of ATP to produce only one molecule of selenophosphate. However, recent studies reported that hydrolysis of the second high energy bond may serve two important functions. The first is to protect the highly reactive selenophosphate product and the second is to provide energy for the release of the reaction products from the binding pocket (Wang et al. 2009).

Studies have shown that the conserved residues cysteine-17 and lysine-20 are essential for catalysis in SPS from *Escherichia coli* and that mutating these residues leads to enzyme inactivation (Veres et al. 1994). These studies also suggest that cysteine-17 may directly coordinate selenium during catalysis and that lysine-20 may form a phosphorylated intermediate, however neither has been isolated nor demonstrated experimentally. In addition, four conserved active site aspartate residues are required for coordinating one or more magnesium ions while a conserved threonine participates in binding a single potassium ion (Wang et al. 2009). While having similar chemical properties to potassium, the presence of sodium or lithium can render SPS inactive, potentially serving a role in the regulation of the enzyme's activity. Interestingly, an unidentified chromophore was reported for SPS having a characteristic absorbance peak at 315 nm. This peak, which may act as a sensor for the formation of a catalytically competent complex, was found to undergo a red shift in the presence of magnesium and ATP, but not in the presence of non-hydrolyzable ATP analogues. It was later reported that mutating any one of the four conserved aspartate residues completely inactivated SPS and also prevented any red shift of the 315 nm peak, even in the presence of both magnesium and ATP (Noinaj et al. 2012).

Recent reports of crystal structures of SPS homologs from *Aquifex aeolicus* (AaSPS), *Escherichia coli* (EcSPS), and humans (hSPS1) have significantly advanced our understanding of the catalytic mechanism of SPS. Importantly, the crystal structures of AaSPS and hSPS1 were solved in transition states with bound ligands and metal cofactors present, providing an invaluable insight into the precise interactions between enzyme and substrate during catalysis by this important enzyme. Providing the first view of the overall structure of SPS, the AaSPS crystal structures (Itoh et al. 2009) provided the molecular details of the catalytic machinery of SPS and its interaction with the non-hydrolyzable ATP analogue AMPCPP. The AMPCPP molecule was found bound along the catalytic site and partially sandwiched along the dimer interface of SPS. It was observed that the glycine-rich N-terminal region of AaSPS underwent a large conformational change upon binding AMPCPP and that four conserved aspartate residues, all essential for enzyme activity, participate in coordinating the metal cofactors which serve both for catalysis and for mediating the binding of the highly electronegative phosphate group of AMPCPP.

Additionally, two crystal structures of hSPS1 were also recently reported (Wang et al. 2009), one in complex with potassium, orthophosphate, and AMPCP and another in a captured intermediate state representing the state immediately following ATP hydrolysis, observing both ADP and $P_i$ products in the crystal structure (Fig. 2). As with AaSPS, both structures were found in a closed conformation. Further, these crystal structures enabled the identification of a conserved threonine residue that serves as a selectivity sensor for monovalent cations, preferring potassium over sodium or lithium for efficient catalysis by SPS and possibly serving a regulatory role in vivo. The essential role of the conserved threonine was further confirmed using mutagenesis studies. Together, the AaSPS and hSPS1 structures provided a detailed description of the molecular interactions involved in cofactor binding, ATP binding, and subsequent hydrolysis.

The majority of what is known about the function of SPS has come from studies investigating SPS from *Escherichia coli* as a model system. The crystal structure of apo EcSPS was also recently reported in an open conformation (Noinaj et al. 2012) and together with the AaSPS and hSPS1 structures helps to provide a more complete view of the steps involved in order for SPS to accomplish its unique and interesting function within the biosynthetic pathway of selenophosphate. However, exactly how selenium is delivered to SPS, whether by direct transfer from a carrier protein or via an inorganic or free state, is still not well understood. Despite this, the current model for the mechanism of selenophosphate synthesis by SPS begins with cofactor binding which is postulated to neutralize the highly electronegative charges of the active site aspartates of SPS and to bridge the binding of the highly electronegative phosphate group of ATP (Fig. 3). Following ATP binding, cysteine-17 and lysine-20 assist in the

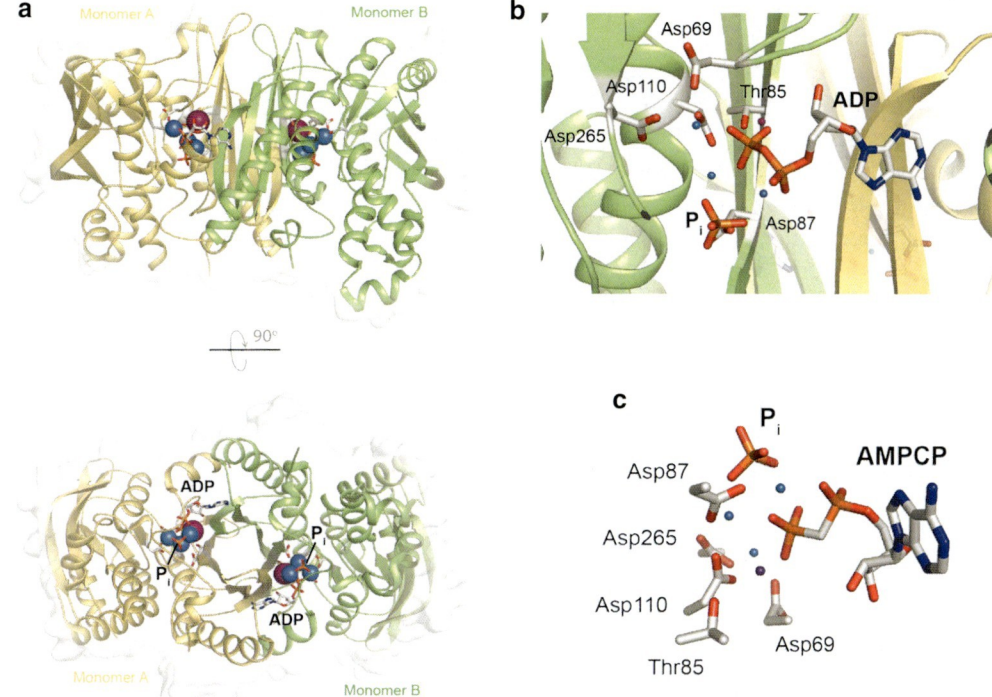

**Selenophosphate Synthetase, Fig. 2** *Crystal structure of human selenophosphate synthetase 1 in complex with sodium, orthophosphate, and ADP.* (**a**) The two monomers of human selenophosphate synthetase 1 (hSPS1, biological dimer) are shown in *gold* and *green*, orthophosphate, ADP, and conserved active site residues are shown in *stick* representation, and magnesium (*blue*) and sodium (*purple*) are shown as *spheres*. The *bottom panel* is rotated 90° along the X-axis relative to the *top panel* (bottom view). (**b**) Zoomed view of the active site of hSPS1 showing conserved residues involved in binding magnesium and sodium in stick representation, magnesium (*blue*) and sodium (*purple*) as *spheres*, and orthophosphate and ADP in stick. (**c**). View of the active site residues from the crystal structure of hSPS1 in complex with potassium, orthophosphate, and AMPCP, following the same representations as in panel (**b**), but at a different angle in order to observe the metal ions mediating ligand binding

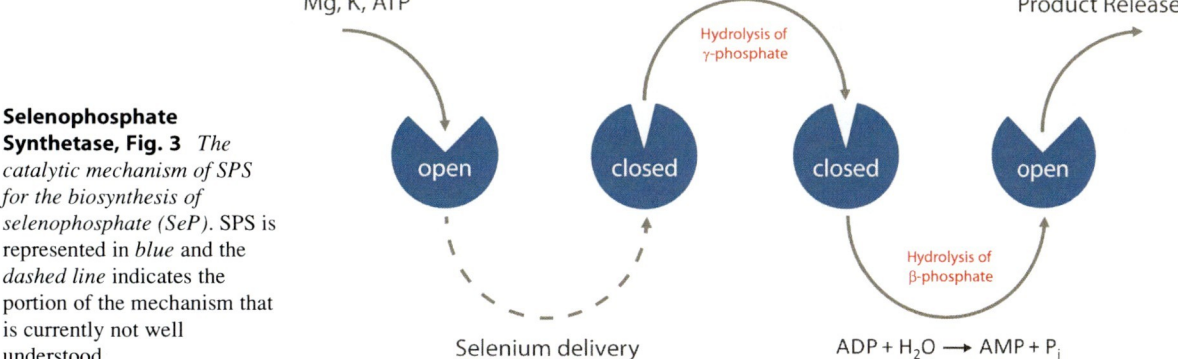

**Selenophosphate Synthetase, Fig. 3** *The catalytic mechanism of SPS for the biosynthesis of selenophosphate (SeP).* SPS is represented in *blue* and the *dashed line* indicates the portion of the mechanism that is currently not well understood

delivery of a selenium donor at the active site, leading to a conformational change and hydrolysis of the γ-phosphate of ATP to form selenophosphate and ADP. To protect the highly reactive selenophosphate, the enzyme remains in a closed conformation preventing immediate release of the products. However, subsequent hydrolysis of the β-phosphate of ADP supplies the energy required for a conformational change back to the open state, thereby releasing the newly formed products from the active site.

SPS plays an integral role within the biosynthetic pathway of selenocysteine by directly forming the essential precursor selenophosphate. Both structural and functional studies of SPS from *Aquifex aeolicus*, *Escherichia coli*, and humans have provided a wealth of information by which to elucidate the mechanistic steps involved in catalysis. However, a longstanding question that remains to be answered asks exactly how selenium is delivered to the active site of SPS. The exact role that cysteine-17 and lysine-20 play within catalysis is still not well understand and more studies are needed in order to provide the last few pieces to complete the mechanistic puzzle of SPS.

## Cross-References

- ▶ Artificial Selenoproteins
- ▶ Selenium and Aquaporins
- ▶ Selenium and Glutathione Peroxidases
- ▶ Selenium in Human Health and Disease
- ▶ Selenium and Iodothyronine Deiodinases
- ▶ Selenium and Muscle Function
- ▶ Selenium and the Regulation of Cell Cycle and Apoptosis
- ▶ Selenium, Biologically Active Compounds
- ▶ Selenium-Binding Protein 1 and Cancer
- ▶ Selenocysteinopathies
- ▶ Selenoprotein Sep15
- ▶ Selenoproteins and the Biosynthesis and Activity of Thyroid Hormones
- ▶ Selenoproteins and Thyroid Gland
- ▶ Selenoproteins in Prokaryotes

## References

Brown KM, Arthur JR (2001) Selenium, selenoproteins and human health: a review. Public Health Nutr 4(2B):593–599, Review

Chen J, Berry MJ (2003) Selenium and selenoproteins in the brain and brain diseases. J Neurochem 86(1):1–12, Review

Itoh Y, Sekine S, Matsumoto E, Akasaka R, Takemoto C, Shirouzu M, Yokoyama S (2009) Structure of selenophosphate synthetase essential for selenium incorporation into proteins and RNAs. J Mol Biol 385(5):1456–1469

Lacourciere GM (1999) Biosynthesis of selenophosphate. Biofactors 10(2–3):237–244, Review

Leinfelder W, Forchhammer K, Zinoni F, Sawers G, Mandrand-Berthelot MA, Böck A (1988) Escherichia coli genes whose products are involved in selenium metabolism. J Bacteriol 170(2):540–546

Noinaj N, Wattanasak R, Lee DY, Wally JL, Piszczek G, Chock PB, Stadtman TC, Buchanan SK (2012) Structural insights into the catalytic mechanism of *Escherichia coli* selenophosphate synthetase. J Bacteriol 194(2):499–508

Rother M, Resch A, Wilting R, Böck A (2001) Selenoprotein synthesis in archaea. Biofactors 14(1–4):75–83, Review

Turanov AA, Xu XM, Carlson BA, Yoo MH, Gladyshev VN, Hatfield DL (2011) Biosynthesis of selenocysteine, the 21st amino acid in the genetic code, and a novel pathway for cysteine biosynthesis. Adv Nutr 2(2):122–128

Veres Z, Kim IY, Scholz TD, Stadtman TC (1994) Selenophosphate synthetase. Enzyme properties and catalytic reaction. J Biol Chem 269(14):10597–10603

Wang KT, Wang J, Li LF, Su XD (2009) Crystal structures of catalytic intermediates of human selenophosphate synthetase 1. J Mol Biol 390(4):747–759

# Selenoprotein

▶ Selenium in Human Health and Disease

# 15kDa Selenoprotein

▶ Selenoprotein Sep15

# Selenoprotein 15 KDa (Sep15)

▶ Selenium and Muscle Function

# Selenoprotein K

Mohamed E. Moustafa
Department of Biochemistry, Faculty of Science, Alexandria University, Alexandria, Egypt

## Synonyms

Dsel; dSelK; G-rich; Sel K; SelG; SelK

## Definition

Selenoprotein K (SelK) is encoded in one of the 25 selenium-containing protein genes in humans. It is present in most organisms in the animal kingdom including mammals, birds, reptiles, and fish, and its homologs also occur in other eukaryotes. It is a small protein of approximately 10 kDa located in the endoplasmic reticulum membrane and possibly involved in retrotranslocation of proteins from the endoplasmic reticulum. The exact function of SelK remains to be determined.

## Introduction

Selenium (Se) is an essential micronutrient that plays important roles in human health, and its severe dietary deficiency is associated with various diseases such as Keshan disease, Kashin-Beck disease, and numerous other disorders including cancer and heart disease (Hatfield et al. 2012). The biological roles of selenium in health and development are exerted in large part through its presence in selenoproteins as selenocysteine (Sec, one-letter code is U), the 21st amino acid in the genetic code (Hatfield et al. 2012). There are 25 selenoprotein genes in humans and 24 in rodents (Kryukov et al. 2003). Various selenoproteins are present in the three domains of life: bacteria, archaea, and eukaryotes.

SelK is a novel selenoprotein identified in human and other mammalian genomes using bioinformatics (e.g., see Kryukov et al. 2003) and has been found to exist in numerous other animals as well as in unicellular eukaryotes.

## Chromosomal Location and Amino Acid Sequence of SelK

Human *SelK* gene is encoded on chromosome 3 at position 3p21.31 and is composed of 5 exons. It is a small selenoprotein composed of 94 amino acids with a calculated molecular weight of 10.5 kDa (Kryukov et al. 2003; Al-Rifai and Moustafa 2012). SelK sequences contain a single Sec residue located at or near the penultimate C-terminal end depending on the length of the protein which varies between 92 and 94 amino acids (Al-Rifai and Moustafa 2012).

All human selenoproteins, with the exception of selenoprotein P, contain a single Sec residue. Sec occurs at the active site of most selenoenzymes including glutathione peroxidases and thioredoxin reductases, where Sec is more reactive than cysteine (Cys) at physiological pH.

SelK is a highly conserved protein in mammals, wherein mammalian SelK shares more than 90% amino acid sequence identity. The calculated isoelectric point of SelK sequences in the animal kingdom is 10.3–11.6 indicating that SelK is a basic protein (Al-Rifai and Moustafa 2012).

Bioinformatic approaches revealed the occurrence of Cys-containing homologs of SelK in some eukaryotes such as insects, nematodes, and plants in which Cys residues have been identified in place of Sec. SelK Cys-containing homologs are also small basic proteins with molecular weights ranging from 10.7 to 11.5 kDa, and they share 36% or more identity with human SelK (Shchedrina et al. 2011a; Al-Rifai and Moustafa 2012).

The Selk/SelS family of proteins in eukaryotes was recently defined by using bioinformatics approaches, based on a set of similar domain organizations. This family of proteins is composed of eukaryotic SelK and SelS, other selenoproteins (e.g., plasmodial Sel1 and Sel4), Cys-homologs of SelK/SelS in insects and plants, as well as other unrelated proteins (Shchedrina et al. 2011a).

## Distribution and Subcellular Localization of SelK

The distribution and subcellular localization of SelK are important in elucidating its biological roles. SelK is expressed in a variety of tissues in humans and rodents including heart, skeletal muscles, pancreas, testes, liver, brain, kidneys, spleen, placenta, and lungs (Lu et al. 2006; Al-Rifai and Moustafa 2012). It is expressed at a relatively high level in a variety of immune tissues and myeloid cells, macrophages, neutrophils, and dendritic cells (Huang et al. 2011; Verma et al. 2011).

SelK is localized in the endoplasmic reticulum (ER) and plasma membrane (Kryukov et al. 2003; Lu et al. 2006; Chen et al. 2006; Verma et al. 2011). Six other selenoproteins are residents in the ER: selenoprotein Sep15, type 2 iodothyronine deiodinase (D2), and selenoproteins M, N, S, and T. Among the ER-resident selenoproteins, SelK is the most widespread selenoprotein as well as the most widespread eukaryotic selenoprotein (Shchedrina et al. 2011a).

SelK is a single spanning integral protein that belongs to type III integral membrane proteins and is located in the ER membrane where a short N-terminal sequence is localized in the ER lumen and the Sec-containing C-terminal region is exposed to the cytosolic side of the ER (Chen et al. 2006). The targeting signal in SelK for localization in the ER remains to be determined.

Selenoproteins are present in vertebrates including mammals and fish, and also in some lower plants such as algae. Some invertebrates, higher plants, and fungi lack selenoprotein genes (Shchedrina et al. 2011b). The insect, *Drosophila melanogaster*, has preserved only three selenoproteins: SelK (G-rich), SelH, and selenophosphate synthetase 2 (SPS2) (Shchedrina et al. 2011b). Drosophila SelK, also called G-rich, dSelK, or SelG, is a homolog of mammalian SelK. These three selenoprotein genes are not essential in *Drosophila*, wherein their functions appear to be active only under certain stress conditions such as starvation. This observation provides a possible explanation for the reason that several species of insects such as *Drosophila willistoni* and the red flour beetle *Tribolium castaneum* have lost all selenoprotein genes (Shchedrina et al. 2011b). G-rich is a type III integral membrane protein localized in the Golgi apparatus of *Drosophila* cells with its C-terminus in the cytosol and its N-terminus in the lumen of Golgi. It was the first selenoprotein found to reside in the Golgi apparatus (Chen et al. 2006).

## Sec Incorporation into SelK and Selenoproteins in General

The Sec incorporation machinery that is responsible for inserting Sec into selenoproteins is conserved among eukaryotes. Sec is incorporated into the nascent polypeptide on ribosomes in response to a specific UGA codon in the mRNA coding for SelK or any other selenoprotein.

Decoding of the UGA codon during translation as Sec requires (1) the Sec insertion sequence (SECIS) element, which is a stem-loop structure located in the 3′-untranslated region of the mRNA coding for SelK or other selenoproteins; (2) a SECIS-binding protein 2 (SBP2) that binds to both the SECIS element and the ribosome; (3) a Sec-specific elongation factor (eEFsec); (4) a unique Sec tRNA, designated Sec tRNA$^{[Ser]Sec}$; (5) a Sec-associated protein (SECp43); (6) a L30 ribosome protein (rpL30); and (7) a SECIS-interacting nucleolin. SBP2 recruits EFsec and Sec tRNA$^{[Ser]Sec}$ then interacts with SecP43 and the ribosomal L protein (Hatfield et al. 2012).

## Roles of SelK in Immunity

SelK expression was decreased in mice fed a selenium-deficient diet containing 0.08 ppm Se, while its expression was increased when the content of selenium in the diet was increased to 1.0 ppm (Verma et al. 2011).

As noted above, SelK is expressed in the spleen and lymph tissues, and also in a variety of immune cells including T and B lymphocytes, neutrophils, and macrophages in humans. SelK was the first selenoprotein found to be expressed in relatively high levels in lymphoid and immune cells.

Immune responses are regulated by m-calpains that are cysteine proteases activated by calcium. m-Calpains mediate the cleavage of SelK as well as other target proteins in resting macrophages. SelK is cleaved at the peptide bond between Arg 81 and Gly 82 by action of m-calpain resulting in a truncated SelK isoform lacking the Sec residue at position 92 (Huang et al. 2011). The cleavage of the C-terminal peptide (82–94) of SelK containing the Sec residue by m-calpain in resting macrophages raises a question of whether the Sec residue at position 92 in full-length SelK may play a role in the immune responses exerted in activated macrophages (Verma et al. 2011).

Stimulation of T cell-like receptors in macrophages by ligands is associated with increased levels of calpastatin that prevent the proteolytic cleavage of SelK by m-calpains resulting in an increased level of full-length SelK in the ER membrane of macrophages. Therefore, the calpain/calpastatin system regulates the distribution of full-length SelK and truncated SelK isoforms in macrophage activation (Huang et al. 2011). It has been shown that SelK is required for efficient calcium flux and optimal Fcγ receptor-mediated activation of macrophage functions such as phagocytosis, ERK phosphorylation, and secretion of cytokines, TNF-α, and IL-6, as well as nitric oxide (Huang et al. 2011; Verma et al. 2011).

The catalytic subunit of m-calpain is oriented to the cytosolic side of the ER in proximity to SelK and the interaction between SelK, m-calpain, and calpastatin occurs on the cytosolic side of the ER membrane (Huang et al. 2011). Deletion of the *SelK* gene (SelK$^{-/-}$) in mice resulted in decreased receptor-mediated calcium flux in the ER of T cells, neutrophils, and macrophages. Calcium-dependent functions were either eliminated or at least partially impaired resulting in altered immune responses in SelK$^{-/-}$ mice. T cell proliferation and neutrophil migration were partially impaired, while T cell migration was eliminated in these mice. In addition, the Fcγ receptor-mediated oxidative burst was reduced in macrophages of SelK$^{-/-}$ mice. These results suggest that SelK promotes calcium-dependent functions in immune cells and SelK is not a limiting factor in regulating calcium-dependent immune cell function.

## Role of SelK in Redox Homeostasis

Selenium is involved in the regulation of the cellular redox balance in mammals through its presence in antioxidant selenoenzymes that include the five glutathione peroxidases (GPx 1–4, GPx 6), three thioredoxin reductases (TR1–TR3), and iodothyronine deiodinases, where Sec is located at the active site of these selenoenzymes (Kryukov et al. 2003). Other selenoproteins such as selenoprotein P and selenoprotein W may also have antioxidant roles.

Reducing SelK levels in HepG2 cells by RNA interference decreased the antioxidant capability of the cells and increased apoptosis after treatment of cells with ER stress agents such as tunicamycin (Du et al. 2010). On the other hand, overexpression of human SelK in cultured cardiomyocytes decreased the level of intracellular reactive oxygen species (ROS) and protected these cells from oxidative stress-induced toxicity after treatment with a high level of hydrogen peroxide. These results indicate that SelK may be involved in the regulation of cellular redox homeostasis (Lu et al. 2006). Further studies are required to demonstrate whether SelK is an antioxidant in vivo and to understand its exact role in reducing the level of ROS.

## Role of SelK as a Chaperone: SelK-Protein Interaction

Proteins that fail to refold properly to their native conformation are degraded through ER-associated degradation (ERAD). Many proteins are involved in the multistep ERAD process which includes recognition of misfolded protein substrates in the ER, and delivery of misfolded proteins to the cytosol through a channel to proteasomes for degradation. It has been shown that SelK is a member of the ERAD complex which is composed of many other proteins such as Derlins, SelS, and p97 in mammals. SelK associates with members of this family during elimination of misfolded proteins. It interacts with Derlins 1, 2, and 3b with a higher affinity for Derlin-1. SelK also associates with SelS to form a SelK-SelS complex that interacts with p97. Derlin-1 or SelK associates with one unique peptide of p97 (Shchedrina et al. 2011a).

SelK is a chaperone that interacts with misfolded proteins and translocates them to the ERAD through the Derlin-selenoprotein-p97 complex (Shchedrina et al. 2011a). It should be noted that many proteins that interact with Derlins and required for ERAD do not interact with SelK suggesting that the interaction of SelK with Derlins and p97 exists at an early stage in ERAD (Shchedrina et al. 2011a).

The promoter region of the *SelK* gene contains the ER stress response element (ERSE) which is also present in the promoter regions of ER chaperons and SelS. This explains why SelK expression is upregulated by conditions that induce accumulation of misfolded proteins in the ER such as inhibition of the proteasome or accumulation of misfolded proteins in the ER (Shchedrina et al. 2011a).

N-linked glycosylation is involved in folding and maturation of membrane and secreted proteins, and it takes place by the oligosaccharyltransferase (OST) complex. It has been shown that SelK is coprecipitated with soluble glycosylated ERAD substrates, such as RPN$_{332}$ and NHK, and is involved in their degradation in the event they do not refold properly. SelK associates with components of the OST complex that include ribophorins, RPNI and RPNII, OST48, and ER chaperones. It is also involved in the Derlin-dependent ERAD of glycosylated misfolded proteins in eukaryotes (Shchedrina et al. 2011a).

In summary, SelK is an interesting, small selenoprotein that exists in the ER and is widely spread in cells and tissues of animals. Homologs of SelK also occur in many other eukaryotes. The exact functions of SelK remain to be elucidated. However, recent studies indicate that SelK could be involved in cellular redox homeostasis, immune responses, and as a chaperone protein.

**Acknowledgments** The author thanks Dr. Dolph L. Hatfield, NCI, NIH, Bethesda, MD, USA and Dr. Vadim N. Gladyshev, Department of Medicine, Brigham and Women's Hospital, Harvard Medical School, Boston, MA, USA for valuable comments on this chapter.

## Cross-References

- Peroxidases
- Scandium, Physical and Chemical Properties
- Selenium and Glutathione Peroxidases
- Selenium and Iodothyronine Deiodinases
- Selenium and Muscle Function
- Selenium and the Regulation of Cell Cycle and Apoptosis
- Selenium in Human Health and Disease
- Selenium, Biologically Active Compounds
- Selenium-Binding Protein 1 and Cancer
- Selenocysteinopathies
- Selenophosphate Synthetase
- Selenoprotein P
- Selenoprotein Sep15
- Selenoproteins and the Biosynthesis and Activity of Thyroid Hormones
- Selenoproteins and Thyroid Gland
- Selenoproteins in Prokaryotes

## References

Al-Rifai MF, Moustafa ME (2012) Analysis of conserved structural features of selenoprotein. EJBMB 30(1):1–18

Chen CL, Shim MS, Chung J, Yoo HS, Ha JM, Kim JY, Choi J, Zang SL, Hou X, Carlson BA, Hatfield DL, Lee BJ (2006) G-rich, a Drosophila selenoprotein, is a Golgi-resident type III membrane protein. Biochem Biophys Res Commun 348(4):1296–1301

Du S, Zhou J, Jia Y, Huang K (2010) SelK is a novel ER stress-regulated protein and protects HepG2 cells from ER stress agent-induced apoptosis. Arch Biochem Biophys 502(2):137–143

Hatfield DL, Berry MJ, Gladyshev VN (2012) Selenium: its molecular biology and role in human health, 3rd edn. Springer, New York

Huang Z, Hoffmann FW, Norton RL, Hashimoto AC, Hoffmann PR (2011) Selenoprotein K is a novel target of m-calpain, and cleavage is regulated by Toll-like receptor-induced calpastatin in macrophages. J Biol Chem 286(40):34830–34838

Kryukov GV, Castellano S, Novoselov SV, Lobanov AV, Zehtab O, Guigó R, Gladyshev VN (2003) Characterization of mammalian selenoproteomes. Science 300(5624): 1439–1443

Lu C, Qiu F, Zhou H, Peng Y, Hao W, Xu J, Yuan J, Wang S, Qiang B, Xu C, Peng X (2006) Identification and characterization of selenoprotein K: an antioxidant in cardiomyocytes. FEBS Lett 580(22):5189–5197

Shchedrina VA, Everley RA, Zhang Y, Gygi SP, Hatfield DL, Gladyshev VN (2011a) Selenoprotein K binds multiprotein complexes and is involved in the regulation of endoplasmic reticulum homeostasis. J Biol Chem 286(50):42937–42948

Shchedrina VA, Kabil H, Vorbruggen G, Lee BC, Turanov AA, Hirosawa-Takamori M, Kim HY, Harshman LG, Hatfield DL, Gladyshev VN (2011b) Analyses of fruit flies that do not express selenoproteins or express the mouse selenoprotein, methionine sulfoxide reductase B1, reveal a role of selenoproteins in stress resistance. J Biol Chem 286(34):29449–29461

Verma S, Hoffmann FW, Kumar M, Huang Z, Roe K, Nguyen-Wu E, Hashimoto AS, Hoffmann PR (2011) Selenoprotein K knockout mice exhibit deficient calcium flux in immune cells and impaired immune responses. J Immunol 186(4): 2127–2137

# Selenoprotein K (SelK)

- Selenium and Muscle Function

# Selenoprotein N (SelN, SepN)

- Selenium and Muscle Function

# Selenoprotein P

Ulrich Schweizer
Institute for Experimental Endocrinology, Charité-Universitätsmedizin Berlin, Berlin, Germany

## Synonyms

SelP (conflict with P-selectin); SeP (old literature); SePP; SEPP1 (human gene name)

## Definition

Selenoprotein P (SePP) is a vertebrate-specific selenoprotein and was first characterized as the major selenoprotein in plasma, hence its name. Selenoproteins are proteins containing the rare amino acid selenocysteine (Sec) in their polypeptide chain. Sec incorporation is specific, in contrast, selenomethionine (SeMet) represents a minor fraction of the methionine pool and is thus randomly inserted into protein in the place of methionine. Sec therefore represents the 21st proteinogenic amino acid and is encoded in mRNA by UGA codons. The cis-acting *Selenocysteine-Insertion-Sequence* (SECIS) in the 3′-untranslated region of selenoprotein mRNAs is required for the re-coding of UGA stop codons as Sec. SePP is the only selenoprotein containing two SECIS elements. SePP contains the highest number of selenocysteines among selenoproteins and plays the central role in body selenium transport and distribution.

## Selenoprotein P is a Plasma Protein and Exhibits Neurotrophic Activity

Mammalian plasma contains 40–200 μg selenium per liter. About 40–60% is contained in SePP, a glycosylated protein of 55 kD (Motsenbocker and Tappel 1982) and most of the remaining selenium is bound in plasma glutathione peroxidase (pGPx, GPX3). Selenoprotein P was discovered and purified based on the incorporation of $^{75}$Se into plasma proteins after metabolic labeling with $^{75}$selenite. Metabolic labeling studies have further suggested that the protein has a plasma half-life of 3–4 h and is subject to rapid turnover (Burk and Hill 2005). Plasma levels of SePP decrease upon dietary selenium restriction and increase rapidly upon selenium supplementation. Based on its response to selenium nutrition and high abundance, it has been speculated that SePP represents a selenium transport protein. In accordance with this notion, *Sepp* mRNA expression is highest in liver and kidney, albeit its widespread expression in most other tissues suggested an additional local function.

SePP was purified as a neurotrophic factor from bovine serum. Serum fractionation revealed a 50–60 kD glycoprotein that was needed to support primary hippocampal neuron cultures in vitro when cultivated with chemically defined medium. Biochemical characterization identified SePP as this neurotrophic activity that was distinct from growth factors (Yan and Barrett 1998). SePP is expressed in the human brain (Scharpf et al. 2007) and mutations in the selenoprotein biosynthetic pathway lead to neurological disease in humans (Schweizer et al. 2011). Recently, available data on the essential roles of selenoproteins in the brain has been reviewed (Schweizer 2012).

## Structure of Selenoprotein P

The amino acid sequence of SePP reveals two domains. The N-terminal domain is predicted to resemble a thioredoxin-like domain carrying the first Sec residue in the loop preceding an α-helix. The position of the Sec as part of a Sec-X-X-Cys motif is conserved among vertebrates and suggests a functional, possibly catalytic, role conferring a weak phospholipid-hyperperoxide peroxidase activity to the protein. The following C-terminal domain is less conserved and contains 9–17 s residues, depending on species. The C-terminal domain is generally believed to act as a selenium carrier for cells. Both domains are separated by a histidine-rich, heparin-binding domain. As a plasma protein, SePP contains several *N*- and *O*-glycosylation sites (Fig. 1). Based on in vitro studies, several researchers have put forward possible antioxidant or heavy metal-binding roles of SePP which still await confirmation in vivo. Selenoprotein P can be cleaved between the two domains by the protease plasma kallikrein (Saito et al. 2004). In zebrafish (*Danio rerio*), two genes exist encoding a short and long form of SePP representing the N-terminal domain and the full-length protein. Interestingly, in rat also several isoforms have been identified, while human and mouse plasma mostly contain full-length SePP. In an attempt, to define the physiological function of the two isoforms, the C-terminus has been genetically ablated. This study has not revealed a function beyond selenium transport (Hill et al. 2007). In a small study in humans, polymorphisms in the human *SEPP* gene were associated with plasma selenium status and body response to selenium supplementation (Meplan et al. 2007), but the results have not yet been confirmed independently.

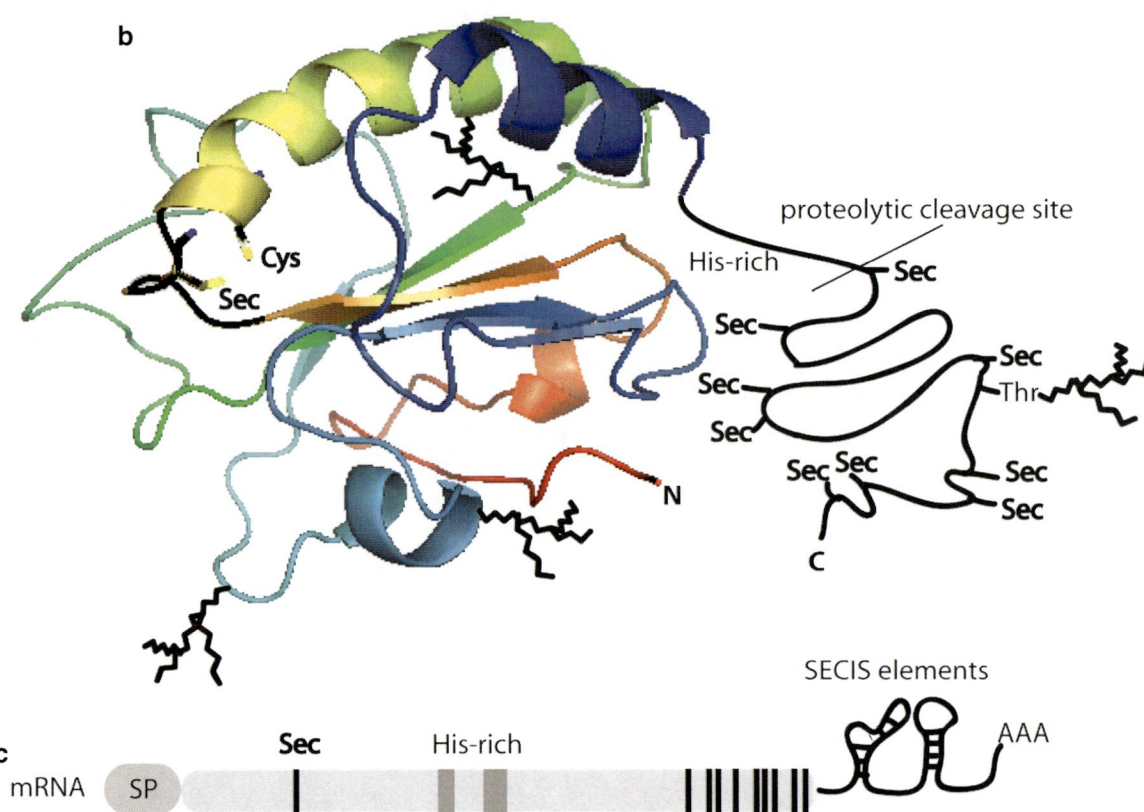

**Selenoprotein P, Fig. 1** Structure of Selenoprotein P. (**a**) Amino acid sequence alignment of the N-terminal domain of selenoprotein P from human, chimpanzee, cattle, mouse, and zebrafish generated with CLUSTALW2 at European Bioinformatics Institute. (**b**) Model structure of the N-terminal domain of human SePP after clipping of the signal peptide. The structure was modeled using PDB 3GL3 as a template and SWISS MODEL at the Swiss Institute of Bioinformatics (Schwede et al. 2003). Glycosylation sites, Sec and conserved Cys are indicated. (**c**) The mRNA sequence of selenoprotein P shows the location of Sec residues, the His-rich regions, and two SECIS elements in the 3′-untranslated region

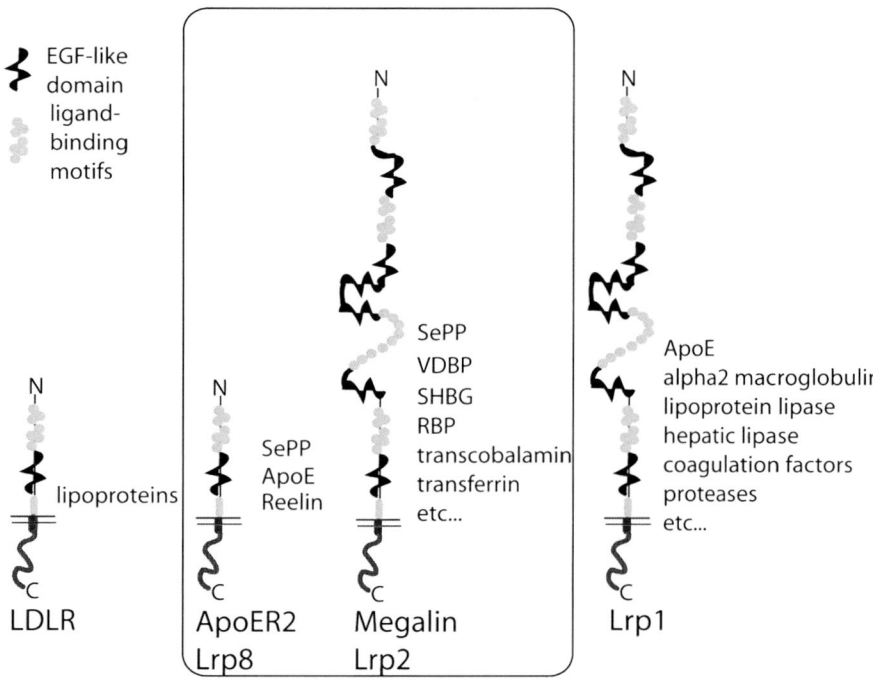

**Selenoprotein P, Fig. 2** Selenoprotein P receptors and related proteins. Members of the lipoprotein receptor-related protein family mediate internalization of multiple ligands and resemble the LDL-receptor. ApoER2 and megalin are SePP receptors

## Role of Selenoprotein P in Selenium Transport

The role of SePP in body selenium distribution and transport was carefully delineated using a number of transgenic mouse models. Targeted inactivation of the *Sepp* gene decreased plasma selenium by more than 70% and resulted in a significant reduction of tissue selenium levels in SePP target tissues, including kidney, brain, and testes (Hill et al. 2003; Schomburg et al. 2003). $Sepp^{-/-}$ mice developed a neurological phenotype with ataxia and seizures, which for the first time definitely established a role of selenium in brain function. The neurological deficits responded to increased dietary selenium intake supporting the role of SePP in selenium transport into the brain. Male $Sepp^{-/-}$ mice sustained infertility (possibly because of impaired expression of the selenoprotein GPx4 in sperm). Targeted ablation of selenoprotein biosynthesis in hepatocytes confirmed the liver as the primary source of plasma SePP, but revealed the independence of the brain from liver-derived SePP (Schweizer et al. 2005). Subsequent studies employing hepatocyte-specific *SEPP1* expression in $Sepp^{-/-}$ mice dissected the roles of systemic versus cerebral SePP expression (Renko et al. 2008).

## Selenoprotein P Receptors

A selenium transport role for SePP would require the expression of SePP receptor molecules in SePP target tissues. Members of the lipoprotein receptor-related protein family (LRP) are multifunctional endocytic receptors which are expressed at the cell surface and mediate internalization of many target molecules including lipoproteins, hormone binding proteins, and SePP. ApoER2/LRP8 is expressed in brain and testis and binds SePP in vitro. $ApoER2^{-/-}$ mice developed neurological dysfunction when exposed to selenium-deficient diet and display the same sperm defects as $Sepp^{-/-}$ mice (Burk et al. 2007). Megalin/LRP2 is highly expressed along kidney tubular epithelium cells. Similarly, *Lrp2*-deficient mice have reduced kidney selenium levels, reduced expression of renal selenoproteins, and significant reductions in brain selenium content upon dietary selenium restriction (Chiu-Ugalde et al. 2010). Several selenium target

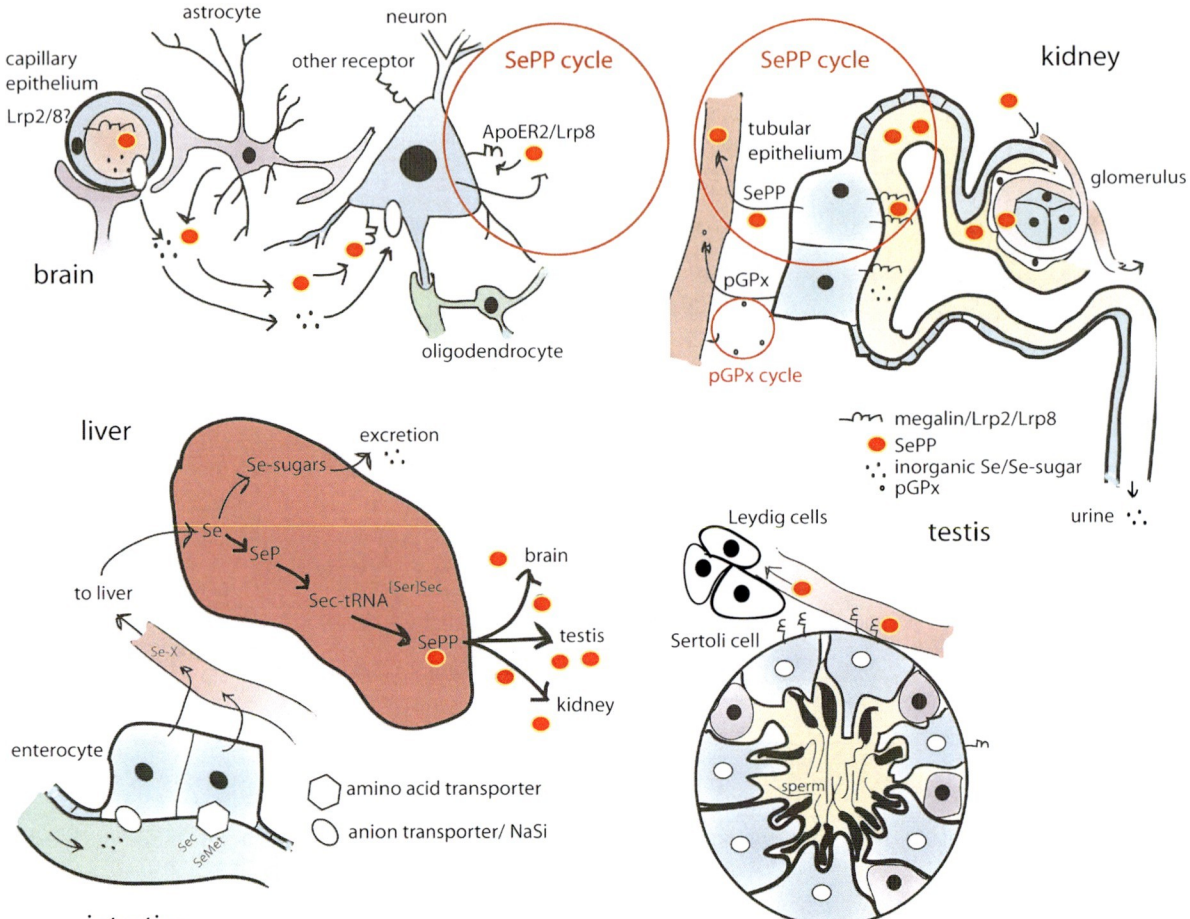

**Selenoprotein P, Fig. 3** Role of Selenoprotein P and its receptors in body selenium distribution. Dietary selenium is primarily taken up along the absorptive epithelium of the small intestine. Inorganic selenium transporters (selenite and selenate uptake) as well as amino acid transporters, which do not distinguish between sulfur- and selenium-containing amino acids, are involved. Selenium travels as an unidentified species (Se-X) to the liver. Within the liver, selenium is converted into Sec-tRNA[Ser]Sec and inserted into selenoproteins. Excess selenium is converted into selenosugars and excreted. The liver is the source of plasma selenoprotein P, which is transported to privileged target tissues expressing cognate receptors. Brain selenium metabolism is complicated, since several cellular membranes must be crossed by selenium to reach the neurons. Megalin may be involved in SePP uptake along the choroid plexus and ependymal epithelium. ApoER2 is a SePP receptor expressed by neurons. Astrocytes express SePP in vitro and may thus contribute to neuronal selenium supply. Neurons express SePP and may store excess selenium extracellularly in the form of SePP. Cerebral SePP expression fixes selenium locally during dietary selenium restriction. Neuronal SePP synthesis and ApoER2-mediated SePP-reuptake have been named the cerebral "SePP cycle." Testis function depends on ApoER2-mediated SePP uptake. Inactivation of either protein leads to decreased GPx4 expression in maturing spermatozoa and infertility. In contrast to brain, selenite does not cross the blood-testis-barrier. Megalin expressed along the kidney tubular epithelium is involved in re-uptake of SePP filtrated in the glomerulus. Accordingly, inactivation of megalin leads to urinary loss of SePP. Megalin-expressing cells express the highest levels of GPx1, GPx3, and SePP within the kidney and inactivation of megalin decreases expression of all three proteins. Plasma GPx3 originates from the kidney epithelium, but most GPx3 is deposited locally within the kidney. SePP secretion by the kidney contributes to another "SePP cycle" involving glomerular filtration, re-uptake, and re-synthesis of SePP. Kidney insufficiency in patients is associated with low selenium status. Tissues expressing selenoproteins, but not expressing ApoER2 and megalin likely operate a still elusive selenium uptake mechanism tissues including skeletal muscle express neither LRP8 nor LRP2 suggesting the presence of an additional receptor (Fig. 2).

## The Current Model of Selenium Distribution in the Body

Taken together, a complex picture of systemic and local selenium distribution by SePP has emerged (Köhrle et al. 2012). The model of selenium metabolism envisages small intestinal uptake of selenium-containing amino acids (selenoproteins from animals, SeMet from plants, inorganic selenium from dietary supplements), and transfer of selenium to the liver where SePP and other selenoproteins are synthesized. SePP is then secreted into plasma and made available to selenium target tissues expressing ApoER2 (brain, testis) and megalin (kidney, brain). SePP is endocytosed with its receptor and proteolytically degraded to yield free Sec, which is further degraded and subsequently activated for Sec biosynthesis via selenophosphate (Fig. 3).

## Summary

SePP is the central selenium transport protein in mammals. It is expressed in many tissues, but most plasma SePP derived from the liver. Plasma levels of SePP represent a good measure of body selenium status (Hollenbach et al. 2008; Xia et al. 2005). SePP interacts with endocytic receptors of the lipoprotein receptor-related protein family which are expressed in SePP target tissues, including testis, kidney, and brain. SePP has been purified from serum as a neuronal survival factor underlining the role of selenium in brain function. Accordingly, *Sepp*-deficient mice represent the first animal model in which brain selenium levels are significantly reduced leading to neurological disease.

## Cross-References

▸ Selenium and Glutathione Peroxidases
▸ Selenium in Human Health and Disease
▸ Selenocysteinopathies

## References

Burk RF, Hill KE (2005) Selenoprotein P: an extracellular protein with unique physical characteristics and a role in selenium homeostasis. Annu Rev Nutr 25:215–235

Burk RF, Hill KE, Olson GE, Weeber EJ, Motley AK, Winfrey VP, Austin LM (2007) Deletion of apolipoprotein E receptor-2 in mice lowers brain selenium and causes severe neurological dysfunction and death when a low-selenium diet is fed. J Neurosci 27:6207–6211

Chiu-Ugalde J, Theilig F, Behrends T, Drebes J, Sieland C, Subbarayal P, Köhrle J, Hammes A, Schomburg L, Schweizer U (2010) Mutation of megalin leads to urinary loss of selenoprotein P and selenium deficiency in serum, liver, kidneys and brain. Biochem J 431:103–111

Hill KE, Zhou J, McMahan WJ, Motley AK, Atkins JF, Gesteland RF, Burk RF (2003) Deletion of selenoprotein P alters distribution of selenium in the mouse. J Biol Chem 278:13640–13646

Hill KE, Zhou J, Austin LM, Motley AK, Ham AJ, Olson GE, Atkins JF, Gesteland RF, Burk RF (2007) The selenium-rich C-terminal domain of mouse selenoprotein P is necessary for the supply of selenium to brain and testis but not for the maintenance of whole body selenium. J Biol Chem 282:10972–10980

Hollenbach B, Morgenthaler NG, Struck J, Alonso C, Bergmann A, Köhrle J, Schomburg L (2008) New assay for the measurement of selenoprotein P as a sepsis biomarker from serum. J Trace Elem Med Biol 22:24–32

Köhrle JS, Schweizer U, Schomburg L (2012) Selenium transport in mammals: selenoprotein P and its receptors. In: Hatfield DL, Berry MJ, Gladyshev VN (eds) Selenium. Its molecular biology and role in human health, vol 16, 3rd edn. Springer, New York, pp 205–220

Meplan C, Crosley LK, Nicol F, Beckett GJ, Howie AF, Hill KE, Horgan G, Mathers JC, Arthur JR, Hesketh JE (2007) Genetic polymorphisms in the human selenoprotein P gene determine the response of selenoprotein markers to selenium supplementation in a gender-specific manner (the SELGEN study). FASEB J 21. doi:10.1096/fj.1007-8166com

Motsenbocker MA, Tappel AL (1982) A selenocysteine-containing selenium-transport protein in rat plasma. Biochim Biophys Acta 719:147–153

Renko K, Werner M, Renner-Muller I, Cooper TG, Yeung CH, Hollenbach B, Scharpf M, Köhrle J, Schomburg L, Schweizer U (2008) Hepatic selenoprotein P (SePP) expression restores selenium transport and prevents infertility and motor-incoordination in Sepp-knockout mice. Biochem J 409:741–749

Saito Y, Sato N, Hirashima M, Takebe G, Nagasawa S, Takahashi K (2004) Domain structure of bi-functional selenoprotein P. Biochem J 383:841–846

Scharpf M, Schweizer U, Arzberger T, Roggendorf W, Schomburg L, Köhrle J (2007) Neuronal and ependymal expression of selenoprotein P in the human brain. J Neural Transm 114:877–884

Schomburg L, Schweizer U, Holtmann B, Flohé L, Sendtner M, Köhrle J (2003) Gene disruption discloses role of selenoprotein P in selenium delivery to target tissues. Biochem J 370:397–402

Schwede T, Kopp J, Guex N, Peitsch MC (2003) SWISS-MODEL: an automated protein homology-modeling server. Nucleic Acids Res 31:3381–3385

Schweizer U (2012) Selenoproteins in nervous system development, function, and degeneration. In: Hatfield DL, Berry MJ, Gladyshev VN (eds) Selenium. Its molecular biology and role in human health, vol 18. Springer, New York, pp 235–248

Schweizer U, Streckfuss F, Pelt P, Carlson BA, Hatfield DL, Köhrle J, Schomburg L (2005) Hepatically derived selenoprotein P is a key factor for kidney but not for brain selenium supply. Biochem J 386:221–226

Schweizer U, Dehina N, Schomburg L (2011) Disorders of selenium metabolism and selenoprotein function. Curr Opin Pediatr 23:429–435

Xia Y, Hill KE, Byrne DW, Xu J, Burk RF (2005) Effectiveness of selenium supplements in a low-selenium area of China. Am J Clin Nutr 81:829–834

Yan J, Barrett JN (1998) Purification from bovine serum of a survival-promoting factor for cultured central neurons and its identification as selenoprotein-P. J Neurosci 18:8682–8691

# Selenoprotein S (SelS)

▶ Selenium and Muscle Function

# Selenoprotein Sep15

Petra A. Tsuji[1] and Bradley A. Carlson[2]
[1]Department of Biological Sciences, Towson University, Towson, MD, USA
[2]Molecular Biology of Selenium Section, Laboratory of Cancer Prevention, National Cancer Institute, National Institutes of Health, Bethesda, MD, USA

## Synonyms

15kDa selenoprotein; SelM homologue; Sep15

## Definition

The 15 kDa selenocysteine-containing protein (selenoprotein) "Sep15" belongs to the family of oxidoreductases and has been shown to be involved in quality control of folding of proteins and in cataract formation (Kasaikina et al. 2011). It is a homologue of selenoprotein M (SelM), which is also a selenocysteine-containing protein with redox activity, thought to be involved in the antioxidant response (Ferguson et al. 2006). The biological function of Sep15 is not completely understood, and it appears to have a strong tissue specificity and split personality in terms of cancer initiation and promotion. Thus, among the many selenoproteins, Sep15 continues to generate interest due to its potential implications in health and disease.

## Sep15 and Selenium

The expression of Sep15, like many other selenoproteins, is regulated by the selenium status of the organism (Ferguson et al. 2006). Whereas the expression of essential, housekeeping selenoproteins, such as thioredoxin reductase-1, is maintained even at low levels of organismal selenium, expression of inducible selenoproteins, such as Sep15, is reduced under conditions of low systemic selenium (Berry 2005). Selenium is a dietary micronutrient and essential trace mineral, necessary for human health. It is found in the soil and its levels in plants and animals that feed on them correspond to soil levels. Seleniferous soils are widespread, and some locations within the North American Great Plains area are known for their high soil selenium content. Selenium can exist stably in the environment as inorganic species, including elemental selenium, selenide, selenite, and selenate. Typically, in humans, grains and animal products are the primary exposure source. Organic species, found in biological systems of living matter, include the methylated selenium compounds, selenoamino acids, and selenocysteine (Tsuji et al. 2012), the amino acid important in selenoproteins such as Sep15.

## Selenocysteine, Selenoprotein Synthesis, and Sep15

One of the major forms of selenium in mammals is selenocysteine (Sec), the 21st amino acid in the genetic code. Sec recently has been shown to be synthesized via several steps on its tRNA (Sec tRNA) after initially being aminoacylated with serine by seryl-tRNA synthetase (Xu et al. 2012). It is encoded by the UGA codon and has been found in bacteria, archaea, and eukaryotes (Gladyshev 2012). UGA codons specify Sec coding rather than translation-termination due to the presence of stem-loop structures called Sec insertion (SECIS) elements in the 3′-untranslated region of mRNAs. It is becoming increasingly evident that the essential roles of selenium in human health are due to its presence in proteins in the form of Sec (Gladyshev 2012). These Sec-containing proteins are commonly referred to as selenoproteins, and nearly always use Sec in redox catalysis. The number of selenoprotein genes varies greatly among organisms. Twenty-five selenoprotein genes have been identified in humans, including the gene encoding for selenoprotein M (SelM) and its distant sequence homologue, the 15 kDa selenoprotein, Sep15.

## Sep15 Structure and Function

Sep15 was first identified and characterized in human tissues by Gladyshev et al. in 1998 (Gladyshev et al. 1998). The Sep15 gene is found on human chromosome 1p31, a location often deleted or mutated in many cancers. The expressed protein consists of 162 residues and the Sec UGA codon is found in exon 3 at position 93. The SECIS element is present in the 3′-untranslated region of mRNA (Fig. 1).

There appear to be two alternative transcripts for Sep15. The longer transcript variant expresses five exons, whereas the second variant lacks an exon in the 3′-coding region. Sep15 was originally isolated and characterized using a human T-cell line and its mRNA found to be expressed in a wide range of tissues. Subsequently, the highest levels of Sep15 expression were reported to occur in human, rat, and mouse organs such as liver, kidney, testes, thyroid, and prostate (Hu et al. 2001). Other animals, such as dogs, cows, chicken, zebra fish, mice, and rats, show intraspecies homologues of Sep15 that appear to be highly conserved in nature. This selenoprotein was also detected in various unicellular eukaryotes.

Similarly to many other selenoproteins, Sep15 belongs to the class of thiol- oxidoreductase selenoproteins, and has been shown to exhibit redox activity (Kumaraswamy et al. 2000; Labunskyy et al. 2007). It contains a thioredoxin-like fold (Marino et al. 2012), with Sec located in the predicted catalytic position, and also contains an endoplasmic reticulum (ER)–targeting peptide in the N-terminal region (Fig. 2).

Sep15 was found to form a strong complex with UDP-glucose:glycoprotein glucosyltransferase (UGT), a 150 kDa enzyme that recognizes misfolded protein domains in the ER of eukaryotic cells and specifically glucosylates these proteins. This suggests that Sep15 may assist UGT function via redox processes and assist in controlling folding or secretion of certain glycoproteins (Ferguson et al. 2006).

Sep15 has also been found to be regulated by ER stress, resulting in increased Sep15 expression in response to mild ER stressors such as the antibiotic drugs tunicamycin and brefeldin A, as well as in rapid proteasomal degradation of Sep15 in response to agents inducing a more robust ER stress, such as dithiothreitol (Labunskyy et al. 2009). Recent studies with a Sep15 knockout mouse model further implicated its role in redox homeostasis as well as its importance in glycoprotein folding, especially in developing eyes (Kasaikina et al. 2011). Although Sep15 knockout mice appeared normal and did not activate ER stress pathways, parameters of oxidative stress were increased in livers of knockout mice. Sep15 knockout mice exhibited a prominent nuclear cataract at an early age, possibly due to an improper folding status of lens proteins (Kasaikina et al. 2011).

## Sep15, Human Polymorphisms and Cancer

Much like with some other selenoproteins, Sep15's role extends into chemoprotective effects. However, based on previous studies, much like thioredoxin reductase-1, this selenoprotein may also harbor a split personality in that it may possess cancer-preventive properties (Kumaraswamy et al. 2000; Diwadkar-Navsariwala and Diamond 2004; Jablonska et al. 2008), but also cancer-promoting roles (Irons et al. 2010; Tsuji et al. 2011a, b).

Higher versus lower Sep15 expression in human populations has been linked to altered cancer risks, such as the observation of a decrease of Sep15 expression in lung cancer patients (Jablonska et al. 2008). Low expression of Sep15 also has been observed in malignant lung, breast, prostate, and liver tissues (Wright and Diamond 2012) as well as in cell lines derived from malignant mesothelioma cells compared to normal lung cells (Apostolou et al. 2004).

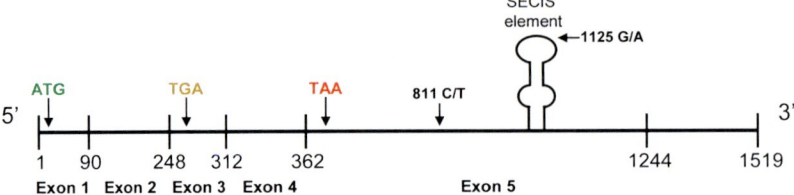

**Selenoprotein Sep15, Fig. 1** Human Sep15 cDNA organization. The relative positions of the ATG initiation site, the TGA Sec codon, the TAA termination signal, and the detected polymorphisms (C/T at position 811, and G/A at position 1125 in the SECIS element) are shown. Alternative 3′-end sequences (position 1244 or 1519) are also indicated (Adapted with modification from Kumaraswamy et al. 2000)

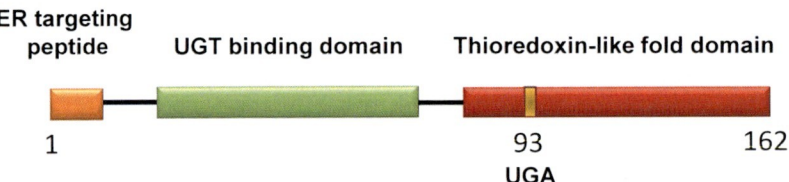

**Selenoprotein Sep15, Fig. 2** Schematic representation of domain organization of Sep15. Protein domains (ER targeting, UGT binding and thioredoxin-like fold) and location of the Sec codon (position 93) within Sep 15 are shown

These observations suggest a role of Sep15 in tumor suppression, whereas mouse in vivo studies suggest lack of Sep15 to be of protective value (Tsuji et al. 2011a). Therefore, even though much of its biological function remains to be elucidated, the importance of Sep15 in health and disease seems supported.

Gene (mRNA) expression analyses suggest that Sep15 expression varies among human populations, with higher expression to be expected in food crops and cattle (and subsequently in inhabitants) of high-selenium soils compared to those in low-selenium soils. Higher versus lower Sep15 expression in populations may also be due to human polymorphisms in the Sep15 gene. The protein expression of Sep15 among various polymorphic populations and the subsequent human tissue distribution remains to be validated. However, two single nucleotide polymorphisms (SNP) with functional consequences have been reported for the Sep15 gene (Hu et al. 2001; Apostolou et al. 2004), resulting in differentially regulated Sep15 protein expression in response to selenium status. These SNPs are located in the 3′-untranslated region in the Sep15 gene (rs5859 and rs5845), and result in C-T substitution at position 811 and a G-A substitution at position 1125, respectively (Fig. 2). Interestingly, the haplotype with a TAT position 811 and an AAT 1125 is relatively rare, only occurring in 7% of Caucasian-Americans, but in about 31% of African-Americans (Hu et al. 2001). Also, small-cell or non-small-cell lung cancers were observed in those individuals with an AA genotype at position 1125 and low-selenium status (Jablonska et al. 2008), suggesting that those with an AA genotype might benefit from increased dietary selenium supplementation. In a Korean patient population, a gender-specific increased rectal cancer risk has been described that associated the minor alleles for rs5845, rs5859, and rs34713741 with increased rectal cancer risk (Wright and Diamond 2012). Based on these epidemiological and case–control studies, a higher expression of Sep15 appeared to have protective effects. However, investigations using preclinical models point to strong tissue specificity as well as evidence to the contrary.

## Sep15 in Preclinical Models

In vitro investigations using RNAi-technology comparing the consequences of loss of Sep15 through targeted downregulation of the gene in mouse lung cancer cells and mouse colon cancer cells showed no effect in lung cancer cells, and an unexpected protective effect of Sep15 loss in colon cancer cells. These murine colon cancer CT26 cells demonstrated an at least partial reversal of the cancer phenotype upon downregulation of Sep15 (Irons et al. 2010). Such protective effects by Sep15-deficiency were subsequently also demonstrated in two human colon cancer cell lines, HCT-116 and HT-29 (Tsuji et al. 2011b). These observations are in stark contrast to the epidemiological literature, and, taken together with previous indications in lung cancer tissues, indicate strong tissue specificity in the response of malignant cells to Sep15 status.

As mentioned, a systemic Sep15 knockout mouse model exists (Kasaikina et al. 2011). Investigations addressing the effects of Sep15 loss in colon cancer in vivo, compared a chemically induced colon cancer model (azoxymethane) in normal and Sep15 knockout mice. Quantitation of the chemically induced aberrant crypt foci development as indicators of preneoplastic lesions in colonic epithelia indicated that – similarly to the results using targeted downregulation of Sep15 in colon cancer cells – a lack of Sep15 was protective against formation of aberrant crypt foci (Tsuji et al. 2011a). Further investigations of molecular targets affected by the loss of Sep15 demonstrate a possible link to expression of inflammation-associated genes and transcription factors (Tsuji et al. 2011a). Possible links between Sep15 expression and inflammation response of the organism continue to be elucidated.

## Summary

Dietary selenium mostly acts at the level of selenoproteins. Many selenoproteins are involved in the redox regulation and play a critical role in maintaining cellular homeostasis. One such selenoprotein is Sep15. Its specific biological remains to be elucidated, but it appears it is involved in quality control of protein folding. A strong tissue specificity and split personality in terms of cancer initiation and promotion has been described. Because of its differential expression among populations due to polymorphisms, further research is needed to evaluate the importance of this gene/protein in health and disease.

**Acknowledgments** The authors would like to thank Drs. Vadim Gladyshev and Dolph Hatfield for their support and critical review.

## Cross-References

▶ Selenium in Human Health and Disease
▶ Selenium, Physical and Chemical Properties
▶ Selenocysteinopathies
▶ Selenoproteins and the Biosynthesis and Activity of Thyroid Hormones
▶ Selenoproteins in Prokaryotes

## References

Apostolou S, Klein JO, Mitsuuchi Y et al (2004) Growth inhibition and induction of apoptosis in mesothelioma cells by selenium and dependence on selenoprotein SEP15 genotype. Oncogene 23:5032–5040

Berry MJ (2005) Insights into the hierarchy of selenium incorporation. Nat Genet 37:1162–1163

Diwadkar-Navsariwala V, Diamond AM (2004) The link between selenium and chemoprevention: a case for selenoproteins. J Nutr 134:2899–2902

Ferguson AD, Labunskyy VM, Fomenko DE et al (2006) NMR structure of the selenoprotein Sep15 and SelM reveal redox activity of a new thioredoxin-like family. J Biol Chem 281:3536–3543

Gladyshev VN (2012) Selenoproteins and selenoproteomes. In: Hatfield DL, Berry MJ, Gladyshev VN (eds) Selenium – its molecular biology and role in human health, 3rd edn. Springer, New York

Gladyshev VN, Jeang KT, Wootton JC et al (1998) A new human selenium-containing protein. Purification, characterization, and cDNA sequence. J Biol Chem 273:8910–8915

Hu YJ, Korotkov KV, Mehta R et al (2001) Distribution and functional consequences of nucleotide polymorphisms in the 3′-untranslated region of the human Sep15 gene. Cancer Res 61:2307–2310

Irons R, Tsuji PA, Carlson BA et al (2010) Deficiency in the 15kDa selenoprotein inhibits tumorigenicity and metastasis of colon cancer cells. Cancer Prev Res (Phila) 3:630–639

Jablonska E, Gromadzinska J, Sobala W et al (2008) Lung cancer risk associated with selenium status is modified in smoking individuals by Sep15 polymorphism. Eur J Cancer 47:47–54

Kasaikina MV, Fomenko DE, Labunskyy VM et al (2011) Roles of the 15-kDa selenoprotein (Sep15) in redox homeostasis and cataract development revealed by the analysis of Sep 15 knockout mice. J Biol Chem 286:33203–33212

Kumaraswamy E, Malykh A, Korotkov KV et al (2000) Structure-expression relationships of the 15-kDa selenoprotein gene. Possible role of the protein in cancer etiology. J Biol Chem 275:35540–35547

Labunskyy VM, Hatfield DL, Gladyshev VN (2007) The Sep15 protein family: roles in disulfide bond formation and quality control in the endoplasmic reticulum. IUBMB Life 59:1–5

Labunskyy VM, Yoo MH, Hatfield DL et al (2009) Sep15, a thioredoxin-like selenoprotein, is involved in the unfolded protein response and differentially regulated by adaptive and acute ER stresses. Biochemistry 48:8458–8465

Marino SM, Gladyshev VN, Dikiy A (2012) Structural characterization of mammalian selenoproteins. In: Hatfield DL, Berry MJ, Gladyshev VN (eds) Selenium – its molecular biology and role in human health, 3rd edn. Springer, New York

Tsuji PA, Carlson BA, Naranjo-Suarez S et al (2011a) Sep15 knockout in mice provides protection against chemically-induced aberrant crypt formation. FASEB J 25:110

Tsuji PA, Naranjo-Suarez S, Carlson BA et al (2011b) Deficiency in the 15 kDa selenoprotein inhibits human colon cancer cell growth. Nutrients 3:805–817

Tsuji PA, Davis CD, Milner JA (2012) Selenium: dietary sources and human requirements. In: Hatfield DL, Berry MJ, Gladyshev VN (eds) Selenium – its molecular biology and role in human health, 3rd edn. Springer, New York

Wright ME, Diamond AM (2012) Polymorphisms in selenoprotein genes and cancer. In: Hatfield DL, Berry MJ, Gladyshev VN (eds) Selenium – its molecular biology and role in human health, 3rd edn. Springer, New York

Xu XM, Turanov AA, Carlson BA et al (2012) Selenocysteine biosynthesis and the replacement of selenocysteine with cysteine in the pathway. In: Hatfield DL, Berry MJ, Gladyshev VN (eds) Selenium – its molecular biology and role in human health, 3rd edn. Springer, New York

## Selenoprotein T (SelT)

▶ Selenium and Muscle Function

## Selenoproteins and the Biosynthesis and Activity of Thyroid Hormones

Roland Gaertner
Department of Endocrinology, University Hospital, Ludwig-Maximilians University Munich, Munich, Germany

## Synonyms

Activation of thyroxine (T4) by deiodination to triiodothyronine (T3); Biosynthesis of selenoproteins; Degradation of thyroid hormones; Enzymes regulating thyroid hormones synthesis

## Definition

Selenium is the only trace element which is incorporated into selenoproteins by a distinct proteogenomic pathway. Selenoproteins play an essential role in development and survival of all known organisms. They are mainly involved in regulation of the nuclear, cytosolic as well as extracellular redox homeostasis of the body, cell proliferation, and immune regulation, and in this context, the selenoenzymes are known for the production as well as degradation of active thyroid hormone. The thyroid is the organ with the highest concentration of iodine and selenium among all human tissues and also the organ which continuously produces free oxygen radicals necessary for the thyroid hormone synthesis.

The main selenoenzymes within the thyroid are the cytosolic glutathione peroxidases 1(GPx-1), the phospholipid GPx (GPx-3), the PH-GPx (GPx-4), the type I 5′deiodinase (D1), type II 5′deiodinase (D2), the cytosolic thioredoxin reductase (TRx-1), and the mitochondrial TRx (TRx-2). Beneath these, also selenoprotein P (SePP), selenoprotein N (SeP15), and several others until now not clearly identified selenoproteins are located within the thyroid. The type III 5 deiodinase (D3) mainly is expressed in extrathyroidal tissues. The thioredoxin reductases are involved in modulation of transcription factor activation, signal transduction (TRx-1), and regulation of cell proliferation and tissue development (TRx-2). The SePP is delivering selenium into the thyroid cell and SeP15 in the degradation of $H_2O_2$.

Within the thyroid follicles, TSH-dependent $H_2O_2$ is continuously synthesized as this is necessary for synthesizing thyroid hormones. Especially, the GPx-3 is an effective scavenger of free oxygen radicals and produced and secreted into the follicular lumen where the thyroid hormone synthesis takes place. Selenium deficiency therefore might initiate cytotoxic effect within the thyroid and also trigger the common thyroid autoimmune diseases.

## Introduction

It is now obvious that selenium is not a toxic substance as previously assumed but an essential trace element for development and maintenance of all known organisms. Selenium is the only trace element which is incorporated by a distinct proteogenomic pathway; 25 genes encoding several selenoproteins had been identified (Kryukov and Gladyshev 2002). For selenium, there is no storage in the body, but the amount of selenoproteins synthesized is dependent on the selenium uptake with the food, mainly selenomethionine from plants and selenocysteine from animal sources. The amount of selenium in nature varies widely, depending on the selenium content in soil. Whereas the soil of some parts of central Asia as well as central and southern Europe is poor in selenium, it is abundant in North America and certain areas of Asia. Main sources of selenium are fish, crabs, kidney, liver, and Brazil nuts (Combs 2001).

## Biosynthesis of Selenoproteins

Selenium is inserted by a unique mechanism as selenocysteine, the 21st amino acid (Sec) into the genetic code of selenoproteins. The cotranslational incorporation into specific proteins is highly regulated. The UGA codon which normally is a stop codon encodes the translational incorporation of Sec into proteins when the mRNA contains a distinct hairpin mRNA downstream of the UGA codon in its 3′ untranslated region. This Sec insertion sequence (SECIS) prevents termination of the translation by competing for release factors that would otherwise lead to disassembly of the mRNA-ribosome complex. The SECIS structure recruits the Sec-binding protein 2 (SBP2) and binds the specific elongation factor EFsec loaded with its tRNA-Sec. Some selenoproteins of the human selenoproteom display multiple genes. The main selenoprotein families, playing a role in the thyroid, are the glutathione peroxidases (seven genes), the thioreductases (three genes), and the iodothyronine deiodinases (three genes) (Köhrle et al. 2005).

In the last years, three families with mutations of SBP2 had been identified, resulting in decreased activity of the deiodinases, GPx, and other selenoproteins (Dumitrescu 2010). These individuals primarily were identified by abnormal thyroid function tests, growth retardation, and several other clinical pictures.

## The Role of Iodothyronine Deiodinases

The action of thyroid hormones is terminated by the amount of circulating hormones, but the individual cells and organs are able to modify the thyroid hormone action by receptor expression, transporter regulation, and alteration of the thyroid hormone activation or degradation (Germain et al. 2009). All the three selenoenzymes D1, D2, and D3 are involved in this local regulation of thyroid hormone metabolism (Bianco 2011). The deiodinases contain selenocysteine at their active site and therefore are dependent on the selenium intake. Interestingly, they are highly preserved compared to other selenoproteins even under mild to moderate selenium deficiency, indicating their importance for the development and maintenance of normal thyroid function. The main role of the deiodinases is to deliver the right concentration of the active thyroid hormone T3 to the nucleus receptor of an individual cell but also inactivate and degradate T4 as well as T3 (Dentrice 2011). The D1 as well as D2 catalyzes the outer ring (5′) deiodination of the prohormone T4 to form the active hormone T3 but also the deiodination of reverseT3 (rT3) to form T2. The D1 as well as D3 catalyzes the deiodination of T4 in the inner ring (5) to form rT3 and also the deiodination of T3 to T2. Thus, D1 is the only deiodinase that can function as either outer or inner ring deiodinase. It was the first to be cloned (Berry et al. 1991). The D1 is located not only within the thyroid but also the liver, kidney, and pituitary and the D2 in pituitary, brain, brown adipose tissue, heart, bone, and skeletal muscle. The D3 is not located in the thyroid but in brain, skin, uterus, placenta, and fetus inactivating T4 to reverse (rT3), the inactive form of T3 and in degradation of T3 into the inactive T2 (Germain et al. 2009).

In the case of iodine deficiency and hypothyroidism, the activity of D1 is increased in the thyroid, heart, muscle, brain, and brown adipose tissue. This results in a higher local T3 delivery and counteracts the low systemic thyroid hormone concentrations. The D3 activity is decreased at the same time to reduce the degradation of T4 and T3. The elevated TSH in hypothyroidism upregulates the D1 activity, thereby increasing preferentially T3 secretion, and preserves iodine consumption and downregulation of D3. During primary hyperthyroidism, the opposite occurs, namely, activation of D3 and downregulation of D1.

The D1 activity is also susceptible to selenium deficiency, and selenium-deficient individuals have mildly elevated T4 and T4 to T3 ratio but normal TSH. Whereas D1 is constitutively expressed in the thyrocytes, D2 is regulated by TSH, but the mechanisms of the changes in the activity of the deiodinases in peripheral organs are not fully understood. The D2 is also mainly involved in deiodination of T4 to T3 in the hypothalamus and essential for normal control of the hypothalamic-pituitary axis and TSH secretion (Maia et al. 2011; Dentrice 2011).

## The Role of Deiodinases in Fasting and Illness

In catabolic situations like fasting or severe illness, not only TSH but also plasma T3 is low but the metabolic inactive rT3 is elevated. The pathophysiological mechanisms are complex and not yet fully understood (Boelen et al. 2011). It seems to be clear that the D1 activity in the liver is decreased, and the D2 activity in the hypothalamus, however, increased, reducing the TSH response to low circulating T3. The D3 activity is also increased in liver and muscle, degrading T3 to rT3. The consequences therefore are low plasma thyroid hormones as well as TSH but elevated rT3. This so called non-thyroidal illness syndrome (NTIS) is therefore caused by distinct changes in the activity of the deiodinases. The latter might be caused by stress hormones like cortisol and cytokines but not by selenium deficiency.

## The Role of Glutathione Peroxidases (GPx)

For the synthesis of thyroid hormones, iodide ($I^-$) is accumulated by the sodium-potassium-symporter (NIS), located at the basolateral thyrocyte membrane. Iodide then is transported through the cell and excreted by the anion transporter pendrin into the follicular lumen. The dual oxidase (Duox) generates $H_2O_2$ at the apical surface which is utilized by the thyroid peroxidase (TPO) to iodinate the tyrosine residues of thyroglobulin. Since $H_2O_2$ is highly cytotoxic, the synthesis of scavenger enzymes is essential to prevent oxidative damage of the cells by $H_2O_2$. These are the selenium-dependent glutathione peroxidases, mainly GPx-3, which is produced by the thyrocytes and secreted into the follicular lumen (Schomburg 2008). The production and secretion of GPx-3 is under negative control of $Ca^{2+}$ signaling pathways. The abundant $H_2O_2$ can be degraded to $H_2O$ by GPx-3 to prevent the thyrocytes from oxidative damage. Depending on the amount of GPx-3, more or less $I^+$ can be produced and coupled via thyroid specific peroxidase (TPO) onto the tyrosine residues of thyroglobulin. In addition, the intracellular GPx-1 and GPx-4 prevent oxidative damage of the thyroid cells from $H_2O_2$ that might diffuse into the cells.

## Selenium and Iodine Deficiency

Iodine deficiency is long known as the cause of neurological cretinism, characterized by mental retardation, deaf-mutation, and goiter. Severe iodine combined with severe selenium deficiency has been shown to be the reason for the development of myxoedematous cretinism, and its pathogenesis first has been clarified in Congolese children. It is characterized by poor development, growth retardation (dwarfism), myxoedema, but no goiter. The thyroid glands of these children are small, fibrotic, with only few overactive thyroid follicles. It turned out that the reason of the damage of the thyroid is low GPx-1 activity due to low selenium intake, followed by an oxidative damage of thyroid cells, necrosis, and inflammation, resulting in fibrosis and atrophy of the thyroid gland, already starting in utero. Therefore, iodine supply after birth is ineffective, and maintenance of hypothyroidism causes the clinical picture of myxoedematous cretinism (Köhrle et al. 2005).

The myxoedematous cretinism might be the extreme picture of severe selenium and iodine deficiency, but also mild to moderate nutritional selenium deficiency combined with iodine deficiency might be responsible for the initiation of autoimmune thyroid disorders in patients with the genetic susceptibility background to develop autoimmune diseases (Beckett 2005; Köhrle 2009). The reduced oxidative defense and decreased

GPx-3 activity might cause thyrocyte damage and contribute to the initiation of an autoimmune reaction. Selenium deficiency is also accompanied by loss of immune competence (Hoffmann 2008). Both cell-mediated immunity and B-cell function can be impaired. This might be related to the fact that the Se-dependent enzymes GPx and thioredoxin reductase (TxR) exhibit not only antioxidative effects (Huang et al. 2012). Selenium-dependent compounds also reduce the nuclear translocation of the transcription factor NF-κB in macrophages by interference with its cytosolic inhibitor IκB and the signaling cascade and hereby decrease inflammatory cytokine expression. Therefore, several prospective, randomized clinical trials were performed in patients with autoimmune thyroiditis (AIT) to examine whether a selenium supplementation might reduce the inflammatory activity within the thyroid and prevent progression of this chronic disease affecting 10–15% of women and 2% of men. Several prospective, double-blinded studies examined the effect of various Se forms on AIT with different outcome, but it seems that especially those performed in mild to moderate selenium-deficient areas, the selenium supplementation was effective to reduce the inflammatory activity within the thyroid (Duntas 2010). This might indicate that mild selenium deficiency is a trigger for the initiation of AIT.

## Summary

The thyroid is the organ with the highest selenium and iodine content. The cause of the high selenium content in the thyroid is the high amount of selenoproteins. The essential trace element selenium is incorporated as selenocysteine, the 21st amino acid into the three iodothyronine deiodinases (D1, D2, and D3) which can both activate or inactivate thyroid hormones. This regulatory process in the individual cells is important for regulation of thyroid hormone action under normal and disease conditions. Thus, selenium is an essential micronutrient for development, growth, and maintenance of the metabolism. In addition, selenocysteine is an essential compound in the catalytic center of several selenoenzymes protecting the thyrocytes from free radical damage, but some of these enzymes are also important for the immune regulation. Severe iodine and selenium deficiency is deleterious for the development of children resulting in myxoedematous cretinism. Whether gradual low iodine and selenium intake or a mismatch of both affects the thyroid function has to be established.

## Cross-References

▶ Selenium and Glutathione Peroxidases
▶ Selenium and Iodothyronine Deiodinases
▶ Selenium in Human Health and Disease
▶ Selenoproteins and Thyroid Gland

## References

Beckett GJ, Arthur JR (2005) Selenium and endocrine systems. J Endocrinol 184:455–465
Boelen A, Kwakkel J, Fliers E (2011) Beyond low plasma T3: local thyroid hormone metabolism during inflammation and infection. Endocr Rev 32:670–693
Combs GF Jr (2001) Selenium in global food. Br J Nutr 85:517–547
Dentrice M, Salbatore D (2011) Deiodinases: the balance of thyroid hormone. Local impact of thyroid hormone inactivation. J Endocrinol 209:273–282
Dumitrescu AM, Refetoff S (2007) Novel biological and clinical aspects of thyroid hormone metabolism. Endocr Dev 10:127–139
Duntas L (2010) Selenium and the thyroid: a close-knit connection. J Clin Endocrinol Metab 95:5180–5188
St Germain DL, Galton VA, Hernandez A (2009) Minireview: defining the roles of iodothyronine deiodinases: current concepts and challenges. Endocrinology 150:1097–1107
Hoffmann PR, Berry MJ (2008) The influence of selenium on immune responses. Mol Nutr Food Res 52:1273–1280
Huang Z, Rose AH, Hoffmann PR (2012) The role of selenium in inflammation and immunity: from molecular mechanisms to therapeutic opportunities. Antioxid Redox Signal 16:705–743
Köhrle J, Gärtner R (2009) Selenium and thyroid. Best Pract Res Clin Endocrinol Metab 23:815–827
Köhrle J, Jakob F, Contempre B, Dumont JE (2005) Selenium, the thyroid, and the endocrine system. Endocr Rev 26:944–984
Kryukov GV, Gladyshev VN (2002) Mammalian selenoprotein gene signature: identification and functional analysis of seleprotein genes using biotransformatic methods. Methods Enzymol 347:84–100
Maia AL, Goemann IM, Meyer EL et al (2011) Deiodinases: the balance of thyroid hormone. Type I iodothyronine deiodinase in human physiology and disease. J Endocrinol 209:283–297
Schomburg L, Köhrle J (2008) On the importance of selenium and iodine metabolism for thyroid hormone biosynthesis and human health. Mol Nutr Food Res 52:1235–1246

# Selenoproteins and Thyroid Gland

Lutz Schomburg
Institute for Experimental Endocrinology,
Charité – University Medicine Berlin,
Berlin, Germany

## Synonyms

3,3′,5′-reverse triodothyronine, reverse T3, rT3; 3,5,3′-triodothyronine, T3; Glutathione peroxidase, GPx; Iodothyronine deiodinase, DIO; Selenocysteine-containing proteins, selenoproteins; Thyroxine, 3,5,3′,5′-tetraiodothyronine, T4

## Definition

Selenium is present in proteins in form of two amino acids, i.e., selenocysteine (Sec) or selenomethionine (SeMet). Sec is incorporated during translation of selenoproteins by a well-conserved highly specific process requiring Sec-loaded tRNA and the decoding of the stop codon UGA as Sec-insertion signal. SeMet is incorporated randomly into newly synthesized proteins at AUG codons specifying methionine insertion as the translational machinery is unable to discriminate Met-tRNA from SeMet-tRNA. Consequently, any given protein can become a SeMet-containing protein by incorporation of SeMet instead of Met during translation, while selenoproteins always contain Sec at genetically predetermined specified positions. Currently, 25 separate human genes are known to encode selenoproteins. The number of actual gene products, i.e., different selenoproteins, is considerably higher in humans due to alternative transcription, translation, splicing, or posttranslational processing events giving rise to a number of isoforms for a given selenoprotein.

## Introduction

The thyroid gland is a very specialized tissue. As a central endocrine gland, it controls growth, development, and energy turnover mainly by synthesizing and secreting thyroid hormones (TH). Tetraiodothyronine (T4) and triiodothyronine (T3) are the major TH. T4 is considered as a prohormone which needs to be converted into the biological active T3 by a deiodination reaction. There are three specific deiodinating enzymes in the mammalian organism, called iodothyronine deiodinases (DIO). They differ by site of expression, regulation, and reaction catalyzed (Bianco et al. 2002). But most importantly, all three DIO isozymes are selenoproteins carrying an active selenocysteine (Sec) residue in their active sites (Köhrle 2005).

## Anatomy, Physiology, and Oddities of the Thyroid Gland

The functional units of the thyroid gland are the thyroid follicles. In the center of the follicles is a protein-dense colloid containing a number of enzymatically active proteins and the high molecular weight protein thyroglobulin (Tg) as the matrix on which TH biosynthesis takes place. The thyrocytes are the differentiated thyroid cells implicated in TH formation. They form a tight and polarized continuous monolayer around the colloid, thereby controlling iodide influx into the colloid and TH biosynthesis and secretion rates. A unique enzymatic reaction is taking place within the follicles needed for iodination of Tg during the course of TH biosynthesis. Iodide becomes oxidized by hydrogen peroxide ($H_2O_2$) actively generated by the follicular cells on their apical membranes facing the colloid (Song et al. 2007). The resulting $H_2O_2$ tone is not found anywhere else in another body fluid explaining the unique anatomic assembly of the thyroid follicles needed for protection of the tissue structure from the aggressive intermediates generated during iodide organification. Similar high concentrations of $H_2O_2$ may only be encountered intracellularly within phagocytic cells of the immune system or hydrolyzing organelles such as lysosomes. It is this unique chemical environment within the thyroid gland which appears to necessitate antioxidative selenoproteins for protecting and maintaining the organ's functional and structural integrity (Schweizer et al. 2008).

Among all of the body tissues, the thyroid gland contains the highest concentration of selenium (Se) and expresses a large number of selenoproteins. There are only two actively secreted selenoproteins within the family of 25 human selenoprotein genes identified, i.e., the Se-transport and Se-storage protein selenoprotein P (SePP) and the extracellular glutathione peroxidases (GPx3). Both proteins are highly expressed in the thyroid gland and their transcript concentrations are among the highest among all the human tissues. GPx3 has been detected in colloid preparations from human thyroid tissue and found to be noncovalently associated with Tg. It is unknown at present whether SePP is also secreted into the colloid. Besides GPx3 and SePP, cDNA profiling database resources maintained at the NCBI indicate high mRNA concentrations of most of the other known selenoproteins in human thyroid. However, little is known on their functional role and physiological importance as no selenoprotein genotypes have been associated with thyroid diseases or malfunctioning. One respective mouse model abrogating selenoprotein biosynthesis specifically in thyrocytes has indicated some functional alterations of the thyroid hormone axis but yielded surprisingly a relatively mild phenotype only (Schomburg 2011).

## Inherited Diseases of Selenoproteins and the Thyroid Gland

Inherited diseases in humans related to impaired expression of specific selenoproteins or inefficient general selenoprotein biosynthesis are rare (Schomburg 2010). Among the 25 human candidate genes, only selenoprotein N (SELN) gene mutations have been unequivocally identified to cause an inherited form of muscular dystrophy now called SELN-related myopathy (Lescure et al. 2009). A thyroid gland phenotype is not described in the respective individuals. Besides SELN, genetic defects in two functional components of the selenoprotein biosynthesis machinery have been identified to cause phenotypes in humans, i.e., the Sec-synthetase (SEPSECS) and an RNA-binding protein implicated in cotranslational Sec-insertion, SECIS-binding protein 2 (SBP2) (Schweizer et al. 2011). Children with function-compromising SEPSECS mutations suffer from a very severe phenotype mainly characterized by cerebrocerebellar-degenerating atrophy. In light of this life-threatening and debilitating progressive severe disease, alterations of the thyroid hormone axis appear of marginal significance for these patients and have not been reported. In contrast, children with mutations in SBP2 have been identified because of their thyroid hormone pattern alterations in an endocrine unit of the University of Chicago (Dumitrescu et al. 2010). SBP2 binds the characteristic stem-loop structure in the 3′-untranslated region of selenoprotein transcripts, the so-called Sec-insertion sequence (SECIS) element (Fletcher et al. 2001). Impaired SBP2 function thus affects not only one specific selenoprotein but reduces selenoprotein biosynthesis in general. Among the experimentally proven selenoproteins affected by impaired expression in children with SBP2 mutations are SePP, GPx1, GPx3, and DIO2. SePP deficiency likely underlies the reduced serum Se concentrations measured in these children, while GPx1 deficiency has been associated with the increased photosensitivity of skin in SBP2 subjects. Low DIO2 expression appears to be responsible for the deranged TH pattern in blood of the mutation carriers while GPx3 deficiency has not been linked to an overt pathology in keeping with the lack of phenotype in Gpx3 knockout mice (Schomburg 2010). Thyroid gland appearance is reported to be normal in children with SBP2 mutations.

## Nutritional Selenium Status and Thyroid Hormones

The notion that both TH activation and inactivation are catalyzed by Se-dependent DIO isozymes raises the question on the importance of dietary Se intake and Se status for TH concentrations. This hypothesis is supported by the TH phenotype of SBP2 children and its analogy to the essentiality of iodide intake and supply for thyroid gland health and TH biosynthesis (Vitti et al. 2001). Se intake and status differ widely among different populations without an obvious correlation to TH patterns in the residents. Only a severe combined deficiency of Se and iodide causes an overt human phenotype described as myxedematous cretinism (Zimmermann and Köhrle 2002). Besides these profound nutritional deficits, the impact of Se intake and Se status on circulating TH concentrations are marginal. This is reflected by a lack of clear association in both epidemiological analyses and supplementation studies. Even in critical illness, when both Se concentrations decline and the TH feedback axis fails, there were no consistent effects of Se supplementation on the TH concentrations in blood (Gärtner 2009).

However, the thyroid gland size and integrity depend on the Se status. A French epidemiological analysis indicated an inverse association of serum Se concentrations with gland volume in adult females. An independent study verified this notion and reported an increased risk of multinodular goiter in Se-deficient females in Denmark. Respective studies linking Se status with thyroid gland cancer have not been reported.

## Selenium and Autoimmune Thyroid Diseases

Besides goiter and nodule risk, Se has been implicated in autoimmune thyroid diseases (AITD). AITD are among the most common autoimmune diseases affecting mainly women (Duntas 2008). Two major AITD account for 1–2% disease prevalence in adult females, i.e., Graves' disease (GD) and Hashimoto's thyroiditis (HT), both partly depending on genetic predisposition (Tomer 2010). A major complication of GD is Graves' ophthalmopathy (GO), an extrathyroidal disease manifestation characterized by eye inflammation and eyeball protrusion (Lindholm and Laurberg 2012). The European Group on GO (EUGOGO) conducted a multicenter placebo-controlled supplementation study on patients with mild GO. Many disease parameters improved after 6 months of daily treatment with 100–200 μg sodium selenite, including inflammation, protrusion, and eye motility. An analysis of changes in quality of life revealed an impressive > 70% improvement in the group of Se-treated patients while most probands in the placebo group showed a decline or remained stable. It is expected that this study will be repeated in populations with different baseline Se status, as the positive supplementation success may be confined to Se-deficient patients thus rather representing a substitution of Se deficit but not a supplementation effect (Gillespie et al. 2012).

There are a growing number of respective Se supplementation trials in HT. Here, supplementation success differs despite similar treatment regimen. Initial reports documented strongly decreased autoantibody concentrations in Se-treated HT patients. Different selenocompounds (selenite or selenomethionine) yielded similar results. However, a growing number of respective supplementation studies are reporting also null results. At present, the selenocompounds chosen, treatment dosages, and application times along with patient-specific characteristics are discussed as potential factors controlling supplementation success. No unifying theory on the most successful adjuvant Se treatment regimen in AITD has been agreed on. Moreover, AITD represent sexual dimorphic diseases. By now, the effects of adjuvant Se supplementation on HT and GD/GO are mainly analyzed for female patients only. It remains to be studied whether Se-deficient men will similarly profit from increased Se intake (Schomburg 2011).

As the general importance of a sufficiently high daily Se intake becomes more widely accepted, so is the acceptance of adjuvant Se supplementation in inflammatory diseases like AITD and as measures for disease prevention (Rayman 2012). In this regard, the thyroid gland appears to represent one of the prime targets being positively affected by avoiding a Se deficit both for prevention in general and as adjuvant treatment especially in AITD. However, Se is an active micronutrient and excessive intake may cause toxic side effects (Lenz and Lens 2009). Therefore, a personal approach in combination with individual Se status assessment is needed in order to safely avoid deficiency without oversupplementation.

## Summary

The thyroid gland represents a unique target tissue of Se as it contains highest relative levels of this trace element and expresses many different selenoproteins. Thyroid selenoproteins are implicated in tissue protection as the gland actively produces a high peroxide concentration needed for thyroid hormone biosynthesis. Selenoproteins of the deiodinase family are responsible for thyroid hormone activation and inactivation. Inherited diseases of selenoprotein biosynthesis have been identified by a disrupted thyroid hormone feedback causing a deranged pattern of circulating thyroid hormones. While insufficient nutritional Se intake predisposes to thyroid goiter and nodules, supplemental Se may improve disease parameters and quality of life in widespread autoimmune thyroid diseases. The number of respective studies is limited but grows fast due to the unexpected strong positive effects observed especially upon supplementing poorly supplied European individuals.

## Cross-References

▶ Peroxidases
▶ Selenium and Glutathione Peroxidases
▶ Selenium and Iodothyronine Deiodinases
▶ Selenium and Muscle Function
▶ Selenium in Human Health and Disease
▶ Selenium, Biologically Active Compounds
▶ Selenoprotein P
▶ Selenoproteins and the Biosynthesis and Activity of Thyroid Hormones

## References

Bianco AC, Salvatore D, Gereben B et al (2002) Biochemistry, cellular and molecular biology, and physiological roles of the iodothyronine selenodeiodinases. Endocr Rev 23(1):38–89

Dumitrescu AM, Di Cosmo C, Liao XH et al (2010) The syndrome of inherited partial SBP2 deficiency in humans. Antioxid Redox Signal 12(7):905–920

Duntas LH (2008) Environmental factors and autoimmune thyroiditis. Nat Clin Pract Endocrinol Metab 4(8):454–460

Fletcher JE, Copeland PR, Driscoll DM et al (2001) The selenocysteine incorporation machinery: interactions between the SECIS RNA and the SECIS-binding protein SBP2. RNA 7(10):1442–1453

Gärtner R (2009) Selenium and thyroid hormone axis in critical ill states: an overview of conflicting view points. J Trace Elem Med Biol 23(2):71–74

Gillespie EF, Smith TJ, Douglas RS (2012) Thyroid eye disease: towards an evidence base for treatment in the 21st century. Curr Neurol Neurosci Rep 12(3):318–324

Köhrle J (2005) Selenium and the control of thyroid hormone metabolism. Thyroid 15(8):841–853

Lenz M, Lens PN (2009) The essential toxin: the changing perception of selenium in environmental sciences. Sci Total Environ 407(12):3620–3633

Lescure A, Rederstorff M, Krol A et al (2009) Selenoprotein function and muscle disease. Biochim Biophys Acta 1790(11):1569–1574

Lindholm J, Laurberg P (2012) Hyperthyroidism, exophthalmos, and goiter: historical notes on the orbitopathy. Thyroid 20(3):291–300

Rayman MP (2012) Selenium and human health. Lancet 379(9822):1256–1268

Schomburg L (2010) Genetics and phenomics of selenoenzymes–how to identify an impaired biosynthesis? Mol Cell Endocrinol 322(1–2):114–124

Schomburg L (2011) Selenium, selenoproteins and the thyroid gland: interactions in health and disease. Nat Rev Endocrinol 8(3):160–171

Schweizer U, Chiu J, Köhrle J (2008) Peroxides and peroxide-degrading enzymes in the thyroid. Antioxid Redox Signal 10(9):1577–1592

Schweizer U, Dehina N, Schomburg L (2011) Disorders of selenium metabolism and selenoprotein function. Curr Opin Pediatr 23(4):429–435

Song Y, Driessens N, Costa M et al (2007) Roles of hydrogen peroxide in thyroid physiology and disease. J Clin Endocrinol Metab 92(10):3764–3773

Tomer Y (2010) Genetic susceptibility to autoimmune thyroid disease: past, present, and future. Thyroid 20(7):715–725

Vitti P, Rago T, Aghini-Lombardi F et al (2001) Iodine deficiency disorders in Europe. Public Health Nutr 4(2B):529–535

Zimmermann MB, Köhrle J (2002) The impact of iron and selenium deficiencies on iodine and thyroid metabolism: biochemistry and relevance to public health. Thyroid 12(10):867–878

# Selenoproteins in Prokaryotes

Michael Rother
Institut für Mikrobiologie, Technische Universität Dresden, Dresden, Germany

## Synonyms

Proteins containing 2-selenoalanine; Selenocysteine-containing proteins in Bacteria and Archaea

## Definition

Selenocysteine, a structural analog of cysteine but synthesized from serine, is found in members of all three domains of life, Eukarya, Bacteria, and Archaea. Selenocysteine is mostly present in the catalytic site of redox-active enzymes with diverse biological functions. The pathway for selenocysteine synthesis and incorporation into proteins during translation differs from that of the 20 canonical amino acids, as it is synthesized bound to a transfer RNA (tRNA) and encoded on the selenoprotein mRNA by UGA, which normally signals termination of translation.

## Introduction

The element selenium (Se) is a member of the chalcogen group in the periodic table, and thus a nonmetal, with properties that are intermediate between those of the neighboring elements sulfur and tellurium. Se was first isolated in lead chambers of a sulfuric acid production factory and named after the Greek Goddess of the Moon, Selene, by the Swedish chemist Jöns Jacob Berzelius in the early nineteenth century (Flohe 2009). Its photovoltaic potential made it a constituent of the first solar cell but today its primary commercial use is in glass manufacturing and pigment production.

Se occurs in four oxidation states, selenate Se(VI), selenite Se(IV), elemental selenium Se(0), and selenide Se(-II), and was for long regarded highly toxic, until its importance for the formate metabolism of *Escherichia coli* was recognized in the 1950s. Extensive scientific effort since then has revealed that Se is an essential trace element for many organisms including humans. Biologically active Se occurs in three forms, (1) as a constituent of a base modification (5-[(methylamino)methyl]-2-selenouridine) in certain tRNAs (Björk 1995), (2) as a non-covalently bound selenium-containing cofactor (Haft and Self 2008), and (3) as the cotranslationally inserted amino acid selenocysteine (Sec, 2-selenoalanine).

## Selenocysteine

As the name suggests, Sec is structurally identical to cysteine (Cys), only with the thiol group replaced by a selenol group. It is mostly found in the catalytic site of various redox-active enzymes. However, for most characterized selenoproteins, the specific functions of Sec are still uncertain because for all but one ($P_A$ of glycine reductase, see below) of the selenoproteins, homologous proteins with Cys at the respective position exist in other organisms. It is, therefore, still a matter of debate why organisms contain this unusual amino acid but probably its physicochemical properties (high nucleophilicity, lower $pk_a$ value than that of Cys) make it more robust and reactive than Cys. Still, the majority of known organisms do not use Sec. However, since selenoproteins are found in members of all three domains of life with fundamental commonalities for synthesis and incorporation conserved, it seems that the Sec-utilizing trait must have been present in the last universal common ancestor, that is, was added to the genetic code before the division of the three domains. Since then, it probably was lost in most lineages and yet in some occasionally retrieved again by lateral gene transfer.

## Synthesis and Incorporation of Selenocysteine in Prokaryotes

During translation, canonical amino acids are attached to their cognate tRNA by their corresponding aminoacyl-tRNA synthetase (aaRS). The resulting aminoacyl-tRNA is subsequently bound by the canonical translation elongation factor (EF-Tu in bacteria, EF-1a in archaea) and brought to the A site of the ribosome where the mRNA triplet is translated via base pairing of the tRNA anticodon and subsequent peptidyl transfer to the nascent polypeptide (Leung et al. 2011).

The pathway of Sec biosynthesis and incorporation in prokaryotes, which was comprehensively reviewed before (Böck et al. 2005; Commans and Böck 1999; Stock and Rother 2009), differs in three fundamental aspects from this paradigm. First, Sec biosynthesis proceeds in a tRNA-bound fashion; a Sec-specific tRNA (tRNA$^{sec}$, the *selC* gene product) is initially acylated with serine by seryl-tRNA synthetase (SerRS), and the seryl-moiety is subsequently converted to a selenocysteyl-moiety. In bacteria, this conversion is catalyzed in a single step by Sec synthase (SS, the *selA* gene product), whereas Archaea employ a two-step mechanism. There, the seryl-moiety is first phosphorylated in an ATP-dependent reaction by *O*-phosphoseryl-tRNA$^{sec}$ kinase (PSTK), and *O*-phosphoseryl-tRNA$^{sec}$ is subsequently converted to a selenocysteyl-moiety by *O*-phosphoseryl-tRNA$^{sec}$: selenocysteine synthase, SepSecS. Both conversion pathways require selenomonophosphate, which is generated from (a still unidentified) reduced selenium-species and ATP by selenomonophosphate synthetase (SPS, the *selD* gene product). It is still not fully understood why Archaea employ the two-step conversion mechanism, which expends an "extra" ATP. In eukaryotes, which also employ the two-step conversion pathway, cysteyl-tRNA$^{sec}$ is formed with thiophosphate more efficiently from *O*-phosphoseryl-tRNA$^{sec}$ than from seryl-tRNA$^{sec}$. It has thus been argued that targeted insertion of Cys instead of Sec could be achieved as an "emergency route" during selenium deprivation. However, how the consequential amino acid ambiguity for one codon is being avoided in such cases is so far unknown.

The second difference to a canonical translation event is that the Sec-encoding triplet on the mRNA is a stop codon, UGA, which normally leads to translation termination. Thus, an obligatory translational recoding event has to occur in selenoprotein synthesis. This recoding of the UGA into a sense codon (for Sec) is brought about by a secondary structure on the selenoprotein mRNA, the SECIS (*se*lenocysteine *i*nsertion *s*equence) element. In bacteria, the SECIS element is located directly downstream of the UGA codon, whereas in Archaea the SECIS element is found in the nontranslated regions (UTR) of the selenoprotein mRNAs (in all but one case in the 3'-UTR).

The third fundamental difference to "conventional" protein synthesis is the presence of a specialized translation elongation factor SelB (the *selB* gene product). It delivers exclusively selenocysteyl-tRNA$^{sec}$ to the A site of the ribosome. SelB is homologous to the canonical elongation factor in its N-terminal part but, depending on the organism, carries a C-terminal extension of various lengths. In bacteria, this extension is responsible for binding the SECIS, which triggers a conformational change in the GTP•SelB•Sec-tRNA$^{sec}$ complex allowing for insertion of the charged tRNA into the ribosomal A site, and, thus, recoding of UGA (Thanbichler and Böck 2001). Archaeal SelB does not bind the SECIS and the mechanism of UGA recoding has not been elucidated there.

## Selenoproteins in Prokaryotes

Despite the fact that the majority of prokaryotes do not contain selenoproteins, nearly all phylogenetic clades harbor members which do. Their unifying feature appears to be an (facultative) anaerobic life style, which possibly reflects the primordial nature of the Sec-utilizing trait. Prominent examples are the proteobacteria (with enterobacteria being most intensively studied), the spirochetes, the deep-rooting *Aquifex*- and *Thermotoga*-clades, and the Gram-positive clostridia. Among the Archaea, selenoproteins have so far been experimentally demonstrated only in members of the orders Methanococcales and Methanopyrales, strictly hydrogenotrophic methanogens. It is intriguing that all but two of the selenoproteins in methanogens (with the exception of selenomonophosphate synthetase and the HesB-like selenoprotein) are involved in the organism's central energy metabolism, hydrogenotrophic methanogenesis. It is even more striking that the mesophilic *Methanococcus* species contain, in addition to these otherwise essential selenoproteins, selenium-independent isoenzymes (with the exception of formate dehydrogenase) which are synthesized during selenium starvation or when the pathway of Sec synthesis and incorporation is disrupted. It appears that these organisms have mostly overcome their selenium-dependence by gene duplication followed by mutation.

In the following, typical selenoproteins of prokaryotes will be introduced. Potential selenoproteins, identified merely through (meta-)sequencing efforts and "validated" only by the fact that orthologous proteins with Cys at the position of the potential Sec are known, are not included here. For potential selenoproteins, the reader is referred to the original literature, for example, Zhang et al. (2006) and Zhang and Gladyshev (2008). The most common selenoproteins in prokaryotes are Sec-containing formate dehydrogenase (FDH) and the Sec-containing Ni-/Fe-hydrogenase. In fact, FDH it is the most widely distributed selenoprotein found in nature and it has been suggested that the genes encoding FDH, together with the genes encoding the system for Sec biosynthesis and incorporation, were extensively transferred laterally.

*Formate dehydrogenase*. FDH catalyzes the reversible reduction of $CO_2$ to formate and can be involved in energy metabolism, carbon fixation, or pH homeostasis (Ferry 1990). The diverse roles FDH employ is reflected by the considerable differences found in subunit composition, kinetic properties, and types of electron acceptors utilized. FDH contains Fe/S clusters and a pterin cofactor coordinating either molybdenum or tungsten. The natural electron acceptor for Sec-containing FDH can be cytochrome (cyt$_b$ for the *Escherichia coli* FdhN), NADP$^+$ (for the *Moorella thermoacetica* enzyme), ferredoxin (for the *Clostridium pasteurianum* enzyme), or $F_{420}$ (a 2-deazaflavin derivative functionally analogous to

NAD$^+$; in the FDH of methanogenic Archaea), where FDH functions in formate oxidation rather than CO$_2$ reduction.

*Hydrogenase.* Hydrogenases catalyze the reversible reduction of protons and are widely distributed among prokaryotes and lower eukaryotes. The electron carriers used vary depending on the organism. Three distinct classes of hydrogenases are known, the Ni-containing (Ni/Fe) hydrogenases, the iron-only (or Fe/Fe) hydrogenases, and the Fe/S cluster-free (formerly metal-free) hydrogenases (Thauer et al. 2010). The vast majority of known hydrogenases fall into the former class which also harbors hydrogenases containing Sec in the active site, the latter class has so far been found exclusively in methanogenic archaea. In the active site of Ni/Fe hydrogenase four Cys residues (or three Cys plus a Sec) coordinate a Ni atom, with two of these Cys also binding the Fe atom. In addition, the Fe possesses three non-protein diatomic ligands, CN, CO, or SO, depending on the source of the enzyme. Sec-dependent hydrogenases are also present in some methanogenic archaea. Hydrogenotrophic (i.e., growing with H$_2$ + CO$_2$) methanogens contain at least two types of hydrogenase, the coenzyme F$_{420}$-reducing hydrogenase, and the coenzyme F$_{420}$-nonreducing hydrogenase. As the names suggest, the physiological electron acceptor for H$_2$ oxidation of the former is F$_{420}$, while electron transfer of the latter was only recently elucidated. F$_{420}$-nonreducing hydrogenase is tightly associated with the heterodisulfide reductase and electrons derived from H$_2$ oxidation are transferred via Fe/S clusters and FAD from the hydrogenase to both the heterodisulfide and ferredoxin by flavin-based electron bifurcation (Thauer et al. 2008). *Methanococcus voltae* contains a Sec-dependent and a Sec-independent set of F$_{420}$-reducing and nonreducing hydrogenases (Sorgenfrei et al. 1997). This situation appears to be identical in all Sec-encoding *Methanococcus* species for which genome sequences are available.

*Glycine reductase.* The glycine reductase system is almost exclusively found in clostridia (Andreesen 2004). It consists of three proteins, P$_A$, P$_B$, and P$_C$. P$_B$ is the substrate-specific glycine-activating (to a Se-carboxymethyl selenoether) Sec-containing protein. Although direct evidence for the involvement of Sec in glycine activation is still lacking, the fact that corresponding subunits of all characterized substrate-specific P$_B$ proteins of glycine, betaine, sarcosine, and proline reductase (see below), respectively, contain Sec, strongly argues for its involvement in catalysis (Andreesen 2004). The small acidic P$_A$ accepts the carboxymethyl group from P$_B$. It is the only selenoprotein for which no Cys-containing homolog is known (Kryukov and Gladyshev 2004). Upon transfer of the carboxymethyl group from P$_A$ to P$_C$, the selenium in P$_A$ becomes oxidized and the carboxymethyl is reduced resulting in a P$_C$-bound acetyl thioester. Oxidized P$_A$ is re-reduced by the thioredoxin/thioredoxin reductase system and phosphorolysis of the acetyl thioester liberates acetyl-phosphate. Sarcosine reductase and betaine reductase share their P$_A$ and P$_C$ components with glycine reductase but contain different substrate-specific, Sec-containing P$_B$ proteins.

*Proline reductase.* The reduction of D-proline to 5-aminovalerate in clostridia seems to proceed by a different mechanism compared to glycine reduction, although in both cases a carbon-nitrogen bond is cleaved and D-proline reductase is similar to P$_B$ of glycine reductase. The enzyme is composed of three different subunits, PrdB (which contains Sec) and two proteins resulting from processing of proprotein PrdA.

*Antioxidant defense proteins.* Methionine sulfoxide reductase (Msr) reduces oxidized methionine residues in proteins, which arise by the unwanted action of reactive oxygen species (Stadtman et al. 2003). MsrA is specific for the *S*-form of methionine sulfoxide, whereas MsrB is specific for the *R*-form. Both MsrA and MsrB can either be selenoproteins or non-selenoproteins, depending on the organism. Furthermore, *Eubacterium acidaminophilum* synthesizes a Sec-containing peroxiredoxin with thiol-dependent peroxidase activity. Metabolic labeling in conjunction with genome analysis further revealed that the organism also produces another, small selenoprotein of unknown function which contains a motif reminiscent of thioredoxin.

*Selenomonophosphate synthetase.* SPS of *E. coli* (encoded by the *selD* gene), which is no selenoprotein, was shown to provide selenomonophosphate for Sec synthesis (see above). The reaction proceeds rather slowly by transfer of the γ-phosphate group of ATP to a reduced selenium species (selenide in vitro) liberating AMP and orthophosphate via a phosphorylated enzyme-intermediate. SPS homologs containing Sec have been identified in eukaryotes, archaea, proteobacteria, and Gram-positive bacteria. What role the Sec residue plays during catalysis in these enzymes is not clear. Also, how Sec synthesis can be initiated employing an enzyme which itself is a selenoprotein, is a classic "chicken or egg" question; *Haemophilus influenzae* apparently solves this problem by initially operating SPS that is either devoid of Sec, or by forming Sec-tRNA$^{sec}$ in a selenophosphate-independent fashion.

*Formylmethanofuran dehydrogenase.* Formylmethanofuran dehydrogenase (FMD) catalyzes the reduction of $CO_2$ and methanofuran to formylmethanofuran, which is the initial step in methanogenesis from $CO_2$ in all methanogenic Archaea. FMD is composed of five subunits of which two share considerable similarity with FDH, and contains either Mo or W, and Fe/S clusters. In *Methanococcus* and *Methanopyrus* species a Sec-containing isoform was found.

*Heterodisulfide reductase.* Obligate hydrogenotrophic methanogens reduce the heterodisulfide of coenzyme M and coenzyme B (CoM-S-S-CoB) by a cytoplasmic multienzyme complex composed of the $F_{420}$-nonreducing hydrogenase (see above) and a soluble heterodisulfide reductase (HDR), an iron-sulfur flavoprotein, which in *Methanococcus* and *Methanopyrus* species also contains Sec. Interestingly, *Geobacter* species capable of anaerobically degrading aromatic compounds may synthesize, deduced from genomic sequences, a tungsten-dependent enzyme complex, benzoyl-CoA reductase BamBCDEFGHI, which resembles the methanogenic HDR/hydrogenase complex and probably also contains Sec (Fuchs et al. 2011). However, the role which Sec might play in this complex remains to be elucidated.

*HesB-like selenoprotein.* The small ca. 11 kDa selenoprotein of *Methanococcus maripaludis* similar to HesB from *Synechococcus* was originally identified through genome analyses and later verified by experimentation (Stock et al. 2010). It is distantly related to IscA, which is involved in Fe/S-cluster assembly. However, except for the fact that it is not required in *M. maripaludis* nothing is known about the function of the HesB-like selenoprotein.

## Summary

The number of potential selenoproteins is ever-increasing through (meta-)genome sequencing projects, which holds the promise that novel and useful enzymatic activities may be discovered. However, experimentation lacks behind sequence data accumulation as biochemical analysis of selenoproteins in not trivial. Still, one needs to understand the (catalytic) role Sec plays in a protein in order to fully appreciate its cellular context, its potential usefulness, and its evolution. It will be interesting to see if glycine reductase shall remain the only protein for which selenocysteine cannot be replaced by another amino acid.

## Cross-References

- Artificial Selenoproteins
- Molybdenum and Ions in Prokaryotes
- Molybdenum Cofactor, Biosynthesis and Distribution
- Molybdenum in Biological Systems
- Nickel in Bacteria and Archaea
- [NiFe]-Hydrogenases
- Selenium and Glutathione Peroxidases
- Selenium and Iodothyronine Deiodinases
- Selenium in Human Health and Disease
- Selenophosphate Synthetase
- Tungsten Cofactors, Binding Proteins, and Transporters in Biological Systems
- Tungsten in Biological Systems

## References

Andreesen JR (2004) Glycine reductase mechanism. Curr Opin Chem Biol 8:454–461

Björk GR (1995) Biosynthesis and function of modified nucleosides. In: Söll D, Rajbhandary UL (eds) tRNA: structure, biosynthesis, and function. ASM Press, Washington, DC

Böck A, Thanbichler M, Rother M et al (2005) Selenocysteine. In: Ibba M, Francklyn CS, Cusack S (eds) Aminoacyl-tRNA synthetases. Landes Bioscience, Georgetown

Commans S, Böck A (1999) Selenocysteine inserting tRNAs: an overview. FEMS Microbiol Rev 23:335–351

Ferry JG (1990) Formate dehydrogenase. FEMS Microbiol Rev 87:377–382

Flohe L (2009) The labour pains of biochemical selenology: the history of selenoprotein biosynthesis. Biochim Biophys Acta 1790:1389–1403

Fuchs G, Boll M, Heider J (2011) Microbial degradation of aromatic compounds – from one strategy to four. Nat Rev Microbiol 9:803–816

Haft DH, Self WT (2008) Orphan SelD proteins and selenium-dependent molybdenum hydroxylases. Biol Direct 3:4

Kryukov GV, Gladyshev VN (2004) The prokaryotic selenoproteome. EMBO Rep 5:538–543

Leung EK, Suslov N, Tuttle N et al (2011) The mechanism of peptidyl transfer catalysis by the ribosome. Annu Rev Biochem 80:527–555

Sorgenfrei O, Müller S, Pfeiffer M et al (1997) The [NiFe] hydrogenases of *Methanococcus voltae*: genes, enzymes and regulation. Arch Microbiol 167:189–195

Stadtman ER, Moskovitz J, Levine RL (2003) Oxidation of methionine residues of proteins: biological consequences. Antioxid Redox Signal 5:577–582

Stock T, Rother M (2009) Selenoproteins in Archaea and Gram-positive bacteria. Biochim Biophys Acta 1790:1520–1532

Stock T, Selzer M, Rother M (2010) In vivo requirement of selenophosphate for selenoprotein synthesis in archaea. Mol Microbiol 75:149–160

Thanbichler M, Böck A (2001) Functional analysis of prokaryotic SELB proteins. Biofactors 14:53–59

Thauer RK, Kaster AK, Goenrich M et al (2010) Hydrogenases from methanogenic archaea, nickel, a novel cofactor, and $H_2$ storage. Annu Rev Biochem 79:507–536

Thauer RK, Kaster AK, Seedorf H et al (2008) Methanogenic archaea: ecologically relevant differences in energy conservation. Nat Rev Microbiol 6:579–591

Zhang Y, Gladyshev VN (2008) Trends in selenium utilization in marine microbial world revealed through the analysis of the global ocean sampling (GOS) project. PLoS Genet 4:e1000095

Zhang Y, Romero H, Salinas G et al (2006) Dynamic evolution of selenocysteine utilization in bacteria: a balance between selenoprotein loss and evolution of selenocysteine from redox active cysteine residues. Genome Biol 7:R94

# Selenoproteome

▶ Selenium in Human Health and Disease

# Self-Assembly

▶ Silicateins

# Self-Subunit Swapping

▶ Nitrile Hydratase and Related Enzyme

# SelG

▶ Selenoprotein K

# SelK

▶ Selenoprotein K

# SelM Homologue

▶ Selenoprotein Sep15

# SelP (Conflict with P-Selectin)

▶ Selenoprotein P

# Senescence

▶ Magnesium and Cell Cycle

# Sensitivity

▶ Chromium and Allergic Reponses

# SeP (Old Literature)

▶ Selenoprotein P

# Sep15

▶ Selenoprotein Sep15

# SEPN1 Gene

▶ Selenium and Muscle Function

# SEPN1-Related Myopathies (*SEPN1*-RM)

▶ Selenium and Muscle Function

# SePP

▶ Selenoprotein P

# SEPP1 (Human Gene Name)

▶ Selenoprotein P

# SGLTs

▶ Sodium/Glucose Co-transporters, Structure and Function

# Sheddase

▶ Zinc Adamalysins

# Siderophilin

▶ Chromium(III) and Transferrin

# Signal Transduction

▶ C2 Domain Proteins

# Signaling by Cadherins and Cadherin-Associated Proteins

▶ Cadherins

# Significance of Zinc in Hemostasis Regulation

▶ Zinc in Hemostasis

# Silica, Immunological Effects

Naoko Kumagai-Takei[1], Suni Lee[1], Hidenori Matsuzaki[1], Hiroaki Hayashi[2], Megumi Maeda[3], Yasumitsu Nishimura[1] and Takemi Otsuki[1]
[1]Department of Hygiene, Kawasaki Medical School, Okayama, Japan
[2]Department of Dermatology, Kawasaki Medical School, Kurashiki, Japan
[3]Department of Biofunctional Chemistry, Division of Bioscience, Okayama University Graduate School of Natural Science and Technology, Okayama, Japan

## Synonyms

Autoimmune diseases as the complication of silicosis; Environmental dysregulation of autoimmunity

## Definition

Patients with silicosis suffer from lung fibrosis causing respiratory dyspnea, complicated chronic bronchitis, pulmonary tuberculosis, emphysema, and other pulmonary diseases and are often complicated by autoimmune/collagen diseases such as RA (rheumatoid arthritis, well known as Caplan syndrome), SLE (systemic lupus erythematosus), SSc (systemic sclerosis), and ANCA (antineutrophil cytoplasmic autoantibody)-related vasculitis/nephritis. It had previously been considered that dysregulation of autoimmunity caused by silica exposure was induced by an adjuvant effect of silica.

However, recent developments in immunology have provided new concepts such as recognition of the danger signal by the NLRP3 (nucleotide-binding domain, leucine-rich-containing family, pyrin-domain-containing 3, Nalp3) inflammasome in antigen-presenting cells, alteration of the CD95/Fas molecule in autoimmune diseases, and the importance of functional and/or changes in the number of CD4 + 25+ FoxP3 (Forkhead box p3, Scurfin) positive regulatory T cells.

## Silica-Induced Dysregulation of Autoimmunity

### Epidemiology

The complication of autoimmune diseases with silicosis was first reported more than half a century ago. Caplan syndrome, which describes silicosis with rheumatoid arthritis, was initially reported in 1953 (Caplan 1953). Since then, many reports and reviews have been published regarding the accompaniment of silicosis with autoimmune diseases, such as RA, SLE, and SSc. These three types of autoimmune connective tissue diseases are typical forms of silica-induced autoimmune diseases (Haustein and Anderegg 1998; Steenland and Goldsmith 1995). However, other forms of these diseases have also been reported such as Wegener's granulomatosis, ANCA (anti-neutrophil cytoplasmic autoantibody)-related vasculitis/nephritis, Sjögren syndrome, and pemphigus vulgaris. Furthermore, many autoantibodies are detected in silicosis patients even in the absence of typical clinical manifestations. In addition to autoantibodies related to the aforementioned autoimmune diseases, antinuclear, ANCA, rheumatoid factor, and anti-topoisomerase I, anti-Fas, anti-caspase-8, and anti-desmoglein antibodies have been reported. These findings indicate that silica possesses immunological effects to induce dysregulation of autoimmunity, in addition to having immunological effects on local immunocompetent cells located at the lung, which can lead to pulmonary fibrosis.

Although several reports have indicated that the incidence of Caplan syndrome is between 20% and 30% in all silicosis patients, the prevalence seemed to differ depending on the report. However, the odds ratio or RA-defined cases with silicosis in South African gold miners indicated a value of 3.79 ($p = 0.0006$) (Sluis-Cremer et al. 1986). Regarding mortality, the SMR (standardized mortality ratio) for RA was reported as 2.01 and 2.29 (Steenland and Goldsmith 1995). While the SMR for SSc was 2.45, the SMR for renal disease with silicosis was 2.22–2.77.

Overall, it is clear that silica can induce immunological effects and cause dysregulation of autoimmunity.

### Role of Inflammasome in the First Signal for the Biological Body

Silica enters the body through the pulmonary region. Following entry, alveolar macrophages acting as antigen-presenting cells are mainly responsible for recognizing silica particles and attempt to exclude it. Although it is known that these alveolar macrophages produce interleukin (IL)-1$\beta$ as a consequence of processing foreign danger signals such as those associated with the presence of silica, the detailed cellular and molecular mechanisms involved have yet to be clarified. However, a recent discovery concerning the role of the NLRP3 inflammasome has assisted in clarifying the cellular response (Dostert et al. 2008). When silica, asbestos fibers, uric acid crystals, or cholesterol encounter antigen-presenting cells, the resulting extrinsic and intrinsic danger signals lead to activation of NLRP3 and recruitment of apoptosis-associated speck-like protein (ASC) with CARD (Caspase activation and recruitment domains)/pyrin-containing adaptor. This complex cleaves pro-caspase-1 to generate active caspase-1. Subsequent caspase-1 activity leads to the production of pro-inflammatory cytokines, and in particular IL-1$\beta$ and IL-18. These cytokines function to serve fibroblasts to form lung fibrosis. This represents the latest scenario to account for the initial response of a body to silica, although it remains undetermined whether these cascaded reactions are closely related to the subsequent dysregulation of autoimmunity.

Moreover, details of the interaction between silica particles and alveolar macrophages have been reported. Based on experimental results, silica particles are toxic to alveolar macrophages as a result of the production of ROS (reactive oxygen species) directly from the particle surface, and indirectly from the cellular response following an encounter with silica. During this encounter, the important role of surface class A scavenger molecules, such as SR-A1/2 (scavenger receptors type 1 and 2), which are trimers with a molecular weight of about 220–250 kDa and preferentially bind modified LDL (Low-density lipoprotein) via acetic acid and oxidized LDL, and MARCO to induce binding and internalization of silica particles has been reported (Hamilton et al. 2008). Thereafter, as a result of interactions with the aforementioned inflammasome, the locus where silica particles first meet with human immunocompetent cells such as alveolar macrophages participate in alternative cytokine production and formation of chronic inflammation, which leads to lung fibrosis and chronic retention of silica particles with subsequent induction of autoimmune dysregulation.

## Status of CD95/Fas and Related Molecules in Silicosis

The CD95/Fas death receptor possesses an important role in the apoptosis of lymphocytes. Furthermore, the molecular alteration of CD95/Fas causes the pathological enhancement of autoimmunity. For example, the genetic mutation of Fas or Fas-ligand in mouse (*lpr* and *gld* genes, respectively) manifests a SLE-like pathological status and genetic mutation of the Fas gene in humans causes ALPS (autoimmune lymphoproliferative syndrome). Furthermore, Fas-mediated apoptosis is inhibited by the soluble form (alternatively spliced form lacking the membrane-bound domain) (sFas), DcR3 (decoy receptor 3), and splicing variants other than sFas, which contain the Fas-ligand-binding domain but lack the membrane-bound domain. Serum sFas and soluble DcR3 are detected in the serum of individuals covering a variety of autoimmune diseases. When silicosis is positioned as a pre-autoimmune disease, this may result in dysregulation and/or altered Fas and Fas-ligand apoptosis system in immunocompetent cells such as lymphocytes and macrophages in peripheral blood or pulmonary lesions. The following findings have been reported in the last decade.

1. In the lung, silicosis progression is related to the expression of Fas-ligand, mast cells, and extracellular matrix remodeling.
2. Fas and Fas-ligand systems may regulate apoptosis in alveolar macrophages and the occurrence of Fas-mediated apoptosis may be a biomarker of silicosis development.
3. Genetic polymorphism of the Fas gene is related to the development of silicosis.
4. Serum sFas and messenger transcript expression of sFas, DcR3, and alternatively spliced variant messenger transcripts of the Fas gene in peripheral blood mononuclear cells derived from silicosis are elevated.
5. Parameters related to Fas-mediated apoptosis are independent of the respiratory parameters in silicosis.
6. Autoantibodies against Fas-related molecules such as Fas and caspase-8 are detected in serum from silicosis.
7. In the animal model, mutation of Fas-ligand prevents the development of acute silicosis.

From aforementioned findings 1–3, it seems that Fas and Fas-ligand apoptosis is related to lung fibrosis caused by silica exposure via apoptosis of alveolar macrophages. The important role of Fas-mediated apoptosis is not restricted to silicosis. Fibrotic interstitial lung diseases such as interstitial pneumonitis, asbestosis, and experimental models of bleomycin-induced lung fibrosis show altered Fas and/or Fas-ligand expression in lung parenchyma or bronchoalveolar lavage.

Furthermore, exposure to silica may influence circulating immunocompetent cells associated with aforementioned findings 4–7 (Otsuki et al. 2006). Additionally, anti-Fas autoantibodies detected in silicosis can induce Fas-mediated apoptosis and mRNA expression of several genes such as I-Flice, sentrin, survivin, and ICAD (inhibitor of caspase-3-activated DNase), all of which can act as inhibitors of Fas-mediated apoptosis, and are reduced in peripheral blood mononuclear cells from silicosis. These findings indicate that Fas-mediated apoptosis is enhanced in circulating lymphocytes under conditions of silicosis. On the other hand, aforementioned findings 4–7 suggest that Fas-mediated apoptosis is protected by extracellular inhibitory molecules such as sFas and DcR3 to facilitate earlier binding with Fas-ligand. From these results, it may be that two populations of circulating T cells are present in silicosis (Otsuki et al. 2006; Lee et al. 2011). One population may comprise T cells which express high levels of membrane Fas, are sensitive to anti-Fas autoantibodies, and undergo Fas-mediated apoptosis with recruitment from bone marrow. The other population of T cells may express lower levels of membrane Fas, produce sFas and DcR3 to escape from Fas-mediated apoptosis, and survive longer. The latter population may contain self-recognizing clones which generate autoimmune diseases. This idea is schematically summarized in Fig. 1, and may account for the role of the Fas and Fas-ligand apoptotic pathway in circulating T cells with respect to the occurrence of autoimmune

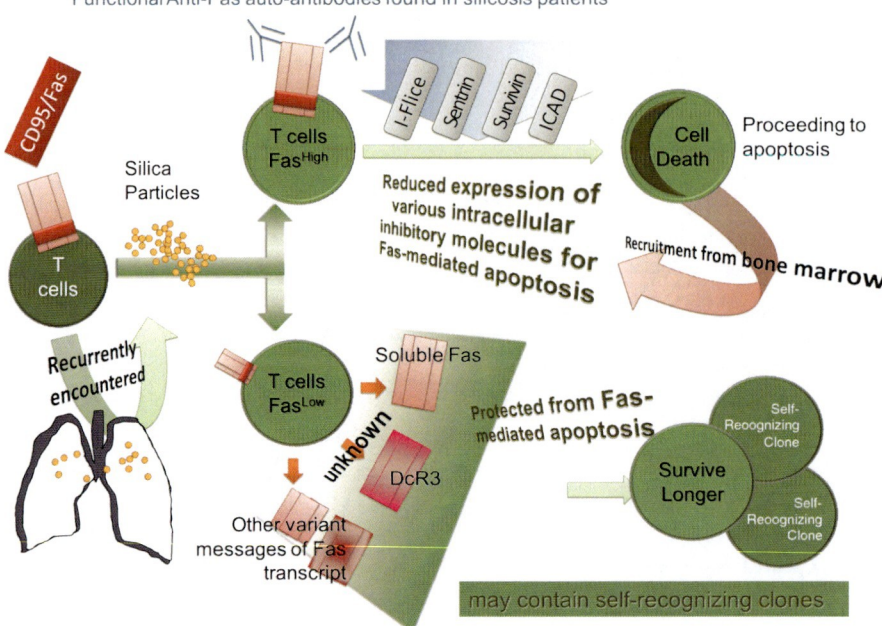

**Silica, Immunological Effects, Fig. 1** Summary of the alteration of Fas and related molecules in silicosis patients. There are two populations of Cd4+ lymphocytes, with one population possessing higher expression of membrane Fas compared with the other. The cell population with higher membrane Fas expression is sensitive to functional anti-Fas autoantibodies found in silicosis (approximately 25%) and progresses to Fas-mediated apoptosis as represented by the reduced expression of various inhibitory molecules for apoptosis. This population may be repeating cell death and involve recruitment from bone marrow. The cell population with lower Fas membrane expression secretes soluble Fas, Decoy Receptor 3, and other variant Fas messenger transcripts, all of which inhibit Fas-mediated apoptosis by binding Fas-ligand at the extracellular area. Avoiding apoptosis, this cell population survives longer and may include self-recognizing clones

dysregulation with silicosis, aside from the biological role of this apoptotic pathway in the development of lung fibrosis from local pulmonary lesions resulting from silica exposure.

### Activation of Responder and Regulatory T Cells by Silica Exposure

The recent discovery of CD4 + 25 + FoxP3 (Forkhead box protein P3) positive regulatory T cells (Tregs) and Th17 cells has contributed toward investigations concerning the cellular and biological development of autoimmune diseases. Tregs comprise a subpopulation of T cells which downregulate the immune system, maintain tolerance to self-antigens, and suppress autoimmune disease. Several autoimmune diseases show reduced function and/or a lower population of peripheral Tregs. Interestingly, manifestation of the CD4+25+ phenotype is also found in antigen-activated responder T cells. Thus, if the T cell population in silicosis is being chronically activated by auto-antigen, the peripheral CD4+25+ fraction may be contaminated by activated responder T cells together with a reduced true Treg (FoxP3 positive) population. This condition may result in reduced inhibitory function of the peripheral CD4+25 + fraction which has been reported. Moreover, several reports have detailed the silica-induced activation of responder T cells.

1. Silica slowly and gradually activates peripheral T cells in vitro as determined by monitoring CD69 surface expression as a marker of early T cell activation.
2. Gene expression of CD69 is higher in the peripheral CD4+25+ fraction of silicosis patients compared with healthy donors.
3. Gene expression of PD-1 (Programmed cell death 1, as an activation marker of T cells) is higher in CD4 +25− and CD4+25+ fractions of silicosis patients compared with healthy donors.

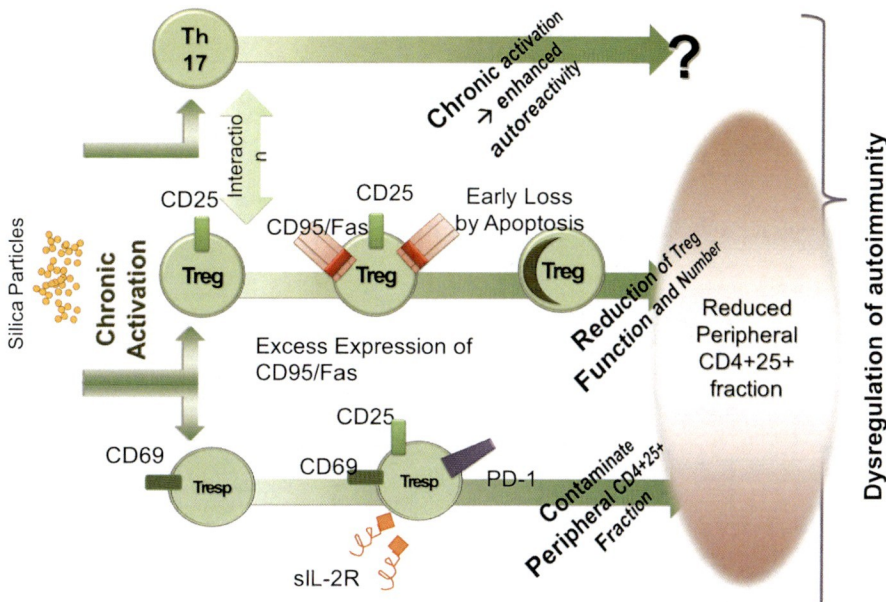

**Silica, Immunological Effects, Fig. 2** Model showing the activating effects of silica on responder and regulatory T cells. Silica can activate both cell types. Activated responder T cells (CD4 + 25- fraction) can express CD69, PD-1 (programmed cell death1), and CD25. Activated regulatory T cells (CD4 + 25+ and FoxP3 positive) can express CD95/Fas death receptor and are sensitive to Fas-mediated apoptosis. Since chronically activated Tregs progress to apoptosis, the inhibitory effects involving the termination of activated responder T cells has been reduced. As a result, chronically activated responder T cells may avoid termination caused by the suppressive function of regulatory T cells, and prolonged activation is manifested. These conditions may result in the longer survival of self-recognizing responder T cell clones and reaction with self-antigens. Furthermore, the inhibitory function of the peripheral CD4 + 25+ fraction is reduced given the earlier loss of regulatory T cells and contamination of activated responder T cells expressing CD4 + 25+

4. The serum level of soluble IIL-2 receptor (sIL-2R) is higher in silicosis patients compared with healthy donors, and the level of sIL-2R is higher in silicosis patients compared with healthy donors, with the highest levels being detected in patients with systemic sclerosis according to a positive correlation.

On the other hand, exposure to silica also activates Tregs. When Tregs are activated, the expression of surface CD95/Fas is enhanced.
1. Expression of surface Fas in peripheral FoxP3 Tregs is higher in silicosis patients compared with health donors.
2. Tregs derived from silicosis patients are sensitive to Fas-mediated apoptosis-inducing antibodies.
3. In vitro exposure of peripheral blood mononuclear cells to silica caused a reduction in FoxP3 expressing cells, but had no effect on the CD25 expressing cell population. These findings indicate that in vitro cultivation facilitates the processing of Tregs toward apoptosis and the activation of originally CD25 negative responder T cells.

Taken together and summarized in Fig. 2, silica exposure chronically activates both responder and regulatory T cells (Hayashi et al. 2010). The development of early loss of Tregs by Fas-mediated apoptosis with contamination of chronically activated responder T cells into the peripheral CD4+25+ fraction results in reduced inhibitory function of this fraction. Chronically activated responder T cells may include self-antigen recognizing clones, and activation of these cells may not be reduced by Tregs (Lee et al. 2011). These conditions may further lead to the dysregulation of autoimmunity.

## Immunological Effects of Asbestos, Mineral Silicate

Although the physical form of asbestos is rigid and linear compared with particulate silica, asbestos is chemically composed of silicon and oxygen, and additional minerals such as magnesium, sodium, and iron. Regarding initiation of the fibrogenicity of asbestos which leads to asbestosis, the important role of

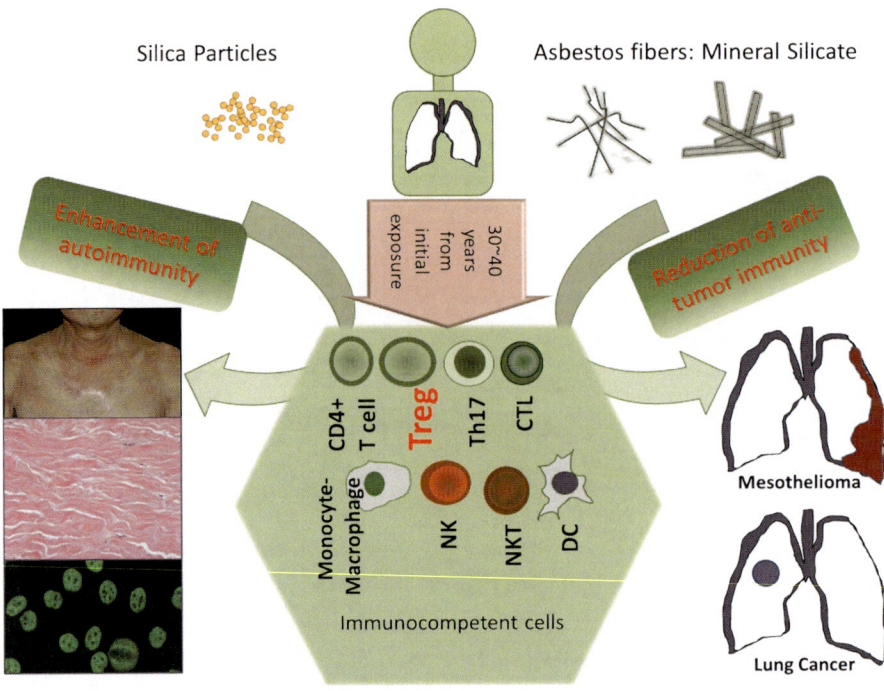

**Silica, Immunological Effects, Fig. 3** Comparison of the immunological effects of silica and mineral silicate (asbestos). If Treg function is placed at the center of these reactions, silica and asbestos are assumed to induce opposite effects. A detailed analysis of the immunological effects of silica and asbestos should be performed to better understand the biological effects of these environmental substances

forming the inflammasome, as mentioned above, has been well recognized. The physical differences between asbestos fibers and silica particles may influence the pathological features of fibrosis which manifest in silicosis and asbestosis. The former forms middle to upper pulmonary areas with small rounded silicotic nodules, while the latter forms middle to lower lesions with linear and irregular progression of fibrosis, which finally develops a honeycomb lung.

Investigations concerning in vitro exposure of immunocompetent cell lines to asbestos, ex vivo exposure of freshly isolated lymphoid cells derived from healthy donors, and circulating peripheral blood lymphoid cells have revealed evidence of reduced antitumor immunity. For example, the functional activation of NK (natural killer) cell receptors such as 2B4, NKG2D, and NKp46 with suppressed intracellular signaling such as the MAPK (mitogen-activated protein kinase) cascade. Additionally, the surface expression of CXCR3 (chemokine (C-X-C motif) receptor 3) on CD4+ T cells was reduced, together with the capacity to produce IFN (interferon)-$\gamma$, both processes being important in the development of antitumor immunity. Moreover, marked secretion of IL-10 and TGF-$\beta$ occurs when the human CD4+ T cell line is continuously exposed to asbestos, with both cytokines possessing immune suppressive effects and also typical soluble factors for the function of Tregs.

Taken together, the immunological effects of asbestos tend to reduce antitumor immunity. This is reasonable according to the complications which arise in patients exposed to asbestos, such as the occurrence of malignant mesothelioma and lung cancers, as summarized in Fig. 3. In particular, if Tregs are placed at the center of the immunological effects of these environmental substances, silica can reduce Treg function and population, although asbestos seemed to enhance Treg function at least from the results pertaining to cytokine production in asbestos-exposed cell lines and patients (Matsuzaki et al. 2012).

## Further Perspectives

The effects of silica on Th17 cells and the role of Th17 cells in silica-induced dysregulation of autoimmunity have yet to be delineated. One report has indicated that IL-17A-producing $\gamma\delta$ T and Th17 lymphocytes mediate lung inflammation but not fibrosis in experimental silicosis. However, further clarification is required concerning cytokine alterations in silicosis and the induction of Th17, the status of Th17 and related cytokines such as IL-17A, IL-21, IL-22, Il-23, IL-6, and TGF (transforming growth factor)-$\beta$, and the effects of silica on generalized antigen-presenting cells such as dendritic cells not located at the local pulmonary area. The difficulties associated with an analysis of the immunological effects of silica relate to the need to consider these effects on the local processing of lung fibrosis and the development of general autoimmune dysregulation. The important players may be similar, such as CD95/Fas and its ligand, Th17, Tregs, and others. The need to distinguish the immunological effects of silica in terms of the local pulmonary area and general immune system is important in contributing toward an understanding of the total effects of silica on the human body. Furthermore, a comparison of the immunological effects of particulate silica and asbestos fibers is also needed to better understand the biological effects of these environmental substances.

## Cross-References

- Chromium(III) and Immune System
- Gallium Nitrate, Apoptotic Effects
- Lead and Immune Function
- Mercury and Immune Function
- Thallium, Cell Apoptosis Inducement
- Zinc and Immunity

## References

Caplan A (1953) Certain unusual radiological appearances in the chest of coal-miners suffering from rheumatoid arthritis. Thorax 8:29–37
Dostert C, Pétrilli V, Van Bruggen R et al (2008) Innate immune activation through Nalp3 inflammasome sensing of asbestos and silica. Science 320:674–677
Hamilton RF Jr, Thalur SA, Holian A (2008) Silica binding and toxicity in alveolar macrophages. Free RadicBiol Med 44:1246–1258
Haustein UF, Anderegg U (1998) Silica induced scleroderma– clinical and experimental aspects. J Rheumatol 25:1917–1926
Hayashi H, Miura Y, Maeda M et al (2010) Reductive alteration of the regulatory function of the CD4(+)CD25(+) T cell fraction in silicosis patients. Int J Immunopathol Pharmacol 23:1099–1109
Lee S, Hayashi H, Maeda M et al (2011) Environmental factors producing autoimmune dysregulation – chronic activation of T cells caused by silica exposure. Immunobiology. 24 Dec 2011. http://www.sciencedirect.com/science/article/pii/S0171298511002853
Matsuzaki H, Maeda M, Lee S et al (2012) Asbestos-induced cellular and molecular alteration of immunocompetent cells and the relationship with chronic inflammation and carcinogenesis. J Biomed Biotechnol. doi:10.1155/2012/492608 (e-Pub: Article ID 492608)
Otsuki T, Miura Y, Nishimura Y et al (2006) Alterations of fas and fas-related molecules in patients with silicosis. ExpBiol Med (Maywood) 231:522–533
Sluis-Cremer GK, Hessel PA, Hnizdo E et al (1986) Relationship between silicosis and rheumatoid arthritis. Thorax 41:596–601
Steenland K, Goldsmith DF (1995) Silica exposure and autoimmune diseases. Am J Ind Med 28:603–608

## Silica-Induced Fibrosis

▶ Silicosis

## Silicateins

Lukmaan Bawazer
School of Chemistry, University of Leeds, Leeds, UK

## Synonyms

Bioinspired materials; Biomimicry; Biomineralization; Biosilica; Cysteine hydrolases; Marine sponge skeletons; Self-assembly

## Definition

Silicateins are a class of biomineralizing enzymes ($\sim$23–27 kDa in their active form) that catalyze and template silica mineralization while directing skeletal growth in a variety of marine sponges. They share a close evolutionary relationship with a well-characterized family of proteolytic digestive enzymes, but are distinguished by unique self-assembling and mineralizing functionalities. Silicateins were first discovered as the majority protein fraction occluded within needle-like biosilica skeletal elements of the marine sponge *Tethya aurantia* (Shimizu et al. 1998; Weaver and Morse 2003), and were later identified in a number of other siliceous sponge species as well (Wang et al. 2012; Schröder et al. 2007). Research elucidating silicatein structure and function has provided insights into the molecular mechanisms of

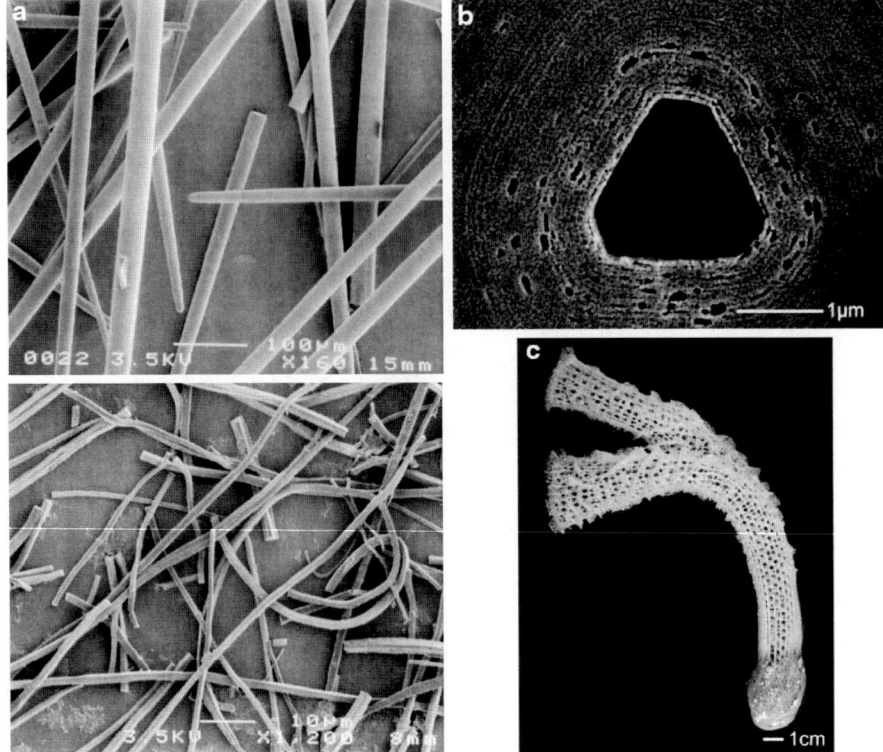

**Silicateins, Fig. 1** Examples of biosilica skeletal structures from which silicateins have been identified. (**a**) Scanning electron micrographs showing needle-like biosilica spicules from the marine demosponge *Tethya aurantia* (*top*). Each spicule contains a polymerized filament of silicatein (*bottom*) running axially through the center of the spicule (Reproduced with permission from Shimizu et al. (1998)). (**b**) A cross section of one of the spicules shown in (**a**), revealing a distorted hexagonal axial canal in which the silicatein filament would be found. Also revealed are layers of condensed silica alternating with layers of less condensed (protein-containing) silica. (**c**) A full silica skeleton of a hexactinellid sponge representing an example of the structural diversity exhibited by siliceous sponges (**b** and **c** are reproduced with permission from Weaver and Morse (2003))

natural silica biomineralization, and has established these proteins as model systems for the development of bioinspired and biogenetic strategies for synthesizing technological materials.

## Introduction

In their natural form, silicateins are predominantly found as polymeric filaments running along the central axis of pseudo-cylindrical skeletal elements, known as spicules, in which they are embedded (Fig. 1a). The exact geometries of spicules and their embedded silicatein filaments vary both by species and by skeletal function. In the sponge *Tethya aurantia*, the majority spicule type (constituting ~75% the dry-weight of the sponge) is up to millimeters in length and tens of micrometers in diameter, and its associated silicatein filaments are ~2 μm in diameter and often appear hexagonal in cross section (e.g., Fig. 1b) though other cross-sectional shapes are also observed (Weaver and Morse 2003; Uriz et al. 2000). Overall, siliaceous sponge skeletons exhibit a remarkable diversity of architectures that have intrigued biologists for more than 100 years (e.g., Fig. 1c) (Weaver and Morse 2003; Wang et al. 2012), but the nature of the mineral-directing silicatein proteins remained unknown until the end of the twentieth century when molecular biology techniques opened paths to gene identification.

Natural isolates of silicatein filaments are obtained by selective dissolution, first by hypochlorite dissolution of the sponge's majority organic matter to yield spicules, and then by selective silica dissolution using either hydrofluoric acid to expose the axial protein filaments (Shimizu et al. 1998) or a buffered glycerol solution (pH 8.8) with physically ground spicules to concomitantly solubilize mineral-occluded silicateins upon extraction (Wang et al. 2012). When originally isolated, mass spectrometry was used to partially determine the amino acid sequences of the filaments' constituent proteins; from this partial information, DNA primers were designed that permitted amplification of the full silicatein-encoding genes from a library of expressed sponge genes (Shimizu et al. 1998). Identification of these genes facilitated a variety of subsequent in vitro silicatein studies, including evolutionary sequence analysis (Shimizu et al. 1998); mineralization studies with natural silicatein isolates (Weaver and Morse 2003; Brutchey and Morse 2008); investigations with recombinant silicateins produced in nonnative hosts such as bacteria (Weaver and Morse 2003; Wang et al. 2012; Brutchey and Morse 2008; Schroder et al. 2007), yeast (Patwardhan et al. 2010), and emulsion-based artificial cells (Bawazer et al. 2012); protein localization studies in loose sponge-cell aggregates (Wang et al. 2012; Muller et al. 2009); and the design of nonbiological molecular analogues that mimic silicatein's mineralizing and templating functionalities (Brutchey and Morse 2008).

## Biochemistry and Molecular Biology

Three isoforms of silicatein, termed silicateins α, β, and γ, were identified in the silicatein filaments of *Tethya aurantia* (Shimizu et al. 1998), and α and β forms were similarly found co-expressed in other species (Wang et al. 2012). Computational sequence comparison with a database of known proteins revealed that the silicateins share a high primary sequence identity with the papain superfamily of protein hydrolases. In particular, cathepsin L is the hydrolase with which silicatein α of *Tethya aurantia* share the most structural similarity (Fig. 2), with these proteins sharing 50% of their amino acids exactly, and sharing 75% sequence identity when considering amino acids representing similar biochemical functionality (Fig. 2) (Shimizu et al. 1998). Importantly, key amino acid residues present at cathepsin's putative enzyme active site are conserved in silicatein, as are six cysteine residues that impart structural stability by forming three disulfide bridges in the folded protein. The main sequence differences between silicateins and their cousin hydrolases include the replacement of cysteine (in cathepsins) with serine (in silicateins) in the well-characterized Cys-His-Asn catalytic triad involved in cathepsin-catalyzed hydrolytic proteolysis. Additionally, silicateins exhibit a higher overall percentage of hydroxylated amino acids (Ser, Thr, and Tyr), including a conspicuous stretch of serine residues proximate to the active site. Both silicatein's serine-based catalytic triad and its surface hydroxyls have been implicated as important for its mechanism of enzymatic mineralization, as discussed below.

The x-ray structure of cathepsin L has been solved to 1.73-Å resolution (Brutchey and Morse 2008), facilitating structural modeling of silicatein's three-dimensional protein fold by computational energy minimization using cathepsin's fold as a starting structural template. The silicatein models show that hydrogen bonding within the relevant Ser-His-Asn triad is conserved, thus reconstituting the catalytic triad configuration found in cathepsin and papain hydrolases. This structural conservation suggested a catalytic mechanism for enzymatic mineralization in which initially nonreactive silicon-based small molecules are catalytically hydrolyzed at the enzyme active site, thereby releasing reactive silanol species that condense to initiate silica polycondensation (Shimizu et al. 1998; Weaver and Morse 2003; Brutchey and Morse 2008). As condensation progresses, hydroxyl groups at the protein surface are expected to act as deposition sites to guide placement of the forming mineral onto the protein template. Other proteins, including non-filamentous silicatein monomers that interact with the growing silica surface, continue to direct mineralization once the filament becomes occluded by mineral (Wang et al. 2012).

When natural isolates of silicatein filaments are reacted with tetraethylorthosilicate (TEOS), a small molecule that is nonreactive until hydrolyzed, the filaments catalyze mineral silica formation from this precursor, with the formed mineral depositing on the silicatein filament surface. When the filaments are heated prior to the mineralization reaction they lose their catalytic function (no silica is produced), thus demonstrating that protein conformation (which degrades upon heating) is critical to the protein's mineralization activity (Brutchey and Morse 2008). In the mechanism first proposed for this silicatein reaction, in analogy with enzymatic proteolysis, the active site Ser's side-chain oxygen (the active site Cys's sulfur

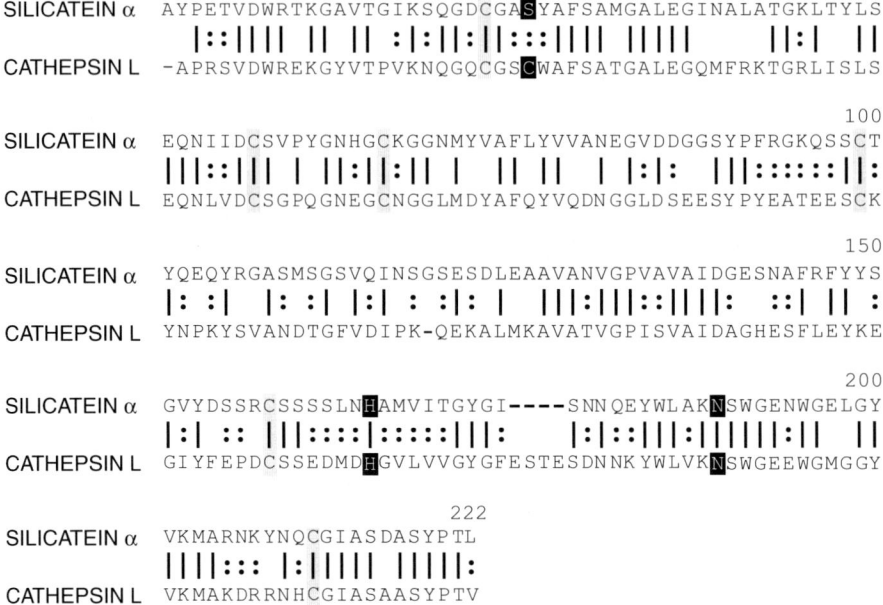

**Silicateins, Fig. 2** Sequence alignment of mature polypeptide sequences (i.e., after cleavage of the propeptide leader sequences) showing the close sequence similarity between silicatein α (from *Tethya aurantia*) and cathepsin L. Catalytic triad residues are highlighted in *white* on *black*. *Black bars* indicate exactly conserved amino acids, while *double dots* indicate different amino acids with shared or similar biochemical functionality. *Grey bars* highlight cysteine residues that form disulfide bridges to stabilize the folded protein (Reproduced from Weaver and Morse (2003) with permission)

in cathepsin) is expected to be activated, through hydrogen bonding with the neighboring His, for nucleophilic attack by silicatein on a TEOS silicon center. Through an $S_N2$-type reaction, the side-chain oxygen forms a covalent bond with the silicon atom with concomitant release of ethanol. Hydrolysis of the silicon-protein intermediate then releases a silanol species while regenerating the Ser-His hydrogen bond within the protein. This mechanism was substantiated through site-directed mutagenesis of recombinant silicatein (Weaver and Morse 2003; Brutchey and Morse 2008). When either the critical Ser or His functionality was converted to a nonreactive Ala residue, the yield of silica produced from the recombinant enzyme was decreased by 90%.

A similar mechanism as that proposed for silicatein's in vitro activity might also operate in vivo, though the exact molecular form of the natural silica precursor is unknown. Silicic acid complexed with sugars, alcohols, or catechols have been suggested as possible forms of the natural substrate (Weaver and Morse 2003), while others have implied that uncomplexed silicic acid ($Si(OH)_4$) may be the substrate (Wang et al. 2012). The latter suggestion stems from the observation that aqueous reaction with silicatein yields solid silica from silicic acid concentrations much lower than those needed for spontaneous condensation (with critical $Si(OH)_4$ concentrations being <100 μM with silicatein versus >1 mM without silicatein). Even in mechanistic variations proposed for silicatein activity with silicic acid (versus those considering an organically complexed silicon species such as TEOS), active site-precursor interactions are still thought to be critical (Wang et al. 2012; Patwardhan et al. 2010). The ability to operate by multiple mechanisms may represent an evolutionary advantage for silicatein, allowing the protein to respond flexibly to environmental stressors affecting the silicon uptake and transport pathways within the sponge.

The filamentous geometry of the silicatein polymer is a major structural director of final needle-like spicules that are mineralized. When naturally isolated silicatein filaments are chemically disassembled in vitro, under certain solution conditions the silicatein monomers oligomerize and proceed to spontaneously reassemble into ordered filaments via intermediate networks characterized by fractal geometry (Murr and Morse 2005). Several hydrophobic patches can be identified on the surface of silicatein's modeled protein fold; these are presumed important to self-assembly, as such patches are absent in protein hydrolase congeners (e.g., cathepsin L) that do not self-assemble. In addition to hydrophobic interactions, intermolecular disulfide bridges may stabilize silicatein oligomers, and electrostatic interactions further contribute to the formation of higher-order structures (Murr and Morse 2005; Muller et al. 2009). The importance of electrostatics is evidenced by the fact that silicateins extracted by hydrofluoric acid self-assemble only above pH 9, a pH at which the protein exhibits a negative charge (Murr and Morse 2005), while fractions isolated without HF, in which posttranslational phosphorylations (i.e., negatively charged groups) are possibly maintained, self-assemble at pH ~7 (Muller et al. 2009). In addition to the roles of silicatein's primary protein sequence and possible posttranslational modifications, other protein partners also support silicatein assembly. Proteins called silintaphins have been shown to bind silicatein and facilitate its assembly into filamentous structures in vitro (Wang et al. 2012). Silicatein-silintaphin binding likely occurs through Plekstrin homology (PH) domains, which have been previously implicated to support protein-protein binding in cell signal transduction and in cytoskeletal networks. The silicatein-silintaphin interaction apparently enhances mineralization activity, perhaps by providing an expanded mineral templating surface. In vitro reaction with silicic acid yields an approximately fivefold increase in silica product when both recombinant proteins (silicatein and silintaphin) are reacted together versus when silicatein alone is reacted with the precursor (Wang et al. 2012).

## Cellular Biology

When first identified, silicatein's primary protein sequence, in addition to revealing aspects of enzymatic action, further suggested possible physiological processing events that the silicateins might undergo during spiculogenesis. This information was again obtained by analogy with the well-understood cysteine hydrolases. Both silicateins and hydrolases such as cathepsin L share similar signaling and transport sequence domains upstream of the sequence region that folds into a mature hydrolytically active enzyme (Shimizu et al. 1998; Weaver and Morse 2003). For cathepsin L, the propeptide leader sequence is known to direct the expressed protein to the lysosome, a membrane-bound vesicle where digestion takes place. Once in the lysosome, active enzyme is obtained through proteolytic cleavage of the leader sequence. This processing strategy permits intracellular enzyme transport while ensuring that hydrolytic activity is expressed only in the appropriate compartmentalized environment. As initially suggested by its leader sequence (Shimizu et al. 1998), silicatein has subsequently been found to be transported to an extracellular space called the silicalemma (Wang et al. 2012), a membrane-bound compartment where the majority of silicatein-catalyzed mineralization occurs (Weaver and Morse 2003; Wang et al. 2012; Uriz et al. 2000; Schroder et al. 2007).

Lessons learned of silicatein's biological processing and cellular localization have stemmed from studies with cell cultures including loose sponge-cell aggregates called primmorphs. Primmorphs are derived from sclerocytes (mineralizing cells) from live sponge tissue that are used to inoculate salt-containing media (including silicic acid) that promote cellular reaggregation and growth of new spicular elements (Wang et al. 2012). Fluorescent protein labeling and transmission electron microscopy (TEM) are used to analyze the primmorphs at different stages of growth. Antibodies raised against silicatein or its interacting proteins are exposed to the primmorph tissue and fluorescently labeled (e.g., with secondary antibodies), allowing cellular or extracellular protein localization to be determined. Morphological characteristics of cellular compartments and growing mineral structures may be observed with TEM imaging on sectioned tissue samples, while electron dispersive x-ray spectroscopy provides chemical information (allowing, e.g., areas of high Si content to be identified).

After initial stages of intracellular spicule growth, the nascent spicules are extruded into the mesohyl (a gelatinous matrix inside the silicalemma) where spicule growth is completed (Wang et al. 2012; Uriz et al. 2000). At very early growth stages, crystalline silicate nanorods appear to direct initial silicatein filament formation but these crystalline species later disappear. Once in the silicalemma, spicules grow both in length, by continued polymerization of the silicatein filament template, and in width, by appositional mineral deposition (Wang et al. 2012; Muller et al. 2009; Schröder et al. 2007). Sclerocytes evaginate into one end of the extruded spicule and secrete new silicatein monomers into the spicule's axial canal. These silicatein monomers then polymerize to lengthen the filament template. This secretion process involves intracellular transport of expressed silicatein proteins in vacuoles known as silicasomes, which also contain a high concentration of elemental silicon (Wang et al. 2012). Ejection of silicasome contents into the extracellular mineralizing space occurs not only at the axial canal but also all around the silicalemma, thereby supplying fresh precursors and proteins for continued silica growth.

Appositional mineralization yields final spicules exhibiting radially layered rings of condensed silica as observed in cross section (Fig. 1c). These condensed rings apparently form as silicatein successively attaches to freshly mineralized spicule surfaces and catalyzes additional mineral growth in the radial direction, ultimately forming layers of protein-silica composite that alternate with layers of more fully condensed silica. A relevant study showed that recombinant silicatein-like cathepsin variants form saturated protein monolayers on silica surfaces (Patwardhan et al. 2010), providing in vitro evidence that unpolymerized silicateins exert surface-binding activity during mineralization. Such surface binding is corroborated by antibody staining revealing that silicateins binds to spicule surfaces in live cell cultures (Muller et al. 2009). In addition to the silintaphins (see above), other proteins have been identified as interaction partners with silicatein, thus contributing to mineral growth and sculpting. Gelactin, a glycoprotein with a galactose-binding domain, forms protein networks in conjunction with collagen fibrils in the mesohyl. Non-filamentous silicateins associate with these networks and are thus spatially organized in advance of the growing spicule surface (Muller et al. 2009). Another enzyme, silicase, catalyzes silica mineral dissolution and may assist in sculpting the final form of the mineralized skeletal elements (Wang et al. 2012; Schröder et al. 2007).

## In Vitro Mineralization

Silicatein filaments isolated from Tethya aurantia retain their catalytic mineralization activity, as demonstrated by their reaction with TEOS (Weaver and Morse 2003). While precipitation of TEOS is typically achieved under acidic or basic conditions, silicatein provides the unique advantage of catalyzing silica formation at neutral pH. When different organic-substituted silicon precursors are used in place of TEOS, silicatein in turn produces solid gels of polysilsesquioxanes (Weaver and Morse 2003).

In addition to producing silicon-based materials, both natural silicatein isolates as well as heterologously expressed silicateins have been found to be catalytically active with a number of water-soluble metallorganic precursors, forming corresponding oxide or oxo-hydroxide minerals (Brutchey and Morse 2008; Schröder et al. 2007). Further, in one case silicatein was used to reductively form gold (Schröder et al. 2007), probably through activity of free sulfhydryl groups at the silicatein surface. In another case, silicatein was shown to catalyze the polymerization of poly(L-lactide) through a ring-opening mechanism (Brutchey and Morse 2008).

Naturally isolated filaments have been shown to catalyze the following mineralization reactions from the indicated water-soluble precursors: $TiO_{2(s)}$ from titanium bis-ammonium-lactatodihydroxide (TiBALDH); $GaOOH_{(s)}$ and $Ga_2O_{3(s)}$ from gallium nitrate ($Ga(NO_3)_3$); and $BaTiOF_{4(s)}$ from barium fluorotitanate ($BaTiF_6$) (Brutchey and Morse 2008). In all cases, mineralization does not occur with heat-denatured filaments, confirming the importance of a stable protein structure for silicatein-catalyzed mineralization. In part from understanding the complexation chemistry of the employed metal precursors at various pHs, each formed mineral class was determined to arise through an analogous hydrolytic activity of silicatein as the one proposed for reaction with TEOS (Brutchey and Morse 2008). Further, in each of these demonstrations, crystalline minerals that normally only form at several hundred degrees celsius (anatase $TiO_2$, $\gamma$-$Ga_2O_3$, and perovskite $BaTiOF_4$, respectively) were found as product minerals formed at the silicatein filament surface after reactions at ambient temperature and pressure.

The ability of silicatein to mineralize technological crystals under such low-temperature conditions has been proposed to be due in part to the relatively slow reaction kinetics that arise from enzymatic hydrolysis of mineral precursors in a vectorially controlled manner (i.e., with hydrolysis occurring only at the filament surface); further, ordered arrays of biochemical hydroxyls are expected to pattern the filament surface, providing a bio-organic structure that supports psuedo-epitaxial templating of the formed crystals. The latter proposition (of psuedo-epitaxial templating) is supported by the finding that, after reaction with Ga (NO$_3$)$_3$, catalyzed $\gamma$-$Ga_2O_3$ crystals exhibited specific crystalline alignment, with (311) crystal faces preferentially orientated relative to the filament surface (Brutchey and Morse 2008). The importance of possible chemical-spatial ordering at the filament surface is further supported by the observation that, except in the case of laboratory-evolved silicatein variants (Bawazer et al. 2012), crystalline products have been observed to form only from reaction with natural silicatein isolates, and not from reaction with recombinant silicateins which lack the defined polymeric structure of the natural filaments (Schröder et al. 2007).

Heterologous silicatein expression, for example in *E. coli* bacteria, often leads to formation of insoluble inclusion bodies, and thus in general the expressed protein must be solubilized and refolded prior to use. Display of recombinant silicatein from organic or inorganic scaffolds has also been used to achieve active silicateins in the form of protein-functionalized surfaces. For example, silicatein sequestered to inorganic surfaces via a polyhistidine tag (after a Ni-NTA polymer was grafted to a gold surface) led to demonstrations of silicatein-catalyzed synthesis of $SiO_{2(s)}$ from TEOS, $TiO_{2(s)}$ from TiBALDH, or $ZrO_{2(s)}$ from hexafluorozirconate ($ZrF_6^{2-}$) at the surface of the gold substrate (Schröder et al. 2007). Other studies, such as the demonstration of biocatalytic synthesis of gold nanoparticles, as well as a study showing $SnO_{2(s)}$ synthesis from sodium hexafluorostannate ($Na_2SnF_6$), were conducted in a system in which silicateins were grafted to titania nanorods via a polyglutamate tag (Schröder et al. 2007). In an alternate display approach that led to the creation of bacteria-based whole-cell biocatalysts, silicateins were expressed as fusions with the OmpA cell membrane protein to achieve silicatein display at the bacterial cell surfaces; these cells were able to catalyze synthesis of titanium phosphate from TiBALDH (Brutchey and Morse 2008).

## Genetic Engineering

In addition to site-directed mutagenesis (Weaver and Morse 2003) or insertion of short peptide tags at one end of the expressed silicatein (Schröder et al. 2007), further studies have explored more extensive mutational diversity across the functional mineral-synthesizing silicatein protein fold. Genetic engineering of cathepsin L, which is readily expressed as a soluble protein, has been used to produce cathepsin-silicatein hybrids that exhibit silicatein-like mineralization functionality, but that require no extra protein refolding step for their production (Patwardhan et al. 2010). This approach, while promising, has not yet achieved protein chimeras with the same hydrolytic activity as silicatein (i.e., yielding mineral product from reaction with substrates such as TEOS). More recent genetic engineering work has utilized strategies of directed enzyme evolution to select new functional silicateins by screening a large biotechnologically derived gene pool of variant silicateins (Bawazer et al. 2012). In this case, recombinant silicatein-encoding genes were anchored to the surfaces of polystyrene microbeads and compartmentalized in droplets of water-in-oil emulsions in the presence of bacterial extracts for in vitro protein expression. The system allowed expressed proteins to be captured on the surfaces of the same microbeads from which their encoding DNA was displayed. All beads were reacted with TEOS or TiBALDH and those beads displaying functional silicateins at their surfaces catalyzed formation of $SiO_{2(s)}$ or $TiO_{2(s)}$, respectively, causing production of mineral oxide coatings on the polymer bead scaffold. Mineralized beads were isolated using a cell-sorting instrument on the basis of their ability to scatter light more intensely than background beads. Once isolated, encoding genes were amplified and sequenced for further study. Two variants were identified that synthesized silica and titania, respectively, which were named silicatein X1 and silicatein XT. Both these variant silicateins exhibited a large degree of hydroxyl-associated mutations on their surfaces and were found to catalyze formation of crystalline materials (e.g., mixed silicates) that were inaccessible through similar in vitro reactions with the parent (naturally derived) silicateins.

## Silicatein Mimicry

After hydrogen bonding between serine and histidine active site residues was identified as a key feature to silicatein-controlled mineralization, a similar biochemical bonding strategy was implemented using a variety of nonbiological platforms, thus creating silicatein mimetic reaction systems for the low-temperature synthesis of semiconductors and other technological materials. The majority of these biomimetics have utilized the display of a nucleophilic group (e.g., -OH or -SH) with a hydrogen bonding acceptor group (e.g., -NH₃ or imidazole) displayed on chemical or structural scaffolds that permit hydrogen bonding between the two functional moieties. Scaffolds presenting the relevant chemical groups have included short-chained organic molecules, block co-polypeptides, non-peptide polymers, self-assembled monolayers (SAMs) displayed on gold nanoparticles, and SAMs patterned onto flat substrates (Brutchey and Morse 2008). In all cases, both the nucleophile and the proton acceptor moieties are required; when either is removed, mineralization does not occur. In the majority of these biomimetic systems, silica mineralization from TEOS was targeted, but in one, using SAMs on flat substrates, $GaOOH_{(s)}$ and $\gamma\text{-}Ga_2O_{3(s)}$ were produced from $Ga(NO_3)_3$. In this case, hydroxyl-terminated SAMs and imidazole-terminated SAMs were patterned in alternating stripes on a gold surface. In a result substantiating the importance of the hydroxyl-imidazole interaction, at early reaction time-points gallium-based minerals formed only on the lines that defined the boundary between hydroxyl- and imidazole-terminated stripes (Brutchey and Morse 2008). In the case of block co-polypeptides the structure-directing ability of silicatein was further mimicked through use of sulfhydryl-presenting cysteine as the nucleophile. When cysteines were oxidized to cystines (i.e., when disulfide bridges were formed), the structure of the polymer backbone was altered, which in turn altered the structure of the synthesized silica (from spherical to columnar microstructure) and demonstrated that the shape of the polymeric scaffold is critical for directing corresponding mineral shape (Brutchey and Morse 2008).

Further translation of silicatein's mechanisms has resulted in a simple but flexible method in which a chemical catalyst is diffused through the vapor phase into a solution of mineral precursor (Brutchey and Morse 2008). Example catalysts include ammonia, if the precursor is dissolved in water, or water vapor, if the liquid precursor is hydroscopic. Both inorganic thin films and nanoparticle products are achievable. Diffusional entry of the catalyst into solution at the air-liquid interface catalyzes precipitation of the inorganic material, which nucleates at the liquid surface and grows into the solution. Thus, vectorially constrained metal hydrolysis in the vicinity of a mineral templating surface (here, the air-liquid interface) mimics mechanisms utilized by silicatein. With this vapor diffusion–based approach, a variety of crystalline materials have been produced at low temperature, including (among others) cobalt hydroxide $(Co(OH)_2)$, zinc oxide (ZnO), and chromium phosphate $(CrPO_4)$ (Brutchey and Morse 2008).

## Future Prospects

The demonstrated ability to convert lessons learned from silicatein mineralization into new, nonbiological strategies for low-temperature mineral synthesis suggests that other mineralizing enzymes, once identified, may serve as additional laboratory models for biomimetic materials engineering. The demonstration that silicateins may be diversified and evolutionarily selected in vitro (Bawazer et al. 2012) further suggests that material-synthesizing enzymes may be obtained, in principle, for any material of interest for which an appropriate selection strategy can be designed. This evolutionary selection approach holds promise for not only developing new enzyme models, but also for developing technological materials and devices that are genetically encoded by DNA, just as the diverse biosilica sponge skeletons observed in nature are so encoded.

Silicatein's close evolutionary relationship with the cathepsin hydrolases suggests that enzymes other than silicatein are amenable to evolutionary selection for mineralizing functionality. This idea is supported by prior work showing that cathepsin and other hydrolase enzymes may exhibit or be imparted with mineralizing behavior (e.g., Patwardhan et al. (2010)). A combination of studies with both silicatein and other enzymes will be necessary to further reveal the complete set of mechanisms by which enzymatic mineralization operates. For example, initial studies have revealed that silicatein's catalytic rate is slow compared to other enzymes (Wang et al. 2012; Brutchey and Morse 2008), but the detailed nature of each step in the mechanistic pathway and the question of whether conventional enzyme descriptors such as $k_{cat}$ and $K_M$ are the most useful parameters for characterizing enzymatic mineralization remain to be determined. Further, an x-ray crystal structure of silicatein has not yet been obtained. Considering silicatein's ability to work on a wide range of chemically diverse metal species, as well as its ability to self-assemble into higher-order polymer structures, additional mechanisms of mineralization could still be discovered that may be novel among known enzyme functions. The discovery of silicatein itself (Shimizu et al. 1998) revealed for the first time that nature utilizes enzymes as direct mineral-synthesizing tools to control biomineralization. It appears that this family of proteins has yet to reveal its full set of mineralizing secrets.

## Cross-References

▶ Silica, Immunological Effects
▶ Silicon Transporters
▶ Titanium as Protease Inhibitor
▶ Titanium, Physical and Chemical Properties

## References

Bawazer LA et al (2012) Evolutionary selection of enzymatically synthesized semiconductors from biomimetic mineralization vesicles. Proc Natl Acad Sci. Accepted (in press)

Brutchey R, Morse D (2008) Silicatein and the translation of its molecular mechanism of biosilicification into low temperature nanomaterial synthesis. Chem Rev 108:4915–4934

Muller WEG et al (2009) Sponge spicules as blueprints for the biofabrication of inorganic-organic composites and biomaterials. Appl Microbiol Biotechnol 83:397–413

Murr M, Morse D (2005) Fractal intermediates in the self-assembly of silicatein filaments. Proc Natl Acad Sci 102:11657–11662

Patwardhan SV et al (2010) Silica condensation by a silicatein α homologue involves surface-induced transition to a stable structural intermediate forming a saturated monolayer. Biomacromolecules 11:3126–3135

Schröder HC et al (2007) Enzymatic production of biosilica glass using enzymes from sponges: basic aspects and application in nanobiotechnology (material sciences and medicine). Naturwissenschaften 94:339–359

Shimizu K, Cha J, Stucky G, Morse D (1998) Silicatein α: cathepsin L-like protein in sponge biosilica. Proc Natl Acad Sci 95:6234

Uriz MXAJ, Turon X, Becerro MA (2000) Silica deposition in Demosponges: spiculogenesis in Crambe crambe. Cell Tissue Res 301:299–309

Wang X et al (2012) Silicateins, silicatein interactors, and cellular interplay in sponge skeletogenesis: formation of glass fiber-like spicules. FEBS J 279:1721–1736

Weaver JC, Morse DE (2003) Molecular biology of demosponge axial filaments and their roles in biosilicification. Microsc Res Tech 62:356–367

## Silicic Acid Transporters in Plants

▶ Silicon Transporters

## Silicon

▶ Silicon Exposure and Vasculitis

## Silicon and Aquaporins

Gerd Patrick Bienert and François Chaumont
Institut des Sciences de la Vie, Universite catholique de Louvain, Louvain-la-Neuve, Belgium

## Synonyms

Aquaporin-mediated silicon transport; Major intrinsic proteins and silicon transport; Silicon channels

## Definition

Aquaporins or major intrinsic proteins (MIPs) are transmembrane channel proteins, which facilitate the passive and bidirectional diffusion of water and a variety of small and noncharged compounds across biological membranes. MIPs are found in organisms of all kingdoms of life and are present in all main subcellular membrane systems. The substrate specificity/spectra of aquaporins are highly isoform-dependent. Certain isoforms of plants were shown to facilitate the transmembrane diffusion of the metalloid silicic acid, a derivative of silicon. All organisms face the challenge to handle considerable variations in the concentration of metalloids to which they are exposed to. This happens when facing the demand to acquire sufficient amounts for their metabolism or, conversely, the necessity to extrude them to prevent toxicity. This is achieved through homeostatic processes that require, among others, aquaporin-facilitated diffusion across membranes at the cellular level.

## Essentiality and Benefits of Silicon for Living Organisms

Silicon (Si) is the second most abundant element in the Earth's crust, after oxygen. Silicon was shown to be essential for animals and diatoms. In humans, silicon is important for bone structure, and it is present in nearly all connective tissues, indicating its major functions in stability and elasticity (Jugdaohsingh 2007). In higher plants, silicon is not essential. However, it was shown to be highly beneficial for plant growth in both monocot and dicot species (Epstein 1999). Silicon represents a structural component in cell walls (physical barrier and stabilizer) that protects plants from various biotic and abiotic stresses. Silicon increases resistance of plants to herbivores and to diseases caused by fungi and bacteria. The benefit of silicon accumulation to some plant species is also seen as an enhanced tolerance to salt stress, metal toxicity, nutrient imbalance, drought, extreme temperature variations, and UV irradiation (Ma 2010). Whether these effects are due to structural or other metabolic functions is not known yet.

## Identification of the First Silicon Transporter in Higher Plants: The Aquaporin OsNIP2;1

In higher plants, especially in gramineous plants like rice (*Oryza sativa*), silicon can account for up to 10% of the shoot dry weight. Silicon was shown to be taken up by the roots in the chemical form of silicic acid $(Si(OH)_4)$, a noncharged molecule at physiological pH ranges. Silicic acid polymerizes at concentrations above 2 mM and forms polysilicic acid. Until recently, it was thought that silicic acid crosses plant membranes by simple passive diffusion, as no transport protein was identified. However, experimental data provided evidences that silicic acid is taken up against chemical concentration gradients suggesting the involvement of active transporters. The first silicon transport protein in higher plants was identified in 2006 by Ma and coworkers (Ma et al. 2006). This silicic acid transporter was shown to be an aquaporin protein, which was a very exciting discovery for both the aquaporin and the plant nutritional research fields. Even though it was known that the movement of small uncharged solutes across plant membranes relied on aquaporins, the aquaporin isoforms characterized so far were shown to mainly fulfill physiological roles in water conductance, gas exchange, and signal transduction but not in metalloid transport (Bienert and Chaumont 2011).

Rice accumulates particularly high amounts of silicon, and both its growth and yield benefit enormously from sufficient silicon supply. Rice seeds from a mutated population were germinated in medium containing germanium dioxide. Germanium and silicon are chemically very similar metalloids that plant roots cannot discriminate in terms of uptake. However, while silicon is beneficial for plant growth, germanium is toxic at elevated concentrations. In a forward genetic screen, the rice mutant *lsi1* (abbreviation of "*low silicon 1*") was isolated and found to be both resistant to elevated germanium levels and defective in silicic acid uptake (Ma et al. 2006). The *lsi1* mutant suffers from reduced grain yield and an increased susceptibility to pests and diseases. Through fine mapping, *lsi1* was cloned and found to encode a protein belonging to the nodulin26-like intrinsic protein (NIP) aquaporin subfamily, namely, *OsNIP2;1*. A single amino acid residue substitution from alanine to threonine in OsNIP2;1 caused the phenotypes of the *lsi1* mutant. A rigorous and detailed molecular, functional, and physiological analysis of OsNIP2;1 clearly confirmed its function as a crucial channel protein responsible for silicon uptake into rice roots (Ma et al. 2006). OsNIP2;1 is expressed in the main and lateral roots, but not in root hairs which is in accordance with the important uptake sites for silicic acid identified in rice. Furthermore, *OsNIP2;1* is transcriptionally upregulated under low silicon conditions. OsNIP2;1 is localized in both the exodermis and endodermis, where the Casparian strips prevent the apoplastic (cell wall) transport of nutrients into the vascular tissue (Ma et al. 2006).

NIPs cluster into three subgroups, NIPI, NIPII, and NIPIII, based on the amino acid residue composition of their aromatic/arginine selectivity filter (see ▶ Aquaporins and Transport of Metalloids). The physicochemical properties of these residues are involved in the determination of the substrate specificity of the pore (Ma et al. 2008). Until now, all silicic acid-permeable NIP aquaporins belong to the NIPIII subgroup whose selectivity filter consists of glycine, serine, glycine, and arginine (Ma 2010). The relative small size of these residues potentially forms a larger constriction region compared with other NIP subgroups or microbial and mammalian aquaglyceroporins. The pore diameter and channel path of NIPIIIs were modeled to be among the widest of all aquaporins. This is in accordance with the ability to channel silicic acid molecules which are larger than typical aquaporin substrates such as water or glycerol (Ma 2010).

NIPI proteins have been reported to transport water, glycerol, lactic acid, and arsenite, while NIPII proteins are permeable to solutes such as urea or the metalloids boric acid and arsenite (Bienert and Chaumont 2011). Si(OH)$_4$ is a tetrahydroxylated metalloid species and a rather big noncharged molecule (about 4.38 Å in diameter) compared to other metalloid species like boric acid and arsenite which are also transported via NIPs. OsNIP2;1 was shown to be also permeable for boric acid, arsenite, monomethylarsonic acid, and dimethylarsinic acid, which are other hydroxylated metalloids present at physiological pH ranges as undissociated species (Ma 2010; Ma et al. 2008).

## NIPs are Important for Silicon Uptake and Allocation in Plants

After OsNIP2;1 had been identified as a silicon influx transporter in rice, Ma and coworkers resolved the silicic acid uptake mechanisms in two other gramineous plant species, maize (*Zea mays*) and barley (*Hordeum vulgare*) (Chiba et al. 2009; Mitani et al. 2009a). Like OsNIP2;1, ZmLsi1/ZmNIP2;1 and HvLsi1/HvNIP2;1 were identified to be responsible for silicic acid uptake in these species. However, their cell-type specificities differ. In addition, the expression level of both NIPs is unaffected by silicon soil concentrations (Chiba et al. 2009; Mitani et al. 2009a), suggesting that silicon homeostasis is differentially regulated in different plant species, a fact that could explain the variable silicon contents (Ma 2010).

OsNIP2;2 (OsLsi6) from rice, which is a homolog of OsNIP2;1 (OsLsi1), also facilitates silicic acid diffusion in the oocyte system. It is expressed both in roots and shoots and localized in the adaxial side of the xylem parenchyma cells of leaf sheaths and leaf blades. *Osnip2;2* knockout does not affect the uptake of silicon by the roots but modifies the deposition pattern of silicon in leaf sheaths and blades. The abaxial epidermis cells are silicified in the *nip2;2/lsi6* mutant, whereas this happens infrequently in wild type plants. Furthermore, the lack of OsNIP2;2 leads to an increase in excretion of silicic acid in the guttation solution. These results suggest that mutation of *OsNIP2;2* results in an alteration of the silicic acid allocation to specific cells. OsNIP2;2 is responsible for unloading of silicic acid out of the xylem and for intervascular transfer of silicic acid at the nodes, which are crucial to ensure silicon allocation to the panicle (Yamaji et al. 2008; Yamaji and Ma 2009). The importance of such transmembrane xylem-associated transporter is best exemplified by the fact that, in rice, more than 90% of the silicic acid taken up by the roots is translocated via the xylem to the shoots.

While NIPI and NIPII isoforms seem to be present in all higher plant species, NIPIII isoforms like *OsNIP2;1*, *OsNIP2;2*, *ZmNIP2;1*, and *HvNIP2;1* are mainly found in gramineous plant species belonging to the *Poaceae* family which shows very high silicon accumulation (>4% Si of their dry weight). All plants contain silicon, but its shoot accumulation level greatly differs amongst them. Whereas plant species belonging to the *Cucurbitales*, *Urticales*, and *Commelinaceae* show an intermediate silicon accumulation (2–4% Si of their dry weight), most other species show a very low accumulation (Ma and Yamaji 2008; Mitani et al. 2011). These differences in silicon accumulation have been attributed to the capacity of the roots to absorb silicon and, therefore, to depend on the presence of specific NIPIII isoforms.

Ma and coworkers also discovered and functionally characterized silicon transporters of "high-silicon" eudicot plants (*Cucurbitales*) such as pumpkin (*Cucurbita moschata*) and zucchini (*Cucurbita pepo*). NIPs from two pumpkin cultivars differing in their silicon accumulation capacity were cloned and analyzed (Mitani et al. 2011). Whereas CmNIP2;1 from the high silicic acid uptake capacity plants transports silicic acid and functionally complements the rice *lsi1* defective mutant (*nip2;1*), the CmNIP2;1 isoform from the low silicic acid uptake capacity plants does not. A single amino acid residue substitution between the two NIP homologs of both cultivars accounted for the different substrate selectivity (Mitani et al. 2011). The impaired silicon uptake capacity of the pumpkin with the nonfunctional NIP channel is of agronomical importance as it is used in Japanese horticultural industry to obtain bloomless cucumber fruits (*Cucumis sativus*). Bloom is a white powder on the surface of cucumber fruits, which is mainly composed of silicon dioxide. Bloomless cucumber is popular in Japan due to its attractive appearance and is generated by grafting cucumber on some specific pumpkin cultivars.

## The Team Play of NIP and Lsi2 Transport Proteins Regulates Plant Silicon Uptake and Distribution

In addition to the crucial role of NIP aquaporins in silicic acid uptake and allocation, a second type of essential silicon transporters, the so-called Lsi2 proteins, were identified (Ma et al. 2007). These proteins are predicted to have 11 transmembrane helices and show homology to citrate anion efflux transporters. Lsi2 proteins are directional transporters as they only mediate the efflux of silicic acid out of the cells driven by the proton gradient (Ma et al. 2007). Lsi2 and silicic acid transporting NIPs were demonstrated to cooperatively function together to ensure the crucial silicon uptake and allocation for plants (Ma 2010). For example, rice OsNIP2;1/OsLsi1 is localized to the distal membrane domains of root endodermis and exodermis cells, whereas OsLsi2 is localized in the opposite plasma membrane domains facing the vascular tissue. While OsNIP2;1 mediates the diffusion of silicic acid along its concentration gradient into these root cells, it provides the substrate for Lsi2 proteins that subsequently transport silicon out of the cells into the vascular tissue, which maintains the concentration gradient across the membrane (Ma 2010).

Similar spatial localization of NIP and Lsi2 proteins was also shown in maize, barley, and pumpkin, all plants requiring high amounts of silicon (Mitani et al. 2009b; Mitani-Ueno et al. 2011). Furthermore, such specific and cooperative coupling of a passive channel and an active transporter was also shown to be crucial to regulate the uptake and allocation of the essential metalloid boron in Arabidopsis (▶ Boron and Aquaporins).

## Concluding Remarks

NIP aquaporins of several higher plant species have been demonstrated to be crucially involved in silicon uptake and distribution. Molecular and physiological characterization of other NIP isoforms in different plant species will provide a more complete understanding of the mechanisms regulating silicon transport in plants. As silicon transport proteins are unknown in mammals, it has to be resolved whether some animal aquaporins might play a similar role to NIPs in silicon transport processes.

## Funding

This work was supported by grants from the Belgian National Fund for Scientific Research (FNRS), the Interuniversity Attraction Poles Programme–Belgian Science Policy, and the "Communauté française de Belgique–Actions de Recherches Concertées". GPB was supported by a grant from the FNRS.

## Cross-References

▶ Aquaporins and Transport of Metalloids
▶ Arsenic and Aquaporins
▶ Boron and Aquaporins
▶ Selenium and Aquaporins

## References

Bienert GP, Chaumont F (2011) Plant aquaporins: roles in water homeostasis, nutrition, and signalling processes. Springer, Berlin

Chiba Y, Mitani N, Yamaji N et al (2009) HvLsi1 is a silicon influx transporter in barley. Plant J 57:810–818

Epstein E (1999) Silicon. Annu Rev Plant Physiol Plant Mol Biol 50:641–664

Jugdaohsingh R (2007) Silicon and bone health. J Nutr Health Aging 11:99–110

Ma JF (2010) Silicon transporters in higher plants. Landes Bioscience-Springer, New York

Ma JF, Yamaji N (2008) Functions and transport of silicon in plants. Cell Mol Life Sci 65:3049–3057

Ma JF, Tamai K, Yamaji N et al (2006) A silicon transporter in rice. Nature 440:688–691

Ma JF, Yamaji N, Mitani N et al (2007) An efflux transporter of silicon in rice. Nature 448:209–212

Ma JF, Yamaji N, Mitani N et al (2008) Transporters of arsenite in rice and their role in arsenic accumulation in rice grain. Proc Natl Acad Sci USA 105:9931–9935

Mitani N, Yamaji N, Ma JF (2009a) Identification of maize silicon influx transporters. Plant Cell Physiol 50:5–12

Mitani N, Chiba Y, Yamaji N et al (2009b) Identification and characterization of maize and barley Lsi2-like silicon efflux transporters reveals a distinct silicon uptake system from that in rice. Plant Cell 21:2133–2142

Mitani N, Yamaji N, Ago Y et al (2011) Isolation and functional characterization of an influx silicon transporter in two pumpkin cultivars contrasting in silicon accumulation. Plant J 66:231–240

Mitani-Ueno N, Yamaji N, Ma JF (2011) Silicon efflux transporters isolated from two pumpkin cultivars contrasting in Si uptake. Plant Signal Behav 6:991–994

Yamaji N, Ma JF (2009) A transporter at the node responsible for intervascular transfer of silicon in rice. Plant Cell 21:2878–2883

Yamaji N, Mitatni N, Ma JF (2008) A transporter regulating silicon distribution in rice shoots. Plant Cell 20:1381–1389

## Silicon and Plant Pathogen

▶ Silicon-Mediated Pathogen Resistance in Plants

## Silicon Channels

▶ Silicon and Aquaporins

## Silicon Exposure and Vasculitis

Jan Willem Cohen Tervaert
Clinical and Experimental Immunology, Maastricht University, Maastricht, The Netherlands

### Synonyms

ANCA; Silicon; Vasculitis

### Definition

Silicon (Si) has a molecular mass of 28 Da. In nature, silicon is found as silicon dioxide (silica, $SiO_2$) or in a variety of silicates (e.g., in talc or asbestos). Furthermore, silicon is present in silicones, a polymer of $[SiO(CH_3)_2]_n$. Silica is abundant in rock, sand, and soil. Exposure to silicon-containing compounds has long been recognized as dangerous for humans, especially because inhalation of crystalline silica can result in serious occupational lung fibrosis (i.e., silicosis). Silicon exposure is associated with different systemic autoimmune diseases such as systemic lupus erythematosus, rheumatoid arthritis, progressive systemic sclerosis, and vasculitis. Within the spectrum of autoimmune diseases that can be induced by silicon-containing compounds, probably the strongest effect is on the induction of vasculitis that is caused by antineutrophil cytoplasmic autoantibodies (ANCA). The association is especially strong with vasculitis in which ANCA with specificity for myeloperoxidase is found. Silica inhalation causes an influx of neutrophils in the lung and subsequently apoptosis and possibly NETosis of these neutrophils resulting in the local release of myeloperoxidase. In the context of the strong adjuvant effect of silica, this may result in the induction of autoimmunity against myeloperoxidase and vasculitis.

### Introduction

Silicon (Si) is the major constituent of the earth's crust. Si has a molecular mass of 28 Da. The crystalline forms of silicon are gray and similar to those in diamond, albeit not as strong as the crystalline forms as found in diamond.

In nature, silicon is found as silicon dioxide (silica, $SiO_2$) or in a variety of silicates (e.g., in talc or asbestos). Furthermore, silicon is present in silicones, a polymer of $[SiO(CH_3)_2]_n$. Silica is abundant in rock, sand, and soil. Exposure to silicon-containing compounds has long been recognized as dangerous for humans, especially because inhalation of crystalline silica can result in serious occupational lung fibrosis (i.e., silicosis). Crystalline silica is present in quartz and flint and is an important constituent of granite, sandstone, and slate. Importantly, silica also occurs in a noncrystalline (amorphous) form, and also this form of silica, e.g., in grain dust-containing silicious fibers, poses a health hazard.

Silicon exposure is associated with different systemic autoimmune diseases such as systemic lupus erythematosus, rheumatoid arthritis, progressive systemic sclerosis, and vasculitis.

Vasculitis is an inflammatory disorder of blood vessels. Vessels of any type in any organ can be affected resulting in a wide variety of signs and symptoms. Vasculitis can have different causes, such as infections, drugs, and/or malignancies. The most severe systemic forms of vasculitis, however, are the so-called primary vasculitides in which no underlying cause is demonstrated. These forms of vasculitis are believed to be autoimmune diseases.

Autoimmune diseases affect approximately 5–10% of the developed world population and are a significant cause of morbidity and mortality. The etiopathogenesis of autoimmune diseases comprises a combination of genetic, immune, hormonal, and environmental factors. During the last decades, accumulating data demonstrate that the burden of autoimmune diseases is rising, possibly due to the increased presence of environmental factors such as silica

(Cohen Tervaert et al. 2006). Primary (autoimmune) vasculitides can be subdivided into large vessel vasculitis, medium-sized vessel vasculitis, and small vessel vasculitis (Wilde et al. 2011). Additionally, this latter form can be subdivided into either small vessel vasculitis due to antineutrophil cytoplasmic autoantibodies (ANCA; ANCA-associated vasculitis, AAV) or immune complex-mediated, non-ANCA-associated vasculitis. Within the spectrum of vasculitides, silicon exposure is mainly related to AAV. In patients with silicosis, i.e., patients with chest X-rays showing rounded opacities and a history of exposure to silica, the prevalence of systemic autoimmune diseases is clearly increased and an up to 25-fold increased risk to develop AAV has been demonstrated in patients with well-documented silicosis (Makol et al. 2011).

## Silicon Exposure

Silicon exposure is associated with silicosis, lung cancer, and chronic renal failure. Furthermore, silicon exposure is related to autoimmune diseases. Silicon exposure, however, does not cause autoimmune diseases itself but may help to set the stage for these diseases to occur.

The research connecting silica dust to autoimmune disease goes back almost 100 years. Initially, research focused on occupational exposure of silica in miners and various construction trade workers since it was demonstrated that different autoimmune diseases were more frequently occurring in these people.

The permissible exposure limit for respirable silica is approximately 0.10 mg/m$^3$. Approximately one to three million American workers are potentially exposed to crystalline silica each year (Table 1). At least 10% of these may have dangerously high silicon exposure levels (Parks et al. 1999). Crystalline silica exposure is especially high among people working in mining, metal services (abrasive blasting), construction (masonry and highway repair), and iron foundries.

Silica is formed from oxygen and silicon. $SiO_2$ occurs in a noncrystalline (amorphous) or a crystalline form. Crystalline silica is found in seven different forms (polymorphisms) of which quartz, cristobalite, and tridymite are the most common (Mulloy 2003). Quartz is an abundant component of soil and rock. When rock and sand is used or processed (i.e., mining, quarrying, drilling, sand blasting, tunneling operations), workers can be exposed to high levels of airborne dust with respirable crystalline silica. Also, nonindustrial silicon exposure may occur since silica is part of the small particulate fraction of air pollution. Moreover, silica content in the air is increased after disasters in which major buildings are destructed.

**Silicon Exposure and Vasculitis, Table 1** Industries and occupations with silica exposure

| |
|---|
| Automobile repair |
| Construction activities (buildings, highways, tunnels) |
| Drilling |
| Farming |
| Iron foundries |
| Mining |
| Metal services |
| Painters (fillers) |
| Production of |
|   Abrasives |
|   Cement |
|   Ceramics |
|   Concrete |
|   Cosmetics |
|   Dental supplies |
|   Detergents |
|   Electric/electronic machinery |
|   Glass |
|   Insulation products |
|   Jewels |
|   Rubber |
|   Soaps |
|   Textiles (cotton or wool) |
| Quarrying |
| Sandblasting |
| Shipbuilding repair |
| Stonecutting |

Interestingly, also exposure to amorphous silica has been related to the occurrence of AAV (Gregorini et al. 1997). Groundwater contains silicon as silicic acid ($H_4SiO_4$). Crystalline silica dissolves in water. The dissolved silica is named (ortho)silicic acid, and this form is readily absorbed by plants and sponges. Soil-grown plants, therefore, may contain high concentrations of Si. The form in which Si is ultimately deposited and accumulated in plant material is mainly amorphous silica, i.e., $SiO_2.nH_2O$ or opal. All soils contain about 2–3% amorphous silica. Deposition of amorphous silica in plants occurs in cell lumens, cell walls, and intercellular spaces or external layers. Hence, farmers can also be exposed to high levels of silica, i.e., agricultural exposure, especially from harvesting crops.

## ANCA-Associated Vasculitis

AAV is a life-threatening autoimmune disease characterized by necrotizing vasculitis of small- and medium-sized vessels. Vasculitis occurs in many different organs. About 70–80% of patients with AAV suffer from renal vasculitis, which is histologically characterized as pauci-immune necrotizing crescentic glomerulonephritis. ANCAs with specificity for either myeloperoxidase (MPO) or proteinase-3 (PR3) are hallmarks of AAV. AAV comprises three disease types: granulomatosis with polyangiitis (GPA, Wegener's), eosinophilic granulomatosis with polyangiitis (EGPA, Churg–Strauss syndrome), and microscopic polyangiitis (MPA) (Cohen Tervaert et al. 2012). The disease types differ with respect to clinical manifestations and histological findings. Granulomatous inflammation is found in GPA and EGPA, but not in MPA. Furthermore, neutrophils are abundantly found in GPA- and MPA-associated inflammation, whereas in EGPA eosinophils dominate the inflammatory infiltrate. MPA and EGPA are mostly associated with MPO-ANCA, whereas in most patients with GPA, PR3-ANCA is found. For classification of patients, ANCA serotype (either MPO-ANCA or PR3-ANCA) is more important than classifying patients according to their clinical subtype since genetics, clinical manifestations, and response to therapy are more related to ANCA serotype than to clinical subtype (Cohen Tervaert et al. 2012).

The pathophysiology of AAV is complex (Fig. 1). Both the humoral and the cellular immune system are involved. ANCAs themselves are pathogenic. In many patients, ANCA levels parallel disease activity. ANCAs bind to neutrophils, monocytes, and endothelial cells. After binding to membrane-bound MPO or PR3, ANCAs promote the degranulation of neutrophils and monocytes facilitating endothelial damage. The endothelium is activated and neutrophil adherence is enhanced. The initial damage results in a cascade of events such as leukocyte tissue infiltration, T cell-driven granuloma formation, and further damage (Fig. 1). ANCA pathogenicity has been demonstrated in animal models (Wilde et al. 2011). When MPO-knockout mice are immunized with murine MPO, MPO-ANCA is induced. Transferring IgG-containing MPO-ANCA to wild-type mice results in AAV. Also, in rats, MPO-ANCA induction results in AAV. Finally, human anti-PR3 antibodies can induce acute vasculitis in mice with a human immune system. This model of PR3-ANCA-associated AAV is useful in dissecting mechanisms of microvascular injury. Granulomatous inflammation, however, is not induced in these animal models.

Genetic, geographic, and seasonal factors are involved in the induction of AAV (de Lind van Wijngaarden et al. 2008). Furthermore, environmental factors such as vitamin D, *Staphylococcus aureus*, cocaine, drugs (especially antithyroid drugs), and silica exposure are postulated to play an important role during disease onset and/or reactivation of the disease.

Since ANCAs are of major importance for disease induction and their interaction with neutrophils are crucial for vascular damage, whereas T cells enhance autoantibody production and drive tissue inflammation (Fig. 1), therapy needs to interfere with these pathogenic mechanisms. The outcome of AAV is fatal if left untreated, but outcome improved dramatically when patients are treated with a combination of high-dose corticosteroids and cyclophosphamide. In severe cases, plasmapheresis to remove pathogenic antibodies is added to this combination. Recently, it has been demonstrated that cyclophosphamide can be replaced by the selective B cell depleting antibody Rituximab, a monoclonal antibody directed against CD20 (Wilde et al. 2011).

## Silicon Exposure and ANCA-Associated Vasculitis

More than 50 years ago, it was demonstrated that pulmonary silicosis was sometimes associated with renal vasculitis (Cohen Tervaert et al. 1998). Furthermore, in 1973, two patients with pulmonary silicosis and a pulmonary–renal syndrome due to small vessel vasculitis were described. These early cases suggested the presence of AAV in an era in which ANCA testing was not yet possible, since testing for MPO-ANCA and PR3-ANCA started in 1988 (Cohen Tervaert et al. 2012). During the third ANCA workshop in Washington, 1990, a group from Angers, France,

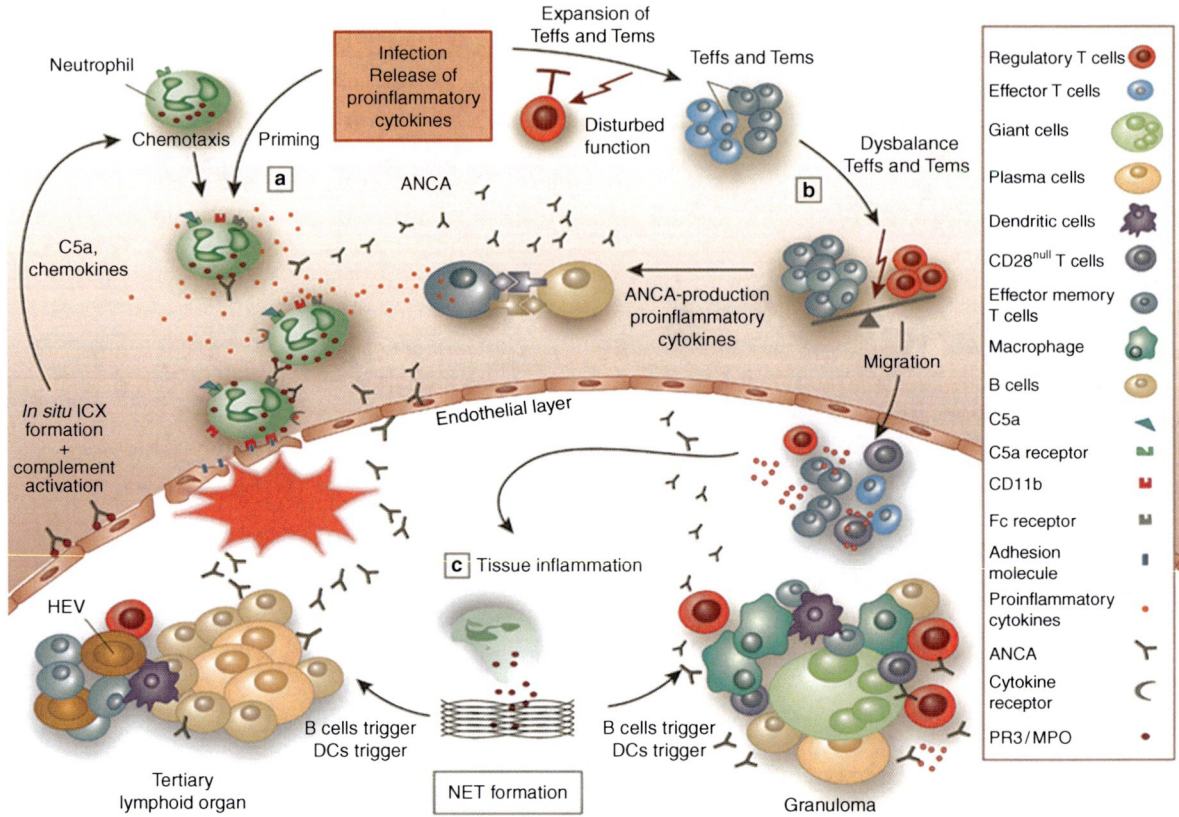

**Silicon Exposure and Vasculitis, Fig. 1** *Pathophysiology of antineutrophil cytoplasmic antibody (ANCA)-associated vasculitis (AAV).* Both neutrophils and T cells promote necrotizing vasculitis in AAV. Silica exposure or infections are the starting point; they trigger priming of neutrophils (*a*), upregulation of adhesion molecules on endothelial cells, and expansion of circulating effector T cells (*b*). Primed neutrophils show increased surface expression of ANCA antigens and adhesion molecules. ANCA binding activates the neutrophil in the following ways: (1) enhancing vessel wall adherence and transmigration capacity, (2) production and release of oxygen radicals, and (3) degranulation and release of enzymes including myeloperoxidase (MPO) and proteinase-3 (PR3) (*a*). Transient immune complexes are formed locally by binding of ANCA to PR3/MPO sticking to endothelial cells. Subsequently, complement is activated, which further promotes neutrophil degranulation. This all adds to the development of necrotizing vasculitis. The expanded effector memory T cells (Tems) are not sufficiently regulated by regulatory T cells (Tregs, *b*). This leads to disbalance in the homeostasis of Tregs and Tems, resulting in further release of proinflammatory cytokines promoting neutrophil priming (*a*); moreover, ANCA production is enhanced by further T cell/B cell interaction. (*c*) Expanded circulating Tems migrate into target organs such as the lungs or the kidney. Within tissues, Tems drive granuloma formation, which is considered an "executioner" of tissue destruction. Granulomas are composed of numerous cell types such as T cells, B cells, giant cells, and dendritic cells (DCs). Moreover, ANCA production occurs in granulomas. Possibly, tertiary lymphoid organs (TLOs) are "local controllers" of tissue inflammation, as induction of Tregs is thought to take place in TLOs. Neutrophil extracellular trap (NET) formation occurs in lesions as a consequence of NETosis. DNA and serine proteases are deployed in these NETs. NET-derived products activate DCs and B cells by sensing via Toll-like receptors (TLRs). *HEV* high endothelial venules, *ICX* immune complexes (Reproduced with permission. From Wilde et al. 2011)

reported on the occurrence of MPO-ANCA-related nephropathy in three patients with pulmonary silicosis. Later on, during the fourth ANCA workshop in Lubeck, 1992, Gregorini reported that MPO-ANCA-related AAV not only occurred in patients with pulmonary silicosis but that MPO-ANCA-associated AAV could also be found in silica-exposed patients without pulmonary silicosis (Gregorini et al. 1997). Furthermore, Gregorini et al. evaluated all their male patients with ANCA-associated renal vasculitis and found that 12 of 37 patients (33%) had significant silica exposure. Otherwise, a group from Sevilla, Spain, reported in 1996 that 14 of 52 (27%) patients chronically exposed to silica at a scouring powder factory developed MPO-ANCA. Similar findings have been reported in workers exposed to asbestos, another silicon-containing mineral. Importantly, ANCA is more often found in patients with pulmonary silicosis compared to patients with silica exposure but without pulmonary silicosis (de Lind van Wijngaarden et al. 2008). Furthermore, MPO-ANCA in these patients may be present without evidence of AAV. Nowadays, many patients with AAV and silica exposure are described.

Most silica-exposed patients have MPO-ANCA, but a few with PR3-ANCA-associated AAV are described as well (Gregorini et al. 1997; Cohen Tervaert et al. 1998). In 1995, silica exposure was related to PR3-ANCA-associated GPA. In a small study from Antwerp, Belgium, it was found that seven of 16 patients with GPA clearly had exposure to silica. The clinical picture of silica-exposed AAV, however, is that of MPA or EGPA (both forms of AAV that are more associated with MPO-ANCA) and only infrequently that of GPA (which is more associated with PR3-ANCA).

Interestingly, an increase of MPO-associated AAV occurred after a cargo plane of the El Al airline crashed into two flats in the Bijlmermeer, Amsterdam, the Netherlands, in October 1992. Later, Yashiro et al. reported a dramatic increase of MPO-ANCA-associated glomerulonephritis after the 1995 earthquake in Kobe, Japan (Yashiro et al. 2000). Both disasters resulted in destruction of large urban buildings with probably a dramatic increase of silica exposure resulting in MPO-ANCA-associated AAV.

To find epidemiological evidence for a relationship between silica exposure and AAV, several case-control studies have been performed and consistently demonstrated more silica exposure in cases than in controls. Among patients with AAV, 20–70% have been exposed to silica (de Lind van Wijngaarden et al. 2008). As expected, the association between silica and MPO-ANCA-associated vasculitis was significantly more stronger than with PR3-ANCA (Hogan et al. 2007). Furthermore, duration of silica exposure was found to be more important than intensity of silica exposure in these cases. Only few patients have pulmonary silicosis at the time of diagnosis of AAV. Moreover, many patients with silica-exposed AAV do not have any pulmonary lesions. In AAV patients, silicon exposure is related either to mineral silica or to biologic forms of silica. Infrequently, however, asbestos exposure is found in the AAV patients. Finally, a patient with AAV occurring 30 years after silicone breast implantation has been described. From these epidemiological studies, it became clear that the most important risk factor for developing AAV due to silica exposure is, however, working as a farmer and/or being exposed to livestock.

## Hypothetical Mechanisms by Which Silicon Exposure Results in AAV

The demonstrated association between AAV and silica exposure does not mean that there is a causal relation between the two. The following scenario, however, could be operative. Inhaled silica particles are captured by alveolar macrophages resulting in entrapment within lysosomes. Subsequently, these macrophages are activated resulting in the production of cytokines, e.g., interleukin-1β, reactive oxygen species (ROS), and reactive nitrogen species. In the end, this leads to apoptosis of alveolar macrophages resulting in the release of silica particles that can be taken up once again by other alveolar macrophages. Exposure to inhaled silica particles also leads to a massive production of interleukin-17 resulting in an influx of neutrophils that are activated and produce ROS and release MPO and PR3. Additionally, the silica particles are transported to the regional lymph nodes resulting in a pronounced adjuvant effect. Silica particles are well-known potent stimulators of lymphocytes. Silica particles induce a type 2 inflammatory response characterized by increases of IgE and IgG1. Furthermore, T cells are chronically activated resulting in chronic elevations of soluble interleukin-2 receptor levels in the circulation. T cells may be chronically activated because negative regulators are dysfunctional. Firstly, significantly lower expression of programmed cell death 1 (PD-1) and cytotoxic T lymphocyte antigen-4 were demonstrated on T cells of silica-exposed workers when compared to nonexposed workers (Rocha et al. 2011). Since dysregulation of the PD-1/PD-1ligand pathway may be involved in the pathogenesis of AAV (Wilde et al. 2012), this may be important. Secondly, also regulatory T cells are dysfunctional after silica particle inhalation (Lee et al. 2012). The combination of a strong adjuvant effect and the presence of the antigens, i.e., MPO and PR3, may result in ANCA production and subsequently AAV. The release of MPO and PR3 via apoptosis may make the enzymes antigenic, and it has been postulated that apoptotic antigens are the natural targets for many autoantibodies (Cohen Tervaert et al. 1998). Recently, however, it has been demonstrated that antigens released from NETotic neutrophils may be an ever stronger inducer of ANCA. Whether silica particles induce NETosis is, however, not yet studied. Other crystals such as monosodium urate crystals, however, have been clearly demonstrated to induce NETosis.

## Conclusion

Exposure to silicon, mainly in its form as silicon dioxide (silica), may result in autoimmunity. Especially, people working in mining, metal services, construction, and iron foundries are at risk since they may be exposed during a long period to high levels of crystalline silica. In addition, however, farmers may be exposed to amorphous silica especially during harvest. Importantly, also during disasters such as an earthquake, silica exposure is increased resulting in increases of autoimmunity. Not only silica may induce autoimmunity but also exposure to other silicon-containing compounds, e.g., asbestos, may be detrimental. Within the spectrum of autoimmune diseases that can be induced by silicon-containing compounds, probably the strongest effect is on the induction of vasculitis that is caused by antineutrophil cytoplasmic autoantibodies (ANCA). The association is especially strong with vasculitis in which ANCA with specificity for myeloperoxidase is found. Silica inhalation causes an influx of neutrophils in the lung and subsequently apoptosis and possibly NETosis of these neutrophils resulting in the local release of ANCA antigens. In the context of the strong adjuvant effect of silica, this combination of events may be responsible for the induction of autoimmunity against these ANCA antigens which finally results in the occurrence of vasculitis.

## Cross-References

▶ Silica, Immunological Effects
▶ Silicon, Biologically Active Compounds
▶ Silicon, Physical and Chemical Properties
▶ Silicosis

## References

Cohen Tervaert JW et al (1998) Silicon exposure and vasculitis. Curr Opin Rheumatol 10:12–17

Cohen Tervaert JW et al (2006) Principles and methods for assessing autoimmunity associated with exposure to chemicals, vol 236, Environmental health criteria. International Programme on Chemical Safety/World Health Organization, Geneva, pp 122–130

Cohen Tervaert JW et al (2012) Antineutrophil cytoplasmic autoantibodies: how are they detected and what is their use for diagnosis, classification and follow-up? Clin Rev Alleg Immunol. doi:10.1007/s12016-012-8320-4

De Lind van Wijngaarden RA et al (2008) Hypotheses on the etiology of antineutrophil cytoplasmic autoantibody-associated vasculitis: the cause is hidden, but the result is known. Clin J Am Soc Nephrol 3:237–252

Gregorini G et al (1997) ANCA-associated diseases and silica exposure. Clin Rev Allerg Immunol 15:21–40

Hogan SL et al (2007) Association of silica exposure with antineutrophil cytoplasmic autoantibody small-vessel vasculitis: a population-based, case-control study. Clin J Am Soc Nephrol 2:290–297

Lee S et al (2012) Environmental factors producing autoimmune dysregulation – chronic activation of T cells caused by silica exposure. Immunobiology 217:743–748

Makol A et al (2011) Prevalence of connective tissue disease in silicosis (1985–2006) – a report from the state of Michigan surveillance system for silicosis. Am J Ind Med 54:255–262

Mulloy KB (2003) Silica exposure and vasculitis. Environ Health Perspect 111:1933–1938

Parks CG et al (1999) Occupational exposure to crystalline silica and autoimmune disease. Environ Health Perspect 107(S5):793–802

Rocha MC et al (2011) Genetic polymorphisms and surface expression of CTLA-4 and PD-1 on T cells of silica-exposed workers. Int J Hyg Environ Health. doi:10.1016/j.ijheh.2011.10.010

Wilde B et al (2011) New pathophysiological insights and treatment of ANCA-associated vasculitis. Kidney Int 79:599–612

Wilde B et al (2012) Aberrant expression of the negative costimulator PD-1 on T cells in granulomatosis with polyangiitis. Rheumatology 51:1188–1197

Yashiro M et al (2000) Significantly high regional morbidity of MPO-ANCA-related angitis and/or nephritis with respiratory tract involvement after the 1995 great earthquake in Kobe (Japan). Am J Kidney Dis 35:889–895

# Silicon Nanowires

Yao He[1], Chunhai Fan[2] and Hong-Yuan Chen[3]
[1]Institute of Functional Nano & Soft Materials, Soochow University, Jiangsu, China
[2]Laboratory of Physical Biology, Shanghai Institute of Applied Physics, Shanghai, China
[3]National Key Laboratory of Analytical Chemistry for Life Science, School of Chemistry and Chemical Engineering, Nanjing University, Nanjing, China

## Synonyms

Biosensors; Nanostructures

## Definition

Silicon nanowires (SiNWs) are a type of one-dimensional silicon-based nanostructures with diameters smaller than 100 nm and length of micrometer or even longer. SiNWs feature many attractive properties including excellent electronic/mechanical properties, favorable biocompatibility, huge surface-to-volume ratios, surface tailorability, improved multifunctionality, as well as their compatibility with conventional silicon technology.

Metal-catalyzed vapor-liquid-solid (VLS) (Wu and Yang 2001) and oxide-assisted growth (OAG) (Zhang et al. 2003) are widely recognized as two classic methods for synthesizing silicon nanowires (SiNWs) (He et al. 2010). In 1998, the groups of Lieber (Morales and Lieber 1998) and Lee (Zhang et al. 1998) independently reported the method of laser ablation-assisted VLS growth for the synthesis of single-crystal SiNWs with diameters of 6–20 nm and lengths ranging from 1 to 30 μm, realizing the first large-quantity fabrication of SiNWs (Fig. 1). Thereafter, the VLS method has been extensively optimized to be a well-established approach for preparing SiNWs. In a typical VLS growth, nanometer-diameter metals or metal compounds (e.g., Fe, Au) act as catalysts that effectively define the diameter of wires. OAG method is considered as another established strategy for preparation of the SiNWs, which is well complementary to and coexistent with the metal-catalyst VLS approach. Instead of metal catalysts used in the VLS method, oxides play an important role in inducing the nucleation and growth of nanowires in the OAG process. Consequently, SiNWs free of metal impurities are readily achieved via the OAG technique. In addition, Peng et al. recently developed a HF-etching-assisted nanoelectrochemical method to produce SiNWs (Peng et al. 2002), which is free of high temperature, vacuum, templates, complex equipment, or hazardous silicon precursors. In this method, noble metal atoms deposited from HF solution on the silicon wafer surface could form nuclei that behave as a cathode, and the area surrounding these nuclei behaves as an anode, and can subsequently be etched and dissolved into the solution by the galvanic cell reaction. As a result, silicon nanowire arrays could be readily grown on silicon wafers in noble metal HF solution via selective etching (Fig. 2), based on the principle of micro-electrochemical redox reaction.

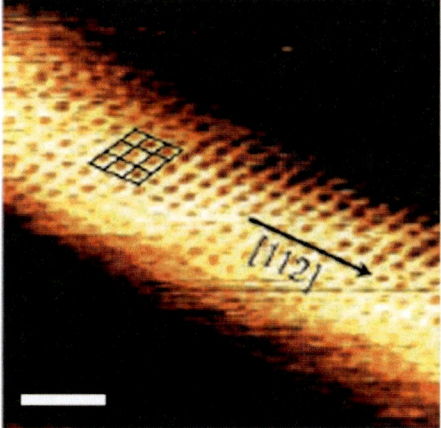
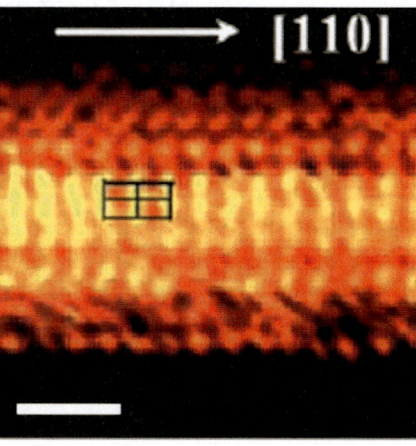

**Silicon Nanowires, Fig. 1** Constant-current scanning tunneling microscopy (STM) image of a SiNW prepared via OAG method. The wire axis is along the [112] (*left*) and [110] (*right*) direction. Scale bar: 1 nm (Ref. Ma et al. 2003)

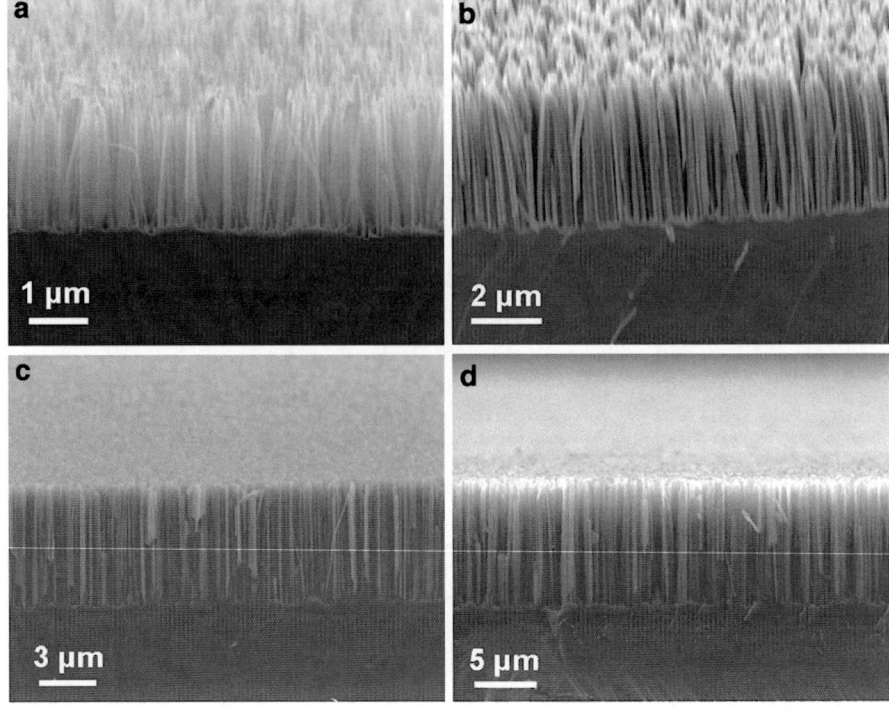

**Silicon Nanowires, Fig. 2** SEM images of silicon nanowires arrays prepared via metal-assisted chemical etching method. SiNWs with controllable lengths of 2 μm (**a**), 4 μm (**b**), 6 μm (**c**), and 10 μm (**d**), could be readily achieved through adjustment of etching time (Ref. He et al. 2010)

SiNWs have drawn intensive attention for their great promise as a platform for various applications ranging from electronics to biology (He et al. 2010). Biological applications of SiNWs are particularly interesting. Lieber and coworkers fabricated SiNWs-based field-effect transistors for high-sensitivity biomolecular detection and stimulation of neuronal signal propagation (Patolsky et al. 2006). Yang and coworkers demonstrated interfacing of SiNWs with mammalian cells without any external force (Kim et al. 2007). Recently, SiNWs were employed as a bioimaging agent with intrinsic 3D spatial resolution, high photostability, and orientation information, providing new opportunities for cellular interaction studies (Jung et al. 2009). Wang et al. presented a SiNWs-based system for quantifying mechanical behavior of cells lines, revealing force information of cancer cells (Li et al. 2009). More recently, SiNW arrays were also designed for isolating special cells (e.g., primary $CD4^+$ T lymphocytes) from the heterogeneous mixture of cell populations, which offers a specific single cell isolation platform and launches a new avenue for the development of advanced biochips allowing simultaneous isolation and activation of cells with concomitant detection of biogenic molecules (Kim et al. 2010). It is also worthwhile to note that, silicon-based nanohybrids have been recently demonstrated to be highly efficient for a variety of bioapplications. For examples, silver nanoparticles (AgNPs)-decorated SiNWs show ultrahigh surface-enhanced Raman scattering (SERS) and antibacterial activity, much better than free AgNPs (He et al. 2011a). As a result, DNA of low concentrations (∼fM) was readily detected by using the SiNWs-based SERS sensors, which is comparable to the lowest concentration ever detected via SERS (He et al. 2011a). Moreover, the AgNPs-decorated SiNWs exhibit strong bacterial properties to both gram-negative and gram-positive bacteria (Lv et al. 2010). In addition, fluorescent quantum dots (QDs)-decorated SiNWs, featuring remarkable anti-photobleaching property, were demonstrated to be superbly suitable for long-term cellular imaging (e.g., SiNWs-labeled cells preserve stable and bright fluorescence during 90-min UV irradiation, whereas CdTe QDs rapidly diminish in fluorescence signals in 20-min observation under the same conditions) (Fig. 3) (He et al. 2011b). Very recently, Gold nanoparticles (AuNPs)-decorated SiNWs were explored as novel near-infrared (NIR) hyperthermia agents due to their

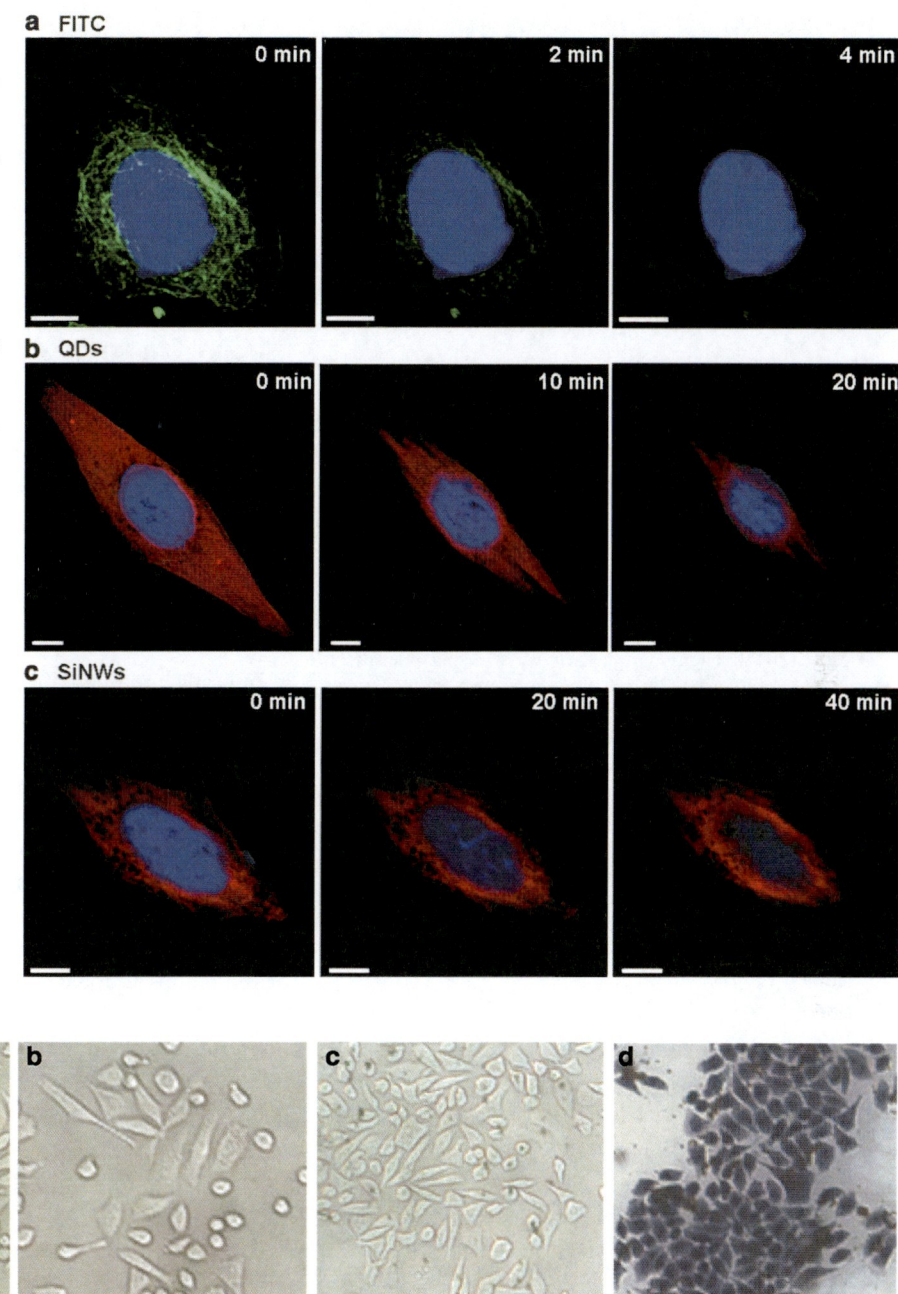

**Silicon Nanowires, Fig. 3** Stability comparison of fluorescence signals of Hela cells imaged by FITC, Hoechst, CdTe QDs, and SiNWs. Scale bar is 5 μm. The *green* fluorescence signals of FITC rapidly diminish in 2 min due to severe photobleaching (**a**). Comparatively, the *red* fluorescence signals of the CdTe QDs almost disappear at 20 min irradiation (**b**). The *blue* signals of Hoechst nearly vanish after 40-min irradiation (**b–c**). In sharp contrast, the SiNWs preserve strikingly stable fluorescence against photobleaching throughout a long-time imaging period. The *red* signals are persistently strong and even slightly brighter during 40-min observation (Ref. He et al. 2011b)

**Silicon Nanowires, Fig. 4** Optical images of KB cells. (**a**) Control cells, (**b**) control cells after laser irradiation, (**c**) cells cultured with SiNWs-based NIR hyperthermia agents, and (**d**) cells cultured with SiNWs-based NIR hyperthermia agents after 3-min 2 W/cm$^2$ 808 nm laser irradiation. *Blue color* indicates dead cells (Trypan *blue* test). Scale bar = 20 μm (Ref. Su et al. 2012)

strong optical absorbance in the NIR spectral window. Significantly, cancer cells treated with the SiNWs-based hyperthermia agents were completely destroyed within 3 min of NIR irradiation, demonstrating the exciting potential of SiNWs for NIR hyperthermia agents (Fig. 4) (Su et al. 2012).

Given that SiNWs can be readily prepared with high reproducibility and low cost, together with the above-mentioned attractive achievements, SiNWs may serve as a promising platform for the designs of high performance with unique merits for myriad bioapplications.

## Cross-References

▶ Silicon Transporters
▶ Silicon-Mediated Pathogen Resistance in Plants

## References

He Y, Fan C, Lee ST (2010) Silicon nanostructures for bioapplications. Nano Today 5:282–295

He Y, Su S, Xu TT, Zhong YL, Zapien JA, Li J, Fan CH, Lee ST (2011a) Silicon nanowires-based highly-efficient SERS-active platform for ultrasensitive DNA detection. Nano Today 6:122–130

He Y, Zhong YL, Peng F, Wei XP, Su YY, Su S, Gu W, Liao LS, Lee ST (2011b) Highly luminescent water-dispersible silicon nanowires for long-term immunofluorescent cellular imaging. Angew Chem Int Ed 50:3080–3083

Jung Y, Tong L, Tanaudommongkon A, Cheng JX, Yang C (2009) In vitro and in vivo nonlinear optical imaging of silicon nanowires. Nano Lett 9:2440–2444

Kim W, Ng JK, Kunitake ME, Conklin BR, Yang PD (2007) Interfacing silicon nanowires with mammalian cells. J Am Chem Soc 129:7728–7729

Kim ST, Kim DJ, Kim TJ, Seo DW, Kim TH, Lee SY, Kim KM, Lee SK (2010) Novel streptavidin-functionalized silicon nanowire arrays for CD4 (+) T lymphocyte separation. Nano Lett 10:2877–2883

Li Z, Song J, Mantini G, Lu MY, Fang H, Falconi C, Chen LJ, Wang ZL (2009) Quantifying the traction force of a single cell by aligned silicon nanowire array. Nano Lett 9:3575–3580

Lv M, Su S, He Y, Huang Q, Hu WB, Li D, Fan CH, Lee ST (2010) Long-term antimicrobial effect of silicon nanowires decorated with silver nanoparticles. Adv Mater 22:5463–5467

Ma DDD, Lee CS, Au FCK, Tong SY, Lee ST (2003) Small-diameter silicon nanowire surfaces. Science 299:1874–1877

Morales AM, Lieber CM (1998) A laser ablation method for the synthesis of crystalline semiconductor nanowires. Science 279:208–211

Patolsky F, Timko BP, Yu GH, Fang Y, Greytak AB, Zheng GF, Lieber CM (2006) Detection, stimulation, and inhibition of neuronal signals with high-density nanowire transistor arrays. Science 313:1100–1104

Peng KQ, Yan YJ, Gao SP, Zhu J (2002) Synthesis of large-scale silicon nanowire arrays via self-assembling nanoelectrochemistry. Adv Mater 14:1164–1167

Su YY, Wei XP, Peng F, Zhong YL, Lu YM, Su S, Xu TT, Lee ST, He Y (2012) Gold nanoparticles-decorated silicon nanowires as highly efficient near-infrared hyperthermia agents for cancer cells destruction. Nano Lett 12:1845–1850

Wu YY, Yang PD (2001) Direct observation of vapor-liquid-solid nanowire growth. J Am Chem Soc 123:3165–3166

Zhang YF, Tang YH, Wang N, Yu DP, Lee CS, Bello I, Lee ST (1998) Silicon nanowires prepared by laser ablation at high temperature. Appl Phys Lett 72:1835–1837

Zhang RQ, Lifshitz Y, Lee ST (2003) Oxide-assisted growth of semiconducting nanowires. Adv Mater 15:635–640

## Silicon Transporters

Jian Feng Ma
Plant Stress Physiology Group, Institute of Plant Science and Resources, Okayama University, Kurashiki, Japan

## Synonyms

Membrane proteins with transport activity for silicic acid; Silicic acid transporters in plants

## Definition

Plants accumulate silicon in the tissues for enhancing tolerance to various stresses. Different transporters are involved in the uptake, xylem loading, and distribution of silicon, including influx transporter and efflux transporter. Influx transporter belongs to a Nod26-like major intrinsic protein (NIP) subfamily of aquaporin-like proteins, while efflux transporter is a member of putative anion transporters.

## Introduction

Silicon (Si) is the second most abundant element after oxygen in the Earth's crust. It is essential for animals and diatoms, but has not been recognized as an essential element for plant growth, because there is no evidence showing that Si is involved in the metabolism, which is required for the essentiality in higher plants (Epstein 1999; Ma and Takahashi 2002; Ma et al. 2011). However, beneficial effects of Si have been observed in a wide range of plant species including dicots and monocots. The effects of Si are characterized by protecting the plant from various biotic and abiotic stresses (Ma 2004; Ma and Yamaji 2006). These effects have been mostly attributed to the deposition of Si at different tissues, which acts as a physical barrier to prevent penetration by fungi and insects, to increase mechanical strength, and to reduce transpiration.

## Influx Transporter of Silicon

Plant roots take up Si in the form of silicic acid [$Si(OH)_4$], a non-charged molecule. The first transporter (Lsi1 for Low Silicon 1) for Si uptake was identified in rice, a typical Si-accumulating species (Ma et al. 2006). Lsi1 comprises 298 amino acids and belongs to a Nod26-like major intrinsic protein (NIP) subfamily of aquaporin-like proteins. The predicted amino acid sequence has six transmembrane domains and two Asn-Pro-Ala (NPA) motifs, features which are well conserved in typical aquaporins.

Lsi1 shows influx activity for silicic acid in *Xenopus* oocyte (Ma et al. 2006). *Lsi1* is constitutively expressed in the roots, but its expression is decreased to one fourth by Si supply. Within a root, the expression of *Lsi1* is much lower in the root tip region between 0 and 10 mm than in the basal regions of the root (>10 mm), which is consistent with the site of Si uptake (Yamaji and Ma 2007). At heading stage with high requirement of Si, there is a transient increase in the expression of *Lsi1*.

In the roots including seminal, lateral, and crown roots, the Lsi1 protein is localized to the plasma membrane of both exodermis and endodermis (Ma et al. 2006), where the Casparian strips prevents apoplastic transport into the root stele. Furthermore, Lsi1 shows polar localization at the distal side of both the exodermis and endodermis cells (Fig. 1).

Lsi1 is also permeable to arsenite and selenite (Ma et al. 2008; Zhao et al. 2010). The aromatic/arginine (ar/R) selectivity filter influences selectivity of transport substrates and the residue at the H5 position of the ar/R filter of OsLsi1 plays a key role in the permeability to Si (Mitani et al. 2011).

Homologs of rice Lsi1 (OsLsi1) have been identified in barley (HvLsi1), maize (ZmLsi1), pumpkin (CmLsi1), and wheat (TaLsi1) (Chiba et al. 2009; Mitani et al. 2009a, b; Montpetit et al. 2012). They are all localized to the plasma membrane and show influx transport activity for silicic acid. However, unlike OsLsi1, HvLsi1 from barley and ZmLsi1 from maize are localized at epidermal, hypodermal, and cortical cells. CmLsi1 from the pumpkin is localized at all root cells (Mitani et al. 2011). In contrast to *OsLsi1*, the expression levels of *HvLsi1*, *ZmLsi1*, and *TaLsi1* are unaffected by Si.

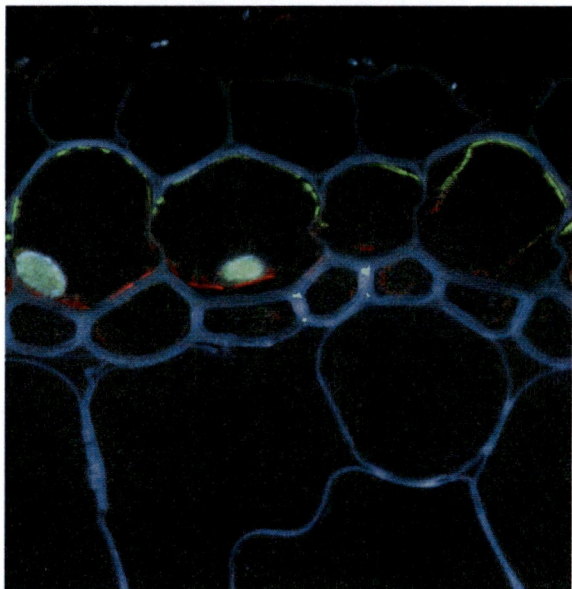

**Silicon Transporters, Fig. 1** Polar localization of influx Si transporter Lsi1 (*green*) and efflux Si transporter Lsi2 (*red*) in rice roots. Nuclei stained by DAPI (*cyan*)

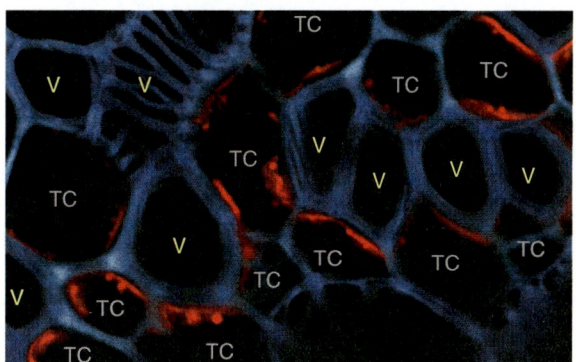

V: xylem vessel, TC: xylem transfer cell

**Silicon Transporters, Fig. 2** Cross section showing localization of Lsi6 (*red*) at the transfer cells surrounding large vascular bundles in rice node I

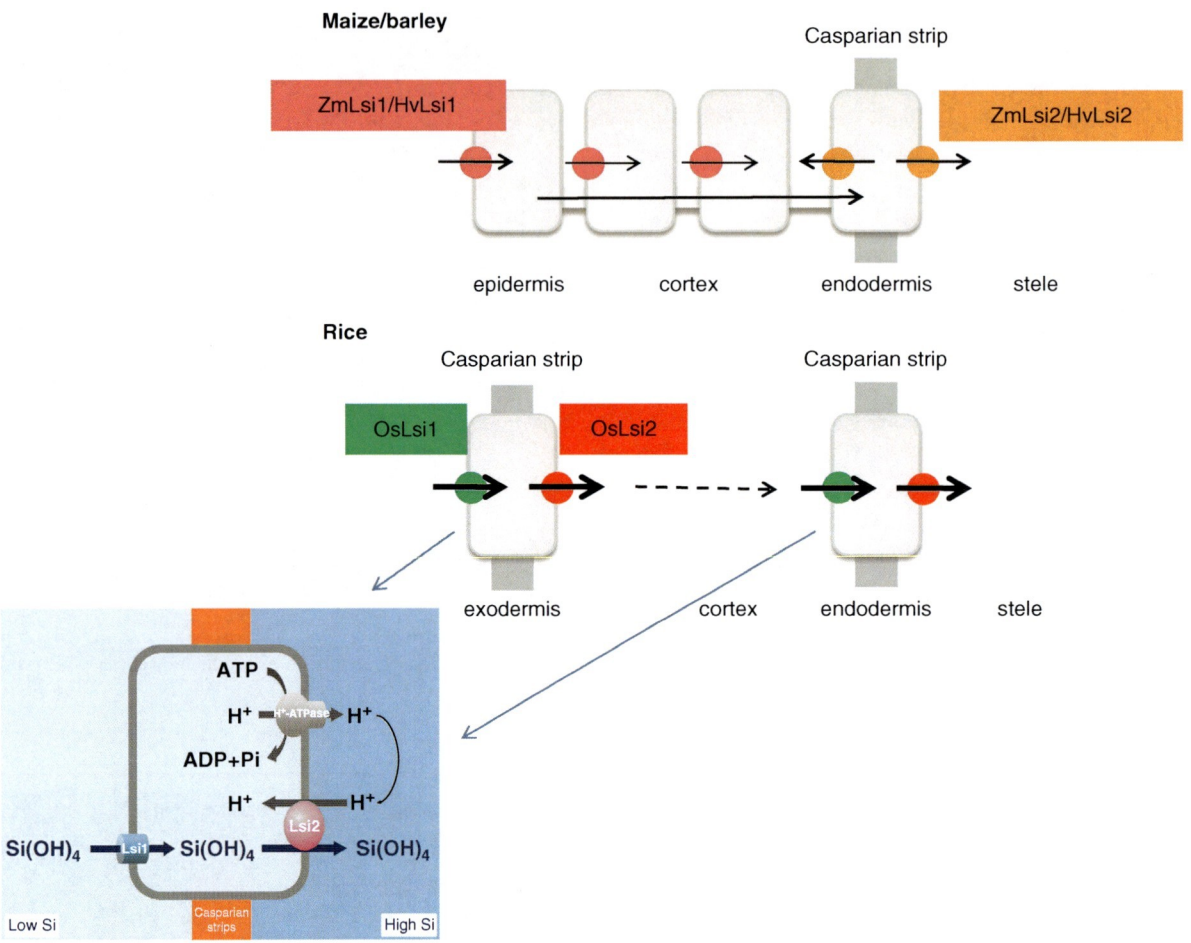

**Silicon Transporters, Fig. 3** Schematic presentation of different Si uptake system in barley/maize (**a**) and rice (**b**) roots

Lsi6, a homolog of Lsi1 in rice is involved in the export of silicic acid from the xylem to other leaf cells. Lsi6 is expressed in the leaf sheaths and leaf blades in addition to the root tips (Yamaji et al. 2008). Lsi6 is localized in the adaxial side of the xylem parenchyma cells in the leaf sheaths and leaf blades (Fig. 2). Knockout of *Lsi6* does not affect the uptake of Si by the roots, but affects the silica deposition pattern in the leaf blades and sheaths and increases excretion of Si in the guttation fluid in rice due to altered pathway.

At reproductive growth stage, Lsi6 is required for the intervascular transfer of Si in node of rice (Yamaji and Ma 2009). Lsi6 is highly expressed at the node I below the panicles and mainly localized at the xylem transfer cells with polarity facing toward the xylem vessel. These cells are located at the outer boundary region of the enlarged vascular bundles and characterized by large surface area due to cell wall ingrowth. Knockout of *Lsi6* decreased Si accumulation in the panicles but increased Si accumulation in the flag leaf. Lsi6 is required for transfer of Si from the large vascular bundles coming from the roots to the diffuse vascular bundles connected to the panicles.

## Efflux Transporter of Silicon

Lsi2 consists of 472 amino-acid residues with 11 transmembrane domains. Lsi2 shows efflux transport activity for silicic acid, but not influx transport activity (Ma et al. 2007). The expression pattern and tissue- and cellular-localization of Lsi2 is the same as that of Lsi1, but in contrast to Lsi1, Lsi2 is localized at the proximal side of the exodermis and the endodermis cells (Fig. 1). Transport of Si by Lsi2 is driven by the proton gradient (Ma et al. 2007).

Similar transporters of Lsi2 have also been identified in barley and maize (Mitani et al. 2009b). However, ZmLsi2 and HvLsi2 are localized only to the endodermis of roots in maize and barley without polarity. These differences result in different pathway of Si from external solution to the xylem between upland crops (barley and maize) and paddy crop (rice). In barley and maize, Si can be taken up from external solution (soil solution) by HvLsi1/ZmLsi1 at different cells including epidermal, hypodermal, and cortical cells (Fig. 3). In contrast, in rice, Si is only taken up at the exodermal cells by OsLsi1 (Fig. 3). After being taken up into the root cells, Si is transported to the endodermis by symplastic pathway and then released to the stele by HvLsi2/ZmLsi2 in maize and barley. By contrast, in rice, Si taken up by OsLsi1 at the exodermal cells is released by OsLsi2 to the apoplast and then transported into the stele by both OsLsi1 and OsLsi2 again at the endodermal cells. The difference in the uptake system may be attributed to the root structures. In rice roots, there are two Casparian strips at the exodermis and endodermis, whereas one Casparian strip is usually present at the endodermis of maize and barley roots under non-stressed conditions. Furthermore, mature roots in rice have a distinct structure, a highly developed aerenchyma, wherein almost all cortex cells between exodermis and endodermis are destructed. Therefore, Si transported into the exodermis cells by the influx transporter, OsLsi1, has to be released by the efflux transporter, OsLsi2, into the apoplast of a spoke-like structure across the aerenchyma. However, in maize and barley roots, there is no such structure or if any; it is developed poorly. These differences in the localization of transporters and polarity may be one of the reasons for the different Si uptake capacities among species.

## Cross-References

▶ Arsenic and Aquaporins
▶ Silicon and Aquaporins
▶ Silicon-mediated Pathogen Resistance in Plants

## References

Chiba Y, Mitani N, Yamaji N, Ma JF (2009) HvLsi1 is a silicon influx transporter in barley. Plant J 57:810–818
Epstein E (1999) Silicon. Annu Rev Plant Physiol Plant Mol Biol 50:641–664
Ma JF (2004) Role of silicon in enhancing the resistance of plants to biotic and abiotic stresses. Soil Sci Plant Nutr 50:11–18
Ma JF, Takahashi E (2002) Soil, fertilizer, and plant silicon research in Japan. Elsevier, Amsterdam
Ma JF, Yamaji N (2006) Silicon uptake and accumulation in higher plants. Trends Plant Sci 11:392–397
Ma JF, Tamai K, Yamaji N, Mitani N, Konishi S, Katsuhara M, Ishiguro M, Murata Y, Yano M (2006) A silicon transporter in rice. Nature 440:688–691
Ma JF, Yamaji N, Mitani N, Tamai K, Konishi S, Fujiwara T, Katsuhara M, Yano M (2007) An efflux transporter of silicon in rice. Nature 448:209–212
Ma JF, Yamaji N, Mitani N, Xu XY, Su YH, McGrath S, Zhao FJ (2008) Transporters of arsenite in rice and their role in arsenic accumulation in rice grain. Proc Natl Acad Sci USA 105:9931–9935
Ma JF, Yamaji N, Mitani-Ueno N (2011) Transport of silicon from roots to panicles in plants. Proc Jpn Acad Ser B 87:377–385
Mitani N, Yamaji N, Ma JF (2009a) Identification of maize silicon influx transporters. Plant Cell Physiol 50:5–12
Mitani N, Chiba Y, Yamaji N, Ma JF (2009b) Identification and characterization of maize and barley Lsi2-like silicon efflux transporters reveals a distinct silicon uptake system from that in rice. Plant Cell 21:2133–2142
Mitani N, Yamaji N, Ago Y, Iwasaki K, Ma JF (2011) Isolation and functional characterization of an influx silicon transporter in two pumpkin cultivars contrasting in silicon accumulation. Plant J 66:231–240
Mitani-Ueno N, Yamaji N, Zhao FJ, Ma JF (2011) Aromatic/arginine selectivity filter of NIP aquaporins plays a critical role in substrate selectivity for silicon, boron and arsenic. J Exp Bot 62:4391–4398
Montpetit J, Vivancos J, Mitani-Ueno N, Yamaji N, Rémus-Borel W, Belzile F, Ma JF, Bélanger RR (2012) Cloning, functional characterization and heterologous expression of TaLsi1, a wheat silicon transporter gene. Plant Mol Biol 79:35–46
Yamaji N, Ma JF (2007) Spatial distribution and temporal variation of the rice silicon transporter Lsi1. Plant Physiol 143:1306–1313
Yamaji N, Ma JF (2009) Silicon transporter Lsi6 at the node is responsible for inter-vascular transfer of silicon in rice. Plant Cell 21:2878–2883
Yamaji N, Mitani N, Ma JF (2008) A transporter regulating silicon distribution in rice shoots. Plant Cell 20:1381–1389
Zhao XQ, Mitani N, Yamaji N, Shen RF, Ma JF (2010) Involvement of silicon influx transporter OsNIP2;1 in selenite uptake in rice. Plant Physiol 153:1871–1877

# Silicon, Biologically Active Compounds

Leonid Breydo
Department of Molecular Medicine, Morsani College of Medicine, University of South Florida, Tampa, FL, USA

## Synonyms

SDVs – silicon deposition vehicles; SITs – silicon transporter proteins

## Definition

Silicon is extremely common in nature where it occurs as silicon dioxide and various silicates. The primary role of silicon-containing compounds in biological environment is a structural one with silicates incorporated into exoskeleton of marine organisms, cell walls of plants, and connective tissues of animals.

## Introduction

Silicon is the second most abundant element in the Earth's crust accounting for 27.7% of it by weight. Silicon is the closest analogue of carbon in periodic table. However, it is more electropositive, resulting in stronger Si-O bonds and much weaker Si-Si and Si-H bonds compared to carbon. High stability of Si-O bond makes formation of silicon dioxide and silicates highly thermodynamically favorable. Silica and most silicates are poorly water soluble and highly chemically stable. Due to their high stability, silicon dioxide and silicates are the only silicon-containing compounds found in the biological environments.

## Silicon in Diatoms and Other Marine Animals

Formation of complex inorganic materials (biominerals) via interaction on minerals with living organisms is a widespread phenomenon in nature. Among the most fascinating examples of biomineral structures are the intricately patterned, silicified cell walls of unicellular algae (diatoms), which contain tightly associated proteins. These structures are made by precipitation of soluble sililic acid ($Si(OH)_4$) as an insoluble silicon dioxide. Under control of the diatom proteins, this process occurs rapidly and results in the material with characteristic particle and pore sizes.

Diatoms are the only group of organisms whose development is completely dependent on the presence of soluble forms of silica. In the absence of silicon sources, their DNA replication stops. Diatoms developed highly sophisticated mechanism of transport of sililic acid to appropriate cellular compartments, its concentration there, and subsequent precipitation. Silicates are transported from the seawater to silica deposition vehicles (SDVs) where synthesis of the silica shell occurs. Diatom silicon transporters (SITs) are species-specific proteins (bicarbonate/silicate cotransporters) able to bind sililic acid and transport across the cell membrane to the interior of the cell.

Several proteins accelerate silica precipitation or play a structural role in the silica shells. Cell walls of diatoms contain three families of proteins: sillafins, frustulins, and pleuralins. Sillafins are small proteins that mediate and control precipitation of silica from a solution of sililic acid (Poulsen et al. 2003; Sumper et al. 2003). They contain a wide variety of modified amino acids and a significant number of posttranslational modifications. These serine-rich proteins are heavily phosphorylated (about 8 phosphates/protein). In addition, sillafins contain several lysines modified with long-chain polyamines. This combination of high concentration of both positive and negative charges results in highly soluble proteins that nonetheless readily self-assemble. It has been shown that self-assembly of sillafins is required for silica precipitation providing silica particles with three-dimensional framework for assembly. Long-chain polyamine side chains of sillafins catalyze the precipitation of silica (Poulsen et al. 2003). Frustulins are large (75–200 kDa) calcium-binding glycoproteins that bind the silica surface via a common acidic, cysteine-rich domain with repetitive structure (Kroger et al. 1997; Rezanka and Sigler 2008). Pleuralins are 150–200 kDa proteins that are involved in formation of new cell wall after cell division.

Sponges also rely on silica for their exoskeleton. In some sponges (e.g., *Tethya aurantia*), silica needles constitute up to 75% of the dry weight of the organism. These silica shells contain protein filaments that catalyze and direct their growth. Primary component of

**Silicon, Biologically Active Compounds, Fig. 1** Catalysis of polymerization of sililic acid by silicatein

these filaments is are silicateins (Wang et al. 2012). Silicateins are 27–29 kDa serine proteases related to cathepsins. They catalyze condensation of sililic acid to insoluble silicates. The reaction starts by forming a silicate ester of the active site serine. Then this ester is deprotonated and the resulting anion acts as a nucleophile reacting with another sililic acid molecule to form a dimer (Fig. 1). Polymerization proceeds further, and eventually, the bond between the active site serine and the growing polymer is hydrolyzed and the polymer is released. Silicateins have similar combination of polylysines with phosphorylated hydroxyamino acids, resulting in their ability to self-assemble into filaments. Enzymes in these filaments are catalytically active. In addition to silicateins, these filaments contain silintaphins (Wang et al. 2012). Silintaphins are structural proteins that promote binding of silicateins to silica, organize them into highly ordered filaments, and enhance their enzymatic activity. In the presence of silintaphins, filaments are maintained at a fixed diameter (20 nm) and direct deposition of silica nanoparticles into fractal-like network. These filaments also contain silicase (an enzyme structurally similar to a carbonic anhydrase) that can hydrolyze the polymeric silicates (Wang et al. 2012). Further maturation of silica shells involves their dehydration by extracellular matrix proteins. Overall, sponge proteins tightly control the structure of silica shells underlining their importance to the organism.

## Silicon in Plants and Animals

Silicon is taken up by plants from soil as sililic acid. Studies that have grown plants in silica-free environments have found that plants do not grow as well. For example, the stems of certain plants will collapse when grown in soil lacking silica (Rezanka and Sigler 2008). In plants, silicon appears to play a protective role providing support to growing shoots and protecting them from environmental stresses. Silicon in plants colocalizes with complex carbohydrates and proteins enriched in hydroxyamino acids. It is thus likely that it forms a part of some type of exoskeleton similar to that found in diatoms and sponges. Two silicon transporters (Lsi1 and Lsi2) have been identified in rice, indicating tightly controlled silicon uptake system in plants

(Ma et al. 2006, 2007). Deficiency in these transporters results in increased susceptibility to diseases and decrease in grain yield by 60–90%.

Silicon is an essential microelement in animals as well. It is absorbed via the intestinal walls as sililic acid. Silicon was shown to perform an important role in connective tissue, especially in bone and cartilage. Silicon deficiency resulted in defects in formation of organic matrix of the connective tissue. Silicon levels are especially high in osteogenic cells, and it is likely to be involved in early stages of bone mineralization and formation of cartilage matrix (Carlisle 1984). These data indicate that silicates play a structural role in bones, cartilage, and other connective tissues.

## Toxicity of Silicon

Toxicity of silicon is primarily associated with lung inflammation. It is a reaction of the body to the small particles of silicon dioxide or silicates (e.g., asbestos) that enter the lungs. This disease occurs primarily in workers who have been exposed to sand or asbestos. Unlike particles of many other minerals, silica or asbestos particles provoke severe fibrogenic reaction. Accumulation of these particles over time causes inflammation and swelling in lungs. Eventually, lungs can fill up with fluid, creating problems with breathing. Asbestos exposure also causes mesothelioma, a form of lung cancer. Inflammation is also believed to be the primary cause of this disease. Toxic effects of silica and asbestos are due to physical disruption of the cells by the particles and not due to chemical reactions of silica.

## Conclusions

Silicon is present in nature primarily as silicon dioxide and various silicates. These compounds are highly stable, and silicon chemistry in biological systems is dominated by transitions between soluble and insoluble silicates. Silicates are important to marine organisms, plants, and likely animals as components of their exoskeleton, cell walls, and connective tissues. Toxic effects of silicon are primarily due to inflammation caused by the particles of silica or silicates.

## Cross-References

▶ Silicateins
▶ Silicon Exposure and Vasculitis
▶ Silicon Transporters
▶ Silicon, Physical and Chemical Properties
▶ Silicosis

## References

Carlisle EM (1984) Silicon. In: Frieden E (ed) Biochemistry of the essential ultratrace elements. Plenum Press, New York, pp 319–340
Kroger N, Lehmann G, Rachel R, Sumper M (1997) Characterization of a 200-kDa diatom protein that is specifically associated with a silica-based substructure of the cell wall. Eur J Biochem 250:99–105
Ma JF, Tamai K, Yamaji N, Mitani N, Konishi S, Katsuhara M, Ishiguro M, Murata Y, Yano M (2006) A silicon transporter in rice. Nature 440:688–691
Ma JF, Yamaji N, Mitani N, Tamai K, Konishi S, Fujiwara T, Katsuhara M, Yano M (2007) An efflux transporter of silicon in rice. Nature 448:209–212
Poulsen N, Sumper M, Kroger N (2003) Biosilica formation in diatoms: characterization of native silaffin-2 and its role in silica morphogenesis. Proc Natl Acad Sci USA 100:12075–12080
Rezanka T, Sigler K (2008) Biologically active compounds of semi-metals. Phytochemistry 69:585–606
Sumper M, Lorenz S, Brunner E (2003) Biomimetic control of size in the polyamine-directed formation of silica nanospheres. Angew Chem Int Ed Engl 42:5192–5195
Wang X, Schlossmacher U, Wiens M, Batel R, Schroder HC, Muller WE (2012) Silicateins, silicatein interactors and cellular interplay in sponge skeletogenesis: formation of glass fiber-like spicules. FEBS J 279:1721–1736

## Silicon, Physical and Chemical Properties

Fathi Habashi
Department of Mining, Metallurgical, and Materials Engineering, Laval University, Quebec City, Canada

Silicon is a metalloid having the diamond structure (Fig. 1). It has the appearance of a metal (Fig. 2) but lacks all properties of metals. Modern electronics is almost exclusively (>95%) based on silicon devices

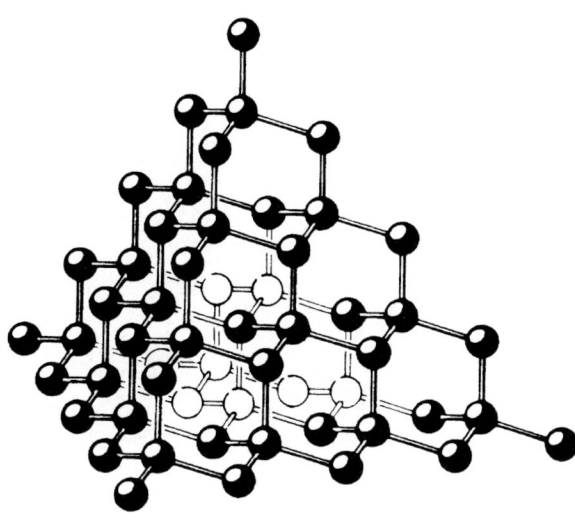

Silicon, Physical and Chemical Properties, Fig. 1 Structure of silicon

and these are made from single-crystal silicon. Silica, $SiO_2$, is the major constituent of rock-forming minerals in magmatic and metamorphic rocks; it accounts for ca. 75% of the Earth's crust. The basic building block of the silicates is the $SiO_4^{4-}$ tetrahedron (Fig. 3). The variety in the structures of the silicates is due to the various possible combinations of these tetrahedral, both with each other and also with other ions.

## Physical Properties

| | |
|---|---|
| Atomic number | 14 |
| Atomic weight | 28.086 |
| Relative abundance in earth's crust, % | 27.72 |
| Density, g/cm³ | |
| At room temperature | 2.329 |
| At melting point | 2.51 |
| Volume increase at transition from liquid to solid, % | +9.1 |
| Melting point, °C | 1,414 |
| Boiling point, °C | 3,231 |
| Specific heat at room temperature, J g$^{-1}$ K$^{-1}$ | 0.713 |
| Thermal expansion (300 K), K$^{-1}$ | $2.6 \times 10^{-6}$ |
| Thermal conductivity (300 K), W cm$^{-1}$ K$^{-1}$ | 1.5 |
| Critical temperature, K | 5,193 |
| *(continued)* | |

| | |
|---|---|
| Critical pressure, MPa | 145 |
| Latent heat of fusion, kJ/mol | 50.66 |
| Heat of evaporation, kJ/mol | 385 |
| Combustion heat, (Si/SiO$_2$), kJ/mol | 850 |
| Bulk modulus (300 K), Pa | 97.84 |
| Vickers hardness, GPa | 10.2 |
| Mohs hardness | 7 |
| Surface tension at melting point, mJ/m² | 885 |
| Bandgap (300 K), eV | 1.126 |

## Chemical Properties

Silicon is produced by reduction of silica with carbon or aluminum:

$$SiO_2 + 2C \rightarrow Si + 2CO$$
$$3SiO_2 + 4Al \rightarrow 3Si + 2Al_2O_3$$

For semiconductor devices impure silica is refined by first converting it into a compound such as $SiHCl_3$, $SiCl_4$, $SiH_2Cl_2$, or $SiH_4$. After purification, the silicon compound is transformed back into elemental, high-purity silicon by chemical vapor deposition. For example,

$$Si + 3HCl \rightarrow SiHCl_3 + H_2$$
$$Si + 4HCl \rightarrow SiCl_4 + 2H_2$$

Monosilane, $SiH_4$, is formed by the hydrolysis of magnesium silicide, $Mg_2Si$, with 10% hydrochloric acid at 50°C. Direct use of silicon hydrides involves their decomposition on surfaces, leading to deposition of elemental silicon or silicon compounds.

Bulk silicon is resistant to acid, including HF or $HNO_3$, but not to a mixture of the two. In this mixture, HF first etches away the $SiO_2$ layer from the silicon surface, then $HNO_3$ oxidizes silicon again, the new $SiO_2$ layer is removed by HF, and so on. Silicon reacts violently with dilute alkaline solutions to generate hydrogen. Heating in an oxygen-containing atmosphere forms an $SiO_2$ layer that inhibits further oxidation. At room temperature, silicon and fluorine produce $SiF_4$ (Habashi 2003; Zulehner et al. 1997).

**Silicon, Physical and Chemical Properties, Fig. 2** Elemental silicon

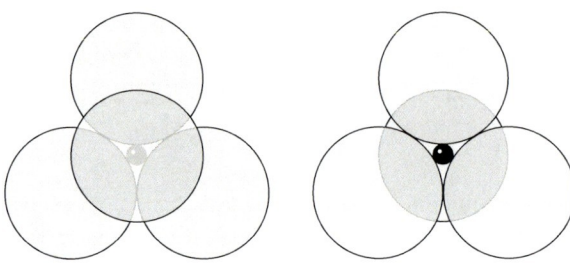

**Silicon, Physical and Chemical Properties, Fig. 3** The $SiO_4$ tetrahedron, the basic building unit of silicates. *Left*: plan, *right*: side view

## References

Habashi F (2003) Metals from ores. An introduction to extractive metallurgy. Métallurgie Extractive Québec, Québec City. Distributed from Laval University Bookstore. www.zone.ul.ca

Zulehner W et al (1997) In: Habashi F (ed) Handbook of extractive metallurgy. Wiley-VCH, Weinheim, pp 1861–1984

# Silicon-Mediated Pathogen Resistance in Plants

Kunzheng Cai
Institute of Tropical and Subtropical Ecology, South China Agricultural University, Guangzhou, China

## Synonyms

Silicon and plant pathogen; The role of silicon in suppressing plant disease

## Definition

Silicon-mediated pathogen resistance in plants refers to an increase in resistance to pathogen and improvement in plant health by silicon amendment. When attacked by plant diseases, silicon treatment may increase Si accumulation and deposition in leaves to form a cuticle-Si double layer to impede pathogen's penetration, and Si also can induce the biochemical defense responses of host and even prime the defense capacity of plant to reduce disease incidence.

## Introduction

Silicon (Si) is the second most prevalent mineral element in soil following oxygen and comprises approximately 28% of the earth's crust (Epstein 1994). Plants uptake Si through the roots in the form of monosilicic acid [Si(OH)4], at a typical concentration of 0.1–0.6 mM in soil water. Once absorbed by the roots, Si is translocated to the shoot via xylem and deposited in the form of amorphous silica ($SiO_2$-$nH_2O$) throughout the plant (Fig. 1), mainly in the cell walls, where it interacts with pectins and polyphenols, and enhances cell wall rigidity and strength, so as to protect plants from multiple environmental stresses. The Si content in plants varies greatly and ranges from 0.1% to 10% of shoot dry weight, an amount equivalent to, or even exceeding, several macronutrients, depending on the species (Epstein 1994). Poaceae, Equisetaceae, and Cyperaceae plants accumulate high level of Si (>4%), the Cucurbitales, Urticales,

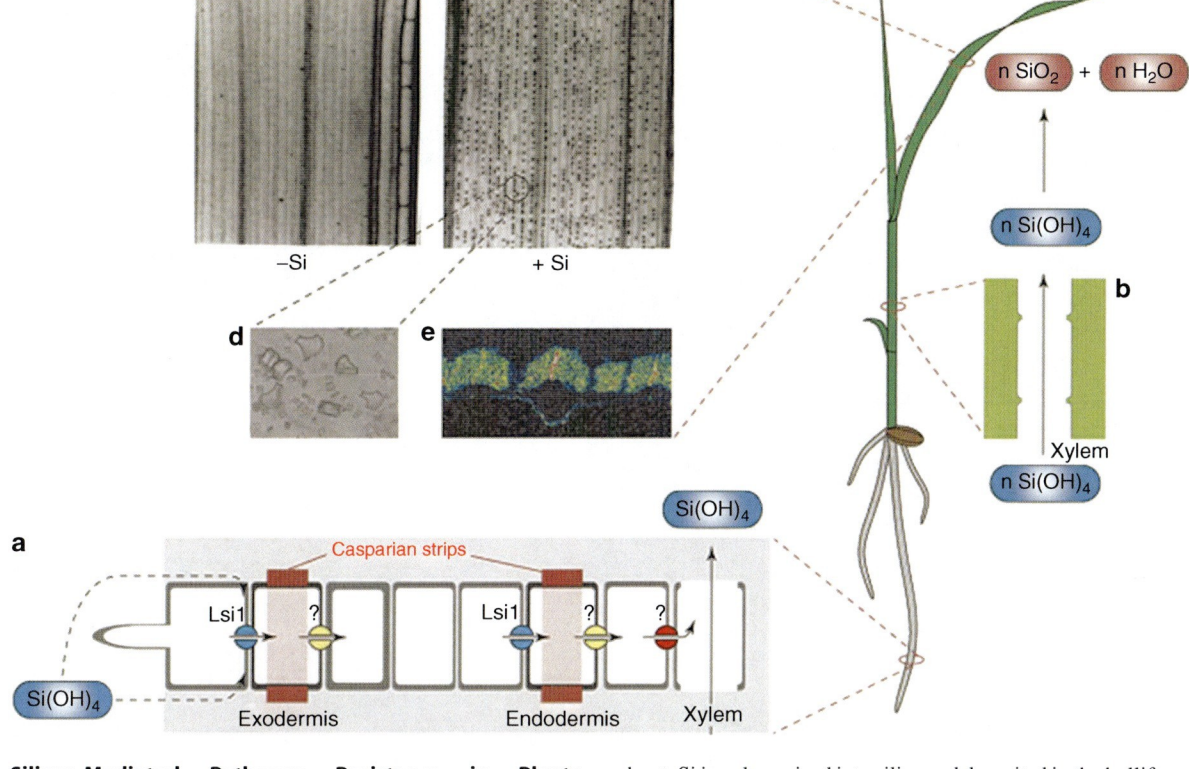

**Silicon-Mediated Pathogen Resistance in Plants, Fig. 1** Uptake, distribution, and accumulation of silicon (Si) in rice. Si is taken up via transporters in the form of silicic acid (**a**) and then translocated to the shoot in the same form (**b**). In the shoot, Si is polymerized into silica and deposited in the bulliform cells (silica body) (**c**, **d**) and under the cuticle (**e**). Silicon detected by soft X-ray (**c**) and by SEM (**e**) (Reprinted from Ma and Yamaji (2006) with permission from Elsevier)

and Commelinaceae show intermediate accumulation of Si (2–4%), while most other species demonstrate low accumulation (Currie and Perry 2007). Differences in Si accumulation among plant species result from the capacity of the roots to absorb Si, and the mechanisms of Si loading into the xylem are also distinct (Ma and Yamaji 2008). Si is not considered as an essential element for higher plants, since there is no evidence that Si is involved in the metabolism of plants. However, the beneficial effects of this element in plant growth, development, yield, and plant resistance to biotic (pathogen and pest) and abiotic stress (UV-B radiation, ion toxicity, drought, salinity, extreme temperature, etc.) have been reported in a wide variety of plant species (Fauteux et al. 2005; Ma and Yamaji 2006; Liang et al. 2007). Furthermore, Si has distinct effects in alleviating plant diseases caused by fungi and bacteria. The in-depth understanding of the mechanisms of silicon-mediated biotic stress will be helpful for the effective use of silicon to enhance pathogen resistance and reduce yield losses to plant pathogens.

## Role of Silicon in Plant Resistance to Pathogen Stress

Plant pathogens pose a serious problem for food security. At least 10% of global crop production is lost by plant diseases annually, either as yield loss or as quality loss (Strange and Scott 2005). Breeding resistant cultivars (high-yielding, less susceptible to pathogens), pesticide application, biological control, and agricultural practices (such as intercropping, irrational fertilizer application) have been traditionally used to reduce yield loss caused by the pathogens. However, cultivar resistance is not durable because of the complicated genetic biodiversity and coevolution between plants and their pathogens. The use of synthetic pesticides will lead to the establishment of races of pathogen resistance to the chemicals and increase the negative effects to the environment and the health of animals and humans. Therefore, there is a need to develop novel environmental-friendly strategies to control plant diseases.

Silicon and its products may provide a viable method to reduce the negative effects of pathogen on crop production. Many studies showed that Si has an important role in protection of plants against different pathogens, such as blast and sheath blight in rice; powdery mildew in wheat, barley, cucumber, and *Arabidopsis thaliana*; ring spot in sugarcane; rust in cowpea; and bacterial wilt in tomato. However, the exact nature of protective effects of silicon in alleviating biotic stress imposed by pathogens is uncertain presently and represents the subject of active debate.

## Mechanisms of Si-Mediated Pathogen Resistance

Regarding the protective role of silicon, numerous studies have been dedicated to explore the possible mechanisms (Fauteux et al. 2005; Cai et al. 2009). One is that Si can be accumulated and deposited beneath the cuticle to form a cuticle-Si double layer. This impedes the pathogen's penetration and colonization through mechanical barrier (Fig. 2). Another explanation is that soluble Si can trigger biochemical defense responses of host, thereby acting as a signal that induces resistance (Fig. 3). Molecular studies show that Si can modulate the expression of defense-related genes and may prime the defense capacity of the plant.

### Mechanical Barrier

Since Si was found to control plant disease, physical barrier was traditionally used to explain its role in enhancing pathogen resistance. For leaves, it was proposed that Si deposition in the cell walls constitutes a mechanical barrier which impedes the penetration of fungal hyphae (Samuels et al. 1991; Bowen et al. 1992). A blast-resistant rice cultivar has larger number of silicified cells than a susceptible cultivar (Kwon et al. 1974). Seebold et al. (2001) indicated that Si manifested its effect to establish blockage to ingress

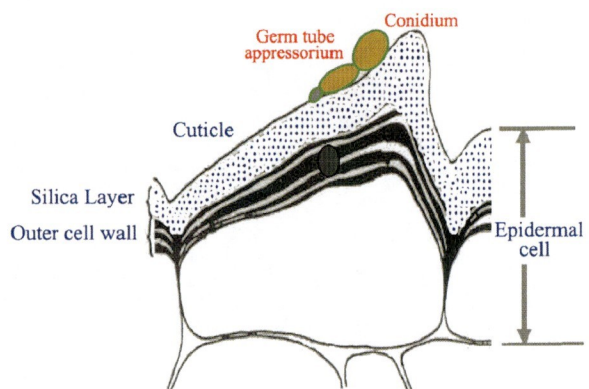

**Silicon-Mediated Pathogen Resistance in Plants, Fig. 2** Physical mechanisms of silicon-mediating pathogen resistance (Courtesy of L. E. Datnoff and F. A. Rodrigues)

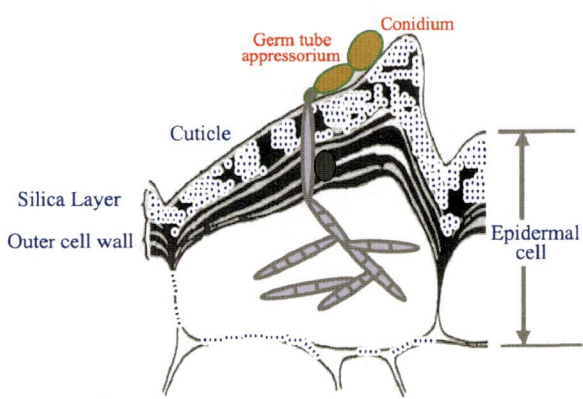

**Silicon-Mediated Pathogen Resistance in Plants, Fig. 3** Induced biochemical defense response by silicon (Courtesy of L. E. Datnoff and F. A. Rodrigues)

by fungus. Cytological analysis showed that the Si-induced resistance to blast in rice was correlated with a specific leaf cell reaction that interfered with the development of *Magnaporthe oryzae* (Rodrigues et al. 2003). Through microscopy and X-ray microanalysis technologies Kim et al. (2002) reported that cell wall reinforcement induced by silicon was closely associated with the reduced blast severity in susceptible and partially resistant cultivars. These results suggested that silicon accumulation might limit fungal penetration and invasion by taking a physical role. Liang et al. (2005) showed that foliar Si applications can effectively control *Podosphaera xanthii* infection in cucumber only through physical action, while root-applied Si leads to systemic acquired resistance in response to infection by the pathogen.

Papilla formation in epidermal cells is recognized to be useful for the enhancement of plant resistance against powdery mildew in wheat and rice (Belanger et al. 2003; Zhang et al. 2006). Cytological observations showed that Si-treated rice plants infected by *Rhizoctonia solani* and *M. oryzae* had much more silica cells and papillae on the guard cell of stoma from the zones where pathogen grew (Zhang et al. 2006; Cai et al. 2008). Furthermore, one study showed that the prevalence of papillae in Si-treated plants could increase penetration resistance of wheat cell against *B. graminis f. sp. Tritici* compared to that without Si treatment (Belanger et al. 2003). Based on light microscopic observation of the adaxial surface of rice leaves amended with or without Si, Hayasaka et al. (2008) further confirmed that Si in the leaf epidermis may confer resistance against appressorial penetration.

Therefore, these results suggested that cell silicification and silicon deposition were associated with silicon-mediated plant resistance to pathogen.

**Induced Biochemical Defense Response**

Although the results of early and recent studies partly supported the role of Si as a physical barrier to inhibit pathogen, the hypothesis has always been strongly disputed. Accumulation and polymerization of insoluble Si at fungal penetration sites or in epidermal cell walls was not associated with the increased resistance (Carver et al. 1987; Chérif et al. 1992). Si-induced resistance against *S. fuliginea* was rapidly declined when the cucumber plants were transferred to Si-free solution (Samuels et al. 1991). Heine et al. (2007) reported that the accumulation of Si in root cell walls did not represent a physical barrier to the spread of *Pythium aphanidermatum* in roots of bitter gourd and tomato.

Based on the above doubts, an alternative mechanism for the silicon protective role was proposed, according to which silicon can modulate and trigger host defense response to pathogens (Fig. 3). Fawe et al. (2001) suggested that defense responses mediated by silicon are functionally similar to systemic acquired resistance (SAR). Si-treated cucumber plants could enhance activity of chitinases, peroxidases, polyphenol oxidases, and flavonoid phytoalexins, all of which may protect against *Pythium*'s infection (Chérif et al. 1994; Fawe et al. 1998). The protective role of silicon in preventing sheath blight and blast attack showed that the lower disease severity was associated with higher activities of peroxidase (POD), polyphenol oxidase (PPO), and phenylalanine ammonia-lyase (PAL) in Si-treated plants (Zhang et al. 2006; Cai et al. 2008). These findings were similar to the results of other studies in *Pythium* cucumber leaves, *B. graminis*-infected wheat leaves, and *P. xanthii*-infected cucumber roots (Chérif et al. 1994; Yang et al. 2003; Liang et al. 2005), showing that Si could enhance activity of POD, PPO, and PAL in leaves coupled with decreased disease infectivity. However, non-inoculated plants exhibited no difference in the protective enzyme activities regardless of Si treatment, which showed that Si alone has apparently no obvious effect on the normal metabolism of plants growing in a controlled environment (Chérif et al. 1992; Yang et al. 2003; Fauteux et al. 2006; Cai et al. 2008).

The protective enzymes (POD, PPO, PAL, etc.) had important roles in regulating the production of antifungal chemicals. Evidence showed that silicon treatment might result in the accumulation of antifungal compounds such as phytoalexins and pathogenesis-related proteins in plants after the pathogens enter the epidermal cell (Chérif et al. 1992, 1994; Remus-Borel et al. 2005; Rodrigues et al. 2003). Fawe et al. (1998) reported that Si-induced resistance to powdery mildew (*Podosphaera xanthii*) in cucumber was associated with the production of a novel flavonol phytoalexin. Further studies by Rodrigues et al. (2004, 2005) showed that Si-induced rice resistance against *M. oryzae* was strongly associated with the production of phytoalexin(s), which were mainly momilactones A and B that were two- to threefold increased in leaf extracts of rice plants amended with Si and inoculated with *M. oryzae*.

Lignin is one of the products of phenolic metabolism and it plays an important role in disease resistance. Si can increase the production of lignin-carbohydrate complexes in the cell wall of rice epidermal cells (Inanaga et al. 1995). Furthermore, the lignin content of the silicon-supplied plants of susceptible rice line CO39 was increased, but the change was not observed in non-Si-treated or *M. oryzae*-infected plants (Cai et al. 2008).

**Molecular Mechanism**

Looking toward the current studies on the role of silicon in pathogen resistance, it is important to investigate the resistance mechanisms of silicon at the molecular level. Differences in gene expression were not appreciable between Si-treated and non-treated rice plants in the absence of stress (Watanabe et al. 2004). By using subtractive cDNA libraries or microarray methods at a genome-wide level, Fauteux et al. (2006) reported that only two genes differentially expressed in silicon-treated plants compared with the control plants. In contrast, inoculation with *E. cichoracearum*, treated or not with Si, altered the expression of a set of nearly 4,000 genes. Of these, many of the up-regulated genes were defense-related, whereas a large proportion of down-regulated genes were involved in primary metabolism. These results suggested that the effect of silicon was limited to attenuating the reaction of the host to pathogen.

A study by Kauss et al. (2003) showed that, during the induction of system-acquired resistance mediated by silicon in cucumber, silicon could enhance the expression of a gene encoding a novel proline-rich protein. This protein was associated with cell wall reinforcement at the site of the attempted penetration of fungi into epidermal cells. Rodrigues et al. (2005) reported that PAL transcripts exhibited variable accumulation at low levels from 0 h to 36 h after *M. oryzae* inoculation in rice, and the accumulation was associated with the limited colonization in epidermal cells of a susceptible rice cultivar supplied with Si. Brunings et al. (2009) found that, compared to the single pathogen-infected plants, Si treatment and pathogen infection resulted in the differential expression of 54 unique genes. This implies that silicon affects the rice response to blast infection at a transcriptional level. By using PCR molecular technology, Ghareeb et al. (2011a, b) showed that silicon can increase the expression of housekeeping genes (PGK, TUB, and ACT) and jasmonic acid/ethylene marker genes. Oxidative stress and the basal defense marker genes were also up-regulated upon challenging the silicon-treated plants with *Ralstonia solanacearum*. These results suggested that Si primed the defense capacity of plant, thereby reducing the disease incidence of pathogen rather than constitutively inducing defended-related genes.

## Future Perspective

Even though significant progress has been made toward the understanding of how silicon modulates plant pathogen resistance, there are still many unanswered questions. Upon pathogen attack, the infected tissue will synthesize antimicrobial compounds together with systemic stress signals such as salicylic acid (SA), jasmonic acid (JA), and ethylene (ET). It is necessary to investigate whether SA/JA/ET signaling plays a role in silicon-mediated resistance. What is the interaction between Si and the stress signal substances and stress hormones? In addition, the targets of the majority of defense-related proteins regarding the Si-mediating stress resistance remain unknown. Identification of these proteins is critical for better understanding of how silicon mediates pathogen resistance. High-throughput technologies, such as transcriptomics and proteomics, have tremendous potential in establishing the Si-responsive genes and protein networks. Since Si-mediating pathogen resistance is a complicated process, it is necessary to use proteomics technology to identify or screen novel Si-responsive proteins and genes in plants exposed to Si treatment. A similar approach can simultaneously target the transcriptome (by DNA microarray) and metabolites (via metabolomics) in plants tolerant or sensitive to Si. Various "OMICS" approaches, including genomics, proteomics, and metabolomics are necessary to investigate the biological function regulated by Si in plant (Zargar et al. 2010).

Finally, the great challenge for the future is applying our increasing knowledge of silicon-mediated plant resistance to develop Si-responsive traits to improve crop productivity, and in particular, to develop suitable Si fertilizers that display durable and significant efficiency of pathogen control across time and space.

**Acknowledgments** The study is financially supported by grants from the National Key Basic Research Funds of China (2011CB100400), Natural Science Foundation of China (31070396), and Doctoral Fund of Ministry of Education of China (20094404110007).

## Cross-References

▶ Silicon Transporters
▶ Silicon and Aquaporins
▶ Silicon, Biologically Active Compounds
▶ Peroxidases
▶ Silicon, Physical and Chemical Properties

# References

Bélanger RR, Benhamou N, Menzies JG (2003) Cytological evidence of an active role of silicon in wheat resistance to powdery mildew (*Blumeria graminis* f.sp. *tritici*). Phytopathology 93:402–412

Bowen P, Menzies J, Ehret D, Samuels L, Glass ADM (1992) Soluble silicon sprays inhibit powdery mildew development on grape leaves. J Am Soc Hortic Sci 117:906–912

Brunings AM, Datnoff LE, Ma JF, Mitani N, Nagamura Y, Rathinasabapathi B, Kirst M (2009) Differential gene expression of rice in response to silicon and rice blast fungus *Magnaporthe oryzae*. Ann Appl Biol 155:1–10

Cai KZ, Gao D, Luo SM, Zeng RS, Yang JY, Zhu XY (2008) Physiological and cytological mechanisms of silicon induced resistance in rice against blast disease. Physiol Plant 134:324–333

Cai KZ, Gao D, Chen JN, Luo SM (2009) Probing the mechanisms of silicon-mediated pathogen resistance. Plant Sig Behav 4:1–3

Carver TLW, Zeyen RJ, Ahlstrand GG (1987) The relationship between insoluble silicon and success or failure of attempted primary penetration by powdery mildew (*Erysiphe graminis*) germlings on barley. Physiol Mol Plant Pathol 31:133–148

Chérif M, Benhamou N, Menzies JG, Bélanger RR (1992) Silicon induced resistance in cucumber plants against *Pythium ultimum*. Physiol Mol Plant Pathol 41:411–425

Chérif M, Asselin A, Bélanger RR (1994) Defense responses induced by soluble silicon in cucumber roots infected by *Pythium* spp. Phytopathology 84:236–242

Currie HA, Perry CC (2007) Silica in plants: biological, biochemical and chemical Studies. Ann Bot 100:1383–1389

Epstein E (1994) The anomaly of silicon in plant biology. Proc Natl Acad Sci USA 91:11–17

Fauteux F, Rémus-Borel W, Menzies JG, Bélanger RR (2005) Silicon and plant disease resistance against pathogenic fungi. FEMS Microbiol Lett 249:1–6

Fauteux F, Chain F, Belzile F, Menzies JG, Bélanger RR (2006) The protective role of silicon in the *Arabidopsis*-powdery mildew pathosystem. Proc Natl Acad Sci USA 103:17554–17559

Fawe A, Abou-Zaid M, Menzies JG, Bélanger RR (1998) Silicon-mediated accumulation of flavonoid phytoalexins in cucumber. Phytopathology 88:396–401

Fawe A, Menzies JG, Chérif M, Bélanger RR (2001) Silicon and disease resistance in dicotyledons. In: Datnoff LE, Snyder GH, Korndörfer GH (eds) Silicon in agriculture. Elsevier Science, New York, pp 159–169

Ghareeb H, Bozsó Z, Ott PG, Repenningc C, Stahlc F, Wydra K (2011a) Transcriptome of silicon-induced resistance against *Ralstonia solanacearum* in the silicon non-accumulator tomato implicates priming effect. Physiol Mol Plant Pathol 75:83–89

Ghareeb H, Bozsób Z, Ott PG, Wydra K (2011b) Silicon and *Ralstonia solanacearum* modulate expression stability of housekeeping genes in tomato. Physiol Mol Plant Pathol 75:176–179

Hayasaka T, Fujii H, Ishiguro K (2008) The role of silicon in preventing appressorial penetration by the rice blast fungus. Phytopathology 98:1038–1044

Heine G, Tikum G, Horst WJ (2007) The effect of silicon on the infection by and spread of *Pythium aphanidermatum* in single roots of tomato and bitter gourd. J Exp Bot 58:569–577

Inanaga S, Okasaka A, Tanaka S (1995) Does silicon exist in association with organic compounds in rice plant? Jpn J Soil Sci Plant Nutr 11:111–117

Kauss H, Seehaus K, Franke R, Gilbert S, Dietrich RA, Kröger N (2003) Silica deposition by a strongly cationic proline-rich protein from systemically resistant cucumber plants. Plant J 33:87–95

Kim SG, Kim KW, Park EW, Choi D (2002) Silicon-induced cell wall fortification of rice leaves: a possible cellular mechanism of enhanced host resistance to blast. Phytopathology 92:1095–1103

Kwon SH, Oh JH, Song HS (1974) Studies on the relationship between chemical contents of rice plants and resistance to rice blast disease. Kor J Plant Prot 13:33–39

Liang YC, Sun WC, Si J, Römheld V (2005) Effects of foliar- and root-applied silicon on the enhancement of induced resistance to powdery mildew in *Cucumis sativus*. Plant Pathol 54:678–685

Liang Y, Sun W, Zhu YG, Christie P (2007) Mechanisms of silicon-mediated alleviation of abiotic stresses in higher plants: a review. Environ Poll 147:422–428

Ma JF, Yamaji N (2006) Silicon uptake and accumulation in higher plants. Trends Plant Sci 11:392–397

Ma JF, Yamaji N (2008) Functions and transport of silicon in plants. Cell Mol Life Sci 65:3049–3057

Rémus-Borel W, Menzies JG, Bélanger RR (2005) Silicon induces antifungal compounds in powdery mildew-infected wheat. Physiol Mol Plant Pathol 66:108–115

Rodrigues FÁ, Benhamou N, Datnoff LE, Jones JB, Bélanger RR (2003) Ultrastructural and cytochemical aspects of silicon mediated rice blast resistance. Phytopathology 93:535–546

Rodrigues FÁ, McNally DJ, Datnoff LE, Jones JB, Labbé C, Benhamou N, Menzies JG, Bélanger RR (2004) Silicon enhances the accumulation of diterpenoid phytoalexins in rice: a potential mechanism for blast resistance. Phytopathology 94:177–183

Rodrigues FÁ, Jurick WM, Datnoff LE, Jones JB, Rollins JA (2005) Silicon influences cytological and molecular events in compatible and incompatible rice-*Magnaporthe grisea* interactions. Physiol Mol Plant Pathol 66:144–459

Samuels AL, Glass ADM, Ehret DL, Menzies JG (1991) Mobility and deposition of silicon in cucumber plants. Plant Cell Environ 14:485–492

Seebold KW, Kucharek TA, Datnoff LE, Correa-Victoria FJ, Marchetti MA (2001) The influence of silicon on components of resistance to blast in susceptible, partially resistant and resistant cultivars of rice. Phytopathology 91:63–69

Strange RN, Scott PR (2005) Plant disease: a threat to global food security. Annu Rev Phytopathol 43:83–116

Watanabe S, Shimoi E, Ohkama N, Hayashi H, Yoneyama T, Yazaki J, Fujii F, Shinbo K, Yamamoto K, Sakata K, Sasaki T, Kishimoto N, Kikuchi S, Fujiwara T (2004) Identification of several rice genes regulated by Si nutrition. Soil Sci Plant Nutr 50:1273–1276

Yang YF, Liang YC, Lou YS, Sun WC (2003) Influences of silicon on peroxidase, superoxide dismutase activity and lignin content in leaves of wheat *Tritium aestivum* L. and its relation to resistance to powdery mildew. Sci Agric Sin 36:813–817

Zargar SM, Nazir M, Agrawal GK, Kim DW, Rakwal R (2010) Silicon in plant tolerance against environmental stressors: towards crop improvement using omics approaches. Curr Proteomics 7:135–143

Zhang GL, Dai QG, Zhang HC (2006) Silicon application enhances resistance to sheath blight (*Rhizoctonia solani*) in rice. J Plant Physiol Mol Biol 32:600–606

# Silicosis

Andrij Holian, Melisa Bunderson-Schelvan and Raymond F. Hamilton Jr.
Department of Biomedical and Pharmaceutical Sciences, Center for Environmental Health, The University of Montana, Missoula, MT, USA

## Synonyms

Matrix metallopeptidases; Silica-induced fibrosis

## Definition

Silicosis is a debilitating and often life-threatening fibrotic disease resulting from exposure to crystalline silica. Silicosis is generally characterized by chronic inflammation and an aberrant fibrotic response to injury. In addition, a role for the matrix metalloproteinases in the etiology of silicosis has been suggested following the observation that MMP-2, -9, and -13 are upregulated in experimental lung silicosis (Perez-Ramos et al. 1999).

## Silica and Silicosis

The dangers of silica exposure have been reported as far back as the ancient Greeks, particularly in relation to pulmonary disorders (Mason and Thompson 2010). While there are various forms of silica, the crystalline structure is known to be the most dangerous, as opposed to the amorphous form. It is also important to note that crystalline silica is the most common type involved in occupational exposures. In general, crystalline silica particles are deposited in the lung, leading to a fibrotic disease known as silicosis, a disabling and sometimes fatal lung disease. Furthermore, respirable crystalline silica has been associated with other respiratory diseases, such as chronic obstructive pulmonary disease (including bronchitis and emphysema) (Cohen et al. 2008) and lung cancer (OSHA 2012).

Extensive work has been done to elucidate the mechanisms leading to the development of silicosis following exposure to particles. The pathophysiology is believed to be fundamentally due to chronic inflammation and accumulation of inflammatory mediators along with aberrant regulation of the fibrogenic pathways (Greenberg et al. 2007). In particular, an upregulation of fibrogenic cytokines such as tumor necrosis factor alpha (TNF-α) (Piguet et al. 1990) and transforming growth factor beta (TGF-β) (Olbruck et al. 1998) have been observed with increased synthesis and remodeling of lung extracellular matrix components (Delgado et al. 2006). In addition, an emerging role for the matrix metalloproteinases (MMPs) and the tissue inhibitors of metalloproteinases (TIMPs) in the etiology of silicosis has been suggested (Pardo et al. 1999).

## Matrix Metalloproteinases

The matrix metalloproteinases (sometimes referred to as matrix metallopeptidases), are a family of matrix-degrading proteinases requiring a zinc ion at the active site for catalysis (Shapiro and Senior 1999). MMPs are generally categorized according to their primary catalytic properties. Categories include the collagenases (MMP-1, -8, and -13), gelatinases (MMP-2 and -9), stromelysins (MMP-3, -10, and -11), matrilysins (MMP-7 and -26), the membrane-type MMPs (MMP-14, -15, -16, -17, -24, and -25) and the ungouped MMPs (MMP-12, -19, -21, and -28), which are extensively reviewed in Sbardella et al. (2012).

The three MMPs most associated with silicosis are MMP-2 (Gelatinase-A, 72 kDa gelatinase), MMP-9 (Gelatinase-B, 92 kDa gelatinase), and MMP-13 (Collagenase 3). MMP-2, as seen in Fig. 1, degrades type IV collagen. MMP-9, as seen in Fig. 2, degrades type IV and type V collagens. Whereas MMP-13, found in Fig. 3, degrades type IV, IX, X, and XIV collagens, fibronectin, laminin, fibrilin 1, and serine protease inhibitors (Amalinei et al. 2010).

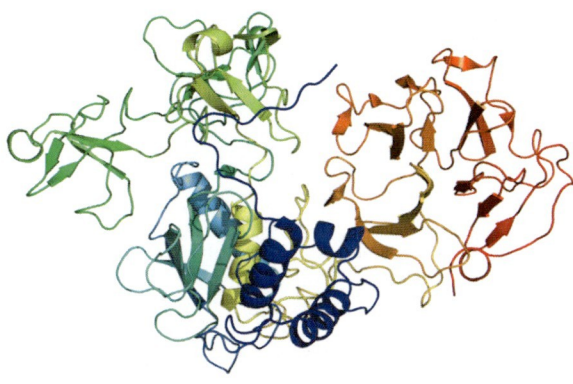

Silicosis, Fig. 1 MMP-2 protein structure by X-ray diffraction (Source: protein data bank)

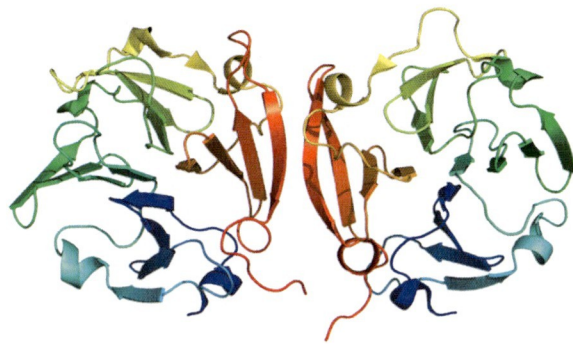

Silicosis, Fig. 2 MMP-9 protein structure by X-ray diffraction (Source: protein data bank)

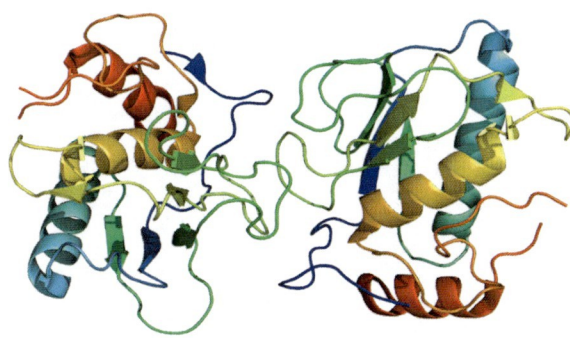

Silicosis, Fig. 3 MMP-13 protein structure by X-ray diffraction (Source: protein data bank)

Expression and inhibition of the MMPs is a tightly regulated process via interactions with growth factors, cytokines, and extracellular matrix components (Crosby and Waters 2010). Under pathological conditions, upregulation of MMPs, resulting in degradation of the extracellular matrix, is believed to play a significant role in acute and chronic diseases such as silicosis. Specifically, aberrant degradation of the extracellular matrix is believed to result in inflammatory cell accumulation and disruption of the epithelial/endothelial cell architecture followed by upregulation of the fibrotic machinery (Crosby and Waters 2010). In fact, pulmonary clara cells, progenitors of bronchial epithelial cells, may play a role in the gene expression of MMP-2 and -9 following silica exposure (Yatera et al. 2001). Additionally, it has been suggested that structural instability resulting from undo expression of MMPs following an environmental insult may have additional effects such as activation of signal transduction pathways that may lead to apoptosis or other cellular events (Shapiro 1998).

## Matrix Metalloproteinases and Silicosis

Changes in the expression of MMPs and TIMPs in experimental silicosis have lead to studies aimed at elucidating the role of MMPs in disease development. For example, in tumor-bearing silicotic mice, there is a decrease in the ratio of TIMPs/MMP-9 compared to tumor-bearing nonsilicotic mice, suggesting that aberrant changes in the ratio of TIMPs to MMPs may play a role in the development of silicosis (Ishihara et al. 2003). Furthermore, while crystalline silica exposure is known to induce fibrotic nodule formation, continued growth of the fibrotic nodules appear to involve remodeling mechanisms similar to wound healing typically orchestrated by MMP-10 (Scabilloni et al. 2005). In addition, there appears to be a time-dependent influence of the MMPs on silicotic granuloma formation. Early stage silicotic granulomas have increased expression of MMP-2, MMP-9, and MMP-13 as well as TIMP-1 and TIMP-2 (Perez-Ramos et al. 1999). In contrast, older granulomas have similar staining for TIMPs, but MMP staining is reduced, suggesting that decreased collagenolytic activity in advanced phases may contribute to collagen accumulation of progressive fibrosis (Pardo et al. 1999). The inflammatory cytokine TNF-α may mediate MMP-13 and TIMP-1 expression, as deletion of two different TNF-α receptors modified protein expression and fibrosis in a C57BL/6 silicosis model (Ortiz et al. 2001). Additionally, the MMP-12 (macrophage metalloelastase) gene was significantly upregulated in a subchronic silicosis rat model, although it was not detectable as a protein increase (Langley et al. 2011). Fragmentation of extracellular collagen may be mediated to a large degree by MMP-14; however, recent evidence from clinical and degradomic studies suggest that the upregulation of MMPs in disease may not always result in a detrimental effect (Butler and Overall 2009; Rodriguez et al. 2010). In fact, recent evidence suggests that macrophage-derived MMPs have anti-fibrotic properties under specific conditions (Wynn and Barron 2010) and have therefore been termed "antidrug targets" (Rodriguez et al. 2010).

## Potential Role of the NLRP3 Inflammasome

The NLRP3 inflammasome is a complex system of bioactive proteins in immune cells resulting in proinflammatory cytokine (IL-1β, IL-18, and IL-33) release, all of which must be tightly regulated. Both asbestos and silica are known to activate the NLRP3 inflammasome (Dostert et al. 2008; Hornung et al. 2008). In fact, the inflammatory response leading to development of a fibrotic condition following silica exposure is dependent on activation of the NLRP3 inflammasome (Cassel et al. 2008). The specific cellular mechanism in alveolar macrophages is believed to involve phagosomal destabilization resulting in lysosomal damage and rupture, which releases cathepsin B and initiates NLRP3 inflammasome assembly (Hornung et al. 2008). Interactions between the NLRP3 inflammasome and MMP activity are suggested by induction of MMP-9 expression by increases in IL-1β, an indicator of NLRP3 inflammasome activation (Cheng et al. 2010). The importance of the NLRP3 inflammasome in chronic cases of silicosis requires further investigation, although one study did find significant increases in cathepsins including cathepsin B in silicosis patient lavage samples (Perdereau et al. 2006). This could be indicative of continuing NLRP3 inflammasome activation. While MMP-2 and -9 are upregulated in both acute and chronic silicosis models, proinflammatory cytokines (including IL-18 and IL-1β) were only upregulated in acute silicosis (Langley et al. 2011). The connection between NLRP3 inflammasome and MMP activity may in fact be tangential, as the inflammasome activation is probably responsible for initial inflammatory cytokines resulting in sustained inflammation and eventually tissue remodeling via MMPs.

## Summary

Silicosis is a debilitating and often times fatal disease of the lungs resulting from environmental exposure to crystalline silica. While the exact etiology of the disease is still under investigation, MMPs are thought to play a role in tissue remodeling resulting in a fibrotic pathology. MMP-2, -9, and -13 are primarily overexpressed in silicotic patients. The initiators of the inflammatory process are probably activated by the NLRP3 inflammasome resulting from damaged alveolar macrophages that take up crystalline silica in the lung. The connection between inflammasome activation and MMP expression is not entirely clear, although MMP expression may simply be a response to sustained inflammation over time.

## Cross-References

▶ Silicon Exposure and Vasculitis
▶ Silicon, Physical and Chemical Properties

## References

Amalinei C, Caruntu ID et al (2010) Matrix metalloproteinases involvement in pathologic conditions. Rom J Morphol Embryol (Revue roumaine de morphologie et embryologie) 51(2):215–228

Butler GS, Overall CM (2009) Updated biological roles for matrix metalloproteinases and new "intracellular" substrates revealed by degradomics. Biochemistry 48(46):10830–10845

Cassel SL, Eisenbarth SC et al (2008) The Nalp3 inflammasome is essential for the development of silicosis. Proc Natl Acad Sci USA 105(26):9035–9040

Cheng CY, Kuo CT et al (2010) IL-1beta induces expression of matrix metalloproteinase-9 and cell migration via a c-Src-dependent, growth factor receptor transactivation in A549 cells. Br J Pharmacol 160(7):1595–1610

Cohen RA, Patel A et al (2008) Lung disease caused by exposure to coal mine and silica dust. Semin Respir Crit Care Med 29(6):651–661

Crosby LM, Waters CM (2010) Epithelial repair mechanisms in the lung. Am J Physiol Lung Cell Mol Physiol 298(6):L715–731

Delgado L, Parra ER et al (2006) Apoptosis and extracellular matrix remodelling in human silicosis. Histopathology 49(3):283–289

Dostert C, Petrilli V et al (2008) Innate immune activation through Nalp3 inflammasome sensing of asbestos and silica. Science 320(5876):674–677

Greenberg MI, Waksman J et al (2007) Silicosis: a review. Disease Month: DM 53(8):394–416

Hornung V, Bauernfeind F et al (2008) Silica crystals and aluminum salts activate the NALP3 inflammasome through phagosomal destabilization. Nat Immunol 9(8):847–856

Ishihara Y, Nishikawa T et al (2003) Expression of matrix metalloproteinase, tissue inhibitors of metalloproteinase and adhesion molecules in silicotic mice with lung tumor metastasis. Toxicol Lett 142(1–2):71–75

Langley RJ, Mishra NC et al (2011) Fibrogenic and redox-related but not proinflammatory genes are upregulated in Lewis rat model of chronic silicosis. J Toxicol Environ Health A74(19):1261–1279

Mason E, Thompson SK (2010) A brief overview of crystalline silica. J Chem Health Safety 17:6–8

Olbruck H, Seemayer NH et al (1998) Supernatants from quartz dust treated human macrophages stimulate cell proliferation of different human lung cells as well as collagen-synthesis of human diploid lung fibroblasts in vitro. Toxicol Lett 96–97:85–95

Ortiz LA, Lasky J et al (2001) Tumor necrosis factor receptor deficiency alters matrix metalloproteinase 13/tissue inhibitor of metalloproteinase 1 expression in murine silicosis. Am J Respir Crit Care Med 163(1):244–252

OSHA. (2012) Safety and Health Topics: Silica, Crystalline. Retrieved 7 Apr 2012, from http://www.osha.gov/dsg/topics/silicacrystalline/index.html

Pardo A, Perez-Ramos J et al (1999) Expression and localization of TIMP-1, TIMP-2, MMP-13, MMP-2, and MMP-9 in early and advanced experimental lung silicosis. Ann N Y Acad Sci 878:587–589

Perdereau C, Godat E et al (2006) Cysteine cathepsins in human silicotic bronchoalveolar lavage fluids. Biochim Biophys Acta 1762(3):351–356

Perez-Ramos J, de Lourdes Segura-Valdez M et al (1999) Matrix metalloproteinases 2, 9, and 13, and tissue inhibitors of metalloproteinases 1 and 2 in experimental lung silicosis. Am J Respir Crit Care Med 160(4):1274–1282

Piguet PF, Collart MA et al (1990) Requirement of tumour necrosis factor for development of silica-induced pulmonary fibrosis. Nature 344(6263):245–247

Rodriguez D, Morrison CJ et al (2010) Matrix metalloproteinases: what do they not do? New substrates and biological roles identified by murine models and proteomics. Biochim Biophys Acta 1803(1):39–54

Sbardella D, Fasciglione GF et al (2012) Human matrix metalloproteinases: an ubiquitarian class of enzymes involved in several pathological processes. Mol Aspects Med 33(2):119–208

Scabilloni JF, Wang L et al (2005) Matrix metalloproteinase induction in fibrosis and fibrotic nodule formation due to silica inhalation. Am J Physiol Lung Cell Mol Physiol 288(4):L709–717

Shapiro SD (1998) Matrix metalloproteinase degradation of extracellular matrix: biological consequences. Curr Opin Cell Biol 10(5):602–608

Shapiro SD, Senior RM (1999) Matrix metalloproteinases. Matrix degradation and more. Am J Respir Cell Mol Biol 20(6):1100–1102

Wynn TA, Barron L (2010) Macrophages: master regulators of inflammation and fibrosis. Semin Liver Dis 30(3):245–257

Yatera K, Morimoto Y et al (2001) Increased expression of matrix metalloproteinase in Clara cell-ablated mice inhaling crystalline silica. Environ Health Perspect 109(8):795–799

# Silver

▸ Silver, Pharmacological and Toxicological Profile as Antimicrobial Agent in Medical Devices

# Silver as Disinfectant

Charles P. Gerba
Department of Soil, Water and Environmental Science, University of Arizona, Tucson, AZ, USA

## Synonyms

Antimicrobial action of silver; Metals as antimicrobials

## Definition

Disinfection refers to the killing or inactivation (in the case of viruses) of microorganisms. Silver is capable of damaging critical sites needed for metabolism and structural integrity of microorganisms by interacting, both with proteins and nucleic acids.

Silver has been known since ancient times for its therapeutic properties. It was first used to treat drinking water over 3,000 years ago and was used to treat burns and chronic wounds. Sodium nitrate, as eye drops have been added to the eyes of newborn to prevent gonococcal infections before the advent of antibiotics. While the rise of antibiotics to treat bacterial infections lessens interest in silver, a resurgence in its use has increased in recent decades because of antibiotic resistant organisms and the development of silver nanoparticles. Advances in material science have created new approaches to the formulations and in optimizing the different mechanisms in which silver can kill pathogenic microorganisms.

Silver-based antimicrobials are used today in several different forms. Silver nitrate has been used the longest to treat a burn, eye, and wound infections in concentrations around 0.5%. Silver sulfadiazine is a combination of silver with the antibiotic sulfadiazine resulting in a synergistic effect against bacteria causing burn wound infections. It is usually applied as a cream. Silver can be combined with zeolites a complex crystal aluminosilicate which acts as an ion exchanger to slowly release the silver over time. The zeolites can be incorporated into material or applied as coatings. Silver nanoparticles are particles in the 5–100 nm size that can be synthesized in the laboratory or by certain microorganisms (referred to as biogenic silver). They have a large surface area to interact with microorganisms and can, if small enough penetrate bacteria. They can also be used to coat surfaces or used in water for killing microorganisms.

## Antimicrobial Action Against Bacteria

While the antibacterial action of silver is not completely known, several mechanisms are generally accepted (Table 1). Silver can be used to stain proteins as it readily binds to the sulfhydryl (−SH) groups (Thurman and Gerba 1989). When this happens it can result in the denaturalization of structural proteins or enzymes resulting in loss of their function. Silver has been found to complex with sulfhydryl groups in the bacterial membrane that are components of enzymes which participate in blocking electron transport and respiration. Membrane transport of metabolites in the bacterial membrane may also be affected. Silver can also effect the expression of important proteins and enzymes within the cell, such as the 30S ribosomal subunits, succinyl co A synthetase, maltose transporter, and fructose bisphopshate adolase (Rai et al. 2012). Within the cell, silver can bind to the DNA by displacing the hydrogen bonds between adjacent nitrogens of purine and pyrimidine bases. This may result in a DNA complex preventing replication of the DNA and subsequent cell division. Metals also bind to phosphate groups, forming a positive dipole, followed by formation of a cyclic phosphate and cleavage of these molecules at the phosphodiester bond.

**Silver as Disinfectant, Table 1** Antimicrobial mechanisms of silver

| |
|---|
| Combines with thio groups in proteins causing denaturalization of enzymes and structural proteins |
| Inhibits cell wall formation in bacteria |
| Damages the cell membrane of bacteria |
| Inhibits cell wall formation in bacteria |
| Attaches to the 30S ribosomal subunit interfering with protein synthesis |
| Interferes with bacterial peptides that can effect cell signaling |
| Intercalates between nucleic acid bases preventing replication |
| Produces free hydroxyl radicals which oxidize organic compounds |

Silver may also act as a semiconductor causing the generation of reactive oxygen species such as hydroxyl groups, which readily oxidize organic molecules. When combined with titanium dioxide and other metals, this effect can be enhanced with visible light serving as an energy source to drive the formation of free radicals. The enhanced activities of these combinations are attributed to increased specific surface area, reduced carrier recombination (recombination of the energized electron within the $TiO_2$ rather than with oxygen), as well as improved electron–hole pair separation. The silver acts as a mediator in storing and transporting photogenerated electrons from the $TiO_2$ surface enhancing the degradation of organic molecules.

The generation of silver nanoparticles has been seen as a new approach in enhancing the antimicrobial effects of silver (Rai et al. 2009). They can be generated in the laboratory or by precipitation of silver within bacterial or fungal cells. Silver nanoparticles are in the range of 5–100 nm in size. The mode of action of silver nanoparticles is similar to that of Ag in solution, but the effective concentrations delivered to the organism are much greater. They have been found to be almost twice as efficient in killing bacteria. The positive charge of silver nanoparticles acts to attract them to the negative charge of cell membrane of bacteria, where damage to its integrity of the bacterial cell readily occurs. The shape and size of the nanoparticle may also affect its antibacterial properties, with small triangle-shaped particle being more effective.

Silver probably first reacts with the bacterial cell membrane and wall causing damage to enzymes and cell wall components followed by penetration of the silver ions into the cytoplasm and reaction with the nucleic acids and other proteins needed for respiration (Lara et al. 2011).

## Silver Action Against Other Microorganisms

Silver is also effective against viruses, fungi, and protozoa, but few studies have been conducted on how it kills/inactivates these organisms (Silvestry-Rodriguez et al. 2007). In general, viruses are more resistant to the action of silver than bacteria, since they are more simply composed. Denaturalization of capsid proteins, combination with the nucleic acid or lipid membrane have all been proposed as responsible for inactivation of viruses. In the case of viruses containing a lipid layer (e.g., human immunodeficiency virus) silver nanoparticles bind with the viral glycoprotein and inhibit the virus binding to host cells. In the case of influenza nanosilver particles may interfere with the fusion of the viral membrane with that of the host cell membrane preventing penetration into the host cell.

## Synergism with Other Disinfectants

One of the advantages of silver as an antimicrobial is that it can be used in combination with antibiotics, other metals, UV light, and other chemical disinfectants resulting in an enhanced killing ability in terms of reduced exposure time and dose. Synergy has been reported for potassium permanganate, chlorine, chloramines, and other oxidizers. It has been suggested that in the case of chemical oxidizers that the cell wall of bacteria is damaged and this enhances the penetration of the silver and allows it to react with other critical components of the bacteria (i.e., nucleic acid replication, metabolic enzymes). Enhanced activity is also seen with natural products that effect membrane function, such as, cinnamonaldehyde. Silver has also been shown to be synergistic with other metals such as copper and zinc. Inactivation of *Legionella pneumophila* with silver and copper is relatively slow, but in combination with a low level of chlorine it proceeds much faster than chlorine or the metals alone. The presence of low levels of silver has been shown to reduce the amount of UV radiation to inactivate the coliphage MS-2.

## Silver Resistance

One of the concerns with any antimicrobial is the development of resistance. This phenomenon is clearly seen with antibiotics, but not with disinfectants like chlorine. Despite the widespread use of chlorine for more than 100 years no selection for resistance has been documented. The reason is the non-site-specific action of chlorine on microorganisms, compared to the site-specific action of antibiotics. Silver has multiple mechanisms of action and sites which it can attack. Many studies have documented the occurrence of resistant of strains of bacteria and fungi. However, in some cases these may be an artifact of how the studies were conducted, and may reflect just an increase in tolerance to higher levels of silver or longer times of contact for the organism to be killed. Resistance is better defined as a level of an antimicrobial that can no longer be used for a particular application. Some bacteria appear to have a natural resistance to silver in areas where they are regularly exposed to silver, such as, silver mines, hospital burns wards, and hospital water distribution systems were silver is used to disinfect the water to control *Legionella* bacteria. Two proposed mechanisms of this resistance are that silver ions are excluded from the cell or immobilized outside of the cell. Plasmid-mediated silver resistance is believed to be the most common means of silver resistance in Gram-negative bacteria and typically involves energy-dependent rejection of the silver from the cell. Silver resistance mediated by plasmids in *Salmonella* involves a total of nine genes and is unusual in that it includes three separate types of resistance mechanisms: a periplasmic metal-binding protein that binds silver to the cell surface, a chemiosmotic efflux pump, and an ATPase efflux pump (Silver 2003).

## Applications

### Water

Silver has been used to treat drinking water for centuries and is currently used in a number of specialized applications (Silvestry-Rodriguez et al. 2007). Both the United States Environmental Protection Agency and the World Health Organization regard silver safe for human consumption. Based on epidemiology and pharmacokinetic studies, a lifetime threshold of 10 g of silver can be considered a No Observable Adverse

Effect Level (NOAEL) for humans. In the United States, silver in drinking water is a secondary (nonenforceable) standard of 100 µl. Silver can be added to drinking water as silver nitrate or by ionization by passing the water between two electrodes of silver or silver/copper. The rate of microbial killing is much slower with silver for some organisms compared to chlorine or other common water disinfectants, but it has seen applications in the developing world and in the control of pathogens in water health-care facilities to control *Legionella* bacteria. An advantage of silver over chlorine is that it is much more persistence in water and its efficacy less influenced by soluble organic matter. Probably the largest use of silver as a disinfectant is to control the presence of *Legionella* in cooling towers and health-care facilities drinking water distribution systems. *Legionella* is difficult to control by chlorine alone, because of its resistance to chlorine and ability to escape interaction with chlorine by being engulfed by protozoa which live in the distribution system. Silver is usually added on a continuous basis in combination with copper via ionization, with the water passing between the electrodes to deliver approximately 200 µl of silver and 1,200 µl of copper. Silver has also been used commercially in swimming pools and hot tubs, but has to be used with chlorine because it does not act fast enough alone. However, because of its synergistic effect, lower levels of chlorine or bromine can be added to achieve a more rapid rate of microbial killing. Silver has been used extensively for decades for retarding biofilms in faucet-mounted activated carbon filters sold for household use.

## Antimicrobial Materials

Silver can be applied as coatings or added to materials during manufacturing. It has been incorporated into a number of consumer products including telephones, toilet seats, toys, infant pacifiers, socks, sports basics, shrubs, towels, bed sheets, drapes to name a few. Silver can be added alone or in combination with $TiO_2$, which acts as a photocatalyst, or with other metals like copper (Xiong et al. 2011). Silver can be added to ceramics that are able to trap metal ions and may be then added to other materials, such as paints, plastics, waxes, polyesters. Its use in fabrics has been suggested as a means of controlling pathogens and reducing odors. How useful silver is in these products under actual use conditions has not been well studied.

Silver added to fabrics is known to kill bacteria and is effective in household hand towels in reducing bacterial numbers after repeated use. Zeolite ceramic (sodium aluminosilicate) has a porous three-dimensional crystalline structure in which silver ions can be retained. Zeolites act as ion exchangers, releasing silver into the environment in exchange for other cations. The amount of silver released is dependent upon the concentration of cations in the environment. Zeolites containing silver/copper/zinc in various combinations have seen extensive use as antimicrobial coatings in recent years.

Another application of silver has been in medical devices including urinary and peritoneal catheters, prosthetic heart values, sutures, and fracture-fixing devices. The coatings of these devices are intended to retard the formation of biofilms and reduce the risk of nosocomial bacterial infections. Clinical studies have shown that silver-coated catheters were more effective in preventing infection. The type of silver may be important as meta-analysis of clinical studies found silver oxide catheters were not as effective as silver-alloyed catheters. Silver is effective against a wide range of antibiotic resistant bacteria. In contrast to silver's benefits in medical devices the use of silver in creams or bandages in preventing infection and reducing the time for healing from burn infections is not clear and clinical studies have provided mixed results.

Silver is clearly an antimicrobial that is valuable in certain applications, such as control of *Legionella* in drinking water distribution systems and certain medical devices. However, its effectiveness in many practical applications is not well documented and needs more study. The development of silver nanoparticles could lead to the more effective use of silver as an antimicrobial, but there are concerns about the nonspecific killing of animal cells (cytotoxicity) by nanoparticles that need to be answered before widespread application.

## Cross-References

▶ Metallothioneins and Silver
▶ Silver, Neurotoxicity
▶ Silver-Induced Conformational Changes of Polypeptides
▶ Zinc and Wound Healing

## References

Lara HL, Garza-Trevino EN, Ixtepan-Turrent L, Singh DK (2011) Silver nanoparticles are broad spectrum bactericidal and virucidal compounds. J Nanobiotechnol 9(30):7–8

Rai M, Yadav A, Gade A (2009) Silver nanoparticles as a new generation of antimicrobials. Biotechnol Adv 27:76–83

Rai MK, Seshmukh SD, Ingle AP, Gade AK (2012) Silver nanoparticles: the powerful nanoweapon against multidrug-resistant bacteria. J Appl Microbiol 112:841–852

Silver S (2003) Bacterial silver resistance: molecular biology and uses and misuses of silver compounds. FEMS Microbiol Rev 27:341–353

Silvestry-Rodriguez N, Sicaritos-Rulelas EE, Gerba CP, Bright KR (2007) Silver as a disinfectant. Rev Environ Contam Toxicol 191:23–45

Thurman RB, Gerba CP (1989) The molecular mechanisms of copper and silver ion disinfection of bacteria and viruses. CRC Crit Rev Environ Contr 18:195–315

Xiong Z, Ma J, Ng WJ, Waite TD, Zhao XS (2011) Silver-modified mesoporous $TiO_2$ photocatalyst for water purification. Water Res 45:2095–2103

## Silver Binding to Polypeptides

▶ Silver-Induced Conformational Changes of Polypeptides

## Silver Impregnation Methods in Diagnostics

Toshiki Uchihara
Laboratory of Structural Neuropathology, Tokyo Metropolitan Institute of Medical Science,
Tokyo, Japan

## Synonyms

Argyrophilia; Epi-molecular; Histological diagnosis; Neuropathology

## Definition

Silver impregnation includes classical techniques, which have revolutionalized histochemistry and neuroscience for over 100 years. Initial probing of the target with silver ion to form silver atom in situ and its subsequent amplification for better visualization are both regulated by catalytic reaction of silver ion, which could be sometimes capricious. Recent progress has achieved sensitive and less capricious visualization of the targets by introducing physical developer. Because the final visualization is uniformly mediated by silver grains, silver-impregnated targets are labeled "argyrophilic" regardless of the method used. In diagnostic neuropathology, however, argyrophilia could be heterogeneous and dependent on the target and method, possibly representing qualitative differences indicative of the nature of the target. This empirical correlation between argyrophilic profiles and tau isoforms is currently in routine use for diagnostic differentiation among Alzheimer disease, Pick body disease, and corticobasal degeneration/progressive supranuclear palsy. Although its mechanism may be too complex to explain based on current knowledge on molecules, its clarification may improve not only the silver impregnation procedures but also our understanding how lesions are generated in disease-specific fashion.

## Introduction

Silver impregnation methods, initially invented more than a century ago, have contributed a great deal in the development of histopathology (Heinz 2005). In the field of neuroscience, it has opened a door to substantiate and expand the concept of "neuron" in relation to its development and functions. Furthermore, various alterations in the brain have been visualized by silver impregnation methods, which defined major hallmark lesions of Alzheimer disease or Pick body disease. Although some silver impregnation methods are capricious and difficult to control to obtain consistent results, recent improvements are sufficiently successful in obtaining more stable results to facilitate their reliable application for histological differentiation in the field of neuropathology. In addition to this improved reliability, recent neuropathological studies demonstrated that "argyrophilia" is not homogeneous and is dependent on the method and the target (Uchihara 2007). If an argyrophilic reaction is obtained through an interaction between a method and a target but not through another, this method/target-dependent argyrophilia may represent some qualitative aspects of the target, which may be potentially useful

**Silver Impregnation Methods in Diagnostics, Table 1** General principles for different staining methods

|  | Probe-reporter-amplification immuno-HC, lectin-HC, ISH | Conventional dyes | Silver impregnations |
|---|---|---|---|
| *Probing Agents* | Antibodies, lectin, nucleic acid | Dyes | Ag+ |
| Mechanism | Specific affinity based | Affinity based (charge etc.) | Catalyic (or affinity based?) |
| *Reporter molecules* | Biotin, peroxidase conjugates | – | Ag+ → Ag |
| Mechanism | Affinity based | | Reduction (catalytic) |
| *Amplifying system* | ABC, PAP, CARD* | – | Ag+ and reducing agents |
| Mechanism | Affinity based or catalytic* | | Autocatalytic |
| *Visualizing Agents* | DAB, fluorochrome, Al-P | Dyes | Enlarged Ag grains |
| Coloration | Various | Intrinsic color (transparent) | Black (blocked) |
| *Interpretation* | Selective to a single molecule by neglecting others | Barely specific | Potentially qualitative |
| | | | Epi-molecular |

*HC* histochemistry, *ISH* in-situ hybridization, *Ag+* silver ion, *Ag* silver atom, *ABC* avidin-biotin peroxidase complex, *PAP* peroxidase anti-peroxidase, *CARD* catalysed reporter amplification, *DAB* diaminobenzidine, *Al-P* alkaline phosphatases
* CARD is a catalytic reaction

for their histological differentiation. Even though the molecular mechanism for argyrophilia and its method/target-dependent nature remain speculative, this empirical knowledge is quite helpful for diagnostic practice. Even if this argyrophilia is not directly related to a molecule, it may represent "epi-molecular" factors not probed otherwise. In this entry, general principles of silver impregnation, different from other staining methods, are outlined first. Although detailed mechanisms of silver impregnation are not fully elucidated, it may be useful and necessary to keep the known principles in mind so that researchers, diagnostic practitioners, and technicians share and enjoy silver impregnations to improve our understanding on neurological diseases for better management and care for the patients.

## Catalytic Chemical Reactions Different from Dye Staining

In general, histological staining is successful when target structures are contrasted enough against the surroundings so that their morphological features are visually recognized (Prento 2001; Horobin 2002). This morphological visualization is based on the relative abundance of what are visualized (molecules, lesions, conformational state, etc.), which are organized into the structures. Pinpointing a molecule by immunohistochemistry (IHC) is mediated by the specific affinity of the primary antibody. This selective affinity is visualized through reporters (biotin, peroxidase, etc.), which is then amplified with avidin-biotin-peroxidase, peroxidase anti-peroxidase method (both affinity based) or catalyzed reporter amplification (catalytic) and is finally visible with diaminobenzidine, alkali-phosphatase, or fluorochromes (Table 1). This separation into probing, reporting, amplifying, and visualizing steps is successful in establishing a specific and sensitive way to identify the molecules in the structures with flexible combinations of probes, reporters, and visualization steps. In contrast, conventional dyes, already visible with its intrinsic color, work both as a probe and a visualizing agent. Although IHC achieves a highly specific, sensitive, and flexible identification of the target, its highly biased attention limited to a known single target is nothing but a neglect of all the others including unknown species (Table 1). In contrast, barely specific nature of conventional dyes may provide a more comprehensive view of the entire organization, while it is not possible to pinpoint the molecules. Silver impregnation is different from these methods because of the following reasons: (1) The same agent silver ion/atom plays multiple roles from probing, reporting, amplifying, and visualization. (2) Underlying mechanisms for these steps are pinpoint catalytic reactions, which transform invisible silver ions into silver grains so that the original localization of what are probed is retained and represented. This catalytic nature could be so capricious that it is sometimes difficult to obtain consistent results even with the same set of histological sections, solutions, and protocol, manipulated by the same technician. However, recent improvements make silver impregnations adequately stable and reliable to be applied for histological diagnosis and research.

## In Situ Probing (Argyrophilia) with Silver Ion

Solutions used for silver impregnation methods are usually colorless and transparent, which contain silver ion. In situ transformation of this silver ion into metallic silver atom is the primary step of silver impregnation. This in situ transformation requires (1) specific attachment of silver ion to the expected targets and (2) transformation of silver ion into atomic silver by reduction (acceptance of electron) (3) without nonspecific deposition of silver (Gallyas 2008). If the intrinsic reducing capacity of the target, as with vitamin C or cathecolamines, is sufficient for direct generation of visible silver grains, it is called "argentaffin reaction." Otherwise, it is sometimes necessary to reduce the attached silver ion to form metallic silver atom, which is called "argyrophilic reaction." Because these metallic silver grains are sometimes too small, it is often necessary to increase their size so that their light microscopic identification is possible. Visualization of attached silver ion/atom by reduction and/or amplification is collectively called "development" as in photography.

Because the attachment of some conventional dyes is guided by coulombic attraction at least partly, elevation of pH may be associated with increased binding of negatively charged dyes and this attachment occurs without delay (Prento 2001; Horobin 2002). In contrast, the initial attachment of silver ion exhibits optimal pH and requires an incubation period followed by a plateau along S-shaped curve (Gallyas 2008). This indicates that some chemical transformations of silver ion through interaction with histological samples rather than simple attraction, as with conventional dyes, are the primary mechanism. Because this catalytic nature of chemical transformation is dependent on the interaction between the target, silver ion, and environment (content of silver solution, pH, temperature, etc.), it is possible that argyrophilia may represent some qualitative nature of the target even though visualized end products are aggregates of silver atoms, regardless of the silver impregnation methods.

## Chemical or Physical Development

Silver ion attached to the target is not yet visible before being transformed into metallic silver after accepting electrons. Electrons may be derived from the target tissue to transform silver ion into silver atom, which is now tightly anchored to the tissue elements. Otherwise, external reducing agents are necessary. If the anchored silver ion is of sufficient quantity, their reduction by external reducing agents may yield metallic silver grain, observable on light microscope. This is called chemical development. Even if initial metallic silver atoms are formed, they are sometimes too small to be visualized with light microscope. It is then necessary to increase their size by way of accumulation of silver ion to grow as silver atom around the primary site of attachment, now anchoring the primary silver atoms. In addition to some reducing agent, silver ions should be provided in parallel so that silver grains accumulate and increase in size around the primary target to the extent adequate for light microscopic examination, a step called "physical development" (Newman and Jasani 1998; Gallyas 2008). This reaction is guided by the autocatalytic nature of silver atom through interaction with silver ion in developer solutions as follows. Once silver atom is anchored to the target, silver ions in touch with this primary silver atom get ready to accept electron from the reducing agent and the reduced silver ions are incorporated into the silver grain as silver atoms. Because these newly integrated silver atoms may serve as a new catalyst for subsequent deposition of silver ion, this autocatalytic reaction follows an exponential increase by order of three. This autocatalytic reaction is the basis how transformation of silver ions into atomic silver is site-directed around the initial attachment of silver so that initial topology is ultimately represented on the tissue section. This silver-based signal amplification is also useful in enhancing the signal of colloidal gold or diaminobenzidine deposits of IHC or in situ hybridization (Danscher et al. 1994).

Although this silver-based development or amplification provides a highly sensitive way of visualization, it is sometimes capricious and unwanted precipitation of silver grains could be troublesome. In order to guide this capricious nature of this autocatalytic silver under control, protective colloids that may protect silver ions apart from reducing agent and catalytic silver grains have been successful. Even in the presence of protective colloid such as arabic gum, however, it is sometimes necessary to avoid light during development. Introduction of tungstosilicate as a protective colloid allows physical development under unprotected light at room temperature, when monitoring of the development over time is possible (Newman and Jasani 1998; Gallyas 2008). This highly sensitive nature of autocatalytic physical development could be a potential source of nonspecific impregnation, not necessarily related to initial formation of silver atom. Even under the control of protective colloid, copresence of silver ions and reducing agents in the physical developer may predispose spontaneous formation of metallic silver atoms to serve as a nidus for subsequent autocatalytic growth of silver grains. This nidus formation could be induced in some remaining focus in the tissue, not probed initially by silver ion but detected only in the copresence of reducing agents. Although several pretreatments may enhance the expected foci and suppress unexpected foci for better contrast, careful monitoring of the section during development is often necessary. Because of this catalytic nature of silver solution and developer, following precautions to avoid non-specific precipitate are mandatory during the entire manipulation. Because even minute precipitates on glassware may induce silver grains, all glassware should be treated with hypochlorous acid and thoroughly washed with double distilled water. Metallic holder and metallic forceps may induce precipitates, which should be avoided. Although some slide-coating material may induce nonspecific silver precipitates, silan and ovalbumin are usually acceptable for coating to our experience.

## Practical Application in Diagnostic Neuropathology

Because silver-impregnated targets exhibit a dense color from copper brown to black that inhibits light transmission, one may have an impression that "argyrophilia" represents some homogeneous characteristics regardless of the silver impregnation methods and their targets. Because a great variety of silver impregnation methods have been invented mainly through experience, and they are more or less specific to different cell types or structures, resultant argyrophilia may be heterogeneous and potentially represent some qualitative nature (Heinz 2005). Their delicate tinctorial difference from copper brown to black may be related to the size of the silver grains rather than qualitative differences between the targets. Although the molecular basis for this specificity remains unknown, it is probable that chemical interaction between target and silver ion, both modifiable by pretreatment or coexistent molecules in silver solution, is responsible for this specificity. In the field of neuropathology (Uchihara 2007), a variety of pathological deposits have been identified initially by silver impregnations, especially Bielschowsky method (Fig. 1D), one of the classic and most popular methods. This method highlights senile plaques (SPs), Pick bodies (PBs), and neurofibrillary tangles (NFTs) as argyrophilic, while their morphologies are quite distinct probably related to their distinct molecular compositions: amyloid-beta protein (Aβ) for SPs, three-repeat (3R) isoform of tau protein for PBs, and combination of 3R and four-repeat (4R) isoforms for NFTs. The fact that this silver impregnation method labels different structures with different molecular compositions indicates that it is not the molecule itself but rather some structural commonalities shared between these lesions as seen with fluorescent probes, thioflavin, or thiazin-red endowed with affinity to β-pleated structure of fibrils. Indeed, treatment with formic acid enhances immunoreactivity for Aβ in SPs while it markedly reduces argyrophilia. This contrast is compatible with the apparent discrepancy between what are probed with antibody (molecules, Aβ) and what exhibits argyrophilia (SPs).

## Quantitative or Qualitative?

Although the exact mechanism of each silver impregnation method and the identity probed by each method remain to be clarified, different methods have been introduced in the field of neuropathology to improve the sensitivity and specificity to visualize argyrophilic structures. Probably because of their experience-based development without solid molecular explanations for the mechanism, most silver impregnation methods have been named after the researchers who established the method, such as Bielschowsky, Bodian, Cross, Gallyas, Campbell-Switzer, Reusche, etc. (Heinz 2005). What are the differences between these methods? (Uchihara 2007) When dealing with

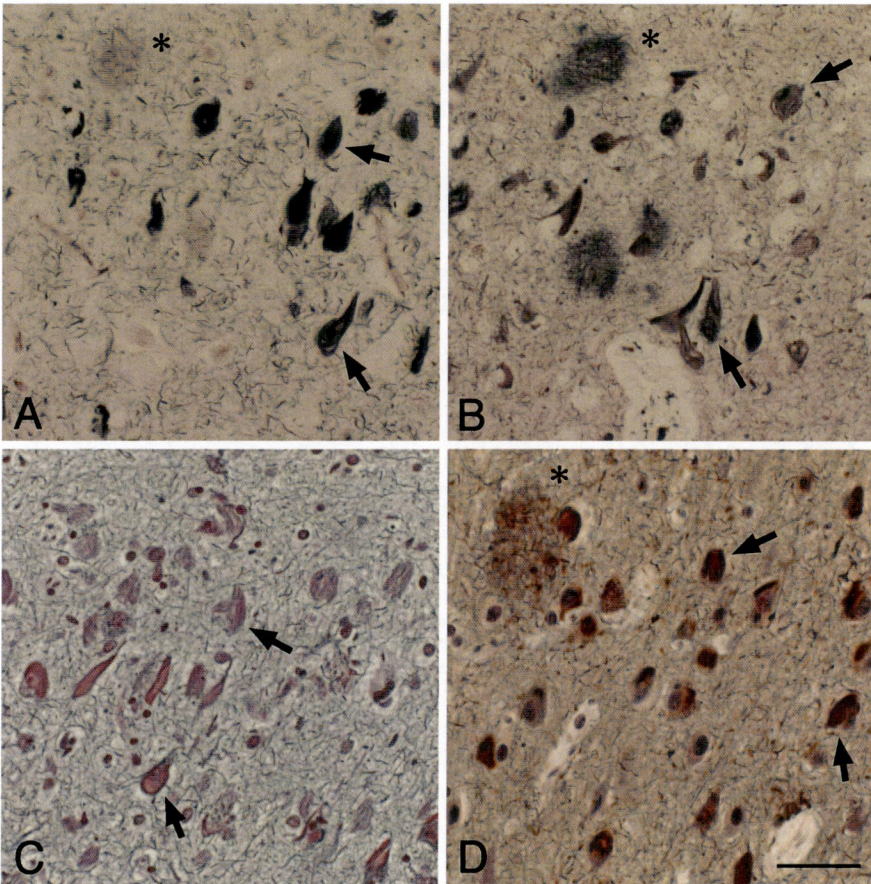

**Silver Impregnation Methods in Diagnostics, Fig. 1** Argyrophilic structures in Alzheimer disease brain by different silver impregnation methods (**A**) Gallyas method, Neurofibrillary tangles (NFTs, *arrows*) are clearly visible while senile plaques (SPs) are recognized through neuritic reactions (*asterisk*). (**B**) Campbell-Switzer method, in addition to NFTs (*arrows*), SPs are well visible (*asterisks*). (**C**) Bodian method, Visualization of NFTs (*arrows*) is less clear. (**D**) Bielschowsky method, SPs (*asterisk*) and NFTs (*arrows*) are readily visible. In addition to SPs and NFTs, Bodian and Bielschowsky methods exhibit nuclear stain and normal axons while these structures are not visible with Gallyas and Campbell-Switzer methods. Bar = 50 μm

Alzheimer disease (AD) brains, for example, argyrophilic structures such as senile plaques (SPs) and neurofibrillary tangles (NFTs) are differently labeled according to the method used. For example, NFTs labeled with Gallyas (Fig. 1A) method are more numerous than those with Bodian (Fig. 1C) or Bielschowsky (Fig. 1D) method, giving an impression that Gallyas method is more sensitive. This apparent higher sensitivity of Gallyas method, however, does not explain why SPs labeled with Gallyas method are less numerous than those with Campbell-Switzer (Switzer) method ((Fig. 1B) or Bielschowsky (Fig. 1D) method. In fact, it is Bielschowsky method that labels more SPs than Gallyas method (Fig. 1D). This quantitative discrepancy is explained if "argyrophilia" is not a homogenous feature, which is engendered through an interaction between the target lesions (what are stained) and the method (how stains).

Major differences between silver impregnation methods are primarily related to silver solution where histological sections are incubated as listed in Table 2. Gallyas method utilizes alkali silver iodate while Switzer method uses pyridine silver as initial silver solution. This simple difference is related to the above-mentioned differential profiles between tau-positive structures. Ammoniacal silver solution is customized for modified Bielschowsky method to detect senile plaques and neurofibrillary tangles on paraffin sections (Kiernan 1996). Although Bielschowsky method is powerful to visualize NFTs and SPs, underlying structures such as normal axons, nucleus, and neurofibrils are also visualized as with Bodian method (Fig. 1). In contrast, Gallyas and Switzer methods are more specific to pathological lesions with less underlying structures, which may facilitate quantification. More practical protocols and detailed principles are available elsewhere (Uchihara 2007).

This quantitative discrepancy of argyrophilia is more striking when various tau-positive lesions are probed with Gallyas or Switzer method as summarized

**Silver Impregnation Methods in Diagnostics, Table 2** Characteristics of different silver impregnation methods

| Method | Gallyas | Campbell-Switzer | Bielschowsky | Bodian |
|---|---|---|---|---|
| Silver solution | Alkali silver iodate | Pyridine silver | Ammoniacal silver | Protein-silver |
| Developer | silver nitrate | Silver nitrate | Ammoniacal silver | – |
| Tau-positive lesions | CBD/PSP, AD | Pick bodies, AD | CBD/PSP, Pick bodies, AD | Pick bodies, AD |
| Related tau isoforms | 4R | 3R | 3R & 4R | ? |
| Senile plaque | Neurites | Diffuse, cored | Diffuse, cored | Cored |
| Background structures | Blood vessel | Axons | Axons, nucleus Neurofibrils | Axon, nucleus Neurofibrils |

*CBD* corticobasal degeneration, *PSP* progressive supranuclear palsy, *AD* Alzheimer disease, *3R* three-repeat, *4R* four-repeat

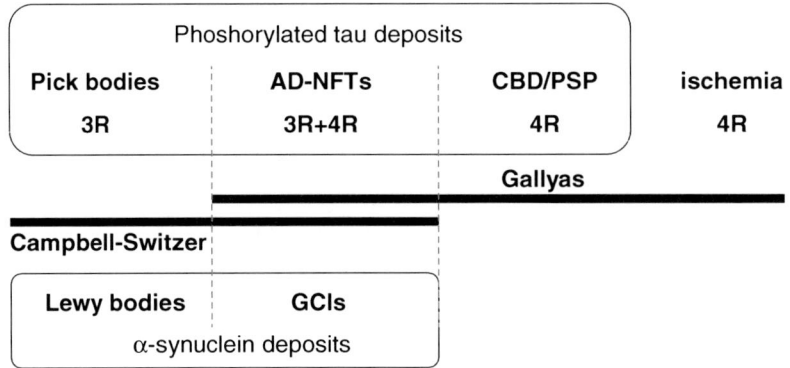

**Silver Impregnation Methods in Diagnostics, Fig. 2** Argyrophilic differentiation by Gallyas and Campbell-Switzer methods. Phosphorylated tau deposits and α-synuclein deposits are classified according to the argyrophilic properties on Gallyas and Campbell-Switzer silver impregnation methods. 4-repeat (4R) tau selective deposition is shared between CBD/PSP and ischemia, both accompanied by selective argyrophilia with Gallyas method. It is of note that 4R tau deposits in ischemia are not phosphorylated. *AD-NFTs* neurofibrillary tangles of Alzheimer type, *CBD/PSP* corticobasal degeneration/progressive supranuclear palsy, *3R* three-repeat tau, *4R* four-repeat tau, *GCIs* glial cytoplasmic inclusions of multiple system atrophy

in Fig. 2. Argyrophilia of Pick body is evident when probed with pyridine silver solution (Switzer method), while it is absent when probed with alkali silver iodate solution (Gallyas method). Because the subsequent step of physical development is essentially the same between these methods, argyrophilic profile of Pick body (Switzer+/Gallyas−) is dependent on the pyridine silver solution of Switzer method. This profile of Pick body (Switzer+/Gallyas−) is completely reversed with tau-positive lesions in corticobasal degeneration/progressive supranuclear palsy (CBD/PSP). Argyrophilia of tau-positive lesions in CBD/PSP is evident with Gallyas method, while it is absent with Switzer method, yielding the complementary profile (Switzer−/Gallyas+). It is known that tau proteins harbor either three or four microtubule-binding domains in tandem, hence named 3R or 4R tau isoform, respectively. Biochemical analyses further clarified a contrast between Pick bodies (Switzer+/Gallyas−) mainly containing 3R tau and tau-positive lesions in CBD/PSP (Switzer−/Gallyas+) containing 4R tau. If this biochemical contrast mirrors their complementary argyrophilic profiles, it is probable that 3R tau deposits exhibit argyrophilia with Switzer method but not with Gallyas method (Fig. 2). Similarly, 4R tau deposits of CBD/PSP exhibit argyrophilia with Gallyas method but not with Switzer method. Because AD brains contain both 3R and 4R isoforms, NFTs of AD exhibit argyrophilia with both Switzer and Gallyas methods (Switzer+/Gallyas+), which confirmed the above-mentioned interpretation. Although these distinctive argyrophilic profiles may be of differential value to distinguish tau deposits, it is not specific to tau. Indeed, α-synuclein-positive deposits exhibit argyrophilia as well. Glial cytoplasmic inclusions of multiple system atrophy (GCIs, deposits of α-synuclein in oligodendroglia) and Lewy bodies (LBs) in Parkinson disease (another α-synuclein deposits in neuron), both exhibit argyrophilia with Switzer method (Fig. 2). Because argyrophilia with Gallyas method is usually limited to GCIs but not with LBs, argyrophilic profile with Switzer/Gallyas methods is useful in differentiating α-synuclein deposits as well, even though their biochemical properties are not fully elucidated for molecular differentiation. Although some studies demonstrated affinity of silver ions in these silver solutions to brain extract containing tau or neurofilament, this intermolecular differentiating potential of silver impregnation is hardly explained by its direct affinity to molecules such as tau or α-synuclein but rather represents epi-molecular organizations, such as conformation or fibril formation (Uchihara 2007).

## Modifications for Silver Impregnations

Methenamine, useful for histological demonstration of glycogen and mucin, may be combined with silver nitrate (methenamine silver) or silver proteinate (methenamine Bodian) to enhance argyrophilia of SPs even in its premature form with scarce fibrillary composition. The methenamine-silver modification does not enhance argyrophilia on neurofibrillary tangles, Kuru plaques of Gerstmann-Sträussler-Scheinker disease. Therefore, its affinity to the targets may be related to some higher structures related to the molecules but not directly related to the molecules themselves. Pretreatment with 0.4% lanthanum nitrate/2% sodium acetate proposed by Gallyas is now the standard of Gallyas silver impregnation method for neurofibrillary tangles. Additional pretreatment with potassium permanganate followed by oxalic acid further enhances argyrophilia of NFTs by Gallyas method. This additional pretreatment with potassium permanganate/oxalic acid greatly enhances immunoreactivity for 4R tau and reduces nonspecific staining for 3R tau, suggesting some commonality between what is probed with 4R tau antibody and those with Gallyas method (Uchihara et al. 2011). In addition to tau-positive deposits in neurodegenerations, some neurons and glia of ischemic foci exhibit argyrophilia with Gallyas method. Interestingly enough, these Gallyas-positive cells exhibit 4R tau IR as well (Fig. 2). Taken together, what are impregnated by Gallyas method share some commonalities with what are proved by 4R tau-specific antibodies. Therefore, argyrophilic properties are not restricted to neurodegeneration but also useful in other conditions to trace neurotoxicity (Switzer 1993), experimental degeneration of axons (Nauta and Gygax 1951), or neurons collectively described as "dark neurons."

## Conclusions

Immunohistochemistry is a sophisticated method that can pinpoint a molecule (what is it?) and its localization (where is it?) by a sharp contrast with other constituents. On the other hand, silver impregnation pinpoints the localization of the target but does not directly reveal its molecular composition. Target/method-dependent nature of silver impregnations may provide disease-specific profiles, possibly representative of some qualitative nature of the target. Because this qualitative nature is not readily reducible to a molecule, it may represent some epi-molecular nature such as conformation or ultrastructure (how is it?), featured by the disease-specific process or the molecule itself. This qualitative interpretation is already useful in histological differentiation of brain lesions in neurodegenerative disorders. Its mechanism, if clarified, may further expand our view by taking this qualitative (epi-molecular) nature into account to decipher how lesions are engendered from molecules to argyrophilic lesions in the human brains.

## Cross-References

▶ Silver in Protein Detection Methods in Proteomics Research
▶ Silver-Induced Conformational Changes of Polypeptides

## References

Danscher G, Stoltenberg M, Juhl S (1994) How to detect gold, silver and mercury in human brain and other tissues by autometallographic silver amplification. Neuropathol Appl Neurobiol 20:454–467

Gallyas F (2008) Physicochemical mechanisms of histological silver staining and their utilization for rendering individual silver methods selective and reliable. Biotech Histochem 83:221–238

Heinz T (2005) Evolution of the silver and gold stains in neurohistology. Biotech Histochem 80:211–222

Horobin RW (2002) Biological staining: mechanisms and theory. Biotech Histochem 77:3–13

Kiernan JA (1996) Review of current silver impregnation: techniques for histological examination of skeletal muscle innervation. J Histotechnol 19:257–267

Nauta WJ, Gygax PA (1951) Silver impregnation of degenerating axon terminals in the central nervous system: (1) Technic. (2) Chemical notes. Stain Technol 26:5–11

Newman GR, Jasani B (1998) Silver development in microscopy and bioanalysis: past and present. J Pathol 186:119–125

Prento P (2001) A contribution to the theory of biological staining based on the principles for structural organization of biological macromolecules. Biotech Histochem 76:137–161

Switzer RC 3rd (1993) Silver staining methods: their role in detecting neurotoxicity. Ann N Y Acad Sci 679:341–348

Uchihara T (2007) Silver diagnosis in neuropathology: principles, practice and revised interpretation. Acta Neuropathol 113:483–499

Uchihara T, Nakamura A, Shibuya K, Yagishita S (2011) Specific detection of pathological three-repeat tau after pretreatment with potassium permanganate and oxalic acid in PSP/CBD brains. Brain Pathol 21:180–188

# Silver in Protein Detection Methods in Proteomics Research

Thierry Rabilloud
CNRS, UMR 5249 Laboratory of Chemistry and Biology of Metals, Grenoble, France
CEA, DSV, iRTSV/LCBM, Chemistry and Biology of Metals, Grenoble Cedex 9, France
Université Joseph Fourier, Grenoble, France

## Synonyms

Post-electrophoretic detection of proteins; Silver staining of proteins

## Definition

Proteomics is defined as the large-scale analysis of proteins and relies, in its final experimental step, on the identification of peptides by mass spectrometry. However, biological samples are so complex that the separating power of the mass spectrometer alone cannot handle this complexity, so that extensive fractionation of the proteins and/or peptides must be carried out prior to the mass spectrometry analysis.

One of the classical setups for proteomic analysis is based on the high-resolution separation of proteins by two-dimensional gel electrophoresis, combining isoelectric focusing in the first dimension and SDS electrophoresis on polyacrylamide gels in the second dimension. In this setup, prior to spot cutting and analysis by mass spectrometry, the proteins are detected and quantitatively analyzed directly on the second dimension polyacrylamide gels by various methods that can use organics dyes, fluorophores or metal ions. Among those detection methods, which must combine a high sensitivity, a high homogeneity from one protein to another and a good linear dynamic range, silver staining, i.e., the detection of proteins by the means of formation of metallic silver at the protein-containing sites in the polyacrylamide gels, offer a high sensitivity at a moderate price, and is thus highly popular.

## Historical Background

Silver staining of proteins in gels is one of the numerous examples of the relations between histology and electrophoresis. Some dyes, such as Light green SF, Eosin Y, and Fast green FCF, have been used both for histology and for protein detection after electrophoresis. This cousinhood is due to similarities in the constraints and aims of the two methods. Both try to detect specific components (varied in the case of histology, proteins in the case of 2D electrophoresis) in a complex gel matrix (the slice for histology, the polyacrylamide gel for electrophoresis), with maximum selectivity and sensitivity. Both use a complex fixation process to insolubilize the components of interest in the matrix, and try to remove components that may interfere with the subsequent detection process, and it is quite clear that gel electrophoresis has copied the concepts and sometimes the recipes used in histology (e.g., some fixatives and some dyes).

Among all the techniques that have been developed in histology, silver staining occupies a special place. It has been totally instrumental in our understanding of the ultrastructure of the nervous system, thanks to the work of Golgi and Ramon y Cajal. There are numerous variants of silver-staining techniques in histology, and their mechanisms are clear for some of these variants and applications, and still unclear for some other ones. It was therefore very tantalizing to try to apply silver staining for protein detection in polyacrylamide gels, and it is therefore no wonder that

1. One of the key authors of the first paper on silver staining for protein detection in polyacrylamide gels (Merril et al. 1979) is also a histologist well-versed into silver staining (Switzer 2000).
2. A blossoming of silver-staining methods appeared for around a decade, with more than a hundred different silver-staining methods published.

## Rationale and Mechanisms of Silver Staining

In order to understand the rationale and mechanisms of silver staining in polyacrylamide gels, two very important chemical facts must be kept in mind.

First, among all the metals, silver and gold share the rare feature to show a strong affinity for almost every protein, both as their ions (Ag + and Au 3+) and as metals (Ag° and Au°) especially in their nanoparticular, colloidal form for the latter.

Second, and opposite to a microscopy slide, a polyacrylamide gel is a very thick matrix with small pores (necessary for the sieving of proteins), which considerably hampers mass transport of molecules, by diffusion and particularly by convection. Thus, reactions that can proceed within minutes in microscopy slides would take hours to perform in polyacrylamide gels, with all the side effects that can be induced by such a delay (e.g., parasite competing reactions).

When taking into account these two phenomena, an obvious staining mechanism emerges at once, which would be to treat the polyacrylamide gel with a colloidal solution of gold or silver. However, this scheme does not work at all for polyacrylamide gels because of the limited penetration of the nanoparticles in the gel. In fact, the nanoparticles deposit at the surface of the gel instead of entering it, resulting in a total absence of contrast and thus of detection. However, this very simple and efficient detection scheme works wonderfully (van Oostveen et al. 1997), but only when the proteins have been transferred (generally electrophoretically) from the polyacrylamide gel onto a porous membrane that is much thinner and with much larger pores (the blotting process).

Although both the blotting process and the colloidal silver staining are very efficient, this scheme has encountered very limited use in proteomics. The bottleneck, in this case, is due to the fact that the silver-staining step occurs roughly halfway in the whole process, and that the downstream part of the process requires an efficient digestion of the separated proteins to produce peptides that must be very efficiently recovered, as they are the final analytes in mass spectrometry. Unfortunately, the porous membranes used in blotting have a strong affinity for proteins and for peptides, so that the recovery of peptides is difficult to carry out under conditions that are compatible with mass spectrometry.

The second obvious scheme that can be proposed would be to form the metal colloid in situ in the polyacrylamide gel, taking advantage of the affinity of proteins for both the ions and the metals. This would lead to a very simple scheme with four steps:

- First the gels would be fixed, to remove all the buffers and chemicals used to carry out SDS electrophoresis and that also have a strong affinity for silver ion
- Second the gels would be washed, to remove the fixative and all the remaining interfering chemicals
- Third the gels would be soaked in the silver (or gold) ion solution
- Fourth the gels would be treated with a reducer to form the opaque metal colloid in situ

Most unfortunately this scheme does not work, neither for silver (Merril 1986), nor for gold (Casero et al. 1985). In both cases, a negative stain is obtained, with the metal colloid formed in the background and NOT in the protein zones within the gels. This negative stain combines the two drawbacks of being poorly sensitive and not quantitative and has thus never been used.

This failure roots itself in the affinity of proteins for silver (and gold) ions. In fact, this affinity is so high that this complexation of the precious metal ions by the proteins decreases considerably their reactivity toward the reducers to form the reduced metal, so that the reaction takes place everywhere but in the protein-containing zones. In other words, this means that it will not be possible to perform the core reaction of silver staining, i.e., the reduction of colorless silver ion into opaque metallic silver at the protein-containing sites, under conditions corresponding to thermodynamic reaction control. This means in turn that successful silver staining will have to be performed under carefully selected kinetic control conditions, i.e., conditions where the silver ion reduction will occur much faster at the protein-containing sites than in the background, despite the complexation phenomenon.

Fortunately for silver staining, it happens that silver ion reduction is extremely sensitive to solid phase catalysis, and that even extremely small deposits of catalysts can induce a massive deposition of metallic silver under carefully selected reaction conditions. These catalysts can have varied structures, but metallic silver itself is one of the most efficient ones. This strong autocatalysis of the silver ion reduction can occur within a silver halide crystal, a process at the very root of photography (Mason 1979) and called chemical development, or can occur in solution in a process called physical development (Mason 1979; Uchihara 2007). In addition to metallic silver itself, various metal sulfides, including silver sulfide and zinc sulfide (Danscher and Rytter Norgaard 1985), can be very efficient catalysts for silver ion reduction.

Thus, a crucial step for achieving a positive, highly sensitive silver stain will be to produce these catalysts exclusively at the protein-containing sites, taking advantage of the numerous reactive functions on proteins, in what can be a called a sensitization step. In all silver-staining protocols published to date, the catalysts formed are either metallic silver or silver sulfide, produced by various reactions, among which the most common are silver ion reduction by aldehydes under alkaline conditions and the acid-catalyzed decomposition of thiosulfate or tetrathionate into sulfide, precipitated in situ as silver sulfide.

Within this frame, the ideally controlled conditions would be to proceed under the conditions prevailing in the Gallyas and Switzer histochemical protocols (described in Uchihara 2007), i.e., formation of the catalyst by contacting the sensitized gel (or slice) with silver nitrate in a first step, and then a real physical development step with an "unlimited" supply of silver ion in the developer, thereby saturating all the silver-binding sites of the proteins and preventing the negative effects of the silver ion complexation.

Unfortunately, this scheme cannot be used with polyacrylamide gels, here again because of their thickness and poor mass transport features. In physical developers, the reaction in solution will eventually take place within the bulk of the solution, and the reaction must be stopped at this stage. This process usually takes place within 15–30 min depending on the exact conditions. In this time frame, the physical developer will have poorly penetrated in the polyacrylamide gel, and stopping the reaction in all the thickness of the gel to avoid general blackening will be equally difficult.

Thus, the only option left for silver staining in polyacrylamide gels is the split reaction, close to the conditions prevailing in the Bielschowsky method (described in Uchihara 2007) in which the gel is first impregnated with silver, then rinsed, then dipped in a reducing solution to perform the silver ion reduction. In this scheme, however, the sole source of silver ion is the silver-impregnated polyacrylamide gel, where the complexation phenomena with proteins take place. Thus, the only way to keep the reaction under the adequate kinetic control is to use a very slow developer, in which the silver ion will have time to leak out of the proteins before being reduced. This process has been described in more details in (Rabilloud et al. 1994), and this very delicate balance explains why so few chemicals can be used as developers in silver staining. Despite all its drawbacks, formaldehyde is almost the only reducer used in silver staining, although aldoses and semicarbazide can also be used, at the expense of sensitivity (Rabilloud 2012).

## Practical Aspects of Silver Staining

The flowchart of all the silver-staining methods described to date is relatively homogeneous, and derives from the principles exposed just above. Basically, a silver-staining protocol is composed of eight main steps:

1. Fixation, just after electrophoresis, in order to insolubilize the proteins in the gels and to remove all the buffers and detergents used in performing the electrophoretic separation. Aldehydes such as formaldehyde were very commonly used in the early days of silver staining. However, the integration of silver staining in a proteomic workflow, where proteins must be digested and peptides extracted, has decreased the use of aldehydes, as they cross-link proteins to too great an extent for adequate proteomic analysis afterward. In the same trend, for both safety reasons and to prevent spurious methylations that could be misinterpreted in the

proteomic analysis afterward, methanol is almost no longer used. Thus, the fixative is now almost always an acetic acid ethanol water bath.
2. Washes, to remove the fixative and decrease the acidity of the gel.
3. Sensitization. Here again, things have evolved since the early days of proteomics because of the additional constraints brought by proteomics. In the early days, completely in line with the histological reactions involving silver, the favorite sensitizing agent was glutaraldehyde, used in conjunction with silver diammine to produce metallic silver microcatalysts. The most common sensitizer nowadays is sodium thiosulfate, used in a weakly acidic environment to induce formation of silver sulfide.
4. Washes to remove the excess sensitizing agent from the bulk of the polyacrylamide gel.
5. Silver impregnation. Here again, there has been changes since the early days. In fact two silvering agents can be used, silver nitrate and silver diammine. Provided that the silver/ammonia ratio is well controlled, silver diammine gives the cleanest and most consistent results. However, it works best with aldehyde reagents that have largely disappeared for proteomic applications, and it does not work with sulfide producing sensitizers such as sodium thiosulfate, as the free silver ion concentration in silver diammine is not high enough to induce formation of silver sulfide precipitates. This explains why almost all the silver-staining methods used for proteomic applications use plain silver nitrate as the silvering agent.
6. Washes to remove the excess silver, and at the very least the silver nitrate liquid film coming on the surface of the polyacrylamide gel after the silvering step.
7. Development, i.e., the formation of the metallic silver precipitate by reduction. When formaldehyde is used, which represents 999 cases out of 1,000, the pH must be close to 11. This explains why citric acid is used in addition to formaldehyde when silver diammine is used as the silvering agent, as the pH of the silver diammine solution is generally close to 13. Oppositely, a carbonate solution is used in conjunction with silver nitrate to raise the pH from 7 (silver bath) to 11. In this case, a low concentration of thiosulfate is added to the developer (Rabilloud et al. 1994; Rabilloud 2012)

to prevent spurious onset of silver reduction in the bulk of the polyacrylamide gel. This keeps the background clean and ensures excellent contrast between the protein-containing zones and the background. This has been a quantum leap in silver staining, as poor contrast and excess background were very commonly encountered in the early days and made silver staining quite erratic in its performances.

In the rare cases where sugars have been used as developing agents, the buffer of the developer is a pH 12 borate buffer (Rabilloud 2012).
8. Stop solution. This solution must at the same time stop the reduction of silver ion, but also prevent the reoxidation of the metallic silver into silver ion by the reverse reaction. Modern stop solutions work by controlling the pH down to 7, and avoid any silver-chelating chemicals that would drive reoxidation favorably.

Experimental details and tips are beyond the scope of this entry but can be easily found in dedicated publications and book chapters, e.g., in (Rabilloud 2012).

## Expected Results

A typical silver-stained two-dimensional gel is shown on Fig. 1.

Control experiments with known proteins have demonstrated that the detection limit of such silver-staining methods is below the ng/mm$^3$ threshold, ten times lower than the best methods using organic dyes and still lower than most of the fluorescence-based methods. In fact, silver staining still offers a very happy compromise between sensitivity, speed, cost, and ease of detection. Its major drawback lies in the protein modification and cross-linking that are induced mostly at the developing stage by the presence of formaldehyde (discussed in Rabilloud 2012 and in references cited therein). These modifications decrease the efficiency of protein identification and the quality of protein coverage by mass spectrometry after silver staining when compared to other detection methods. However, adequate performance can still be achieved (i.e., each and every protein spot visible on the gel can be identified) using modern silver-staining methods, modern MS/MS mass spectrometers and by destaining the excised silver-stained spots on the very same day as staining (Rabilloud 2012).

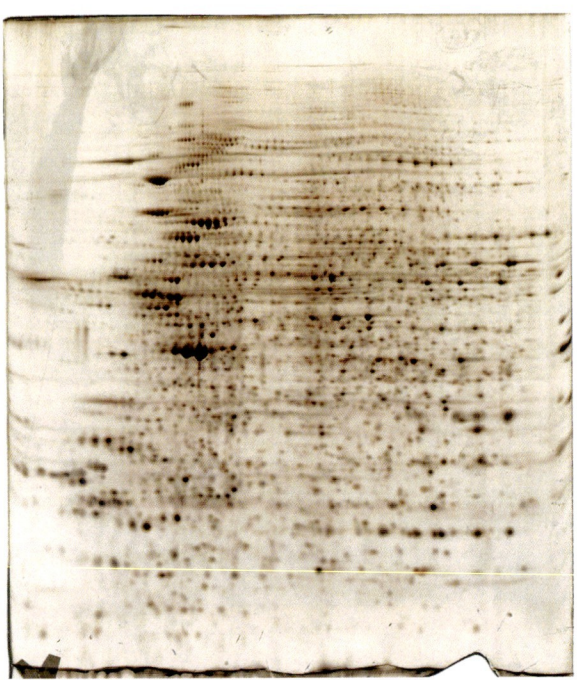

**Silver in Protein Detection Methods in Proteomics Research, Fig. 1** Separation of total cellular proteins by two-dimensional electrophoresis and silver staining. Total cellular proteins (150 micrograms) have been loaded on the two-dimensional gel, which has been stained by a silver-staining method using thiosulfate as the sensitizing agent, silver nitrate as the silvering agent, and formaldehyde/carbonate/thiosulfate as the developer. More than 2,000 protein spots can be detected by the image analysis software

## Future Directions

As most protein detection methods after electrophoresis, silver staining is now quite a mature technique. In the early days image development had to be stopped prematurely because of background appearance. In the modern silver-staining methods, gels can be left in the developer until nothing more happens, which suggests that all the silver ion that can be reduced has been reduced. Common early days such as "hollow" or "doughnut" spots also belong to the past.

This situation means in turn that little can be expected in the area of sensitivity of detection, as the constraints driving silver staining are very tight and interconnected.

At equal sensitivity, an open area for progress would be to replace formaldehyde as the developing agent, and the aldose-based developers (Rabilloud 2012) can probably be optimized to reach the same sensitivity as formaldehyde.

On the very long run, the only way to increase sensitivity would be to be able to deposit more silver on proteins. This will require to go back to physical development to have a real long life physical developer available. This has never been developed in the other fields using silver staining (e.g., histology), as a 30 min life of the developer is very convenient in these fields. This will need lengthy development, as such a dimension has never been investigated before.

## References

Casero P, Delcampo GB et al (1985) Negative aurodye for polyacrylamide gels – the impossible stain. Electrophoresis 6:367–372

Danscher G, Rytter Norgaard JO (1985) Ultrastructural autometallography: a method for silver amplification of catalytic metals. J Histochem Cytochem 33:706–710

In selecting the references cited in this chapter, priority has been given to references where the full text can be freely downloaded, either from the PubMed Central website (references with a *) or from the HAL institutional website (references with a $). The references in HAL must be retrieved by use of the search function at the following url: http://hal.archives-ouvertes.fr/index.php?halsid=69un2cd9njil7ousvo4fjric25&action_todo=search&s_type=simple

Mason LFA (1979) Photographic processing chemistry. Focal Press, London

Merril CR (1986) Development and mechanisms of silver stains for electrophoresis. Acta Histochem Cytochem 19:655–667

*Merril CR, Switzer RC et al (1979) Trace polypeptides in cellular extracts and human body fluids detected by two-dimensional electrophoresis and a highly sensitive silver stain. Proc Natl Acad Sci USA 76:4335–4339

$Rabilloud T (2012) Silver staining of 2D electrophoresis gels. Methods Mol Biol 893:61–73

$Rabilloud T, Vuillard L et al (1994) Silver-staining of proteins in polyacrylamide gels – a general overview. Cell Mol Biol 40:57–75

*Switzer RC 3rd (2000) Application of silver degeneration stains for neurotoxicity testing. Toxicol Pathol 28:70–83

*Uchihara T (2007) Silver diagnosis in neuropathology: principles, practice and revised interpretation. Acta Neuropathol 113:483–499

van Oostveen I, Ducret A et al (1997) Colloidal silver staining of electroblotted proteins for high sensitivity peptide mapping by liquid chromatography-electrospray ionization tandem mass spectrometry. Anal Biochem 247:310–318

## Silver In Situ Hybridization

▶ Enzyme Metallography and Metallographic In Situ Hybridization

## Silver Nanoparticles (AgNPs)

▶ Nanosilver, Next-Generation Antithrombotic Agent

## Silver Nanoparticles and Alterations in Drug Effects

▶ Silver, Neurotoxicity

## Silver Nanoparticles and Astrocytic Activation

▶ Silver, Neurotoxicity

## Silver Nanoparticles and Biological Functions

▶ Silver, Neurotoxicity

## Silver Nanoparticles and Blood–Brain Barrier Breakdown

▶ Silver, Neurotoxicity

## Silver Nanoparticles and Cerebrolysin Treatment

▶ Silver, Neurotoxicity

## Silver Nanoparticles and Gene Expression in the Brain

▶ Silver, Neurotoxicity

## Silver Nanoparticles and Heat Shock Protein Expression

▶ Silver, Neurotoxicity

## Silver Nanoparticles and Hypertension and Diabetic Disorders

▶ Silver, Neurotoxicity

## Silver Nanoparticles and Hyperthermia

▶ Silver, Neurotoxicity

## Silver Nanoparticles and Nitric Oxide Synthase Upregulation

▶ Silver, Neurotoxicity

## Silver Nanoparticles and Oxidative Stress

▶ Silver, Neurotoxicity

## Silver Nanoparticles Neurotoxicity

▶ Silver, Neurotoxicity

## Silver Staining of Proteins

▶ Silver in Protein Detection Methods in Proteomics Research

# Silver Sulfadiazine

▶ Silver, Pharmacological and Toxicological Profile as Antimicrobial Agent in Medical Devices

# Silver, Burn Wound Sepsis and Healing

Bishara S. Atiyeh[1], Amir Ibrahim[2], Saad A. Dibo[3] and Marwan S. Rizk[4]

[1]Division of Plastic Surgery, Department of Surgery, American University of Beirut Medical Center, Beirut, Lebanon

[2]Plastic and Reconstructive SurgeryBurn Fellow, Massachusetts General Hospital / Harvard Medical School & Shriners Burn Hospital, Boston, USA

[3]Division Plastic and Reconstructive Surgery, American University of Beirut Medical Center, Beirut, Lebanon

[4]Deptartment of Anesthesiology, American University of Beirut Medical Center, Beirut, Lebanon

## Synonyms

Burn dressing; Burn wound infection; Burn wound reepithelialization; Burn wound treatment

## Definition

Topical treatment and burn wound dressings are an essential component of burn wound management. Several topical agents are being used to reduce surface colonization and prevent burn wound sepsis. Silver-based products have been most widely used over the last few decades and since the introduction of the silver-containing formulation, silver sulfadiazine (SSD) (Flamazine®, Silvadene®), by Fox shortly, silver became the main antibacterial prophylactic agent in burns and a viable treatment option for burn wounds, open wounds, and chronic ulcers. Nevertheless, the use of the many silver-containing products for burn wounds is still controversial and opinions in clinical practice conflicting.

Silver does possess antimicrobial properties. However, evidence of its effectiveness remains poorly defined and does not support its use for topical treatment. Silver does not appear to promote wound healing or prevent wound infection. It can only be concluded from the available evidence that silver-containing dressings and topical silver are not better or worse than other topical modalities in preventing wound infection and promoting healing of burn wounds.

## Introduction

Burns frequently become infected. Burn wound sepsis further aggravates morbidity and affects survival of severely burnt victims. Although types and actual numbers of bacteria are critical in the development of wound sepsis, it is recognized nowadays that the presence of biofilms, identified in animal burns and more recently in human burn wounds, increases risks of wound sepsis and is a major additional factor in delayed healing.

Burn treatment methods have changed in recent decades. Increasingly, aggressive surgical approach with early tangential excision and wound closure by a variety of modalities ranging from simple autologous skin grafts to keratinocyte cultures and bioengineered materials is being applied. Despite these major advances, prevention of burn wound infection remains a challenge. Moreover, local burn wound care continues to be a significant component of the overall burn management and sometimes, due to lack of proper facilities or adequate resources, may be the major and only possible management modality (Atiyeh et al. 2007).

Local burn wound care involves application of topical creams, ointments, or solutions with or without an overlying dressing until excision and grafting is performed or reepithelialization of superficial burns is completed. Not infrequently, local burn wound care is continued until the eschar of deeper burns separates uncovering granulation tissue that is subsequently grafted. One of the most used topical chemotherapeutic products to prevent or treat bacterial infection and burn wound sepsis, are noble metal antimicrobials the most prevalent of which are silver-based products. The past few decades have seen an increasing clinical application of ointments, gels, and foams or fibrous silver-impregnated wound dressings for the prevention and management of burn infections.

Silver compounds have been exploited for their medicinal properties for centuries. A detailed historical review about the medicinal usage of silver has been published by Klasen HJ. As early as 1000 B.C., the properties of silver in rendering water potable were appreciated. Phoenicians have noticed that storing wine, water, and vinegar in silver containers prevented their spoiling and Hippocrates had recognized its antidisease and healing properties. Even in the early twentieth century, silver coins were put in milk bottles to "prolong the milk's freshness." Its modern usage as a medical agent began in 1881 with Crede's silver nitrate solution as prophylaxis for gonorrhea opthalmicum. Silver compounds became popular remedies for tetanus, rheumatism, colds, and gonorrhea in the nineteenth century before the advent of antibiotics. Silver in the form of hardened silver nitrate (lunar caustic) was used as early as the eighteenth century to remove granulation tissue from slowly healing chronic wounds. Silver nitrate in the solid form as a rod or pencil or as a solution in concentrations varying from 0.2% to 2% was applied to fresh burns early in the nineteenth century and during the first few decades of the twentieth century, silver foil and silver nitrate solution became more commonly used on fresh and infected burn wounds. The first recorded use of silver in surgery was by Halsted in 1913. He reported the use of silver foil as the initial dressing for fresh surgical incisions. In 1919, Lotichius advised application of silver foil to second-degree burns after removal of blisters, until healing is complete. The application of silver foil to clean third-degree burns was also proposed. In 1935, a combination of tannic acid (5%) and silver nitrate (10%) was proposed in the belief that tanning would bind the toxins present in the burnt area. As penicillin and sulfa drugs became available for topical treatment around the Second World War, interest in silver salts or silver salt solutions for the treatment of burn patients, completely disappeared. It took many years to revive topical treatment of burns with silver (nitrate) by Moyer et al. On the basis of in vitro and in vivo studies, he concluded that the lowest silver nitrate concentration for antibacterial action is a 0.5% solution. A silver-containing formulation, silver sulfadiazine (SSD) (Flamazine®, Silvadene®), was introduced by Fox shortly after. Since that time, silver has reemerged as the main antibacterial prophylactic agent in burns and as a viable treatment option for burn wounds, open wounds, and chronic ulcers infections (Becker 1999; Klasen 2000a, b).

Surprisingly, the use of the many silver-containing products for burn wounds is still controversial and opinions in clinical practice are conflicting. Some strongly believe that their use in burn wounds is justified while others claim that routine use of silver-containing dressings particularly for superficial and partial thickness wounds is not warranted because of limited evidence on their clinical effectiveness and mostly is not cost effective (Aziz et al. 2012; Storm-Versloot et al. 2010).

## Silver Antibacterial Mechanism of Action

Among all metals, silver is the most toxic element to microorganisms. It is also effective against fungi and some viruses. However, as a metal, silver is relatively inert and is poorly absorbed by mammalian or bacterial cells. For antimicrobial efficacy ionization is essential. Silver cations, Ag+, are the active antimicrobial entity (Atiyeh et al. 2007).

In presence of wound fluids, silver readily ionizes. Ag+ cations exhibit broad antimicrobial action at low concentrations. They are highly reactive and bind readily to negatively charged cytosolic proteins, RNA, DNA, chloride ions, as well as other anions. Thus, they probably are not capable of passive diffusion across biological membranes. Instead they are believed to traverse membranes primarily by binding ion transporters. Ag+ bactericidal mechanisms may occur via a number of routes probably mediated by deposition on the microbial cell wall and/or by interaction with cysteine residues. A major contributor to their antibacterial efficacy is also their binding affinity to thiol (−SH) groups of the bacterial cell respiratory enzymes generating reactive oxygen species (ROS). Upon binding to other biological molecules they also inhibit activities that are vital to the bacteria's regulatory processes and cause bacterial inactivation and subsequent death. They also bind preferentially to DNA thereby reducing the ability of all known bacteria and bacterial spores to replicate. In vitro studies suggested that the bactericidal effect of Ag+ ions can also be due to inactivation of the phosphomannose isomerase (PIM) enzyme", alternately mannose-6 phosphate isomerase (MPI), by binding to free sulfhydryl groups.

As ionization lies at the heart of silver antibacterial properties, it complicates delivery and maintenance of optimal antibacterial concentrations in the wound bed. When bound to wound fluid proteins Ag+ ions become unavailable to exert their therapeutic potential (Atiyeh et al. 2007).

Because Gram-negative bacteria and Gram-positive bacteria have different cellular structures, the mechanism of action of silver on these two classes of bacteria may be dissimilar. Gram-negative bacteria have relatively thin cell walls and thus may be more prone to death at lower Ag+ concentrations (Jung et al. 2008; Liu et al. 2010).

Silver nanoparticles (AgNPs) have unique physicochemical properties due to a crystallographic surface structure and a high ratio of surface area to mass. With an increase in atom numbers on the surface AgNPs have greater potential to generate Ag+ ions. Silver nanoparticles penetrate deeper into tissues and are able to penetrate the bacterial cell wall via pinocytosis and endocytosis. Properties related to the nanostructure may thus give rise to intrinsic antimicrobial activity over and above the release of Ag+ ions (Fidel Martínez-Gutierrez et al. 2012; Thomas et al. 2011).

When AgNPs deposit on the surface they affect cell wall permeability and normal transport of electrolytes and other metabolites. Because of their high affinity for sulfur and phosphorous they affect also the activity of cell wall sulfur/phosphorous-containing components. Studies conducted on *Staphylococcus aureus* have concluded that AgNP causes the cell DNA to become condensed to a tension state probably indicating loss of their replicating abilities. Subsequently the cell wall breaks down resulting in cellular collapse and release of intracellular contents (Li et al. 2011). AgNPs could reduce also the enzymatic activity of respiratory chain dehydrogenase and inhibit cellular respiration. Furthermore, they alter expression abundance of some proteins; that of formate acetyltransferase increases while that of aerobic glycerol-3-phosphate dehydrogenase, ABC transporter ATP-binding protein, and recombinase A protein decreases. Bactericidal effect of AgNPs is size dependent. It is greatest with smaller AgNPs size. Their increased reactivity raises concerns for potential greater toxicity (Li et al. 2011; Liu et al. 2010).

Studies with biofilms suggest that the Ag+ ions can destabilize the biofilm matrix. By binding to electron donor groups on biological molecules Ag+ decrease the number of sites for hydrogen bonding and electrostatic and hydrophobic interactions. This effect on biofilms is of critical importance. Silver could be an effective alternative to antimicrobials which are needed to eradicate biofilms in dosages often up to 100 times higher than that required for standard bactericidal action (Chaw et al. 2005; Thomas et al. 2011).

Antibiotic resistance of opportunistic and strict pathogenic bacteria in burn wounds is also a growing concern. Ag+ bactericidal efficacy on these problematic bacteria has been demonstrated. However, when applied in low doses, silver resistance has been encountered in a number of microorganisms. Most importantly, even when applied adequately, bacteria often may regrow at a rapid rate even with continued silver treatment (Atiyeh et al. 2007).

Bioavailability of active free Ag+ ions is of paramount importance. It is affected by numerous factors including cationic exchange, ability to form complexes, precipitation, and adsorption. Fluctuating physiological and biochemical conditions in infected wounds, which are in a state of dynamic biological flux, result in pH fluctuations. This inevitably affects activity and performance of Ag+ which is greatest at low pH. To ensure that maximum antimicrobial performance of ionic silver can be achieved, wound pH monitoring could be a clinically relevant component of a burn wound management strategy. This, however, needs to be determined and investigated by future research (Percival et al. 2011).

## Silver Products and Delivery Modalities

Over the past few years, there has been a rapid increase in the number of silver-containing products available for topical treatment of burns and other problematic chronic wounds. Inorganic salts (silver nitrate, silver calcium phosphate, silver chloride), organic compounds (silver sulfadiazine-SSD), and more recently nanocrystalline silver (AgNP) have been incorporated in creams, ointments, hydrocolloids, hydrogels, and foam or fibrous dressings (Atiyeh et al. 2007).

Colloidal silver solutions in which charged pure silver particles (3–5 ppm) were held in suspension by small electric currents were most commonly used prior to 1960. Silver proteins are more stable in solution. These, however, proved to possess much less antibacterial action and were rapidly replaced by silver salts. Effective salts include silver nitrate, lactate, tartrate, and citrate. Silver nitrate of 0.5% is the standard and most popular silver salt solution used for topical burn wound therapy and is highly bactericidal. It is however unstable; when exposed to light it produces typical black stains. Moreover, nitrate is toxic to wounds and the reduction of nitrate to nitrite causes oxidant-induced cell damage. SSD combines the inhibitory action of silver with the antibacterial effect of sulfadiazine. This silver complex acts on the bacterial wall in contradistinction to the silver ions which act more generally on the bacterial energy system.

Sustained nanocrystalline silver releasing systems have been the latest innovation in wound care products. Silver is incorporated within the dressing. Widely used silver-containing dressings include hydrocolloids (SSD/hydrocolloid, Contreet-H®), silver alginates (Silvercel®), foams (Avance®, Contreet Ag®), hydrofibers (e.g., Aquacel® Ag), and polymeric films and meshes (Arglaes®) including metallic nanocrystalline silver with sustained release of silver into the wound (Acticoat®) or ionic silver (Aquacel® Ag) and dressings with silver bound to activated charcoal (Actisorb Silver®). When silver present in the dressing comes into contact with body fluids, biologically active Ag+ ions are produced. Except very few which do not release silver, these dressings exert bactericidal action within the dressing itself by absorbing exudates as well as in the wound bed by releasing Ag+. AgNPs are assumed to ensure a more controlled, sustained, and prolonged release of Ag+ particles into the wound theoretically ensuring increased antimicrobial activity (Aziz et al. 2012).

Silver-containing products and dressings currently available differ regarding the characteristics of the "carrier" and the delivery modality of Ag+ onto the wound bed at constant optimal bactericidal concentration and minimal toxicity. The difficulties with many of these products lie in their variable and low levels of silver release, limited number of silver species released, lack of penetration, rapid consumption of silver ions, and the presence of pro-inflammatory nitrate, or cream bases that negatively affect wound healing. Some may cause also staining, electrolyte imbalance, and patient discomfort (Atiyeh et al. 2007).

Silver compounds that produce low levels of silver ions in the wound environment, or that do not require frequent applications and dressing changes or paradoxically have characteristics permitting long-term application with fewer dressing changes as a cost saving measure favor the development of silver resistant strains that must be avoided considering the rapid increase of resistant strains of bacteria to the available most potent antibiotics. It is possible that in the near future Ag+ ions may soon be one of the few effective topical antibacterial agents remaining for clinical use in burn patients.

## Silver Absorption and Toxicity

Except some historical studies associating high doses of silver nitrate with damage to the gastrointestinal tract and occasional fatality, there is little in the literature to suggest significant silver toxicity. There is no evidence to date that silver in any form causes significant toxicity to the immune, cardiovascular, nervous, or reproductive systems in humans. Mild allergic hypersensitivity reactions have, however, been encountered and chronic ingestion, inhalation, or dermal exposure in sufficient quantities may give rise to disfiguring usually irreversible argyria (Atiyeh et al. 2007; Wilkinson et al. 2011). This depends on the degree to which silver is absorbed, metabolized, and ultimately excreted from the body. Organic compounds are better absorbed by the body and present a slightly heightened risk of toxicity. However, organic silver, particularly SSD, may have faster metabolic rates. Efficacy and toxicity of silver products are determined by the type and extent of the wound, duration of application, and mechanism of delivery. Some patients may even be more vulnerable to silver toxicity than others. Normal physiologically acceptable levels of silver are very low. Patients treated with SDD may have increased renal background silver concentration by 4,000%. Silver levels, though, return quickly to normal upon cessation of topical application.

Possible mechanisms of silver toxicity on various mammalian cells, including fibroblasts and keratinocytes, have been investigated in vitro involved in wound healing. Nanoparticles appear to be the most toxic causing oxidative stress via a number of pathways. Even though these adverse effects occur in vivo only when the natural antioxidant defense mechanisms of the body are overwhelmed, special consideration must be given to a liver toxicity case reported in a young male with 30% second-degree burns, though his liver function returned to normal after cessation of nanocrystalline silver application (Elliot 2010).

## Effect of Silver on Wound Healing

Silver has been considered for many years to improve wound healing and reduce infections. By producing a variety of toxins and proteases bacterial colonization and burn wound sepsis delay healing. It can thus be assumed that the antibacterial properties of topical silver products have a positive effect on healing and reepithelialization of second-degree burns. Even though there has been a direct link between higher doses of silver and faster microbial kill-time, there is no definitive and clinical evidence to suggest that this is an appropriate therapeutic measure and further to determine whether the antimicrobial kill-time had any relevance to the effectiveness of silver in wound healing.

The potent anti-inflammatory properties of silver have been recognized for centuries. Silver-based technologies in particular downregulate matrix metalloproteinases (MMPs) and the inflammatory response to levels that would facilitate wound healing. It must be stressed, however, that not all silver is anti-inflammatory. This property is largely dependent on the delivery vehicle, the available concentration and species of silver, and the duration of release. Increased inflammation observed with SSD is caused by the water soluble cream base itself. Similarly, silver nitrate application results in more than tenfold increase in MMP levels indicating an exaggerated inflammatory response.

Despite their beneficial bactericidal effects, all silver-based dressings result in a significant delay of reepithelialization with delayed eschar separation and healing often observed clinically. Cytotoxicity is related to the concentration of Ag+ released from each dressing. Epithelial regeneration is inhibited whenever fluid evaporation from the dressing causes the concentration of AgNO3 to exceed 1%. SSD delays healing as well as wound contracture of full thickness burns. Healing may be delayed also following overexposure to nanocrystalline silver. In his original report, Moyer recognized silver cytotoxicity on keratinocytes. Laboratory studies have demonstrated that both keratinocytes and fibroblasts are susceptible to lethal damage when exposed to silver concentrations lethal for bacteria. Silver-based products cannot discriminate between healthy cells involved in wound healing and pathogenic bacteria. Among 17 wound care products recently tested on human keratinocyte cultures, silver-containing products were found to be among the most toxic (Atiyeh et al. 2007; Burd et al. 2007; Fidel Martínez-Gutierrez et al. 2012; Kempf et al. 2011; Maghsoudi et al. 2010).

## Conclusion

Silver-containing ointments, creams, and wound dressings are widely used to treat burn wounds on the assumption that silver prevents burn wound sepsis and may promote wound healing. They are widely used as well to treat all types of chronic infected, nonhealing wounds. Undoubtedly, silver does possess antimicrobial properties. However, evidence of its effectiveness remains poorly defined. Two recently published systematic reviews concluded that available evidence does not support the use of silver-containing dressings or creams for topical burn treatment since generally these treatments did not promote wound healing or prevent wound infection. One small trial of a silver-containing dressing showed significantly better healing time of second-degree burns while all other reviewed studies suggested that topical silver had significantly worse healing time and was not effective in preventing wound infection. SSD in particular has no effect on infection prevention, and actually slows down healing in patients with partial thickness burns. Even Fox has mentioned that SSD does not offer sufficient protection against the regrowth of Gram-negative bacteria in patients with more than 50% TBSA burns (Atiyeh et al. 2007; Aziz et al. 2012; Storm-Versloot et al. 2010; Wasiak et al. 2008).

Despite proven antibacterial properties, it can only be concluded from the available evidence that silver-containing dressings and topical silver are neither significantly better nor worse than other topical modalities in preventing wound infection and promoting healing of burn wounds.

## Cross-References

▶ Colloidal Silver Nanoparticles and Bovine Serum Albumin
▶ Metallothioneins and Silver
▶ Silver as Disinfectant
▶ Silver in Protein Detection Methods in Proteomics Research
▶ Silver, Pharmacological and Toxicological Profile as Antimicrobial Agent in Medical Devices
▶ Silver-Induced Conformational Changes of Polypeptides

## References

Atiyeh BS, Costagliola M, Hayek SN, Dibo SA (2007) Effect of silver on burn wound infection control and healing: review of the literature. Burns 33:139–148

Aziz Z, Abu SF, Chong NJ (2012) A systematic review of silver-containing dressings and topical silver agents (used with dressings) for burn wounds. Burns 38:307–318

Becker RO (1999) Silver ions in the treatment of local infections. Met Based Drugs 6:311–314

Burd A, Kwok CH, Hung SC, Chan HS et al (2007) A comparative study of the cytotoxicity of silver-based dressings in monolayer cell, tissue explant, and animal models. Wound Repair Regen 15:94–104

Chaw KC, Manimaran M, Tay FE (2005) Role of silver ions in destabilization of intermolecular adhesion forces measured by atomic force microscopy in *staphylococcus epidermidis* biofilms. Antimicrob Agents Chemother 12:4853–4859

Elliot C (2010) The effects of silver dressings on chronic and burns wound healing. Br J Nurs (Tissue Viability Suppl) 19:S32–S36

Fidel Martínez-Gutierrez F, Thi EP, Silverman JM, de Oliveira CC et al (2012) Antibacterial activity, inflammatory response, coagulation and cytotoxicity effects of silver nanoparticles. Nanomedicine 8:328–336

Jung WK, Koo HC, Kim KW, Shin S et al (2008) Antibacterial activity and mechanism of action of the silver Ion in *staphylococcus aureus* and Escherichia coli. Appl Environ Microbiol 74:2171–2178

Kempf M, Kimble RM, Cuttle L (2011) Cytotoxicity testing of burn wound dressings, ointments and creams: a method using polycarbonate cell culture inserts on a cell culture system. Burns 37:994–1000

Klasen HJ (2000a) Historical review of the Use of silver in the treatment of burns. I. Early uses. Burns 26:117–130

Klasen HJ (2000b) A historical review of the use of silver in the treatment of burns. II. Renewed interest for silver. Burns 26:131–138

Li WR, Xie XB, Shi QS, Duan SS, Ouyang YS et al (2011) Antibacterial effect of silver nanoparticles on *Staphylococcus aureus*. Biometals 24:135–141

Liu HL, Dai SA, Fu KY, Hsu SH (2010) Antibacterial properties of silver nanoparticles in three different sizes and their nanocomposites with a new waterborne polyurethane. Int J Nanomedicine 5:1017–1028

Maghsoudi H, Monshizadeh S, Mesgari M (2010) A comparative study of tthe burn wound healing properties of saline-soaked dressing aand silver sulfadiazine in rats. Indian J Surg 73:24–27

Percival SL, Thomas J, Linton S, Okel T et al (2011) The antimicrobial efficacy of silver on antibiotic-resistant bacteria isolated from burn wounds. Int Wound J. doi:10.1111/j.1742-481X.2011.00903.x (Epub ahead of print)

Storm-Versloot MN, Vos CG, Ubbink DT, Vermeulen H (2010) Topical silver for preventing wound infection. Cochrane Database Syst Rev (3):Art. No: CD006478. doi:10.1002/14651858.CD006478.pub2

Thomas JG, Slone W, Linton S, Corum L et al (2011) A comparison of the antimicrobial efficacy of two silver-containing wound dressings on burn wound isolates. J Wound Care 20:580–586

Wasiak J, Cleland H, Campbell F (2008) Dressings for superficial and partial thickness burns. Cochrane Database Syst Rev (4). Art. No.: CD002106. doi:10.1002/14651858.CD002106.pub3

Wilkinson LJ, White RJ, Chipman JK (2011) Silver and nanoparticles of silver in wound dressings: a review of efficacy and safety. J Wound Care 20:543–549

## Silver, Neurotoxicity

Hari Shanker Sharma and Aruna Sharma
Laboratory of Cerebrovascular Research, Department of Surgical Sciences, Anesthesiology & Intensive Care medicine, University Hospital, Uppsala University, Uppsala, Sweden

## Synonyms

Nanowired cerebrolysin and superior neuroprotective effects in silver neurotoxicity; Silver nanoparticles and alterations in drug effects; Silver nanoparticles and astrocytic activation; Silver nanoparticles and biological functions; Silver nanoparticles and blood–brain barrier breakdown; Silver nanoparticles and Cerebrolysin treatment; Silver nanoparticles and gene expression in the brain; Silver nanoparticles and heat shock protein expression; Silver nanoparticles and Hypertension and diabetic disorders; Silver nanoparticles and hyperthermia; Silver nanoparticles and nitric oxide synthase upregulation; Silver nanoparticles and oxidative stress; Silver nanoparticles neurotoxicity

## Definition

### Silver (Ag)

The metal silver, known as *Argentum* (in Latin) and the chemical symbol of which is Ag, has atomic number 47 and atomic mass 107.8682. Ag is a soft metal with the highest electrical and thermal conductivity and the lowest contact resistance as compared to any other metals. Due to these properties, it is used in electrical contacts and conductors. Ag compounds are used in photographic film, disinfectants, and antimicrobial creams alone or with other antibiotics. Ag ions and compounds induce a toxic effect on bacteria, viruses, algae, and fungi, like any other heavy metals, but their toxicity to humans is not well characterized.

Different silver compounds are often used in traditional medicine for various diseases. However, excessive use of Ag preparations leads to argyria in humans followed by toxicity. Ag toxicity could result in coma, pleural edema, and hemolysis. Thus, exposure to Ag or its compounds in humans requires further investigation.

### Silver Nano or Silver Nanoparticles (AgNPs)

Silver nano denotes an antibacterial technology used in washing machines, refrigerators, air conditioners, air purifiers, and vacuum cleaners in which silver nanoparticles (AgNPs) are used. Silver nano coating on the inner surfaces of these machines is given for their antibacterial and antifungal effects. However, it is still uncertain whether use of AgNPs in household machines could have any toxic effect on humans.

### Silver Nanowires (AgNWs)

In recent years, Ag nanowires (AgNWs) are being prepared using chemical synthesis to make them a soft biocompatible for drug delivery. AgNWs are stretchable conductors and have greater electrical conductivity than carbon nanotubes (CNTs). The biocompatible AgNWs are used as floating or swimming devices in the blood stream that could be monitored and controlled from the outside to deliver drugs at specific cells or organs using bionanotechnology. However, the adverse effects of AgNWs or AgNPs are still not known well.

## Human Exposure of Silver (Ag)

Ag is present naturally in the environment in mines, silver-containing rocks, and/or soil. From these natural sources, wearing down of silver-containing rocks and soil by the rain or wind often releases large amounts of silver into the environment (ATSDR 1990). The released silver is carried to long distances through the air or water routes. Rain-washing of silver compounds from the soil leads to their transport to the groundwater.

However, the major source of silver in the environment are human activities, for example, processing of ores, cement manufacturing, steel refining, waste incineration, fossil fuel combustion, and cloud seeding (ATSDR 1990). A rough estimate indicates that in the USA alone about 2 million kilograms of Ag is released into the environment due to human activity, out of which 77 % comes from land disposal of solid waste, 17 % from surface waters, and 6 % from the atmosphere. The human population is exposed to Ag mainly through the ingestion of water and food contaminated with Ag. The food sources of Ag include seafood from areas near sewage or disposal of industrial waste materials and/or crops grown in regions with high levels of ambient Ag in the water or in air (ATSDR 1990; Sharma and Sharma 2007). It has been suggested that about 70,000 people are exposed to Ag at workplace environments every year largely due to inhalation. Industrial sources in USA alone could release large amounts of Ag annually into the ambient air due to metal production (30 k kg), electrical contact and conductors manufacturing and usage (22 k kg), coal and petroleum combustion (9 k kg), iron and steel production (7 k kg), and cement factories (2 k kg). Moreover, release of huge amounts of Ag in USA annually into water is from photographic film developing (65 k kg), photographic material manufacturing (5 k kg), sewage treatment plants (70 k kg), and soil erosion (72 k kg).

Other sources of Ag in water are textile plants and related waste disposal (see ATSDR 1990).

### Average Concentration of Ag Exposure

Apart from Ag exposure to humans from the air, food, and water, dental or medicinal usage also results in high amounts of Ag intoxication. The average public water supply in USA contains 0.1–10 μg/L Ag, whereas in some areas the Ag content may be as high as 30 μg/L. In extreme situations, the water content of Ag may go up to 50 μg/L (ATSDR 1990). On the other hand, average intake of Ag from food and water could be in the range of 70–80 μg/day. In a typical US diet, average consumption of Ag varies from 3 to 4.5 μg/day. Inhalation of Ag from air may account for 0.23 μg/day. Excessive exposure to Ag from water, food, and air could vary depending on the level of Ag concentration in the particular region's environment and could reach a level of 100 μg/day (ATSDR 1990).

### Entry of Ag into the Body and Excretion

Ag could enter the human body through oral, dermal, or nasal routes. Ingestion of contaminated food, water, or medicine results in entry of Ag into the blood stream. Handling of photographic materials or industrial waste, Ag powders, and/or breathing air containing Ag will result in Ag entry into the body. Inhalation of air and ingestion of food and water containing Ag induces large amount of the metal into the body than the dermal route. Most of the Ag that has entered the body is excreted in the feces within a week (Chang et al. 2005; Elder et al. 2006; Oberdorster et al. 2005). However, Ag is not excreted in the urine. Ag that has entered the body may not be completely eliminated and thus remains localized in blood or in different organs extra- or intracellularly including the brain (Oberdorster 1996). As silver is not metabolized by the body enzymes, Ag deposits could stay in the body fluid environment for a very long time (ATSDR 1990; Sharma and Sharma 2007; Sharma 2009a, b).

### Adverse Health Effects of Ag Exposure

Short- or long-term exposure to Ag leads to various adverse health effects in humans ranging from dermal, cardiovascular, respiratory, gastrointestinal, neurological, hematological, immunological, renal, and reproductive disorders (Takenaka et al. 2001). Excessive exposure of Ag depending on the duration and dose could also induce developmental effects, cancer, or even death. In this entry, neurological effects following Ag exposure are discussed in detail.

### Ag Deposits in Brain Areas After Exposure

Human exposure to Ag following use of nasal drops containing silver nitrate resulted in deposition of Ag granules in the brain of certain subjects together with argyria in a woman (Landas et al. 1985). Autopsy studies showed that silver-containing granules were present in the circumventricular organs lacking a blood–brain barrier (BBB) as well as in the periventricular nucleus and supraoptic nucleus in the hypothalamus (ATSDR 1990). In this study, unfortunately, other brain areas were not examined. Thus, it is difficult to establish whether entry of Ag into the brain was due to breakdown of the BBB or it happened only through the circumventricular organs that lack an effective BBB regulation.

In animal studies, exposure of silver nitrate through drinking water for 120 days to 20 female mice made them hypoactive after 4 months of exposure than the saline-treated control group. Postmortem studies revealed massive Ag granules in the brain areas involved in motor control, for example, deep cerebellar nuclei, motor nuclei, and red nucleus of the brain stem, as well as other brain areas such as the cerebral cortex, basal ganglia, and anterior olfactory nucleus. However, in this study, no relationship was observed between Ag deposition in the brain and behavioral activity (Rungby and Danscher 1983, 1984).

### Human Exposure to Ag Nanoparticles (AgNPs)

Due to the advancement in nanobiotechnology, nanomaterials are now present in almost all our day-to-day activities (Sharma 2009 PBR). Accordingly, Ag nanoparticles (AgNPs) are used in several preparations for human applications ranging from topical antibacterial creams to nanomedicine. AgNPs have more powerful effects on the cell and tissues as compared to the parent Ag because of their small size. Any material could be considered as nanoparticles (NPs) if they are smaller than 100 nm in size (Sharma 2009; Sharma and Sharma 2012a, b). These NPs, whether occurring as by-products of some industrial process or engineered material, affect the cells or organisms in an identical fashion (Sharma 2009a, b). The effects of AgNPs are far more superior in inducing bactericidal or preservative effects than Ag alone. However, studies in cell culture suggest that toxic effects of AgNPs are also very higher than parent Ag.

Apart from use of AgNPs in broad-spectrum antibacterial, antifungal, or antiviral agents for the treatment of wound, they are also used in wound dressing to contain infection and in coating of textiles for military uniform to protect against chemical or biological warfare (Chen and Schluesener 2008; Thilagavathi et al. 2008). Extensive use of AgNPs is also common in weapon technology and explosive detection procedures. Thus, increasing use of AgNPs necessitates investigation of their possible neurotoxicity in humans using in vivo animal studies.

## AgNPs Affect Central Nervous System (CNS) Function

Several in vitro studies have shown that AgNPs are capable of inducing cytotoxic or pro-inflammatory effects probably through an increase in oxidative stress, release of various cytokines, and/or upregulation of stress proteins, for example, heat shock proteins (HSP) (Kruszewski et al. 2011). In addition, AgNPs may also induce apoptosis, necrosis, altered gene expression, and DNA damage (Bouwmeester et al. 2011). Few in vivo studies have demonstrated that AgNPs induce loss of body weight, alterations in hematological or biochemical parameters, and several organ damages or dysfunction (Kim et al. 2010). Systemic administration of AgNPs (size range 80–115 nm) either through subcutaneous, intraperitoneal, or intravenous route (5–60 mg/kg) over a period of 4–13 weeks resulted in Ag accumulation in different organs, for example, kidney, liver, spleen, heart, lungs, and brain in various quantities (Lankveld et al. 2010; Loeschner et al. 2011; Takenaka et al. 2011; Powers 2010). However, studies on neurotoxicity of brain or spinal cord following AgNPs intoxication in vivo is still not investigated in depth.

### Essential Criteria for Neurotoxicity of AgNPs

Excessive exposure to AgNPs from the environment, food, water, medicine, or any other household materials containing Ag results in accumulation of Ag granules in various cells and tissues. An increased level of blood AgNPs is also known to occur in these circumstances. When blood with a high level of AgNPs is circulated through the brain microvessels, it is quite likely that Ag may enter into the brain microenvironment either through circumventricular organs or due to breaking down of the BBB. In addition, AgNPs could also enter the brain fluid microenvironment through the choroid plexus or by disrupting the blood-cerebrospinal fluid barrier (BCSFB). Since the BCSFB is less tight than the BBB, it appears that AgNPs could easily gain access into the CSF and consequently into the brain through the BCSFB (Sharma and Johansson 2007a, b). However, further studies are needed to explore precise routes of AgNPs entry into the brain fluid microenvironment (Rapoport 1976).

However, it is still unclear whether high levels of AgNPs in blood could induce neurotoxicity either directly or through inducing breakdown of the BBB or BCSFB. Obviously, breakdown of the blood-CNS barrier (BCSNB) is instrumental in neurotoxicity (Sharma 2009a; Sharma and Sharma 2010; Sharma and Westman 2004a).

### Blood–Brain Barrier is the Gateway to Neurological Diseases

The brain and spinal cord are protected from external disturbances to their fluid microenvironment by the BBB and blood-spinal cord barrier (BSCB) (Fig. 1). The anatomical site of the BBB and BSCB resides in the endothelial cells of the brain or spinal cord microvessels (Rapoport 1976). Anatomically the BBB and BSCB are very similar in nature as the endothelial cells of the cerebral or spinal capillaries are connected with the tight junctions making them function as extended plasma membranes (Sharma and Westman 2004). Furthermore, these endothelial cells lack microvesicular transport and their albuminous cell membranes are surrounded by a thick basement membrane or basal lamina (Fig. 1). The basal lamina in itself does not constitute the barrier but somehow regulates the BBB or BSCB functions (Sharma and Westman 2004). In addition, the endothelial cells also have direct contact with neurons and glia. Almost 85 % of their surfaces are covered by the astrocytic end feet (Fig. 1). Thus, the endothelial cells complex collectively constitutes the BBB and BSCB in the CNS. Any breach of the cerebral or spinal cord endothelial cells, either caused directly by chemicals and metals toxicity or indirectly induced by oxidative stress and/or release of free radicals, cytokines, or neurochemicals, results

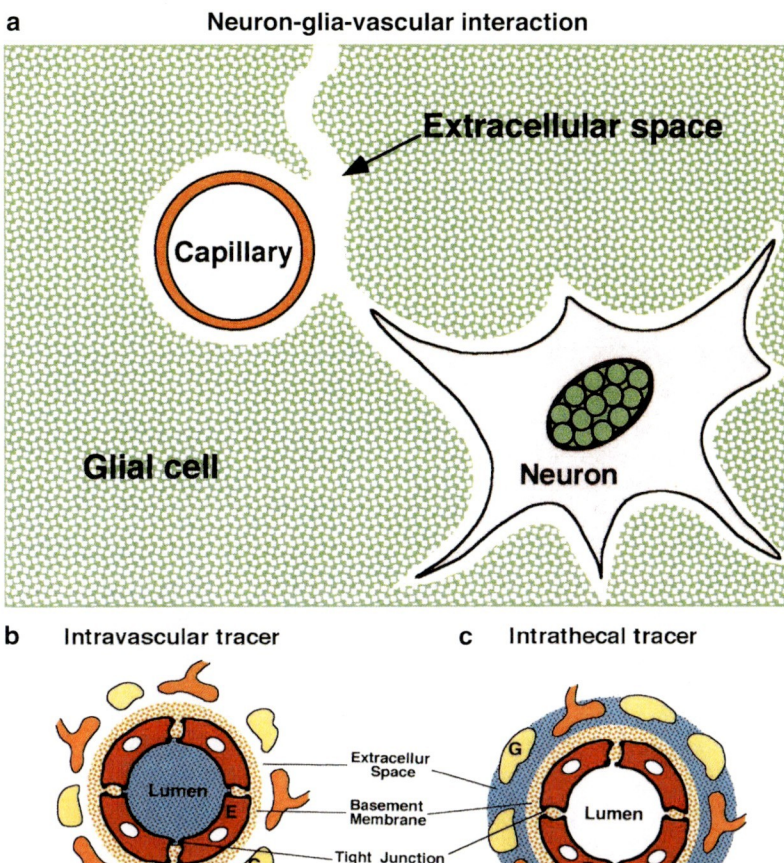

**Silver, Neurotoxicity, Fig. 1** Anatomical features of the blood–brain barrier (BBB). (**a**). Neurons are suspended in a pool of glial cells and microvessels supply nutrients and other essential elements for neuronal and glial cell survival. The cerebral capillaries however maintain a strict regulation on the transport of substances between the blood and brain due to the presence of the BBB. The cerebral endothelium is connected by tight junctions so that the cerebral capillary containing one or more endothelial cells behaves like an extended plasma membrane (Sharma and Westman 2004a; Sharma 2009a). The endothelial cells also lack vesicular transport (Rapoport 1976). Any leakage of the BBB to proteins will alter the composition of the extracellular fluid microenvironment leading to edema formation and cell injuries. The blood-spinal cord barrier (BSCB) is also anatomically identical and strictly regulates the exchange between blood and the spinal cord like the BBB (Sharma 2009a). Likewise, passage of transport from brain to blood is also limited as the permeability properties of the abluminal side of the cerebral endothelium (brain–blood barrier, bbb, C) are quite comparable to that of the luminal membrane (BBB) permeability (**b**). The brain or spinal cord microvessels also contain a thick basement membrane and are surrounded by astrocytic end feet that cover more than 85 % of the capillary surface (**b, c**). Neurons also make contact with both the astrocytes and the microvessels (**b, c**). Thus, the neuronal-glia-vascular complex constitutes the effective BBB and bbb that regulates the fluid compartments s of the CNS strictly within a narrow limit. Any breach to these barrier system leads to central nervous system (CNS) diseases and cell injuries. Thus, the blood-CNS barrier (BCSNB) could be regarded as a gateway to neurological disease (The data modified after Sharma 1982, 2009a)

in the breakdown of the BBB or BSCB. Depending on the magnitude and intensity of the BBB or BSCB disruptions, large molecules such as endogenous proteins like albumin, globulins, or exogenous protein molecules, for example, horse radish peroxidase (HRP), ferritin, and microperoxidse (MP) could enter into the brain or spinal cord microenvironment (Rapoport 1976). Entry of unwanted substances into the brain, especially proteins, results in alterations in the fluid microenvironment of the CNS leading to abnormal brain functions. Thus, the normal maintenance of the BBB or BSCB is essential to keep our CNS healthy. On the other hand, breakdown of these barriers will result in brain or spinal cord pathology. Taking these facts into consideration, the BBB and BSCB could be considered as the gateway to neurological diseases (see Sharma 1982, 1999, 2009). It appears that AgNPs affect the BBB or BSCB function to induce neurotoxicity (see below).

### Blood-CSF Barrier a Regulatory System for Brain Function

The CNS fluid environment apart from the tight BBB or BSCB is also regulated by the blood-CSF barrier (BCSFB). The BCSFB is anatomically located within the choroid plexus epithelial cells that are connected by tight junctions (Fig. 2). The endothelial cells in the choroid plexus are leaky as they do not possess tight junctions (Rapoport 1976; Sharma and Johanson 2007a, b). Thus, the intravenously administered tracer substances could easily reach up to the choroidal plexus epithelial cells. However, their passage across the choroidal epithelium into the CSF is restricted by the tight junctions located between the apical sides of the choroidal epithelial cells (Rapoport 1976; Sharma and Westman 2004a; Stalberg et al., 1998). The BCSFB is considered to be relatively less tight than the BBB (Fig. 2). This is because of the fact that the main function of the choroid plexus is to regulate the microenvironment of the CSF within the ventricular system (Sharma and Johansson 2007a, b) that is secreted by the choroidal epithelial cells (Sharma and Westman 2004). Thus, under normal circumstances, the BCSFB strictly regulates the entry of proteins or large molecules within the CSF. However, under disease conditions, or due to a direct toxic effect of metals or indirect action of cytokines, free radicals, release of neurochemicals, and other adverse stimuli, a breakdown of the BCSFB results in marked alterations in CSF composition. An altered CSF composition could adversely affect brain function and/or result in neurotoxicity. There are reasons to believe that AgNPs could also affect the BCSFB disruption similar to the BBB or BSCB.

### Breakdown of the BCNSB Leads to Brain Pathology

Breakdown of the BCNSB comprising BBB, BSCB, or BCSFB to plasma proteins leads to brain pathology (Sharma 1982, 1999, 2009). This is also evident in clinical situations where abnormally high levels of proteins are detected in the CSF in almost all kinds of neurological diseases (Sharma and Westman 2004a, b; Sharma and Johanson 2007a, b). In addition, leakage of serum proteins into the brain microenvironment causes neuronal, glial, and axonal damages in several neurological diseases, for example, stroke, ischemia, trauma, or other neurodegenerative diseases including Alzheimer's, Parkinson's, or Huntington's Diseases (Sharma 2012 IRN; Sharma et al. 2012a, b; Sharma and Sharma 2012a, b). However, it is still unclear whether the disease manifestation or progression could be further aggravated by AgNPs exposure or intoxication (Sharma et al. 2009a, b; Sharma and Sharma 2012a, b).

### Extravasation of Serum Proteins into the Brain Leads to Brain Edema Formation

One of the most adverse events following breakdown of the BBB or BSCB is the formation of brain or spinal cord edema (Sharma et al. 1998a, b) that is primarily responsible for CNS pathology (Sharma et al. 1998a; Sharma 1999, 2009). When serum proteins enter the brain fluid microenvironment, several biochemical, immunological, or neurochemical reactions take place leading to water intoxication into the CNS microenvironment from the vascular compartment (Rapoport 1976). Water from the vascular compartment enters the CNS extracellular environment due to altered osmotic balance between the plasma and brain across the endothelial cell membrane (Rapoport 1976; Sharma and Westman 2004a, b). Accumulation of water within the CNS microenvironment causes swelling of the brain or spinal cord. Excessive swelling of brain in the closed cranial compartment results in compression of the vital centers leading to instant death. It appears that AgNPs intoxication in vivo leads to brain edema formation by disrupting the BBB function. However, additional studies are needed to confirm this hypothesis.

### Alterations in the Brain Fluid Microenvironment Alters Gene Expression

Leakage of serum proteins and edema fluid within the CNS compartment after breakdown of the BCNSB leads to alteration in neuronal, glial, and myelin functions. Spread of edema fluid into different brain or spinal cord regions with advancing time exposes local cells or tissues to adverse fluid

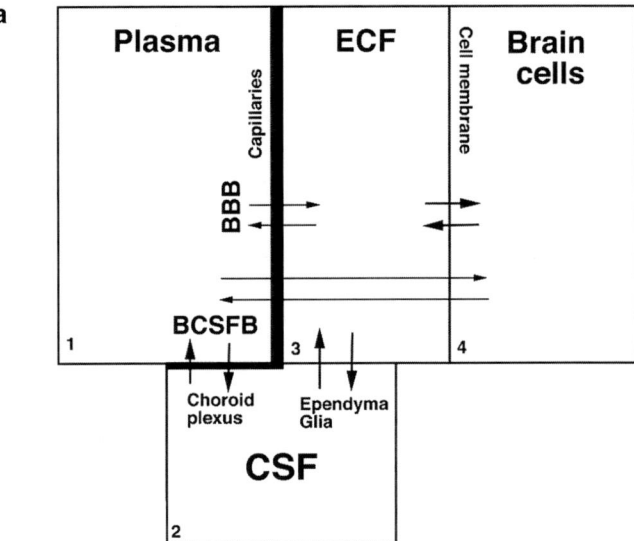

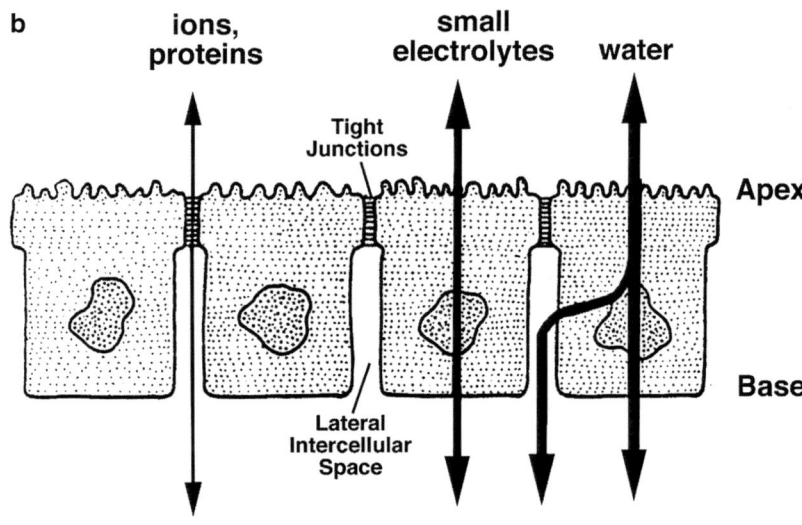

**Silver, Neurotoxicity, Fig. 2** The blood-cerebrospinal fluid (CSF) barrier (BCSFB). The composition of the CSF is entirely different from the plasma infiltrate and is devoid of any protein fragments (Sharma and Johanson 2007a). This active secretion CSF and strict maintenance of its composition is achieved by the choroid plexuses that hang on the roof of the cerebral ventricles. The choroidal epithelium (**b**) has specialized cells and the apical side of these cells is connected with tight junctions (**b**). The microvessels that supply blood to the choroid plexuses are fenestrated and lack any BBB (A1) or bbb (Sharma and Westman 2004a). However, the BCSFB (A2) is not as tight as the BBB (A1). Relatively, greater exchange of materials takes place routinely across the BCSFB interface (A2) as compared to the BBB compartment (A1). The extracellular fluid (ECF) compartment (A3) and brain cells (A4) can exchange materials readily. Thus, any substance that has entered into the ECF compartment can affect the neuronal, glial, and axonal environment (A4). There are also regular exchanges between CSF and brain cells through the ependymal lining that constitute the Brain CSF Barrier, (*bcsfb*, A2). In general, the *bcsfb* is not very restrictive and drugs or materials upon entering the CSF could affect brain cells readily. However, the exchanges across the *bcsfb* compartments are also regulated up to a certain extent (Rapoport 1976; Sharma 2009a; Sharma and Westman 2004a) (Data modified after Sharma 1982, 2009a)

microenvironment. As a result, these cells and tissues feel threatened resulting in severe cellular stress. When cells are stressed, they react by abnormal production of different kinds of proteins to counteract those external influences or to survive, resulting in altered gene expression. Altered gene expression could be either meant for their self-defense or survival strategies or simply as a marker of heightened cellular stress or activity (Sharma 2012 IRN). In these circumstances, altered expressions of several genes, for example, c-fos, c-jun, HSP, glial fibrillary acidic protein (GFAP), and myelin basic protein (MBP), are seen in the CNS (Sharma 2004; Cervós-Navarro et al. 1998; Westman and Sharma 1998; Sharma and Westman 2004a, b). Interestingly, these genes are also altered in various neurodegenerative diseases or brain pathologies following trauma or ischemia. There are reasons to believe that AgNPs could also enhance these gene expression in cells and tissues in the brain after acute or chronic intoxication. However, this idea requires additional investigation using in vivo models of AgNPs exposure.

### Abnormal Exposure of Serum Constituents Leads to Apoptotic or Necrotic Cell Death

Exposure of serum components to the brain cells after breakdown of the BCNSB results in altered gene expression reflecting changes in genetic material of the cells and/or damage to their DNA (Sharma 2004a). In such situations, cells or tissues feel threatened due to exposure of serum elements and edema fluid. When cell membranes are porous enough due these insults, intracellular accumulation of water and/or edema fluid occurs. Intracellular accumulation of edema fluid is cytotoxic (Sharma et al. 1998a; Rapoport 1976) and leads to cell death either due to necrotic or apoptotic processes. This necrotic and/or apoptotic cell death is seen following almost all kinds of CNS insults in experimental situations or in neurodegenerative diseases in clinical cases. Metal toxicity or AgNPs exposure in cell culture also results in necrotic and/or apoptotic cell death. This suggests that AgNPs in vivo could also induce cell death by necrotic or apoptotic processes. However, detailed studies on cell death following AgNPs in brain in vivo situations are not well known and require further investigations. It is also unclear whether AgNPs may have different effects in different brain regions with regard to neuronal, glial, or axonal damages or death in a specific and precise manner.

## AgNPs Induces CNS Toxicity in Animal Models

Keeping the above views in consideration, our laboratory has initiated a series of investigations on AgNPs-induced neurotoxicity in animal models in in vivo situations (Sharma 2009a, b). Effects of AgNPs dose and route of administration on neurotoxicity in rats and mice under normal healthy conditions was explored. Furthermore, influence of AgNPs neurotoxicity was also examined in animals subjected to environmental stressful situations, that is, hyperthermia (Sharma and Sharma 2007, 2011, 2012a, b; Sharma et al. 2009a, b, 2011) or alterations in internal milieu, for example, hypertension or diabetes (Sharma and Sharma 2012a, b; Lafuente et al. 2012; Sharma et al. 2012a; Sharma and Sharma 2012c). Our neurotoxicity studies are focused on AgNPs-induced BBB dysfunction; brain edema formation; neuronal, glial, and axonal damages; as well as alterations in various gene expressions within the CNS.

*Methodological Consideration.* Experiments were carried out on male Sprague–Dawley Rats (200–250 g) and C57 male Balb mice (30–35 g) for controlled AgNPs exposure under the Guidelines of National Institute of Health (NIH) for Care of the Experimental Animals and approved by the Local Institutional Ethics Committee.

Engineered AgNPs obtained from US Air Force Research Lab and/or commercially procured from Denzlingen, Germany, in the size range of 50–60 nm were used (Sharma et al. 2009a, b). These AgNPs were either suspended in sterile water or mixed in 0.05 % Tween 80 in 0.7 % NaCl solution for in vivo administration (See Sharma et al. 2009b). For control group, equimolar NaCl solution or carbonized microspheres (15 ± 0.6 μm in diameter) suspended in Tween 80 or in sterile saline were used for comparison. AgNPs dose was adjusted in such a manner that each animal (rats or mice) received 50 mg/kg through intraperitoneal (ip) route, 30 mg/kg doses for intravenous (iv) route, 2.5 mg/kg for intracarotid (ica) route, or 20 μg in 10 μl through intracerebroventricular (icv) route (Sharma et al. 2009a). To have almost identical effects of AgNPs given from any route, the intravenous dose was reduced to 1/2 of the intraperitoneal dose, the intracarotid dose was reduced to 1/20th of the intraperitoneal dose, and the intracerebroventricular infusion was reduced to 1/250th of the intravenous doses

(Sharma et al. 2009a, b). The dose and duration of exposure of AgNPs neurotoxicity in rats and mice was also compared to understand the species differences, if any, in our model.

The vehicle used for AgNPs suspension was equivalent to 0.1–0.3 M NaCl or 0.5 mg/ml whole blood concentrations (see Sharma et al. 2009a). Thus, the hyperosmolality of the vehicle (without AgNPs) was also examined on the BBB function. Our results showed that the vehicle alone was unable to open the BBB to protein tracers (see Table 1). Our calculations on the hyperosmolality are based on the assumption that rats or mice blood volume is ca. 7 ml/100 g body weight (roughly 3 ml blood volume per mice). Accordingly, a dose of 50 mg/kg Ag for a 500 g rat will be 25 mg total amount distributed in 35 ml of blood plus 0.5 ml solvent. Assuming Ag mol wt. 107 g in 1 L solvent will be 1 M Ag solution (Sharma et al. 2009a). Thus, the AgNPs concentration in the whole blood was approximately 0.5 mg/ml. Accordingly, the molarity of AgNPs may range from 0.1 to 0.3 M. Infusion of 0.1–0.3 M NaCl did not affect the BBB breakdown (Sharma et al. 2009a). All animals in this experimental group were allowed to survive either 4 h or 24 h after the administration of AgNPs (see Table 1).

## AgNPs Induces Breakdown of the BBB and BSCB

When AgNPs were administered intraperitoneally (50 mg/kg) into rats or mice, a mild leakage was observed at 4 h but not at 24 h after its administration (Table 1), whereas, intravenous administration of AgNPs (30 mg/kg) disrupted the BBB to protein tracers, for example, Evans blue albumin or radioiodine in rats and mice that was most pronounced after 24 h administration as compared to 4 h (Table 1). The brain and spinal cord in mice showed greater extravasation of tracers than rats following identical doses of AgNPs administration (Table 1). Intracarotid AgNPs (2.5 mg/kg) resulted in extravasation of Evans blue albumin or radioiodine in the ipsilateral side of the brain mainly in rats or mice (Table 1) that was most pronounced after 24 h as compared to 4 h after injection (Table 1). Intracerebroventricular (icv) superfusion of AgNPs (20 μg in 10 μl) into the right lateral ventricle resulted in extravasation of Evans blue and radioiodine into the perfused side of the cerebral hemisphere 24 h after administration (Table 1). In all these groups, the intensity of BBB breakdown was higher in mice than rats irrespective of the route of administration or observation period (Table 1).

A representative example of Evans blue leakage in the brain and spinal cord in AgNPs-treated mice is shown in Fig. 3. It is apparent from Fig. 1 that AgNPs-treated mice showed profound leakage of Evans blue at 24 h in the brain and spinal cord after intravenous, intracarotid, or intracerebroventricular AgNPs administration (Fig. 3). In most cases, cerebellum, hypothalamus, piriform cortex, brain stem, and ventral surface of the brain showed pronounced leakage of Evans blue (Fig. 3). In case of right ica or icv administration of AgNPs, although leakage of Evans blue was confined to the ipsilateral side, contralateral side of the brain also exhibited some blue staining (Fig. 3). This blue staining was seen on the dorsal and ventral surfaces of the brain and spinal cord (Fig. 3).

Marked blue staining was also seen in rats 24 h after AgNPs treatment intravenously (Sharma et al. 2009a). Moderate blue staining was seen in many parts of the dorsal and ventral surfaces of the brain. Deeper structures as seen in the coronal sections showed profound blue staining in the cortex, hippocampus, thalamus, and hypothalamus (Fig. 4a).

### Leakage of Serum Albumin in the Neuropil

Leakage of endogenous albumin in AgNPs-treated rats or mice is evident in the neuropil following 24 h after administration through systemic or intracerebroventricular routes (See Table 2). This albumin leakage was most prominent after intravenous, intracarotid, and intracerebroventricular administration of AgNPs (Table 2). The magnitude and intensity of albumin leakage was much higher in mice than rats (see Table 2). A representative example of albumin immunostaining in AgNPs-treated rats is shown in Fig. 4b. As evident in the figure, cerebral cortex shows several albumin-labeled neurons 24 h after intravenous administration of AgNPs (Fig. 4b). These albumin-positive neurons are largely localized in the edematous areas of the brain (Fig. 4b).

### Exudation of Lanthanum ($La^3$) Across Endothelial Cells

The above-mentioned results suggest that AgNPs are capable of inducing BBB breakdown. However, it is unclear whether AgNPs affect the BBB dysfunction by widening the tight junctions or by enhancing the endothelial cell membrane transport. To answer this question, the passage of lanthanum ($La^3$), an electron dense tracer, was examined across the cerebral endothelial

**Silver, Neurotoxicity, Table 1** Effect of saline, vehicle, or nanoparticles on the blood–brain barrier permeability and brain edema formation

| Type of Experiment | n | Blood–brain barrier permeability (Whole Brain) | | | | $^{131}$Iodine % | | | |
|---|---|---|---|---|---|---|---|---|---|
| | | EBA mg % | | | | Rats | | Mice | |
| | | Rats | | Mice | | 4 h | 24 h | 4 h | 24 h |
| | | 4 h | 24 h | 4 h | 24 h | | | | |
| **A. Control group** | 6 | | | | | | | | |
| *a. Intraperitoneal* | | | | | | | | | |
| Saline | | 0.24 ± 0.04 | 0.26 ± 0.08 | 0.28 ± 0.06 | 0.30 ± 0.05 | 0.30 ± 0.04 | 0.34 ± 0.08 | 0.33 ± 0.05 | 0.34 ± 0.08 |
| Tween 80 | | 0.22 ± 0.08 | 0.30 ± 0.06 | 0.32 ± 0.06 | 0.33 ± 0.09 | 0.34 ± 0.07 | 0.36 ± 0.08 | 0.36 ± 0.06 | 0.038 ± 0.10 |
| *b. Intravenous* | | | | | | | | | |
| Saline | | 0.26 ± 0.05 | 0.24 ± 0.06 | 0.28 ± 0.04 | 0.30 ± 0.08 | 0.30 ± 0.05 | 0.34 ± 0.06 | 0.36 ± 0.10 | 0.34 ± 0.09 |
| Tween 80 | | 0.30 ± 0.10 | 0.28 ± 0.09 | 0.34 ± 0.11 | 0.36 ± 0.12 | 0.29 ± 0.08 | 0.33 ± 0.08 | 0.38 ± 0.12 | 0.38 ± 0.11 |
| *c. Intracarotid* | | | | | | | | | |
| Saline | | 0.28 ± 0.08 | 0.26 ± 0.08 | 0.30 ± 0.08 | 0.34 ± 0.06 | 0.29 ± 0.10 | 0.26 ± 0.14 | 0.34 ± 0.08 | 0.36 ± 0.09 |
| Tween 80 | | 0.30 ± 0.06 | 0.32 ± 0.10 | 0.33 ± 0.12 | 0.36 ± 0.14 | 0.32 ± 0.08 | 0.33 ± 0.09 | 0.35 ± 0.12 | 0.35 ± 0.09 |
| *d. Intracerebroventricular* | | | | | | | | | |
| Saline | | 0.28 ± 0.08 | 0.32 ± 0.09 | 0.33 ± 0.10 | 0.36 ± 0.12 | 0.36 ± 0.11 | 0.36 ± 0.08 | 0.38 ± 0.14 | 0.36 ± 0.10 |
| Tween 80 | | 0.32 ± 0.12 | 0.34 ± 0.08 | 0.36 ± 0.10 | 0.38 ± 0.13 | 0.37 ± 0.10 | 0.39 ± 0.14 | 0.36 ± 0.10 | 0.38 ± 0.12 |
| **B. Hyperosmolar saline group** | 8 | | | | | | | | |
| *a. Intraperitoneal* | | 0.28 ± 0.08 | 0.30 ± 0.07 | 0.30 ± 0.04 | 0.36 ± 0.08 | 0.33 ± 0.08 | 0.34 ± 0.06 | 0.36 ± 0.08 | 0.36 ± 0.10 |
| *b. Intravenous* | | 0.30 ± 0.06 | 0.34 ± 0.08 | 0.32 ± 0.10 | 0.35 ± 0.09 | 0.36 ± 0.10 | 0.37 ± 0.05 | 0.37 ± 0.08 | 0.35 ± 0.05 |
| *c. Intracarotid* | | 0.26 ± 0.07 | 0.28 ± 0.06 | 0.30 ± 0.06 | 0.34 ± 0.06 | 0.32 ± 0.07 | 0.34 ± 0.08 | 0.34 ± 0.06 | 0.36 ± 0.08 |
| *d. Intracerebroventricular* | | 0.32 ± 0.10 | 0.34 ± 0.10 | 0.36 ± 0.09 | 0.34 ± 0.12 | 0.37 ± 0.13 | 0.38 ± 0.14 | 0.36 ± 0.09 | 0.39 ± 0.12 |
| **B. AgNPs group** | 8 | | | | | | | | |
| *a. Intraperitoneal 50 mg/kg* | | 0.33 ± 0.06* | 0.23 ± 0.05 | 0.48 ± 0.08* | 0.23 ± 0.03 | 0.38 ± 0.08* | 0.32 ± 0.07 | 0.43 ± 0.07* | 0.28 ± 0.12 |
| *b. Intravenous 30 mg/kg in saline* | | 0.46 ± 0.08* | 0.67 ± 0.12** | 0.50 ± 0.08* | 0.78 ± 0.08** | 0.55 ± 0.08* | 0.89 ± 0.11** | 0.60 ± 0.07* | 0.89 ± 0.12** |
| *c. Intravenous 30 mg/kg in Tween 80* | | 0.48 ± 0.10 | 0.70 ± 0.08** | 0.50 ± 0.10 | 0.78 ± 0.10** | 0.50 ± 0.08* | 0.87 ± 0.08** | 0.55 ± 0.08* | 0.90 ± 0.08** |
| *d. Intracarotid 2.5 mg/kg* | | 0.48 ± 0.06* | 0.81 ± 0.12** | 0.52 ± 0.08* | 0.94 ± 0.08** | 0.48 ± 0.07* | 0.93 ± 0.21** | 0.60 ± 0.06* | 0.99 ± 0.08** |
| *e. Intracerebroventricular 20 μg/10 μl* | | 0.38 ± 0.04* | 0.68 ± 0.11** | 0.48 ± 0.06* | 0.76 ± 0.21** | 0.40 ± 0.06* | 0.74 ± 0.12** | 0.56 ± 0.08* | 0.89 ± 0.22** |

Values are Mean ± SD from five to six rats at each point

\* = $P < 0.05$, \*\* = $P < 0.01$, compared from respective controls

\# = $P < 0.05$, \#\# = $P < 0.01$, compared from experimental group,

ANOVA followed by Dunnet's test

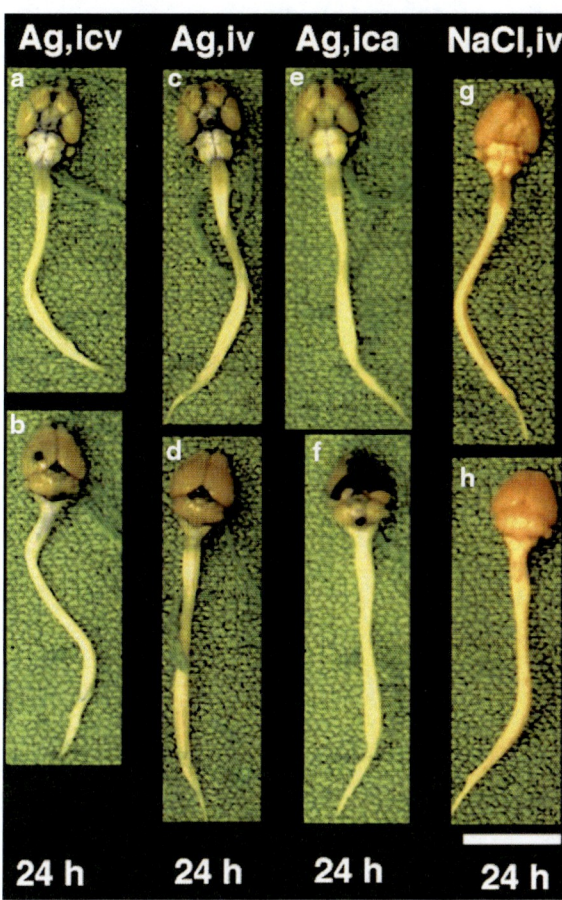

**Silver, Neurotoxicity, Fig. 3** Silver (Ag) neurotoxicity in mouse CNS. Intracerebroventricular (icv), intravenous (iv), or intracarotid (ica) administration of AgNPs induces profound leakage of Evans *blue* albumin (EBA) over the dorsal (b, d, f) and ventral (a, c, e) surfaces of the brain and spinal cords selectively and specifically (for details see text) 24 h after administration. On the other hand, saline treatment in an identical manner did not result in EBA leakage in the mouse brain or spinal cord (g, h). Bar = 3 mm (Data modified after Sharma 2009a)

cells in selected cortical areas in rats or mice after AgNPs treatment using transmission electron microscopy (TEM) (see Table 3). The results have shown that AgNPs resulted in profound exudation of $La^3$ across the cerebral microvessels by transcellular routes without modifying the tight junctions (Table 3). Thus, extravasation of lanthanum was confined within endothelial cell cytoplasm and in some cases in the basal lamina, whereas the tracer was stopped at the tight junctions (See Fig. 3c). Several microvesicular profiles containing $La^3$ were located within the endothelial cell cytoplasm of the large superficial microvessels of the brain or spinal cord in rats or mice (Sharma et al. 2009). This suggests that AgNPs are able to somehow directly or indirectly influence endothelial cell membrane transport without opening of the tight junctions (see Sharma et al. 2009a).

### AgNPs Induces Breakdown of the BCSFB

In our experiments, the lining of the cerebral ventricular walls underlying cortical structures, for example, caudate nucleus, hippocampus, and roof of fourth ventricle, were stained with Evans blue following intravenous or intracarotid administration of AgNPs in rats or mice. This suggests that the BCSFB was also disrupted by AgNPs in addition to the BBB and BSCB. To further confirm these observations, in separate groups of rats, albumin concentration in the CSF collected from cisterna magna was analyzed biochemically (Sharma HS unpublished observations). There was a 20-fold increase in albumin concentration (8.23 ± 0.56 ng/ml) in AgNPs-treated rats as compared to those treated with saline (0.04 ± 0.02 ng/ml). This observation confirmed that the AgNPs are able to induce breakdown of the BCSFB as well (Sharma HS unpublished observations).

### AgNPs Induces CNS Edema Formation

Administration of AgNPs induces profound brain edema formation at the time of the BBB breakdown to proteins (See Fig. 4, Table 3). Thus, intravenous, intracarotid, or intracerebroventricular administration of AgNPs increased the brain or spinal cord water content and volume swelling. Roughly, an increase in 1 % brain or spinal cord water content is equal to ca. 3 % volume swelling (% $f$) (Rapoport 1976; Sharma et al. 1998). The most marked increase in % $f$ was seen in mice (% $f$ = 3–14 %) as compared to rats (% $f$ = 2–8 %) after AgNPs administration (Table 3). On the other hand, intraperitoneal administration of AgNPs resulted in much less volume swelling and edema (<1 % $f$) in mice or rats (Table 3). This increase in brain water content is tightly correlated with Evans blue albumin leakage in the CNS after AgNPs treatment irrespective of the route of administration or the observation period (See Fig. 4).

### AgNPs Induce Brain Pathology

AgNPs administration induces profound brain pathology after intravenous, intracarotid, or intracerebroventricular administration in the areas exhibiting protein

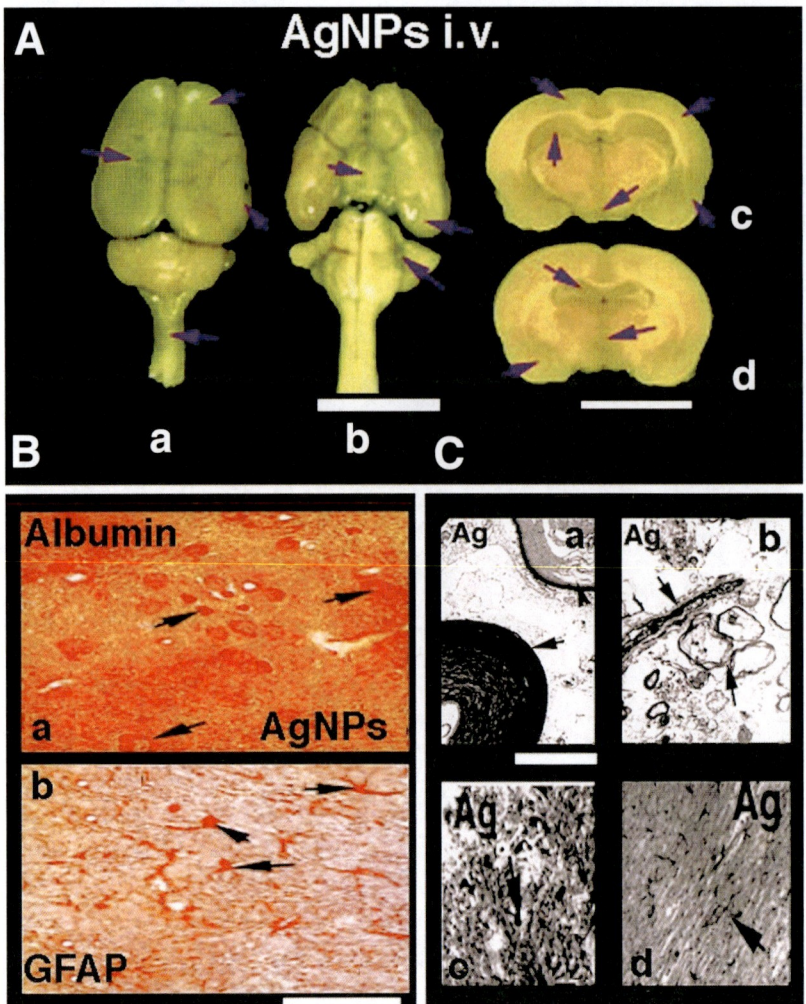

**Silver, Neurotoxicity, Fig. 4** Silver (Ag) neurotoxicity. A. Intravenous (iv) administration of silver nanoparticles (AgNPs) resulted in profound leakage of Evans blue albumin (EBA) on the dorsal (A.a) and ventral (A.b) surfaces of the rat brain (*arrows*). Coronal sections passing through hippocampus (A.c,d) also showed EBA leakage into the deep brain tissue (*arrows*). B. Immunocytochemical studies showing albumin leakage in the cerebral cortex (B.a, *arrows*) and overexpression of astrocytic protein and glial fibrillary acidic protein (GFAP, B.b, *arrows*). In the neuropil, marked expansion and edema are also visible (B.a, b). C. Ultrastructural changes showing myelin vesiculation (C.a, *arrowhead*) after AgNPs intoxication (iv) in the rat. However, the adjacent myelin (*arrow*) appears normal. Decreased dosage of lanthanum across the cortical microvessels and myelin vesiculation (*arrows*) is evident in the rat (C.b). Perivascular edema and structural damage is also evident (C.b). Immunohistochemical studies showing heat shock protein (HSP 72) upregulation (C.c), and GFAP overexpression (C.d) in the spinal cord of AgNPs-treated rats (*arrows*). Bar: A = 3 mm; B = 40 μm, C.a = 600 nm; C.b = 1 μm; C.c, d = 30 μm (Data modified after Sharma et al. 2009a)

leakage across the BBB or BSCB. All cellular components, for example, neurons, glia, endothelial cells, and axons, are affected by AgNPs treatment in both rats and mice (Tables 1 and 3; Figs. 4 and 5). Several damaged or distorted nerve cells in the cerebral cortex, hippocampus, cerebellum, thalamus, hypothalamus, and brain stem in AgNPs-treated rats or mice were seen at light microscopy (Table 3, Fig. 5). These neuronal and nonneural changes in the brain were most marked following AgNPs administration via intravenous, intracarotid, or intracerebroventricular routes (see Table 3). In terms of neuronal loss and/or nerve cell damage, hippocampus appeared to be the most sensitive structure for AgNPs neurotoxicity (Fig. 5)

**Silver, Neurotoxicity, Table 2** Effect of nanoparticles on physiological variables and expression of albumin, GFAP, HSP, and MBP in rats and mice at 24 h after administration

| Type of Expt. | | n | Physiological variables | | | | Immunoreactivity in cortex (nr. of positive cells) | | | |
|---|---|---|---|---|---|---|---|---|---|---|
| | | | MABP torr | Arterial pH | PaO$_2$ torr % | PaCO$_2$ torr | Albumin + cells/section | GFAP + cells/section | MBP + cells/section | HSP + cells/section |
| **1. Control Group (saline treatment equimolar concentration)** | | | | | | | | | | |
| Rats | iv | 6 | 120 ± 8 | 7.38 ± 0.04 | 80.34 ± 0.12 | 34.56 ± 0.21 | nil | 4 ± 2 | ++++ | 2 ± 3 |
| | ip | 6 | 123 ± 7 | 7.38 ± 0.06 | 80.54 ± 0.23 | 34.53 ± 0.22 | nil | 2 ± 2 | ++++ | 2 ± 1 |
| | ica | 6 | 118 ± 6 | 7.38 ± 0.06 | 80.34 ± 0.23 | 34.56 ± 0.08 | nil | 4 ± 2 | ++++ | 3 ± 2 |
| | icv | 5 | 110 ± 8 | 7.36 ± 0.08 | 80.29 ± 0.12 | 34.76 ± 0.23 | 3 ± 2 | 7 ± 5 | ++++ | 8 ± 2 |
| Mice | iv | 8 | 118 ± 6 | 7.40 ± 0.02 | 81.78 ± 0.23 | 34.68 ± 0.21 | nil | nil | ++++ | nil |
| | ip | 5 | 120 ± 9 | 7.42 ± 0.08 | 81.83 ± 0.34 | 34.89 ± 0.31 | nil | 4 ± 2 | ++++ | 3 ± 2 |
| | ica | 8 | 120 ± 9 | 7.43 ± 0.04 | 81.78 ± 0.23 | 34.87 ± 0.13 | nil | 3 ± 2 | ++++ | 2 ± 2 |
| | icv | 6 | 124 ± 12 | 7.41 ± 0.08 | 80.32 ± 0.43 | 35.32 ± 0.54 | +/−? | 4 ± 2 | ++++ | 4 ± 3 |
| **2. AgNPs intraperitoneal (50 mg/kg)** | | | | | | | | | | |
| Rats | Ag | 5 | 108 ± 8* | 7.36 ± 0.07 | 80.56 ± 0.47 | 34.46 ± 0.23 | 18 ± 1* | 98 ± 13**1 | ++ | 68 ± 14** |
| Mice | Ag | 5 | 106 ± 8* | 7.39 ± 0.09 | 81.46 ± 0.43 | 34.44 ± 0.33 | 110 ± 12** | 135 ± 16** | ++ | 160 ± 15** |
| **3. AgNPs intravenous (30 mg/kg)** | | | | | | | | | | |
| Rats | Ag | 5 | 90 ± 8** | 7.34 ± 0.08 | 81.56 ± 0.67 | 33.45 ± 1.23 | 120 ± 13** | 168 ± 13** | ++ | 167 ± 14* |
| Mice | Ag | 5 | 80 ≈ 12* | 7.30 ± 0.08 | 82.56 ± 0.89 | 33.56 ± 0.54 | 140 ± 10** | 155 ± 16** | ++ | 187 ± 25** |
| **#4. AgNPs intracarotid (2.5 mg/kg)** | | | | | | | | | | |
| Rats | Ag | 5 | 96 ± 8** | 7.32 ± 0.08 | 81.76 ± 0.37 | 33.45 ± 1.23 | 80 ± 13** | 88 ± 13**1 | ++ | 87 ± 14** |
| Mice | Ag | 5 | 80 ≈ 12* | 7.30 ± 0.08 | 81.86 ± 0.48 | 34.67 ± 0.45 | 98 ± 8** | 105 ± 16** | ++ | 117 ± 15** |
| **°5. AgNPs intracerebroventricular administration (20 μg/10 μl)** | | | | | | | | | | |
| Rats | Ag | 8 | 90 ± 10* | 7.29 ± 0.08 | 82.44 ± 0.18* | 32.56 ± 0.35* | 120 ± 19** | 160 ± 18** | ++ | 189 ± 16** |
| Mice | Ag | 8 | 98 ± 9* | 7.30 ± 0.06 | 82.46 ± 0.11** | 32.56 ± 0.37* | 150 ± 19** | 160 ± 18** | + | 189 ± 16** |

Values are Mean ± SD of five to eight animals
*icv* intracerebroventricular administration, *ica* internal carotid artery, *nil* absent
+ = faint, ++ = mild, +++ = moderate, ++++ = extensive, +/− = unclear, ? = border line, ° = cells are counted in half brain only
* $P < 0.05$; ** $P < 0.01$, ANOVA followed by Dunnet's test for multiple group comparison from one control group.
# Semiquantitative data were analyzed according to nonparametric Chi-square test.
Data in rats and mice were compared to rats and mice groups, respectively.
Data from Sharma et al. 2009.

**Silver, Neurotoxicity, Table 3** Effect of nanoparticles on the blood–brain barrier (BBB) permeability, brain edema formation, and cell changes in rats and mice at 24 h after administration

| Type of Expt. | | n | BBB permeability Evans blue mg % | [131]Iodine % | Brain edema Water content % | %f | Cell changes (LM) Neuron Nissl | glia GFAP | Myelin Luxol Fast Blue | EM La+++ TEM |
|---|---|---|---|---|---|---|---|---|---|---|
| **1. Control Group (saline treatment equimolar concentration)** | | | | | | | | | | |
| Rats | iv | 6 | 0.23 ± 0.08 | 0.34 ± 0.03 | 76.34 ± 0.28 | – | nil | +/– | ++++ | nil |
| | ip | 6 | 0.24 ± 0.08 | 0.30 ± 0.08 | 76.12 ± 0.32 | – | nil | +/– | ++++ | nil |
| | ica | 6 | 0.28 ± 0.06 | 0.34 ± 0.05 | 76.21 ± 0.24 | – | nil | +/– | ++++ | nil |
| | icv | 5 | 0.22 ± 0.04 | 0.28 ± 0.10 | 76.08 ± 0.18 | – | nil | +/– | ++++ | nil |
| Mice | iv | 8 | 0.21 ± 0.10 | 0.28 ± 0.14 | 74.56 ± 0.32 | – | nil | +/– | ++++ | nil |
| | ip | 5 | 0.28 ± 0.08 | 0.34 ± 0.12 | 74.72 ± 0.32 | – | nil | +/– | ++++ | nd |
| | ica | 7 | 0.21 ± 0.04 | 0.32 ± 0.06 | 74.68 ± 0.13 | – | nil | +/– | ++++ | nd |
| | icv | 6 | 0.24 ± 0.10 | 0.30 ± 0.08 | 74.65 ± 0.41 | – | nil | nil | ++++ | nd |
| **2. AgNPs intraperitoneal (50 mg/kg)** | | | | | | | | | | |
| Rats | Ag | 5 | 0.23 ± 0.05 | 0.32 ± 0.07 | 76.23 ± 0.12 | <1 | +/– | +/– | +++/? | nd |
| Mice | Ag | 5 | 0.23 ± 0.03 | 0.28 ± 0.12 | 74.54 ± 0.21 | <1 | +/– | +/– | +++/+? | nd |
| **3. AgNPs intravenous (30 mg/kg)** | | | | | | | | | | |
| Rats | Agg | 8 | 0.67 ± 0.12** | 0.89 ± 0.11** | 77.34 ± 0.21** | 4 | +++ | ++ | ++ | ++ |
| Mice | Ag | 8 | 0.78 ± 0.08** | 0.89 ± 0.12** | 76.89 ± 0.23** | 8 | +++ | +++ | ++ | ++ |
| **4. AgNPs intracarotid (2.5 mg/kg)** | | | | | | | | | | |
| Rats | Ag | 8 | 0.81 ± 0.12** | 0.93 ± 0.21** | 77.94 ± 0.09** | 5 | +++ | ++ | ++ | ++ |
| Mice | Ag | 8 | 0.94 ± 0.08** | 0.99 ± 0.08** | 77.19 ± 0.08** | 9 | +++ | +++ | ++ | ++ |
| **5. AgNPs intracerebroventricular administration (20 μg/10 μl)** | | | | | | | | | | |
| Rats | Ag | 8 | 0.68 ± 0.11** | 0.74 ± 0.12** | 77.38 ± 0.11** | 4 | ++ | +++ | ++ | ++ |
| Mice | Ag | 8 | 0.76 ± 0.21** | 0.89 ± 0.22** | 77.48 ± 0.21** | 14 | ++++ | ++++ | + | +++ |

Values are Mean ± SD of five to eight animals

*c.s.* cortical superfusion, *nil* absent, *nd* not done, *LM* light microscopy, *EM* electron microscopy, *Na* sodium, *K* potassium, *La* lanthanum ion, *% f* volume swelling calculated from changes in water content

+ = faint, ++ = mild, +++ = moderate, ++++ = extensive, +/– = unclear, ? = boarder line (for details see Sharma et al. 2009)

* $P < 0.05$; ** $P < 0.01$, ANOVA followed by Dunnet's test for multiple group comparison from one control group

Data in rats and mice were compared to rats and mice groups, respectively

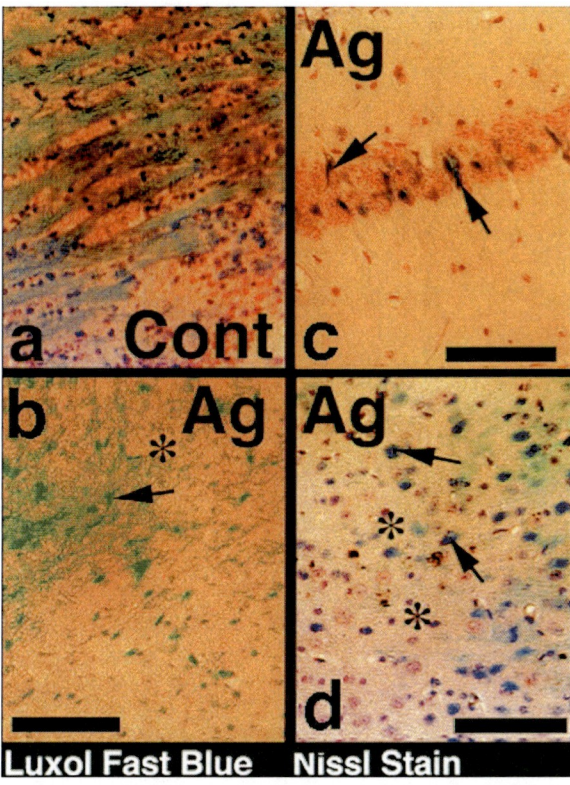

**Silver, Neurotoxicity, Fig. 5** Silver neurotoxicity of myelin and nerve cells. Intravenous administration of AgNPs resulted in severe myelin loss as seen using Luxol fast *blue* (LFB) histochemical staining (**a**, **b**) and neuronal loss in the hippocampus (**c**) and in cortex (**d**) using Nissl staining on paraffin section in the rat. In normal rats, LFB-stained myelin fibers are abundant in the cortex (**a**). AgNPs treatment resulted in drastic loss of these LFB-stained myelinated fibers (**b**, *arrow*). General sponginess and edema (*) are also seen in the neuropil together with loss of LFB-stained fibers. AgNPs toxicity resulted in massive loss of neurons in the hippocampus that was most marked in the CA-3 region (**c**, *arrows*), whereas loss of nerve cells are also seen in the cortex (**d**, *arrows*). Edema and sponginess of the neuropil in the cortex (*) are clearly seen. Bar: 50 μm (Data modified after Sharma et al. 2009a)

followed by cerebellum, cerebral cortex, brain stem, thalamus, and hypothalamus (Sharma HS unpublished observations). A close correlation between albumin-positive cells and neuronal damages in AgNPs-treated animals irrespective of routes of administration suggests that breakdown of the BBB is instrumental in Ag neurotoxicity (see Fig. 5).

In AgNPs-treated animals, distortion of microvessels with perivascular edema is also prominent (Fig. 4). These endothelial cell changes are most pronounced in mice brain than the rats (Sharma HS unpublished observations).

### Ultrastructural Changes

Studies using TEM further confirmed that AgNPs induced neurotoxicity, causing neuronal and glial cell damages in mouse and rat brains (Sharma et al. 2009a). In these AgNPs-treated rats, $La^3$ showed pronounced infiltration across the endothelial cells in the brain or spinal cord (Table 2, Fig. 4). These ultrastructural changes were most frequent in the areas showing BBB disruption to $La^3$ (Sharma et al. 2009) (see Table 2). Membrane damage, myelin vesiculation, collapse of endothelial cells and microvessels with perineuronal or perivascular edema are frequent in AgNPs-treated animals (Fig. 4c). Interestingly, one nerve cell or axon in a particular brain area is damaged, whereas the adjacent neurons or axons are quite normal in appearance (Sharma et al. 2009). This suggests that AgNPs induce selective vulnerability to cells and tissues directly or through a cascade of adverse cellular reactions, for example, oxidative stress or release of neurochemicals (Sharma et al. 2009).

### AgNPs Induce Gene Expression

Systemic administration of AgNPs altered expression of several genes regulating glial cell function, myelin stability, stress response of cells and tissues, as well as oxidative stress in the CNS.

### Glial Fibrillary Acidic Protein (GFAP) Gene Expression

Treatment with AgNPs induced astrocytes reaction as evidenced with activation of glial fibrillary acidic protein (GFAP) gene expression in the areas showing extravasation of serum proteins (Sharma et al. 2009). The GFAP is a specific marker of astrocytic proteins and overexpression of this gene in the brain or spinal cord denotes glial cell pathology (Cervós-Navarro et al. 1998) (Figs. 4 and 5; Tables 2 and 3). The normal astrocytes are star-shaped structures with elongated astrocytic foot processes that are in contact with neurons and or microvessels. This shape of astrocytes was altered severely in AgNPs-treated animals (Figs. 4 and 5). Accordingly, the astrocytes were distorted in appearance and overexpress the GFAP gene. These activated astrocytes are largely located in the edematous brain or spinal cord regions (Figs. 4 and 6). Perivascular astrocytes often show profound swelling and their end feet are either damaged or contain water-filled channels around the microvessels (Figs. 4–6, Table 2). These changes in astrocytes were most

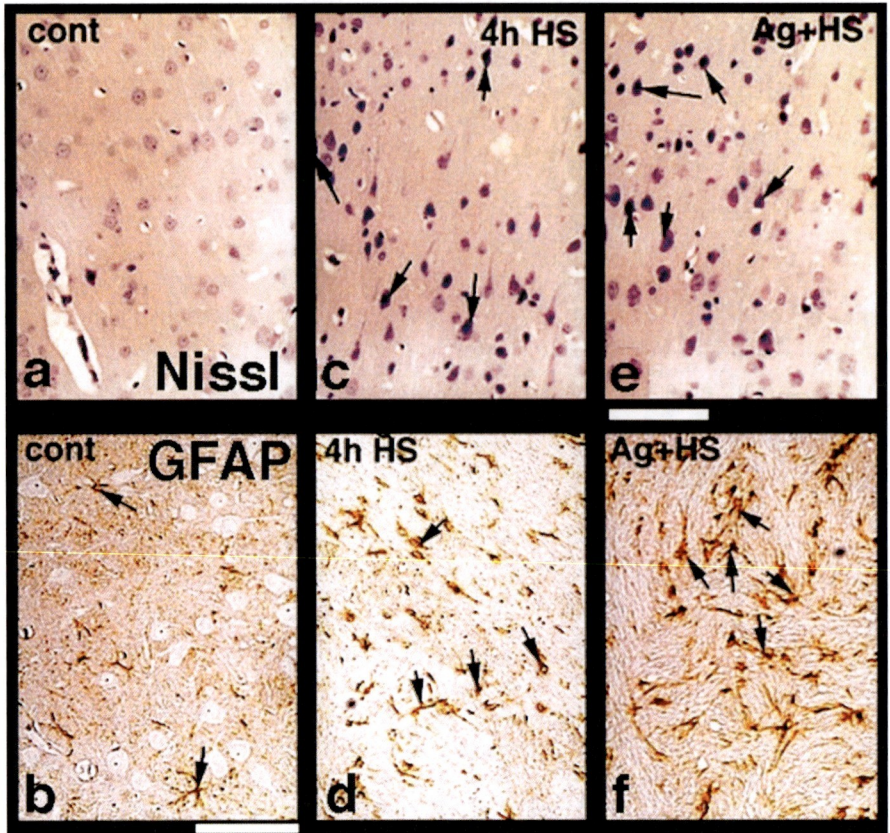

**Silver, Neurotoxicity, Fig. 6** Silver neurotoxicity is exacerbated following heat stress–induced hyperthermia. Nissl-stained cerebral cortical neurons (**a, c, e**) and immunostaining of glial fibrillary acidic protein (GFAP, **b, d, f**) in normal rats and following 4 h heat stress at 38 °C in a biological oxygen demand (BOD) incubator. Control cerebral cortex of normal rats did not show neuronal damage (**a**) or marked expression of GFAP (**b**). However, when normal rats were subjected to 4 h heat stress, neuronal damage and loss (**c**, *arrows*) and activation of GFAP (**d**, *arrows*) were prominent in the cortex. These neuronal (**e**) and glial cell (**f**) changes were further exacerbated after heat stress in AgNPs-treated rats. Thus, the neuronal damage and loss (*arrows*) and overexpression of GFAP gene (*arrows*) are much more aggravated in AgNPs-treated, heat-stressed rats (**e, f**). This suggests that AgNPs neurotoxicity is aggravated in hyperthermia (Data modified from Sharma et al. 2009b)

pronounced in mice than rats after intravenous or intracarotid injection of AgNPs (Tables 2 and 3).

Myelin Basic Protein (MBP) Gene Expression
Myelin is an important part of the CNS (Givogri et al. 2000; Sharma 2004). Myelinated nerve fibers carry electrical impulses much faster than unmyelinated nerve fibers in the brain or in periphery (Stålberg et al. 1998-book). Loss of myelin in disease conditions impairs the ability of nerve transmission and results in neuropathic pain or other long-term neural diseases. AgNPs treatment induces marked loss of myelin as seen using Luxol-fast blue (LFB) histochemical staining in both rats and mice (Tables 2 and 3, Fig. 7). Myelin fibers in the normal brain or spinal cord are stained deep blue using LFB that were severely damaged after AgNPs treatment indicating myelin degeneration (Sharma et al. 2009). To further confirm the myelin damage, myelin basic protein (MBP) gene expression was examined using immunohistochemistry in AgNPs-treated group. The results have shown a significant downregulation of MBP gene expression in several brain and spinal cord areas in mice or rats after intravenous, intracarotid, or intracerebroventricular administration of AgNPs (see Tables 2 and 3). Mice brain showed greater loss

of MBP gene expression as compared to rats (Tables 2 and 3). These observations clearly point out that AgNPs induce myelin damage (Sharma et al. 2009a, b).

Heat Shock Protein 72 kD Gene Expression

Heat shock protein (HSP) 72 kD genes constitute the basic response of all cells and organisms against any kind of stress or adverse stimuli (Sharma and Westman 2004a, Sharma 2005b; BBB book). Thus, upregulation of HSP gene in brain or spinal cord denotes cellular stress (Westman and Sharma 1998) and often precedes cellular injury. Since AgNPs induced profound neuronal damage, HSP gene expression in rats and mice in brain or spinal cord is quite likely. Our studies have shown a profound upregulation of HSP gene expression in neurons, glial cells, and in endothelial cells in the areas exhibiting BBB breakdown and cell injuries selectively in the brain or spinal cord areas of rats and mice after AgNPs administration (Fig. 4, Table 2). The HSP upregulation was prominent in the cell cytoplasm of neurons, glial cells, and in endothelia in the cortex, hippocampus, cerebellum, thalamus, hypothalamus, brain stem, and the spinal cord (Sharma et al. 2009). These areas also show extensive leakage of albumin (Fig. 4). A very strong correlation between number of albumin and HSP-positive cells is seen in AgNPs-treated animals irrespective of their route of administration and duration of exposure (Fig. 4c). This indicates that BBB breakdown is instrumental in upregulation of HSP gene expression. Interestingly, the magnitude and intensity of HSP expression in mice brains was very high after AgNPs treatment than rats. This indicates that AgNPs-induced stress response is species dependent (Sharma et al. 2009).

c-fos and c-jun Gene Expression

Cellular stress affects neuronal activity resulting in overexpression of proto-oncogenes c-fos and c-jun in the CNS (Hughes and Dragunow 1995; Sharma 2004, chapter 15). Since AgNPs exert cellular stress, it is quite likely that c-fos and c-jun gene expressions are altered selectively in the rat or mice brain. Thus, experiments were conducted in our lab to investigate these proto-oncogene expressions in the CNS using immunohistochemistry after AgNPs treatment. Our results have shown that AgNP enhances c-fos and c-jun expression in the areas of HSP expression and cell injuries at 4 or 24 h after intracarotid, intravenous, or intracerebroventricular administration. Expression of these proto-oncogenes is seen in the brain areas perfused by AgNPs. Thus, intracarotid administration of AgNPs showed overexpression of c-fos and c-jun in the ipsilateral hemisphere only at 4 h and a slight expression in the contralateral side was seen after 24 h survival. On the other hand, intravenous administration resulted in pronounced expression of c-jun and c-fos in the hippocampus, cerebellum, and cortex at and 4 and 24 h after AgNPs intoxication. Intracerebroventricular application of AgNPs induced c-fos and c-jun upregulation in superficial cortex and cerebellum only. This indicates that AgNPs are able to induce profound expression of proto-oncogenes selectively.

nNOS Gene Expression

Nitric oxide (NO) is a free radical gas and is synthesized within the CNS by enzyme nitric oxide synthase (NOS) that normally present in some neurons and in glial cells (Calabrese et al. 2007; Sharma 1999, 2009; Sharma et al. 1998b; Sharma and Alm 2004). Any noxious stimulus to the brain that is sufficient enough to disrupt normal functioning of the neuronal or glial cells will activate neuronal NOS (nNOS) expression in the brain or spinal cord. There are reports indicating that NO derived from nNOS is involved in c-fos regulation (Chan et al., 2004; Samuel and Guggenbichler 2004). Since AgNPs induced c-fos overexpression, nNOS upregulation was also examined following AgNPs intoxication in rats or mice. The results have shown that AgNPs significantly enhanced nNOS expression in the areas showing neuronal damage. Interestingly, nNOS expression was observed in cells that are also expressing c-fos and HSP upregulation. These cells are normally located within the regions exhibiting albumin leakage. This suggests that AgNPs-induced breakdown of the BBB leads to disturbances in the cellular microenvironment resulting in a variety of cellular or molecular responses, for example, oxidative stress, release of free radicals, and several neurochemicals within the CNS. These factors either alone or in combination lead to alteration in gene expression, nNOS upregulation, and cell damage. Upregulation of nNOS releases NO, the free radical molecule that can induce direct cell membrane damage (Sharma and Alm 2004). It is likely that breakdown of the BBB caused by AgNPs may be one of the leading factors in overexpression of nNOS, c-fos, c-jun, and/or HSP gene expression in the brain. Obviously, alterations in these genes within the CNS may represent either injured neurons or intensive activity of nerve cells that could lead to even cell death.

## AgNPs Exacerbates Hyperthermia-Induced Brain Pathology

AgNPs acute intoxication induces neurotoxicity in naive animals (Sharma et al. 2009a). However, whether chronic exposure of AgNPs may alter the response of additional stress or traumatic injuries to the CNS is still not well known (Sharma and Sharma 2007). Military personnel are often exposed to long-term AgNPs intoxication and they have to work under extreme environmental conditions and/or stressful situations (Sharma and Hoopes 2003). Thus, it is important to understand whether AgNPs could alter the magnitude or intensity of stress-induced brain dysfunction. For combat operation or peace keeping tasks, military personnel are often exposed to either extreme hyperthermia, for example, in Middle East or extreme cold environment, that is, Afghanistan, Canada, or other high-altitude countries. When these military personnel are exposed to various nanoparticles including AgNPs, whether they are adversely affected under extreme environmental conditions is still not well known (Sharma 2005a, 2006a, b; Sharma et al. 2009b). To answer this question, rats were chronically treated with a mild dose of AgNPs (50–60 nm in Tween 80 in a dose of 50 mg/kg, ip) daily for 7 days. This dose of AgNPs did not induce BBB breakdown or brain edema formation (Sharma and Sharma 2007). However, when these animals were subjected to 4 h heat exposure at 38 °C in a biological oxygen demand incubator, the BBB disruption to protein tracers or lanthanum was exacerbated than saline-treated groups after identical heat stress (Sharma and Sharma 2007). In these animals, hyperthermia-induced brain pathology was also exacerbated as compared to saline-treated rats (Sharma et al. 2009b; Fig. 7). This observation suggests that AgNPs exacerbate brain dysfunction in hyperthermia.

## AgNPs Exacerbated Hypertension- and Diabetes-Induced Brain Pathology

Apart from external heat stress, hypertension or diabetes that induces profound cellular or system stress to the body is very common in large number of human populations. However, whether hypertensive and/or diabetic people are more vulnerable to nanoparticles exposure with regard to brain dysfunction is still not well known. Since AgNPs exacerbated hyperthermia-induced brain pathology, it appears that nanoparticles exposure in hypertension or diabetes may further aggravate brain pathology. This hypothesis was examined in our laboratory in rat models. Rats were made either hypertensive or diabetic using standard protocol (Sharma and Sharma 2012). The two-kidney one-clip (2K1C) method was used to induce chronic hypertension in rats (Muresanu and Sharma 2007), whereas streptozotocin (75 mg/kg, ip for 3 days) injection in rats was used to develop diabetes mellitus (Sharma et al. 2010; Muresanu et al. 2010; Lafuente et al. 2012). These hypertensive (mean arterial blood pressure, MABP $186 \pm 8$ mmHg) or diabetic (blood glucose $18 \pm 0.6$ mM/L) rats were intoxicated with AgNPs (50 mg/kg, ip daily for 7 days) and BBB permeability and brain edema was examined on the 8th day. Normotensive (MABP $108 \pm 4$ mmHg) rats with normal blood glucose ($6 \pm 0.8$ mM/L) level were used as controls. Our studies have shown that AgNPs resulted in 300–450 % increase in Evans blue or radioiodine leakage in the brain and spinal cord of hypertensive and diabetic rats, respectively, as compared to normal animals (Sharma HS, unpublished observation). The brain swelling in these rats were increased by 24 % and 30 % (volume swelling $f$) in hypertensive and diabetic rats, respectively, after AgNPs treatment as compared to normal controls (Sharma HS, unpublished observations). These observations suggest that AgNPs neurotoxicity is further aggravated in hypertension or diabetes. Thus, special precaution may be taken by hypertensive or diabetic people to avoid AgNPs exposure to prevent further deterioration of their mental or physical health.

## AgNPs Alter Neuroprotective Effect of the Drug

Since AgNPs exacerbate hyperthermia-, hypertension-, or diabetes-induced brain pathology, it seems likely that in such situations, neuroprotective agents that are able to attenuate brain damage in healthy situations require certain adjustment in their dose for effective therapy. To test this hypothesis, effects of Cerebrolysin

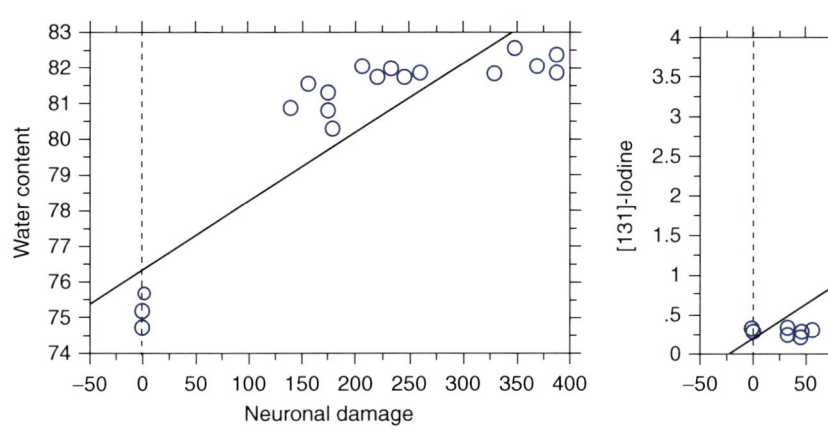
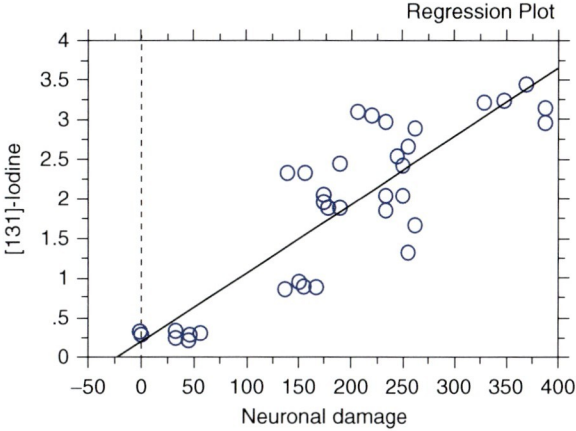

**Silver, Neurotoxicity, Fig. 7** Silver neurotoxicity and breakdown of the blood–brain barrier and edema formation. AgNPs induce neurotoxicity by inducing significant breakdown of the blood–brain barrier (BBB) to protein tracers, for example, radioiodine ([131]-I). Radioiodine when administered into the circulation binds to serum proteins and forms iodine-protein complex in the circulation. Leakage of iodine across the brain suggests protein leakage (Rapoport 1976; Sharma 2004a). Entry of serum proteins into the fluid microenvironment of the brain causes edema formation that leads to subsequent brain damage. A tight correlation (r2 = 0.788) between increase in brain water (*left*) with neuronal damage in AgNPs-treated rats (iv,) is in line with this hypothesis. Leakage of iodine is related to brain edema formation and/or neuronal damage is evident from an equally tight correlation between iodine leakage and neuronal damage (r2 = 0.78) (Data modified after Sharma et al. 2011)

(Ever NeuroPharma, Austria) were examined on hyperthermia-induced brain pathology in AgNPs-intoxicated rats (Sharma and Sharma 2207, 2012a, b; Sharma et al. 2007, 2011; 2012a, b). Cerebrolysin is a combination of several neurotrophic factors and active peptide fragments that are capable to enhance neuroregeneration, neuroplasticity, and neuroprotection (Menon et al. 2012; Sharma et al. 2012a, b, c). Cerebrolysin is known to attenuate hyperthermia-induced brain damage in a dose of 2.5 ml/kg given intravenously in rats (Sharma et al. 2007). Thus, it was interesting to examine whether cerebrolysin in the same dose could attenuate AgNPs-induced exacerbation of hyperthermia-induced brain damage as well (Sharma et al. 2011). Our results have shown that cerebrolysin in standard doses of 2.5 ml/kg, iv, was unable to attenuate AgNPs-induced exacerbation of brain pathology (Sharma et al. 2011). However, a double dose of cerebrolysin (5 ml/kg, iv) was able to induce appreciable neuroprotective effects in AgNPs-treated, heat-stressed rats (Sharma et al. 2011). Alternatively, when cerebrolysin was delivered through TiO2 nanowires, 2.5 ml/kg iv dose was enough to attenuate brain damage in these AgNPs-treated, heat-stressed animals (Sharma et al. 2011, 2012; Menon et al. 2012; Muresanu et al. 2012). This suggests that AgNPs not only enhance brain pathology after stress, but also lower the effectiveness of the drug treatment in standard doses. To overcome these effects, either the drug dose is enhanced further or the compound is administered using nanodrug delivery to achieve considerable neuroprotection in AgNPs intoxicated group (Sharma et al. 2012b).

## Summary and Conclusion

Evidences in human and animals data presented above clearly demonstrate that AgNPs depending on the dose and duration of exposure are able to induce selective neurotoxicity. This effect of AgNPs appears to be species dependent. Thus, mice are more vulnerable than rats to AgNPs neurotoxicity. There appears to be a selective vulnerability of AgNPs on regional brain toxicity. This is evident with marked neuronal damages and profound gene expressions in the hippocampus followed by cerebellum, cerebral cortex, thalamus, hypothalamus, brain stem, and spinal cord. This AgNPs neurotoxicity appears to be due to breakdown of the BCNSB to large molecules, for example, proteins. Extravasation of proteins into the CNS leads to edema formation and subsequently to cell injuries. Selective expressions of proto-oncogenes and stress proteins in the areas of albumin leakage are in line with this hypothesis. Interestingly, various external or internal factors also affect the magnitude of AgNPs-induced neurotoxicity. Thus, stress induced by high ambient temperature enhances AgNPs-induced brain damage. Likewise, hypertension and diabetes aggravate neurotoxicity of AgNPs. Obviously, enhanced brain damage by AgNPs either alone or in combination with external (hyperthermia) or internal (hypertension or diabetes) factors could alter neuroprotective effects of drugs, such as cerebrolysin. Accordingly, intoxication of AgNPs and resulting brain damage require either double doses of Cerebrolysin or its nanodelivery, for example, nanowired-cerebrolysin to achieve good neuroprotection. These observations suggest that humans exposed to AgNPs for a long time require additional care with regard to drugs dosage and treatment schedule following stress, trauma, or other diseases for effective therapy.

## Future Perspectives

Further research is necessary to find out whether AgNPs neurotoxicity may also be exacerbated following cold exposure. Moreover, whether the magnitude and intensity of cardiac arrest, chronic neuropathic pain, and psychostimulants' abuse induced brain damage is also altered by AgNPs intoxication. Is this exacerbation of AgNPs-induced brain pathology under various conditions affected by the size of nanoparticles and/or age of animals? These are the potential new features of AgNPs neurotoxicity that require further investigation.

**Acknowledgments** This investigation is partially supported by the Air Force Office of Scientific Research (London), Air Force Material Command, USAF, under grant number FA8655-05-1-3065. The US government is authorized to reproduce and distribute reprints for government purpose notwithstanding any copyright notation thereon. The views and conclusions contained herein are those of the authors and should not be interpreted as necessarily representing the official policies or endorsements, either expressed or implied, of the Air Force Office of Scientific Research or the US Government. We express sincere gratitude to several laboratories where a part of the work was done or some data was recorded and evaluated. Financial support from Swedish Medical Research Council (Grant Nr. 2710, HSS); Astra-Zeneca, Mölndal, Sweden (HSS), Alexander von Humboldt Foundation, Germany (HSS); IPSEN Medical, France (HSS); Ever NeuroPharma, Austria (AS); University Grants Commission, New Delhi, India (HSS); Indian Council of Medical Research, New Delhi, India (HSS/AS), is gratefully acknowledged. The authors have no conflict of interest with any of the financial agencies mentioned above.

## Cross-References

▶ Colloidal Silver Nanoparticles and Bovine Serum Albumin
▶ Magnesium, Physical and Chemical Properties
▶ Silver as Disinfectant
▶ Silver in Protein Detection Methods in Proteomics Research
▶ Silver-Induced Conformational Changes of Polypeptides

## References

Agency for Toxic Substances and Disease Registry (ATSDR) (1990) Toxicological profile for silver. U.S. Department of Health and Human Services, Public Health Service, Atlanta, GA. http://www.atsdr.cdc.gov/toxprofiles/tp.asp?id=539%26tid=97. Accessed 30 June 2012

Bouwmeester H, Poortman J, Peters RJ, Wijma E, Kramer E, Makama S, Puspitaninganindita K, Marvin HJ, Peijnenburg AA, Hendriksen PJ (2011) Characterization of translocation of silver nanoparticles and effects on whole-genome gene expression using an in vitro intestinal epithelium coculture model. ACS Nano 5(5):4091–4103

Calabrese V, Mancuso C, Calvani M, Rizzarelli E, Butterfield DA, Stella AM (2007) Nitric oxide in the central nervous system: neuroprotection versus neurotoxicity. Nat Rev Neurosci 8(10):766–775

Cervós-Navarro J, Sharma HS, Westman J, Bongcam-Rudloff E (1998) Glial reactions in the central nervous system following heat stress. Prog Brain Res 115:241–274, Review

Chan SH, Chang KF, Ou CC, Chan JY (2004) Nitric oxide regulates c-fos expression in nucleus tractus solitarii induced by baroreceptor activation via cGMP-dependent protein kinase and cAMP response element-binding protein phosphorylation. MolPharmacol 65(2):319–325

Chang CC, Tsai SS, Ho SC, Yang CY (2005) Air pollution and hospital admissions for cardiovascular disease in Taipei, Taiwan. Environ Res 98:114–119

Chen X, Schluesener HJ (2008) Nanosilver: a nanoproduct in medical application. Toxicol Lett 176(1):1–12, Epub 2007 Oct 16. Review

Elder A, Gelein R, Silva V, Feikert T, Opanashuk L, Carter J, Potter R, Maynard A, Ito Y, Finkelstein J, Oberdorster G (2006) Translocation of inhaled ultrafine manganese oxide particles to the central nervous system. Environ Health Perspect 114(8):1172–1178

Givogri MI, Bongarzone ER, Campagnoni AT (2000) New insights on the biology of myelin basic protein gene: the neural-immune connection. J Neurosci Res 59(2):153–159

Hughes P, Dragunow M (1995) Induction of immediate-early genes and the control of neurotransmitter-regulated gene expression within the nervous system. Pharmacol Rev 47(1):133–178

Kim YS, Song MY, Park JD, Song KS, Ryu HR, Chung YH, Chang HK, Lee JH, Oh KH, Kelman BJ, Hwang IK, Yu IJ (2010) Subchronic oral toxicity of silver nanoparticles. Part Fibre Toxicol 7:20

Kruszewski M, Brzoska K, Brunborg G, Asare N, Dobrzyńska M, D_usinská M, Fjellsbø LM, Georgantzopoulou A, Gromadzka-Ostrowska J, Gutleb AC, Lankoff A, Magdolenová Z, Pran ER, Rinna A, Instanes C, Sandberg W, Schwarze P, Stępkowski T, Wojewódzka M, Refsnes M (2011) Toxicity of silver nanomaterials in highereukaryotes. In: Fishbein JC (ed) Advances in molecular toxicology, vol 5. Elsevier, Amsterdam, pp 179–218

Lafuente JV, Sharma A, Patnaik R, Muresanu DF, Sharma HS (2012) Diabetes exacerbates nanoparticles inducedbrainpathology. CNS Neurol Disord Drug Targets 11(1):26–39

Landas S, Fischer J, Wilkin LD, Mitchell LD, Johnson AK, Turner JW, Theriac M, Moore KC (1985) Demonstration of regional blood–brain barrier permeability in human brain. Neurosci Lett 57(3):251–256

Lankveld DP, Oomen AG, Krystek P, Neigh A, Troost-de Jong A, Noorlander CW, Van Eijkeren JC, Geertsma RE, De Jong WH (2010) The kinetics of the tissue distribution of silver nanoparticles of different sizes. Biomaterials 31(32):8350–61

Loeschner K, Hadrup N, Qvortrup K, Larsen A, Gao X, Vogel U, Mortensen A, Lam HR, Larsen EH (2011) Distribution of silver in rats following 28 days of repeated oral exposure to silver nanoparticles or silveracetate. Part Fibre Toxicol 8:18

Menon PK, Muresanu DF, Sharma A, Mössler H, Sharma HS (2012) Cerebrolysin, a mixture of neurotrophic factorsinducesmarked neuroprotection in spinal cord injuryfollowingintoxication of engineered nanoparticles from metals. CNS Neurol Disord Drug Targets 11(1):40–49

Muresanu DF, Sharma HS (2007) Chronic hypertension aggravates heat stress induced cognitive dysfunction and brainpathology: an experimental study in the rat, using growth hormone therapy for possible neuroprotection. Ann N Y Acad Sci 1122:1–22

Muresanu DF, Sharma A, Sharma HS (2010) Diabetes aggravates heat stress-induced blood-brain barrier breakdown, reduction in cerebral bloodflow, edema formation, and brain pathology: possible neuroprotection with growth hormone. Ann N Y Acad Sci 1199:15–26

Muresanu DF, Sharma A, Tian ZR, Smith MA, Sharma HS (2012) Nanowired drug delivery of antioxidant compound H-290/51 enhances neuroprotection in hyperthermia-induced neurotoxicity. CNS Neurol Disord Drug Targets 11(1):50–64

Oberdorster G (1996) Significance of particle parameters in the evaluation of exposure-dose–response relationships of inhaled particles. Inhal Toxicol 8(Suppl):73–89

Oberdorster G, Maynard A, Donaldson K, Castranova V, Fitzpatrick J, Ausman K, Carter J, Karn B, Kreyling W, Lai D, Olin S, Monteiro-Riviere N, Warheit D, Yang H, ILSI Research Foundation/Risk Science Institute Nanomaterial Toxicity Screening Working Group (2005) Principles for characterizing the potential human health effects from exposure to nanomaterials: elements of a screening strategy. Part Fibre Toxicol 2:8

Powers CM (2010) Developmental neurotoxicity of silver and silver nanoparticles modeled in vitro and in vivo. PhD thesis, Duke University, Durham

Rapoport SI (1976) Blood–brain barrier in physiology and medicine. Raven, New York

Rungby J, Danscher G (1983) Neuronalaccumulation of silver in brains of progeny from argyric rats. Acta Neuropathol 61(3–4):258–262

Rungby J, Danscher G (1984) Hypoactivity in silver exposedmice. Acta PharmacolToxicol (Copenh) 55(5):398–401

Samuel U, Guggenbichler JP (2004) Prevention of catheter-related infections: the potential of a new nano-silver impregnated catheter. Int J Antimicrob Agents 23(Suppl 1):S75–S78

Sharma HS (1982) Blood–brain barrier in stress. PhD thesis, Banaras Hindu University, Varanasi, pp 1–85

Sharma HS (1999) Pathophysiology of blood–brain barrier, brain edema and cell injury following hyperthermia: new role of heat shock protein, nitric oxide and carbon monoxide. An experimental study in the rat using light and electron microscopy. Acta Universitatis Upsaliensis 830:1–94

Sharma HS (2004) Blood–brain and spinal cord barriers in stress. In: Sharma HS, Westman J (eds) The blood-spinal cord and brain barriers in health and disease. Elsevier/Academic, San Diego, pp 231–298

Sharma HS (2004a) Pathophysiology of the blood – spinal cord barrier in traumatic injury. In: Sharma HS, Westman J (eds) Blood – spinal cord and brain barriers in health & disease, Elsevier,/Academic Press, Oxford/Boston/San Diego, pp. 437–518

Sharma HS (2005a) Heat-related deaths are largely due to brain damage. Indian J Med Res 121(5):621–623

Sharma HS (2005b) Selective neuronal vulnerability, blood–brain barrier disruption and heat shock protein expression in stress induced neurodegeneration. Invited review. In: Sarbadhikari SN (ed) Depression and Dementia: progress in brain research, clinical applications and future trends. Nova Science Publishers, New York, pp 97–152

Sharma HS (2006a) Hyperthermia induced brain oedema: current status and future perspectives. Indian J Med Res 123(5):629–652

Sharma HS (2006b) Hyperthermia influences excitatory and inhibitory amino acid neurotransmitters in the central nervous system. An experimental study in the rat using behavioural, biochemical, pharmacological, and morphological approaches. J Neural Transm 113(4):497–519

Sharma HS, Ali SF, Dong W, Tian ZR, Patnaik R, Patnaik S, Sharma A, Boman A, Lek P, Seifert E, Lundstedt T (2007) Drug delivery to the spinal cord tagged with nanowire enhances neuroprotective efficacy and functional recovery following trauma to the rat spinal cord. Ann N Y Acad Sci 1122:197–218

Sharma HS (2009a) Blood–central nervous system barriers: the gateway to neurodegeneration, neuroprotection and neuroregeneration. In: Lajtha A, Banik N, Ray SK (eds) Handbook of neurochemistry and molecular neurobiology:

brain and spinal cord trauma. Springer, Berlin/Heidelberg/New York, pp 363–457

Sharma HS (2009b) Nanoneuroscience and Nanoneuropharmacology. Prog Brain Res 180:1–286

Sharma HS (2012) New perspectives of central nervous system injury and neuroprotection. Int Rev Neurobiol 102: 1–395

Sharma HS, Alm P (2004) Role of nitricoxide on the blood–brain and the spinal cord barriers. In: Sharma HS, Westman J (eds) The blood-spinal cord and brain barriers in health and disease. Elsevier/Academic, San Diego, pp 191–230

Sharma HS, Hoopes PJ (2003) Hyperthermiainduced pathophysiology of the central nervous system. Int J Hyperthermia 19(3):325–354, Review

Sharma HS, Johanson CE (2007a) Intracerebroventricularly administered neurotrophins attenuate blood cerebrospinal fluid barrier breakdown and brain pathology following whole-body hyperthermia: an experimental study in the rat using biochemical and morphological approaches. Ann N Y Acad Sci 1122:112–129

Sharma HS, Johanson CE (2007b) Blood-cerebrospinal fluid barrier in hyperthermia. Prog Brain Res 162:459–478, Review

Sharma HS, Sharma A (2007) Nanoparticles aggravate heat stress induced cognitive deficits, blood–brain barrier disruption, edema formation and brain pathology. Prog Brain Res 162:245–273, Review

Sharma HS, Sharma A (2010) Breakdown of the blood–brain barrier in stress alters cognitive dysfunction and induces brain pathology. New perspective for neuroprotective strategies. In: Ritsner M (ed) Brain protection in Schizophrenia, mood and cognitive disorders. Springer, Berlin/New York, pp 243–304

Sharma HS, Sharma A (2011) New strategies for CNS injury and repair using stem cells, nanomedicine, neurotrophic factors and novel neuroprotective agents. Expert Rev Neurother 11(8):1121–1124

Sharma HS, Sharma A (2012a) Neurotoxicity of engineered nanoparticles from metals. CNS Neurol Disord Drug Targets 11(1):65–80

Sharma HS, Sharma A (2012b) Nanowired drugdelivery for neuroprotection in central nervous system injuries: modulation by environmental temperature, intoxication of nanoparticles, and comorbidity factors. Wiley Interdiscip Rev Nanomed Nanobiotechnol 4(2):184–203

Sharma HS, Sharma A (2012c) Recent perspectives on nanoneuroprotection & nanoneurotoxicity. CNS Neurol Disord Drug Targets 11(1):5–6

Sharma HS, Westman J (2004a) The blood-spinal cord and brain barriers in health and disease. Academic, San Diego, pp 1–617 (Release date: Nov. 9, 2003)

Sharma HS, Westman J (2004b) The heat shock proteins and hemeoxygenase response in central nervous system injuries. In: Sharma HS, Westman J (eds) The blood-spinal cord and brain barriers in health and disease. Elsevier/Academic, San Diego, pp 329–360

Sharma HS, Westman J, Nyberg F (1998a) Pathophysiology of brain edema and cell changes following hyperthermic brain injury. In: Sharma HS, Westman J (eds) Brain functions in hot environment. Progr Brain Res, 115: 351–412

Sharma HS, Alm P, Westman J (1998b) Nitricoxide and carbonmonoxide in the brain pathology of heat stress. Prog Brain Res 115:297–333, Review

Sharma HS, Ali SF, Tian ZR, Hussain SM, Schlager JJ, Sjöquist PO, Sharma A, Muresanu DF (2009a) Chronic treatment with nanoparticles exacerbate hyperthermia induced blood–brain barrier breakdown, cognitive dysfunction and brain pathology in the rat. Neuroprotective effects of nanowired-antioxidant compound H-290/51. J Nanosci Nanotechnol 9(8):5073–5090

Sharma HS, Ali SF, Hussain SM, Schlager JJ, Sharma A (2009b) Influence of engineered nanoparticles from metals on the blood–brain barrier permeability, cerebral blood flow, brain edema and neurotoxicity. An experimental study in the rat and mice using biochemical and morphological approaches. J Nanosci Nanotechnol 9(8):5055–5072

Sharma HS, Patnaik R, Sharma A (2010) Diabetes aggravates nanoparticles induced breakdown of the blood-brainbarrier permeability, brain edema formation, alterations in cerebral bloodflow and neuronal injury. An experimental study using physiological and morphological investigations in the rat. J Nanosci Nanotechnol 10(12):7931–7945

Sharma HS, Muresanu DF, Patnaik R, Stan AD, Vacaras V, Perju-Dumbrav L, Alexandru B, Buzoianu A, Opincariu I, Menon PK, Sharma A (2011) Superior neuroprotective effects of cerebrolysin in heat stroke following chronic intoxication of Cu or Ag engineered nanoparticles. A comparative study with other neuroprotective agents using biochemical and morphological approaches in the rat. J Nanosci Nanotechnol 11(9):7549–7569

Sharma HS, Sharma A, Mössler H, Muresanu DF (2012a) Neuroprotective effects of cerebrolysin, a combination of different active fragments of neurotrophic factors and peptides on the whole body hyperthermia-inducedneurotoxicity: modulatoryroles of co-morbidityfactors and nanoparticleintoxication. Int Rev Neurobiol 102:249–276

Sharma HS, Castellani RJ, Smith MA, Sharma A (2012b) The blood–brain barrier in Alzheimer's disease: noveltherapeutic targets and nanodrug delivery. Int Rev Neurobiol 102:47–90

Sharma A, Muresanu DF, Mössler H, Sharma HS (2012c) Superior neuroprotective effects of cerebrolysin in nanoparticle-induced exacerbation of hyperthermia-induced brain pathology. CNS Neurol Disord Drug Targets 11(1):7–25

Stalberg E, Sharma H, Sharma HS, Olsson Y (1998) Spinal cord monitoring. Basic principles, regeneration, pathophysiology and clinical aspects. Springer, Wien/New York, pp 1–525

Takenaka S, Karg E, Roth C, Schulz H, Ziesenis A, Heinzmann U, Schramel P, Heyder J (2001) Pulmonary and systemic distribution of inhaled ultrafine silver particles in rats. Environ Health Perspect 109(Suppl 4):547–551

Takenaka K, Saidoh N, Nishiyama N, Inoue A (2011) Fabrication and nano-imprintabilities of Zr-, Pd- and Cu-based glassy alloy thin films. Nanotechnology 22(10):105302

Thilagavathi G, Raja ASM, Kannaian T (2008) Nanotechnology and protective clothing for defence personnel. Def Sci J 58(4):451–459

Westman J, Sharma HS (1998) Heat shock protein response in the central nervous system followinghyperthermia. Prog Brain Res 115:207–239, Review

# Silver, Pharmacological and Toxicological Profile as Antimicrobial Agent in Medical Devices

Alan B. G. Lansdown
Faculty of Medicine, Imperial College, London, UK

## Synonyms

Colloidal silver; Metal-binding proteins; Metallothioneins; Nanocrystalline silver; Silver; Silver sulfadiazine

## Definition

Silver is a white lustrous transitional metallic element found widely in the human environment. Low concentrations of silver accumulate in the human body through inhalation of particles in the air and contamination of the diet and drinking water. Silver serves no trace metal value in the human body. Increasing use of silver as an antibacterial and antifungal agent in wound care products, medical devices (bone cements, catheters, surgical sutures, cardiovascular prostheses, and dental fillings), textiles, cosmetics, and even domestic appliances in recent years has lead to concern as to the safety aspects of the metal and potential risks associated with absorption of the biologically active $Ag^+$ into the human body.

Silver exists in its elemental form in natural deposits and in complexes with lead, copper, arsenic, and mercury. Metallic silver is inert in the human body but ionizes readily in the presence of water and body fluids to release biologically active $Ag^+$ ions. These ions bind sulfhydryl and other ligands on amino acids and proteins (notably albumins, cyto-keratins, and macroglobulins) to form stable complexes. Ag(I) induces and binds cysteine-rich metallothioneins (MTI and MTII) which play a fundamental role in mobilization and cytoprotection of silver and certain other metals (Lansdown 2002a). Silver metalloprotein complexes in mammalian and microbial flora underlie the pharmacology, antimicrobial action, and cytoprotection of the body against Ag(I) ion overload. Silver is one of the safest metals known to man. Occupational and clinical exposure to silver residues, environmental silver, and silver-related antimicrobial therapies (i.e., colloidal silver) are a cause of disfiguring and psychologically disturbing dark discolorations of the skin and eye: they are not life threatening. Silver induces contact sensitization and delayed hypersensitivity but other major adverse effects are not seen (Lansdown 2010).

## Silver as an Antimicrobial Agent and Its Application in Medical Devices

The antimicrobial properties of silver have been appreciated for many years and its value in water purification, wound care, and personal hygiene is recognized (Lansdown 2002b, c). Drugs like silver arsphenamine, colloidal silver proteins (Argyrol, Protargyrol, Collargol, etc.), silver nitrate, and silver sulfadiazine (Flamazine) have a long history in treating venereal and fungal diseases, Legionella sp., *Pseudomonas aeruginosa*, *Staphylococcus aureus*, and many other life-threatening infections. Ag(I) is an efficacious broad range antimicrobial agent; silver-resistant organisms are not widespread (Lowbury 1977).

The antimicrobial efficacy of silver is directly related to the ionization of the silver source (metallic silver, inorganic salt) in moisture and body fluids and the availability of bioactive silver ion ($Ag^+$). Silver ion binds readily to proteins, cell membranes, and tissue debris to form stable precipitates. Ionization of metallic silver is proportional to the surface area of particles; release of $Ag^+$ from nanocrystalline particles of <20 nm being more than 100-fold higher than from silver foil or other metallic silver forms (Fig. 1). Nanocrystalline silver exhibits a sixfold or higher log reduction in *Pseudomonas aeruginosa* in culture (Burrell 2003). Experimental studies suggest that concentrations of 60 ppm $Ag^+$ are sufficient to control the majority of bacterial and fungal pathogens. $Ag^+$ binds protein residues on cell membranes of sensitive bacteria, fungi, and protozoa and is absorbed intracellularly (Fig. 2). Subsequent denaturation and inactivation of intracellular proteins and essential enzymes including RNA and DNA-ases forms the basis of the genetically regulated antimicrobial action of silver (Silver 2003). Silver-sensitive bacteria and fungi are known to absorb and concentrate $Ag^+$ from dilute solutions (1 ppm) by an oligodynamic action.

The following are the characteristics of silver-resistant bacteria or fungi:

1. Do not respond to the antibacterial action of $Ag^+$ in culture media
2. Exhibit a stable mutational change within populations
3. Resistance is transferable to other susceptible strains of bacteria by in vitro mating or conjugation (Silver 2003)

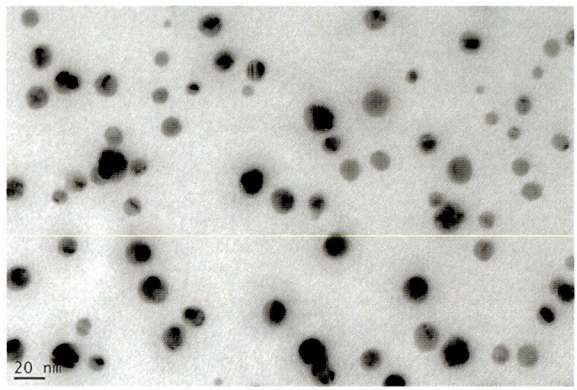

**Silver, Pharmacological and Toxicological Profile as Antimicrobial Agent in Medical Devices, Fig. 1** Nanocrystals of metallic silver <20 nm diameter. The high surface to volume ratio greatly increases their capacity to ionize in moisture and body fluids to release biologically active $Ag^+$ (Grain boundary phenomenon)

Silver-resistance was first identified in a strain of *Salmonella typhimurium* in the burns clinic in Massachusetts General Hospital (Boston, USA) (McHugh et al. 1975). The bacterium was isolated from patients subject to a fatal septicemia. It was resistant to 0.5 % silver nitrate and other antibiotics. This silver-resistant strain of *Salmonella typhimurium* (pMGH100) has provided fundamental information on the genetical and molecular basis of silver-resistance in bacteria and the role of the Sil-gene complex (Silver 2003). Silver-resistant strains of *Escherichia coli* and *Pseudomonas stutzeri* isolated from burn wound patients in Canada were used to study patterns of resistance conferred by intracellular bodies (plasmids). Silver-resistant bacteria accumulate and retain silver more than nonresistant strains which show up to 63 % silver efflux. Silver complexes and denatures bacterial enzymes and structural proteins leading to cell death.

## Silver Technology and Antimicrobial Efficacy

### Water Treatment

Silver-based technologies have proved superior to other purifications including heating and chloride methods in water preservation and sterilization up to

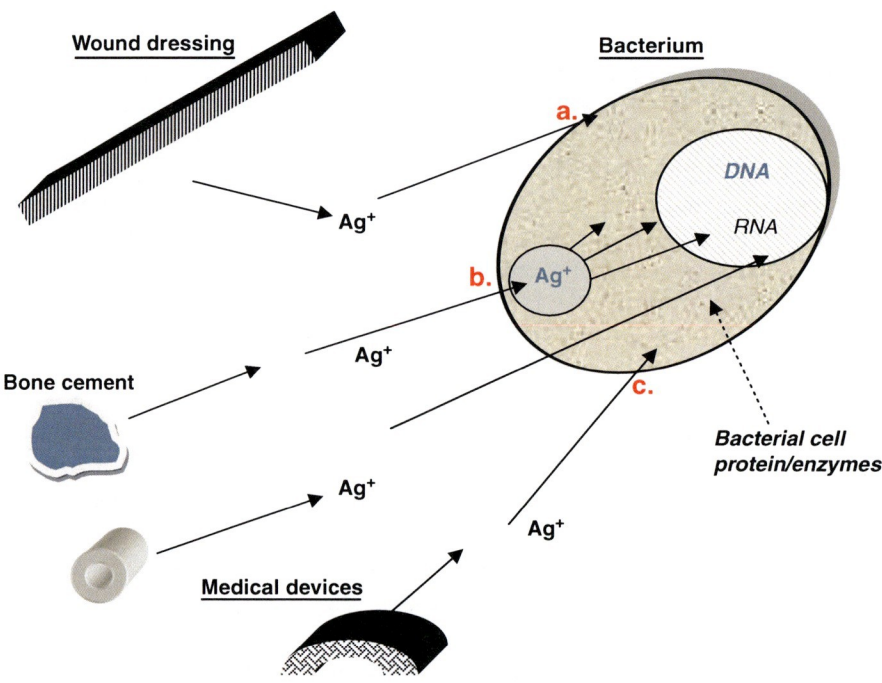

**Silver, Pharmacological and Toxicological Profile as Antimicrobial Agent in Medical Devices, Fig. 2** Antimicrobial action of silver ($Ag^+$): attachment to cell membrane, absorption or transmembrane uptake into the cell (pinocytic vesicle), coagulation and denaturation of bacterial enzymes and intracellular metalloprotein precipitation

the present time. They are controllable and of minimal human health risk, yet efficacious in controlling outbreaks of Legionnaire's disease and cleansing of domestic water systems, hospital water supplies, and swimming baths. Technological advances since the 1920s include introduction of the Katadyn process (a spongy preparation of 0.006–0.5 ppm metallic silver or silver coating on sand, or impregnated into filter material) (Katadyn Products Inc, Switzerland), development of silver-copper ionization filters, electrolytically modulated silver ionization (EMSI), silver-carbon filters, and zeolite ceramic silver-zinc coatings (Sykes 1958). Small quantities of other metals like palladium or gold may be present to accentuate the bioactivity of $Ag^+$. Pathogenic bacteria are more sensitive to Katadyn, but spores, molds, and protozoa are seemingly resistant. Bacillus mesentericus spores were shown to survive in Katadyn (300 ml water + 20 g silver) for more than 5 weeks. The EMSI process introduced later included electrodes to stimulate the ionization of the silver in the purification of swimming baths. The method used either two silver electrodes or a silver electrode and an electrode of another metal.

The World Health Organization (1996) does not regulate silver in drinking water presently as available information is "insufficient to identify a health standard." United States (1985) guidelines specify maximal levels of silver in drinking water at 0.1 mg/L in the long term. EEC Drinking Water Directive (1980) specified a maximal admissible concentration of 10 µg/L standard. Colloidal silver products which were used earlier to improve water quality are not legally permitted in the European Union, USA, and many other countries for reasons of safety and lack of efficacy.

## Wound Care and Management

Silver has a history of use in wound care extending back over more than 400 years. The development of silver as an antibiotic in wound care and its central role in wound management reflects the state of knowledge of microbiology at the time, technological advances in biomaterials and the chemistry of silver, and understanding of the biological events comprising the principal phases of wound healing (Lansdown 2010). Early wound therapies were based upon 1 % silver nitrate to treat pathogenic strains of *Staphylococcus aureus* and *Pseudomonas aeruginosa*, but it is an irritant and has the propensity of staining the skin black. Silver nitrate is still used in burn wound therapy but has been largely superseded by less irritant silver sulfadiazine (SSD). SSD, which is widely available today, was developed as a prophylactic therapy for *Pseudomonas aeruginosa* infections in burn wounds. It combines the antibiotic efficacy of silver nitrate with sodium sulfonamide. Modern wound dressings embrace controlled patterns $Ag^+$ release and are designed according to the type and levels of infection of wounds. They comply with current international guidelines on wound bed preparation."

Sustained silver-release wound dressings are now well established in the care and management of acute and chronic skin wounds and burns. They comprise:

- An ionizable silver source to provide antimicrobial action over a defined period
- Absorbent material to control exudates and moisture control
- Fibers, fabrics, and materials with minimal bioactivity to provide a matrix or structural component for the dressing or means of application

They are safe, readily applied, and efficacious in controlling a wide range of infections. The dressings differ in total silver content, the nature of the ionizable silver source, patterns of $Ag^+$ release in the presence of wound exudates and moisture, and in their mode of pharmacological and antibacterial action. Exposure to wound fluid or moisture triggers release of the so-called bioactive silver ion for bactericidal and fungicidal action, neutralization of any toxins produced, and elimination of antisocial odors. Modern sustained silver ($Ag^+$)-release wound dressings tailored to control infections and to improve the clinical condition of indolent and difficult-to-heal wounds include:

1. High silver content – rapid ionization and $Ag^+$ release
2. Modest to high silver content but with more sustained ionization and $Ag^+$ release
3. Low silver content – designed for wounds with low risk of infection or as barrier dressings for postoperative wounds and acute lesions

The dressings release more silver than is expected for antibiotic action (10–40 ppm). Some excess silver precipitates in the wound bed and some as nontoxic silver metalloprotein complexes.

## Catheters

Catheters are available for intravenous, intraperitoneal, intracerebral, or intrauterine access for the administration of therapeutics, fluids, and electrolytes or food supplements, blood transfusion, and withdrawal of body fluids (Tobin and Bambauer 2003). Catheter technology is complex and silver-impregnated or coated polymers, silicones, etc., are now available for long-term insertion and control of bacterial/fungal adhesions and biofilm formation. The severity of catheter-related infections is usually related to the duration of placement, the number of procedures performed, and catheter design. Infections like Proteus, Providencia, and Morganella colonize catheter tips and evoke calcium phosphate deposition leading to catheter blockage. Silver coating (<1 μm thick on inner and outer surfaces) by ion beam-assisted deposition (IBAD) or sputter coating technology has been developed to mitigate microbial adhesions and biofilm formation in in-dwelling catheters. It is efficacious in controlling catheter-related bloodstream infections which arise following the migration of infections from bacterial or fungal cell contaminations at catheter hubs, colonies at insertion sites. Other technology (Bacti-Guard®, BARD® hydrogel, BARDEX® I.C. Foley catheter) involves dispersion of silver particles within a hydrogel matrix on catheter surfaces. This matrix absorbs moisture to form a close interface between the catheter and the urethral wall, the moisture triggering release of bioactive $Ag^+$ from the silver dispersant. Other metals including gold and palladium alloy may be included to potentiate bactericidal action. Silver-metalloprotein complexes mitigate the local toxic effects of excess silver.

Polyurethane or silicone polymer tubing impregnated with 2 % nanoparticulate silver and an insoluble silver salt provides a means for external ventricular drainage and measurement of intracranial pressure in cases of hydrocephaly (Silverline, Spiegelberg, Hamburg). The combination of a fast-acting silver salt with the slow ionization of metallic silver is held to optimize control of catheter-related cerebrospinal infections and catheter dysfunction.

## Orthopedic Devices

Metallic silver and silver-containing alloys are used in surgical instruments and devices where insertion into the body is accompanied by risk of infections and biofilm formation (Lansdown 2010). Implantable devices using silver-coating technology include wires, pins, and screws; external fixation pins, fracture fixation plates, and total joint prostheses for hips, knee, shoulders, and ankles; and bone cements. Silver coatings are biocompatible and do not deteriorate with prolonged usage and provide rigidity of function. Clayton Parkhill in 1897 used silver-coated metals to fix fractures and to "secure the antiseptic action of the device." He believed that the silver coating leading to production of silver-protein complexes created an unfavorable environment on the surface of his fixation devices for bacterial adhesion and colonization. IBAD technology has proved beneficial in aiding anchorage of orthopedic pins and decreasing the incidence of biofilm formation. Stainless steel coated with a silver zeolite complex (2.5 % silver and 14 % zinc) reduced bacterial adhesion. A 2 mm thick titanium/silver coating by a physical vapor deposition process releasing 0.5–2.ppb $Ag^+$ was efficacious against *Staphylococcus aureus* and *Klebsiella pneumoniae* without evidence of cytotoxicity in osteoblasts or epithelial cells. Other technologies designed to provide an antimicrobial effect in orthopedic devices include Bioglass® technology (Novabone, Florida) and bone cements containing an ionizable silver source. These devices are generally safe in use and blood silver levels have rarely exceeded permissible levels.

## Cardiovascular Devices

Silver is included into synthetic prosthetic material, heart valves, vascular grafts, and sewing cuffs to control associated infections and biofilm formation and reduce risks of morbidity and increased mortality. Silver metalloprotein in blood, liver, and kidney are raised without toxic changes.

## Hygiene Textiles

Silver-coated textiles have achieved widespread acclaim in the production of hygiene clothing and in reducing health risks associated with pathogenic strains of bacteria in patients with atopic eczema. New technology enables silver to be impregnated or coated on a wide range of natural and synthetic fibers. $Ag^+$ release is controlled, low level, and sustained for the expected life of garments. Textile fibers serve as vehicles for delivery of bioactive silver ion to eliminate or otherwise protect against bacterial imbalances in the skin, microbial overgrowth accompanied by excessive odor, or local discomfort. Silver coated or impregnated textile fibers are safe in use but delayed hypersensitivity and allergy to silver may be a health risk to some people (Lansdown 2010).

## Silver Absorption, Metabolism, Deposition, and Excretion

The toxicity of any xenobiotic material relates to the amount absorbed into the body, its metabolism and accumulation in target organs, and cellular vulnerability to supraphysiological levels. Silver is absorbed into the human body through ingestion, inhalation, implantation, or insertion of medical devices, and through dermal contact (Lansdown 2010). It competes for binding sites on carrier proteins and MBP in the circulation and intracellularly. Where protective mechanisms afforded by MT, cyto-keratins, and MBP become saturated, toxic changes occur. Metabolic pathways are similar irrespective of route of uptake. Blood levels of silver in workers exposed to silver occupationally ranged from 0.1 to 23.0 $\mu g.L^{-1}$, with highest levels in silver reclamation (Armitage et al. 1998). DiVincenzo et al. (1985) reported mean silver levels in blood, urine, and feces as 11 $\mu g.L^{-1}$, <0.0005 $\mu g.g^{-1}$, and 15 $\mu g.g^{-1}$, respectively, in silver smelting and refining. Hair concentrations are a good monitor of occupational silver exposure and have been reported in the range 130 ± 160 $\mu g.g^{-1}$ (placebo 0.57 ± 0.56 $\mu g.g^{-1}$) following chronic exposure.

### Oral and Intranasal

Buccal or gastrointestinal absorption of silver may result from:

1. Ingestion of contaminated food, or drinking water purified through silver: copper filters (mean levels of silver in drinking water <5 $\mu g.L^{-1}$) permitted in the USA)
2. Occupational exposure to metallic silver dust, nanocrystalline silver, silver oxide, and silver nitrate, silver in aerosols
3. Silver nitrate or colloidal silver therapies for oral hygiene or gastrointestinal infection
4. Silver acetate antismoking therapies
5. Silver amalgams used in dentistry
6. Accidental consumption of silver nitrate or other colorless silver compounds

Silver absorption through buccal membranes and gastrointestinal mucosae is determined by the ionization of the silver source and availability of "free" $Ag^+$ to interact with protein receptors on cell membranes. Passive uptake is not indicated on account of the high reactivity of $Ag^+$ and its capacity to bind sulfhydryl, carboxyl, hydroxyl groups and protein ligands on mucosal surfaces and cell debris. Biologically active $Ag^+$ readily binds and precipitates with organic constituents of food (phytate, fibers, etc.) and precipitates inorganic cations like chloride and phosphate, thereby reducing absorption. Current estimates suggest that <10 % of the silver ingested by humans is absorbed into the circulation, but this is influenced by age, health, and nutritional status and composition of an individual's diet. Absorption of silver from drinking water ranges from 7 to 80 $\mu g/day$ and that total lifetime intake is estimated to be about 10 g. Absorption of silver from silver acetate antismoking remedies is low but in a patient suffering from argyria for 2 years total body silver was 6.4 g and tracer studies showed that of that 18 % administered was retained (East et al. 1980). Silver has a biological half-life in the human body of about 5 days but may be retained in bone, liver, and kidney for 30, 15, and 10 days, respectively.

### Inhalation

Argyria and argyrosis provide unequivocal evidence for silver absorption following long-term inhalation of colloidal silver preparations or occupational exposures to silver or silver oxide dusts or silver nitrate. Inhalation of silver residues or silver nitrate is a cause of respiratory distress, squamous metaplasia, and pigmentation of the respiratory tract (Lansdown 2010). Hygienists place the minimal risk of nasal and pharyngeal inflammation at exposures of 0.1 $mg.m^{-1}$ with particles of grain size of <20 nm diameter. Detailed study of the uptake of silver through nasal and respiratory membranes has not been seen, but it is expected that inhaled silver or silver compounds will ionize in mucoid secretions or alveolar pulmonary surfactants allowing $Ag^+$ to be absorbed through alveolar epithelia. Silver metalloprotein precipitates in lungs are absorbed by alveolar macrophages. It is unclear to what extent $Ag^+$ interacts with or is precipitated by the phospholipid content of pulmonary surfactants, or whether this secretion acts as a protective barrier to absorption. Workers exposed chronically to >0.01 $mg.m^{-3}$ environmental silver exhibited raised blood (>0.27 $\mu g.100\ ml^{-1}$) and urine (>1.91 $\mu g.L^{-1}$) silver levels. Blood and urine silver-metalloprotein levels are not accurate measurements of occupational silver exposure.

### Dermal and Percutaneous Absorption

The majority of products containing silver or silver compounds for antibiotic purposes come into contact

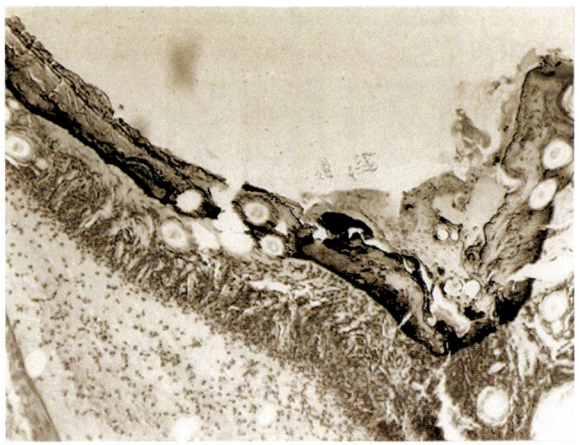

**Silver, Pharmacological and Toxicological Profile as Antimicrobial Agent in Medical Devices, Fig. 3** Black silver sulfide and silver metalloprotein complex deposition in the exudates and debris of an experimental skin wound

with human skin at some time, but percutaneous absorption of silver ($Ag^+$) is exceedingly low. Exposed sulfhydryl groups of keratin and phospholipids of the epidermal barrier function are effective in irreversibly binding free $Ag^+$ (Lansdown 2010). Tracer studies using $^{111}Ag$ indicate that <4 % of the total $Ag^+$ released from topically applied silver nitrate solution is absorbed through intact skin (Nørgaard 1954). Low but perceptible (0.43–11.6 ng/cm$^2$) percutaneous absorption was noted in vitro in skin exposed to nanoparticulate silver. Penetration of silver through guinea pig skin (with a thickness similar to human skin) was estimated to be <1 % in 5 h.

In generalized argyria, most $Ag^+$ is absorbed through inhalation or ingestions of contaminated food or drinking water, and use of colloidal silver therapies (Bleehen et al. 1981). Risks of argyria or increased blood silver through contact with silver antibiotics in textiles and hygiene clothing are negligible (Lansdown 2010). Blood levels of silver were not significantly raised in a patient treated with a wound dressing containing 85 mg.Ag.100 cm$^{-1}$ for 4 weeks. Experimental studies have demonstrated that silver precipitates in epidermal wound debris, proteins (albumins and macroglobulins) in wound exudates, or as relatively insoluble silver chloride in the skin surface exudates to be lost in normal repair processes (Fig. 3).

SSD (Flamazine) is an amphiphilic preparation widely used in burns therapy (Coombs et al. 1992). It is not noticeably absorbed through intact skin but in burned patients percutaneous absorption increases according to the severity of the wounds, their depth, vascularity, and the amount applied. Patients with >20 % whole-body burns exhibited blood silver levels exceeding 200 µg.L$^{-1}$ with the highest concentrations at 310 µg.L$^{-1}$, without argyria. Blood silver levels increased 20-fold within 6 h to at least 40 µg/L, rising to plateau after 4–7 days. Silver nitrate is appreciably more astringent than silver sulfadiazine and ionizes more rapidly when applied topically as strong silver nitrate (75 %), silver nitrate sticks/pencils, or douches to remove warts, callus, or undesirable granulations, but $Ag^+$ penetration is very low. $Ag^+$ binds to epidermal keratin and blackens on exposure to solar radiation to give characteristic brown-black discoloration. Local skin discolorations rarely occur following application of sustained silver-release wound dressings and occupational contact with silver oxide and other ionizable silver compounds.

## Intraparenteral Uptake and Absorption from Implanted Medical Devices

Silver is employed as an antimicrobial and antibiofilm treatment in a variety of catheters orthopedic devices as a means of protection against pathogenic infections. All achieve a sustained release of $Ag^+$ to achieve antimicrobial efficacy and, as these devices are in long-term contact with the human circulation and body issues, induction of Ag-MT and Ag-other MBP complexes is expected. Schierholz, in 1999, considered that most of this irreversibly bound silver has no toxicological, physiological, or antimicrobial significance and that silver coating or impregnation of medical devices is only effective clinically when the concentration of free $Ag^+$ is increased and the effect of contact with serum proteins and inorganic anions minimized.

## Silver Toxicity

### Cellular Uptake and Tissue Management

Silver is absorbed into the body as $Ag^+$ binds strongly to intracellular proteins, notably serum albumins and macroglobulins, for metabolism to bone and soft tissues. $Ag^+$ or silver-protein complexes do not pass the blood-brain barrier into neurological tissues (Lansdown 2010). $Ag^+$ released through ionization of silver nitrate or silver sulfadiazine induces and binds the cysteine-MT-I and MT-II in metabolically active cells (Lansdown 2002). Controversies exist on the predominant routes of silver metabolism in the human body; its transitory or longer-term accumulation in kidney, liver, and bone; and its excretion patterns in bile, urine, hair, and nail as seen in occupational health studies (Wan et al. 1981; East et al. 1980). The biliary route of excretion predominates over the urinary route but urinary silver measurement provides a convenient index of silver absorption by all routes and serves as a guide to the total body silver content, particularly at blood levels of $<100$ $\mu g.L^{-1}$. Urinary excretion of silver is irregular at higher concentrations. Fecal silver includes that excreted in bile protein-complexes and that ingested and not absorbed gastrointestinally ($>90$ %). Estimates suggest that exposures to environmental concentrations of 0.1 $mg/m^3$ (TLV) resulted in fecal excretion of silver of about 1 mg daily (Di Vincenzo et al. 1985).

**Manifestations of Toxicity**

Metallic silver is inert in the presence of biological materials. It ionizes in moisture and body fluids to release bioactive $Ag^+$ which binds sulfhydryl, carboxyl, hydroxyl, and protein ligands. It readily precipitates with chloride, selenide, and phosphate moieties (Fig. 3). Silver-MBP complexes absorbed pinocytically by living cells actively induces and binds MTs to form stable complexes (Lansdown 2010).

The principal adverse effects associated with silver exposure are:
- Argyria and argyrosis
- Contact allergy and delayed hypersensitivity

Argyria, Argyrosis

Argyria is characterized by deposition of inert silver sulfide and/or silver selenide in the vascular connective tissues of the dermis and around hair follicles and sweat ducts (Bleehen et al. 1981). Argyrosis more specifically refers to deposition of insoluble silver precipitates in the cornea and retina of the eye (Lansdown 2011; Moss et al. 1979). Argyria and argyrosis result from a chronic overload of silver in the circulation and incapacity of the liver and kidney to eliminate the silver-protein complexes in bile or urine (Rosenman et al. 1979). The black deposits are intracellular (lysosomal) or intracellular in distribution and evoke a long-lasting blue-gray discoloration of the skin and nail-beds in light exposed areas. Solar irradiation promotes manifestation of argyria which can be profoundly disfiguring and socially embarrassing. It is psychologically disturbing and not readily removed by chemical treatment or dermabrasion. There is no evidence to suggest that argyria is a cause of, or is associated with cellular damage in any tissue. Mechanisms of argyria are not known but thought to involve imbalances in soluble and insoluble silver compounds, local pH, and lysosomal enzymes. Selenium seems to be contributory to argyria, with patients exhibiting tenfold increase in the element. Silver selenide is a highly insoluble inert compound. Melanin granules may protect against argyria and argyrosis by absorbing UV irradiation, but silver ($Ag^+$) is not associated with melanogenesis or changes in melanocyte function (Bleehen et al. 1981).

Argyrosis is defined as a dusky gray/blue pigmentation of the cornea and conjunctiva resulting from deposition of inert silver precipitates following chronic occupational, therapeutic, or environmental exposure to silver or soluble silver salts (Moss et al. 1979; Lansdown 2011). It occurs with and is a more sensitive outward sign of silver exposure than argyria. It is unrelated to pathological damage in any tissue. Argyrosis is a common manifestation of chronic occupational exposure to silver and older colloidal silver therapies (Moss et al. 1979).

Contact Allergy and Delayed Hypersensitivity

Allergy is a known adverse effect of silver exposure in coinage, cosmetics, and therapy with silver nitrate, SSD, and sustained silver-release wound dressings to control wound infections (Fisher 1987). Aged solutions of silver nitrate with greater ionization are more allergenic than freshly prepared reagents. Predisposed metal workers, jewelers, photographers, and other persons exposed to silver or silver salts occupationally may exhibit symptoms of "silver-workers finger," "silver-fulminate itch" (in explosives industry), "silver coat dermatitis," and delayed contact hypersensitivity. Silver allergy is identified through standard patch tests using 2 % aq. silver nitrate or using scratch tests or intradermal injection of 0.05 ml of 1 % silver nitrate.

Silver in Bone and Soft Tissues

The skin, eye, brain, liver, kidney, bone and bone marrow, and spleen are listed as sites for silver precipitation following systemic absorption. Silver is not absorbed into neurological tissues and is not a cause of neuropathic change (Lansdown 2007). Although biochemical evidence shows that silver interacts with calcium in hydroxyapatite binding in bone, clinical and experimental studies have failed to link chronic silver exposure with osteoporosis, even though blood silver levels were raised.

Clinical and experimental studies list the liver as the principal organ for silver accumulation and elimination. Apart from transitory changes in metabolizing enzymes, there is no evidence of irreversible pathological hepatic damage in patients with blood silver of >200 $\mu g.L^{-1}$ or advanced argyria (Coombs et al. 1992). Daily administration of 50 mg silver leaf to 30 healthy volunteers for 20 days led to transitory increases in blood phospholipid, triglycerides, cholesterol, glycemia, and related enzymes, but no functional changes. Silver ingestion promotes hepatic MT-synthesis. Urinary silver excretion is lower than the biliary route and provides an inaccurate measure of silver uptake. Functional renal impairment has not been reported in patients subject to long-term SSD or occupational silver exposure. Inflammatory damage is reported in rare cases where <3 % silver nitrate has been injected intraurethrally or into the renal pelvis to control severe renal infections or chyluria.

Silver nitrate and SSD therapy in burns cases have been associated with leucopenia and reduced granulocyte counts. These and other blood dyscrasias attributed to silver exposure are equivocal. Current views are that they are attributable to by-products of thermal injury and not silver uptake (Choban and Marshall 1987). Alternatively, the sulfadiazine moiety may be responsible.

Methemoglobinemia is a rare complication of 0.5 % silver nitrate therapy in burn clinics, particularly in children. This is not due to silver toxicity. It occurs through the action of the nitrate cation on the oxygen-carrying capacity of hemoglobin. Normal hematopoiesis is restored through intravenous injection of methylene blue.

### In Vitro Toxicity

In vitro studies in human cell lines have demonstrated that cytotoxic metallic silver, SSD, and soluble silver compounds are toxic to keratinocytes and fibroblasts (Lansdown 2010). The tests have been designed also to study molecular toxicity of $Ag^+$, target cell sensitivity, cytoprotective patterns afforded by silver protein complexes, and the action of $Ag^+$ on intracellular membranes, growth factors, and enzymes. Thus, $Ag^+$ is shown to react with sulfhydryl groups, protein residues, and enzymes associated with cell membranes leading to denaturation, structural damage, and mitochondrial dysfunction, in much the same way as that seen in bacterial and fungal cells. Fibroblasts are more sensitive to $Ag^+$ than keratinocytes, but silver toxicity in cultured cells may be influenced by the age of the donor and the composition of the culture medium (Hidalgo et al. 1998). Human diploid fibroblasts and dermal fibroblasts were inhibited by short-term exposure to SSD, and apoptosis was associated with cytoplasmic deterioration and degeneration in nuclei and cell organelles, and alterations in growth factor sensitivity (McCauley et al. 1994).

Cytokinetic and morphological studies have demonstrated that human fibroblasts and glioma cells are more vulnerable to nanocrystalline silver during the G2 and M phases of the cell cycle, and that cell injury and death is principally due to effects of the $Ag^+$ on mitochondrial function, ATP-synthesis, production of oxygen-reactive species, and oxidative stress. Other experiments with a cancer cell line demonstrated that nuclear-targeting silver nanoparticles cause DNA double-strand breaks with a subsequent increase in the sub-G1 (apoptotic) population at much lower concentrations than expected (Austin et al. 2011). Intracellular silver accumulated in cancer cells in the M and G2 phases of the cell cycle. Dark field imaging showed that these cells failed to undergo cell division and ultimately underwent programmed cell death.

### Mutagenicity, Carcinogenicity, and Teratogenicity

Silver is not documented by the US Environmental Protection Agency (EPA) (2010) as a human carcinogen. In vitro studies conducted so far have not shown $Ag^+$ to be mutagenic or clastogenic. No cancers have been reported in humans exposed to metallic silver, its alloys, or inorganic or organic compounds (Lansdown 2010). Carcinogenic changes have been reported in a small number of experimental studies in animals exposed to metallic silver, silver nitrate, or colloidal silver but the reports are inconsistent and uncorroborated.

There is no evidence to show that ingestion of metallic silver or inorganic or organic silver compounds in human pregnancy or during lactation is a cause of impaired fetal development, postnatal growth, or survival. Intrauterine injection of 1 % silver nitrate in mid-gestation led to vaginal bleeding and termination of pregnancy in cynomolgus monkeys, but the silver absorbed was not detrimental to the outcome of pregnancies following remating.

## Discussion

Silver is an efficacious antimicrobial agent with a broad spectrum of action against most common human bacterial and fungal pathogens. Experience has emphasized its value in wound care and burns therapy where potentially lethal pathogens like *Pseudomonas aeruginosa* and *Staphylococcus aureus* are prevalent. Additionally, silver coating and impregnation technologies have proved beneficial in the production of hygiene textiles, and possibly catheter production, but limited evidence is available to show that silver is effective against bacterial adhesion and biofilm formation.

Silver as $Ag^+$ is absorbed into the human body mainly through inhalation and through ingestion of contaminated food and water. The uptake and toxicity of silver in body tissues is mitigated by the strong affinity of $Ag^+$ for protein residues in the circulation, exudates, and cell surface membranes. Silver strongly induces and binds cysteine-rich MT.

Silver and silver compounds are of low toxicity in the human body (Lansdown 2010a, b). Argyria and argyrosis manifest by silver precipitates in dermal connective tissue are the principal adverse effects of chronic silver exposure but these are not life threatening. The discolorations of the skin and eyes are socially unacceptable and psychologically disturbing. Silver allergy and delayed hypersensitivity are well documented and represent the most important of toxic effects associated with silver in occupational and therapeutic medicine (Fisher 1987).

Permitted exposure limits for silver and soluble silver compounds in air and drinking water are 0.01 mg.m$^3$ and 0.10 mg.L$^{-1}$, respectively (U.S. EPA 1985). Under normal circumstances, a human can be expected to consume up to 10 g from food and water in his lifetime. This is below the expected Adverse Effect Level which has yet to be established by the World Health Organization.

## Cross-References

▶ Metallothioneins and Silver
▶ Nanosilver, Next-Generation Antithrombotic Agent
▶ Silver as Disinfectant
▶ Silver, Burn Wound Sepsis and Healing
▶ Silver, Neurotoxicity
▶ Zinc and Wound Healing
▶ Zinc Metallothionein

## References

Armitage SA, White MA, Wilson HK (1996) The determination of silver in whole blood and its application to biological monitoring of occupationally exposed groups. Ann Occup Hyg 40:331–338

Bleehen SS, Gould DJ, Harrington CI, Durrant TE, Slater DN, Underwood JC (1981) Occupational argyria; light and electron microscopic studies and X-ray microanalysis. Br J Dermatol 104:19–26

Burrell RE (2003) A scientific perspective on the use of topical silver preparations. Ostomy Wound Manage 49:19–24

Choban PS, Marshall WJ (1987) Leukopenia secondary to silver sulfadiazine: frequency, characteristics and clinical consequences. Am Surg 53:515–517

Coombs CJ, Wan AT, Masterton JP, Conyers RAJ, Pedersen J, Chia YT (1992) Do burn patients have a silver lining. Burns 18:179–184

Di Vincenzo GD, Giordano CJ, Schriever LS (1985) Biologic monitoring of workers exposed to silver. Int Arch Occup Environ Health 56:207–215

East BW, Boddy K, Williams ED, Macintyre D, McLay AL (1980) Silver retention, total body silver and tissue silver concentrations in argyria associated with exposure to an anti-smoking remedy containing silver acetate. Clin Exp Dermatol 5:305–311

Fisher AA (1987) Contact dermatitis. Lea and Febiger, Philadelphia

Hidalgo E, Bartolome R, Barroso C, Moreno A, Domınguez C (1998) Silver nitrate: antimicrobial activity related to cytotoxicity in cultured human fibroblasts. Skin Pharmacol Appl Skin Physiol 11:140–151

Lansdown ABG (2002) Metallothioneins: potential therapeutic aids for wound healing in the skin. Wound Repair Regen 10:130–132

Lansdown ABG (2007) Critical observations on the neurotoxicity of silver. Crit Rev Toxicol 37:237–250

Lansdown ABG (2010) Silver in healthcare: its antimicrobial efficacy and safety in use. Royal Society of Chemistry, Cambridge

Lansdown ABG (2011) Metal ions affecting the skin and eyes. In: Sigel A, Sigel H, Sigel RO (eds) Metal ions in life sciences. Royal Society of Chemisty, Cambridge

Lowbury EJL (1977) Problems of resistance in open wounds and burns. In: Mouton RP, Brumfitt W, Hamilton-Miller JMT (eds) The Rational Choice of Antibacterial Agents. Kluwer, London

McCauley RL, Li YY, Chopra V, Herndon DN, Robson MC (1994) Cytoprotection of human dermal fibroblasts against silver sulfadiazine using recombinant growth factors. J Surg Res 56:378–384

McHugh GL, Mollering RC, Hopkins CC, Swartz MN (1975) *Salmonella typhimurium* resistant to silver nitrate, chloramphenicol and ampicillin. Lancet 1:235–240

Moss AP, Sugar A, Hargett NA (1979) The ocular manifestations and functional effects of occupational argyrosis. Arch Ophthalmol 97:906–908

Silver S (2003) Bacterial silver resistance, molecular biology and uses and misuses of silver compounds. FEMS Microbiol Rev 27:341–353

Tobin EJ, Bambauer R (2003) Silver- coating of dialysis catheters to reduce bacterial colonisation and infection. Ther Apher Dial 7:504–509

U.S. EPA (1985) Drinking water criteria document for silver, Final Draft ECAO-CIN-026, The Office of Health and Environmental Assessment, Environmental Criteria and Assessment Office, Cincinnati, for the Office of Drinking Water, Washington, DC

Wan AT, Conyers RA, Coombs CJ, Masterton JP (1991) Determination of silver in blood, urine, and tissues of volunteers and burn patients. Clin Chem 37:1683–1687

World Health Organisation (1996) Silver in drinking water: background document for the development of WHO guideline for drinking water quality. WHO, Geneva, WHO/SDE/WSH/03.04/14

# Silver, Physical and Chemical Properties

Fathi Habashi
Department of Mining, Metallurgical, and Materials Engineering, Laval University, Quebec City, Canada

Silver occurs in native state and therefore has been known since ancient times. In the second millennium BC, the Phœnicians obtained large quantities of silver from deposits on the Iberian Peninsula. The first large-scale production of silver took place when the Greeks exploited the mines of Laurion, which reached their peak productivity in ca. 500 BC. The Greeks also worked silver mines in Thrace and in Asia Minor. The Greek Empire was based to a large extent on the role of silver for coinage.

## Physical Properties

| | |
|---|---|
| Atomic number | 47 |
| Atomic weight | 107.87 |
| Relative abundance in earth's crust, % | $2 \times 10^{-6}$ |
| Naturally occurring isotopes | $^{107}$Ag (51.8%) $^{109}$Ag (48.2%). |
| Crystal structure | Face-centered cubic |
| Lattice constant at 20°C, nm | 0.40774 |
| Atomic radius (12- coordination), nm | 0.144 |
| Ionic radius of Ag$^+$ (6-coordination), nm | 0.137 |
| Density, g/cm$^3$ | |
| At 20°C | 10.49 |
| Of liquid at melting point | 9.30 |
| Of liquid at 1,250°C | 9.05 |
| Melting point, °C | 961.9 |
| Boiling point, °C | 2,210 |
| Vapor pressure, Pa | |
| At 1,030°C | 1.33 |
| At 1,190°C | 13.3 |
| At 1,360°C | 133 |
| At 1,580°C | 1,330 |
| At 1,870°C | 13,300 |
| Specific heat capacity at 25°C, J kg$^{-1}$ K$^{-1}$ | 0.23 |
| Thermal conductivity, W m$^{-1}$ K$^{-1}$ | 418 |
| Brinell hardness | 26 |
| Modulus of elasticity, MPa | 82,000 |
| Tensile strength, MPa | 140 |
| Resistivity at 0°C, Ω cm | 1.50 |

(*continued*)

Silver has the highest electrical conductivity and the highest thermal conductivity of all metals.

## Chemical Properties

Molten silver absorbs oxygen and releases it during solidification, with bubbling of the metal surface. Hydrogen is slightly soluble in molten silver, whereas nitrogen, carbon monoxide, carbon dioxide, and the noble gases are insoluble. Halogens react violently with silver at red heat. Moist chlorine gas corrodes silver even at low temperature. Ozone blackens the surface of silver due to oxide formation. Hydrogen sulfide (gaseous or in solution) and aqueous solutions of sulfides immediately form a black coating of silver sulfide on the surface of the metal.

Silver dissolves readily in hot, concentrated nitric and sulfuric acids:

$$3Ag + 4HNO_3 \rightarrow 3AgNO_3 + 2H_2O + NO$$

$$2Ag + 3H_2SO_4 \rightarrow 2AgHSO_4 + SO_2 + 2H_2O$$

The initially vigorous reaction with aqua regia

$$2Ag + 3HCl + HNO_3 \rightarrow 2AgCl + 2H_2O + NOCl$$

is rapidly slowed by the formation of a coating of insoluble silver chloride.

Most common oxidizing aqueous media attack silver, including chromic acid, permanganate solutions, persulfuric acid, selenic acid, and aqueous solutions of free halogens. The following also attack silver: hydrochloric acid, phosphoric acid, bromine water, solutions of alkali-metal chlorides, copper chloride, and iron chloride. The reactions are often strongly dependent on temperature and in some cases on the formation of protective coatings. Silver is resistant to aqueous solutions of organic acids and aqueous alkali.

Silver is resistant toward fused sodium hydroxide at ca. 550°C in the absence of atmospheric oxygen and moisture. Potassium hydroxide is more aggressive. Silver is attacked by fused salts such as sodium peroxide, potassium nitrate, sodium carbonate, potassium hydrogen sulfate, and potassium cyanide. It dissolves in aqueous solutions of sodium cyanide:

$$2Ag + 4CN^- + O_2 + 2H_2O \rightarrow 2[Ag(CN)_2]^- + H_2O_2 + 2OH^-$$

For centuries, amalgamation was the most important process for the treatment of silver ores. Today, it is rarely used because of the toxicity of mercury. The Patera process of leaching silver ores with thiosulfate solution was developed. Halide-containing silver ores or chloridized roasted ores undergo the same reactions that form the basis of the photographic fixing:

$$AgCl + 2Na_2S_2O_3 \rightarrow Na_3[Ag(S_2O_3)_2] + NaCl$$

Silver is precipitated from solutions by the addition of sodium sulfide. Silvering of glass, mainly for mirrors and vacuum flasks is performed with an ammoniacal silver nitrate solution to which a solution of formaldehyde and sodium hydroxide is added as reducing agent. Silver is deposited on the surface of the glass on warming. Almost all photographic technology is based on the light sensitivity of silver halides Renner (1997).

## References

Renner H (1997) Silver. In: Habashi F (ed) Handbook of extractive metallurgy. WILEY-VCH, Weinheim, pp 1215–1268

# Silver-Induced Conformational Changes of Polypeptides

Gabi Drochioiu
Alexandru Ioan Cuza University of Iasi, Iasi, Romania

## Synonyms

Silver binding to polypeptides; Silver-peptide complexes

## Definition

Polypeptides are polymers of amino acids linked by peptide bonds with various levels of organization. The alternative structures of the same polypeptide are referred to as different conformations, and transitions between them are called conformational changes. Silver binding to polypeptides is known to induce conformational changes, which are different from those produced by other heavy metals, such as copper, mercury, nickel, etc. The toxicity degree of silver ions seems to be modulated by their ability to change peptide and protein conformations. Glycine-rich polypeptides undergo severe conformational changes depending on the environmental conditions and especially the presence of silver ions. Contrary to other metal ions, such as copper, iron, nickel, or mercury ones, silver binding may reduce β-sheet or β-turn conformations, with the corresponding increase in the α-helical ones.

## Conformation of Polypeptides

Depending on the linear sequence of the amino acids of the polypeptide chain, it may adopt a highly regular local arrangement of atoms, such as α-helix and the β-strand spatial geometries (Branden and Tooze 1999). In principle, the α-helix, the β-strand, or the β-sheet secondary structures represent ways of accommodating all of the hydrogen-bond donors (always more acceptors than donors in proteins) in the polypeptide backbone (Pauling et al. 1951). The unfolded polypeptide chain lacking any fixed three-dimensional structure is referred as a random coil or unordered secondary structure. The three-dimensional structure (tertiary structure) of a single protein molecule may contain various proportions of α-helices and β-sheets, which are folded into a compact globule.

The polypeptide and protein structures can be determined by X-ray crystallography. This method determines the three-dimensional density distribution of electrons in the crystallized polypeptides, and thereby gives the three-dimensional coordinates of all the atoms to be determined to a certain resolution. Nuclear Magnetic Resonance (NMR) techniques and several other techniques such as circular dichroism (CD) or vibrational (infrared and RAMAN) spectroscopy (Arrondo and Goni 1999) have been developed to provide structural information on polypeptides. Cryo-electron microscopy is an alternative way of determining protein structures to high resolution. Nevertheless, the secondary structure composition of polypeptides can easily be determined via far-ultraviolet (far-UV, 170–250 nm) circular dichroism. The double minimum at 208 and 222 nm indicate α-helical structure, whereas a single minimum at 204 nm or 217 nm is characteristic for random-coil or β-sheet structure, respectively.

## Silver Binding to Polypeptides

Silver ions form aqueous phase complexes with both sulfur and non-sulfur containing peptides and proteins. The structures of silver ion complexes of polypeptides are dependent on the sites of the silver ion attachment. Silver binding by polypeptides or even simple peptides is often assisted by a dramatically conformational change (Murariu et al. 2011). Silver clusters also induce a strong enhancement of the optical absorption of the peptide (Tabarin et al. 2008). Their binding to a peptide can reduce the conformational flexibility and induce transitions between secondary structures.

Circular dichroism spectropolarimetry indicates that polypeptides often undergo a random coil to α-helical conformational change upon binding of one equivalent of silver(I) ion, but not zinc(II), copper(II), or nickel(II) ions. Under normal conditions, silver exists in solutions in only the +1 valence state. Moreover, transmission electron microscopy (TEM) revealed that some histidine-containing polypeptides self-assemble into long helical fibers in the presence of silver(I) ions.

The structural and sequence motifs refer to short segments of protein three-dimensional structure or amino acid sequence that are found in a large number of different proteins. Consequently, several peptides with various residues in their sequence have been synthesized to investigate silver-induced conformational changes at various pH values by CD spectroscopy (Murariu et al. 2011). Uniquely, the glycine-based, histidine-containing peptide showed a severe change from a random coil and β-turn conformation to large α-helices during silver binding. When comparing the effect of silver ions on the conformation of bradykinin, a similar tendency was found. Besides, silver ions reduce the tendency of amyloid-β peptides to aggregation.

## Silver-Peptide Complexes

Polypeptide complexes with silver ions are easily obtained by mixing the aqueous peptide solutions (1 mg/mL peptide) with silver nitrate solutions for 1 or 2 h. The molar ratios may vary from 1:1 to 1:10, and the pH could be 7.4 or higher. Several small peptides, such as a nine amino acid residues peptide (P9; CHQYHHNRE), a ten amino acid residues peptide (P10; RCHQYHHNRE), and Bradykinin (Brd), which is a nine amino acid peptide chain, RPPGFSPFR, have been used to investigate silver ions to polypeptides (Murariu et al. 2011). Besides, some histidine-containing, 19 amino acid residues peptides, with various sequence motifs, amyloid-β peptides (involved in Alzheimer's disease) as well as their fragments, or even large polypeptides, such as casein, have been used to form silver ion complexes at various pH conditions (Drochioiu et al. 2009).

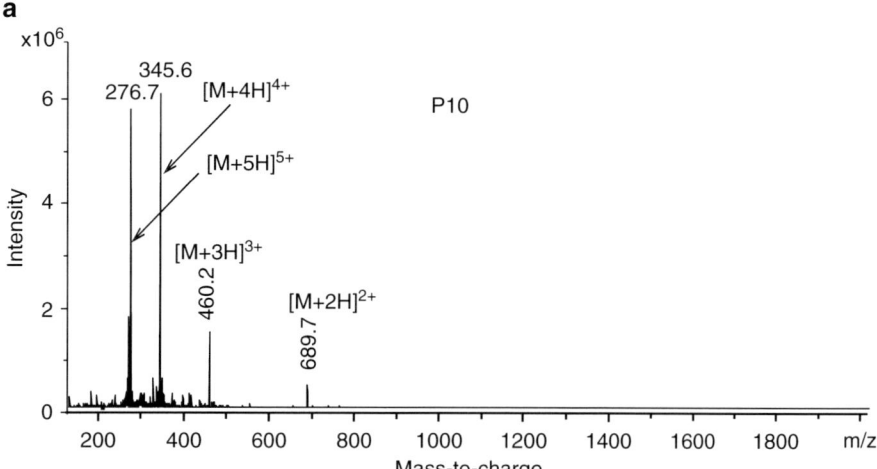
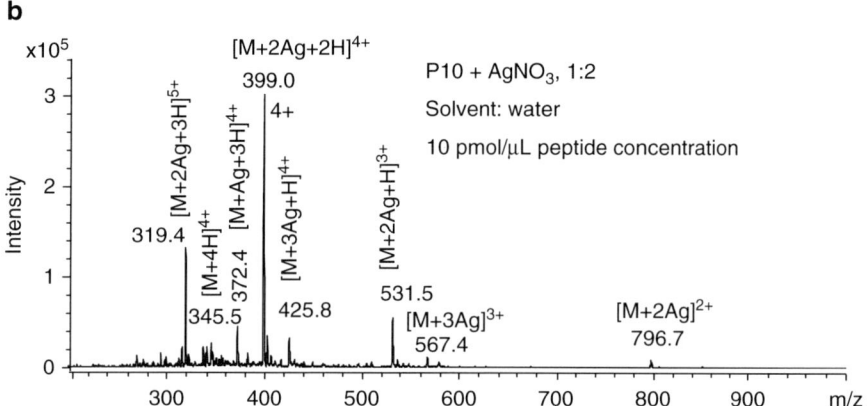

**Silver-Induced Conformational Changes of Polypeptides, Fig. 1** Electrospray ionization ion trap mass spectra of (**a**) decapeptide RCHQYHHNRE (P10) and (**b**) its complexes with silver ions. Most molecules of peptide bound two silver ions each, whereas some of them made the complex with three silver ions

## Mass Spectrometric Measurements

The silver polypeptide complex solutions are regularly diluted to 10 pmol/μL because the electrospray ionization mass spectrometers (ESI-MS) are sensitive to extremely low peptide concentrations. The formation of silver polypeptide complexes of a cysteine-containing peptide, with MW 1377.5, and the following sequence RCHQYHHNRE, is shown in Fig. 1. Silver ions bind strongly to this decapeptide to form coordination compounds containing only one molecule of peptide and one, two, or even three silver ions, without favoring dimerization. The signals for the unbound molecules were the lowest in the spectrum, whereas that for $[M+2Ag+2H]^{4+}$ was the highest.

Amyloid-β peptide containing 40 amino acid residues proved also to bind strongly silver ions at physiological pH, as shown in Fig. 2. Although Aβ(1–40) does not contain cysteine, which has high affinity toward heavy metals, each molecule of Aβ may bind one or two silver ions to form multicharged molecular ions, such as $[M+Ag+6H]^{7+}$, $[M+Ag+5H]^{6+}$, $[M+Ag+4H]^{5+}$, $[M+Ag+3H]^{4+}$, which correspond to m/z 634.6, 740.1, 888.0, and 1109.9, respectively. Aβ may also form adducts with two silver ions, in which sodium ions stabilized the molecular ions $[M+2Ag+Na+3H]^{6+}$ and $[M+2Ag+Na+2H]^{5+}$ (m/z 762.0 and 914.2).

The physiological pH may vary from around 2.0 (stomach) to 8.5 (mitochondria); some experiments have shown that the proportion of silver ions bound to polypeptides increase significantly with increasing pH (Drochioiu et al. 2009). Moreover, on mixing both silver and nickel ions with amyloid-β peptide at higher pH, a competition among these ions has been observed. In contrast to spectra of peptides or metal ion, peptide

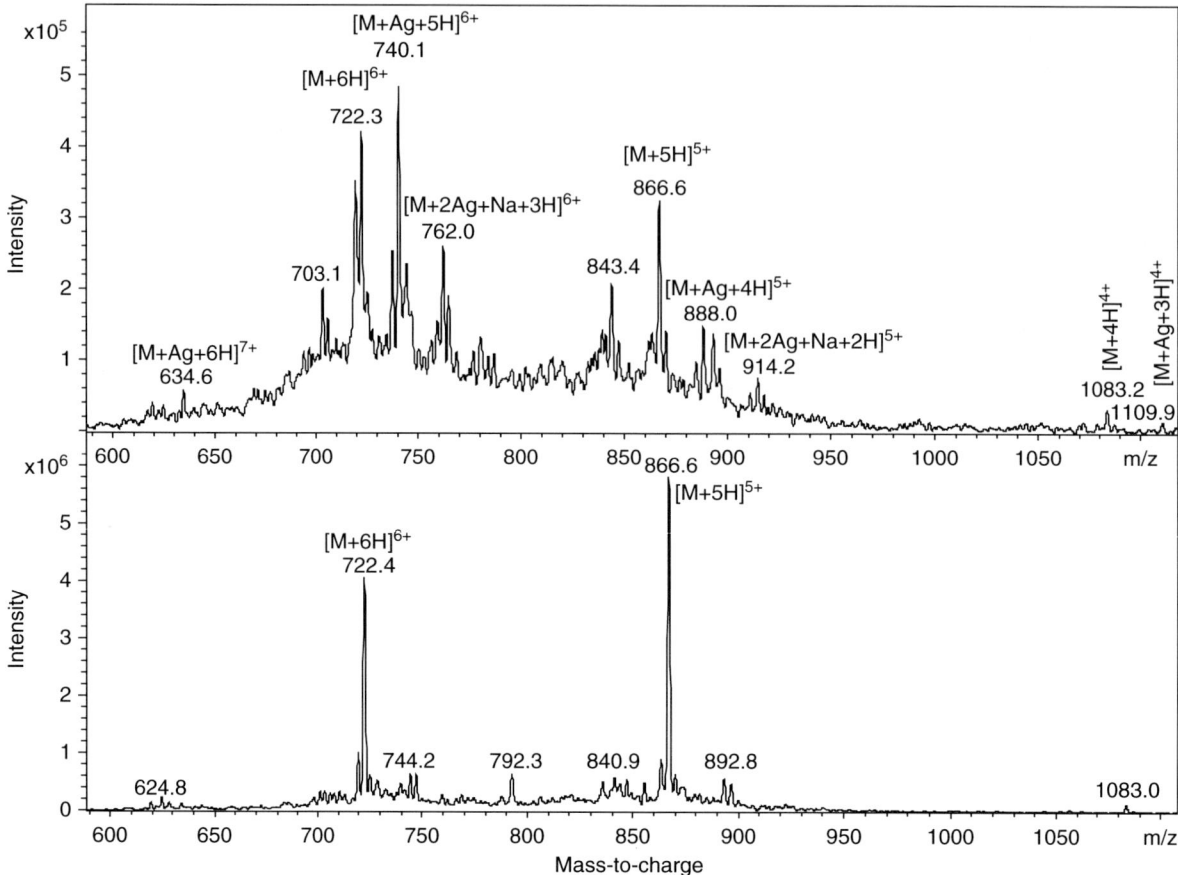

**Silver-Induced Conformational Changes of Polypeptides, Fig. 2** Electrospray ionization ion trap mass spectra of Amyloid-β peptide (M = 4327.0) and its complexes with silver ions at physiological pH. The solution contains both unbound peptide molecules as revealed by the peaks at m/z 722.3 (the molecular ion $[M + 6H]^{6+}$) and m/z 866.6 ($[M + 5H]^+$), and complexes containing one molecule of Aβ(1–40) and one or two silver ions (peaks at m/z 740.1 and m/z 762.0). No dimmers or oligomers were found

complexes at neutral pH, each group of signals in a given area of the spectrum of metal ion, peptide complexes could not be assigned to an isotopic distribution but many individual complexes. For example, the major signal at m/z 355.0 was assigned to the $[M+6Ag+8H]^{14+}$ ion, that at m/z 356.0 to $[M+11Ag+3Ni-H]^{16+}$, whereas the signal at m/z 357.0 (calcd. m/z 357.2) to $[M+8Ag+3Ni+H]^{15+}$. The peak at m/z 358 could be assigned to $[M+3Ag+10H]^{13+}$. All these ions appearing in the same area of the spectrum are charged differently and contain various ratios between the metal ions.

The main signal in the area from m/z 201.9 to 209.9 was assigned to $[M+11Ag+3Ni+11H]^{28+}$, corresponding to a silver-nickel-Aβ1–40 complex with the molecular weight of 5678.0 (calcd. 5678.6). Its presence was also sustained by the peaks at m/z 710.9 (8+), 632.0 (9+), 356.0 (16+), and 247.9 (23+), respectively.

## CD Spectropolarimetry

Typically, CD measurements are made on polypeptides at concentrations around 0.2 mM and peptide/silver ion ratios in the range from 1:1 to 1:10. The direct CD measurements (θ, in millidegrees) are then converted to molar ellipticity, $\Delta\varepsilon$ ($M^{-1} \cdot cm^{-1}$).

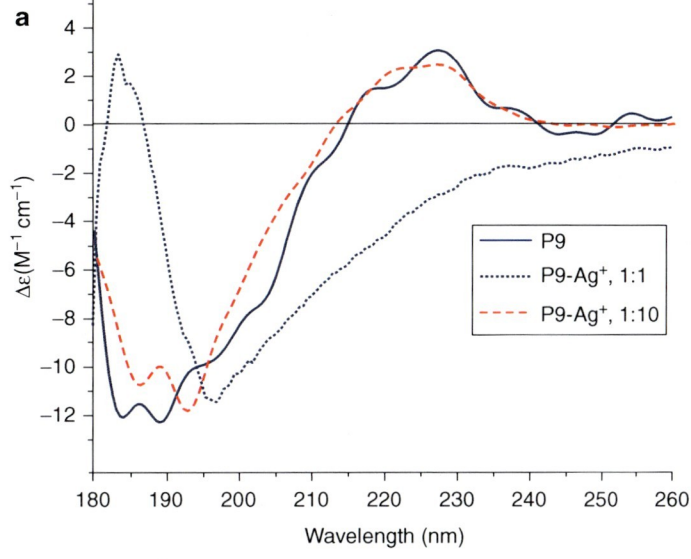
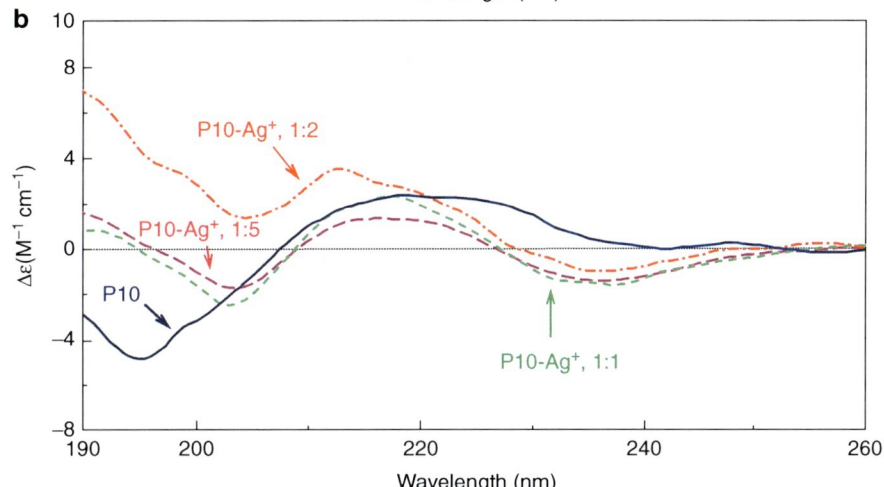

**Silver-Induced Conformational Changes of Polypeptides, Fig. 3** CD spectra of (**a**) peptide P9 in water (pure peptide, 0.2 mM) and in the presence of 1:1 and 1:10 molar ratios of silver ions (P9-Ag$^+$, 1:1, and P9-Ag$^+$, 1:10), and (**b**) CD spectra of P10 and its complexes with silver ions

Proportions of each secondary structure type are obtained by spectral deconvolution of the peptide CD spectrum as a linear sum of predetermined basis spectra.

It was shown that the introduction of a single amino acid residue in a peptide sequence may affect drastically the conformation, whereas the addition of silver ions to the solutions of these peptides (P9 and P10) results in unexpected conformational changes (Fig. 3).

The far UV CD spectrum of P9 has two negative maxima at 184 nm and 189 nm suggesting a mixture of β-turn and unordered populations. Silver ions severely changed the peptide conformation, supporting the idea that the β-turn proportion of conformers decreased with an increase in the α-helix one. A very sharp band at 197 nm was found in the negative region of absorption spectrum, which seems to be a characteristic for a combination of both α-helix and random coil spectra, with a diminished contribution from β-turn populations. Besides, the relatively intense peak at 183.5 nm has confirmed the tendency of silver ions to increase α-helical proportions when bound to peptides under the 1:1 molar ratio. It was supposed that silver ions bound first to amino terminal group, the sulfhydryl group SH, and a nitrogen atom in the histidine moiety and was coordinated by more that two atoms in the peptide backbone. The next silver ions bound also to peptide backbone, each one at a different site, generating a conformation similar to that of unbound peptide. However, the high affinity of polypeptides toward silver ions can also be explained by the tendency of this ion to form complexes with the amino nitrogen and carboxyl oxygen, which are present in all amino acids and peptides. As a result of binding of more silver ions than one, the proportion of unordered populations increases significantly. In the case of 1:10 P9: $Ag^+$ complex, the two bands in the negative region are found to be shifted to 186 nm and 192.5 nm, respectively.

P10 in distilled water solutions is found as a mixture of random coil (46.1%) and β-sheet (40.5%) forms. Silver ions induce an increased absorption in the positive region around 193 nm as well as in the negative one at 222 nm. Hence, the 196 nm negative minimum of pure peptide shifted toward 203 nm in the case of 1:1 P10-$Ag^+$ complex. The 1:2 P10-$Ag^+$ has a minimum in the positive region at 204.5 nm. Such minima have been assigned to a mixture of α-helical and unordered conformers, with an increased proportion of α-helical conformers in the case of 1:2 molar ratio P10: $Ag^+$. A large proportion of α-helical conformers was calculated for 1:2 molar ratio adduct of P10 with silver ions, whereas an increase in the concentration of $Ag^+$ resulted in the accumulation of unordered populations. Possibly, the first silver ion is bound to the guanidine group of the N-terminal arginine residue, increasing the proportion of unordered conformers. The second one is bound to the SH group of cysteine and nitrogen atoms in amino groups and imidazol moiety to form a chelate with silver ion wrapped by several atoms of peptide backbone. The next silver ions interfere one to another inducing a random coil conformation for P10.

The CD spectra of polypeptide $H_2N$-GGGGHGGGGHGGGGHGGGG-COOH (P19) in aqueous solutions are characteristic to 55.7% β-turn populations and 44.3% unordered populations (Murariu et al. 2011). Since glycine is a flexible amino acid residue, it is also expected that some parts of the same polypeptide molecule be β-turn and others unordered. CD measurements estimate the average proportions of conformers. On adding silver ions to P19 in aqueous solutions at pH 7.4, the peptide conformation changed severely. Silver ions transformed unstructured peptide molecules into α-helical conformers (from 0% to 34.8%, and from 0% to 45%, depending on peptide: silver ions ratio) and stabilized β-turn structures. Indeed, a negative minimum has been observed at 222.5 nm, suggesting the presence of α-helical conformers. On increasing the silver-peptide ratio up to 10, the shape of spectrum changed dramatically, being similar to that of α-helical peptides. The deconvoluted CD spectrum for the 1:10 P19-$Ag^+$ complex was characterized by a positive peak at 197.5 nm, with the intensity of the negative band at 208 nm being less negative than that of the 222 nm band in comparison with the regular α-helix, with a positive band at 190 nm and two negative bands at 208 and 222 nm.

The amyloid-β(1–40) and amyloid-β(1–42) peptides are mainly α-helical in their native conformation, but undergo an α-helix to β-strand conversion before or during fibril formation associated with Alzheimer's disease. Due to their high affinity to such polypeptides and the tendency to induce α-helical conformers, silver ions could dissociate amyloid plaques and Aβ aggregates. Figure 4 shows the CD spectrum of an aqueous solution of Aβ1–40 with a large negative absorption band having a maximum at 197.5 nm. Aβ1–40 in the fresh water solutions is found as a mixture of random coil (49.01%), β-sheet (33.7%), and β-turn (17.3%) forms. On adding silver ions, the peptide conformation changes severely from β-sheet to α-helix. Even a large molar ratio $Ag^+$: Aβ1–40 (10:1) induces a significant decrease in the negative absorption around 198.0 nm, suggesting an increase in the proportion of α-helical conformers. Because β-turn conformers absorb significantly at 202.0 nm and 225.0 nm, the characteristic absorption of α-helical populations at 222.0 nm was relatively low. The large molecule of Aβ accommodated easier silver ions than the other peptides investigated here, and silver ions interacted with more atoms in the peptide backbone to induce an α-helix-like structure. Sodium dodecyl sulfate (SDS) is known to induce an increase in α-helix populations of polypeptides. On adding SDS to Aβ, the conformational equilibrium changed from β-sheet and unordered populations to α-helical and β-turn structures. SDS induced important conformational changes even at 1:10 molar ratio (α-helix: 28.0%, β-turn: 35.0%, and unordered: 36.9%). Therefore, both silver ions and SDS induced the abolition of β-sheet structures of Aβ1–40 peptide, which are related to Aβ neurotoxicity in AD. Because SDS is toxic, sodium stearate and sodium palmitate have also been used to diminish β-sheet proportion of Aβ1–40 peptide with rather similar results. An increase in α-helical populations from zero to 15.8%,

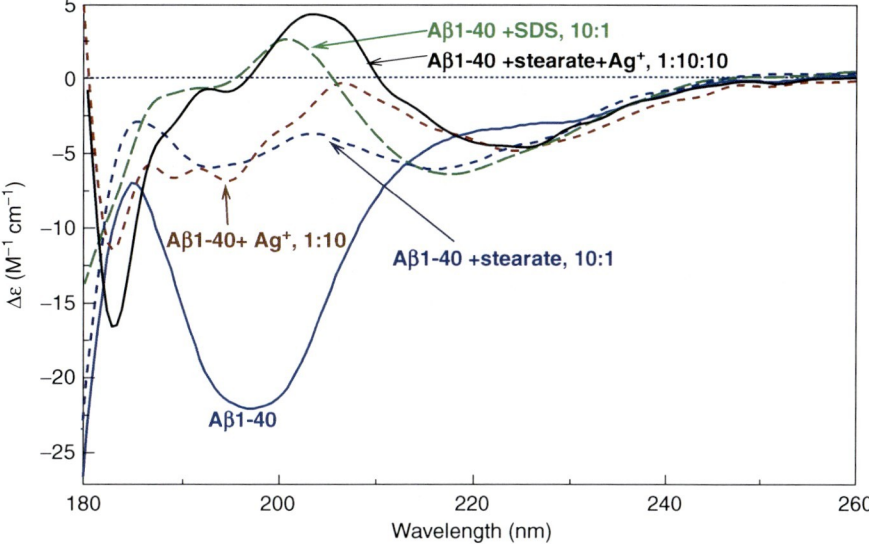

Silver-Induced Conformational Changes of Polypeptides, Fig. 4 CD spectra of amyloid-β peptide 1–40 (Aβ, 0.2 mM), and in the presence of silver ions, SDS, and stearate (1:10 and 10:1 molar ratio)

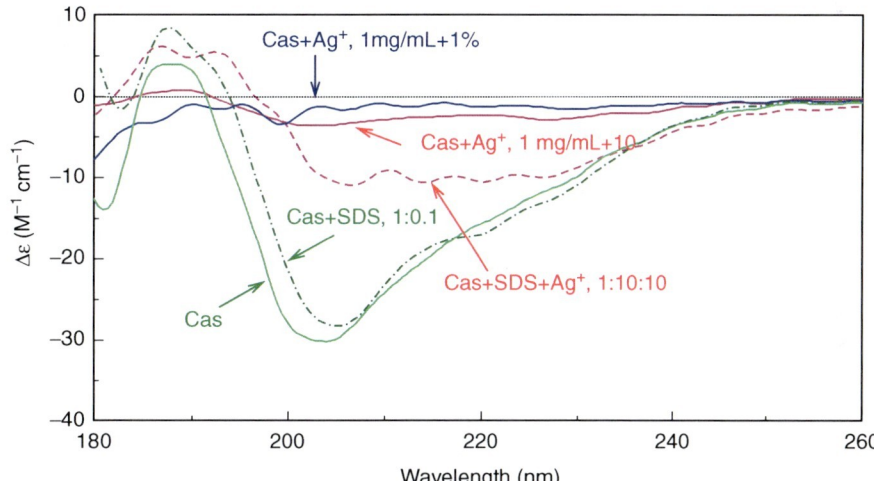

Silver-Induced Conformational Changes of Polypeptides, Fig. 5 Effect of silver ions (Ag⁺), sodium dodecyl sulfate (SDS), and their mixture on the conformation of casein (Cas)

and β-sheet ones from 33.7% to 13.6% could provide a motivation to investigate the synergic effect of such tensio-active agents and silver ions in neurodegeneration.

## Conformational Changes of Large Polypeptides

On treating a larger polypeptide, such as casein with silver ions, the conformation changed dramatically (Fig. 5). Casein, a polypeptide with a molecular mass MW = 23,600, has a large proportion of β-sheet conformers (42.9%) and unordered forms (around 41.0%). On adding 1% SDS, the proportion of α-helical conformers increased little, up to only 5.7%, whereas a tenfold increase in SDS concentration produced an additional increase in the proportion of α-helical populations up to 8.0%. Moreover, 1% SDS induced a slight increase in β-sheet population (48.2%). An unexpected increase in the proportion of α-helical conformers, from 4.6% to 20.7%, was observed on adding small volumes of silver nitrate solution, whereas β-sheet populations have completely disappeared. As expected, β-turn forms increased from 5.2% up to 26.6%. A tenfold increase in the concentration of silver ions slowly decreased the proportions of α-helical and β-turn forms, with the corresponding increase in the proportion of unordered molecules. However, a mixture of SDS and silver ions produced the highest proportion of α-helix conformers (28.4%) and β-turn ones (27.2%). There was direct evidence for the synergism between SDS and silver ions in inducing α-helical conformers starting from β-sheet ones.

## Biological and Medical Implications

Since most proteins are required to adopt a specific three-dimensional structure to be biologically active, interactions with metals which change the conformation are expected to alter their function. The role of silver ions in various pathologies, as well as their effect on peptide conformation and properties are less well understood. However, it is known that silver ions bind tightly to the cysteine residues preventing normal dynamics and catalysis of cellular enzymes. Treatment with silver or other heavy metals in genetically susceptible mice induces a systemic autoimmune condition characterized by antinuclear antibodies targeting the 34-kDa nucleolar protein fibrillarin, as well as lymphoproliferation and systemic immune-complex (IC) deposits (Havarinasab et al. 2009; Hultman et al. 1994). The effects of short-term exposure of cells to extracellular $Ag^+$ ions have also been investigated. Ionic silver (1 μM $Ag^+$) acting on peptide moieties significantly increases the inward flow of cations.

After exposure of the yeast, *Saccharomyces cerevisiae*, to ionic silver or nanosilver containing plasma coating, the main common feature was the presence of electron dense nodules all over the cell (Despax et al. 2011). The formation of these nodules was related to the $Ag^+$ reactivity toward sulfur-containing compounds to form clusters with $Ag_2S$-like structures, together with the production of a few silver nanocrystals, mainly at the cell wall periphery. Following nanosilver-based treatment, some sulfur containing silver clusters preferentially located at the cell wall periphery have been detected, together with nodules composed of silver, sulfur, and phosphorus all over the cell. In these treatments, nitrogen and silver signals overlapped, confirming the affinity of silver entities for polypeptides.

Phycobilisomes (PBSs) absorb light energy and transfer it to chlorophyll of photo systems (Liu et al. 2005). Only silver ions induce disassembly of the cores of PBSs, which serve as a specific target of $Ag^+$ in vivo, binding to proteins followed by reduction to free silver or formation silver sulfide with sulfur containing amino acids.

A specific and protective role for silver ions in brain pathologies (Murariu et al. 2011) was suggested, contrary to silver toxicity on developing brains, which could be related to their high affinity toward physiologically and pharmacologically active peptides. Due to its tendency to induce α-helical conformations, the silver ion could display a low toxicity toward the nervous system, in direct relationship with its binding to amyloid-β peptides. The polypeptides or single amino acids have a high affinity for silver ion and chelate it by using the nitrogen atoms. The silver ion adopts a collinear coordination, without any deprotonation of peptides at physiological pH. This finding may be of paramount importance to understand the behaviors of several peptides and proteins involved in the neurodegenerative pathologies such as prion protein or amyloid-β peptide. It was suggested that silver ions might have both a toxic effect on binding to SH groups of biologically active proteins and peptides and a protective role due to their folding capability.

## Perspectives

All the findings presented here aim at suggesting that silver binding to polypeptides could change dramatically their structure and properties. It was hypothesized that the level of toxicity of silver ions can be modulated by their ability to change peptide and protein conformations (Murariu et al. 2011). Since silver binding to polypeptides changes their conformations, and consequently their properties, further research is still needed. Because of their capability to change the β-sheet forms into the α-helical ones, alone or in the presence of tensioactive agents, the role of silver ions in neurodegenerative pathologies should also be investigated.

## Cross-References

- ► CusCFBA Copper/Silver Efflux System
- ► Metallothioneins and Silver
- ► Nanosilver, Next-Generation Antithrombotic Agent
- ► Silver as Disinfectant
- ► Silver in Protein Detection Methods in Proteomics Research
- ► Silver, Burn Wound Sepsis and Healing
- ► Silver, Neurotoxicity
- ► Silver, Pharmacological and Toxicological Profile as Antimicrobial Agent in Medical Devices
- ► Zinc, Metallated DNA-Protein Crosslinks as Finger Conformation and Reactivity Probes

## References

Arrondo JL, Goni FM (1999) Structure and dynamics of membrane proteins as studied by infrared spectroscopy. Prog Biophys Mol Bio 72:367–405

Branden C, Tooze J (1999) Introduction to protein structure, 2nd edn. Garland Publishing, New York

Despax B, Saulou C, Raynaud P, Datas L, Mercier-Bonin M (2011) Transmission electron microscopy for elucidating the impact of silver-based treatments (ionic silver versus nanosilver-containing coating) on the model yeast *Saccharomyces cerevisiae*. Nanotechnology 22:175101

Drochioiu G, Manea M, Dragusanu M, Murariu M, Dragan ES, Petre BA, Mezo G, Przybylski M (2009) Interaction of β-amyloid(1-40) peptide with pairs of metal ions: an electrospray ion trap mass spectrometric model study. Biophys Chem 144:9–20

Havarinasab S, Pollard KM, Hultman P (2009) Gold- and silver-induced murine autoimmunity – requirement for cytokines and CD28 in murine heavy metal-induced autoimmunity. Clin Exp Immunol 155:567–576

Hultman P, Johansson U, Turley SJ, Lindh U, Eneström S, Pollard KM (1994) Adverse immunological effects and autoimmunity induced by dental amalgam and alloy in mice. FASEB J 8:1183–1190

Liu X-G, Zhao J-J, Wu Q-Y (2005) Oxidative stress and metal ions effects on the cores of phycobilisomes in *Synechocystis* sp. PCC 6803. FEBS Lett 579:4571–4576

Murariu M, Dragan ES, Adochitei A, Zbancioc G, Drochioiu G (2011) Silver binding to peptides: a CD study. J Pept Sci 17:512–519

Pauling L, Corey RB, Branson HR (1951) The structure of proteins; two hydrogen-bonded helical configurations of the polypeptide chain. Proc Natl Acad Sci USA 37:205–211

Tabarin T, Kulesza A, Antoine R, Mitrić R, Broyer M, Dugourd P, Bonačić-Koutecký V (2008) Absorption enhancement and conformational control of peptides by small silver clusters. Phys Rev Lett 101:213001

## Silver-Peptide Complexes

► Silver-Induced Conformational Changes of Polypeptides

## SISH

► Enzyme Metallography and Metallographic In Situ Hybridization

## Site: Cleft

► Monovalent Cations in Tryptophan Synthase Catalysis and Substrate Channeling Regulation

## SITs – Silicon Transporter Proteins

► Silicon, Biologically Active Compounds

## Skin

► Chromium and Allergic Reponses

## *SLC24* Gene Family

► Potassium Dependent Sodium/Calcium Exchangers

## Slc30a = ZnT

► Zinc Regulation in Fish Biology

## Slc39a = Zip

► Zinc Regulation in Fish Biology

## SMADIP1gene and SIP1 Gene

▶ Zinc and Mowat-Wilson Syndrome

## SMF

▶ Sodium as Primary and Secondary Coupling Ion in Bacterial Energetics

## Smoking

▶ Polonium and Cancer

## SmtB

▶ Zinc Sensors in Bacteria

## Snake-Venom Metallopeptidase

▶ Zinc Adamalysins

## SOD

▶ Zinc in Superoxide Dismutase

## SOD-1

▶ Zinc in Superoxide Dismutase

## SOD1. UniProtKB/Swiss-Prot P00441 (SODC_HUMAN)

▶ Copper-Zinc Superoxide Dismutase and Lou Gehrig's Disease

## SOD1/CRS4

▶ Zinc Storage and Distribution in S. cerevisiae

## Sodium (Na$^+$)

▶ Magnesium and Vessels

## Sodium and Potassium Transport in Mammalian Mitochondria

Evgeny Pavlov
Department of Physiology & Biophysics, Faculty of Medicine, Dalhousie University, Halifax, NS, Canada

### Synonyms

Mitochondrial sodium and potassium channels and exchangers; Sodium and potassium uniporter

### Definition

Mitochondrial sodium and potassium transport is a process of these ions movement between mitochondrial inner compartment (matrix) and cytoplasm of the cell. This process is highly regulated and provided by a number of molecular structures localized in the mitochondrial inner membrane. This transport can occur by the uniporter mechanism when either sodium or potassium is transported through corresponding selective channel or by the exchanger mechanism when sodium or potassium transport occurs in exchange of calcium or protons.

### Introduction

Mitochondria are intracellular organelles found in nearly all types of eukaryotic cells. Morphologically, mitochondria can be described as membrane vesicles composed of two membranes. Outer mitochondrial membrane contains a number of large weakly selective channels (pores). Outer membrane regulates transport of small organic molecules (most notably ATP) but is freely permeable to various metal ions. On the other hand mitochondrial inner membrane is largely impermeable for charged molecules and possesses a number of specialized and highly selective systems which allow tightly regulated transport of certain types of ions between mitochondrial inner compartment (matrix) and cytoplasm. Among other transporting systems mitochondria possess several specialized mechanisms for import and export of potassium and sodium ions. $Na^+$ and $K^+$ uptake into the mitochondria occurs through uniport mechanisms when these ions enter mitochondria using the energy of electrical potential gradient of the inner membrane (Bernardi 1999). This gradient is generated and maintained at $-180$ mV under normal physiological conditions due to the activity of mitochondrial respiratory chain. Another pathway for $Na^+$ and $K^+$ transport across mitochondrial membrane is provided through exchanger mechanism when transport of $Na^+$ or $K^+$ is coupled with the transport of other positively charged ions – specifically $H^+$ or $Ca^{2+}$ (Castaldo et al. 2009; Murphy and Eisner 2009).

## Physiological Roles of Mitochondria Ion Transport

Very early studies of mitochondrial physiology have established that these organelles are capable of transporting various ions including $Na^+$ and $K^+$. However, initially these ion-transporting systems were studied largely in the context of their relationship to oxidative phosphorylation process (Bernardi 1999). According to the classical Mitchell's chemiosmotic model, electrochemical gradient of the mitochondrial inner membrane is the major source of energy for generation of ATP and as such it was postulated that this membrane should be largely impermeable to the ions other than protons. Indeed, any active transport of ions would cause energy dissipation and disruption of ATP production which would make this transport unfavorable. However, later it was recognized that mitochondrial ion transport plays an essential role in intracellular signaling and is critical for regulation of many processes occurring in mitochondria and cell under both physiological and pathological conditions. To date, several roles for mitochondrial $Na^+$ and $K^+$ transport systems have been proposed and are subject of intensive investigation.

## Physiological Roles of K-Transporting Systems

Mitochondria can accumulate $K^+$ through electrogenic uniporter mechanism most likely due to the presence of a number of $K^+$ channels in their inner membrane. Under physiological conditions, this $K^+$ uptake is balanced by its release in exchange of $H^+$ through KH-exchanger (KHE). Under physiological conditions, the main role of $K^+$ uptake is linked to the regulation of mitochondrial volume. As discussed in the following review (Bernardi 1999), existence of such mechanisms might play a critical role during mitochondrial biogenesis. Furthermore, changes in the volume of mitochondria induced by $K^+$ transport might play a very important role in the regulation of mitochondrial metabolism. It is also proposed that mitochondrial $K^+$ uptake plays a critical role in cell response to the pathological conditions. For example, opening of $K^+$ channels is involved in cardioprotection against ischemia reperfusion (IR) injury. Ischemia reperfusion injury is a phenomenon of high clinical importance which is observed during patient treatment following heart attack. The main treatment strategy for patients during heart attack or stroke (ischemia) is reestablishment of blood flow to provide oxygen supply to damaged heart or brain tissue (reperfusion). However, paradoxically, process of reperfusion causes even further damage compared to the ischemia. Thus, prevention of reperfusion damage is an important and currently not well-resolved problem. The idea about the involvement of mitochondrial $K^+$ transport in protection from the reperfusion injury is based on the number of experimental data, which indicate that activation of mitochondrial $K^+$ uptake causes protective effect against the cell damage caused by reoxygenation following the hypoxia (Nishida et al. 2010). Although the exact mechanism of how this protection occurs is not fully understood it has been proposed that most likely activation of electrogenic $K^+$ uptake causes partial depolarization of the mitochondrial inner membrane, which reduces the net $Ca^{2+}$ accumulation inside mitochondria. Excessive $Ca^{2+}$ accumulation inside mitochondria is known to be the major cause of the injury and thus its prevention by increased $K^+$ uptake may indeed be protective. Another very important role which this "mild depolarization" may play is in reduction of the amount of reactive oxygen species (ROS) produced during the process of reperfusion (Szewczyk et al. 2009). In addition to $Ca^{2+}$, ROS are other major contributors to the tissue damage during IR injury. Reduction of membrane potential is expected to affect the activity of the mitochondrial respiratory chain and thus cause decreased rate of ROS production. It should be mentioned here that although evidence of the involvement of mitochondrial $K^+$-transporting system in stress response is overall strong, it should be taken with some degree of caution. As it was pointed out in a number of recent reviews, most of the drugs which affect mitochondrial $K^+$ transport also affect the activity of plasma membrane $K^+$ channels. Taking into account that cell membrane $K^+$ channels are closely involved in stress response it is possible that some of the effects attributed to the changes of mitochondrial $K^+$ transport might in fact be related to the changes induced by activation of cell membrane $K^+$ transport.

**Physiological Role of Mitochondrial Na$^+$ Transport**

The two main $Na^+$-transporting mechanisms are recognized to exist in mitochondria. $Na^+$ can be transported through the membrane either in exchange of protons (Na-H-exchanger or NHE) or in exchange of $Ca^{2+}$ (NCE) (Murphy and Eisner 2009). NCE is the main mechanism of $Na^+$ influx into mitochondria. One of the major roles of $Na^+$ transport in mitochondria is in regulation of the levels of mitochondrial and cytoplasmic $Ca^{2+}$. $Ca^{2+}$ is the major signaling ion in the cell involved in regulation of a large number of biochemical pathways including those which involve mitochondria. For example, $Ca^{2+}$ is an activator of a number of mitochondrial enzymes involved in energy metabolism. Under normal conditions mitochondrial $Ca^{2+}$ is constantly recycled through well-coordinated uptake and release mechanisms. It enters mitochondria through electrogenic $Ca^{2+}$ uniporter and is released through NCE. NCE activity largely determines the concentrations of free $Ca^{2+}$ in mitochondrial matrix as well as in cytoplasm. Thus, NCE plays an important role in regulation of many signaling processes in both physiological and pathological conditions. Under physiological conditions, NCE is mostly involved in regulation of the activity of mitochondrial metabolism (Bernardi 1999). This occurs through regulation of mitochondrial $Ca^{2+}$ levels, which in turn regulated three key dehydrogenases involved in energy metabolism: isocitrate, pyruvate, and oxoglutarate dehydrogenases. In addition, $Ca^{2+}$ can directly regulate the activity of ATP synthase – major enzyme responsible for production of ATP. The role of NCE in pathology can be best illustrated for ischemic conditions in cardiac cells found during heart attack. Under these conditions, due to the lack of oxygen, ATP production is decreased and mitochondrial membrane potential is reduced. These changes cause the reversal mode of NCE when $Na^+$ is now extruded from mitochondria and $Ca^{2+}$ is accumulated in mitochondrial matrix. This contributes significantly to $Ca^{2+}$ overload of mitochondria and development of $Ca^{2+}$-induced permeability transition pore (Zoratti and Szabo 1995). Opening of permeability transition pore causes irreversible damage to the mitochondria, and eventually leads to the cell death. Similar mechanism also contributes to ischemia-induced neuronal damage during stroke.

## Kinetic Parameters of K$^+$ and Na$^+$ Ion-Transporting Systems

### Methods for Investigation of K$^+$ and Na$^+$ Ion Transport

One of the major challenges in studies of mitochondrial ion transport is the fact that, unlike cell membrane ion transport, it cannot be studied directly. Indeed, cell membrane ion transport can be directly investigated by patch clamp technique. This technique gives direct information about nature, kinetics, and regulation of certain ion-transporting system under conditions closely related to ones observed in vivo. In order to apply similar type of technique to the mitochondrial inner membrane prior to the assay, mitochondria need to be isolated from the cell and osmotically swollen to make their inner membrane accessible to the patch pipette. Although a number of ion channels have been detected using this approach the question remains whether these channel activities reflect the behavior of ion-transporting systems of the inner membrane in native environment. In addition to patch clamp technique, ion fluxes measurements can be performed at the level of intact isolated mitochondria (Bernardi 1999). In this case, kinetic parameters of ion fluxes can be assessed indirectly by optical measurements of changes in mitochondrial volume. The classical model system which uses swelling assay approach can be illustrated by the method which uses protonated form of acetic acid. When NHE is active, it will induce accumulation of $Na^+$ inside mitochondria in exchange of protons which will be transported into the recording media. This creates proton gradient and if recording media contains acetate which is highly permeable to the mitochondrial inner membrane it will induce its uptake. Eventually, this results in accumulation of $Na^+$ acetate which induces swelling of mitochondria and can be detected optically as change in light absorbance. Similar strategy can be used for assay of the kinetics of KHE as well as in studies of $K^+$ and $Na^+$ uptake by uinporter mechanism. In the case of uniporter studies, swelling can also be measured in the presence of acetate but it requires presence of substrates to energize mitochondria. Mitochondria are needed to be energized in order to ensure extrusion of protons and buildup of proton gradient. In addition to $K^+$ and $Na^+$ flux measurements by electrophysiological (patch clamp) and optical (isolated mitochondria) methods, several $Na^+$-sensitive fluorescent probes have been developed recently (Murphy and Eisner 2009). These probes allow, to some extent, to estimate changes in mitochondrial $Na^+$ transport at the level of intact cells and add important information about $Na^+$ transport in mitochondria in their native environment.

## Kinetic Parameters and Ionic Regulation of $K^+$ and $Na^+$-Transporting Systems

### Exchangers

The following rates and regulation of mitochondrial $Na^+$ and $K^+$ transport have been estimated (Bernardi 1999). The maximal velocity of KHE is estimated to be 300 nmol $K^+$/mg protein/min. The main properties of the KHE are the following: its activity is increased with the decrease in matrix $Mg^{2+}$ concentration; it is inhibited by matrix protons and inhibited by quinine. The maximal velocity of NCE is 18 nmol $Na^+$/mg protein min. The stochiometry of the NCE is estimated to be close to 3:1 ($Na^+:Ca^{2+}$) suggesting that this transporting mode is electrogenic and its activity depends on the presence of mitochondrial membrane potential. NCE activity is strongly regulated by protons and is decreased during matrix alkalinization. NCE is also strongly regulated by extramitochondrial $Ca^{2+}$ and some other divalent cations including $Ni^{2+}$, $Ba^{2+}$, and $Mg^{2+}$. Finally, $La^{3+}$ is a potent blocker of NCE.

### Uniporters (Channels)

Although the principal mechanism of $Na^+$ entry into mitochondria is provided NCE it was also suggested that it can occur through the channel mechanism (Bernardi 1999). The velocity of such uptake is 0.2 nmol $Na^+$/mg protein min mV. This putative channel can be inhibited by $Mg^{2+}$ and Ruthenium Red, and is regulated by protons with optimum pH in the range of 7.5–8. The estimated maximal velocity of electrogenic $K^+$ uptake is 22.5 nmol $K^+$/mg protein min mV. This activity is strongly inhibited by $Mg^{2+}$ and it is considered that it can never reach under normal conditions. $K^+$ conductance in energized mitochondria in the presence of physiological concentrations of $Mg^{2+}$ is estimated at the level of 0.11 $K^+$/mg protein min mV which is well below its maximal velocity.

## Molecular Identities of $K^+$ and $Na^+$-Transporting Systems

### Exchangers

Identification of the molecular structures responsible for mitochondrial ion transport remains to be the major challenge. The molecular identity of most of the mitochondrial ion-transporting systems remains to be conclusively proven. To date, a number of proteins have been suggested to participate in this transport. For example, NHE transport might be mediated by 59 kDa protein which has been partially purified from beef heart mitochondria (Bernardi 1999). Putative NCE has been identified as a 110 kDa protein which was found in protein extracts from heart, liver, and kidney mitochondria (Castaldo et al. 2009). Alternatively, it has been proposed that NCE transport might be mediated by expression of one of the cellular NCE proteins in mitochondria. In experiments with neuronal and astrocytes cultured cells, it was demonstrated that plasmalemmal NCE exchangers (NCX1, NCX2, and NCX3) can be found localized in mitochondria suggesting that at least in part NCE activity in mitochondria might be mediated by these proteins. Finally, novel NCE-related protein (NCLX) has been recently found in mitochondria (Hajnoczky and Csordas 2010). Overexpression of NCLX significantly enhanced NCE activity of mitochondria whereas its silencing with RNAi reduced this activity suggesting that NCLX might be the major contributor to NCE transport in mitochondria.

## Channels

Over the past two decades, several putative $K^+$ channels have been identified in the mitochondrial inner membrane (reviewed in Szewczyk et al. (2009, 2010)). The first mitochondrial $K^+$ channel was identified in 1991 by the group of Inoue. In these experiments, patch clamp of mitoplasts (mitochondria with their outer membrane removed by osmotic swelling) reveal channel activity which was selective to potassium and inhibited by ATP. This channel (named mitoKatp channel) has conductance of 10 pS and can be inhibited by ATP, ADP, and by glibenclamide and activated by diazoxide. The protein responsible for the mitoKatp channel activity has been partially purified from mitochondria and has molecular weight of 54 kDa. It was demonstrated that this protein when reconstituted into artificial bilayer system has properties and regulation similar to those found in native mitochondrial membranes. Pharmacological profile and immunoreactivity suggest that this channel likely belongs to the family of inward-rectifier channels. The second $K^+$ channel which has been found in mitochondrial inner membrane has been described as large-conductance $Ca^{2+}$-activated $K^+$ channel. The conductance of this channel is 300 pS in the 150 mM KCl recording solution. This channel activity is stimulated by addition of $Ca^{2+}$. Importantly, this channel can be blocked by charybdotoxin – highly specific blocker of plasmalemmal BKca channel. This suggests close similarity between mitochondrial and plasmalemmal BKca channels. The idea about close structural relationship between these two channels is further supported by the finding that mitochondria demonstrate immunoreactivity with antibodies against beta subunit of plasma membrane BKca channel. It has been suggested that mitochondrial and plasma membrane BKca channels are splice variants of the protein encoded by the same gene; however, this suggestion remains to be confirmed experimentally. In addition, two other candidates to the role of mitochondrial $K^+$ channels have been proposed. It was demonstrated that mitochondria might contain 17 pS voltage-gated $K^+$ channel similar to Kv 1.3 channel (mito Kv1.3). Finally, very recently, $K^+$-transporting properties were found to be linked to the activity of twin-pore domain $K^+$ channel TASK-3 (Szewczyk et al. 2009). This channel has been found in melanoma and keratinocyte cells. However, its detailed functional properties remain to be established.

## Conclusion

In conclusion, past three decades of intensive investigation have established the critical role of mitochondrial $K^+$- and $Na^+$-transporting systems in cellular physiology. Activity of these systems has been linked to a number of diseases suggesting that these systems can be promising targets for drug development. Further, more detailed characterization of the molecular identities of these systems remains the major challenge in studies of mitochondrial ion-transporting systems.

## Cross-References

- Calcium and Mitochondrion
- $K^+$-Dependent $Na^+/Ca^{2+}$ Exchanger
- Potassium in Health and Disease
- Sodium Channel Blockers and Activators
- Sodium-Hydrogen Exchangers, Structure and Function in Human Health and Disease

## References

Bernardi P (1999) Mitochondrial transport of cations: channels, exchangers, and permeability transition. Physiol Rev 79(4):1127–1155

Castaldo P, Cataldi M, Magi S, Lariccia V, Arcangeli S, Amoroso S (2009) Role of the mitochondrial sodium/calcium exchanger in neuronal physiology and in the pathogenesis of neurological diseases. Prog Neurobiol 87(1):58–79

Hajnoczky G, Csordas G (2010) Calcium signalling: fishing out molecules of mitochondrial calcium transport. Curr Biol 20(20):R888–R891

Murphy E, Eisner DA (2009) Regulation of intracellular and mitochondrial sodium in health and disease. Circ Res 104(3):292–303

Nishida H, Matsumoto A, Tomono N, Hanakai T, Harada S, Nakaya H (2010) Biochemistry and physiology of mitochondrial ion channels involved in cardioprotection. FEBS Lett 584(10):2161–2166

Szewczyk A, Jarmuszkiewicz W, Kunz WS (2009) Mitochondrial potassium channels. IUBMB Life 61(2):134–143

Szewczyk A, Kajma A, Malinska D, Wrzosek A, Bednarczyk P, Zablocka B, Dolowy K (2010) Pharmacology of mitochondrial potassium channels: dark side of the field. FEBS Lett 584(10):2063–2069

Zoratti M, Szabo I (1995) The mitochondrial permeability transition. Biochim Biophys Acta 1241(2):139–176

## Sodium and Potassium Uniporter

▸ Sodium and Potassium Transport in Mammalian Mitochondria

## Sodium Arsenate

▸ Arsenic in Pathological Conditions
▸ Arsenic, Free Radical and Oxidative Stress

## Sodium Arsenite

▸ Arsenic in Pathological Conditions
▸ Arsenic, Free Radical and Oxidative Stress

## Sodium as Primary and Secondary Coupling Ion in Bacterial Energetics

Pavel Dibrov and Judith Winogrodzki
Department of Microbiology, University of Manitoba, Winnipeg, MB, Canada

## Synonyms

*SMF*; Sodium cycle; Sodium pumps; Sodium-dependent transporters; Sodium-motive force

## Definition

*Primary coupling ion* – ion whose transmembrane electrochemical gradient could directly be generated at the expense of the external energy source and used to perform chemical (ATP synthesis), osmotic (active transport), or mechanical (motility) work. *Secondary coupling ion* – can be actively transported across the membrane only at the expense of the transmembrane electrochemical gradient of the *primary coupling ion*.

## Introduction

Perhaps the most prominent biological role of sodium-binding/transporting proteins is their involvement into the transformation of energy in biomembranes. All mitochondria and chloroplasts, as well as most bacteria on our planet, transform a plethora of available external energy sources into the transmembrane electrochemical gradient of $H^+$ ions (proton-motive force, or *PMF*). The *PMF* could be directly used to drive a variety of endergonic processes, such as ATP synthesis, active solute transport (in organelles and bacteria), motility and reverse electron transport (in bacteria). Thus, $H^+$ acts as a *primary coupling ion* in membrane energetics (Skulachev 1988). The *PMF* is generated by *primary $H^+$ pumps* ranging from relatively simple individual polypeptides (bacteriorhodopsin) to elaborate multienzyme complexes (respiratory chains) residing in energy-transforming membranes. Consumers of *PMF* vary in structure and composition according to their function. Working together, primary $H^+$ pumps and *PMF* consumers constitute a $H^+$ *cycle* that was postulated by Peter Mitchell in 1961–1969 (Skulachev 1988).

One of universal biological functions that are directly energized by the *PMF* in bacteria, mitochondria, and chloroplasts is active $Na^+$ transport. In other chapters, the role of $Na^+$ as an important secondary messenger and the role of $Na^+$ homeostasis in health and disease are considered in detail. Here, the role of $Na^+$ in energetic exchange in bacteria will be discussed.

## $Na^+$ as a Secondary Coupling Ion: $Na^+/H^+$ Antiport and SMF Consumers

$Na^+/H^+$ antiporters are found in virtually all prokaryotic organisms. Despite their great phylogenetic diversity (Saier 2000), they all are single (with the only exception of *mrp*-encoded ion exchangers), cofactor-less, membrane polypeptides that catalyze counter-transport of $1Na^+$ with $nH^+$ ($n$ is stoichiometry of antiport) thus using the *PMF* to build up a sodium-motive force, or *SMF*. When an antiporter is at equilibrium, partial components of *PMF* and *SMF* obey the equation

$$\Delta pNa = n\Delta pH + (n-1)\Delta \psi,$$

where $\Delta pNa$ and $\Delta pH$ are chemical concentration gradients of $Na^+$ and $H^+$, respectively, that are maintained on the membrane, and $\Delta\psi$ is the transmembrane difference of electric potentials. As one can see from the above equation, $Na^+/H^+$ antiporters can amass considerable energy in the form of asymmetric distribution of $Na^+$ on the membrane: magnitude of *SMF* maintained by an electrogenic ($n > 1$) antiporter operating close to its equilibrium may considerably exceed that of the primary energetic resource, *PMF*.

It should be mentioned here that in bacteria, as well as in unicellular eukaryotes such as yeast, primary $H^+$ pumps are expelling protons from the cytoplasm, so that $Na^+/H^+$ antiporters allow these protons to return back in exchange for intracellular $Na^+$ ions, thus not only converting *PMF* into *SMF* but also detoxifying the cytoplasm (Padan 2008). In alkaline environments, acidification of the cytoplasm resulting from the activity of $Na^+/H^+$ antiporters is one of major mechanisms of bacterial pH homeostasis (Slonczewski et al. 2009). Under the conditions of energetic stress, when the activity of primary proton pumps is for any reason arrested, $Na^+/H^+$ antiporters, being perfectly reversible, may operate in the reverse mode: they would expel intracellular protons in exchange for external sodium ions, therefore spending the accumulated *SMF* to prevent an immediate dissipation of the *PMF*. This function of $Na^+/H^+$ antiport as *an energetic buffer of PMF* is clearly evident in halophylic species such as *Halobacterium salinarium*, where inverted $Na^+/H^+$ antiport could use large $\Delta pNa$ existing on the membrane to feed *PMF*-driven flagellar motors for many hours after an abrupt stoppage of primary proton pumps (Skulachev 1988). Given fundamental importance of the $Na^+/H^+$ exchange for bacterial metabolism briefly outlined above, it is no wonder that $Na^+/H^+$ antiporters are ubiquitous components of the membrane proteomes in prokaryotes.

Recent success in crystallization and X-ray analysis of Ec-NhaA, archetypical enterobacterial $Na^+/H^+$ antiporter from *Escherichia coli*, yielded a number of structural features that are important for the understanding of function and regulation in this class of membrane ion transporters. Antiporters of the NhaA type contribute to salt resistance, pH, and volume homeostasis in many bacteria. Ec-NhaA catalyzes the electrogenic ($n = 2$) exchange at a very high turnover rate ($10^5$ $min^{-1}$), which is comparable to the conductance of ion channels. The activity of Ec-NhaA is highly pH-dependent: from zero at cytoplasmic pH 6.5 to the maximum at pH 8.5 (Padan 2008). Crystallographic analysis revealed that the cation-binding site of Ec-NhaA is formed by negatively charged side chains of amino acid residues belonging to a pair of non-neighboring transmembrane segments, TMS IV and XI. Both these TMSs are *discontinuous*: their α-helical halves are interrupted by the non-helical oligopeptide segments which cross in the middle of the membrane, forming a pocket for translocated cations ($Na^+$, or $Li^+$, or $2H^+$). While discontinuous TMSs *per see* have been reported in such primary ion pumps as $Ca^{2+}$-motive ATPase or $Na^+$-$K^+$ ATPase, as well as in diverse secondary $Na^+$-coupled transporters (see below), the "TMS IV-XI assembly" represents a unique structural fold characteristic for NhaA-type antiporters (Padan 2008). Two hydrophilic "funnels" formed by bended TMSs connect the cation-binding site to the aqueous compartments separated by the membrane. Funnels, however, are isolated from one another by a densely packed group of nonpolar residues so that the passive transmembrane ion leakage through the protein is prevented. Based on the Ec-NhaA structure, Dr. Padan and her colleagues suggested that the $Na^+/H^+$ exchange catalyzed by this antiporter proceeds by the *alternate-accessibility mechanism* (Padan 2008). According to this model, delicate electrostatic balance of the unoccupied cation-binding site (achieved by mutual compensation of electric charges associated with the TMS IV-XI assembly) is perturbed when a substrate cation (or protons) binds to it. This triggers a swift reorientation of the active site, which can now be unloaded at the opposite side of the membrane. Transferred alkali cation is then replaced by two $H^+$ ions (or vice versa), provoking another reorientation completing the catalytic cycle (Padan 2008). A ratio of the *SMF* relative to the *PMF* will solely determine the direction of net ion fluxes *via* Ec-NhaA. It also is worth mentioning that the translocation of small inorganic cations requires only small conformational movements within the TMS IV-XI assembly in the course of the Ec-NhaA catalytic cycle. This apparently accounts for amazingly fast turnover rates of NhaA-type antiporters (Padan 2008).

The *SMF* generated by $Na^+/H^+$ antiport can directly energize a number of $Na^+$-substrate symports, such as widespread $Na^+$-dependent uptake of melibiose, proline, and glutamate that have been documented in many bacteria, including *E. coli* and *Bacillus subtilis* which both rely on the complete $H^+$ cycle in their energetics (Häse et al. 2001; Mulkidjanian et al. 2008). $Na^+$-substrate symports are typical for microorganisms inhabiting sodium-rich environments. For example, in marine *Vibrios* the uptake of all amino acids and some other substrates is $Na^+$-dependent (Häse et al. 2001; Mulkidjanian et al. 2008).

Recent crystallographic analyses of a number of $Na^+$-substrate symporters, including glutamate transporter GltPh from *Pyrococcus horikoshii* and leucine transporter LeuT from *Aquifex aeolicus*, while revealing two individual folds distinct from that of Ec-NhaA, have at the same time highlighted some recurrent structural themes, in particular discontinuous TMSs forming the substrate-binding site that is located roughly in the middle of the lipid bilayer and is able to alternate between inward-facing and outward-facing state. This noticeable structural conservation between fylogenetically unrelated NhaA-type antiporter and $Na^+$-coupled symporters indicates the *alternating-access* as a possible universal mechanism of the secondary sodium-coupled transport (Krishnamurthy et al. 2009). Structural studies of well-diffracting crystals of LeuT yielded a wealth of structural information, including the identities and precise positions of the coordination ligands for $Na^+$ ions and co-transported substrates. It has been found that $Na^+$ ion in the primary ion-binding site, Na1, is octahedrally coordinated by six ligands. Most importantly, only five of them are oxygens provided by the residues of discontinuous TMSs, while the last one comes from the carboxylate of the co-transported substrate, leucine, which is bound nearby. This elegant structural arrangement underlies the strict coupling in binding/translocation of alkali cation and organic co-substrate (Krishnamurthy et al. 2009). Of note, there also is an additional ion-binding site, Na2, $\sim 6$ Å away from the bound leucine. The ion coordination shell in Na2 has a trigonal bipyramidal geometry. This site supposedly has lower affinity for $Na^+$ and readily releases it to the bulk water phase, thus facilitating the dissociation of co-transported amino acid and converting the thermodynamic force (*SMF*) into the osmotic work of substrate accumulation (Caplan et al. 2008; Krishnamurthy et al. 2009).

Either *PMF* or *SMF* could directly power bacterial *flagellar motors*, remarkable rotary nano-engines that traverse the entire cell envelope and transfer mechanical rotation to the long helical *flagella*, locomotory organelles propelling bacteria through liquid media (Skulachev 1988; Häse et al. 2001; Thormann and Paulick 2010). Inwardly directed $H^+$ or $Na^+$ flux creates a torque between the flagellum-bearing rotor and the cell-wall-anchored stator, formed by 11–12 individual structural units surrounding the rotor. Multimer of a rotor subunit, FliG, forms a ring that contacts stator near the cytoplasmic face of the membrane. Stator units determine the specificity of motor for coupling ion. Although the immense complexity of the complete motor so far forbade its isolation and X-ray analysis, combination of such approaches as cryo-electron tomography and extensive site-directed mutagenesis suggests that MotAB subunit complexes of $H^+$-stators in *E. coli* as well as PomAB complexes of $Na^+$- stators in *Vibrio* sp. form two selective ion channels per unit. Presumably, the C-terminal domain of highly conserved rotor subunit FliG and the loop between TMSs A2 and A3 of Mot/PomA in each stator unit interact electrostatically; passing through the motor, coupling ions induce periodic conformational distortions of the A2-A3 loop thus pushing FliG and resulting in continuous rotation of the rotor-flagellum assembly (Thormann and Paulick 2010).

Such a surprisingly "minimalistic" coupling mechanism has very serious biological consequences. In particular, it allows a given motor to function with multiply stators using different coupling ions. Indeed, chimeric motors constructed by expressing the $Na^+$-motive PomAB complexes in a MotAB-deficient mutant of *E. coli* use *SMF* instead of *PMF*, while in *V. alginolyticus* substitution of native PomAB by the heterologous $H^+$-motive MotAB converted their motors from the *SMF*-driven into *PMF*-driven ones (Häse et al. 2001; Thormann and Paulick 2010). More importantly, growing body of experimental data suggests that in vivo motors of some species can perform an "in-flight swapping" of stator units by the removal of old ones from the stator ring and replacing them by new units recruited from an intramembrane pool. "Prefabricated" stator units of the intramembrane pool are presumably inactive until properly incorporated into the stator ring; measurements of rotational dynamic indicate that the exchange of stator units may proceed in the running motor without interruption of rotation. In general, it seems that considerable number of bacterial species respond to environmental challenges that affect magnitudes of

*PMF/SMF* (pH, [Na$^+$]) or efficiency of motility (microviscosity) by "fine-tuning" of their flagellar motors through the exchange of stator power-transducing modules and, if needed, coupling ions. For example, the single flagellar motor of *Shewanella oneidensis* MR-1 uses Na$^+$-motive PomAB-based stator in Na$^+$-rich environments, but as Na$^+$ concentration decreases, gradually replaces PomAB complexes by H$^+$-motive MotABs (Thormann and Paulick 2010). However, Na$^+$-motors may co-exist with H$^+$-motors in the same membrane, as in the case of some marine *Vibrios* that constitutively express a single polar Na$^+$-motive flagellum and induce the assembly of additional *PMF*-dependent lateral flagella when swarming in viscous environments. In addition, swapping may occur between two sets of stator units using the same coupling ion apparently as a response to changing dynamic load (Thormann and Paulick 2010).

Often overlooked is biological significance of the fact that PomAB-based motors mediate huge inwardly directed Na$^+$ fluxes – providing a return route for the Na$^+$ ions expelled from the cell by primary Na$^+$ pumps and/or Na$^+$/H$^+$ antiporters and thus completing the sodium cycle in membrane energetics. The highest flagellar rotation speed ever measured is 1,700 Hz for the Na$^+$-dependent polar flagellum of *V. alginolyticus*, which, at estimated 1,000 Na$^+$ ions crossing the membrane per revolution, amounts to ~5% of cytoplasmic Na$^+$ (see (Thormann and Paulick 2010) and references therein). This enormous sodium "leakage" explains why Na$^+$-motors are as a rule found in organisms that have, in addition to the battery of Na$^+$/H$^+$ antiporters, at least one type of primary Na$^+$ pumps (Häse et al. 2001; Mulkidjanian et al. 2008).

In all the above cases Na$^+$ acted as a *secondary coupling ion*, linking membrane-stored energetic resource to a certain physiological function (osmotic or mechanical work): the Na$^+$ gradients that served as energy sources for these functions have initially been generated by primary H$^+$ pumps (*PMF*) and then converted to *SMF* by Na$^+$/H$^+$ antiporters.

## Na$^+$ as a Primary Coupling Ion: Complete Sodium Cycle

However, certain extremophiles, such as alkalophilic and thermophilic bacteria, can use Na$^+$ as *primary coupling ion* in a Na$^+$ cycle instead of, or in addition to, the H$^+$ cycle (Häse et al. 2001; Mulkidjanian et al. 2008). Similarly to H$^+$ cycle, a complete Na$^+$ cycle should include not only a number of Na$^+$-coupled transporters, Na$^+$/H$^+$ antiport machinery, and an *SMF*-dependent flagellar motor, but also a primary Na$^+$ pump(s) that directly couple(s) a chemical reaction to Na$^+$ translocation, as well as – last but not least – the Na$^+$-transporting ATP synthase that can use the generated *SMF* for the ATP synthesis under physiological conditions. Of course, an actual repertoire of primary sodium pumps and *SMF* consumers in any given species reflects both its evolutionary history and current environmental challenges shaping the vector of the selective pressure, but a minimal set, defining Na$^+$ as a primary coupling ion, must include a functional Na$^+$-ATP synthase.

There are at least five major types of primary Na$^+$ pumps documented to date (Dimroth 1997; Mulkidjanian et al. 2008; von Ballmoos et al. 2009): (1) biotin-dependent Na$^+$-motive decarboxylases that couple export of Na$^+$ ions to the decarboxylation of oxaloacetate, malonate, methylmalonyl-CoA, or glutaconyl-CoA; (2) respiratory Na$^+$-motive NADH: ubiquinone oxidoreductases (NQRs) and closely related Na$^+$-translocating ferredoxin:NAD oxidoreductase (RNF); (3) Na$^+$-motive N$^5$-methyltetrahydromethanopterin: coenzyme M methyltransferase from methanogenic archaea; (4) Na$^+$-pyrophosphatases (Na$^+$-PPases), coupling hydrolysis of pyrophosphate to the Na$^+$ transfer across the membrane; and (5) Na$^+$-transporting ATPases belonging to the ABC (ATP-binding cassette)-type or P-type ATPase families. Despite the significant progress in understanding of the mechanisms of primary Na$^+$ transport (Mulkidjanian et al. 2008), primary Na$^+$ pumps still await their detailed crystallographic characterization.

Fortunately, structures of both H$^+$-motive and Na$^+$-motive ATP synthases of F type (also known as F$_1$F$_0$-type) are rather well documented, allowing the comparative analysis at relatively high resolution (von Ballmoos et al. 2009). Similar to bacterial flagellar motors, F-type ATP synthases are complex rotary nano-machines comprised of a rotor and stator modules, F$_0$ and F$_1$. They transform chemiosmotic driving force (*PMF* or *SMF*) into the $\Delta$G of phosphorylation of ADP, yielding ATP, predominant "soluble" form of cellular energy. (Working in the opposite directions, these enzymes can hydrolyze ATP to build up *PMF* or *SMF*.) The synthesis of ATP is catalyzed by the hydrphilic $F_1$ portion of the enzyme protruding into the cytoplasm. $F_1$ is an $\alpha_3\beta_3$ hexamer that has a threefold symmetry. It carries three nucleotide-binding sites and through the $\delta$ subunit is attached to a "stalk," $b_2$-dimer, which is in turn anchored to the massive membrane-embedded subunit a. The $ab_2$ assembly together with attached $\alpha_3\beta_3$ hexamer comprises the stator part of ATP synthase. The membrane module of prokaryotic enzyme, $F_0$, contains an oligomeric ring of c subunits, $c_{10-15}$, in addition to the $ab_2$ complex that flanks it laterally. The c ring is attached by $\varepsilon$ subunit to $\gamma$ subunit, a long eccentric shaft that traverses a central cavity of $\alpha_3\beta_3$ hexamer and can rotate in it without friction. The $\gamma\varepsilon c_{10-15}$ assembly is the rotor part of the enzyme. Coupling ions cross the membrane through subunit a and its interface with c-ring, generating the torque and forcing c-ring together with the attached $\gamma$ to rotate in steps. Rotating inside $F_1$, $\gamma$ subunit induces synchronous changes in nucleotide-binding affinity of its catalytic sites, thus facilitating ATP synthesis (von Ballmoos et al. 2009 and references therein).

The torque-generating mechanism in $F_0$ and especially its transmission to the catalytic module through the rotation of $\gamma$ are rather similar in $Na^+$- and $H^+$-motive enzymes. In either case, coupling ion must be temporarily bound by the c subunit; an elementary rotational movement of c-ring is needed to complete the ion translocation across the membrane (see (von Ballmoos et al. 2009) for detailed discussion). However, structural requirements for the coordination of coupling ion in these two cases are expected to differ. While it usually takes six protein ligands to coordinate $Na^+$ ion, a single side-chain carboxylate would be sufficient to accommodate tranlsocated proton. Indeed, in widely popular models of the $H^+$-motive F-type ATPases, the conserved negative residue of c subunit (Asp61 in *E. coli* enzyme) that is located in the middle of lipid bilayer and undergoes reversible protonation, is implicated in the mechanism of $H^+$ translocation (see Mulkidjanian et al. 2008 and references therein).

Structural data confirm that in the $Na^+$-motive enzyme from *Ilyobacter tartaricus*, $Na^+$ is directly coordinated by four protein ligands: side chain oxygens of Gln32 and Glu65 of one c subunit, the hydroxyl oxygen of Ser66, and the backbone carbonyl oxygen of Val63 of the neighboring c subunit. Further, hydrogen bonds between the side chain oxygen of Glu65 and the $NH_2$ group of Gln32 as well as between the other oxygen of Glu65 and OH groups of Ser66 and Tyr70 contribute to the formation of the coordination sphere by keeping Glu65 deprotonated at neutral pH so that it is available as a ligand for $Na^+$ binding (von Ballmoos et al. 2009). In the absence of $Na^+$, the *I. tartaricus* enzyme translocates protons; probing of the protonation state of c-ring with the specific inhibitor N,N'-dicyclohexylcarbodiimide (DCCD) at different pHs indicated that the $H^+$ transfer here, as well as in the $H^+$-ATP synthase from *H. salinarium*, occurs via protonation-deprotonation of a carboxylate within the ion-binding site of c subunit. However, the same approach applied to the $H^+$-ATP synthase from *E. coli* and organellar eukaryotic enzymes indicates the coordination of a hydronium ion, $H_3O$ (requiring three ligands for its efficient coordination) rather than protonation of a single carboxylate as a probable mechanism of $H^+$ translocation (von Ballmoos et al. 2009). Unfortunately, a crucial fragment of the puzzle – structural information about $H^+$-binding sites in the c ring of a *bona fide* $H^+$-translocating enzyme – is still missing.

Of note, V-type ATPases, being phylogenetically quite distinct from F-type enzymes, display not only a striking similarity in overall architecture but, most curiously, in the geometry of $Na^+$-coordinating spheres associated with c/K subunits comprising the membrane-imbedded rotor of $Na^+$-motive V-type enzymes (Mulkidjanian et al. 2008). This might have unexpected evolutionary implications (see below).

## Evolution of $Na^+$ and $H^+$ Energetics: $Na^+$ as a Primordial Coupling Ion

The ability to use the *SMF* instead of *PMF* for all major biochemical functions, including ATP synthesis, allows microorganisms with $Na^+$ cycle to conquer ecological niches where it is impossible to maintain sufficiently high *PMF* on the membrane for various reasons (alkaline pH, high temperatures). This is one of the reasons why initially $Na^+$ cycle has been considered as a relatively late evolutionary acquisition (Skulachev 1988). However, metagenomic analysis in conjunction with recently obtained structural data suggests that the opposite is probably correct.

Several lines of reasoning summarized in (Mulkidjanian et al. 2008) support the notion about $Na^+$ cycle as the evolutionary predecessor of $H^+$ cycle: First, the nearly identical $Na^+$-binding sites in F- type and V-type ATPases, as well as multiple appearance of $Na^+$-motive V/F-ATPases in distinct branches of the phylogenetic tree strongly suggest that their common ancestorial enzyme translocated $Na^+$ rather than $H^+$; further divergent evolution apparently resulted in ligand loss in several clades. These unrelated simplification events should invariably lead to the loss of the ability to coordinate $Na^+$, thus yielding a plethora of enzymes selective for $H^+$. An alternative scenario, i.e., independent gain of identical sets of ligands, seems implausible. Further, considering co-evolution of ion-transporting membrane proteins in conjunction with the evolution of biological membranes themselves, one should take into account that, being structurally less demanding, sodium-impermeable (but proton-leaky) membranes probably preceded proton-tight membranes (Skulachev 1988). Of note, contemporary membranes are still much less permeable for $Na^+$ than for $H^+$, especially at higher temperatures; the difference reaches several orders of magnitude (Mulkidjanian et al. 2008). On the other hand, both $Na^+$-decarboxylases and $Na^+$-PPases are present in bacteria as well as in archaea, apparently antedating the divergence of the three major domains of life. In addition, a number of modern $Na^+$-substrate symporters in organisms with $H^+$ cycle do not have "simplified" $H^+$-dependent counterparts. Taken together, these three observations point toward ancestral status of the $Na^+$-based membrane energetics.

Within this conceptual framework, transition from $Na^+$ to $H^+$ as a coupling ion brought the benefit of direct mechanistic coupling of scalar redox reactions to vectorial translocation of proton across the membrane, allowing the organization of primary redox $H^+$ pumps with very different mid-point potentials into electron-transfer chains covering the redox span of $\sim$1.2 eV from organic substrates to oxygen (Mulkidjanian et al. 2008). This thermodynamic gain was accompanied by a radical simplification of ion-binding sites with three ($H_3O^+$) or even one ($H^+$) coordination ligands instead of six ($Na^+$). Robust proton pumps and efficient electron-transfer chains in combination with progressive $H^+$-tight membranes eventually made the $H^+$ cycle a predominant type of energy transformation in microbial membranes (and, ultimately, in mitochondria and chloroplasts). The $Na^+$ cycle remained operative only in niches where the high-potential electron acceptors, such as oxygen, nitrate, sulfate, or sulfite are not available and the huge thermodynamic advantage of $H^+$-coupled electron transfer could not be realized. Therefore, bacterial and archaeal alkaliphiles and hyperthermophiles that rely exclusively upon fermentations of different kinds retained the $Na^+$ cycle, while respiring species eventually switched to more progressive $H^+$ circulation despite of the necessity to overcome massive proton leakage. Another case of the operational $Na^+$ cycle is the plasma membrane of animal cells, where it plays regulatory but not energetic role: the plasmalemmal $Na^+/K^+$ ATPase energizes $Na^+/H^+$ antiporters that operate "backward," allowing $Na^+$ to enter the cytoplasm in exchange for internal protons thus preventing over-acidification at active glycolysis.

## $Na^+$ Cycle and Microbial Pathogens

Disturbingly, cross-genome comparisons suggest that the complete $Na^+$ cycle or at least individual primary $Na^+$ pumps (together with *SMF* consumers and $H^+$-motive ATPsynthase) may be common among human and animal pathogens (Häse et al. 2001). One could try to extend the above evolutionary logic to rationalize this finding. Indeed, innards of animal body are typically anaerobic and poor in alternative electron donors, but often sodium-rich (blood). On the other hand, certain pathogens (such as *Vibrio cholerae*) during the free-living stages of their life cycles may confront conditions where *PMF* on the membrane is severely diminished, forcing them to use $Na^+$ as a primary coupling ion (see (Mulkidjanian et al. 2008) and references therein). But whatever the concrete reasons might be in each specific case, the very fact that $Na^+$ pumping systems of different types are "overrepresented" in microbial pathogens makes these systems an attractive target for prospective development of novel anti-microbial remedies.

## Cross-References

▶ Sodium-Binding Site Types in Proteins
▶ Sodium-Coupled Secondary Transporters, Structure and Function
▶ Sodium-Hydrogen Exchangers, Structure and Function in Human Health and Disease

## References

Caplan DA, Subbotina JO, Noskov SY (2008) Molecular mechanism of ion-ion and ion-substrate coupling in the $Na^+$-dependent leucine transporter LeuT. Biophys J 95:4613–4621

Dimroth P (1997) Primary sodium ion translocating enzymes. Biochim Biophys Acta 1318:11–51

Häse CC, Fedorova N, Galperin MY, Dibrov P (2001) Sodium ion cycle in bacterial pathogens: evidence from cross-genome comparisons. Microbiol Mol Biol Rev 65:353–370

Krishnamurthy H, Chayne L, Piscitelli CL, Gouaux E (2009) Unlocking the molecular secrets of sodium-coupled transporters. Nature 459:347–355

Mulkidjanian AY, Dibrov P, Galperin MY (2008) The past and present of sodium energetics: may the sodium-motive force be with you. Biochim Biophys Acta 1777:985–992

Padan E (2008) The enlightening encounter between structure and function in the NhaA $Na^+/H^+$ antiporter. Trends Biochem Sci 33:435–443

Saier MH Jr (2000) A functional-phylogenetic classification system for transmembrane solute transporters. Microbiol Mol Biol Rev 64:354–411

Skulachev VP (1988) Membrane bioenergetics. Springer, Berlin

Slonczewski JL, Fujisawa M, Dopson M, Krulwich TA (2009) Cytoplasmic pH measurement and homeostasis in bacteria and archaea. Adv Microb Physiol 55:1–80

Thormann KM, Paulick A (2010) Tuning the flagellar motor. Microbiology 156:1275–1283

von Ballmoos C, Wiedenmann A, Dimroth P (2009) Essentials of ATP synthesis by $F_1F_0$ ATP synthases. Annu Rev Biochem 78:649–672

## Sodium Channel Agonists

▶ Sodium Channel Blockers and Activators

## Sodium Channel Antagonists

▶ Sodium Channel Blockers and Activators

## Sodium Channel Blockers

▶ Sodium Channel Blockers and Activators

## Sodium Channel Blockers and Activators

Boris S. Zhorov
Department of Biochemistry and Biomedical Sciences, McMaster University, Hamilton, ON, Canada
Sechenov Institute of Evolutionary Physiology and Biochemistry, Russian Academy of Sciences, St. Petersburg, Russia

### Synonyms

Antiarrhythmics; Local anesthetics; Sodium channel agonists; Sodium channel antagonists; Sodium channel blockers; Sodium channel toxins

### Definition

The term "sodium channel blockers" covers a wide range of naturally occurring and synthetic compounds that block the ion permeation through sodium-selective ion channels, large transmembrane proteins that control rapid membrane depolarization in electrically excitable cells. Due to their key roles in the cell physiology, sodium channels are targets for deadly toxins, which are synthesized by poisonous organisms as the attack or defense weapons, and for many therapeutic drugs. The latter are used to manage pain and treat various health disorders, including cardiac arrhythmias, heart attack, stroke, migraine, and epilepsy. Most of these compounds block the sodium channels. A broader term "sodium channel ligands" covers sodium channel blockers and compounds that modify gating of sodium channels. The gating modifiers include scorpion toxins and activators such as batrachotoxin, aconitine, and pyrethroid insecticides, which increase the ion permeation though sodium channels.

Ion channels control the ion flow across cell membranes. The human genome encodes over 400 ion channels. Among them are nine voltage-gated sodium channels (designated $Na_v1.1$ $Na_v1.9$), which play key physiological roles (Hille 2001; Yu and Catterall 2003). These channels open from the resting, ion-impermeable state in response to membrane depolarization, then transit to ion-impermeable inactivated state(s), and return to the resting state upon membrane repolarization. Channels $Na_v1.1$, $Na_v1.2$, $Na_v1.3$ are predominantly expressed in the central nervous system. Channels $Na_v1.4$ and $Na_v1.5$ are expressed, respectively, in the skeletal and cardiac muscle cells. Channels $Na_v1.6$ are expressed in both the central and peripheral nervous systems. Nociception-related sodium channels $Na_v1.7$, $Na_v1.8$, and $Na_v1.9$, which are expressed in peripheral neurons, are targets for pain management. Thousands of small-molecule ligands, which modulate sodium channels, have been synthesized with the aim to develop drugs with improved potency and selectivity toward specific channel subtypes. Practically all drugs, which target sodium channels, have been developed using empirical try-and-error approach.

## Potassium Channels as Templates to Model Sodium Channels

In the absence of x-ray structures of sodium channels, atomistic mechanisms of their modulation by ligands are poorly understood. Homology models of sodium channels, which are based on x-ray structures of potassium channels, are used to rationalize experimental data and design new experiments. The x-structure of a prototype voltage-gated potassium channel Kv1.2 in the open state shows four subunits circumferentially arranged around the pore axis (Fig. 1). Each subunit contains a voltage-sensing domain S1-S4, a linker-helix L45, an outer helix S5, an inner helix S6, and an extracellular membrane-diving P-loop between helices S5 and S6. The P-loop consists of a P-helix, a P-turn, an ascending limb, and an extracellular linker to helix S6. Helices S6s and P-turns line the inner pore. Four ascending limbs expose backbone carbonyls of the signature-sequence motif TVGYG into the outer pore forming the selectivity filter that distinguishes potassium ions from other monovalent and divalent cations. Potassium channels have several potassium-binding sites along the ion permeation pathway. Four sites are located in the selectivity filter where the energetically unfavorable dehydration of potassium ions is compensated by their attraction to the backbone carbonyls. Another site is seen in the center of a large water-filled cavity within the inner pore where the hydrated potassium ion is stabilized by macrodipoles from the four P-helices. In the x-ray structure of a closed bacterial potassium channel KcsA with a tetraalkylammonium blocker, the latter binds at the focus of P-helices.

Sodium channels have evolved from calcium channels and the pore domains of these channels are thought to share similar folding (Hille 2001). Sodium and calcium channels fold from a single-polypeptide chain of four homologous repeats that are analogous to potassium channel subunits (Fig. 1). Initial interpretation of experimental data on accessibility of engineered cysteines to sulfhydryl reagents (MTSET) questioned similarity of the inner pore region between potassium and calcium channels (Zhen et al. 2005) and therefore reliability of homology models of calcium and sodium channels, which are based on the x-ray structures of potassium channels. However, the experimental data by (Zhen et al. 2005) were explained in a Kv1.2-based homology model of the calcium channel in which multiple conformations of long side chains of MTSET-substitutes engineered cysteines were analyzed (Bruhova and Zhorov 2010). The results of this analysis imply that helices S5 and S6 in calcium and sodium channels fold like in potassium channels.

## Marine Toxins Block the Outer Pore of Sodium Channels

The ion selectivity of sodium channels is governed by side chains of four residues, one residue from each ascending limb. These are Asp, Glu, Lys, and Ala residues that form the so-called DEKA ring. Glutamate and aspartate residues three to four positions downstream from the DEKA residues form the so-called outer ring EEDD. Marine toxins tetrodotoxin and saxitoxin are the hallmark blockers of the outer vestibule of sodium channels. A large body of mutational, electrophysiological, and structure-activity data on the action of these toxins is available, including effects of mutations in the outer pore on the potency of marine toxins. However, interpretation of these data in structural terms is difficult. Indeed, it is generally accepted that the outer-pore region of sodium channels, which includes the DEKA and EEDD rings, folds not like in potassium channels. Two competitive homology models of the sodium channel P-loop region have been proposed (Lipkind and Fozzard 2000; Tikhonov and Zhorov 2005a). These models have different backbone conformations, but both of them explain key experimental data on interactions of the DEKA and EEDD rings with tetrodotoxin and saxitoxin. The possibility to explain the same set of experimental data with

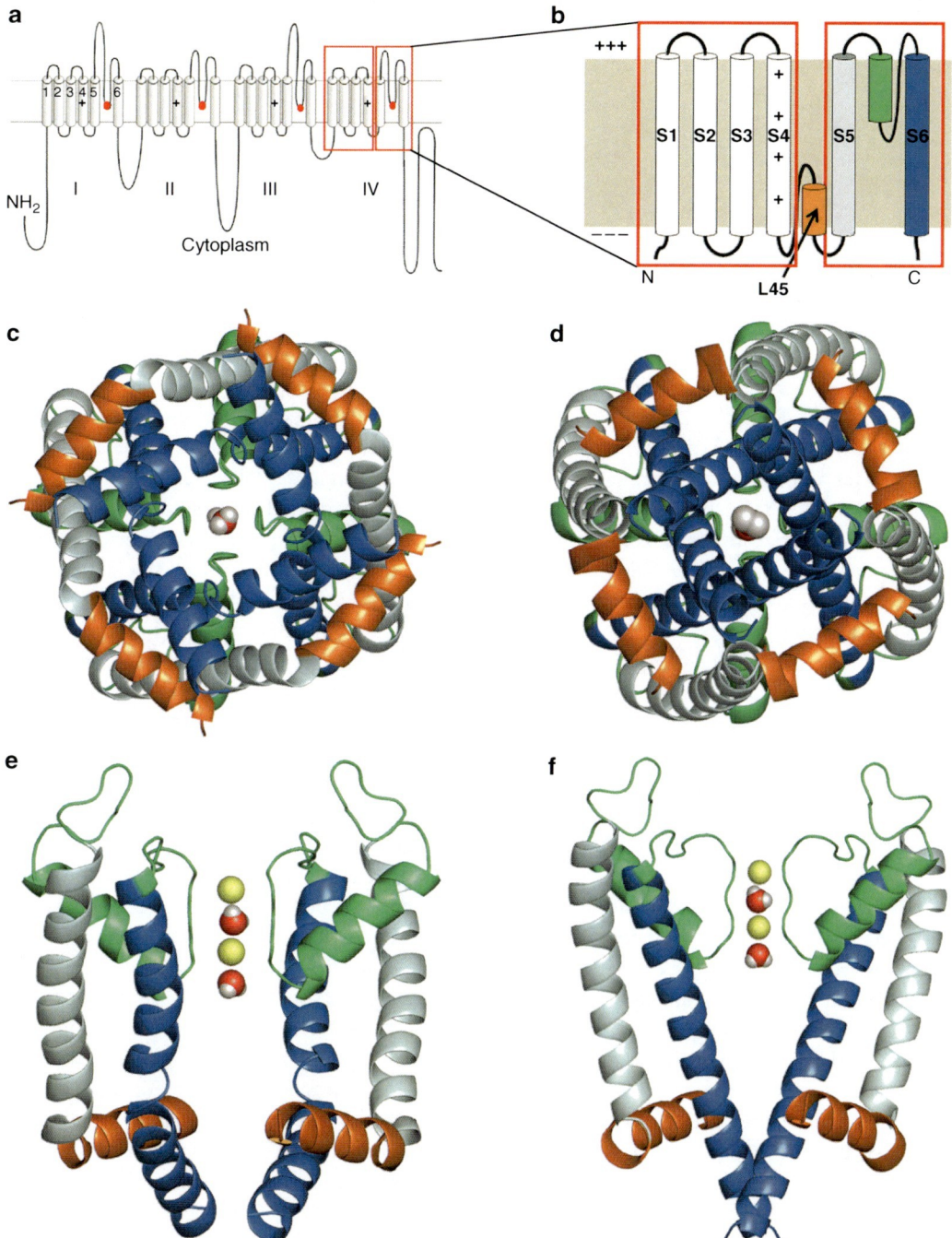

**Sodium Channel Blockers and Activators, Fig. 1** *Voltage-gated ion channels*. (**a**) Calcium and sodium channels are assembled of four homologous repeats. *Red dots* indicate approximate positions of the selectivity-filter residues. (**b**) The enlarged view of a single repeat, which is equivalent of a potassium channel subunit, shows the voltage-sensing domain S1-S4 (*white*), the linker-helix L45 (*orange*), the outer helix S5 (*gray*), the pore helix P (*green*), and the inner helix S6 (*blue*). Helices L45, S5, P, and S6 contribute one fourth to the pore domain. (**c**) Cytoplasmic view of the x-ray structure of the open Kv1.2 channel. Only the pore domain is shown for clarity. (**d**) Cytoplasmic view of a closed-state Kv1.2 model. (**e** and **f**) Side views of (**c**) and (**d**) in which only two subunits are shown for clarity. The selectivity filter is occupied by two potassium ions and two water molecules

**Sodium Channel Blockers and Activators, Fig. 2** *Some ligands* of sodium channels, which are mentioned in the text

Lidocaine   Benzocaine   Batrachotoxin

two different models is due to the flexibility of long side chains in the DEKA and EEDD rings, which can adjust around rigid toxins and compensate for dissimilarity of the backbone folding. The predictive potential of the models can be characterized by their ability to explain experimental data, which have not been used at the stage of the model building. The backbone conformation of model (Lipkind and Fozzard 2000) has been modified to explain experimental data on the action of mu-conotoxin GIIIA (Choudhary et al. 2007). No backbone modification of model (Tikhonov and Zhorov 2005a) was necessary to explain cross-linking experiments involving the outer-pore residues (Tikhonov and Zhorov 2007) and adaptation of garter snakes to tetrodotoxic prey (Tikhonov and Zhorov 2011).

*Cationic blockers of the inner pore of sodium channels.* The inner-pore region of sodium channels harbors binding sites for many small-molecule blockers of medical importance. Effect of some ligands increases upon repetitive membrane depolarizations. Such use-dependent drugs are desirable to treat pain and hyperexcitability conditions including the heart attack, stroke, arrhythmias, and neurological disorders. For most of ligands, the binding sites and binding modes (the ligand orientation in the protein) are unknown. A local anesthetic and antiarrhythmic lidocaine is a well-studied blocker of the inner pore (Fig. 2). Models of lidocaine binding have been elaborated to explain mutational, electrophysiological, and structure-activity data (Tikhonov and Zhorov 2007; Lipkind and Fozzard 2005). In the open-channel model shown in Fig. 3a, lidocaine interacts with two aromatic lidocaine-sensing residues, which were identified in mutational studies, whereas the ammonium group at the focus of P-helices approaches the selectivity filter and would destabilize binding of the sodium ion through electrostatic repulsion (Tikhonov et al. 2006). This destabilization of the permeant ion in the selectivity filter may explain action of some other small-molecule cationic drugs that block the inner pore of sodium channels. However, further experimental and theoretical studies are necessary to understand atomistic mechanisms of the sodium channel block by more complex cationic blockers of cardiac sodium channels such as ranolazine and mibefradil.

*Competition of metal ions with organic-cation ligands.* Metal ions in the pore of cationic channels would compete with the pore-targeting cationic ligands and thus destabilize respective ligand-channel complexes. Therefore, the deficiency of permeant cations in the pore should increase binding of the organic-cation ligands. This may explain why many cationic ligands exhibit higher affinity to non-permeating slow inactivated states than to the closed or open states (Tikhonov and Zhorov 2007). Action of some ligands is affected by mutations of residues, which are far from the ligand-binding sites. For example, ingress/egress of cationic local anesthetics into/from the closed sodium channel is affected by mutations in the outer pore. Furthermore, the binding of tetrodotoxin and mu-conotoxin in the outer pore also affects the access of local anesthetics into the closed channel. Computer modeling suggests that local anesthetics are unlikely to enter the inner pore of the closed channel through the selectivity filter or the closed activation gate, whereas a lipid-exposed domain interface is wide enough to let through a local anesthetic molecule (Bruhova et al. 2008). However, this access rout seems inconsistent with the above-mentioned experimental data. To explain this controversy, a model was proposed in which the

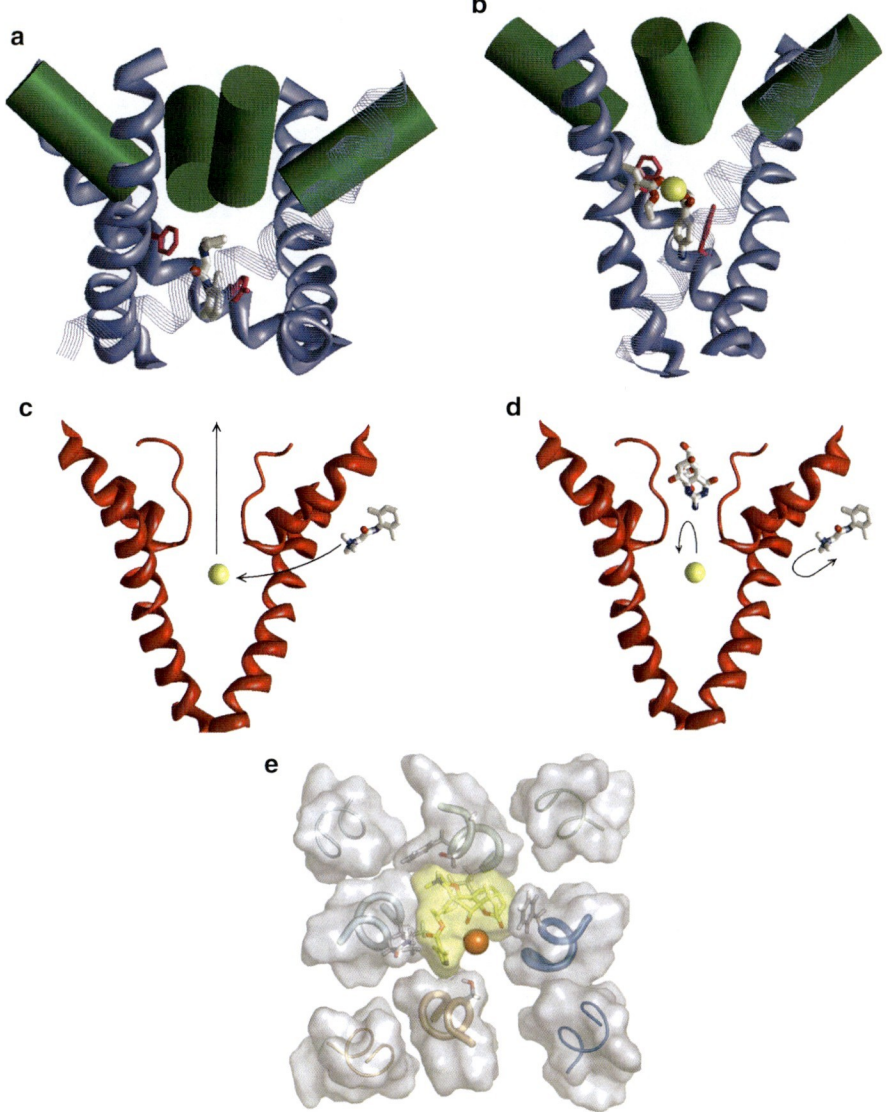

**Sodium Channel Blockers and Activators, Fig. 3** *Models of sodium channel Nav1.4 with ligands and sodium ions.* (**a**) A local anesthetic lidocaine is a hydrophobic cation, which blocks the channel in the 1:1 stoichiometry by an ammonium group stabilized at the focus of P-helices (*green cylinders*). (**b**) A cationophilic local anesthetic benzocaine lacks an ammonium group. Two benzocaine molecules jointly chelate a sodium ion at the focus of P-helices, thus forming a hydrophobic tripartite cation, which blocks the channel. (**c** and **d**) A scheme of coupled movement of a sodium ion and a permanently charged local anesthetic in the closed channel. (**c**) The local anesthetic molecule, which enters the central cavity through a membrane pathway, repels the resident sodium ion that moves away via the unblocked outer pore and releases the nucleophilic central cavity for the ligand binding. (**d**) When the outer pore is blocked by tetrodotoxin, the sodium ion in the central cavity lacks a hydrophilic escape route. It remains in the central cavity and prevents the local anesthetic access. (**e**) Extracellular and side views of the open channel with an agonist batrachotoxin (*yellow carbons*). Batrachotoxin-sensing residues are shown with *gray carbons*. For clarity, only parts of transmembrane helices around batrachotoxin are shown. The van der Waals shape of batrachotoxin approximately matches that of the central cavity. A sodium ion binds to the hydrophilic inner surface of the agonist and can permeate though it. Figure 3a, b (Tikhonov et al. 2006) was originally published in *FEBS Letters*. Figure 3c, d (Bruhova et al. 2008) is reproduced with the permission of *the American Society for Pharmacology and Experimental Therapeutics*. Figure 3e (Du et al. 2011a) was originally published in the *Journal of Biological Chemistry* © the American Society for Biochemistry and Molecular Biology focus of P-helixes in the central cavity is occupied by a sodium ion in both the open and closed states of the ligand-free channel. A local anesthetic molecule entering the central cavity of the closed channel should expel the resident sodium ion (Fig. 3c). Furthermore, since a vacant state of the cation-binding site is energetically unfavorable, the egress of the local anesthetic molecule from the central cavity should be coupled with the cavity reoccupation by a sodium ion (Bruhova et al. 2008). This model also explains the effects of tetrodotoxin (Fig. 3d) and mu-conotoxin on the close-channel block by local anesthetics.

## Cationophilic Blockers of Sodium Channels

Intriguingly, some blockers of sodium channels are cationophilic molecules, which lack cationic groups. Examples include lacosamide, which blocks sodium channels involved in nociception, and a local anesthetic benzocaine (Fig. 2). Importantly, while cationic ligands block the channel with the 1:1 stoichiometry, binding of at least two benzocaine molecules is necessary to block the channel. To explain these data, a model was proposed in which two benzocaine molecules bind a sodium ion to form a cationic tripartite blocking particle (Tikhonov et al. 2006) with the sodium ion located at the focus of P-helices, the position where the ammonium group of lidocaine binds (Fig. 3b). Stability constants of alkaline and alkaline earth metals with organic ligands have long been available, but physiological importance of such interactions is underappreciated due to low concentration of metal-ligands complexes in the bulk solvent. However, local concentrations of metal ions and ligands inside ion channels may be higher than in the bulk solvent. A ternary ligand-metal-channel complex can be more stable than respective binary metal-ligand and metal-channel complexes due to two reasons. First, the ligand and the channel would provide more ion-coordinating groups than either the ligand or the channel. Second, ligand-channel interactions immobilize the ligand inside the channel.

## Steroidal Toxins as Sodium Channel Activators

Batrachotoxin, veratridine, aconitine, and grayanotoxin are steroidal sodium channel activators, which cause sodium channels to activate easier and stay open longer than normal channels. The activators, which are lipid-soluble bulky molecules, are traditionally believed to bind at the channel-lipid interface and increase the ion permeation by allosteric mechanisms (Hille 2001). More recently, batrachotoxin-sensing residues were found in all the four pore-lining helices S6s. These data are inconsistent with the concept that batrachotoxin binds at the lipid-protein interface. In a modeling study, batrachotoxin, veratridine, and aconitine were hypothesized to bind in the inner pore and preclude the activation-gate closure (Tikhonov and Zhorov 2005b). In a recent model (Du et al. 2011a), batrachotoxin is proposed to adopt a horseshoe conformation with the horseshoe plane normal to the pore axis (Fig. 3e). Oxygen atoms at the horseshoe inner surface constitute a transient binding site for permeation cations, while the bulky batrachotoxin molecule would resist the pore closure, thus causing persistent channel activation. This model of "stent in the channel" explains why batrachotoxin-modified channels have different ion selectivities and longer open states than normal channels.

Effects of steroidal toxins, which target sodium channels, are sensitive to small chemical modifications of the ligand and channel. Thus, a point mutation of a pore-facing hydrophilic residue switches the channel-activating action of batrachotoxin to the channel-blocking action (Wang et al. 2007a). Bulleyaconitine A, an analogue of the activator aconitine, is used in China for the treatment of chronic pain and rheumatoid arthritis. This naturally occurring compound blocks neuronal sodium channels, which are crucial for nociception (Wang et al. 2007b).

## Insecticides

Pyrethroid insecticides, which are exemplified by deltamethrin, are lipid-soluble sodium channel activators. Results of intensive experimental studies of pyrethroid insecticides are summarized in a model where the ligand binds at the lipid-exposed cavity formed by the short intracellular linker-helix IIS4-S5 and transmembrane helices IIS5 and IIIS6 (O'Reilly et al. 2006). A recent study of a long flexible insecticide BTG 502 predicts that this partial antagonist wraps around helix IIIS6, exposes its hydrophobic linker to the inner pore to partially preclude the ion permeation (Du et al. 2011b). The binding site for BTG 502 appears to overlap with the binding sites of batrachotoxin and pyrethroids.

## Conclusion

The ion permeation through sodium channels is modulated by many naturally occurring and synthetic molecules, which have dramatically different chemical structures and mechanisms of action. This entry describes only few sodium channel blockers and activators for which 3D models of interaction with sodium channels have been proposed, in particular models that propose direct interaction of sodium channel ligands with metal ions. In the absence of experimental structures of ligand-metal-channel complexes, these models can be used to design new experiments.

**Acknowledgments** This work was supported by grant GRPIN/238773-2009 from the Natural Sciences and Engineering Council of Canada.

## Cross-References

▶ Bacterial Calcium Binding Proteins
▶ Sodium Channels, Voltage-Gated

## References

Bruhova I, Zhorov BS (2010) A homology model of the pore domain of a voltage-gated calcium channel is consistent with available SCAM data. J Gen Physiol 135:261–274
Bruhova I, Tikhonov DB, Zhorov BS (2008) Access and binding of local anesthetics in the closed sodium channel. Mol Pharmacol 74:1033–1045
Choudhary G, Aliste MP, Tieleman DP, French RJ, Dudley SC Jr (2007) Docking of mu-conotoxin GIIIA in the sodium channel outer vestibule. Channels 1:344–352
Du Y, Garden DP, Khambay B, Zhorov B, Dong K (2011a) Batrachotoxin, pyrethroids and BTG 502 share overlapping binding sites on insect sodium channels. Mol Pharmacol 80:426–433
Du Y, Garden D, Wang L, Zhorov BS, Dong K (2011b) Identification of new batrachotoxin-sensing residues in segment IIIS6 of sodium channel. J Biol Chem 286:13151–13160
Hille B (2001) Ion channels of excitable membranes, 3rd edn. Sinauer Associates, Sunderland, p 814
Lipkind GM, Fozzard HA (2000) KcsA crystal structure as framework for a molecular model of the Na(+) channel pore. Biochemistry 39:8161–8170
Lipkind GM, Fozzard HA (2005) Molecular modeling of local anesthetic drug binding by voltage-gated sodium channels. Mol Pharmacol 68:1611–1622
O'Reilly AO, Khambay BP, Williamson MS, Field LM, Wallace BA, Davies TG (2006) Modeling insecticide-binding sites in the voltage-gated sodium channel. Biochem J 396:255–263
Payandeh J, Scheuer T, Zheng N, Catterall WA (2011) The crystal structure of a voltage-gated sodium channel. Nature 475:353–358
Tikhonov DB, Zhorov BS (2005a) Modeling P-loops domain of sodium channel: homology with potassium channels and interaction with ligands. Biophys J 88:184–197
Tikhonov DB, Zhorov BS (2005b) Sodium channel activators: model of binding inside the pore and a possible mechanism of action. FEBS Lett 579:4207–4212
Tikhonov DB, Zhorov BS (2007) Sodium channels: ionic model of slow inactivation and state-dependent drug binding. Biophys J 93:1557–1570
Tikhonov DB, Zhorov BS (2011) Possible roles of exceptionally conserved residues around the selectivity filters of sodium and calcium channels. J Biol Chem 286:2998–3006
Tikhonov DB, Bruhova I, Zhorov BS (2006) Atomic determinants of state-dependent block of sodium channels by charged local anesthetics and benzocaine. FEBS Lett 580:6027–6032
Wang CF, Gerner P, Wang SY, Wang GK (2007a) Bulleyaconitine A isolated from aconitum plant displays long-acting local anesthetic properties in vitro and in vivo. Anesthesiology 107:82–90
Wang SY, Tikhonov DB, Mitchell J, Zhorov BS, Wang GK (2007b) Irreversible block of cardiac mutant Na+ channels by batrachotoxin. Channels 1:179–188
Yu FH, Catterall WA (2003) Overview of the voltage-gated sodium channel family. Genome Biol 4:207
Zhen XG, Xie C, Fitzmaurice A, Schoonover CE, Orenstein ET, Yang J (2005) Functional architecture of the inner pore of a voltage-gated $Ca^{2+}$ channel. J Gen Physiol 126:193–204

# Sodium Channel Toxins

▶ Sodium Channel Blockers and Activators

# Sodium Channels, Voltage-Gated

Tsai-Tien Tseng
Center for Cancer Research and Therapeutic Development, Clark Atlanta University, Atlanta, GA, USA

## Synonyms

SCN; VGSC

## Definition

Voltage-gated sodium channels refer to members of the voltage-gated ion channel (VIC) superfamily, which are responsible for energy-independent transport of sodium ions. The opening of these channels is regulated by voltage, ligand, or receptor-binding. Voltage-gated sodium channels participate in transmission and propagation of action potentials in excitable tissues.

## Basic Biological Roles of Voltage-Gated Ion Channels

Transmembrane transport is essential for all aspects of life, including uptake and excretion of nutrients, nucleic acids, ions, and drugs. Excitability of tissues accounts for the transmission of chemical and electrical signals in a wide range of organisms. Over 50 years ago, the participation of plasma membrane ion channels in cellular electrogenesis and excitability was described by the Hodgkin-Huxley model (1952). Such transmission and propagation of signals rely on the voltage-gated ion channels (VICs), which are known to transport potassium, sodium, and calcium ions. These channels are generally associated with one of the following states: deactivated (closed), activated (open), and inactivated (closed). Most importantly, these channels are responsible for the generation of action potentials in nerve cells. They are represented in the Hodgkin-Huxley model to provide the mathematical foundation on how action potentials in neurons initiated and propagated.

The principal pore-forming unit, also known as α subunit, of voltage-gated sodium channel does not function alone in the lipid plasma membrane. Fine-tuning of the kinetics and biogenesis is carried out by auxiliary subunits, which form a complex with the principal unit in the membrane. The associations with other proteins have also been found to participate in cytoskeleton anchoring, cell adhesion, and signal transduction. In addition to cellular excitability, VICs play important roles in motility, cellular homeostasis in various organisms. Subsequent studies have implicated ion channels in the contribution of nearly all basic cellular behaviors, including tissue homeostasis such as apoptosis, differentiation, and proliferation. The mechanism which ion channels contribute to the above processes is by regulating cell volume, maintaining membrane potential, and providing influx of essential signaling ions. Several aspects of voltage-gated sodium channels are discussed here, including basic structures of the pore-forming and auxiliary subunit complex, relationships with cancer, and sodium channels in bacteria.

## Structural Features of Eukaryotic Sodium Channels

As stated above, the voltage-gated ion channels (VICs) are a superfamily of energy-independent transmembrane transporters with selectivity toward potassium, calcium, and sodium ions. While individual members of this superfamily are selective toward one of the above three ions, they have a wide variety of topologies as a result of multiple duplication and fusion events. Voltage-gated sodium channel is also known as a putative ancestor to the last three transmembrane segments of certain members from the Major Facilitator Superfamily (MFS) (Saier 2003). One of the simplest and smallest configuration belong to the potassium channel, with only two transmembrane segments in each peptide and 97 amino acids in length, forming a homotetrameric structure in membrane with a total of eight (2 + 2 + 2 + 2) transmembrane segments. The first solved structure of this family came from a potassium channel, KcsA, encoded in *Streptomyces lividans*. The rudimentary structure of KcsA has been used for structural modeling of the pore-forming helices for sodium channels.

Proteins forming eukaryotic voltage-gated sodium channels are generally more than 2,000 amino acids in length. Each pore-forming subunit commonly contains 24 transmembrane segments (TMSs) with hydrophilic N and C termini encoded by genes in the human genome, sharing the same topology with a majority of mammalian calcium channels. Each 24-TMSs sodium channel is consist of four 6-TMSs repeats, forming a heterotetrameric structure. A single repeat is composed of one voltage sensor with four TMSs and one pore-forming domain with two TMSs. The fourth transmembrane segment, S4, contains positively charged residues to serve as gating charges (Catterall 2010). Individual repeats are connected by large hydrophilic loops. Between second and third repeats of a human sodium channel, the hydrophilic loop is much larger than any of the inter-helix loops from other members of the VIC superfamily.

These loops also serve as interaction sites with other peptides or subjected to posttranslational modification. It is also known that both posttranscriptional and posttranslational modifications play important roles in the diversity of VICs in eukaryotes. For example, alternative splicing provides localized modifications on tissue-specific channels. Other modifications such as phosphorylation and glycosylation are usually found in expressed VICs.

## Bacterial Sodium Channel (Na$_v$Bac)

*Bacillus halodurans,* an extremophile that grows in high salt and high pH, encodes a 6-TMS sodium channel, NaChBac, which can be expressed in mammalian cells after transfection and form a 12-pS channel. The fourth transmembrane segment for NaChBac is similar to many voltage-gated sodium channels, containing arginine residues every third amino acid apart. It is likely that the arginines from the fourth transmembrane segment are neutralized by the acidic residues of second and third transmembrane segments.

NaChBac is activated by depolarization, followed by inactivation and deactivation. In terms of pharmacological properties, it can be blocked by dihydropyridines, nifedipine, and nimodipine, similar to L-type calcium channels. While this channel is sensitive to calcium channel blockers, its specificity is solely toward sodium ions (Martinac et al. 2008). Mutational analyses established that an increase in the number of aspartic substitutions in the filter sequence will increase calcium selectivity. This selectivity is possibly due to the lack of flexible tertiary structure around the acidic residues in four nonidentical repeats, commonly found in animal calcium channels.

While NaChBac is the first bacterial sodium channel characterized, its homologues have been found in at least 11 other species of bacteria. This family of sodium channels has been named Na$_v$Bac. Several members, including that from *Paracoccus zeaxanthinifaciens* and *Silicibacter pomeroyi,* have been expressed in mammalian cells for functional studies. It has been proposed that members of this family are responsible for controlling sodium-driven flagellar motility. For its unique structure among all voltage-gated sodium channels, Na$_v$Bac might prove to be a good source for crystallography studies in determining the differences between sodium and calcium filtration (Martinac et al. 2008).

## Auxiliary Subunits

While the conductance of sodium ion always relies on the principal subunit, the kinetics and biogenesis of the complete complex depend heavily on the three families of auxiliary subunits, β1/β3, β2/β4, and temperature-induced paralytic E (TipE). These auxiliary subunits are associated with the principal pore-forming subunit of the voltage-gated sodium channel complex through covalent and noncovalent interactions.

TipE subunits are found to be encoded exclusively in insect genomes as revealed by similarity searches, for examples, mosquito, houseflies, and fruit flies. The temperature sensitive phenotype associated with TipE was first discovered in fruit flies which were paralyzed at a temperature above 38°C. Further characterization of the TipE subunit has revealed its ability in regulating the expression of the sodium channel para α-subunit with its chaperon-like properties. Sodium currents cannot be measured when para is expressed in the absence of TipE. TipE was also observed to express transiently in the fruit fly during plural development in order to rescue the adult paralysis. It is most likely that TipE does not serve as a stable component of the channel complex at the cell surface. Although TipE has been cloned, little studies have been done to further elucidate its functions. Phylogenetic analysis was used to divide homologues of TipE into multiple potential functional types. The phylogeny of this family also indicated that TipE arose prior to the speciation events giving rise to housefly and fruit fly and after the divergence from other animals. Along with the fact that TipE homologues are found exclusively in insect genomes, it can be suggested that this family of proteins arose relatively recently, possibly within the past 570 million years (Tseng et al. 2007).

In comparison, TipE is the functional equivalent of the β-subunit for the shaker potassium channels. TipE also shares functional similarities with the β-subunits in mammalian neuronal and cardiac sodium channels. However, structurally, TipE is quite dissimilar to β-subunits from other voltage-gated ion channels.

TipE contains two transmembrane helices instead of one in the β-subunit of a voltage-gated sodium channel. One of these transmembrane helices is located at the N-terminus of TipE from *Drosophila*, followed by a hydrophilic region of 200 amino acids in length and another transmembrane helix located near the hydrophilic C-terminal tail. The hydrophilic region between these two transmembrane helices is about 200 amino acids in length. These two putative transmembrane helices also correspond to highly conserved regions among TipE homologues (Tseng et al. 2007). Hydrophilic loops between these two helices from various TipE homologues can have varied lengths. Like the principal pore-forming subunit, auxiliary subunits can also be subjected to posttranslational modifications.

In addition to TipE, two other families have been studied with functional and phylogenetic approaches. These two families, β1/β3 and β2/β4, share limited degree of similarity, suggesting the possibility that all β subunits arose from the same ancestor (Tseng et al. 2007). The first family, β1 (SCN1B) and β3 (SCN3B), can modulate the gating kinetics of voltage-sensitive sodium channels based on tissue specificity. β1 (ca. 36 kDa), a type I transmembrane protein, associates with the pore-forming α-subunit via noncovalent interactions. The expression of β subunits might be important in the development of mammals. It has been found that abnormalities, including epilepsy, ataxia, neuronal pathfinding errors, and premature death, are associated with beta-1 null mice (Fraser and Pardo 2008). The surface expression of α subunits can be increased two- to fourfold via the regulation of the trafficking mechanism by β1 subunits. β1-mediated modulation of $Na_v1.2$ surface density is dependent on interaction with contactin and ankyrin. A version of β1, β1a, is known to express early in embryonic brain development. β3 was first discovered in human, characterized by expression in *Xenopus* oocytes, demonstrating a slower inactivation for sodium channels. Differential distributions in brain tissues of β1 and β3 are accompanied by distinct kinetic properties.

Similarity searches found that all characterized β subunits are exclusively mammalian in origin, with all exhibiting sequence similarity with immunoglobulin (Ig) folds. In β1 and β3 subunits, the Ig-like fold is in the extracellular portion of the proteins. However, the degree of primary sequence similarity between β1/β3 and Ig is low and insufficient to establish homology. This Ig-like fold is located in a hydrophobic region near the N-terminus as demonstrated with Kyte-Doolittle hydropathy plot. This region, similar to Ig, might be involved in interaction with extracellular matrix, which may result in immobilization to nodes of Ranvier and axon hillocks with high concentration of β subunits. While the similarity between β subunits and Ig is the most pronounced, a limited degree of sequence similarity can also be observed between β1/β3 and myelin (Tseng et al. 2007).

A putative signal peptide (~25 aa) on the N-terminus of β1 is followed by a hydrophilic region with four N-glycosylation sites and a C-terminal transmembrane helix. Truncation of 36 residues on the C-terminal showed no impact to the inactivation toward the complete channel complex. This serves as an implication that modulation of inactivation kinetics relies on the hydrophilic loop or the N-terminal hydrophobic region. The second hydrophobic region likely serves as a transmembrane segment for anchoring the auxiliary subunit within the membrane. While all members of β1/β3 share a similar topology, the loop of β3 is more hydrophobic in comparison with the hydrophilic region from β1. This generally hydrophilic region between two hydrophobic segments might be of medical importance. Febrile seizure can be caused by a mutation at cysteine 121.

Phylogenetic analysis suggested that β1/ β3 and β2/ β4 should not be placed in the same family due to low sequence similarity (Tseng et al. 2007). β2 subunits modulate gating kinetics of neuronal sodium channels exclusively and are not found in other tissue types. All β subunits have the same topology, as seen in hydrophobicity plots. β2 and β4 subunits also contain a hydrophobic signal peptide, an Ig-like domain (an extracellular segment), a TMS for membrane anchoring, and a C-terminal tail. The domain in β2/β4 with an Ig-like fold is similar to ones found in contactin, a neuronal cell adhesion molecule. Phylogenetic analysis also revealed that β2 and β4 are distantly related with Ig light chains. While β2 and β4 subunits share a very high degree of sequence similarity, they also form disulfide linkages with α-subunit, not found in β1/β3. This formation of a stable complex likely relies on one of the five extracellular cysteine residues. Two of these cysteine residues are in the Ig-like domain. Further interaction with ankyrin and tenascin-R also allows β2 to target sodium channel complex to node of Ranvier.

## Voltage-Gated Sodium Channels in Cancer

Ion channels have been implicated in the progression of many types of cancer and their progressions. Voltage-gated sodium channels are upregulated in a wide variety of cancers, including prostate, breast, lung (both small-cell and non-small-cell), cervical cancer, leukemia, and mesothelioma as well as highly metastatic cancer-derived cell lines. In leukemia cells, voltage-gated sodium channels are not only upregulated on the transcriptional level, but also associated with multidrug resistance (Yamashita et al. 1987). Expression of sodium channels has been observed on both transcription and translation levels in several prostate cancer cell lines. Enhanced metastasis was found to correlate with the appearance of membrane channels, similar to ones in excitable membranes with inward currents characteristics of sodium and calcium ions. The expression of voltage-gated sodium channels can also influence cell proliferation, transformation, apoptosis, and cellular behaviors for metastatic cell spread, including motility and invasion.

The enhanced electrophysiological properties in prostate cancer epithelial cells were suggested to increase the perturbation of ionic homeostasis and/or secretory activities in addition to motility. In metastatic prostate cancer cells, the shift to a "more excitable" phenotype during the increasing malignancy is associated with tetrodotoxin (TTX)-sensitive $Na_v1.7$, expressed at approximately 20-fold along with a decrease in potassium current, as determined by RT-PCR analysis (Prevarskaya et al. 2007). Channel blocker, TTX, was found to directly reduce the invasiveness of cancer cells, suggesting voltage-gated sodium channels as a potential therapeutic target. Other channel blockers, flunarizine and riluzole, also cause dose-dependent growth inhibition to prostate cancer cell lines. Such observation with channel blockers is not limited to prostate cancer cells. In non-small-cell lung cancer cell lines, TTX was also found to prevent aggressive and invasive phenotype, including migration of highly metastatic cells (Prevarskaya et al. 2010). An antiepileptic, phenytoin is also known to reduce the migration of metastatic human cell line, PC-3 without changing the motility of weakly metastatic human LNCaP cells. In contrast, veratrine, a channel opener, has been shown to increase growth of both androgen-insensitive (PC-3 and DU-145) and androgen-sensitive (LNCaP and MDA-PCA-2B) prostate cancer cell lines. Other channel openers, aconitine and ATX II, was also found to enhance the migration of metastatic PC3 cell line from human and MAT-LyLu from rat without changing the motility of LNCaP (weakly metastatic) and AT-2 from rat. This facilitation of channel opening by agonists also enhances migration with no impairment to cell proliferation or viability.

Due to the association with various cardiac phenotypes and generation of cardiac sodium current, cardiac channel $Na_v1.5$ is one of the best studied voltage-gated sodium channels. Mutation in SCN5A, the gene encoding $Na_v1.5$, can cause congenital and acquired long QT syndrome, sick sinus syndrome, atrial fibrillation, Brugada syndrome, and conduction slowing. Most recently, $Na_v1.5$ has been implicated in cancer development (Fraser and Pardo 2008). The invasiveness of breast cancer cells is known to associate with $Na_v1.5$, which was found to be approximately 1,000-fold overexpressed in highly metastatic cells. This association was by an increased cysteine cathepsin activity. As a proteolytic enzyme, cysteine cathepsins are known to influence in carcinomas and play an important role in the metastasis of many other types of cancer as well, including prostate, non-small-cell cancer and leukemia. An increase in $Na_v1.5$ amplitude was also found after acute estrogen application. Estrogen is a steroid hormone known to be important in the progression of breast cancer. Activation of protein kinase A mediates a recently characterized G-protein-coupled receptor, GPR30, which increases the $Na_v1.5$ amplitude. Furthermore, the action of estrogen might also be important for breast cancer metastasis due to its ability to reduce cell adhesiveness in a sodium channel-dependent manner (Fraser and Pardo 2008).

While studies have focused on the principal pore-forming unit, nonconduction function carried out by other components of the sodium channel complex may also play a role in cancer. Auxiliary subunits of voltage-gated sodium channels are known to participate in the pathogenesis of cancer (Brackenbury et al. 2008). These include transport proteins, kinases, phosphatases, and extracellular matrix proteins as well as the β subunits. All of these above components can contribute to the disease conditions of cancer by having abnormal functionalities and modulation of the complex. It has been suggested that the involvement of β-subunits might be complex, as they are capable of functioning separately from the α-subunit. For example, an increased amount of intracellular domain of β2, after cleavage by the β-site of APP cleaving enzyme (BACE1) and γ-secretase, was found in neuroblastoma cells along with increased mRNA and protein expression for $Na_v1.1$ (Fraser and Pardo 2008).

It is important to point out that, for certain cancer types, while the overexpression of voltage-gated sodium channel does not necessarily contribute to the metastasis, the presence of sodium current is required. It has been found that multiple types of non-voltage-gated sodium channels are expressed in high-grade glioma cells, responsible for the sodium current. Such functional diversity for transporting sodium highlights the importance of sodium current rather than sodium channel type in determining the malignancy. In essence, the overexpression of either non-voltage gated, such as ENaC, and voltage-gated can contribute to altered electrophysiological properties of cancers or perturb the intracellular ionic homeostasis, representing a crucial factor in promoting cellular motility and metastasis.

Highly metastatic cancers mostly express embryonic isoforms of voltage-gated sodium channels, as established by structural and functional studies (Prevarskaya et al. 2010). Such observation supports the idea that human embryonic genes could be reexpressed in cancer cells. The pro-invasive and upregulation mechanisms are not well understood. However, the expression of voltage-gated sodium channels might be under the control of steroid receptors in hormone-responsive tissues, such as uterus, ovaries, breast, and prostate. As shown for many highly malignant cancer cell types, growth factors such as epidermal growth factor (EGF) and nerve growth factor (NGF) might exert additional control on the expression of sodium channels.

## Cross-References

▶ Sodium Channel Blockers

## References

Brackenbury WJ, Djamgoz MBA, Isom LL (2008) An emerging role for voltage-gated $Na^+$ channels in cellular migration: regulation of central nervous system development and potentiation of invasive cancers. Neuroscientist 14(6):571–583

Catterall WA (2010) Ion channel voltage sensors: structure function, and pathophysiology. Neuron 67(6):915–928

Fraser SP, Pardo LA (2008) Ion channels: functional expression and therapeutic potential in cancer. EMBO Rep 9(6):512–515

Hodgkin AL, Huxley AF (1952) A quantitative description of membrane current and its application to conduction and excitation in nerve. J Physiol 117(4):500–544

Martinac B, Saimi Y, Kung C (2008) Ion channels in microbes. Physiol Rev 88(4):1449–1490

Prevarskaya N, Skryma R, Bidaux G, Flourakis M, Shuba Y (2007) Ion channels in death and differentiation of prostate cancer cells. Cell Death Differ 14(7):1295–1304

Prevarskaya N, Skryma R, Shuba Y (2010) Ion channels and the hallmarks of cancer. Trend Mol Med 16(3):107–121

Saier MH Jr (2003) Tracing pathways of transport protein evolution. Mol Microbiol 48(5):1145–1156

Tseng T-T, McMahon AM, Johnson VT, Mangubat EZ, Zahm RJ, Pacold ME, Jakobsson E (2007) Sodium channel auxiliary subunits. J Mol Microbiol Biotechnol

Yamashita N, Hamada H, Tsuruo T, Ogata E (1987) Enhancement of voltage-gated $Na^+$ channel current associated with multidrug resistance in human leukemia cells. Cancer Res 47(14):3736–3741

## Sodium Cycle

▶ Sodium as Primary and Secondary Coupling Ion in Bacterial Energetics

## Sodium Formate, Formic Acid

▶ Chromium(VI), Oxidative Cell Damage

## Sodium Glucose Transporters

▶ Sodium/Glucose Co-transporters, Structure and Function

## Sodium Pump

▶ Sodium/Potassium-ATPase Structure and Function, Overview

## Sodium Pumps

▶ Sodium as Primary and Secondary Coupling Ion in Bacterial Energetics

# Sodium, Na

▶ Sodium, Physical and Chemical Properties

# Sodium, Physical and Chemical Properties

Sergei Yu. Noskov
Institute for Biocomplexity and Informatics,
Department of Biological Sciences, University of Calgary, Calgary, AB, Canada

## Synonyms

Alkali cations; Alkali metals; Monovalent cations; Sodium, *Na*

**Sodium, Physical and Chemical Properties, Table 1** Physical properties of sodium

| | |
|---|---|
| Atomic number | 11 |
| Atomic mass | 22.98977 g mol$^{-1}$ |
| Relative abundance on Earth | 2.6% |
| Oxidation states | +1, 0, −1 |
| Ionization energy (1, 2, 3) | 495.7, 4,562, and 6,910.3 kJ mol$^{-1}$ |
| Ionic radius | 0.097 nm |
| Van der Waals radius | 0.227 nm |
| Atomic radius | 0.186 nm |
| Electronic shell | [Ne]3 s$^1$ |
| Standard potential | −2.71 V |
| Electronic shell | [Ne]3 s$^1$ |
| Phase | Solid |
| Molar heat capacity | 28.23 J mol$^{-1}$ K$^{-1}$ |
| Melting point | 97.8°C |
| Boiling point | 883°C |
| Hydration free energy ($\Delta G^0$) | −371 kJ/mol |
| CAS registry number | 7440-23-5 |

## Definitions

*CAS*: Chemical Abstracts Service
  *Alkali Metal*: Group 1 elements
  *Physical Properties*: Sodium (Na) is a member of the group 1 of alkali cations, sometimes also referred to as group IA. Its atomic number is 11 and atomic weight is 22.98 g mol$^{-1}$. The metal is highly reactive and cannot be found in a free form. Being highly soluble in water (reacts with it exothermically), a monovalent (Na$^+$) sodium cation is abundantly present in salt- and freshwater basins, soil, and earth crust. Sodium is also among the most prominent in the solar spectrum. It is also the most abundant type of alkali metals. It is most commonly found in complex with chloride (ordinary or table salt) but also occurs in other minerals such as natural zeolites, sodalities, and many others. A list of key physical properties of Na is collected in Table 1. Like the other alkali metals, Na has NMR-active nuclei (Na$^{23}$).
  *Chemical and Biochemical Properties*: As all other alkali metals, sodium is highly reactive comparing to metals from other groups. A pure metal is usually obtained by electrolysis of their molten salts. It is generally established that sodium was first isolated by Davy in 1807 with electrolytic extraction. Sodium can be easily oxidized, as its first ionization energy is low compared to other elements. Because of its solubility, it is mostly found as an ion. Sodium standard potential is also very electronegative, which makes sodium excellent reducing agent. All of the alkali metals are soluble as well in liquid NH$_3$, forming a metastable solution.

When Na is heated with oxygen or in an excess of air, a sodium peroxide is formed:

$$2Na + O_2 \rightarrow Na_2O_2.$$

The sodium oxide (Na$_2$O) can be obtained by thermal decomposition of the peroxide. Na salts and hydroxides are highly soluble in water. The Na$^+$ ions are relatively weekly complexed by simple anions in solution and hardly form any complexes with monodentate neutral ligands. The average hydration number for Na$^+$ is 6, but other coordination states were also reported. The hydration enthalpy ($\Delta_{hyd}H^0(298\ K)$) for Na$^+$ is −404 kJ mol$^{-1}$, hydration entropy ($\Delta_{hyd}S^0(298\ K)$) is −110 JK$^{-1}$ mol$^{-1}$, and the corresponding free energy of hydration ($\Delta_{hyd}G^0(298\ K)$) is −371 kJ mol$^{-1}$. The hydration data is reported for the following reaction:

$$Na^+_{(gas)} \rightarrow Na^+_{(aqueous)}.$$

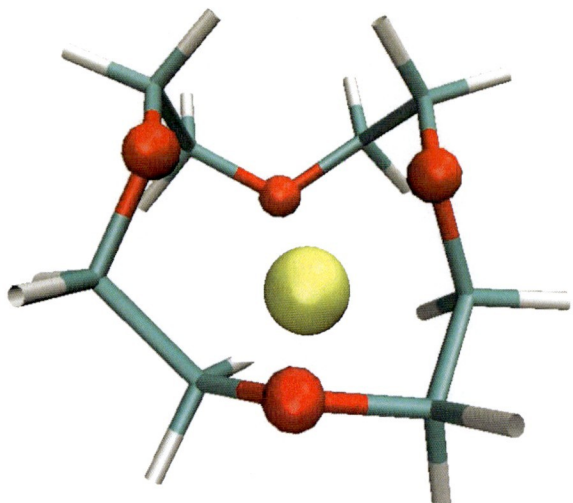

**Sodium, Physical and Chemical Properties, Fig. 1** Ball-and-stick representation of a 12-crown-4 complex with bound Na⁺. Oxygen atoms shown in *red*

The organosodium compounds are characterized by high polarity, and essentially all of them are ionic and highly reactive. Some of the most important compounds are ones formed by acidic hydrocarbons such indene or acetylenes. They are obtained by the reaction of sodium in liquid ammonia:

$$3C_6H_6 + 2Na \rightarrow 2C_5H_5^- Na^+ + C_5H_8$$

$$RC \equiv CH + Na \rightarrow RC \equiv C^- Na^+ + \frac{1}{2}H_2$$

The sodium-naphtalene complex is often used as a powerful reducing agent.

One of the most important and stable complexes of Na⁺ with organic ligands is the one with macrocyclic polyethers and cryptates. Na complexes with cyclic polyethers are commonly known as crown-ethers. The nomenclature of crown-ethers usually provides the total number of C and O atoms and the number of O in the ring. Figure 1 shows the structure of a 12-crown-4 sodium complex.

It is generally accepted that the radius of the binding cavity determines preference for a particular cation. It should be noted that ion binding to a crown-ether may lead to conformational change and thus alteration of the cavity radius. The crown-ether complexes of sodium are large and hydrophobic molecules, which are primarily soluble in organic solvents.

Sodium ions are a critical component of proper biochemical and physiological function in biological systems. Na⁺ transport is at the heart of most transport and signaling processes in normal cell. The intracellular concentration of Na⁺ is lower than the extracellular one. Therefore, Na⁺ can enter the cell, following an electrochemical gradient. The energy produced by the movement of Na⁺ ions is used to fuel secretion of a number of neurotransmitter molecules, K⁺ and H⁺ ions, as well to uptake of glucose molecules against their concentration gradient. Movement of the Na⁺ ion across cellular membranes is utilized by the cell to generate the resting and action potentials in excitable membranes (nerves, muscle cells containing gated Na⁺ and K⁺ channels) and to control of cell volume.

## Cross-References

▶ Cellular Electrolyte Metabolism
▶ Sodium-Binding Site Types in Proteins

## References

Dean JA, Lange NA (1998) Lange's handbook of chemistry. McGraw-Hill, New York
Hille B (2001) Ion channels of excitable membranes, 3rd edn. Sinauer, Sunderland
Housecroft CE, Sharpe AG (2008) Inorganic chemistry, 3rd edn. Pearson-Prentice Hall, Harlow

# Sodium/Glucose Co-transporters, Structure and Function

Louis J. Sasseville and Jean-Yves Lapointe
Groupe d'étude des protéines membranaires (GÉPROM) and Département de Physique, Université de Montréal, Montréal, QC, Canada

## Synonyms

Cotransporters; Glucose transporters; Na⁺/sugar symporter; Na⁺-dependent glucose transporter; Secondary-active transporters; SGLTs; Sodium glucose transporters; Solute carrier family 5 (SLC5) genes

## Definition

Na$^+$/glucose cotransporters (SGLTs) are responsible for the "secondary-active" transport of glucose and other substrates across cellular membranes. They concentrate glucose inside the cell using electrochemical energy from the transmembrane Na$^+$ gradient, employing an alternating access cotransport mechanism. Their most important roles are to mediate the absorption of glucose in the intestine and its reabsorption in the proximal tubules of the kidney.

## Introduction

The general mechanism for active glucose cotransport through epithelial cells was first proposed by Robert Crane in the early 1960s. It described the secondary-active pumping of glucose across the brush border membrane of the intestinal epithelium using the Na$^+$ gradient generated by the basolateral Na$^+$/K$^+$-ATPase. Subsequent studies validated this model, which has been generally accepted (Wright and Turk 2004). In the 1980s, it was established that the renal early proximal tubule exhibits a low-affinity/high-capacity cotransporter while the late proximal tubule has a high-affinity/low-capacity cotransport system.

Some 30 years after the initial cotransport hypothesis, cloning of the human Na$^+$/glucose cotransporters SGLT1 and SGLT2 followed by the cloning of homologous cotransporters allowed them to be grouped into the solute carrier gene family 5 (SLC5). The SLC5 family is comprised of 12 members and includes Na$^+$-dependent transporters for glucose, galactose, *myo*-Inositol, and a few other substrates. Cloning allowed expression in different heterologous expression systems including *Xenopus laevis* oocytes. This lead to the determination of the transport stoichiometry for the different Na$^+$/glucose cotransporters and refinement of their kinetic models using electrophysiology and radioactive substrate uptake (Wright and Turk 2004).

Recent crystallisation of the *Vibrio parahaemolyticus* Na$^+$/galactose cotransporter (vSGLT) allowed atomic level investigations of the cotransport mechanism. Surprisingly, it also revealed that Na$^+$/glucose cotransporters share the same basic architecture with the prokaryote Leucine Transporter LeuT and several other members of unrelated transporter gene families (Abramson and Wright 2009; Boudker and Verdon 2010; Forrest et al. 2010).

## The SLC5 Family

SGLTs are encoded by genes which belong to the solute carrier family SLC5, which is comprised of 12 genes in humans. The functions of 11 of those 12 proteins are known, only 3 of them are strictly Na$^+$/sugar cotransporters (SGLT1, 2, and 4) and one (SGLT3) would be a glucose sensor. We will briefly review the members of the SLC5 family found in humans.

### SGLT1

SGLT1 (SLC5A1) is by far the best studied member of the SGLTs. This is partly because it was the first to be cloned in 1987, but also because of its strong expression in a variety of expression systems (*X. laevis* oocytes, mammalian and insect cell lines, and because of its physiological importance.

SGLT1 is a ~75 kDa membrane protein. It is primarily expressed in the small intestine and the proximal renal tubules of the kidney, where it is responsible for the transport of glucose and galactose (Wright and Turk 2004; Sabino-Silva et al. 2010), though it is also found in the heart and in salivary glands (Sabino-Silva et al. 2010). As a secondary-active transporter, SGLT1 uses the sodium gradient generated by the Na$^+$/K$^+$-ATPase to pump one sugar along with two Na$^+$ ions across the membrane toward the intracellular milieu (Bergeron et al. 2000; Sabino-Silva et al. 2010). SGLT1 transports glucose ($K_m = 0.4$ mM) and galactose, and is inhibited by a natural product called phlorizin. SGLT1 is a high-affinity/low-capacity transporter, in comparison to SGLT2, which displays a lower affinity and a higher transport capacity (Chao and Henry 2010). The complementary work of these transporters in the kidney will be detailed later.

SGLT1 is an electrogenic transporter, which means that each cotransport event (i.e., pumping of 1 sugar across the membrane) cause a net movement of two elementary charges (i.e., the two Na$^+$ ions). This feature, combined with strong expression of SGLT1 in expression systems, has allowed extensive electrophysiological studies of the cotransport kinetics, leading to formulation of the basic cotransport mechanism.

## SGLT2

SGLT2 (SLC5A2) is expressed primarily in renal proximal tubules where it is responsible, with SGLT1, for the near complete reabsorption of the glucose filtered at the glomerulus. It displays low affinity for glucose ($K_m = 2$ mM) (Chao and Henry 2010) and a stoichiometry of 1 $Na^+$:1 glucose. This means that the electromotive force of only 1 $Na^+$ is available during cotransport, reducing the concentrative power of SGLT2 compared to SGLT1, which has a stoichiometry of 2 $Na^+$:1 glucose.

Despite its cloning in the early 1990s, very low expression in heterologous expression systems (Bergeron et al. 2000; Wright and Turk 2004) has limited investigation of this cotransporter. SGLT2 has recently been the subject of considerable attention as it has emerged as a new target for therapeutic strategies aimed at controlling glycemia of diabetic patients (Chao and Henry 2010).

## SGLT3-5

SGLT3 (SLC5A4) has been described as a glucose sensor in humans, as it transports very little glucose while generating a strong inward current that contributes to depolarizing the host cell whenever external glucose is present. In sharp contrast to this, pig SGLT3 tightly couples $Na^+$ and glucose transport. Human SGLT3 is expressed in muscle and neurons (Wright and Turk 2004; Sabino-Silva et al. 2010).

SGLT4 (SLC5A9) is believed to be a $Na^+$/Mannose cotransporter. The presence of SGLT4 messenger RNA in the small intestine and kidney suggests expression of SGLT4 in those tissues (Tazawa et al. 2005).

Little is known about SGLT5 (SLC5A10) apart from the fact that it is expressed in the kidney (Wright and Turk 2004).

## SMIT1-2, NIS, SMCT1-2, SMVT, and CHT1

SMIT1 (SLC5A3) is a $Na^+$/*myo*-Inositol cotransporter, but is also known to transport D-glucose with low affinity (Wright and Turk 2004). *Myo*-inositol is one of a few compatible osmolytes that are accumulated in the cells of the kidney medulla to reach a high intracellular osmolarity without increasing cytosolic ionic strength (Lahjouji et al. 2007).

SMIT2 (SLC5A11) is a $Na^+$/*myo*-Inositol cotransporter. It also weaky cotransports D-glucose, thus its previous name of $Na^+$/glucose cotransporter 6 (SGLT6). It has a stoichiometry of 2 $Na^+$:1 substrate. SMIT2 is a 75 kDa protein expressed in kidney, heart, skeletal muscle, liver, and brain, and is responsible for luminal transport of *myo*-Inositol in the intestine and in the renal proximal tubules of the kidney (Lahjouji et al. 2007).

NIS (SLC5A5) is a $Na^+$/iodide transporter. It has a stoichiometry of 2 $Na^+$:1 iodide. It is expressed in the thyroid, the salivary glands, gastric mucosa, and the lactating mammary gland (Bizhanova and Kopp 2009).

SMCT1 (SLCA8) is a $Na^+$/monocarboxylate transporter. It has a 2 $Na^+$:1 monocarboxylates stoichiometry (Coady et al. 2007). It is involved in absorption of short-chain fatty acids in the colon and small intestine, reabsorption of lactate and pyruvate in the kidney, and cellular uptake of lactate and ketone bodies in neurons. Nonsteroidal anti-inflammatory drugs such as ibuprofen, ketoprofen, and fenoprofen have been shown to act as inhibitors (Ganapathy et al. 2008).

SMCT2 (SLC5A12), another $Na^+$/monocarboxylate transporter, is a low-affinity transporter. It is localized in the intestinal tract and in the proximal tubule of the kidney, but is also found in astrocytes of the brain and Müller cells of the retina. Cotransport by SMCT2 is electroneutral, suggesting a stoichiometry of 1 $Na^+$: 1 monocarboxylate (Ganapathy et al. 2008).

SMVT (SLC5A6) is a sodium-dependent multivitamin transporter. It is responsible for the transport of pantothenate, biotin, and lipoate, three physiologically important molecules. The reported affinity for pantothenate is 1–3 μM, while it is 8–20 μM for biotin. There is a certain controversy about the stoichiometry of cotransport, regarding the possibility of one or two $Na^+$ being cotransported for each substrate. SMVT is ubiquitously expressed in mammalian tissues, but higher messenger RNA concentrations are present in the intestine, kidney, and placenta (Prasad and Ganapathy 2000).

CHT1 (SLC5A7) is the high-affinity choline transporter. It is responsible for the reuptake of choline from the synaptic cleft into presynaptic neurons. It only shares a 20–26% sequence identity with the other members of the SLC5 gene family (Ribeiro et al. 2006).

| Protein name | Human gene name | Preferred substrates |
|---|---|---|
| SGLT1 | *SLC5A1* | Glucose, galactose |
| SGLT2 | *SLC5A2* | Glucose |
| SGLT3 | *SLC5A4* | Glucose |

(*continued*)

| Protein name | Human gene name | Preferred substrates |
|---|---|---|
| SGLT4 | *SLC5A9* | Mannose |
| SGLT5 | *SLC5A10* | |
| SMIT1 | *SLCA3* | *myo*-Inositol |
| SMIT2 or SGLT6 | *SLC5A11* | *myo*-Inositol |
| NIS | *SLC5A5* | Iodide |
| SMCT1 | *SLC5A8* | Monocarboxylate |
| SMCT2 | *SLC5A12* | Monocarboxylate |
| SMVT | *SLC5A6* | Pantothenate, biotin, and lipoate |
| CHT1 | *SLC5A7* | Choline |

### *v*SGLT

vSGLT is a bacterial member of the SLC5 family. vSGLT shows $Na^+$-dependent cotransport of galactose, fucose, and glucose and is inhibited by phlorizin. The stochiometry of cotransport is 1 $Na^+$:1 sugar (Leung et al. 2002).

Despite its bacterial origin and a lack of functional data, in part due to its low level of expression in heterologous expression systems such as *X. laevis* oocytes, vSGLT is of the utmost importance to the comprehension of the human SLC5 proteins as it is the only member of the SLC5 family whose structure has been resolved at the atomic level.

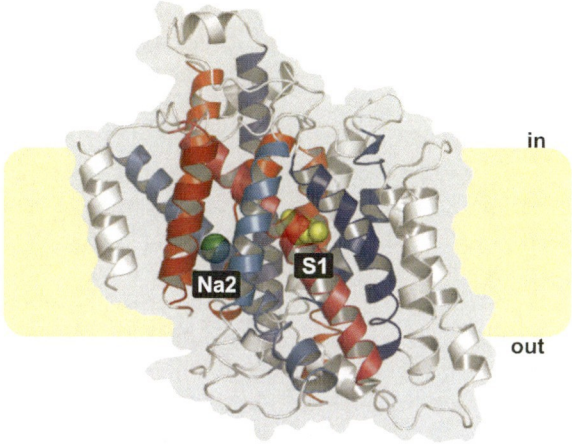

**Sodium/Glucose Co-transporters, Structure and Function, Fig. 1** *General structure of vSGLT*. Structural repeats forming the inverted symmetry are identified by shades of *red* (TM1-5) and *blue* (TM6-10). S1 is shown with bound galactose (*yellow*), and Na2 with bound sodium (*green*). Extraneous TMs (TM -1, 11–13) are in *white*, silhouette of the volumetric representation of vSGLT in *light gray*, and the cellular membrane in *light yellow*. Based on PDB 3DH4

### Structure of Na$^+$/Glucose Cotransporters

Currently, only two crystal structures of vSGLT have been resolved (Protein Databank accession codes (PDBs) 3DH4 and 2XQ2). They provide the first structural pictures of a member of the SLC5 family at an atomic level. These show 14 transmembrane segments (TMs), with extracellular N- and C-terminals. Both structures show an intracellular vestibule, suggesting an inward-facing conformation (Fig. 1).

### LeuT Structural Family

Unexpectedly, the core helices of vSGLT showed a strong structural homology with the core helices of a member of the unrelated Neurotransmitter/$Na^+$ Symporter family (NSS), the Leucine Transporter LeuT. It was subsequently found that this overall fold is also shared by members of the Major Facilitator superfamily (MFS) (agmatine/arginine exchanger AdiC), the nucleobase-cation-symport-1 family (NCS1) (the nucleobase/cation symporter Mhp1), the betaine/choline/carnitine transporters family (BCCT) (the $Na^+$/betaine symporter BetP and the L-carnitine/γ-butyrobetaine antiporter CaiT) and the amino acid, polyamine, and organocation (APC) transporters family (the proton-dependent amino acid transporter ApcT) (Boudker and Verdon 2010). These transporters, i.e., *vSGLT, LeuT, Adic, Mhp1, BetP, CaiT*, and *ApcT*, form a new *structural* family, coined the "LeuT structural family."

vSGLT shows sequence homology to members of the SLC5 family (e.g., 32% sequence identity with hSGLT1). This allows use of the crystal structure of vSGLT as a reliable template for the structure of the SLC5 family. While there is no significant *sequence* similarity between the members of the LeuT structural family, strong *structural* similarities allow relevant comparisons. This is particularly useful since members of the LeuT structural family have been resolved in a number of conformational states (inward or outward facing, occluded or non-occluded) and with different bound substrates, notably a second $Na^+$ ion (Forrest et al. 2010).

While shared architecture between members of unrelated families is interesting, the particularities of this architecture are both surprising and fascinating. Because this architecture was first identified in

LeuT, the 10 helices forming the core domains are numbered from one to ten. Thus, in vSGLT, the N-terminal helix becomes transmembrane segment -1 (TM -1), the next becomes TM1, and so forth. The "LeuT architecture" is characterized by an inverted symmetry. This means that TM1-5 and TM6-10 are superimposable following a ~180° rotation around an axis set parallel to the membrane plane. This is unexpected as there is no significant sequence similarity between TM1-5 and TM6-10 (Abramson and Wright 2009; Forrest et al. 2010). It is worth noting that other architectures found in a number of different transporters, such as LacY, ATP/ADP carrier, NhaA, and Glt$_{Ph}$ also present structural repeats and/or inverted symmetry (Boudker and Verdon 2010; Forrest et al. 2010).

Analysis of the members of the Leut structural family gives valuable insight into the structural characteristics of Na$^+$/glucose cotransporters.

### Substrate-Binding Site

The sugar-binding site S1 was directly identified in the vSGLT crystal (PDB 3DH4) and its relative location is conserved in other members of the LeuT structural family. It is located approximately halfway across the membrane, in the center of the core domain. This corresponds to the unwound segments in the first TMs of each repeat (TM1 in repeat TM1-5, and TM6 in repeat TM6-10). These unwound segments are believed to play an important role in the binding of substrates, as they generate a local polar environment. While the relative location of the substrate-binding site is conserved, the specific interactions vary considerably in order to accommodate the diverse substrates. In vSGLT, the hydroxyl groups of the galactose molecule are coordinated by hydrogen atoms from polar side chains of TM1, 6, 7, and 10 (Abramson and Wright 2009).

Secondary substrate-binding sites have been reported in LeuT and in CaiT. In LeuT, an extracellular, secondary, low-affinity-binding site called S2 is seen between the main binding site S1 and the extracellular side of the membrane. Two roles have been postulated for it. First, S2 appears to be involved in the binding of noncompetitive inhibitors which restrict the closure of the extracellular vestibule, and thus inhibit translocation of the substrate to the intracellular side (Boudker and Verdon 2010). Second, substrate binding to S2 is believed to allosterically promote intracellular release of substrate from S1 (Abramson and Wright 2009; Boudker and Verdon 2010). In CaiT, the substrate is seen to bind to the main binding site S1, but also to two different secondary-binding sites. A first one is located between S1 and the *intra*cellular side of the membrane, while a second one is seen on the *extra*cellular side, at the top of the extracellular vestibule (Forrest et al. 2010). As no secondary-binding site have been identified in the crystal structure of vSGLT, it is difficult to assess the relevance of such findings for Na$^+$/glucose cotransporters.

### Na$^+$-Binding Site

All sodium-dependent members of the LeuT structural family (LeuT, vSGLT, Mhp1, and BetP) present a Na$^+$-binding site 7–10 Å away from the main substrate-binding site S1. This site, named Na2 for historical reasons, is located at the intersection of TM1 and TM8. The requirement of Na$^+$ for sugar binding suggests allosteric regulation of S1 by Na2. In members of the LeuT structural family with a 2 Na$^+$:1 substrate stoichiometry (LeuT and BetP), the second Na$^+$ ion has been found to interact directly with the substrate bound in S1 (Boudker and Verdon 2010).

Also, it is worth noting that molecular dynamic simulations suggests that the crystal structure of vSGLT with Na$^+$ assigned in Na2 (PDB 3DH4) is believed to be in a ion-releasing state, as the Na$^+$ ion was found to spontaneously unbind (Abramson and Wright 2009; Boudker and Verdon 2010) in the first few nanoseconds of simulation.

## Cotransport Cycle and Alternating Access

The mechanism of cotransport in Na$^+$/glucose cotransporters has been the subject of extensive work since the initial identification of secondary active glucose transport more than 50 years ago (Wright and Turk 2004). Experimental work in situ, in vivo, and in vitro provided information on the *kinetic* of cotransport that allowed elaboration of a general cotransport mechanism for Na$^+$/glucose cotransporters. Crystal structures of members of the LeuT structural family provided *static* "snapshots" of the transporters in different conformations along the cotransport cycle, confirming and further refining the previously proposed mechanism.

**Sodium/Glucose Co-transporters, Structure and Function, Fig. 2** *Cotransport cycle of $Na^+$/glucose cotransporters.* The cotransporter is in state $S_1$, where the binding sites are accessible from the extracellular side only. The binding of a $Na^+$ ion (*green circle*) leads to state $S_2$, which has a greater affinity for sugars. Binding of the sugar (*yellow hexagon*) (state $S_3$) induces a closing of the extracellular gates (*orange sticks*) (state $S_4$), which in turn allows closing of the extracellular vestibule and opening of the intracellular vestibule by the tilting of the four-helix bundle (*gray wedge*) (state $S_5$). This is followed by the opening of the intracellular gate (*orange stick*) (state $S_6$), allowing intracellular release of the substrate. The empty transporter (state $S_7$) is then believed to undergo a similar process in order to return to its initial state $S_1$, reexposing the substrate-binding sites to the extracellular side

## Alternating Access

The alternating access mechanism states that, in order to allow stoichiometric cotransport of substrates, the binding sites are exposed to one side other of the membrane at a time during a sequence of conformational changes.

Historically, two distinct mechanisms have been proposed to account for alternating access: the "rocking-bundle" and the "gated pore" models. In the rocking-bundle model, whole domains of the proteins would "rock" about the binding site, while in the gated pore model, small local gates would alternatively open/close to allow access to the inside/outside of the cell. Crystallographic structures suggest that $Na^+$/glucose cotransporters incorporate both gated pore and rocking-bundle characteristics (Abramson and Wright 2009; Boudker and Verdon 2010; Forrest et al. 2010). In vSGLT, side chains from TMs 1, 2, 6, and 10 form intra- and extracellular hydrophobic plugs which can occlude the substrate-binding site and prevent any interaction with the corresponding intra/extracellular vestibule. Small movements of those side chains modulate access to the substrate-binding site, in a way reminiscent of the gated-pore mechanism. Analysis of the crystal structures of members of the LeuT structural family in various conformational states suggests that, during cotransport, vSGLT undergoes large conformational changes. The so-called four-helix bundle, comprised of the first two TMs of each repeats (TMs 1, 2, 6, and 7) could tilt by as much as 25°, opening/closing the intra/extracellular vestibules, in a way that is reminiscent of the rocking-bundle mechanism (Abramson and Wright 2009; Forrest et al. 2010).

Furthermore, a strong correlation is seen between the alternating access mechanism and the inverted symmetry found in the architecture of the transporter, as it has been shown that alternating access can be achieved by simply swapping the conformations of each repeat (Forrest et al. 2010).

## Cotransport Mechanisms

Based on the general concept of alternating access, the basic cotransport cycle of the LeuT structural family would work as follows. In the initial state, the binding sites are accessible from the extracellular side only. A $Na^+$ ions first bind to the cotransporter, producing an increase in sugar affinity. Binding of the sugar (with or without a second $Na^+$, depending on the stoichiometry) induces the closing of the extracellular gates, triggering the tilting of the four-helix bundle, resulting in the closing of the extracellular vestibule and opening of the intracellular vestibule. This is followed by the opening of the intracellular gate, allowing intracellular release of $Na^+$ ions and substrate. The empty transporter is then believed to experience a similar process in order to return to its initial state, reexposing the substrate-binding sites to the extracellular side (Fig. 2) (Abramson and Wright 2009). Because dissimilar cotransporters share similar architectures, the homologous members of the SGLT family are believed to share this general cotransport mechanism.

While the binding order in $Na^+$/glucose cotransporters is generally accepted as being the $Na^+$

ions first and then the substrate, the release order remains to be clearly established. Nevertheless, spontaneous unbinding of the Na$^+$ found in the crystal structure of vSGLT during molecular dynamic simulation (Abramson and Wright 2009; Boudker and Verdon 2010) suggests that Na$^+$ ions are first released, followed by sugar release.

## Physiological Relevance of Na$^+$/Glucose Cotransporters

The most important physiological role of Na$^+$/glucose cotransporters is transepithelial transport of glucose in the intestine and the kidney. Essential to the function of the Na$^+$/glucose cotransporters is the presence of a Na$^+$ gradient across the apical membrane. This Na$^+$ gradient is generated by the Na$^+$/K$^+$-ATPase. The Na$^+$/K$^+$-ATPase is located on the basolateral side of epithelial cells and actively pumps three Na$^+$ ions out of the cell in exchange for two K$^+$ ions transported in, using energy from the hydrolysis of one ATP molecule. The energy of this gradient allows Na$^+$/glucose cotransporters to pump glucose across the apical membrane. Glucose molecules will then passively permeate through the basolateral membrane through facilitated diffusion glucose transporters (GLUTs), which are part of the SLC2 family. Taken together, these three groups of proteins (Na$^+$/K$^+$-ATPases, SGLTs and GLUTs) allow transepithelial absorption of glucose (Fig. 3). This general mechanism is followed in both the intestine and kidney (Sabino-Silva et al. 2010).

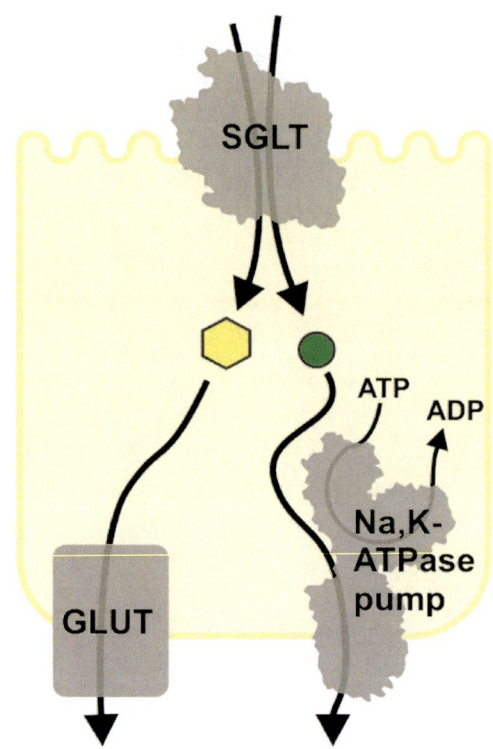

**Sodium/Glucose Co-transporters, Structure and Function, Fig. 3** *Transepithelial absorption of glucose.* The Na$^+$/K$^+$-ATPase uses energy from the hydrolysis of ATP to actively pump Na$^+$ outside the cell, effectively lowering intracellular Na$^+$ concentration, and creating a transmembrane Na$^+$ gradient. Na$^+$/glucose cotransporters use this gradient to achieve secondary-active cotransport of sugar (*yellow hexagon*) and Na$^+$ (*green circle*) across the apical membrane. Sugars can then cross the basolateral membrane through the passive transporters GLUTs

### Renal Reabsorption of Glucose

SGLT2 is a low-affinity/high-capacity cotransporter, while SGLT1 is a high-affinity/low-capacity transporter. This allows them to play complementary roles in the reabsorption of glucose in the kidney. SGLT2 is expressed in the early proximal convoluted tubule of the kidney. Because of its high capacity, it accounts for up to 90% of total renal glucose reabsorption. Expressed in the late proximal straight tubule, SGLT1 is able to mediate the remaining 10% of total renal glucose reabsorption due to its high affinity. This combination accounts for the nearly complete glucose reabsorption which corresponds to approximately 180 g of glucose per day under normal circumstances (Chao and Henry 2010).

### Regulation of Na+/Glucose Cotransporter

The physiological functions of SGLTs are known to be regulated through two major mechanisms: transcriptional regulation and controlling the distribution of the transporter.

The hepatocyte nuclear factors 1α and 1β (HNF-1α et HNF-1β) have been shown to be essential regulators of SGLT1 expression, and it was shown that HNF-1α directly controls SGLT2 expression in mouse and man (Sabino-Silva et al. 2010).

Protein kinase A and C (PKA and PKC) have been shown to regulate SGLT1. They are believed to control the distribution of cotransporters between the plasma membrane and intracellular compartments by regulating the rates of insertion into or retrieval from the plasma membrane (Sabino-Silva et al. 2010).

Finally, SGK1 and SGK3 have been shown to promote the activity of SGLT1 in *X. laevis* oocytes via phosphorylation of Nedd4-2 (Dieter et al. 2004).

## Glucose-Galactose Malabsorption and Familial Renal Glycosuria

Defects in the expression of SGLT1 in the intestine is known to cause glucose-galactose malabsorption syndrome (GGM). GGM is caused by the lack of absorption of glucose and galactose in the intestine, causing life-threatening diarrhea in newborn infants (Wright and Turk 2004). More than 30 different GGM-causing mutations have been identified in the SLC5A1 gene. Most of these mutations cause a defect in the targeting of SGLT1 to the apical membrane but, in a few cases, SGLT1 is thought to be present in the plasma membrane but nonfunctional (Bergeron et al. 2000).

Defect in the expression of SGLT2 is associated with familial renal glycosuria (FRG). FRG causes increased loss of glucose through urine, but otherwise does not cause significant pathological problems (Chao and Henry 2010).

## Pharmaceutical Relevance

Oral rehydration therapy (ORT) is a remarkably straightforward treatment preventing the dehydration associated with infectious diarrhea. Oral administration of an aqueous solution containing $Na^+$ and glucose stimulate cotransport by SGLT1 in the intestine, which in turn promotes water absorption through osmosis (Sabino-Silva et al. 2010).

Suppression of SGLT2-mediated cotransport results in increased excretion of glucose, lowered plasma glucose levels, and little adverse effects. This provides a novel therapeutic strategy for the treatment of type 2 diabetes. Inhibiting SGLT2 would help diabetic patients to regulate their glycemia and prevent adverse effect of hyperglycemia on several organs including the kidney itself (diabetic nephropathies). Use of phlorizin, the classic SGLT inhibitor, is prohibited due to its low bioavailability, its lack of specificity for SGLT2, and the blocking effect of one of its metabolites (phloretin) on GLUT1. This has encouraged the development of novel SGLT2-specific inhibitors based on phlorizin. Some of these drugs are currently in Phase III trials (Chao and Henry 2010).

## Summary

$Na^+$/glucose cotransporters are responsible for the secondary-active transport of a number of substrates, using the electrochemical energy of the transmembrane $Na^+$ gradient. Recently resolved atomic structures show conservation of a core architecture between members of unrelated gene families, suggesting similar underlying transport mechanisms. Being responsible for the absorption of glucose, $Na^+$/glucose cotransporters represent a novel and promising therapeutic approach to type 2 diabetes.

## Cross-References

▶ Sodium, Physical and Chemical Properties
▶ Sodium-Binding Site Types in Proteins
▶ Sodium-Coupled Secondary Transporters, Structure and Function

## References

Abramson J, Wright EM (2009) Structure and function of Na + -symporters with inverted repeats. Curr Opin Struct Biol 19(4):425–432

Bergeron M, Goodyer PR, Gougoux A, Lapointe JY (2000) Pathophysiology of renal hyperaminoacidurias and glucosuria. In: Giebish DWSG (ed) The Kidney, vol 2, 3rd edn. Lippincott Williams & Wilkins, New-York, pp 2211–2233

Bizhanova A, Kopp P (2009) The sodium-iodide symporter NIS and pendrin in iodide homeostasis of the thyroid. Endocrinology 150(3):1084

Boudker O, Verdon G (2010) Structural perspectives on secondary active transporters. Trend Pharmacol Sci 31(9): 418–426

Chao EC, Henry RR (2010) SGLT2 inhibition – a novel strategy for diabetes treatment. Nat Rev Drug Discov 9(7):551–559

Coady MJ, Wallendorff B et al (2007) Establishing a definitive stoichiometry for the Na+/monocarboxylate cotransporter SMCT1. Biophys J 93(7):2325–2331

Dieter M, Palmada M et al (2004) Regulation of glucose transporter SGLT1 by ubiquitin ligase Nedd4-2 and kinases SGK1, SGK3, and PKB&ast; &ast. Obesity 12(5):862–870

Forrest LR, Krämer R et al (2010) The structural basis of secondary active transport mechanisms. Biochim Biophys Acta (BBA) Bioenergetics 1807:167–188

Ganapathy V, Thangaraju M et al (2008) Sodium-coupled monocarboxylate transporters in normal tissues and in cancer. AAPS J 10(1):193–199

Lahjouji K, Aouameur R et al (2007) Expression and functionality of the Na+/myo-inositol cotransporter SMIT2 in rabbit kidney. Biochim Biophys Acta (BBA)-Biomembranes 1768(5):1154–1159

Leung D, Turk E et al (2002) Functional expression of the vibrio parahaemolyticus Na+/galactose (vSGLT) cotransporter in xenopus laevis oocytes. J Membr Biol 187(1):65–70

Prasad PD, Ganapathy V (2000) Structure and function of mammalian sodium-dependent multivitamin transporter. Curr Opin Clin Nutr Metab Care 3(4):263

Ribeiro FM, Black SAG et al (2006) The "ins" and "outs" of the high affinity choline transporter CHT1. J Neurochem 97(1):1–12

Sabino-Silva R, Mori R et al (2010) The Na+/glucose cotransporters: from genes to therapy. Braz J Med Biol Res 43(11):1019–1026

Tazawa S, Yamato T et al (2005) SLC5A9/SGLT4, a new Na + -dependent glucose transporter, is an essential transporter for mannose, 1, 5-anhydro-D-glucitol, and fructose. Life Sci 76(9):1039–1050

Wright EM, Turk E (2004) The sodium/glucose cotransport family SLC5. Pflügers Archiv Eur J Physiol 447(5):510–518

# Sodium/Potassium-ATPase Structure and Function, Overview

Dwight W. Martin
Department of Medicine and the Proteomics Center, Stony Brook University, Stony Brook, NY, USA

## Synonyms

ATP phosphohydrolase ($Na^+,K^+$, exchanging); EC 3.6.3.9; Na(+)/K(+)-exchanging ATPase; Na, K-activated ATPase; $Na^+/K^+$ pump; Sodium-potassium pump; Sodium pump; Sodium/potassium-exchanging ATPase; Sodium/potassium-transporting ATPase; Sodium-potassium adenosine triphosphatase; Sodium-potassium-dependent adenosine triphosphatase

## Definition

The $Na^+,K^+$-ATPase is a transmembrane protein found in nearly all animal cells. It couples the energy released from the hydrolysis of the gamma phosphate of ATP to the transport of sodium and potassium ions across the cell membrane in a process called active transport. It is responsible, to a large extent, for the maintenance of transmembrane sodium and potassium concentration gradients and transmembrane electrical potential gradients. These gradients are in turn utilized by numerous biological processes, many of which are essential for higher-order life.

## Historical Perspective

Most cells and particularly animal cells establish low cytoplasmic sodium concentrations. In the animal kingdom, it can be argued that the linchpin of sodium homeostasis is the $Na^+,K^+$-ATPase (a.k.a. $Na^+$-pump). This singular transport protein is responsible for establishing and maintaining an electrochemical gradient that is utilized by numerous cellular, physiological, muscular, and neurological functions. In essence, it is pivotal for much of higher-order life. To place the biological importance of the $Na^+,K^+$-ATPase in perspective, >25% of the ATP in the cells of a resting animal is consumed by this one enzyme. The need for $Na^+$ and $K^+$ exchange and maintenance of $Na^+$ gradients in neuronal tissue was recognized over 100 years ago (Glynn 2002). Molecular insight into the mechanism of $Na^+/K^+$ homeostasis began to take hold in the early 1950s with the development of the red blood cell as an experimental tool for transport studies and later in the same decade Jens Skou identified a $Na^+$- and $K^+$-dependent enzyme that hydrolyzed ATP in neuronal membranes. It was soon realized that the same enzyme was present in both red blood cells and nerve tissue and the ubiquitous nature of this transport system was rapidly established. For the last 50 years, the $Na^+,K^+$-ATPase has been the subject of intense biochemical investigations culminating in atomic resolution crystal structures in recent years. Yet, much of what we know about the structure and function of this enzyme was obtained and predicted in studies using biological systems. Adding to the experimental tissue repertoire, it was discovered that $Na^+,K^+$-ATPase existed naturally in high concentrations in such diverse tissues as shark rectal gland, mammalian kidneys, and avian nasal salt glands. These tissues provided rich sources of $Na^+,K^+$-ATPase needed for biochemical investigations.

## The $Na^+,K^+$-ATPase Transport Reaction Cycle

Early in the investigations of the $Na^+,K^+$-ATPase, it was discovered that ouabain, a steroid originating from plants, was a potent inhibitor of the transporter (Lingrel 2010). The availability of this inhibitor proved to be a valuable tool in the dissection of the transport reaction cycle (analogously, the use of other specific inhibitors

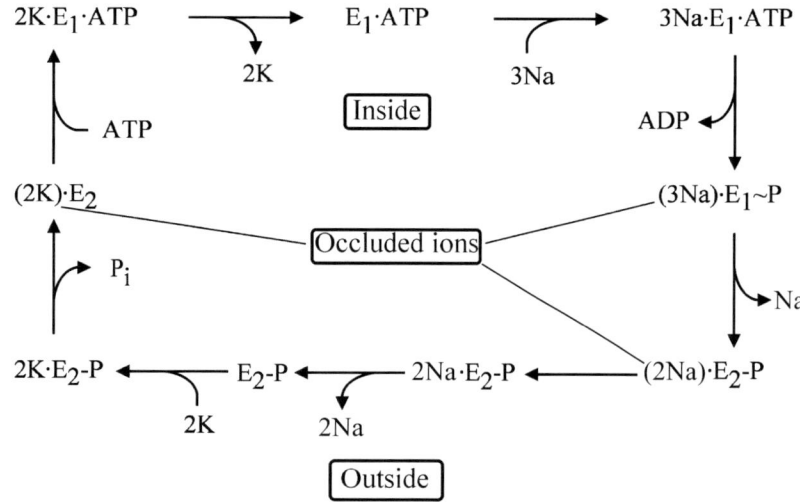

Sodium/Potassium-ATPase Structure and Function, Overview, Fig. 1 The $Na^+$, $K^+$-ATPase transport reaction cycle (Albers–Post Model). The above reaction cycle contains the major steps in the transport process. The *arrows* indicate the normal direction of the cycle; however, all steps are reversible. Inside the reaction *rectangle* represents the cytoplasmic space and outside the rectangle is extracellular space

has aided in the investigation of other transport/channel systems). As early as the late 1950s it was established that the $Na^+,K^+$-ATPase transports 3 $Na^+$ out of the cell in exchange for transporting 2 $K^+$ into the cell. A distinguishing and class-defining feature of the $Na^+$, $K^+$-ATPase is that the enzyme, in the presence of $Na^+$, autophosphorylates by transferring the high energy phosphate from bound ATP to an aspartyl residue of the protein forming a covalently bound phosphorylated reaction cycle intermediate. It is through this intermediate that the enzyme obtains the necessary energy for active transport. This phosphorylation step is characteristic of the family of P-type ATPases of which there are over 300 members widely distributed throughout biology (Thever and Saier 2009; Palmgren and Nissen 2011). In mammalian biology, some of the more recognizable members, in addition to the $Na^+,K^+$-ATPase, include the $Ca^{2+}$-ATPase found in abundance in muscle tissue and present in most cells and the gastric $H^+,K^+$-ATPase (these three are part of the $P_2$ subfamily). This phosphorylated intermediate is one step in a multistep transport reaction cycle that has been fine-tuned over the last few decades but the essence of which was originally proposed in the late 1960s and is commonly known as the Albers–Post (or Post–Albers) reaction cycle (named after the pioneering work of Robert Post and Wayne Albers). While developed from studies of the $Na^+,K^+$-ATPase, there are substantial analogies between the Albers–Post reaction cycle and the reaction cycle of other P-type ATPases. Another key feature of the Albers–Post model is that the enzyme must exist in at least two conformations corresponding to altered states of affinity for $Na^+$ and $K^+$.

Figure 1 presents the major steps in the $Na^+,K^+$-ATPase reaction cycle (Kaplan 2002; Jorgensen et al. 2003). Following the convention of early models, the two conformations are designated as $E_1$ and $E_2$. This should not be confused with enzyme stoichiometry (i.e., monomer-dimer), an issue I will discuss below. The $E_1$ conformation can be characterized as having a high affinity for $Na^+$ ($K_D$ approx. 0.6 mmol/L) and a lower affinity for $K^+$ ($K_D$ approx. 10 mmol/L), with ion-binding sites facing the cytosol. Upon undergoing the conformation change to $E_2$ the ion-binding sites face the extracellular space and now have a lower affinity for $Na^+$ ($K_D > 300$ mmol/L) and a higher affinity for $K^+$ ($K_D < 1.3$ mmol/L). The details of the reaction cycle have been the object of many clever and sophisticated biochemical, biophysical, and molecular biological investigations. Starting at the top, center step, $Na^+,K^+$-ATPase in the $E_1$ conformation binds 3 $Na^+$ ions. One of the $Na^+$ binds with somewhat lower affinity than the other two. The binding of this "third $Na^+$" is voltage sensitive and necessary for the transfer of the $\gamma$-phosphate of ATP to the ATPase. This high energy state of the enzyme is indicated by the $\sim$P. In this state, the ions bond to the enzyme are "occluded" and not accessible to either intra- or extracellular space. During the transition from $E_1$ to $E_2$, the third $Na^+$ is released to the extracellular space followed by the deocclusion and release of the other two $Na^+$ ions from the phosphorylated form of the $E_2$ conformation.

The $E_2$-P state binds 2 $K^+$, the $E_2$ form releases the phosphate and produces an occluded state with two bound $K^+$. ATP binds to this state with low affinity, accelerating the transition to the $E_1$ state and deocclusion and release of the $K^+$ ions into the cytosol producing $E_1$ with ATP bound with high affinity. The above description and arrows in Fig. 1 indicate the normal direction of the reaction cycle. However, all steps are reversible and under the proper conditions it is possible to synthesize ATP from ADP and free inorganic phosphate ($P_i$). Under optimal conditions, the ATPase can cycle in the normal direction with maximum rates near 150 cycles/s.

## Na$^+$,K$^+$-ATPase Structure

Major advances in our understanding of the structure of Na$^+$,K$^+$-ATPase have occurred over the past two decades, culminating in recent atomic resolution crystal structures. The ATPase is composed of $\alpha$ and $\beta$ subunits, both of which are transmembrane proteins. The $\alpha$ subunit has a mass of about 112,000 u. The $\beta$ subunit has a mass based upon amino acid composition of approximately 35,000 u, but it is highly glycosylated and appears as a broad band ranging from $M_r$ 45,000 to 58,000 when analyzed by gel electrophoresis due to the heterogeneous makeup of the attached N-linked oligosaccharides. All known transport and enzymatic activities occur through the $\alpha$ subunit. Although the $\beta$ subunit does not appear to have catalytic activity, removal of the $\beta$ subunit results in loss of Na$^+$,K$^+$-ATPase activity. Studies suggests that the $\beta$ subunit may be involved in trafficking the ATPase to the plasma membrane and may also influence $K^+$ interaction with the $\alpha$ subunit. Recent investigations also implicate the $\beta$ subunit in functions that are unrelated to Na$^+$,K$^+$-ATPase activity (Geering 2008). A third membrane protein ("$\gamma$ subunit") is also closely associated with the $\alpha\beta$ complex in some, but not all, tissues and is part of the FXYD family of proteins (Garty and Karlish 2006). The FXYD proteins are small (6,500–7,500 u) highly hydrophobic, single-span membrane proteins, so named because of the signature sequence on their extracellular N-terminus. Also known as phospholemman in cardiac and skeletal muscle, they are the target of several kinases (e.g., protein kinase A and protein kinase C). Phosphorylation of the $\gamma$ subunit is believed to perform a regulatory role by affecting Na$^+$, K$^+$, and ATP-binding affinities. There are multiple isoforms of the ATPase subunits (four $\alpha$ and three $\beta$). Isoforms $\alpha_1$ and $\beta_1$ are the most ubiquitous, found in most tissues, and the predominant isoform in kidney, red blood cells, and organs involved in ion homeostasis. The other isoforms are more tissue specific: $\alpha_2$ is found in skeletal, cardiac and smooth muscle, brain, lung, and adipose tissue; $\alpha_3$ is expressed in neurons and cardiac cells; $\alpha_4$ has only been found in testes; $\beta_2$ is expressed in neurons and cardiac cells; and $\beta_3$ has been found in testes. In addition to minor sequence variations, the isoforms have different affinities for Na$^+$, K$^+$, and ouabain and can yield altered reaction kinetics compared to $\alpha_1$.

Although the $\alpha\beta$ heterodimeric unit has long been recognized as a required unit for activity, there was a long-standing debate as to whether a higher order of oligomerization (e.g., $(\alpha\beta)_2$) was needed for activity. The arguments favoring higher orders of oligomerization were supported by the common observation that Na$^+$,K$^+$-ATPase units were often found aggregated in tissues with high abundance of units and in preparations solubilized in detergents. Additionally, higher-order oligomerization provided a mechanism to explain complex kinetic observations. Many graduate level biochemical text books continue to illustrate the pump as an aggregate of multiple $\alpha\beta$ units. However, highly purified Na$^+$,K$^+$-ATPase preparations and the application of sophisticated analytical techniques have established that the $\alpha\beta$ heterodimer is the minimal functional unit for ATPase activity and transport. It is also clear from the high-resolution structural data (discussed below) that each $\alpha$ subunit has all the necessary structural elements for ATPase activity and transport. The strong tendency of ATPase units to aggregate remains an observation that continues to simulate investigations into possible regulatory functions. Yet, to this date, no clearly defined function can be identified that requires an oligomeric structure.

## Subunit Structural Detail

The primary structure of the $\alpha$ subunit consists of about 1,016 amino acids. There is considerable sequence conservation throughout the animal kingdom. As an example, comparing the sequences of $\alpha$ subunits from duck nasal salt glands and human kidney reveals

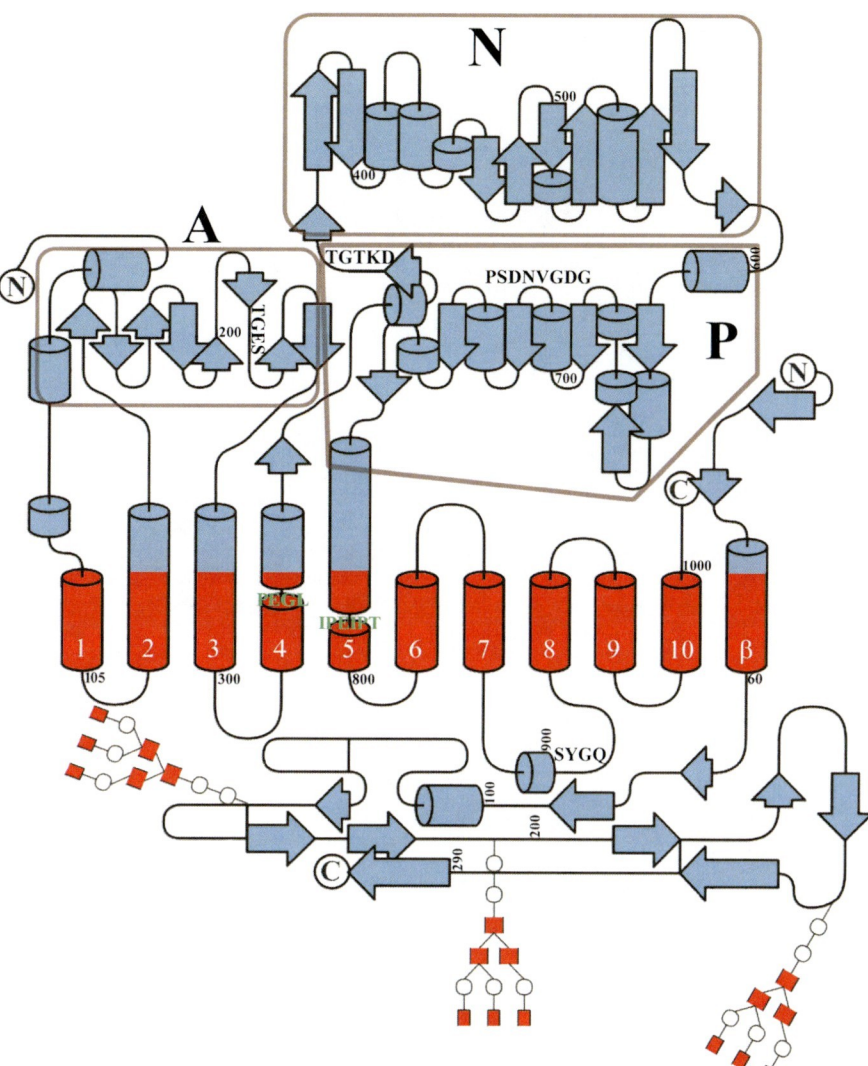

Sodium/Potassium-ATPase Structure and Function, Overview, Fig. 2 Two-dimensional topology representation of the Na⁺,K⁺-ATPase structure. The *red helices* represent hydrophobic transmembrane segments. The more hydrophilic segments are in *blue*. The *small black numbers* indicate approximate location in the primary sequence. The segments that comprise the N, P, and A domains are indicated by the *gray borderlines*. The amino acid motif residues discussed in the text are indicated at their approximate locations along the primary sequence. The motifs in the P domains read *right* to *left* consistent with the direction of the primary sequence illustrated in that domain. The PEGL and IPEIPT motifs, indicated in *green*, in TM4 and TM5, respectively, cause breaks (kinks) in those transmembrane segments. The β chain disulfide bridges, and branched N-linked oligosaccharides are positioned at approximate locations (Modified from Martin (2005))

95.5% similarity and 93.5% identity. The amino acids are folded into α-helices and β-sheets resulting in ten transmembrane helices (TMs) and a large cytoplasmic segment comprised mostly of three domains that are formed, in the main, from two inter-TM peptide loops which comprise about 70% of the total mass of the α-subunit. The three cytoplasmic domains, A (actuator), N (nucleotide binding), and P (phosphorylation) contain the machinery for ATP hydrolysis and substantial domain movement is coupled to affect TM position and the orientation of the ion-binding sites located in the TM residues. Within the primary structure are amino acid sequence patterns (motifs) that are conserved in many P-type ATPases and can be related to functional steps in the transport reaction cycle. Figure 2 shows a two-dimensional topology cartoon model of the Na⁺,K⁺-ATPase which illustrates the major structural domains and the approximate location of six important motifs in the context of the linear amino acid sequence. The motif TGES is in the inter-TM2-TM3 loop of the A domain and is believed to be involved in stabilizing the $E_2P$ conformation. The cytoplasmic loop between TM4 and TM5 contain the other two important domains (N and P). Within the P domain is the DKTGT motif (read right to left in Fig. 2 because of direction of amino acid sequence). This is the P-Type ATPase phosphorylation site signature motif. Additionally, the P domain contains the

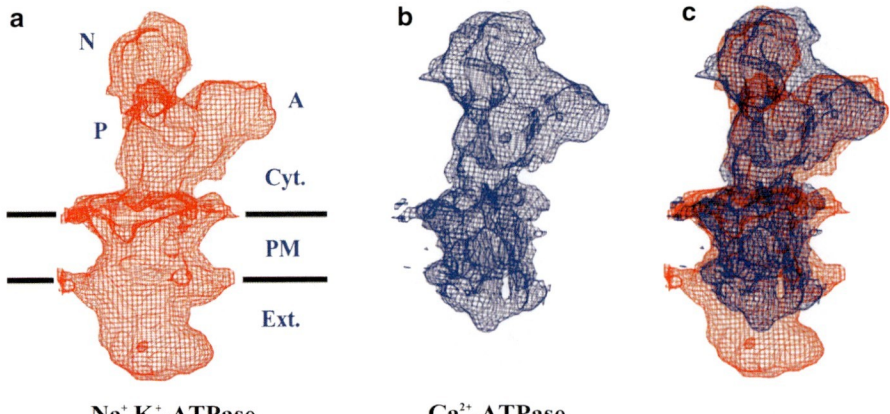

**Sodium/Potassium-ATPase Structure and Function, Overview, Fig. 3** Comparison of $Na^+,K^+$-ATPase and $Ca^{2+}$-ATPase structures obtained from electron micrographs of tubular crystals. Both ATPases were crystallized in the $E_2$ conformation. The $Na^+,K^+$-ATPase has an 11 Å resolution and the $Ca^{2+}$-ATPase is at 8 Å. The location N, P, and A domains are indicated, as is the orientation relative to the cytoplasm (Cyt), plasma membrane (PM), and extracellular (Ext). Panel C shows an overlap of the two structures, clearly highlighting domain structural similarity as well as the distinguishing β subunit of the $Na^+,K^+$-ATPase (Adapted from the data of Rice et al. (2001))

GDGVNDSP motif (also read right to left in figure) which is involved with the stabilization of phosphoenzyme intermediates. TM4 contains the motif PEGL, which resides just above the metal ion-binding sites creating a somewhat unwound section in the helix. TM5 contains the IPEIPT sequence segment which also causes some unwinding and kinking of this helix. Together, TM1-TM6 make up the core of the membrane transport domain and are the minimal segments needed for ion transport by P-type pumps.

Only about 6% of the $Na^+,K^+$-ATPase mass resides in the inter-TM loops that are located in the extracellular space. Of that, the loop between TM7 and TM8 contains most of the mass. Within this loop is the SYGQ sequence motif which is believed to be important for the interaction between the α and β chains. The β subunit has a single transmembrane helix (TMβ) which, as seen in recent high-resolution three-dimensional crystal structures, is somewhat detached from the α subunit but situated closest and parallel to TM7. The extracellular C-terminal segment (~244 amino acids) accounts for about 80% of the β subunit mass. The β subunit primary structure is not as well conserved as that of the α subunit and there is considerable interspecies variability. Generally, there are about three disulfide bonds and three to four sites of N-glycosylation in the extracellular segment. The extracellular segment makes a close association with the SYGQ motif in the TM7-TM8 loop and completely covers the TM5-TM6 and TM7-TM8 loops of the α subunit. This is believed to be consistent with its role in ion access and $K^+$ occlusion during the reaction cycle.

The two-dimensional structure given in Fig. 2 can be instructional in the context of the primary structure; however, protein function is determined by three-dimensional ternary and quaternary interactions. Early success in obtaining three-dimensional structures was obtained through studying specially prepared samples in which the $Na^+,K^+$-ATPase units had formed ordered arrays either in a planar, membrane-like structure or in tubular structures where the units form an ordered helical array. These structures are sometimes collectively called two-dimensional crystals. The arrays are diffracted using an electron microscope and the three-dimensional structure of the ordered units can be reconstructed. Using tubular crystals that provide more angels of diffraction, the entire structure of the $Na^+,K^+$-ATPase was obtained at a resolution of 11 Å. Figure 3 shows a comparison of structures obtained from tubular crystals of both the $Na^+,K^+$-ATPase and the closely related $Ca^{2+}$-ATPase. The structural similarity is obvious and consistent with the many regions of sequence similarity. The additional mass and location of the β subunit in the $Na^+,K^+$-ATPase is also apparent.

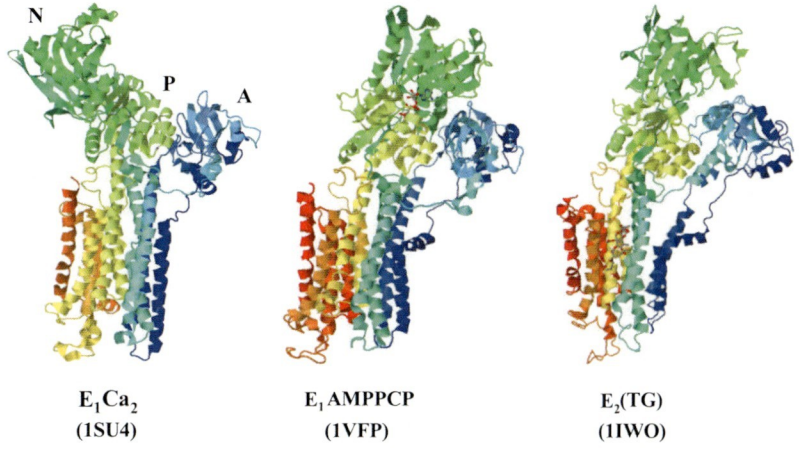

**Sodium/Potassium-ATPase Structure and Function, Overview, Fig. 4** High resolution structures of the $Ca^{2+}$-ATPase crystallized in conformations corresponding to different steps in the ATPase reaction cycle. The structures are aligned with TM10 in approximately the same location. See text for discussion about the conformations. The characters in parentheses are the structure accession numbers. The images were created using JMOL and downloaded from the RCSB PDB (www.pdb.org) from the following PDB ID entries: 1SU4 (Toyoshima et al. 2000); 1VFP (Toyoshima and Mizutani 2004); 1IWO (Toyoshima and Nomura 2002)

Over the past decade, the $Ca^{2+}$-ATPase has yielded to efforts to crystallize it in three-dimensional structures that diffract to produce structures with atomic resolution of a few Å. Using varied substrate conditions it has been possible to generate crystal structures of the ATPase in a number of the conformations associated with specific reaction steps in the $Ca^{2+}$-ATPase reaction cycle that are analogous steps in the $Na^+,K^+$-ATPase reaction cycle. Three conformational structures of the $Ca^{2+}$-ATPase are displayed in Fig. 4. With these structures, it is possible for us to visualize the structural alterations which occur during the energy transduction and transport process (Toyoshima and Inesi 2004). Briefly, starting at the $E_1Ca_2$ conformation ("open conformation") the N, P, and A domains are widely separated. Upon binding ATP (the crystals for the center structure were generated using the ATP analog AMPPCP which fixes the structure to the transition state before transfer of the γ P to the aspartate of the DKTGT motif of the P domain), the N domain, essentially as a rigid body, makes a nearly 90° tilt toward the P domain and the A domain rotates about 30° on an axis parallel to the membrane making contact with the N domain. This leads to the compact cytoplasmic region in the middle structure. Concomitant to the large domain movements are substantial and complex movements in the TM1-TM3 helices (linked to the A domain, Fig. 2) leading to essentially the closing of the gate preventing bound $Ca^{2+}$ from diffusing back into the cytoplasm. The final structure in Fig. 4, $E_2$(TG) was crystallized with the aid of thapsigargin which is a $Ca^{2+}$-ATPase inhibitor that locks the ATPase in the $E_2$ conformation. In this conformation, the domains have separated slightly and the A domain has rotated about 110° around an axis approximately perpendicular to the membrane relative to its position in the $E_1Ca_2$ structure. Associated with these movements are additional complex tilts and rotations of the TM1-TM6. Given the close structural homology between $Na^+,K^+$-ATPase and $Ca^{2+}$-ATPase, it is suspected that the $Na^+,K^+$-ATPase goes through similar domain movements and TM rearrangements and the first opportunities to test this supposition is at hand.

Over the last few years, the $Na^+,K^+$-ATPase has also been crystallized to yield detailed structures at a few Å resolution. At this time, all of the crystal structures that have been analyzed are of essentially the same conformation ($K_2E_2$-P) with and without bound ouabain. A high resolution structure is shown in Fig. 5. Panel 1 displays the ATPase in an orientation similar to that of the structures in Fig. 4. The ATPase is rotated approximately 90° in panel 2 to give a clearer picture of the positions of the β and γ subunits. At atomic resolution, there are also many detailed

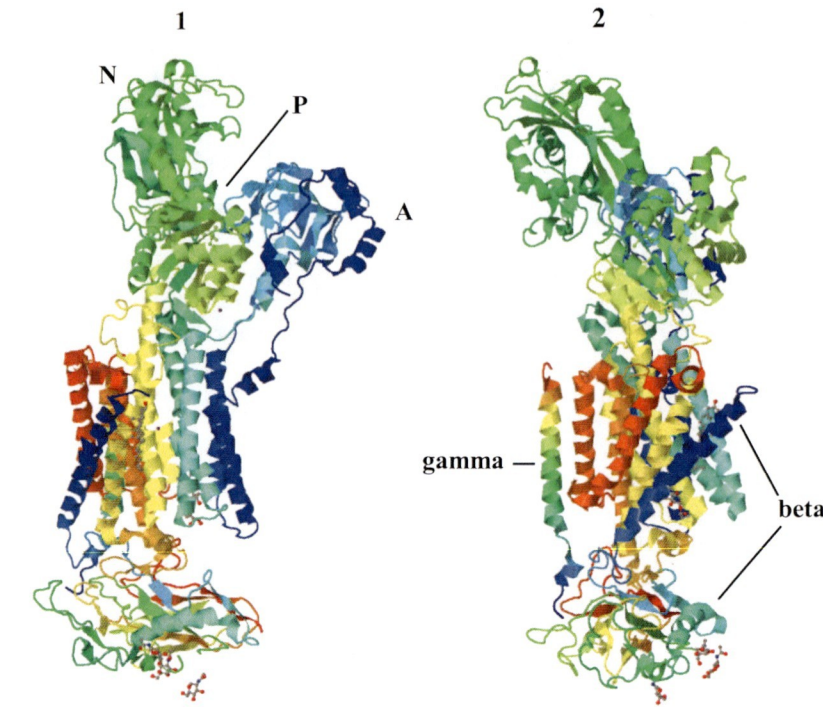

Sodium/Potassium-ATPase Structure and Function, Overview, Fig. 5 High resolution of the structure of the $Na^+,K^+$-ATPase. The crystal was formed of a structure analogous to $K_2E_2$-P using a phosphate analog ($MgF_4^{2-}$) in the presence of bound ouabain. In view 1, the ATPase is rotated to a position analogous to those of the $Ca^{2+}$-pump structures in Fig. 4. In view 2, the structure is rotated approximately 90° on its vertical axis to present a better view of the gamma and beta subunits. The images were created using JMOL and downloaded from the RCSB PDB (www.pdb.org) from PDB ID 3A3Y (Ogawa et al. 2009)

differences in structure between the two pumps, for example, the A and N domains do not interact as closely as they do in the $Ca^{2+}$-ATPase structures and the N domain is about 22º further from the P domain. A more detailed discussion can be found in the references cited. Earlier studies on the $Ca^{2+}$-ATPase crystals prompted investigators to use a technique of homology modeling to predict the $Na^+$- and $K^+$-binding in the $Na^+,K^+$-ATPase from the $Ca^{2+}$-binding site of the $Ca^{2+}$-ATPase structure. The new $Na^+,K^+$-ATPase structures have, in the main, confirmed the results of the homology model. With the new information, a more comprehensive model of ion binding can now be constructed, still drawing upon the information from the multiple structures of the $Ca^{2+}$-ATPase. Figs. 4 and 5 displayed structures which were downloaded from the Protein Data Bank web site (http://www.rcsb.org/pdb/home/home.do). The reader is encouraged to visit this site and view the structures in greater detail with full rotation.

Figure 6 presents the current modeling of the $Na^+$- and $K^+$-binding sites viewed perpendicularly to the plane of the membrane (Morth et al. 2011). This view allows us to address the question of how the changes in ion specificity from $Na^+$ to $K^+$ may be actuated and provides reasonable speculative locations for the third $Na^+$-binding site. As of this writing, only the two $K^+$-binding sites have been identified in crystal structures. A strong rationale based on the homology modeling and mutagenesis studies argues that the two $K^+$ sites are converted into two $Na^+$ sites upon reorientation of the transmembrane helices. The binding site for the third $Na^+$, a unique feature of the $Na^+,K^+$-ATPase, is inferred from structural details of the $K_2E_2$-P crystal, and mutagenesis data.

The ion-binding pocket created by TM1-TM6 is shown by two large blue spheres in both panels of Fig. 6. Panel A shows the $E_2$ form of the enzyme with two bound $K^+$ as small pink spheres labeled K1 and K2. K2 is slightly closer to the extracellular surface. Six of the nine residues associated with binding $K^+$ were found to be the same as those used in binding $Ca^{2+}$ in $Ca^{2+}$-ATPase crystals. Panel B shows the $E_1$ form with two $Na^+$ as small blue spheres labeled Na1 and Na2. The large pink spheres are the regions believed to be involved in binding the third $Na^+$. This region is controlled, in the main, by residues that are part of TM7-TM10, outside the main transport core. The binding of the third $Na^+$, which is voltage sensitive, may actually be stabilized by two binding regions (small yellow spheres, Na3a and Na3b). The $Na^+$-binding sites are created from the

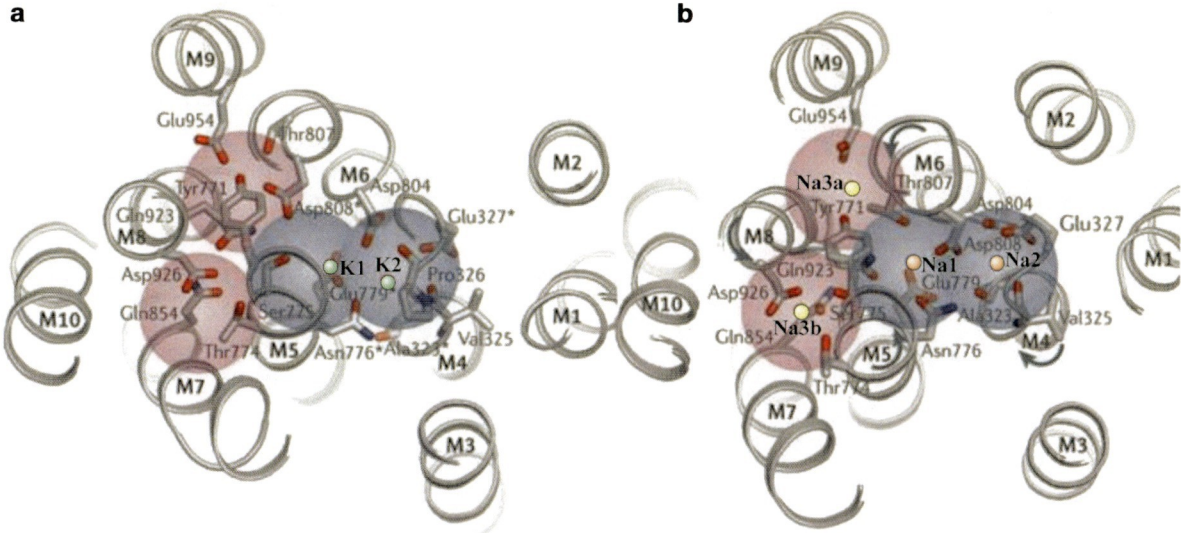

**Sodium/Potassium-ATPase Structure and Function, Overview, Fig. 6** Ion-binding sites of the Na$^+$,K$^+$-ATPase. Panel A: K$^+$-binding sites observed in crystal structures; ions K1 and K2 (small *pink spheres*) interact with residues represented by *sticks*, residues marked with *asterisks* are identical to those of the Ca$^{2+}$-ATPase. The proline (Pro 326), which unwinds TM4 is also indicated by a *stick* structure. Panel B: Homology model of two Na$^+$-binding sites (Na1 and Na2, small *blue spheres*) based on the Ca$_2$E$_1$-P structure of Ca$^{2+}$-ATPase. The *large blue spheres* represent a binding cavity for the two ions. The *large pink spheres* represent proposed binding pockets for the third Na$^+$ (Na3a and Na3b, small *yellow spheres*). The proposed model calls for a counterclockwise rotation of M5, M6, and M8 and a clockwise rotation of M4 (see *arrows*) in going from the E$_2$-P to the E$_1$ conformation (Reprinted by permission from Macmillan Publishers Ltd: Nat Rev Mol Cell Bio. Morth JP, Petersen BP, Buch-Pedersen MJ et al. (2011) A structural overview of the plasma membrane Na$^+$,K$^+$-ATPase and H$^+$-ATPase ion pumps. 12:60–70)

K$^+$-binding sites by rotation of TM4, TM5, TM6, and TM6. The kink created in TM4 by the proline discussed in Fig. 2 plays an important role in this transition. The helical rotations are linked to the large domain movements in the cytoplasmic region discussed above. The actual movements that occur within the binding pocket are subtle, but enough to change the coordinate valences and affinities from that which favors K$^+$ to having a higher affinity for Na$^+$ and lower affinity for K$^+$. That the same residues bind K$^+$ and Na$^+$ within a very tight binding pocket is consistent with consecutive binding models of transport in which K$^+$ is released to the cytoplasmic side before Na$^+$ is bound and transported to the extracellular side of the membrane.

The Na$^+$,K$^+$-ATPase and related P-type ATPases are exquisite machines that convert the stored energy of ATP to transport ions against steep concentration gradients. They accomplish this feat by using large movements of cytoplasmic domains that leverage rotational, lateral, and vertical movements of complex transmembrane helices. The process, in the case of the Na$^+$,K$^+$-ATPase, is made all the more impressive by the fact that, unfettered, it can accomplish this transport cycle at rates of over 150 cycles/s.

## Cross-References

▶ Calcium ATPase
▶ Cellular Electrolyte Metabolism
▶ K$^+$-Dependent Na$^+$/Ca$^{2+}$ Exchanger
▶ Potassium Channels, Structure and Function
▶ Potassium in Biological Systems
▶ Potassium in Health and Disease
▶ Potassium, Physical and Chemical Properties
▶ Sodium Channels, Voltage-Gated
▶ Sodium, Physical and Chemical Properties
▶ Sodium/Glucose Co-transporters, Structure and Function
▶ Sodium-Binding Site Types in Proteins
▶ Sodium-Coupled Secondary Transporters, Structure and Function
▶ Sodium-Hydrogen Exchangers, Structure and Function in Human Health and Disease

## References

Garty H, Karlish SJD (2006) Role of FXYD proteins in ion transport. Annu Rev Physiol 68:431–459

Geering K (2008) Functional roles of Na, K-ATPase subunits. Curr Opin Nephrol Hypertens 17:526–532

Glynn IM (2002) A hundred years of sodium pumping. Annu Rev Physiol 64:1–18

Jorgensen PL, Hakansson KO, Karlish SJ (2003) Structure and mechanism of $Na^+$, $K^+$-ATPase functional sites and their interactions. Annu Rev Physiol 65:817–849

Kaplan JH (2002) Biochemistry of $Na^+$, $K^+$-ATPase. Annu Rev Biochem 71:511–535

Lingrel JB (2010) The physiological significance of the cardiotonic steroid/ouabain binding site of the $Na^+$, $K^+$-ATPase. Annu Rev Physiol 72:395–412

Martin DW (2005) Structure–function relationships in the $Na^+$, $K^+$-pump. Semin Nephrol 25:282–291

Morth JP, Pedersen BP, Buch-Pedersen MJ et al (2011) A structural overview of the plasma membrane $Na^+$, $K^+$-ATPase and $H^+$-ATPase ion pumps. Nat Rev Mol Cell Biol 12:60–70

Ogawa H, Shinoda T, Cornelius F, Toyoshima C (2009) Crystal structure of the sodium-potassium pump ($Na^+$, $K^+$-ATPase) with bound potassium and ouabain. Proc Natl Acad Sci USA 106:13742–13747

Palmgren MG, Nissen P (2011) P-type ATPases. Annu Rev Biophys 40:243–266

Rice WJ, Young HS, Martin DW et al (2001) Structure of $Na^+$, $K^+$-ATPase at 11-Å resolution: comparison with $Ca^{2+}$-ATPase in E1 and E2 states. Biophys J 80:2187–2197

Thever MD, Saier MH Jr (2009) Bioinformatic characterization of P-Type ATPases encoded within the fully sequenced genomes of 26 eukaryotes. J Membr Biol 229:115–130

Toyoshima C, Inesi G (2004) Structural basis of ion pumping by $Ca^{2+}$-ATPase of the sarcoplasmic reticulum. Annu Rev Biochem 73:269–292

Toyoshima C, Mizutani T (2004) Crystal structure of the calcium pump with a bound ATP analog. Nature 430:529–535

Toyoshima C, Nomura H (2002) Structural changes in the calcium pump accompanying the dissociation of calcium. Nature 418:605–611

Toyoshima C, Nakasako M, Nomura H, Ogawa H (2000) Crystal structure of the calcium pump of sarcoplasmic reticulum at 2.6 Å resolution. Nature 405:647–655

## Sodium/Potassium-Exchanging ATPase

▶ Sodium/Potassium-ATPase Structure and Function, Overview

## Sodium/Potassium-Transporting ATPase

▶ Sodium/Potassium-ATPase Structure and Function, Overview

## Sodium-Binding Site Types in Proteins

Bogdan Lev[1], Benoît Roux[2] and Sergei Yu. Noskov[1]
[1]Institute for Biocomplexity and Informatics, Department of Biological Sciences, University of Calgary, Calgary, AB, Canada
[2]Department of Pediatrics, Biochemistry and Molecular Biology, The University of Chicago, Chicago, IL, USA

## Synonyms

Cation binding sites and structure-function relations; $Na^+$-dependent proteins

## Definitions

*Sodium*: An alkali metal (Na) with atomic number 11. Sodium ion ($Na^+$) is essential for numerous physiological functions including participation in the action potential generation, regulation of body fluid volume, and acid-base balance.

*Sodium-binding protein*: Any protein or enzyme that requires the binding of a sodium ion to its structural stability or functional activity.

*Primary Active Transport*: Substrate transport against its concentration gradient coupled to chemical energy, usually from hydrolysis of adenosine triphosphate.

*Secondary Active Transport*: Substrate transport against its concentration gradient coupled to electrochemical gradient created by ion concentration difference between extra- and intracellular milieus.

$Na^+$ is among the most abundant elements on the planet. Several studies regarding the evolution of $Na^+$-binding proteins suggest that its abundance in the water ensures its involvement into biophysical and physiological processes in virtually all-living organisms (Mulkidjanian et al. 2008). It seems likely that $Na^+$ binding played a critical role in stabilizing flexible elements during the early evolution of protein sequences able to adopt stable folds. The binding of $Na^+$ ions to proteins is one of the key regulators of enzyme function, stability, and cell signaling. For example, the binding of $Na^+$ is known to regulate the folding of proteins and nucleic acids and activate enzymatic catalysis. $Na^+$ is also involved in the generation and propagation of electrical signals by excitable cells, where the ionic currents across the cell membrane give rise to a variety of intercellular and intracellular communications. The goal in this section is to give an overview of some of the basic principles of $Na^+$-binding site organization, including common coordination geometries, the types of ligating atoms, and the relation to biochemical function. Important concepts are illustrated using information on $Na^+$ complexes with small organic molecules and several well-characterized $Na^+$-binding proteins.

## General Principles in the Organization of $Na^+$ Sites from a Survey of Known Binding Sites

A comprehensive analysis of ion-ligand distances, geometries, and preferred coordination numbers in $Na^+$-binding proteins has been provided by Harding (2002). Statistics regarding the frequency of the coordination numbers extracted from the Cambridge Structural Database (CSD) and Protein Data Base (PDB) are shown in Fig. 1. The analysis of small molecules in the CSD has shown that $Na^+$ is predominantly coordinated by oxygen ligands (over 90% of all sites), with a mean ion-oxygen distance of 2.42 Å. The coordination distances vary from 2.01 to 2.80 Å, with six being the most common coordination number (~50% of all sites). Penta-coordinated $Na^+$ represents approximately 28% of all sites, while sites with $n = 4$ or 7 represent ~13% and 8% of all reported sites, respectively. A higher numbers of coordinating ligands are observed as well (Fig. 1). It has been hypothesized that partial hydration and the presence of charged ligands could potentially enhance the ability of a protein site to bind $Na^+$ with high affinity and specificity.

Different geometries of the binding site were observed in the CSD and PDB databases. Most of the neutral sites may have distorted tetrahedral, bipyramidal, or octahedral coordinations. Binding sites formed by six or more coordinating ligands contain at least one negative charge and tend to display covalent connectivity (covalently linked cage forming a tight and well-defined cavity). This can be achieved by the engagement of the backbone and side chain atoms from the same amino acid ($i$) or the nearest amino acid neighbor along the sequence ($i$ with $i - 1$ or $i + 1$) forming the binding site. Because backbone atoms are often involved in the coordination of $Na^+$ ions, an a priori prediction of the $Na^+$ site location in a protein with an unknown structure represents a formidable challenge.

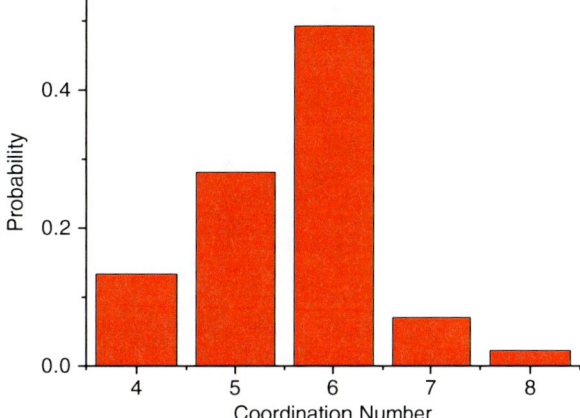

**Sodium-Binding Site Types in Proteins, Fig. 1** Distribution of the coordination numbers for $Na^+$-binding molecules deposited to the CSD and PDB databases

## Examples of the $Na^+$-Binding Sites in Proteins

The list of proteins with selective $Na^+$-binding sites includes a large number of soluble enzymes, such as serine proteases and thrombins as well as membrane proteins such as ATP-driven pumps, channels, and transporters, all of which are of great medical importance. In many cases, proteins contain multiple $Na^+$-binding sites, but do not display any ion-dependent catalytic activity. It has been suggested that ions can stabilize surface-exposed flexible elements, thus enhancing the overall stability under the conditions of high temperatures, pressures, or ionic strengths. Structural organization of $Na^+$-binding sites in proteins and its functional implications will be reviewed.

*$Na^+$-coupled secondary transporters*: Sodium-dependent secondary transporters utilize the electrochemical gradient of $Na^+$ to drive the uphill translocation of a substrate molecule. In mammals, these transporters prevent neurotoxic levels of neurotransmitters, transport nutrients and osmolytes (Abramson and Wright 2009). Despite a low sequence homology, multiple studies have identified striking

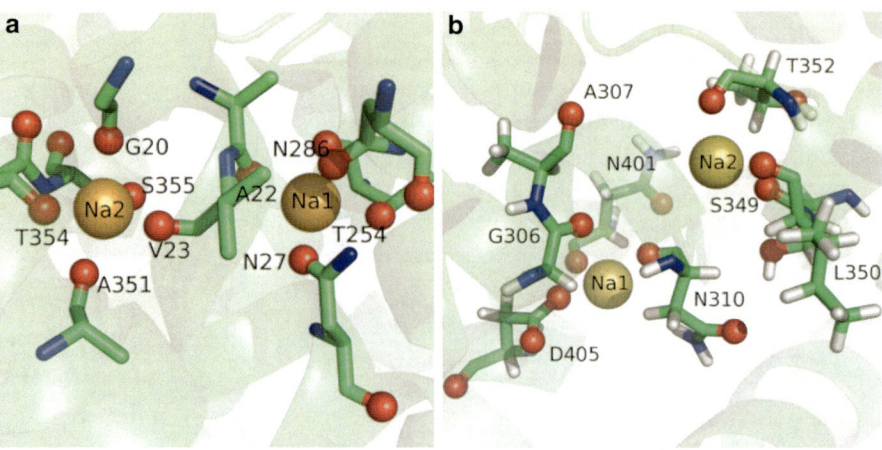

**Sodium-Binding Site Types in Proteins, Fig. 2** Two representative structures of ion-binding sites Na1 and Na2 in the Leucine Transporter LeuT (PDB: 2A65) (**a**) and Aspartate Transporter Glt$_{Ph}$ (**b**) (PDB:2NWX). Bound Na$^+$ ions (*yellow*) and coordinating oxygens (*red*) are shown with CPK representation

similarities in the topological organization of proteins from different families, such as the sodium-coupled leucine transporter LeuT of the NSS family, the sodium-galactose symporter vSGLT of the SSS family, the Mhp1 of the NCS1 family, the BetP of the BCCTs family, and two unrelated proton-coupled amino acid transporters, AdiC and ApcT (Abramson and Wright 2009). The shared structural features among these transporters include a twofold symmetry with inverted repeats and a break in the TM helices that form the substrate and ion-binding pocket, with at least one Na$^+$-binding site preserved across the different families. While transporters can have many distinct conformational states with different ion occupancies, the focus will be on the states for which high-resolution X-ray crystal structures are available.

## Leucine Transporter (LeuT) from the Neurotransmitter Sodium Symporter (NSS) Family

For LeuT, the state with available high-resolution structure (Yamashita et al. 2005) has two Na$^+$ ions bound at sites labeled as Na1 and Na2 with the cotransported solute directly coupled to one of the sites (Na1). It has been suggested that it is the second solute-uncoupled Na$^+$-binding site (Na2) that is conserved among different transporters with known crystal structures. While the exact sequence of events in the transport cycle is a hotly debated topic, there is little doubt that sodium plays central role by supplying the energy that fuels conformational changes.

It has also been suggested that the binding preference (selectivity) of these two sites may be controlled via different mechanisms. This selectivity may be translated into different functional roles for these two binding sites (Noskov and Roux 2008). The organization of the binding pocket is shown in Fig. 2a. A number of factors appear to determine the preferential binding of sodium to LeuT. The high-affinity/high-specificity binding of an ion to the Na1 site relies on the presence of charged carboxylate group donated by the substrate and on the cage of oxygens provided by the protein residues. In molecular dynamics simulations, this site can exhibit significant flexibility while remaining highly selective for Na$^+$. The second site (Na2) provides a rigid cage of ligands coordinating the bound ion, a type of host-guest mechanism leading to selective binding that is common to many ionophores. In this case, the local rigidity could be achieved through a combination of local connectivity with inter-residue hydrogen bonding. It has also been suggested that the presence of a two ion-binding motif plays a significant role in increasing the Na$^+$/K$^+$ selectivity of the protein, which is advantageous for a thermophilic protein such as LeuT.

## Aspartate Transporter from the Excitatory Amino-Acid Transporters (EAAT) Family

Transporters from EAAT family use Na$^+$ gradients as a driving force to transport glutamate against its concentration gradient. The transport event is coupled to the co-transport of three sodium ions and one proton and/or the counter-transport of one potassium ion (Boudker et al. 2007). A bacterial homologue of EAAT transporters, aspartate transporter Glt$_{Ph}$, has been crystallized in three different conformational states with two ion-binding sites exposed in the structure. Although Tl$^+$ was used to label ion-binding sites in the crystal structure, these sites are thought to be Na$^+$ selective. The binding pocket organization is shown in Fig. 2b. The side chains of Asp405, Gly306, Asn401, and Asn310 are involved in the formation of the Na1 sodium-binding site, while the main chains of Ala307, Ser349, Thr352, and Leu350 form Na2 in the protein's closed state (residue numbering is from EAAT). The Na2 site does not exist in the glutamate-free open state of the transporter. It has been proposed that Na$^+$ binding to the transporter leads to the stabilization/destabilization of the extracellular and intracellular hairpin loops, thus controlling the accessibility of substrates, solvents and ions to the binding pocket. A third binding site formed by residues Thr314 and Asn401 with the solute has also been proposed based on a combination of electrophysiological recordings and molecular modeling, which suggests that Na$^+$ may be directly coupled with the substrate. This mode of Na$^+$ coordination has been observed in LeuT from unrelated NSS family of proteins. Detailed mechanisms regarding the Na$^+$ activation of the transporter conformational dynamics have yet to be established.

## Voltage-Gated Na$^+$ Channels

Signaling within nervous networks relies on the electrical signals produced by sustainable electrochemical flows from the concentration gradients of ions across a cell membrane. Such concentration gradients are maintained by combination of active and passive ion transport powered by a number of membrane proteins, including ion channels, pumps, and transporters. The generation of an action potential and the propagation of the nerve impulse arise from the opening and closing of Na$^+$ and K$^+$ channels. A major breakthrough in our understanding of the Na$^+$-binding site organization in voltage-gated Na$^+$ channels was recently made when the first crystal structure of one of the voltage-gated sodium channels (NavAb) of the NaChBac family was reported (Payandeh et al. 2011). Channels from this family are most likely the ancestors of Nav in vertebrate channels and therefore provide a great

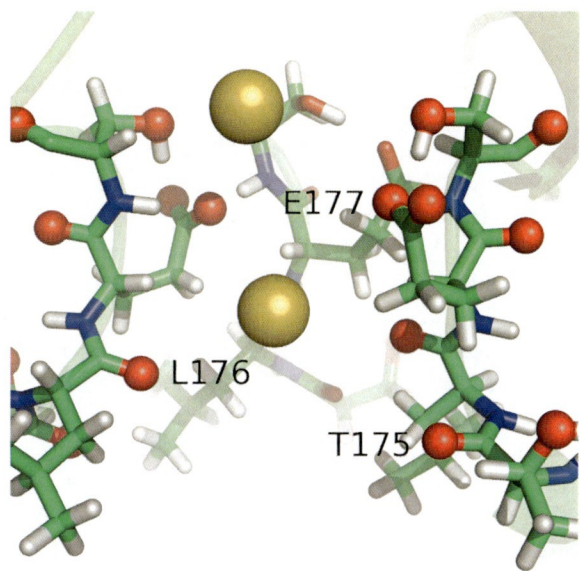

**Sodium-Binding Site Types in Proteins, Fig. 3** Proposed organization of the Na$^+$-binding sites in the selectivity filter of the Nav$_{Ab}$ channel (PDB: 2RVY). Bound Na$^+$ ions (*yellow*) and coordinating oxygens (*red*) are shown with CPK representation

structural example for the organization of the Na$^+$ binding site. It is important to emphasize that ion channels evolved to combine the selective binding of transported cations with rapid permeation, which imposes additional restraints on the structure of the binding pocket. The crystal structure of this channel shows that a pathway for ion entry from the bulk solution is created by a line of electronegative residues that funnel an ion into the narrowest part of the pore: the so-called selectivity filter showed in Fig. 3b.

Residues Thr175, Leu176, Glu177, Ser178, and Trp179 from four identical monomers form the Nav$_{Ab}$ selectivity filter. Glu177 appears to be well conserved among other sodium/calcium channels. The region from residue 175 to residue 179 forms tight turns, exposing the backbone carbonyls of Thr175 and Leu176 to ions. Most of the other residues are involved in additional hydrogen bonding that adds to the stability of the filter. Although the overall fold of the channel, with a "pore loop" located between two transmembrane α-helices, is similar to that of potassium channels, the selectivity filter of Nav$_{Ab}$ is much wider, which allows the passage of Na$^+$ in its hydrated or partially hydrated forms. In order to access the filter, hydrated ions should partially dehydrate in order to go through. The size of the pore allows direct interactions between the ion and one of the glutamates supplied by a number of hydrogen bonds with waters from the first solvation shell of the ion. By providing a line of polar and negatively charged residues, the channel is able to compensate for the partial dehydration penalty. It is worth mentioning that the selectivity of the channel for $Na^+$ over $Ca^{2+}$ can be switched by a single mutation that introduces additional negative charge into the selectivity filter. This structure reveals that there are important differences between the selectivity filters of $Na^+$ and $K^+$ channels. The canonical selectivity filter in $K^+$ channels provides an almost complete dehydration for a permeating ion as they are coordinated by backbone carbonyl oxygen. In $Na^+$ channels, the filter is much wider, allowing for partial ion hydration and creating an asymmetric coordination sphere for a bound ion. Analysis suggests that there may be up to two binding sites in the selectivity filter of $Nav_{Ab}$, suggesting the possibility that permeation may rely on a multi-ion knock-on mechanism as in $K^+$ channels.

## ATP-Synthase

One of the most important molecular systems evolved is a class of proteins known as ATP-synthases. ATP-synthases couple the electrochemical gradient from the directional movement of ions to the synthesis of ATP. ATP molecules can later be used for a variety of cellular processes, ranging from the maintenance of an ionic balance to efficient signaling and volume control. In most biological systems ATP synthesis is coupled to proton transport, but several bacteria evolved to couple it to $Na^+$ transport instead. Available crystal structures revealed that ATP-synthases are proteins with a relatively complex structure. Two major domains can be defined: the $F_1$ region, which is located outside of the membrane, and the $F_0$ region, which is a rotary mechanism embedded in the membrane. The membrane-embedded rotors of $Na^+$-dependent F-ATP-synthases consist of 11 c-subunits that form a c-ring, forming 11 $Na^+$-binding sites in between adjacent subunits. An ion is bound between the N-terminal helix and two C-terminal helices, thus ion binding stabilizes so-called locked conformation of the pump. The ion-binding site is located on the outer surface at the narrowest part of the ring around the middle of the membrane, therefore exposed to hydrocarbon core of a lipid bilayer. It has been hypothesized that the site may be accessed by $Na^+$ ions via two channels created by the c-ring in complex with a subunit, although this mechanism has not been demonstrated. The binding site is formed by the side chains of Glu65 and Gln32 from one of the subunits and side chain of Ser66 and the carbonyl oxygen of Val63 donated by another subunit. The strictly conserved glutamate residue Glu65 is required for favorable $Na^+$ binding to this site. This mode of coordination (involvement of the main-chain, polar side chains and a charged carboxylate group) is observed in other proteins discussed in this entry.

## $Na^+$-Activated Enzymes

$Na^+$-activated enzymes are ubiquitous in the plant and animal kingdoms. $Na^+$ binding plays a very important role in the regulation of enzymatic activity for many proteases, including the well-studied examples of thrombin and the ion-dependent regulation of allostery (Page and Di Cera 2006). While $Na^+$ itself is not catalytically active, its binding to the protein may have several consequences with regard to the enzymatic activity. The structural and functional studies of the FXa protein have allowed us to understand this regulation in greater detail. The binding site for an ion is adjacent to the enzyme's active site. The bound ion is coordinated by six oxygen ligands, in which two ligands are carbonyl oxygens and four ligands are structural water molecules. One of the waters bridges the bound ion to an aspartate residue, which is part of the active center. Ion binding has been suggested to attenuate the protonation ($pK_a$ value) of aspartate, increasing the substrate dissociation rate and thus controlling enzymatic activity of FXa protease.

One of the most remarkable examples of enzymes activated and by $Na^+$ binding is the allosteric regulation of thrombin, which is a member of the blood coagulation cascade (Page and Di Cera 2006). The $Na^+$-bound form of the protein has procoagulant and signaling functions, whereas the $Na^+$-free form is an anticoagulant. The binding of $Na^+$ enhances both the substrate binding and catalytic activity of an enzyme. It has been shown that thrombin activity decreases in the presence of other monovalent cations, such as $Li^+$ or $K^+$, suggesting that the enzyme structure

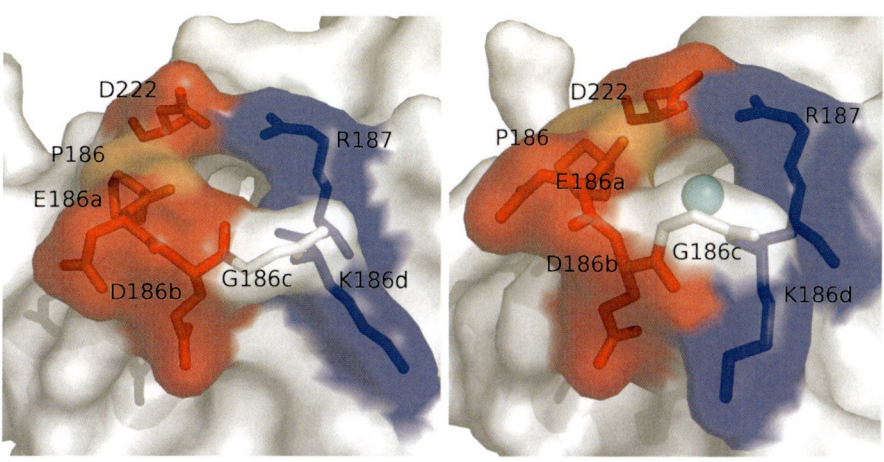

**Sodium-Binding Site Types in Proteins, Fig. 4** Examples of the structural effects of Na$^+$ binding on the stability of the salt-bridges between Lys186d/Asp186b and Arg187/Asp222 (PDB entries are for Na$^+$-free structure:1SGI and Na$^+$-bound structure:1SG8, respectively). The residue numbering corresponds to one used by Page and Di Cera (Page and Di Cera 2006)

is uniquely optimized for Na$^+$ (Page and Di Cera 2006). How is this rather complex biological mechanism achieved? Detailed comparisons of two crystal structures for the Na$^+$-bound and Na$^+$-free forms of thrombin suggest that ion binding induces several critical interactions. The presence of a bound ion leads to the stabilization of a salt-bridge between Arg187 and Asp222 (Fig. 4).

This, in turn, causes a reorientation of Asp186 and breaking of essential salt-bridge, which affects substrate-binding affinity. Na-free form of an enzyme displays considerable widening of the active center, where carbonyl group of Gly186c adopts different orientation. It was also proposed that Na$^+$ binding is essential for retention of the highly ordered water network, which is a requirement for effective catalysis. This is similar to the FXa protease, in which a wire of structural water molecules connects an ion to residues forming the active center. It has been shown that de-wetting of the protein surface in the Na$^+$-free form causes disintegration of the active center.

## Concluding Remarks

The proteins in the cell have evolved to possess a broad diversity of Na$^+$-binding proteins. Unlike many of the physiological cations (Zn$^{2+}$, Ca$^{2+}$, Mg$^{2+}$, etc.), Na$^+$ displays limited ability to act as a catalyst. However, Na$^+$ binding is at the heart of normal physiological function, participating in the finely tuned regulation of Na$^+$/K$^+$ transport across the membrane, attenuating the enzymatic activity of critical proteases and acting as a structural and allosteric agent that affects the activity of enzymes.

## Cross-References

▶ Cellular Electrolyte Metabolism
▶ Potassium-Binding Site Types in Proteins
▶ Sodium as Primary and Secondary Coupling Ion in Bacterial Energetics
▶ Sodium Channels, Voltage-Gated
▶ Sodium-Coupled Secondary Transporters, Structure and Function

## References

Abramson J, Wright EM (2009) Structure and function of Na$^+$-symporters with inverted repeats. Curr Opin Struct Biol 19:425–432

Boudker O et al (2007) Coupling substrate and ion binding to extracellular gate of a sodium-dependent aspartate transporter. Nature 445:387–393

Harding MM (2002) Metal-ligand geometry relevant to proteins and in proteins: sodium and potassium. Acta Cryst D 58:872–874

Mulkidjanian AY et al (2008) Evolutionary primacy of sodium bioenergetics. Biol Direct 3:13

Noskov SY, Roux B (2008) Control of ion selectivity in LeuT: two Na$^+$ binding sites with two different mechanisms. J Mol Biol 377:804–818

Page MJ, Di Cera E (2006a) Is Na$^+$ a coagulation factor? Thromb Haemost 95:920–921

Page MJ, Di Cera E (2006b) Role of Na$^+$ and K$^+$ in enzyme function. Physiol Rev 86:1049–1092

Payandeh J, Scheuer T et al (2011) The crystal structure of a voltage-gated sodium channel. Nature 475:353–358

Yamashita A et al (2005) Crystal structure of a bacterial homologue of Na$^+$/Cl–dependent neurotransmitter transporters. Nature 437:215–223

# Sodium-Coupled Secondary Transporters, Structure and Function

Chunfeng Zhao and Sergei Yu. Noskov
Institute for Biocomplexity and Informatics and Department of Biological Sciences,
University of Calgary, Calgary, AB, Canada

## Synonyms

LeuT; LeuT$_{Aa}$; Secondary active transporter; Secondary transporter

## Definition

Secondary active transporters are membrane proteins that move a substrate (main substrate) across the cell membrane against its concentration gradient, utilizing the free energy stored in the downhill concentration gradient of one or more coupled substrates, usually ions. *Active*, in contrast to passive, indicates that the movement of the main substrate is against its concentration gradient, in other words, uphill. *Secondary*, in contrast to primary, indicates that the energy source of the active transport is not directly from the hydrolysis of ATP (adenosine triphosphate), instead the transporter utilizes a gradient established by a primary transporter, which is often ATP requiring.

Na$^+$-coupled secondary transporters refer to secondary active transporters that utilize the free energy stored in the electrochemical potential difference of sodium ions in and out of the cell membrane. This electrochemical potential difference is primarily established by the Na$^+$/K$^+$ATPase, which pumps sodium (Na$^+$) out of the cell and potassium (K$^+$) into the cell.

## Introduction

Na$^+$-coupled secondary transporters include integral membrane proteins that actively transport various solutes across the cell membrane. For example, Na$^+$/neurotransmitter transporters couple the uptake of a neurotransmitter with one or more Na$^+$ ions, removing neurotransmitters from the synaptic cleft (Krishnamurthy et al. 2009). Another family of secondary transporters, the Na$^+$/glucose transporters transport glucose molecule coupled with one or more Na$^+$ ions, playing an important role in small intestines. Malfunction of Na$^+$-coupled transporters are implicated in various diseases and syndromes, including depression, epilepsy, and some severe congenital diseases (Abramson and Wright 2009). As a result, Na$^+$-coupled secondary transporters are designated targets for drugs in the treatment of depression, diabetes, and obesity (Krishnamurthy et al. 2009; Abramson and Wright 2009).

The general accepted model explaining the mechanism of secondary active transport is the alternating access model proposed by Jardetzk in 1966 (Jardetzk 1966). Briefly, the transporter protein changes conformations between the outward-facing and inward-facing conformations, allowing the main substrate and ion(s) to bind from one side and to unbind from the other side. However, microscopic mechanisms of Na$^+$-coupled active transport of solutes remains largely unknown until the recent surges in 3D (three-dimensional) structure determination of many bacterial Na$^+$-coupled secondary transporters (Krishnamurthy et al. 2009; Abramson and Wright 2009; Forrest et al. 2011). This entry focuses on the structures and transport mechanisms of two transporters: the aspartate transporter Glt$_{ph}$, a homologue of the human glutamate transporters, and the leucine transporter (LeuT$_{Aa}$), a homologue of the human serotonin transporter. These structures represent two known main folds (essential topological arrangements) of Na$^+$-coupled secondary transporters. They provide extensive insights on the mechanisms of Na$^+$-coupled active transport, relating structure to function.

## Structure of the Na$^+$-Coupled L-Aspartate Acid Transporter Glt$_{ph}$

One of the first Na$^+$-coupled secondary transporters that a crystal structure has been determined for is the Na$^+$-coupled aspartate transporter from *Pyrococcus horikoshii* (Glt$_{ph}$) (Yernool et al. 2004). Glt$_{ph}$ is a homologue of the excitatory amino acid transporters (EAATs). The EAAT family includes the Na$^+$-coupled

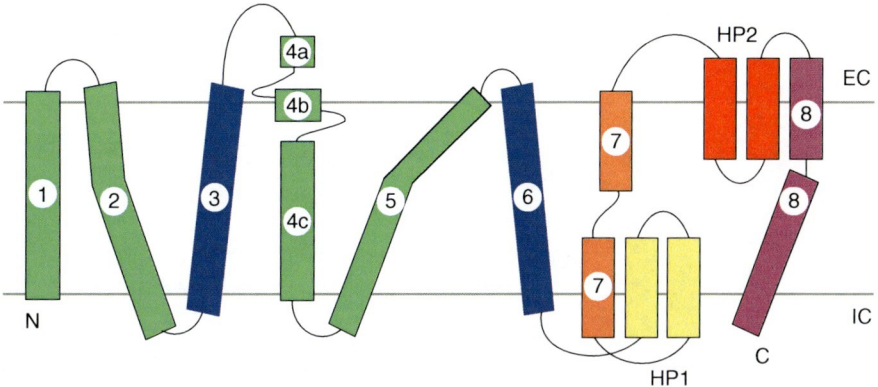

**Sodium-Coupled Secondary Transporters, Structure and Function, Fig. 1** Illustration of the topology of a protomer of $Na^+$-coupled L-aspartate acid transporter $Glt_{ph}$. The reported crystal structure (Yernool et al. 2004) is a trimer formed by three protomers. Each single protomer has eight helical transmembrane segments (TMs 1–8) and two helical hairpins (HPs 1–2). TMs 1,2,4,and 5 (*in green*) form the trimerization domain, while TMs 3,6,7, and 8, and the two HPs 1–2 (not in green) form the transport domain (Reyes et al. 2009)

glutamate transporter, which is responsible for the removal of glutamate from glutamatergic synapses in the human nervous system. Three structures have been determined for $Glt_{ph}$: an outward-facing occluded structure with an L-aspartate substrate and two $Tl^+$ ions (i.e., thallium, as sodium replacement) bound (pdb entry 2NWX), an outward-facing structure with a non-transportable substrate analogue L-threo-β-benzyloxyaspartate (TBOA) and just one $Tl^+$ bound (pdb entry 2NWW), and an inward-facing occluded structure with an L-aspartate substrate and 2 $Na^+$ bound captured for a $Hg^{2+}$-cross-linked double cysteine mutant $Glt_{ph}$ K55C/A364C (PDB entry 3KBC) (Reyes et al. 2009). In all structures, the transporters are in a homotrimeric structure (i.e., a trimer formed by three single protomers), but the transport function is most likely exerted independently by each protomer (Yernool et al. 2004).

A protomer of $Glt_{ph}$ has 425 amino acid residues. The topology of $Glt_{ph}$ is illustrated in Fig. 1. Each protomer has eight helical transmembrane segments (TMs 1–8) and two helical hairpins (HPs 1–2). These elements can be partitioned to two domains: the trimerization domain consisting of TMs 1, 2, 4, and 5, and the transport domain consisting of TMs 3, 6, 7, and 8 and HPs 1–2. Within the transport domain, TMs 7–8 and HPs 1–2 bind to the substrate and two sodium ions directly, forming the substrate-binding core (Yernool et al. 2004; Reyes et al. 2009).

## Transport Mechanism of $Glt_{ph}$: Shuttled Alternating Access

The transport mechanism of $Glt_{ph}$ can be described as "shuttled" alternating access. When the transporter change from outward-facing occluded to inward-facing occluded state, the trimerization domain (TMs 1, 2, 4, and 5) remains largely unchanged. However, within the transport domain (TMs 3, 6, 7, and 8 and HPs 1–2), the peripheral lipid-facing hydrophobic TM3 and TM6 traverse the lipid bilayer directly, moving toward the intracellular side by ~12Å. These two TMs serve as two arms extended from the trimerization domain, holding the substrate-binding core of the transport domain (TMs 7–8 and HPs 1–2). As a result, HPs 1–2 move as much as 20 Å across the lipid bilayer toward the intracellular side, facilitated by the intra-protein track provided by the trimerization domain. Other than the descent toward the intracellular side, the transport domain also rotates about 37° around an axis passing through the transport domain center of mass and loops TM2-TM3 and TM5-TM6 achieved by the hinge movements in these loops (Figs. 2, 3). Accompanying these large-scale conformational changes of the transport domain, the substrate and $Na^+$-binding sites are translocated. In the outward-facing occluded structure, the L-aspartate substrate is bound ~5Å beneath the extracellular surface and occluded from the extracellular solution by HP2.

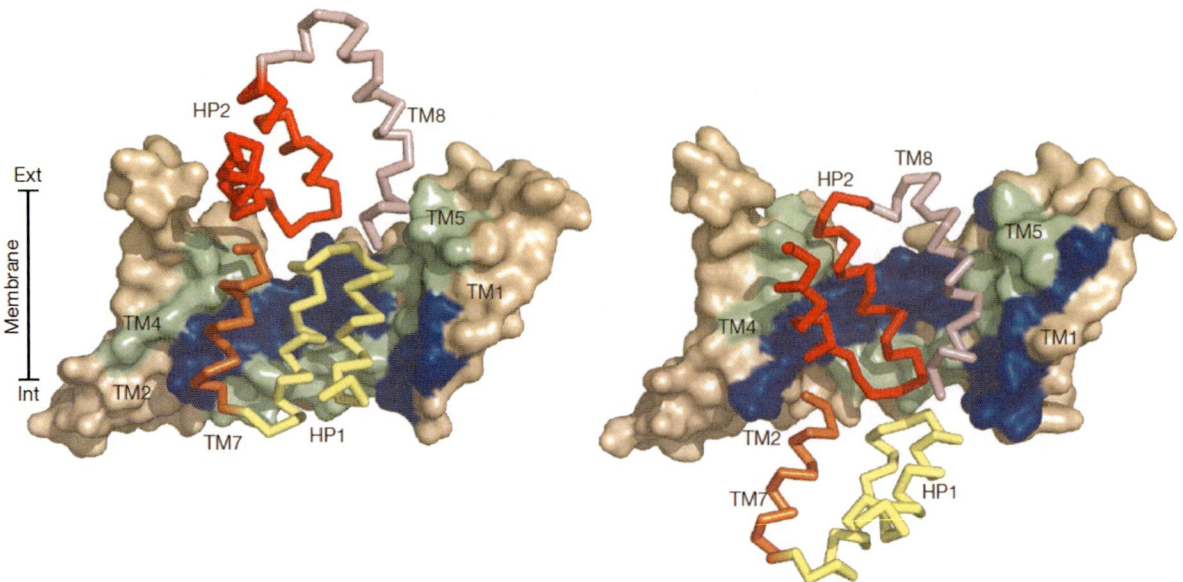

**Sodium-Coupled Secondary Transporters, Structure and Function, Fig. 2** Illustration of the movement of the substrate-binding core of the transport domain (TMs 7–8 and HPs 1–2) for the transition from the outward-facing occluded conformation (*left figure*, Glt$_{ph}$) to the inward-facing occluded conformation (*right figure*, Glt$_{ph}$ K55C/A364C Hg$^{2+}$). The substrate-binding core of the transport domain is shown in *ribbon* representation, and the trimerization domain (TMs 1–2 and 4–5) are shown in *surface* representation. For the conformational transition, the trimerization domain stays relatively unchanged while the transport domain experiences substantial movements toward the intracellular side, accompanied by a rotation of the substrate-binding core (Reprinted by permission from Macmillan Publishers Ltd: Nature, (Reyes et al. 2009), copyright 2009)

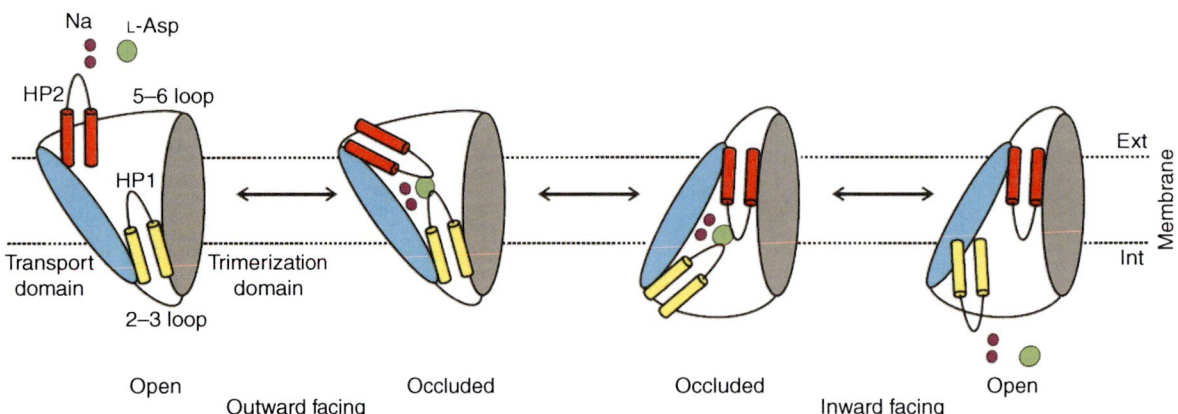

**Sodium-Coupled Secondary Transporters, Structure and Function, Fig. 3** Proposed (Reyes et al. 2009) transport cycle for the Glt$_{ph}$ transporter. A single protomer is shown. The trimerization domain is illustrated in *gray*. TMs 3,6,7, and 8 of the transport domain are illustrated in *blue*. HP1 and HP2 are shown in *yellow* and *red*, respectively. In the outward-facing and the inward-facing occluded states, the extracellular and intracellular gates, HP2 and HP1, respectively, are closed. Transition between the outward- and the inward- facing occluded states occurs upon a movement of the transport domain relative to the stable trimerization domain, consisting of a "descendent" toward the intracellular side and a "rotation" (Reprinted by permission from Macmillan Publishers Ltd: Nature, (Reyes et al. 2009), copyright 2009)

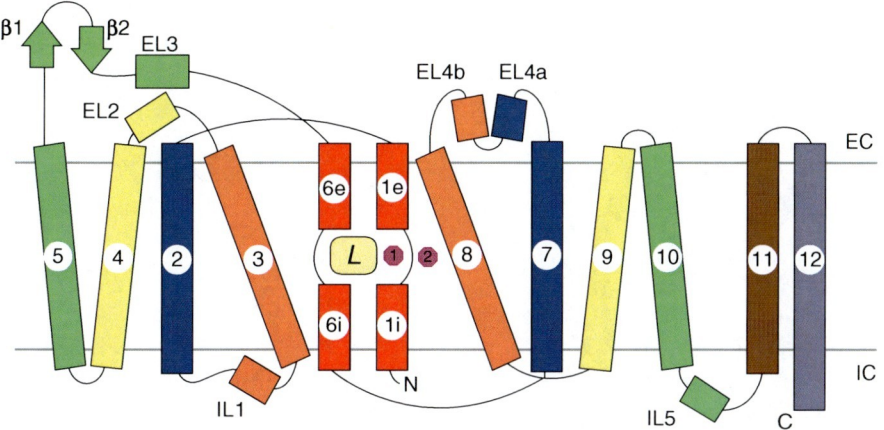

**Sodium-Coupled Secondary Transporters, Structure and Function, Fig. 4** Illustration of the topology of the Na$^+$-coupled leucine transporter LeuT$_{Aa}$ (Yamashita et al. 2005). It consists of 12 helical transmembrane segments (TMs 1–12), for which 10 are essential (TMs 1–10). TMs 1–5 and TMs 6–10 form an inverted repeat. There are unwounded regions in the middle of TMs 1 and 6, which form part of the substrate and Na$^+$-binding sites

In the inward-facing occluded structure, the substrate is bound ~5Å beneath the intracellular surface and occluded from the intracellular solution by HP1. In summary, the transport domain acts like a "shuttle bus" that moves between the extracellular and the intracellular side against a stable "track" formed by the trimerization domain. Coupled to this process is the translocation of the L-aspartate substrate and Na$^+$ ions from the extracellular to the intracellular side.

To complete the cycle, HP2 serves as an extracellular (EC) gate for the binding of the substrate and Na$^+$ ions, and HP1 may serve as an intracellular (IC) gate for the releasing of the substrate and Na$^+$ ions. Coupling between the transport cycle and the energy stored in the Na$^+$ gradient is proposed to be via synergistic binding of the substrate and ions. In other words, binding of Na$^+$ ions facilitates the binding of L-aspartate substrate and vice versa. Figure 3 illustrates the complete transport cycle based on the current understanding (Reyes et al. 2009): Starting from the outward-facing open state, the substrate and Na$^+$ ions bind to the protein and induce the closure of the extracellular gate HP2; The conversion from the outward-facing occluded state to the inward-facing occluded state, featuring the shuttling and rotation of the transport domain, is proposed to be driven by thermal fluctuation. At the inward-facing occluded state, the intracellular gate, HP1 is open and the substrate and ions are released. The apo-transporter is converted back from the inward-facing open state to the outward-facing open state, again possibly driven by thermal energy, and thus restarts the cycle (Reyes et al. 2009).

## Structure of the LeuT-Fold Na$^+$-Coupled Secondary Transporters

Starting from the leucine transporter LeuT from *Aquifex aeolicus* in 2005 (Yamashita et al. 2005), series of crystal structures of Na$^+$-coupled secondary transporters with the same fold, in other words, essential topological organization, as LeuT have been reported (Krishnamurthy et al. 2009; Abramson and Wright 2009; Forrest et al. 2011). These transporters include the Na$^+$/galactose transporter vSGLT from *Vibrio parahaemolyticus*, the Na$^+$/benzyl-hydantoin transporter Mhp1 from *Microbacterium liquefaciens*, and the Na$^+$/glycine betaine transporter BetP from *Corynebacterium glutamicum*. Although these transporters belong to different protein families, they all have the essential LeuT-fold topology features: First, they all have ten essential transmembrane (TM) α- helices, and there is an internal inverted symmetry within these ten essential helices (5 + 5). Second, there are unwounded regions (helix breaks) in two of the essential helices that form part of the substrate and Na$^+$-binding pockets.

The topology for LeuT is illustrated in Fig. 4. The crystal structure of LeuT is shown in Fig. 5 (Yamashita et al. 2005). The structure is in an occluded state with a main substrate, leucine, and two Na$^+$ bound

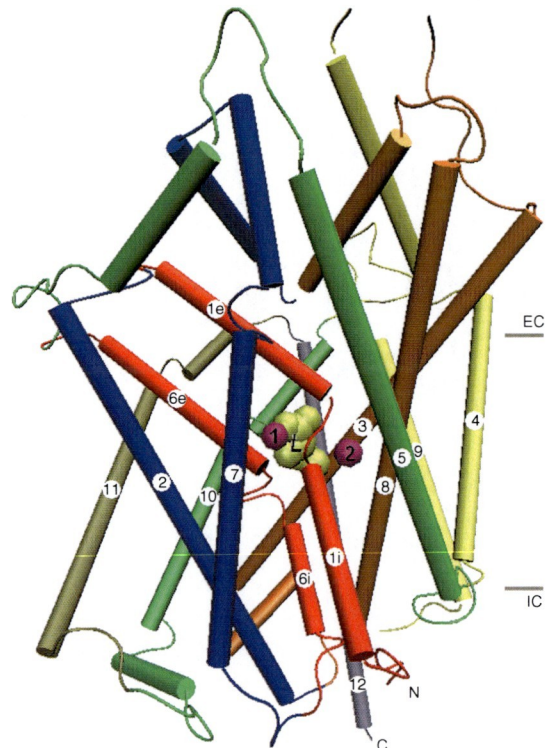

**Sodium-Coupled Secondary Transporters, Structure and Function, Fig. 5** Crystal structure of the $Na^+$-coupled leucine transporter LeuT (pdb entry 2A65) (Yamashita et al. 2005). The protein is shown in *carton* representation. The color scheme is kept consistent as Fig. 4. The leucine substrate is shown in yellow ball representation. It is directly coordinated by residues from TMs 1, 6, and 3 as well as the $Na^+$ in $Na_1$. $Na^+$ in the $Na_1$ and $Na_2$ sodium-binding sites are shown in magenta balls. Other than the carboxyl group of the leucine substrate, $Na^+$ in $Na_1$ is also coordinated by residues from TMs 1, 6, and 7. $Na^+$ in $Na_2$ is coordinated by residues from TMs 1 and 8. EC (*extracellular*) and IC (*intracellular*) roughly indicate the position of the membrane lipid bilayers. The figure is made with the molecular visualization software VMD

in the core of the protein, roughly halfway across the membrane. The protein has 519 residues that form 12 transmembrane helices. However, only 10 of the 12 TMs (TM1–TM10) are essential for the function and there is an internal structural repeat relating TM1–TM5 and TM6–TM10 by a pseudo-twofold axis located in the plane of the membrane (Yamashita et al. 2005). There are helix breaks in TM1 and TM6, roughly in the middle of the membrane bilayers. Residues in the helix breaks adopt extended, non-helical conformation and expose their main chain carbonyl oxygen and nitrogen for hydrogen bonding and substrate/ion binding.

The single leucine-binding site is formed by TMs 1, 6, 3, and 8. For LeuT, two sodium ions, named $Na_1$ and $Na_2$, were identified. $Na_1$ directly binds to the carboxyl group of the leucine substrate. It is also coordinated by residues from TMs 1, 6, and 7. $Na_2$ is roughly 6 Å away from the α-carbon of leucine substrate, and is coordinated by residues from TM1 and TM8.

### Ion/Substrate Stoichiometry and the Conserved Sodium-Binding Site

Unlike its human homologue, the human serotonin transporter (hSERT), which is sodium and chloride dependent, LeuT and other crystallized LeuT-fold transporters (LeuT, Mhp1, vSGLT, BetP) are chloride independent. The mechanism for chloride coupling in hSERT has been elucidated nicely elsewhere by Forrest et al. in 2007 and is beyond the scope of this entry.

The $Na^+$/substrate binding stoichiometries are different among the crystallized LeuT-fold $Na^+$-coupled secondary transporters. LeuT and BetP have a stoichiometry of 2:1 $Na^+$: substrate, while vSGLT and Mhp1 have a 1:1 stoichiometry. Despite the differences in binding stoichiometry, the $Na_2$ sodium-binding site of LeuT is found to be conserved in vSGLT, Mhp1, and potentially BetP (Abramson and Wright 2009). Figure 6 shows the conserved $Na^+$-binding site in LeuT, vSGLT, and Mhp1. The conservation of this site indicates some general principals of coupling between this $Na^+$ with the main substrate.

### Alternating Access Model for $Na^+$-Coupled Secondary Transporters

For the LeuT-fold $Na^+$-coupled secondary transporters, crystal structures have been reported on different conformational states (Fig. 7) (Zhao and Noskov 2011). These structures, together with many other physiological and computational studies (Krishnamurthy et al. 2009; Abramson and Wright 2009; Forrest et al. 2011), provide a picture of the alternating access mechanism for transporters in this structural family. Starting from the outward-facing "ion" state with the conserved $Na^+$ bound, the binding of the main substrate helps to form the occluded state in which the $Na^+$ ion and the substrate are largely "occluded" from the bulk of either side of the membrane bilayers. Next, the $Na^+$ is released to form the inward-facing "sub" state and the substrate is then

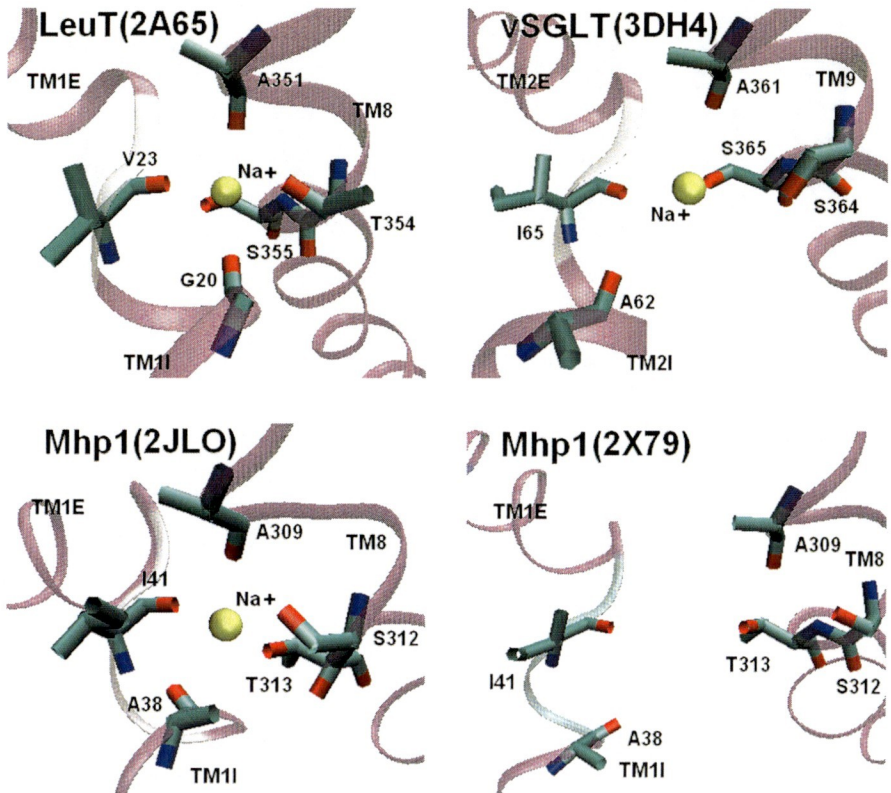

**Sodium-Coupled Secondary Transporters, Structure and Function, Fig. 6** The conserved Na$^+$-binding site (Na$_2$ of LeuT) for the LeuT-fold Na$^+$-coupled secondary transporter. Pdb entry names are shown in the *brackets* after each transporter. LeuT (2A65) (Yamashita et al. 2005) and Mhp1(2JLO) (Shimamura et al. 2010) are in the occluded state with the Na$^+$ site intact, while vSGLT (3DH4) (Abramson and Wright 2009) and Mhp1 (2X79) (Shimamura et al. 2010) are in inward-facing conformations and the Na$^+$ site is (partially) disgruntled. The proteins are shown in *ribbon* representation and colored by the secondary structure: *magenta* for α-helices and *white/blue* for the breaks in the helices. The residues that form the Na$^+$-binding site are shown in stick representation and the Na$^+$ is shown as a yellow ball. Reprinted from Biochimica et Biophysica Acta (BBA) - Biomembranes, 1818, Igor Zdravkovic, Chunfeng Zhao, Bogdan Lev, Javier Eduardo Cuervo, Sergei Yu. Noskov, Atomistic models of ion and solute transport by the sodium-dependent secondary active transporters, 337–347, © (2012), with permission from Elsevier

released to form the inward-facing open state. The transporter will then revert to an outward-facing conformation for a new cycle. Notably, the revealing of the Mhp1 structures at the outward-facing "ion" state, occluded state, and inward-facing open state provides valuable insights of the mechanisms of transport (Shimamura et al. 2010). It is also worthwhile to mention that the conformational difference between the outward-facing "ion" state and the occluded state are minimal compared to the conformational difference between the occluded state and the inward-facing open state. Shimamura et al. (2010) described this version of alternating access model using a combination of extracellular thin gates, intracellular thin gates, and the intracellular thick gate. For the Mhp1 transporter, the extracellular thin gate is a single helix, namely, TM10, which occludes the substrate from extracellular solvent. The intracellular thick gate is proposed to be TM5. In LeuT, the extracellular and intracellular thin gates are proposed to be some salt bridges between charged residues (Yamashita et al. 2005). The intracellular thick gate, both in Mhp1 and LeuT, is ~20Å-thick, tightly packed protein moieties of the transporter. To change from the outward-facing "ion" state to the occluded state, the transport only closes its extracellular thin gate.

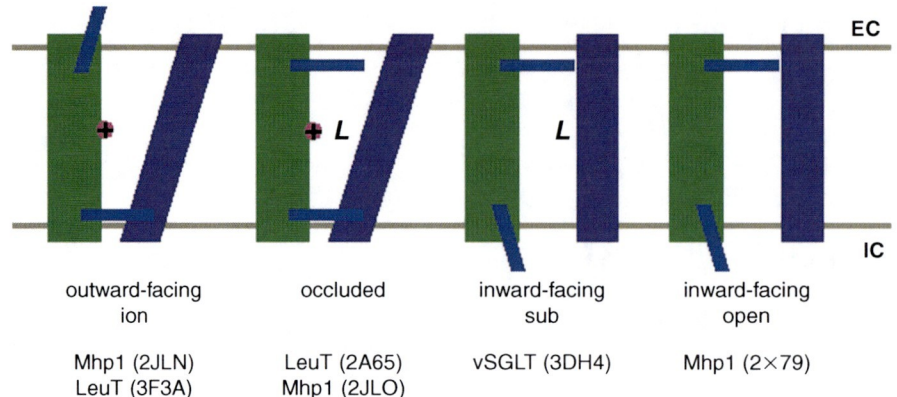

**Sodium-Coupled Secondary Transporters, Structure and Function, Fig. 7** Proposed transport cycle of LeuT-fold Na$^+$-coupled secondary transporters (Zhao and Noskov 2011; Shimamura et al. 2010). The protein is simplified by two *thick sticks* and two *thin blue sticks*, representing the extracellular (EC) and intracellular (IC) *thin gates*. The conserved Na$^+$ is shown as a *magenta ball* with a "+" sign in the *middle*. The main substrate is represented by an "*L*." Crystal structures from LeuT, Mhp1, and vSGLT with pdb entry name in the *brackets*, are indicated underneath the states which they were crystallized in

To change from the occluded state to any inward-facing conformation, the transporter needs to undergo major conformational changes to open both the intracellular thin gate, and the intracellular thick gate (Fig. 7).

### The Role of the Conserved Na$^+$ in the Transport Cycle

So, what is the role of the conserved Na$^+$ in the transport cycle of LeuT-fold Na$^+$-coupled secondary transporters? How is the free energy stored in the downhill concentration gradient (from extracellular to intracellular) utilized to facilitate the active transport of the main substrate? Zhao and Noskov (2011) showed, in a computational study of LeuT, that the binding of a Na$^+$ to Na$_2$ effectively locks its coordinating residues from TM1i and TM8 in a close distance, preventing the relative away-movement of TM1i and TM8, which is evident in the inward-facing structures of vSGLT and Mhp1 (Fig. 6). As a result, the binding of the conserved Na$^+$ encourages outward-facing/occluded conformations, while the removal of Na$_2$ ion encourages inward-facing conformations. These results are consistent with single molecular spectroscopy studies on LeuT (Zhao et al. 2010). Thus, Zhao and Noskov (2011) further proposed that the binding of the conserved Na$^+$ may play an essential role in reverting the inward-facing open conformation to the outward-facing ion conformation and helps to form the binding site for the main substrate.

### Cross-References

▶ Sodium, Physical and Chemical Properties
▶ Sodium/Glucose Co-transporters, Structure and Function
▶ Sodium-Binding Site Types in Proteins

### References

Abramson J, Wright EM (2009) Structure and function of Na$^+$-symporters with inverted repeats. Curr Opin Struct Biol 19:425–432

Forrest LR, Kramer R, Ziegler C (2011) The structural basis of secondary active transport mechanisms. Biochim Biophys Acta 1807:167–188

Forrest LR, Tavoulari S, Zhang YW, Rudnick G, Honig B (2007) Identification of a chloride ion binding site in Na$^+$/Cl–dependent transporters. Proc Nat Acad Sci USA 104:12761–12766

Jardetzk O (1966) Simple allosteric model for membrane pumps. Nature 211:969–970

Krishnamurthy H, Piscitelli CL, Gouaux E (2009) Unlocking the molecular secrets of sodium-coupled transporters. Nature 459:347–355

Reyes N, Ginter C, Boudker O (2009) Transport mechanism of a bacterial homologue of glutamate transporters. Nature 462:880–885

Shimamura T, Weyand S, Beckstein O, Rutherford NG, Hadden JM, Sharples D, Sansom MSP, Iwata S, Henderson PJF, Cameron AD (2010) Molecular basis of alternating access membrane transport by the sodium-hydantoin transporter Mhp1. Science 328:470–473

Yamashita A, Singh SK, Kawate T, Jin Y, Gouaux E (2005) Crystal structure of a bacterial homologue of Na$^+$/Cl$^-$-dependent neurotransmitter transporters. Nature 437:215–223

Yernool D, Boudker O, Jin Y, Gouaux E (2004) Structure of a glutamate transporter homologue from *Pyrococcus horikoshii*. Nature 431:811–818

Zhao CF, Noskov SY (2011) The role of local hydration and hydrogen-bonding dynamics in ion and solute release from Ion-coupled secondary transporters. Biochemistry 50:1848–1856

Zhao YF, Terry D, Shi L, Weinstein H, Blanchard SC, Javitch JA (2010) Single-molecule dynamics of gating in a neurotransmitter transporter homologue. Nature 465:188–U173

## Sodium-Dependent Transporters

▶ Sodium as Primary and Secondary Coupling Ion in Bacterial Energetics

## Sodium-Hydrogen Exchangers, Structure and Function in Human Health and Disease

Larry Fliegel
Department of Biochemistry, University of Alberta, Edmonton, AB, Canada

## Synonyms

Membrane proteins; Na$^+$/H$^+$ exchanger; pH regulation

## Definition

Na$^+$/H$^+$ exchangers are a family of membrane proteins that exchange sodium for protons across lipid bilayers. They are widely distributed in all living cell types and are critical in several human diseases. The best known type of Na$^+$/H$^+$ exchanger in multicellular species is the mammalian Na$^+$/H$^+$ exchanger isoform 1 (NHE1). It is a plasma membrane protein that regulates intracellular pH by removing one intracellular hydrogen ion in exchange for one extracellular sodium ion. NHE1 regulates intracellular pH, but is also involved several diseases in the myocardium and in cancer, in addition to its role in cell growth, movement and differentiation. This review summarizes current knowledge of Na$^+$/H$^+$ exchangers, with emphasis on the most well characterized human Na$^+$/H$^+$ exchanger, the NHE1 isoform.

## Introduction, Na$^+$/H$^+$ Exchangers

Na$^+$/H$^+$ exchangers are a class of membrane proteins that exchange Na$^+$ for H$^+$'s across lipid bilayers. The prokaryotic and eukaryotic genes that encode this monovalent cation proton antiporter (CPA) superfamily were recently reviewed (Brett et al. 2005). Briefly, the superfamily includes the CPA1, CPA2, and NaT-DC (Na$^+$-transporting carboxylic acid decarboxylase) families, each of which has unique bacterial ancestors. The CPA1 family includes bacterial NhaP transporters. It also includes many well studied Na$^+$/H$^+$ exchangers including from fungi, plants and mammals, and the human NHE1–NHE10 isoforms, the SLC9A paralogous genes. The CPA2 family shares its origins with prokaryotic NhaA. The *E. coli* form of NhaA is structurally, the most well studied Na$^+$/H$^+$ exchanger and its crystal structure has been deduced. However, many other genes of this family are not well studied. This family also includes HsNHA1 and HsNHA2 which are two relatively recently characterized forms of human Na$^+$/H$^+$ exchanger that may be involved in hypertension (Brett et al. 2005). The NaT-DC family is a smaller family that mediates transmembrane export of 1–2 Na$^+$ in exchange for an extracellular H$^+$ (Brett et al. 2005).

## Human NHE1–10

1. *General* – Human NHE1–NHE10 are membrane proteins that transport hydrogen ions in exchange for sodium ions. Human NHE1 is ubiquitously expressed in mammalian cells and is widespread throughout the animal kingdom. NHE1 is critical in intracellular pH (pH$_i$) regulation, protecting cells from acidification, as well as regulating cell volume and sodium fluxes (reviewed in). Of the family of ten isoforms, the NHE1 isoform is ubiquitous and is on the plasma membrane of mammalian cells. NHE1 is important in the myocardium as it is implicated in the diseases ischemia/reperfusion injury, and in heart hypertrophy (see below and

reviewed in Fliegel 2008). NHE1 has a 500 amino acid membrane domain that transports ions and a cytosolic regulatory domain of approximately 315 amino acids that modulates activity of the membrane domain and are a target of regulation by proteins and phosphorylation (Fliegel 2008).

2. *Subtypes* – Human NHE1 is an 815 amino acid protein. The family of mammalian NHE-like proteins includes ten isoforms (NHE1–10). Each are the product of a different gene and with different tissue distributions and physiological roles. The first type cloned was named NHE1. NHE2–4 are principally expressed in the kidney and gastrointestinal tract. NHE5 is located in the brain. NHE6–9 are found in intracellular organelle membranes, such as endosomes, mitochondria, and the golgi apparatus, NHE10 is found in osteoclasts. The identity of the various isoforms varies from 25% to 70%; however, all have a similar predicted secondary structure (see below) (Fliegel 2008).

## Structure

1. *Topological models of NHE1* – Currently, the high-resolution structure of NHE1 has not yet been solved, due to the difficulty in expressing and crystallizing membrane proteins. The first 500 residues are predicted to be 12 transmembrane spanning segments, and the remaining 315 residues constitute a cytosolic intracellular regulatory domain (Fliegel 2008). Most of the information on structure of the membrane domain comes from topology models of NHE1, some biochemical studies, a low resolution electron diffraction envelope, a crystal structure of a bacterial homologue and NMR studies of fragments of the NHE1 protein. Our laboratory produced and purified the full length NHE1 protein (Fliegel 2008). The EM structure showed that NHE1 exists as a homodimers which was confirmed by intermolecular cross-linking (Fliegel 2008; Lee et al. 2011).

Two initial studies investigated the topology of the membrane domain (Landau et al. 2007; Wakabayashi et al. 2000) and vary somewhat between each other and are reviewed in detail in (Lee et al. 2011). The topology was first examined in detail by Wakabayashi et al. (Wakabayashi et al. 2000) and using substituted cysteine accessibility they suggested that NHE1 contains 12 TM helices with both the N- and C-terminal in the cytoplasm, and the N-linked glycosylation site on the extracellular side (Fig. 1). They also suggested that two intracellular loops and one extracellular loop may be pore or membrane associated.

A more recent model for the topology of NHE1 was developed by Landau et al. (Landau et al. 2007) and was based on computational methods, including evolutionary conservation analyses and fold alignment methods with *E. coli* NhaA. This model has some significant differences from that of Wakabayashi et al. (Wakabayashi et al. 2000). The model of Landau suggests that the first two helices predicted by the Wakabayashi et al. (2000) are removed and function only as a signal sequence (Fig. 1). The next 6 TM segments of amino acids 127–315, have the same topology in both models. Amino acids 328–398, vary in their assignments between the two models. The last three transmembrane segments, TM 10–12 (amino acids 410–500) are very similar in both models.

More recently, Nygaard et al. (Nygaard et al. 2010) followed up with further structural modeling of NHE1, based on the structure of NhaA. Their 3D model principally was similar to the model of Wakabayashi et al. (Wakabayashi et al. 2000) in basic transmembrane topology. However, their conclusions were not in agreement with some biochemical data and their EPR results did not distinguish between the two earlier models, so that the structure of the NHE1 protein remains unresolved and controversial (Lee et al. 2011).

2. *Structural Studies on Human NHE1* – Our laboratory has used a "divide and conquer" approach to study the NHE1 protein (Lee et al. 2011). We studied individual TM (transmembrane) segments, which are components of the larger multi-TM domain. This avoids the difficulties that have plagued researchers in producing multi-TM spanning proteins. Studies on individual TM helices and loops have shown that the structures are often very similar to the structures of helices found in crystal structures of entire TM proteins (Lee et al. 2011). We determined the structures of several isolated TM helices of NHE1 using NMR. These were TM

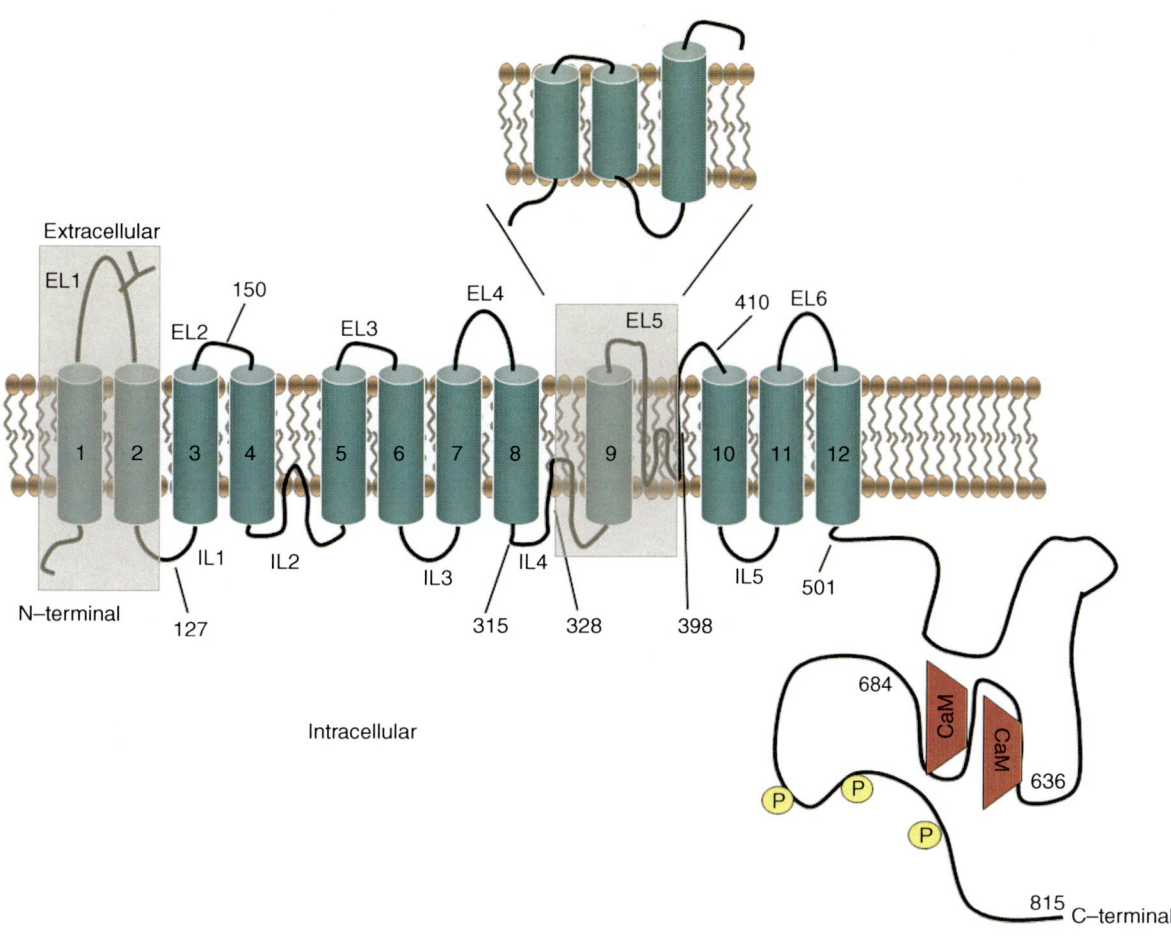

**Sodium-Hydrogen Exchangers, Structure and Function in Human Health and Disease, Fig. 1** *Schematic diagram of different versions of the two dimensional topology of NHE1.* Alternative two dimensional models of NHE1 topology according to Wakabayashi et al. (2000) and Landau et al. (2007). TM segments according to Wakabayashi et al. (2000) are indicated by 1–12. The carboxyl and amino termini are indicated. Shaded TM's 1–2 are proposed to be deleted by Landau et al. (2007). The region with varying predicted topology of amino acids 328–398 is indicated by shading and the alternative topology proposed by Landau et al. (2007) is shown above. *P* indicates putative regulatory phosphorylation sites. CaM indicates regulatory calmodulin binding region. *EL* and *IL* refer to intracellular and extracellular loops according to Wakabayashi et al. (2000)

IV (residues 155–180), TM VI (229–236), TM VII (250–275), TM IX (338–365), and TM XI (447–472). Each putative TM segment had unique structures and none were complete, perfect helices. Often, discontinuity in the helices was in the middle of the TM segments and near or containing functionally important residues. Very briefly, TM IV was irregularly structured, not resembling a canonical alpha-helix. Residues 165–168, were an extended region containing two proline residues. For TM VI, residues 229–236 contained two helical regions oriented at approximately right angles to each other (residues 229–236 and 239–250) surrounding a central unwound region. TM VII (residues 250–275) was helical over residues 255–260 and 264–272, with the central residues at 261–263 non helical. TM IX (residues 338–365) contained two regions containing alpha helix structure at residues 340–344 and 353–359, with a 90° kink at Ser$^{351}$ between the two regions which could provide flexibility. TM XI also contains two helical regions, residues 447–454 and 460–471 that are linked by an extended region of residues 455–459 (Lee et al. 2011).

## NHE1 Physiological Roles

NHE1 has many roles in different cell types (Brett et al. 2005; Fliegel 2008). Knockout of NHE1 from cells demonstrates a role in growth, which is pronounced in more acidic media. Similarly, NHE1-knockout mice demonstrated decreased growth and also ataxia and epileptic-like seizures. NHE1 is additionally permissive in cell differentiation and in cell cycle progression. NHE1 modifies apoptosis. NHE1 is activated by trophic factor withdrawal in mouse β-cells, leading to cellular alkalinization and progression of apoptosis. This activation is through p38 dependent phosphorylation of the NHE1 tail (Malo and Fliegel 2006).

## Pathological Roles

1. *NHE1 in transformed cells* – The role of NHE1 in cellular transformation and metastasis is significant. NHE1-dependent alkalinization plays a pivotal role in the development of a transformed phenotype of malignant cells, and inhibition of NHE1 prevents or reduces such development (Stock et al. 2008). This is particularly well studied in breast cancer. Metastasis is the leading cause of fatality in breast cancer. In breast cancer cells, NHE1 activation contributes to cell invasion (Stock et al. 2008). Early work showed that serum deprivation stimulated NHE1 activity in breast epithelial cell lines, in contrast to inhibiting it in non-tumor cells. Reshkin and coworkers championed the idea that a RhoA/p160ROCK/p38MAPK signaling pathway is responsible for mediating activation of NHE1 through serum deprivation (Stock et al. 2008).

   The mechanism by which NHE1 activity enhances invasion by breast cancer cells is not only by raising $pH_i$, but also by acidification of the extracellular microenvironment of tumor cells. The extracellular acidification is thought to be necessary for protease activation which facilitates the digestion and remodeling of the extracellular matrix (Stock et al. 2008), critical in metastasis.

2. *NHE1 and heart disease* – NHE is important in $pH_i$ regulation in the myocardium. However, it also has a key role to play in several myocardial pathologies. The best known of these is the role of NHE1 in ischemia/reperfusion damage in the myocardium. During ischemia, anaerobic glycolytic metabolism is elevated resulting in increased proton production. This decreases $pH_i$ and serves to activate NHE1, which leads to a rapid accumulation of sodium in the cell (Fliegel 2008). The high sodium concentration drives an increase in $Ca^{2+}$ via reversal of the activity of the $Na^+/Ca^{2+}$ exchanger and this triggers various deleterious pathways leading to cell death. Many preclinical studies have shown that NHE1 inhibition protects the myocardium from this calcium overload (Fliegel 2008). Activation of NHE1 regulatory pathways is also important in NHE1-mediated damage to the myocardium and this results in further detrimental activity of the NHE1 protein (Fliegel 2008).

3. *NHE1 in cardiac hypertrophy* – NHE1 is also important in cardiac hypertrophy (Fliegel 2008). Studies have shown that NHE1 inhibition prevents cardiac hypertrophy in several different models of hypertrophy. We also demonstrated that the effect of the hypertrophic agonist aldosterone can be blocked by NHE1 inhibition. A possible mechanism by which NHE1 inhibition prevents hypertrophy is by prevention of increases in intracellular $Na^+$. The expression level and activity of NHE1 are elevated in a variety of cardiovascular diseases, including in hypertensive, hypertrophied, or diabetic myocardium (Fliegel 2008) which may accentuate the effects of NHE1 in these diseases.

*Other cardiovascular conditions* – NHE1 inhibitors have been useful in inhibiting diabetic vascular hypertrophy and they also prevent alterations in coronary endothelial function in streptozotocin-induced diabetes. NHE1 inhibitors also have antifibrillatory and antiarrhytmic effects in dogs and rats. Inhibition of NHE1 has also been shown to be protective in cardiac resuscitation models and NHE1 inhibition was beneficial in improving the outcome in a canine transplantation model. NHE1 inhibition may additionally be useful in treatment of circulatory. Overall, NHE1 inhibition has many and varied beneficial cardiovascular effects (Karmazyn et al. 2003).

## Clinical Use of NHE1 Inhibitors

Trials with NHE1 inhibitors have not been very successful. Large-scale studies with various inhibitors have given mostly disappointing results (see (Avkiran et al. 2008) for review). In a small trial of patients with myocardial infarction who received coronary angioplasty, the NHE1 inhibitor cariporide had some beneficial effects on ejection fraction, wall-motion abnormalities and enzyme release. The NHE1 inhibitor eniporide was tested in a larger scale two-stage trial of myocardial infarction. It showed a dose dependent effect to reduce enzyme release, indicating reduced infarction. However, in a second later stage of the trial there was no beneficial effect and an overall negative effect. The GUARDIAN trial tested patients undergoing coronary artery bypass graft surgery. Analysis of subgroups of the trial showed that cariporide was beneficial possibly because treatment with the inhibitor in this trial was early (Avkiran et al. 2008). The EXPEDITION trial tested if the inhibitor cariporide reduces myocardial injury in patients with coronary artery bypass graft surgery. It was successful for this purpose; however, it also had adverse side effects and increasing cerebrovascular events significantly (Avkiran et al. 2008).

## Summary

$Na^+/H^+$ exchangers are an important superfamily of cation transporting membrane proteins. There are several families of the monovalent cation proton antiporter (CPA) superfamily. Mammalian NHE1-NHE10 are part of the CPA1 family. They comprise ten distinct isoforms of NHE protein. NHE1 is the best studied. It is a plasma membrane pH regulatory protein that extrudes one intracellular proton in exchange for an extracellular $Na^+$. It is comprised of a 500 amino acid membrane domain and a 315 amino acid cytosolic regulatory domain. The topology of NHE1 is in dispute, though the structures of regions of the transmembrane domain have been elucidated. NHE1 is involved heart disease and cancer and is critical in cell growth and differentiation. To date, NHE1 inhibitors have shown great promise in preclinical studies on the myocardium, but these have not translated to beneficial clinical results. Future improvements in the design and use of inhibitors may result in clinically useful treatments.

**Acknowledgments** Research by LF in this area is supported by the Canadian Institute of Health Research. LF is supported by an Alberta Ingenuity Medical Scientist award.

## Cross-References

▶ Calmodulin
▶ Sodium, Physical and Chemical Properties
▶ Sodium-Binding Site Types in Proteins

## References

Avkiran M, Cook AR, Cuello F (2008) Targeting $Na^+/H^+$ exchanger regulation for cardiac protection: a RSKy approach? Curr Opin Pharmacol 8:133–140

Brett CL, Donowitz M, Rao R (2005) Evolutionary origins of eukaryotic sodium/proton exchangers. Am J Physiol Cell Physiol 288:C223–C239

Fliegel L (2008) Molecular biology of the myocardial $Na^+/H^+$ exchanger. J Mol Cell Cardiol 44:228–237

Karmazyn M, Avkiran M, Fliegel L (2003) The sodium-hydrogen exchanger. From molecule to its role in disease. Kluwer, Boston/Dordrecht/London, p 318

Landau M, Herz K, Padan E et al (2007) Model structure of the $Na^+/H^+$ exchanger 1 (NHE1): functional and clinical implications. J Biol Chem 282:37854–37863

Lee BL, Sykes BD, Fliegel L (2011) Structural analysis of the $Na^+/H^+$ exchanger isoform 1 (NHE1) using the divide and conquer approach. Biochem Cell Biol 89:189–199

Malo ME, Fliegel L (2006) Physiological role and regulation of the $Na^+/H^+$ exchanger. Can J Physiol Pharmacol 84:1081–1095

Nygaard EB, Lagerstedt JO, Bjerre G et al (2010) Structural modeling and electron paramagnetic resonance spectroscopy of the human Na+/H + exchanger isoform 1, NHE1. J Biol Chem 286:634–648

Stock C, Cardone RA, Busco G et al (2008) Protons extruded by NHE1: digestive or glue? Eur J Cell Biol 87:591–599

Wakabayashi S, Pang T, Su X et al (2000) A novel topology model of the human $Na^+/H^+$ exchanger isoform 1. J Biol Chem 275:7942–7949

# Sodium-Motive Force

▶ Sodium as Primary and Secondary Coupling Ion in Bacterial Energetics

# Sodium-Potassium Adenosine Triphosphatase

▶ Sodium/Potassium-ATPase Structure and Function, Overview

## Sodium-Potassium Pump

▶ Sodium/Potassium-ATPase Structure and Function, Overview

## Sodium-Potassium-Dependent Adenosine Triphosphatase

▶ Sodium/Potassium-ATPase Structure and Function, Overview

## sodN

▶ Nickel Superoxide Dismutase

## Sol

▶ Gold Nanoparticle Platform for Protein-Protein Interactions and Drug Discovery

## Solute Carrier Family 5 (SLC5) Genes

▶ Sodium/Glucose Co-transporters, Structure and Function

## Solute Channels

▶ Aquaporins and Transport of Metalloids

## Speciation Analysis

▶ Aluminum Speciation in Human Serum

## Specific Binding

▶ Mercury and DNA

## Spectroscopic Properties of Bioinorganic Moieties

▶ Iron-Sulfur Cluster Proteins, Ferredoxins

## Spermatogenesis

▶ Selenium and Glutathione Peroxidases

## Spirogermanium

▶ Germanium-Containing Compounds, Current Knowledge and Applications

## SPR

▶ Gold Nanoparticles, Biosynthesis

## $Sr^{2+}$

▶ Strontium, Calcium Analogue in Membrane Transport Systems

## Sterol

▶ Chromium and Membrane Cholesterol

## Stress

▶ Arsenic-Induced Stress Proteins
▶ Chromium(III) and Immune System

## Strontium

▶ Strontium, Calcium Analogue in Membrane Transport Systems

# Strontium and DNA Aptamer Folding

Vladimir N. Uversky
Department of Molecular Medicine, University of South Florida, College of Medicine, Tampa, FL, USA

## Synonyms

d($G_2T_2G_2TGTG_2T_2G_2$); $G_4$-DNA; G-quartet; G-tetrad; Intramolecular G-quadruplex structure; Telomere-like structure; Thrombin aptamer

## Definition

In addition to classical double-stranded helical structure, some DNA sequences can form some unusual structures. Such unusual nucleic acid structures are illustrated by aptamers that consist of short synthetic DNA or RNA sequences that adapt well-defined three-dimensional structures and bind to specific biomolecules. For example, the 15-mer oligonucleotide with a d($G_2T_2G_2TGTG_2T_2G_2$) sequence is one of the thrombin-binding aptamers (TBA) that binds to the serine protease thrombin with high affinity and inhibits the thrombin-catalyzed fibrin-clot formation. The unique spatial structure of TBA, the intramolecular G-quadruplex, is stabilized by the specific coordination of metal ions. Although various cations can stabilize aptamers, the highest stability was ascribed to $Sr^{2+}$-aptamers.

## Unusual DNA Structures: Telomeres, G-Quadruplexes, and Aptamers

Among various unusual DNA structures are telomeres, which are repetitive nucleotide sequences located at the termini of linear chromosome. Telomeres are known to stabilize the ends of chromosomes (Blackburn 1991a, b), protecting them from deterioration or from fusion with neighboring chromosomes, and control the proper replication and segregation of eukaryotic chromosomes (DePamphilis 1993). In fact, since the eukaryotic DNA replication enzymes cannot replicate the DNA sequences present at the ends of the chromosomes, these sequences and the information they carry may get lost. The telomeres, that "cap" the end-sequences, serve as disposable buffers at the ends of the chromosomes, get lost in the process of DNA replication, being consumed during cell division, and are replenished by an enzyme, telomerase reverse transcriptase.

Telomere length varies greatly between species, ranging from ~300 base pairs in yeast to many kilobases in humans and is typically composed of arrays of guanine-rich, six- to eight-base-pair-long repeats. The telomere ends of 12–16 bases usually have G and T repeats in one strand that overhangs at the 3' end, which is essential for telomere maintenance and capping (Henderson et al. 1987; Henderson and Blackburn 1989). Telomeres are known to form large loop structures, telomere loops (T-loops), where the single-stranded DNA curls around in a long circle stabilized by telomere-binding proteins (Griffith et al. 1999). At the very end of the T-loop, the single-stranded telomere DNA is held onto a region of double-stranded DNA by the telomere strand disrupting the double-helical DNA, base pairing to one of the two strands, thus forming a specific triple-stranded structure, displacement loop or D-loop (Burge et al. 2006).

Nucleic acid sequences enriched in guanine are capable of forming a four-stranded structure, G-quadruplexes, also known as G-quartets, G-tetrads, or $G_4$-DNA, which can be formed by DNA and RNA, be parallel or antiparallel (depending on the direction of the strands or parts of a strand that form the tetrads), and can represent intramolecular, bimolecular, or tetramolecular structures. Since telomeres in all vertebrates consist of many repeats of the d(GGTTAG) sequence, they can form G-quadruplexes. In the G-tetrad structure, four guanine bases associate through Hoogsteen hydrogen bonding to form a square planar structure called a guanine tetrad, and two or more guanine tetrads can stack on top of each other to form a G-quadruplex. The quadruplex structure is further stabilized by the presence of a cation, such as $Na^+$ or $K^+$, which sits in a central channel between each pair of tetrads (see Fig. 1) (Williamson et al. 1989; Miura et al. 1995).

Besides being found in telomeres, G-quadruplexes can be found in promoter regions of some genes (Simonsson et al. 1998; Siddiqui-Jain et al. 2002).

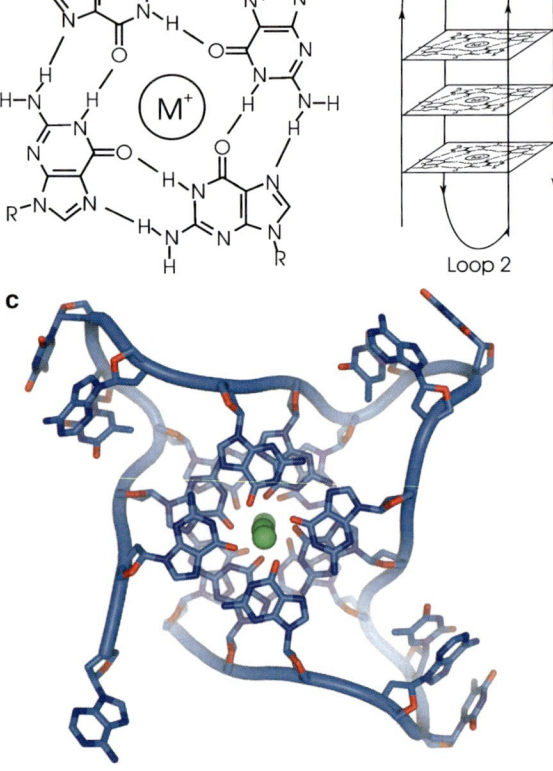

**Strontium and DNA Aptamer Folding, Fig. 1** Structure of a G-quadruplex. (**a**) A G-tetrad. (**b**) An intramolecular G-quadruplex. (**c**) Structure of parallel quadruplexes that can be formed by human telomeric DNA (PDB ID: 1KF1)

Quadruplex folding rules have been elaborated and used for the genome-wide analysis of the abundance of quadruplexes. These analyses revealed that human genome contains 376,000 Putative Quadruplex Sequences (Huppert and Balasubramanian 2005). Furthermore, quadruplexes are also common in the bacterial genomes, where they are predominantly located within promoters of genes pertaining to transcription, secondary metabolite biosynthesis, and signal transduction (Rawal et al. 2006).

G- quadruplexes are also commonly found in DNA aptamers, which have been identified by some selection process to bind specific targets (Ellington and Szostak 1990). The remarkable kinetic and thermodynamic stability of G-quadruplexes is determined by the tight association of cations with guanine residues (Kang et al. 1992).

## Strontium and Thrombin Aptamers

The ability of nucleic acids to fold into a wide array of different structures was utilized in the development of techniques for the generation of aptamers (Ellington and Szostak 1990). Aptamers are DNA or RNA oligonucleotides that have been screened from a randomly generated population of sequences for their ability to bind a desired molecular target (Ellington and Szostak 1990).

The aptamer isolation process consists of repeated cycles of selection for, and enrichment of, oligonucleotides with an affinity to a specific target (e.g., with the ability to inhibit the activity of a target protein), followed by the PCR-based amplification of these sequences. One of the most studies aptamers is the thrombin-binding aptamer (TBA) selected for its ability to bind to thrombin and inhibit thrombin-catalyzed fibrin-clot formation in vitro (Bock et al. 1992). TBA is the d($G_2T_2G_2TGTG_2T_2G_2$) oligonucleotide, which is able to form a complex structure that is comprised of two G-quartets connected by a TGT loop at the center and two $T_2$ loops (see Fig. 2), and stabilized by the formation of an intramolecular complex with $K^+$ (Macaya et al. 1993).

Detailed structural characterization revealed that $K^+$, $Rb^+$, $NH_4^+$, $Sr^{2+}$, and $Ba^{2+}$ are able to form stable intramolecular cation – aptamer complexes at temperatures above 25°C, whereas the cations $Li^+$, $Na^+$, $Cs^+$, $Mg^{2+}$, and $Ca^{2+}$ were shown to form weaker complexes at very low temperatures (Kankia and Marky 2001). Rationalization for these observations is based on the consideration of ionic radii. Here, metal cations with an ionic radius in the range 1.3 – 1.5 Å fit well within the two G-quartets of the complex, whereas the other cations cannot (Kankia and Marky 2001). Comparison of the G-quadruplexes with $K^+$ and $Sr^{2+}$ revealed that the $Sr^{2+}$ aptamer complex unfolds with a higher transition temperature and lower endothermic heat, but its favorable formation (in terms of $\Delta G°$) is comparable to that of the $K^+$ aptamer complex. Furthermore, hydration/dehydration plays a similar role in the formation of both $K^+$ and $Sr^{2+}$ aptamers, where there are two major hydration-related contributions, dehydration of both cations and guanine O6 atomic groups and the water uptake upon folding of a single strand into a G-quadruplex structure (Kankia and Marky 2001).

**Strontium and DNA Aptamer Folding, Fig. 2** Structure of the thrombin aptamer. (**a**) Sequence of the thrombin aptamer. (**b**) Schematic structure of TBA. (**c**) TBA structure stabilized by interaction with $Sr^{2+}$ (PDBID: 1RDE)

## Cross-References

▶ Calcium, Physical and Chemical Properties
▶ Strontium, Physical and Chemical Properties

## References

Blackburn EH (1991a) Telomeres. Trends Biochem Sci 16:378–381
Blackburn EH (1991b) Structure and function of telomeres. Nature 350:569–573
Bock LC, Griffin LC, Latham JA, Vermaas EH, Toole JJ (1992) Selection of single-stranded DNA molecules that bind and inhibit human thrombin. Nature 355:564–566
Burge S, Parkinson GN, Hazel P, Todd AK, Neidle S (2006) Quadruplex DNA: sequence, topology and structure. Nucleic Acids Res 34:5402–5415
DePamphilis ML (1993) Eukaryotic DNA replication: anatomy of an origin. Annu Rev Biochem 62:29–63
Ellington AD, Szostak JW (1990) In vitro selection of RNA molecules that bind specific ligands. Nature 346:818–822
Griffith JD, Comeau L, Rosenfield S, Stansel RM, Bianchi A, Moss H, de Lange T (1999) Mammalian telomeres end in a large duplex loop. Cell 97:503–514
Henderson ER, Blackburn EH (1989) An overhanging 3′ terminus is a conserved feature of telomeres. Mol Cell Biol 9:345–348
Henderson E, Hardin CC, Walk SK, Tinoco I Jr, Blackburn EH (1987) Telomeric DNA oligonucleotides form novel intramolecular structures containing guanine-guanine base pairs. Cell 51:899–908
Huppert JL, Balasubramanian S (2005) Prevalence of quadruplexes in the human genome. Nucleic Acids Res 33:2908–2916
Kang C, Zhang X, Ratliff R, Moyzis R, Rich A (1992) Crystal structure of four-stranded Oxytricha telomeric DNA. Nature 356:126–131
Kankia BI, Marky LA (2001) Folding of the thrombin aptamer into a G-quadruplex with Sr(2+): stability, heat, and hydration. J Am Chem Soc 123:10799–10804
Macaya RF, Schultze P, Smith FW, Roe JA, Feigon J (1993) Thrombin-binding DNA aptamer forms a unimolecular quadruplex structure in solution. Proc Natl Acad Sci USA 90:3745–3749
Miura T, Benevides JM, Thomas GJ Jr (1995) A phase diagram for sodium and potassium ion control of polymorphism in telomeric DNA. J Mol Biol 248:233–238
Rawal P, Kummarasetti VB, Ravindran J, Kumar N, Halder K, Sharma R, Mukerji M, Das SK, Chowdhury S (2006) Genome-wide prediction of G4 DNA as regulatory motifs: role in Escherichia coli global regulation. Genome Res 16:644–655
Siddiqui-Jain A, Grand CL, Bearss DJ, Hurley LH (2002) Direct evidence for a G-quadruplex in a promoter region and its targeting with a small molecule to repress c-MYC transcription. Proc Natl Acad Sci USA 99:11593–11598
Simonsson T, Pecinka P, Kubista M (1998) DNA tetraplex formation in the control region of c-myc. Nucleic Acids Res 26:1167–1172
Williamson JR, Raghuraman MK, Cech TR (1989) Monovalent cation-induced structure of telomeric DNA: the G-quartet model. Cell 59:871–880

# Strontium Binding to Proteins

Vladimir N. Uversky
Department of Molecular Medicine, University of South Florida, College of Medicine, Tampa, FL, USA

## Synonyms

Alkaline earth metal; Tracer for calcium

## Definition

Due to the general similarity between the strontium and calcium ions (both metals are members of the alkaline earth series (Group IIB of the Periodic Table) that have many common properties, such that both have two positive charges in their ionic forms, similar ionic radii, and the ability to form complexes of various binding strengths), strontium can replace calcium to some extent in various biochemical processes in the body, for example, being able to interact with many calcium-binding proteins. Being heavier than calcium, strontium is frequently used as a tracer for calcium. It is also an important and useful analog of calcium in clinical research. Strontium has been safely used as a medicinal substance for more than a 100 years.

### Some Physicochemical Properties of Strontium

The chemical element strontium is the 38th element in the chemical periodic table with the symbol of Sr, atomic number of 38, and atomic mass of the 87.62. Strontium is a soft silver-white or yellowish metallic element that is softer than calcium and even more reactive in water. As all alkaline earth metals, strontium is highly reactive and does not occur in nature as a free metal, being incorporated in various minerals. Strontium is one of the most abundant elements on earth, ranking approximately 15th in natural abundance in the Earth's crust and comprising about 0.04% of the Earth's crust. In fact, there is more strontium in the Earth's crust than carbon, sulfur, or chromium. At a concentration of 8.1 ppm, strontium is also the most abundant trace element in seawater. The human body contains about 320 mg of strontium, nearly all of which is located in bones and connective tissues.

There are four stable isotopes in the naturally occurring strontium, $^{88}$Sr (83%), $^{86}$Sr (9.9%), $^{87}$Sr (7.0%), and $^{84}$Sr (0.6%). Furthermore, 16 radioactive isotopes can be produced synthetically. Of these, $^{89}$Sr and $^{90}$Sr, both beta emitters, got the most attention due to the fact that these radionuclides of strontium are formed as by-products of nuclear fission reactions and atomic bomb detonation. Among different radioactive isotopes of strontium, $^{90}$Sr is the most hazardous, due to its ability to be deposited in the skeleton, its long biological half-life ($\sim$10 years), and a decay half-life of $\sim$30 years.

Chemically, strontium is rather similar to barium, magnesium, and calcium. It always exhibits the oxidation state of +2 and easily and exothermally reacts with water. It is also highly reactive with oxygen, and finely powdered strontium metal is pyrogenic and will ignite spontaneously in air at room temperature. Due to its ability to react with oxygen, air, and water, strontium is kept under a liquid hydrocarbon such as mineral oil or kerosene to prevent oxidation.

### Physiological Effects of Strontium

As already mentioned both $Ca^{2+}$ and $Sr^{2+}$ belong to the class of alkaline earth metals, have a valency of 2, highly reactive, and are able to be involved in formation of various complexes and chelates of various solubilities and various binding strengths with a number of anionic compounds. This similarity of major physicochemical properties of these two metals determines their remarkable physiological similarity. In fact, strontium and calcium are absorbed in the gastrointestinal tract, concentrated in bone, and excreted primarily in urine. Furthermore, one of the mechanisms of strontium incorporation into bone involves ionic exchange with bone calcium. Because of these similarities, strontium is currently used in bone therapies. For example, $^{89}$Sr and $^{85}$Sr are used for treating the scattered painful bone metastases that affect two-thirds of patients suffering from advanced and metastatic cancers (Giammarile et al. 1999). Stable strontium isotope has a pronounced dose-dependent multiphasic effect on bone formation, where high doses induce decreased resorption and mineral density of bone (Cabrera et al. 1999). However, low doses of strontium were shown to possess some beneficial effects on bone formation, leading to the increase in the number of bone forming sites and vertebral bone volume, and did not have detectable adverse effects on the mineral profile, bone mineral chemistry, or bone matrix mineralization (Grynpas et al. 1996). Furthermore, stable strontium in the form of strontium ranelate is used in the treatment of osteoporosis (Brandi 1993; Wasserman 1998; Reginster et al. 2003).

Although $Ca^{2+}$ and $Sr^{2+}$ have many similar properties, these two cations are not identical. The atomic, covalent, and van der Waals radii of strontium are of 215, 195 $\pm$ 10, and 249 pm, respectively. These values somewhat exceed those of calcium (197, 176 $\pm$ 10, and 231 pm, respectively). Furthermore, $Sr^{2+}$ with its atomic mass of 87.62 Da, is more than twofold heavier than $Ca^{2+}$ (whose standard atomic weight is of 40.078 Da). Likely, the interplay between these similarities and differences defines the fact that different anionic compounds possess different binding affinities for $Ca^2$ and $Sr^2$. As a result, some anionic compounds prefer $Ca^2$, whereas others bind preferentially to $Sr^2$. For example, the binding affinity of $Sr^2$ to alginates is 1.5- to 4.3-fold greater than that of $Ca^2$. On the other hand, calcium is clearly preferred in other interactions, such as controlling negative charges on membrane, binding to collagen, and interaction with G-actin (Wasserman 1998). Although in the majority of biological systems, preference is generally given to $Ca^2$ over $Sr^2$, a marine organism, *Acantharia* (radiolaria), possesses a strontium sulfate-made internal skeleton, that consists of 10 or 20 spicules and defines the characteristic star-shaped morphology of this ubiquitous and abundant rhizarian protists in the world ocean (Decelle et al. 2012). However, the bony skeleton of vertebrates is mostly composed of the highly insoluble calcium phosphate complex, hydroxyapatite. Some strontium is present in bone too, but it is usually considered a contaminant, with no physiological function at the trace values present therein (Wasserman 1998).

## Interaction of Strontium with Proteins
### Strontium and Human Alkaline Phosphatase
The aforementioned dose-related multiphasic effects of strontium on bone formation with high strontium doses promoting osteomalacia, a disease characterized by the impairment of bone mineralization, leading to accumulation of unmineralized matrix or osteoid in the skeleton (Schrooten et al. 1998), showed that strontium is not fully equivalent to calcium and indicated that there could be a difference in how these two metal ions interact with their environment. Since the defective bone mineralization symptoms of strontium-induced osteomalacia are quite similar to those of hypophosphatasia, a rare inherited disorder associated with mutations in the tissue-nonspecific alkaline phosphatase (AP), some effects of strontium on bone mineralization could occur through a direct interaction with AP, through the AP pathway, decrease in AP expression, or through a decrease in AP catalytic activity (Llinas et al. 2006).

APs are highly conserved dimeric enzymes. There is a noticeable polymorphism in human APs which are differentiated into four isozymes: tissue-nonspecific AP (TNAP, found in bone, liver, and kidney) and three tissue-specific isozymes, placental (PLAP), germ cell (GCAP), and intestinal (IAP). Although the three tissue-specific isozymes are 90–98% homologous, and their genes are clustered on Chromosome 2, TNAP is only 50% identical with the other three, and its gene is located on Chromosome 1 (Harris 1990).

APs are known to catalyze the hydrolysis of phosphomonoesters to generate inorganic phosphate and alcohol. Being metalloenzymes, APs contain one magnesium and two zinc ions in their catalytic site and also have a fourth metal binding site, 10 Å away from the active site (Llinas et al. 2006). This fourth metal binding site contains a calcium ion which stabilizes a large area comprising of loops 210–228 and 250–297 and which can be substituted for $Sr^{2+}$. Structural analysis of human PLAP cocrystallized with strontium chloride revealed that strontium efficiently substitutes the calcium ion in the fourth metal binding site. Binding of strontium to this site was characterized by noticeable changes in the metal coordination, suggesting that substitution of $Ca^{2+}$ for $Sr^{2+}$ might affect the stability of this important region (Llinas et al. 2006).

### Vitamin D-Induced Calcium-Binding Protein
The calbindins are a family of vitamin D-dependent calcium-binding proteins found in the intestine (calbindin D-9 k) and kidney (calbindin D-28 k) that play a number of important roles in calcium transport and homeostasis. Detailed structural and functional analysis revealed that the mostly α-helical vitamin D-induced calcium-binding protein (also known as calbindin-D, CaBP) possesses two maxima in its pH-calcium-binding curve, one at about pH 6.6 and the other at about pH 9.6 and is able to interact with other alkali earth metals. Based on the competitive inhibition studies, it was also shown that the binding affinities of this protein to the alkali earth metals can be ranged in the following order $Ca^{2+} > Sr^{2+} > Ba^{2+} > Mg^{2+}$, which is not predicted based on the energy of hydration or the crystal radius of the metal cations (Ingersoll and Wasserman 1971).

### Strontium and $Ca^{2+}$-Sensitive ATPase
In addition to be activated by $Ca^{2+}$ ions, adenosine triphosphatase of human erythrocyte membranes (also known as the $Ca^{2+}$-sensitive ATPase or ATP

phosphohydrolase, a protein which is responsible for active $Ca^{2+}$ transport outward across the erythrocyte membrane) can be regulated by other bivalent metal ions, such as $Sr^{2+}$, $Ba^{2+}$, $Mn^{2+}$, $Ni^{2+}$, $Co^{2+}$, $Cd^{2+}$, $Cu^{2+}$, $Zn^{2+}$, and $Pb^{2+}$. Contrarily to the calbindin-D discussed above, the degree of activation of the $Ca^{2+}$-sensitive ATPase was shown to be dependent on the radius of the ion rather than on its nature, with $Ca^{2+}$ and $Sr^{2+}$ possessing the highest activation potency among all bivalent ions analyzed (Pfleger and Wolf 1975). Since $Ca^{2+}$ and $Sr^{2+}$ have rather similar atomic radii (see above), it was proposed that interaction of only these two ions with the protein is able to induce an optimal conformational changes enhancing enzymatic activity of ATP phosphohydrolase, whereas any deviation from the optimum atomic radius value resulted in a considerable decrease in the catalytic power of the enzyme (Pfleger and Wolf 1975). Interestingly, the dissociation constants of the enzyme-metal ion complex were independent on the atomic radii of studied bivalent metal ions. The fact that $Ca^{2+}$-sensitive ATPase can be activated by a wide range of bivalent metal ions of different nature and rather different size suggested that this protein possessed an unusually high tolerance to the activating ion radius (Pfleger and Wolf 1975).

### The Extracellular Calcium-Sensing Receptor

Strontium ranelate (a $Sr^{2+}$ salt of ranelic acid, or distrontium 5-[bis(2-oxido-2-oxoethyl)amino]-4-cyano-3-(2-oxido-2-oxoethyl)thiophene-2-carboxylate) is composed of an organic moiety (ranelic acid) capable of binding two stable strontium atoms. Strontium ranelate is marketed as Protelos® and Protos® and was shown to be a dual action bone agent (DABA), which has opposite effects on bone resorption and formation, resulting in beneficial outcomes on trabecular bone mass in vivo (Marie 2005).

The extracellular parathyroid calcium-sensing receptor (CaSR), a member of family C of G-protein-coupled receptors (GPCRs), is modulated by physiological $Ca^{2+}$ and $Mg^{2+}$ and regulates the secretion of calciotropic hormones in thyroid C-cells and parathyroid cells, whereas in the distal tubule of the kidney, activation of the CaSR mediates inhibition of tubular $Ca^{2+}$ reabsorption (Brown and MacLeod 2001). This receptor represents a potential target for divalent cations' actions in tissues since intracellular responses associated with the CaSR can be activated with other divalent cations such as $Sr^{2+}$, $Ba^{2+}$, $Mn^{2+}$, and $Ni^{2+}$ (Coulombe et al. 2004). The CaSR response to change in the concentration of the bivalent metal ions stimulates the phospholipases C (PLC) and A2 (PLA2), the mitogen-activated protein kinase (MAPK), and phosphoinositide 3-kinase (Pi 3-kinase), ultimately leading to the production of major second messengers including inositol phosphates (IP), intracellular $Ca^{2+}$, and arachidonates (Coulombe et al. 2004). The maximal effect of $Sr^{2+}$ on CaSR activation was shown to be comparable to that of $Ca^{2+}$.

Importantly, CaSR might mediate some of the effects of strontium ranelate, since in cell lines, $Sr^{2+}$ and strontium ranelate displayed comparable agonist activity for the CaSR (Coulombe et al. 2004). Modulation of the CaSR activity in bone cells by strontium ranelate may contribute to its reported antiosteoporotic effects (Coulombe et al. 2004).

### Strontium and Calmodulin

Calmodulin (CaM) is a small ubiquitous calcium binding protein, which is highly abundant, is known to bind four calcium ions into four EF-hand motifs and is shown to exist in equilibrium between the apo- and holo-forms wherein four calcium ions are bound. Among its many functions, CaM modulates the calcium concentration in the cell. Furthermore, CaM is able to bind to a very wide range of different proteins in a calcium-dependent manner. Since the functional properties of this protein clearly depend on calcium, other metal ions can compete with $Ca^{2+}$ for regulation of CaM functionality. Analysis of the selectivity of the calcium-trigger protein CaM for divalent metal ions by electrospray ionization mass spectrometry revealed that different metal ions have different properties with respect to the calmodulin binding and competition with calcium. The order of affinity for apo-CaM was shown to follow sequence $Ca^{2+} \gg Sr^{2+} \approx Mg^{2+} > Pb^{2+} \approx Cd^{2+}$. In the presence of calcium the affinity alters to $Pb^{2+} > Ca^{2+} > Cd^{2+} > Sr^{2+} > Mg^{2+}$ (Shirran and Barran 2009).

### Engineering Strontium Binding Affinity in an EF-Hand Motif

An interesting approach for designing a protein with the ability to specifically bind strontium was proposed (Rinaldo et al. 2004). The need for such protein is determined by the facts that the existence of a natural protein that is able to sense radioactive isotopes of strontium is questionable, but such strontium sensor/detoxifier could be of great potential use in the field of nuclear waste management. The starting point for the design of such a strontium sensor is a 33 amino acids long peptide which corresponds to the helix-turn-helix (EF-hand) motif of the calcium binding site I of *Paramecium tetraurelia* calmodulin. Then, some residues were mutated to change the motif's specificity for calcium into one for strontium. Next, the dynamic and binding behaviors of the resulting mutant sequences of the EF-hand motif were analyzed using molecular dynamics simulations and free-energy calculations. Based on the results of these analyzes several characteristics that could lead to mutant peptides with enhanced strontium affinity were proposed (Rinaldo et al. 2004).

### Strontium and α-Lactalbumin

α-Lactalbumin is the non-catalytic subunit of the lactose synthetase enzyme complex containing one calcium-binding site with an affinity constant of $2.5 \times 10^6$ M$^{-1}$. This high binding affinity indicates that $Ca^{2+}$ is playing a structural role and does not readily dissociate from the protein under the physiological conditions. In fact, removal of $Ca^{2+}$ from human and bovine α-lactalbumin was shown to transform these proteins into the molten globule-like conformation (Permyakov and Berliner 2000). Being added to the α-lactalbumin apo-form, $Sr^{2+}$ is able to bind to the $Ca^{2+}$-binding site of this protein, possessing the affinity constant of $5.1 \times 10^5$ M$^{-1}$ (Schaer et al. 1985). Importantly, binding of both metal ions to the apoprotein was shown to be accompanied by a very large enthalpy change, supporting the idea that metal binding induced the global conformational changes from the molten globule-like conformation to a folded state with well-developed tertiary structure (Schaer et al. 1985).

### Strontium and Serum Albumin

An analysis of interaction between strontium and bovine serum albumin (BSA) by a combination of gel chromatography and inductively coupled plasma atomic emission spectrometry (ICP-AES, a method that is particularly sensitive to the presence of strontium in solution allowing detection of strontium at concentrations as low as 1 μg/L) revealed that BSA does not have any high affinity site for $Sr^{2+}$ ions, but possesses one medium affinity site (with $K_1$ of the order of $3 \times 10^4$ M$^{-1}$) and ten low affinity sites ($K_2 = 800$ M$^{-1}$) (Sandier et al. 1999).

## Cross-References

▶ α-Lactalbumin
▶ Calbindin D$_{28k}$
▶ Calcium and Extracellular Matrix
▶ Calcium in Health and Disease
▶ Calcium-Binding Protein Site Types
▶ Calcium-Binding Proteins, Overview
▶ Calmodulin
▶ EF-Hand Proteins
▶ Strontium, Physical and Chemical Properties

## References

Brandi ML (1993) New treatment strategies: ipriflavone, strontium, vitamin D metabolites and analogs. Am J Med 95:69S–74S

Brown EM, MacLeod RJ (2001) Extracellular calcium sensing and extracellular calcium signaling. Physiol Rev 81:239–297

Cabrera WE, Schrooten I, De Broe ME, D'Haese PC (1999) Strontium and bone. J Bone Miner Res 14:661–668

Coulombe J, Faure H, Robin B, Ruat M (2004) In vitro effects of strontium ranelate on the extracellular calcium-sensing receptor. Biochem Biophys Res Commun 323:1184–1190

Decelle J, Suzuki N, Mahe F, de Vargas C, Not F (2012) Molecular phylogeny and morphological evolution of the Acantharia (Radiolaria). Protist 163:435–450

Giammarile F, Mognetti T, Blondet C, Desuzinges C, Chauvot P (1999) Bone pain palliation with 85Sr therapy. J Nucl Med 40:585–590

Grynpas MD, Hamilton E, Cheung R, Tsouderos Y, Deloffre P, Hott M, Marie PJ (1996) Strontium increases vertebral bone volume in rats at a low dose that does not induce detectable mineralization defect. Bone 18:253–259

Harris H (1990) The human alkaline phosphatases: what we know and what we don't know. Clin Chim Acta 186:133–150

Ingersoll RJ, Wasserman RH (1971) Vitamin D3-induced calcium-binding protein. Binding characteristics, conformational effects, and other properties. J Biol Chem 246:2808–2814

Llinas P, Masella M, Stigbrand T, Menez A, Stura EA, Le Du MH (2006) Structural studies of human alkaline phosphatase in complex with strontium: implication for its secondary effect in bones. Protein Sci 15:1691–1700

Marie PJ (2005) Strontium ranelate: a novel mode of action optimizing bone formation and resorption. Osteoporos Int 16(Suppl 1):S7–S10

Permyakov EA, Berliner LJ (2000) α-Lactalbumin: structure and function. FEBS Lett 473:269–274

Pfleger H, Wolf HU (1975) Activation of membrane-bound high-affinity calcium ion-sensitive adenosine triphosphatase of human erythrocytes by bivalent metal ions. Biochem J 147:359–361

Reginster JY, Deroisy R, Jupsin I (2003) Strontium ranelate: a new paradigm in the treatment of osteoporosis. Drugs Today (Barc) 39:89–101

Rinaldo D, Vita C, Field MJ (2004) Engineering strontium binding affinity in an EF-hand motif: a quantum chemical and molecular dynamics study. J Biomol Struct Dyn 22:281–297

Sandier A, Amiel C, Sebille B, Rouchaud JC, Fedoroff M (1999) A study of strontium binding to albumins, by a chromatographic method involving atomic emission spectrometric detection. Int J Biol Macromol 24:43–48

Schaer JJ, Milos M, Cox JA (1985) Thermodynamics of the binding of calcium and strontium to bovine alpha-lactalbumin. FEBS Lett 190:77–80

Schrooten I, Cabrera W, Goodman WG, Dauwe S, Lamberts LV, Marynissen R, Dorrine W, De Broe ME, D'Haese PC (1998) Strontium causes osteomalacia in chronic renal failure rats. Kidney Int 54:448–456

Shirran SL, Barran PE (2009) The use of ESI-MS to probe the binding of divalent cations to calmodulin. J Am Soc Mass Spectrom 20:1159–1171

Wasserman RH (1998) Strontium as a tracer for calcium in biological and clinical research. Clin Chem 44:437–439

---

# Strontium, Calcium Analogue in Membrane Transport Systems

Kim L. Hein and J. Preben Morth
Centre for Molecular Medicine Norway (NCMM), Nordic EMBL Partnership, University of Oslo, Oslo, Norway

## Synonyms

$Sr^{2+}$; Strontium

## Definition

▶ Strontium is a chemical element with symbol Sr and one of the alkaline earth metals. It was discovered in 1790 near the Scottish village Strontian and was eventually isolated in 1808. ▶ Strontium is found in group 2 of the periodic system, which also includes the biological important divalent metals magnesium (Mg) (▶ Magnesium in Biological Systems) and calcium (Ca) (▶ Calcium in Biological Systems). It is never found as a pure metal in nature since the metallic Sr is easily oxidized to form strontium oxide (SrO). ▶ Strontium is, however, often found trapped in minerals like celestite ($SrSO_4$) and strontianite ($SrCO_3$).

Strontium has attracted less attention than the more common divalent ions $Mg^{2+}$ and $Ca^{2+}$; important for human biology and pathology, however, the biological role of Sr has received increased interest since the introduction of strontium ranelate. This compound was the first antiosteoporotic agent reported to both stimulate bone formation and decrease bone resorption.

## The Biological Role of Strontium

The chemical properties of Sr in biological systems are highly similar to Ca, and the protein binding capacity in the serum or plasma is in the same order of magnitude (Pors Nielsen 2004). The radioactive isotopes Sr, mainly $^{85}$Sr and $^{89}$Sr, have also been used as tools for studying Ca kinetics (Cohn and Gusmano 1966; Litvak et al. 1967). However, a biological difference between the two elements is detectable and can in part be explained by the larger radii of the hydrated $Sr^{2+}$ ion. $Sr^{2+}$ and $Ca^{2+}$ utilize common transport paths across the plasma membrane, where Sr competes directly with Ca for absorption in the intestines, and during renal tubular reabsorption (Pors Nielsen 2004). Intracelluar $Sr^{2+}$ has been shown to be transported by a highly selective mitochondrial calcium uniporter with a relative divalent conductances equivalent to $Ca^{2+}$ (Kirichok et al. 2004); however, $Sr^{2+}$-specific transporter is yet to be found. The unicellular green algae (Chlorella) is able to completely replace the $Ca^{2+}$ requirement with $Sr^{2+}$ throughout the entire life cycle, and furthermore, $Sr^{2+}$ have also been shown to act as a growth stimulant for some plant species. Sr is not an essential element for biological function and total absence has also no apparent effect (Pors Nielsen 2004).

In general, the toxicity of Sr compounds to humans is low. Only strontium chromate ($SrCrO_4$) is known to be highly carcinogenic; the toxicity is however caused by the chromate. Additionally, the radioactive isotope $^{90}$Sr (half-life: 28.8 years) can be very harmful for the human body because of its bone-seeking properties and potentially cause bone cancer or leukemia. But because $^{90}$Sr from the environment and nuclear waste pose a considerable health risk, ways to sequester strontium in the presence of excess $Ca^{2+}$ are of considerable interest. However, as mentioned above, $Sr^{2+}$ and $Ca^{2+}$ utilize common transport paths across the plasma membrane which makes it difficult for organisms to discriminate and sequester strontium specifically. However, despite this, a small number of organisms with the ability to form the biominerals barite ($BaSO_4$) and celestite ($SrSO_4$) have been identified (Wilcock et al. 1989). The desmid green alga *Closterium moniliferum* have been shown to be able to sequester $Sr^{2+}$ by forming Sr-substituted barite crystals, but there is little evidence for the involvement of selective transport proteins in uptake, transport, or efflux of $Sr^{2+}$ vs. $Ca^{2+}$. Instead, the mechanism seems to depend on undersaturation and supersaturation of $SrSO_4$ and $CaSO_4$ in intracellular organelles, but the exact mechanism remains to be elucidated (Krejci et al. 2011).

## Clinical Applications

Until recently, Sr compounds have not been used in any clinical applications. In the early 1950s, it was discovered that strontium lactate given in moderate dose could increase Ca deposition in bones (Shorr and Carter 1952), and a few years later, it was shown that strontium lactate could reduce bone pain in patients with osteoporosis. Despite these early observations, the pharmacological properties of strontium compounds were ignored for years, which may be the result of the reported rachitogenic effect observed for Sr when present in high doses.

In the 1980s and 1990s, a renewed interest for Sr compounds developed, which eventually lead to a new treatment procedure for osteoporosis. Osteoporosis is considered a worldwide public health concern, estimated to effect 40% of all woman aged over 60 years one or multiple times in their lifetime (Reginster and Neuprez 2010) causing high personal and human society costs. Strontium ranelate (Fig. 1) is today the only compound that concomitantly decrease bone resorption and stimulate bone formation and is today considered the first line of defense in the prevention and the treatment of osteoporosis. The radioactive isotope $^{89}Sr$ has been used as pain relief in terminal cancer patients with bone metastases and refractory bone pain; however, the exact mechanism of pain relief by administration of Sr still remains elusive (Giammarile et al. 2001).

**Strontium, Calcium Analogue in Membrane Transport Systems, Fig. 1** Strontium complex structure of strontium ranelate

## Scientific Uses

The use of Sr in scientific setups is rather limited; however, because of the chemical similarities to Ca, Sr has been used successfully to study the Ca metabolism and kinetics during bone accretion and bone reabsorption in humans. Here have strontium-85 proven to be a good calcium tracer due to the longer half-life and lower production cost compared to other calcium isotopes, like calcium-45 and calcium-47 (Litvak et al. 1967). Sr has also been used to study neurotransmitter release in neurons and the substitution of Ca with Sr and allows the researcher to measure the effects of a single vesicle fusion event, because Sr only affects the fusion event by causing asynchronous vesicle fusion, while the facilitation of the synaptic vesicle fusion with the synaptic membrane is retained (Miledi 1966) (Cordeiro et al. 2011).

Several protein, protein–nucleotide, and nucleotide structures have been reported with Sr bound, including a few membrane proteins (Esser et al. 2008; Berman et al. 2000). In most cases, Sr originates from experimental conditions and is found in known or putative $Ca^{2+}$ ligand binding sites and with a similar structural coordination (Fig. 2). There are, however, small deviations in the binding coordination of $Sr^{2+}$ compared to that of $Ca^{2+}$; these deviations have been shown to alter the enzyme kinetics (Llinas et al. 2006) and also shown to stabilize ▶ DNA resulting in better resolved structures of ▶ DNA and DNA–protein complexes (Thorpe et al. 2003). The origin of these small deviation in binding coordination of Sr can in large be attributed to the larger ionic radius of Sr, approximately 0.15 Å

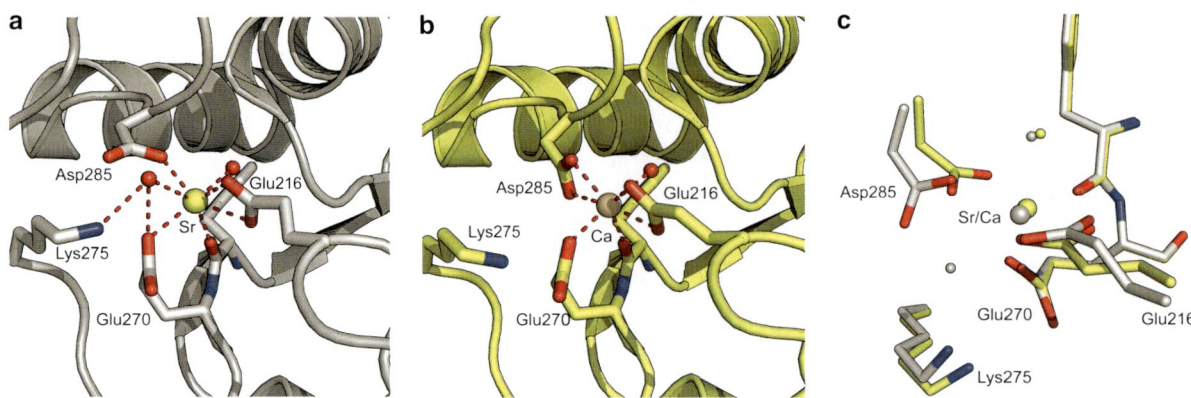

**Strontium, Calcium Analogue in Membrane Transport Systems, Fig. 2** Interacting residues in the fourth metal binding site of alkaline phosphatase showing the coordination network of Sr (**a**) or Ca (**b**). *Red spheres* illustrate water. (**c**) Superimposition of the binding sites, showing the deviation between the Sr and Ca structures. Sr structure draw in *gray* and Ca structure draw in *yellow*. Redrawn from (Llinas et al. 2006). The figure was generated with protein databank identification code: 2GLQ and 1ZED. Using the PyMOL Molecular Graphics System, Version 1.3, Schrödinger, LLC

**Strontium, Calcium Analogue in Membrane Transport Systems, Table 1** Solubility of various strontium compounds in water, at 1 atm pressure, units of solubility in g/100 g $H_2O$

| Substance | Formula | 0°C | 10°C | 20°C | 30°C | 40°C | 50°C | 60°C | 70°C | 80°C | 90°C | 100°C |
|---|---|---|---|---|---|---|---|---|---|---|---|---|
| Strontium acetate | $Sr(C_2H_3O_2)_2$ | 37 | 42.9 | 41.1 | 39.5 | 38.3 | | 36.8 | | 36.1 | 36.2 | 36.4 |
| Strontium bromate | $Sr(BrO_3)_2 \times H_2O$ | | | 30.9 | | | | | | | | 41 |
| Strontium bromide | $SrBr_2$ | 85.2 | 93.4 | 102 | 112 | 123 | | 150 | | 182 | | 223 |
| Strontium chlorate | $SrClO_3$ | | | 175 | | | | | | | | |
| Strontium chloride | $SrCl_2$ | 43.5 | 47.7 | 52.9 | 58.7 | 65.3 | | 81.8 | | 90.5 | | 101 |
| Strontium chromate | $SrCrO_4$ | | | 0.085 | | 0.090 | | | | | | |
| Strontium fluoride | $SrF_2$ | | | 0.00012 | | | | | | | | |
| Strontium formate | $Sr(HCO_2)_2$ | 9.1 | 10.6 | 12.7 | 15.2 | 17.8 | | 25 | | 31.9 | 32.9 | 34.4 |
| Strontium hydroxide | $Sr(OH)_2 \times 8H_2O$ | 0.91 | 1.25 | 1.77 | 2.64 | 3.95 | | 8.42 | | 20.2 | 44.5 | 91.2 |
| Strontium iodate | $Sr(IO_3)_2$ | | | 0.19 | | | | | | | | 0.35 |
| Strontium iodide | $SrI_2$ | 165 | | 178 | | 192 | | 218 | | 270 | 365 | 383 |
| Strontium molybdate | $SrMoO_4$ | | | 0.01107 | | | | | | | | |
| Strontium nitrate | $Sr(NO_3)_2$ | 39.5 | 52.9 | 69.5 | 88.7 | 89.4 | | 93.4 | | 96.9 | 98.4 | |
| Strontium selenate | $SrSeO_4$ | | | 0.656 | | | | | | | | |
| Strontium sulfate | $SrSO_4$ | 0.0113 | 0.0129 | 0.0132 | 0.0138 | 0.0141 | | 0.0131 | | 0.0116 | 0.0115 | |
| Strontium thiosulfate | $SrS_2O_3 \times 5H_2O$ | | | 2.5 | | | | | | | | |
| Strontium tungstate | $SrWO_4$ | | | 0.0003957 | | | | | | | | |

longer than calcium. For human alkaline phosphatase, this increases the coordination number, allowing interaction with an additional residue which is in close vicinity to the active site, thereby affecting enzyme activity. The molecular differences between Sr and Ca ligand binding have been reported in human alkaline phosphatase (Fig. 2). The calcium coordination involves the carboxylate of Glu216, Glu270, and

Asp285, backbone carbonyl of Phe269, and one water molecule. The interaction of Glu270 is monodentate with Ca–O distances of 2.2 Å and 4.2 Å; that of Glu216 is bidentate with Ca–O distances of 2.4 Å and 2.4 Å. The Ca–O distance is 2.3 Å and 2.5 Å between the calcium and Asp285, respectively. The strontium coordination involved an additional water molecule, and the coordination pattern is altered. The carboxylate of Asp285 has moved away from Sr altering binding distances to 2.5 Å and 3.9 Å making the character of the interaction more monodentate. Minor changes to the binding network induced by Sr include the carbonyls of the Tyr269 main chain, Glu270, and the first water molecule which moves only by 0.05, 0.1, and 0.3 Å, respectively. More substantial changes are observed in the Cδ of Glu216 which shifts by 0.7 Å, but the interaction remains bidentate, and the Cγ of Asp285 which moves 1.1 Å and becomes a monodentate interaction. Furthermore, an asdditional water molecule adds further stabilization through interaction with Lys275, whereas it plays no role in the presence of calcium in the native enzyme (Llinas et al. 2006).

Even though the biological significance of substitution Ca with Sr might be minor in respect to biological function, the use of Sr to aid structural determination can be very useful.

The K absorption edge for Sr is 16.1046 keV (0.7699 Å) and is well within the spectral range of modern synchrotrons. Considering the relatively low toxicity and relatively high solubility of most of the strontium compounds (see Table 1), Sr could be useful for obtaining phase information, especially if Ca binding already has been established.

## Cross-References

▸ Calcium Ion Selectivity in Biological Systems
▸ Calcium-Binding Proteins, Overview
▸ Magnesium in Biological Systems
▸ Strontium and DNA Aptamer Folding
▸ Strontium, Physical and Chemical Properties

## References

Berman HM, Westbrook J, Feng Z et al (2000) The protein data bank. Nucleic Acids Res 28:235–242. doi:10.1093/nar/28.1.235

Cohn SH, Gusmano EA (1966) Kinetics of strontium and calcium skeletal metabolism in the rat. Riv Patol Nerv Ment 87:79–83

Cordeiro JM, Gonçalves PP, Dunant Y (2011) Synaptic vesicles control the time course of neurotransmitter secretion via a $Ca^{2+}/H+$ antiport. J Physiol 589:149–167. doi:10.1113/jphysiol.2010.199224, London

Esser L, Elberry M, Zhou F, Yu CA, Yu L, Xia D (2008) Inhibitor-complexed structures of the cytochrome bc1 from the photosynthetic bacterium *Rhodobacter sphaeroides*. J Biol Chem 283(5):2846–2857

Giammarile F, Mognetti T, Resche I (2001) Bone pain palliation with strontium-89 in cancer patients with bone metastases. Q J Nucl Med 45:78–83

Kirichok Y, Krapivinsky G, Clapham DE (2004) The mitochondrial calcium uniporter is a highly selective ion channel. Nature 427(6972):360–364

Krejci MR, Wasserman B, Finney L, McNulty I, Legnini D, Vogt S, Joester D (2011) Selectivity in biomineralization of barium and strontium. J Struct Biol 176(2):192–202, Epub 2011 Aug 17

Litvak J, Oberhauser E, Riesco J et al (1967) Strontium-85 kinetics in hypoparathyroidism at different levels of calcium intake. J Nucl Med 8:60–69

Llinas P, Masella M, Stigbrand T et al (2006) Structural studies of human alkaline phosphatase in complex with strontium: implication for its secondary effect in bones. Protein Sci 15:1691–1700. doi:10.1110/ps.062123806

Miledi R (1966) Strontium as a substitute for calcium in the process of transmitter release at the neuromuscular junction. Nature 212:1233–1234

Pors Nielsen S (2004) The biological role of strontium. Bone 35:583–588. doi:10.1016/j.bone.2004.04.026

Reginster J-Y, Neuprez A (2010) Strontium ranelate: a look back at its use for osteoporosis. Expert Opin Pharmacother 11:2915–2927. doi:10.1517/14656566.2010.533170

Shorr E, Carter AC (1952) The usefulness of strontium as an adjuvant to calcium in the remineralization of the skeleton in man. Bull Hosp Joint Dis 13:59–66

Thorpe JH, Teixeira SCM, Gale BC, Cardin CJ (2003) Crystal structure of the complementary quadruplex formed by d (GCATGCT) at atomic resolution. Nucleic Acids Res 31:844–849

Wilcock J, Perry C, Williams R, Brook A (1989) Biological minerals formed from strontium and barium sulphates. II. Crystallography and control of mineral morphology in desmids. Proc R Soc Lond B Biol Sci 238:203–221

# Strontium, Physical and Chemical Properties

Timothy P. Hanusa
Department of Chemistry, Vanderbilt University, Nashville, TN, USA

## Synonyms

Alkaline-earth metal; Atomic number 38

## Definition

A heavy member of the alkaline-earth elements (atomic number 38), strontium is a soft, silvery, highly reactive metal. Its compounds have various uses, particularly in pyrotechnics, as phosphors, and in sugar refining. Strontium salts are generally nontoxic, and the human body treats strontium much as it does calcium; strontium supplementation can be used to increase bone density and growth.

## Background

Strontium is a member of the alkaline-earth family of metals (Group 2 in the periodic table). It is a relatively common element, ranking approximately 15th in natural abundance in the Earth's crust; it is more plentiful than, for example, sulfur or chromium. The free metal does not occur in nature, but is found combined in minerals such as celestite (strontium sulfate, $SrSO_4$) (Fig. 1), or less commonly, in strontianite (strontium carbonate, $SrCO_3$). A mineral from a lead mine near the village of Strontian, in Scotland, was originally misidentified as a type of barium carbonate, but in 1790 the physician and chemist Adair Crawford confirmed that it was a different substance. The chemist Thomas Charles Hope named the new mineral *strontites*, after the village, and the corresponding "earth" (strontium oxide, SrO) was accordingly referred to as *strontia*. When Sir Humphry Davy first isolated the metal by electrolysis of a mixture of strontium oxide and mercuric oxide in 1808, he used the stem of the word strontia to form the name of the element. The most important commercial source of strontium is celestite; over two thirds of the world's supply comes from China, with Spain and Mexico supplying much of the remainder.

**Strontium, Physical and Chemical Properties, Fig. 1** Strontium sulfate in the mineral form celestite often displays pale blue crystals

**Strontium, Physical and Chemical Properties, Table 1** Selected isotopes of strontium

| Nuclide | Nat. abund (%) | Nuclear spin | Half-life |
|---|---|---|---|
| $^{79}Sr$ | – | −3/2 | 2.25 min |
| $^{80}Sr$ | – | 0 | 106.3 min |
| $^{81}Sr$ | – | −1/2 | 22.3 min |
| $^{82}Sr$ | – | 0 | 25.4 days |
| $^{83}Sr$ | – | +7/2 | 32.4 h |
| $^{84}Sr$ | 5.6% | 0 | Stable |
| $^{85}Sr$ | – | +9/2 | 64.9 days |
| $^{86}Sr$ | 9.9% | 0 | Stable |
| $^{87}Sr$ | 7.0% | +9/2 | Stable |
| $^{88}Sr$ | 82.6% | 0 | Stable |
| $^{89}Sr$ | – | +5/2 | 50.6 days |
| $^{90}Sr$ | – | 0 | 28.9 year |
| $^{91}Sr$ | – | +5/2 | 9.6 h |
| $^{92}Sr$ | – | 0 | 2.7 h |
| $^{93}Sr$ | – | +5/2 | 7.4 min |

## Isotopes

Naturally occurring strontium comprises four stable isotopes; $^{88}Sr$ is the most common (83%), followed by $^{86}Sr$ (9.9%) and $^{87}Sr$ (7.0%) (Table 1). The $^{87}Sr$ isotope stems both from primordial sources and from the radioactive decay of $^{87}Rb$ (half-life $= 4.9 \times 10^{10}$ years). Depending on the location, therefore, it is possible for $^{87}Sr/^{86}Sr$ ratios to differ substantially, sometimes by more than a factor of 5. This variation is used in radiodating geological samples and in identifying the provenance of skeletons and clay artifacts. Of the 16 synthetic isotopes, all are radioactive, and $^{89}Sr$ and $^{90}Sr$, both beta emitters, are of particular importance. $^{90}Sr$ (half-life of 29 years) is produced in nuclear fission reactions, and was a component of fallout from above ground atomic explosions. Owing to its chemical resemblance to calcium, it is assimilated in bones, and is a cause of osteosarcoma, although it is also used for treatment of some superficial cancers. It also finds use as a source in thickness gauges, and was one of the nuclides used in NASA's SNAP radioisotope thermoelectric generators. $^{89}$Sr is employed in the treatment of bone cancer, as it targets bone tissues, delivers its beta radiation, and then decays in a few month's time (half-life of 51 days).

## Properties of the Metal

Various physical and chemical properties of strontium are listed in Table 2. Strontium is a lustrous, silvery or silver-yellow metal that is relatively soft (Mohs hardness of 1.5). It crystallizes in a cubic close-packed lattice with $a = 6.085$ Å. It readily oxidizes when exposed to air, and must be protected from oxygen during storage. One of the few uses for elemental strontium is as an alloying agent for aluminum or magnesium in cast engine blocks and wheels; the strontium improves the machinability and creep resistance of the metal.

## General Properties of Compounds

Strontium exclusively displays the +2 oxidation state in its compounds. The metal is highly electropositive ($\chi = 0.95$ on the Pauling scale; 0.99 on the Allred-Rochow scale; cf. 0.93 and 1.01, respectively, for sodium), and with a noble gas electron configuration for the Sr$^{2+}$ ion ([Kr]$5s^0$), the metal-ligand interactions are usually viewed as electrostatic. To a first approximation, the bonding can be considered as nondirectional, and strongly influenced by ligand packing. The presence of multidentate and sterically bulky ligands can produce highly irregular geometries.

As with the halides of several other heavy alkaline-earth metals, the difluoride and dichloride of strontium display nonlinear geometries in the gas phase (108° and 120°, respectively) (Hargittai 2000). An argument based on the "reverse polarization" of the metal core electrons by the ligands has been used to explain their geometry, an analysis that makes correct predictions about the ordering of the bending for the dihalides (i.e., Ca < Sr < Ba; F > Cl > Br > I). The "reverse polarization" analysis can be recast in molecular

**Strontium, Physical and Chemical Properties, Table 2** Atomic and physical properties of strontium

| Atomic Number | 38 | $E$ for M$^{2+}$(aq) + 2e$^-$ → M(s) | −2.89 V |
|---|---|---|---|
| Number of naturally occurring isotopes | 4 | Melting point | 768°C |
| Atomic mass | 87.62 | Boiling point | 1,381°C |
| Electronic configuration | [Kr]$5s^2$ | Density (20°C) | 2.63 |
| Ionization energy (kJ mol$^{-1}$) | 549.2 (1st); 1,064(2$^{nd}$) | $\Delta H_{fus}$ (kJ mol$^{-1}$) | 8.2 |
| Metal radius | 2.15 Å | $\Delta H_{vap}$ (kJ mol$^{-1}$) | 137 |
| Ionic radius (6-coordinate) | 1.18 Å | Electrical resistivity (20°C)/ μohm cm | 13.5 |

orbital terms; that is, bending leads to a reduction in the antibonding character in the HOMO.

An alternative explanation for the bending in ML$_2$ species has focused on the possibility that metal $d$ orbitals might be involved. Support for this is provided by calculations that indicate a wide range of small molecules, including MH$_2$, MLi$_2$, M(BeH)$_2$, M(BH$_2$)$_2$, M(CH$_3$)$_2$, M(NH$_2$)$_2$, M(OH)$_2$, and MX$_2$ (M = Ca, Sr, Ba) should be bent, at least partially as an effect of metal d-orbital occupancy (Kaupp and Schleyer 1992). Spectroscopic confirmation of the bending angles in most of these small molecules is not yet available, however.

The standard classification of alkali and alkaline-earth ions as hard (type a) Lewis acids leads to the prediction that ligands with hard donor atoms (e.g., O, N, halogens) will routinely be preferred over softer (type b; P, S, Se) donors. This is generally true, but studies have demonstrated that the binding of s-block ions to "soft" aromatic donors can be quite robust; hydrated strontium (and barium) clusters rapidly undergo exchange reactions with benzene molecules in the gas phase, for example (Rodriguez-Cruz and Williams 2001).

The Sr$^{2+}$ ion is relatively large (the 6-coordinate radius is 1.18 Å) and approximately the same size as lead (1.19 Å) or the largest of the divalent lanthanides (Sm$^{2+}$, Eu$^{2+}$). Coordination numbers in strontium compounds are often high; a 12-coordinate strontium center exists in the {Sr[Ni(L)]$_6$}$^{2+}$ (L = 2,2'-(propane-1,3-diylbis(azanylylidene))dipropanoate) ion (Fig. 2) (Lin et al. 2003), for example. In the presence of sterically compact ligands (e.g., −NH$_2$, −OMe,

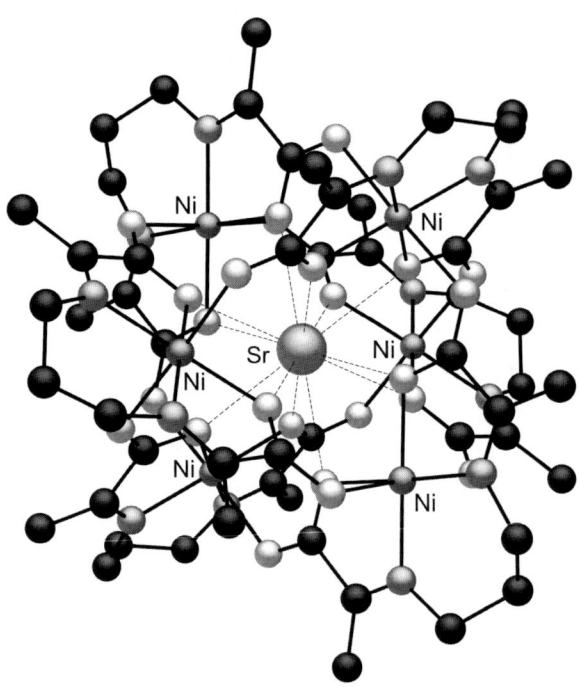

**Strontium, Physical and Chemical Properties, Fig. 2** Solid state structure of the $\{Sr[Ni(L)]_6\}^{2+}$ (L = 2,2'-(propane-1,3-diylbis(azanylylidene))dipropanoate) ion. Bonds from strontium to oxygen are dotted

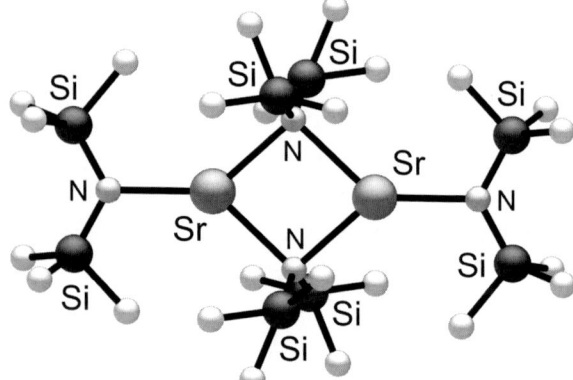

**Strontium, Physical and Chemical Properties, Fig. 3** Solid state structure of $\{Sr[N(SiMe_3)_2]_2\}_2$

**Strontium, Physical and Chemical Properties, Table 3** Bond energies in diatomic strontium molecules

| Sr–E | Enthalpy (kJ mol$^{-1}$) |
|---|---|
| Sr–H | 163 ± 8 |
| Sr–O | 426 ± 17 |
| Sr–F | 542 ± 7 |
| Sr–Cl | 406 ± 13 |
| Sr–Br | 333 ± 9 |
| Sr–I | 270 ± 6 |

halides), extensive oligomerization or polymerization will occur, leading to the formation of nonmolecular compounds of limited solubility or volatility. However, sterically bulky ligands can be used to produce compounds with low coordination numbers and improved solubility; strontium is only 3-coordinate in $\{Sr[N(SiMe_3)_2]_2\}_2$, for example (Fig. 3) (Westerhausen 1998).

The strengths of common strontium-element bonds are listed in Table 3. These are for gas-phase species, and so must be used with caution in comparison to the solid state. Bonds to oxygen and fluorine are the strongest that strontium forms, and are comparable to those of some early transition metals (e.g., Cr–O = 429 kJ mol$^{-1}$; W–F = 548 kJ mol$^{-1}$).

## Common Reactions and Compounds of Strontium

### With Strontium-Oxygen Bonds

In addition to its reaction with oxygen at room temperature, strontium burns in air to yield a mixture of white strontium oxide, SrO, and strontium nitride, $Sr_3N_2$ (1, 2). The peroxide, $SrO_2$, is formed directly from the metal and high-pressure oxygen (3). Strontium oxide is cleanly made by heating strontium carbonate, $SrCO_3$ (4), or strontium nitrate, $Sr(NO_3)_2$ (5). The oxide is a high melting solid (2,530°C) with a rock salt lattice that reacts with $CO_2$ in the reverse of the reaction used to prepare it. At one time, a major use of strontium oxide was as an additive to glass to prevent X-ray emission from the faceplates of color television and other cathode ray tubes. Strontium oxide was preferred over the equally effective lead oxide, as the latter would discolor over time. As flat panel display technologies have come to dominate the commercial marketplace, this use of strontium oxide has declined sharply.

$$2Sr(s) + O_2(g) \rightarrow 2SrO(s) \quad (1)$$

$$3Sr(s) + N_2(g) \rightarrow Sr_3N_2(s) \quad (2)$$

**Strontium, Physical and Chemical Properties, Fig. 4** Strontium salts are responsible for the intense red color of fireworks

$$Sr(s) + O_2(g) \rightarrow SrO_2(s) \quad (3)$$

$$SrCO_3(s) \rightarrow SrO(s) + CO_2(g) \quad (4)$$

$$2Sr(NO_3)_2(s) \rightarrow 2SrO(s) + 4NO_2(g) + O_2(g) \quad (5)$$

Some strontium carbonate is used in glass manufacturing, but an important use for it and strontium nitrate (and to a lesser extent the chlorate, oxalate, and sulfate) are in pyrotechnics. There is currently no substitute for the brilliant red color produced by strontium salts in fireworks, flares, and tracer ammunition (Fig. 4); the color from lithium compounds, for example, is less intense. About 5–10% of all strontium production is consumed in pyrotechnics.

Strontium and strontium oxide react with water to form the hydroxide, $Sr(OH)_2$ (6, 7). The reaction involving elemental strontium also releases hydrogen gas ($H_2$). Alternatively, the hydroxide can be formed by precipitation reactions (e.g., (8)), owing to its limited solubility in water at room temperature (18 g L$^{-1}$); it is much more soluble in hot water (218 g L$^{-1}$ at 100°C). Superheated steam will convert strontium carbonate to the hydroxide (9). A large-scale use of the hydroxide has been in sugar refining; the reaction of strontium hydroxide with crude sucrose liquors (molasses) forms strontium saccharate, which has low solubility ($K_{sp} = 5.6 \times 10^{-9}$) and can be separated from impurities. Treatment of the saccharate with carbon dioxide releases the sucrose and generates the poorly soluble and readily removed strontium carbonate.

$$Sr(s) + 2H_2O(g) \rightarrow Sr(OH)_2(aq) + H_2(g) \quad (6)$$

$$SrO(s) + H_2O(l) \rightarrow Sr(OH)_2(aq) \quad (7)$$

$$Sr(NO_3)_2(aq) + 2KOH(aq) \rightarrow Sr(OH)_2(s) \downarrow + 2KNO_3(aq) \quad (8)$$

$$SrCO_3(s) + H_2O(>100°C) \rightarrow Sr(OH)_2(s) + CO_2(g) \quad (9)$$

Strontium sulfate (celestite) is the main commercial source of strontium, although the metal will react with sulfuric acid to form the sulfate (10), or it can be formed from an exchange reaction (e.g., (11)). The latter works because of the sulfate's low solubility in water ($K_{sp} = 3.4 \times 10^{-7}$), which causes it to precipitate from solution. Strontium sulfate itself is not the direct source of most other strontium compounds, but it is converted to the carbonate and nitrate for further use. The sulfate finds some use as a pigment in paints and inks.

$$Sr(s) + H_2SO_4(l) \rightarrow SrSO_4(s) + H_2(g) \quad (10)$$

$$SrCl_2(aq) + Na_2SO_4(aq) \rightarrow SrSO_4(s) + 2NaCl(aq) \quad (11)$$

The technique of chemical vapor deposition (CVD; sometimes abbreviated as MOCVD [metalorganic chemical vapor deposition]) has been under intensive development for the s-block elements, and particularly the alkaline-earth metals, since the late 1980s (Pierson 1999). The production of complex oxides of strontium, such as the perovskite-based titanate $SrTiO_3$ and superconducting cuprates (e.g., BSCCO, $Bi_2Sr_2Ca_{n-1}Cu_nO_{2n+4+x}$), has been one focus of this research. Strontium titanate is a paraelectric ceramic that is used as a gate dielectric in electronics, in capacitors, as a substrate for oxide-based films, and as a diamond simulant. The high-temperature superconductors have a variety of applications, including use in current-carrying cable.

The strontium ferrites comprise a family of compounds of general formula $SrFe_xO_y$, often formed from the high-temperature (1,300°C) reaction between $SrCO_3$ and $Fe_2O_3$, although solution routes to the material are available. Permanent ceramic magnets can be formed from the ferrimagnetic $SrFe_{12}O_{19}$ material, and are used when high coercivity, thermal and electrical resistivity, and chemical inertness are important. Although brittle, such magnets find use in loudspeakers and other electronic devices, and in applications as diverse as motors for automobile windshield wipers and children's toys.

### With Strontium-Sulfur Bonds

Strontium sulfide can be formed from the high-temperature (1,100°C) reaction of strontium sulfate with carbon (coal) as the reducing agent (12). The resulting soluble sulfide is dissolved in water and filtered. In the "black ash" process, the sulfide solution is treated with $CO_2$ or $Na_2CO_3$ to form the carbonate and sulfur byproducts (13, 14). Like the oxide, strontium sulfide possesses a rock salt crystal structure, and is soluble in water. It finds use as a depilatory and as an ingredient in phosphors for electroluminescent devices. Europium-activated strontium sulfide-calcium sulfide phosphor, $Ca_xSr_{1-x}S$:Eu, and cesium-doped strontium sulfide, SrS:Ce, are of interest in this regard (Braithwaite and Weaver 1990). Phosphorescent materials in demand for watches and instrument dials also fall into this category. Among the best phosphors currently available for these purposes is the mixed oxide $M[Al_2O_4]$ (M = Ca, Sr, Ba), activated by 0.001–10% of Eu, and co-activated by 0.001–10% of at least one element selected from several of the lanthanide elements, tin or bismuth. The phosphor $Sr[Al_2O_4]$:(Eu, Dy), for example, is visible for up to 15 h after exposure to light.

$$SrSO_4(s) + 2C(s) \rightarrow SrS(s) + 2CO_2(g) \quad (12)$$

$$SrS(aq) + H_2O + CO_2(g) \rightarrow SrCO_3(s) + H_2S(g) \quad (13)$$

$$SrS(aq) + Na_2CO_3(s) \rightarrow SrCO_3(s) + Na_2S(aq) \quad (14)$$

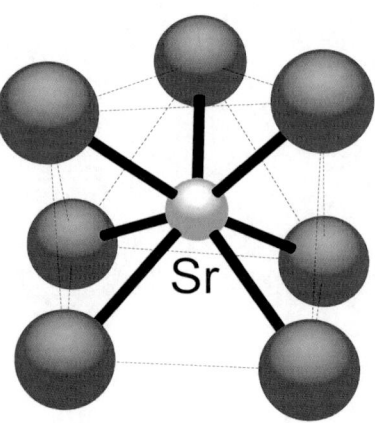

**Strontium, Physical and Chemical Properties, Fig. 5** Coordination environment around strontium in solid $SrI_2$

### With Strontium-Halogen Bonds

All four halides of strontium are known; $SrF_2$ has the $CaF_2$ lattice type, $SrCl_2$ has a distorted $TiO_2$ type lattice, $SrBr_2$ displays a distorted $PbCl_2$ lattice and $SrI_2$ has its own unique nonmolecular type. In the latter, strontium is 7-coordinate (Fig. 5), with Sr–I contacts ranging from 3.26 to 3.42 Å. It is related to the lattice found for $EuI_2$, which has similarly sized ions.

The strontium halides are not formed commercially from the metal and halogens, but rather from displacement reactions. $SrF_2$, for example, is formed from strontium carbonate; the insolubility of the fluoride ($K_{sp} = 7.9 \times 10^{-10}$) facilitates separation of the product (15). Strontium chloride is made by the neutralization of strontium hydroxide with HCl in water (16); the dihydrate product $SrCl_2 \cdot 6H_2O$ can be heated to leave the anhydrous $SrCl_2$. Alternatively, the chloride, bromide, and iodide can be formed analogously to the fluoride, by reaction of the carbonate with the appropriate hydrohalide acid (17).

$$SrCO_3(aq) + 2HF(aq) \rightarrow SrF_2(s) + H_2O + CO_2(g) \quad (15)$$

$$Sr(OH)_2 + 2HCl(aq) \rightarrow SrCl_2(aq) + 2H_2O \quad (16)$$

$$SrCO_3 + 2HX(aq) \rightarrow SrX_2(aq) + H_2O + CO_2(g)$$
$$(X = Cl, Br, I) \quad (17)$$

Although brittle, crystalline strontium fluoride (Mohs hardness = 3.5) is used to make optical

**Strontium, Physical and Chemical Properties, Fig. 6** Portion of the solid-state structure of $[Sr(NH_3)_8]_3[C_{70}]_2 \cdot 19NH_3$ (the ammonia molecules of crystallization are omitted for clarity)

windows and coatings on lenses, as it is suitably transparent from roughly 150 nm (ultraviolet) to 11 μm (infrared). Strontium fluoride has also been used as the carrier compound for strontium-90 in thermoelectric generators. Strontium chloride has been incorporated into toothpastes to reduce sensitivity; it fills tubules in the dentin that contain nerve endings, and blocks access to the nerve. Strontium iodide has been used as a nutritional source of iodide.

Strontium hydride, which is made by the high-temperature reaction of elemental strontium and hydrogen (18), can be fused with the strontium halides to produce the strontium hydride halides SrHX (X = Cl, Br, I). An alternative method of preparation is to heat elemental strontium and a strontium halide in a hydrogen atmosphere at 900°C (19). The SrHX compounds have the PbClF crystal structure, and melt without decomposition at high temperatures (e.g., mp of SrHCl is 840°C).

$$Sr(s) + H_2(g) \rightarrow SrH_2(s) \quad (18)$$

$$Sr(s) + SrX_2(s) + H_2(g) \rightarrow 2SrHX(s) \quad (19)$$

### With Strontium-Nitrogen Bonds

Strontium dissolves in liquid ammonia to give a deep blue-black solution that yields a copper colored ammoniate, $Sr(NH_3)_6$, on evaporation. With time, the ammoniate will decompose to form the amide (20). In the presence of the carbon fullerene $C_{70}$, however, the $[Sr(NH_3)_8]^{2+}$ cation is generated from strontium in liquid ammonia. The X-ray crystal structure of $[Sr(NH_3)_8]_3[C_{70}]_2 \cdot 19NH_3$ reveals a distorted tetragonal antiprism around the metal (Sr–N = 2.63–2.80 Å); the fullerene units are linked in slightly zigzagging linear chains by single C–C bonds (1.59 Å) (Fig. 6) (Panthöfer et al. 2004).

$$Sr(NH_3)_6(s) \rightarrow Sr(NH_2)_2(s) + 4NH_3(s) + H_2(g) \quad (20)$$

Whereas the parent amide $Sr(NH_2)_2$ possesses an ionic lattice, replacement of the hydrogen atoms with groups of increasing size leads to molecular complexes. Such amido complexes are versatile reagents, and can be used to prepare other main-group and transition metal complexes. The chemistry of $Sr[N(SiMe_3)_2]_2$ (Fig. 3) has been reviewed (Westerhausen 1998).

### With Strontium-Carbon Bonds

Strontium compounds with bonds to carbon (organostrontium complexes) are relatively rare. Owing to the largely ionic character of the bonding in strontium compounds, all are reactive species, and decompose in the presence of air and moisture. They are formed by a variety of methods, often involving exchange reactions (e.g., (21)) (Hanusa et al. 2007). Some organostrontium species can function as initiators for polymerization reactions and serve as precursors to oxides under CVD conditions. The bis

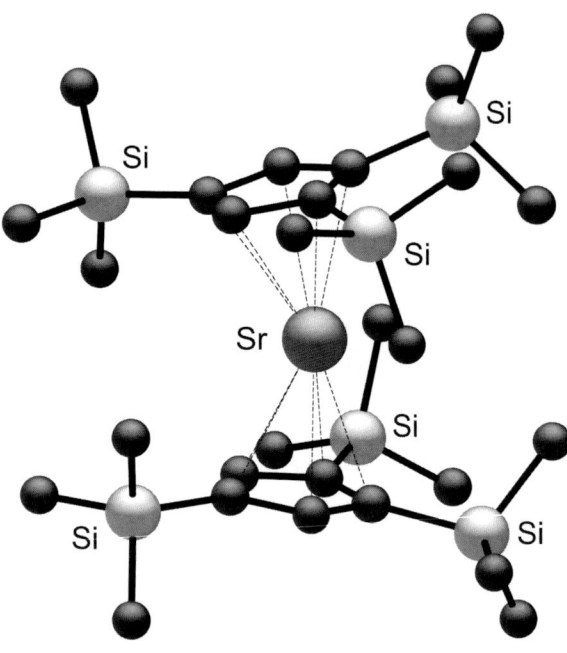

**Strontium, Physical and Chemical Properties, Fig. 7** Solid state structure of $[C_5(SiMe_3)_3H_2]_2Sr$. The molecule is "bent," with an angle between the cyclopentadienyl ring planes of 161°

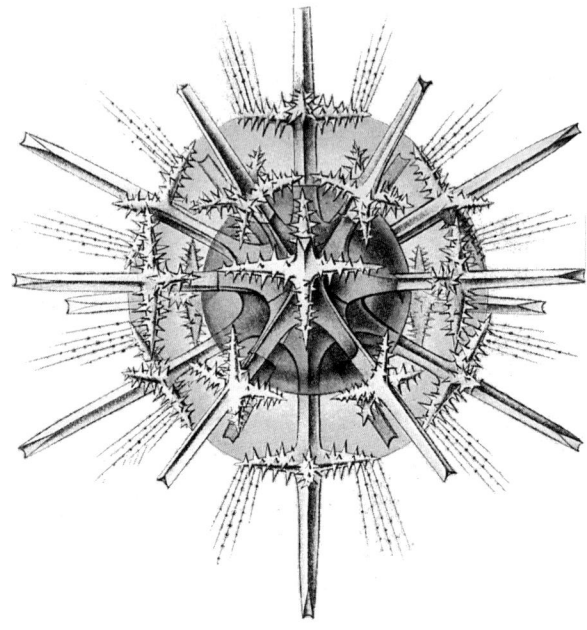

**Strontium, Physical and Chemical Properties, Fig. 8** Schematic drawing of strontium ranelate, used to treat osteoporosis

(cyclopentadienyl) metallocenes usually display "bent" (nonlinear) geometries (Fig. 7), similar to the bent gas-phase dihalides; related issues in their metal-ligand bonding (e.g., reverse polarization effects, d-orbital participation) may be involved.

$$2K[C_5R_5] + SrI_2(s) \rightarrow (C_5R_5)_2Sr + 2KI(s) \quad (21)$$

## Biological Features of Strontium Chemistry

The $Sr^{2+}$ ion is sufficiently close in size to $Ca^{2+}$ that the human body treats the ▶ strontium as a calcium analogue. Strontium is considered a "bone seeker," and although this is dangerous in the case of radioactive $^{90}Sr$, strontium supplementation can be used to increase bone density and growth. Strontium ranelate (Fig. 8) is a chelated version of strontium that has been shown both to promote bone growth by contributing to the production of osteoblasts and to slow bone resorption by osteoclasts. It is available as a prescription drug in many countries, although not in Canada or the USA, for treatment of postmenopausal osteoporosis. Other salts of strontium, including the malonate and citrate, may offer comparable benefits.

Although strontium is not an essential element for higher life forms, a type of large (up to ca. 0.5 mm length) marine protozoa belonging to the class *Acantharia* constructs its skeleton out of strontium sulfate (Fig. 9). It is possible that the bioaccumulation of strontium by *Acantharia* protozoa may locally affect Sr/Ca ratios in seawater (typically listed as 8.1 ppm $Sr^{2+}$, 411 ppm $Ca^{2+}$) (De Deckker 2004).

**Strontium, Physical and Chemical Properties, Fig. 9** Drawing of the *Acantharia* protozoa *Stauracantha spinulosa*, whose skeleton is composed of $SrSO_4$ (Drawing from Ernst Haeckel's *Kunstformen der Natur* (1904))

## Cross-References

- Strontium and DNA Aptamer Folding
- Strontium, Calcium Analogue in Membrane Transport Systems

## References

Braithwaite N, Weaver G (1990) Electronic materials. Butterworth, London

De Deckker P (2004) On the celestite-secreting Acantharia and their effect on seawater strontium to calcium ratios. Hydrobiologia 517:1–13

Hanusa TP et al (2007) Alkaline-earth metals: beryllium, magnesium, calcium, strontium, and barium. In: Crabtree RH, Mingos DMP (eds) Comprehensive organometallic chemistry-III, vol 2. Elsevier, Oxford, pp 67–152

Hargittai M (2000) Molecular structure of metal halides. Chem Rev 100:2233–2301

Haeckel E (1904) Kunstformen der Natur. Bibliographisches Institut, Leipzig, plate 21

Kaupp M, Schleyer PvR (1992) The structural variations of monomeric alkaline earth $MX_2$ compounds (M = Ca, Sr, Ba; X = Li, BeH, $BH_2$, $CH_3$, $NH_2$, OH, F). An ab initio pseudopotential study. J Am Chem Soc 114:491–497

Lin X, Doble DMJ et al (2003) Cationic assembly of metal complex aggregates: structural diversity, solution stability, and magnetic properties. J Am Chem Soc 125:9476–9483

Panthöfer M, Wedig U et al (2004) Geometric and electronic structure of polymeric $C_{70}$-fullerides: the case of $^1_\infty[C_{70}^{3-}]$. Solid State Sci 6:619–624

Pierson HO (1999) Handbook of chemical vapor deposition: principles, technology, and applications. Noyes, Norwich

Rodriguez-Cruz SE, Williams ER (2001) Gas-phase reactions of hydrated alkaline earth metal ions, $M^{2+}(H_2O)_n$ (M = Mg, Ca, Sr, Ba and n = 4-7), with benzene. J Am Soc Mass Spectrom 12:250–257

Westerhausen M (1998) Synthesis, properties, and reactivity of alkaline earth metal bis[bis(trialkylsilyl)amides]. Coord Chem Rev 176:157–210

## Structural Zinc

- Zinc Structural Site in Alcohol Dehydrogenases

## Structural Zinc Site

- Zinc Structural Site in Alcohol Dehydrogenases

## Structure, Spatial Arrangement

- NMR Structure Determination of Protein-Ligand Complexes using Lanthanides

## Structures and Functions of Extracellular Calcium-Binding Cadherin Repeats

- Cadherins

## Substrate: Reactant

- Monovalent Cations in Tryptophan Synthase Catalysis and Substrate Channeling Regulation

## Superoxide:superoxide Oxidoreductase

- Zinc in Superoxide Dismutase

## Superoxides Anions

- Arsenic, Free Radical and Oxidative Stress

## Suzuki Biology

- Palladium-Mediated Site-Selective Suzuki-Miyaura Protein Modification

## Suzuki-Miyaura Couplings at Unnatural Amino Acids

- Palladium-Mediated Site-Selective Suzuki-Miyaura Protein Modification

## Synaptic Transmission

▶ Zinc-Secreting Neurons, Gluzincergic and Zincergic Neurons

## Synaptic Zinc

▶ Zinc-Secreting Neurons, Gluzincergic and Zincergic Neurons

## Synergistic Toxicity of Arsenic and Ethanol

▶ Arsenic and Alcohol, Combined Toxicity

## Synthetic Metalloenzymes

▶ Palladium-catalysed Allylic Nucleophilic Substitution Reactions, Artificial Metalloenzymes

## 5f Systems – Actinide Elements

▶ Actinide and Lanthanide Systems, High Pressure Behavior

## 4f Systems – Lanthanide Elements

▶ Actinide and Lanthanide Systems, High Pressure Behavior

# T

## T & B Lymphocytes

▶ Chromium(III) and Immune System

## T:T Mismatched Base Pair

▶ Mercury and DNA

## Tacrolimus

▶ Calcineurin

## Tantalum, Physical and Chemical Properties

Fathi Habashi
Department of Mining, Metallurgical, and Materials Engineering, Laval University, Quebec City, Canada

Tantalum is a transition metal like the other members of the group. They are less reactive than the typical metals but more reactive than the less typical metals. The transition metals are characterized by having the outermost electron shell containing two electrons and the next inner shell an increasing number of electrons. Tantalum is closely related to niobium with which they are associated in their ores that are mainly complex oxides, $(Nb,Ta)_2O_5$.

## Physical Properties

| | |
|---|---|
| Atomic number | 73 |
| Atomic weight | 180.95 |
| Relative abundance in Earth's crust, % | $2.1 \times 10^{-4}$ |
| Crystal structure | Body-centered cubic |
| Lattice constant, nm | $a = 0.33025$ |
| Density, g/cm$^3$ | 16.6 |
| Melting point, °C | 2,996 |
| Boiling point, °C | 5,425 |
| Heat of fusion, kJ/mol | 28.5 |
| Heat of vaporization at 3,273 K, kJ mol$^{-1}$ | 78.1 |
| Specific heat capacity at 20°C, J mol$^{-1}$ K$^{-1}$ | 25.41 |
| Linear coefficient of thermal expansion at 20°C, K$^{-1}$ | $6.5 \times 10^{-6}$ |
| Thermal conductivity at 20°C, W m$^{-1}$ K$^{-1}$ | 54.4 |
| Specific electrical conductivity at 20°C, $\Omega^{-1}$ cm$^{-1}$ | 0.081 |
| Superconductivity, K | 4.3 |
| Temperature coefficient (0–100°C) | 0.00383 |

The mechanical properties of the metal are strongly dependent on its purity, structure, and crystal defects, as is the case with almost all refractory metals. Even low concentrations of interstitial impurities increase the hardness and reduce the ductility. The yield strength is strongly temperature dependent, the value at 200°C being ca. 30% of that at 20°C.

## Chemical Properties

Tantalum and niobium are closely associated because of the lanthanide contraction (Fig. 1). Like niobium,

V.N. Uversky et al. (eds.), *Encyclopedia of Metalloproteins*, DOI 10.1007/978-1-4614-1533-6,
© Springer Science+Business Media New York 2013

**Tantalum, Physical and Chemical Properties, Fig. 1** The similarity of atomic radii of Zr-Hf, Nb-Ta, Mo-W, etc., due to the lanthanide contraction

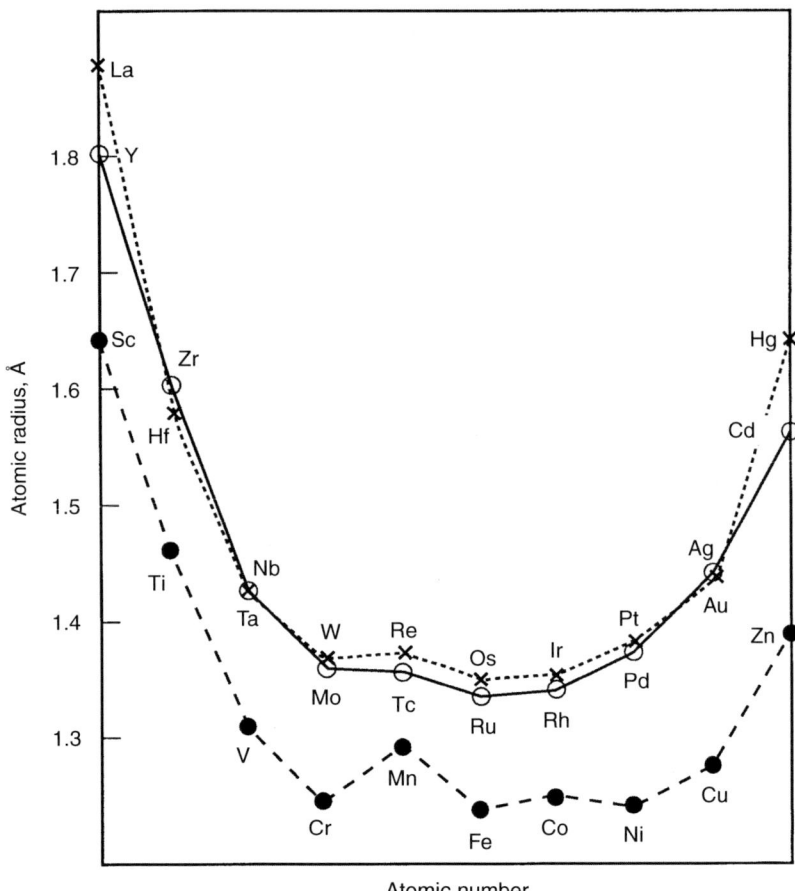

tantalum has principal oxidation states +2 and +5. Below 100°C, the metal is extremely resistant to corrosion by most organic and inorganic acids, with the exception of hydrofluoric acid. This is due to a dense adherent film of tantalum oxide, a characteristic which is utilized in the manufacture of electrolytic capacitors. Metallic tantalum is produced almost exclusively by reduction of potassium heptafluorotantalate with sodium:

$$K_2TaF_7 + 5Na \rightarrow Ta + 2KF + 5NaF$$

Tantalum pentoxide, $Ta_2O_5$, occurs in two thermodynamically stable modifications. The transition temperature from the orthorhombic modification to the tetragonal a modification is ca. 1,360°C. Tantalum pentafluoride, $TaF_5$, is produced by passing fluorine over metallic tantalum, or from tantalum chloride by adding anhydrous hydrogen fluoride. Both tantalum pentafluoride and niobium pentafluoride are used in the petrochemical industry as catalysts.

Potassium tantalum fluoride, $K_2TaF_7$, is an industrially important intermediate in the production of tantalum metal. It crystallizes in colorless, rhombic needles when $K^+$ ions are added to solutions of tantalum in hydrofluoric acid from the extraction process. Its solubility in hydrofluoric acid solution decreases from 60 g/100 ml at the boiling point to <0.5 g/100 ml at room temperature. As the corresponding stable niobium salt $K_2NbOF_5$ is significantly more soluble, tantalum can also be separated from niobium by fractional crystallization (Marignac process).

Tantalum pentachloride, $TaCl_5$, forms colorless, hygroscopic, needle-shaped crystals produced on an industrial scale by chlorination of metallic tantalum, ferrotantalum, or tantalum metal scrap. It is soluble in absolute alcohol, with formation of the corresponding alkoxide. Ultrafine tantalum oxide powder can be produced by hydrolysis of this alkoxide. Tantalum pentachloride and tantalum alkoxides are suitable for use in the chemical vapor deposition of tantalum metal or tantalum oxide.

## References

Andersson K, Reichert K, Wolf R (1997) Tantalum and tantalum compounds. In: Habashi F (ed) Handbook of extractive metallurgy. Wiley-VCH, Weinheim, pp 1417–1430

Habashi F (2003) Metals from ores. An introduction to extractive metallurgy. Métallurgie extractive Québec, Quebec City, Canada. Distributed by Laval University Bookstore, www.zone.ul.ca

# Targeted Drug Delivery

▶ Gold Nanomaterials as Prospective Metal-based Delivery Systems for Cancer Treatment

# Targeted Mechanism of Gold Complex

▶ Gold Complexes as Prospective Metal-Based Anticancer Drugs

# Technetium, Physical and Chemical Properties

Fathi Habashi
Department of Mining, Metallurgical, and Materials Engineering, Laval University, Quebec City, Canada

Technetium is a radioactive metal that does not occur in nature; first obtained by irradiation of molybdenum with deuterons at the Berkeley cyclotron. The isotope $^{93}$Tc ($t_{1/2} = 4.2 \times 10^6$ years) is the longest-lived isotope but it cannot be obtained in weighable amounts. The isotope $^{99}$Tc ($t_{1/2} = 2.14 \times 10^5$ years) is obtained by fission of $^{235}$U with thermal neutrons in about 6% yield and used for chemical and physical studies of the metal. This corresponds to formation of 2.8 g technetium per day and 100 MW$_{th}$ power of a nuclear reactor (i.e., about 30 kg/year in a modern 1,300 MW$_{el}$ nuclear power reactor). $^{99}$Tc can be recovered from high active waste solutions after reprocessing of spent fuel by the PUREX process.

The most important technetium isotope, however, is $^{99m}$Tc ($t_{1/2} = 6$ h) which is currently the most frequently used radionuclide in nuclear medicine. Due to its short half-life, it must be separated from its mother nuclide $^{99}$Mo ($t_{1/2} = 66$ h) by the use of a generator at the place of application. Several types of $^{99}$Mo/$^{99m}$Tc generators exist which can supply sterile $^{99m}$Tc in high purity and good yield, e.g., by elution of technetium with NaCl solution from an alumina column loaded with $^{99}$Mo. $^{99}$Mo itself can be produced by two processes:

- Irradiation of $^{98}$Mo-enriched molybdenum in a nuclear reactor via neutron capture $^{98}$Mo(n, γ)$^{99}$Mo
- Irradiation of highly $^{235}$U-enriched (≥20%) uranium in a nuclear reactor and separation of the fission product $^{99}$Mo (fission yield ~ 6%)

The second type of production is preferred because it yields $^{99}$Mo with a high specific activity so that the eluted $^{99m}$Tc is obtained in high concentration. In contrast, the product of $^{98}$Mo irradiation only delivers $^{99}$Mo with a much lower specific activity and, therefore, requires larger generators. Nowadays, an adequate supply of $^{99m}$Tc is provided by the so-called $^{99m}$Tc generators. These generators are efficient and reliable and differ principally from common generators in the technique of separating technetium from molybdenum.

## Physical Properties

| | |
|---|---|
| Atomic number | 43 |
| Atomic weight | 97.91 |
| Relative abundance in Earth's crust | 0 |
| Melting point, °C | $2,250 \pm 50$ |
| Density, g/cm$^3$ | 11.47 |
| Crystal structure | Hexagonal close-packed |
| Lattice dimensions at 25°C, pm | $a = 274.07$ |
| | $c = 439.80$ |
| Critical temperature (superconducting), K | $7.73 \pm 0.02$ |
| Debye temperature, K | 454 |
| Standard entropy of formation, J mol$^{-1}$ K$^{-1}$ | $31 \pm 0.8$ |
| Magnetic susceptibility at 25°C | $270 \times 10^{-6}$ |
| Specific heat $C_p$ (1,000 K), J g$^{-1}$ K$^{-1}$ | 0.29 |

## Chemical Properties

The chemical properties of technetium are more similar to rhenium than to manganese. Valency states between 1− and 7+ are known, the most stable ones being Tc (IV) and Tc (VII). Solid technetium is resistant to oxidation, but in sponge or powder form it is readily oxidized to the volatile heptaoxide when heated in air. The metal dissolves in dilute or concentrated nitric acid, concentrated sulfuric acid, and bromine water, but not in hydrochloric acid. It also dissolves in aqueous acid hydrogen peroxide solutions. It is oxidized by these solvents to pertechnetate, but in the case of bromine water $[TcBr_6]^{2-}$ is also formed. Contrary to rhenium, no heptafluoride of technetium is known – only $TcF_5$ and $TcF_6$ are obtained. Sulfur reacts with technetium at elevated temperature to give the disulfide and carbon to give TcC.

Oxidation of technetium metal by oxygen at 500°C yields pale yellow $Tc_2O_7$ (m.p. 119.5°C, b.p. 310.6°C) which reacts with water vapor to form dark red $HTcO_4$.

## References

Griebel J (1997) Technetium. In: Habashi F (ed) Handbook of extractive metallurgy. Wiley-VCH, Weinheim, pp 1589–1591

Habashi F (2003) Metals from ores. An introduction to extractive metallurgy. Métallurgie extractive Québec, Quebec City, Canada. Distributed by Laval University Bookstore, www.zone.ul.ca

## Tellurite Tolerance

▶ Bacterial Tellurite Processing Proteins
▶ Bacterial Tellurite Resistance

## Tellurite=Chalcogen

▶ Tellurite-Resistance Protein TehA from Escherichia coli

## Tellurite-Detoxifying Protein TehB from *Escherichia coli*

Hassanul Ghani Choudhury and
Konstantinos Beis
Division of Molecular Biosciences, Imperial College London, London, South Kensington, UK
Membrane Protein Lab, Diamond Light Source, Harwell Science and Innovation Campus, Chilton, Oxfordshire, UK
Research Complex at Harwell, Harwell Oxford, Didcot, Oxforsdhire, UK

## Synonyms

Chalcogen family; Selenium; Tellurium

## Definition

Tellurite detoxification by methylation is a process in which the metal oxyanion is modified by a methyltransferase. Methyltransferases catalyze the transfer of a methyl group from their S-adenosyl methionine (SAM) cofactor to an acceptor; the SAM acts as the methyl donor. Studies have shown that methylated tellurite is less toxic than tellurite.

## Introduction

Micromolar concentrations of selenium and tellurite in the environment can cause deformities and even death of wildlife (Ranjard et al. 2002). Tellurium can be found at higher concentrations near waste discharge sites (Sabaty et al. 2001). Oxyanion metals are able to enter bacterial cells by utilizing pathways for other similar chemical compounds. Some bacteria are resistant to potassium tellurite and have evolved two main mechanisms for detoxification of oxyanion metals (1) reduction to the elemental state and (2) methylation (Walter and Taylor 1992). Reduction of tellurite to its elemental state results in precipitation of tellurium within cells and it is seen as black deposits. Methylation occurs in a SAM dependent methylation mechanism and results in volatile species.

When volatile tellurite species are formed (dimethyltellurite, DMTe) a garlic-like odor can often be smelt, this has been claimed "the most abominable odour of all organometallics" (Chasteen and Bentley 2003). The smelling of this odor was the original means of detection and was used to identify dimethyltellurite in *S. brevicaulis* cultures (Bird and Challenger 1939), which was later confirmed by other methods. *P. fluorescens* K27, a facultative anaerobe, was shown to produce DMTe analyzed by headspace chromatography (Burton et al. 1987; Basnayake et al. 2001). However the production of DMTe is dependent on the amount and oxidation state of the tellurium oxyanions added. The resulting methylated compounds (e.g., dimethyl selenide) have lower toxicity than inorganic species. Bioremediation of oxyanion metals by volatilisation from contaminated areas within the environment offers an alternative to the use of selenate-respiring bacteria (Ranjard et al. 2003).

A strain of *Penicillium* isolated from sewage has been identified by GC-MS headspace chromatography as producing DMTe (Basnayake et al. 2001). However DMTe was only produced when both tellurium and selenium compounds were introduced to the culture. This may be due to selenium activating certain components of the methylation pathway that tellurium could not.

Bacterial methylation of tellurite can occur in a SAM dependant mechanism to produce volatile methylated tellurite compounds. The *Corynebacterium* enzyme system has been shown to be capable of the SAM dependant methylation of inorganic selenium to produce dimethyl selenide. Enzymatic SAM dependant methylation of sodium selenide in mammals has also been shown to form methane selenol and dimethyl selenide (Doran and Alexander 1977). Bacterial thiopurine methyltransferase (bTPMT) from *Pseudomonas syringae* has been shown to catalyze S-adenosylmethylation. The overexpression of bTPMT in *Escherichia coli* cultures conferred resistance to high concentrations of sodium biselenite and tellurite. bTPMT detoxifies oxyanion metals by the methylation of organic and inorganic to dimethylselenide and dimethyldiselenide by a SAM dependant mechanism. Volatile methylated compounds from *E. coli* harboring bTPMT were also detected using GC-MS headspace chromatography. bTPMT is also capable of methylating tellurite to dimethyltelluride, conferring tellurite resistance. *Pseudomonas syringae* cultures spiked with sodium selenite, selenate, and (methyl)selenocysteine produced volatile methylated

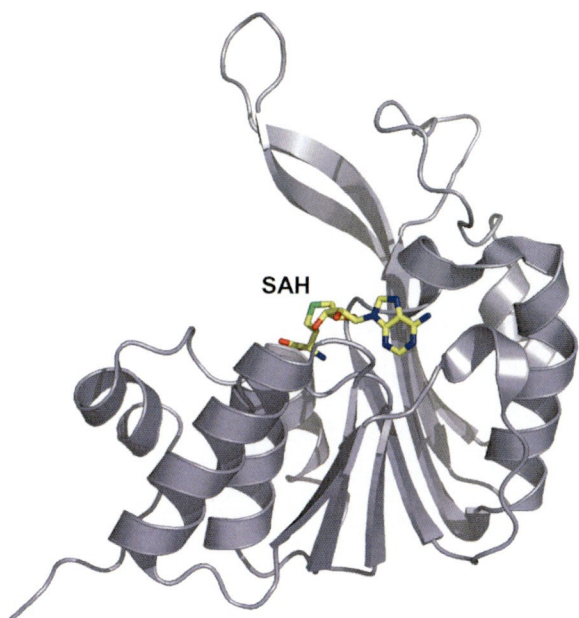

**Tellurite-Detoxifying Protein TehB from *Escherichia coli*, Fig. 1** Ribbon representation of the TehB crystal structure. The SAH cofactor is shown in stick with carbon in *gray*, nitrogen in *blue*, oxygen *red*, and sulfur in *green*

selenium oxyanions (Cournoyer et al. 1998). In *E. coli* detoxification of tellurite occurs by the methyltransferase TehB.

## TehB Structure

TehB belongs to the SAM methyltransferase superfamily; it has three conserved motifs that are involved in the binding of the cofactor SAM. Motif I consists of a glycine rich region containing a consensus sequence h(D/E)h GxGxG(x represents any residue and h represents hydrophobic residues). Motif I is followed by an acidic aspartate residue, 16 residues downstream of motif I. Motif II consists of 8 residues and is located 34 residues downstream of motif I. Motif II has a central aspartate residue surrounded by a high number of aromatic residues. Motif III consists of 9 residues located 22 residues downstream of motif II. Motif III is consistent with the consensus sequence L(R/K)PGG (R/I/J)(L/I)(L/F/I)(I/L). The two central glycine residues of motif III are highly conserved (Liu et al. 2000).

TehB contains seven β-strands and six α-helices. The first six β-strands are flanked by α-helices and adopt a Rossmann-like fold (Fig. 1). The first six

**Tellurite-Detoxifying Protein TehB from *Escherichia coli*, Fig. 2** Reaction mechanism of the tellurite methylation

β-strands are parallel and β7 sits antiparallel (Choudhury et al. 2011). The TehB structure has the cofactor S-adenosyl homocysteine (SAH) bound in its active site since SAM can easily be hydrolyzed; the SAH is bound in a cleft formed by α4, β1, β4 and a loop connecting β4 to α5. The SAM adenosine ring nitrogen N1 forms hydrogen bonds with the main chain of Leu87 and the exocyclic N6 with the side chain carboxyl oxygen of Asp86. The ribose moiety oxygens O2' and O3' are hydrogen bonded to the conserved Asp59. The rest of the cofactor molecule is hydrogen bonded to Gly38, Arg43, Asn44, and Thr102.

## Kinetics and Reaction Mechanism

TehB methylates tellurite in a SAM dependant manner at 350 nmol of $TeO_3^{-2}$/mg of TehB/min (Liu et al. 2000). Further kinetic analysis of the purified TehB have shown that the protein does not discriminate between chalcogens and it has the ability to methylate both tellurium and selenium oxyanions (Choudhury et al. 2011). The specificity is much higher for tellurite and lower for selenium dioxide and selenite ($TeO_3^{-2}$ > $SeO_2$ > $SeO_3^{-2}$). For selenate ($SeO_4^{-2}$) to be methylated, it requires its reduction to selenite ($SeO_3^{-2}$). TehB is not capable of the reduction step and it would require the coupling with a reductase prior to methylation. TehB is very specific for chalcogens as assays in the presence of arsenic compounds did not show any activity.

The active site is lined by conserved charged residues, which are able to bind and neutralize the negative charge of the oxyanions. The cavity is quite large, enabling it to accommodate metal oxyanions of different sizes. $TeO_3^{-2}$, $SeO_2$, and $SeO_3^{-2}$ are good nucleophiles with a lone pair for attack. Sequence alignments and structural analysis of the active site have shown that there are three conserved residues near to the binding site that are capable of binding and orientating the metals for nucleophilic attack, His176, Arg177, and Arg184. These residues are found in a loop region close to substrate binding cleft and mutagenesis studies showed that mutating Arg177 to Ala can abolish almost 75% of the protein's activity, whereas the other mutants only showed decreased activity.

The general mechanism of TehB is (1) SAM and tellurite bind to the enzymes active site. The Arg177 binds the tellurite, neutralizes the oxygen negative charges, and positions it for a nucleophilic attack. (2) The tellurium lone pair attacks the methyl group at the positively charged sulfonium group in an $S_N2$ dependent manner. (3) The methylated tellurium and SAH molecules leave the active site and the enzyme is regenerated (Fig. 2).

## Summary

The *E. coli* TehB protein detoxifies toxic chalcogens by SAM dependent methylation. It has higher efficiency for tellurite oxyanions than for those of selenium. The key residues responsible for catalysis and binding of the oxyanion metals are His176, Arg177, and Arg184 as identified by mutagenesis studies. Mutagenesis and kinetic analysis studies have shown that methylation of chalcogens proceeds via an $S_N2$ nucleophilic attack to the methyl group of SAM with Arg177 as the key residue.

## Cross-References

- ▶ Arsenic Methyltransferases
- ▶ Arsenic, Mechanisms of Cellular Detoxification
- ▶ Bacterial Tellurite Resistance
- ▶ Scandium, Physical and Chemical Properties
- ▶ Selenium, Biologically Active Compounds
- ▶ Tellurite-Resistance Protein TehA from *Escherichia coli*
- ▶ Tellurium and Oxidative Stress
- ▶ Tellurium in Nature

## References

Basnayake RST, Bius JH, Akpolat OM et al (2001) Production of dimethyl telluride and elemental tellurium by bacteria amended with tellurite or tellurate. Appl Organomet Chem 15(6):499–510

Bird ML, Challenger F (1939) 39. The formation of organo-metalloidal and similar compounds by micro-organisms. Part VII. Dimethyl telluride. J Chem Soc (Resumed) 1939:163–168

Burton GA, Giddings TH, Debrine P et al (1987) High-incidence of selenite-resistant bacteria from a site polluted with selenium. Appl Environ Microbiol 53(1):185–188

Chasteen TG, Bentley R (2003) Biomethylation of selenium and tellurium: microorganisms and plants. Chem Rev 103(1):1–25

Choudhury HG, Cameron AD, Iwata S et al (2011) Structure and mechanism of the chalcogen-detoxifying protein TehB from *Escherichia coil*. Biochem J 435:85–91

Cournoyer B, Watanabe S, Vivian A (1998) A tellurite-resistance genetic determinant from phytopathogenic pseudomonads encodes a thiopurine methyltransferase: evidence of a widely-conserved family of methyltransferases. Bba-Gene Struct Expr 1397(2):161–168

Doran JW, Alexander M (1977) Microbial transformations of selenium. Appl Environ Microbiol 33(1):31–37

Liu MF, Turner RJ, Winstone TL et al (2000) *Escherichia coli* TehB requires S-adenosylmethionine as a cofactor to mediate tellurite resistance. J Bacteriol 182(22):6509–6513

Ranjard L, Prigent-Combaret C, Nazaret S et al (2002) Methylation of inorganic and organic selenium by the bacterial thiopurine methyltransferase. J Bacteriol 184(11):3146–3149

Ranjard L, Nazaret S, Cournoyer B (2003) Freshwater bacteria can methylate selenium through the thiopurine methyltransferase pathway. Appl Environ Microbiol 69(7):3784–3790

Sabaty M, Avazeri C, Pignol D et al (2001) Characterization of the reduction of selenate and tellurite by nitrate reductases. Appl Environ Microbiol 67(11):5122–5126

Walter EG, Taylor DE (1992) Plasmid-mediated resistance to tellurite – expressed and cryptic. Plasmid 27(1):52–64

# Tellurite-Resistance Protein TehA from *Escherichia coli*

Hassanul Ghani Choudhury and Konstantinos Beis
Division of Molecular Biosciences, Imperial College London, London, South Kensington, UK
Membrane Protein Lab, Diamond Light Source, Harwell Science and Innovation Campus, Chilton, Oxfordshire, UK
Research Complex at Harwell, Harwell Oxford, Didcot, Oxforsdhire, UK

## Synonyms

Tellurite=chalcogen

## Definition

Tellurium and its oxyanions are highly toxic. Bacteria have developed resistance mechanisms to overcome the tellurium and tellurite toxic effects. These include reduction to elemental state, production of volatile species via methylation, and efflux from the cell by membrane proteins.

## Introduction

Tellurium belongs to the 16th column of the Periodic Table of elements which are called chalcogens. Tellurium can be found in trace quantities (Avazeri et al. 1997) within the environment. Tellurium exists as elemental tellurium ($Te^0$), inorganic telluride ($Te^{2-}$), the oxyanion forms tellurite ($TeO_3^{2-}$), and tellurate ($TeO_4^{2-}$) and organic dimethyl telluride (DMTe); $TeO_3^{2-}$ and $TeO_4^{2-}$ are the more common and soluble oxyanions compared to elemental tellurium. There are no reports that tellurium is an essential metal, but its toxicity to organisms is well studied. Tellurium is referred to as toxic metalloid with the oxyanion tellurite being highly soluble in water and highly toxic to biological systems. Tellurite is more toxic than cadmium, chromium, cobalt, mercury, and zinc, to living organisms (Nies 1999; Chasteen et al. 2009). Tellurite toxicity affects microbial growth and ranges from 1 to 1,000 μg/ml (Rathgeber et al. 2002; Coral et al. 2006). Microorganisms have developed three main mechanisms of dealing with tellurite toxicity; these are, (1) increased efflux, (2) decreased uptake, (3) chemical modification via methylation to a volatile nontoxic form or reduction to the less toxic elemental form.

Tellurite toxicity can be attributed partly as a result of the generation of reactive oxygen species that act as strong oxidizing agents and can lead to the oxidation of many cellular thiols, disrupting and causing the stoppage of protein/DNA synthesis and many reductases (Perez et al. 2007). Tellurium compounds have been shown to inhibit the selenocysteine containing thioredoxin reductases and flavoenzyme (Chasteen and Bentley 2003; Chasteen et al. 2009). Tellurite has also been suggested to replace sulfur in various biological reactions, with fatal effects on the cell.

Many bacteria are capable of reducing tellurite oxyanions to elemental tellurium. Elemental tellurium appears as black deposits within cells and even though it is less toxic than tellurite oxyanions it still retains some of its toxicity (Perez et al. 2007).

Multiple organisms have been identified to expel methylated oxyanion compounds. These range from bacterial systems where the volatile compounds have been identified by GC-MS headspace chromatography (Ranjard et al. 2002, 2003) to plant systems where an increase in methylated oxyanions has been observed from tellurite polluted ground.

In some bacteria the Ter determinant *tehAB* has been identified as the tellurite resistant operon (Taylor 1999); TehA encodes for an inner membrane protein and TehB encodes for a methyltransferase. The closest homologues of these two genes are found in *Haemophilus influenzae*. Cloning of these genes on

**Tellurite-Resistance Protein TehA from *Escherichia coli*, Fig. 1** Ribbon representation of the HiTehA trimer viewed from outside the membrane. Each protomer is shown in different color

a multicopy plasmid and expression in *Escherichia coli* give resistance to tellurite. *E. coli* TehA (EcTehA) is attributed to provide tellurite resistance to the cells. EcTehA expression has also been shown to provide resistance to most lipophilic cationic dyes and related compounds and single quaternary cations (e.g., ethidium). Hypersensitivity to compounds with two quaternary cations is also conferred by the expression of TehA. EcTehAB determinant was found to continuously remove tellurite from the growth medium. This is consistent with the occurrence of a modification of tellurite within the cell (Turner et al. 1997). EcTehA is believed to provide resistance by allowing passage of substrates out of the cell. It is unclear whether EcTehA detoxifies the cell by allowing passage for tellurite or methyl tellurite or both out of the cell.

## TehA Structure

The crystal structure of the *H.influenzae* (HiTehA) is the closest structural homologue to the EcTehA (Chen et al. 2010). HiTehA forms a homotrimer with each monomer made of ten transmembrane helices (Fig. 1). The TehA protomers exhibit a novel fold of ten transmembrane helices that are tandemly repeated helical hairpins arranged with a quasi-five-fold symmetry.

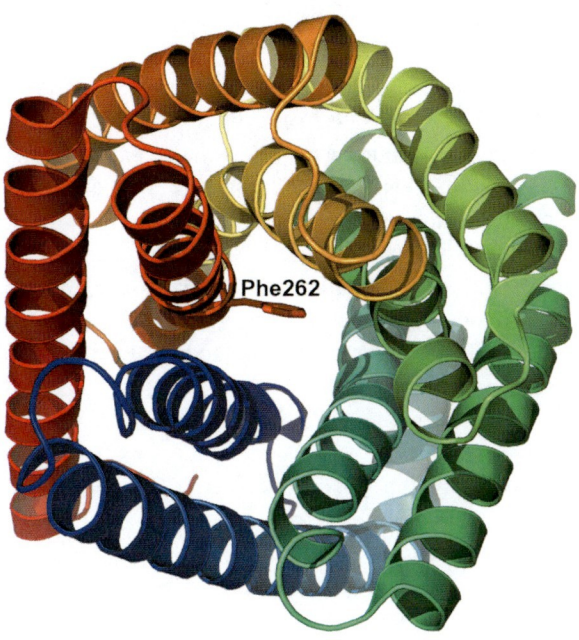

**Tellurite-Resistance Protein TehA from *Escherichia coli*, Fig. 2** Ribbon representation of one HiTehA protomer viewed from the *top*. The conserved Phe262 that constricts the pore is shown as a stick representation

Each of the HiTehA protomer has a central pore, formed by the five odd numbered helices outwardly directed. The five even numbered helices linked via the hairpins to the inner helices surround the inner pore to make the outer layer. The pore of each monomer has a relatively consistent diameter across the five helical turns of approximately 5 Å; centrally located proline residues form kinks in four of the five central helices which maintain a consistent diameter across the membrane. The five outer helices are straighter and longer in form and are more slanted to the membrane plane in comparison to the five central helices. TM6 is the only proline kinked outer helix and is located at the trimer threefold axis (Chen et al. 2010).

There are numerous conserved hydrophobic residues, which line the interior of the pore. The side chain of a highly conserved Phe262 constricts the pore (Fig. 2). The electrostatic potential of the pore surface is polarized whereas the pore itself is highly hydrophobic. This electropositive charge of the cytoplasmic surface may enhance efflux (Chen et al. 2010).

Electrophysiological experiments of membrane currents across voltage-clamped *Xenopus* oocytes in the presence of wild type HiTehA only resulted in low-signal currents but when Phe262 was mutated to an alanine or glycine a large influx of current across the membrane could be measured. Mutation of Phe262 to a threonine, valine, or leucine increased the current across the membrane but not as significantly when it was mutated to glycine or alanine. The increases in current are related to other Phe262 mutants that are consistent with a less constricted pore. The crystal structures of HiTehA and the electrophysiological experiments demonstrate the crucial role of Phe262 and demonstrate evidence of channel gating of this residue. The physiological stimuli that activates and regulates the HiTehA channel is unknown. HiTehA has also yet to be shown to have any known function in vitro (Chen et al. 2010).

## Summary

EcTehA can detoxify tellurite and other compounds by extruding them from the cell. The homologous TehA structure from *H.influenza* exists as a homotrimer. Each of the HiTehA protomers is made of ten transmembrane helices that form a central pore. A key phenylalanine, Phe262, in the middle of the pore controls passage of substrates through the channel. This is supported by voltage current measurements of the wild type and mutated protein. It is thought that the tellurite will transverse through the pore and reduce its toxic levels.

## Cross-References

▸ Arsenic, Mechanisms of Cellular Detoxification
▸ Bacterial Tellurite Resistance
▸ Tellurium, Physical and Chemical Properties
▸ Tellurium and Oxidative Stress
▸ Tellurium in Nature
▸ Tellurite-Detoxifying Protein TehB from *Escherichia coli*

## References

Avazeri C, Turner RJ, Pommier J et al (1997) Tellurite reductase activity of nitrate reductase is responsible for the basal resistance of *Escherichia coli* to tellurite. Microbiol UK 143:1181–1189

Chasteen TG, Bentley R (2003) Biomethylation of selenium and tellurium: microorganisms and plants. Chem Rev 103:1–25

Chasteen TG, Esteban Fuentes D, Carlos Tantalean J et al (2009) Tellurite: history, oxidative stress, and molecular mechanisms of resistance. FEMS Microbiol Rev 33(4): 820–832

Chen Yh, Hu L, Punta M et al (2010) Homologue structure of the SLAC1 anion channel for closing stomata in leaves. Nature 467:1074–U1157

Coral G, Arikan B, Coral MNU (2006) A preliminary study on tellurite resistance in Pseudomonas spp. isolated from hospital sewage. Pol J Environ Stud 15:517–520

Nies DH (1999) Microbial heavy-metal resistance. Appl Microbiol Biotechnol 51:730–750

Perez JM, Calderon IL, Arenas FA et al (2007) Bacterial toxicity of potassium tellurite: unveiling an ancient enigma. PLoS One 2(2):e211

Ranjard L, Prigent-Combaret C, Nazaret S et al (2002) Methylation of inorganic and organic selenium by the bacterial thiopurine methyltransferase. J Bacteriol 184(11):3146–3149

Ranjard L, Nazaret S, Cournoyer B (2003) Freshwater bacteria can methylate selenium through the thiopurine methyltransferase pathway. Appl Environ Microbiol 69: 3784–3790

Rathgeber C, Yurkova N, Stackebrandt E et al (2002) Isolation of tellurite- and selenite-resistant bacteria from hydrothermal vents of the juan de fuca ridge in the pacific ocean. Appl Environ Microbiol 68:4613–4622

Taylor DE (1999) Bacterial tellurite resistance. Trends Microbiol 7:111–115

Turner RJ, Taylor DE, Weiner JH (1997) Expression of *Escherichia coli* TehA gives resistance to antiseptics and disinfectants similar to that conferred by multidrug resistance efflux pumps. Antimicrob Agents Chemother 41:440–444

# Tellurium

▶ Tellurite-Detoxifying Protein TehB from *Escherichia coli*
▶ Tellurium in Nature

# Tellurium and Oxidative Stress

Alex Elías and Claudio C. Vásquez
Laboratorio de Microbiología Molecular,
Departamento de Biología, Universidad de Santiago de Chile, Santiago, Chile

## Synonyms

Bacterial tellurite resistance; Oxidative stress; Reactive oxygen species

## Definition

The shadowy element Tellurium (Te) belongs to the group of chalcogens, the same group of other elements of biological importance such as oxygen and sulfur. It was described early in 1782 by Franz Joseph Mueller. Curiously, although its name derives from the Latin word *tellum*, meaning "earth," tellurium abundance in the Earth's crust is rather low ($\sim$0.027 ppm). On the other hand and unlike its closely related element selenium, tellurium seems unrelated to biological systems to date (Cunha et al. 2009).

## Generalities

Elemental tellurium ($Te^0$) is insoluble in water and is usually found as black deposits inside bacteria grown on tellurite-containing, selective, growth media. In nature, Te is found commonly associated with other metal forming alloys as calaverite ($AuTe_2$), sylvanite ($AgAuTe_4$) and nagyagita ($AuPb(Sb,Bi)Te_{2-3}\,S_6$), among others. However, Te forms mainly copper- or sulfur-bearing ores; thus, tellurium production is usually related to copper refining (Chasteen et al. 2009).

Tellurium's major use is as an alloying additive in steel, and copper to improve machining characteristics. In the chemical industry, it is used as a vulcanizing agent and accelerator in the processing of rubber, and as a catalyst for synthetic fiber production. Other uses include those in photoreceptor and thermoelectric electronic devices, thermal cooling, as an ingredient in blasting caps, and as a pigment to produce various colors in glass and ceramics. Currently, tellurium is increasingly used in the production of cadmium-tellurium-based solar cells.

## Elemental Tellurium in Biology

There is no known tellurium toxicity for higher mammals. In fact, tellurium poisoning is a very rare event although a typical human being exhibits $>$ 0.5 g of Te mostly in bone. With the exception of iron, zinc, and rubidium, this amount exceeds largely the levels of all other trace elements in humans (Chasteen et al. 2009).

Tellurium found historical applications in the treatment of microbial infections prior to the discovery of antibiotics. Early documentation in 1926 reports its use

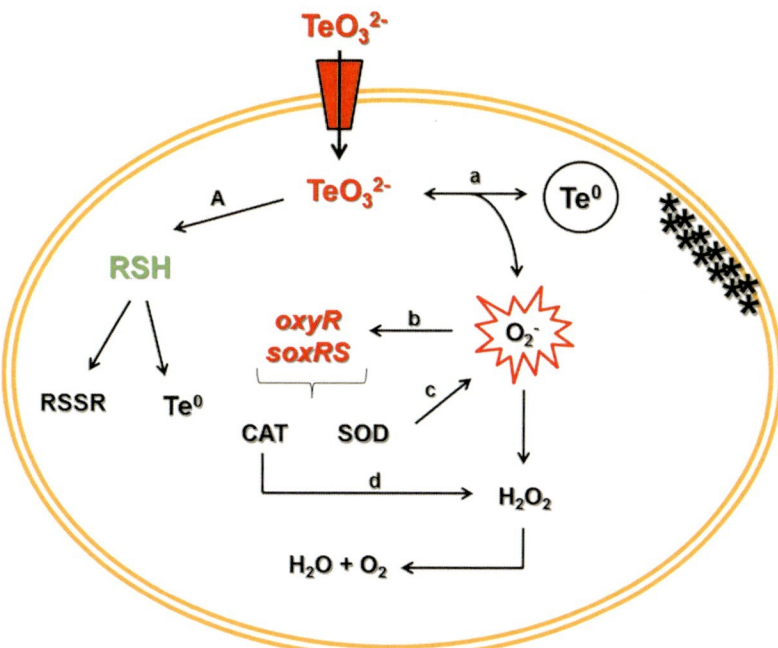

**Tellurium and Oxidative Stress, Scheme 1** Current model for tellurite toxicity. Once inside the cell, tellurite is reduced enzymatically (a) or via RSH (A) generating elemental tellurium, which forms characteristics *black* deposits in the cell membrane. Cell-reduced thiols (RSH), mainly GSH, are directly affected by tellurite, depleting them and generating a first redox imbalance. Superoxide is generated along with tellurite reduction, imposing a second redox imbalance. This ROS activates regulatory genes related to its detoxification (b). The picture shows the two main antioxidant enzymes produced catalase (CAT) and superoxide dismutase (SOD). Superoxide is dismutated by SOD into peroxide (c), which in turn is transformed into water and oxygen by CAT (d)

in the treatment of syphilis and leprosy (Chasteen et al. 2009; Cunha et al. 2009). However, there was also an association with unpleasing effects in humans: ingesting tellurium compounds, such as $TeO_2$ or tellurite, generated a "disagreeable garlic-like breath" (De Meio 1946).

It is important to note that the toxic effect associated with tellurium depends on its chemical form: tellurium has a low solubility at physiological pH and the oxidation to tellurite ($TeO_3^{2-}$), tellurate ($TeO_4^{2-}$), or $TeO_2$ occurs easily. However, tellurium dioxide is practically insoluble in water at physiological pH (Cunha et al. 2009), so one could assume that the oxyanions are primarily responsible for the observed toxic effects. In fact, tellurite has been used in microbiology since the 1930s, when Alexander Fleming reported its antibacterial properties along with penicillin (Fleming 1932). Recently, a rough ranking of tellurium toxicity for bacteria has been published being tellurite > tellurate > elemental tellurium (Basnayake et al. 2001). This ranking also seems to apply to humans.

Uptake routes of tellurium compounds into an organism differ: tellurium poisoning can occur due to oral ingestion as tellurium-containing dust may be inhaled. Once inside, the tellurium compound in question may exert its toxicity in several ways.

## Tellurium and Oxidative Stress

The tellurium oxyanion $TeO_3^{2-}$ is highly toxic for most bacteria at concentrations as low as 1 μg/ml (Taylor 1999). This is especially important when compared with the toxicity exhibited by other metals and metalloids of public health and environmental concern such as cobalt, zinc, and chromium.

However, several bacterial species display tolerance to this metalloid and in fact this trait has been exploited to elaborate selective media for isolating these microorganisms. A number of genetic tellurium resistance determinants ($Te^R$) have been identified in different

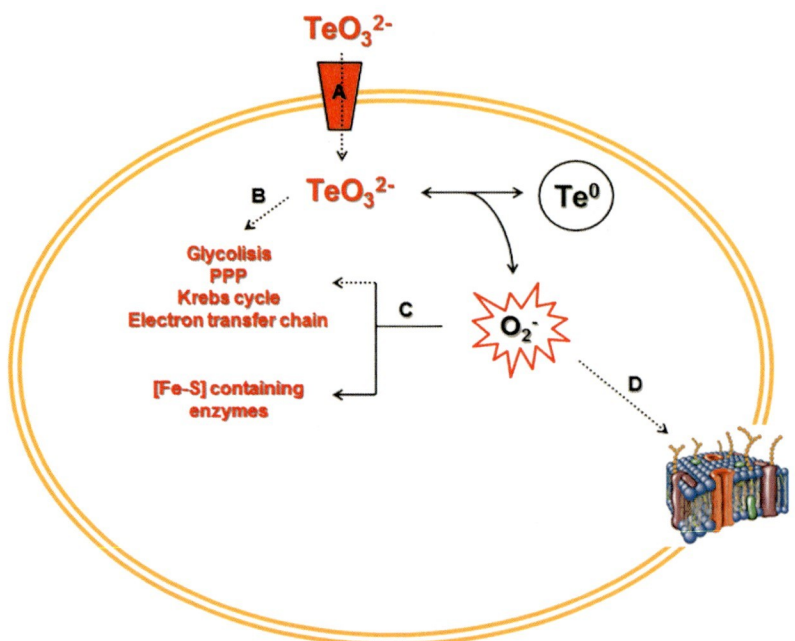

**Tellurium and Oxidative Stress, Scheme 2** Some new features of the tellurite toxicity model currently under study (*doted arrows*), **A**, uptake of some tellurite takes place through an acetate transporter (ActP), but it occurs mainly associated to the PitA phosphate-transport system. **B**, tellurite may have a direct effect on the main metabolic pathways of the cell, becoming a metabolic poison. **C**, superoxide generated by tellurite reduction uncouples the iron-sulfur clusters of many metabolically important enzymes as fumarase and aconitase, among others. **D**, lipid peroxidation may be a direct cause of toxicity generated by the $O_2^-$ generated by $TeO_3^{2-}$ reduction

species of bacteria. These mediate $TeO_3^{2-}$ resistance by as yet unknown mechanisms. Analysis of the nucleotide/amino acid sequence of $Te^R$ determinants shows a very high variation, making it difficult to propose a universal $TeO_3^{2-}$ resistance mechanism. In general, it is accepted that resistance mechanisms are related to direct extrusion of the toxicant, its conversion to volatile (alkylated) forms, and/or enzymatic or nonenzymatic $TeO_3^{2-}$ reduction to the insoluble elemental form, $Te^0$. Actually, most microorganisms share the ability to reduce $TeO_3^{2-}$ to the less toxic form $Te^0$, which results in the formation of black Te deposits inside cells (Taylor 1999). Note that there is a difference between microbial tellurite resistance and reduction. Several tellurite-sensitive microorganisms, for example, *Escherichia coli* K12, are able to reduce $TeO_3^{2-}$.

The observed differences between tellurite-sensitive and tellurite-resistant organisms may be related to oxyanion extrusion mechanisms or by different chemical modifications that do not include reduction. Generation of methylated forms of Te, for example, has been detected in the headspace of recombinant *E. coli* strains. Methyl telluride is volatile and easily eliminated from the cell, generating the classic "garlic-like" odor associated with this toxicant (Chasteen et al. 2009).

In the past 10 years, a number of research groups have reported that the overexpression of certain genes involved in basal metabolism results in an increased tolerance to $TeO_3^{2-}$. Most importantly, *cysK* genes (encoding cysteine synthase) from several different species are capable of generating $Te^R$ when expressed in heterologous hosts. The most possible explanation for these findings may be related with the toxicant's primary targets: it is well known that Te (and Se) oxyanions strongly react with cellular thiols (RSH), especially with glutathione (GSH), so it was postulated (and later demonstrated) that GSH can reduce $TeO_3^{2-}$ to $Te^0$, thus causing a decrease of the oxyanion's toxic effect (Turner et al. 2001). Closely related with this phenomenon, it has been shown that superoxide ($O_2^-$), a reactive oxygen specie (ROS), is generated during tellurite reduction (Tremaroli et al. 2007; Pérez et al. 2007; Zannoni et al. 2008).

ROS are highly reactive molecules (such superoxide and peroxide) due to the presence of unpaired valence shell electrons. ROS form as a natural by-product of the normal metabolism of oxygen and under certain environmental stress situations (UV or heat exposure), ROS levels increase dramatically resulting in significant damage to cell structures and/or biomolecules, a situation known as oxidative stress. Several mechanisms exist to protect the microorganism from this kind of stress. ROS increase leads to thiol oxidation, among other effects. Some of these thiols form part of cellular proteins such as the OxyR transcriptional regulator. This factor mediates the expression of several peroxide-inducible genes encoding for key enzymes related to the bacterial response to oxidative stress. Furthermore, several *E. coli* genes, regulated by the *soxRS* regulatory system, results activated. This regulon specifically responds to superoxide. So, bacteria seem to cope with tellurite toxicity through general adaptation mechanisms like those used when they face other environmental stressors.

The current model of tellurite toxicity proposes that, once inside the cell, tellurite quickly depletes the GSH pool, generating a "first" redox imbalance. Associated with the reduction of tellurite by GSH and other molecules as enzymes like catalase and pyruvate dehydrogenase, among others, $O_2^-$ is generated, thus sharpening the redox imbalance and finally generating an oxidative stress situation. The cell responds activating all the systems related to cope with the oxidative stress situation imposed by tellurite detoxification (Schemes 1 and 2).

## Cross-References

▶ Arsenic, Free Radical and Oxidative Stress
▶ Arsenic, Mechanisms of Cellular Detoxification
▶ Bacterial Tellurite Resistance
▶ Cadmium and Oxidative Stress
▶ Catalases as NAD(P)H-Dependent Tellurite Reductases
▶ Chromium(VI), Oxidative Cell Damage
▶ Nickel Superoxide Dismutase
▶ Peroxidases
▶ Tellurite-Detoxifying Protein TehB from *Escherichia coli*
▶ Tellurite-Resistance Protein TehA from *Escherichia coli*
▶ Tellurium in Nature
▶ Zinc in Superoxide Dismutase

## References

Basnayake RST, Bius JH, Akpolat OM, Chasteen TG (2001) Production of dimethyl telluride and elemental tellurium by bacteria amended with tellurite or tellurate. Appl Organomet Chem 15:499–510

Chasteen TG, Fuentes DE, Tantaleán JC, Vásquez CC (2009) Tellurite: history, oxidative stress, and molecular mechanisms of resistance. FEMS Microbiol Rev 33:820–832

Cunha RL, Gouvea IE, Juliano L (2009) A glimpse on biological activities of tellurium compounds. An Acad Bras Cienc 81:393–407

De Meio RH (1946) Tellurium: the toxicity of ingested elementary tellurium for rats and rat tissue. J Ind Hyg Toxicol 28:229–231

Fleming A (1932) On the specific antibacterial properties of penicillin and potassium tellurite. Incorporating a method demonstrating some bacterial antagonisms. J Pathol Bacteriol 35:831–842

Pérez JM, Calderón IL, Arenas FA, Fuentes DE, Pradenas GA, Fuentes EL, Sandoval JM, Castro ME, Elías AO, Vásquez CC (2007) Bacterial toxicity of potassium tellurite: unveiling an ancient enigma. PLoS ONE 14:e211

Taylor DE (1999) Bacterial tellurite resistance. Trends Microbiol 7:111–115

Tremaroli V, Fedi S, Zannoni D (2007) Evidence for a tellurite dependent generation of reactive oxygen species and absence of a tellurite-mediated adaptive response to oxidative stress in cells of *Pseudomonas pseudoalcaligenes* KF707. Arch Microbiol 187:127–135

Turner RJ, Aharonowitz Y, Weiner JH, Taylor DE (2001) Glutathione is a target in tellurite toxicity and is protected by tellurite resistance determinants in *Escherichia coli*. Can J Microbiol 47:33–40

Zannoni D, Borsetti F, Harrison JJ, Turner RJ (2008) The bacterial response to the chalcogen metalloids Se and Te. Adv Microb Physiol 53:1–71

# Tellurium in Nature

Torsten Burkholz and Claus Jacob
Division of Bioorganic Chemistry, School of Pharmacy, Saarland State University, Saarbruecken, Germany

## Synonyms

Materials; Novel therapeutics; Redox regulation; Tellurium; Toxicity

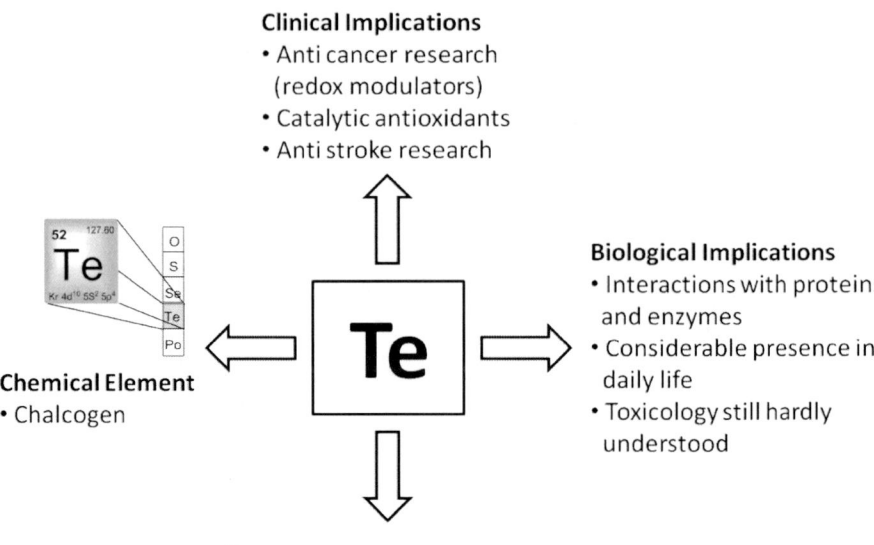

**Tellurium in Nature, Fig. 1** Despite the fact that tellurium is often considered a rather "shadowy" element, it is present in various alloys, photochemical, and photovoltaic devices. Furthermore, its (organic) compounds exhibit a rather interesting spectrum of biological activities. Ultimately, tellurium is of great interest in the context of material science, renewable energies, toxicology, and (redox modulating) drug design

## Definition

Tellurium is a rather interesting chemical element which together with oxygen, sulfur, selenium, and polonium forms the group of chalcogens. The element is not extremely rare in nature, and has already been discovered in the eighteenth century. Nonetheless, its presence in daily life is limited, and often goes unnoticed. It is used, for instance, in specific alloys and, as CdTe, in (photo-)optical devices. At first sight, tellurium is an alien to biology. Nonetheless, recent studies have shown that (a) most organisms are able to metabolize tellurium agents along sulfur and selenium pathways and (b) tellurium-based compounds exhibit a wide spectrum of interesting biological activities which can be explained by the unique chemistry of inorganic, complex-like, and organic tellurium agents. Together, these findings place tellurium and its compounds at the center of biochemical, toxicological, and pharmaceutical research, especially in the context of antimicrobial, anti-inflammatory, and anticancer drug development.

## Introduction

Chemically speaking, tellurium belongs to the group of chalcogens, which also includes the elements oxygen, sulfur, selenium, and polonium. It was discovered in 1782 by Franz-Joseph Mueller von Reichenstein (1742–1825), and is named after the Latin word *tellus*, which means "Earth." Despite its early discovery in the eighteenth century and down-to-earth name, tellurium is still a rather "shadowy" element, with apparently little importance in daily life. There are exceptions, of course, such as the increasing presence of CdTe particles in photochemical devices, for example, in solar panels and photovoltaic systems (Contreras-Puente et al. 2000). Nonetheless, there is a notable absence of tellurium from most areas of science, including biology: Tellurium is not an essential biological trace element and together with its various inorganic, complex, and organic compounds has so far escaped wider consideration in medicine, toxicology, pharmacology, and drug development.

When considering the various aspects of tellurium chemistry and the known biological activities of different tellurium compounds, this apparent lack of interest in tellurium in the Life Sciences is actually rather surprising (Fig. 1). Tellurium is not particularly exotic as far as its chemistry is concerned, and in many ways its unique reactivity is well suited to convey biological activity, for instance in the context of drug design. Here, the following arguments come to mind: First of all, tellurium and most of its compounds are not

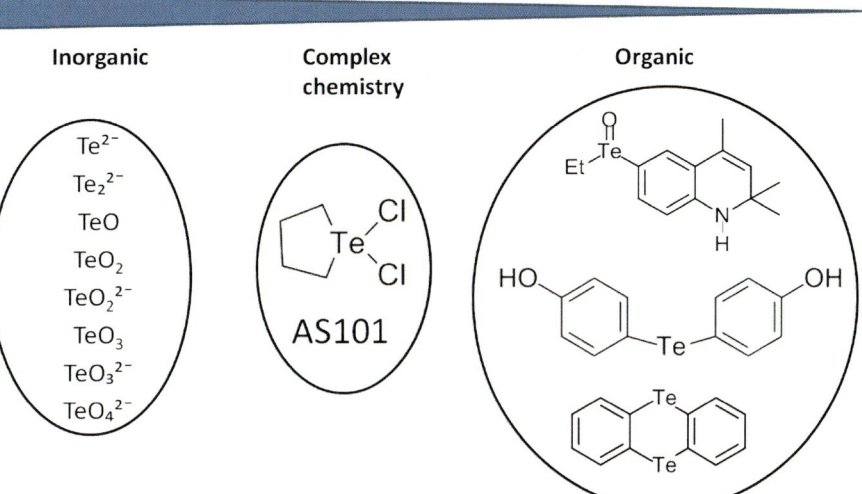

**Tellurium in Nature, Fig. 2** Tellurium is a metalloid (semimetallic element) and as such exhibits a rather unique chemistry. The latter includes various inorganic tellurium anions and oxoanions, complex-like structures with a tellurium central atom and (mostly oxygen- or halide-based) ligands and typical organic molecules where tellurium is bound covalently to carbon. In the context of biological activity, inorganic, organic and complex-like structures of tellurium are of almost equal interest

particularly toxic, radioactive, or otherwise "dangerous." Secondly, tellurium provides a multifaceted chemistry which is rich in inorganic and organic substances, including various inorganic salts (e.g., tellurides $Te^{2-}$, tellurites $TeO_3^{2-}$ and tellurates $TeO_4^{2-}$), complexes, and a wide range of diverse organotellurium compounds. Many of these compounds are stable and may be used in the context of drug development. Thirdly, tellurium and its various compounds are indeed reactive, and also interact readily with important biomolecules, such as cysteine- and selenocysteine-based proteins and enzymes. Here, highly selective inhibition of certain key enzymes by specific tellurium compounds may provide a basis for therapeutic intervention. The fact that tellurium is not an essential trace element does not necessarily imply that it is also useless in the context of drug design, as the immense success of gold- and platinum-based drugs illustrates.

Despite the notable absence of tellurium in proteins, enzymes, and indeed all known biomolecules, it is therefore still worth considering its appearance, emergence, and possible roles in biochemistry. Within this context, one needs to consider the following aspects of tellurium in nature: (a) its occurrence in nature, and especially in biology, (b) its special reactivity which may allow selectivity in a therapeutic context, (c) its chemistry and hence (bio-) availability of suitable tellurium agents, and last but not least (d) already existing evidence of specific biological and suspected pharmacological activities associated with currently available tellurium compounds which are known to date and may set tellurium apart from the other elements of the Periodic Table.

## Natural Occurrence of Tellurium in the Soil, Plants, and Animals

Despite its apparent absence from most aspects of daily life, tellurium is not an extremely rare element. An average abundance of 0.027 ppm tellurium in the soil has been estimated based on samples from Australia, China, Europe, New Zealand, and North America, which means that tellurium is as rare as silver or gold. In contrast to silver and gold, however, tellurium does not occur naturally in its elemental form but in the form of inorganic tellurium oxides and salts (Fig. 2). Oxides include, for instance, $TeO_2$ (oxidation state +4) and $TeO_3$ (oxidation state +6). These oxides form acids, namely, tellurous acid ($H_2TeO_3$) and telluric acid ($H_2TeO_4$). The corresponding salts are known as tellurites $TeO_3^{2-}$ and tellurates $TeO_4^{2-}$. Tellurates are generally more stable than tellurites. In addition, the anion $TeO_2^{2-}$ (oxidation state of +2), the telluride anion $Te^{2-}$ (oxidation state -2), and the ditelluride dianion $Te_2^{2-}$ (oxidation state -1) also exist, for instance, in $Ag_2Te$ and in $Na_2Te_2$.

Like most elements, tellurium is not distributed equally throughout the soil, but is enriched in some

## Biosyntethic pathway

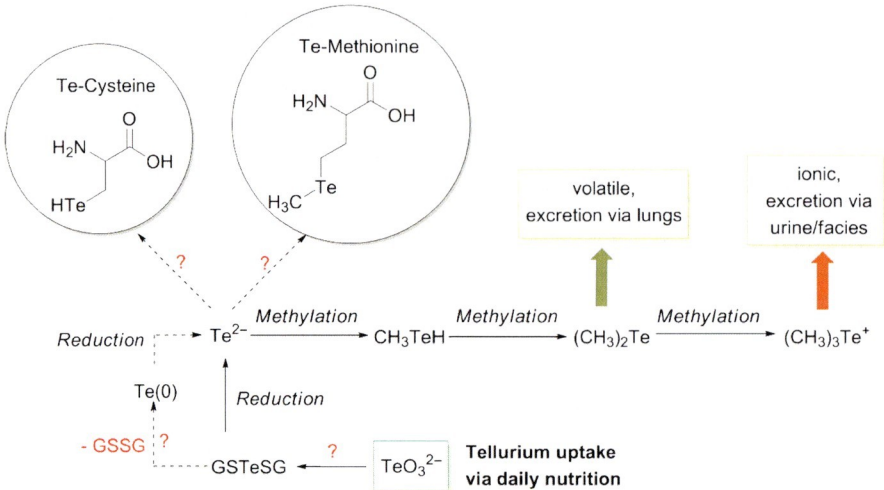

**Tellurium in Nature, Fig. 3** To the best of our current knowledge, tellurium is not an essential trace element in living organisms. Nonetheless, many of these organisms are able to process tellurium and subsequently excrete various, often methylated forms of this element. The underlying metabolism differs from organism to organism, yet seems to rely mostly on pathways normally used for the metabolism of sulfur or selenium. In fact, some lower organisms are even able to actively and purposefully use tellurium instead of sulfur under extreme conditions. In these cases, tellurium may well be considered as "essential." In the case of humans, tellurium is metabolized via the selenium pathways and excreted via the lung and kidneys

regions, while others are tellurium "deficient." The tellurium content in Chinese soils, for instance, may range from 0.007 to 0.113 ppm (Wenqi et al. 1992). Varying concentrations of tellurium in nature are also found in certain geothermal waters, such as in the New Zealand geothermal water, or at polluted sites, such as in wastewater, sewage sludge, and landfills (Dopp et al. 2004).

Not surprisingly, the level of tellurium in plants depends on the plant species, yet it also reflects the level of tellurium in the soil or the surrounding environment. In the late 1980s, Cowgill and colleagues examined over 1,000 samples from trees, shrubs, and flowering plants from the Ely mining region in Nevada, USA, and from western Colorado (Cowgill 1988). These studies indicate that plants known to accumulate selenium are also able to accumulate tellurium up to levels of approximately 1 ppm. Despite the fact that tellurium has no (obvious) biological function or role in these plants, accumulation seems to occur via the metabolic pathway of its close relative, selenium. As a consequence, tellurium uptake (mostly as tellurite) by many organisms is followed by the generation of the tellurium analogs of well-known selenium metabolites. In the case of tellurium, these species often include $H_2Te$, $(CH_3)_2Te$ and $(CH_3)_3Te^+$, which can then be excreted via the lung (volatile species) or kidney (ionic species). A basic metabolic pathway of tellurium which combines many of the known aspects of this metabolism is outlined in Fig. 3.

A distinct tellurium metabolism has been found in cases where microorganisms grow in the presence of (elevated levels of) tellurium salts and appear to be resistant against tellurium toxicity. Many of these organisms have developed a bioreductive strategy to cope with and even to remove tellurium (e.g., by generating elemental tellurium or by biovolatization). Many microorganisms able to grow in the presence of tellurate and tellurite, for instance, reduce these anions to elemental, insoluble $Te^0$, which does no longer pose any danger (Fig. 3). At the same time, reduction of tellurate and tellurite to $Te^0$ enables some of these microorganisms to sustain anaerobic growth by driving respiration with these anions as electron acceptors (instead of dioxygen). Bacteria able to employ tellurate and tellurite in this manner are rather exotic, though, and include *Bacillus selenitireducens*, *Sulfurospirillum barnesii*, (Baesman et al. 2007) and *Bacillus beveridgei* sp. nov., which have been isolated from deep ocean hydrothermal vent worms (Csotonyi et al. 2006; Baesman et al. 2009).

When grown with tellurate or tellurite as terminal electron acceptors, these microbes produce uniformly sized nanoparticles of Te(0), such as nanorods of 10 nm × 200 nm in size (*Bacillus selenitireducens*) or irregular spheres of less than 50 nm in diameter (*Sulfurospirillum barnesii*). These tellurium (nano)particles resemble similar selenium (nano)particles, generated by bacteria such as *Klebsiella pneumoniae* (Fesharaki et al. 2010). Since the production, physicochemical properties, and ultimate fate of these particles has not been fully investigated so far, they represent an exciting and rather novel field of current nano-bio research.

Alternatively, reduction of tellurium salts in certain organisms proceeds further to $H_2Te$ which is methylated to $(CH_3)_2Te$ or to ionic $(CH_3)_3Te^+$, both of which can subsequently be excreted (see above) (Ogra 2009). Here, the formation of $(CH_3)_2Te$ by certain fungi is particularly interesting. Although $(CH_3)_2Te$ is toxic, it is also volatile and is continuously removed from the system via "evaporation" (Chasteen and Bentley 2003). This kind of biovolatization may actually play a role in the removal of tellurium, not only from the fungus itself but also from the environment the fungus is living in. Indeed, biovolatization of tellurium (by fungi) and accumulation of tellurium in plants may have a practical use in the context of bioremediation (e.g., removal of tellurium from contaminated soil).

Interestingly, $H_2Te$, $(CH_3)_2Te$, $(CH_3)_3Te^+$ are not the only organotellurium species produced by living organisms. In rare cases, when fungi are grown without a sulfur source but in the presence of $Na_2TeO_3$, tellurium-containing amino acids (tellurocysteine, tellurocystine, and telluromethionine) and even tellurium-containing proteins are formed (Chasteen and Bentley 2003). This more or less random incorporation of the heavier (Te) for the lighter (S) chalcogen analog seems to follow a pattern similar to the one already known for selenium: If present at elevated levels (e.g., in the soil, growth medium), selenium is frequently incorporated into the amino acid methionine (instead of sulfur) by lower organisms.

While it is questionable that such tellurium-containing amino acids fulfill a major biochemical function in these fungi, their presence nonetheless proves two important points: Firstly, tellurium is not entirely alien to biology but is taken up and metabolically processed by various organisms which apparently know how to handle tellurium-based agents. Secondly, under extreme conditions, such as during severe deprivation of sulfur, tellurium may even "jump into the breach" and substitute for sulfur in some biomolecules. In these cases, it would certainly not be entirely wrong to talk about an apparent "essential role" of tellurium in biochemistry, even if this role is ultimately forced upon the organism (for instance, by deprivation of essential sulfur sources).

## The "Special Relationship"

One of the most surprising and, at the same time, interesting facts about tellurium in (living) nature is the high reactivity of tellurium compounds toward certain sulfur and selenium-based biomolecules – and the high selectivity for these targets. These interactions often (but not always) occur in the form of redox processes; alternatively, complex formation is also possible (Fig. 4). Many tellurium-based substances, including tellurides (RTeR), telluroxides (RTe(O)R), tellurols (RTeH), and ditellurides (RTeTeR), are redox-active, with formal oxidation states of tellurium in these compounds ranging from −2 to +6. Their redox behavior differs considerably from the "classical" one of metal ions: Direct electron transfer from and to the Te atom is rare, at least under physiological conditions (aqueous media, ambient temperature, pH around 7). Instead, oxidation and reduction at the Te atom often proceeds in concert with (nucleophilic) substitution reactions. Within this context, the close relationship with the other chalcogen elements, in particular with oxygen, sulfur, and selenium, plays a major role. Reactive oxygen species (ROS), such as hydrogen peroxide ($H_2O_2$), for instance, are able to oxidize tellurides to telluroxides and tellurols to ditellurides. Here, $H_2O_2$ acts as an electrophile. In contrast, thiols often act as reducing agents for tellurium agents, transforming telluroxides to tellurides and ditellurides to tellurols. In this case, the thiol(ate) acts as a nucleophile.

This particularly close connection between the redox chemistries of oxygen, sulfur, selenium, and tellurium also forms the basis for highly specific redox interactions involving biomolecules. Many tellurium compounds exhibit an exquisite glutathione peroxidase (GPx)-like activity, that is, they catalyze the reduction of $H_2O_2$ in the presence of reduced glutathione (GSH) (Fig. 4a). GPx itself is a selenocysteine-containing antioxidant enzyme, and its mimics often contain

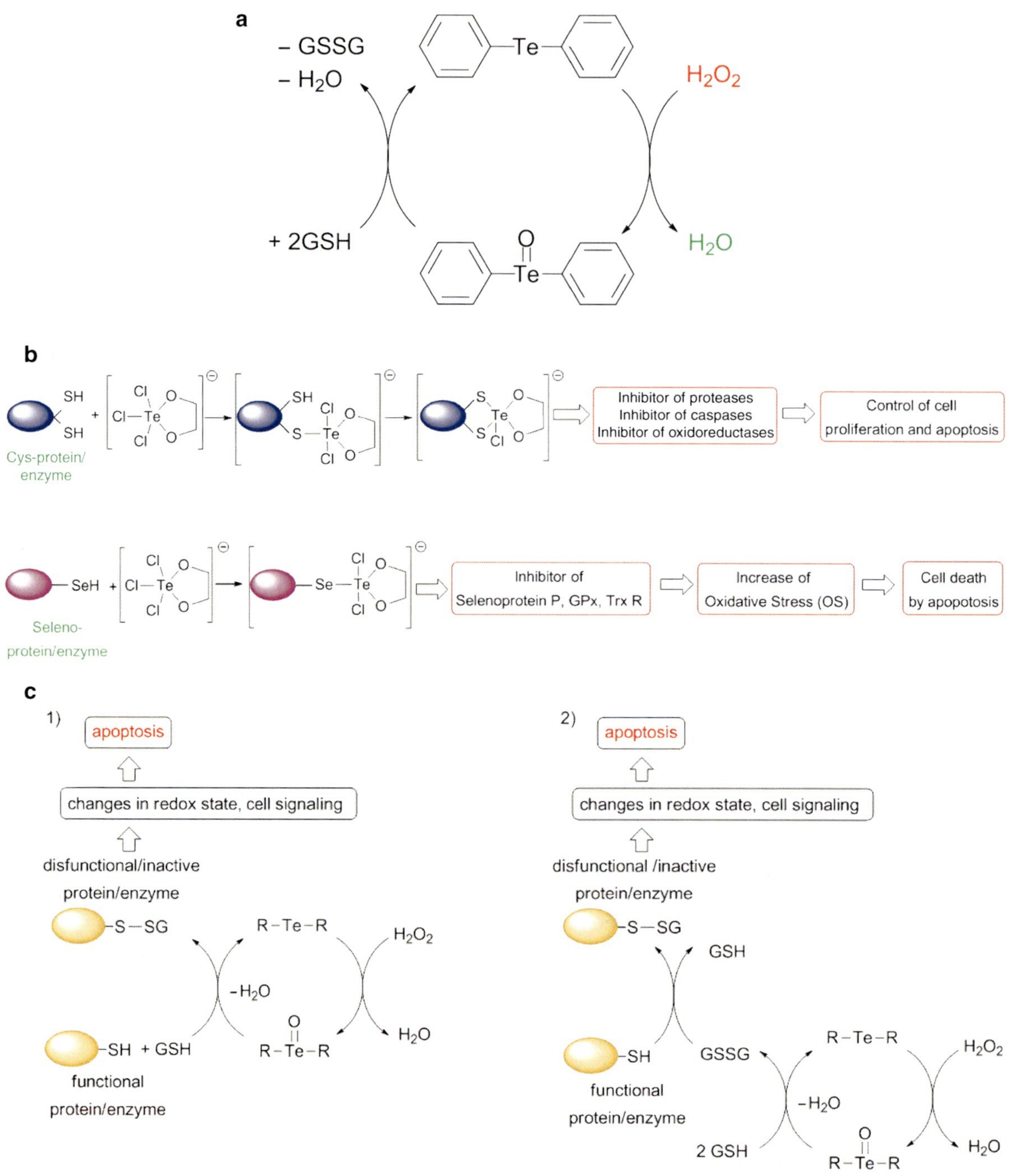

**Tellurium in Nature, Fig. 4** The perhaps most interesting biochemical action associated with tellurium compounds is the modulation of the (intracellular) redox state which explains many of the biological activities observed for these compounds, including antioxidant activity and selective cytotoxicity. Some antioxidant tellurium compounds mimic the selenoenzyme GPx by catalytically removing peroxides in the presence of reduced glutathione (GSH) (Panel a). In contrast, other tellurium compounds, such as AS101, form fairly strong complexes with active site cysteine or selenocysteine residues of enzymes (the thiol and selenol groups serve as "ligands" for the Te atom). This kind of complexation may result in the inhibition of essential, often antioxidant enzymes (such as the human selenoenzyme TrxR) and in an increase of Oxidative Stress which may subsequently result in cell death (Panel b). Indeed, a pro-oxidant activity of tellurium compounds is also observed when the GPx-like catalytic cycle fails to use GSH but consumes protein-based thiols instead. In this case, widespread oxidation of proteins and enzymes affects the cellular thiolstat and may also ultimately lead to cell death via the induction of apoptosis (Panel c). Figure in part adapted from Ba et al. (2011)

redox-active selenium or indeed tellurium atoms. Other organotellurides have been tested as chain-breaking antioxidants, whose (biochemical) activity resembles the one of vitamin E. Such "vitamin E mimics" appear to reduce peroxyl radicals more efficiently than α-tocopherol and are also able to decompose hydroperoxides in the presence of stoichiometric amounts of thiol-based reducing agents (Kumar et al. 2010).

In contrast, there is also a more sinister side to this special relationship which involves tellurium agents interacting with cysteine residues in proteins and enzymes (Fig. 4b, c). Such interactions are often detrimental to protein function and enzyme activity and ultimately interfere with the "cellular thiolstat," which ultimately may result in cell death (often via apoptotic pathways) (Jacob et al. 2012; Jacob 2011). Several in vivo studies with tellurite performed in rats and mice, for instance, indicate that ingestion of $TeO_3^{2-}$ causes a transient demyelination of peripheral nerves due to inhibition of squalene epoxidase (Abe et al. 2007) which appears to be due to binding of methyltellurium compounds to the thiol groups of the vicinal cysteine residues at the active site (complex chemistry, Fig. 4b). Similarly, Nogueira and colleagues have recently performed toxicological studies with organotellurium compounds in mice and rats. The toxicity of tellurium compounds in these studies is, at least in part, due to the interaction of tellurium with the thiol groups of cysteine-containing proteins and enzymes, such as δ-ALA-D and $Na^+$, $K^+$-ATPase (Nogueira et al. 2001; Stangherlin et al. 2009; Meotti et al. 2003; Borges et al. 2005).

While the special tellurium-sulfur chemistry may explain certain biological effects associated with tellurium compounds, one should not ignore the high affinity of tellurium to selenium, either. Prominent selenium-containing targets may include selenoprotein P and various GPx enzymes (Garberg et al. 1999). The activity of the human selenoenzyme thioredoxin reductase (TrxR) is also affected by tellurium compounds, possibly as a result of such tellurium-selenium interactions (McNaughton et al. 2004). By interacting with – mostly antioxidant – selenium proteins and enzymes, certain tellurium compounds may therefore indirectly induce Oxidative Stress – rather than acting as GPx-like antioxidants (see above). Not surprisingly, such effects may have severe consequences, from liver damage to the impairment and possible damage of hippocampal and prefrontal cortex neurons (Widy-Tyszkiewicz et al. 2002). Indeed, a possible pro- rather than antioxidant activity of tellurium compounds has recently raised considerable interest in the context of redox modulation and drug design, and is discussed separately and in more detail in Section "Aspects of an Emerging Pharmaceutical Chemistry of Tellurium".

## Chemical Diversity and Availability of Tellurium Agents for Medical Applications

The special relationship between oxygen, sulfur, and selenium on the one hand, and tellurium on the other, obviously raises the question if tellurium-based agents, by interfering with other chalcogen-based cellular systems, may provide the basis for effective and possibly also selective modulation of certain (intra-)cellular processes. Besides the notable absence of tellurium in biological systems, and its often suspected (but probably largely unfounded) toxicity, one major argument traditionally employed against the use of tellurium in biological systems (and therefore also in medicine/pharmacology) is the lack of a broad range of chemically stable "designer" tellurium compounds which may be used for such purposes. Indeed, from a chemistry perspective, tellurium is a niche element with only a couple of research groups worldwide currently focusing on the synthesis and characterization of such agents. So far, it seems that the rather vicious cycle of lack of interest among biochemists and pharmaceutical researchers on the one side, and hence the lack of interest among chemists to come up with new and more sophisticated tellurium agents on the other side, has somewhat held back the development and evaluation of such agents in the past. In many aspects, this is rather unfortunate, as the metalloid ("semi-metal") tellurium provides colorful chemistry including inorganic compounds, organometallic complexes, and organotellurium agents (Fig. 2). In fact, many of these agents are highly stable chemically as well as metabolically, exhibit extraordinary chemical and biochemical properties, and may well be of interest in biochemical and medical research.

In order to better understand the special chemistry and biochemistry/biological activity of tellurium compounds, we briefly need to consider its chemical properties. Tellurium is located in the *p*-block of the Periodic

Table (electron configuration [Kr] $4d^{10}\ 5s^2\ 5p^4$) and as such exhibits certain metallic and nonmetallic properties. It occurs in two elemental modifications, one of them metallic, silver-shiny and one of them brownish-black. Not surprisingly, its primary industrial use is as a metal, for instance, as part of alloys (Fig. 1) (Bouad et al. 2003). Tellurium also features in the development of innovative new materials. Here, the (highly toxic) salt cadmium telluride (CdTe) may serve as an example. Because of its unique photochemical properties, CdTe is increasingly used in photovoltaic panels, and also may ultimately serve as part of novel probes in biological detection (Deng et al. 2007).

In the medium term, the currently anticipated rapidly increasing use of CdTe in solar panels raises a number of issues, including questions related the toxicity, bioavailability, degradation, and the impact of such materials on the environment and ultimately on humans. At the same time, disposal and recycling of CdTe-containing materials is slowly becoming a major issue, and clearly should stimulate more research into tellurium, its various compounds, their biological activity, and ultimately also toxicity (Fig. 1).

Besides CdTe, a number of other tellurium-containing agents are of similar interest to (biological) chemistry. As already mentioned, the chemistry of tellurium as a metalloid is very diverse and includes inorganic tellurium salts, tellurium-containing complex-like structures, as well as a wide range of organotellurium agents. Despite a few notable differences, the "inorganic" chemistry of tellurium resembles the one of sulfur and selenium, and includes a range of tellurium oxides, tellurides, tellurites, and tellurates (see Section Natural Occurrence of Tellurium in the Soil, Plants, and Animals). As the field of inorganic tellurium chemistry is rather extensive and diverse, several excellent reviews on this topic exist which may be consulted for more detailed information (Bureau et al. 2009).

The somewhat "metallic" character of tellurium obviously raises the rather naïve but certainly not unreasonable question if this element can also participate in a kind of "complex chemistry." Indeed, some of the tellurium compounds currently studied in biological test systems, including experimental animals, actually resemble metal-ligand complexes as far as their structure, bonding, stability, and reactivity are concerned. The widely studied and biologically active substance trichloro(dioxoethylene-O,O')-tellurate (often referred to as AS101), and its relative (3E)-3-[dichloro(4-methoxyphenyl)-1$^4$-tellanyl]-2-methylpent-3-en-2-ol (often referred to as RT-04), structurally as well as in the context of reactivity remind one of a metal-ligand complex rather than a typical organotellurium compound.

These structures contain a central tellurium atom, which is bound to a range of – often multidentate – ligands. The latter coordinate to tellurium via an oxygen atom. Coordination numbers vary, but tend to be either 4, 5, or 6, which results in mostly square-planar, trigonal bipyramidal, or octahedral structures. The chemical behavior of these complexes differs from the one of inorganic tellurium salts and organotellurium compounds – and hence provides the basis for a rather distinct biochemical reactivity and biological activity. The inhibitory action of AS101 against various cysteine proteases, for instance, seems to be due to ligand exchange at the tellurium atom, whereby one of the original chloride ligands is exchanged for the thiol(ate) group of the active site cysteine residue of the protease (as observed for cathepsin B). Formation of this Te-S complex results in inactivation of the protease, whose inhibition subsequently triggers a range of wider biochemical events.

Last but not least, the field of organotellurium chemistry includes a vast and diverse range of different organotellurium compounds with characteristic tellurium-carbon bonds. These compounds may be the most interesting as far as chemical diversity, biological activity, and possible applications in the area of drug design are concerned. There are several excellent and also rather recent reviews on the synthesis of organotellurium compounds, such as the book entitled "Tellurium in Organic Synthesis" by Petragnani and Stefani (2007) and reviews by Petragnani and Stefani (2005) and by Zeni et al. (2006) It should therefore be sufficient to highlight a few key aspects of this chemistry here. First of all, tellurium compounds are considerably more reactive but also less stable when compared to their selenium analogs. Secondly, most organotellurium agents are deeply colored and unfortunately therefore also often light sensitive. This is partially due to the rather weak tellurium-carbon bond: the trend within Group 16, which proceeds from 358 kJ mol$^{-1}$ for a typical C-O-bond to 272 kJ mol$^{-1}$ for a C-S-bond and 234 kJ mol$^{-1}$ for the C-Se-bond places the C-Te-bond at around 200 kJ mol$^{-1}$ (or even lower).

Fortunately, Te-aryl or Te-alkylaryl compounds are generally more stable when compared to Te-alkyl compounds – and indeed, most organotellurium compounds evaluated in biological systems are aromatic tellurides (Jacob et al. 2000).

Thirdly, the tellurium atom is mostly di- or tetravalent in organic compounds and also redox active. Here, two distinct tellurium-based redox pairs play a major role, and should never be confused with each other. The rather stable telluroxides (RTe(O)R, R $\neq$ H, R often aryl) are formed from equally stable tellurides (RTeR) by oxidation – and this telluride/telluroxide redox pair plays an important role as far as the catalytic properties of tellurium (e.g., for biological redox modulation) are concerned. Besides the better known tellurides (RTeR), various tellurols (RTeH) are also redox active. These compounds are reasonably acidic (the p$K$a of $H_2Te$ is 2.6, compared to 3.8 for $H_2Se$ and 7.0 for $H_2S$), (Ba et al. 2010) and the tellurolate (RTe$^-$) anion is readily oxidized to a ditelluride (RTeTeR). The tellurolate/ditelluride redox pair is, in theory, also able to participate in catalytic conversions (similar to the thiol/disulfide pair), yet instability of the tellurolate is often a major drawback. Importantly, higher oxidation states of tellurium, maybe in analogy to the selenium-based selenenic (RSeOH) and seleninic acids (RSe(O)OH), do not seem to play a major role in the redox behavior of tellurium compounds, at least not under physiological conditions. This is probably due to the inherent instability or reactivity of these tellurium species at ambient temperature and in aqueous solution. Without intending to ignore tellurols entirely, the following section therefore focuses on the more stable and more readily accessible Te-aryl and Te-alkylaryl compounds.

## Aspects of an Emerging Pharmaceutical Chemistry of Tellurium

Traditionally, tellurium and its various compounds have been considered as alien and even toxic to humans. Indeed, the element has long been associated with toxic and otherwise "disagreeable" effects. Since the nineteenth century, it has been known, for instance, that humans or animals ingesting tellurium compounds, such as $TeO_2$ or tellurite, breathe out a "disagreeable garlic-like odour" (see also Fig. 3). More severe clinical manifestations of acute tellurium toxicity in patients include a metallic taste, nausea, and vomiting (Yarema and Curry 2005). Nonetheless, it is perhaps rather disappointing that such findings should have prevented a more widespread research into the possible uses of tellurium agents in medicine and drug development. First of all, these examples represent classical cases of metal/metalloid poisoning, that is, situations involving the exposure to an exceptionally high amount of the element or its oxides – and in an "inappropriate" form (such as dust) and manner (e.g., via inhalation). Other metals (and nonmetals) may result in similar symptoms when taken up in excessive amounts or in toxic modifications (e.g., inhalation of zinc dust, swallowing of large amounts of selenium oxide, etc.). Secondly, it is natural that some compounds of a given element may be toxic, while others are not. One only needs to consider aggressive agents associated with oxygen, such as singlet oxygen or ozone, to underline this point. Hence, the argument that tellurium is generally toxic because there have been some instances of $TeO_2$ poisoning is clearly not sustainable. Indeed, there is some very basic evidence which points toward a particular toxicity of tellurite, while other tellurium compounds are less toxic. Chasteen et al. have recently published a rough ranking of (inorganic) tellurium toxicity in *Pseudomonas fluorescens* K27, where tellurite is more toxic than tellurate which in turn is more toxic than elemental tellurium (Basnayake et al. 2001). And finally, the fact that a specific chemical element (and its various compounds) do not occur naturally in a given organism does not inevitably mean that such substances are also irrelevant in the context of drug design (see, for instance, gold- and platinum-based drugs).

In fact, there is even some convincing evidence, which counts against the "uselessness" of tellurium in biochemical and pharmacological research. First of all, the "disagreeable garlic-like odour" mentioned above is probably the result of a metabolic conversion of $TeO_2$ to volatile $H_2Te$ or $(CH_3)_2Te$. The occurrence of such a conversion *in humans* is interesting as it seems to confirm that the human body can actually metabolize tellurium compounds and also detoxify and excrete them in a controlled manner.

At the same time, there has been some rather early evidence of "useful" biological activities associated with tellurium compounds, such as tellurite. In the pre-penicillin era, tellurite was used to inhibit the growth of many microorganisms, yet its action on bacteria was highly variable and it was not employed to treat infections in humans (Turner et al. 1999).

Nonetheless, tellurite often served as a "benchmark antibiotic." In the 1930s, for instance, Sir Alexander Fleming compared the antibacterial activities of penicillin to the ones of tellurite – and in nearly all cases, penicillin-insensitive bacteria were tellurite-sensitive and vice versa (Fleming 1932).

Based on these arguments it may be worthwhile to revisit the various tellurium compounds and to evaluate their possible uses as antibiotics, antifungal, or even anticancer agents. Many synthetic organotellurium compounds have turned out to be toxic when tested in cell culture or animal models, yet the selectivity of such toxicity, or the underlying biochemical mechanisms, have hardly been explored. Such knowledge would be essential, however, if one were to narrow down possible diseases which could be treated with tellurium compounds, if one were to identify cellular targets or if one were to derive certain structure-activity relationships.

There is some evidence which points into the direction of tellurium-based redox modulation in cells, and hence to a particularly strong activity against certain (antioxidant impaired) microorganisms and against cancer cells under Oxidative Stress. It is known, for instance, that *Plasmodium falciparum* – which causes malaria – appears to be particularly sensitive against certain tellurium and selenium compounds (Mecklenburg et al. 2009). As plasmodia lack certain antioxidant defense systems – and hence are particularly sensitive toward elevated levels of Oxidative Stress, the activity of these tellurium agents may be due to intracellular redox modulation.

Indeed, several organotellurium compounds exhibit distinct redox activity in vitro, such as GPx-like catalytic redox properties, or the ability to sequester radicals, to reduce peroxynitrite ($ONOO^-$) in the presence of RSH, and to protect metallothionein proteins against certain ROS (Jacob et al. 2000; Giles et al. 2003). Not surprisingly, such compounds have been considered as potential antioxidants, in line with the benchmark GPx-mimic ebselen, which is already in clinical trials for acute ischemic stroke (see Section The Special Relationship and Fig. 4a) (Yamaguchi et al. 1998).

Interestingly, redox catalysis may not only result in antioxidant events. Depending on the substrates consumed, GPx-like catalysis can also damage cells by consuming essential thiol groups, for example, in proteins and enzymes. This has recently led to a more thorough investigation of the redox-modulating activities of various organotellurium agents in cell culture and in animal models. Indeed, as the intracellular redox state of some cancer cells is rather close to the critical oxidative threshold, such redox-modulating "sensor/effector" agents are able to recognize cancer cells under Oxidative Stress and kill these cells quite selectively, while leaving normal cells vastly unaffected. Here, a fairly selective activity of 2,3-bis(phenyltellanyl)naphthoquinone has been observed in various cancer cell lines, such as HT29 and CT26 human colon cancer cells, which are more affected by this tellurium compound when compared to normal cultured NIH 3T3 fibroblast cells. Likewise, when CLL B-cells isolated from the blood of patients suffering from chronic lymphocytic leukemia (CLL) where treated with 2,3-*bis*(phenyltellanyl)naphthoquinone, a clear reduction in cell numbers was observed, while healthy B-cells from the same patients and control peripheral blood mononuclear cells (PBMC) were considerably less affected (Doering et al. 2010). A similar biochemical outcome, despite a different underlying mode of enzyme inhibition, has been observed for AS101 and several vinyl tellurium agents, all of which inhibit cysteine proteases, first by complexation and subsequently possibly by oxidation (see Section Chemical Diversity and Availability of Tellurium Agents for Medical Applications) (Cunha et al. 2009). RT-04, for instance, inhibits cathepsin B and induces apoptosis in HL60 cells with no significant toxic effects observed in normal bone marrow cells, possibly via regulation of Bcl-2 proteins in these cancer cells (Abondanza et al. 2008).

These examples may suffice to underline the great potential tellurium compounds may carry in the field of medicine and drug development. At the same time, tellurium has also attracted the attention of medical researchers working in the field of diagnostics. The unique physical, chemical, and spectroscopic properties of tellurium imply that tellurium-containing substances may serve as effective biological markers. It is possible, for instance, to synthesize the unnatural amino acid telluromethionine, which is taken up by certain organisms and incorporated "naturally" into proteins and enzymes. Such a formal exchange of sulfur for the heavy atom tellurium is of considerable advantage in the context of protein structural studies by X-ray crystallography. Furthermore, the radioactive $^{123m}Te$ isotope can be used to radiolabel certain molecules and monitor their distribution within cells, tissues, and organisms. Knapp, Kirsch, and colleagues, for instance, have synthesized a range of fatty acids containing tellurium, including radioactive $^{123m}$Te, and subsequently have studied the distribution and biological activity of these fatty acid derivatives in rats and dogs. These Te-containing fatty acids were hardly metabolized, but apparently "trapped" in the myocardium, pointing toward a possible use in the diagnosis of heart (and pancreatic) diseases, or in nuclear medicine (Knapp et al. 1980; Kirsch et al. 1983).

## Conclusions and Outlook

In summary, tellurium may appear as a "forgotten" element in the context of practical uses, yet this impression is becoming increasingly unsustainable. In the field of material sciences, CdTe-containing products are at the forefront of research and development, primarily in optical devices and as part of solar cells (Chen et al. 2009). Similar considerations apply to biology: Despite the fact that tellurium is apparently not a trace element, most organisms, including humans, are able to metabolize tellurium-containing compounds. Tellurium is therefore not really an "alien" to biology. Quite on the contrary, several interesting, therapeutic properties of tellurium compounds have recently emerged, ranging from antioxidant activity to redox modulation and (selective) cysteine enzyme inhibition. Here, the special relationship between the sulfur (and selenium)-based cellular redox and signaling systems (cellular thiolstat) on the one hand, and tellurium-based compounds on the other, may hold the key to a very selective, target-specific action of some of these agents.

It is therefore rather unfortunate that the pharmaceutical chemistry (and toxicology) of tellurium is still in its infancy. Hence, it is high time that we face up to the tasks ahead: The rather difficult synthetic chemistry of tellurium needs to be explored further, and we need to evaluate the chemical reactivity, biochemical mode(s) of action, and subsequent biological activity of these substances in earnest. Here, questions of metabolic stability, tissue selectivity, and also outright toxicity of certain compounds and their metabolites need to be addressed. In some instances, such studies need to evolve considerably from the current status quo in order to deliver satisfactory answers. Issues such as neurotoxicity and effects of tellurium compounds on behavior cannot easily be answered in cell culture or in rodent models and demand wider and more extensive studies. The challenges are certainly there for synthetic chemists as well as for biochemists and pharmacologists.

## Cross-References

▶ Bacterial Tellurite Resistance
▶ Scandium, Physical and Chemical Properties
▶ Selenium and the Regulation of Cell Cycle and Apoptosis
▶ Selenium, Biologically Active Compounds
▶ Tellurite-Resistance Protein TehA from *Escherichia coli*
▶ Tellurium and Oxidative Stress
▶ Tellurium, Physical and Chemical Properties

## References

Abe I, Abe T, Lou WW et al (2007) Site-directed mutagenesis of conserved aromatic residues in rat squalene epoxidase. Biochem Biophys Res Commun 352:259–263

Abondanza TS, Oliveira CR, Barbosa CMV et al (2008) Bcl-2 expression and apoptosis induction in human HL60 leukaemic cells treated with a novel organotellurium(IV) compound RT-04. Food Chem Toxicol 46:2540–2545

Ba LA, Doring M, Jamier V et al (2010) Tellurium: an element with great biological potency and potential. Org Biomol Chem 8:4203–4216

Ba LA, Doring M, Jamier V et al (2011) Tellurium: an element with great biological potency and potential. Org Biomol Chem 8:4203–4216

Baesman SA, Bullen TD, Dewald J et al (2007) Formation of tellurium nanocrystals during anaerobic growth of bacteria that use te oxyanions as respiratory electron acceptors. Appl Environ Microbiol 73:2135–2143

Baesman SM, Stolz JF, Kulp TR et al (2009) Enrichment and isolation of Bacillus beveridgei sp nov., a facultative anaerobic haloalkaliphile from Mono Lake, California, that respires oxyanions of tellurium, selenium, and arsenic. Extremophiles 13:695–705

Basnayake RST, Bius JH, Akpolat OM et al (2001) Production of dimethyl telluride and elemental tellurium by bacteria amended with tellurite or tellurate. Appl Organomet Chem 15:499–510

Borges VC, Rocha JBT, Nogueira CW (2005) Effect of diphenyl diselenide, diphenyl ditelluride and ebselen on cerebral Na+, K + -ATPase activity in rats. Toxicology 215:191–197

Bouad N, Marin-Ayral RM, Nabias G et al (2003) Phase transformation study of Pb-Te powders during mechanical alloying. J Alloys Compd 353:184–188

Bureau B, Boussard-Pledel C, Lucas P et al (2009) Forming glasses from Se and Te. Molecules 14:4337–4350

Chasteen TG, Bentley R (2003) Biomethylation of selenium and tellurium: microorganisms and plants. Chem Rev 103:1–25

Chen LY, Chen CL, Li RN et al (2009) CdTe quantum dot functionalized silica nanosphere labels for ultrasensitive detection of biomarker. Chem Commun 2670–2672

Contreras-Puente G, Vigil O, Ortega-Lopez M et al (2000) New window materials used as heterojunction partners on CdTe solar cells. Thin Solid Films 361:378–382

Cowgill UM (1988) The tellurium content of vegetation. Biol Trace Elem Res 17:43–67

Csotonyi JT, Stackebrandt E, Yurkov V (2006) Anaerobic respiration on tellurate and other metalloids in bacteria from hydrothermal vent fields in the eastern Pacific Ocean. Appl Environ Microbiol 72:4950–4956

Cunha RLOR, Gouvea IE, Feitosa GPV et al (2009) Irreversible inhibition of human cathepsins B, L, S and K by hypervalent tellurium compounds. Biol Chem 390:1205–1212

Deng ZT, Zhang Y, Yue JC et al (2007) Green and orange CdTe quantum dots as effective pH-sensitive fluorescent probes for dual simultaneous and independent detection of viruses. J Phys Chem B 111:12024–12031

Doering M, Ba LA, Lilienthal N et al (2010) Synthesis and selective anticancer activity of organochalcogen based redox catalysts. J Med Chem 53:6954–6963

Dopp E, Hartmann LM, Florea AM et al (2004) Environmental distribution, analysis, and toxicity of organometal(loid) compounds. Crit Rev Toxicol 34:301–333

Fesharaki PJ, Nazari P, Shakibaie M et al (2010) Biosynthesis of selenium nanoparticles using *Klebsiella pneumoniae* and their recovery by a simple sterilization process. Braz J Microbiol 41:461–466

Fleming A (1932) J Pathol Bacteriol 35:831–842

Garberg P, Engman L, Tolmachev V et al (1999) Binding of tellurium to hepatocellular selenoproteins during incubation with inorganic tellurite: consequences for the activity of selenium-dependent glutathione peroxidase. Int J Biochem Cell Biol 31:291–301

Giles NM, Watts AB, Giles GI et al (2003) Metal and redox modulation of cysteine protein function. Chem Biol 10:677–693

Jacob C (2011) Redox signalling via the cellular thiolstat. Biochem Soc Trans 39:1247–1253

Jacob C, Arteel GE, Kanda T et al (2000) Water-soluble organotellurium compounds: catalytic protection against peroxynitrite and release of zinc from metallothionein. Chem Res Toxicol 13:3–9

Jacob C, Battaglia E, Burkholz T et al (2012) Control of oxidative posttranslational cysteine modifications: from intricate chemistry to widespread biological and medical applications. Chem Res Toxicol 25:588–604

Kirsch G, Goodman MM, Knapp FF (1983) Organotellurium compounds of biological interest – unique properties of the N-chlorosuccinimide oxidation-product of 9-telluraheptadecanoic acid. Organometallics 2:357–363

Knapp FF, Ambrose KR, Callahan AP (1980) Tellurium-123 m-labeled 23-(isopropyl telluro)-24-nor-5alpha-cholan-3beta-Ol – new potential adrenal imaging agent. J Nucl Med 21:251–257

Kumar S, Johansson H, Kanda T et al (2010) Catalytic chain-breaking pyridinol antioxidants. J Org Chem 75:716–725

McNaughton M, Engman L, Birmingham A et al (2004) Cyclodextrin-derived diorganyl tellurides as glutathione peroxidase mimics and inhibitors of thioredoxin reductase and cancer cell growth. J Med Chem 47:233–239

Mecklenburg S, Shaaban S, Ba LA et al (2009) Exploring synthetic avenues for the effective synthesis of selenium- and tellurium-containing multifunctional redox agents. Org Biomol Chem 7:4753–4762

Meotti FC, Borges VC, Zeni G et al (2003) Potential renal and hepatic toxicity of diphenyl diselenide, diphenyl ditelluride and Ebselen for rats and mice. Toxicol Lett 143:9–16

Nogueira CW, Rotta LN, Perry ML et al (2001) Diphenyl diselenide and diphenyl ditelluride affect the rat glutamatergic system in vitro and in vivo. Brain Res 906:157–163

Ogra Y (2009) Toxicometallomics for research on the toxicology of exotic metalloids based on speciation studies. Anal Sci 25:1189–1195

Petragnani N, Stefani HA (2005) Advances in organic tellurium chemistry. Tetrahedron 61:1613–1679

Stangherlin EC, Ardais AP, Rocha JBT et al (2009) Exposure to diphenyl ditelluride, via maternal milk, causes oxidative stress in cerebral cortex, hippocampus and striatum of young rats. Arch Toxicol 83:485–491

Turner RJ, Weiner JH, Taylor DE (1999) Tellurite-mediated thiol oxidation in *Escherichia coli*. Microbiol Uk 145:2549–2557

Wenqi Q, Yalei C, Yiping Z et al (1992) Tellurium contents in soils in China. Int J Environ Stud 41:263–266

Widy-Tyszkiewicz E, Piechal A, Gajkowska B et al (2002) Tellurium-induced cognitive deficits in rats are related to neuropathological changes in the central nervous system. Toxicol Lett 131:203–214

Yamaguchi T, Sano K, Takakura K et al (1998) Ebselen in acute ischemic stroke: a placebo-controlled, double-blind clinical trial. Ebselen study group. Stroke 29:12–17

Yarema MC, Curry SC (2005) Acute tellurium toxicity from ingestion of metal-oxidizing solutions. Pediatrics 116:E319–E321

Zeni G, Ludtke DS, Panatieri RB et al (2006) Vinylic tellurides: from preparation to their applicability in organic synthesis. Chem Rev 106:1032–1076

# Tellurium, Physical and Chemical Properties

Fathi Habashi
Department of Mining, Metallurgical, and Materials Engineering, Laval University, Quebec City, Canada

Tellurium is a metalloid found in close association with sulfur and selenium, has the appearance of a metal but lacks all properties of metals (Fig. 1). It is recovered from the anodic slimes obtained during the electrorefining of copper and nickel during processing of their sulfide ores. It also occurs in nature in the form of gold telluride. Tellurium is a $p$-type semiconductor, demonstrates the phenomenon of piezoelectricity and becomes

Tellurium, Physical and Chemical Properties, Fig. 1 Tellurium

superconductive at 3.3 K. Tellurium is used in cadmium telluride, CdTe, solar panels. On melting, the specific volume increases by ca. 5%. Tellurium vapor is yellow-gold and consists mainly of $Te_2$ molecules up to 2,000°C.

## Physical Properties

| | |
|---|---|
| Atomic number | 52 |
| Atomic weight | 127.60 |
| Relative abundance in Earth's crust, % | $1.8 \times 10^{-7}$ |
| Density, g/cm$^3$ | |
| At room temperature | 6.245 |
| At melting point | 5.70 |
| Melting point, °C | 449.51 |
| Boiling point, °C | 988 |
| Heat of fusion, kJ·mol$^{-1}$ | 17.49 |
| Heat of vaporization, kJ·mol$^{-1}$ | 114.1 |
| Molar heat capacity, J·mol$^{-1}$·K$^{-1}$ | 25.73 |
| Atomic radius, pm | 140 |
| Crystal structure | Hexagonal |
| Thermal conductivity, W·m$^{-1}$·K$^{-1}$ | 1.97–3.38 |
| Standard electrode potentials, V | |
| $Te + 2e^- \rightarrow Te^{2-}$ | −0.92 |
| | *(continued)* |

| | |
|---|---|
| $Te^{4+} + 4e^- \rightarrow Te$ | 0.63 |
| Young's modulus, GPa | 43 |
| Shear modulus, GPa | 16 |
| Bulk modulus, GPa | 65 G |
| Mohs hardness | 2.25 |
| Brinell hardness, MPa | 180 |

## Chemical Properties

Tellurium burns in air to form tellurium dioxide, $TeO_2$. Tellurium dioxide can also be obtained by treating tellurium powder with nitric acid, followed by heating the nitrate.

Tellurium dioxide dissolves in water to form tellurous acid, $H_2TeO_3$. Sodium tellurite can be oxidized to tellurate:

$$Na_2TeO_3 + {}^1/_2 O_2 \rightarrow Na_2TeO_{4(insoluble)} \quad (>550°C)$$

Tellurium is obtained by electrolysis of a tellurite solution:

$$Anode: 4OH^- \rightarrow 2H_2O + O_2 + 4e^-$$
$$Cathode: TeO_3^{2-} + 3H_2O + 4e^- \rightarrow Te + 6OH^-$$

Tellurium obtained by electrolysis may be further refined by vacuum distillation and zone refining to yield the higher-purity grades 99.99–99.9999%.

Other oxides of tellurium are tellurium monoxide, TeO, tellurium trioxide, $TeO_3$, and tellurium pentoxide $Te_2O_5$. Hydrogen telluride, $H_2Te$, is a colorless, highly toxic gas which decomposes above 0°C. In humans, tellurium is partly metabolized into dimethyl telluride, $(CH_3)_2Te$, a gas with a garlic-like odor which is exhaled in the breath of victims of tellurium toxicity or exposure (Knockaert 1997).

## References

Knockaert G (1997) In: Habashi F (ed) Handbook of extractive metallurgy. Wiley-VCH, Weinheim, pp 1571–1583

# Telomere-Like Structure

▶ Strontium and DNA Aptamer Folding

# Terbium

Takashiro Akitsu
Department of Chemistry, Tokyo University of Science, Shinjuku-ku, Tokyo, Japan

## Definition

A lanthanoid element, the eighth element (yttrium group) of the f-elements block, with the symbol Tb, atomic number 65, and atomic weight 158.92534. Electron configuration [Xe] $4f^9 6s^2$. Terbium is composed of stable ($^{159}$Tb, 100%) and two synthetic radioactive ($^{157}$Tb; $^{158}$Tb) isotopes. Discovered by C. G. Mosander in 1843. Terbium exhibits oxidation states III and IV; atomic radii 177 pm, covalent radii 195 pm; redox potential (acidic solution) $Tb^{3+}/Tb$ $-2.391$ V; $Tb^{4+}/Tb^{3+}$ $-2.7$ V; electronegativity (Pauling) 1.1. Ground electronic state of $Tb^{3+}$ is $^7F_6$ with S = 3, L = 3, J = 6 with $\lambda = -290$ cm$^{-1}$. Most stable technogenic radionuclide $^{158}$Tb (half-life 180 years). The most common compounds: $Tb_2O_3$, $TbO_2$, and $Tb_4O_7$. Biologically, terbium is of low to moderate toxicity, and after intravenous infusion, accumulated in the liver and the kidney and discharged to choluria and biliuria (Atkins et al. 2006; Cotton et al. 1999; Huheey et al. 1997; Oki et al. 1998; Rayner-Canham and Overton 2006).

## Cross-References

▶ Lanthanide Ions as Luminescent Probes
▶ Lanthanide Metalloproteins
▶ Lanthanides and Cancer
▶ Lanthanides in Biological Labeling, Imaging, and Therapy
▶ Lanthanides in Nucleic Acid Analysis
▶ Lanthanides, Physical and Chemical Characteristics

## References

Atkins P, Overton T, Rourke J, Weller M, Armstrong F (2006) Shriver and Atkins inorganic chemistry, 4th edn. Oxford University Press, Oxford/New York
Cotton FA, Wilkinson G, Murillo CA, Bochmann M (1999) Advanced inorganic chemistry, 6th edn. Wiley-Interscience, New York
Huheey JE, Keiter EA, Keiter RL (1997) Inorganic chemistry: principles of structure and reactivity, 4th edn. Prentice Hall, New York
Oki M, Osawa T, Tanaka M, Chihara H (1998) Encyclopedic dictionary of chemistry. Tokyo Kagaku Dojin, Tokyo
Rayner-Canham G, Overton T (2006) Descriptive inorganic chemistry, 4th edn. W. H. Freeman, New York

## Terminal Oxidases

▶ Heme Proteins, Cytochrome c Oxidase

## Ternary Crosslinks

▶ Zinc, Metallated DNA-Protein Crosslinks as Finger Conformation and Reactivity Probes

## Terpyridine Platinum(II) Complexes as Cysteine Protease Inhibitors

▶ Platinum(II), Terpyridine Complexes, Inhibition of Cysteine Proteases

## TET, Tetrahedral-Shaped Aminopeptidase

▶ Zinc Aminopeptidases, Aminopeptidase from Vibrio Proteolyticus (Aeromonas proteolytica) as Prototypical Enzyme

## Thallium

▶ Thallium, Distribution in Animals
▶ Thallium, Effects on Mitochondria

## Thallium and Programmed Cell Death

▶ Thallium, Cell Apoptosis Inducement

## Thallium Myocardial Perfusion Imaging

▶ Thallium-201 Imaging

## Thallium Myocardial Perfusion Scintigraphy

▶ Thallium-201 Imaging

## Thallium Myocardial Perfusion SPECT (Single Photon Emission Computed Tomography)

▶ Thallium-201 Imaging

## Thallium Stress Test

▶ Thallium-201 Imaging

## Thallium Tumor Imaging

▶ Thallium-201 Imaging

## Thallium Tumor Scintigraphy

▶ Thallium-201 Imaging

## Thallium, Cell Apoptosis Inducement

Sandra Viviana Verstraeten
Department of Biological Chemistry, IQUIFIB (UBA-CONICET), School of Pharmacy and Biochemistry, University of Buenos Aires, Argentina, Buenos Aires, Argentina

## Synonyms

Thallium and programmed cell death

## Definitions

*Apoptosis*: Apoptosis (from the Greek *apo*: from/off, *ptosis*: falling) is a physiological programmed cell death that is highly conserved in nature. This process is composed of a tightly controlled chain of events. The hallmark of apoptosis is nuclear condensation, DNA fragmentation, membrane blebbing, and cell fragmentation into the apoptotic bodies. The latter are removed by phagocytes but without affecting the neighboring cells.

*Necrosis*: In opposition to apoptosis, necrosis (from the Greek νεκρός: dead) constitutes a traumatic form of cell death that results from the acute injury of cells by certain toxins, trauma, or infection. Necrosis is characterized by the abrupt rupture of the cell with the consequent release of the intracellular contents into the surroundings. As a consequence, an inflammatory process is triggered that may affect relatively large portions of tissue.

*Caspases*: They are a group of enzymes that belongs to the cysteine proteases family and that plays essential roles in the progress of apoptosis. These enzymes are classified into two groups: the initiator caspases (caspases 2, 8, 9, and 10) and executioner caspases (caspase 3, 6, and 7). The first group has the executioner caspases as substrates and activates them, and the second group acts on diverse target proteins to trigger apoptosis.

*Reactive oxygen species (ROS)*: They are a group of oxygen-containing molecules, with moderate to high reactivity toward biological macromolecules. From a chemical point of view, some ROS are free radicals since they bear one or more unpaired electrons that confer reactivity to these molecules. The simplest ROS are superoxide anion ($O_2^{\bullet-}$), hydrogen peroxide ($H_2O_2$), and hydroxyl radical ($^{\bullet}OH$), which are intermediates in the reduction of molecular oxygen ($O_2$) to water in the electron transfer chain. The reaction of these molecules with lipids, proteins, and nucleic acids alters their chemical properties and affects their biological functions.

## Introduction

Thallium (Tl) is a nonessential heavy metal that is present in the earth's crust at very low concentrations, mainly in the form of salts and minerals. However, mining activity causes the mobilization of this metal to water and soils, thus increasing its bioavailability and reaching the living organisms (ATSDR 1999). Along with lead, cadmium, and mercury, Tl is considered as a priority pollutant by the USA Environmental Protection Agency. Tl accumulation affects several human tissues and systems, including the central, peripheral, and autonomic nervous systems, as well as the epidermal, gastrointestinal, cardiovascular, reproductive, and renal systems (▶ Thallium, Distribution in Animals).

Tl has two ionic species, the monovalent or thallous cation (Tl(I)) and the trivalent or thallic cation (Tl(III)). While the clinical symptoms of Tl(I) poisoning has been described in the initial decades of the twentieth century, the cellular mechanisms involved in its toxicity are still poorly understood. Very few information is available on Tl(III). However, the increasing use of these cations in many industries stresses the importance of studying their toxic effects at a biochemical level. By unraveling the mechanisms involved, it will be possible in the future to design strategies to prevent and/or to revert the damage to the organs that are the targets of Tl noxious effects.

Apoptosis is an evolutionarily conserved and highly organized form of programmed cell death. It is formed by a tightly controlled sequence of events that ultimately leads to nuclear changes, chromatin shrinkage and DNA fragmentation, membrane blebbing, and the formation of apoptotic bodies. The latter are removed by phagocytes. Thus, and in opposition to necrosis where the intracellular content of the affected cells is released causing inflammation and damage to their neighbors, apoptosis allows the removal of dying cells without affecting its surroundings. Traditionally, apoptosis is triggered by either the stimulation of death receptors in cell surface (extrinsic pathway) or by the damage to mitochondria with the release of pro-apoptotic proteins and enzymes into the cytosol (mitochondrial pathway). In the recent years, a third mechanism for apoptosis initiation has been recognized and progressively accepted: the destabilization of lysosomal compartment with the release of hydrolytic enzymes into the cytosol. The following sections will review the evidence showing that both Tl(I) and Tl(III) induce apoptosis in different experimental models as a possible mechanism to explain the toxicity of these cations.

## Tl and the Mitochondrial Pathway of Apoptosis

This pathway is initiated by the destabilization of mitochondrial membrane due to a variety of stimuli, such as certain toxins, UV radiation, and others. Increased steady-state concentrations of reactive oxygen species (ROS) or reactive nitrogen species (RNS) generated in the plasma membrane, endoplasmic reticulum, or mitochondria can damage these organelles and initiate apoptosis.

It has been demonstrated that in rat pheochromocytoma (PC12) cells, both Tl(I) and Tl(III) (10–100 µM) increase the steady-state concentration of hydrogen peroxide ($H_2O_2$) within mitochondria. $H_2O_2$ is a ROS, and results from the partial reduction of oxygen to water by the electron transfer chain. Also, it can be generated from the reduction of another ROS, the superoxide anion ($O_2^{\bullet-}$) produced by the enzyme NADPH oxidase (NOX) at the plasma membrane level. $O_2^{\bullet-}$ is metabolized by the enzyme superoxide dismutase (SOD), producing $H_2O_2$. This molecule is highly diffusible and can cross intracellular membranes, reaching mitochondrial matrix. In PC12 cells, Tl(I) and Tl(III) do not significantly affect NOX activity. Thus, the increased $H_2O_2$ found in this model must be generated within the mitochondria. In accordance with the oxidative stress status due to the high $H_2O_2$ content, the concentration of glutathione in its reduced form (GSH) is significantly decreased (Hanzel and Verstraeten 2006). The tripeptide GSH (Glu-Cys-Gly) is the main intracellular nonprotein thiol and it has a dual effect on the toxicity of the metals. Firstly, GSH chelates metals and facilitates their removal from the cell. Secondly, GSH has an antioxidant action that helps to prevent ROS-mediated damage to cell components. A strong correlation between $H_2O_2$ concentration in Tl-exposed PC12 cells and the decrease in cells viability supports the hypothesis that Tl-mediated oxidative stress may be related to cell death in this experimental model. Similar observations were reported in cultured rat hepatocytes, where both Tl(I) and Tl(III) increased ROS production and altered the GSSG/GSH balance toward an oxidized state (Pourahmad et al. 2010). The pro-oxidant effect of Tl(I) was prevented by incubating hepatocytes with a variety of antioxidants, mitochondrial transition pores (MTP) sealers, and other compounds that favor a reductant state within cells that compensates Tl(I) pro-oxidant effects. Although Tl(III)-mediated ROS generation was also prevented by these compounds, their effect was limited and only attenuated the pro-oxidant effect of this cation. Interestingly, in this experimental model, Tl-mediated ROS production was also partially prevented by inhibitors of cytochrome P450, indicating that endoplasmic reticulum stress might be another source of ROS upon Tl exposure (Pourahmad et al. 2010).

Working with isolated mitochondria, Korotkov et al. demonstrated that millimolar concentrations of Tl(I) alter mitochondrial permeability and open MTP (▶ Thallium, Effects on Mitochondria). This observation was extended to cultured cells (hepatocytes, PC12 cells) where micromolar concentrations of either Tl(I) or Tl(III) decrease mitochondrial membrane potential (Fig. 1). As a consequence of mitochondrial depolarization, MTP open and allow the release of different pro-apoptotic proteins into the cytosol. One of these proteins is cytochrome c, a component of the electron transfer chain that is anchored to mitochondrial internal membrane through the binding to the phospholipid cardiolipin. In vitro, the interaction of Tl(III) with cardiolipin-containing membranes not only alters the biophysical properties of the bilayer, but also causes the oxidative cleavage of this phospholipid, producing phosphatidylglycerol (Puga Molina and Verstraeten 2008). From a chemical point of view, it is important to consider that the couple Tl(III)/Tl(I) has a redox potential ($\varepsilon^0$) of +1.25 mV. Therefore, Tl(III) is considered a strong oxidant, being able to oxidize a variety of macromolecules per se and hence alter their biological functions. As a result of cardiolipin loss by Tl(III), a decreased binding of cytochrome c to the bilayer was found. Until today, there are no reported evidences showing that Tl(III) – in its trivalent valence – reaches mitochondria in vivo, but in the case that this phenomenon occurs, this might be a plausible mechanism to explain the detachment of cytochrome c from mitochondrial membranes, an effect that triggers cell apoptosis.

As mentioned before, cytochrome c exits the mitochondria though the MTP. Supporting that, cells' preincubation with cyclosporine A which blocks these pores prevents cytochrome c release in PC12 cells exposed to Tl(I) or Tl(III) (Fig. 1). A second mechanism for mitochondrial pores formation has been proposed: This involves members of Bcl-2 superfamily composed of over 30 members. This family comprises both pro-apoptotic (Bax) and anti-apoptotic (Bcl-2, Bcl-XL) proteins. In PC12 cells, Tl(III) – and to a lesser extent Tl(I) – induces Bax oligomerization into mitochondrial membrane (Fig. 2). In addition, the expression of Bax increases while Bcl-2 expression remains constant (Hanzel and Verstraeten 2009). A misbalance between the expression of pro- and anti-apoptotic members of the Bcl-2 superfamily was also observed in C6 glioma cells exposed for 24 h to Tl(I) (10 μM to 3 mM) (Chia et al. 2005) indicating that this must be a common mechanisms of action in Tl toxicity, regardless of the cell type involved.

Once inside the cytosol, cytochrome c together with pro-caspase 9, Apaf-1 (apoptosis protease activating factor 1), and dATP forms a multi-protein complex, the apoptosome, that activates caspase 9. In turn, this caspase acts on pro-caspase 3, and the now active caspase 3 will inactivate the DNA-repairing enzyme poly(ADP-ribose) polymerase (PARP), among other proteins. Supporting that, both Tl(I) and Tl(III) increase the activity of caspase 9 and caspase 3 (Fig. 3), and promote the cleavage – and therefore, inactivation – of PARP (Hanzel and Verstraeten 2009). In cultured hepatocytes Tl(I)- and Tl(III)-supported caspase 3 activation was partially or fully prevented by cells' pretreatment with antioxidants such as mannitol and α-tocopherol, the iron chelator deferoxamine, and the MTP sealer, carnitine (Pourahmad et al. 2010).

In addition to Tl-dependent cytochrome c release, other proteins can exit the mitochondria and exert their pro-apoptotic effect in a caspase-independent manner. To this group belongs the apoptotic inducing factor (AIF) and endonuclease G (EndoG). Once mitochondrial membrane is permeabilized by members of the Bcl-2 family, AIF is released into the cytosol, translocates to the nucleus, and participates in chromatin condensation in chromatinolysis. It has been proposed that AIF causes stage I of DNA condensation and the large-scale DNA fragmentation that precedes intranucleosomal cleavage by caspase-activated DNAase (CAD) during stage II of DNA condensation in apoptosis. In addition to AIF, EndoG is also released and translocates to the nucleus and participates in DNA fragmentation. Both AIF and EndoG have been detected in PC12 cells nuclei upon Tl(I) or Tl(III) exposure (Hanzel and Verstraeten 2009). Together, the imbalance between DNA damaging (AIF and EndoG) and repairing (PARP) enzymes in Tl-treated cells results in a decrease in DNA integrity that leads to the nuclear fragmentation that constitutes the hallmark of apoptosis.

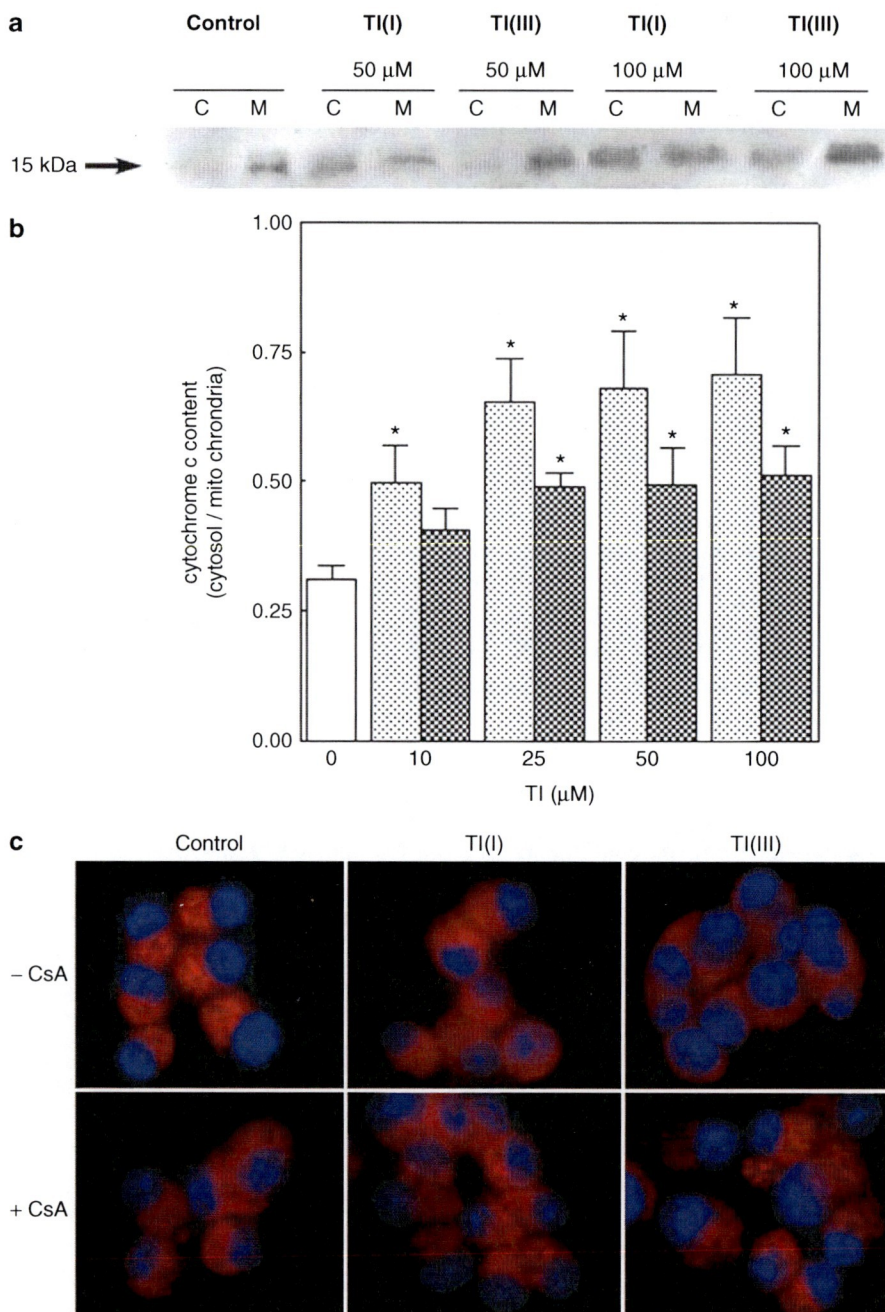

**Thallium, Cell Apoptosis Inducement, Fig. 1** *Tl(I) and Tl(III) cause cytochrome c release.* Rat pheochromocytoma (PC12) cells were incubated for 24 h in the presence of Tl(I) or Tl(III) (10–100 μM), and cytochrome c release was measured. (**a**) A representative Western blot showing cytochrome c content in cells cytosol (C) and mitochondria (M). (**b**) Quantitation of cytochrome c release in control (☐), Tl(I)- (▦), and Tl(III)-treated cells (▦). *Significantly different from the value measured in control cells ($P < 0.05$, ANOVA). (**c**) Cytochrome c immunofluorescence in PC12 cells exposed for 18 h to Tl either in the absence or presence of 1 μM cyclosporine A (CsA). *Red*: cytochrome c; *blue*: cell nuclei stained with the DNA probe Hoechst 322258. Magnification: 600X. (Reprinted from Hanzel and Verstraeten 2009, with permission from Elsevier)

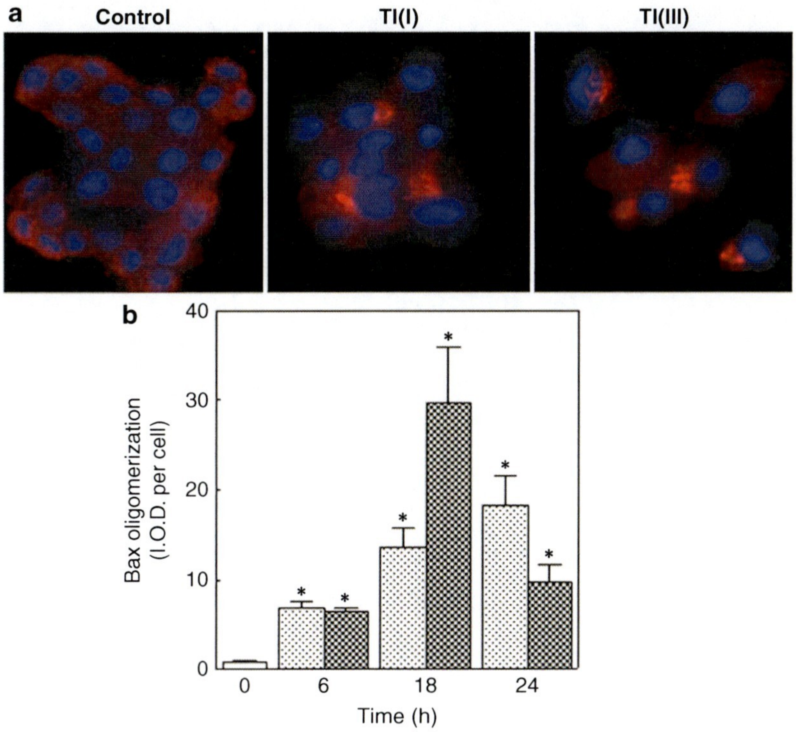

**Thallium, Cell Apoptosis Inducement, Fig. 2** *Tl(I) and Tl(III) cause Bax oligomerization.* Rat pheochromocytoma (PC12) cells were incubated for 1–24 h in the presence of Tl(I) or Tl(III) (100 μM), and Bax was detected by immunofluorescence. (**a**) Photograph showing Bax's presence in the cytoplasm of control cells and its oligomerization in Tl(I)- and Tl(III)-treated cells. *Red fluorescence*: Bax; *blue fluorescence*: cell nuclei stained with the DNA probe Hoechst 322258. Magnification: 600 X. (**b**) Quantitation of Bax oligomerization in control (☐), Tl(I)- (▦), and Tl(III)-treated cells (▩). Results are shown as the integrated optical density (IOD.) per cell, and are the mean ± SEM of three independent experiments. *Significantly different from the value measured in control cells ($P < 0.001$, ANOVA) (Reprinted from Hanzel and Verstraeten 2009, with permission from Elsevier)

## Tl and the Extrinsic Pathway of Apoptosis

The extrinsic pathway of apoptosis is initiated by the action of external signals including certain hormones, the members of the tumor necrosis factor (TNF) family, Fas ligand (FasL), and TRAIL. All these signals bind to the denominated "death receptors" present in the extracellular portion of the plasma membrane, and trigger a cascade of events that leads to apoptosis. Once these receptors are activated by their corresponding ligands, a number of proteins interact with their intracellular domain forming the death-inducing signaling complex, or DISC. This complex recruits pro-caspase 8 and/or pro-caspase 10 which in turn activate the executioner caspases (mainly caspase 3), and proceeds with the events that lead to DNA fragmentation described in the previous section.

Interestingly, Tl(III) but not Tl(I) is capable of activating this pathway in PC12 cells (Fig. 4). After 1 h of exposure to 100 μM Tl(III), PC12 cells increase the expression of FasL receptor. The kinetics of FasL receptor expression is similar to that observed for the activation of mitochondrial pathway of apoptosis, suggesting that both mechanisms equally contribute to the apoptotic effect of Tl(III). In agreement with the activation of FasL receptor, caspase 8 is rapidly activated, an effect that reaches a maximum after 12 h of Tl(III) exposure (Fig. 4). On the other hand, Tl(I) is incapable of inducing FasL receptor expression and

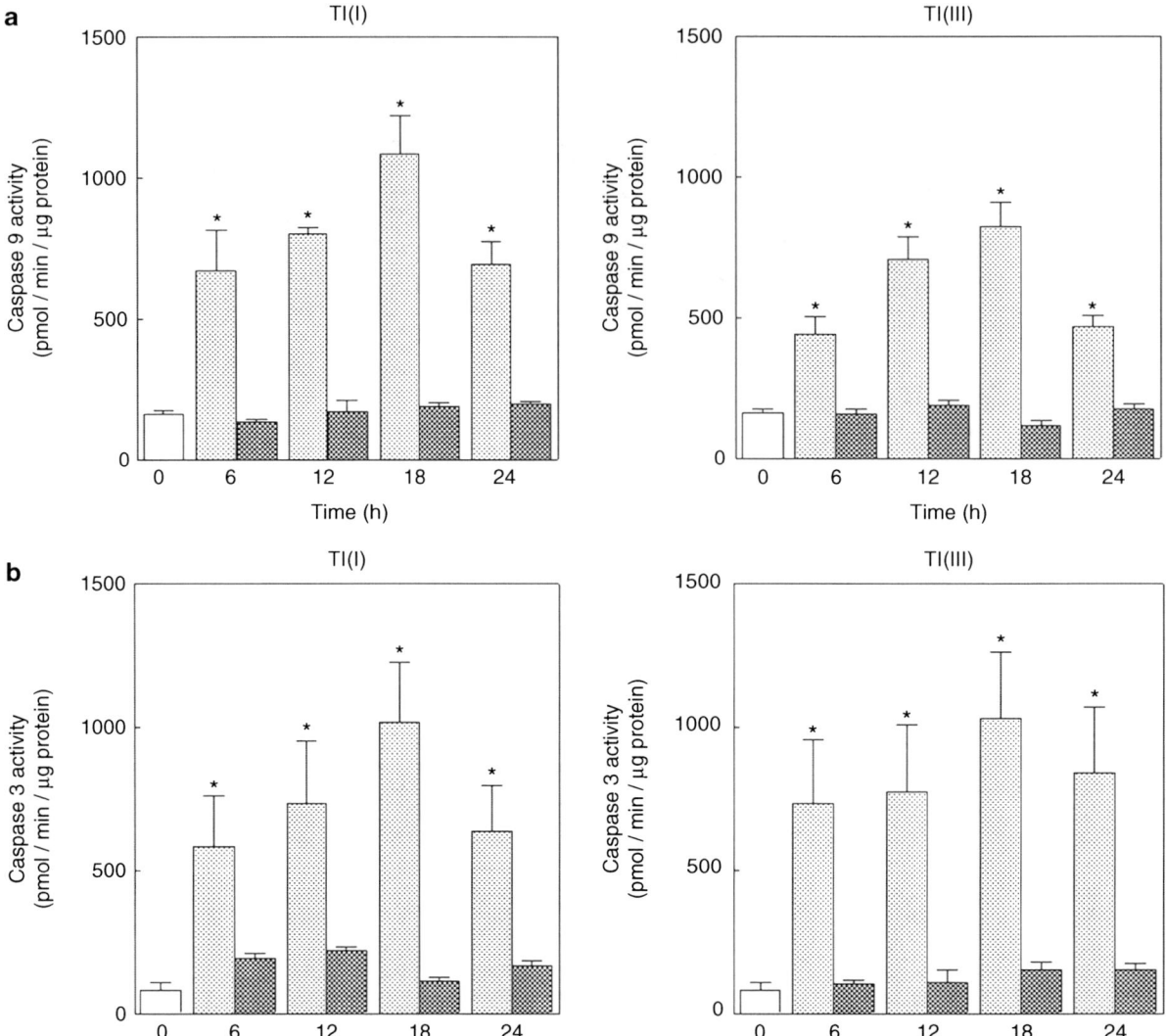

**Thallium, Cell Apoptosis Inducement, Fig. 3** *Tl(I) and Tl(III) activate caspases 9 and 3*. Rat pheochromocytoma (PC12) cells were incubated for 6 to 24 h either in the absence or in the presence of Tl(I) or Tl(III) (100 μM), and the activities of (**a**) caspase 9 and (**b**) caspase 3 were evaluated. Caspases activities in Tl(I)- (*left panel*) and Tl(III)- (*right panel*) treated cells were measured in the absence (▨) or in the presence (▨) of the corresponding specific inhibitors (caspase 9: Ac-LEHD-CHO, caspase 3: Ac-DEVD-pNA). Results are shown as mean ± SEM of six independent experiments. *Significantly different from the value measured in control cells (☐) ($P < 0.05$, ANOVA) (Reprinted from Hanzel and Verstraeten 2009, with permission from Elsevier)

caspase 8 activation. The cause of the activation of the extrinsic pathway of apoptosis by Tl(III) remains to be elucidated, but it seems to be a common mechanism for multivalent heavy metals, since Cd(II), Hg(II), As(III), and Co(II) also increase FasL receptor and promote caspase 8 activation in diverse experimental models (Liu et al. 2003, Eichler et al. 2006).

It has been proposed that the depolarization of mitochondria and enhanced ROS generation may induce the expression of FasL receptor independently of the binding to FasL, and via c-Jun terminal kinase (JNK) activation. This mechanism seems to be unlikely since both Tl(I) and Tl(III) induce mitochondrial depolarization and ROS production to a similar extent and with similar kinetics but only Tl(III) activates the extrinsic pathway of apoptosis.

Regardless of the activation of FasL receptor, caspase 8 may be activated by another mechanism.

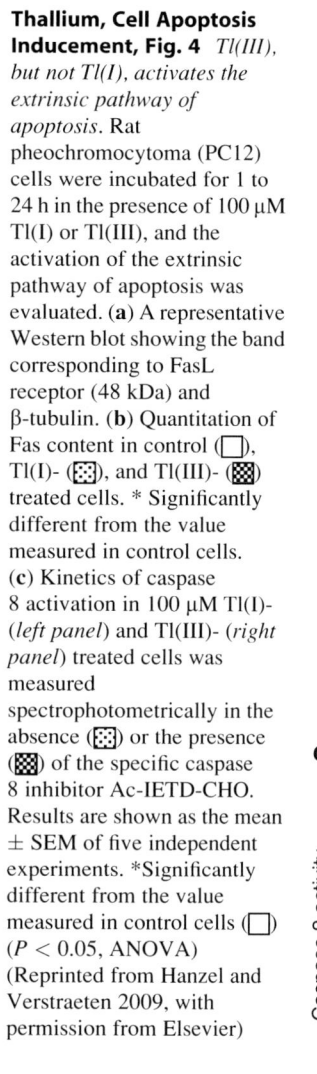

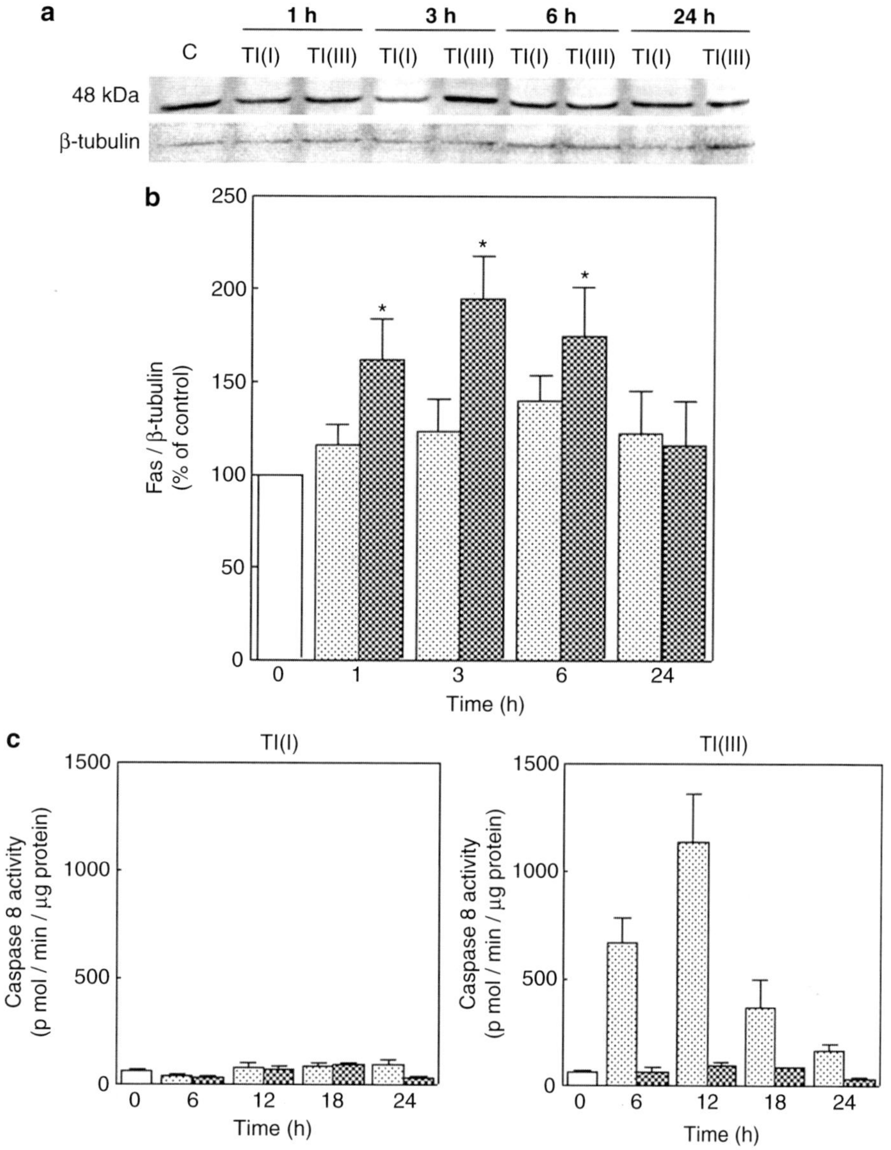

**Thallium, Cell Apoptosis Inducement, Fig. 4** *Tl(III), but not Tl(I), activates the extrinsic pathway of apoptosis*. Rat pheochromocytoma (PC12) cells were incubated for 1 to 24 h in the presence of 100 μM Tl(I) or Tl(III), and the activation of the extrinsic pathway of apoptosis was evaluated. (**a**) A representative Western blot showing the band corresponding to FasL receptor (48 kDa) and β-tubulin. (**b**) Quantitation of Fas content in control (□), Tl(I)- (▨), and Tl(III)- (▨) treated cells. * Significantly different from the value measured in control cells. (**c**) Kinetics of caspase 8 activation in 100 μM Tl(I)- (*left panel*) and Tl(III)- (*right panel*) treated cells was measured spectrophotometrically in the absence (▨) or the presence (▨) of the specific caspase 8 inhibitor Ac-IETD-CHO. Results are shown as the mean ± SEM of five independent experiments. *Significantly different from the value measured in control cells (□) ($P < 0.05$, ANOVA) (Reprinted from Hanzel and Verstraeten 2009, with permission from Elsevier)

The lysosomal hydrolytic enzyme cathepsin D activates caspase 8 in the cytosol. In order to be this mechanism operative, lysosome's integrity must be compromised, allowing the exit of their internal content. The list of agents that cause lysosomal destabilization is growing, and includes both exogenous factors, such as certain xenobiotics and photodamage, and endogenous, such as sphingosine and ROS, among others. To notice, the massive destruction of lysosomes leads to cell necrosis, while a discrete damage to these organelles results in cell apoptosis. It has been recently demonstrated that both Tl(I) and Tl(III) compromise lysosomal membrane integrity in cultured rat hepatocytes (Pourahmad et al. 2010). In a model of rats' intoxication with $TlCl_3$, in addition to mitochondria and endoplasmic reticulum damage, an enlargement of lysosomal compartment was observed in hepatocytes that are consistent with an autophagic process (Woods and Fowler 1986). The involvement of Tl-mediated damage to lysosomes in the initiation and/or progression of cell apoptosis remains to be elucidated at the molecular level.

**Thallium, Cell Apoptosis Inducement,**

**Fig. 5** *Proposed sequence of events leading to Tl(I)- and Tl(III)-mediated apoptosis.* For further details, see text. *Dashed arrows* indicate hypothetical steps. Abbreviations: AIF apoptosis inducing factor, Apaf 1 apoptotic protease activating factor 1, casp caspase, cath cathepsin, cyt c cytochrome c, DISC death-inducing signaling complex, EndoG endonuclease G, FADD Fas-associated protein with death domain, FasL Fas ligand, FasL rec Fas ligand receptor, MTP mitochondrial transition pores, PARP poly(ADP-ribose) polymerase, iPARP inactive PARP

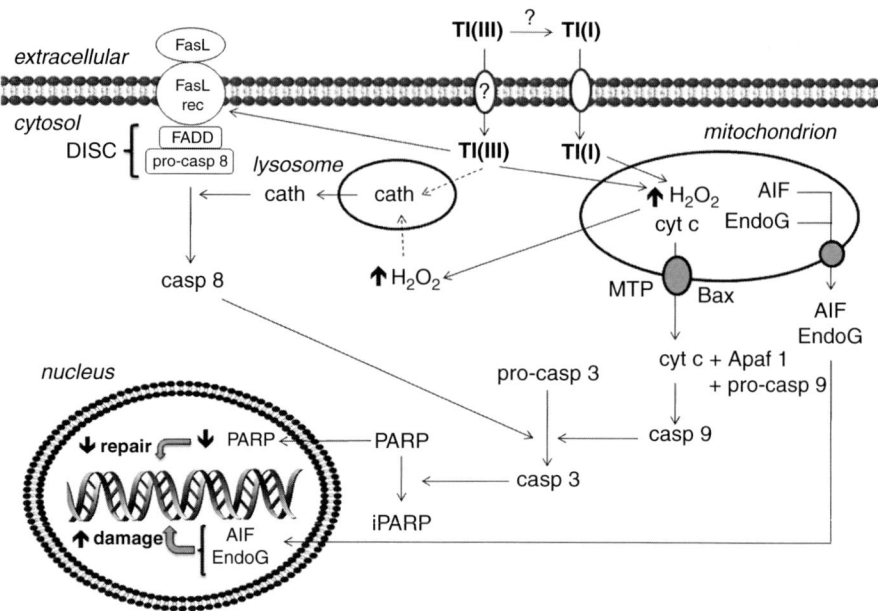

## Conclusions

In different experimental models, including cultured cells and intoxicated animals, Tl was found to cause cell apoptosis. The activation of mitochondrial and/or extrinsic pathways by this metal relies on the valency of the cation assessed, which suggests that Tl(I) and Tl(III) interact with dissimilar extra- and intracellular components (Fig. 5). The induction of apoptosis by these cations reduces cell viability, an effect that may account for the neurodegeneration observed in Tl intoxication, as well as for the alterations found in other human organs and systems.

## Cross-References

▶ Apoptosis
▶ Thallium, Effects on Mitochondria
▶ Oxidative Stress
▶ Thallium, Distribution in Animals

## References

ATSDR (1999) Thallium. Agency for toxic substances and disease registry eds. Prepared by Clement International Corp., under contract 205-88-0608: Atlanta

Chia CF, Chen SC, Chen CS, Shih CM, Lee HM, Wu CH (2005) Thallium acetate induces C6 glioma cell apoptosis. Ann N Y Acad Sci 1042:523–530

Eichler T, Ma Q, Kelly C, Mishra J, Parikh S, Ransom RF, Devarajan P, Smoyer WE (2006) Single and combination toxic metal exposures induce apoptosis in cultured murine podocytes exclusively via the extrinsic caspase 8 pathway. Toxicol Sci 90: 392–399

Hanzel CE, Verstraeten SV (2006) Thallium induces hydrogen peroxide generation by impairing mitochondrial function. Toxicol Appl Pharmacol 216:485–492

Hanzel CE, Verstraeten SV (2009) Tl(I) and Tl(III) activate both mitochondrial and extrinsic pathways of apoptosis in rat pheochromocytoma (PC12) cells. Toxicol Appl Pharmacol 236:59–70

Liu Q, Hilsenbeck S, Gazitt Y (2003) Arsenic trioxide-induced apoptosis in myeloma cells: p53-dependent G1 or G2/M cell cycle arrest, activation of caspase-8 or caspase-9, and synergy with APO2/TRAIL. Blood 101:4078–4087

Pourahmad J, Eskandari MR, Daraei B (2010) A comparison of hepatocyte cytotoxic mechanisms for thallium (I) and thallium (III). Environ Toxicol 25: 456–467

Puga Molina LC, Verstraeten SV (2008) Thallium(III)-mediated changes in membrane physical properties and lipid oxidation affect cardiolipin–cytochrome c interactions. Biochim Biophys Acta 1778:2157–2164

Woods JS, Fowler BA (1986) Alteration of hepatocellular structure and function by thallium chloride: ultrastructural, morphometric, and biochemical studies. Toxicol Appl Pharmacol 83:218–229

# Thallium, Distribution in Animals

Laura Osorio-Rico[1], Sonia Galván-Arzate[1] and Abel Santamaría[2]
[1]Departamento de Neuroquímica, Instituto Nacional de Neurología y Neurocirugía Manuel Velasco Suárez, Mexico City, DF, Mexico
[2]Laboratorio de Aminoácidos Excitadores, Instituto Nacional de Neurología y Neurocirugía Manuel Velasco Suárez, Mexico City, DF, Mexico

## Synonyms

Accumulation: Concentration, content, uptake
Thallium: Thallium acetate, Tl$^+$, Tl (I), Tl (III), Tl$_2$SO$_4$

## Definition

*Thallium*: Toxic metalloid with oxidation states from +1 and +3. It is highly toxic and was used in rat poisons and insecticides. This metal is actually used in electronic and pharmaceutical industry and glass manufacture.

*Accumulation*: Content of metal in exposed animals estimated by different analytical techniques.

*Distribution*: Metal levels quantified in tissues and organs at different times.

*Blood–Brain Barrier (BBB)*: Separation of circulating blood and the brain extracellular fluid (BECF) in the central nervous system (CNS). It occurs along all capillaries and consists of tight junctions around the capillaries that do not exist in normal circulation. Endothelial cells restrict the diffusion of hydrophilic molecules into the cerebrospinal fluid (CSF), while allowing the diffusion of small hydrophobic molecules. This barrier also includes a thick basement membrane and astrocytic endfeet.

## General Issues

Thallium (Tl$^+$) is a highly toxic metal and is present in the environment. Tl$^+$ is released to the air from coal-burning power plants, cement factories, and smelting operations (Repetto et al. 1998). Accumulation and distribution of this metal in animal tissues is relevant for a better understanding of the toxic nature of this agent, although this topic has only been modestly studied using a few different animal species, times of administration, and dosages. Acute exposure to Tl$^+$ is known to produce damage to different organs, both at central and peripheral levels, whereas chronic exposure induces direct toxicity to brain, spinal cord, and peripheral nervous system (Galván-Arzate et al. 2000, 2005).

The toxic features of Tl$^+$ include its complexation with protein sulfhydryl groups, inhibition of cellular respiration, interaction with riboflavin cofactors, and disruption of calcium homeostasis (Díaz and Moreal 1994). However, the best-known mechanism of this metal to date is probably the one related with the interference of vital potassium-dependent processes when replacing K$^+$ in Na$^+$/K$^+$-ATPases. Tl$^+$ and K$^+$ are both univalent ions with similar ionic radii, so Tl$^+$ is able to interfere in K$^+$-dependent processes, then mimicking its flux and intracellular accumulation in mammal cells. Tl$^+$ also replaces physiological K$^+$ during the activation of several enzymes (including pyruvate kinase and aldehyde dehydrogenase), the stabilization of ribosomes, the modulation of enzyme production and amino acid synthesis, and the modification of transport mechanisms (Peter and Viraraghavan 2005). Tl$^+$ salts are absorbed by the gastrointestinal tract, the respiratory system, and the skin. This ion is rapidly accumulated in different tissues and widely distributed into the brain, kidney, liver, heart, skeletal muscle, testis, intestine, and salivary glands. Its lethal dose 50 (LD$_{50}$) reported for rodents is 32 mg/kg i.p. (Ríos and Monroy-Noyola 1992). Moreover, some recent reports provide valuable information on Tl$^+$ distribution and concentrations in wild animals that have been exposed to this metal upon conditions of environmental pollution.

## Brain

The content of Tl$^+$ in exposed animals has been estimated for the first time by Ríos et al. (1989) in the following brain regions of adult Wistar rats: cortex (Cx), hippocampus (Hc), corpus striatum (S), hypothalamus (Ht), and midbrain (Mb), following an acute i.p. dose of 32 mg/kg. Brain Cx exhibited significantly less Tl$^+$ than other regions when estimated 24 h post-injection. The content of Tl + in Ht was almost three times higher than in Cx. In addition, there was a dose-dependent accumulation of $Tl^+$ in brain regions when different doses (16, 32, and 48 mg/kg) and time courses (3, 6, 18, 24, and 48 h) were assessed, achieving a peak concentration at 24 h after its administration.

Another study (Galván-Arzate and Ríos 1994) explored the distribution of $Tl^+$ in the rat brain in direct relation to the development of the blood–brain barrier (BBB) and its role in modulating $Tl^+$ transport from the periphery. Newborn and 1, 5, 7, 15 and 20 days-old rats were i.p. injected with $Tl^+$ at a dose of 16 mg/kg. $Tl^+$ homogeneously accumulated among the studied regions (Cx, cerebellum (Ce), Hc, Ht, and Mb). Time course of $Tl^+$ distribution in brain regions of newborn rats showed a peak in Cx at 1–2 days after its administration. All other regions reached peak concentrations between 0.75 and 1 day, thus indicating a slower entrance into the cortical tissue when compared with the rest of the brain. $Tl^+$ accumulation was also evaluated in Cx, Ce, and Ht of 24-h-exposed rats at ages 1, 5, 10, 15, and 20 days. The metal content was low in Cx and Ce at 5 days after birth, and remained the same at 20 days of age. Hypothalamic $Tl^+$ accumulation remained almost the same than other regions at all ages tested, except for rats of 20 days of age that exhibited the lowest levels. Altogether, these data suggests that the functional development of the BBB defines the patterns of differential distribution of the metal in the central nervous system.

A subchronic administration of sublethal doses of $Tl^+$ (0.8 and 1.6 mg/kg) to adult rats was found to increase its accumulation in brain regions in a dose-dependent manner. However, with the dose of 1.6 mg/kg i.p., Hc and Ht exhibited a higher accumulation of the metal when compared to all the other regions (Galván-Arzate et al. 2000).

In a more recent study (Galván-Arzate et al. 2005), i.p. doses of 8 and 16 mg/kg of $Tl^+$ given as single administrations to adult rats caused an homogeneous and dose-dependent accumulation in different brain regions (Ht, Ce, S, Hc and Cx) at day 7 post-injection.

Considering the evidence together, the pattern of $Tl^+$ distribution in the brain of rodents, and by extension, of mammals, could be caused by a selective transport across the BBB, as Ht exhibited the highest $Tl^+$ accumulation. This finding could be likely due to the limited action of the protective mechanisms exerted by the BBB in this region (Willis and Grossman 1981).

## Other Organs

Using radioactive thallium sulfate, Thyresson found the highest content in the kidneys, followed (in decreasing order) by intestinal mucosa, thyroid, testis, pancreas, skin, bone, and muscle. On the contrary, liver and nerve tissue were homogeneously low in $Tl^+$ content (Thyresson 1951).

In studies using a field desorption mass spectrometry technique, $Tl^+$ concentrations were established in stomach, heart, kidney, liver, and brain of mice after feeding 80, 130, and 160 mg/kg $Tl^+$, as a time-dependent process. It was found that the stomach is the organ with the maximal metal concentration, followed by kidney, heart, liver, and brain. Brain $Tl^+$ uptake was comparatively low and constant during the first 12 h. In the terminal stage (24 h) all organs, including the brain, contained increased $Tl^+$ levels of the same order of magnitude (Achenbach et al. 1980).

In 1989, Aoyama demonstrated that $Tl^+$ administration to hamsters (at a dose of 10 and 50 mg/kg; orally) produced the highest concentrations in brain, testis, liver, and kidney at a 50 mg/kg dose, when compared with the 10 mg/kg dose. The kidney showed the highest content of $Tl^+$ at 1 and 3 days after its administration, whereas the brain was the organ exhibiting the lower content.

Noteworthy, kidney has been shown to be the organ that accumulates the highest amounts of $Tl^+$ among all body organs, at 24 h post-injection of 16 mg/kg to adult rats (Ríos et al. 1989). In contrast, the metal levels in newborn rats at 1 day after its systemic infusion followed the order: testis > heart > kidney > liver > brain. These findings could be the result of different rates of absorption of the metal into the tissues of immature animals modulated by the blood-testicular barrier (Galván-Arzate and Ríos 1994).

Leung and Ooi (2000) found uneven levels of $Tl^+$ in different organs of rats sacrificed 4 days after the injection of 30 mg/kg of $Tl_2SO_4$. The content of the metal was higher in the kidney, followed by ileum, stomach, and liver, all in a range from 12 to 66 μg/g of tissue.

In dogs exposed to $Tl^+$ throughout the ingestion of metal-tainted food and beverage, the metal concentrations in liver and kidney were 18 and 26 μg/g, respectively, whereas $Tl^+$ was absent in other organs (Volmer et al. 2006).

## Placental Distribution

Placental absorption and distribution of $Tl^+$ constitutes another relevant issue that has been addressed. In a comparative study, a group of rats were fed on the tenth day of pregnancy with 8 mg/kg of $Tl_2SO_4$. Later on, animals were killed at different times (10, 20, 30, 40, or 50 min, as well as 1, 6, 12, 20, or 30 h) after the metal administration. In addition, mice were fed on the ninth day with the same dose and sacrificed at 0.5, 1, 2, 4, 6, 8, 27, or 50 h. The quantitative determinations were done by field desorption mass spectrometry after dilution of the homogenized tissue samples with enriched stable isotopes of thallium. The kidneys of pregnant rats and mice showed very high $Tl^+$ concentrations at short times (ranking a minimum of $7 \times 10^{-5}$ and maximum $4 \times 10^{-4}$ M). The brain, however, presented amounts of $Tl^+$ which were lower ($3 \times 10^{-5}$ M) by a factor of 10–20-fold in comparison with other organs. Interestingly, no differences between rats and mice were found (Ziskoven et al. 1983). After 1 h, the fetal tissue of rats reached maximum values up to $7 \times 10^{-5}$ M, whereas in mice the maximal $Tl^+$ concentration ($4 \times 10^{-5}$ M) was achieved 4 h after the metal administration. Fetal and maternal organ levels were well correlated.

## Thallium as Environmental Pollutant

To monitor pollutants, bio-monitoring through measurement of chemical concentrations in wild animals offers greater benefits than measuring inorganic environmental material. In 2007, Suzuki and coworkers reported the accumulation of $Tl^+$ and other elements considered as factory wastes in semiconductor-producing facilities in the Taiwan area. They found the highest $Tl^+$ concentrations in kidney of Formasan squirrels, being even twofold to threefold higher than those of Japanese squirrels. The levels found in kidney were followed in magnitude by liver, lung, and muscle, maintaining the same relation between Taiwan and Japanese specimens. The values ranged from 0.061 to 0.009 μg/g dry tissue.

## Conclusion

Characterizing $Tl^+$ distribution in organs in animals is relevant for designing experimental strategies to counteract its toxic effects based on previously acquired knowledge on where and in which proportion this metal is preferentially accumulated. This is critical for toxicological and clinical research, especially if it is considered that this toxic heavy metal is capable of producing damage at several levels in different organs and body systems. Here were presented a brief compilation on those studies detailing the $Tl^+$ distribution in organs of different species in an attempt to bring a tool to better understand $Tl^+$ toxicity.

## References

Achenbach C, Hauswirth O, Heindrichs C et al (1980) Quantitative measurement of time-dependent thallium distribution in organs of mice by field desorption masss spectrometry. J Toxicol Environ Health 6:519–528

Aoyama H (1989) Distribution and excretion of thallous after oral and intraperitoneal administration of thallous malonate and thallous sulfate in hamster. Bull Environ Contam Toxicol 42:456–463

Díaz R, Moreal J (1994) Thallium mediates a rapid chloride/hydroxyl ion exchange through myelin lipid bilayers. Mol Pharmacol 46:1210–1216

Galván-Arzate S, Ríos C (1994) Thallium distribution in organs and brain regions of developing rats. Toxicol 90:63–69

Galván-Arzate S, Martínez A, Medina E et al (2000) Subchronic administration of sublethal doses of thallium to rats: effect on distribution and lipid peroxidation in brain regions. Toxicol Lett 116:37–43

Galván-Arzate S, Pedraza-Chaverri J, Medina-Campos O et al (2005) Delayed effects of thallium in rat brain: regional changes in lipid peroxidation and behavioral markers, but moderate alterations in antioxidants, after a single administration. Food Chem Toxicol 43:1037–1045

Leung KM, Ooi VEC (2000) Studies of thallium toxicity, its tissue distribution and histopathological effects in rats. Chemosphere 41:155–159

Peter A, Viraraghavan T (2005) Thallium: a review of public health and environmental concerns. Environ Int 31:493–501

Repetto G, Del Peso A, Repetto M (1998) Human thallium toxicity. In: Nriagu JO (ed) Thallium in the environment. Wiley, New York, pp 167–199

Ríos C, Monroy-Noyola A (1992) D-Penicillamine and Prussian blue as antidotes against thallium intoxication in rats. Toxicol 74:69–76

Ríos C, Galván-Arzate S, Tapia R (1989) Brain regional thallium distribution in rats acutely intoxicated with $Tl_2SO_4$. Arch Toxicol 63:34–37

Suzuki Y, Watanabe I, Oshida T et al (2007) Accumulation of trace elements, used in semiconductor industry in Formosan squirrel, as a bio-indicator on their exposure, living in Taiwan. J Chemosphere 68:1270–1279

Thyresson N (1951) Experimental investigation on thallium poisoning in the rat; distribution of thallium, especially in the skin, and excretion of thallium under different experimental conditions, a study with the use of the radioactive isotope Tl204. Acta Derm Venereol 31:3–27

Volmer P, Merola V, Osborne T et al (2006) Thallium toxicosis in a Pit Bull Terrier. J Vet Diagn Invest 18:134–137

Willis W, Grossman R (1981) Medical neurobiology. Mosby, New York

Ziskoven R, Achenbach C, Schulten HR et al (1983) Thallium determinations in fetal, tissues and maternal brain and kidney. Toxicol Lett 19:225–231

## Thallium, Distribution in Plants

Bernd Markert[1], Stefan Fränzle[2], Simone Wünschmann[1], Ana Maria Figueiredo[3], Astrid van der Meer[4] and Bert Wolterbeek[4]
[1]Environmental Institute of Scientific Networks, in Constitution, Haren/Erika, Germany
[2]Department of Biological and Environmental Sciences, Research Group of Environmental Chemistry, International Graduate School Zittau, Zittau, Germany
[3]Instituto de Pesquisas Energeticas e Nucleares, IPEN-CNEN/SP, Sao Paulo, Brazil
[4]Interfaculty Reactor Institute, Delft University of Technology, Delft, The Netherlands

## Synonyms

*Thallium*: Tl; *Distribution*: Partition; *Plants*: Plantae; Photoautotroph metazoan

## Definition

Till today, only an overseeable number of publications deal with the distribution of Tl in plants (f.e. Xiao et al. 2004; Sasmaz et al. 2007). Mainly highly polluted areas such as mining areas were examined, and significantly higher Tl-concentrations could be found in soil and plant samples there. Thallium (Tl, Z = 81) is a soft, ductile, lowly melting, and highly toxic heavy metal discovered by Crookes in 1861 (in lead chamber sludge residues, containing As, Se, Te, and Os besides). Thallium was discovered in the very first days of optical-emission spectroscopy and like with the similar cases of rubidium, cesium, and indium, was named according to its flame color (Greek thallos = *green as a young twig*; $\lambda$ = 535 nm).

There are hardly any minerals which contain Tl as a principal component, except for lorandite $TlAsS_2$ (first spotted in Macedonia [FYROM]) and manganese nodules in deep ocean. For now, thallium is mainly obtained from pyrite and larger amounts are released during roasting $CaCO_3$ for cement production. With the former uses of Tl (rodenticide, gas discharge lamps, epilation agent) being outphased (actually, banned) now all over the developed world, thallium consumption did significantly decrease to <15 t/a, comparable to those of Ga, or of the rarer platinum-group metals, with the neat metal totaling <1 t.

## Chemical Features

Chemically, thallium is the heaviest nonradioactive member of group 13 (third main group), possessing two stable isotopes, $^{203}$Tl [29.5% of natural mixture] and $^{205}$Tl [70.5%]. There are two principal oxidation states of Tl: +I and +III. The intermediate oxidation state $Tl_{aq}^{2+}$ which can, for example, be photogenerated is an extremely strong oxidant ($\epsilon \approx$ 2.2 V vs NHE). Thallium-based anions such as $Tl_4^{4-}$ are well known in solid or liquid alloys (Zintl phases) or certain solvents, but do not form or persist in water. It is kind of ambiguous in its chemical behavior, resembling partly alkali metals, partly silver, in forming readily soluble, strongly alkaline Tl(OH), Tl(CN), Tl(NO$_2$), and pyrophosphate $Tl_4P_2O_7$, $Tl_2CO_3$ (like Na, K, Cs) while monovalent halides other than TlF, molybdate $Tl_2MoO_4$, and chalcogenides such as $Tl_2S$ are insoluble in water – like with Ag. Like silver [or lead(II)] chalcogenides and halides, the latter compounds are photosemiconductors but – unlike AgBr or AgI – not sensitive to photodecomposition. Unlike the lighter homologues aluminum, gallium, and indium, its trivalent state is strongly oxidizing toward both inorganic and organic substrates although some Tl(III) complexes thereof like $[Tl(CN)_4]^-$ or $TlBr_4^-$ or even $[TlI_2(bipy)]^+$ are stable toward decay (i.e., reductive elimination) into Tl(I) and coupled ligand oxidation product (in this case, cyanogen or free halogens). A valuable organic transformation involving Tl(III) nitrate oxidant is *oxythallation*, converting alkenes into those aldehydes or their corresponding ketals, which would be formed upon hydration of the respective alkynes otherwise, cycloalkenes into ring-contrived aldehydes, and producing shortened or rearranged carboxylic acids from alkynes. The lower oxidation state Tl$^+$ forms rather readily from Tl(0); yet, although the formal potential Tl/Tl$^+$ is well below that of hydrogen and Tl(OH) is highly soluble in water, like most other Tl(I) salts, thallium metal does dissolve neither in pure, if aerated, water nor in dilute nonoxidizing acids. Whereas the *alkyl* compounds of the lighter homologues, like AlR$_3$, violently react with water, let alone oxidizing acids, and sometimes spontaneously ignite in air (AlR$_3$), most Tl(III) alkyl cations are remarkably stable, some even withstanding prolonged boiling in concentrated nitric acid. In fact, given this and additional properties (resistance toward solar UV radiation in both salts and aqueous solution), and since it is readily formed by biomethylation by various organisms, it is remarkable that (CH$_3$)$_2$Tl$^+$ ion is not the only form of thallium which occurs in the biosphere (it is present, making up to some 40% of marine thallium [about 1 ng/l] but far from being the only speciation form around [Schedlbauer and Heumann 2000]). Possibly the very redox transalkylation by which Tl trialkyls can be prepared from alkyl compounds of more oxidizing ions like Hg(II), Au(III), or Pt(IV), for example, according to

$$3HgR_2 + 2Tl(\text{metal}) \rightarrow 2TlR_3 + 3Hg(\text{redox transalkylation}),$$

forms a pathway of cleavage if there is "super-reduced" [Co(I)] cobalamine (vitamin B$_{12}$), namely:

$$(CH_3)_2Tl^+ + [Co^I(\text{cobalamine})] + H_2O \rightarrow$$
$$[CH_3 - Co^{III}(\text{cobalamine})]^+ + Tl^+ + CH_4 + OH^-$$

## Toxicological Actions

Tl$^+$ differs from alkali metal ions such as K$^+$, Rb$^+$ by having a free electron pair additionally, which was taken up after reduction of Tl$^{3+}$. Having a free electron pair at the metal ion, complex formation is poor with Tl (I) (mutual repulsion with ligand electron pairs). Due to that free valence electron pair, Tl ions are shaped rather like an egg than being spherical. The net charge is positive while the negative valence electron pair is located on one end, making the entire atomic ion a dipole (one positive [shorter] end of the "egg"'s long axis opposing a negative [extended] end), and thus, although complex formation is feeble, much less pronounced than in monovalent coinage metals, will cause havoc when competing with K$^+$ ions (of similar net diameter) in, for example, nerve membranes: they would slip into these nerve cell membrane channels but turn a little, thereafter blocking passage for K$^+$ (or, Rb, Ag) ions while slowly accumulating until nerve function breaks down altogether. The typical aspects of thallium toxicity, other than the conspicuous but reversible total body hair loss some 12 days after ingestion, are thus all related to neurotoxicity: first hyperesthesia, then loss of tactile sense, blindness, paralysis, unconsciousness (sometimes irreversible coma), and finally failure of the sine node of the heart, that is, cardiac arrest (Tl poisoning cannot be treated by infunding K$^+$ to shift equilibrium since this would stop heartbeat also. Rather, it is absorbed by Prussian Blue administered orally). Most likely, the first effect of Tl ingestion is also related to K/Tl antagonism in both sensory and motor nerves (of the guts and colon): obstipation, which of course increases the share of Tl absorbed from a poisoned meal, this situation being aggravated by Tl reabsorption from guts via the enterohepatic cycle, which however can be blocked by Tl absorbers orally given (Prussian Blue).

In the central nervous system, there is enhanced production and release of catecholamines after Tl ingestion, probably related to both tremor symptoms and psychical illnesses associated with Tl toxicity, namely, depression but also psychotic episodes.

Tl(I) sulfate Tl$_2$SO$_4$ formerly was used as a rodenticide (not arising any suspicion by rats because it does neither taste nor smell anyhow but being frequently abused for homicide for these very reasons also; 1 g will kill an average adult human), with the bright green emission of Tl employed to correct the emission spectra of neon or argon gas discharge lamps into a more whitish overall appearance. While there was the distinctive green band in the spectrum of such lamps only 30 years ago, this kind of Tl use meanwhile was outphased also.

Table 1 represents a so-called ecotoxicological "identity card" for Tl. Based on the ecotoxicological identity cards of As and Sn from Fränzle and Markert (2002) this card is meant to give a first-hand description of properties relevant to biological and toxicological features of a certain chemical element and its geobiochemically plausible speciation form.

**Thallium, Distribution in Plants, Table 1** Ecotoxicological "identity card" for Tl species (explanations in the text)

| | | | | | | |
|---|---|---|---|---|---|---|
| Formal oxidation state | +I | +I | +III | +III | +III | Negative |
| Principal speciation form | $Tl_{aq}^{+}$ | Complexes of Tl(I) | $Tl_{aq}^{3+}$ | Complexes of Tl(III) | Organometallic, e.g., $(CH_3)_2Tl^+$ | Zintl ions like $Tl_4^{4-}$ |
| Remarkable chemical features pertinent to biochemistry and toxicology | Nonspherical ion, tends to "lock" in ion channels | Usually unstable | Strong oxidant, attacking, e.g., alkenes, cycloalkenes, styrenes like cinnamic acid and alkynes (oxythallation) | Branching in relative stabilities between O donor and other complexes[a] | | Do not occur in biomass, hardly stable in aqueous media |
| Origins, fates | | | Possibly formed by very strongly oxidizing enzymes or environmental peroxides ($H_2O_2$; PAN), $NO_2$, $NO_3$ radicals; reduction by alkenes, phenols, e.g., humic acids or photolysis | | Biomethylation by unidentified marine and sludge organisms, very resistant toward all oxidation, hydro- and photolysis; pathway of cleavage in marine environment must exist but is unknown | Direct reduction of Tl or Tl alloys by solvated electrons in solvents like $NH_3(l)$, amines or THF |
| In environmental compartments | Present, also in minerals lorandite $TlAsS_2$, crookesite $Cu_7TlSe_4$; volatile Tl compounds (TlCl, Tl(OH)) released in cement production | Some are abundant | Not identified, presumably short-lived | Not detected so far | Common in ocean (up to 48% of total Tl), others like $TlR_3$ or Tl(I) organics would be far less stable | Absent |
| In animals | Neurotoxic, tends to accumulate in nerves, reversible hair loss, also resorbed transdermally | | Fast cleavage by oxythallation or thiolate reduction would be likely | Not detected so far | | Would be rapidly oxidized |
| In plants | Accumulated by rape leaves, pine trees, certain flowers (up to >1%!), less so by grapes and cabbage; root absorption is common and occurs readily | | Fast cleavage by oxythallation or thiolate reduction would be likely | Not detected so far | | Would be rapidly oxidized |
| In bacteria | | | Like with plants, animals | | | |
| Other, unidentified or unspecified organisms | | | | | See above for oxidative (Challenger-type) methylation of Tl(I) | |

[a]Unlike most other transition-group, main-group, REE, or actinoide metal ions, stabilities of thallium (both common oxidation states, and Hg(II)) for chelator – ligand complexes cannot be calculated directly from the equation
$-\log k_{diss} = x * E_L(L) + c$
Fränzle (2010) introducing the electrochemical ligand parameter $E_L(L)$ for whatever neutral or anionic ligand, but different donor atoms produce unlike regression equations (one of which holds for oxygen donors [ligands like oxalate], the other for N, S, Se, C [such as cyanide])

The starting point is the environmental levels of different oxidation states and speciation forms, modified by bioenrichment (BCF) factors.

With other elements, known to be essential at last for certain species of organisms (Fränzle 2010; Fraenzle et al. 2012), such identity cards define speciation forms involved in biochemistry or outright toxic ones, without removing the differences caused by mutual conversion (e.g., Cr(III) is assumed to be essential for some vertebrates, and can be produced in vivo by reduction of chromate(VI); nevertheless the latter remains toxic, carcinogenic, by no means useful for "effective," acceptable Cr supply). Things are a little different with elements for which no essential role is known; here the focus rests with probable modes of interaction among different speciation forms and components of biomass, and (also with Tl) with the possible feature of biomethylation. Of course, this also covers differences in toxicity and modes of toxic action attributed to the different speciation states. As a rule, except for alkylated species, or Cr toxicity, differences among speciation forms of heavy metals are much less pronounced than with nonmetals.

Elements which do no good to any living being yet have some bioinorganic chemistry, usually displaying ion–biomass interactions even if only destructive ones. With thallium, Tl(III) is likely to do the same kind of oxythallation transformations of many biogenic organics in vivo, which is known to be useful in preparative organic chemistry.

## Thallium in Plants

There are no creatures known for which Tl is established or purported to be essential (Fig. 1).

Given the specific mechanism of toxic action toward nerve membranes, it comes as no surprise that plants – having no nerves – tolerate Tl much better than animals would; some plants and fungi even accumulate it. Due to the strongly oxidizing properties toward all alkenes (e.g., unsaturated lipid-forming carboxylic acids such as oleic or linoleic acid), alkynes, and certain amino acids, $Tl^{3+}$ would not last long after ingestion; hence the biochemistry and toxicology deal with Tl(I) only. While commonly Tl levels in plants are rather low ($\geq 0.1$ mg/kg), plants can readily absorb it owing to the similarity with K and Ag, and the fact that K can be completely and reversibly exchanged also with Rb in hydroponic cultures. Rape leaves have levels of several 100 ppm, as do certain parts of Scots pine *Pinus sylvestris*, and there are even higher concentrations in certain flowers, prompting studies to do not only phytoremediation but actually biomining of Tl, like with Ni, Au, and U. Known hyperaccumulator plants for Tl are among brassicaceae, cp. the tremendously high levels observed in green rape leaves. Table 2 gives some Tl concentrations found in various kinds of plants.

In the past, the content of Tl in lower or epigeic plants could be used as so-called bioindicators of observing the pollution status of the atmospheric environment (i.a. Djingova et al. 2001; Markert et al. 2003, 2011; Wolterbeek et al. 2003). Wappelhorst (1999) reported the Tl distribution in moss samples of *Polytrichum formosum* collected in the triangle of Germany, Czech Republic, and Poland. Table 2 represents some Tl-concentrations given by Wappelhorst (1999) in various kinds of plants.

Wierzbicka et al. (2004) determined Tl concentrations in four plant species (*Plantago lanceolata, Biscutella laevigata, Dianthus carthusianorum, Silene vulgaris*) that were collected in a highly polluted area (metal mining) near by Olkusz (southern Poland). Wierzbicka et al. (2004) report that *P. lanceolata* accumulated extremely large amounts of thallium (average, 65 mg Tl/kg dry wt.; maximum 321 mg Tl/kg dry wt. in roots). *S. vulgaris* and *D. carthusianorum* accumulated much less (averages, 10 and 6.5 mg Tl/kg dry wt., respectively). On the other hand, *B. laevigata* accumulated negligible amounts of thallium in its tissues. The concentration of thallium in plants (shoots, roots) from the calamine waste heap was 100–1,000 times the level normally found in plants (0.05 mg Tl/kg dry wt.).

## Resume

Regardless of all environmental abundance, catalytic versatility (platinum group metals!), or existence of very different chemical features (e.g., the possibility to "switch" from cation or complex formation to ligand properties upon simple alkylation [Sn, Pb, Bi]) among the speciation forms, no chemical element with $Z > 42$ (Mo) is essential for a really large number of living beings, the rare examples being Cd, I, Ba, and W.

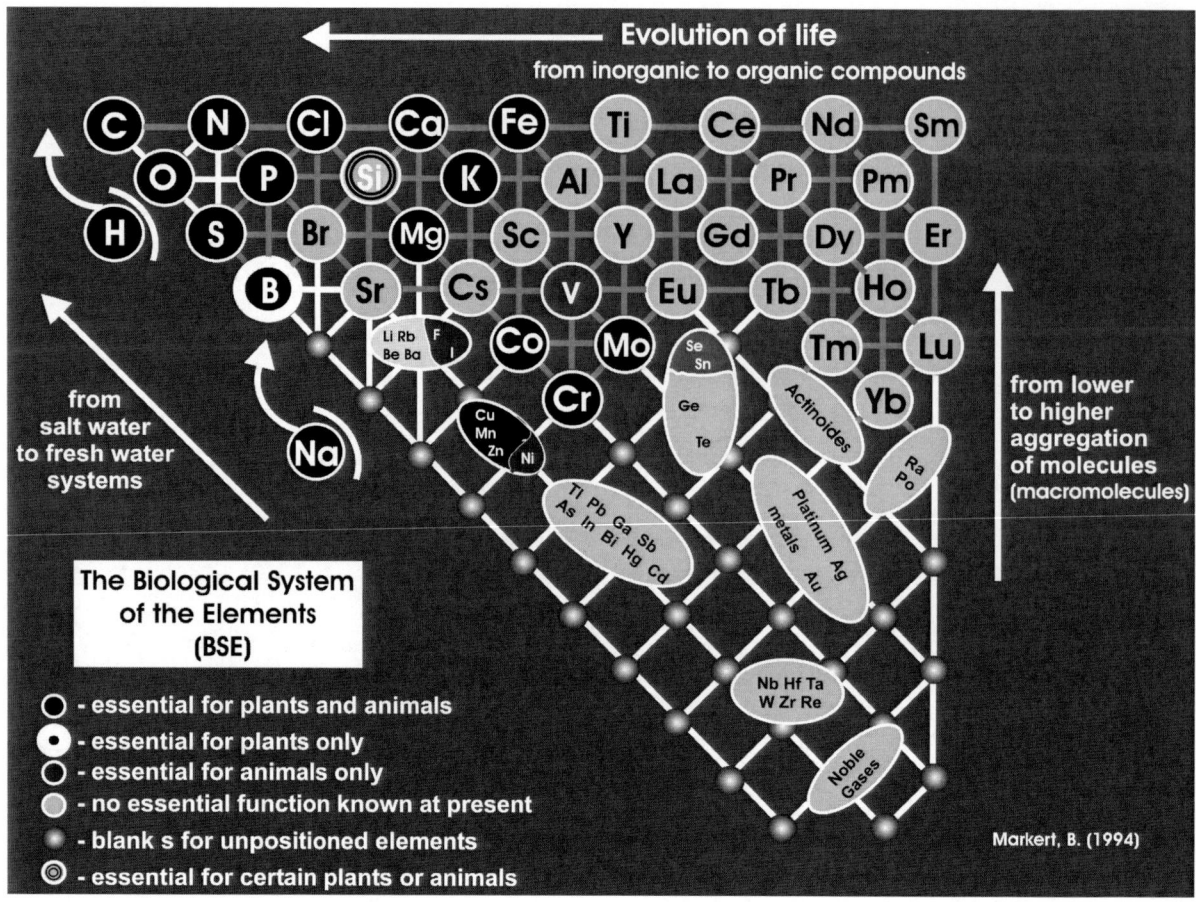

**Thallium, Distribution in Plants, Fig. 1** The Biological System of the Elements (BSE) from Markert (1994). The BSE compiled from data on correlation analysis, physiological function of the individual elements in the living organisms, evolutive development out of the inorganic environment, and with respect to their uptake form by the plant organism as a neutral molecule or charged ion. The elements H and Na exercise various functions in the biological system so that they are not conclusively fixed. The ringed elements can at present only be summarized as groups of elements with a similar physiological function since there is a lack of correlation data. The element Tl belongs to the category with "no essential function known at present"

Given this, the biological or biochemical behavior of Tl is fairly typical. Most notable are the differences concerning impact of Tl to different kingdoms, which are mainly due to the K/Tl antagonism hitting nerve activities (action potentials) which have no correlates in either plants, fungi, or bacteria (even though both green plants and bacteria need K). The peculiarities of heavy-element chemistry ($Z \approx 80$), which are due to very pronounced relativistic orbital contraction here, are also related to toxicology (while the other elements Pt through Bi, differently ready to form M-M bonds, show extreme differences in toxicity but are all susceptible to biomethylation at least in certain circumstances and organisms) and may be involved in bioaccumulation.

**Thallium, Distribution in Plants, Table 2** Tl concentrations in various kinds of plants

| Plants | Tl (dry matter) | References |
|---|---|---|
| Reference content in plants | 0.03–0.3 mg/kg | Markert (1992, 1996) |
| Moss *Pleurozium schreberi* | 0.05 µg/g | Wappelhorst (1999) |
| | 0.03 µg/g | Berg and Steinnes (1997) |
| Moss *Polytrichum formosum* | 0.02 µg/g | All data in Wappelhorst (1999) |
| Spruce needles (*Picea*) first year | 0.008 µg/g | |
| Spruce needles (*Picea*) second year | 0.02 µg/g | |
| Pine needles (*Pinus*) first year | 0.003 µg/g | |

## Cross-References

- Potassium in Biological Systems
- Metals and the Periodic Table
- Thallium, Distribution in Animals
- Thallium, Effects on Mitochondria

## References

Berg T, Steinnes E (1997) Use of mosses (*Hylocomium splendens* and *Pleurozium schreberi*) as biomonitors of heavy metal deposition: from relative to absolute deposition values. Environ Poll 98(1):61–71

Djingova R, Ivanova J, Wagner G, Korhammer S, Markert B (2001) Distribution of lanthanoids, Be, Bi, Ga, Te, Tl, Th and U on the territory of Bulgaria using *Populus nigra* 'Italica' as an indicator. Sci Total Environ 280:85–91

Fraenzle S, Markert B (2002) The Biological System of the Elements (BSE) – a brief introduction into historical and applied aspects with special reference to "ecotoxicological identity cards" for different elements (f.e. As and Sn). Environ Pollut 120:27–45

Fraenzle S, Markert B, Wuenschmann S (2012) Introduction to environmental engineering. Wiley-VCH, Weinheim

Fränzle S (2010) Chemical elements in plants and soil. Springer, Berlin

Markert B (1992) Multi-element analysis in plants – analytical tools and biological questions. In: Adriano DC (ed) Biogeochemistry of trace metals. Lewis, Boca Raton, pp 401–428

Markert B (1994) The Biological System of the Elements (BSE) for terrestrial plants (glycophytes). Sci Total Environ 155:221–228

Markert B (1996) Instrumental element and multielement analysis of plant samples – methods and applications. Wiley, Chichester

Markert B, Breure A, Zechmeister H (2003) General aspects and integrative approaches. In: Markert B, Breue T, Zechmeister H (eds) Bioindicators & Biomonitors. Priciples, Concepts & Applications. Elsevier, pp 3–39

Markert B, Wuenschmann S, Fraenzle S, Figueiredo A, Ribeiro AP, Wang M (2011) Bioindication of trace metals – with special reference to megacities. Environ Pollut 159:1991–1995

Sasmaz A, Sen O, Kaya G, Yaman M, Sagiroglu A (2007) Distribution of thallium in soil and plants growing in the Keban mining district of Turkey and determined by ICP-MS. Atomic Spectrosc 28(5):157–163

Schedlbauer OF, Heumann KG (2000) Biomethylation of thallium by bacteria and first determination of biogenic dimethylthallium in the ocean. Appl Organomet Chem 14:330–340

Wappelhorst O (1999) Charakterisierung atmosphärischer Depositionen in der Euroregion Neiße durch ein terrestrisches biomonitoring. PhD thesis, IHI Zittau

Wierzbicka M, Szarek-Łukaszewska G, Grodzińska K (2004) Highly toxic thallium in plants from the vicinity of Olkusz (Poland). Ecotoxicol Environ Saf 59(1):84–88

Wolterbeek HT, Garty J, Reis MA, Freitas MC (2003) Biomonitors in use: lichens and metal air pollution. In: Markert B, Breure AM, Zechmeister HG (eds) Bioindicators and biomonitors, principles, concepts and applications. Elsevier, Amsterdam, pp 377–419

Xiao T, Guha J, Boyle D, Liu CQ, Chen J (2004) Environmental concerns related to high thallium levels in soils and thallium uptake by plants in southwest Guizhou, China. Sci Total Environ 318(1–3):223–244

# Thallium, Effects on Mitochondria

Sergey M. Korotkov
Sechenov Institute of Evolutionary Physiology and Biochemistry, The Russian Academy of Sciences, St. Petersburg, Russia

## Synonyms

*Thallium*: Thallium(I), Tl(I), $Tl^+$; *Effects*: Biological action, Influence; *Mitochondria*: Intracellular organelles, Organoids

## Definitions

*Thallium*: Thallium is a metal with atomic number of 81 and two oxidation states of +1 and +3. Thallium belongs to the group of trace elements and is used in the electronics industry, in the pharmaceutical industry (with care), and in glass manufacturing. Thallium is highly toxic and was used in rat poisons and insecticides.

*Mitochondria*: Mitochondria are two-membrane granular or prolate organelles of 0.5 μm size which are located between the cytoplasm and nuclear membranes of the majority eukaryotic cells of both autotroph (photosynthesizing plants) and heterotroph (mushrooms and animal) organisms. The basic function of mitochondria is oxidation of organic compounds following use of energy, released at their disintegration, in creation of both the proton gradient and electrochemical potential on the inner mitochondrial membrane for the purpose of ATP synthesis which occurs by the mitochondrial H(+)-ATP-synthase.

## Additional Definitions

*Mitochondrial swelling:* Mitochondrial swelling was evaluated as a decrease in $A_{540}$ at 20°C using a SF-46 spectrophotometer (LOMO, St. Petersburg, Russia).

*Mitochondrial respiration:* Mitochondrial respiration (oxygen consumption rate) was measured polarographically by using LP-7 (Czechoslovakia) in a 1.5-ml closed thermostatic chamber with magnetic stirring at 26°C.

*Respiratory states:* Respiratory states have been defined by Chance and Williams (1956) according to a protocol for oxygraphic experiments with isolated mitochondria. State 3 is defined as the state after addition of ADP and state 4 as one after phosphorylation of all ADP to ATP.

*Mitochondrial membrane potential:* Mitochondrial membrane potential ($\Delta\Psi_{mito}$) was evaluated as an intensity of safranin fluorescence (arbitrary units) in the mitochondrial suspension with magnetic stirring at 20°C using a Shimadzu RF-1501 spectrophotofluorimeter (Shimadzu, Germany) at 485/590 nm wavelength (excitation/emission).

## Effects of Thallium on Cells and Living Organisms

Thallium is a highly toxic metal which belongs to a group of trace elements. Human toxicity was found in use of Tl compounds as a human depilatory (thallium acetate), a component in manufacturing of optical glasses (thallium oxide), a homicidal agent, and a rodenticide (thallium sulfate) to kill rats, mice, and other animals. After penetrating into an organism, Tl damages cardiovascular, central nervous, and renal systems as well as the gastrointestinal system and skin, and it results in hair loss (Goel and Aggarwal 2007 and references herein). It has been shown that the essence of the harmful effects of $Tl^+$ on living organisms lies in its ability both to easily penetrate the inner mitochondrial membrane (Saris et al. 1981; Korotkov et al. 2008 and #3 ref. herein) and to substitute $K^+$ in $K^+$-dependent enzymes and biochemical processes (Douglas et al. 1990). These effects of $Tl^+$ resulted in both the proximity of crystal-chemical radii of $K^+$ and $Tl^+$, and easy thallium polarizability that allows $Tl^+$ to form manifold chemical bonds with reactive groups of molecules which constitute living organisms.

Experiments with rats, exposed to chronic thallium intoxication in vivo, showed that $Tl^+$ stimulated massive mitochondrial swelling which was followed by disruption of mitochondrial and other intracellular membranes of kidney, liver, brain, and other organs (Korotkov et al. 2008, refs. of #9 and #10 herein). $Tl^+$ has triggered apoptosis in Jurkat and PC12 cells (Bragadin et al. 2003; Hanzel and Verstraeten 2009). It was postulated earlier that $Tl^+$ can stimulate release of $Ca^{2+}$ from intracellular compartments (Korotkov et al. 2008, #9 ref. herein). Increase of cytoplasmic concentration of $Ca^{2+}$, $Na^+$, or $P_i$ and decrease of one of $K^+$ were both found in experiments with isolated rat hepatocytes in medium containing TlCl. Interaction of $Tl^+$ with SH groups of mitochondrial and cellular membranes, glutathione depletion, and the increased production of reactive oxygen species can be among other reasons of the thallium toxicity (Zierold 2000; Korotkov et al. 2008, #9 ref. herein; Hanzel and Verstraeten 2009). It was quite recently shown that $Tl^+$ could injure isolated rat hepatocytes that resulted in hepatocyte proteolysis, in glutathione depletion, in decline of the inner mitochondrial membrane potential ($\Delta\Psi_{mito}$), in ROS formation, and in lipid peroxidation (Pourahmad et al. 2010).

## Influence of $Tl^+$ on Isolated Mitochondria

The study of swelling of isolated mitochondria in nitrate media showed that the inner mitochondrial membrane (IMM) is poorly penetrated by univalent cations such as $H^+$, $K^+$, and $Na^+$. However, visible swelling of the mitochondria in $TlNO_3$ medium (Fig. 1) exposed substantial permeability of the membrane to $Tl^+$ (Saris et al. 1981; Korotkov et al. 2008). Subsequent energization of the mitochondria stimulated their massive contraction which occurred by means of a $Tl^+/H^+$ exchange mechanism (Saris et al. 1981; Korotkov et al. 2008). Nonenergized mitochondria in Tl acetate medium showed massive swelling that was realized by means of $Tl^+/H^+$ exchange (Saris et al. 1981; Korotkov et al. 2007). Further energization of the mitochondria stimulated additional swelling due to an electrophoretic uniport of $Tl^+$ into the matrix (Saris et al. 1981; Korotkov et al. 2007; Korotkov et al. 2008, #3 ref. herein). The participation of the mitochondrial $K_{ATP}$-dependent channel in the electrophoretic uniport of $Tl^+$ in the matrix was demonstrated

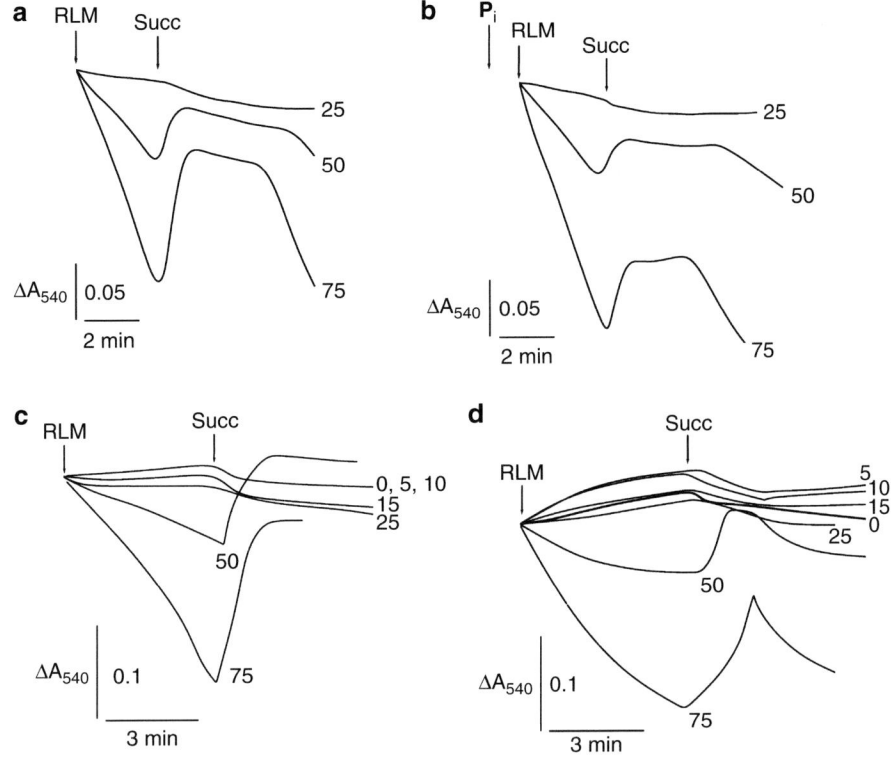

**Thallium, Effects on Mitochondria, Fig. 1** Effects of Tl$^+$ on swelling of rat liver mitochondria in a 160 mOsm nitrate medium. Mitochondria (1.5 mg protein/ml) were added to the medium containing 5–75 mM TlNO$_3$, as well as 5 mM Tris-NO$_3$ (pH 7.3), 1 mM Tris-PO$_4$ (**b**), 10$^{-8}$ M nonactin (**d**), 4 μM rotenone, and 3 μg/ml of oligomycin. To obtain 160 mOsm, sucrose is added to the medium. Additions of mitochondria (RLM) and 5 mM succinate (Succ) are shown by *arrows*. Typical traces for three different mitochondrial preparations are presented. Concentrations of TlNO$_3$ (mM) are shown on the *right* of the traces (This material is reproduced with the courteous permission of John Wiley & Sons, Inc. See Korotkov et al. 2008 and #6 ref. herein)

in both swelling and fluorescent experiments (Wojtovich et al. 2010 and #45 ref. herein). Tl$^+$ resulted in both the increase of state 4 respiration of rat liver mitochondria (Fig. 2), and potent futile cycling of Tl$^+$ via the IMM (Bragadin et al. 2003; Korotkov et al. 2007; Korotkov et al. 2008 and #3 ref. herein). Nonactin, a cyclic ionophore, and inorganic phosphate (Saris et al. 1981; Korotkov et al. 2007; Korotkov et al. 2008 and #12 ref. herein) have both facilitated transport of Tl$^+$ into mitochondria and increased both state 4 and swelling of mitochondria (Figs. 1 and 2). With lower affinity to molecular SH groups, Tl$^+$ unlike bivalent heavy metals (Cd$^{2+}$, Zn$^{2+}$, Hg$^{2+}$, Cu$^{2+}$, and Pb$^{2+}$) has not inhibited state 3 or 2,4-dinitrophenol (DNP)-stimulated respiration in medium containing TlNO$_3$ or Tl acetate (Figs. 1 and 2) owing to the lack of inhibition of mitochondrial respiratory enzymes by thallium (Korotkov et al. 2007; Korotkov et al. 2008 and refs. of #3 and #10 herein; Korotkov 2009).

## Tl$^+$ Increases the Permeability of the Inner Mitochondrial Membrane to Univalent Cations (H$^+$, K$^+$, Na$^+$, and Li$^+$)

It was quite recently shown in experiments using media containing nitrate salts of univalent cations and TlNO$_3$ (Korotkov 2009; Korotkov and Saris 2011) that Tl$^+$ similar to the bivalent heavy metals increased the permeability of the inner membrane of rat liver mitochondria to univalent cations (H$^+$, K$^+$, Na$^+$, and Li$^+$). The Tl$^+$-induced increase of the permeability (for more details, see Korotkov 2009 and

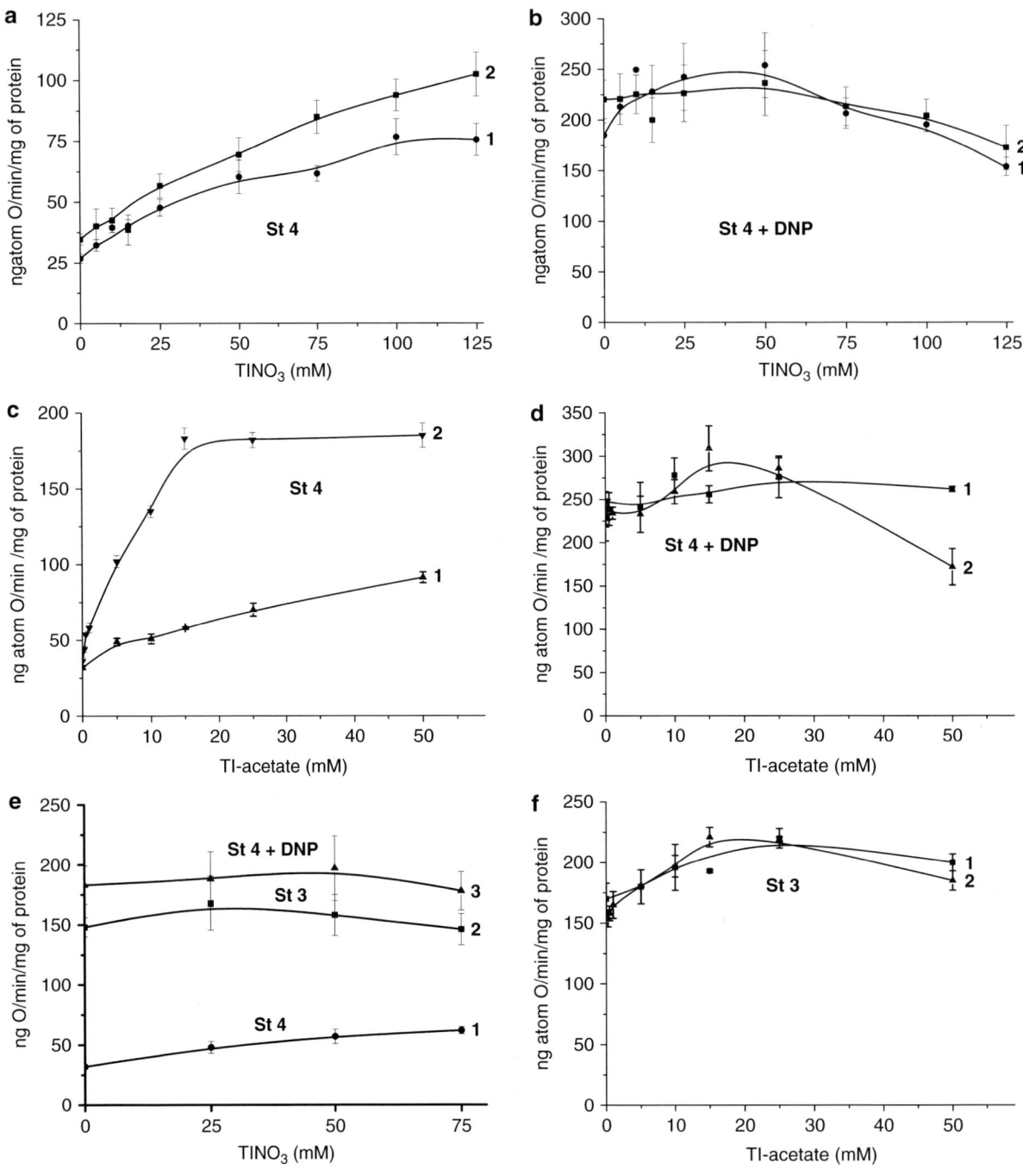

**Thallium, Effects on Mitochondria, Fig. 2** Effects of Tl$^+$ on oxygen consumption rates (ng atom O min/mg of protein) of rat liver mitochondria in a 290 mOsm medium. Mitochondria (1.5 mg protein/ml) were suspended in the medium containing 5 mM Tris-NO$_3$ (pH 7.3) and 0–125 mM TlNO$_3$ (**a, b, e**) or 5 mM Tris acetate (pH 7.3) and 0–50 mM Tl acetate (**c, d, f**) as well as 3 mM Mg(NO$_3$)$_2$ (**c–f**), 5 mM succinate, and 4 μM rotenone. Additionally the medium was supplemented with 3 mM Tris-PO$_4$ (**a–b** [trace 2], and **e–f**) and 10$^{-8}$ M nonactin (**c–d, f** [trace 2]). To obtain 290 mOsm, sucrose was added to the medium; 130 μM ADP or 30 μM DNP was added to the media after 2 min recording of state 4 to induce state 3 or DNP-stimulated respiration. Error bars were calculated by the Muller formula from rates found for three different mitochondrial preparations (see Korotkov 2009) (This material is reproduced with the courteous permission of both John Wiley & Sons, Inc. (**a–d, f**) and Springer (**e**). See Korotkov et al. 2008 and #6 ref. herein; Korotkov 2009)

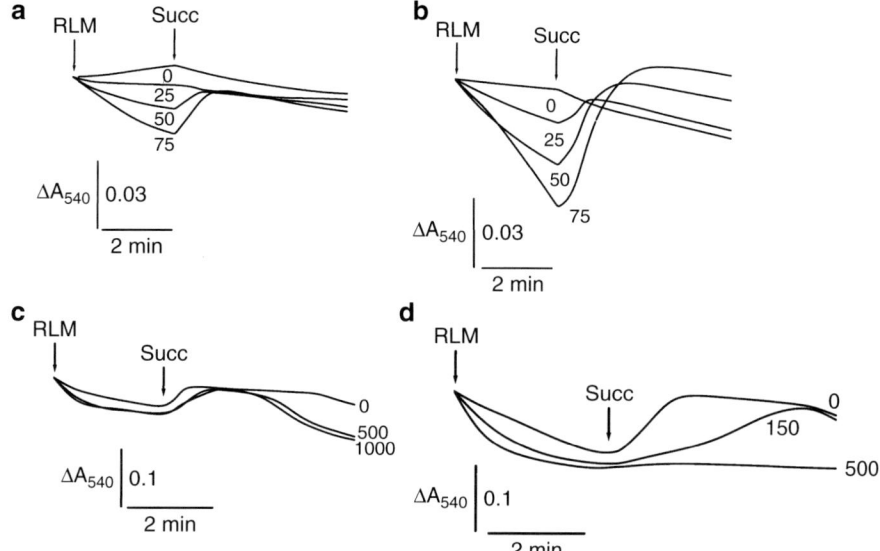

**Thallium, Effects on Mitochondria, Fig. 3** Effects of Tl$^+$ on swelling of rat liver mitochondria in a 400 mOsm medium. Mitochondria (1.5 mg protein/ml) were added to the medium containing 0–75 mM TlNO$_3$ (**a–b**) or 75 mM TlNO$_3$ and 50–1,000 μM quinine (**c–d**) as well as 5 mM Tris-NO$_3$ (pH 7.3), 250 mM sucrose (**a, c**), 125 mM of KNO$_3$ (**b, d**), 4 μM rotenone, and 3 μg/ml of oligomycin. To obtain 400 mOsm, sucrose was added additionally to the medium. Numbers near the traces show concentrations of TlNO$_3$ (mM) or quinine (μM) on panels A and B or C and D, correspondingly. Additions of mitochondria (RLM) and 5 mM succinate (Succ) are shown by *arrows*. Typical traces for three different mitochondrial preparations are presented (This material is reproduced with the courteous permission of Springer. Korotkov 2009)

Korotkov and Saris 2011) was showed by the risen swelling of both non-energized (Fig. 3) and energized mitochondria, and by the accelerated dissipation of $\Delta\Psi_{mito}$ (Fig. 4). Contraction of succinate-energized mitochondria, swollen in the nitrate medium (Fig. 3d) but not sucrose one (Fig. 3c), was inhibited by quinine, which blocks mitochondrial K$^+$/H$^+$ exchange (Korotkov 2009). The participation of the exchanger in extruding the Tl$^+$-induced excess concentration of the univalent cations from the matrix was early hypothesized (Saris et al. 1981; Korotkov et al. 2008). The Tl$^+$-induced swelling of succinate-energized mitochondria resulted in the decrease of state 3 and 2,4-dinitrophenol (DNP)-stimulated respiration (Fig. 4b) (more detail see Korotkov 2009).

## Thallium Induces the Permeability Transition in the Inner Membrane of Ca$^{2+}$-Loaded Rat Liver Mitochondria

It is known that binding of Ca$^{2+}$ with matrix calcium-specific trigger sites, located near the adenine nucleotide translocase (ANT), with following fall of $\Delta\Psi_{mito}$ triggers opening of the mitochondrial permeability transition pore (MPTP) in the inner membrane which becomes permeable to molecules up to 1,500 kDa. The pore opening results in massive mitochondrial swelling, lowering of the matrix concentration of Ca$^{2+}$ and ATP, and dissipation of the $\Delta\Psi_{mito}$. If the Ca$^{2+}$ sites are insufficiently saturated, MPTPs are opened in the low conductance state and small molecules up to 300 kDa or ions (H$^+$, K$^+$, and Ca$^{2+}$) may penetrate easily the IMM. In the earlier times, it was believed that the MPTP is formed by ANT, cyclophilin D (CyP-D), and the voltage-dependent anion channel (VDAC). In modern times, many researchers consider the mitochondrial phosphate carrier and CyP-D to be primary components of the MPTP, whereas ANT is viewed as a regulatory part of the MPTP.

It was recently showed that Tl$^+$ has induced opening of the MPTP in Ca$^{2+}$-loaded rat liver mitochondria (CaRLM) energized by glutamate *plus* malate or succinate which are substrates of I and II respiratory complexes, respectively (Figs. 5–7) (for more details,

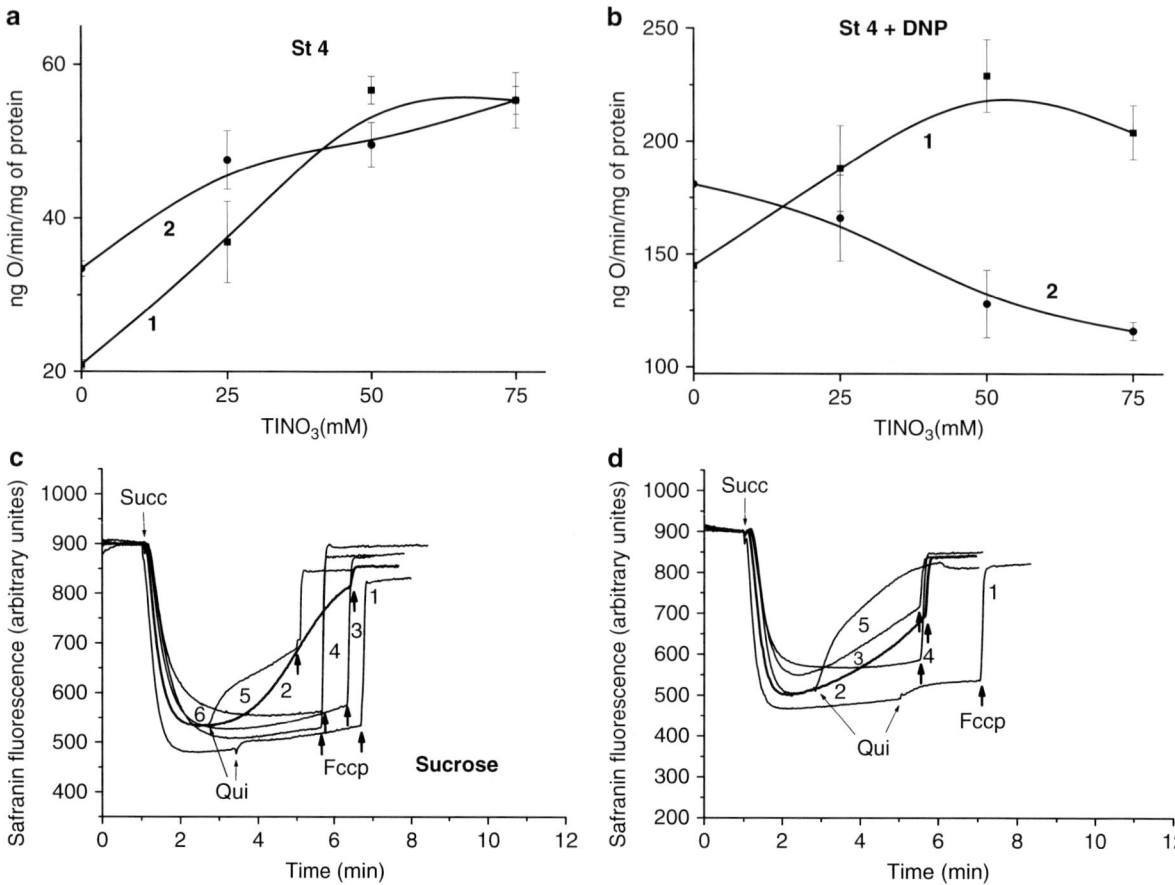

**Thallium, Effects on Mitochondria, Fig. 4** Effects of TlNO₃ on oxygen consumption rates (ng atom O min/mg of protein) and the inner membrane potential ($\Delta\Psi_{mito}$) of rat liver mitochondria in a 400 mOsm medium. Mitochondria (0.5 (**c–d**) or 1.5 (**a–b**) mg protein/ml) were suspended in the medium containing: 0–75 mM (**a–b**) or 30 mM (**c–d**) of TlNO₃, 5 mM Tris-NO₃ (pH 7.3), 250 mM sucrose (**a–b** [trace 1], and **c**), 125 mM KNO₃ (**a–b** [trace 2], and **d**), 5 mM succinate (**a–b**), 4 μM rotenone, and 3 μg/ml of oligomycin, as well as 1 mM Tris-PO₄ (**c–d**) and 3 μM safranin (**c–d**). To obtain 400 mOsm, sucrose was added additionally to the medium (Figs. 4–7). DNP of 30 μM (**b**) was added to the medium to trigger DNP-stimulated respiration after 2 min recording of state 4 (**a**). Error bars (**a–b**) were calculated by the Muller's formula from rates found for three different mitochondrial preparations. Additions before mitochondria (**c–d**) were as follows: 1, none (free of TlNO₃); 2, none (marked in *bold*); 3, 5 mM Mg(NO₃)₂; 4, 2 mM ADP; 5, 500 μM (**d**) or 1,000 μM (**c**) quinine [traces 1 and 5]; 6, 0.5 mM ADP, 1 mM Mg(NO₃)₂, and 1 μM CsA. Typical traces for three different mitochondrial preparations are presented (This material is reproduced with the courteous permission of Springer. Korotkov 2009)

see Korotkov and Saris 2011). This effect of Tl⁺ in the nitrate media resulted in both the increased swelling (Figs. 5 and 6), as well as the accelerated dissipation of $\Delta\Psi_{mito}$ and the decreased state 4, or state 3, or DNP-stimulated respiration (Fig. 7). It should be stressed that the swelling of nonenergized mitochondria was stimulated by both Tl⁺ and Ca²⁺ in the media free of rotenone (Fig. 5). However, the swelling was increased by Tl⁺ but not Ca²⁺ in the presence of rotenone (Fig. 6).

It is possible that these differences may be due to the induction of energy-dependent uptake of Ca²⁺ together with the participation of mitochondrial endogenous substrates under conditions in which pyridine nucleotides are in a more oxidized state. One can see on Figs. 5 and 6 that Tl⁺ distinct from the bivalent heavy metals can stimulate opening of the MPTP only in the presence of Ca²⁺. The most probable reason is: Tl⁺ reacts poorly with the Ca²⁺-binding sites as it has

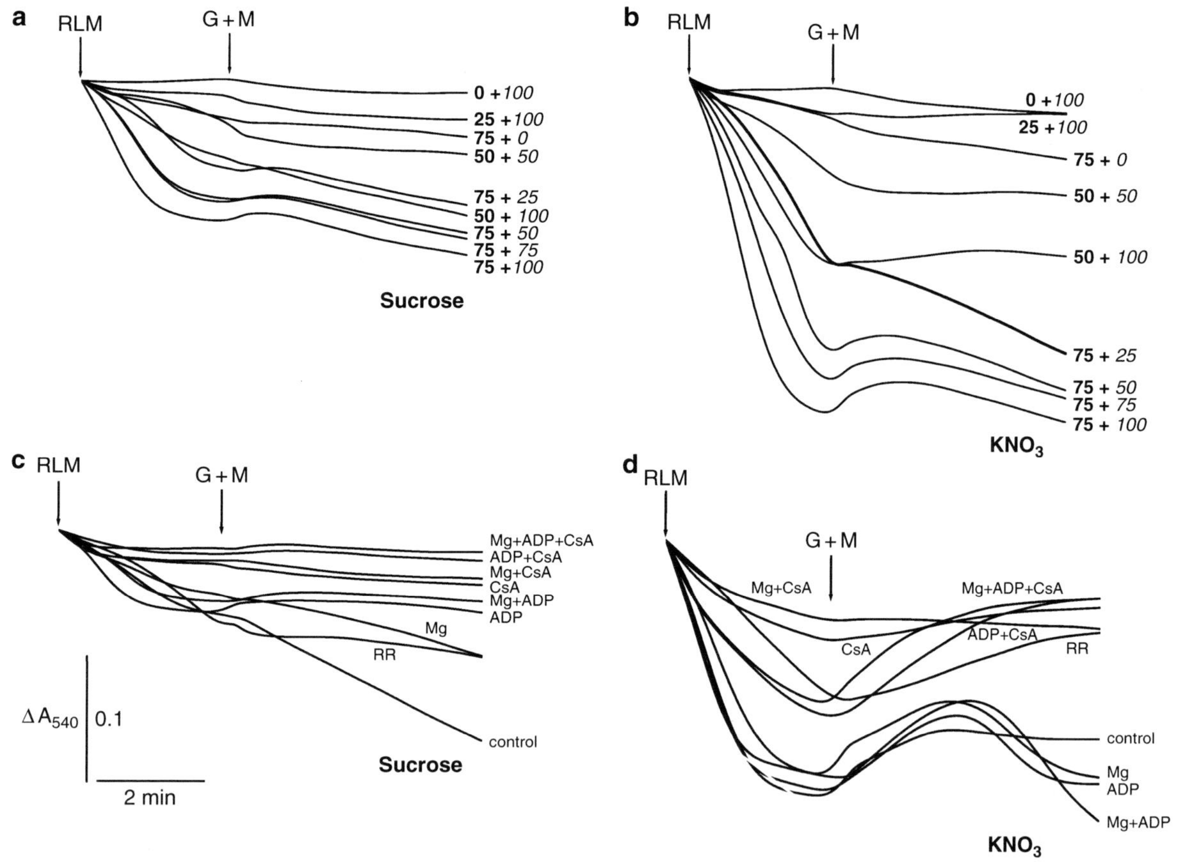

**Thallium, Effects on Mitochondria, Fig. 5** Effects of Tl$^+$ and Ca$^{2+}$ on swelling of rat liver mitochondria in the presence of glutamate and malate. Mitochondria (1.5 mg protein/ml) were added to the 400 mOsm medium containing 0–75 mM (**a–b**) or 50 mM (**c–d**) of TlNO$_3$, 5 mM Tris-NO$_3$ (pH 7.3), 0–100 μM (**a–b**) or 100 μM (**c–d**) of Ca$^{2+}$, and 1 μg/ml of oligomycin, as well as 250 mM sucrose (**a**) or 125 mM KNO$_3$ (**b**). The numbers on the right of the traces (**a–b**) show concentrations of TlNO$_3$, (mM) [in *bold*] or CaCl$_2$ (μM) [in *italics*] in this medium. Additions before mitochondria are indicated on the right of the traces (**c–d**): none (control); 1 μM CsA (CsA); 0.5 mM ADP (ADP); 3 mM Mg$^{2+}$ (Mg); 7 μM RR (RR). Injections of mitochondria (RLM) and 5 mM of glutamate and malate (G + M) are shown by *arrows*. Typical traces for three different mitochondrial preparations are presented (This material is reproduced with the courteous permission of Springer. Korotkov and Saris 2011)

a single charge and shows comparatively low affinity to molecular SH groups. The phenomenon of the substrate specificity manifested in the fact that the total concentrations of Ca$^{2+}$ stimulated the maximum swelling of mitochondria energized by glutamate *plus* malate (Fig. 5), but not by succinate *plus* rotenone (Fig. 6). The possible roles of Ca$^{2+}$-binding sites, located near the respiratory complex I, and the ANT in inducing opening of the MPTP are discussed (Korotkov and Saris 2011).

It was discovered that the Tl$^+$-induced MPTP in CaRLM was potentiated by inorganic phosphate and diminished by the MPTP inhibitors (ADP, CsA, Mg$^{2+}$, Li$^+$, rotenone, EGTA, and ruthenium red) (Korotkov et al. 2008; Korotkov 2009; Korotkov and Saris 2011). The Tl$^+$-induced swelling of CaRLM, energized by glutamate *plus* malate (Fig. 5) or succinate in the presence of rotenone (Fig. 6), was reduced by ADP, CsA, Mg$^{2+}$, or ruthenium red, an inhibitor of the mitochondrial Ca$^{2+}$ uniporter. The MPTP inhibitors prevented both the decreased DNP-stimulated respiration and the fall of $\Delta\Psi_{\text{mito}}$ in succinate-energized CaRLM (Fig. 7). Maximal effect was found in the simultaneous presence of ADP and CsA in the

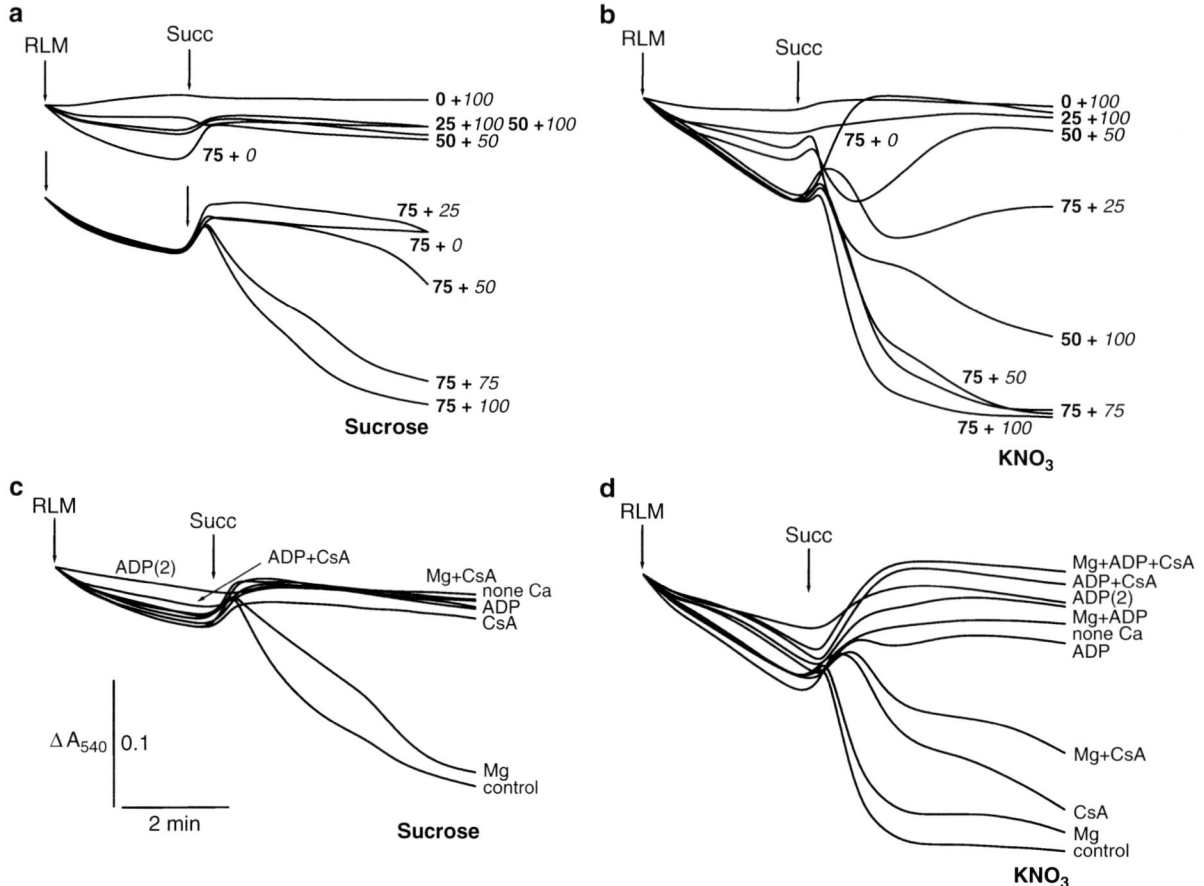

**Thallium, Effects on Mitochondria, Fig. 6** Effects of Tl$^+$ and Ca$^{2+}$ on swelling of rat liver mitochondria in the presence of succinate. The medium, additions, and designations are as shown in the Fig. 5. Exceptions from Fig. 5 are that 5 mM succinate and 2 μM rotenone were used there instead of 5 mM of glutamate *plus* malate; 75 mM TlNO$_3$ was used in experiments with the MPTP inhibitors (**c–d**). Typical traces for three different mitochondrial preparations are presented (This material is reproduced with the courteous permission of Springer. Korotkov and Saris 2011)

media (Figs. 5–7). It was suggested that swelling of energized mitochondria in the nitrate media containing Ca$^{2+}$ and Tl$^+$ may be caused by opening of CsA-inhibited and ADP-dependent pores in the IMM (for more details, see Korotkov and Saris 2011). The Tl$^+$-induced swelling was inhibited by both Li$^+$ (Korotkov and Saris 2011) and Mg$^{2+}$ (Figs. 5–7) that is possibly related to the involvement of Ca$^{2+}$ sites on the outer surface of the IMM in opening the MPTP. The decrease of state 4 and DNP-stimulated respiration of CaRLM (Fig. 7) is related to the increased swelling of the mitochondria (Fig. 6) that may be associated with the reduced activity of the respiratory enzymes because the mitochondrial structure can be disturbed by the more massive swelling of CaRLM (Korotkov and Saris 2011). Comparing experiments with the sucrose and KNO$_3$ media (Figs. 5 and 6), one can conclude that Tl$^+$, similar to Cd$^{2+}$ and Ca$^{2+}$, induces opening of the MPTP less actively in the high conduction states and furthermore, that this effect of Tl$^+$ occurs in the presence of the substrates of respiratory complex I (glutamate plus malate) and II (succinate). Thus, it

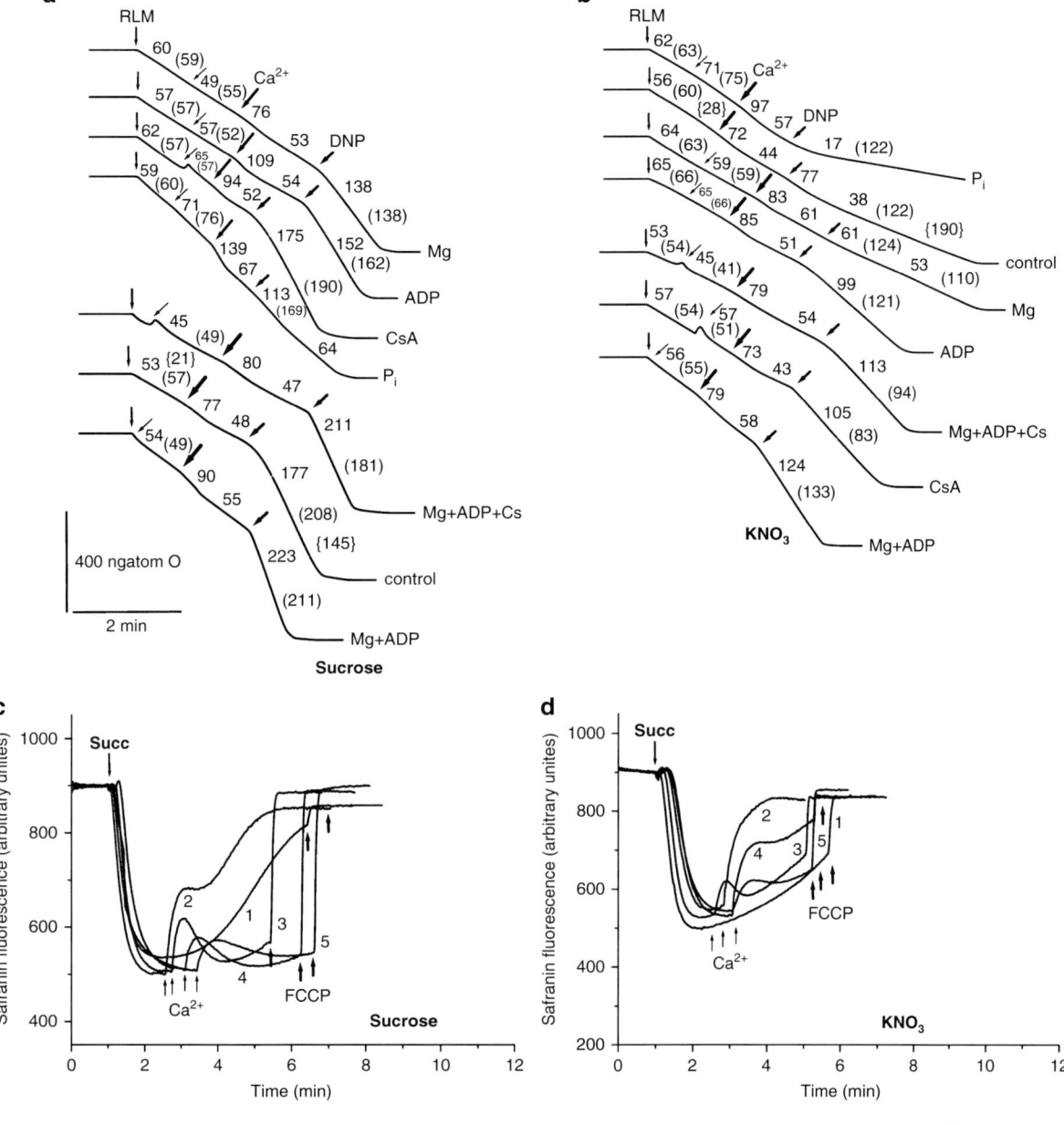

**Thallium, Effects on Mitochondria, Fig. 7** Effects of TlNO₃ and Ca²⁺ on oxygen consumption rates (ng atom O min/mg of protein) and the inner membrane potential (ΔΨ$_{mito}$) of rat liver mitochondria in a 400 mOsm medium. The medium, additions, and designations are as in the Fig. 4. Exception from Fig. 4 is that additions of trace 5 on Fig. 7 are the same ones of trace 6 on Fig. 4. The rates are presented as numbers placed above experimental traces. Numbers in parentheses were obtained from experiments with Ca²⁺-free media. The numbers in braces were calculated from experiments with the Ca²⁺-free media, where 75 mM TlNO₃ was substituted by 150 mM sucrose (Korotkov 2009). Accessorial additions of 30 μM (**d**) and 50 μM (**c**), or 100 μM (**a–b**) [traces 2–5] of Ca²⁺ (Ca²⁺) are correspondingly shown by short (**c–d**) or long *bold* (**a–b**) *arrows*. Typical traces for three different mitochondrial preparations are presented (This material is reproduced with the courteous permission of Springer. Korotkov and Saris 2011)

can be postulated that opening of the MPTP can play an important role in the development of toxic processes during thallium intoxication in living organisms (Bragadin et al. 2003; Korotkov et al. 2008; Korotkov 2009; Pourahmad et al. 2010; Korotkov and Saris 2011).

## Cross-References

▶ $Ca^{2+}$
▶ Sodium-Hydrogen Exchangers, Structure and Function in Human Health and Disease
▶ Tobacco

## References

Bragadin M, Toninello A, Bindoli A et al (2003) Thallium induces apoptosis in Jurkat cells. Ann N Y Acad Sci 1010:283–291
Chance B, Williams GR (1956) The respiratory chain and oxidative phosphorylation. Adv Enzymol 17:65–134
Douglas KT, Bunni MA, Baindur SR (1990) Thallium in biochemistry. Int J Biochem 22:429–438
Goel A, Aggarwal P (2007) Pesticide poisoning. Natl Med J India 20:182–191
Hanzel CE, Verstraeten SV (2009) Tl(I) and Tl(III) activate both mitochondrial and extrinsic pathways of apoptosis in rat pheochromocytoma (PC12) cells. Toxicol Appl Pharmacol 236:59–70
Korotkov SM, Glazunov VV, Yagodina OV (2007) Increase in toxic effects of $Tl^+$ on isolated rat liver mitochondria in the presence of nonactin. J Biochem Mol Toxicol 21:81–91
Korotkov SM (2009) Effects of $Tl^+$ on ion permeability, membrane potential and respiration of isolated rat liver mitochondria. J Bioenerg Biomembr 41:277–287
Korotkov SM, Saris NE (2011) Influence of $Tl^+$ on mitochondrial permeability transition pore in $Ca^{2+}$-loaded Rat liver mitochondria. J Bioenerg Biomembr 43:149–162
Korotkov SM, Emel'yanova LV, Yagodina OV (2008) Inorganic phosphate stimulates the toxic effects of $Tl^+$ in rat liver mitochondria. J Biochem Mol Toxicol 22:148–157
Pourahmad J, Eskandari MR, Daraei B (2010) A comparison of hepatocyte cytotoxic mechanisms for thallium (I) and thallium (III). Environ Toxicol 25:456–467
Saris NE, Skulskii IA, Savina MV et al (1981) Mechanism of mitochondrial transport of thallous ions. J Bioenerg Biomembr 13:51–59
Wojtovich AP, Williams DM, Karcz MK et al (2010) A novel mitochondrial K(ATP) channel assay. Circ Res 106:1190–1196
Zierold K (2000) Heavy metal cytotoxicity studied by electron probe X-ray microanalysis of cultured rat hepatocytes. Toxicol In Vitro 14:557–563

# Thallium, Physical and Chemical Properties

Fathi Habashi
Department of Mining, Metallurgical, and Materials Engineering, Laval University, Quebec City, Canada

Thallium is a silvery white metal, considered a less typical metal because when it loses its outermost electrons it will not achieve the inert gas electronic structure. It occurs as a trace element in pyrite, and is extracted as a by-product of roasting this mineral for the production of sulfuric acid. Thallium can also be obtained from the smelting of lead and zinc sulfide ores. The fresh surface of the metal has a bluish-white luster. In many of its physical properties, thallium resembles its neighbor, lead. It exists in two allotropic forms: α-thallium, hexagonal close-packed, stable at room temperature; and β-thallium, body-centered cubic, stable above 226°C. A volume increase of 3.23% takes place on solidification.

Thallium and its compounds are extremely toxic. The odorless and tasteless thallium sulfate was once widely used as rat poison and ant killer. Thallium (I) sulfide's electrical conductivity changes with exposure to infrared light therefore making this compound useful in photoresistors. Thallium selenide has been used in a bolometer for infrared detection.

## Physical Properties

| Atomic number | 81 |
|---|---|
| Atomic weight | 204.38 |
| Relative abundance in Earth's crust, % | $3 \times 10^{-5}$ |
| Density, g cm$^{-3}$ | 11.85 |
| Melting point, °C | 303 |
| Boiling point, °C | 1,473 |
| Atomic radius, pm | 170 |
| Heat of fusion, kJ mol$^{-1}$ | 4.14 |
| Heat of vaporization, kJ mol$^{-1}$ | 165 |
| Molar heat capacity, J mol$^{-1}$ K$^{-1}$ | 26.32 |
| Electrical resistivity at 20°C, μΩ m | 0.18 |
| Thermal conductivity, W m$^{-1}$ K$^{-1}$ | 46.1 |
| Thermal expansion at 25°C, μm m$^{-1}$ K$^{-1}$ | 29.9 |
| Young's modulus, GPa | 8 |

(continued)

| | |
|---|---|
| Shear modulus, GPa | 2.8 |
| Bulk modulus, GPa | 43 |
| Poisson ratio | 0.45 |
| Mohs hardness | 1.2 |
| Brinell hardness, MPa | 26.4 |

## Chemical Properties

Thallium is a reactive metal. It oxidizes slowly in air, even at room temperature forming thallium (I) and thallium (III) oxides. In the presence of water, a hydroxide is formed. Thallium is therefore stored under petroleum spirit, glycerine, or de-aerated water. It combines with fluorine, chlorine, and bromine at room temperature, and with iodine, sulfur, phosphorus, selenium, and tellurium on heating. The metal is dissolved slowly by hydrochloric and dilute sulfuric acids, but more rapidly by nitric and concentrated sulfuric acids. The two main oxidation states of thallium are +1 and +3.

$Tl_2O$, is formed by heating thallium in air at <100°C. It is reddish-black to black, volatile in air at higher temperatures, being converted to $Tl_2O_3$, and is hygroscopic, forming a colorless, alkaline solution of TlOH with water. Thallium (III) oxide, $Tl_2O_3$, can also be formed by treating thallium (III) salts with KOH or $NH_3$, or treating thallium(I) salts with oxidizing agents such as NaOCl. It is brown to black, forms cubic crystals. Thallium is precipitated from solutions by zinc amalgam, and the relatively pure metal is obtained by an electrolytic process. Metal of the highest purity (99.9999%) can be obtained by multistage electrorefining.

Thallium (I) chloride, TlCl, is formed when HCl or alkali metal chlorides are added to neutral thallium (I) salt solutions. The precipitate is white, becoming gray-brown to blackish-brown on exposure to light. It reacts with ammonium sulfide to form $Tl_2S$:

$$2TlCl + (NH_4)_2S \rightarrow Tl_2S + 2NH_4Cl$$

Thallium (I) hydroxide reacts with carbon dioxide forming water-soluble thallium carbonate. This is the only water soluble heavy metal carbonate. Thallium (I) bromide is a photosensitive yellow compound similar to silver bromide.

## References

Habashi F (2001) Cadmium, indium, and thallium production. In: K. H. Jürgen Buschow, Robert W. Cahn, Merton C. Flemings, Bernard Ilschner (print), Edward J. Kramer, Subhash Mahajan, & Patrick Veyssière (Eds.), Encyclopedia of materials: science and technology, pp 879–880

Micke H, Wolf HU (1997) In: Habashi F (ed) Handbook of extractive metallurgy. Wiley-VCH, Weinheim, pp 1543–1556

## *Thallium*: Tl

▶ Thallium, Distribution in Plants

## Thallium-201 Imaging

Hakan Demir
Department of Nuclear Medicine, School of Medicine, Kocaeli University, Umuttepe, Kocaeli, Turkey

## Synonyms

Thallium myocardial perfusion imaging; Thallium myocardial perfusion scintigraphy; Thallium myocardial perfusion SPECT (single photon emission computed tomography); Thallium stress test; Thallium tumor imaging; Thallium tumor scintigraphy

## Definition

201 Thallium (Tl) imaging is a nuclear medicine diagnostic imaging method, called scintigraphy. There are two types of thallium imaging in nuclear medicine: tumor and myocardial imaging. Primary and/or metastasis of a tumor are investigated in thallium tumor imaging. In myocardial imaging, which is a more known type of thallium imaging, blood supply and viability of myocardium (heart muscle) is evaluated.

## Principles and Methodology

### 201 Thallium (Tl-201)

201 Thallium (Tl-201) is an unstable isotope of thallium element. It is used for thallium imaging

in nuclear medicine. Tl-201 decays by electron capture to mercury-201, emitting mainly X-rays of energy 67–82 keV (88% abundance) and gamma photons of 135 and 167 keV (12% abundance) (Hesse et al. 2005). Tl-201 has 73 h of physical half-life (t ½). It is produced in the cyclotron. It is administered intravenously to the patients as thallous chloride. After intravenous injection, about 88% is cleared from the blood after the first circulation. Almost 4% of the injected activity localizes in the myocardium. Although thallium belongs to group IIIA, thallous ion behaves like $K^+$ because they are both monovalent and have similar ionic radii (Saha 2010a, b). Approximately 60% enters the cardiac myocytes using the sodium–potassium ATPase-dependent exchange mechanism, and the remainder enters passively along an electropotential gradient (Hesse et al. 2005). Viable myocardial cells have normal $Na^+/K^+$ ion exchange pumps. Thallium binds the $K^+$ pumps and is transported into the cells. The amount of thallium that enters the tissues is proportional to the tissue blood supply.

## History

About 30 years ago, in early studies, K-43 was used in animal models and in man for myocardial perfusion imaging. In the mid-1970s, with the availability of Tl-201, planar myocardial imaging became widely utilized for the diagnosis of coronary artery disease. In the mid-1980s, SPECT was introduced and became the predominant method of acquisition of myocardial perfusion imaging. Initially, exercise stress was used for all patients. After the introduction of dipyridamole, adenosine, and dobutamine for pharmacologic stress in the 1980s and 1990s, perfusion studies could be performed in patients in whom exercise was contraindicated. In the mid-1990s ECG gating of the perfusion data methods was introduced. With this technique, called gated SPECT, global and regional functional ventricular information could be obtained from perfusion data sets. Today, almost all myocardial SPECT perfusion studies are performed with ECG gating as gated SPECT (Cerqueira and Vidigal Ferreira 2007). Since more than 5 years SPECT/CT systems have been used for myocardial perfusion imaging. SPECT/CT systems consist of two imaging modalities: SPECT and computed tomography (CT). With these hybrid systems attenuation correction could be performed more easily and precisely. Not only calcium scoring of coronary arteries but also noninvasive coronary angiography could be performed using CT portion of device at the same session of myocardial perfusion imaging.

Indications of myocardial thallium imaging (Strauss 2002):
1. Diagnosis of coronary artery disease (presence, location, and severity)
2. Assessment of the impact of coronary stenosis on regional perfusion
3. Help distinguish viable ischemic myocardium from scar
4. Risk assessment and stratification
    (a) Post-myocardial infarction
    (b) Preoperative for major surgery in patients who may be at risk for coronary events
5. Monitor treatment effect
    (a) After coronary revascularization
    (b) Medical therapy for congestive heart failure or angina
    (c) Lifestyle modification

## Myocardial Thallium Imaging

Myocardial thallium imaging (perfusion imaging, scintigraphy) consists of two parts:
1. Stress imaging
2. Rest or redistribution imaging

After intravenous injection of Tl-201 at stress, the radiotracer is distributed in the myocardium according to myocardial perfusion and viability. Thallium ions undergo continuous exchange between extracellular and intracellular compartments, leading to what is called the redistribution of thallium. After a certain time, there is equilibrium between the intracellular and interstitial compartments of activity and the defect seen in the hypoperfused areas on the stress image will disappear on the 3–4 h redistribution images. However, at equilibrium phase, the defect does not disappear on the redistribution image because of very low initial uptake (Saha 2010a, b). Comparison between the stress and redistribution images distinguishes between the reversible defect of inducible hypoperfusion (ischemia) and the fixed defect of myocardial necrosis (myocardial infarction). Ischemic defects are mostly reversible and viable, and the patient is likely to benefit from revascularization procedures (bypass surgery or angioplasty) (Saha 2010a, b).

## Stress Myocardial Perfusion Imaging

Stress imaging is provided by either exercise or pharmacological stress.

Exercise Stress Test

Exercise is performed using treadmill or bicycle ergometers. Tl-201 is injected intravenously at peak exercise. Numerous treadmill exercise protocols are available. They have different speed and inclination of the treadmill. The Bruce and modified Bruce protocols are the most widely used. Before starting exercise test the clinical history of the patient should be recorded. The indication for the test, symptoms, risk factors, medication, and prior diagnostic or therapeutic procedures must be noted. The patient should be stable for a minimum of 48 h prior to the test. Cardiac medications which may interfere with the stress test should, if possible, be interrupted. Before exercise, an intravenous cannula should be inserted for Tl-201 injection. The electrocardiogram should be monitored continuously during the exercise test and for at least 3–5 min of recovery. The blood pressure should be checked at least every 3 min during exercise. Exercise should be symptom limited, with patients achieving $\geq 85\%$ of their age-predicted maximum heart rate (220 − age). The Tl-201 should be injected close to the peak exercise. The patients should be encouraged to continue the exercise for at least 1 min after the injection (Hesse et al. 2005).

There are some absolute contraindications for exercise stress test (Hesse et al. 2005):
- Acute coronary syndrome
- Acute pulmonary embolism
- Uncontrolled, severe hypertension (blood pressure $\geq 200/110$ mmHg)
- Severe pulmonary hypertension
- Acute aortic dissection
- Symptomatic aortic stenosis and hypertrophic, obstructive cardiomyopathy
- Uncontrolled cardiac arrhythmias

If possible, exercise is terminated when peak heart rate is reached. On the other hand, according to electrocardiogram, blood pressure alterations or symptoms of the patient exercise has to be terminated early.

Absolute indications for early termination of exercise are (Hesse et al. 2005):
- Marked ST segment depression ($\geq 3$ mm)
- Ischemic ST segment elevation of $>1$ mm in leads without pathological Q waves
- Appearance of ventricular tachyarrhythmia
- A decrease in systolic blood pressure of $>20$ mmHg, despite increasing work load, when accompanied by other evidence of ischemia
- Markedly abnormal elevation of blood pressure (systolic blood pressure $\geq 250$ mm or diastolic blood pressure $\geq 130$ mmHg)
- Angina sufficient to cause distress to the patient
- Central nervous system symptoms (e.g., ataxia, dizziness, or near-syncope)
- Peripheral hypoperfusion (cyanosis or pallor)
- Sustained ventricular tachycardia or fibrillation
- Inability of the patient to continue the test
- Technical difficulties in monitoring ECG or blood pressure

Relative indications for early termination of exercise are (Hesse et al. 2005):
- ST segment depression $>2$ mm horizontal or downsloping
- Arrhythmias other than sustained ventricular tachycardia (including multifocal premature ventricular contractions (PVCs), triplets of PVCs, supraventricular tachycardia, heart block, or bradyarrhythmias), especially if symptomatic
- Fatigue, dyspnea, cramp, or claudication
- Development of bundle branch block or intraventricular conduction defect that cannot be distinguished from ventricular tachycardia

Pharmacological Stress Test

Stress imaging could be performed using pharmacological stress when exercise is contraindicated or in patients not be able to achieve maximal, age-predicted heart rate during exercise. Adenosine, dipyridamole, dobutamine, and regadenoson are most known pharmacological stress drugs.

Caffeine-containing beverages (coffee, tea, cola, etc.), foods (chocolate, etc.), and medications (some pain relievers, stimulants, and weight-control drugs) and methylxanthine-containing medications should be avoided for at least 12 h prior to pharmacological stress testing since they could interfere with vasodilator tests.

## Adenosine/Dipyridamole

Adenosine is a direct coronary arteriolar dilator. It causes three- to fourfold increase in myocardial blood flow in a normal coronary artery. Dipyridamole is an indirect coronary arteriolar dilator. It increases the tissue levels of adenosine by preventing the intracellular reuptake and deamination of adenosine (Hesse et al. 2005). Myocardium supplied by a diseased coronary artery has a reduced perfusion reserve and this leads to heterogeneity of perfusion during vasodilatation. Because myocardial tracer uptake is proportional to perfusion, this results in heterogeneous uptake of tracer in myocardium (Hesse et al. 2005).

The following list includes the absolute contraindications for adenosine or dipyridamole stress test (Hesse et al. 2005):
- Acute coronary syndrome
- Severe bronchospasm
- Greater than first-degree heart block or sick sinus syndrome, without a pacemaker
- Symptomatic aortic stenosis and hypertrophic obstructive cardiomyopathy
- Systolic blood pressure <90 mmHg
- Unstable angina
- Cerebral ischemia

Relative contraindications to adenosine or dipyridamole stress tests are (Hesse et al. 2005):
- Severe sinus bradycardia (heart rate <40/min)
- Severe atherosclerotic lesions of extracranial artery
- Use of dipyridamole during the last 24 h

### Adenosine/Dipyridamole Stress Test Procedure and Dosage

Adenosine should be given as a continuous infusion at 140 µg/kg/min over 4–6 min with injection of the tracer at 3–4 min. The infusion should be continued for 1–2 min after radiotracer injection (Hesse et al. 2005).

Dipyridamole should be given as a continuous infusion intravenously at 140 µg/kg/min over 4 min. The radiotracer is injected 3–5 min after the completion of dipyridamole infusion. Severe hypotension (systolic blood pressure <80 mmHg), persistent second-degree or sign of third-degree atrioventricular or sinoatrial block, wheezing, severe chest pain are the indications of early termination of adenosine or dipyridamole (Hesse et al. 2005).

### Side Effects of Adenosine/Dipyridamole Stress Test

Minor side effects are common and occur in approximately 80% of patients after adenosine infusion. Because of the very short half-life of adenosine (<10 s), most side effects resolve rapidly after discontinuing the infusion (Hesse et al. 2005).

The common side effects are listed below (Hesse et al. 2005):
- Flushing (35–40%)
- Chest pain (25–30%)
- Dyspnea (20%)
- Dizziness (7%)
- Nausea (5%)
- Symptomatic hypotension (5%)
- High-degree atrioventricular (AV) and sinoatrial (SA) blocks (7%)
- ST segment depression $\geq 1$ mm (15–20%)
- Fatal or nonfatal myocardial infarction (rare, <1/1,000)

More than 50% of patients develop side effects (flushing, chest pain, headache, dizziness, or hypotension) after dipyridamole infusion. The frequency of these side effects is less than that seen with adenosine, but they last longer (15–25 min). Aminophylline (125–250 mg, i.v.) is required usually. The incidence of high-degree AV and SA block with dipyridamole is 2% (Hesse et al. 2005).

## Dobutamine

Dobutamine is a secondary pharmacological stressor. It is indicated in patients who cannot tolerate exercise stress and have contraindications to other pharmacological stress drugs. Since dobutamine increases heart rate, blood pressure, and myocardial contractility, myocardial oxygen demand is raised. These effects result in a secondary coronary vasodilatation. Because of reduced coronary flow reserve in areas supplied by critically stenosed coronary arteries, vasodilatation could not be seen (Hesse et al. 2005).

### Procedure and Dosage of Dobutamine

Dobutamine is infused incrementally, starting at a dose of (5 to) 10 µg/kg/min and increasing at 3-min intervals to 20, 30, and 40 µg/kg/min. The Tl-201 should be injected when the heart rate is $\geq 85\%$ of the age-predicted maximum heart rate (220 − age). Dobutamine infusion should be continued for 2 min after the radiotracer injection. It is customary to use atropine in patients if heart rate does not reach 85% of age-predicted maximal heart rate. The indications for early termination of dobutamine are similar to those for exercise stress (Hesse et al. 2005).

### Side Effects
Control of side effects is easy because of short plasma half-life (120 s). Side effects occur in about 75% of patients. The common side effects are palpitation (29%), chest pain (31%), headache (14%), flushing (14%), dyspnea (14%), and significant supraventricular or ventricular arrhythmias (8–10%). Ischemic ST segment depression occurs in approximately one third of patients undergoing dobutamine infusion (Hesse et al. 2005).

### Regadenoson
Regadenoson is an A2A adenosine receptor agonist that is a coronary vasodilatator. Activation of the A2A adenosine receptor by regadenoson produces coronary vasodilatation and increases coronary blood flow (Holly et al. 2010).

#### Procedure and Dosage of Regadenoson
The recommended intravenous dose of regadenoson is 5 mL (0.4 mg regadenoson). Rapid injection is advised (10 s). And also, 5-mL saline flush immediately after the injection of regadenoson is suggested. Ten to twenty seconds later Tl-201 could be injected. A 12-lead electrocardiogram will be recorded every minute during the infusion. Blood pressure should be monitored every minute during infusion and 3–5 min into recovery (Holly et al. 2010).

#### Side Effects of Regadenoson
Shortness of breath, headache, and flushing are the most common side effects of regadenoson. Chest discomfort, angina pectoris or ST, dizziness, chest pain, nausea, abdominal discomfort, dysgeusia, and feeling hot are less common side effects. First and second degree AV blocks were detected in 3% and 0.1%, respectively. Aminophylline may be administered to attenuate severe and/or persistent adverse reactions to regadenoson (Holly et al. 2010).

Contraindications of regadenoson (Holly et al. 2010):
- Patients with second- or third-degree AV block or sinus node dysfunction without artificial pacemaker.
- The safety profile of regadenoson has not yet been definitively established in patients with bronchospasm. Inadequate data exists to confidently use regadenoson in patients with bronchospasm.
- Systolic blood pressure <90 mmHg.
- Known hypersensitivity to regadenoson.

#### Relative Contraindications of Regadenoson
Profound sinus bradycardia (heart rate\40/min) is considered for relative contraindications of regadenoson (Holly et al. 2010).

Indications for reversal of regadenoson infusion (Holly et al. 2010):
- Severe hypotension (systolic blood pressure <80 mmHg)
- Development of symptomatic, persistent second degree or complete heart block
- Wheezing
- Severe chest pain associated with ST depression of 2 mm or greater
- Signs of poor perfusion (pallor, cyanosis, and cold skin)
- Technical problems with the monitoring equipment
- Patient's request to stop

1. Combination with low-level exercise with pharmacological stress:
   Low-level exercise could be added to pharmacological stress test to reduce the side effects. And also this procedure improves the image quality because of lower bowel activity. However, low-level exercise is not recommended for patients with left bundle branch block or ventricular paced rhythm.
2. Redistribution or rest myocardial perfusion imaging:
   After 3–4 h of stress imaging redistribution or rest, myocardial imaging is performed without second radiopharmaceutical injection since thallium has redistribution effect.
3. Reinjection imaging:
   In some cases, a second injection of Tl-201 (1 mCi) before rest imaging can be given for a more accurate assessment of myocardial viability.
4. Late redistribution imaging:
   In some cases, redistribution after 3–4 h of injection of Tl-201 is not enough. In these cases, late redistribution imaging (after 24–72 h of injection) can also be performed for a more accurate assessment of myocardial viability.

### Administration and Activity
The usual activity of Tl-201 is 74 MBq (2 mCi) for stress and redistribution imaging. An additional

**Thallium-201 Imaging, Fig. 1** Reconstructed SPECT slices. The stress and rest or images are displayed in two rows of images to facilitate comparison. The short-axis images are displayed from apex (*left*) to base (*right*). The vertical long-axis slices from septum (*left*) to lateral wall (*right*). The horizontal long-axis slices are displayed from inferior wall (*left*) to anterior wall (*right*). In this patient, normal perfusion is not only in stress but also in rest images

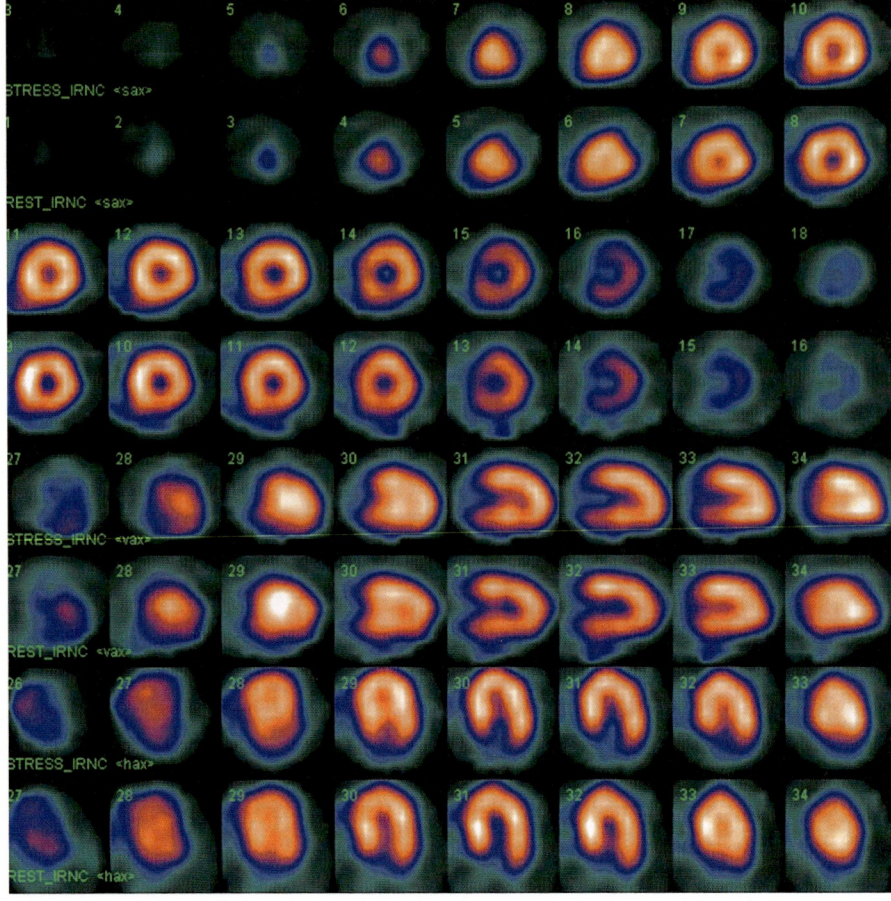

37 MBq (1 mCi) can be given at rest for reinjection imaging if redistribution is thought to be incomplete at the time of redistribution imaging or if redistribution is predicted to be slow. Tl-201 should be administered through a secure intravenous line. Paravenous injection must be avoided due to risk of local tissue necrosis. After injection of thallium, the line can be flushed with either saline or glucose to ensure that the full dose is given.

## Myocardial Thallium Imaging Technique

Tl-201 myocardial imaging is performed using SPECT (with gamma camera) in recent years. Tl-201 emits emitting mainly X-rays of energy 67–82 keV (88% abundance) and gamma photons of 135 and 167 keV (12% abundance). X or gamma rays could not be visible. Gamma camera is a kind of medical imaging device that converts gamma rays to visible light (scintillation photons). Gamma camera that has a rotating detector is called SPECT. Patient, who is injected radiopharmaceutical, emitting gamma rays, is put into the bed under the detector of camera in supine position. Detector of SPECT is rotated around the patient in a circular or an ellipsoid orbit in 180°–360°. Camera acquires multiple planar images (projections) from multiple angles over 180°–360°. Projections are acquired at defined points during the rotation, typically every 3°–6°. A computer is then used to apply a tomographic reconstruction algorithm to the multiple projections, yielding a 3-D dataset. This dataset may then be manipulated to show thin slices along any chosen axis of the body. Thus, cross-sectional slices could be obtained with SPECT. The time taken to obtain each projection is also variable, but 20–35 s is typical. This gives a total imaging time of 15–20 min for myocardial thallium imaging in modern gamma cameras (with two or triple detectors).

## Assessment of Myocardial Thallium Imaging

Myocardial SPECT imaging performed after stress and rest reveals the distribution of the radiopharmaceutical,

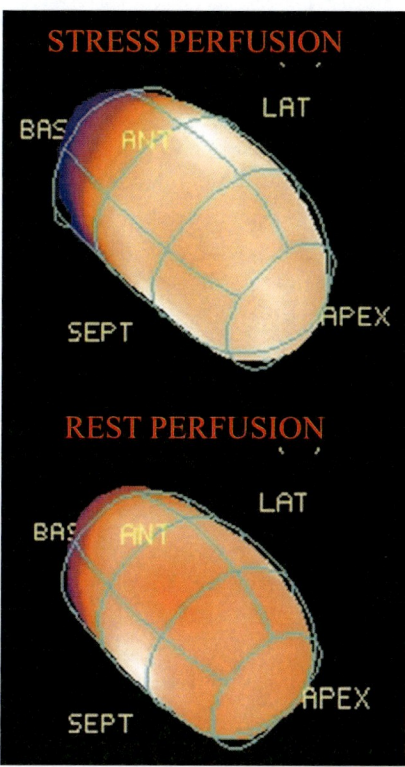

**Thallium-201 Imaging, Fig. 2** Three-dimensional rendering of myocardial perfusion of the left ventricle

and therefore the relative blood flow to the different regions of the myocardium. Head to head comparison of stress and rest image sets is performed for assessment of myocardial perfusion. The display of SPECT myocardial perfusion images is standardized. Reconstructed slices of the heart are displayed as short-axis, horizontal long-axis, and vertical long-axis slices (Fig. 1).

Images are assessed both visually and semiquantitatively. Stress and rest reconstructed slices are often divided in 17–20 segments according to standards developed by the American Heart Association (AHA), American Colleges of Cardiology (ACC), and American Society of Nuclear Cardiology (ASNC). Because of the many reconstructed SPECT images available for analysis, it is useful to compress all information into one image. This can be done by either displaying a three-dimensional rendering of myocardial perfusion of the left ventricle (Fig. 2), or by generating color-coded polar maps or "bull's-eye" images (Fig. 3). A standardized segmentation and nomenclature for bull's eye display myocardial perfusion images for 17 segments model is shown in Fig. 4.

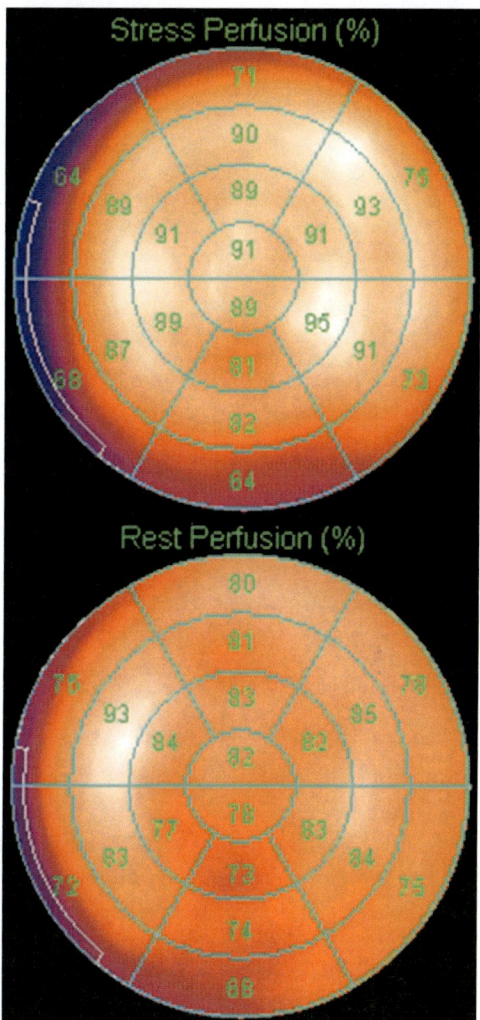

**Thallium-201 Imaging, Fig. 3** Color-coded polar maps or "bull's-eye" images of the left ventricle. Myocardial perfusion image data are projected onto one plane. Image data of the apex are projected in the center of the bull's eye. Image data of the base of the left ventricle are projected on the periphery of the bull's eye. Mid-ventricular image data are projected between these two areas. Polar maps could be evaluated with either myocardial segmental models (17–20 segments) or coronary artery anatomy pattern. Nomenclature of 17 segments model is determined in Fig. 4. However, present polar map is divided into 20 segments. In this model, apical region of left ventricle is divided into 6 segments (anterior, anteroseptal, inferoseptal, inferior, inferolateral, and anterolateral) instead of 4 segments in 17 segments model. Also, apex is represented in one segment, instead of 2 segments in 17 segments model

## Visual Analysis

Myocardial perfusion images are assessed for relative differences in radiotracer accumulation. Head to head comparison of stress and rest image sets is performed.

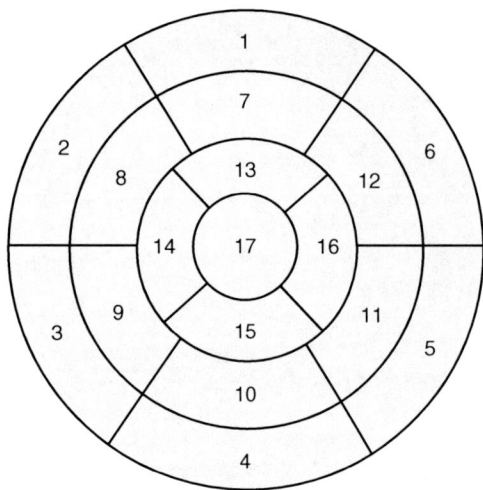

**Thallium-201 Imaging, Fig. 4** Segmentation of left ventricle myocardium (17 segments model): *1* basal anterior, *2* basal anteroseptal, *3* basal inferoseptal, *4* basal inferior, *5* basal inferolateral, *6* basal anterolateral, *7* mid anterior, *8* mid anteroseptal, *9* mid inferoseptal, *10* mid inferior, *11* mid inferolateral, *12* mid anterolateral, *13* apical anterior, *14* apical septal, *15* apical inferior, *16* apical lateral, *17* apex

The following defines the various image patterns (Germano and Berman 2002):

- Normal: Homogeneous uptake of the radiopharmaceutical throughout the myocardium is observed (Fig. 1).
- Fixed defect: A localized myocardial area with decreased radiotracer uptake is observed both in stress and rest imaging. This pattern generally indicates myocardial infarction and scar tissue.
- Reversible defect: A defect that is present on the stress images and is no longer present or present to a lesser degree on the rest images indicates myocardial ischemia (Fig. 5).
- Reverse reversible defect: The stress images are either normal or show a defect, whereas the rest images show a more severe defect. This pattern is frequently observed in patients who have undergone thrombolytic therapy or percutaneous coronary angioplasty. This phenomenon is thought to be caused by initial excess of tracer uptake in a reperfused area with a mixture of scar tissue and viable myocytes. Initial accumulation is followed by rapid clearance from scar tissue. Although the significance of this finding is controversial, it does not represent evidence of exercise-induced ischemia.
- Lung uptake: Normally no, or very little, radiotracer is observed in the lung fields on post-stress images. Increased lung uptake is an important abnormal finding that indicates exercise-induced ischemic left ventricular dysfunction.
- Transient left ventricular dilation: Occasionally, the left ventricle appears to be larger following exercise than on the rest images. This pattern also has been associated with exercise-induced left ventricular dysfunction and greater amount of myocardium at risk.
- Transient right ventricular visualization: The right ventricle is more clearly visualized on the post-stress images than on the rest images. This pattern indicates transient left ventricular dysfunction during stress.

### Semiquantitative Analysis

Myocardial perfusion images could be interpreted semiquantitatively by using a scoring system. Generally, a 5 points scoring system is used (0 = normal perfusion, 1 = mild hypoperfusion, 2 = moderate hypoperfusion, 3 = severe hypoperfusion, 4 = absent uptake). By applying the scoring system for each segment of the 17 segments model to both rest and stress images, a summed stress score (SSS), a summed rest score (SRS), and a summed difference score (SDS = SRS-SSS) can be derived. A normal image thus has a score of "0," whereas the maximal abnormal score is "68" (no heart visualized) in 17 segments model. A summed score of <8 is considered small, 9–13 moderate, and >13 large defect. These semiquantitative scores have been shown to provide important prognostic information.

### Gated SPECT

Gated SPECT imaging allows the simultaneous examination of left ventricular myocardial perfusion and function. With this technique not only calculation of end-diastolic, end-systolic volumes, and ejection fraction but also analysis of left ventricular wall motion and thickening are possible. Since being introduced in the early 1990s, ECG gating is now a critical element of all SPECT studies. More than 90% of all myocardial perfusion studies are performed with gating. Gated SPECT studies are acquired using 8 or 16 time frames in the heart cycle. It has been demonstrated that 16 frames are more accurate for ejection fraction calculation. However, image quality of 16 frames is poorer than 8 frames

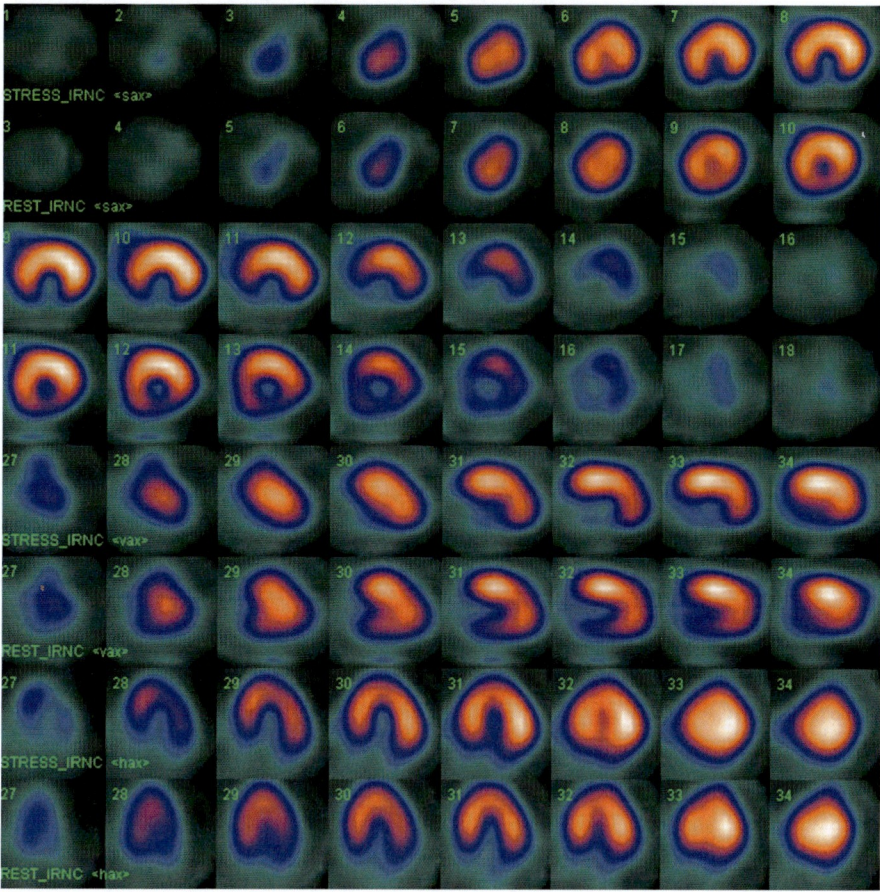

**Thallium-201 Imaging, Fig. 5** Moderate perfusion defect is seen in inferior wall of the left ventricle in stress images. Redistribution is noticed in the same region in rest images. Inferior wall ischemia is diagnosed according to myocardial perfusion SPECT

because of low counts. On the other hand, 16 frame acquisition allows analysis of filling and ejection rates that can be used for assessment of diastolic function (Cerqueira and Vidigal Ferreira 2007).

Left ventricular systolic and diastolic functional parameters (end-diastolic, end-systolic volumes, ejection fraction, transient ischemic dilatation value) can be obtained from gated SPECT studies. Also, visual and semiquantitative analysis of wall motion and wall thickening of left ventricle is possible with this technique.

In ischemic region of left ventricle wall, motion and thickening are usually decreased during peak exercise. Since gated studies are acquired a minimum of 5–10 min following peak exercise stress, regional wall motion in most patients with ischemia usually returns to normal by the time of gated SPECT imaging. If ischemic, sufficiently profound ischemic areas of myocardium may continue to have persistent wall motion abnormalities that are detected during gated acquisition. One of the most important benefits of gated SPECT acquisition is helping to differentiate attenuation due to the diaphragm or breast tissue from areas of old myocardial infarction or scar tissue. Areas of scar have absent or diminished wall motion and thickening while areas with decreased perfusion due to attenuation will show normal wall motion and thickening (Cerqueira and Vidigal Ferreira 2007).

### Radiation Dosimetry of Myocardial Thallium Imaging

The absorbed radiation doses to various organs in healthy subjects following administration of Tl-201 chloride are given in Table 1.

### Thallium Tumor Imaging

Primary and/or metastasis of a tumor are investigated in thallium tumor imaging. Particularly, this technique is useful in soft tissue and bone sarcoma, lymphoma, lung cancer, cerebral tumors, and differentiated

**Thallium-201 Imaging, Table 1** Absorbed dose* per unit activity administered (mGy/MBq) for adults mGy/patient examination

| | |
|---|---|
| Bone surfaces | 27 |
| Gall-bladder | 6 |
| Small intestine | 11 |
| Colon | 18 |
| Kidneys | 38 |
| Urinary bladder | 3 |
| Heart | 16 |
| Ovaries | 58 |
| Testes | 26 |
| Effective Dose | 17.6 mSv |

*The absorbed doses (mGy/MBq) are adopted from EANM/ESC procedural guidelines for myocardial perfusion imaging in nuclear cardiology (Hesse et al. 2005)

*mGy* miliGray, *MBq* Megabecquerel, *mSv* miliSievert

thyroid carcinoma. After radiotherapy, differential diagnosis of residual viable tumor or radiation necrosis is very difficult using anatomical diagnostic techniques such as CT or magnetic resonance imaging (MR). Because Tl-201 uptake is possible in only viable cells, it is useful for differentiation of residual viable tumor from radiation necrosis, especially in cerebral tumors. After 10 min of intravenous injection of 1–3 mCi Tl-201, whole body and regional SPECT images are obtained using gamma camera. Except physiological uptake and uptake in clearance organs (myocardium, skeletal muscles, salivary glands, thyroid, liver, spleen, colon, kidneys, bladder) any retention of Tl-201 is accepted pathological.

In the last two decades tumor imaging using thallium has been declining. Positron emission tomography (PET) and, recently, positron emission tomography-computed tomography (PET-CT) substituted scintigraphic techniques in diagnostic oncology because of easy availability of these systems. Sensitivity, specificity, and accuracy of PET and PET-CT are higher than thallium tumor imaging. Also, resolution of PET or PET-CT is better than scintigraphic techniques. The smallest detectable lesion is about 1 cm for Tl-201 imaging. However, with modern PET-CT systems, lesions of 3–4 mm could be detectable.

## Cross-References

▸ Potassium, Physical and Chemical Properties
▸ Thallium, Distribution in Animals

## References

Cerqueira MD, Vidigal Ferreira MJ (2007) Heart. In: Biersack HJ, Freeman LM (eds) Clinical nuclear medicine. Springer, Berlin/Heidelberg/New York

Germano G, Berman DS (2002) Cardiovascular. In: Sandler MP, Coleman RE, Patton AJ et al (eds) Diagnostic nuclear medicine. Lippincott Williams & Wilkins, Philadelphia

Hesse B, Tägil K, Cuocolo A, EANM/ESC Group et al (2005) EANM/ESC procedural guidelines for myocardial perfusion imaging in nuclear cardiology. Eur J Nucl Med Mol Imaging 32:855–897

Holly TA, Abbott BG, Al-Mallah M et al (2010) Single photon-emission computed tomography. J Nucl Cardiol 17:941–973 (ASNC Imaging Guidelines for Nuclear Cardiology Procedures)

Saha GB (2010a) Diagnostic uses of radiopharmaceuticals in nuclear medicine. In: Saha GB (ed) Fundamentals of nuclear pharmacy, 6th edn. Springer, New York

Saha GB (2010b) Characteristics of specific radiopharmaceuticals. In: Saha GB (ed) Fundamentals of nuclear pharmacy, 6th edn. Springer, New York

Strauss HW, Miller DD, Wittry MD et al (2002) Society of Nuclear Medicine Procedure Guideline for Myocardial Perfusion Imaging version 3.0 (Society of Nuclear Medicine Procedure Guidelines Manual)

## The Effect of Selenium on Cell Proliferation

▸ Selenium and the Regulation of Cell Cycle and Apoptosis

## The Molecular Effects of Gallium on Bacterial Physiology

▸ Gallium in Bacteria, Metabolic and Medical Implications

## The Molecular Mechanisms of the Therapeutic Action of Boron-Containing Compounds

▸ Boron-containing Compounds, Regulation of Therapeutic Potential

## The Resistance Mechanism of the Platinum Anticancer Drugs

▶ Platinum-Containing Anticancer Drugs and proteins, interaction

## The Role of ATR1 Paralogs in Boron Stress Response

▶ Boron Stress Tolerance, YMR279c and YOR378w

## The Role of Silicon in Suppressing Plant Disease

▶ Silicon-Mediated Pathogen Resistance in Plants

## Therapeutic Targeting of Iron-Dependent Tumor Growth with Iron-Mimetic Metals

▶ Gallium Nitrate, Apoptotic Effects

## Thermolysin

Olayiwola A. Adekoya[1] and Ingebrigt Sylte[2]
[1]Pharmacology Research group, Department of Pharmacy, Institute of Pharmacy, University of Tromsø, Tromsø, Norway
[2]Medical Pharmacology and Toxicology, Department of Medical Biology, University of Tromsø, Tromsø, Norway

### Synonyms

Thermolysin-like enzymes

### Definition

These are zinc-containing enzymes that hydrolyze proteins into smaller polypeptides and/or amino acids by cleavage of their peptide bonds. They appear to be key factors in the pathogenesis of various diseases. Most of them are secreted to degrade proteins and peptides for bacteria nutrition.

### Thermolysin

Thermolysin from *Bacillus thermoproteolyticus* is the type-example of the M4 family of peptidases, which also is termed the thermolysin family. All peptidases in the family bind a single, catalytic zinc ion. The zinc ion is tetrahedrally coordinated by the three amino acid ligands and a water molecule. The water molecule is activated and forms the nucleophile in catalysis. As of February 2010 there are 493 sequences in the Merops database that belong to the thermolysin family (http://merops.sanger.ac.uk/) (Rawlings 2010; Rawlings et al. 2010). The family members are found in bacteria, fungi, and archaea. Typically, they consist of a presequence (signal sequence), cleaved of during export, a propeptide sequence with inhibitory and chaperone functions (facilitating folding), and a peptidase unit. The propeptide acting as an inhibitor remains attached until the peptidase is secreted and can be safely activated (Yeats et al. 2004). All peptidases of the thermolysin family are members of the clan MA, while those belonging to the subclan MA(E) of the thermolysin family are termed thermolysin-like peptidases. Thermolysin-like peptidases are zinc-dependent metallopeptidases in which two of the zinc ligands are the histidines in the motif: His-Glu-Xaa-Xaa-His (HEXXH). The Glu is located at the active site. The catalytic zinc ion is also bound by a glutamic acid located C-terminal to the HEXXH motif and a water molecule. The amino acids spacing between the third zinc ligating amino acid and an aspartic acid are identical in subclan MA(E) (four amino acids), and this Glu-$(Xaa)_3$-Asp motif in addition to the HEXXH motif are motifs used for identifying peptidases that may have the thermolysin fold. Furthermore, the residue which is the third zinc ligand allows the clan to be divided into two subclans: subclan MA(E) and subclan MA(M). In the subclan, MA(E), known as "Glu-zincins," the third ligand is a Glu located 18–72 residues C-terminal to the HEXXH motif. The HEXXH motif and the third amino acid ligating zinc are located in α helices connected by a turn. The turn is required to bring the amino acids together forming the zinc-binding site (Fig. 1).

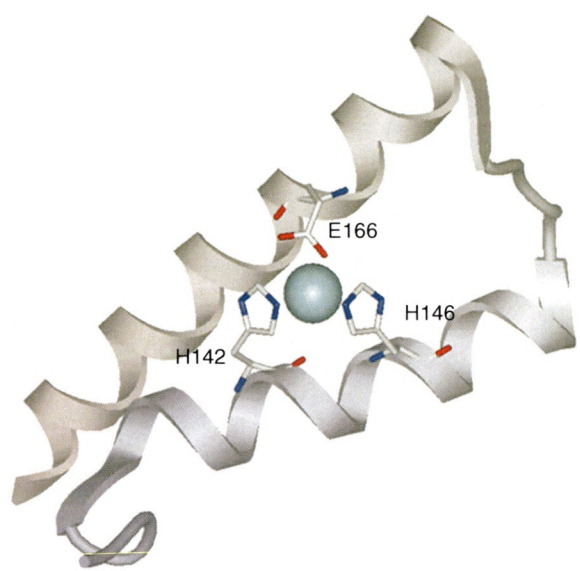

**Thermolysin, Fig. 1** The thermolysin active site. The zinc ion and the coordinating amino acids are shown

In the subclan, MA(M), known as the "Metzincins," the third zinc ligand is a histidine or an aspartic acid occurring in the extended motif **HEXXHXXGXX(H/D)** (X is any amino acid residue). The presence of the conserved Gly is needed for forming the β turn that brings the three zinc ligating amino acids together (Bode et al. 1992). The region C-terminal to the third histidine forms a loop that ends at the catalytic zinc in a unique turn. The turn contains a strictly conserved methionine and is designated the "Met-turn," thus the name of the subclan.

## Tertiary Structures of the Thermolysin Family M4

Thermolysin from *Bacillus thermoproteolyticus* was the first metalloendopeptidase whose 3D structure was solved by X-ray crystallography (Matthews et al. 1972, 1974; Colman et al. 1972). Thermolysin consists of a two-domain structure with the active site between the domains. The N-terminal domain includes both α-helices and β-sheets, and carries the HEXXH motif. The C-terminal domain is predominantly helical and carries the third zinc ligating amino acid. The 3D structure has been determined for six M4 family members. These are thermolysins from *Bacillus cereus* (1npc) and *Bacillus thermoproteolyticus* (1kei), pseudolysin from *Pseudomonas aeruginosa* (1ezm), aureolysin from *Staphylococcus aureus* (1bqb), vibriolysin MCP-02 from *Pseudoalteromonas* sp. SM9913 (3nqx), and protealysin precursor from *Serratia* sp. (2vqx). The structures of the two thermolysins contain four calcium ions suggested to be responsible for the thermostability. The aureolysin structure has three calcium ions, pseudolysin has only one calcium ion, but contains two disulfide bridges; vibriolysin MCP-02 has one calcium ion; while the protealysin has no calcium ion. The structures of at least 9 mature thermolysins have also been solved at resolutions of 1.7–2.2 Å (Matthews et al. 1972; Holland et al. 1995; Holmes and Matthews 1982; English et al. 1999, 2001; Hausrath and Matthews 2002). Furthermore, the structure of thermolysin in complex with at least 25 ligands has been solved (Gaucher et al. 1999; Hausrath and Matthews 1994; Holland et al. 1994; Monzingo and Matthews 1984; Shen and Wendoloski 1995; Tronrud et al. 1986, 1987; Holmes et al. 1983; Holmes and Matthews 1981). Two X-ray structures of pseudolysin (1u4g and 3bdk unpublished) in complex with small molecule inhibitors are also deposited in the PDB database. Structures of the autoprocessed complex of MCP-02 (E346A) (3nqy), *and of MCP-02 (*E369A) (3nqz) are also found in the PDB database.

## Specificity of Peptidases in the Thermolysin Family

Very little is known about the substrate specificities of the enzymes of the thermolysin family. Most of what is known about the family is based on numerous studies on thermolysin. As a result, there is large amount of data available on thermolysins but the information about other family members is limited. Thermolysin preferentially cleaves at the N-terminal side of hydrophobic or bulky amino side chains such as Leu, Phe, Ile, and Val. Thermolysin also cleaves bonds of Met, His, Tyr, Ala, Asn, Ser, Thr, Gly, Lys, Glu, or Asp at the $P1'$ site (Ligne et al. 1997; Heinrikson 1977). Like thermolysin, pseudolysin and griselysin favor hydrophobic residues at the $P1'$ position, but show higher preference for aromatic residues than for other hydrophobic residues. Hydrophobic residues are also favored in P1 and $P2'$ (Morihara 1974, 1995; Kajiwara et al. 1991; Tsuyuki et al. 1991). The specificity of aureolysin, coccolysin, and lambda toxin (λ-Toxin) is similar to that of thermolysin with preference for hydrophobic $P1'$ residues (leucine, valine, tyrosine, isoleucine, and phenylalanine) and alanine. Zinc is required for aureolysin catalytic activity, but it can be substituted with cobalt producing a proteinase that is

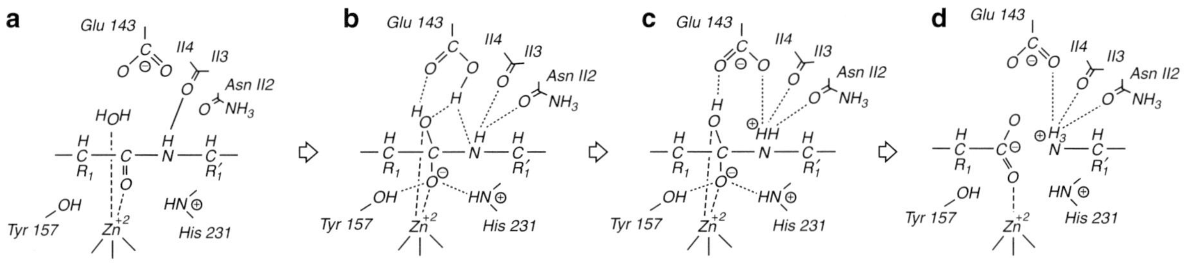

**Thermolysin, Fig. 2** The first proposed mechanism in which the glutamate (E) of the HEXXH motif acts as a proton acceptor during catalysis. The incoming substrate displaces the water molecule toward Glu143 of the HEXXH motif, the water molecules still ligate the zinc ion (**a**). The nucleophilicity of the water molecule is increased; the lone pair of the water oxygen is directed toward the substrate and the water attacks the carbonyl carbon and forms an intermediate (**b**). A proton is transferred to the leaving nitrogen. The side chain oxygen of Asn112 and the backbone carbonyl of Ala113 accept hydrogen bonds from the doubly protonated tetrahedral nitrogen of the scissile bond (**c**). A second proton transfer through Glu143 to the leaving nitrogen forms the product (**d**) (Permission received from ACS publishing)

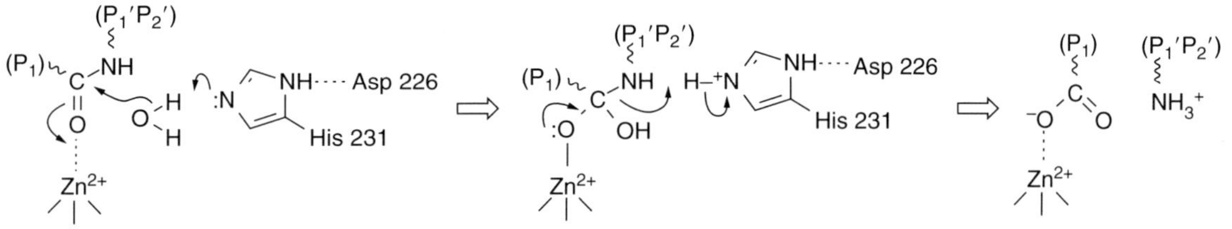

**Thermolysin, Fig. 3** The second proposed mechanism in which the active-site histidine acts as the general base instead of the glutamate residue of the HEXXH motif. $P_1$, $P_1'$, and $P_2'$ are side chains of peptides to be cleaved (Permission received from ACS publishing)

more active than native enzyme (Drapeau 1978). Specificity of vimelysin is different from that of thermolysin. Vimelysin specifically recognizes phenylalanine at the P1′ position, whereas thermolysin specifically recognizes phenylalanine at the P1′ position (Takahashi et al. 1996).

## Catalytic Mechanism of the Thermolysin Family

The zinc ion in native thermolysin is tetrahedrally coordinated by three amino acid ligands (His142 and His146 of the HEXXH motif, and Glu166 of the Glu-(Xaa)$_3$-Asp motif). The fourth coordination is provided by a water molecule. Other amino acid residues presumed to be involved in catalysis are Asn 112, Ala113, Glu143 of the HEXXH motif, Tyr 157, Asp226 and His 231. The catalytic mechanism of thermolysin is the best understood of all family members, even though it is not completely understood. Two mechanisms of action have been proposed, one in which glutamate of the HEXXH motif acts as a proton acceptor during catalysis (Hangauer et al. 1984; Matthews 1988) and the second in which an active-site histidine acts as the general base instead of the glutamate residue(Mock and Stanford 1996; Mock and Aksamawati 1994). In the first proposed mechanism, the catalytic cleavage of peptide bonds proceeds by a general base type mechanism with the attack of a water molecule or hydroxide ion on the carbonyl carbon of the scissile bond. The incoming substrate is presumed to displace the water molecule toward Glu143 of the HEXXH motif, such that both hydrogen atoms of the water molecules are hydrogen bonded to Glu143, while the oxygen still ligates the zinc ion (Fig. 2). In that way the nucleophilicity of the water molecule is enhanced, and the remaining lone pair of the water oxygen is directed toward the substrate being aligned for nucleophilic attack. The water attacks the carbonyl carbon and forms an intermediate (Fig. 2b). Tyr 157 and His231 help to stabilize the intermediate by hydrogen bonding with the carbonyl oxygen of the substrate. The proton accepted by Glu143 is transferred to the leaving nitrogen. The side chain oxygen of Asn112 and the backbone carbonyl of Ala113 accept hydrogen bonds from the doubly protonated tetrahydral nitrogen of the scissile bond (Fig. 2c). A second proton transfer via Glu143 to the leaving nitrogen forms the product (Fig. 2d).

In the second proposed mechanism, three other residues are suggested to be important for catalysis besides the activated water molecule, the zinc ion, and the zinc ligands. These three residues are Glu143, which acts as an electrophile, Asp226, and His231. Asp226 orientates the imidazolium ring of His231, and His231 acts as proton donor and general base (Fig. 3).

## Inhibition of Enzymes of the Thermolysin Family

Small molecule inhibitors of enzymes in the thermolysin family are phosphoramidon, from *Streptomyces tanashiensis*, and metal chelating inhibitors like 1,10-phenanthroline and EDTA. Chelation causes removal of the zinc ion, resulting in inhibition of the metallopeptidase. The zinc ion is removed to yield an inactive apoenzyme. 1,10-phenanthroline has less affinity for calcium ions than EDTA. Similarities of the active site region and structure of zinc-containing enzymes provide both a huge possibility for design of inhibitors as well as the challenge of achieving specificity. The widespread development of resistance to antibiotics in clinical use today calls for urgent development of anti-infectives that could be used alongside antibiotics in clinical practice. This will curtail the incidence of resistance to these antibiotics. Development of small molecule inhibitors of thermolysin-like enzymes that are virulence factors could indeed prove ground breaking in that direction.

The naturally occurring protein inhibitors, Streptomyces metalloproteinase inhibitor (SMPI) and the inducible metalloproteinase inhibitor (IMPI) from the moth*Galleria mellonella* caterpillar, are protein inhibitors of the family. SMPI, from *Streptomyces nigrescens*TK-23, is a protein inhibitor of metalloproteinases.It is a specific inhibitor of the gluzincin family of metalloproteinases.The NMR structure shows that the protein(102 amino acids) contains two disulfide bridges. Studies have shown that the loop Arg60-Ala73 is involved in inhibition, and the inhibitor function has been connected to Cys64-Val65 segment in the loop. This loop also contains a disulfide bridge between Cys64 and Cys69, resulting in a rigid loop structure. The X-ray crystal structure of SMPI has not been determined. Experimentally derived binding affinity of SMPI is several hundredfold higher than all available small molecule inhibitors of the thermolysin family. This indicates a huge potential in designing inhibitors mimicking the peptidase-binding region of SMPI. Such inhibitors are of immense industrial as well as therapeutic value.

IMPI, from *Galleria mellonella*, was identified in the larvae of the greater wax moth. IMPI shares no sequence similarity with known vertebrate or invertebrate proteins or other natural inhibitors of metalloproteinases. IMPI inhibits microbial metalloproteinases, such as bacterial thermolysin, but is not active against matrix metalloproteinases (MMPs). Its molecular mass is 8.6 kDa. IMPI contains ten cysteine residues which are likely to form five disulfide bridges. These five intermolecular disulfide bonds probably explain its pronounced stability in high temperatures and for the treatment with trichloroacetic acid (TCA) (Wedde et al. 1998). The 3D structure of IMPI has not been determined. The low sequence similarity between IMPI and known vertebrate or invertebrate proteins or other natural inhibitors of metalloproteinases also precludes the construction of a 3D homology model. Certainly, the understanding of the inhibition interactions between IMPI and the peptidases in the thermolysin family could be important for the development of effective anti-infectives.

## Therapeutic and Biotechnological Potentials

A large number of these enzymes are implicated as virulence factors of the microorganisms producing them and are potential drug targets. The similarity in the active site structure and architecture of metalloproteinases is also important and could be exploited. In fact, the discovery of the mechanistic similarities between thermolysin, carboxypeptidase A, and angiotensin-converting enzyme was a key factor in the development of antihypertensives in the present clinical use. Some enzymes in the family are able to function at the extremes of temperatures, and some function in organic solvents.

Thereby, enzymes of the thermolysin family have an innovative potential for biotechnological and therapeutic applications.

## Diseases in Which Peptidases in the Family M4 Are Implicated

Thermolysin-like metalloproteinases are key factors in the pathogenesis of various diseases (Adekoya and Sylte 2009). They are secreted to degrade host proteins and peptides for bacteria nutrition. This family of enzymes has great potentials as putative drug targets for therapeutic uses. Pseudolysin, the extracellular elastase of *Pseudomonas aeruginosa* plays an important role in the pathogenesis of *P. aeruginosa* infections. Pseudolysin causes tissue destruction, disrupts some cell functions, interferes with host-defense mechanisms, and impairs the normal function of host proteases. All these are achieved because pseudolysin cleaves casein, elastin, human IgG, collagen types III and IV, serum $\alpha$1-proteinase inhibitor, and human bronchial mucosal proteinase inhibitor. Pseudolysin has also been implicated in various lung infections, chronic ulcers by degradation of human wound fluids and human skin proteins, and corneal infections, causing corneal liquefaction which could be sight-threatening. *P. aeruginosa* elastase-specific inhibitors have been used to treat rabbit cornea infections as adjuncts to antibiotics. Coccolysin from *Enterococcus faecalis* has been implicated in several opportunistic infections caused by *E. faecalis*, such as soft tissue and urinary tract infections, intra-abdominal abscesses and root canal infections, and secondary bacteremia and food poisoning. Coccolysin also hydrolyzes casein, gelatin, and hemoglobin; clots milk and inactivates human endothelin-1; and plays important roles in the pathogenesis of these organisms. $\lambda$-Toxin, the thermolysin-like peptidase from *Clostridium perfringens*, degrades various host proteins and also activates the precursors of clostridial potent toxins. $\lambda$-Toxin can degrade immunoglobulin G, complement C3 component, fibrinogen, fibronectin and $\alpha_2$-macroglobulin, which contribute to innate or adaptive immune defense against infection. The thermolysin-like peptidase from *Legionella* cleaves $\alpha_1$-antitrypsin, tumor necrosis factor $\alpha$, interleukin 2 and CD4 on human T-cell surfaces. It has been suggested that it may have a role in the virulence of Legionnaire's disease and pneumonia. Hemolysin from *Renibacterium salmoninarum* possess hemolytic properties observed in the temperature range from 6°C to 37°C for mammalian and fish erythrocytes. Hemolysin is implicated in kidney disease of wild and farmed salmon fish. The thermolysin-like peptidase from *Helicobacter pylori* is implicated as causative agent of gastritis, peptic ulcer, and gastric carcinoma (Smith et al. 1994), while the hemagglutinin/proteinase from *Vibrio cholerae* is implicated in cholera, characterized by severe vomiting and watery diarrhea. Furthermore, studies have shown that the thermolysin-like peptidase from *Vibrio cholerae* affects intracellular tight junctions by degrading occluding. The thermolysin-like peptidase, hemagglutinin/proteinase, is a bifunctional molecule capable of mediating hemagglutination and proteolysis. The hemagglutinin activity is still present at 4°C, while the protease activity is lost. Interestingly, autodigestion of the C-terminus results in increase in proteolytic activity and decrease in hemagglutinating activity. Aureolysin from *Staphylococcus aureus* targets human fibrinolytic system for spread and invasion. Studies indicate that Aureolysin activates plasminogen, degrades plasminogen activator inhibitor, and stops the inhibitory activity of $\alpha_2$ –antiplasmin. All these effects result in increased fibrinolytic effects. The thermolysin-like peptidase, neutral protease from *Bacillus nematocida*, is suggested to possess nematotoxic activity, and was proposed to synergize with serine protease to kill nematodes. *Aeromonas hydrophila* exhibits prothrombin activator activity, and is capable of cleaving prothrombin into active thrombin. It was suggested that this thermolysin-like peptidase could be an alternative to several snake venoms for the analysis of prothrombin content in plasma as it is easily obtained in large amounts from cultured bacteria. The thermolysin-like peptidases, ZmpA and ZmpB from *Burkholderia cenocepacia*, cleave several proteins important for host defence. ZmpB had proteolytic activity against alpha-1 proteinase inhibitor, alpha(2)-macrogobulin, type IV collagen, fibronectin, lactoferrin, transferrin, and immunoglobulins. Studies suggest that these play important roles in the resistance of *B. cenocepacia* to host antimicrobial peptides as well as alter the host protease/anti-proteasebalance in chronic respiratory infections. Furthermore, both ZmpA and ZmpB cleaved protamine, a fish antimicrobial peptide.

## Temperature Adaptation of the Thermolysin Family M4

In their 1974 article, Mattews et al. stated : "it is suggested that the enhanced stability of thermostable proteins relative to thermolabile ones cannot be attributed to a common determinant such as metal ion or hydrophobic stabilization, but in a given instance may be due to rather subtle differences in the hydrophobic character, metal binding, hydrogen bonding, ionic interactions, or a combination of all of these." (Matthews et al. 1974) Thermostable proteins usually have similar 3D structures to their less-stable counterparts, and the differences in stability are due to a combination of different effects which can be more or less important from protein to protein. The same can also be said for cold-adapted enzymes. The "core" residues are usually conserved but subtle differences in certain regions of the structures exit that make the temperature adaptation possible.

Thermolysins from *Alicyclobacillus acidocaldarius, Bacillus* sp. *Strain EA1, Bacillus megaterium* have $T_{50}$ of 82°C, 85°C, and 58°C respectively. $T_{50}$ is defined as the temperature where an enzyme loses 50% of its activity in a 30-min period of incubation. $T_{50}$ of 82°C, 76.7°C, and 68.5°C has been reported for *B. thermoproteolyticus, B. caldolyticus, and B. stearothermophilus CU21*, respectively. Mutation of six residues situated near the binding site of the third calcium ion of *B. stearothermophilus* CU21 led to the production of an enzyme with a $T_{50}$ of 96.9°C. Bacillolysin from *B. cereus* was reported to be less temperature stable than thermolysin. It has fewer hydrogen bonds in the 3D structure, but the four calcium-binding sites are present (Stark et al. 1992; Pauptit et al. 1988). The higher thermal stability of thermolysin was proposed to be due to a combination of rigidification by proline residues, hydrogen bonding or salt-bridges (Stark et al. 1992). Vimelysin from *Vibrio* str. T1800 has an optimum temperature of 50°C, and has high activity in the presence of organic solvents. The peptidase from *Serratia marcescens* has a $T_{50}$ of 50°C. Metalloprotease from ***Listeria monocytogenes*** is active at temperatures up to 80°C, and is the most heat-stable natural thermolysin-like peptidase known with a $T_{50}$ of 88°C. Addition of high salt concentrations to growing cultures increased its protease activity. Some members of the thermolysin family are enzymes from cold-adapted bacteria. These are Vibriolysin from *Antarctic bacterium str. 643,VAB*, vibriolysin MCP-02 from a deep sea bacterium and vibriolysin E495 from an Arctic sea ice bacterium (Xie et al. 2009; Adekoya et al. 2006a). While vibriolysin MCP-02 and vibriolysin E495 have optimum temperature of 57°C, 5°C less than mesophlic pseudolysin, VAB's cold adaptation has so far not been proven biochemically. Rigorous analysis of its modeled 3D structure, amino acid sequence analysis of the whole thermolysin family, and molecular dynamics simulation studies show features of cold adaptation. Amino acid sequence analysis of 44 thermolysin enzymes showed that VAB compared to the other enzymes has the following characteristics: (1) Fewer arginines; (2) A smaller Arg/(Lys + Arg) ratio; (3) A lower fraction of large aliphatic side chains, expressed by the (Ile + Leu)/(Ile + Leu + Val) ratio; (4) More methionines; (5) More serines; and (6) More of the thermolabile amino acid asparagine. MD simulations for 3 nanoseconds (ns) of VAB, pseudolysin, and thermolysin showed that VAB has a more flexible 3D structure than their thermophilic and mesophilic counterparts, especially in some loop regions. The structural analysis indicated that VAB has fewer intramolecular cation–π electron interactions and hydrogen bonds than its mesophilic (pseudolysin) and thermophilic (thermolysin) counterparts. Lysine is the dominating cationic amino acids involved in salt bridges in VAB, while arginine is the dominating in thermolysin and pseudolysin. VAB had a greater volume of the inaccessible cavities than pseudolysin and thermolysin. The electrostatic potentials at the surface of the catalytic domain were also more negative for VAB than for thermolysin and pseudolysin (Adekoya et al. 2006a). Molecular dynamics studies comparing pseudolysin, MCP-02, and E495 showed increased flexibility from pseudolysin to MCP-02 and E495 resulting from the decrease of hydrogen-bond stability in the dynamic structure, which was due to the increase in the number of asparagine, serine, and threonine residues (Xie et al. 2009).

## Industrial and Therapeutic Use of Peptidases of Family M4

Members of the thermolysin family of enzymes are stable in the extremes of environmental conditions, such as high and low temperatures, pH, organic solvents, etc., and have, therefore, huge industrial and biotechnological potentials.

Vibriolysin from *Vibrio proteolyticus* is used in the production of the precursor to the sweetener, aspartame. It is used in mediating the coupling of *N*-protected aspartic acid and phenylalanine methyl ester to yield *N*-protected aspartylphenylalanine methyl ester, a precursor to the sweetener aspartame. It is also used for the removal of necrotic tissue from wounds such as burns or cutaneous ulcers. It has been reported that it stimulated the healing of partial-thickness burn wounds.

Vibriolysin from *Vibrio proteolyticus*, therefore, has several industrial as well as biomedical applications (Durham 1990; Durham et al. 1993). Thermolysin is also used as a peptide and ester synthetase and in the production of the artificial sweetener aspartame. It is also widely used as a nonspecific proteinase to obtain fragments for peptide sequencing in the biotechnology industry. Due to its higher activity in organic solvents, Vimelysin from *Vibrio* str. T1800 showed great potentials in peptide condensation reactions. The optimum pH of vimelysin was 8.0 when casein was used as substrate, while the optimum pH was 6.5 when furylacryloyl-glycyl-leucine amide (FAGLA) was used. The optimum temperature of vimelysin was 50°C when casein was used as a substrate, while it was 15°C when FAGLA was used as a substrate(Oda et al. 1996; Kunugi et al. 1996). Bacillolysin from *Paenibacillus polymyxa* (*Bacillus polymyxa*) is suggested to be involved in the generation of α and β-amylases from a large amylase precursor (Takekawa et al. 1991).

## Experimental and Computational Studies of Inhibition of the Thermolysin Family

Most experimental and computational studies of inhibition of the enzymes in the thermolysin family have focused on thermolysin and pseudolysin. There is very scanty information on the inhibition studies on other members of the family. However, there is a patent grant for the use of aureolysin inhibitors in the treatment of inflammatory skin conditions characterized by colonization with *Staphylococcus aureus* (Layton and Chandler 2007). Most experimental inhibition studies reported are those done during initial characterization of the peptidases using general small molecule inhibitors like phosphoramidon from *Streptomyces tanashiensis* (Suda et al. 1973) and metal chelators like Ethylenediaminetetraacetic acid, EDTA, and 1,10-phenanthroline. The metal chelators target the catalytic zinc ion. Experimental inhibition studies using SMPI mutants with mutations in the reactive site loop indicated that pseudolysin is inhibited 10–100-fold stronger than thermolysin (Hiraga et al. 1999). Furthermore, whereas $ClCH_2CO$-HOLeu- Ala-Gly-$NH_2$ can irreversibly inhibit both pseudolysin and thermolysin, only thermolysin is irreversibly inhibited by $ClCH_2CO$-HOLeu-$OCH_3$ (HOLeu: *N*-hydroleucine), while pseudolysin is not. These findings suggest that the binding pocket of pseudolysin is bigger than the binding pocket of thermolysin, since an increase in the length of the ligand by two amino acids (Ala and Gly) results in inhibition of both pseudolysin and thermolysin. Most of the amino acids involved in active site interactions and ligand binding are highly conserved between thermolysin and pseudolysin. (The corresponding amino acids in pseudolysin are given in brackets): Asn112 (Asn112), Ala113 (Ala113), Trp115 (Trp115), Arg203 (Arg198), Tyr157 (Tyr155), Phe130 (Phe129), Leu133 (Leu132), Val139 (Val137), Glu143 (Glu141), Ile188 (Ile186), Val192 (Ile190), Leu202 (Leu197), and His231 (His223). Studies of thermolysin have indicated that these amino acids residues are important for substrate/inhibitor binding or catalyses (Matthews 1988). Molecular dynamics simulation studies have, however, indicated that some amino acids not conserved between pseudolysin and thermolysin are also important for SMPI binding to pseudolysin. These amino acids are Tyr114, Asp206, and His224. The corresponding amino acids of thermolysin are Phe114, Tyr206, and Ile224, respectively. These differences may contribute to the stronger affinities of SMPI to pseudolysin than to thermolysin (Adekoya et al. 2005; 2006b).

Experimentally derived binding affinities of SMPI against enzymes of the thermolysin family are 100-folds more favorable than for available small molecule inhibitors, indicating a huge potential in designing inhibitors mimicking the peptidase-binding region of SMPI. Such inhibitors are of immense industrial as well as therapeutic value. The experimentally determined inhibition constant ($K_i$) of SMPI is $2.54 \times 10^{-12}$ M for pseudolysin, and $1.14 \times 10^{-10}$ M for thermolysin. Mutational analysis of the SMPI-thermolysin complex showed that Arg60, Arg61, and Arg66 within the rigid active site loop are important for thermolysin inhibition. The double mutant, R60/61A, had 5 times weaker inhibition than the wild type, while the single mutant R66A showed 100 times reduced inhibition, whereas triple mutant R60/61/66A showed 200 reduced inhibition compared to the wild type _ENREF_82. Computational alanine scanning of the amino acid residues at the interaction interfaces of pseudolysin-SMPI complexes has suggested hot spot residues responsible for the strong-binging affinities observed. Auxiliary binding sites away from the active site contributing to better interactions have also been suggested by computational studies of SMPI-thermolysin and SMPI-pseudolysin complexes. These auxiliary binding sites were also subjected to computational alanine scanning which predicted hot spot residues that might contribute to the high binding free energies.

## Cross-References

▶ Calcium-Binding Protein Site Types
▶ Calcium-Binding Proteins, Overview
▶ Zinc and Zinc Ions in Biological Systems
▶ Zinc Matrix Metalloproteinases and TIMPs
▶ Zinc-Binding Proteins, Abundance
▶ Zinc-Binding Sites in Proteins

## References

Adekoya OA, Sylte I (2009) The thermolysin family (M4) of enzymes: therapeutic and biotechnological potential. Chem Biol Drug Des 73(1):7–16

Adekoya OA, Willassen NP, Sylte I (2005) The protein-protein interactions between SMPI and thermolysin studied by molecular dynamics and MM/PBSA calculations. J Biomol Struct Dyn 22(5):521–531

Adekoya OA, Helland R, Willassen NP, Sylte I (2006a) Comparative sequence and structure analysis reveal features of cold adaptation of an enzyme in the thermolysin family. Proteins 62(2):435–449

Adekoya OA, Willassen NP, Sylte I (2006b) Molecular insight into pseudolysin inhibition using the MM-PBSA and LIE methods. J Struct Biol 153(2):129–144

Bode W, Gomis-Ruth FX, Huber R, Zwilling R, Stocker W (1992) Structure of astacin and implications for activation of astacins and zinc-ligation of collagenases. Nature 358(6382):164–167

Colman PM, Jansonius JN, Matthews BW (1972) The structure of thermolysin: an electron density map at 2–3 A resolution. J Mol Biol 70(3):701–724

Drapeau GR (1978) Role of metalloprotease in activation of the precursor of staphylococcal protease. J Bacteriol 136(2):607–613

Durham DR (1990) The unique stability of vibrio proteolyticus neutral protease under alkaline conditions affords a selective step for purification and use in amino acid-coupling reactions. Appl Environ Microbiol 56(8):2277–2281

Durham DR, Fortney DZ, Nanney LB (1993) Preliminary evaluation of vibriolysin, a novel proteolytic enzyme composition suitable for the debridement of burn wound eschar. J Burn Care Rehabil 14(5):544–551

English AC, Done SH, Caves LS, Groom CR, Hubbard RE (1999) Locating interaction sites on proteins: the crystal structure of thermolysin soaked in 2% to 100% isopropanol. Proteins 37(4):628–640

English AC, Groom CR, Hubbard RE (2001) Experimental and computational mapping of the binding surface of a crystalline protein. Protein Eng 14(1):47–59

Gaucher JF, Selkti M, Tiraboschi G, Prange T, Roques BP, Tomas A, Fournie-Zaluski MC (1999) Crystal structures of alpha-mercaptoacyldipeptides in the thermolysin active site: structural parameters for a Zn monodentation or bidentation in metalloendopeptidases. Biochemistry 38(39):12569–12576

Hangauer DG, Monzingo AF, Matthews BW (1984) An interactive computer graphics study of thermolysin-catalyzed peptide cleavage and inhibition by N-carboxymethyl dipeptides. Biochemistry 23(24):5730–5741

Hausrath AC, Matthews BW (1994) Redetermination and refinement of the complex of benzylsuccinic acid with thermolysin and its relation to the complex with carboxypeptidase A. J Biol Chem 269(29):18839–18842

Hausrath AC, Matthews BW (2002) Thermolysin in the absence of substrate has an open conformation. Acta Crystallogr D Biol Crystallogr 58(Pt 6 Pt 2):1002–1007

Heinrikson RL (1977) Applications of thermolysin in protein structural analysis. Methods Enzymol 47:175–189

Hiraga K, Seeram SS, Tate S, Tanaka N, Kainosho M, Oda K (1999) Mutational analysis of the reactive site loop of Streptomyces metalloproteinase inhibitor, SMPI. J Biochem (Tokyo) 125(1):202–209

Holland DR, Barclay PL, Danilewicz JC, Matthews BW, James K (1994) Inhibition of thermolysin and neutral endopeptidase 24.11 by a novel glutaramide derivative: X-ray structure determination of the thermolysin-inhibitor complex. Biochemistry 33(1):51–56

Holland DR, Hausrath AC, Juers D, Matthews BW (1995) Structural analysis of zinc substitutions in the active site of thermolysin. Protein Sci 4(10):1955–1965

Holmes MA, Matthews BW (1981) Binding of hydroxamic acid inhibitors to crystalline thermolysin suggests a pentacoordinate zinc intermediate in catalysis. Biochemistry 20(24):6912–6920

Holmes MA, Matthews BW (1982) Structure of thermolysin refined at 1.6 A resolution. J Mol Biol 160(4):623–639

Holmes MA, Tronrud DE, Matthews BW (1983) Structural analysis of the inhibition of thermolysin by an active-site-directed irreversible inhibitor. Biochemistry 22(1):236–240

Kajiwara K, Fujita A, Tsuyuki H, Kumazaki T, Ishii S (1991) Interactions of Streptomyces serine-protease inhibitors with *Streptomyces griseus* metalloendopeptidase II. J Biochem (Tokyo) 110(3):350–354

Kunugi S, Koyasu A, Kitayaki M, Takahashi S, Oda K (1996) Kinetic characterization of the neutral protease vimelysin from vibrio sp. T1800. Eur J Biochem 241(2):368–373

Layton, TG, Chandler RS (2007) Use of an aureolysin inhibitor for the treatment of inflammatory skin conditions characterised by colonisation with *Staphylococcus aureus*. Patent No. PCT/EP2006/065863

Ligne T, Pauthe E, Monti JP, Gacel G, Larreta-Garde V (1997) Additional data about thermolysin specificity in buffer- and glycerol-containing media. Biochim Biophys Acta 1337(1):143–148

Matthews BW (1988) Structural basis of the action of thermolysin and related zinc peptidases. Acc Chem Res 21(9):333–340

Matthews BW, Colman PM, Jansonius JN, Titani K, Walsh KA, Neurath H (1972) Structure of thermolysin. Nature New Biol 238:41–43

Matthews BW, Weaver LH, Kester WR (1974) The conformation of thermolysin. J Biol Chem 249(24):8030–8044

Mock WL, Aksamawati M (1994) Binding to thermolysin of phenolate-containing inhibitors necessitates a revised mechanism of catalysis. Biochem J 302(Pt 1):57–68

Mock WL, Stanford DJ (1996) Arazoformyl dipeptide substrates for thermolysin. Confirmation of a reverse protonation catalytic mechanism. Biochemistry 35(23):7369–7377

Monzingo AF, Matthews BW (1984) Binding of N-carboxymethyl dipeptide inhibitors to thermolysin determined by X-ray crystallography: a novel class of transition-state analogues for zinc peptidases. Biochemistry 23(24):5724–5729

Morihara K (1974) Comparative specificity of microbial proteinases. Adv Enzymol Relat Areas Mol Biol 41(0):179–243

Morihara K (1995) Pseudolysin and other pathogen endopeptidases of thermolysin family. Methods Enzymol 248:242–253

Oda K, Okayama K, Okutomi K, Shimada M, Sato R, Takahashi S (1996) A novel alcohol resistant metalloproteinase, vimelysin, from vibrio sp. T1800: purification and characterization. Biosci Biotechnol Biochem 60(3):463–467

Pauptit RA, Karlsson R, Picot D, Jenkins JA, Niklaus-Reimer AS, Jansonius JN (1988) Crystal structure of neutral protease from *Bacillus cereus* refined at 3.0 A resolution and comparison with the homologous but more thermostable enzyme thermolysin. J Mol Biol 199(3):525–537

Rawlings ND (2010) Peptidase inhibitors in the MEROPS database. Biochimie 92(11):1463–1483

Rawlings ND, Barrett AJ, Bateman A (2010) MEROPS: the peptidase database. Nucleic Acids Res 38(Database issue):D227–D233

Shen J, Wendoloski J (1995) Binding of phosphorus-containing inhibitors to thermolysin studied by the Poisson-Boltzmann method. Protein Sci 4(3):373–381

Smith AW, Chahal B, French GL (1994) The human gastric pathogen *Helicobacter pylori* has a gene encoding an enzyme first classified as a mucinase in Vibrio cholerae. Mol Microbiol 13(1):153–160

Stark W, Pauptit RA, Wilson KS, Jansonius JN (1992) The structure of neutral protease from *Bacillus cereus* at 0.2-nm resolution. Eur J Biochem 207(2):781–791

Suda H, Aoyagi T, Takeuchi T, Umezawa H (1973) Letter: a thermolysin inhibitor produced by actinomycetes: phospholamidon. J Antibiot (Tokyo) 26(10):621–623, 0021-8820 (Print)

Takahashi S, Okayama K, Kunugi S, Oda K (1996) Substrate specificity of a novel alcohol resistant metalloproteinase, vimelysin, from *Vibrio* sp. T1800. Biosci Biotechnol Biochem 60(10):1651–1654

Takekawa S, Uozumi N, Tsukagoshi N, Udaka S (1991) Proteases involved in generation of beta- and alpha-amylases from a large amylase precursor in *Bacillus polymyxa*. J Bacteriol 173(21):6820–6825 (0021-9193 (Print))

Tronrud DE, Monzingo AF, Matthews BW (1986) Crystallographic structural analysis of phosphoramidates as inhibitors and transition-state analogs of thermolysin. Eur J Biochem 157(2):261–268

Tronrud DE, Holden HM, Matthews BW (1987) Structures of two thermolysin-inhibitor complexes that differ by a single hydrogen bond. Science 235(4788):571–574

Tsuyuki H, Kajiwara K, Fujita A, Kumazaki T, Ishii S (1991) Purification and characterization of *Streptomyces griseus* metalloendopeptidases I and II. J Biochem (Tokyo) 110(3):339–344

Wedde M, Weise C, Kopacek P, Franke P, Vilcinskas A (1998) Purification and characterization of an inducible metalloprotease inhibitor from the hemolymph of greater wax moth larvae, *Galleria mellonella*. Eur J Biochem 255(3):535–543

Xie BB, Bian F, Chen XL, He HL, Guo J, Gao X, Zeng YX, Chen B, Zhou BC, Zhang YZ (2009) Cold adaptation of zinc metalloproteases in the thermolysin family from deep sea and arctic sea ice bacteria revealed by catalytic and structural properties and molecular dynamics: new insights into relationship between conformational flexibility and hydrogen bonding. J Biol Chem 284(14):9257–9269

Yeats C, Rawlings ND, Bateman A (2004) The PepSY domain: a regulator of peptidase activity in the microbial environment? Trends Biochem Sci 29(4):169–172

# Thermolysin-Like Enzymes

▶ Thermolysin

# Thiols and Lanthanides

▶ Lanthanides, Rare Earth Elements, and Protective Thiols

# Thioneins

▶ Metallothioneins and Copper
▶ Metallothioneins and Lead
▶ Metallothioneins and Mercury

# Thorium, Physical and Chemical Properties

Fathi Habashi
Department of Mining, Metallurgical, and Materials Engineering, Laval University, Quebec City, Canada

Thorium is a silvery-white ductile metal, with low strength properties and chemical resistance. The most

**Thorium, Physical and Chemical Properties, Fig. 1** Decay scheme of thorium

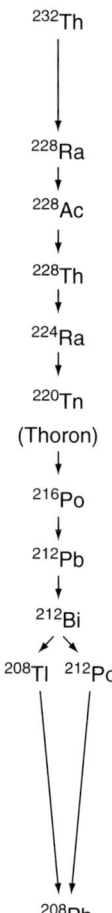

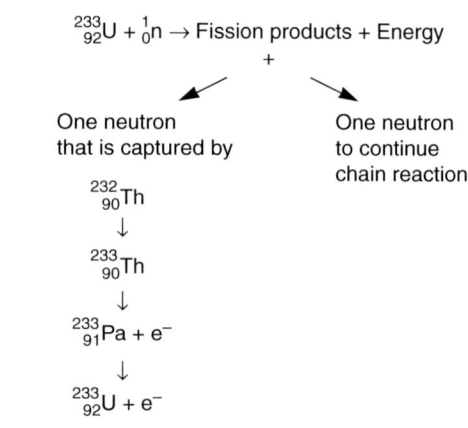

widely occurring and important mineral of thorium is monazite, a phosphate of rare earths and thorium. Thorium is radioactive with a half-life of $2.2 \times 10^{10}$ years (Fig. 1). It absorbs neutrons and is transformed to uranium 233:

$$^{232}_{90}\text{Th} + ^{1}_{0}\text{n} \rightarrow ^{233}_{90}\text{Th} \rightarrow ^{233}_{91}\text{Pa} + e^-$$
$$\downarrow$$
$$^{233}_{92}\text{U} + e^-$$

Uranium 233 undergoes fission and can be used to operate a breeder reactor in presence of a thorium blanket according to the scheme:

Thorium exists in two solid phases; the α-phase (lattice constant 0.5068 nm) is stable at room temperature and is transformed to the β-phase (lattice constant 0.411 nm) at $1{,}360 \pm 10$°C. Heat of transformation $3{,}600 \pm 125$ J/mol.

Mechanical, thermal, and electrical properties of thorium metal are sensitive to the level of impurities, especially the carbon content. The level of impurities typically varies according to the method of production. Thorium metal is superconducting below 1.368 K.

## Physical Properties

| | |
|---|---|
| Atomic number | 90 |
| Atomic weight | 232.04 |
| Relative abundance in Earth's crust, % | $1.1 \times 10^{-3}$ |
| Density, g/cm$^3$ | 11.72 |
| Melting point, °C | 1,750 |
| Boiling point, °C | 4,702 |
| Electrical resistance at room temperature, Ω/cm | $14 \pm 1 \times 10^{-6}$ |
| Temperature coefficient of resistivity, K$^{-1}$ | $(3.6\text{--}4) \times 10^{-3}$ |
| Standard electrode potential, Th $\rightarrow$ Th$^{4+}$, V | $-1.90$ |

## Chemical Properties

In moist air, thorium oxidizes slowly at its surface even at room temperature. Oxidation proceeds in two stages. First, a surface layer of the dark gray hydride is formed by reaction with atmospheric moisture. The hydride is then oxidized, with formation of a strongly bonded oxide layer. Above 670°C, thorium reacts with nitrogen to form Th$_2$N$_3$. It reacts with hydrogen at 300–400°C, with the evolution of heat, to form Th$_4$H$_{15}$, which itself decomposes above 1,100°C, probably forming ThH$_2$, a gray-black powder that ignites spontaneously. Thorium can therefore be melted only in an inert-gas atmosphere or in a vacuum.

Thorium reacts violently with the halogens and with sulfur on heating to 450°C, producing a flame, and it reacts with fluorine even in the cold. Thorium reacts at 800–900°C with water vapor to form ThO$_2$. On heating with phosphorus, finely divided boron, or carbon, tetravalent compounds are formed, as in the reaction with nitrogen. Thorium is stable toward dissolved or molten alkali and is attacked only slowly by dilute mineral acids. It is passivated in HNO$_3$. Thorium dissolves in aqua regia and concentrated HCl. Thorium salts have only a slight tendency to hydrolyze:

$$Th^{4+} + H_2O \rightarrow Th(OH)^{3+} + H^+$$

The thorium ion forms anionic complexes with a large number of ions in acidic or neutral solution, e.g., CO$_3^{2-}$, C$_2$O$_4^{2-}$, SO$_4^{2-}$, HSO$_4^-$, citrate, and tartrate. Thorium forms positively charged complexes with F$^-$ if the molar ratio of F$^-$ to Th$^{4+}$ is <2. Complexes are also formed with an excess of thiocyanide and nitrate. Thorium has a strong tendency to form double salts such as K$_2$ThF$_6$·4H$_2$O, NaTh$_2$(PO$_4$)$_3$, MnTh(SO$_4$)$_3$·7H$_2$O, and ThSO$_4$HPO$_4$·4H$_2$O. It also forms basic salts such as ThOCO$_3$·2H$_2$O.

Thorium is nearly always tetravalent. It cannot be deposited electrolytically from aqueous solutions.

## References

Habashi F (2003) Metals from ores. An introduction to extractive metallurgy. Métallurgie Extractive Québec, Québec City. Distributed by Laval University Bookstore, www.zone.ul.ca

Stoll W (1997) Thorium. In: Habashi F (ed) Handbook of extractive metallurgy. Wiley, Weinheim, pp 1649–1684

# Thrombin Aptamer

▶ Strontium and DNA Aptamer Folding

# Thulium

Takashiro Akitsu
Department of Chemistry, Tokyo University of Science, Shinjuku-ku, Tokyo, Japan

## Definition

A lanthanoid element, the 12th element (yttrium group) of the f-elements block, with the symbol Tm, atomic number 69, and atomic weight 168.93421. Electron configuration [Xe] 4f$^{13}$6s$^2$. Thulium is composed of stable ($^{169}$Tm, 100%) and four synthetic radioactive ($^{167}$Tm; $^{168}$Tm; $^{170}$Tm; $^{171}$Tm) isotopes. Discovered by P. T. Cleve in 1879. Thulium exhibits oxidation state III; atomic radii 176 pm, covalent radii 190 pm; redox potential (acidic solution) Tm$^{3+}$/Tm −2.278 V; Tm$^{3+}$/Tm$^{2+}$ −2.3 V; electronegativity (Pauling) 1.25. Ground electronic state of Tm$^{3+}$ is $^3$H$_6$ with S = 1, L = 5, J = 6 with λ = −1290 cm$^{-1}$. Most stable technogenic radionuclide $^{171}$Tm (half-life 1.92 years). The most common compounds: Tm$_2$O$_3$, TmCl$_3$ and Tm(OH)$_3$. Biologically, thulium is of low to moderate toxicity, and has high affinity for tumor cells (Atkins et al. 2006; Cotton et al. 1999; Huheey et al. 1997; Oki et al. 1998; Rayner-Canham and Overton 2006).

## Cross-References

▶ Lanthanide Ions as Luminescent Probes
▶ Lanthanide Metalloproteins
▶ Lanthanides and Cancer
▶ Lanthanides in Biological Labeling, Imaging, and Therapy
▶ Lanthanides in Nucleic Acid Analysis
▶ Lanthanides, Physical and Chemical Characteristics

## References

Atkins P, Overton T, Rourke J, Weller M, Armstrong F (2006) Shriver and Atkins inorganic chemistry, 4th edn. Oxford University Press, Oxford/New York

Cotton FA, Wilkinson G, Murillo CA, Bochmann M (1999) Advanced inorganic chemistry, 6th edn. Wiley-Interscience, New York

Huheey JE, Keiter EA, Keiter RL (1997) Inorganic chemistry: principles of structure and reactivity, 4th edn. Prentice Hall, New York

Oki M, Osawa T, Tanaka M, Chihara H (1998) Encyclopedic dictionary of chemistry. Tokyo Kagaku Dojin, Tokyo

Rayner-Canham G, Overton T (2006) Descriptive inorganic chemistry, 4th edn. W. H. Freeman, New York

## Thyroxine Deiodinase (TXD)

▶ Selenium and Iodothyronine Deiodinases

## Thyroxine, 3,5,3′,5′-Tetraiodothyronine, T4

▶ Selenoproteins and Thyroid Gland

## Timothy Syndrome (TS)

▶ Calcium Sparklets and Waves

## Tin

▶ Tin, Toxicity

## Tin (IV)

▶ Tin Complexes, Antitumor Activity

## Tin Complexes, Antitumor Activity

Farukh Arjmand
Department of Chemistry, Aligarh Muslim University, Aligarh, UP, India

## Synonyms

Antineoplastic; Antiproliferative compounds; Apoptosis; Cytotoxicity; Organotin; Tin (IV)

## Definition

The prototypical success story of cisplatin, cis-diamminedichloroplatinum(II), an archetypical inorganic drug (as it contains not a single carbon atom) in the treatment of several types of solid tumors, in particular, testicular and ovarian tumors has fostered a growing interest in metal-based pharmaceuticals. However, the clinical effectiveness of cisplatin is greatly limited by drug resistance and significant side effects which are considered to be the prime factors responsible for recurrence of the disease and morbidity. This has spurred considerable attention to non-platinum chemotherapeutics with the aim to optimize the efficiency of platinum drugs and to avoid their toxic side effects. Among the noteworthy non-platinum metal-based therapeutics, tin and organotins has emerged as potentially active antitumor antiproliferative compounds. Tin compounds and their therapeutic potentials have been reviewed (Gielen and Tiekink 2005; Hadjikakou and Hadjiliadis 2009), while a number of early reviews have recorded advances in the screening for antitumor potential of organotin complexes (Saxena and Huber 1989; Crowe 1987; Yang and Guo 1999). It seems that although tin may exist in either in $Sn^{2+}$ or in $Sn^{4+}$ oxidation state, almost all organotins have a tetravalent structure because tin(II) compounds are readily oxidized to tin (IV). Recent reports in literature (Hadjikakou and Hadjiliadis 2009) have established organotin(IV) compounds are very important in cancer chemotherapy as they display another spectrum of antitumor activity, may show non-cross-resistance with platinum drugs and may possess less or different toxicity as compared to platinum compounds (de Vos et al. 1998). Furthermore, tin compounds are strong apoptotic directors.

A plethora of organotin derivatives have been prepared and tested in vitro and in vivo, firstly against murine leukemia cell lines and after that, against different panels of human cancer cell lines (Crowe 1987). Most of the compounds tested early on, exhibited interesting activity in specific cancer models but often lacked activity against a broad spectrum of tumors. Nevertheless, the possibilities for variation of the organotin moieties and donor ligands linked to the metal has resulted in several diorganotin and triorganotin(IV) compounds with high antiproliferative activity in vitro against a panel of solid and hematologic cancers (Hadjikakou and Hadjiliadis 2009).

**Tin Complexes, Antitumor Activity, Scheme 1** Irreversible binding to the peripheral phosphate groups of phosphoribose residues by Sn(IV)

## Mechanistic Pathway of DNA Inhibition

The diorganotins are considered to be the ultimate cytotoxic agents in solution, triorganotin compounds may undergo spontaneous disproportionation with the corresponding diorganotin and tetraorganotin derivatives while, in vivo, the loss of the alkyl or aryl group occurs through the intervention of enzymes such as aromatase. Inhibition of macromolecular synthesis, mitochondrial energy metabolism, and reduction of DNA synthesis, as well as direct interaction with the cell membrane (increase in cytosolic $Ca^{2+}$ concentration) has been implicated in organotin-induced cytotoxicity. In addition, promotion of oxidative and DNA damage in vivo has been detected and these are the major factors which contribute to tin-induced apoptosis in many cell lines (Alama et al. 2009). Apoptosis also named as programmed cell death is a form of cell deletion aimed at the control of cell differentiation and proliferation. Morphologically, apoptosis is characterized by chromatin margination around the nuclear membrane followed by its condensation in cup-shaped dense masses. A number of metal compounds able to activate apoptosis directly involved in the apoptotic pathway, such as p53 tumor suppressor, TRAIL receptor, caspases, and the Bcl-2 family of proteins have been recently developed (Chauhan et al. 2007). Since, there are two primary modes of apoptosis, i.e., extrinsic and intrinsic, metal-induced apoptosis is thought to be initiated intracellularly, the mitochondria being most pertinent in mediating apoptosis via metal-induced reactive oxygen species. Tributyl compounds have shown to induce apoptosis by causing extracellular $Ca^{2+}$ influx and generating reactive oxygen species ROS. Several, in vitro and in vivo studies with different organotin compounds have reported that these compounds have differential cytotoxic effects on various cell types depending on the length of the alkyl chain. The induction of apoptosis (Sarma 2008; Ray et al. 2000) by three tri-n-butylstannyl benzoate (TBSB), tri-n-butylstannyl-2,6-difluorobenzoate (TBSDFB), and tri-n-butyl-stannyl-2-iodobenzoate (TBSIB) in human leukemia K562 cells was analyzed to study differential cytotoxic effects of halogen groups present in the phenyl ring of tributyltin halobenzoates. Since the induction of apoptosis by these compounds in K562 cells resulted from the extracellular $Ca^{2+}$ influx and generation of ROS, the initial amount of extracellular $Ca^{2+}$ was greater

**Tin Complexes, Antitumor Activity, Table 1** The most active organotin compound against various common cancerous cell lines

| Compound | Cancerous cell line | ID$_{50}$ (μg/ml) | Coordination mode | ligand | Cisplatin ID$_{50}$ (μg/ml) |
|---|---|---|---|---|---|
| [Sn(C$_6$H$_5$)$_3${OOC$_6$H$_3$-3,4-(NH$_2$)$_2$}] | HCV29T | 0.004 | 4 | Carboxylic acid | 0.7 |
| {{4-[(ClCH$_2$CH$_2$)$_2$N]C$_6$H$_4$COOSnBu$_2$}$_2$O}$_2$ | A549 | 0.2–0.02 | 5 | Carboxylic acid | 3.3 |
| [(Bn)$_3$SnOCOR] | H226 | 0.005 | 5 | Carboxylic acid | 3.269 |
| 2-PhC$_2$N$_3$CO$_2$Sn(C$_6$H$_5$)$_3$ | HeLa | 0.005 | 5 | Carboxylic acid | 0.433 |
| {[(C$_6$H$_5$)$_3$Sn]$_2$(MNA).[(CH$_3$)$_2$CO]} | Leiomyosarcoma | 0.005 | 5 | Thioamide | 1.2-1.5 |
| (n-Bu)$_2$Sn-(Gly-Leu) | A498 | 0.03 | 4 | Amino acid | 2.253 |
| | IGROV | 0.006 | 4 | Amino acid | 0.169 |

in TBSB-treated K562 cells than the cells treated with TBSDFB or TBSIB. Similarly, DNA fragmentation by endonucleases was observed as an early event in case of TBSB-treated K562 cells.

The design of improved organotin (IV) antitumor agents has imposed limitations concerning the cellular targets of these compounds and their mechanism of action although inhibition of mitochondrial oxidative phosphorylation appears to be an important mode of toxicity. The binding ability of organotin compounds, toward DNA depends on the coordination number and nature of groups bonded to the central tin atom (Tabassum and Pettinari 2006). The phosphate group of DNA sugar backbone usually acts as anchoring site as depicted in Scheme 1. Nitrogen of DNA base binding is extremely effective, stabilizing the tin center as an octahedral stable species. A novel antitumor agent triphenyltin benzimidazolethiol (TPT) (Tabassum and Pettinari 2006) was developed and was found to be more potent apoptosis-inducing agent in HeLa cells than cisplatin. The upregulation of Bak at the transcriptional level resulted in the release of cyto C from mitochondria to cytosol and subsequently proceeded by activation of procaspases 9,3 suggesting that TPT-induced apoptosis signaling proceeds by an intrinsic mitochondrial pathway.

## Factors Governing the Antiproliferative Activity

Organotins exhibit significant in vitro antiproliferative activity which in some cases is higher than the corresponding activity of cisplatin or other clinically approved drugs. While the organotin moiety is crucial for cytotoxicity, nevertheless, the ligand plays a key role in transporting and addressing of the molecule to the molecular target, resisting untimely exchange with biomolecules. This results in their structural diversity and therein a broad therapeutic activity. Consequently, a large number of organotin (IV) complexes derived from different ligands, namely, carboxylates, amino acids, oxamates containing oxygen donor ligands, organotin(IV) complexes with sulfur donor ligands (thione, thiol) were tested for in vitro cytotoxicity, in addition to organotins with Schiff bases derived from salicylaldehyde and aminopyridines which were screened against human myelogenous leukemia K562, cervix (HeLa), and murine L929 fibrosarcoma cell lines and results were compared with those of the anticancer drugs, cisplatin, carboplatin, and oxaliplatin. Some of the organotins were found to be more active in vitro than the standard cisplatin (Table 1) and have been described below.

In 2006, first attempts in developing the quantitative structure/activity and structure/property relationship for organotin compounds were reviewed (Song et al. 2006). The prediction of anticancer activity based on mixture of the organic group [R], halide or pseudohalide [X$_n$] and the donor ligand, L in the R$_2$SnL$_2$ moiety of diorganotin carboxylates and R$_2$SnX$_2$L$_2$ (L = bidentate ligand with O and/or N donor atoms) as shown in Fig. 1. Di (n-butyl) compounds were found to be the most active. Examination of the structures of tin compounds containing N-donor atoms which were tested for antitumor activity revealed that in active tin complexes, the average Sn–N bond lengths were >239 pm, whereas the inactive complexes had Sn–N bonds <231 pm which imply that predissociation of the ligand may be an important step in

**Tin Complexes, Antitumor Activity, Fig. 1** (a) Some of the organotin-carboxylate structures in literature. (b) Possible structures of five-coordinated organotin(IV)

the mode of action of these complexes. The coordinated ligand may favor transport of the active species to the site of action in the cells, where they are released by hydrolysis.

For the design of an efficacious antitumor compound, the selection of other ligands attached to the dibutyl tin (IV) moiety is of great significance. Triorganotin (IV) derivatives of N-maleoyl-protected trenaxamic acid HOCOR with formulae [(CH$_3$)$_3$SnOCOR], [(Et)$_3$SnOCOR], [(n-Bu)$_3$SnOCOR], [(Ph)$_3$SnOCOR], and [(Bn)$_3$SnOCOR] were tested in vitro for their bioactivity, against several tumor cell lines of human origin. These compounds are polymers with Sn(IV) five-coordinated and bridging carboxylate groups in the solid state as characterized by spectroscopy (Scheme 2), while five-coordinated with trigonal bipyramidal geometries in solution (Hadjikakou and Hadjiliadis 2009). These complexes display significant in vitro activities in comparison to N-maleoyl-protected tranexamic acid and the reference drugs (doxorubicin, cisplatin, 5-fluorouracil, methotrexate, and etoposide) (Table 2). The nature (alkyl/phenyl/aryl) and size of covalently attached R′ groups of Sn(IV) atom and partition coefficients plays a key role in the toxicities of the reported complexes. It was demonstrated that in vitro toxicity is enhanced by the bulkiness of the functional groups R′ attached to Sn (IV), against the tumor cell lines used. Hydrophobicity increases with the bulkiness of R′ group and boost up the bioactivity of these complexes.

Solubility in water is second important factor concerning the in vivo testing of compounds exhibiting promising in vitro properties. Therefore, several water-soluble organotin(IV) carboxylate compounds were synthesized by the condensation reaction of carboxylic acids with di-n-butyl oxide, triphenyltin hydroxide, and tri-n-butyl tin acetate (Gielen et al. 2000). The carboxylic acids involved were steroid carboxylic acids, terebic acid, 2,3:4,6-di-isopropylidene-2-keto-L-gulonic acid, gibberellic acid, and 3 S,4 S-3-[(R)-1-(tertiobutyl-dimethylsilyloxy)-ethyl]-4-[(R)-1-carboxyethyl]-2-azetidinone as well as 3,6-dioxaheptanoic acid, 3,6,9-trioxadecanoic acid, 4-carboxybenzo-18-crown-6. These complexes were tested on a panel of 60 different human tumor cell lines and in vitro primary screen with

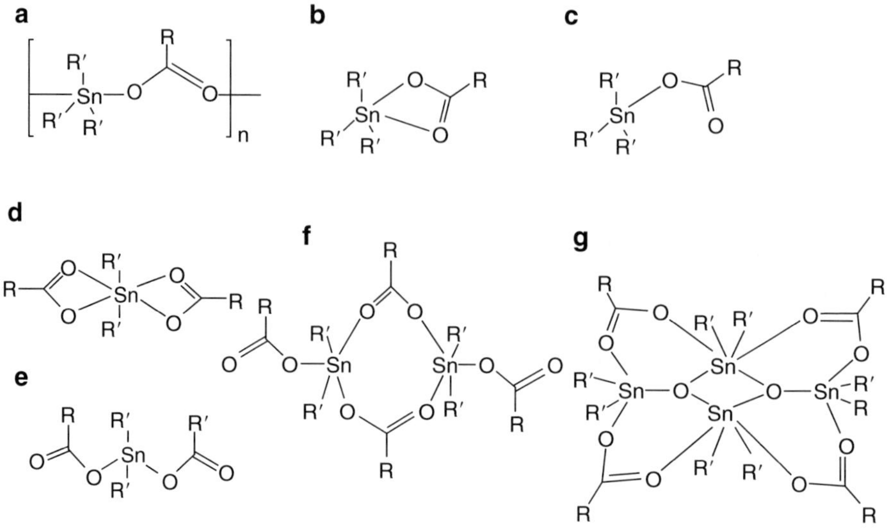

**Tin Complexes, Antitumor Activity, Scheme 2** Common structural motifs found for triorganotin (IV) and diorganotin (IV) complexes with carboxylic acids

**Tin Complexes, Antitumor Activity, Table 2** In vitro inhibitory dose for the 50% of various cancer cell lines ($ID_{50}$) in μg/ml of the organotin(IV) compounds with carboxylic acids

| Compound | Coordination mode | $ID_{50}$(μg/ml) against cancer cell lines | | | | | | |
|---|---|---|---|---|---|---|---|---|
| | | A498 | EVSA-T | H226 | IGROV | M19 | MCF7 | WiDr |
| [$(CH_3)_3$SnOCOR] | Five | 0.093 | 0.082 | 0.066 | 0.112 | 0.105 | 0.096 | 0.101 |
| [$(Et)_3$SnOCOR] | Five | 0.086 | 0.08 | 0.077 | 0.101 | 0.1 | 0.03 | 0.068 |
| [$(n-Bu)_3$SnOCOR] | Five | 0.077 | 0.081 | 0.03 | 0.083 | 0.101 | 0.024 | 0.052 |
| [$(Ph)_3$SnOCOR] | Five | 0.07 | 0.042 | 0.021 | 0.8 | 0.093 | 0.017 | 0.023 |
| [$(Bn)_3$SnOCOR] | Five | 0.05 | 0.033 | 0.005 | 0.12 | 0.89 | 0.006 | 0.017 |

seven human tumor cell lines provided by NCI was studied. In the next step, testing of promising new derivatives in human tumor xenografts of nude mice was also studied. An in vivo murine tumor model was selected initially the mouse L1210 leukemia, later on mouse Colon 26 being expected to possess higher predictive value than L1210 were selected. Though some of the compounds studied were toxic to tumor-bearing mice so that a second infection could not be administered. After administration weekly at a dose of 15 mg/Kg, compound **1** (Fig. 2) gave T/C (relative tumor size of the treated (T) mice divided by the relative tumor size of the control (C) mice expressed in%) of 77% and an ILS (increase of median life span) of 30%.

In order to make organotin compounds more soluble, another structure was designed which contains a five-membered ring moiety as well as polar substituents, organotin terebates (compound **2–4**) owing to their high in vitro antitumor activity. These complexes were tested in vivo in the mouse Colon 26. Two injections of compound **3** resulted in 3/5 toxic deaths in 1 week. The first series of polyoxyalkylene carboxylates related to water-soluble organotin carboxylate compounds (**5–12**) and benzocrown carboxylate compounds (Fig. 3) were found to be more active than doxorubicin (Table 3). The development process led to compounds with definite pharmacological activity, antitumor activity in vivo, but toxicity prevailed in some cases.

Amino acids ligands are a source of chiral stereogenic center which is another deciding factor to enhance the pharmacological behavior of the metal complexes by adopting specific conformation and target selective binding affinity for inherently chiral DNA. Amino acids are the basic structural units of proteins which recognize a specific base sequence of DNA. Many amino acid derivatives are known to exhibit impressive cytotoxic activity. Among these are diorganotin(IV) complexes of N-(5-halosalicylidene) α-amino acid of the formulations $R_2$Sn(5-X-2-O$C_6H_3$CH = NH(i-Pr)COO) (where X = Cl and R = n-Bu, Ph, and Cy) were tested against the three

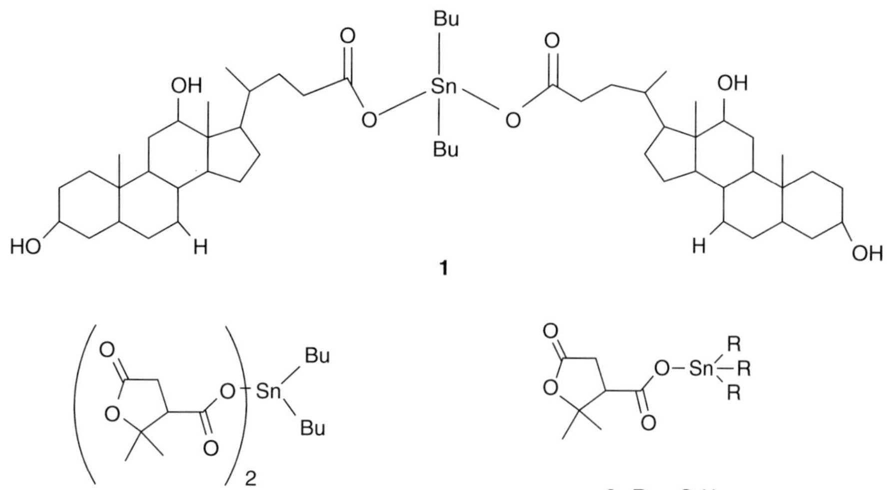

1

2

3 R=n-$C_4H_4$

4 R=$C_6H_5$

**Tin Complexes, Antitumor Activity, Fig. 2** Structure of organotin steroidcarboxylates, **1** and organotin terebates, 2–**4**

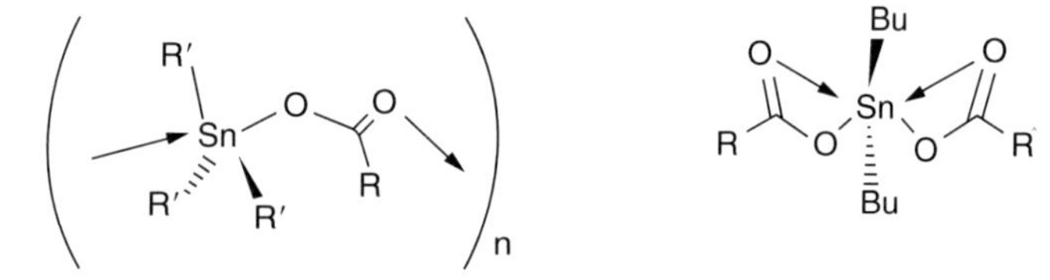

5 R′= $C_6H_5$, R= $CH_3OCH_2CH_2OCH_2$

6 R′= $C_4H_9$, R= $CH_3OCH_2CH_2OCH_2$

9 R′= $C_6H_5$, R= $CH_3O(CH_2CH_2O)_2CH_2$

10 R′= $C_4H_9$, R= $CH_3O(CH_2CH_2O)_2CH_2$

7 R= $CH_3OCH_2CH_2OCH_2$

11 R= $CH_3O(CH_2CH_2O)_2CH_2$

8 R= $CH_3OCH_2CH_2OCH_2$

12 R= $CH_3O(CH_2CH_2O)_2CH_2$

**Tin Complexes, Antitumor Activity, Fig. 3** Structure of organotin polyoxaalkanecarboxylates, **5–12**

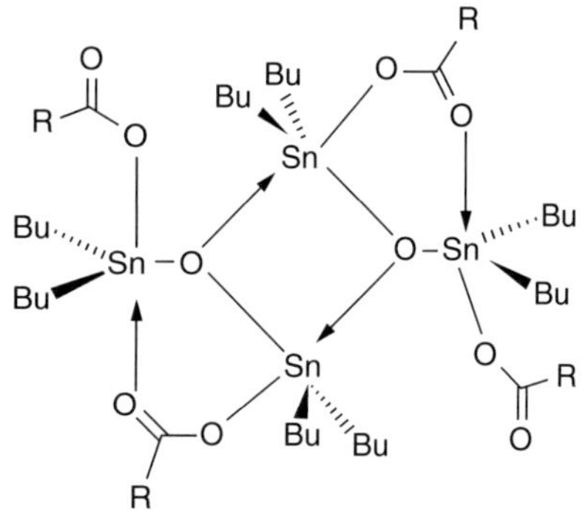

Tin Complexes, Antitumor Activity, Table 3 $IC_{50}$ values (in μM) of the organotin steroid carboxylate 1, organotin terebates 2–4, and organotin polyoxaalkanecarboxylates 5–12

| Compound | MCF7 | EVSA-T | WiDr | IGROV | M19MEL | A498 | H226 |
|---|---|---|---|---|---|---|---|
| 1 | 163 | 61 | 400 | 160 | 122 | 220 | 430 |
| 2 | 49 | 46 | 240 | 33 | 112 | 112 | 190 |
| 3 | 6.7 | 6.7 | 25 | 8.9 | 25 | 34 | 18 |
| 4 | 34 | 5.9 | 34 | 37 | 83 | 83 | 77 |
| 5 | 27 | 25 | 70 | 77 | 64 | 66 | 68 |
| 6 | 74 | 93 | 190 | 190 | 200 | 350 | 141 |
| 7 | 120 | 124 | 760 | 260 | 230 | 270 | 320 |
| 8 | <1 | <1 | 3.9 | <1 | <1 | <1 | 3.3 |
| 9 | 17 | 17 | 36 | 63 | 46 | 40 | 47 |
| 10 | 76 | 53 | 84 | 187 | 160 | 200 | 118 |
| 11 | 147 | 112 | 840 | 300 | 280 | 250 | 480 |
| 12 | <1 | <1 | <1.8 | <1 | <1 | <1 | <1 |

Tin Complexes, Antitumor Activity, Table 4 In vitro inhibitory dose for the 50% of various cancer cell lines ($ID_{50}$) in μg/ml of some organotin(IV) compounds with amino acids of the formulation, $R_2Sn(5\text{-}X\text{-}2\text{-}OC_6H_3CH = NCH(i\text{-}Pr)COO)$

| | | $ID_{50}$ (μg/ml) against cancer cell lines | | |
|---|---|---|---|---|
| Compound | Coordination mode | MCF7 | HeLa | CoLo205 |
| R=n-Bu, X=Cl | Five | 2.1 | 0.31 | 1.4 |
| R=Ph, X=Cl | Five | 2.97 | 2.9 | 2.72 |
| R=Cy, X=Cl | Five | 0.21 | 0.25 | 0.61 |
| R=Ph, X=Br | Five | 2.68 | 0.77 | 3.15 |

Tin Complexes, Antitumor Activity, Table 5 In vitro inhibitory dose for the 50% of various cancer cell lines ($ID_{50}$) in μg/ml of some organotin(IV) compounds with amino acids of the formulation, $(n\text{-}Bu)_2Sn\text{-}L$

| | | $ID_{50}$ (μg/ml) against cancer cell lines | | | | | | |
|---|---|---|---|---|---|---|---|---|
| Compound | Coordination mode | A498 | EVSA-T | H226 | IGROV | M19 | MCF7 | WiDr |
| L=Gly-Tyr | Five | 0.138 | 0.021 | 0.057 | 0.025 | 0.073 | 0.04 | 0.284 |
| L=Gly-Trp | Five | 0.196 | 0.064 | 0.133 | 0.072 | 0.182 | 0.093 | 0.424 |
| L=Leu-Tyr | Five | 0.336 | 0.074 | 0.177 | 0.118 | 0.205 | 0.15 | 0.42 |
| L=Leu-Leu | Five | 0.155 | 0.056 | 0.108 | 0.09 | 0.15 | 0.096 | 0.295 |
| L=Val-Val | Five | 0.134 | 0.032 | 0.78 | 0.046 | 0.089 | 0.051 | 0.265 |
| L=Ala-Val | Five | 0.332 | 0.084 | 0.105 | 0.119 | 0.139 | 0.478 | 0.332 |

human tumor cell lines HeLa, CoLo205, and MCF7 and results demonstrated that these complexes were efficient cytotoxic agents and their cytotoxic activities were higher than those of the clinically widely used cisplatin (Hadjikakou and Hadjiliadis 2009). Literature reports support that both the organotin moiety (R′) and the ligand (L) play an important role in cytotoxicity and results of these studies further corroborate well with structure activity relationship which decreases in the order CyHex>n-Bu>Ph for the R′ group bound to tin. Tables 4 and 5 compares 50% inhibitory dose ($ID_{50}$) against various cancer cell lines (μg/mL) of the organotin(IV) compounds of amino acids with their coordination modes around tin(IV) atom.

Some organotin (IV) derivatives of general formulae $R_2SnL$ (R = Bu, L is the dianion of glycyltyrosine, glycyltryptophane, leucylcytosine, leucylleucine, valylvaline, and alanylvaline) and $Ph_3Sn(HL\text{-}7)$ (HL-7 = glycylleucinate) were characterized. All the complexes were tested against cancer cell lines of

**Tin Complexes, Antitumor Activity, Fig. 4** Structure of triorganotin (IV) lupinylthiolate hydrochlorides

**Tin Complexes, Antitumor Activity, Table 6** Cytotoxic activity of organotin thaloamide complexes at 50% inhibitory dose (ID$_{50}$) µg/ml values against various cancer cell lines

| Compound | Leiomiosarcoma |
|---|---|
| [(CH$_3$)$_2$Sn(PMT)$_2$] | 7.4–22.3 |
| [(n-C$_4$H$_9$)$_2$Sn(PMT)$_2$] | 0.30 |
| [(C$_6$H$_5$)$_2$Sn(PMT)$_2$] | 0.5–1.0 |
| [(C$_6$H$_5$)$_2$Sn(PMT)] | 0.05 |

human origin, namely, MCF7, EVSA-T, WiDr, IGROV, M19 MEL, A498, and H226. It was observed that Ph$_3$Sn(HL) exhibited the lowest ID$_{50}$ values of the tin compounds tested and its activity was comparable to those of methotrexate and 5-fluorouracil.

Appending a carbohydrate to a metal-binding ligand has the ability to reduce toxicity, and improve solubility and molecular targeting. Carbohydrate ligands with chelating functions for metal ions have been developed to generate a well-defined binding environment as well as increase the stability of the resultant metalcomplexes. The antitumor activity of 3-C-[(triphenylstannyl) methyl]-1,2:5,6-di-O-isopropylidene-D-allofuranose (Ph$_3$-SnCH$_2$carbohydrate) and 3-C-[(triphenylstannyl) methyl]-1,2:5,6-di-O-isopyropropylidene-D-allofuranose (Ph$_3$-SnCH$_2$carbohydrate) were reported (Caruso et al. 1993). It was concluded that Sn–C bonded triphenyltin carbohydrates are less active than the Ph$_3$Sn Cl in vitro; Ph$_3$Sn-carbohydrate is more active than Ph$_3$SnCH$_2$–carbohydrate, and this may be related to the long Sn–C carbohydrate bond distance (2.225 Å) in the former compound that shows a striking biological activity in contrast to the normal inactivity of tetraorganotins. A potential benefit of this approach is that the carbohydrate can remain pendent, thereby being freely available to interact with carbohydrate transport and metabolic pathways in the body.

A large number of compounds have been tested against different cancer cell lines via gene-mediated pathway. Organotin compounds of flufenamic acid and flufenamates (flu) were evaluated for antiproliferative activity in vivo. [Bu$_2$(flu)SnOSn(flu)Bu$_2$] and [Bu$_2$Sn (flu)$_2$] (Hflu = N-[(3-trifluoromethyl)-phenyl]-anthranilic acid) exhibited high cytotoxic activity against the cancer cell lines A549 (non-small cell lung carcinoma). Because of interesting activity and better stability, the triethyl and tributyltin(IV) lupinylthiolates, IST-FS 29, IST-FS 35 (Fig. 4) were studied more thoroughly against a number of human cancer cell lines, exhibiting IC$_{50}$ values in the range of 0.6–12.4 µM for IST-FS 29 and 0.6–1.8 µM for IST-FS 35. Moreover, IST-FS 29 was very active in vivo in murine tumor models (P338 myelomonocytic leukemia, B16–F10 melanoma, and 3LL Lewis lung carcinoma) and its cytotoxic effects seem consistent with necrosis or delayed cell death rather than apoptosis. Mild and reversible signs of acute toxicity such as behavioral symptoms, weight loss, and histological alterations were mainly reported at the highest single dose of 28 mg/kg. On the other hand, lower concentrations of the compound ranging from 7 to 21 mg/kg did not result in major toxic effects, even after repeated administration. The antitumor activity studies showed that fractional dosing, rather than single bolus administration, over 1 week, might prove more active and better tolerated by allowing the achievement of the highest therapeutic total dose of IST-FS 29 (42 mg/kg). Indeed, repeated administration of IST-FS 29 resulted in marked significant improvement of antitumor activity against B16F10 (50% of tumor vol. inhibition, $P = 0.0003$) and, to a greater extent, 3 LL (90% of tumor vol. inhibition, $P = 0.0001$) tumors. These results indicate that IST-FS 29 might be a suitable candidate as an orally administrable anticancer drug and support its further development in human tumor xenografts. After repeated oral administration, IST-FS 35 was able to inhibit the tumor growth of implanted P338 and B16–F10 cells, upto 96%, after a single i.v. injection. Since IST–FS 35 produced peritoneal irritation via the i.p. route and did not appear to be well absorbed orally, i.v. administration was considered as the treatment of choice.

The organotin (IV) complexes of heterocyclic thioamides 2-mercapto-pyrimidine (PMTH), of formulae [(CH$_3$)$_2$Sn(PMT)$_2$], [(n-C$_4$H$_9$)$_2$Sn(PMT)$_2$], [(C$_6$H$_5$)$_2$Sn(PMT)$_2$], and [(C$_6$H$_5$)$_2$Sn(PMT)], were used to study their cytotoxicity against sarcoma cancer

L¹, R = R' = H

L², R = H, R' = 3-CH$_3$

L³, R = H, R' = 4-CH$_3$

L⁴, R = OCH$_3$, R' = H

L⁵, R = OCH$_3$, R' = 3-CH$_3$

**Tin Complexes, Antitumor Activity, Fig. 5** Structure of organotin complexes derived from Schiff base ligands, 13–17

cells. The antiproliferative effects of these complexes are tabulated in Table 6. The in vitro cytotoxicity against the cancer cell line of sarcoma cells (mesenchymal tissue) from the Wistar rat, polycyclic aromatic hydrocarbons (PAH, benzo[α]pyrene) carcinogenesis of the organotin(IV) complexes with the heterocyclic thioamides; 2-mercaptobenzothiazole (Hmbzt), and 2-mercapto-nicotinic acid (H$_2$MNA) of formulae [(n-C$_4$H$_9$)$_2$Sn(MBZT)$_2$] and {[(C$_6$H$_5$)$_3$Sn]$_2$(MNA).[(CH$_3$)$_2$CO]} have been studied. The very strong cytotoxic activity exhibited by {[(C$_6$H$_5$)$_3$Sn]$_2$(MNA).[(CH$_3$)$_2$CO]} complex is most probably due to the availability of the free coordination positions around tin(IV) atoms since both Sn(IV) are five-coordinated with trigonal bipyramidal geometry. One additional reason for this behavior should be the presence of two tin (IV) atoms in this compound.

Previously, organotin (IV) complexes with Schiff bases ligands have been studied, not only due to their novel structural features caused by the multidenticity of these ligands but also in view of their pharmacological and antitumor activity. Organotin(IV) complexes with Schiff bases derived from salicylaldehyde and aminopyridines of general formulae Me$_2$SnCl$_2$.2L (L = L¹ (**13**), L² (**14**), L³ (**15**)) or Me$_2$SnCl$_2$.L (L = L⁴ (**16**), L⁵ (**17**)) and the ionic compounds [H$_2$NpyN-H⁺]$_2$ [Ph$_2$SnCl$_4$]$_2$⁻ (H$_2$NpyN-H⁺ = aminopyridines used for the preparation of Schiff bases L¹–L⁵), (Fig. 5) respectively, were screened against the human myleogenous leukemia K562, cervix (HeLa) and murine L929 fibrosarcoma cell lines, and the results were compared with those of the anticancer drugs, cisplatin, carboplatin, and oxaliplatin. The free Schiff bases L¹–L⁵ and their Me$_2$SnCl$_2$ complexes have no cytotoxic activity against any of the three tumor lines used. The most cytotoxic compounds among all the compounds studied were the ionic compounds whose ID$_{50}$ values ranged between 0.2 and 4.7 μg/ml. The cytotoxicity of the ionic compounds is stronger than those of the reference standards, cisplatin (ID$_{50}$ value ranged between 2.8 and 28.3 μg/ml) and carboplatin (ID$_{50}$ value ranged between 12.2 and >50 μg/ml) and even higher than that of oxaliplatin (ID$_{50}$ value is 33.2 μg/ml) against the HeLa cell line. Further, the ID$_{50}$ values of these compounds against the K562 cell line, i.e., 0.2–0.3 μg/ml, are very similar to that of oxaliplatin (ID$_{50}$ = 0.3 μg/ml) against the same cell lines.

## Conclusion

Organotin compounds exhibit significant in vitro and in vivo antiproliferative activity and in most of the cases, was found higher than cisplatin and the other standard drugs used for clinical treatment in cancer chemotherapy. The antiproliferative activity of active organotin compounds is mainly dependent on (1) the availability of vacant coordination positions at tin (IV) atoms, (2) nature of ligand L (alkyl/phenyl/aryl) and size of covalently attached organotin moiety R' groups of Sn(IV) atoms, and (3) the occurrence of relatively stable ligand-Sn bonds e.g., Sn–N and Sn–S and their slow hydrolytic decomposition.

## Cross-References

▶ Calcium ATPase
▶ Tin, Toxicity

## References

Alama A, Tasso B, Novelli F, Sparatore F (2009) Organometallic compounds in oncology: implications of novel organotins as antitumor agents. Drug Discov Today 14:500–508

Caruso F, Bol-Schoenmakers M, Penninks AH (1993) Crystal and molecular structure and in vitro antiproliferative and antitumor activity of two organotin(IV) carbohydrate compounds. J Med Chem 36:1168–1174

Chauhan M, Banerjee K, Arjmand F (2007) DNA binding studies of novel copper(II) complexes containing L-tryptophan as chiral auxiliary: In vitro antitumor activity of Cu–Sn2

complex in human neuroblastoma cells. Inorg Chem 46:3072–3082
Crowe AJ (1987) The chemotherapeutic properties of tin compounds. Drugs Future 12:255–275
de Vos D, Willem R, Gielen M, Wingerden KE, Nooter K (1998) The development of novel organotin anti-tumor drugs: structure and activity. Metal Based Drugs 5:179–188
Gielen M, Tiekink ERT (2005) Tin compounds and their therapeutic potential. In: Gielen M, Tiekink ERT (eds) Metallotherapeutic drugs and metal-based diagnostic agents: the use of metals in medicine. Wiley, Chichester
Gielen M, Biesemans M, Willem R (2000) Synthesis, characterization and in vitro antitumor activity of di- and triorganotin of polyoxa- and biologically relevant carboxylic acids. J Bioinorg Chem 79:139–145
Hadjikakou SK, Hadjiliadis N (2009) Antiproliferative and antitumor activity of organotin compounds. Coord Chem Rev 253:235–249
Ray D, Sarma KD, Antony A (2000) Differential effects of tri-n-butylstannyl benzoates on induction of apoptosis in K562 and MCF-7 cells. IUBMB Life 49:519–525
Saxena AK, Huber F (1989) Organotin compounds and cancer chemotherapy. Coord Chem Rev 95:109–123
Song X, Zapata A, Eng G (2006) Organotins and quantitative-structure activity/property relationships. J Organomet Chem 691:1756–1760
Tabassum S, Pettinari C (2006) Chemical and biotechnological developments in organotin cancer chemotherapy. J Organomet Chem 691:1761–1766
Yang P, Guo M (1999) Interactions of organometallic anticancer agents with nucleotides and DNA. Coord Chem Rev 185–186:189–211

# Tin, Physical and Chemical Properties

Fathi Habashi
Department of Mining, Metallurgical, and Materials Engineering, Laval University, Quebec City, Canada

Tin is an ancient metal. It is obtained chiefly from the mineral cassiterite, $SnO_2$. It is alloyed with copper to form bronze containing 12% Sn, which is used to cast bells, statues, and cannons easily. Pewter, an alloy of 85–90% tin with the remainder of copper, antimony, and lead, was used for cutlery. Tin has long been used as a solder in the form of an alloy with lead, tin accounting for 5–70%. Such solders are used for joining pipes or electric circuits. Tin bonds readily to iron and is used for coating steel to prevent corrosion. Tin-plated steel containers are used for food preservation. Tin is a malleable, ductile silvery-white metal. It is nontoxic but certain organotin compounds are highly toxic.

**Tin, Physical and Chemical Properties, Fig. 1** Transformation of tin into the gray amorphous form after cooling for 10 h at $-38\,^\circ C$

Tin exists in two modifications: β-tin (the metallic form, or white tin) is stable at room temperature and above room temperature is malleable. In contrast, α-tin (nonmetallic form, or gray tin) is stable below 13.2 °C, is brittle with no metallic properties. It is a gray powdery material and has a diamond cubic structure. The transformation from white to gray tin is known as tin pest. The rate of transformation is however very slow. For all practical purposes β-tin is the metal in common use (Fig. 1).

## Physical Properties

| | |
|---|---|
| Atomic number | 50 |
| Atomic weight | 118.71 |
| Relative abundance in Earth's crust,% | $4 \times 10^{-3}$ |
| Crystal structure | |
| α-Sn (gray tin) | Face centered cubic |
| β-Sn (white tin) | Tetragonal |
| Transformation temperature, °C | 13.2 |
| Enthalpy of transformation, J/mol | 1,966 |
| Lattice constants at 25°C, pm | |
| α-Sn | $a = 648.92$ |

(continued)

| | |
|---|---|
| β-Sn | $a = 583.16, c = 318.13$ |
| Density of α − Sn, g/cm$^3$ | 5.765 |
| Density of β-tin at 20°C, g/cm$^3$ | 7.286 |
| Density of liquid tin, g/cm$^3$ | |
| At 240°C | 6.992 |
| At 400°C | 6.879 |
| At 800°C | 6.611 |
| At 1,000°C | 6.484 |
| Melting point, °C | 232.06 |
| Boiling point, °C | 2,603 |
| Enthalpy of fusion, J/mol | 7,029 |
| Enthalpy of vaporization, J/mol | 295,763 |
| Molar heat capacity of β-tin, J mol$^{-1}$ K$^{-1}$ | |
| At 25°C | 27.0 |
| At 230°C | 30.7 |
| Liquid | 28.5 |
| Vapor pressure, Pa | |
| At 1,000 K | $9.8 \times 10^{-4}$ |
| At 1,800 K | 750 |
| At 2,100 K | 8,390 |
| At 2,400 K | 51,200 |
| Cubic coefficient of expansion, K$^{-1}$ | |
| α-Sn at −130°C to +10°C | $(14.1–4.7) \times 10^{-6}$ |
| β-tin at 0°C | $59.8 \times 10^{-6}$ |
| β-tin at 50°C | $69.2 \times 10^{-6}$ |
| β-tin at 100°C | $71.4 \times 10^{-6}$ |
| β-tin at 150°C | $80.2 \times 10^{-6}$ |
| Molten tin at 700°C | $105.0 \times 10^{-6}$ |
| Coefficient of thermal conductivity of β-tin at 0°C, W cm$^{-1}$ K$^{-1}$ | 0.63 |
| Surface tension, N/m | |
| At 232°C | 0.53–0.62 |
| At 400°C | 0.52–0.59 |
| At 800°C | 0.51–0.52 |
| At 1,000°C | 0.49 |
| Dynamic viscosity, Pa·s | |
| At 232°C | $2.71 \times 10^{-3}$ |
| At 400°C | $1.32 \times 10^{-3}$ |
| At 1,000°C | $0.80 \times 10^{-3}$ |
| Specific electrical resistivity, Ωm | |
| α-Sn at 0°C | $5 \times 10^{-6}$ |
| β-Sn at 25°C | $11.15 \times 10^{-6}$ |

(continued)

| | |
|---|---|
| Transition temperature for superconductivity, K | 3.70 |
| Magnetic susceptibility of β-Sn, m$^3$/kg | $2.6 \times 10^{-11}$ |
| Yield strength at 25°C, N/mm$^2$ | 2.55 |
| Ultimate tensile strength, N/mm$^2$ | |
| At −120°C | 87.6 |
| At 15°C | 14.5 |
| At 200°C | 4.5 |
| Brinell hardness (10 mm, 3,000 N, 10 s) | |
| At 0°C | 4.12 |
| At 100°C | 2.26 |
| At 200°C | 0.88 |
| Modulus of elasticity, N/mm | |
| At −170°C | 65,000 |
| At −20°C | 50,000 |
| At 0°C | 52,000 |
| At 40°C | 49,300 |
| At 100°C | 44,700 |
| At 200°C | 26,000 |

## Chemical Properties

Tin is amphoteric, reacting with both strong bases and strong acids with evolution of hydrogen. With sodium hydroxide solution, tin forms Na$_2$[Sn(OH)$_6$]. The reaction with acids is slow in the absence of oxygen. Vigorous reactions occur with nitric acid. While hydrogen fluoride does not attack tin, hydrochloric acid reacts even at a concentration of 0.05% and temperatures below 0°C. Tin is not attacked by sulfurous acid or by <80% sulfuric acid. The most important use of tin and tin-plated materials is in the preserved food industry. For this reason, the possibility of reactions of tin with certain organic acids is important. Lactic, malic, citric, tartaric, and acetic acids either do not react at normal temperatures or do so to a negligible extent, especially in the absence of atmospheric oxygen. It has two oxidation states: +2 and +4.

## References

Graf GG (1997) Molybdenum. In: Habashi F (ed) Handbook of extractive metallurgy. Wiley-VCH, Weinheim, pp 683–714

# Tin, Toxicity

Elke Dopp and Albert W. Rettenmeier
Institute of Hygiene and Occupational Medicine,
University of Duisburg-Essen, Essen, Germany

## Synonyms

Organotin; Tin; Toxicity

## Definition

*Tin* (Sn, from *stannum*) is the 49th most abundant element in the Earth's crust, where it occurs primarily as a mineral containing tin dioxide ($SnO_2$). In its elemental form, tin is a soft, silver-white ▸ metal which is relatively inert toward air and water. Tin is a main group metal (atomic number 50) and appears in its inorganic and organic compounds in two oxidation states: +2 and the more stable +4. While elemental tin and the inorganic tin compounds are considered relatively nontoxic, the more lipid-soluble organic tin species exhibit a variety of distinct toxic reactions. Generally, tri-substituted ($R_3SnX$) and disubstituted ($R_2SnX_2$) organotins are more toxic than monosubstituted ($RSnX_3$) tin species. Toxicity decreases with increasing alkyl chain length independent on the counter ions (Gajda and Jancsó 2010).

## Exposure

Human exposure to tin arises from the release of the metal and its compounds from natural and anthropogenic sources. Elemental tin is used in alloys such as brass or bronze, in pewter and in soldering materials. Inorganic tin compounds are largely employed as intermediates for the synthesis of organotin compounds and in electrolytes for plating tin and tin alloys. They are also found in pigments and glazes, in cosmetics, dental care products, coloring agents, and food additives. Organotin compounds are widely utilized as stabilizers for PVC, as homogenous catalysts for polyurethane foam formation and silicone vulcanization, for glass coating, as pesticides (fungicides, insecticides, acaricides), antifouling agents, wood preservatives, disinfectants, and rodent repellants. Some of the applications of organotin compounds have been discontinued because of their high toxicity, particularly in aquatic environments (ATSDR 2005; Gajda and Jancsó 2010). Naturally occurring organotin compounds are methylated tin species formed from inorganic tin by the action of bacteria present in soil and marine sediments (Dopp et al. 2004; Hirner and Rettenmeier 2010). The most important source for human exposure to tin is the uptake from food contaminated by tin compounds (seafood, food products in tin-lined cans) and from contact with household products. Data on human exposure to inorganic and organotin compounds is however limited. Evidently, tin is not an essential element for humans (ATSDR 2005).

## Toxicokinetics

Inorganic and organic compounds can be absorbed to some extent by inhalation, ingestion, or dermal penetration. Organotin compounds are more readily taken up than inorganic compounds by any of these routes with absorption increasing with higher degrees of alkylation. Absorbed tin is distributed throughout the body. In postmortem samples, tin has been found in kidney, liver, lung, brain, and bone tissue with the highest concentration measured in the latter tissue. The metabolism of inorganic tin compounds have not been investigated as yet. Organic tin compounds are successively dealkylated/dearylated according to animal and in vitro studies. Dealkylation products may be hydroxylated and/or conjugated with glutathione and further metabolized to mercapturic acid derivates. Feces and, to a lesser extent, urine are the major routes of tin excretion. Most tin compounds, particularly the inorganic species, are eliminated rather rapidly, but small amounts are stored in bones for up to a few months. While inorganic and organic tin compounds can be transferred across the placenta, the uptake of these compounds through breastfeeding has not yet been demonstrated (Appel 2004; ATSDR 2005).

## Mechanisms of Action

The toxic effects of organotin compounds result from interactions of the alkyl and aryl moieties with cell membranes and the intracellular reactivity of the alkyltin cation. Basically, the effects can be categorized in $Ca^{2+}$-dependent and $Ca^{2+}$-independent reactions. Organotin compounds disturb ▶ $Ca^{2+}$ homeostasis by increasing free intracellular $Ca^{2+}$ concentration, which among others affects signaling pathways, induces apoptosis, and promotes depolymerization and disintegration of cytoskeletal and nuclear proteins. $Ca^{2+}$-independent are coordinative and covalent binding to proteins leading to inhibition of enzymes involved in energy production and drug metabolism and in the regulation of transmembrane gradients. Increased neuronal release of and/or decreased neuronal uptake of neurotransmitters and decreased expression of neural cell adhesion molecules induced by trimethyltin, and suppressed T-cell-mediated immune response by butyltin compounds are also among the toxic reactions caused by organotin compounds.

Although there are many reports describing the potential toxicity of organotins in human and mammals, the critical target molecules for the toxicity of organotin compounds remain unclear. Recently, organotin compounds including TBT and TPT were identified as nanomolar agonists for retinoid X receptor (RXR) and peroxisome proliferator-activated receptor (PPAR) gamma, which are members of the nuclear receptor superfamily. TBT and TPT are potent activators of these nuclear hormone receptors (RXR, PPARγ) and they promote adipocyte differentiation, suggesting that these organotins might contribute to the development of metabolic diseases (Nakanishi 2008).

## Acute Toxicity

The acute toxicity of inorganic tin compounds is rather low. Following ingestion of canned food contaminated by tin, nausea, vomiting, and diarrhea have been reported. Stomach aches, anemia, and liver and kidney disorders may occur if larger amounts of inorganic tin are ingested. In contrast, organotin compounds exhibit a considerably more pronounced toxicity following acute or subacute exposure. Reported effects include skin and eye irritation, respiratory irritation, gastrointestinal effects, and neurological sequelae. Lethal cases have occurred both after acute inhalation exposure to a mixture of trimethyltin and dimethyltin vapors and after acute oral ingestion of trimethyltin. Treatment of *Staphylococcus* infections with a drug contaminated with triethyltin iodide (the so-called Stalinon® affair) caused about 100 deaths in France in 1954. Neurological symptoms (headache, photophobia, altered consciousness, and convulsions) occurred about 4 days after the intoxication and partially continued in the surviving patients for several years.

## Chronic Toxicity of Inorganic Tin

There are only few reports on chronic toxic effects of inorganic tin compounds. Chronic inhalation exposure to stannic oxide dust or fumes may cause stannosis, a benign form of pneumoconiosis. Gastrointestinal symptoms may occur after repeated ingestion of inorganic tin compounds. Signs of anemia and gastrointestinal distension are effects observed after chronic exposure of animals to inorganic tin compounds.

## Chronic Toxicity of Organotin Compounds

### Neurotoxicity

As indicated by the accidental, sometimes lethal intoxication cases in France in 1954, neurotoxicity is an important toxicological endpoint of certain organotin compounds. Particularly trimethyl and triethyltin are potent neurotoxins, however, symptoms of neurotoxicity (encephalopathy) have also been found after exposure to dimethyltin (Michalke et al. 2009). In addition to the clinical signs and symptoms such as headaches, photophobia, altered consciousness, and convulsions, reported in the intoxication cases, morphological alterations have been demonstrated in various brain regions. Following exposure to trimethyltin, neuronal necrosis have been found in areas of the limbic system, particularly in the hippocampus. Behavioral changes such as aggression, memory loss, and unresponsiveness may result. As observed in the French cases, triethyltin ingestion caused brain and cord swelling characterized by fluid accumulation between myelin layers, splitting of the myelin sheets, and formation of intramyelin vacuoles. Similar findings were obtained in animal studies when the respective short-chain alkyltin compounds were administered. The mechanisms of neurotoxic action of trimethyl and triethyltin have not yet been elucidated in detail. One hypothesis is that small amounts of trimethyltin may inhibit the ability of astrocytes to maintain a transmembrane K+ gradient causing an imbalance between neuronal inhibition/excitation.

## Immunotoxicity

The immune system is a primary target of the butyl- and octyltin-induced toxicity as shown in numerous animal studies, however, adverse immunological effects have not been reported in exposed humans. Immunological alterations observed in rats after administration of tributyltin oxide involve the depletion of lymphocytes in the thymus and a reduced size and weight of this organ. It seems to be a direct and selective action of tributyltin oxide on the lymphocytes, whereas dialkyltins appear to interfere with the proliferation of the thymocytes. Acute immunotoxic effects require daily doses as high as >2 mg/kg in the rats, the impairment of specific and nonspecific resistance to infections may occur already at daily doses as low as 0.25 mg/day if applied long term.

## Hematological Effects

Anemia, as indicated by a decreased hemoglobin concentration, has been observed after subchronic or chronic oral exposure of rats to dibutyltin dichloride, tributyltin oxide, or dioctyltin dichloride. Based on the finding that tributyltin oxide administration led to increased reticulocyte counts and reduced iron concentrations in serum, it has been suggested that the organotin compound interferes with hemoglobin synthesis, either by suppressing iron uptake or by fostering iron loss. It appears unlikely that environmental levels of organotin compounds are high enough to cause hematological effects in humans.

## Reproductive and Developmental Toxicity

Effects of organotin compounds on human reproduction and development have not been investigated yet. In rats and mice, administration of di- and tributyl- and of triphenyltin compounds in daily doses of >3 mg/kg during gestation days 7–9 induced pregnancy failure, pre- and postimplantation losses, resorptions, and stillbirths. In male rats, exposure to tributyltin in daily doses of 10 mg/kg affected reproduction by causing histologic alterations in seminal vesicles and epididymis and reduced sperm counts. Embryotoxic and teratogenic effects, mainly cleft palate and facial malformations, have also been observed after exposure of rat embryos to organotins. There is still uncertainty whether these effects only occur secondary to maternal toxicity.

The observed effects on reproduction and development have been related to the interference of organotin compounds with the synthesis of sex hormones, resulting in an androgen/estrogen imbalance which affects sexual maturation. In in vitro studies in human choriocarcinoma cells, tributyl- and triphenyltin compounds markedly enhanced estradiol biosynthesis along with the increase of both aromatase activity and 17-beta-hydroxysteroid dehydrogenase type I (17beta-HSD I) activity, which converts low-activity estrogen estrone to the biologically more active form estradiol. It is unclear whether these hormonal alterations are relevant to human organotin exposure. Contrary to the enhancement of aromatase activity observed in mammalian cells, tributyl- and triphenyltin compounds appeared to be potential competitive inhibitors of aromatase in gastropods. As this enzyme converts androgen to estrogen, its inhibition leads to increased androgen levels causing the irreversible sexual abnormalities called "imposex" and "intersex," respectively. "Imposex" is a masculinization process in female neogastropod snails, involving the development of male sex organs. The imposition of a vas deferens disrupts the oviducal structure and function preventing normal breeding activity and causing population disappearance. Oogenesis is supplanted by spermatogenesis in some species. "Intersex" is a related condition observed in littorinid mesogastropoda which too become unable to lay eggs. Field evidence clearly associates these syndromes with the use of tributyltin compounds as an antifouling agent, chiefly on boat hulls. Dose-related effects can be replicated in laboratory exposures to environmentally relevant concentrations of these compounds (Matthiessen and Gibbs 1998). After organotin compounds were recognized as persistent organic pollutants with an extremely high endocrine-disrupting activity in some marine organisms, the use of these compounds was banned by the European Union in 2003 and worldwide by the International Maritime Organization.

No data are available on endocrine-related effects of short-chain alkyltins.

## Genotoxicity/Carcinogenicity

Genotoxicity studies with inorganic tin compounds in vitro provided some positive and some negative results. Stannous chloride ($SnCl_2$) gave mainly negative results in bacterial tests, but caused DNA damage and sister chromatid exchanges in Chinese hamster ovary cells (CHO), and chromosomal aberrations in CHO cells and human peripheral lymphocytes. Stannic chloride ($SnCl_4$) was negative in bacterial test systems and in CHO cells, but induced chromosomal aberrations in human peripheral lymphocytes. The rather weak genotoxic effect of the stannous ion has been attributed to the formation of reactive oxygen species. Overall, a similar picture was obtained from in vitro and in vivo genotoxicity studies with organotin compounds. Weakly positive results were obtained in a few tests (mainly in mammalian test systems), but most investigations in bacterial and mammalian test systems turned out to be negative.

There are no studies that evaluated whether inorganic or organic tin compounds cause cancer in humans. A few animal studies showed, however, that some organotin compounds may have a carcinogenic potential which may also be of some relevance to humans. A mixture of mono-*n*-octyltin trichloride and di-*n*-octyltin dichloride induced statistically significant thymus lymphomas in female rats and generalized malignant lymphomas in rats of either sex in the highest dose-groups (Ciba-Geigy 1986). Di-*n*-butyltin diacetate caused a significant increase of hepatocellular adenomas and carcinomas in male mice, and tri-*n*-butyltin oxide significantly induced benignant hypophyseal tumors, adrenal pheochromocytomas, parathyroid adenomas, and a rare anaplastic tumor of the exocrine pancreas in male and/or femala rats. Based on these studies and in view of the negligible genotoxic potential of the organotin compounds, the German Commission for the Investigation of Health Hazards of Chemical Compounds in the Work Area has classified *n*-butyltin and *n*-octyltin compounds as category 4 carcinogens. (Substances with carcinogenic potential for which a non-genotoxic mode of action is of prime importance and genotoxic effects play no or at most a minor part provided the MAK values (Maximum concentration values) are observed.) (DFG 2010). Under these conditions no contribution to human cancer risk is expected. According to the American Conference of Governmental Industrial Hygienists (ACGIH®) are all tin compounds "not classifiable as human carcinogens" (category A4).

A basically identical assessment/evaluation regarding the carcinogenicity of tributyltin oxide has been made by the US Environmental Protection Agency (USEPA) ("not classifiable as to human carcinogenicity").

## Regulations and Advisories

Exposure to tin and its inorganic and organic compounds has been regulated by national and international regulatory authorities to protect human health. The US Food and Drug Administration (FDA) has set limits for the use of stannous chloride as a food additive and of some organic tin compounds in coatings and plastic food packaging. The use of certain organotin compounds in paints has been limited by the US EPA. Several agencies have established workplace exposure limits for tin and tin compounds. The US Occupational Safety and Health Administration (OSHA), the National Institute for Occupational Safety and Health (NIOSH), and the American Conference of Governmental Industrial Hygienists (ACGIH®) recommend workplace exposure limits of 2 mg/m$^3$ for inorganic tin compounds (except tin oxides (OSHA and NIOSH) and tin hydride (ACGIH®)) and of 0.1 mg/m$^3$ for organotin compounds (except tricyclohexyltin hydroxide (NIOSH)). In contrast, the German Commission for the Investigation of Health Hazards of Chemical Compounds in the Work Area has stated that there is insufficient information to establish a maximum concentration value (MAK) for tin and inorganic tin compounds. For n-butyltin and n-octyltin compounds, a MAK value of 0.02 mg/m$^3$ was established, for phenyltin compounds a MAK value of 0.002 mg/m$^3$ and for the other organic tin compounds a MAK value of 0.1 mg/m$^3$ (DFG 2010).

Based on No observed adverse effect levels (NOAEL) for immunological effects of tributyltin oxide in rats ATSDR derived intermediate- and chronic-duration oral Minimal risk levels (MRL) of 0.0003 mg/kg per day. For intermediate duration, exposure to dibutyltin chloride and inorganic tin MRLs of 0.005 mg/kg per day and 0.3 mg Sn/kg per day, respectively, have been derived. There are no MRLs available for oral exposure to methyl-, ethyl-, octyl-, and phenyltin compounds and for any inhalation and dermal exposure to tin compounds. An MRL is defined as an estimate of daily human exposure to a substance that is likely to be without an appreciable risk of adverse effects over a specified duration of exposure (ATSDR 2005).

## Cross-References

▶ Bacterial Calcium Binding Proteins
▶ Calcium Homeostasis: Calcium Metabolism
▶ Metals and the Periodic Table

## References

Appel KE (2004) Organotin compounds: toxicokinetic aspects. Drug Metab Rev 36:763–786
Aschner M, Gannon M, Kimelberg HK (1992) Interactions of trimethyl tin (TMT) with rat primary astrocyte cultures: altered uptake and efflux of rubidium, L-glutamate and D-aspartate. Brain Res 582:181–185
ATSDR (Agency for Toxic Substances and Disease Registry) (2005) Toxicological profile for tin and tin compounds. U.S. Department of Health and Human Services, Atlanta
Ciba-Geigy Ltd (1986) 24 month carcinogenicity study in rats. Final report TK127000/1 Basel, Ciba-Geigy Ltd., 2. January (GU project No. 800218)
Cooke CM (2002) Effect of organotins on human aromatase activity in vitro. Toxicol Lett 126:126–130
Dopp E, Hartmann LM, Florea A-M, Rettenmeier AW, Hirner AV (2004) Environmental distribution, analysis, and toxicity of organometal(loid) compounds. Crit Rev Toxicol 34:301–333
DFG (Deutsche Forschungsgemeinschaft) (2010) List of MAK and BAT Values 2010, Report No. 46. Commission for the Investigation of Health Hazards of Chemical Compounds in the Work Area (ed), Wiley-VCH, Weinheim
Gajda T, Jancsó A (2010) Organotins. formation, use, speciation, and toxicology. In: Sigel A, Sigel H, Sigel RKO (eds) Metal ions in life sciences, vol 7. Royal Society of Chemistry, Cambridge, pp 111–151
Hirner AV, Rettenmeier AW (2010) Methylated metal(loid) species in humans. Met Ions Life Sci 7:465–521
Koczyk D (1996) How does trimethyltin affect the brain: facts and hypotheses. Acta Neurobiol Exp (Warsz) 56:587
Matthiessen P, Gibbs PE (1998) Critical appraisal of the evidence for tributyltin-mediated endocrine disruption in mollusks. Environ Toxicol Chem 17:37–43
Michalke B, Halbach S, Nischwitz V (2009) Metal speciation related to neurotoxicity in humans. J Environ Monit 11:939–954
Mosinger M (1975) Centre D'Exploitations et de Recherches Medicales, Marseille, Final Report Advastab TM 181 FS (Cincinnati-Milacron)
Nakanishi T (2008) Endocrine disruption induced by organotin compounds; organotins function as a powerful agonist for nuclear receptors rather than an aromatase inhibitor. J Toxicol Sci 33(3):269–276

# Titanium and Low Molecular Mass Substances

Jiangxue Wang, Ying Hou and Jiawei Ma
Key Laboratory for Biomechanics and Mechanobiology of the Ministry of Education, School of Biological Science and Medical Engineering, Beihang University, Beijing, China

## Definition

The dialkyl silyl-bridged alkylimido cyclopentadienyl (Cp) ligand gained much concern in metallocenes and catalysts because of its application in the polymerization process and organic substrate processing. In organometallic chemistry, metallocenes and related cylopentadienyl derivatives represent a cornerstone. Titanocene dichloride ($Cp_2TiCl_2$), a metallocene derivative, is a common reagent that behaves as the source of $Cp_2Ti^{2+}$ in organometallic and organic synthesis. Due to the four ligands around the metal center, the structure of $Cp_2TiCl_2$ is not a typical sandwich structure like ferrocene, but a distorted tetrahedral shape where the aromatic rings are not co-planar (Fig. 1). Each of the two Cp rings is attached as $\eta5$ ligands. Viewing the Cp ligands as tridentate, the complex has a coordination number of 8. In 1979, Köpf and Köpf-Maier synthesized $Cp_2TiCl_2$ because of its antitumor activity (Köpf and Köpf-Maier 1979). $Cp_2TiCl_2$ is a precursor to many Ti(II) derivatives. Titanocene, $TiCp_2$, is highly reactive and is not well known.

Because of the short Cp–M–N bond angle (less than 115°), the metal center of metallocene displays stronger Lewis acid behavior, and its coordination sphere is more accessible to reactants. To achieve the desired catalytic behavior, the Cp-amido ligand could be modified.

For titanium, a short bridge ($SiMe_2$) and sterically demanding cyclopentadienyl (Cp) ligand and amido functionalities ($C_5Me_4$, N-t-Bu) appear to give the highest activity, the highest molecular weight, and the highest incorporation of co-monomer. Using Grignard reagent and alkyl lithium compounds, the reduction of $Cp_2TiCl_2$ could result in the fulvalene complex. In the presence of ethylene, the Cp compound cannot be made and $(C_5Me_5)_2TiCl_2$ was derived (Buchwald and Nielsen 1988).

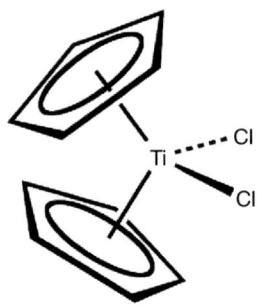

**Titanium and Low Molecular Mass Substances, Fig. 1** Titanocene dichloride

By metathesis reaction of dilithium salts or Grignard reagent, Bertolasi et al. (2007) synthesized the titanium and zirconium complexes [M($\eta^5$-C$_5$Me$_4$SiMe$_2$-$\eta^1$-N-2-R)(NMe$_2$)$_2$] (R = pyridine, thiazole; M = Zr, Ti) containing tetramethylcyclopentadienyl-dialkylsilyl bridged amidinato as pendant ligand. The complexes show a general distorted tetrahedral geometry structure typically of complexes with *ansa*-monocyclopentadienyl-amido ligands acting in a bidentate mode, which is regardless of in solution and the solid state. After activated with methylaluminoxane (MAO), they catalyzed the polymerization of ethylene, even though the titanium complexes present lower activity than the homologous zirconium complexes.

Among the various catalytic processes, the titanium and zirconium compounds catalyze the polymerization of alkenes. In addition, the reaction of zirconium and titanium complexes catalyzes the inter- and intramolecular hydroamination of alkynes and allenes. Thus, they are active precursors in regioselective catalytic hydroamination operating with an anti-Markovnikov mechanism.

## Cross-References

▶ Mercury and Low Molecular Mass Substances
▶ Titanium(IV) Intake by Apotransferrin
▶ Titanium, Biological Effects

## References

Bertolasi V, Boaretto R, Chierotti MR, Gobetto R, Sostero S (2007) Synthesis, characterization and reactivity of new complexes of titanium and zirconium containing a potential tridentate amidinato-cyclopentadienyl ligand. Dalton Trans 28(44):5179–5189

Buchwald SL, Nielsen RB (1988) Group 4 metal complexes of benzynes, cycloalkynes, acyclic alkynes, and alkenes. Chem Rev 88(7):1047–1058

Köpf H, Köpf-Maier P (1979) Titanocene dichloride–the first metallocene with cancerostatic activity. Angew Chem Int Ed Engl 18(6):477–478

# Titanium and Nucleic Acid

Jiangxue Wang, Ying Hou and Jiawei Ma
Key Laboratory for Biomechanics and Mechanobiology of the Ministry of Education, School of Biological Science and Medical Engineering, Beihang University, Beijing, China

## Definition

The titanocene complexes with inorganic Ti(IV) species are the new class of organometallic antitumor activity. Since Ti(IV) is highly charged, it is considered as a hard acid. Ti(IV) is mainly accumulated in the cellular nucleic acid-rich regions, particularly in the nucleic chromatin of tumor cells (Köpf-Maier and Martin 1989). It possibly interacts with the oxygen (hard base) of the phosphoesters and with the nitrogen of the nucleobase in DNA, inhibiting the cell cycle.

Formation of titanocene-DNA complexes has been regarded as the main mechanism of antitumor properties of the titanocenes. At physiological pH, titanocene interacts weakly with nucleotides through the phosphoesters. DNA is the prime cellular target of Ti(IV). The interaction of Ti(IV) complexes with increasing amounts of calf thymus DNA yielded nonlinear and noncooperative binding behaviors with inaccessible four binding site and binding constants of $K_b \sim 10^5$ (Vera et al. 2004). Using UV-vis and fluorescence spectroscopy, it was analyzed that at low concentration of Cp$_2$TiCl$_2$ the binding interaction is dominated by phosphate (O) coordination, while at high concentration both phosphate (O) and N (DNA bases) coordinations are important. Using inductively coupled plasma-atomic emission spectroscopy (ICP-AES) combined with a sample preparation by dialysis, the amount of adduct formation increased with time.

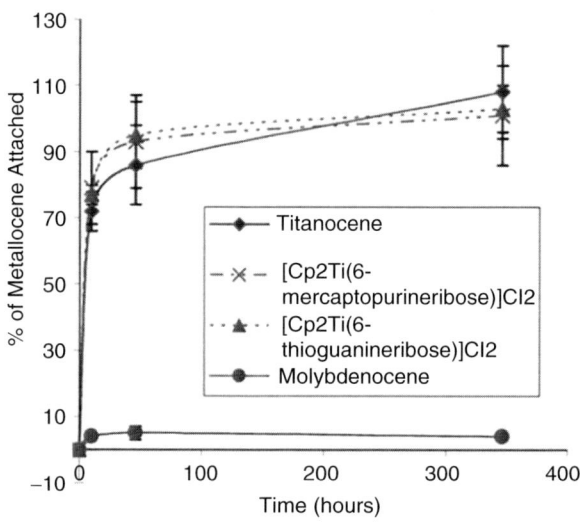

**Titanium and Nucleic Acid, Fig. 1** Comparative interactions between metallocenes and calf thymus DNA at different times (Vera et al. 2004)

After 0–10 h, about 72% titanium formed adducts with DNA. After 36–46 h, 86% titanium is attached to DNA, and 2 weeks after 100% titanium remained bound to DNA under physiological conditions (pH 7.4 and buffer solution). For $Cp_2TiCl_2$, two distinct adducts were formed with tritium labeled DNA. At pH 5.3 two Cp rings are associated with the DNA adduct ($Cp_2Ti$-DNA) while at pH 7.0 only one Cp ring is associated with the DNA adduct (CpTi-DNA), which is the predominant species. The Ti-DNA interaction is strong and irreversible and does not dissociate upon dialysis. In general, more than 90% of the titanium drugs bind irreversibly to calf thymus DNA in a 2 weeks period. However, molybdenocene does not bind DNA constituents at physiological pH (Fig. 1).

## Cross-References

▸ Magnesium
▸ Mercury and DNA
▸ Scandium, Interactions with Nucleotide and Nucleic Acids
▸ Titanium(IV) Intake by Apotransferrin
▸ Yttrium, Interactions with Proteins and DNA

## References

Köpf-Maier P, Martin R (1989) Subcellular distribution of titanium in the liver after treatment with the antitumor agent titanocene dichloride. A study using electron spectroscopic imaging. Virchows Arch B Cell Pathol Incl Mol Pathol 57

Vera JL, Román FR, Meléndez E (2004) Study of titanocene-DNA and molybdenocene-DNA interactions by inductively coupled plasma-atomic emission spectroscopy. Anal Bioanal Chem 379:399–403

# Titanium as Protease Inhibitor

Jiangxue Wang, Ying Hou and Jiawei Ma
Key Laboratory for Biomechanics and Mechanobiology of the Ministry of Education, School of Biological Science and Medical Engineering, Beihang University, Beijing, China

## Definition

In coordination chemistry, metal coordinated ligands are used to indirect binding of metal ions to the active sites of serine proteases. The metal ion charge interferes with the catalytic cleavage of peptide bonds. In the first row transition metal ions $Sc^{3+}$, $Ti^{4+}$, $V^{5+}$, $Cr^{6+}$, $Mn^{2+}$, $Fe^{3+}$, $Co^{2+}$, $Cu^{2+}$, and $Zn^{2+}$, only the aqueous Ti(IV) is a potent inhibitor of trypsin subclass in the case of serine proteases, but not the chymotrypsin subclass.

Serine proteases are found in a wide variety of organisms and in some cases show unique structural variations. It is used to develop clinically protease inhibitors for its importance as targets of drug. Duffy et al. (1998) studied that the aqueous Ti(IV) reduced the enzymatic activity of trypsin.

In water at physiologic pH, Ti(IV) forms five coordinate complexes $TiO(SO_4)(H_2O)$ that is coordinatively distinct from the octahedral form $TiO(SO_4)(H_2O)_2$. The geometry around Ti(IV) is a square-pyramid distorted octahedron. Upon this, the direct binding of Ti(IV) to trypsin was likely to occur at a free carboxyl group of Asp-189 at the bottom of the substrate binding pocket and the five coordinate geometry of $TiO(SO_4)(H_2O)$. That is to say a readily available ligand offered by the active site, and access by the metal ion to square-pyramidal geometry.

During initial contact between Ti(IV) and trypsin, enzyme activity was reduced by −25% and displayed steady-state kinetics. After 15 min, loss of enzyme activity increased and no longer displayed steady-state kinetics, which indicate Ti(IV) consolidates its binding to trypsin with time by irreversible inhibition. Aqueous Ti(IV) form colloidal solutions, which result in the decreased solubility as the total Ti(IV) burden in solution increases. In the initial (competitive) inhibition of trypsin by Ti(IV), a hyperbolic dose response was determined. This nonlinear response is characteristic of a self-limiting phenomena which is caused by decreasing Ti(IV) solubility. Therefore, Ti(IV) binds at the active site of the enzyme, and not at distal sites associated with protein stabilization or noncompetitive inhibition. Three mechanisms by which Ti(IV) may influence trypsin activity are possible: (1) interaction with the $Ca^{2+}$ binding site; (2) interaction with a noncompetitive binding site; or (3) interaction directly with the active site.

## Cross-References

▶ Actinides, Interactions with Proteins

## References

Duffy B, Schwietert C, France A, Mann N, Culbertson K, Harmon B, McCue JP (1998) Transition metals as protease inhibitors. Biol Trace Elem Res 64(1–3):197–213

# Titanium Dioxide as Disinfectant

Charles P. Gerba
Department of Soil, Water and Environmental Science, University of Arizona, Tucson, AZ, USA

## Synonyms

Antimicrobial action of titanium dioxide

## Definition

Disinfection refers to the killing or inactivation (in the case of viruses) of microorganisms. While several metals (most notably copper and silver) are noted for their antimicrobial activity by being toxic (by combining with proteins and nucleic acids), titanium dioxide is believe to be unique in that it primarily acts as a photocatalyst using UV light and is not consumed in the reaction. Titanium dioxide can be used in many formulations that exhibit antimicrobial activity against all types of microorganisms.

## Titanium Dioxide as a Photocatalyst

Titanium dioxide application as an antimicrobial has been recognized for several decades and has been used for this purpose in fabricated materials and in air and water purification. Because its action is believed to be primarily due to photocatalysis, it has been seen as a benign, environmentally friendly means of destruction of both chemical and microbial contaminates. Titanium dioxide is not consumed in the process of photocatalysis. Eventually, the microbes and organics will be mineralized to carbon dioxide and water.

Titanium dioxide occurs in nature in three forms: anatase, rutile, and brookite, with anatase being the most useful for photocatalytic processes. It always occurs as small, isolated, and sharply developed crystals and can be easily synthesized in the laboratory. In the presence of water and light energy (photons), photocatalysis results in the formation of reactive oxygen species which react with organic molecules in microorganisms resulting in their death or inactivation (in the case of viruses). Titanium dioxide, a solid semiconductor, has many advantages including its low cost, chemical stability, nontoxicity, and ability to use ultraviolet light energy (wavelengths of 300–400 nm).

The process of photocatalysis begins when photon energy ($h\nu$) of greater than or equal to the bandgap energy of $TiO_2$ is illuminated onto its surface, usually 3.2 eV (anatase) or 3.0 eV (rutile) (Fujishima et al. 2000). The photo excitation leaves behind an empty unfilled valance band, creating an electron–hole pair ($e^- h^+$) as the electron moves to a higher energy level. The band gap, in semiconductor theory, is the void energy region which separates the valence band from the conduction band (Fig. 1). The band gap is overcome with energy from photons. The absorption of energy and the subsequent generation of the electron–hole pair are the first steps (1).

**Titanium Dioxide as Disinfectant, Fig. 1** Schematic of photocatalytic process on TiO$_2$ surfaces

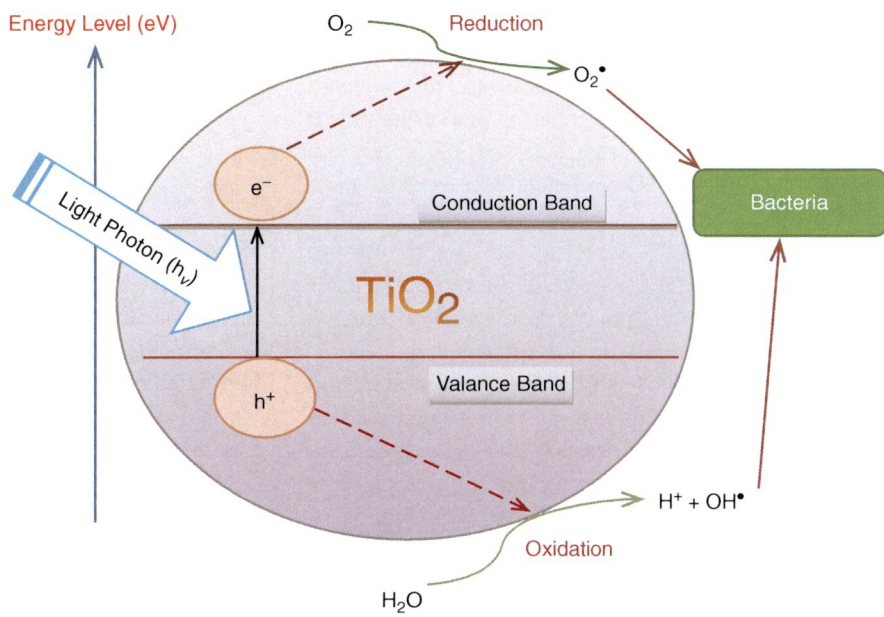

Photoexcitation: $TiO_2 + h\nu \rightarrow e^-_{cb}(TiO_2) + h^+_{vb}(TiO_2)$ (1)

Where $e^-_{cb}$ is the conduction band electron and $h^+_{vb}$ is the valence band hole.

The interaction of the hole with water molecules or hydroxide ions produces very reactive hydroxyl radicals. Equation 2 depicts how the presence of oxygen allows for the formation of superoxide radicals ($O_2^{\bullet-}$).

Photoexcited: $e^-$ scavenging: $(O_2)_{ads} + e^-_{cb} \rightarrow O_2^{\bullet-}$ (2)

This radical can be further pronated to form hydroperoxyl radical ($HO_2^\bullet$) and subsequently hydrogen peroxide ($H_2O_2$) as shown in Equations (3–5).

Protonation of superoxides: $O^\bullet_2 + OH^+ \rightarrow HOO^\bullet$ (3)

Co – scavenging of $e^-$: $HOO^\bullet + e^-_{cb} \rightarrow HO^-_2$ (4)

Formation of $H_2O_2$: $HOO^- + H \rightarrow H_2O_2$ (5)

It is important to recognize that for these processes to precede both dissolved oxygen and water molecules have to be present. Without the presence of water molecules the highly reactive hydroxyl radicals (OH$^\bullet$) could not be formed. Thus, the availability of both moisture (in the air or on the surface of the treated surface) and oxygen are necessary for titanium dioxide to be an effective antimicrobial. Light intensity is one of the few parameters that greatly affect the degree of photocatalytic reaction on reaction with organic molecules.

Some residual photocatalytic activity may occur in the dark after exposure to UV light, but this depends on the light intensity before placement in the dark.

## Photocatalytic Mechanisms Responsible for Antimicrobial Activity of TiO$_2$

Most studies have shown that Gram-positive bacteria are more resistant to photocatalytic disinfection than Gram-negative bacteria. The difference has been ascribed to the differences in the cell wall structure between the two groups of bacteria. Gram-negative bacteria have a triple-layer cell wall with an inner membrane, a thin peptidoglycan layer, and an outer membrane (lipopolysaccharide), whereas Gram-positive bacteria have a thicker peptidoglycan layer and no outer membrane. The lipopolysaccharide layer of the Gram-negative bacteria is believed to be the initial site of attack by the reactive oxygen species. This is followed by the attack on the peptidoglycan layer, peroxidation of the lipid membrane and eventual oxidation of the proteins in the membrane resulting in rapid leakage of potassium ions from the bacterial cells, causing a loss of cell viability. The cell wall eventually becomes broken resulting in morphological changes and rupture of the cell wall. Studies with different strains of *Legionella pneumophila* have shown that the greater the fatty acid composition of the membrane lipids the greater the susceptibility to photocatalysis by $TiO_2$. Eventually, the process leads to the complete destruction of the organism resulting in the formation of carbon dioxide and water.

The mechanisms by which other groups of organisms are killed by photolysis have not been as well studied, but oxidation process can be expected to denature protein molecules in non-lipid-containing viruses and eventually leading to the destruction of the nucleic acid.

One of the limitations of the use of titanium dioxide as an antimicrobial is the need for UV light to activate the process. While only low levels of UV light are needed, the rate of microbial killing is related to energy available from UV light. This is especially a problem in indoor environments, where only low levels of UV are emitted by most artificial lighting sources. To overcome this limitation, doping (addition) of other materials with $TiO_2$ have been assessed to expand the action of these composites to use light in the visible range. Examples are the use of copper, silver, and nitrogen (Khataee and Mansoori 2012). For example, CuO, also a type of semiconductor, requires a band gap energy of only 1.7 eV, which can be activated with a wavelength of ~720 nm. Titanium dioxide doped with either CuO or Ag has been shown to exhibit antimicrobial activity in the presence of fluorescent lighting. They also exhibit killing activity in the dark after exposure to fluorescent lighting. Both copper and silver are antimicrobial, but they can also enhance the production of reactive oxygen species when combined with $TiO_2$.

## Non-photocatalytic Mechanisms Responsible for Antimicrobial Activity

One of the unique aspects of titanium dioxide is photoinduced "superhydrophilicity" or enhanced wettability of the surface. When $TiO_2$ is prepared with a certain percentage of $SiO_2$, it acquires superhydrophilic properties after it is exposed to UV light. Electrons and holes are still produced, but they react in a different way. The electrons reduce the Ti cations from the IV state to the III state, and the holes oxidize the $O^{2-}$ anions. The oxygen atoms are ejected, creating oxygen vacancies. Water molecules can then occupy these oxygen vacancies, producing adsorbed OH groups, which tend to make the surface hydrophilic. The longer the surface is exposed to UV light, the smaller the contact angle (less hydrophobic the surface) for water becomes and on prolonged exposure approaches zero. Titanium dioxide coatings maintain their hydrophilic properties as long as they are illuminated. This property has seen application for antifogging surfaces (mirrors, widows) and self-cleaning building materials.

It has been theorized that the non-UV light-induced germicidal properties of $TiO_2$ surfaces may be due to hydrophobicity of its surface. When freshly prepared, preparations of $TiO_2$ have been reported to rapidly inactivate yeast cells in the absence of UV light. This was explained by the existence of hydrophilic and hydrophobic areas on the surface of $TiO_2$ which caused the stretching of the yeast cells. The hydrophilic cell wall touches the $TiO_2$ surface more and loses its round shape and flattens more resulting in deformation. This eventually results in pressure buildup in the cells resulting in bursting of the cell and release of internal contents (Lifen et al. 2009). The degree of cell deformation is directly related to the wetting property of the $TiO_2$. A greater antimicrobial effect is observed with Gram-negative bacteria, which have thinner cell wall than the thicker walled Gram-positive bacteria.

## Effectiveness Against Microorganisms

Titanium dioxide treated materials have been shown to be effective against a wide range of microorganisms including protozoa algae, fungi, bacteria, and viruses. The degree of effectiveness depends on how the material was prepared and if it has been doped with other substances to enhance its antimicrobial capabilities. Unfortunately, there are no standard test procedures for evaluating and comparing different formulations or materials for their antimicrobial activity by $TiO_2$. Standard test protocols are available for common chemical disinfectants, such as chlorine or quaternary compounds that allow for their registration by government agencies. The assessment is also complicated because it involves a treated surface that requiring a light source for maximum effectiveness. In this aspect, it may be considered more of a reactor than a true chemical disinfectant. It exhibits the nature of a "self-sanitizing surface," which has continuous antimicrobial activity. Recently, the United States Environmental Protection Agency has allowed the term "self-sanitizing surfaces with a residual" to be used for copper surfaces under a defined test protocol that requires a 99.9% reduction in 2 h of both a Gram-negative and Gram-positive test bacterium. Titanium dioxide does have the ability to meet these requirements. Given a long enough contact time $TiO_2$ is capable of being a "self-disinfecting surface." A disinfectant is defined as the ability to reduce the starting microbial numbers by 99.999% or greater in a defined period of time. The rate of kill is determined by the formulation, light intensity, and temperature (lower rate of kill at lower temperatures).

The general effectiveness of $TiO_2$ formulations is a measure of contact time. More resistant organisms will take a longer period of time to kill. Thus, Gram-positive bacteria are generally more resistant than Gram-negative bacteria, i.e., Gram-positive bacteria require a longer time of contact to achieve the same amount of kill as Gram-negative bacteria. Some of the more difficult to kill environmentally stable stages of some pathogens such as *Cryptosporidium parvum* oocysts, *Giardia* cysts, and bacterial endospores have been shown to be killed by $TiO_2$-treated materials (Foster et al. 2011). Hydroxyl radicals were shown to be the major reactive oxygen species involved in the killing of *C. parvum* oocysts, although other species were also believed to play a role. Both lipid and non-lipid-containing viruses including influenza, vaccine, and norovirus are also sensitive to inactivation. Fungal molds (including spores) and yeast are also susceptible.

Close contact between the organism and $TiO_2$ increases the extent of oxidative damage. Thus, disinfection of water occurs at a greater rate with suspended $TiO_2$ particles than when $TiO_2$ is immobilized on a surface. Incorporation of hydroxyapatite with $TiO_2$ has been used to enhance the adsorption of microorganisms to the reactive surface in fabrics (Fujishma et al. 2004). All microorganisms have a net negative charge and are attracted to the positive surface charge of the hydroxyapatite. In addition, it has been suggested the hydroxyapatite acts to encourage the longer persistence of reactive oxygen species.

**Titanium Dioxide as Disinfectant, Table 1** Antimicrobial applications of $TiO_2$ (Data from Fujishma and Zhang 2006)

| |
|---|
| Fabrics |
| Water and wastewater treatment |
| Surgical face masks |
| Paint |
| Food packing films |
| Dental implants |
| Tiles |
| Plastics |
| Coated wood |
| Coated paper |
| Concrete |
| Air filters |
| Window blinds |
| Purification systems for pools and spas |

## Applications

The incorporation of $TiO_2$ into materials to render them antimicrobial has been extensively explored and has resulted in a number of commercially available products. Table 1 lists some of the potential and commercially used applications. Unfortunately, data on the performance of products is often limited, and much of the existing data is limited to laboratory studies with little performance data under in use conditions.

Use of $TiO_2$ for disinfection of water and wastewater has been studied since the solar energy could be used or decreased levels of UV light for the killing of pathogenic microorganisms (Chong et al. 2010). Reactors for water treatment are ones in which the photocatalyst is in the form of suspended particles or with the photocatalyst immobilized onto an inert surface. The $TiO_2$ can be introduced as slurry of particles fed at a continuous basis into a water stream. However, too many particles lead to light attenuation and decreased effectiveness of the disinfection process. With immobilized $TiO_2$ reactors water is passed over the treated surface in the presence of solar energy or a UV light emitting light source. While such systems have been shown to be highly effective in killing microorganisms, they suffer from a number of significant limitations. If particles are used, they have to be removed after treatment, which is a significant cost. With immobilized reactors the amount of water which can be treated is limited by the penetration of UV light through the water and the amount of contact between $TiO_2$ and contaminate.

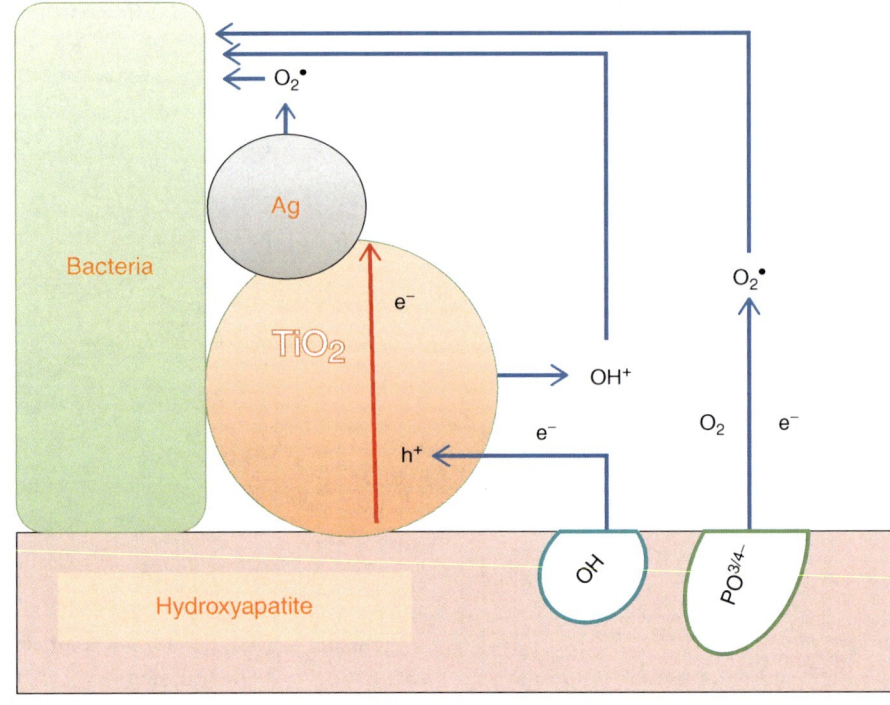

**Titanium Dioxide as Disinfectant, Fig. 2** Mechanisms of combined action of TiO$_2$, Ag, and hydroxyapatite against bacteria. Ag with TiO$_2$ allows the use of visible light energy to generate free radicals, while the hydroxyapatite increases the lifetime of the radicals and adsorbs the bacteria allowing for closer interaction with the generated radicals

Air purification has been a widely explored application of TiO$_2$ for destruction of both chemical contaminates, as well as, bacteria and molds (including their toxins) (Chen et al. 2010). The most effective systems incorporate the TiO$_2$ into the filter, which is then exposed to a low level UV emitting light source. The organisms must come into contact with the treated surface to be effective. A number of factors control the effectiveness of such systems such as air flow rate passing the reactive surfaces, reaction time, temperature, and relative humidity. Humidity plays an important role in the effectiveness of the process. At a given temperature, higher relative humidity (RH) results in more water molecules for the production of reactive oxygen species. However, very high RH does not help much to increase the killing of organisms. In general the filters perform best at typical relative indoor humilities of 30–50%, with reduced performance at RH levels above 75%.

Self-disinfecting surfaces have potential major application in the health care industry, in the reduction of hospital-acquired infections (Highashino and Kamiya 2011). Incorporation of TiO$_2$ and hydroxyapatite and or Ag has been shown to be effective against bacteria in fabrics, both in visible light and in the dark (Fig. 2). They not only kill the bacteria, but also suppress the growth when the fabrics become wet. In addition, to the previous benefits of incorporating hydroxyapatite can also serve to prevent photocatalytic degradation of the treated material. Treatment of ceramic tiles has also been shown to be successful in killing common bacterial pathogens.

The increasing utilization of TiO$_2$ as an antimicrobial are promising and further development of new chemistries and doping with enhancers to increase the performance of this technology should guarantee its future role in the control of pathogenic microorganism and other prevention of the microbial degradation of materials.

## Cross-References

▶ Titanium and Nucleic Acid
▶ Titanium as Protease Inhibitor
▶ Titanium, Biological Effects
▶ Titanium, Physical and Chemical Properties

## References

Chen F, Yang X, Mak HKC, Chan DWT (2010) Photocatalytic oxidation for antimicrobial control in built-environment: a brief literature review. Build Environ 45:1747–1754

Chong MN, Jin B, Chow CWK, Saint C (2010) Recent developments in photocatalytic water treatment. Water Res 44:2997–3027

Foster HA, Ditta IR, Varhese S, Steele A (2011) Photocatalytic disinfection using titanium dioxide: spectrum and mechanism of antimicrobial activity. Appl Microbiol Biotechnol 90:1847–1868

Fujishima A, Rao TN, Tryk DA (2000) Titanium dioxide photocatalysis. J Photochem Photobiol C: Photochem Rev 1:1–21

Fujishma A, Zhang X (2006) Titanium dioxide photocatalysis: present situation and future approaches. CR Chime 9:7450–7760

Fujishma A, Zhang T, Hasse H, Funakoshi K (2004) Apatite-coated titanium dioxide photocatalyst for air purification. Catal Today 96:113–118

Higashino T, Kamiya K (2011) Survival of MRSA attached to materials used in medical facilities of different humidity levels. Med Biol 155:456–460

Khataee A, Mansoori GA (2012) Nanostructured titanium dioxide materials. World Scientific, London

Lifen L, Bradford J, Yeung KL (2009) Non-UV light germicidal activity of fresh $TiO_2$ and $Ag/TiO_2$. J Environ Sci 21:700–706

# Titanium in Cells, Tissues, Organs, and Organisms

Jiangxue Wang, Ying Hou and Jiawei Ma
Key Laboratory for Biomechanics and Mechanobiology of the Ministry of Education, School of Biological Science and Medical Engineering, Beihang University, Beijing, China

## Definition

Titanium as a biomaterial is widely used for orthopedic implants such as the replacement of teeth, bone, and the retention and stabilization of prostheses. At the titanium-tissue interface, the proteoglycan/glycosaminoglycan complexes exist and play a role in establishing the adhesion property. This adhesion property is important in the process of bone-implant contact, osseointegration, and mineralization.

It is well known that proteoglycans, comprised of a core protein and glycosaminoglycans (GAGs), play a role in the attachment and adhesion between osteoblastic cells and the extracellular matrix. Using electron microscopic techniques, an amorphous layer of thickness about 50–400 nm was detected between the bone and titanium, which is abundant in osteocalcin and rarely found in collagen (Ayukawa et al. 1998; Sennerby et al. 1991). This layer exhibits the characteristics of ground substance and a high percentage of sulfur. Several studies implied that the amorphous layer is proteoglycans which permeate the interfacial layer of the tissue facing titanium (Linder et al. 1983; Squire et al. 1996). Using the osteoblastic cells to mineralize titanium surface, Nakamura et al. (2007) reported that the various GAG/proteoglycan complexes were localized at the mineralized tissue/titanium interface, and the expression of bone-related proteoglycan genes were up-regulated including osteoadhesion, fibromodulin, syndecan, biglycan, aggrecan, and all of the GAGs, but these phenomena were not detected at the tissue/polystyrene interface. By addition of the GAGs degrading enzymes, the biomechanical properties of the bone/titanium interface changed. Only the interfacial strength was reduced at the tissue-titanium implant, while the hardness and elastic modulus of the mineralized tissue were not altered (Nakamura et al. 2006). This suggested that the interfacial strength of tissue-titanium is mediated by proteoglycan/GAG complexes. The proteoglycan/GAG complexes provide a novel tool to assess the osteoconductive potential of titanium implant surface, and are also a valuable target to improve the bonding capability of implant surfaces to bone.

## Cross-References

▶ Magnesium in Biological Systems
▶ Molybdenum in Biological Systems
▶ Scandium, Biological Effects
▶ Titanium, Biological Effects
▶ Yttrium, Biological Effects

## References

Ayukawa Y, Takeshita F, Inoue T, Yoshinari M, Shimono M, Suetsugu T, Tanaka T (1998) An immunoelectron microscopic localization of noncollagenous bone proteins (osteocalcin and osteopontin) at the bone-titanium interface of rat tibiae. J Biomed Mater Res 41(1):111–119

Linder L, Albrektsson T, Brånemark PI, Hansson HA, Ivarsson B, Jonsson U, Lundström I (1983) Electron microscopic analysis of the bone-titanium interface. Acta Orthop Scand 54:45–52

Nakamura H, Shim J, Butz F, Aita H, Gupta V, Ogawa T (2006) Glycosaminoglycan degradation reduces mineralized tissue-titanium interfacial strength. J Biomed Mater Res A 77(3):478–486

Nakamura HK, Butz F, Saruwatari L, Ogawa T (2007) A role for proteoglycans in mineralized tissue-titanium adhesion. J Dent Res 86:147–152

Sennerby L, Ericson LE, Thomsen P, Lekholm U, Astrand P (1991) Structure of the bone-titanium interface in retrieved clinical oral implants. Clin Oral Implants Res 2:103–111

Squire MW, Ricci JL, Bizios R (1996) Analysis of osteoblast mineral deposits on orthopaedic/dental implant metals. Biomaterials 17:725–733

# Titanium surfaces, biochemical modification by peptides and ECM proteins

Jiangxue Wang, Ying Hou and Jiawei Ma
Key Laboratory for Biomechanics and Mechanobiology of the Ministry of Education, School of Biological Science and Medical Engineering, Beihang University, Beijing, China

## Definition

Owing to the specific property, titanium is especially suited to medical and dental applications, such as hip and knee replacements and dental implant. However, titanium is inert and completely corrosion-resistant. To improve tissue response at the bone-implant interface and for further clinical use, the surface modification of titanium was broadly designed (Morra 2006). Traditionally, the surface modification of titanium is based on surface topography, on ceramic coatings, on physiochemical or inorganic approaches. Recently, the biochemical surface modification is utilized to immobilize peptides, ExtraCellular Matrix (ECM) proteins, Bone Morphogenetic Proteins (BMPs), enzymes, and growth factors on implants for the purpose of inducing specific cell and tissue responses.

For peptides, the simple sequence Arg-Gly-Asp (RGD) has been identified as mediating attachment of cells to plasma and extracellular matrix proteins. The property of cellular recognition implies that peptides could convey particular cell adhesion on biomaterial surfaces and enhance cell interactions with biomaterials. By the method of silanization, cross-linked action, amination, and polymerization, titanium substrates were modified with different functional groups such as 3-aminopropyltriethoxylsilane (APTES), polypyrole (PPY), PLL-g-PEG, etc. To promote particular cell adhesion, the specific linear- or cyclo-RGD peptides were synthesized and coupled to the functionalized titanium surface. Experiments in vitro showed that RGD-coated titanium surfaces promoted osteoblasts adhesion, proliferation, and increased expression of bone matrix mRNAs. Studies in vivo presented that new bone formed on the titanium implants coated with RGD peptides. Peptides provided by the clever synthetic work are able to stimulate specific cell behavior. Importantly, surface properties, such as hydrophobicity/hydrophilicity and roughness, are an aspect for peptide immobilization and osteoblasts adhesion and proliferation, and always at work over and above specific effects. Much basic work focused on the point to find a sequence specifically targeted to bone cells and overcome the "universality" and lack of selectivity of the RGD signal.

For ECM proteins, many studies look collagen as the basis to research ECM proteins. Based on passive adsorption, collagen type I, type III, fibrillar collagen, tropocollagen, fibronectin, and laminin were studied as the coating on titanium surfaces to evaluate the ability of cell-tissue interaction. Collagen covalent-linked with gamma-aminopropyltriethoxysilane (APS) or poly(ethylene-co-vinyl alcohol) (EVA) functionalized titanium substrate was also assessed. Collagen, containing cell-binding domains, can both recruit more osteogenic precursor cells, and play an important role in osteoblasts behavior, promoting not only cell adhesion but also osteoblastic differentiation of bone marrow cells and controlling their progression along the osteogenic pathway.

Besides collagen, the ultimate protein-based coating on titanium surfaces, a native extracellular matrix coating, showed the highest increase of cell density than RGDS- or fibronectin-immobilized titanium. The bone sialoprotein–coated implants showed osteoinductive activity in rat femora. It is interesting to observe that whenever the peptide approach is compared to the "whole protein" approach, the latter gives better or equivalent results.

## Cross-References

▶ Actinides, Interactions with Proteins
▶ Strontium Binding to Proteins
▶ Titanium in Cells, Tissues, Organs, and Organisms
▶ Yttrium, Interactions with Proteins and DNA

## References

Morra M (2006) Biochemical modification of titanium surfaces: peptides and ECM proteins. Eur Cell Mater 12:1–15

# Titanium(IV) Intake by Apotransferrin

Jiangxue Wang, Ying Hou and Jiawei Ma
Key Laboratory for Biomechanics and
Mechanobiology of the Ministry of Education, School
of Biological Science and Medical Engineering,
Beihang University, Beijing, China

## Definition

Transferrin, an iron-transport protein present in vertebrates, carries iron(III) in blood at pH of 7.4 and delivers to cells and releases it at pH of 5.5. Under normal conditions, its transferring is only 30% iron-saturated, and also acts as a specific carrier for other tripositive, tetrapositive, and bipositive metal ions. Ti(IV) forms a strong complex with human serum transferrin (hTF) by binding to each of the two specific Fe (III) binding sites in this protein.

In general terms, Ti(IV) complexes, such as titanocene dichloride ($Cp_2TiCl_2$), appear to offer a different alternative for cancer chemotherapy. Upon administration, Ti(IV) complexes are transported into cells to perform the antitumor action. However, the extensive hydrolysis at physiological pH occurs in the absence of biomolecules to give insoluble precipitates. The transport mechanism of Ti(IV) complexes in vivo raised much attention in the antitumor action. Analysis of human blood plasma taken from a patient infused with $Cp_2TiCl_2$ solution showed no evidence for free titanocene species, and following separation of the plasma proteins, all titanium was associated with a single protein. Pharmacokinetic studies confirmed that >70% Ti-protein interactions bound in plasma after intravenous injection $Cp_2TiCl_2$ into patients. Serum fractionation studies and $^{45}$Ti radiolabeling experiments confirmed that Ti(IV) is associated only with transferrin both in vivo and in vitro.

Once titanocene dichloride is administered into the human body, the Cp and Cl ligands are released to form $Ti_2$-hTF (Fig. 1), which is stable at neutral pH. However, at pH < 5.5, Ti(IV) is released and preferentially binds ATP (Guo et al. 2000). Thus, following delivery of Ti (IV) by transferrin to the cancer cells which are known to over-express transferrin receptors and have a lower intracellular pH, Ti(IV) could be released and become available for complexation with DNA.

**Titanium(IV) Intake by Apotransferrin, Fig. 1** Ti site in transferring (Guo et al. 1999)

Using UV-vis spectrum, the ability of $Cp_2TiCl_2$ to deliver Ti(IV) to apotransferrin (apo-Tr) was elucidated (Gao et al. 2007). The titration of apo-Tr with increasing amounts of $Cp_2TiCl_2$ yielded three absorption bands, at 240, 295, and 321 nm in about 2–10 min. The band at 321 nm, which is attributed to a ligand-to-metal charge transfer (LMCT) band, was utilized as an indicator to monitor Ti(IV)-transferrin binding. In saturation point, the spectrum reach a plateau at r = 2([Ti]/[Tr]) which demonstrated that the two metal binding sites are being occupied by Ti(IV) and that both N- and C-lobes were loaded. All titanocene complexes are able to transfer Ti(IV) into the N- and C-lobes of transferring because the ring substituent do not change or delay the ligand stripping process and subsequent donation of Ti(IV) to apo-Tr, such as the complexes $(Cp-R)_2TiCl_2$ and $(Cp-R)CpTiCl_2$, where R is $CO_2CH_3$ and $CO_2CH_2CH_3$. Surprisingly, the reaction of apo-hTf with Ti(IV) nitrilotriacetate (NTA) does not lead to formation of appreciable amounts of $Ti_2$-hTf, even after long incubation times (Messori et al. 1999).

## Cross-References

▶ Iron Homeostasis in Health and Disease
▶ Iron Proteins, Transferrins and Iron Transport
▶ Magnesium Binding Sites in Proteins
▶ Nickel-Binding Sites in Proteins
▶ Scandium, Interactions with Transferrin
▶ Titanium, Biological Effects

## References

Gao LM, Hernandez R, Matta J, Melendez E (2007) Synthesis, Ti(IV) intake by apotransferrin and cytotoxic properties of functionalized titanocene dichlorides. J Biol Inorg Chem 12:959–967

Guo ML, Sun H, Sadler PJ (1999) Uptake and release of a titanium anticancer complex by human transferring. J Inorg Biochem 74:150

Guo ML, Sun H, McArdle HJ, Sadler PJ (2000) $Ti^{IV}$ uptake and release by human serum transferrin and recognition of $Ti^{IV}$-transferrin by cancer cells: understanding the mechanism of action of the anticancer drug titanocene dichlorid. Biochemistry 39:10023–10033

Messori L, Orioli P, Banholzer V, Pais I, Zatta P (1999) Formation of titanium(IV) transferrin by reaction of human serum apotransferrin with titanium complexes. FEBS Lett 442:157–161

## Titanium, Biological Effects

Jiangxue Wang, Ying Hou and Jiawei Ma
Key Laboratory for Biomechanics and Mechanobiology of the Ministry of Education, School of Biological Science and Medical Engineering, Beihang University, Beijing, China

## Definition

Titanium has historically maintained the reputation of being an inert and relatively biocompatible biomaterial. In medicine, it is widely used for hip and knee replacements, pace-makers, bone-plates and screws, and orthopedic applications. Synthes Stratec (Oberdorf, Switzerland) produced novel commercially pure titanium foam to promote the orthopedics implant development. Its porosity is 65–70% which improves the scaffold to interlock with the host bone and increases the early implant stability, thereby avoiding the development of local stress peaks, subsidence or migration of the implant. The surface roughness is quite high which promotes the adhesion and differentiation of primary human osteoblasts under perfusion, such as the mRNA expression of osteogenic genes like ALP and col-1. This osteoconductive porosity material might reduce the need for autologous bone.

The surface property including topological structure, porosity, composition, and mechanical stability of implants are important for the contact with bone and host tissue. The different approaches to improve fixation of commercial titanium or titanium alloys implants are studied, that is, surface roughing, coating, or surface immobilization of growth factors or drugs. Recently, nanomaterials are potential coating agents to modulate protein adsorption and promote cell adhesion, osseointegration and bone mineralization at the bone-biomaterial interface both in vitro and in vivo because of nanometric surface topography, high specific surface area, roughness, and quantum size effect. In addition, the surface-immobilized bisphosphonates also could improve the mechanical fixation of $TiO_2$ and hydroxylapatite-coated titanium surfaces and increased bone-to-implant contact.

Ti(IV) complexes offer an alternative agent for cancer chemotherapy which certainly do not follow the rationale and mechanism of platinum complexes (Melendez 2002). Novel Ti(IV) drug development may come from the polynuclear species such as the derivatives of budotitane and titanocene dichloride, and the more recent $[(Q^B)_2Ti(\mu\text{-}O)]_4$. These complexes function as model compounds for the interaction of Ti(IV) with biologically important macromolecules, such as DNA.

Budotitane, cis-diethoxy-bis(1-phenylbutane-1,3-dionato)titanium(IV) $[(bzac)_2Ti(OEt)_2]$, the first non-Pt metal antitumor compound was produced and reached clinical trials against a wide variety of ascites and solid tumors. The ligand, bzac = benzoylacetonato, is an asymmetric β-diketone chelator useful for establishing one key structural feature for activity. The bis (β-diketonato)titanium complexes bind either to macromolecules via coordinative covalent bonds or via intercalation between nucleic acids strands by the aromatic ring of the butadione. The major mechanism might be the fact that bis(β-diketonato)titanium complexes hydrolyze extensively in water to yield oligomeric $[(bzac)_2TiO]_2$. Clinical trials on $[(bzac)_2Ti(OEt)_2]$ showed that the drug has a maximum tolerable dose of 230 mg/m$^2$ on a 2-week schedule with the side effect of cardiac arrhythmia. Higher doses would result in liver and kidney toxicity.

Titanocene dichloride, $(Cp)_2TiCl_2$, is the first metallocene dihalide with antitumor activity and is in phase II clinical trials now. The 2 Cl cleaving groups are analogous to the two OEt cis in budotitane, which are the key active structures. $(Cp)_2TiCl_2$ has a larger spectrum of antitumor activity than Pt complexes with no nephrotoxicity or myelotoxicity due to the good solubility in physiological medium and the interaction with transferrin (See entry ▶ Titanium(IV) Intake by Apotransferrin).

## Cross-References

▶ Titanium and Nucleic Acid
▶ Titanium(IV) Intake by Apotransferrin

## References

Melendez E (2002) Titanium complexes in cancer treatment. Crit Rev Oncol Hematol 42:309–315

# Titanium, Physical and Chemical Properties

Fathi Habashi
Department of Mining, Metallurgical, and Materials Engineering, Laval University, Quebec City, Canada

## Physical Properties

Titanium is one of the most abundant metals in the Earth's crust, was discovered in 1791 but was produced on industrial scale only in the 1950s because of the difficulties in obtaining it in a pure form due to its reactivity. It is a light metal of high melting point and excellent corrosion resistance. Its oxide is an important white pigment.

| | |
|---|---|
| Atomic number | 22 |
| Atomic weight | 47.88 |
| Relative abundance in Earth's crust, % | 0.44 |
| Density at 25°C, g/cm$^3$ | 4.5 |
| Atomic radius for coordination number six in the crystal, nm | 0.145 |
| Melting point, °C | 1,668 |
| Boiling point, °C | 3,287 |
| Phase transformation temperature, °C    Hexagonal ⇌ body centered cubic | 882 |
| Lattice constants of α-Ti at room temperature, nm | $c = 0.4679$ $a = 0.2951$ |
| Heat of transformation, kJ/mol | 3.685 |
| Coefficient of linear expansion at 25°C, K$^{-1}$ | $8.5 \times 10^{-6}$ |
| *(continued)* | |

| | |
|---|---|
| Latent heat of fusion, kJ/mol | 20.9 |
| Latent heat of sublimation, J/mol | 464.7 |
| Latent heat of vaporization, kJ/mol | 397.8 |
| Specific heat capacity at 25°C, J g$^{-1}$ K$^{-1}$ | 0.523 |
| Thermal conductivity at 20–25°C, W cm$^{-1}$ K$^{-1}$ | 0.221 |
| Surface tension at 1,600°C, N/m | 1.7 |
| Diffusion coefficient (self-diffusion) at 750°C, cm$^2$/s       α-Ti       β-Ti | $4 \times 10^{-13}$ $2.4 \times 10^{-9}$ |
| Modulus of elasticity at 25°C, GPa | 100–110 |
| Electrical resistivity, μ Ω cm       At 25°C       At 600°C | 42 140–150 |
| Superconductivity transition temperature, K | $0.40 \pm 0.04$ |
| Magnetic susceptibility of α-Ti at 25°C, cm$^3$/g | $3.2 \times 10^{-6}$ |

Increasing contents of oxygen, nitrogen, and carbon, which are incorporated interstitially in titanium, slightly elongate the $a$-axis of α-Ti and considerably elongate its $c$-axis. These effects are mostly marked with carbon and are very slight with hydrogen. Many properties of titanium at a given temperature show variations that depend on the composition and condition of the metal (purity, alloying elements, and thermal and mechanical pretreatment). Variations in properties such as electrical and thermal conductivity and plastic behavior are caused by lattice defects. Cold working increases hardness and strength by producing dislocations, and reduces modulus of elasticity and electrical conductivity, for example, by producing holes in the lattice. The presence of foreign elements, even at low concentrations, has often a marked effect. In general, electrical resistivity, hardness, and strength increase with decreasing purity.

## Chemical Properties

The corrosion resistance of titanium metal is due to the formation of a thin, dense, stable, adherent surface film of oxide, which immediately reforms after mechanical damage if oxygen is present in the surrounding medium. Titanium is less resistant to corrosion in strongly reducing media. Titanium dioxide is of outstanding importance as a white pigment because of its superior scattering properties, its chemical stability, and lack of toxicity. Titanium dioxide is amphoteric with weak acidic and basic character.

Titanium is usually quadrivalent in its compounds, but may also function as trivalent and in a few compounds as bivalent, but these are unstable in aqueous solution. Titanium tetrachloride, $TiCl_4$, is obtained by passing chlorine over heated titanium, titanium carbide, or over a mixture of titanium dioxide and carbon. In the pure state, it forms a colorless liquid, boiling at 136.5°C. It has a pungent smell, fumes strongly in moist air, and is rapidly hydrolyzed by water:

$$TiCl_4 + 2H_2O \rightarrow TiO_2 + 4HCl$$

When titanium dioxide is fumed with concentrated sulfuric acid, titanyl sulfate, $TiOSO_4$, is formed, as a white powder, soluble in cold water and decomposed by hot water, with the deposition of gelatinous titanium dioxide:

$$TiOSO_4 + H_2O \rightarrow TiO_2 + H_2SO_4$$

Alkali titanates correspond to the formulas $M_2^I TiO_3$ and $M_2^I Ti_2O_5$. They may be obtained by evaporating solutions of α-titanic acid in concentrated alkali hydroxide solutions (β-titanic acid is insoluble) or by fusing titanium dioxide with alkali carbonates.

### References

Habashi F (2003) Metals from ores. An introduction to extractive metallurgy, Métallurgie Extractive Québec, Quebec City. Distributed by Laval University Bookstore, www.zone.ul.ca

Siebum H et al (1997) Titanium. In: Habashi F (ed) Handbook of extractive metallurgy. Wiley-VCH, Weinheim, pp 1129–1180

## Tobacco

▶ Polonium and Cancer

## Tobacco Filters

▶ Polonium and Cancer

## Tobacco Leaves

▶ Polonium and Cancer

## Tobacco Plants

▶ Polonium and Cancer

## Tobacco Smoke

▶ Polonium and Cancer

## *TOM:* Transporter of MAs

▶ Iron Proteins, Plant Iron Transporters

## Topoisomerase IB

▶ Gold(III), Cyclometalated Compound, Inhibition of Human DNA Topoisomerase IB

## Total Internal Reflection Fluorescence (TIRF)

▶ Calcium Sparklets and Waves

## Toxicity

▶ Tellurium in Nature
▶ Tin, Toxicity

## Toxicity of Gold Complex

▶ Gold Complexes as Prospective Metal-Based Anticancer Drugs

## Toxicity of Gold Nanomaterials

▶ Gold Nanomaterials as Prospective Metal-based Delivery Systems for Cancer Treatment

## Toxicity of Gold(III) Complexes

▶ Gold(III) Complexes, Cytotoxic effects

## Trace Metal

▶ Copper, Biological Functions

## Trace Metals

▶ Chromium(III), Cytokines, and Hormones

## Tracer for Calcium

▶ Strontium Binding to Proteins

## Transcription Factor Hijacking

▶ Zinc, Metallated DNA-Protein Crosslinks as Finger Conformation and Reactivity Probes

## Transfer RNA

▶ Lead and RNA

## Transferrin

▶ Chromium(III) and Transferrin

## Transferrin as a Gallium Mediator

▶ Gallium Uptake and Transport by Transferrin

## Transient Receptor Potential Channels TRPV1, TRPV2, and TRPV3

Michael X. Zhu
Department of Integrative Biology and Pharmacology, The University of Texas Health Science Center at Houston, Houston, TX, USA

### Synonyms

Growth factor regulated channel (GRC)

### Definition

Vanilloid subfamily of transient receptor potential (TRP) protein isoforms 1, 2, and 3. The prototypical TRP protein was first identified from the fruit fly, *Drosophila melanogaster*, because of the short-lived or transient response of the insects' eyes to light. The light response of retina as recorded using electroretinograph is referred to as "receptor potential," which is typically sustained during the light stimulation. Mutations at the *trp* gene caused the transient receptor potential phenotype in the flies. The TRP superfamily now includes all proteins with sequences similar to the *Drosophila* TRP protein. All TRP proteins form ion channels as tetramers and conduct cations such as $Na^+$, $K^+$, and $Ca^{2+}$ with varying selectivities. In extreme cases, a few TRP channels conduct only $Ca^{2+}$ (e.g., TRPV5 and TRPV6), while a few others conduct only monovalent cations (e.g., TRPM4 and TRPM5). In vertebrates, the TRP members are further categorized into TRPA (ankyrin), TRPC (canonical), TRPM (melastatin), TRPML (mucolipin), TRPP (polycystin), and TRPV (vanilloid) subfamilies (Montell et al. 2002). These channels play diverse functions, but none of them appears to be tightly associated with phototransduction like in the case of the *Drosophila* TRP channel. Therefore, TRP represents an acronym given to proteins or genes of homologous sequences with no particular implications on function or mechanism of regulation. TRPV1, TRPV2, and TRPV3 are three members of the mammalian TRPV subfamily, which has a total of six members. They are closely related to TRPV1, the founding member of this subfamily, which was originally named as vanilloid

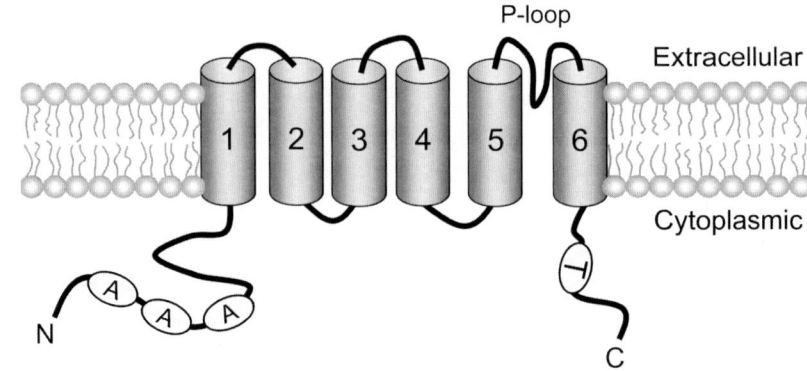

**Transient Receptor Potential Channels TRPV1, TRPV2, and TRPV3, Fig. 1** Membrane topology of TRP. *A* ankyrin-like motif, *C* carboxyl terminus, *N* amino terminus, *P-loop* pore loop, *T* TRP motif

receptor 1 (VR1) because of its activation by compounds containing a vanilloid moiety, such as capsaicin, the pungent ingredient of hot chili peppers (Caterina et al. 1997).

## Introduction

TRPV1 was discovered by expression cloning of receptors for capsaicin using a cDNA library prepared from rat dorsal root ganglia. Transfection of the cDNA into human embryonic kidney-derived HEK293 cells allowed for detection of capsaicin-induced elevation of intracellular $Ca^{2+}$ concentration and subsequent isolation of a single cDNA clone coding for a transmembrane protein of 838 amino acids. The predicted membrane topology resembled that of voltage-gated potassium channels and the *Drosophila* TRP protein characterized by six transmembrane segments and an intervening pore loop situated in between the 5th and 6th transmembrane segments (Fig. 1). Both the amino (~430 aa) and carboxyl (~150 aa) termini are located at the cytoplasmic side. A short stretch (~70 aa) encompassing a part of the putative pore loop, the 6th transmembrane segment and the beginning of the carboxyl terminus, was found to be highly homologous to *Drosophila* TRP and its closely related homologues found in mammals and *C. elegans*, but the rest of the sequences are quite distant despite the apparent conservation in membrane topology and the presence of multiple ankyrin-like motifs at the amino terminus. Nonetheless, in 1997, the TRP channels (later on referred to as TRPCs) were the only known proteins found to be related to the then newly cloned capsaicin receptor (Caterina et al. 1997).

Functionally, the capsaicin receptor acts as a nonselective cation channel, mediating mainly $Na^+$ and $Ca^{2+}$ influx into cells under physiological conditions, upon activated by hot temperatures, low pH, and several exogenous as well as endogenous chemical substances (Jordt et al. 2003). Both the initial characterization and subsequent studies have all demonstrated the importance of the capsaicin receptor (TRPV1) in pain sensation induced by noxious heat, acid, and many inflammatory factors. More recent studies have also implicated the role of TRPV1 in regulating body temperature (Gavva 2008). The TRPV1 protein is mainly expressed, and its channel function is detected in the nociceptive unmyelinated C fibers of small to medium diameter dorsal root ganglion neurons.

TRPV2 was discovered by two independent groups: reported by one as a close homologue of TRPV1 (Caterina, et al. 1999) and by another as a channel involved in the insulin-like growth factor-induced $Ca^{2+}$ entry (Kanzaki et al. 1999). TRPV2 is activated by temperatures higher than 52°C and by membrane stretch and chemical substances, such as probenecid, 2-aminoethoxydiphenyl borate (2-APB), and cannabidiol. TRPV2 proteins are found in neurons, including large diameter sensory neurons of dorsal root ganglia and neurons in the brain, as well as muscle, endocrine, liver, epithelial, and immune cells.

TRPV3 was first reported to be a warm temperature activated cation channel by three independent groups (Peier et al. 2002; Xu et al. 2002; Smith et al. 2002). In mammals, the TRPV3 gene is located adjacent to that of TRPV1 on the same chromosome. TRPV3 is highly expressed in the epithelial layers of the skin, oral, and nasal cavities and that of

**Transient Receptor Potential Channels TRPV1, TRPV2, and TRPV3,**
**Fig. 2** Structure of 2-aminoethoxydiphenyl borate (2-APB) and its analogues

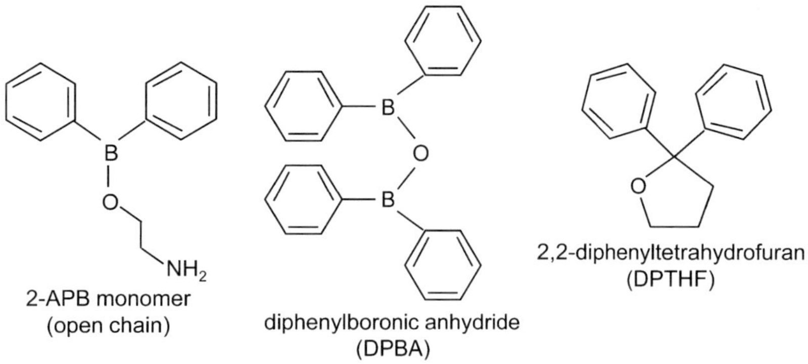

gastrointestinal tract. In addition to warm to hot temperatures, the TRPV3 channel is also activated by some flavor enhancers, chemical allergens, and skin sensitizers, such as, camphor, carvacrol, menthol, eugenol, and thymol, typically at high concentrations. Ablation of the TRPV3 gene in mice caused defects in hair morphogenesis and skin-barrier formation (Cheng et al. 2010). While gain-of-function mutations of TRPV3 in mice and rats resulted in hair loss and susceptibility to dermatitis and itchiness, equivalent mutations in humans have been linked to Olmsted syndrome, a rare genetic disease showing palmoplantar and periorificial keratoderma, alopecia, and severe itching (Lin et al. 2012).

## Common Features

All three channels are $Ca^{2+}$-permeable nonselective cation channels activated by temperature increases, but the temperature activation thresholds ($T_h$) are different. The reported $T_h$ values for TRPV1, TRPV2, and TRPV3 are 43°C, 52°C, and 31–39°C, respectively (Jordt et al. 2003). Temperature activation is characterized by a sharp increase of the activity upon temperature change, assessed using $Q_{10}$ temperature coefficient. Typically, the $Q_{10}$ value for a temperature-gated channel is higher than 10. Critical domains involved in temperature sensing of TRPV1-3 have been located to the N-terminal ankyrin-like repeats (Yao et al. 2011), a turret at the pore loop (Cui et al. 2012), or key residues located near the pore region of TRPV3 (Grandl et al. 2008).

All three channels are activated by a boron-containing compound, 2-APB (Fig. 2) (Hu et al. 2004). This feature is quite unique among TRP channels as most of them are inhibited by 2-APB.

A boron-containing 2-APB analogue, diphenylboronic anhydride (DPBA), also activated TRPV3, but a boron-free analogue, 2,2-diphenyltetrahydrofuran (DPTHF), inhibited TRPV3 (Chung et al. 2005). H426 at the amino terminus and R696 at the carboxyl terminus of TRPV3 appear to be critical to confer the stimulatory effect of 2APB (Hu et al. 2009).

TRPV1 may form heterotetramers with TRPV2 or TRPV3. The heteromertic channels are likely different from homotetrameric channels in pharmacology, mechanism of regulation, and relative permeabilities to different cations. Whether and where these heterotetramers exist in different tissues and cell types are unclear.

## Unique Features

TRPV1 shows strong voltage dependence intrinsic to the channel itself (Voets et al. 2004). At positive membrane potentials, TRPV1 is activated at room temperature or even lower temperatures. This voltage-dependent activation is shifted more negative to physiological membrane voltages upon a temperature increase to higher than the $T_h$ of TRPV1, a pH decrease, or binding by a chemical agonist. However, it is debated whether all stimuli activate the TRPV1 channel by changing its voltage dependence.

When activated, both TRPV2 and TRPV3 show voltage-dependent changes in whole-cell conductance, displaying double rectification at both negative and positive potentials. The reduced conductance at voltages near zero mV is likely a result of blockage due to divalent cation ions, namely, $Ca^{2+}$ and $Mg^{2+}$, rather than intrinsic voltage gating. In this case, the passage of the cations through the channel pore is facilitated by strong driving force at more negative as well as more positive potentials.

TRPV1 is the best known, and also the most frequently studied, TRP channel because of its role in nociception (pain sensation) and therefore the promise that TRPV1 blockers might be useful for pain management. However, side effects on body temperature regulation will likely limit the clinical use of TRPV1 blockers (Gavva 2008). Nevertheless, the knowledge gained from studying the mechanisms of regulation of this channel is very rich and quite informative to the understanding of other TRP channels. A central theme has emerged that TRP channels are polymodal sensors of environmental changes (Clapham 2003). Thus, TRPV1 activities can be initiated and/or modulated by a broad range of modalities, including changes in temperature, membrane potential, pH, concentrations of chemical ligands, kinase and phosphatase activities, cell signaling pathways, and so on. A large number of endogenous and exogenous ligands have been described for the TRPV1 channel. Compared to TRPV1, information on the mechanisms of regulation of TRPV2 and TRPV3 is less abundant.

TRPV1 has been reported to be activated by extracellular divalent cations ($Mg^{2+}$ and $Ca^{2+}$) at higher than 10 mM concentrations. Both divalent and monovalent (e.g., $Na^+$) cations, through increasing extracellular ionic strength, also strongly potentiate ligand-evoked TRPV1 activity. Two glutamate residues (E600 and E648) located at putative extracellular loops of TRPV1 are shown to be critical for these effects via electrostatic interactions with the cations (Ahern et al. 2005).

## Summary

TRPV1, TRPV2, and TRPV3 are closely related nonselective cation channels important for sensing warm to hot temperatures, membrane stretch, and/or extracellular acidosis and other chemical environmental changes. Under physiological conditions, they mediate mainly $Na^+$ and $Ca^{2+}$ influxes, leading to membrane depolarization and increases in intracellular $Ca^{2+}$ concentrations. These cellular events are key contributors to pain sensation, body metabolism/homeostasis regulations, and immune responses. Altered expression and function of TRPV1–3 channels can lead to diseases. Pharmacological and genetic interventions of the expression and channel activities of TRPV1–3 have important implications in disease prevention and treatment.

## Cross-References

▶ Boron, Biologically Active Compounds
▶ Boron-Containing Compounds, Regulation of Therapeutic Potential
▶ Calcium in Biological Systems
▶ Calcium in Health and Disease
▶ Calcium in Heart Function and Diseases
▶ Calcium in Nervous System
▶ Calcium in Vision
▶ Calcium Ion Selectivity in Biological Systems
▶ Calcium Signaling
▶ Calcium, Neuronal Sensor Proteins
▶ Calcium-Binding Protein Site Types
▶ Calcium-Binding Proteins, Overview
▶ Calmodulin
▶ Magnesium and Inflammation
▶ Magnesium Binding Sites in Proteins
▶ Magnesium in Biological Systems
▶ Magnesium in Eukaryotes
▶ Magnesium in Health and Disease

## References

Ahern GP, Brooks IM, Miyares RL, Wang XB (2005) Extracellular cations sensitize and gate capsaicin receptor TRPV1 modulating pain signaling. J Neurosci 25(21):5109–5116

Caterina MJ, Schumacher MA, Tominaga M, Rosen TA, Levine JD, Julius D (1997) The capsaicin receptor: a heat-activated ion channel in the pain pathway. Nature 389(6653):816–824

Caterina MJ, Rosen TA, Tominaga M, Brake AJ, Julius D (1999) A capsaicin-receptor homologue with a high threshold for noxious heat. Nature 398(6726):436–441

Cheng X, Jin J, Hu L, Shen D, Dong XP, Samie MA, Knoff J, Eisinger B, Liu ML, Huang SM, Caterina MJ, Dempsey P, Michael LE, Dlugosz AA, Andrews NC, Clapham DE, Xu H (2010) TRP channel regulates EGFR signaling in hair morphogenesis and skin barrier formation. Cell 141(2):331–343

Chung MK, Güler AD, Caterina MJ (2005) Biphasic currents evoked by chemical or thermal activation of the heat-gated ion channel, TRPV3. J Biol Chem 280(16):15928–15941

Clapham DE (2003) TRP channels as cellular sensors. Nature 426(6966):517–524

Cui Y, Yang F, Cao X, Yarov-Yarovoy V, Wang K, Zheng J (2012) Selective disruption of high sensitivity heat activation but not capsaicin activation of TRPV1 channels by pore turret mutations. J Gen Physiol 139(4):273–283

Gavva NR (2008) Body-temperature maintenance as the predominant function of the vanilloid receptor TRPV1. Trends Pharmacol Sci 29(11):550–557

Grandl J, Hu H, Bandell M, Bursulaya B, Schmidt M, Petrus M, Patapoutian A (2008) Pore region of TRPV3 ion channel is specifically required for heat activation. Nat Neurosci 11(9):1007–1013

Hu HZ, Gu Q, Wang C, Colton CK, Tang J, Kinoshita-Kawada M, Lee LY, Wood JD, Zhu MX (2004) 2-aminoethoxydiphenyl borate is a common activator of TRPV1, TRPV2, and TRPV3. J Biol Chem 279(34):35741–35748

Hu H, Grandl J, Bandell M, Petrus M, Patapoutian A (2009) Two amino acid residues determine 2-APB sensitivity of the ion channels TRPV3 and TRPV4. Proc Natl Acad Sci USA 106(5):1626–1631

Jordt SE, McKemy DD, Julius D (2003) Lessons from peppers and peppermint: the molecular logic of thermosensation. Curr Opin Neurobiol 13(4):487–492

Kanzaki M, Zhang YQ, Mashima H, Li L, Shibata H, Kojima I (1999) Translocation of a calcium-permeable cation channel induced by insulin-like growth factor-I. Nat Cell Biol 1(3):165–170

Lin Z, Chen Q, Lee M, Cao X, Zhang J, Ma D, Chen L, Hu X, Wang H, Wang X, Zhang P, Liu X, Guan L, Tang Y, Yang H, Tu P, Bu D, Zhu X, Wang K, Li R, Yang Y (2012) Exome sequencing reveals mutations in TRPV3 as a cause of Olmsted syndrome. Am J Hum Genet 90(3):558–564

Montell C, Birnbaumer L, Flockerzi V, Bindels RJ, Bruford EA, Caterina MJ, Clapham DE, Harteneck C, Heller S, Julius D, Kojima I, Mori Y, Penner R, Prawitt D, Scharenberg AM, Schultz G, Shimizu N, Zhu MX (2002) A unified nomenclature for the superfamily of TRP cation channels. Mol Cell 9:229–231

Peier AM, Reeve AJ, Andersson DA, Moqrich A, Earley TJ, Hergarden AC, Story GM, Colley S, Hogenesch JB, McIntyre P, Bevan S, Patapoutian A (2002) A heat-sensitive TRP channel expressed in keratinocytes. Science 296(5575):2046–2049

Smith GD, Gunthorpe MJ, Kelsell RE, Hayes PD, Reilly P, Facer P, Wright JE, Jerman JC, Walhin JP, Ooi L, Egerton J, Charles KJ, Smart D, Randall AD, Anand P, Davis JB (2002) TRPV3 is a temperature-sensitive vanilloid receptor-like protein. Nature 418(6894):186–190

Voets T, Droogmans G, Wissenbach U, Janssens A, Flockerzi V, Nilius B (2004) The principle of temperature-dependent gating in cold- and heat-sensitive TRP channels. Nature 430(7001):748–754

Xu H, Ramsey IS, Kotecha SA, Moran MM, Chong JA, Lawson D, Ge P, Lilly J, Silos-Santiago I, Xie Y, DiStefano PS, Curtis R, Clapham DE (2002) TRPV3 is a calcium-permeable temperature-sensitive cation channel. Nature 418(6894):181–186

Yao J, Liu B, Qin F (2011) Modular thermal sensors in temperature-gated transient receptor potential (TRP) channels. Proc Natl Acad Sci USA 108(27):11109–11114

# Transition Element

▶ Vanadium in Biological systems

# Transition Metal Uptake

▶ Cobalt Transporters

# Transition-Metal Uptake

▶ Nickel Transporters

# *trans*-L-Dach (1R, 2R-Diaminocyclohexane) Oxalatoplatinum (L-OHP)

▶ Oxaliplatin, Clinical Use in Cancer Patients

# Transmissible Spongiform Encephalopathy

▶ Copper and Prion Proteins

# Treatment of Depression in Bipolar Disorders

▶ Lithium as Mood Stabilizer

# Treatment of Mania in Bipolar Disorders

▶ Lithium as Mood Stabilizer

# Triacylglycerol Hydrolases

▶ Lipases

# 3,5,3′-Triodothyronine, T3

▶ Selenoproteins and Thyroid Gland

# Tris, Tris(Hydroxymethyl)aminomethane

▶ Zinc Aminopeptidases, Aminopeptidase from Vibrio Proteolyticus (Aeromonas proteolytica) as Prototypical Enzyme

# Trivalent Chromium

- Chromium and Diabetes
- Chromium and Nutritional Supplement

# tRNA

- Lead and RNA

# Troponin

Nikolai B. Gusev
Department of Biochemistry, School of Biology, Moscow State University, Moscow, Russian Federation

## Definition

Troponin is an equimolar heterotrimeric complex consisting of $Ca^{2+}$-binding (troponin C), inhibitory (troponin I), and tropomyosin-binding (troponin T) subunits. Troponin is located on actin filaments and, together with tropomyosin, provides for $Ca^{2+}$-dependent regulation of interaction of actin and myosin in skeletal and cardiac muscles.

In the early 1960s in the laboratory of S. Ebashi, it was found that a special protein fraction isolated from muscle extract and called "native tropomyosin" is able to provide $Ca^{2+}$-sensitivity to Mg-ATPase and superprecipitation of desensitized actomyosin. In 1968, S. Ebashi and coworkers (1968) published the first method of separation of native tropomyosin and troponin purification. At the beginning, troponin was assumed to be a single protein; however, fundamental investigations performed in laboratories of S. Ebashi, S.V. Perry and J. Gergely and many others showed that troponin consists of three different components, each of which performs specific functions. The largest subunit of troponin (troponin T) binds the troponin complex to actin and tropomyosin and assures the proper distribution of the troponin complex along the thin filament so that in vertebrate skeletal muscle, the stoichiometry of troponin-tropomyosin-actin is 1/1/7. The inhibitory subunit (troponin I) inhibits Mg-ATPase activity of actomyosin in the absence of $Ca^{2+}$. Finally, the $Ca^{2+}$-binding subunit (troponin C) provides for regulation of the troponin complex by calcium. Some properties of human troponin subunits are presented in Table 1. Let us analyze properties of troponin components in more detail.

Troponin C is a rather small (molecular weight less than 20 kDa) acidic protein belonging to the family of EF-hand $Ca^{2+}$-binding proteins. The human genome contains two genes coding for slow (cardiac) and fast (skeletal) muscle isoforms of troponin C. According to the crystal structures, troponin C has a dumbbell-like structure, two globular domains of which each contain two cation-binding sites and are connected with each other by a "handle" formed by extended α-helix (Herzberg and James 1985). In the case of skeletal troponin C, the "handle" seems to be rather rigid, whereas in the crystal structure of cardiac troponin C, the central part of the "handle" seems to be partially melted. The C-terminal globular domain usually contains two high-affinity calcium-binding sites ($K_d$ $Ca^{2+}$ ~ 6 $10^{-8}$ M), which are also able to bind magnesium ($K_d$ $Mg^{2+}$ ~ 3 $10^{-4}$ M). These sites are permanently saturated by cations and play important structural role in the interaction of troponin C with the other components of troponin complex. Removal of cations from these sites destabilizes the troponin complex and provides for selective extraction of troponin C from myofibrils. The N-terminal domain of human skeletal troponin C contains two low-affinity ($K_d$ $Ca^{2+}$ ~ 3 $10^{-6}$ M) $Ca^{2+}$-specific sites. The first $Ca^{2+}$-binding site of human cardiac troponin C contains nonconservative replacements and is unable to bind calcium; therefore, cardiac troponin C has only one $Ca^{2+}$-specific site. The cation-binding sites of the N-terminal domain play a regulatory role. $Ca^{2+}$ binding to these sites induces conformational changes, which lead to the strengthening of troponin C-troponin I interaction and to elimination of the inhibitory effect of troponin I on the interaction of actin and myosin. Conformational changes are based on calcium-induced reorientation of α-helices flanking $Ca^{2+}$-binding loops (so-called opening of $Ca^{2+}$-binding site). In the absence of $Ca^{2+}$, α-helices flanking the $Ca^{2+}$-binding site are nearly antiparallel to each other (Fig. 1), whereas after calcium binding, these helices become almost perpendicular to each other (Grabarek et al. 1990). This leads to exposure of certain hydrophobic

**Troponin, Table 1** Some properties of human troponin components

| Troponin components | Gene number | Uni-Prot KB number | Molecular weight (kDa) and pI | Functions | Posttranslational modifications, alternative splicing |
|---|---|---|---|---|---|
| Troponin C | 2 | | | | N-acetylation |
| Fast skeletal | | P02585 | 18.022/4.06 | Binding of 4 $Ca^{2+}$/mole protein | |
| Slow skeletal/cardiac | | P63316 | 18.402/4.05 | Binding of 3 $Ca^{2+}$/mole protein | |
| Troponin I | 3 | | | | N-acetylation |
| Fast skeletal | | P48788 | 21.207/8.88 | Inhibition actomyosin ATPase activity | Phosphorylation (?) |
| Slow skeletal | | P19237 | 21.561/9.61 | | Phosphorylation (?) |
| Slow cardiac | | P19429 | 23.876/9.87 | | Phosphorylation |
| Troponin T | 3 | | | | N-acetylation, Phosphorylation |
| Fast skeletal | | P45378 | 31.693/5.71 | Binding to actin and tropomyosin | 7 isoforms by alternative splicing |
| Slow skeletal | | P13805 | 32.817/5.86 | | 3 isoforms by alternative splicing |
| Slow cardiac | | P45379 | 35.792/4.94 | | 10 isoforms by alternative splicing, restrictive proteolysis |

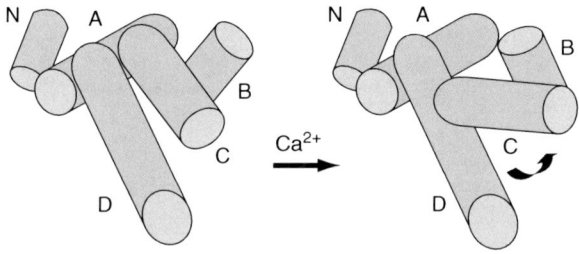

**Troponin, Fig. 1** Oversimplified scheme of calcium-dependent movement of α-helices in the N-terminal domain of skeletal troponin C. Five helices of the N-terminal domain of troponin C are marked N, A, B, C, and D and two calcium-binding loops connect helices A and B and C and D, respectively. Calcium binding induces reorientation of helices A–B and C–D and "opening" of the corresponding sites leading to exposure of hydrophobic sites

sites and promotes interaction of the N-terminal part of troponin C with certain sites in troponin I structure.

Troponin C undergoes N-terminal acetylation; however, the physiological significance of this posttranslational modification remains poorly understood. Troponin C plays crucial role in regulation of cardiac contraction, therefore certain mutation of troponin C correlates with development of hypertrophic or dilated cardiomyopathies (Willott et al. 2010). Since troponin C is responsible for $Ca^{2+}$-dependent regulation of cardiac contractions, it is desirable to find special compounds which will be able to modulate its Ca-binding properties. Therefore, attempts were made to find special drugs that would affect (usually specifically increase) the affinity of the Ca-binding sites of cardiac troponin C (Li et al. 2008;). These investigations might be useful for developing of new approaches in the field of cardiovascular medicine.

Troponin I is small (molecular weight less than 25 kDa), strongly basic, and highly susceptible to proteolysis. The human genome contains three different genes responsible for the expression of the fast and slow skeletal, as well as the cardiac, isoforms of troponin I (Table 1). All three isoforms have similar amino acid sequences and contain (1) two long, conserved α-helices (H1 and H2) involved in formation of triple coil with the corresponding α-helix of troponin T, (2) an inhibitory region (IR) that binds both troponin C and actin-tropomyosin, (3) the so-called switch or triggering region (SW) interacting either with actin-tropomyosin or with the N-terminal domain of troponin C saturated with $Ca^{2+}$, and (4) the mobile C-terminal region (MCR) interacting with actin-tropomyosin (Fig. 2). Troponin I isoforms differ in the length and structure of their variable N-terminal parts, which in the case of cardiac isoform contains ~30 additional amino acid residues. It is supposed that in the relaxed state, the C-terminal globular domain of troponin C is bound to the N-terminal part of helix H1 of troponin I and to the coiled-coil formed by α-helices of troponin I and troponin T, whereas the N-terminal domain of troponin C is free of calcium and does not interact with any other troponin components (Fig. 2). In this state, troponin I is tightly bound to troponin T through its long H1 and H2 helices and the inhibitory and switch regions as well as the mobile C-terminal

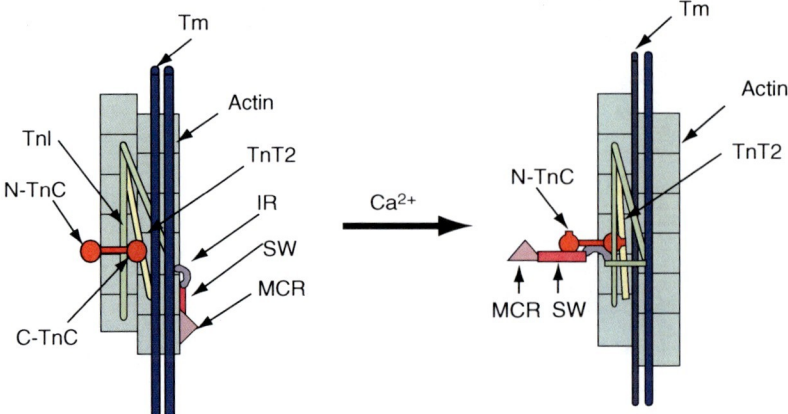

**Troponin, Fig. 2** Proposed model of regulation of actin-myosin interaction by troponin-tropomyosin complex. In relaxed state (*left panel*), troponin C (*red dumbbell* with $Ca^{2+}$ bound to the C-terminal domain (C-TnC) and N-terminal domain (N-TnC) free of $Ca^{2+}$ are bound to the N-terminal part of helix 2 of troponin I (*green cylinder*) and to the TnT2 part of troponin T (*yellow cylinder*). The inhibitory region (IR), switch region (SW), and mobile C-terminal region (MCR) of troponin I are fixed on actin (*squares*) and through troponin T fix tropomyosin in a position outside of the groove on the surface of actin filament, thus hindering productive interaction of myosin with actin. Saturation of $Ca^{2+}$-specific sites located in the N-terminal domain of troponin C (N-TnC) (*right panel*) provides for its interaction with SW domain of troponin I and for binding of IR of troponin I to the central helix of troponin C and to dissociation of these domains and MCR from actin. Structural rearrangements are transmitted to troponin T; these in turn push tropomyosin into the depth of the actin groove, thus eliminating inhibition of productive interaction of actin and myosin

domains of troponin I tightly interact with actin (Vinogradova et al. 2005; Solaro 2010). Being tightly fixed on actin through its inhibitory, switch and mobile C-terminal regions troponin I, via troponin T, pulls tropomyosin outside of the groove of the actin filament and by this means fixes tropomyosin in a position preventing productive interaction of actin with myosin. Contraction is initiated by increase of the $Ca^{2+}$ concentration in the cell. Saturation of $Ca^{2+}$-specific binding sites induces reorientation of α-helices and exposure of certain hydrophobic sites in the N-terminal domain of troponin C (Fig. 1). This makes possible interaction of the switch domain of troponin I with the N-terminal domain of troponin C and provides for interaction of the inhibitory region of troponin I with the central helix of troponin C. All of these interactions destabilize binding of the mobile C-terminal domain of troponin I with actin. Therefore, troponin I can no longer fix troponin T and tropomyosin, bound to troponin T, in the blocking position on the actin filament. Tropomyosin rolls back into the groove of the actin filament thus eliminating hindrances for productive interaction of actin with myosin. This is the general way of regulation of actin-myosin interaction by troponin-tropomyosin. There are certain peculiarities in the functioning of cardiac troponin. For instance, the inhibitory domain of cardiac troponin I is less ordered than the corresponding domain of skeletal troponin I, and in the absence of calcium, the central helix of cardiac troponin C is partially disordered. Therefore, the cardiac troponin complex seems to be less rigid than its skeletal counterpart. The most important difference concerns the regulatory N-terminal domain of cardiac troponin C, which contains only one $Ca^{2+}$-binding site and which, upon binding calcium, is not switched to a completely "open" conformation. Another important difference between skeletal and cardiac troponin is the structure of N-terminal part of troponin I which in the case of the cardiac isoform contains ~ 30 additional amino acids. The detailed location of this fragment inside the troponin complex remains unknown; however, phosphorylation of Ser23 and Ser24 of cardiac troponin I by cAMP-dependent protein kinase significantly decreases the affinity of Ca-specific sites of cardiac troponin C. This modulation of calcium binding seems to be important for fine regulation of cardiac contraction since decreased affinity to $Ca^{2+}$ can lead to increased rate of relaxation and can at least partially explain the positive chronotropic effect of epinephrine. Many other sites of cardiac and skeletal troponin I can be phosphorylated in vitro by a number of different protein kinases; however, the physiological importance of this phosphorylation has yet to be established since phosphorylation of only two of the abovementioned Ser residues of cardiac troponin I were confirmed in the living cell by modern proteomic approaches. Troponin I also undergoes *N*-acetylation. Although this type of posttranslational modification is common for many contractile proteins, its exact physiological effect remains poorly understood.

Troponin I plays a crucial role in the functioning of the troponin complex and regulation of muscle contraction. A large number of point mutations have been found to correlate with development of hypertrophic, dilated, or restrictive cardiomyopathies (Willott et al. 2010). As already mentioned, the human genome contains three genes coding for troponin I, and a special cardiac isoform of troponin I is expressed in the heart. This makes possible development of sensitive and specific immunochemical methods for detection of cardiac troponin I in the blood serum. At present, troponin I is one of the best markers for diagnosis of myocardial infarction (Omland 2010). Utilization of troponin I as a biochemical marker provides prognostic information in stable coronary artery disease and chronic heart failure and improves diagnostic accuracy in the case of acute coronary syndromes. Measurements of cardiac troponin I are also routinely used for diagnosis of acute myocardial infarction. Intense and diverse investigations in this field have lead to organization of a special international working group for standardization of an assay for troponin I.

Troponin T is the largest tropomyosin-binding component of the troponin complex. The human genome contains three genes encoding fast and slow skeletal and cardiac isoforms of troponin T (Table 1). Limited proteolysis leads to formation of two large fragments, namely the N-terminal TnT1 and the C-terminal TnT2 fragments. TnT1 fragment contains the hypervariable N-terminal region with unknown protein partners and unknown detailed location in troponin-tropomyosin complex and the so-called middle conserved region, the N-terminal part of which contains the first tropomyosin-binding site. This first tropomyosin-binding site (~40 amino acid residues) is located in the region of tail-to-head overlap of two neighboring tropomyosin dimers and seems to bridge them. The TnT2 fragment contains the second tropomyosin-binding site (about 25 residues near the beginning of TnT2 fragments) as well as the sites responsible for troponin I and troponin C binding. Troponin T and tropomyosin have an antiparallel orientation, and therefore, the second tropomyosin-binding site of troponin T is located close to conserved Cys190 of α-tropomyosin, whereas the fist tropomyosin-binding site of troponin T is located close to the tail-to-head (C-to-N) junction of two tropomyosin dimers. The three-dimensional structure of troponin complex (Vinogradova et al. 2005) was obtained with the utilization of truncated fragments of troponin T and, therefore, detailed three-dimensional structure of entire TnT1, of part of TnT2, and of a short-conserved C-terminal end of troponin T inside of the whole troponin complex remains unknown (Wei and Jin 2011). Interaction of troponin T with the other troponin components and proposed reorganization of troponin complex induced by calcium binding to the $Ca^{2+}$-specific sites of troponin C was described earlier (see Fig. 2). The main function of troponin T is now considered to be transmission of a conformational signal from troponin C and troponin I to tropomyosin. Troponin T not only transmits the signal from troponin C/troponin I to tropomyosin but is involved in reorganization of the tropomyosin strand on actin filament leading to translocation of tropomyosin *in* or *out* of the actin groove. Moreover, as mentioned earlier, the first tropomyosin-binding site of troponin T is located close to the tail-to-head junction of two neighboring tropomyosin dimers and, therefore, one infers that it can participate in cooperative interaction between the so-called regulatory units, each consisting of seven actin monomers, one tropomyosin dimer, and one troponin complex. This means that rotation of tropomyosin and "switching on" of one regulatory subunit will affect position of tropomyosin and "switching on" of the neighboring regulatory subunit and by this means will provide cooperative regulation of the whole actin filament.

The function of troponin T is regulated in different ways. Firstly, each mRNA derived from the three troponin T genes can undergo very complicated alternative splicing. There are a number of a very short exons located in the hypervariable N-terminal part and, as a rule, one mutually exclusive exon in the C-terminal part of troponin T. In the case of mammalian fast skeletal troponin T gene, this can theoretically lead to formation of 256 splicing variants. Not all of these variants are expressed; however, alternative splicing seems to be dependent on the stage of development and on the type of skeletal fiber. There is strict correlation between isoforms of troponin T and tropomyosin expressed in a given cell. As already mentioned, the protein partners and exact location of the N-terminal, hypervariable part of troponin T still remain unknown; however, it is known that calcium affinity of the whole troponin complex is dependent on the length and exon composition of the hypervariable N-terminal part of troponin T. Secondly, troponin T is very susceptible to proteolysis. It was shown that calpain I associated with cardiac myofilaments can cut the hypervariable N-terminal end of troponin T and, by this means, can affect ATPase activity and myofibril force generation. Thirdly, it was shown that troponin T can be phosphorylated by a number of different proteins kinases in vitro. This phosphorylation affects interactions of troponin T with the other troponin components and, by this means, modifies the regulatory activities of troponin. However, modern proteomic research revealed that under basal in vivo conditions, troponin T is 100% monophosphorylated at Ser1. The earlier published results (Gusev et al. 1980) suggested that casein kinase 2 is responsible for phosphorylation of this site of troponin T; however, physiological significance of this phosphorylation remains enigmatic.

Troponin T plays an important role in fixation of troponin complex on tropomyosin and actin and in transmission of a conformational signal from troponin C/troponin I to tropomyosin. Therefore point mutations which are usually (but not exclusively) located in the C-terminal part of troponin T, i.e., in the region responsible for interaction with the other troponin components, often correlate with development of different kinds of hypertrophic, dilated or restrictive cardiomyopathies (Willott 2010). As already mentioned, there are three genes coding for different isoforms of troponin T, and a special isoform of troponin T is expressed in the heart. This made it possible to develop a special assay for measuring the cardiac specific isoform of troponin T in the blood serum (Omland 2010). The use of cardiac troponin T as a biochemical marker of different cardiovascular diseases is less popular than cardiac troponin I; however, use of troponin T as a biomarkers also provides important and valuable diagnostic information (Omland 2010).

The mechanism of troponin-tropomyosin regulation of muscle contraction seems to be very complicated (Solaro 2010). It is supposed that the thin (actin) filaments can be in three different states, namely blocked (in the absence of $Ca^{2+}$), closed (in the presence of $Ca^{2+}$), and open (in the presence of $Ca^{2+}$ with bound myosin heads). It is still difficult to correlate the structural changes of troponin complex described earlier, with these three states of actin filaments. However, it is supposed that in the blocked state, the regulatory sites of troponin C are free of calcium and that tropomyosin occupies a position on actin that does not prevent binding of myosin heads, but makes impossible liberation of phosphate from the ATPase site of myosin and generation of the power stroke. In the blocked state, $Ca^{2+}$ specific sites of troponin C are saturated with calcium, inhibitory action of troponin I is eliminated, and tropomyosin occupies a position on actin that does not prevent binding of myosin heads to actin filaments and generation of the power stroke. In the open state, myosin heads bound to actin push tropomyosin in the depth of actin groove completely "switching on" the regulatory subunit. Translocation of tropomyosin in the given regulatory subunit affects the position of tropomyosin in the neighboring regulatory subunits (or affects the structure of actin monomers), increases $Ca^{2+}$ affinity of troponin, and, by this means, might cooperatively "switch on" many closely located regulatory subunits.

All attempts to detect and isolate troponin from smooth muscles were unsuccessful. However, these investigations resulted in discovering of two new regulatory proteins which functionally are similar to troponin. Caldesmon and calponin are actin-binding proteins of smooth muscles. In the absence of $Ca^{2+}$, these proteins are able to inhibit productive interaction of myosin with actin and, by this means, prevent contraction. This inhibition is reversed by interaction of caldesmon and/or calponin with $Ca^{2+}$-saturated calmodulin (or other $Ca^{2+}$-binding proteins, i.e., S100) or by phosphorylation of caldesmon and/or calponin by different protein kinases. In this respect, caldesmon (and to a lesser extent calponin) can be considered as a functional analogue of troponin I and troponin T, whereas calmodulin (or other $Ca^{2+}$-binding proteins, such as S100) plays the role of troponin C.

In conclusion, troponin is the main $Ca^{2+}$-dependent regulator of actin-myosin interaction in skeletal and cardiac muscle. The functioning of the troponin complex is delicately regulated by a number of genes, alternative splicing, and different kinds of posttranslational modifications. Mutations of troponin components often correlate with development of different forms of cardiomyopathies, and troponin I and troponin T are successfully used as important biomarkers of different forms of cardiovascular diseases.

## Cross-References

▶ Calmodulin

## References

Ebashi S, Kodama A, Ebashi F (1968) Troponin. I. Preparation and physiological function. J Biochem (Tokyo) 64:465–477

Grabarek Z, Tan RY, Wang J, Tao T, Gergely J (1990) Inhibition of mutant troponin C activity by an intra-domain disulphide bond. Nature 345:132–135

Gusev NB, Dobrovolskii AB, Severin SE (1980) Isolation and some properties of troponin T kinase from rabbit skeletal muscle. Biochem J 189:219–226

Herzberg O, James MN (1985) Structure of calcium regulatory muscle protein troponin-C at 2.8A resolution. Nature 313:653–659

Li MX, Robertson IM, Sykes BD (2008) Interaction of cardiac troponin with cardiotonic drugs: a structural perspective. Biochem Biophys Res Commun 369:88–99

Omland T (2010) New features of troponin testing in different clinical settings. J Int Med 268:207–217

Solaro RJ (2010) Sarcomere control mechanisms and the dynamics of cardiac cycle. J Biomed Biotechnol 2010:105648

Vinogradova MV, Stone DB, Malanina GG, Karatzaferi C, Cooke R, Mendelson RA, Fletterick RJ (2005) $Ca^{2+}$-regulated structural changes in troponin. Proc Natl Acad Sci USA 102:5038–5043

Wei B, Jin J-P (2011) Troponin T isoforms and posttranscriptional modification: evolution, regulation and function. Arch Biochem Biophys 505:144–154

Willott RH, Gomes AV, Chang AN, Parvatiyar M, Pinto JR, Potter J (2010) Mutations in troponin that cause HCM, DCM and RCM: what can we learn about thin filament function?. J Mol Cell Cardiol 48:882–892

## Troponin C

▶ Calcium in Biological Systems
▶ Calcium-Binding Proteins, Overview

## Tungsten

▶ Tungsten Cofactors, Binding Proteins, and Transporters in Biological Systems
▶ Tungsten in Biological Systems

## Tungsten Cofactors, Binding Proteins, and Transporters in Biological Systems

Biswajit Mukherjee, Saikat Ghosh, Samrat Roy Chowdhury and Anandamoy Rudra
Department of Pharmaceutical Technology, Jadavpur University, Kolkata, West Bengal, India

## Synonyms

Heavy stone; Scheelite; Tungsten; Wolframite

## Introduction

Tungsten, "scheelite" in Sweden and "heavy stone" in Nordic language, was obtained from wolframite (wolf soot or cream) in Germany, and the element is symbolized accordingly as "W." Tungsten (atomic weight 183.85; atomic no.74; atomic radius 1.4 Å) belongs to group 6d of period VI with electronic configuration $[Xe] 4f^{14} 5d^4 6s^2$. Transitional element tungsten when bound to an enzyme for providing biological activity acts as a metal cofactor. Tungsten cofactor in tungsten-containing enzyme generally involves in redox reaction but has low redox potential. The metal is generally transported by specific ATP-binding cassette (ABC) transporter.

## Cofactors

Enzymes are generally globular proteins that catalyze chemical reactions. A cofactor is a nonprotein chemical bound to a protein to provide the biological activity of the protein. Enzymes catalyze the substrates into products. All living organisms need enzymes to work for their survival. Many enzymes have a protein part called apoenzyme and a nonprotein part called cofactor, for their activities. Both the parts together form a complete enzyme called holoenzyme. The protein part refers to apoenzyme, and the nonprotein part is either tightly bound (called prosthetic group) or a loosely bound (known as coenzyme or cofactor). Such nonprotein part may be organic or inorganic in nature, for example, biotin (organic), FAD (organic), and different metals (inorganic).

## Metal as Cofactors

Many metals serve as cofactors of various enzymes. In humans, metals, such as Fe, Mg, Co, Zn, Se, Mo, etc., are already detected as cofactors for various enzymes.

## Examples of Other Cofactors

Pterin is a heterocyclic compound consisting of a pteridine ring, a keto group at position 4, and an amino group at position 2. Two important derivatives of this system are folates and pterin. Pterins were first obtained from pigments of butterfly wings. They play important role in coloration of the biological world. They also act as cofactors for different enzymes in catalysis reactions.

Folates (known as conjugated pterins) contain para-amino benzoic acid and L-glutamate with methyl group at the position 6 of the pteridine ring. They play important role in biological group transfer reactions.

## Tungsten as Cofactor

Tungsten as cofactor often catalyzes reactions with low redox potential (about −420 mV). However, the tungsten enzymes also catalyze reactions with positive redox potentials. In case of many enzymes containing molybdenum, molybdenum is replaced by tungsten. Examples of replacement of tungsten by molybdenum are also available. *Methanobacterium thermoautotrophicum* contains tungsten in enzyme formylmethanofuran dehydrogenase in which tungsten is substituted by molybdenum when the organism is grown on molybdenum-rich media. In tungsten-rich medium, when *Methanobacterium wolfei* is grown, the molybdenum in formylmethanofuran dehydrogenase is substituted by tungsten. However, some Mo/W enzymes can function only with a specific ion. Such examples are W-formate dehydrogenase found in hyperthermophilic archaea and Mo-aldehyde oxidoreductase from sulfate-reducing bacteria. In general, in many of the molybdenum-containing enzymes, it may be replaced by tungsten. However, same is not true in case of most of the enzymes containing tungsten. Thus, it suggests that in the process of evolution, in relation with molybdenum, tungsten of tungsten enzymes might have been substituted by molybdenum.

## Metalloproteins

Conjugated proteins, which have metal ions as cofactors, are generally referred to as metalloproteins (e.g., acetylene hydratase which contains tungsten). They serve various functions in the cells such as enzymes, proteins for storage and transport, and proteins involved in signal transduction. In fact, sizable numbers of enzymes require metals for their activities. The metal ions are coordinated to nitrogen, sulfur, oxygen atoms of the amino acids of the polypeptide chains, or macrocytic ligand. The metal facilitates the role of metalloproteins in functions such as redox reaction.

### Tungsten-Containing Proteins

Tungsten is the heaviest element known to occur in biomolecules synthesized by living organisms. Further, the element has been proved to be essential for life processes. Tungsten-containing enzymes have been found in the organisms living in both extreme and non-extreme conditions. There are good numbers of tungsten-containing enzymes found in microorganisms: hyperthermophilic archaea (*Pyrococcus furiosus* and *Thermococcus litoralis*), methanogens (*Methanobacterium thermoautotrophicum* and *Methanobacterium wolfei*), Gram-positive bacteria (*Clostridium thermoaceticum, C. formicoaceticum,* and *Eubacterium acidaminophilum*), Gram-negative anaerobes (*Desulfovibrio gigas, Pelobacter acetylenicus*), and Gram-negative aerobes (*Methylobacterium* sp. *RXM*). The biological role of the element has thus been established in the prokaryotes but not yet in eukaryotes.

*Tungsten-containing proteins at a glance*: The various tungsten-containing proteins along with their sources (Refer to Table 1), their metal cofactors with their metal content (Refer to Table 2), and protein structure and their molecular weight (Refer to Table 3) are given below for a glance.

### Tungsten-Containing Enzymes: An Overview

There are four different families of enzymes that use tungsten as cofactors. They are (1) aldehyde oxidoreductase family, (2) formate dehydrogenase family, (3) dimethyl sulfoxide reductase family, and (4) Benzoyl-coenzyme A reductase family (Hille 2002).

### Aldehyde Oxidoreductase (AOR) Family of Enzymes

This family of tungstoenzymes (Refer to Fig. 1) contains (1) aldehyde oxidoreductase (AOR)

**Tungsten Cofactors, Binding Proteins, and Transporters in Biological Systems, Table 1** Tungsten-containing proteins and their sources

| Number | Tungsten-containing proteins | Source |
|---|---|---|
| 1. | Acetylene hydratase | *Pelobacter acetylenicus* |
| 2. | Aldehyde ferredoxin oxidoreductase | *Desulfovibrio gigas*, *Pyrococcus furiosus* |
| 3. | Aldehyde oxidoreductase | *Eubacterium acidaminophilum* |
| 4. | Benzoyl-coenzyme A reductase | *Geobacter metallireducens* |
| 5. | Carboxylic acid reductase (Form I) | *Clostridium thermoaceticum* |
| 6. | Carboxylic acid reductase (Form II) | *Clostridium thermoaceticum* |
| 7. | Formaldehyde ferredoxin oxidoreductase | *Thermococcus litoralis* |
| 8. | Formate dehydrogenase | *Clostridium thermoaceticum, Methylobacterium extorquens*, *Desulfovibrio gigas*, *Syntrophobacter fumaroxidans*, *Eubacterium acidaminophilum* |
| 9. | Formylmethanofuran dehydrogenase | *Methanobacterium thermoautotrophicum, Methanobacterium wolfei* |
| 10. | Glyceraldehyde-3-phosphate ferredoxin reductase | *Pyrococcus furiosus* |
| 11. | Tungsten oxidoreductases 4(WOR4) | *Pyrococcus furiosus* |
| 12. | Tungsten oxidoreductases 5 (WOR5) | *Pyrococcus furiosus* |

(Roy and Adams 2002; Bevers et al. 2005), (2) ferredoxin oxidoreductase (FOR) (Roy and Adams 2002), (3) glyceraldehyde-3-phosphate oxidoreductase (GAPOR) (Roy and Adams 2002), (4) carboxylic acid reductase (CAR) (Rauh et al. 2004), (5) tungsten-containing oxidoreductase 4 (WOR4) (Roy and Adams 2002) and tungsten-containing oxidoreductase 5 (WOR5) (Bevers et al. 2005). Each enzyme has a single 4Fe:4S cluster with a single tungsten atom present in the nucleus of the cluster and a bis-pterin cofactor. The size of the subunits of these enzymes is nearly similar (approximately 70 kDa) with a high similarity in sequence (more than about 50%). These enzymes bring about the conversion of aldehydes to carboxylic acids with 4Fe:4S ferredoxin as acceptor of the reduced equivalents. Active sites of these enzymes contain two equivalents of pterin cofactors coordinately bound to tungsten. The oxidized enzyme has two tungsten atoms ($W^{VI}$–O and one $W^{VI}$–OH) while the reduced form has a single $W^{VI}$–OH (Hille 2002).

AOR shows broad substrate specificity being more active with amino acid–derived aldehydes by transamination and decarboxylation. AOR plays a vital role in peptide fermentation. The 2-keto oxidoreductases (four types), present in *Pyrococcus furiosus*, generate aldehydes which are oxidized by AOR (Roy and Adams 2002).

FOR shows excellent substrate specificity for semi- and di-aldehyde (4–6 carbon atoms) with a role in catabolism of peptides (Roy and Adams 2002).

The enzymes AOR and FOR have 231 identical sequences, showing 38% identity and 59% similarity with a higher incidence of similarities being observed from residues 1–210 (48% identity). The four cysteine residues which are coordinated to 4Fe:4S cluster along with the Asn-93, Ala-183, and coordinated to magnesium are same in AOR and FOR, but they vary in the Glu-313, and Mis-488 (missense point, $\lambda_{max}$ 488 nm) residues coordinated to pterin which are present near the substrate-binding site (Roy and Adams 2002).

**Aldehyde Ferredoxin Oxidoreductase (AOR)** The homodimer protein aldehyde ferredoxin oxidoreductase (AOR) obtained from *Pyrococcus furiosus* consists of 4Fe:4S cluster with dithiolene groups from two pterin molecules coordinated with mononuclear tungsten that serves as cofactor. Further, iron links the two subunits of the AOR dimer (Refer to Fig. 2) (Roy and Adams 2002). Organisms growing at high temperatures, as found in marine volcanoes, contain this enzyme. Acetaldehyde, isovaleraldehyde, phenylacetaldehyde, and indoleacetaldehyde, the aldehyde derivatives of some of the common amino acids, act as the substrates for AOR.

1. Aldehyde Oxidoreductase

   This enzyme is obtained from *Eubacterium acidaminophilum* which is a monomer with apparent molecular mass of 67 kDa and contains 6.0 mol iron, 1.1 mol tungsten, 0.6 mol each of zinc and pterin cofactor per mole of protein (Refer to Fig. 3) (Rauh et al. 2004). AOR brings about the oxidation

**Tungsten Cofactors, Binding Proteins, and Transporters in Biological Systems, Table 2** Tungsten-containing proteins, metal and metal cofactors, and their content

| Number | Tungsten-containing proteins | Metal and metal cofactors present | Metal and metal cofactors content (Mol/mol protein) |
|---|---|---|---|
| 1. | Acetylene hydratase | W, Fe | 0.5, 3 |
| 2. | Aldehyde ferredoxin oxidoreductase (D. gigas) | W, Fe | 0.68, 4 (as FeS) |
| 3. | Aldehyde ferredoxin oxidoreductase (P. furiosus) | W, Fe | 2, 4 (as FeS) |
| 4. | Aldehyde oxidoreductase | W, Fe, Zn | 1.1, 6, 0.6 |
| 5. | Benzoyl-coenzyme A reductase | W, Fe | 0.9, 15 |
| 6. | Carboxylic acid reductase (Form I) | W, Fe | 1, 29 |
| 7. | Carboxylic acid reductase (Form II) | W, Fe | 3, 82 |
| 8. | Formaldehyde ferredoxin oxidoreductase | W, Fe | 4, 4 (as FeS) |
| 9. | Formate dehydrogenase (C. thermoaceticum) | W, Se, Fe | 2, 2, 20 |
| 10. | Formate dehydrogenase (D. gigas) | W, Fe | 0.9, 7 |

**Tungsten Cofactors, Binding Proteins, and Transporters in Biological Systems, Table 3** Tungsten-containing proteins, their structure, and molecular weight

| Number | Tungsten-containing proteins | Protein structure | Molecular weight (in kDa) |
|---|---|---|---|
| 1. | Acetylene hydratase | $\alpha_1$ | $\alpha_1 = 85$ |
| 2. | Aldehyde ferredoxin oxidoreductase (D. gigas) | $\alpha_2$ | $\alpha_2 = 62$ |
| 3. | Aldehyde ferredoxin oxidoreductase (P. furiosus) | $\alpha_2$ | $\alpha_2 = 67$ |
| 4. | Aldehyde oxidoreductase | Monomer | 67 |
| 5. | Benzoyl-coenzyme A reductase | $\alpha_2\beta_2$ | $\alpha_2 = 73, \beta_2 = 20$ |
| 6. | Carboxylic acid reductase (Form I) | $\alpha\beta$ | $\alpha = 64, \beta = 14$ |
| 7. | Carboxylic acid reductase (Form II) | $\alpha_3\beta_3\gamma$ | $\alpha_3 = 64, \beta_3 = 14, \gamma = 43$ |
| 8. | Formaldehyde ferredoxin oxidoreductase | $\alpha_4$ | $\alpha_4 = 69$ |
| 9. | Formate dehydrogenase (C. thermoaceticum) | $\alpha_2\beta_2$ | $\alpha_2 = 96, \beta_2 = 76$ |
| 10. | Formate dehydrogenase (D. gigas) | $\alpha\beta$ | $\alpha = 96, \beta = 76$ |

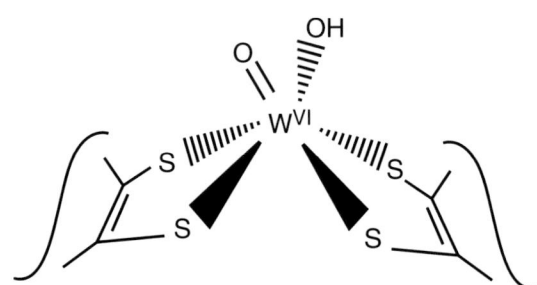

**Tungsten Cofactors, Binding Proteins, and Transporters in Biological Systems, Fig. 1** The general structure of aldehyde ferredoxin oxidoreductase family (Hille 2002) (The figure has been reproduced with the prior approval of Dr. Russ Hille)

(two-electron) of aldehyde substrates (both aliphatic and aromatic) to acids with ferredoxin acting as the acceptor of electron. The activity of the enzyme is induced in presence of high molar concentration of electron donors such as serine and/or formate when supplied in the growth medium. Activity of the enzyme with acetaldehyde, propionaldehyde, butyraldehyde, isovaleraldehyde, and benzaldehyde differs by a factor of less than 2. Further, the deduced sequence of aldehyde oxidoreductase exhibits great similarities with the other tungsten-containing aldehyde oxidoreductases from archaea.

2. Formaldehyde Ferredoxin Oxidoreductase (FOR)
There are three domains in each monomer of the enzyme having tungsten and iron-sulfur clusters acting as cofactor, located at the interface as the domains. Two similar folded halves, each with a six-stranded β-sheet and three α-helices, are present in the N-terminal domain (1st domain). Further, there are 14 α- and 11 α- helices in the second and third domains, which bear symmetry to

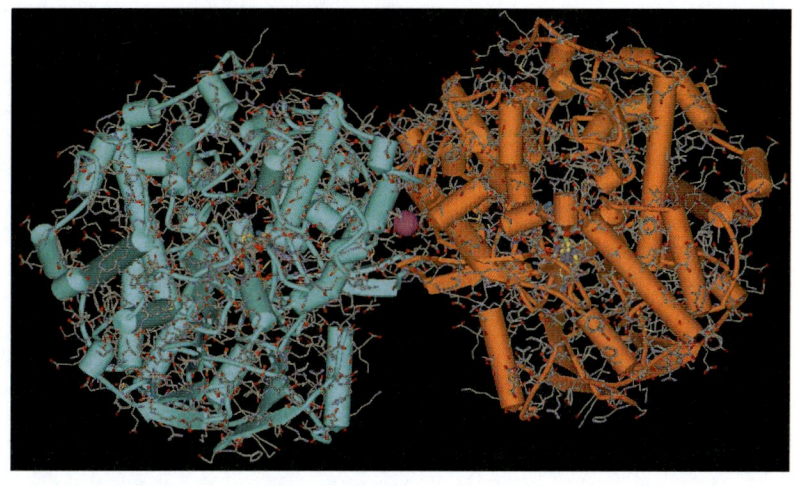

**Tungsten Cofactors, Binding Proteins, and Transporters in Biological Systems, Fig. 2** 3D schematic structure representation of the AOR dimer, showing the two subunits and the associated metallocenters

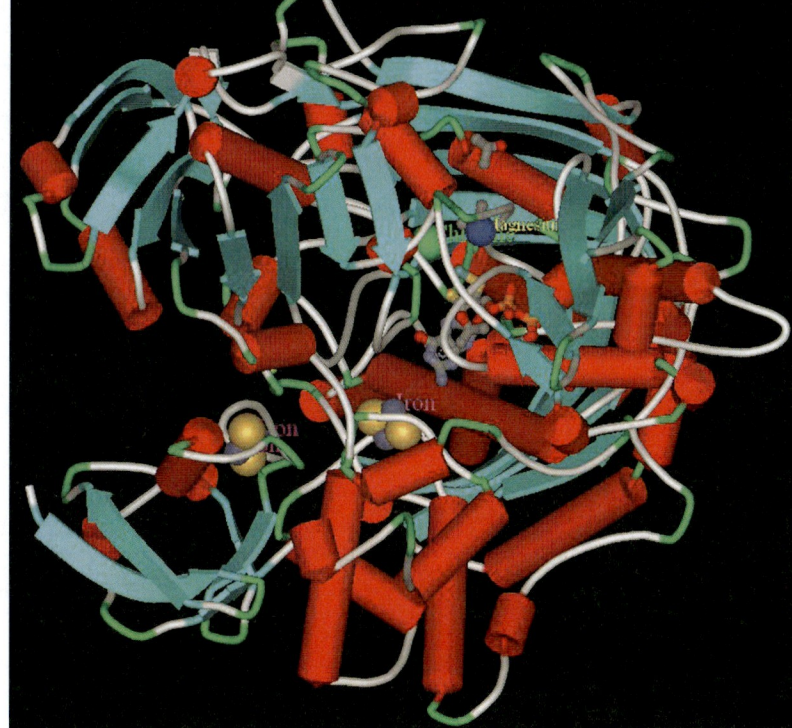

**Tungsten Cofactors, Binding Proteins, and Transporters in Biological Systems, Fig. 3** 3D schematic representation of the structure of one monomer of the aldehyde oxidoreductase from *D. gigas*. The domains (FebondSproximal; FebondSdistal; Mo1, and Mo2) are represented in different colors, and the metal cofactors are shown in *ball-and-stick* mode; tungsten (*blue*), iron (*gray*), chlorine (*green*)

each other (Hille 2002). The second domain interacts with pterin in which the third domain interacts with Q-pterin. A unique short three-stranded sheet present in the second domain has CxxCxxxC motif for Fe:S cluster. The Q-pterin plays an important role in electron movement to and from the enzyme metal centre. This fact is indicated by the presence of Fe:S cluster near Q-pterin (Hille 2002).

(a) Formaldehyde Ferredoxin Oxidoreductase from *Pyrococcus furiosus*

This is a homotetrameric protein that contains one tungstodipterine with one 4Fe:4S cubane as cofactor per 69 kDa subunit (Refer to Fig. 4) (Roy and Adams 2002; Bevers et al. 2005). The substrate concentration, as well as electron acceptor, may have a role for the activity of

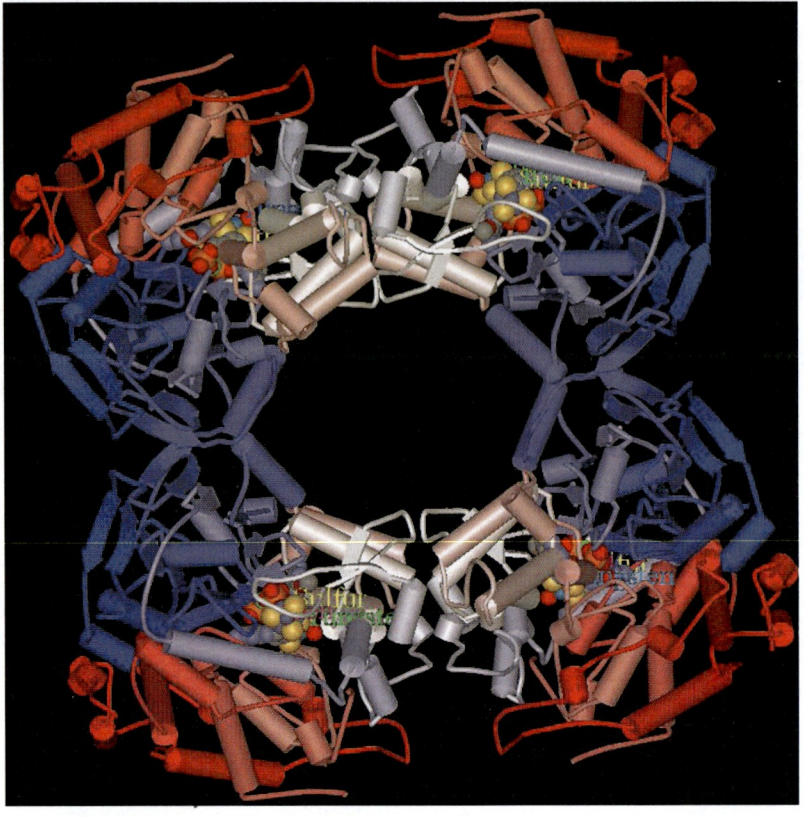

**Tungsten Cofactors, Binding Proteins, and Transporters in Biological Systems, Fig. 4** 3D schematic structure of tungsten-containing formaldehyde ferredoxin oxidoreductase from *P. furiosus* domains 1, 2, 3, and 4 of each subunit contains tungsten (*green*), sulfur (*yellow*), and iron (*gray*). The docking position of *P. furiosus* ferredoxin to one FOR subunit, as observed in the crystal structure of the FOR-ferredoxin complex, is indicated with the *green* trace for the ferredoxin and a CPK model for 4Fe:4S cluster

the enzyme. Steady state condition for the enzyme kinetics studied at 80 °C and pre-steady-state condition of the enzyme kinetics studied at 50 °C showed that two $K_m$ values for formaldehyde with two molecules of ferredoxin as electron acceptor were three times (approximately) lower than those with benzyl viologen as electron acceptor. Denaturing and/or inhibitory effect of the substrate when present in high concentration may be responsible for the variations.

(b) Formaldehyde Ferredoxin Oxidoreductase from *Thermococcus litoralis*

Archae bacterium *Thermococcus litoralis* can grow at up to 98 °C by fermenting peptides. Tungsten-containing protein ferredoxin oxidoreductase (FOR) stimulates its growth. It is a homotetramer containing 4 Fe, 4 acid-labile sulfides, and one tungsten per subunit (69 kDa). Tungsten appears as a pterin cofactor with Fe/S forming an unusual 4Fe:4S cluster existing in reduced state. This metalloprotein works at very high temperatures (90 °C and even above) using ferredoxin as an electron acceptor. However, it neither oxidizes aldehyde and phosphates, utilizing CoA, nor it reduces NAD (P).

FOR has similarity at N-terminal sequence with W-Fe-S–containing enzyme aldehyde FOR of *P. furiosus*. It has a role in pyroglycolytic pathway.

3. Glyceraldehyde-3-Phosphate Ferredoxin Reductase (GAPOR)

It is a tungsten-containing protein known for its potential role in pyroglycolytic pathway (Refer to Fig. 5) (Roy and Adams 2002). This metalloprotein is generally found in *Pyrococcus abyssi*, *Thermococcus kodakarensis*, and *Thermococcus celer* (hydrogen-producing microorganisms) and in *Desulfurococcus amylolyticus*, *Thermoproteus tenax*, *Methanocaldococcus jannaschii*, and *Methanococcus maripaludis* (non-hydrogen-producing microorganisms). The growth of many of these bacteria (*P. furiosus*) is dependent on tungsten. The enzyme contains 0.85 g W atom per mole. Glyceraldehyde-3-phosphate is the only substrate for GAPOR in the unusual pyroglycolytic pathways

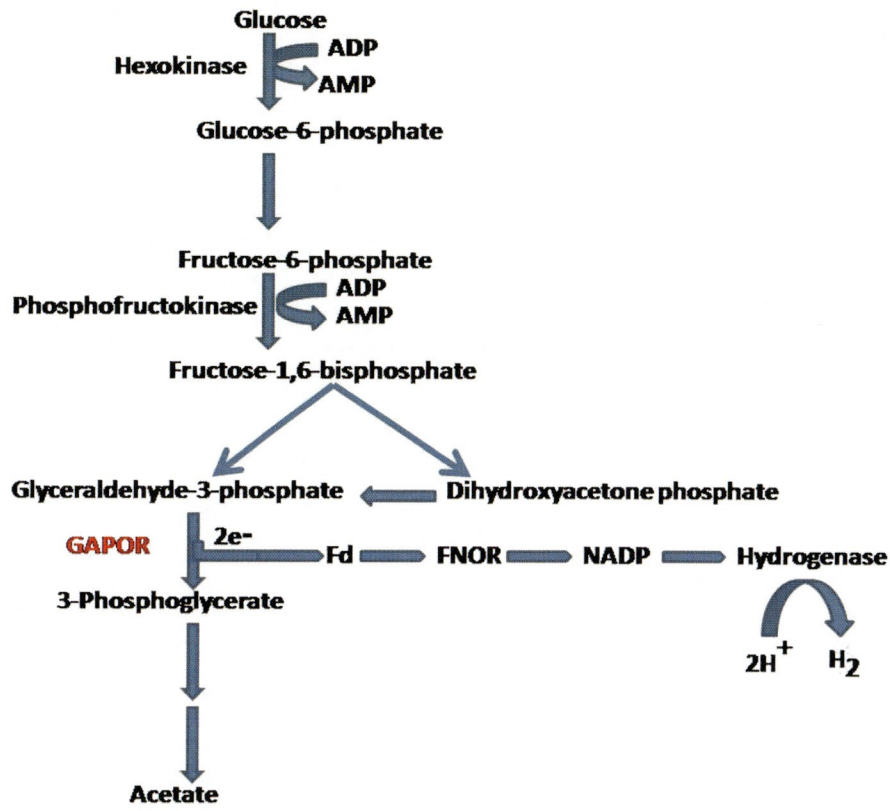

**Tungsten Cofactors, Binding Proteins, and Transporters in Biological Systems, Fig. 5** Proposed role of GAPOR in the conversion of glucose to acetate in *P. furiosus*

(Roy and Adams 2002). It forms glyceraldehyde-3-phosphoglycerate in the pyroglycolytic pathway. The enzyme is distinct from other tungsten-containing enzymes in terms of their oxidative role. It catalyzes glyceraldehyde-3-phosphate oxidation to yield 3-phosphoglycerate and reduction of oxidized ferredoxin by a dual substrate electron transfer reaction. It uses ferredoxin as an electron acceptor since ferredoxin is more stable at high temperatures.

4. Carboxylic Acid Reductase (CAR)

   This enzyme belongs to the AOR family and is also known as aldehyde dehydrogenase. It is obtained from *Clostridium thermoaceticum* (Rauh et al. 2004). This tungsten-containing enzyme reduces nonactivated carboxylic acids to aldehydes in the presence of reduced viologens. The oxidized viologens are able to dehydrogenate aldehydes to carboxylic acid about 20 times faster than the reduced viologens. Carboxylic acid reductase has two forms: One has molecular mass of about 240 kDa, while the other has molecular mass of about ?60 kDa.

5. Tungsten-Containing Oxidoreductases 4 and 5 (WOR4 and WOR5)

   These enzymes belong to family oxidoreductases containing tungsten (Bevers et al. 2005). Tungsten plays an important role in the growth of *Pyrococcus furiosus* with optimal growth at 100°C, with peptides and carbohydrates as carbon sources (Roy and Adams 2002; Bevers et al. 2005). The homodimer WOR4 contains tungsten, three Fe, and 3–4 acid-labile sulfides with a $Ca^{++}$ atom per subunit. Generally, tungsten enzymes oxidize various types of aldehydes having a significant role in catabolism of sugars and amino acids. However, WOR4 neither oxidizes aliphatic or aromatic aldehydes or hydroxyl acids nor reduces the keto acids (Roy and Adams 2002). The homodimeric protein WOR5 has one 4Fe:4S cluster and one tungstobispterin cofactor per subunit (65 kDa). The enzyme shows broad substrate specificity having

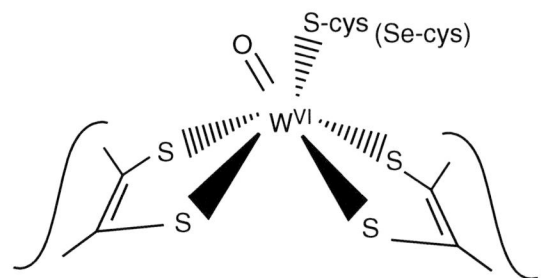

**Tungsten Cofactors, Binding Proteins, and Transporters in Biological Systems, Fig. 6** The general structure of formate dehydrogenase family (Hille 2002) (The figure has been reproduced with the prior approval of Dr. Russ Hille)

higher affinity for substituted and non-substituted aliphatic and aromatic aldehydes. The enzymes WOR5, formaldehyde oxidoreductase, and aldehyde oxidoreductase together catalyze the oxidation of various types of aldehydes in *P. furiosus* (Bevers et al. 2005).

Formate Dehydrogenase Family
This family of enzymes (Refer to Fig. 6) contains two enzymes, formate dehydrogenases (1 and 2 types), and N-formylmethanofuran dehydrogenase, which are involved in the physiological fixation of $CO_2$ into acetate and N-formylmethanofuran by reduction, respectively (Hille 2002). The structural analysis shows that the oxidized enzyme exists in $L_2W^{VI}OX$ coordination sphere (Hille 2002).

**Formate Dehydrogenase 1 and Formate Dehydrogenase 2** Heterodimeric enzymes with two subunits ($\alpha_2\beta_2$) of 107 kDa and 61 kDa, respectively, are obtained from the α-probacterium *Methylobacterium extorquens*. The purified enzymes contain (approximately 5 mol) non-heme iron and acid-labile sulfur, 0.6 mol of non-covalently linked FMN, and tungsten (approximately 1.8 mol). The catalytic subunit of these tungsten-containing enzymes was characterized from *Eubacterium acidaminophilum* and *Moorella thermoacetica*. The β-subunit of the enzymes contains putative motifs for binding with FMN and NAD. The enzymes also have iron-sulfur cluster-binding motif. The β-subunit acts as a fusion protein with N-terminal domain of the enzymes, which is related to NuoE-like subunits. The C-terminal domain of the enzymes is related to NuoE-like subunits known as NADH-ubiquinone oxidoreductase.

Two genes encoding formate dehydrogenases were identified in *Eubacterium acidaminophilum*. Each gene cluster consists of two coding regions: one encoding for catalytic subunit (FDH-1) and the other encoding for electron-transferring subunit (FDH-1, FDH-2).

Two different types of formate dehydrogenases (FDH-1, FDH-2) are also obtained from *Syntrophobacter fumaroxidans*. These two purified enzymes exhibit high formate oxidation and $CO_2$ reduction rates (Moura et al. 2004).

Formate dehydrogenases catalyze the oxidation reaction of formate and produces $CO_2$ (Moura et al. 2004):

$$HCOO^- \rightarrow CO_2 + H^+ + 2e^- \quad E^\circ = -420 \text{ mV}$$

The tungsten-containing formate dehydrogenase, obtained from *Desulfovibrio gigas*, is a heterodimeric enzyme with 92 and 29 kDa subunits, respectively (Moura et al. 2004).

**Formylmethanofuran Dehydrogenase** Methanogenic archaea such as *Methanobacterium thermoautotrophicum* and *Methanobacterium wolfei* contain formylmethanofuran dehydrogenase (a tungsten iron-sulfur protein with pterin cofactor, EC 1.2.99.5). N-formylmethanofuran is reversibly dehydrogenated to $CO_2$ and methanofuran (Bertram et al. 1994). Analysis of the structure of the enzyme shows that it contains tungsten, two iron-sulfur clusters, and molybdoprotein guanine dinucleotide.

Unlike the molybdenum-containing formylmethanofuran dehydrogenases, the tungsten-containing formylmethanofuran dehydrogenases are not inactivated by the cyanide (Bertram et al. 1994). Further, the thermophilic methanogenic archaeon, *Methanobacterium wolfei*, requires tungsten for its growth (Bertram et al. 1994). The enzyme has an apparent molecular mass of 130 kDa with three subunits (35, 51, 64 kDa). Amino acid sequence of N-terminal shows 0.3–0.4 mol W per mol enzyme with the molybdoprotein guanine dinucleotide as pterin cofactor. The enzyme has optimum activity at 65°C and requires potassium ions for thermostability.

This isoenzyme with tungsten obtained from *Methanobacterium thermoautotrophicum* generally consists of four subunits with molecular masses of 65 kDa (FwdA), 53 kDa (FwdB), 31 kDa (FwdC), and 15 kDa (FwdD). Each mole of this enzyme has 0.4 mol of tungsten, 0.6 mol molybdoprotein guanine dinucleotide, and 8 mol of non-heme iron and acid-labile sulfur.

**Tungsten Cofactors, Binding Proteins, and Transporters in Biological Systems, Fig. 7** 3D structure of acetylene hydratase from *Pelobacter acetylenicus*. Tungsten is shown as a *blue sphere*; iron and sulfide atoms of the 4Fe:4S cluster are shown as *gray* and *yellow spheres*. The two MGD ligands of tungsten are shown as blue sticks sulfur is represented as *yellow*

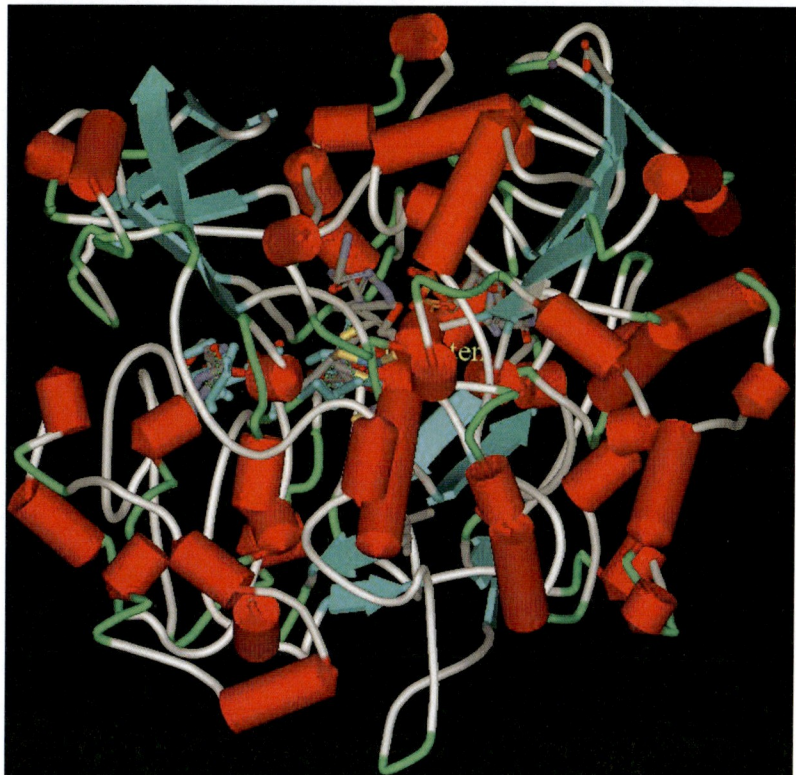

### Dimethyl Sulfoxide Reductase Family

**Acetylene Hydratase** This W-Fe-S enzyme (Refer to Fig. 7) belongs to dimethyl sulfoxide reductase family and is obtained from *Pelobacter acetylenicus* (Hille 2002). The enzyme contains bis-molybdopterine guanine dinucleotide–ligated tungsten atom and a cubane-type 4Fe:4Se cluster (Refer to Fig. 8). The tungsten center requires binding with a water molecule for its activity. The cubane-type 4Fe:4Se cluster causes a predominant shift of the $pK_a$ of the aspartate residue (of the enzyme) which in turn activates a water molecule to bind with the tungsten center of the enzyme. It then attacks the acetylene bound in the hydrophobic pocket of the enzyme. It undergoes non-redox reaction and catalyzes acetylene to acetaldehyde by hydration (Seiffert et al. 2007).

### Benzoyl Coenzyme A Reductase

The enzyme benzoyl coenzyme A reductase (BCoAR) obtained from *Geobacter metallireducens* has a molecular mass 185 kDa (Kunga et al. 2009). The enzyme is used by anaerobes to catalyze aromatic compounds into

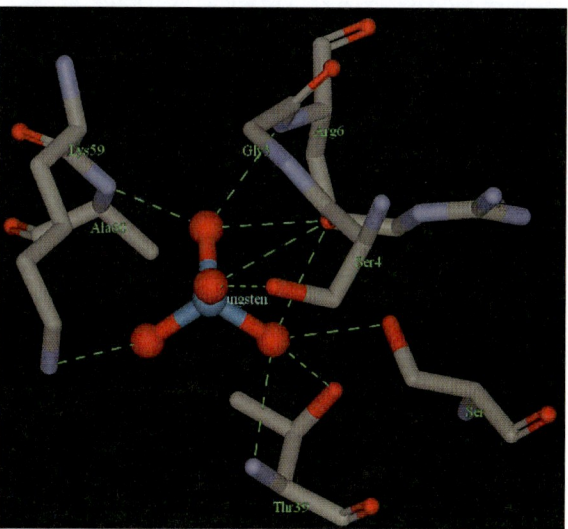

**Tungsten Cofactors, Binding Proteins, and Transporters in Biological Systems, Fig. 8** X-ray crystal structure of tungstate bound in a carrier protein. Tungstate is visible as the tetrahedral anion in the center of the picture. The *green dotted lines* represent hydrogen bonding, *blue* color represents tungsten, and *red* color represents oxygen

dearomatized products. In aerobic bacteria, BCoA is a center which intervenes in the degradation pathways of various aromatic compounds, and this BCoA is used as substrate by BCoAR to dearomatize the aromatic ring driven by a stoichiometric ATP hydrolysis. In facultative anaerobes, this reduction yields dienyl CoA generally catalyzed by a soluble 3Fe:4S cluster containing the enzyme BCoAR (Kunga et al. 2009). The oxygen-sensitive enzymes have $\alpha_2\beta_2$ protein chain composition with 73 kDa and 20 kDa subunits. One 3Fe:4S cluster and three 4Fe:4S clusters exist per $\alpha\beta$ unit. Presence of benzoid chemicals induces Bam BC gene, which transcribes the enzyme (Kunga et al. 2009).

## Other Metalloproteins Whose Activity Is influenced by Tungsten

(a) *Nitrate reductase*: Substitution of tungsten for molybdenum in nitrate reductase enzyme complex (EC 1.7.1.1) results in an inactive enzyme.
(b) *A Novel Iron Hydrogenase Whose Activity Is Dependent upon Tungsten*

The thermophilic eubacterium *Thermotoga maritima* can grow up to 90°C by using a fermentation metabolism having $H_2$, $CO_2$, and organic acids as the end products. The novel cytoplasmic enzyme iron hydrogenase catalyzes the production of $H_2$, and addition of tungsten to the growth medium increases both the concentration of the hydrogenase and its in vitro activity. However, structural analysis of the enzyme shows that tungsten is not the part of the enzyme. Possibly, tungsten has a role in modulating $H_2$-activating FeS center.

## Binding Proteins and Transport of Tungsten

*Ion transporter*: An ion transporter is a transmembrane protein, which moves ions across plasma membrane against the concentration gradient. It differs from the ion channels, where ions move by passive transport process. Transporters are proteins, which convert energy from ATP, sunlight, and other redox reactions, to the potential energy. This energy is then used by secondary transporters, which are generally known as ion carriers. They drive vital cellular processes, such as ATP synthesis. One of the examples of ion transporters is $Na^+/K^+$ ATPase (sodium-potassium adenosine triphosphatase), also called the $Na^+/K^+$ pump (sodium-potassium pump), and is located in the plasma membrane in almost all animals.

*Tungsten-binding protein*: Tungsten enters the biological systems as the tetrahedral oxoanions. Certain proteins have been isolated (known as carrier proteins), which bind and transport tungsten (Refer to Fig. 8) (Sugio et al. 2004). Tungsten-binding protein AP19-3 is a heterodimeric protein isolated from iron-oxidizing bacterium *Acidithiobacillus ferrooxidans* and consists of two subunits with molecular masses of 12 kDa and 20.7 kDa (Sugio et al. 2004). Tungsten shows predominantly much higher binding efficiency with AP19-3, the purified tungsten-binding protein, as compared to albumin, aldolase, catalase, chymotrypsinogen A, ferritin, and ferredoxin (at pH 3.0) (Sugio et al. 2004).

*Tungsten transportation*: Transportation of tungsten is generally done by a specific ATP-binding cassette (ABC) transporter found in *Eubacterium acidaminophilum* (Makdessi et al. 2001). It is a Gram-positive anaerobe that contains at least two tungsten-dependent enzymes: one is formate dehydrogenase, and the other is aldehyde dehydrogenase. The *ABC* transporter, specific for tungstate, is coded by genes *tupABC*. The substrate-binding protein, TupA, is overexpressed in *E. coli*, and tungstate induces shifting of TupA mobility, representing that this anion is bound by TupA. $K_d$ value for tungstate is 0.5 µM. *ABC* transporter exhibits high similarities with putative transporters from *Methanobacterium thermoautotrophicum*, *Haloferax volcanii*, *Vibrio cholerae*, and *Campylobacter jejuni*. When *tupABC* genes are undergoing downstream, the genes *moeA, moeA-1, moaC*, and a truncated *moaC* have been identified by comparison of the sequence of the deduced amino acid sequences. These genes participate in the biosynthesis of the pterin cofactor, which is present in molybdenum- and tungsten-containing enzymes except nitrogenase. The extracytoplasmic binding portion of the TupA transporter is overexpressed in *Escherichia coli*, and the anion-binding characteristics of the protein show specificity for the tungstate (Makdessi et al. 2001).

## Role of Tungsten in Biological System

(a) Effects of Tungsten on the Vascular Responses
The surface oxidation properties of tungsten wire generally used in wire myography significantly and adversely affects vascular responses to vasodilators. The effect is likely due to the paratungstate anion

$[W_{12}O_{42}]^{-12}$ and is not associated with free radical generation or $K^+$ channel inhibition.

(b) Tungsten, as Essential Heavy Metal Element for Prokaryotes

Tungsten serves as the heaviest element in biological systems. It acts as an inhibitory element/anion, in iron-molybdenum cofactor containing enzyme nitrogenase in dinitrogen fixation, and also in many "metal-binding pterins," such as tricyclic pyranopterin, as in molybdoenzymes, and in the sulfite oxidase and the xanthine dehydrogenase families of enzymes. They are involved in the transformation of different types of carbon-, nitrogen-, and sulfur-containing compounds. Tungstate serves as cofactors for enzymes such as dimethyl sulfoxide (DMSO) reductase family, especially in case of $CO_2$-reducing formate dehydrogenases (FDHs), formylmethanofuran dehydrogenases, and acetylene hydratase (catalyzing only an addition of water, but no redox reaction). Tungsten also serves as an essential element for the aldehyde oxidoreductase (AOR) family.

(c) Effect of Tungsten Cofactor on Molybdenum Cofactor

The RNA motif, located upstream of genes encoding molybdate transporters, molybdenum cofactor (Moco) biosynthesis enzymes, and proteins, utilizes Moco as a coenzyme. Variants of this RNA are likely to be activated by the related tungsten cofactor (Tuco) that carries tungsten in place of molybdenum.

(d) Antidiabetic Effect of Tungsten

Oxyanion derivative of tungsten has similar biological activity like vanadium. Tungsten derivative has been shown to have antidiabetic effect on animals. When sodium tungstate was administered in the drinking water of diabetic rats for 8 months, there is a significant reduction of blood glucose level. There was no evidence of intolerance after prolonged use of the tungsten derivative.

(e) Other Effects

In soil, tungsten metal oxidizes to tungstate anion. The soil chemistry determines how the tungsten polymerizes. Generally, alkaline soil is responsible for monomeric tungstates, while acidic soil is responsible for polymeric tungstates. Sodium tungstate has less toxic effect on earthworms, but it completely inhibits their reproductive ability. Tungsten is found as a biological copper metabolic antagonist, and its role is similar to that of molybdenum. Tetrathiotungstates are generally used as biological copper chelation chemicals.

## Conclusions

Tungsten as the heaviest element in biological system has been established as an essential metal in microbes. Tungsten cofactors in tungsten-containing enzymes generally involve in redox reaction. Anti-diabetic potential of tungsten derivatives in animals and use of the tungsten wire in myography in humans claim further research to understand the importance and to utilize the benefit of the element in biomedical sciences. However, it is yet to identify any tungsten protein in higher animals or plants.

## Cross-References

▶ Formate Dehydrogenase
▶ Nitrite Reductase
▶ Tungsten in Biological Systems

## References

Bertram PA, Karrasch M, Schmitz RA, Bocher R, Albracht SPJ, Thauer RK (1994) Formylmethanofuran dehydrogenases from methanogenic Archaea substrate specificity, EPR properties and reversible inactivation by cyanide of the molybdenum or tungsten iron-sulfur proteins. Eur J Biochem 220:477–484

Bevers LE, Boll E, Hagedoorn PL, Hagen WR (2005) WOR5, a Novel Tungsten-containing aldehyde oxidoreductase from *Pyrococus furiosus* with a broad substrate specificity. J Bacteriol 187:7056–7061

Hille R (2002) Molybdenum and tungsten in biology. Trends Biochem Sci 27:360–367

Kunga JW, Lfflera C, Dörnerb K, Heintzc D, Galliend S, Van Dorsselaerd A, Friedrichb T, Boll M (2009) Identification and characterization of the tungsten-containing class of benzoyl-coenzyme A reductases. Proc Natl Acad Sci USA 106:17687–17692

Makdessi K, Andreesen JR, Pich A (2001) Tungstate uptake by a highly specific ABC transporter in *Eubacterium acidaminophilum*. J Biol Chem 276:24557–24564

Moura JJ, Brondino CD, Trincão J, Romão MJ (2004) Mo and W bis-MGD enzymes: nitrate reductases and formate dehydrogenases. J Biol Inorg Chem 9:791–799

Rauh D, Graentzdoerffer A, Granderath K, Andreesen JR, Pich A (2004) Tungsten-containing aldehyde oxidoreductase of *Eubacterium acidaminophilum*: isolation, characterization and molecular analysis. Eur J Biochem 271:212–219

Roy R, Adams MWW (2002) Characterization of a fourth tungsten containing enzyme from the hyperthermophilic archaeon Pyrococcus furiosus. J Bacteriol 184:6952–6956

Seiffert GB, Ullmann GM, Messerschmidt A, Schink B, Kroneck PMH, Einsle O (2007) Structure of the non-redox-active tungsten/4Fe:4S enzyme acetylene hydratase. Proc Natl Acad Sci USA 104:3073–3077

Sugio T, Kuwano H, Hamago Y, Negishi A, Maeda T, Takeuchi F, Kamimura K (2004) Existence of a tungsten-binding protein in Acidithiobacillus ferrooxidans AP19-3. J Biosci Bioeng 97:378–382

# Tungsten in Biological Systems

Partha Basu
Department of Chemistry and Biochemistry,
Duquesne University, Pittsburgh, PA, USA

## Synonyms

Biology; Living system and organisms; Tungsten; Wolfram

## Definition

Cofactor – A cofactor is a small molecule that is generally non-covalently bonded to a protein and is required for the function of the protein.

Michaelis-Menton kinetics – This kinetic model is for enzymes that describes the relation between the rate of substrate conversion and its concentration. $K_m$ is the apparent affinity of the enzyme for the substrate, $V_{max}$ is the maximum rate of conversion, and $k_{cat}$ is the turnover number.

Stable isotopes – Nonradioactive chemical isotopes and those radioisotopes with a very long half-life.

Tungsten cofactor – Pterin dithiolene cofactor coordinated to tungsten.

Tup ABC – ABC transporter for tungstate.

## Introduction

Among the group 6 elements, molybdenum is recognized to be essential in all phyla of life, and its biological role has been investigated at a greater detail. Of them, the pterin-containing molybdenum enzymes are the most diverse where the molybdenum is coordinated by a specially designed cofactor called pyranopterin cofactor, also known as molybdopterin (Basu and Burgmayer, 2011). Tungsten enzymes, while discovered recently, has the same pyranopterin cofactor but coordinated to W. The only natural biological role of tungsten is to catalyze substrate transformation, and unlike molybdenum enzymes, the W-containing enzymes have been isolated exclusively from microorganisms.

## Basic Considerations

### Molecular Properties

W compounds exhibit a wide variety of stereochemistries in addition to different oxidation states, the highest oxidation state being +6. Tungsten enzymes catalyze substrate transformation concomitant with a two electron redox process, changing where the W oxidation state shuttles between +6 to +4. In the earth's crust, W is present at a low concentration, $\sim 1$ ppm, and in oceanic water it is even lower, $\sim 0.1$ ppb. The higher concentration of Mo in water may have provided an evolutionary pressure resulting in a larger family of enzymes incorporating Mo than W. The concentration of W is, however, significantly higher in alkaline brine lakes, hot spring waters, and marine hydrothermal vents. In hydrothermal vent fluids, the concentration of W is $\sim 1,000$ times higher than that found in seawater. The dominant species of naturally occurring tungsten in water is tungstate ($WO_4^{2-}$), which is present over a wide pH range. At a higher pH and in the presence of sulfur, dominant species is still tungstate; even though a significant amount of tungsten disulfide is produced (Fig. 1). Organisms harboring W-containing enzymes generally grow in hydrothermal vents, and have the means to transport tungstate into the cell.

### W in Growth Media

The parallel chemistry of Mo and W led to the hypothesis that replacement of Mo with W in molybdenum enzyme may provide insights into the catalytic role of the metal. However, when different organisms were grown on W or were exposed to W, the resulting enzymes exhibited an attenuated activity or complete loss of activity. Therefore, W was considered to be an antagonist of Mo-enzymes. Indeed, a routine test for

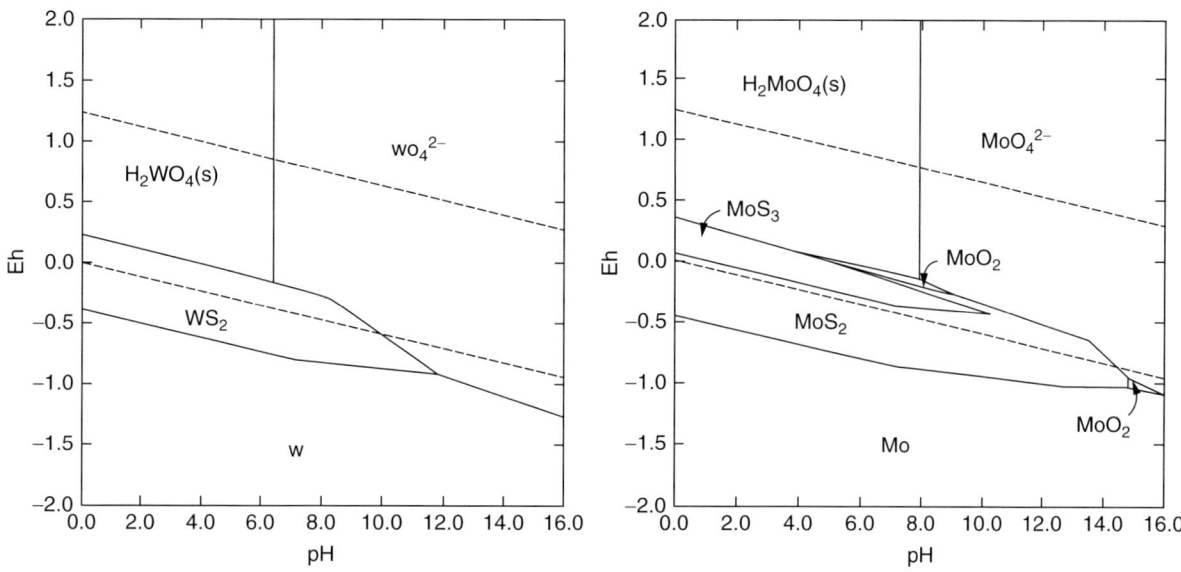

**Tungsten in Biological Systems, Fig. 1** The Eh-pH diagram for M-S-H$_2$O system at 25°C; concentration of M ~10$^{-3}$, S 1.0 (*Left*: M = W; *right*: M = Mo) (Adopted from Osseo-Asare (1982))

Mo-enzymes is the inhibition by W. Stimulated growth of certain *Clostridia* (*Clostridium thermoaceticum* and *Clostridium formicoaceticum*) was reported in 1970s. These organisms fermentatively grow on sugar producing acetate as the sole product. When tungstate is supplemented in the growth media, it increased the NADP-dependent activity which indicated a positive role of tungsten. A positive growth effect of tungstate on another microorganism, *Methanobacterium wolfei*, was demonstrated by incorporation of radio-labeled tungstate (with $^{185}$W). Later it was shown that tungstate impacts the activity of formylmethylfuran dehydrogenase (FMDH). FMDH reduces CO$_2$ to formaldehyde.

## W-Containing Enzymes

To date, all W-containing enzymes have a mononuclear active site where catalysis takes place. The W atom is coordinated by the dithiolene sulfur donors from a pterin cofactor. This cofactor is also present in pterin containing Mo enzymes, and the coordinated form is called molybdenum cofactor. In this entry, the tungsten-bound form is referred as the tungsten cofactor, Wco (Fig. 2). Degradative studies on molybdenum enzymes led to the proposed basic form of this cofactor. The molybdenum cofactor (Moco)

**Tungsten in Biological Systems, Fig. 2** Schematic representation of Wco

is synthesized biochemically in conserved steps, and the same is likely for the tungsten cofactor (Leimkuehler et al. 2011). The structure of the cofactor was confirmed by X-ray crystallographic study of a W-containing protein, aldehyde oxidoreductase (Chan et al. 1995). In addition to the W-containing active site they also harbor other redox components, in most cases an iron sulfur cluster (Table 1). To date, all pterin-containing W-enzymes were isolated from strict anaerobic organisms and are oxygen sensitive. The enzymes were isolated from the cytosolic fraction of the cell and a membrane-bound W-enzyme is yet to be reported. In contrast, Mo enzymes have been isolated from the cytosolic, periplasmic, and membrane fractions.

**Tungsten in Biological Systems, Table 1** Representative tungsten enzymes and their selected properties

| Organism | Enzyme | Family | Subunit | Mass (kDa) | Metal units |
|---|---|---|---|---|---|
| *Pyrococcus furiosus*[a] | AOR | AOR | α2 | 67 | 2Wco, 2[4Fe4S], 1Fe |
| *P. furiosus*[b] | FOR | | α4 | 69 | 4Wco, 4[4Fe4S] |
| *P. furiosus*[c] | GAPOR | | α | 73 | Wco, [4Fe4S], |
| *P. furiosus*[d] | WOR-4 | | α2 | 69 | 2Wco, ~6Fe |
| *P. furiosus*[e] | WOR-5 | | α2 | 65 | 2Wco, 2[4Fe4S], |
| *Clostridium thermoaceticum*[f] | CAR-I | | αβ | 64,14 | Wco, 29Fe |
| *C. thermoaceticum*[g] | CAR-II | | α3β3γ | 64,14,43 | 3Wco, 82Fe |
| *C. formicoaceticum*[h] | CAR | | α2 | 67 | 2Wco, 11 Fe |
| *Desulfovibrio gigas*[i] | ADH | | α2 | 62 | 2Wco, 2[4Fe4S] |
| *C. thermoaceticum*[j] | FDH | FDH | α2β2 | 96,76 | 2Wco, 20–40 Fe |
| *Methanobacterium wolfei*[k] | FMDH | | αβγ | 64,51,35 | Wco, 2–5 Fe |
| *D. gigas*[l] | FDH | | αβ | 110,24 | Wco, [4Fe4S] |
| *M. thermoautotrophicum*[m] | FMDH | | αβγδ | 65,53,31,15 | Wco, 8Fe |
| *Pelobacter acetylenicus*[n] | AH | AH | αβ | 73 | Wco, [4Fe4S] |

[a]Chan et al. (1995); Mukund and Adams (1990), (1991)
[b]Hu et al. (1999); Kletzin et al. (1995)
[c]Mukund and Adams (1995)
[d]Roy and Adams (2002)
[e]Bevers et al. (2005)
[f]White et al. (1989); White and Simon (1992)
[g]White and Simon (1992); White et al. (1989)
[h]White et al. (1991)
[i]Hensgens et al. (1995)
[j]Yamamoto et al. (1983)
[k]Schmitz et al. (1992)
[l]Raaijmakers et al. (2002)
[m]Bertram et al. (1994); Hochheimer et al. (1998)
[n]Seiffert et al. (2007)

## Classification

All pterin-containing W enzymes can be divided into three evolutionarily distinct families: aldehyde oxidoreductase (AOR), formate dehydrogenase (FDH), and acetylene hydratase (AH). The classification is based on the sequence homology, coordination about the metal center, and nature of the cofactor (Table 1) (Johnson et al. 1996).

## The AOR Family

The AOR family is the largest of all and is represented by the prototypical member, aldehyde ferredoxin oxidoreductase (AOR). This family also includes formaldehyde ferredoxin oxidoreductase (FOR), glyceraldehyde-3-phosphate ferredoxin oxidoreductase (GAPOR), carboxylic acid reductase (CAR), and aldehyde dehydrogenase (ADH). With the exception of CAR, enzymes of this family are similar in size with a single subunit of ~70 kDa. However, they differ in the quaternary structure from monomeric (GAPOR) to dimeric (AOR, ADH) to tetrameric (FOR). AOR, FOR, and GAPOR have been isolated from a hyperthermophilic archeon, *Pyrococcus furiosus* that optimally grows at 100°C using sugars as carbon and energy source. Two other W-containing enzymes, WOR4 and WOR5, have also been isolated from *P. furiosus*. Two forms of CAR enzymes have been purified from *Clostridium themoaceticum*.

The enzymes of this family catalyze the basic reaction of oxidation of aldehyde to carboxylic acids, a two-electron process, in many cases using ferredoxin (Fd) to shuttle electrons. The overall reaction is shown in (1).

$$RCHO + H_2O \rightleftharpoons RCOOH + 2H^+ + 2e^- \quad (1)$$

Of all members, AOR and WOR5 show a broad substrate transforming reactivity. AOR is very active

**Tungsten in Biological Systems, Table 2** Selected kinetic parameters for WOR and AOR (From Bevers et al. (2009), Heider et al. (1995))

|  | WOR5 | | | AOR | | |
|---|---|---|---|---|---|---|
|  | $K_m$ (mM) | $K_{cat}$(/s) | $K_{cat}/K_m$ (/M s/) | $K_m$, mM | $K_{cat}$ (/s) | $K_{cat}/K_m$ (/M/s) |
| Acetaldehyde | 1.5 ± 0.2 | 0.315 | 210 | 0.016 | 343 | $21.43 \times 10^6$ |
| Formaldehyde | 45 ± 12 | 7.65 | 170 | 1.42 | 950 | $0.66 \times 10^6$ |
| Crotonaldehyde | 46 ± 6 | 1.01 | 22 | 0.136 | 269 | $1.98 \times 10^6$ |
| Benzaldehyde |  |  |  | 0.057 | 720 | $12.63 \times 10^6$ |
| Phenylacetaldehyde |  |  |  | 0.076 | 960 | $12.63 \times 10^6$ |
| Isovalerylaldehyde |  |  |  | 0.028 | 272 | $9.71 \times 10^6$ |
| Propionaldehyde |  |  |  | 0.150 | 1,100 | $7.33 \times 10^6$ |
| Indoleacetaldehyde |  |  |  | 0.050 | 55 | $1.1 \times 10^6$ |
| Salicylaldehyde |  |  |  | 0.065 | 13 | $0.2 \times 10^6$ |
| Hexanal | 0.18 ± 0.02 | 14.4 | $0.080 \times 10^6$ |  |  |  |
| Hydratropaldehyde | 0.12 ± 0.04 | 8.58 | $0.072 \times 10^6$ |  |  |  |
| 2-methylvaleraldehyde | 0.27 ± 0.03 | 11.72 | $0.043 \times 10^6$ |  |  |  |
| 2-ethylhexanal | 0.17 ± 0.02 | 7.67 | $0.045 \times 10^6$ |  |  |  |
| 3-phenylbutyraldehyde | 0.42 ± 0.12 | 7.4 | $0.018 \times 10^6$ |  |  |  |
| 2-methylbutyraldehyde | 0.43 ± 0.09 | 7.095 | $0.017 \times 10^6$ |  |  |  |
| Isobutyraldehyde | 0.79 ± 0.03 | 10.9 | $0.014 \times 10^6$ |  |  |  |
| 2-naphthaldehyde | 1.3 ± 0.1 | 7.15 | $0.006 \times 10^6$ |  |  |  |
| Cinnamaldehyde | 1.6 ± 0.1 | 7.36 | $0.005 \times 10^6$ |  |  |  |
| 2-methoxybenzaldehyde | 4.8 ± 0.6 | 13.9 | $0.003 \times 10^6$ |  |  |  |

with aldehydes derived from amino acids presumably produced during peptide fermentation or pyruvate when grown on carbohydrate. Aldehydes such as acetaldehyde (derived from alanine), isovaleraldehyde (derived from valine), indole acetaldehyde (derived from tryptophan), and phenylacetaldehyde (derived from phenylalanine) exhibit a higher rate of transformation as determined by $k_{cat}/K_m$. The AOR proteins are oxygen sensitive, and contain a dimeric catalytically competent subunit. The AOR isolated from *P. furiosus* was the first to be crystallographically characterized of any pterin-containing Mo or W enzymes.

Tungsten containing oxidoreductase number four (WOR4) and tungsten containing oxidoreductase number five (WOR5) also belong to the same family. Like AOR, WOR5 shows a broad substrate specificity; however, it exhibits a higher affinity for aliphatic and aromatic aldehydes. The mRNA levels of WOR4 have been shown to increase when *P. furiosus* is was grown on peptides and both the WOR4 and WOR5 mRNA levels were upregulated by cold shock. This suggests that both WOR4 and WOR5 may play a role in stress response and WOR4 may be involved in peptide fermentation. Kinetic data suggest that WOR5 is more active than AOR (Table 2).

FOR exhibits a higher activity with smaller aldehydes (C1-C3) and semi- and di-aldehydes. Various semi-aldehydes (C4-C6) are presumably involved in the metabolism of amino acids such as arginine, lysine, and proline, implying a role of FOR in these cases. GAPOR is specific for glyceraldehyde-3-phosphate yielding 3-phosphoglycerate. Like FOR and AOR, it uses Fd as its electron transfer partner, and does not use NAD or NADP. This is an important enzyme in the sugar fermentation pathway, where it bypasses glyceraldehyde-3-phosphate GAP dehydrogenase (GAPDH) and phosphorglycerate kinase (PGK). In this way, glyceraldehyde-3-phosphate (GAP) is converted to 3-phosphoglycerate directly. Support of this idea stems from lower GAPDH and PGK activities observed in cells grown on maltose in the presence of W. Thus, GAPOR is a key enzyme in transformation of glucose to pyruvate in *P. furiosus* using an unusual glycolytic pathway.

The AOR family also includes the W-containing carboxylic acid reductase (CAR) isolated from acetogenic *C. formicoaceticum*. The enzyme catalyzes the reduction of carboxylic acid to aldehyde, as well as the reverse reaction, that is, the oxidation of aldehyde to carboxylic acid. The latter, being a thermodynamically more favorable process, is catalyzed at a faster rate. The molecular properties of this enzyme are very similar to that of *P. furiosus* AOR sharing broad substrate specificity. However, the exact physiological role of this enzyme is yet to be understood. Two forms of CAR have been isolated form *C. thermoaceticum*; and in both cases, the enzyme has higher Fe content and has additional subunits. In the presence of Mo, a larger (100 kDa) analogous Mo-containing protein was isolated, called Mo-AOR. Sequence-wise, this enzyme is distinct from CAR.

Another enzyme of this family is the aldehyde dehydrogenase (ADH) from sulfate-reducing bacterium, *Desulfovibrio gigas*. The N-terminal sequence of this enzyme is similar to that of *P. furiosus* AOR. A molybdenum containing aldehyde oxidizing protein, called MOP, was isolated from the same organism. This enzyme however is different from *P. furiosus* AOR.

### The FDH Family

There are two enzymes in this class: formate dehydrogenase (FDH) and N-formylmethanofuran dehydrogenase (FMDH). Both enzymes catalyze the activation of $CO_2$ in a reversible manner ((2) and (3), respectively). The catalytic center of these enzymes has a W-center coordinated by two pyranopterin cofactors; the enzymes also harbor other electron transfer partners, for example, [4Fe-4S] or [2Fe-2S] cluster. Interestingly, the FDH from anaerobic organisms contain a Mo center. Several crystal structures of the FDH have been reported, including one from *D. gigas* that harbors the W enzyme. W-containing FMDH have been isolated from *Methanobacterium thermoautotropicum* and *M. wolfei*. Like, FDH, the Mo congener of FMDH has also been reported. In FMDH the reduced $CO_2$ is added to the organic cofactor, methanofuran. This step is believed to be the first of $CO_2$ reduction to methane in methanogens.

$$HCOO^- \rightleftharpoons CO_2 + H^+ + 2e^- \quad (2)$$

$$\text{furfurylformamide} + H_2O \rightleftharpoons \text{furfurylamine} + CO_2 + H^+ + 2e^- \quad (3)$$

While molybdenum containing FDH and FMDH are known, organisms that show W-dependency do not contain W-substituted Mo enzymes. The W-containing enzymes are encoded by different sets of genes than those that encode Mo enzymes, but shows amino acid sequence similarity. Unlike the members of the AOR family, the primary amino acid sequences of these two enzymes are not very similar. Because the crystal structure of FMDH is unavailable, direct structural comparison is not possible. However, FDH has a selenocysteine coordinated W center; FMDH does not have a Se-containing amino acid, which is likely to impact the functions. Both *M. wolfeii* and *M. thermoautotrophicum* can express two FMDHs: one Mo-containing and the other W-containing. In the presence of Mo, *M. wolfeii* expresses only Mo-containing FMDH and in the presence of W both enzymes are expressed containing only W. A different behavior is seen with *M. thermoautotrophicum*; both enzymes are expressed in the presence of Mo and only W-containing FMDH is expressed in the presence of W. The catalytic properties are retained when one metal is substituted with another leading to an isoenzyme.

### Acetylene Hydratase Family

Acetylene hydratase (AH) from acetylene-utilizing bacterium, *Pelobacter acetylenicus* is the only member of this family, and is the only W enzyme that does not catalyze a redox transformation. The enzyme catalyzes the hydration of acetylene to acetaldehyde. The reaction is catalyzed in the presence of a strong reducing agent that reduces the W from +6 to +4 states. The 4Fe4S cluster is proposed to be activating this step. The reactive species has a hydroxo or water molecule coordinated to the W center, which is deprotonated by a neighboring aspartate residue.

## W-Substituted Mo Enzymes

The chemical similarities of tungsten and molybdenum compounds led to the belief that W-substituted Mo enzymes function similarly. In most cases, substitution of Mo with W, leads to an inactive enzyme, and thus was thought to be an antagonist of the biological function of Mo. In enzymes where W substitution shows activity, comparisons provide insight into the chemistry of the Mo/W enzymes.

Sulfite oxidase (SO), a molybdoenzyme that oxidizes sulfite to sulfate, isolated from livers of rats fed with 400 ppm of tungstate, resulted in W-substituted SO that can no longer oxidize sulfite but can transfer electrons. However, unlike the Mo center, the W center is reduced only after the partner electron transfer group, heme, is reduced, and the W center exhibited spectroscopic properties similar to the Mo center. The redox behavior indicates the W center has a lower redox potential than that of the Mo center. Similarly, nitrate reductase (NR), a molybdoenzyme that reduces nitrate to nitrite, isolated from *Neurosporacrassa* incubated with tungstate, was unable to reduce nitrate even though it could transfer electrons.

Another Mo-enzyme, trimethyl amine oxide reductase (TMAOR), that reduces trimethyl amine N-oxide (TMAO) to trimethyl amine (TMA), has been isolated as W enzyme from *Escherichia coli* grown with tungstate. The strategy was to mutate the molybdenum transport protein, ModA, which results in the transport of molybdenum by a sulfate transporter, leading to the formation of the molybdenum enzyme. ModA binds to both molybdate and tunsgtate with near equal affinity. If in the growth medium, tungsten is present in place of molybdate, a tungsten containing protein is produced. Both Mo-TMAOR and W-TMAOR were produced in this way and compared with the wild type TMAOR. The W-TMAOR exhibited a higher kinetic efficiency, was more heat resistant, more sensitive to pH, and less sensitive to salt concentration. The W-TMAOR exhibited a broad substrate specificity including different sulfoxides, while the Mo-TMAOR was inactive with these substrates.

The dimethylsulfoxide reductase (DMSOR) from *Rhodobacter capsulatus*, a Mo-enzyme, has been characterized using crystallography. The structure of the W-substituted enzyme, W-DMSOR, isolated from cells grown on $Na_2WO_4$, has also been determined (Stewart et al. 2000). The structures of Mo-DMSOR and W-DMSOR are very similar, and the spectral features indicate a similar electronic structure. The W-DMSOR was more active in reducing DMSO, but in contrary to the Mo-DMSOR, it was found inactive toward the oxidation of DMS. This suggests that W (IV) is a stronger reductant than Mo (IV); and W (VI) is a weaker oxidant than Mo (VI). At pH 7.0, the W (VI/V) and W (V/IV) couples appear at −194 and −134 mV, respectively; whereas the Mo (VI/V) and Mo (V/IV) couples appear at +26 and +200 mV, respectively, at pH 8.0. The lower potentials of the W-DMSOR are consistent with the efficient reduction of DMSO to DMS but not the oxidation of DMS to DMSO.

## Structural Features of the W-Enzymes

The first crystal structure of any pyranopterin-containing enzyme was that of AOR from *P. furiosus*, which demonstrated ring closed pyran moiety of the cofactor (Chan et al. 1995). The protein is a homodimer with each monomer harboring a W-center and a 4Fe4S cluster. The W center was symmetrically coordinated by the two dithiolene moieties from the two pyranopterin cofactors, and the phosphate groups coordinate to a $Mg^{2+}$ ion; no oxo or hydroxo ligand was found by crystallography. However, the spectroscopic investigation indicated the presence of a W-O distance of 1.7 Å, representing a W = O unit and another longer W-O/N may also be present. The closest distance between the 4Fe4S cluster and W-center is ∼9 Å, well within the range for efficient electron transfer. A cysteine residue is hydrogen bonded to one of the pyranopterin cofactors and the 4Fe4S cluster, potentially providing a pathway for electron transfer. The W center is deeply buried (∼15 Å) from the protein surface.

The crystal structure of a homologous protein, FOR from *P. furiosus*, has also been determined in different forms such as native, complex with an inhibitor (glutarate) and complex with its physiological electron transfer partner, ferredoxin (Hu et al. 1999). The overall structure of FOR was a tetramer, but the structure of the monomer was very similar to that of the AOR. The W center was coordinated by four sulfur donors from the two pyranopterin cofactors with no coordination from any amino acid residue. An oxygen atom was located at 2.10 Å from the W atom, indicating the presence of a coordinated water or hydroxo ligand. The 4Fe4S cluster is 13 Å away from the W-center, again within the range of efficient electron transfer.

The crystal structure of acetylene hydratase, isolated from *P. acetylenicus*, was determined at a high resolution of 1.26 Å. The W-center is constituted similarly – coordinated by four sulfur donors from the pyranopterin cofactors, a cysteine sulfur from the protein backbone, and a water molecule. The water molecule is in hydrogen bonding distance from a nearby aspartate residue rendering a partial positive charge on the oxygen, making it susceptible to nucleophilic attack on the substrate. The W-center is ∼17 Å deep inside the protein, and the 4Fe4S cluster is disposed close to one of the pyranopterin cofactors.

## W Homeostasis

Organisms that need W for cellular viability use regulatory machinery for maintaining the availability of the metal ion, although the minimal intracellular concentrations of W are yet to be determined. Different aspects, for example, the uptake, intracellular distribution, the extrusion of W, and storage need to work in concert to maintain the homeostasis. The minimal cellular requirement of W is unknown, but W functions as a trace metal; thus, its requirement is likely to be small. In oxic environments the predominant species is tungstate, and tungstate is transported inside the cell (Andreesen and Makdessi 2008). The cells can sequester tungstate from natural sources where nanomolar concentration lead to monomeric species rather than polymeric.

W transport. Tungstate is transported by inner membrane ABC (adenosine triphosphate–binding cassette) transporters. In most cases, these transporters consist of three proteins: A, B, and C proteins. The A protein, localized in the periplasmic space, recognizes tungstate and binds it; the B protein forms the transmembrane channel through which tungstate is transported; and the C protein that lies in the cytoplasmic side of the membrane is involved in ATP hydrolysis. A tungsten uptake protein system called TupABC has been identified in *Eubacterium acidaminophilum*. The gene for this transporter is located close to the genes responsible for the cofactor biosynthesis. The binding constant is $\sim 2 \times 10^6 \, M^{-1}$ for $WO_4^{2-}$ and that for binding $MoO_4^{2-}$ is several orders of magnitude smaller. TupA binds to $WO_4^{2-}$ in a 1:1 ratio, and 100-fold excess of $MoO_4^{2-}$ is required to get saturation of TupA (Makdessi et al. 2001). A second distinctly different W-specific ABC transporter has been reported in *P. furiosus*, which is called W-transport protein A or WtpA. It binds to $WO_4^{2-}$ very tightly with a binding constant $\sim 10^{11} \, M^{-1}$. These three proteins, ModA, TupA, and WtpA, are distinct with different binding motifs, with significant differences in sequence (Hagen 2011).

*Campylobacter jejuni* encodes two ABC transporters similar to ModABC and TupABC. While the former cannot distinguish between $WO_4^{2-}$ and $MoO_4^{2-}$ the latter binds to $WO_4^{2-}$ 50,000-fold more tightly than $MoO_4^{2-}$. The TupABC is suggested to supply W to W-containing FDH in *C. jejuni*. Interestingly, in *C. jejuni* a ModE-like protein lacking molybdenum-binding domain can repress ModABC in the presence $WO_4^{2-}$ and $MoO_4^{2-}$. An oxyanion permease (PerO) has also been described to transport $MoO_4^{2-}$ in *R. capsulatus*, which is the first known non-ABC-type $WO_4^{2-}$ and $MoO_4^{2-}$ transporter.

The size of the transported anion is often used in understanding the specificity. For example, the sulfate-binding protein binds sulfate with high affinity (binding constant, $K = 8.33 \times 10^6 \, M^{-1}$) and molybdate with low affinity, and the molybdate transport protein, ModA, binds $MoO_4^{2-}$ with high affinity ($K = 5.0 \times 10^7 \, M^{-1}$) but not sulfate or phosphate. Interestingly, both proteins are very similar with respect to the binding site including the hydrogen-bonding network. Structurally, W–O, Mo–O, and S–O bond lengths are 1.78–1.79, 1.75–1.78, and 1.47–1.49 Å, respectively. The latter distance being shorter than the prior two provides a rationale for the observed difference in the specificity in binding. The same principle cannot be applied for explaining the difference between the molybdenum transport proteins and tungsten transport proteins.

Alternatively, the transport proteins can differentiate oxyanions based on their protonation states. The $pK_a$s of these oxo acids have been difficult to determine due to the formation of multinuclear species as well as a change in the geometry. However, there are subtle differences in the protonation states as a function of pH (Fig. 3), which may provide a clue how the transporters are tuned for specific metal ion and where the specificity is less prevalent. Such an assertion would also depend of the nature of the binding site not only the hydrogen bonding network, but also the overall electrostatic interactions.

W storage. Cells utilize specialized machinery to store metal ions, which can be released on demand at a particular place. In some cases, metal ions are stored in a protein as a protective mechanism, for example, ferritin that stores Fe(II), which otherwise can carry out Fenton chemistry increasing the oxidative stress. Thus far, two types of proteins have been identified to store Mo and W: the "molbindin" or "mop" and the "oxo-anion" storage protein. The "molbindin" proteins have one (e.g., in *C. pasteurianum*) Mop from *Sporomusaovata* or more (two) mop domains in a single protein as in *Azotobacter vinelandii* molybdate-binding domain that can bind both molybdate and tunsgtate presumably with near equal affinity. Crystallography showed that the mono-mop proteins

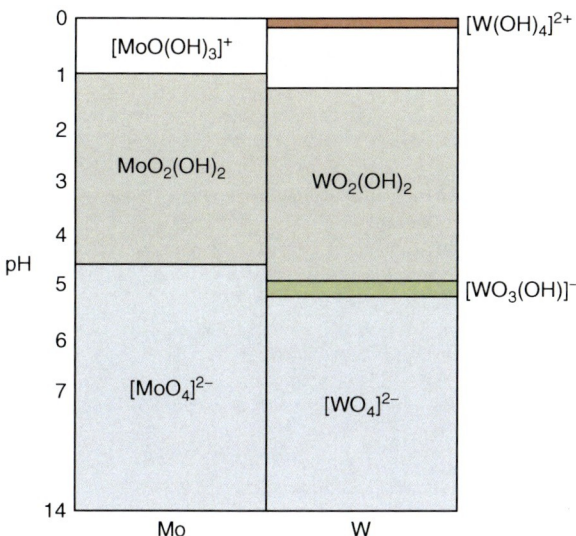

**Tungsten in Biological Systems, Fig. 3** Protonation states of Mo and W oxo groups as a function of pH (Reprinted with permission from Pershina et al. (2001). Copyright 2001 American Chemical Society)

form a trimer of dimers and di-mop proteins form trimers resulting in equal (eight) binding in each case. It is interesting to note that *Eubacterium acidaminophilum* has two W-containing proteins (AOR and FDH), a dedicated tungsten uptake system, and a Mop protein. This Mop protein may be specific for tungsten storage. The oxo-anion storage protein has been isolated from *A. vinelandii*; it binds both molybdate and tungstate and can store up to 100 Mo/W atoms. Thus, it has a much higher binding capacity for the metal ions. Because, no W enzyme has yet been identified in *A. vinelandii*, the physiological storage of W by this protein is questionable.

## Spectroscopic Techniques in Studying W-Centers in Enzymes

Tungsten has five stable isotopes ($^{180}$W, 0.13%; $^{182}$W, 26.3%; $^{183}$W, 14.3%; $^{184}$W, 30.7%; $^{186}$W, 28.6%). Isotopes with nonzero nuclear spin have been used for spectroscopic detection in enzymatic samples. The large number of isotopes provides a useful tool for mass spectrometric analyses. The $^{183}$W isotope is ~14% abundant and has a nuclear spin of one half, and is amenable to nuclear magnetic resonance (NMR) spectroscopy. The typical spectral range of $^{183}$W is ~4,000 ppm and in small molecules such signals have been detected either by polarization transfer or by direct detection. However, sensitivity is very low and thus far no NMR data for any W proteins have been reported. However, the same isotope has provided useful electron paramagnetic resonance (EPR) information. W enzymes catalyze a two electron process (with the exception of AH), and during the regeneration step they pass through W(V) state, which is EPR active. The W(V) state has a $d^1$ configuration which gives an EPR spectrum with $g_{av} < 2$ and with hyperfine splitting originating from the $^{183}$W. The W(V) signals are slow relaxing and can be observed at higher (>100 K) temperatures. In general, a larger deviation from the free electron g–value is observed for W centers than for the corresponding Mo centers due to a much larger spin-orbit coupling constant for W, which also increases the g-anisotropy. EPR spectroscopy has been used in characterizing different natural W-enzymes (e.g., AOR, FOR), as well as W-substituted Mo enzymes. Magnetic circular dichroism (MCD) spectroscopy has also been used in investigating biological W(V) centers enriched with $^{183}$W, which indicated different redox forms of the enzyme.

X-ray absorption spectroscopy (XAS) remains an extremely valuable tool for investigating W centers, especially with newer generations of synchrotron radiation sources. In most metalloproteins, investigation is done at the metal K-edge; however, for W the energy is very high, and data are generally taken at the $L_{III}$-edge. Because the X-ray absorption looks at W of all oxidation states, homogeneous preparation of samples is crucial, which is particularly challenging due to high sensitivity of the W centers. This technique has been applied in investigating W center in AOR, FDH, and GAPOR (Bevers et al. 2009). With the development of well-defined model compounds, interpretation of the spectroscopic data from enzymes will continue to improve.

## Conclusion

Defining the natural biological role of W remains an exciting area of research. While much progress has been made in isolating and characterizing such enzymes, the detailed mechanisms of such enzymes remain unclear. The transport and homeostasis of W is an even newer field than the enzymes themselves, and here also functional details remain to be clarified.

Future studies with more sensitive tools will see an in-depth understanding of these interesting biomolecules, and one wonders whether there is a W-enzyme in eukaryotic system, which remains to be discovered.

**Acknowledgments** The research carried out in the author's laboratory was supported by grants from the National Institutes of Health. The author gratefully acknowledges his coworkers and collaborators.

## Cross-References

▸ Formate Dehydrogenase
▸ Iron-Sulfur Cluster Proteins, Ferredoxins
▸ Metals and the Periodic Table
▸ Molybdenum and Ions in Living Systems
▸ Molybdenum and Ions in Prokaryotes
▸ Molybdenum Cofactor, Biosynthesis and Distribution
▸ Molybdenum in Biological Systems
▸ Molybdenum-enzymes, MOSC Family
▸ Tungsten Cofactors, Binding Proteins, and Transporters in Biological Systems

## References

Andreesen JR, Makdessi K (2008) Tungsten, the surprisingly positively acting heavy metal element for prokaryotes. Ann NY Acad Sci 1125:215

Basu P, Burgmayer SNJ (2011) Pterin chemistry and its relationship to the molybdenum cofactor. Coord Chem Rev 255:1016–1038

Bertram PA, Schmitz RA, Linder D, Thauer RK (1994) Tungstate can substitute for molybdate in sustaining growth of Methanobacterium thermoautotrophicum. Identification and characterization of a tungsten isoenzyme of formylmethanofuran dehydrogenase. Arch Microbiol 161:220–228

Bevers LE, Bol E, Hagedoorn P-L, Hagen WR (2005) WOR5, a Novel Tungsten-Containing Aldehyde Oxidoreductase from Pyrococcus furiosus with a Broad Substrate Specificity. J Bacteriol 187:7056–7061

Bevers LE, Hagedoorn P-L, Hagen WR (2009) The bioinorganic chemistry of tungsten. Coord Chem Rev 253:269

Chan MK, Mukund S, Kletzin A, Adams MWW, Rees DC (1995) Structure of a hyperthermophilic tungstopterin enzyme, aldehyde ferredoxin oxidoreductase. Science 267:1463

Hagen WR (2011) Cellular uptake of molybdenum and tungsten. Coord Chem Rev 255:1117

Heider J, Ma K, Adams M (1995) Purification, characterization, and metabolic function of tungsten- containing aldehyde ferredoxin oxidoreductase from the hyperthermophilic and proteolytic archaeon Thermococcus strain ES-1. J Bacteriol 177:4757

Hensgens CMH, Hagen WR, Hansen TH (1995) Purification and characterization of a benzyl viologen-linked, tungsten-containing aldehyde oxidoreductase from Desufovibrio gigas. J Bacteriol 177:6195–6200

Hochheimer A, Hedderich R, Thauer RK (1998) The formylmethanofuran dehydrogenase isoenzymes in Methanobacterium wolfei and Methanobacterium thermoautotrophicum: induction of the molybdenum isoenzyme by molybdate and constitutive synthesis of the tungsten isoenzyme. Arch Microbiol 170:389–393

Hu Y, Faham S, Roy R, Adams MWW, Rees DC (1999) Formaldehyde ferredoxin oxidoreductase from *Pyrococcus furiosus*: the 1.85 Å resolution crystal structure and its mechanistic implications. J Mol Biol 286:899

Johnson MK, Rees DC, Adams MWW (1996) Tungstoenzymes. Chem Rev (Washington, DC) 96:2817–2839

Kletzin A, Mukund S, Kelley-Crouse TL, Chan MK, Rees DC, Adams MW (1995) Molecular characterization of the genes encoding the tungsten-containing aldehyde ferredoxin oxidoreductase from Pyrococcus furiosus and formaldehyde ferredoxin oxidoreductase from Thermococcus litoralis. J Bacteriol 177:4817–4819

Leimkuehler S, Wuebbens MM, Rajagopalan KV (2011) The history of the discovery of the molybdenum cofactor and novel aspects of its biosynthesis in bacteria. Coord Chem Rev 255:1129–1144

Makdessi K, Andreesen JR, Pich A (2001) Tungstate uptake by a highly specific ABC transporter in *Eubacterium acidaminophilum*. J Biol Chem 276:24557–24564

Mukund S, Adams MW (1990) Characterization of a tungsten-iron-sulfur protein exhibiting novel spectroscopic and redox properties from the hyperthermophilic archaebacterium Pyrococcus furiosus. J Biol Chem 265:11508–11516

Mukund S, Adams MW (1991) The novel tungsten-iron-sulfur protein of the hyperthermophilic archaebacterium, Pyrococcus furiosus, is an aldehyde ferredoxin oxidoreductase. Evidence for its participation in a unique glycolytic pathway. J Biol Chem 266:14208–14216

Mukund S, Adams MW (1995) Glyceraldehyde-3-phosphate ferredoxin oxidoreductase, a novel tungsten-containing enzyme with a potential glycolytic role in the hyperthermophilic archaeon Pyrococcus furiosus. J Biol Chem 270:8389–8392

Osseo-Asare K (1982) Solution chemistry of tungsten leaching systems. Metallurg Trans B 13B:555–564

Pershina V et al (2001) Solution chemistry of element 106: theoretical predictions of hydrolysis of Group 6 cations Mo, W, and Sg. Inorg Chem 40:776–780

Raaijmakers H, Macieira S, Dias JM, Teixeira S, Bursakov S, Huber R, Moura JJG, Moura I, Romao MJ (2002) Gene Sequence and the 1.8 Å Crystal Structure of the Tungsten-Containing Formate Dehydrogenase from Desulfovibrio gigas. Structure 10:1261–1272

Roy R, Adams MWW (2002) Characterization of a fourth tungsten-containing enzyme from the hyperthermophilic archaeon Pyrococcus furiosus. J Bacteriol 184:6952–6956

Schmitz RA, Albracht SPJ, Thauer RK (1992) Properties of the tungsten-substituted molybdenum formylmethanofuran dehydrogenase from Methanobacterium wolfei. FEBS Lett 309:78–81

Seiffert GB, Ullmann GM, Messerschmidt A, Schink B, Kroneck PMH, Einsle O (2007) Structure of the non-redox-active tungsten/[4Fe:4S] enzyme acetylene hydratase. Proc Natl Acad Sci USA 104:3073–3077

Stewart LJ, Bailey S, Bennett B, Charnock JM, Garner CD, McAlpine AS (2000) Dimethyl sulfoxide reductase: an enzyme capable of catalysis with either molybdenum or tungsten at the active site. J Mol Biol 299:593

White H, Simon H (1992) The role of tungstate and/or molybdate in the formation of aldehyde oxidoreductase in Clostridium thermoaceticum and other acetogens; immunological distances of such enzymes. Arch Microbiol 158:81–84

White H, Strobl G, Feicht R, Simon H (1989) Carboxylic acid reductase: a new tungsten enzyme catalyses the reduction of non-activated carboxylic acids to aldehydes. Eur J Biochem 184:89–96

White H, Feicht R, Huber C, Lottspeich F, Simon H (1991) Purification and some properties of the tungsten-containing carboxylic acid reductase from *Clostridium formicoaceticum*. Biol Chem Hoppe Seyler 372:999–1005

Yamamoto I, Saiki T, Liu SM, Ljungdahl LG (1983) Purification and properties of NADP-dependent formate dehydrogenase from Clostridium thermoaceticum, a tungsten-selenium-iron protein. J Biol Chem 258:1826–1832

# Tungsten, Physical and Chemical Properties

Fathi Habashi
Department of Mining, Metallurgical, and Materials Engineering, Laval University, Quebec City, Canada

Tungsten has a silvery white luster, a high density, and the highest melting point of all metals. Typical application includes incandescent lamp filaments (Fig. 1). It is the most important refractory metal used in powder metallurgy. Pure metallic tungsten is brittle at room temperature and is therefore not suitable for cold forming. At elevated temperatures (ca. 100–500°C) it is transformed into the ductile state. Other applications are in high-speed steels, high-temperature steels, and tool steels.

## Physical Properties

| | |
|---|---|
| Atomic number | 74 |
| Atomic weight | 183.84 |
| Relative abundance in Earth's crust, % | $1 \times 10^{-4}$ |
| Density, g/cm$^3$ | 19.3 |

*(continued)*

| | |
|---|---|
| Crystal form | Body-centered cubic |
| Lattice parameter, nm | 0.31648 |
| Atomic radius, nm | 0.139 |
| Melting point, °C | 3,410 |
| Boiling point, °C | 5,900 |
| Heat of fusion, kJ/mol | 35.17 |
| Heat of sublimation, kJ/mol | 850.8 |
| Specific heat capacity, kJ kg$^{-1}$ K$^{-1}$ | |
| At 25°C | 0.135 |
| At 1,000°C | 0.17 |
| At 2,000°C | 0.20 |
| Molar heat capacity at 20°C, J mol$^{-1}$ K$^{-1}$ | 24.28 |
| Coefficient of thermal expansion, mm/m | |
| At 500°C | 2.3 |
| At 1,500°C | 7.6 |
| At 3,000°C | 19.2 |
| Thermal conductivity, W m$^{-1}$ K$^{-1}$ | |
| At 0°C | 129.5 |
| At 1,000°C | 112.5 |
| At 2,000°C | 96.0 |
| Vapor pressure, Pa | |
| At 1,700°C | $1 \times 10^{-10}$ |
| At 2,300°C | $4 \times 10^{-1}$ |
| At 3,000°C | $2 \times 10^{-3}$ |
| Specific electrical resistivity, mW m | |
| At 20°C | 0.055 |
| At 1,000°C | 0.330 |
| At 2,000°C | 0.655 |
| At 3,000 | 1.40 |
| Absorption cross section, barn for thermal neutrons | 19.2 |
| First ionization potential, eV | 8.0 |
| Magnetic susceptibility at 25, cm$^3$/g | $0.32 \times 10^{-6}$ |
| Hardness (Vickers) HV 30 | |
| At 0°C | 450 |
| At 400°C | 240 |
| At 800°C | 190 |
| Modulus of elasticity, kN/mm$^2$ | |
| At 0°C | 407 |
| At 1,000°C | 365 |
| At 2,000°C | 285 |
| Minimum compression strength (sintered), N/mm$^2$ | 1,150 |
| Shear modulus at 20°C, kN/mm$^2$ | 177 |

## Chemical Properties

Tungsten is stable in air up to 350°C, but begins to oxidize above 400°C. A thin film of blue tungsten oxides forms on the surface, and this at first prevents further attack. As the temperature increases, cracks develop in the oxide film, and these favor oxidative

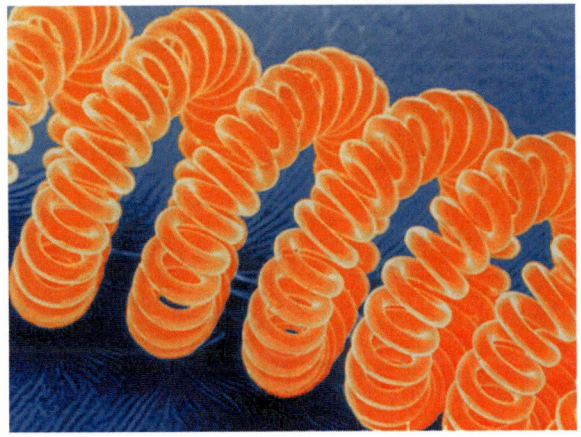

Tungsten, Physical and Chemical Properties, Fig. 1 Incandescent tungsten filament

attack. Above 800°C, rapid oxidation takes place with formation of $WO_3$, which sublimes. Tungsten is stable to mineral acids in the cold, and is only slightly attacked at higher temperatures. It is stable in cold and hot hydrofluoric acid. Mixtures of hydrofluoric and nitric acids dissolve tungsten readily. Alkaline solutions do not attack tungsten, which is also stable to molten alkalis. In the presence of oxidizing agents ($Na_2O_2$ or $NaNO_3$), rapid dissolution takes place.

Tungsten reacts at high temperatures with B, C, Si, P, As, S, Se, Te, and the halogens. It reacts with fluorine at lower temperatures. It is stable to chlorine up to 250°C, and to bromine and iodine up to 500°C. It is stable to gaseous nitrogen and ammonia up to 1,400°C. There is no reaction with hydrogen. With carbon monoxide, the hexacarbonyl is formed at low temperatures, and the carbide above 800°C.

Tungsten occurs in all oxidation states from 2+ to 6+ inclusive, the most important being 6+. This chemistry is extremely complex as the tungstate anion exists in monomeric form only in strongly alkaline solutions. In mildly alkaline solutions, they polymerize. Below pH 1, sparingly soluble tungsten oxide hydrate (tungstic acid) precipitates.

## References

Habashi F (2003) Metals from ores. An introduction to extractive metallurgy, Métallurgie Extractive Québec, Québec City, Canada. Distributed by Laval University Bookstore, www.zone.ul.ca

Lassner E et al (1997) Tungsten. In: Habashi F (ed) Handbook of extractive metallurgy. Wiley-VCH, Weinheim, pp 1329–1360.

# Type 1 Copper Proteins

▶ Monocopper Blue Proteins

# Type I (II, III)-Iodothyronine Deiodinase

▶ Selenium and Iodothyronine Deiodinases

# Type-1 Copper Protein

▶ Plastocyanin

# U

## Unfolded Protein Response

▶ Cadmium and Stress Response

## Uptake of Actinides by Proteins

▶ Actinides, Interactions with Proteins

## Uptake of Cadmium

▶ Cadmium Absorption

## Uranium, Physical and Chemical Properties

Fathi Habashi
Department of Mining, Metallurgical, and Materials Engineering, Laval University, Quebec City, Canada

Uranium is a radioactive element that is transformed into the stable element lead via many intermediate stages involving the emission of α- and β-radiation with a very long half-life (Fig. 1). The most important end product is $UO_2$, which is the only nuclear fuel used in power station reactors, although uranium carbide, UC, is sometimes used in sodium-cooled fast breeder reactors because of its higher density and superior thermal conductivity. Uranium metal is used as the fuel in materials testing reactors. U–Zr alloys are used as nuclear fuel in marine reactors.

Uranium consists of three isotopes whose concentrations in different resources are exactly the same – $^{238}U$, 99.2762%; $^{235}U$, 0.7182%; and $^{234}U$, 0.0056% – except at the Oklo mine in Gabon, where the proportion of $^{235}U$ is 0.7078%. This was attributed to the occurrence of a fission process $1.8 \times 10^9$ years ago. Uranium occurs in three modifications:

α-Uranium, $T < 942$ K, orthorhombic. Lattice constants: $a = 0.2858$ nm, $b = 0.5876$ nm, $c = 0.4955$ nm at 298.15 K

β-Uranium, $T = <942–1049$ K, tetragonal. Lattice constants: $a = 1.0759$ nm, $c = 0.5656$ nm at $< 942$ K (lower transformation point)

γ-Uranium, $T = 1049–1408$ K, body-centered cubic. Lattice constants: $a = 0.3524$ nm at 1,049 K (lower transformation point)

Specific, $cp$, and molar, $Cp$, heat capacities of uranium modifications are given below:

|  | $c_p$, J kg$^{-1}$ K$^{-1}$ | $C_p$, J mol$^{-1}$ K$^{-1}$ |
|---|---|---|
| α-Uranium (298.15 K) | 116 | 27.666 |
| β-Uranium (< 942 K) | 180 | 42.928 |
| γ-Uranium (1,049 K) | 161 | 38.284 |

The hardness of the metal depends strongly on the annealing temperature.

### Physical Properties

| | |
|---|---|
| Atomic number | 92 |
| Atomic weight | 238.03 |
| Relative abundance in Earth's crust,% | $4 \times 10^{-4}$ |
| Density, g/cm$^3$ | 19.13 |

(continued)

V.N. Uversky et al. (eds.), *Encyclopedia of Metalloproteins*, DOI 10.1007/978-1-4614-1533-6,
© Springer Science+Business Media New York 2013

| | |
|---|---|
| Melting point, °C | 1,125–1,135 |
| Boiling point, °C | 4,234 |
| Thermal conductivity at room temperature, W m$^{-1}$ K$^{-1}$ | 27.60 |
| Specific electrical conductivity, $\Omega^{-1}$ cm$^{-1}$ | 3.0–3.38 × 10$^4$ |
| Temperature coefficient of specific electrical resistance in the range 0–100°C | 2.61–2.76 × 10$^{-3}$ |
| Specific magnetic susceptibility at room temperature, cm$^3$/g | 1.66 × 10$^{-6}$ |

Uranium 235 undergoes fission when bombarded by thermal neutrons; it breaks apart into two smaller elements and, at the same time, emits several neutrons and a large amount of energy:

$$^{235}_{92}U + ^{1}_{0}n \rightarrow ^{89}_{36}Kr + ^{144}_{56}Ba + 3^{1}_{0}n + 200 \text{ MeV}$$
$$\downarrow \qquad \downarrow$$
$$^{89}_{39}Y + 3e^- \quad ^{144}_{60}Nd + 4e^-$$

Uranium 238 absorbs neutrons forming uranium 239 which is a beta emitter with a short half-life; its daughter neptunium 239 also emits an electron to form plutonium 239.

$$^{238}_{92}U + ^{1}_{0}n \rightarrow ^{239}_{92}U \rightarrow ^{239}_{93}Np + e^-$$
$$\downarrow$$
$$^{239}_{94}Pu + e^-$$

## Chemical Properties

The freshly exposed surface of massive uranium metal has a silvery luster, but it is oxidized relatively rapidly in air. Within a few hours, the metal becomes covered with a thin oxide layer of iridescent colors and eventually turns black. Uranium combines readily with hydrogen at 225°C to form UH$_3$, which decomposes again at 436°C. This behavior is utilized in the manufacture of the powdered metal from its compact form. Uranium reacts vigorously with halogens and hydrogen halides. It burns in fluorine gas to form UF$_6$. The reaction with HF gas proceeds only to the UF$_4$ stage; reaction with aqueous HF leads to hydrated UF$_4$. Powdered uranium reacts vigorously with water and with dilute acids. Strong solutions of alkali attack the metal to form uranates.

Anionic complexes of uranium are present in both acid and alkaline uranium solutions. Strongly basic ion exchangers are generally used to extract these complexes. The hydrate UO$_4 \cdot x$H$_2$O is precipitated from aqueous uranyl nitrate solutions at room temperature by H$_2$O$_2$:

$$UO_2(NO_3)_2 + H_2O_2 + xH_2O \rightarrow$$
$$UO_4 \cdot xH_2O + 2HNO_3.$$

Calcium, magnesium, and sodium are suitable agents for the reduction of UF$_4$ to uranium metal, e.g.,

$$UF_4 + 2Mg \rightarrow U + 2MgF_2.$$

Three industrial processes are currently used for enrichment of uranium: the diffusion process, the ultracentrifuge process, and the jet nozzle process. They employ gaseous UF$_6$.

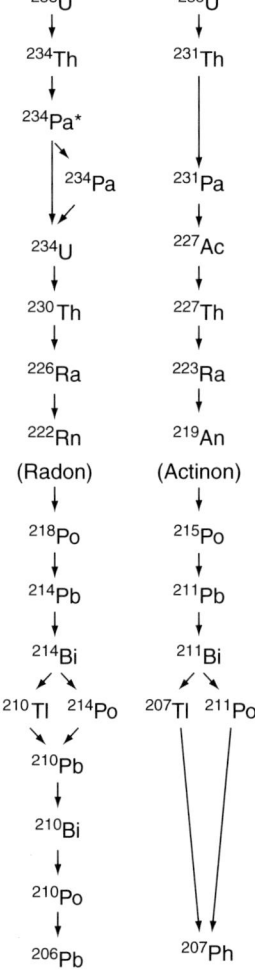

**Uranium, Physical and Chemical Properties, Fig. 1** Decay chain of uranium isotopes. Half-life of $^{238}$U is 4.5 × 10$^9$ and of $^{235}$U is 7.13 × 10$^8$

## References

Habashi F (2003) Metals from ores. An introduction to extractive metallurgy. Métallurgie Extractive Québec, Québec City. Distributed by Laval University Bookstore, www.zone.ul.ca

Peehs M, Walter S, Walter T (1997) Uranium. In: Habashi F (ed) Handbook of extractive metallurgy. Wiley, Weinheim, pp 1599–1648

# Urea Amidohydrolase

▶ Urease

# Urease

Stefano Benini[1], Francesco Musiani[2] and Stefano Ciurli[2]
[1]Faculty of Science and Technology, Free University of Bolzano, Bolzano, Italy
[2]Department of Agro-Environmental Science and Technology, University of Bologna, Bologna, Italy

## Synonyms

Urea amidohydrolase

## Definition

Urease is a nickel-dependent enzyme that catalyzes the reaction of water with urea. The reaction initially produces ammonia and carbamate, which then spontaneously decomposes to yield another molecule of ammonia and hydrogen carbonate.

## Biological Significance of Urease

Two important records characterize the scientific history of urease: in 1926, urease from *Canavalia ensiformis* (jack bean) was the first enzyme to be crystallized, proving that enzymes are proteins, and in 1975, the requirement for nickel in urease catalysis was established, providing the first example for the biological essentiality of this element as an enzyme cofactor (Dixon et al. 1975). Since then, a significant improvement in the understanding of the chemistry and biochemistry of nickel in the urease system has been achieved (Zambelli et al. 2011).

Urease (urea amidohydrolase E.C. 3.5.1.5) is an enzyme produced by plants, algae, fungi, and bacteria, where it catalyzes the reaction of water with urea. The reaction initially produces ammonia and carbamate, which then spontaneously decomposes to yield another molecule of ammonia and hydrogen carbonate causing a pH increase (Ciurli 2007; Carter et al. 2009) (Scheme 1):

The main function of plant urease is to salvage urea nitrogen during germination for nutrition purposes (▶ Nickel in Plants). Bacterial urease activity is widespread in the environment, especially in agricultural soils fertilized with urea, a worldwide used soil nitrogen fertilizer and a substance excreted by many vertebrates as the catabolic product of nitrogen-containing compounds (▶ Nickel in Bacteria and Archaea). In soil, urease is found both as an intracellular enzyme as well as extracellular protein in organo-mineral soil particles. Without the ureolytic activity by soil urease, urea would rapidly accumulate in the environment because of its stability toward hydrolysis. The efficiency of soil nitrogen fertilization with urea is decreased by the urease activity itself because large amounts of ammonia nitrogen can be lost in the atmosphere, while plant damage can be caused by ammonia toxicity and soil pH increase, thereby causing major environmental and economic problems.

Ureolytic bacteria are important in recycling nitrogenous substances in the rumens of domestic livestock. Urease has also implications in human and animal health: it is involved as a virulence factor in a number of bacterial infections of the urinary and gastrointestinal tract. The increase of pH caused by urease activity is the strategy used by *Helicobacter pylori* to survive the very acidic environment of the host's stomach mucosa, thereby causing stomach ulcers, eventually leading to cancer. The enzyme is localized into the cytoplasm in all sources so far examined with the exception of *H. pylori*, for which the urease released by autolysis and adsorbed onto the walls of viable cells accounts for 75% of the total activity.

**Urease,**
**Scheme 1** Enzymatic urea hydrolysis catalyzed by urease

## The Biochemistry of Urease

The enzymatic decomposition pathway catalyzed by urease features $k_{cat}/K_m$ approximately $10^{15}$ times higher than the rate of the uncatalyzed reaction (Callahan et al. 2005). The observed values of $K_m$ fall in the 1–100 mM range and are largely independent of pH, while $k_{cat}$ strongly depends on this parameter. Typical bell-shaped pH profiles for urease activity are observed, with $pK_a$ values of 6.5 for a general base that must be deprotonated, and ca. 9.0 for a general acid that must be protonated. The maximum of activity is detected at pH 7.5–8, with maximal values of $k_{cat}$ and $k_{cat}/K_m$ of $\sim 3,000$ s$^{-1}$ and $\sim 1,000$ s$^{-1}$ mM$^{-1}$, respectively. Enzyme inactivation at pH < 5 is due to irreversible protein denaturation and loss of the essential nickel ions. Urea is not the only substrate for the enzyme, and hydrolytic activity also involves formamide, acetamide, $N$-methylurea, $N$-hydroxyurea, $N,N'$-dihydroxyurea, semicarbazide, and different kinds of phosphoric acid amides. The values of $k_{cat}$ for these alternative substrates are ca. two orders of magnitude lower than that observed for urea.

## Structural Biology of Urease Architecture and Mechanism

The majority of bacterial ureases have a quaternary structure composed of a trimer of trimers $(\alpha\beta\gamma)_3$, with each $\alpha$ subunit featuring an active site containing two Ni$^{2+}$ (Fig. 1a) (Jabri et al. 1995; Benini et al. 1999). In some other cases, the $\beta$ and $\gamma$ subunits are fused together to form $(\alpha\beta)_3$ trimers, which sometimes oligomerize to build a tetrahedral assembly such as $((\alpha\beta)_3)_4$ (Fig. 1b) (Ha et al. 2001). Plant enzymes are generally made up of a dimer of homotrimers $(\alpha_3)_2$, where the $\alpha$ subunit is derived from the fusion of the corresponding bacterial $\alpha\beta\gamma$ subunits (Fig. 1c) (Balasubramanian and Ponnuraj 2010) (▶ Nickel-Binding Proteins, Overview).

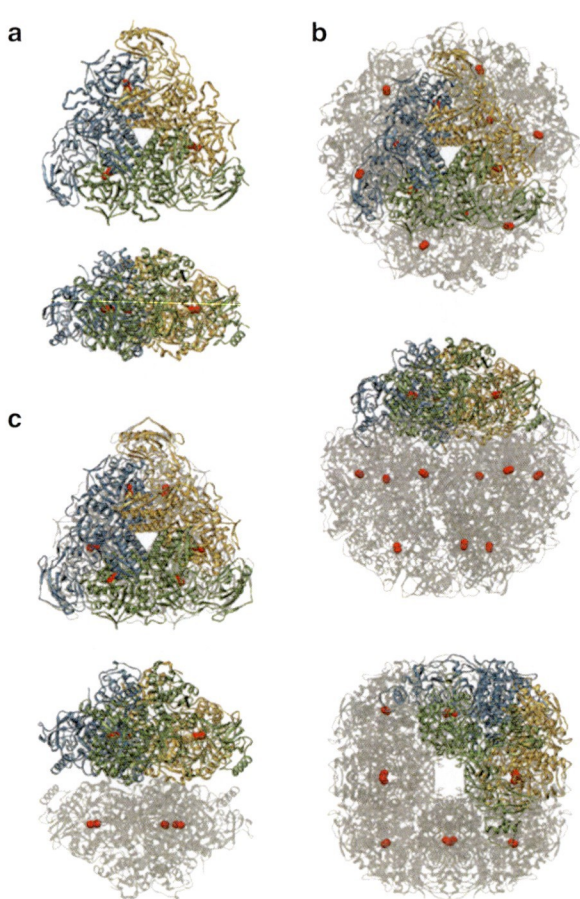

**Urease, Fig. 1** Ribbon diagram of urease from (**a**) *Bacillus pasteurii*, (**b**) *H. pylori*, and (**c**) jack bean. Ribbons are colored to evidence the chains composing the protomers. Ni$^{2+}$ are reported as *red spheres*. The proteins are seen through the ternary axis (*top* panels) and the binary axis (**c**, *bottom* panel). The *bottom* panels of (**a**) and (**b**), as well as the *central* panel of (**c**), are rotated by 90° around the horizontal axis versus the *top* panels

The structure of the urease active site (Fig. 2a) is very conserved (▶ Nickel-Binding Sites in Proteins). The catalytic core features a binuclear nickel center characterized by a Ni–Ni distance of about 3.5–3.7 Å, with each of the two Ni$^{2+}$ ions bound to two histidine residues. The carbamate group of a carbamylated

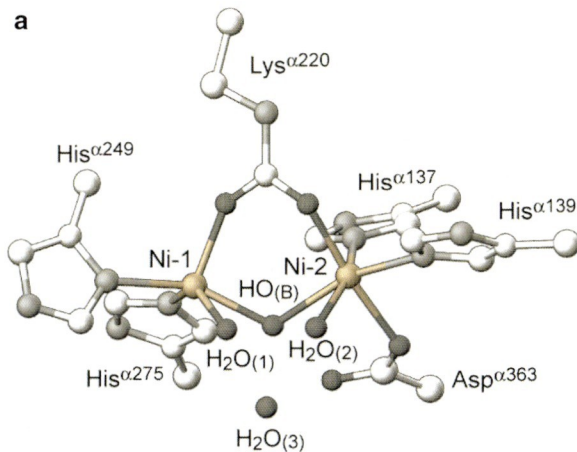

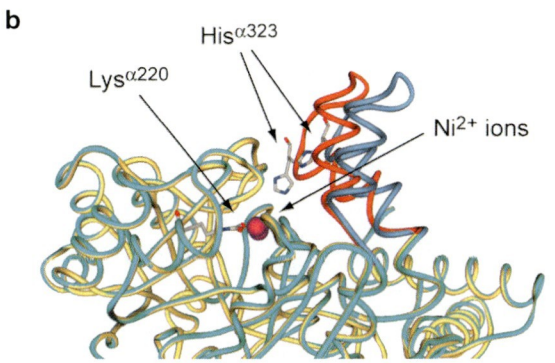

**Urease, Fig. 2** (**a**) Coordination geometry of $Ni^{2+}$ in native urease active site. (**b**) Detail of the flexible flap shown in the open (*blue*) and closed (*orange*) conformations. The $Ni^{2+}$ ions are shown as *red spheres*

lysine bridges the two $Ni^{2+}$ ions, while one of the two metal centers is bound to the carboxylate oxygen of an aspartic acid. The coordination geometry of the metal ions is completed by a terminally bound water molecule and by a nickel-bridging hydroxide. Therefore, one $Ni^{2+}$ is pentacoordinated with a distorted square-pyramidal geometry, and the other is hexacoordinated with a distorted octahedral geometry. An additional water molecule is part of a hydrogen-bonding network that makes up a tetrahedral cluster of four water/hydroxide molecules in the close proximity of the metal centers, suggesting the existence of an active site cavity able to stabilize a tetrahedral transition state and/or intermediate. The access to the active site from solution is gated by a mobile flap, made of a helix-turn-helix motif, which can move from an open conformation to allow substrate entrance, to a closed conformation to allow the reaction to proceed, and back again to an open conformation in order to release the products of the reaction (Fig. 2b). This flap bears a catalytically essential histidine residue, able to move by about 5 Å to and from the active site metal center changing from the open to the closed conformation.

It is from the comparison between the structures of urease in the native form and in complex with selected inhibitors that a proposal of a structure-based reaction mechanism for urease could be formulated (Scheme 2, *Bacillus pasteurii* urease residue numbers) (Ciurli 2007).

Upon urea entrance, with the flap in the open conformation, the structurally characterized hydrated active site of the resting enzyme (Scheme 2a) evolves toward an initial substrate-bound intermediate where urea replaces the three water molecules bound to the $Ni^{2+}$ ions (Scheme 2b). Binding of urea initially involves Ni-1 via the urea carbonyl oxygen, an event followed by further displacement of the water molecules from the active site. Docking and density-functional quantum chemistry calculations suggest and support the hypothesis that flap closure facilitates urea coordination to the second $Ni^{2+}$ via its $-NH_2$ group. This mode of binding of urea into urease active site is stabilized by a specific H-bonding network (H-bond donors on one side and H-bond acceptor on the other side) that locates and steers the substrate to a precise orientation thus allowing hydrolysis to occur (Scheme 2c). The model for a bidentate urea coordination mode is supported by the crystal structure of urease in complex with boric acid, $B(OH)_3$. Boric acid has a similar triangular shape and dimension as urea, is isoelectronic with it, and has the same neutral charge, so it can be considered an inert analog of the substrate. In the structure of urease in complex with $B(OH)_3$, two of the borate hydroxide moieties replace the water molecules terminally bound to Ni-1 and Ni-2, while the third borate hydroxide replaces the distal solvent molecule. According to this bidentate-binding mode, the nucleophilic attack on the urea carbon atom leading to the tetrahedral intermediate is carried out by the Ni-bridging hydroxide (Scheme 2d). This event is favored by the weakening of the bonds between the hydroxide and the metal ions following the bidentate binding of the substrate. Moreover the bridging urea-binding mode is the most efficient method to render the urea central carbon atom electron poor and therefore prone to the nucleophilic attack from the activated hydroxide. The possible formation of a tetrahedral

**Urease, Scheme 2** Structure-based mechanism of urea hydrolysis catalyzed by urease

intermediate located between the two $Ni^{2+}$ ions is supported by calculations and by the structure of urease crystallized in the presence of the inhibitor phenylphosphorodiamidate (PPD). In this structure, the product of the hydrolysis of PPD, diamidophosphate (DAP), is trapped in the active site, replacing all four water/hydroxide molecules found in the resting state of the enzyme. The tetrahedral DAP bound to the dinickel center represents an ideal transition state or intermediate analog of the enzymatic reaction.

After the formation of the tetrahedral intermediate, the nickel-bridging –OH group, now part of a diamino (hydroxy)methanolate moiety and therefore very acidic, loses its hydrogen atom, which is consequently transferred to the distal urea –$NH_2$ group. This proton transfer could be mediated by a dihedral rotation along the C$\alpha$–C$\beta$ bond of the aspartate bound to Ni-2, a movement known to bring the carboxylic oxygen atom not involved in metal binding alternatively close to the bridging hydroxide or to the distal –$NH_2$ group.

As the catalytic histidine residue moves nearer the active site upon flap closure, its neutral imidazole side chain stabilizes the nascent C–$NH_3^+$ group (Scheme 2e). The formation of the C–$NH_3^+$ moiety after proton transfer, and its stabilization by the catalytic histidine, causes the breakage of the distal C–N bond, with the subsequent release of ammonia. The resulting carbamate decomposes into another molecule of ammonia and hydrogen carbonate. These steps possibly occur in a concerted manner together with the flap opening, which could both facilitate products release and return of the enzyme to the ground native state by allowing bulk water to rehydrate the active site (Scheme 2a).

This structure-based mechanism is in agreement with all available kinetics data. In particular, the pH-dependence of the enzyme activity and the noncompetitive inhibition by fluoride thought to replace the bridging hydroxide and therefore strengthening the hypothesis of a bridging nucleophile. The alternative reaction mechanism proposed for KAU involves as a proton donor the catalytic histidine, which should be protonated at the optimal pH for the reaction (pH 8.0) although its p$K_a$ of ~6.5. To overcome the limits of such mechanism, a so-called reverse protonation hypothesis has been proposed, which would imply that only 0.3% of the active sites in the urease functional molecule would be in the correct protonation state for reaction, resulting in a very low enzymatic efficiency (Karplus et al. 1997).

## The Role of Nickel in Urease Catalysis

The presence of nickel as an enzyme metal cofactor came as a surprise when its essentiality for urease activity was discovered. The less toxic $d^{10}$ closed-shell $Zn^{2+}$ is indeed commonly found in hydrolytic enzymes (▶ Zinc-Binding Sites in Proteins). Zinc is usually chosen because of its resilience to undergo deleterious redox state changes and because of its large positive charge density, which renders it able to act as a Lewis acid, by polarizing substrates and preparing them to nucleophilic attack by hydroxide (▶ Zinc, Physical and Chemical Properties). However, these properties are also applicable to $Ni^{2+}$, which additionally features an open-shell $d^8$ electronic configuration that imposes stereoelectronic restrains not available in the $Zn^{2+}$ case (▶ Nickel, Physical and Chemical Properties). This property is exploited in the case of urease, in which the $Ni^{2+}$ ions drive the two substrates, urea and water, into the optimal spatial topology necessary for catalysis. This rationalization was made possible by the determination of the urease crystal structures in the resting state and complexed with substrate and transition state/intermediate analogs: in all steps of the mechanism described above, the two $Ni^{2+}$ ions are either pentacoordinated in a distorted square-pyramid for Ni-1 or hexacoordinated in a distorted octahedron for Ni-2. These geometries are commonly observed for nickel but not for zinc, which is usually found in tetracoordinated environments in enzyme catalytic sites. Considering the protein-based ligand arrangement observed in the active site of urease, a $Zn^{2+}$ ion in place of Ni-1 would be four-coordinated and unable to further bind a molecule of urea, while a $Zn^{2+}$ ion in place of Ni-2 would be five-coordinated and therefore unable to additionally bind the urea $-NH_2$ group in the step that leads to the bidentate substrate-binding/activation mode. Therefore, the two $Ni^{2+}$ ions provide enough binding sites for coordinating to the protein while retaining the capacity of binding the two waters and the bridging hydroxide.

As an example of the difference in catalytic activity of $Ni^{2+}$ and $Zn^{2+}$, it is worth comparing the binuclear metal centers in the resting state of the enzymes urease ($Ni^{2+}$), dihydroorotase ($Zn^{2+}$) and phosphotriesterase ($Zn^{2+}$). In these latter two zinc enzymes, the metal ions are coordinated by the same type of amino acids as in urease, but have different coordination geometry. In dihydroorotase, no terminal water is bound to the metal ions: one of the two $Zn^{2+}$ ions is pentacoordinated with distorted trigonal bipyramidal geometry, and the other is tetracoordinated with distorted tetrahedral geometry. In phosphotriesterase, one water molecule is terminally bound to one of the two metal ions, and both $Zn^{2+}$ are pentacoordinated with distorted trigonal bipyramidal geometry. On the other hand, urease contains one water molecule terminally bound to each $Ni^{2+}$ in the bimetallic center. This configuration allows the presence of two closely spaced binding sites on the two adjacent $Ni^{2+}$ ions, thus allowing urea to bind the binuclear center with a bidentate coordination.

An interesting exception to the requirement of urease for $Ni^{2+}$ is the alternative urease observed in *Helicobacter mustelae*, a bacterium that colonizes the stomach of carnivores. This enzyme is characterized by the absence of $Ni^{2+}$, by inactivation in the presence of oxygen and by $Fe^{2+}$-induced expression, observations that suggest the presence of $Fe^{2+}$ ions in its active site. As compared to $Ni^{2+}$, $Fe^{2+}$ has the same charge, similar radius, is much more sensitive to oxidation, and like $Ni^{2+}$, stabilizes a penta- or hexacoordinated geometry, thus adhering to the rationale described above for its suitability to be a cofactor in this enzyme.

## Urease as a Plant Defense Factor

Plant ureases from *Glycine max* (soybean) and *C. ensiformis* (jack bean) display insecticidal activity that is unrelated to the enzyme activity (Carlini and Grossi-de-Sá 2002). One of the urease isoforms found in jack bean, termed canatoxin, appears to generate an entomotoxic peptide, pepcanatox, in insects possessing cysteine and aspartate protease activity in their digestive systems. A recombinant analog of this peptide, jaburetox, also features insecticidal activity. This peptide represents a loop that interconnects two domains. These domains correspond to the α and β chains in bacterial ureases, where this link is absent, explaining the absence of entomotoxic activity of these enzyme forms. The crystal structure of urease from jack bean has revealed that this region consists of a helix-loop-helix-beta hairpin motif (Balasubramanian and Ponnuraj 2010). The latter is thought to be responsible for the toxicity of this peptide through the inhibition of a membrane ion channel or through the formation of a pore in the cell membrane of the digestive system of the plant-attacking insects.

## Cross-References

▶ Nickel in Bacteria and Archaea
▶ Nickel in Plants
▶ Nickel, Physical and Chemical Properties
▶ Nickel-Binding Proteins, Overview
▶ Nickel-Binding Sites in Proteins
▶ Zinc, Physical and Chemical Properties
▶ Zinc-Binding Sites in Proteins

## References

Balasubramanian A, Ponnuraj K (2010) Crystal structure of the first plant urease from jack bean: 83 years of journey from its first crystal to molecular structure. J Mol Biol 400:274–283

Benini S, Rypniewski WR, Wilson KS et al (1999) A new proposal for urease mechanism based on the crystal structures of the native and inhibited enzyme from *Bacillus pasteurii*: why urea hydrolysis costs two nickels. Structure 7:205–216

Callahan BP, Yuan Y, Wolfenden R (2005) The burden borne by urease. J Am Chem Soc 127:10828–10829

Carlini CR, Grossi-de-Sá MF (2002) Plant toxic proteins with insecticidal properties. A review on their potentialities as bioinsecticides. Toxicon 40:1515–1539

Carter EL, Flugga N, Boer JL et al (2009) Interplay of metal ions and urease. Metallomics 1:207–221

Ciurli S (2007) Urease. Recent insights in the role of nickel. In: Sigel A, Sigel H, Sigel RKO (eds) Nickel and its surprising impact in nature. Wiley, Chichester

Dixon NE, Gazzola C, Blakeley R et al (1975) Jack bean urease (EC 3.5.1.5). A metalloenzyme. A simple biological role for nickel? J Am Chem Soc 97:4131–4132

Ha N-C, Oh S-T, Sung JY et al (2001) Supramolecular assembly and acid resistance of *Helicobacter pylori* urease. Nat Struct Biol 8:505–509

Jabri E, Carr MB, Hausinger RP et al (1995) The crystal structure of urease from *Klebsiella aerogenes*. Science 268:998–1004

Karplus PA, Pearson MA, Hausinger RP (1997) 70 years of crystalline urease: what have we learned? Acc Chem Res 30:330–337

Zambelli B, Musiani F, Benini S et al (2011) Chemistry of $Ni^{2+}$ in urease: sensing, trafficking, and catalysis. Acc Chem Res 44:520–530

# Urease Activation

▶ Nickel in Plants

# UV–vis, Ultraviolet–visible Spectroscopy

▶ Zinc Aminopeptidases, Aminopeptidase from Vibrio Proteolyticus (Aeromonas proteolytica) as Prototypical Enzyme

# V

## V₂O₅

▸ Vanadium Pentoxide Effects on Lungs

## Vanadium

▸ Vanadium Pentoxide Effects on Lungs

## Vanadium – Essential Element for Life

▸ Vanadium in Live Organisms

## Vanadium Compounds

▸ Vanadium in Biological Systems

## Vanadium in Biological Systems

Malay Chatterjee, Subhadeep Das, Mary Chatterjee and Kaushik Roy
Division of Biochemistry, Department of Pharmaceutical Technology, Jadavpur University, Kolkata, West Bengal, India

## Synonyms

Metavanadate; Transition element; Vanadium compounds; $VOSO_4$

## Definition

Vanadium is a transitional metal. It is widely distributed but less abundant element in earth's crust. It can mimic and potentiate the effect of various growth factors such as insulin and epidermal growth factor. Vanadium can affect many cellular processes regulated by cAMP and is implicated in disease of diverse etiology such as diabetes, cancer, chlorosis, anemia, and tuberculosis.

## Background

Vanadium, the twenty-third member of the periodic table, is a soft ductile, silver-gray metal. The transitional element has occupied a place, which belongs to group VB and fourth period in the periodic table. Vanadium is widely distributed throughout the planet but low in abundance. Almost all the living beings inclusive of plants and animals contain vanadium. Deficiency of this micronutrient results in disturbances in metabolism in thyroid glands or bones, growth inhibition, mineralization, and disturbances of metabolic pathways of lipids and carbohydrates. This dietary micronutrient is not accumulated by biota. Only some mushrooms, tunicates, and sea squirts are found to bioaccumulate vanadium to a significant extent. Not only as dietary micronutrient, this transition element has some noteworthy pharmacological potential, too. A large number of scientists have devoted themselves to investigate its physiological relevance and pharmacological importance in human population over the last 30 years. Thus, study of vanadium constitutes a field of interesting research in health and disease.

## Vanadium Ions in Biological System

Vanadium oxidizes NADH (reduced nicotinamide adenine dinucleotide) to its active oxygenated form and thus acts as prooxidant. Free radical generation is one of the mechanisms through which the element exerts its cellular effects. Extracellular and intracellular messengers are responsible for signal transduction from cell to cell. Vanadium compounds regulate the intracellular signaling via second messengers. Adenylate cyclase acts as catalyst in the formation of cAMP (cyclic adenosine monophosphate) from ATP (adenosine triphosphate) and is activated by the action of vanadium at a concentration $>10^{-5}$ M. The most plausible mechanism for adenylate cyclase activation is probably the complex formation with GDP (guanosine diphosphate). Orthovanadate-GDP complex imitating GTP (guanosine triphosphate) binds with protein G and thus activates adenylate cyclase. Other than cAMP, vanadium also regulates the activity of other secondary messengers such as inositol-1,4,5-triphosphate ($IP_3$), $Ca^{2+}$ ion, etc. $IP_3$ synthesis in different cells is stimulated by activation of phospholipase C (PLC)–coupled G-protein regulated by vanadium. Vanadium also helps in mobilization of $Ca^{2+}$ from intracellular stores and subsequent rise in the intracellular $Ca^{2+}$ level. Thus, vanadium performs various roles in regulating many cellular processes. It also helps accumulation of phosphotyrosine in many cellular proteins by inhibiting phosphotyrosine phosphatase activity and stimulating tyrosine kinase. Predominant oxidation state of vanadium is +5 in body fluid, and it enters through nonspecific anion channel. The bioreduced pentavalent state is transformed into tetravalent vanadyl. The pool of oxidant to reductant ratio determines intracellular form of vanadium. Protein phosphorylation and dephosphorylation in cell are regulated by the action of vanadium (Nriagu 1998).

## Vanadium in Different Organs

### Brain and Nervous System

Though brain is least accumulator of this transition metal, it possesses some physiological responsibility there. Vanadium promotes Akt signaling and inhibits forkhead box protein O (FOXO)–dependent and forkhead box protein O (FOXO)–independent death signals which finally prevent neurons from delayed neuronal death. Oral administration of sodium orthovanadate normalizes blood glucose level leading to the prevention of cholinergic neural dysfunction, an alarming diabetic complication. Moreover, a vanadium (IV) compound, namely, [bis-(1-N-oxide-pyridine-2-thiolato) oxovanadium(IV)], has been found to be effective therapeutic in the neurodegeneration of septo-hippocampal cholinergic neurons in a dose-dependent manner. Vanadium, on the other hand, at its +5 oxidation state ($V_2O_5$), acts as a potent neurotoxic agent.

### Heart

Heart is the second least accumulator of vanadium. Vanadium administration in an in vivo model improves cardiac performance, regulates blood pressure level, and thereby imparts cardioprotection along with cardiac function recovery possibly through Fas-associated death domain protein (FADD)–like interleukin-1 beta-converting enzyme (FLICE) inhibitory protein expression.

### Bone

Bone is an organ where vanadium accumulates most. Therefore, it is very much true that vanadium possesses some important physiological relevancy here. The therapeutic efficacy of vanadium absorbed by *Coprinus comatus* has been tested successfully on the bones of streptozotocin (STZ)-induced rats. Trabecular bone volume fraction was found to increase in the treatment along with significant increase in osteoblast surface. Moreover, bone energy separation, osteoclast number, bone stiffness, and trabecular separation were not affected in the treatment.

### Colon

The biological activity of vanadium complex [VO($O_2$) NTA]$K_2$ (where NTA = nitrilotriacetate anion) has been successfully tested in vitro on Caco-2 cell lines, derived from human adenocarcinoma. In a DMH (dimethyl hydrazine)-induced rat colon carcinogenesis model, ammonium monovanadate together with vitamin $D_3$ has shown its pharmacological potentiality. This is shown by reduction in tumor incidence, size and volume, inhibition of cell proliferation, downregulation of Bcl-2 (B-cell lymphoma 2), and upregulation of p53 proteins. $O_6$-methylguanine ($O_6$-Meg) is a potent genotoxic agent in DMH-induced colon cancer. Vanadium supplementation is also capable of lowering $O_6$-Meg level to some extent in experimental rat model (Chatterjee et al. 2008). By means of its antioxidant properties, VO(IV) hesperidin complex plays an important role in cancer chemoprevention. In Caco-2 cell lines, the antiproliferative effect of hesperidin moiety is enhanced when it forms complex with VO(IV), and necessary morphological changes for apoptosis also take place.

### Kidney

Sodium orthovanadate in a time- and dose-dependent manner inhibits autocrine growth of HTB44 (kidney) cell lines. In a spectroscopic study, the pharmacological potential related to vanadium accumulation in kidneys and renal clearance of [bis(ethylmaltolato) oxovanadium (IV)] (BEOV) over $VOSO_4$ has been established in experimental rat model.

### Liver

Liver is one of the major organs to be mentioned for vanadium accumulation with some important physiological significance. Dietary vanadium at a dose of 45–60 mg/kg is capable of causing cell cycle arrest and apoptosis of broiler hepatocytes. Chemopreventive potential of vanadium is also successfully tested against diethylnitrosamine (DENA)-induced hepatocarcinoma in rats. It helps in lowering hepatic 8-OHdGs (8-hydroxy-2′-deoxyguanosine), DNA protein cross-links, single-strand breaks, chromosomal aberrations, and metallothionein immunoreactivity. Anticlastogenic potential of this trace element is also established in rat hepatocarcinogenesis model.

### Pancreas

Pancreas is also not deprived of beneficiary role of vanadium compounds. Oral $VOSO_4$ restores the structure and ultrastructure of pancreatic islets β-cells and thus reverses diabetic symptoms in streptozotocin-induced rat model. A newly synthesized oxovanadium(IV) chelate has also been found to exert same protective effect against pancreatic β-cells.

### Spleen

Using antioxidant property of vanadium, vanadyl sulfate improves the oxidative damage caused by diabetes in spleen (Tunali and Yanardag 2006).

### Lungs

Vanadium accumulation in lungs occurs through inhalation rather than ingestion. Most of the inhaled vanadium complexes are toxic. But there are few complexes which turn beneficiary in maintaining our normal physiology. As for example, azosalophen oxovanadium complex is found to cause apoptotic cell death of human A549 lung cancer cells. Other vanadium salts such as sodium orthovanadate and vanadyl sulfate are also capable of reducing the growth of A549 cell lines.

## Vanadium: Health and Diseases

Biphasic effect of vanadium, beneficiary at low concentration and toxic at higher doses, is a very important matter to discuss with respect to physiological aspect. The wide applications of vanadium compounds as a tonic, as an antiseptic, or as a spirocheticide in the treatment of chlorosis, dental caries, anemia, tuberculosis, and diabetes mellitus are well evident from the past. Well tolerance of vanadium supplementation at pharmacological doses, 100 times greater than dietary intake, was established from a series of human vanadium trials. At higher doses, slight toxicity (diarrhea, cramp) was found to produce on prolonged treatment (6–10 weeks) with ammonium vanadyl tartrate (4.5–18 mg V/day). Continued administration (16 months) of oxytartarovanadate (4.5 mg V/day) to the patients showed no sign of toxicity throughout the study (Nriagu 1998). Bis(ethylmaltolato)oxovanadium (IV) is one of the few noteworthy vanadium complexes which has already gone through phase I and II clinical trials for the treatment of type II diabetes. In phase I of clinical trials, all the volunteers did not face any adverse health effects; bilirubin and hemoglobin levels, kidney, liver, and gastrointestinal functionality were normal. In the phase II trial, well tolerance of bis (maltolato)oxovanadium (BMOV) has successfully been tested. BMOV is another oral glucose-lowering drug successfully employed in lowering plasma glucose level of diabetic rats. In diabetic rats, BEOV takes an important role in increasing bone mineral density (BMD), bone crystal length, and mineralization and thus improves bone dysfunction. A daily dose of vanadyl sulfate (100 mg/kg) has been found to be therapeutically capable of reducing blood glucose, uric acid level, serum catalase, LDH (lactate dehydrogenase) activities, and brain lipid peroxidation in STZ-induced diabetic rats. Thus, vanadyl sulfate may be able to prevent brain damage in STZ-induced diabetic model of rats. Though vanadyl sulfate has some pharmacokinetic potential, it appears toxic at higher doses as stated earlier. Combinatorial effect of vanadyl sulfate with taurine has been investigated in diabetic rat model. The duo was found to normalize blood glucose level and lipid levels in STZ-NA (streptozotocin-nicotinamide) diabetic rats, with minimized potential toxicity of vanadium. In DENA-induced hepatocarcinogenesis model of rats, daily supplementation of ammonium monovanadate (0.5 ppm) was found to be capable of preventing early DNA damage, and further proliferating cell nuclear antigen expression was also limited. Vanadocene dichloride and vanadocene acetylacetonate are considered as potent antiproliferative agents. The plausible mechanism of their antiproliferative action is disruption of bipolar spindle formation which caused cell cycle arrest of cancer cells at G2/M phase. But there are some other evidences which account for toxicity of vanadium. Many attempts are performed in inhibiting toxicity of vanadium without altering its therapeutic efficacy. In a recent study, vanadium administration to diabetic rats was attempted using black tea decoction as vehicle rather than water. Lower vanadium accumulation in liver and rise in blood glucose level indicate to the success of the attempt. Toxicity of vanadium is more intense when inhaled rather ingested because of its poor absorption from the gastrointestinal tract.

## Vanadium and Nucleic Acid

Many investigations have been performed on therapeutic efficacy of vanadium on degenerative diseases, like diabetes and cancer, which arise due to abnormal modification of DNA chain. Thus, the therapeutic efficacy of vanadium could be somewhat related to the prevention of regular wear and tear of DNA chain. At physiological pH, the interactions between calf thymus DNA with $VO^{2+}$ and $VO_3^-$ in aqueous solution also maintain the DNA stability against any structural and functional alteration induced by chemical carcinogen (Ahmed Ouameur et al. 2006). According to Hamilton and Wilker (Hamilton and Wilker 2004), oxoanion of vanadium ($VO_4^{3-}$) acts as a nucleophile against electrophilic DNA alkylating agent and thereby prevents DNA alkylation following "carcinogen interception" mechanism. The most deleterious form of oxidative DNA damage is the double-strand break which is believed to link with induction of chromosomal mutation with tumorigenic potential. Vanadium, with vitamin $D_3$, possibly has synergistic role on detoxification of dimethylhydrazine-induced DNA adduct formation leading to the prevention of DNA damage. Again, vanadium complexes photochemically cleave DNA (Sam et al. 2004). DNA cleavage activity, i.e., nuclease activity of vanadium complex, was first reported for $VOSO_4$. In presence of $H_2O_2$, vanadyl sulfate leads to the generation of Fenton-like hydroxyl radicals that are responsible for the DNA cleavage by vanadium. In most of the cases, DNA cleavage activity of vanadium compounds is triggered either by an oxidizing agent (oxone, $H_2O_2$) or by photoirradiation (Costa Pessoa et al. 2007).

## Vanadium-Binding Protein

Therapeutic efficacy of vanadium has been discussed so far. Now, the question comes how vanadium is transported to site of action after administration into our body. There are metal ion binders in blood plasma. Serum albumin and transferrin are the mostly discussed binding proteins for vanadium (Pessoa and Tomaz 2010). Apart from these two, glutathione is an important metal ion binder which is responsible for the transformation of vanadate to $VO^{2+}$ ion in erythrocytes. The binding of vanadium complex to transferrin can be explained by two possible ways, specific and nonspecific. In the specific mode of binding, the metal carrier binds $VO^{2+}$ ion through $COO^-$ functionality.

## Conclusions

A pool of scientific data has confirmed that vanadium, once considered a potent toxic element, has sufficient potential to be an effective therapeutic of today.

A number of chemical, clinical, and epidemiological studies are still needed to determine its most effective dose in the most effective form and obviously in the most effective route, too. Pharmacological application of vanadium compounds in various diseases needs consideration of chelating agents and dietary factors, including other trace elements and added antioxidants. Ease of administration and options for lessening toxicities warrant further development to yield effective and safe pharmacological formulations. The biological and biochemical basis of this dietary micronutrient needs intense study in order to understand its physiological role in biological system.

## Cross-References

▶ Antioxidant Enzymes
▶ Cancer
▶ Diabetes
▶ Micronutrient
▶ Vanadium

## References

Ahmed Ouameur A, Arakawa H, Tajmir-Riahi HA (2006) Binding of oxovanadium ions to the major and minor grooves of DNA duplex: stability and structural models. Biochem Cell Biol 84(5):677–683

Chatterjee M, Samanta S, Chatterjee M et al (2008) Vanadium and 1, 25 $(OH)_2$ vitamin D3 combination in inhibitions of 1,2, dimethylhydrazine-induced rat colon carcinogenesis. Biochim Biophys Acta 1780(10):1106–1114

Costa Pessoa J, Cavaco I, Correia I et al (2007) In: Kustin K, Costa Pessoa J, Crans DC (eds) Vanadium: the versatile metal, ACS symposium series 974, ACS, pp 340–351

Hamilton EE, Wilker JJ (2004) Inhibition of DNA alkylation damage with inorganic salts. J Biol Inorg Chem 9:894–902

Nriagu JO (1998) Vanadium in the environment. Part 2: health effects. Wiley, New York

Pessoa JC, Tomaz I (2010) Transport of therapeutic vanadium and ruthenium complexes by blood plasma components. Curr Med Chem 17(31):3701–3738

Sam M, Hwang JH, Chanfreau G et al (2004) Hydroxyl radical is the active species in photochemical DNA strand scission by bis(peroxo)vanadium(V) phenanthroline. Inorg Chem 43(26):8447–8455

Tunali S, Yanardag R (2006) Effect of vanadyl sulfate on the status of lipid parameters and on stomach and spleen tissues of streptozotocin-induced diabetic rats. Pharmacol Res 53(3):271–277

# Vanadium in Live Organisms

Alexey N. Antipov
A.N. Bach Institute of Biochemistry Russian Academy of Sciences, Moscow, Russia

## Synonyms

Biological effect of vanadium; Vanadium – essential element for life

## Definition

The element of vanadium was discovered in 1801 by Andrés Manuel del Río when analyzing a new lead-bearing mineral and was rediscovered again in 1831 by Nils Gabriel Sefström, who named it vanadium after the Scandinavian goddess of beauty and fertility, Vanadis (Freya).

For a long time the biological role of vanadium was not known at all. On the one hand vanadium is a very toxic element for any live organisms. It inhibits some biochemical processes and leads to some serious illnesses, for example cancer. But, on the other hand, vanadium is an essential element for life. *Ascidiacea* can concentrate vanadium in their blood. Vanadium-reducing bacteria use vanadium oxides as a terminal acceptor of electrons in respiration. Some enzymes, such as haloperoxidases, alternative nitrogenase and nitrate reductase contain vanadium in their active sites. Several species of macrofungi, namely, *Amanita muscaria* and the related species, are able to accumulate vanadium (up to 500 mg/kg of dry weight). The rats and chickens tested are also known to require vanadium in very small amounts and its deficiency results in reduced growth and impaired reproduction. Vanadium is a relatively controversial dietary supplement, primarily for increasing insulin sensitivity and bodybuilding. Vanadyl sulfate can improve glucose control in the blood of people suffering from type 2 diabetes.

## Basic Information About the Element

Vanadium is the fifth most abundant transition metal in the Earth's crust, which is often found along with titanium and iron in their ores and some coal and oil deposits, such as crude and shale oils contain significant amounts (Emsley 1991). It is discovered in the enzymes of seaweed in nature and is used for catalysts in oxidation chemistry. All oxidation states from −2 to +5 are known in inorganic chemistry, and give numerous beautiful colors to their substances and so are often associated with transition metal compounds. Its multiple oxidation states, ready hydrolysis, and polymerization bestow upon vanadium a chemistry far richer and more complex than that of many elements, formation of aggregated oxyanions and sulfur complexes being just two examples. The highest three oxidation states (III, IV, and V) are of significant importance in water and so are the oxidation states found in the more than 100 known vanadium minerals.

For living organisms vanadium plays a dual role. On the one hand it is an essential element; on the other hand it is toxic. This situation is quite normal, since one substance can be a poison and a "medicine remedy."

## Vanadium in Higher Plants

Vanadium content in plants is of 0.001% (by weight) on average. Plants absorb it from soil. Still in the nineteenth century, vanadium was detected in some plants, and afterward its presence in coal, peat, and shale no longer seemed so strange. However, data on its physiological role in plants has been nearly absent. So far no compound containing vanadium in its structure has been allocated from higher plants. There is little evidence that vanadium in higher plants stimulates photosynthesis and nitrogen metabolism.

## Vanadium in Mushrooms

Amavadin is a vanadium-containing anion found in three species of poisonous *Amanita* mushrooms: *A. muscaria*, *A. regalis*, and *A. velatipes*. Its molar mass is of 398.94 g/mol. Molecular formula is $[V\{NO[CH(CH_3)CO_2]_2\}_2]^{2-}$. Appearance – it is light blue in solution. Amavadin was first isolated and identified by Kneifel and Bayer in 1972. The anion represents an eight-coordinate vanadium complex (Fig. 1). A $Ca^{2+}$ cation is often used to crystallize amavadin to obtain X-ray diffraction of high quality. The oxidized amavadin can be isolated in the form of its phosphoric acid derivative. The oxidized form contains vanadium(V), which can be used to obtain an NMR spectrum. The formation of amavadin begins with the formation of two tetradentate ligands.

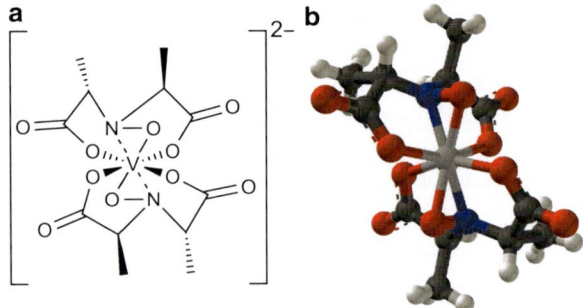

**Vanadium in Live Organisms, Fig. 1** The structure of amavadin (**a**) planar model and (**b**) 3D model

$$2HON(CH(CH_3)CO_2H)_2 + VO^{2+}$$
$$\rightarrow [V\{NO[CH(CH_3)CO_2]_2\}_2]^{2-} + H_2O + 4H^+$$

The ligand found in amavadin was first synthesized in 1954. Amavadin contains vanadium(IV). Initially, amavadin was thought to have a vanadyl, $VO^{2+}$, center. In 1993, it was discovered by crystallographic characterization that amavadin is not a vanadyl ion compound. Instead, it is an octacoordinated vanadium (IV) complex. This complex is bonded to two tetradentate ligands derived from *N*-hydroxyimino-2,2′-dipropionic acid, $H_3(HIDPA)$, ligands. The ligands coordinate through the nitrogen and the three oxygen centers.

Amavadin is a $C_2$-symmetric anion with a 2− charge. The twofold axis bisects the vanadium atom perpendicular to the two NO ligands. The anion features five chiral centers, one at vanadium and the four carbon atoms having S stereochemistry. There are two possible diastereomers for the ligands, (*S*,*S*)-(*S*,*S*)-Δ and (*S*,*S*)-(*S*,*S*)-Λ.

The biological function of amavadin is still unknown, yet it has been thought that it uses $H_2O_2$ and acts as a peroxidase to aid the regeneration of damaged tissues. Amavadin may serve as a toxin to protect the mushroom (Garner et al. 2000).

## Vanadium in Ascidia

About a 100 years ago, the German chemist Martin Henze discovered high levels of vanadium in the blood (coelomic) cells of the ascidian *Phallusia mammillata*, collected from the Bay of Naples. His discovery attracted the interdisciplinary attention of chemists, physiologists, and biochemists, in part because of considerable interest in vanadium as a possible prosthetic group, in addition to iron and copper, in respiratory pigments. This would have implied a role for vanadium in oxygen transport, a hypothesis that later proved to be false. After Henze's finding, many chemists looked for vanadium in other species of ascidians. The highest amount of vanadium was found in blood cells of the ascidian *Ascidia gemmata*. The vanadium concentration in this species can reach 350 mM, which is $10^7$ times as much as the concentration found in the seawater (35 nM). This is thought to be the highest degree of accumulation of a metal in any living organism. The accumulation of vanadium is also found in the fan worms *Pseudopotamilla occelata* and *Perkinsiana littoralis*. In these fan worms, the concentration of vanadium is as high as 60 mM. Fan worms belong to the phylum *Polychaeta*, which is phylogenetically distant from *Chordata*. Unlike the chordates, in fan worms, the highest level of vanadium is found not in blood (coelomic) cells but in the epithelial cells of the branchial crown.

In living blood cells of *P. mammillata* and *Ascidia sydneiensis samea*, vanadium is localized in the vacuoles of signet ring cells and vacuolated amoebocytes.

During the accumulation process, vanadium in the +5 oxidation state ($HVO_4^{3-}$ or $H_2VO_4^{2-}$ at physiological condition; $V^V$) is reduced to +3 oxidation state ($V^{3+}$; $V^{III}$) via +4 oxidation state ($VO^{2+}$; $V^{IV}$). NADPH is a strong reductant of $V^V$ to $V^{IV}$ as it was shown by the fact that enzymes for the pentose phosphate pathway are exclusively expressed in the cytoplasm of vanadocytes, and detailed in vitro studies suggest that chelating agents are necessary for this reaction.

Vanadium-binding proteins (Vanabins) carry out the function of metallochaperones or chelating agents in the accumulation and reduction of vanadium. Vanabins are small cysteine-rich proteins distantly related to metallothioneins, but the repetitive patterns of cysteines in vanabins are different from that of metallothioneins (Ueki and Michibata 2011).

The Vanabin family has at least five homologous members in *Ascidia sydneiensis samea*. Vanabin1, Vanabin2, and Vanabin3 are exclusively localized in the cytoplasm of signet ring cells, while Vanabin4 is

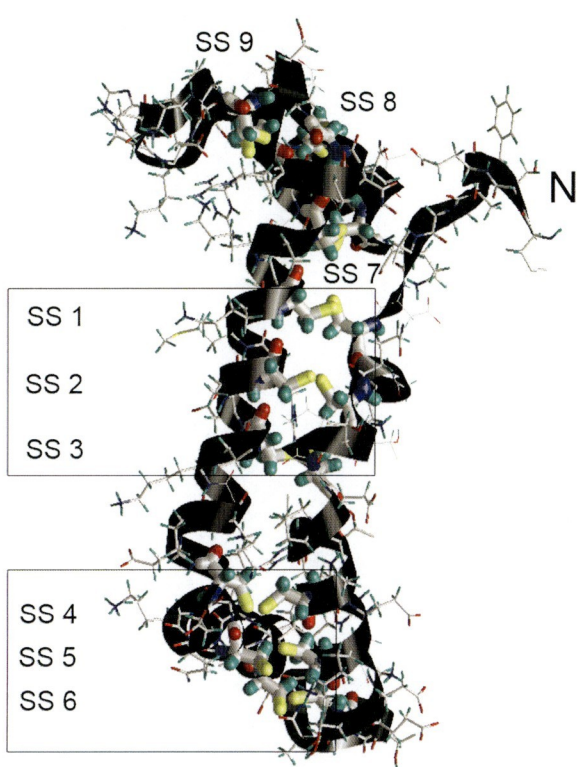

**Vanadium in Live Organisms, Fig. 2** 3D structure of the Vanabin2 (PDB ID: 1 VFI). Nine disulfide bonds are indicated by SS1–9

loosely associated with the cytoplasmic membrane of signet ring cells. VanabinP is exclusively localized in the blood plasma. In these five Vanabins, the 18 cysteine residues are conserved, and the intervals between these cysteine residues are conserved as well. A conserved amino-acid motif described by the consensus sequence {C}–{X2–5}–{C} (C: cysteine, X: any amino acids except for cysteine). In general, residues that function in maintaining structural stability and in metal-binding sites are tightly conserved.

Vanabin1 and vanabin2 bind 10 and 20 vanadium (IV) ($VO^{2-}$) ions with dissociation constants of $2.1 \times 10^{-5}$ and $2.3 \times 10^{-5}$ M, respectively. In spite of their cysteine-rich nature, vanadium (IV) ions are shown not to be coordinated by thiolates but by nitrogen and oxygen atoms. VanabinP bound a maximum of 13 vanadium (IV) ions per molecule with a $K_d$ of $2.8 \times 10^{-5}$ M.

Vanabin2 has a bow-shaped conformation consisting of four α-helices connected by nine disulfide bonds (Fig. 2), and $VO^{2+}$ is mostly coordinated to side-chain nitrogen atoms of amino acids such as lysine and arginine (Hamada et al. 2005).

Although the unusual phenomenon whereby some ascidians accumulate vanadium to the levels more than ten million times higher than those in seawater has attracted researchers in various fields, the physiological role(s) of vanadium still have to be explained. Endean and Smith proposed that the cellulose of the tunic might be produced by vanadocytes. Carlisle suggested that vanadium-containing vanadocytes might reversibly trap oxygen under the conditions of low oxygen tension. However, no direct evidence exists in support of either hypothesis. Also, vanadium was proposed to protect ascidians from fouling or predation, or act as an antimicrobial agent. Antifouling and anti-predation are the most relevant physiological functions of acid contained in the tunic, but the contribution of vanadium to these functions does not sound so evident. Michibata suggested that Vanabin2 could catalyze a redox cascade connecting NADPH to $V^V/V^{IV}$, one should consider whether the reverse reactions, accompanying the oxidations of $V^{III}$ to $V^{IV}$ and $V^{IV}$ to $V^V$, occur in the proposed cascade, perhaps resulting in the release of energy, just like the vanadium redox flow battery. Thus, the ascidians may accumulate the metal ions as an energy source.

## Nitrogen Fixation

Vanadium nitrogenase (Rehder 2000), an enzyme found in the nitrogen-fixing bacterium, is known as a secondary pathway for nitrogen fixation when molybdenum is unavailable to the primary molybdenum nitrogenase pathway. An important component in nature's nitrogen cycle, vanadium nitrogenase in the presence of dinitrogen, converts the nitrogen gas to ammonia making inaccessible nitrogen available to plants. Vanadium nitrogenases are found in some members of the bacteria genus *Azotobacter*, *Anabaena*, *Rhodopseudomonas*, and *Methanosarcina*. Most of the functions of vanadium nitrogenase match that of more common molybdenum nitrogenases and serve as an alternative pathway for nitrogen fixation in molybdenum-deficient conditions. Like molybdenum nitrogenase, dehydrogen functions as a competitive inhibitor and carbon monoxide functions as a noncompetitive inhibitor to the fixation. Vanadium nitrogenase has an $\alpha_2\beta_2\gamma_2$ subunit structure, while molybdenum nitrogenase has an $\alpha_2\beta_2$ structure. Though the structural genes encoding vanadium nitrogenase show only about 15% conservation with molybdenum nitrogenases, the two nitrogenases share the same type of redox centers. At room temperature, vanadium nitrogenase is less active in nitrogen fixing than moldybdenum nitrogenases because it converts more $H^+$ to $H_2$. However, at low temperatures vanadium nitrogenases have been found to be more active than the molybdenum type, and it can fix nitrogen at temperatures as low as 5 °C and its low-temperature activity is ten times as high as that of Mo-Fe nitrogenase. Unlike the molybdenum nitrogenase, small amounts of hydrazine, isonitriles, and acetylene which can be converted to ethylene and ethane can be produced during catalytic activity. Vanadium nitrogenase is easily oxidized and can, thus, only work under anaerobic conditions and has complex protection mechanisms to avoid oxygen.

Chi Chung Lee, Yilin Hu, and Markus Ribbe of the University of California (Irvine, USA) discovered the ability of the enzyme vanadium nitrogenase to convert carbon monoxide into trace amounts of propane by chance while they were running different experiments on the enzyme (Lee et al. 2010). One of their experiments composed of exposing carbon monoxide to the vanadium nitrogenase in an environment that completely lacked nitrogen. From the carbon monoxide exposure, the enzyme produced trace amounts of propane, ethylene, and ethane. Though the genes of enzyme have been known for over 20 years, only recently the new technology allowed for mass production to run the extensive experiments. Vanadium nitrogenase produces alkanes through the reduction of carbon monoxide and the hydrolysis of ATP and dithionite. The process of forming these hydrocarbons is carried out through proton and electron transfer where short carbon chains are formed.

## Vanadium-Dependent Haloperoxidases

Haloperoxidases catalyze the oxidation of halides in the presence of hydrogen peroxide and they are named according to the most electronegative halide that they can oxidize, that is, chloroperoxidases (ClPOs) can catalyze the oxidation of chloride as well as of bromide and iodide, bromoperoxidases (BrPOs) react with bromide and iodide, whereas iodoperoxidases (IPOs)

**Vanadium in Live Organisms, Table 1** Some properties of vanadium-containing haloperoxidases

| Value for | Property — Mol. mass, kDa | No. and mass (kDa) of subunits | Type of peroxidase | $K_m$ Haloid | $K_m$ H$_2$O$_2$ | pH optimum | Thermostability |
|---|---|---|---|---|---|---|---|
| *Laminaria saccharina* | 108 | 66, 64 | Br | 1 mM | | 6.0 | + |
| *Corallina pilulifera* | 790 | 12/64 | Br | | | 6.0 | + |
| *Corallina officinalis* | 740 | 12/64 | Br | 1 mM | 60 μM | 6–7 | + |
| *Ascophyllum nodosum* | 250 | 6/40 | Br | 12.7 mM | | 4.5–6.5 | + |
| *Ceratium rubrum* | 240 | 50 | Br | 2 mM | 17 μM | 6.0–8.3 | – |
| *Saccorhiza polychides* | 125 | 2/64 | I | | | 6.1 | + |
| *Xanthoria parietina* | | 64 | Br | 28 μM | 870 μM | 5.5 | + |
| *Pelvetia canaliculata* | 166 | 66 | I | 2.1 mM | 110 μM | 6.0 | + |
| *Pelvetia canaliculata* | 416 | 66, 72, 157 and 280 | I | 2.4 mM | 20 μM | 6.5 | + |
| *Kappaphycus alvarezii* | | | I | 2.5 mM | 85 μM | 6.5 | n.d. |
| *Embellisia didymospora* | | 67 | Cl | 1.2 mM | 60 μM | 5.2 | + |
| *Curvularia inaequalis* | | 67 | Cl | 0.25 mM | | 5.5 | |

\+ activity at 50°C is stable for 1–2 h
– inactivated at 50°C for a few minutes

are specific of iodide. The first vanadium-dependent haloperoxidase (vHPO) was discovered in *Ascophyllum nodosum*, a brown alga belonging to the *Fucales*. Since vHPO activities have been detected in a very large number of red and brown macroalgae and fungi (Table 1). Most of them have been identified as bromoperoxidases. The values of specific activity reported for these enzymes are extremely variable, that is, from 12 units mg$^{-1}$ of proteins in *Laminaria saccharina* to 1730 units mg$^{-1}$ in *Macrocystis pyrifera*. Whereas the molecular masses of denatured proteins are quite similar, that is, around 64 kDa, the native enzymes consist of dimeric assemblies, ranging from 100 to 800 kDa. For instance, in the red alga *Corallina pilulifera*, vBrPOs are organized in dodecamers. In contrast to heme-haloperoxidases, vHPO activities are relatively resistant to high-temperature exposures and the enzymes are still active in oxidative conditions and in the presence of different organic solvents such as acetone, methanol, or ethanol. These properties as well as the ability to halogenate a broad range of organic compounds of both commercial and pharmaceutical interests make vHPOs good candidates for use in industrial biotransformations.

The global dimeric structure of algal vBrPOs (Weyand et al. 1999) is extremely conserved, folding into α-helices with a few short β-strands (Fig. 3). The main tertiary structural motif of two four-helix bundles

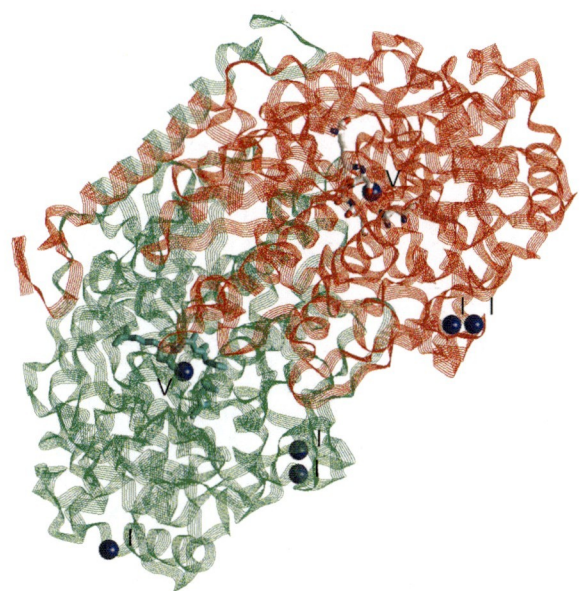

**Vanadium in Live Organisms, Fig. 3** The structure of BrPO from *Ascophyllum nodosum* (PDB ID: 1 QI9). The selected amino acid residues belong to the active site of the enzyme

is similar to the reported X-ray crystal structure of the vClPO from a terrestrial fungi, *Curvularia inaequalis*.

The surfaces of the two monomers are tightly bound one to another by hydrophobic interactions and in the case of *A. nodosum* vBrPOs, two additional intermolecular disulfide bridges are also involved in the interface. Just recently, EXAFS and MS studies have led to reassess partially the electron density map for the brown algal vBrPO structure, showing posttranslational bromination and iodination of tyrosine residues at the surface of *A. nodosum* vBrPO.

The vanadium-binding site motif, HP[S/A]Y[P/G][S/A]GHA, is relatively well conserved in all algal vHPO protein sequences. The vanadium cofactor is coordinated by four nonprotein oxygen atoms and one nitrogen (Nε2) atom from a histidine residue into trigonal bipyrimidal geometry. Based on X-ray data, the linkage between the metal and the histidine ligand was first assigned as a direct coordinating bond and later as a covalent bond. The hydrogen-vanadate ($HOVO_3^{2-}$) is stabilized into a hydrogen-bonding network, involving six highly conserved residues (one apical histidine, two arginine, a lysine, a glycine, and a serine). Kinetics and structural studies of single site mutants of vClPO from *C. curvularia* have shown the importance of these residues in binding vanadate, but also the rigidity of the active site, resulting from a large number of hydrogen bonding interactions around.

Detailed steady-state kinetic analysis has suggested a bi-bi ping-pong two-substrate mechanism, in which hydrogen peroxide first binds the vanadate-active site (Fig. 4), followed by halide oxidation (Wever and Wieger 2001). Coordination of hydrogen peroxide to the vanadium center is the first step in catalysis, and $His^{404}$, which activates the axial water molecule, must be deprotonated for $H_2O_2$ to bind. A crystal structure of the peroxo-V-ClPO shows that peroxide is coordinated in a side-on manner in the equatorial plane and distorts the vanadium site to a tetragonal bipyramidal geometry. After the binding of peroxide, $His^{404}$ is no longer hydrogen-bonded to any oxygen atoms of the cofactor, $Lys^{353}$ makes direct contact with one of the oxygen atoms of the bound peroxide, and further activation the bound peroxide which undergoes through charge separation. A halide ion is the second substrate that binds to the enzyme through nucleophilic attack on the partially positive oxygen atom. This binding breaks the peroxide bond and creates the nucleophilic $OX^-$ group, which leaves the coordination sphere as hypohalous acid after protonation by an incoming water molecule. If an appropriate nucleophile is present, the generated hypohalous acid intermediate will react with the organic substrate, giving rise to a halogenated compound. If not, the oxidized halogen intermediate will react with another equivalent of hydrogen peroxide to generate dioxygen in the singlet state and the halide. A new fluorescence microscopy–based method, which uses a fluorogenic derivative to monitor the formation and migration of HOBr from a V-BrPO-active site, will shed light on where the actual halogenation of organic substrates occurs. Active-site residues assisting with the selection and binding of the halide are still being elucidated. On the basis of crystal structure analysis, $Phe^{397}$ and $Trp^{350}$ (numbering from *C. inaequalis*) may participate in halide binding through their δ + ring edge. VBrPOs alternatively contain His and Arg residues at these positions, respectively. Recently, chlorinating activity was created in the *Corallina pilulifera* vBrPO through a single amino acid substitution. The substitution of the Arg residue at position 350 with Trp or Phe increased the affinity of vBrPO for chloride. The native vBrPO from *A. nodosum* also contains a Trp residue corresponding to position 350, and it, too, was shown to exhibit chloroperoxidase activity. Therefore, this Trp residue is believed to participate in chloride binding. Additional mutagenesis experiments on the His residue corresponding to $Phe^{397}$ showed the importance of the residue in the oxidation of halides. The His residue was mutated to Ala in the vBrPO from *C. officinalis*, which resulted in the enzyme's inability to efficiently oxidize bromide; however, the mutant could still oxidize iodide. Interestingly, like the vBrPOs, the three putative vClPOs from the napyradiomycin biosynthetic cluster also contain a His residue in the position corresponding to $Phe^{397}$, and this residue may also play a role in substrate binding or selectivity.

In the absence of halides, vBPOs catalyze the enantioselective oxidation of sulfides into sulfoxides. These sulfoxidation reactions occur at the vanadium-binding site through a direct oxygen transfer from the peroxovanadium intermediate to the sulfide.

In seaweeds, vBrPOs are involved in the biosynthesis of bromoform, dibromomethane, and dibromochloromethane. These compounds have antimicrobial activity. The enzyme may prevent the seaweeds from fouling by microorganisms or may act as an antifeeding system. The vClPOs of hyphomycetes may participate in some pathogenic processes or in lignin degradation.

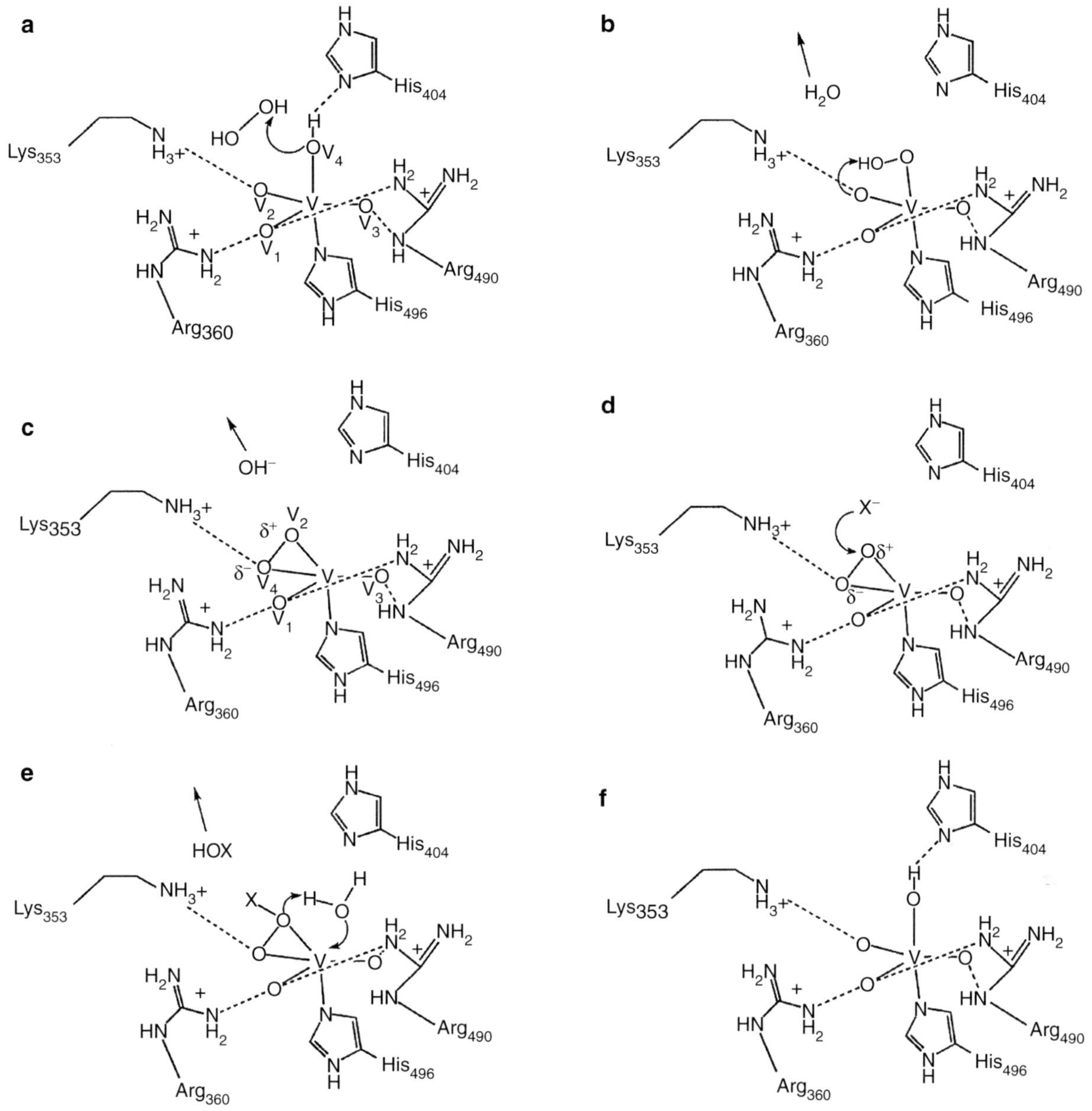

**Vanadium in Live Organisms, Fig. 4** Minimal reaction scheme for vanadium ClPO catalysis (Wever and Wieger 2001)

## Vanadate-Reducing Bacteria

In literature, a few vanadate-reducing microorganisms are described. First the bacteria of the genus *Pseudomonas* reducing vanadate have been described by N.N. Lyalikova (Yurkova and Lyalikova 1991), their biochemistry, for the first time, has been studied in our laboratory. There is also evidence of vanadate-reducing microbial activity in some types of *Shewanella oneidensis*, *Acidithiobacillus ferrooxidans*, *Acidithiobacillus thiooxidans*, and *Geobacter metallireducens*. These bacteria use vanadate as a terminal acceptor of electrons during respiration.

The *Pseudomonas* species can conserve energy using hydrogen, carbon monoxide, various sugars, and organic acids as an electron donor to support their growth. *Pseudomonas isachenkovii* strain A-1 was isolated from an ascidian worm from the Bay of

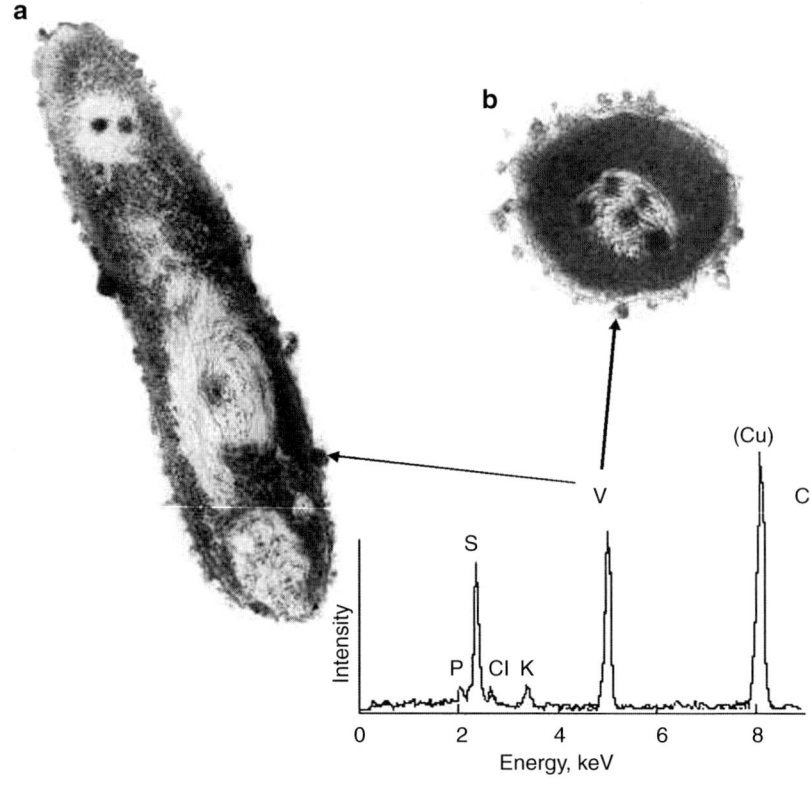

**Vanadium in Live Organisms, Fig. 5** Thin section of the *P. isachenkovii* cells: (**a**) lengthwise; (**b**) cross-section; (**c**) X-ray microanalysis spectrum of the *P. isachenkovii* cells. The copper band (Cu) refers to the greed

Kraternaya (the Kuril Islands in the Pacific Ocean). *P. isachenkovii* can express two catalytically distinct molybdenum-free and molybdenum cofactor-free dissimilatory nitrate reductases, one of which contains vanadium (Antipov et al. 1998). The vanadium-containing nitrate reductase has a mass of 220 kDa and consists of four subunits, which can be independent catalytic units under special physiological conditions of cell's growth. Both enzymes participate in vanadate reduction. The reducing vanadium is bonded with the vanadium-binding protein (molecular mass of peptide chain is 14 kDa) up to 20 atoms per molecule of protein and is excreted into growth medium. Electronmicroscopic studies of the *P. isachenkovii* cells showed that cultivation of the bacterium on a vanadate-containing medium (in contrast to a nitrate-containing one) resulted in the formation of a large number of vanadium-accumulating swells on the surface of the membranes of the cell wall (Fig. 5). While the bacterial growth and vanadate reduction take place, these containers seem to separate from the cell wall, as a result, vanadium-binding protein is accumulated in the culture medium. The synthesis of vanadium-binding protein by the cells of the vanadate-reducing bacterium, followed by its excretion into the culture medium, appears to be physiologically important. For the bacterium, it is a way of vanadate detoxication, whereas the ascidians in which the bacteria cells live go on accumulating the intermediate tetravalent vanadium that is subsequently reduced to the trivalent state in the blood vanadocytes. Perhaps there are symbiotic relations between the ascidians and the vanadate-reducing bacteria living in them.

## Vanadium and Health

Human muscle tissue contains $2 \times 10^{-6}\%$ vanadium, but bone tissue does about $-0.35 \times 10^{-6}\%$, in the blood is less than $2 \times 10^{-4}$ mg/l. The average human body (body weight is about 70 kg) contains 0.11 mg of vanadium altogether. The amount of vanadium in food is unknown. Daily vanadium requirement is 6–63 mg per day (WHO 2000). Only 1% of vanadium from external sources is absorbed in the body, and the rest is excreted through the urine. The main food sources of vanadium are rice, oats, beans, radishes, barley, buckwheat, lettuce, peas, potatoes, dill, parsley, black pepper, shellfish, meat, mushrooms, soy, wheat, olives, unsaturated fats and vegetable oils, some quantity comes from water as well.

All vanadium compounds should be considered to be toxic. Tetravalent $VOSO_4$ has been reported to be over five times more toxic than trivalent $V_2O_3$. The Occupational Safety and Health Administration (OSHA) has set an exposure limit of 0.05 mg/m$^3$ for vanadium pentoxide dust and 0.1 mg/m$^3$ for vanadium pentoxide fumes in workplace air for an 8-hour workday, 40-hour work week. The National Institute for Occupational Safety and Health (NIOSH) has recommended that 35 mg/m$^3$ of vanadium be considered an immediate danger to life and health. This is the exposure level of a chemical that is likely to cause permanent health problems or death.

Vanadium compounds are poorly absorbed through the gastrointestinal system. Inhalation exposures to vanadium and vanadium compounds result primarily in adverse effects on the respiratory system. Quantitative data are, however, insufficient to derive a subchronic or chronic inhalation reference dose. Other effects have been reported after oral or inhalation exposures on blood parameters, on liver, on neurological development in rats, and other organs.

There is little evidence that vanadium or vanadium compounds are reproductive toxins or teratogens. Vanadium pentoxide was reported to be carcinogenic in male rats and male and female mice by inhalation in a National Toxicology Program (USA) study, although the interpretation of the results has recently been disputed. Vanadium has not been classified as carcinogenic by the United States Environmental Protection Agency.

Vanadium compounds exert preventive effects against chemical carcinogenesis on animals, by modifying, mainly, various xenobiotic enzymes, inhibiting, thus, carcinogen-derived active metabolites. Studies on various cell lines reveal that vanadium exerts its antitumor effects through inhibition of cellular tyrosine phosphatases and/or activation of tyrosine phosphorylases. Both effects activate signal transduction pathways leading either to apoptosis and/or to activation of tumor suppressor genes. Furthermore, vanadium compounds may induce cell-cycle arrest and/or cytotoxic effects through DNA cleavage and fragmentation and plasma membrane lipoperoxidation. Reactive oxygen species generated by Fenton-like reactions and/or during the intracellular reduction of V(V) to V(IV) by, mainly, NADPH participate to the majority of the vanadium-induced intracellular events. Vanadium may also exert inhibitory effects on cancer cell metastatic potential through modulation of cellular adhesive molecules, and reverse antineoplastic drug resistance.

Several genes are regulated by this element or by its derivatives, which include genes for tumor necrosis factor-alpha (TNF-$\alpha$), Interleukin-8 (IL-8), activator protein-1 (AP-1), ras, c-raf-1, mitogen-activated protein kinase (MAPK), p53, and nuclear factors – $\kappa$B.

Vanadyl sulfate is a very controversial dietary supplement, popular with bodybuilders and is often purchased in gym shops where it is legal. The vanadate anion is a phosphate mimic that has been used as a probe of the enzymes that transfer phosphates in cell signaling – the phosphatases and kinases. Not surprisingly vanadium shows many interesting biological properties resulting from this activity, not the least of which is its ability to enhance, but not mimic, the action of insulin, the key hormone in diabetes mellitus. This property was first shown in three diabetic humans in France and was published in 1899 in *La Presse Médicale*. Vanadium does not work in the complete absence of insulin – hence, it is an enhancer rather than a mimic of insulin. Significant efforts over the last 25 years, since John McNeill of the University of British Columbia showed that vanadate was effective in a diabetic rat model, have led to a number of vanadium compounds now being clinically investigated in humans as potential agents for the treatment of diabetes.

Deficiency of vanadium is represented as single cases of vanadium-dependent schizophrenia, and is also associated with pathology of carbohydrate metabolism.

## Cross-References

▶ Molybdenum and Ions in Living Systems
▶ Molybdenum in Biological Systems
▶ Vanadium in Biological Systems
▶ Vanadium Ions and Proteins, Distribution, Metabolism, and Biological Significance
▶ Vanadium Pentoxide Effects on Lungs

## References

Antipov AN, Lyalikova NN, Khijniak TV, L'vov NP (1998) Molybdenum-free nitrate reductases from vanadate-reducing bacteria. FEBS Lett 441:257–260

Emsley J (1991) The elements. Clarendon, Oxford

Garner CD, Armstrong EM, Berry RE, Beddoes RL, Collison D, Cooney JJA, Ertok SN, Helliwell M (2000) Investigations of Amavadin. J Inorg Biochem 80:17–20

Hamada T, Asanuma M, Ueki T, Hayashi F, Kobayashi N, Yokoyama S, Michibata H, Hirota H (2005) Solution structure of Vanabin2, a vanadium(IV)-binding protein from the vanadium-rich ascidian *Ascidia sydneiensis samea*. J Am Chem Soc 127:4216–4222

Lee CC, Hu Y, Ribbe MW (2010) Vanadium nitrogenase reduces CO. Science 329:642

Rehder D (2000) Vanadium nitrogenase. J Inorg Biochem 80:133–136

Ueki T, Michibata H (2011) Molecular mechanism of the transport and reduction pathway of vanadium in ascidians. Coord Chem Rev 255:2249–2257

Wever R, Wieger H (2001) Vanadium haloperoxidases. In: Messerschmidt A et al (eds) Handbook of metalloproteins. Wiley, Chichester, pp 1417–1428

Weyand M, Hecht H, Kiess M, Liaud M, Vilter H, Schomburg D (1999) X-ray structure determination of a vanadium-dependent haloperoxidase from *Ascophyllum nodosum* at 2.0 A resolution. J Mol Biol 293:595–611

Yurkova NA, Lyalikova NN (1991) New vanadate-reducing facultative chemolithotrophic bacteria. Microbiology 59:672–677

# Vanadium Ions and Proteins, Distribution, Metabolism, and Biological Significance

Biswajit Mukherjee, Sefali Halder, Miltu Kumar Ghosh and R. Manasadeepa
Department of Pharmaceutical Technology, Jadavpur University, Kolkata, West Bengal, India

## Definition

Vanadium is an element of atypical biological significance. Vanadium ions act in oxidation-reduction reactions in the organism. Vanadium proteins such as vanadium haloperoxidase, vanadium nitrogenase, and vanabins have profound role in biology. The oxidation states of vanadium influence on its absorption and metabolism. The metabolism of this multivalent metal and its ions in humans is yet obscure.

## Vanadium and Vanadium ions

Vanadium is a silvery gray, ductile, soft transitional metal (density 6 g/cm$^3$) (Mukherjee et al. 2004; Lide 2004). Vanadium was named after Old Norse Vanadis (the goddess of beauty and fertility), as salts of vanadium are generally and widely colorful. It does not exist in free form in nature. The element is very prone to oxidation, and pure vanadium when exposed to atmosphere forms an oxide layer on it, which protects vanadium from further oxidation. About 65 vanadium compounds are known so far in nature. Vanadium is not available in nature as pure metal. It exists in combination with different minerals in bound forms. Some of those are bauxite, patronite ($VS_4$), vanadinite ($Pb_5(VO_4)_3Cl$), and carnotite ($K_2(UO_2)_2(VO_4)_2 \cdot 3H_2O$). The minerals are mostly obtained from South Africa, Northwestern China, and Eastern Russia. Vanadium is also obtained from carbon-containing fossil fuel deposits such as crude oil, coal, shale oil, and tar sands. In crude oil, vanadium concentration is up to 1,200 ppm. About 110,000 t of vanadium is released into the atmosphere by burning fossil fuels per year.

Vanadium is abundant in soil in different forms. The amount which is taken up by plants is very little. The humans and other animals get vanadium from the foodstuffs. The foods with relatively high vanadium content (12–15 μg/100 g) are buckwheat, soya beans, olive oil, sunflower oil, apples, eggs, crabs, mushrooms, shellfish, black pepper, parsley, dill seed, beverages (Beer), wine, grain products, and artificial sweeteners. Most fats and oils also contain vanadium but in a very low amount ($<0.3$ μg/100 g).

Vanadyl sulfate is the most common chemical form available in all food supplements. In addition, decavanadate and oxovanadates potentially have many biological activities and have been used as biochemical supplements.

## Physical Properties

In periodic table, vanadium is present in group 5, period 4, and block d and has atomic number 23, the electron configuration [Ar] $3d^3\ 4s^2$ with electron affinity 50.7 kJ/mole and electronegativity 1.63 (Pauling scale) (Lide 2004; Crans et al. 2004). The standard atomic weight is 50.94 with atomic volume 8.3 cm$^3$/mole and atomic radius 134 picometers.

It has electrical conductivity 39.37 S.m$^{-1}$, melting point 2,183 K, and its thermal conductivity is 30.7 Wm$^{-1}$ K$^{-1}$.

## Isotopes

Vanadium has two natural isotopes and 25 artificial isotopes. The stable natural isotope of vanadium is $^{51}$V, and natural radioactive isotope of the element is $^{50}$V (half-life, 1.5 × 10$^{17}$ years) (Lide 2004; Crans et al. 2004). About 24 artificial radioisotopes have been identified. The most stable isotope among them is $^{49}$V (half-life, 330 days). This is followed by $^{48}$V (half-life, 16 days). Some of the isotopes have half-lives even in the range of few seconds. Four metastable isotopes of vanadium are also available. Vanadium isotopes undergo beta decay reaction.

## Chemical Properties of Vanadium Ions

The oxidation state of vanadium exists right from +2 to +5 numbers. The most common and highest oxidation state is +5. Vanadium +5 compounds are oxidizing agents, and +2 compounds are reducing agents, whereas vanadium +3 and +4 numbers exist in different vanadium complexes as tetrahedral form. To observe the different oxidizing states of vanadium, ammonium vanadate can be reacted with elemental zinc, a reducing agent. This gives different colors due to different oxidative states of vanadium, i.e., yellow for +5, blue for +4, green for +3, and pale violet color for +2.

Vanadium +3 makes complexes with amino acids such as alanine, aspartate, etc. In biological fluid, vanadium is available as pentavalent state(V$^{5+}$) as vanadate or monovanadate state (VO$_3^-$, VO$_4^{3-}$, H$_2$VO$_4^-$) or in tetravalent state (V$^{4+}$) as vanadyl ion (VO$^{2+}$) in acidic pH, whereas in basic pH, it exists in the state of orthomonovanadate ion VO$_4^{3-}$ (e.g., sodium orthovanadate). Vanadium ion is present as a regular tetrahedron with four equivalent oxygen atoms with a single double bond, which is conventionally a resonance form (Fig. 1).

Different oxo compounds of vanadium ions are present in different buffer solutions. In a strong basic solution of vanadium pentoxide (V$_2$O$_5$), it gives the colorless VO$_4^{3-}$ ion. If the pH changes to acidic, these solutions gradually darken through orange to red at around pH 7. Brown hydrated V$_2$O$_5$ precipitates at pH 2. On redissolving in water, it forms a light yellow solution of [VO$_2$(H$_2$O)$_4$]$^+$ ion. Different numbers of the

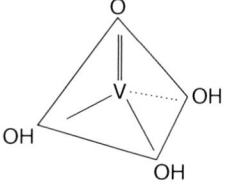

**Vanadium Ions and Proteins, Distribution, Metabolism, and Biological Significance, Fig. 1** Structure of vanadium ion (H$_3$VO$_4$)

oxyanions exist between pH 2 and pH 13. In low pH, HVO$_2$ and H$_2$VO$_4^-$ are produced. The acid dissociation of vanadium has the similarity with that of phosphate. In different pH media of vanadate, vanadium produces various polyoxovanadate ions such as H$_2$V$_{10}$O$_{28}^{4-}$, H$_2$VO$_4^-$ (pH 2–4), H$_2$VO$_4^-$, V$_4$O$_{12}^{4-}$, HV$_{10}$O$_{28}^{5-}$ (pH 4–9), HVO$_4^{2-}$, V$_2$O$_7^{4-}$ (pH 9–12), etc.

In more concentrated solutions, many polyvanadate ions (Lide 2004; Crans et al. 2004) are formed resulting of different chains, rings, and clusters (Fig. 2) by involving tetrahedral vanadium. A wide variety of polyoxovanadate ions formed by vanadium include both discrete ions and infinite polymeric ions. Some examples of discrete ions are:

| Bond type | Orbital configuration |
|---|---|
| VO$_4^{3-}$ "orthovanadate" tetrahedral | $sp^3$ |
| V$_2$O$_7^{4-}$ "pyrovanadate" tetrahedral | $sp^3$ |
| V$_3$O$_9^{3-}$ cyclic tetrahedral | $sp^3$ |
| V$_4$O$_{12}^{4-}$ cyclic tetrahedral | $sp^3$ |
| V$_5$O$_{14}^{3-}$ tetrahedral | $sp^3$ |
| V$_{10}$O$_{28}^{6-}$ "decavanadate" VO$_6$ octahedral | $d^2sp^3$ |
| V$_{13}$O$_{34}^{3-}$ fused VO$_6$ octahedral | $d^2sp^3$ |

Some examples of polymeric "infinite" ions are
- [VO$_3$]$_n^{n-}$ in, e.g., NaVO$_3$, sodium metavanadate
- [V$_3$O$_8$]$_n^{n-}$ in CaV$_6$O$_{16}$

In these ions, vanadium exhibits tetrahedral and octahedral (Fig. 3) coordination due to variable hybridization. Each ligand donates a pair of electrons to form a coordinate covalent link between itself and vanadium, the central atom with an incomplete electron shell. Since sufficient bonding orbital is not available in vanadium ions, it undergoes hybridization to form coordination compounds.

Vanadium has a few important features in coordination chemistry due to its early position in the transition metal series. First, metallic vanadium, electronic

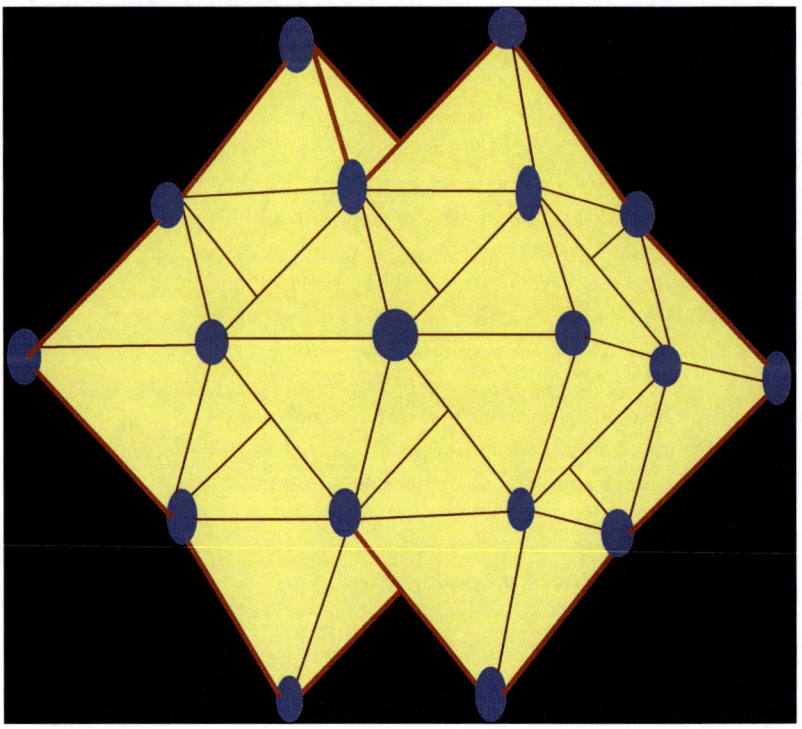

**Vanadium Ions and Proteins, Distribution, Metabolism, and Biological Significance, Fig. 2** Decavanadate ions

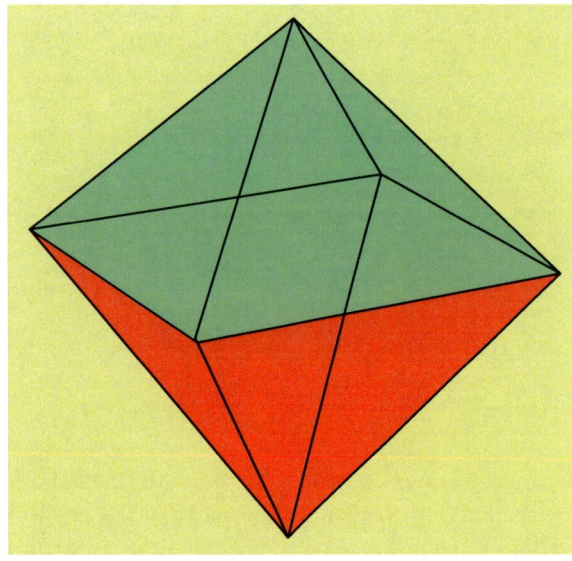

**Vanadium Ions and Proteins, Distribution, Metabolism, and Biological Significance, Fig. 3** Octahedral structure of acetyl acetovanadate

configuration $[Ar]3d^34s^2$, is relatively electron-poor compound. Vanadyl ion, $VO^{2+}$ in many complexes of vanadium such as vanadyl acetylacetonate, is 5 coordinate and can attach to a sixth ligand. Thus, many 5-coordinate vanadyl complexes have a trigonal bypyramidal geometry with $dsp^3$ hybridization, such as $VOCl_2[N(CH_3)]_2$.

Vanadium ion also forms various peroxo complexes. When yellow-colored oxovanadium is treated with acidic hydrogen peroxide, it forms red peroxovanadium $VO(O_2)_2$.

Vanadium ions form complexes with halides (Lide 2004; Crans et al. 2004) as vanadium has different oxidation states, +2, +3, +4, and +5. Vanadium(II, III) chlorides, vanadium oxytrichloride(V), vanadium tetrachloride, vanadium(III) bromide, vanadium(III) fluoride, vanadium(III) iodide, vanadium(IV) fluoride, vanadium(V) oxytrifluoride, vanadyl(IV) oxalate, vanadyl(IV) sulfate, and vanadyl(IV) sulfate hydrate are the important examples of these category.

Organovanadium chemistry is the chemistry of organometallic compounds containing a carbon to vanadium chemical bond. Organovanadium compounds have importance to organic synthesis and polymer chemistry as reagents and catalysts. Low-valency vanadium gets stabilized with carbonyl ligands. When valency increases oxo ligands of vanadyl, ions generally form. Common ligands for organovanadium are carbonyl, phosphine, and cyclopentadienyl.

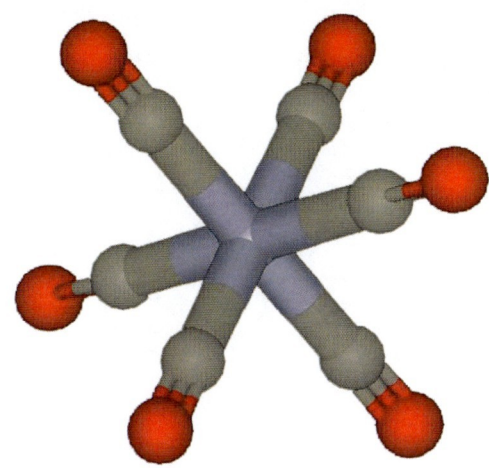

Vanadium Ions and Proteins, Distribution, Metabolism, and Biological Significance, Fig. 4 Structure of vanadium carbonyl $V(CO)_6$

For example, vanadium carbonyl $V(CO)^{6+}$ (Fig. 4) is an important moiety of this class.

## Vanadium Proteins

Metalloproteins consist of protein scaffolds holding metal ion(s) either as individual ion or in a combination with different metal ions. Among these several ions, only few ions bind strongly and exchange slowly to the metalloproteins. These metalloenzymes are essential for catalysis of metabolism of small molecules, such as $H_2$, $CH_4$, CO, $N_2$, and $O_2$, as well as for long-range electron transfer. Some vanadium-containing metalloproteins such as vanadium haloperoxidases, vanadium nitrogenases, and vanabins play vital role in biochemical processes in biological systems.

## Vanadium Haloperoxidases

Vanadium-containing haloperoxidase (V-HPO) was first discovered in 1984 from a brown alga *Ascophyllum nodosum*.

### Structure

The enzyme has helical structure with two numbers of four α-helix bundles and three small β-sheets. Holoenzymes have two active centers, and they contain trigonal bipyramidal coordinated vanadium atoms. Different known haloperoxidases are named after the presence of halide (Cl/Br/I) in their molecule. Cl-containing haloperoxidase is called vanadium chloroperoxidase (V-CPO). Likewise, bromine- and iodine-containing vanadium enzymes are called vanadium bromoperoxidase (V-BPO) and vanadium iodoperoxidase (V-IPO), respectively. Vanadium-binding centers remain on N-terminal site of α-helix, and the centers have catalytic histidine residue ($His^{418}$, for V-BPO/$His^{404}$, for V-CPO). Vanadium is coordinated to three oxygen species in the equatorial plain, to one apical OH group and to nitrogen of a histidine at another apical position. Another histidine close to the active site (e.g., $His^{411}$ for V-BPO) modulates redox potential through protonation and deprotonation reactions of the catalytically active $V_2O_2\text{-}O_2$ species. Although there is significant overlap of the active sites of V-BPO and V-CPO, these enzymes are distinct from each other. $Lys^{353}$, $Arg^{360}$, $His^{404}$, $Arg^{490}$, and $His^{496}$ are the key residues in V-CPO (Winter and Moore 2009). Recombinant V-BPO (*Corallina pilulifera*) contains a second histidine near active site ($His^{478}$) instead of phenylalanine in V-CPO (*Curvularia inaequalis*). This suggests that the oxidation state of vanadium remains in highest oxidation state that does not change during catalysis.

### Reaction Mechanism of Vanadium Haloperoxidases

Vanadium-containing haloperoxidases contain vanadate ion and can oxidize halide ($Cl^-/Br^-/I^-$). The enzymes catalyze two-electron oxidation of a halide using $H_2O_2$. V-CPO can oxidize both $Cl^-$ and $Br^-$, whereas V-BPO oxidizes only $Br^-$ in presence of $H_2O_2$. Common reaction is given below where R is an alkyl group and X stands for Cl/Br/I. The enzyme undergoes catalytic reaction in presence of $H_2O_2$ by protonation of the bound peroxide and addition of halide ($Cl^-/Br^-/I^-$) (Fig. 5).

$$X^- + H_2O_2 \xrightarrow{V-HPO} R - X + 2H_2O_2$$

At an early stage of the reaction, addition of $H_2O_2$ assists to proceed the reaction. Then protonation of the bound peroxide and incorporation of halide take place. It leads to the formation of a mixture of protein peroxovanadium complex and protonated protein

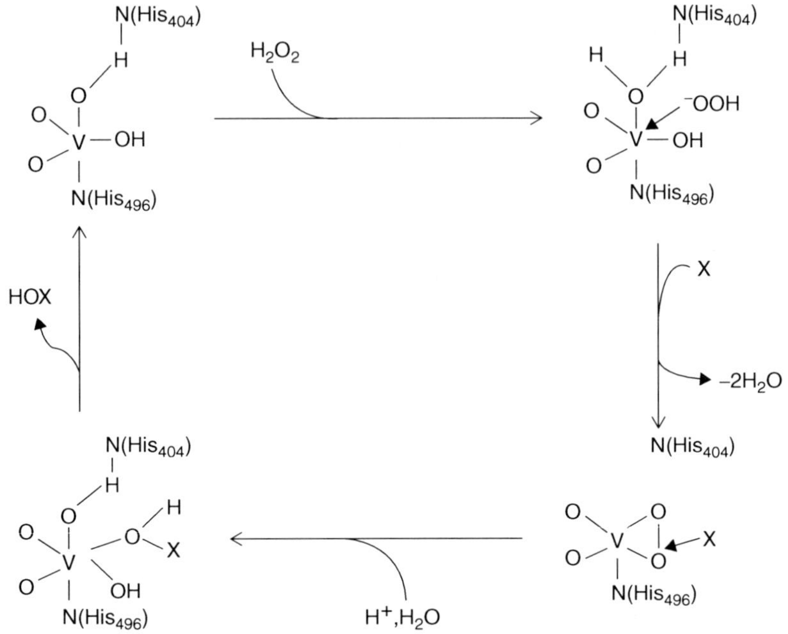

**Vanadium Ions and Proteins, Distribution, Metabolism, and Biological Significance,**
**Fig. 5** Proposed mechanism for the vanadium haloperoxidases – catalyzed oxidation of halide by hydrogen peroxide

peroxovanadium complex. However, halide (e.g., bromide) does not bind to the vanadium atoms in those complexes. Due to the nucleophilic attack by halide (X) on the protonated protein complex, it generates $X^+$ species, which halogenate organic substrate to produce singlet oxygen, a very reactive oxygen species. Distinct mechanistic differences exist between the enzymes, V-CPO and V-BPO. Presence of cosolvents such as 25% tert-butanol or methanol and variation of pH generally influence the stereoselectivity of product of V-BPO. Possibly, due to the presence of excess $H_2O_2$, presence of electron-donating substituent enhances stereoselectivity, whereas electron-withdrawing substituent reduces stereoselectivity of the reaction. V-BPO catalyzes oxidation reaction with $H_2O_2$ to form peroxovanadium, while V-CPO forms a racemic mixture of small amounts of sulfoxide. Thus, their reaction mechanism may be distinct. During oxidation of sulfide by V-HPO, peroxide is protonated before sulfide attack. The sulfide undergoes a one-electron oxidation to yield radical sulfur-containing cation, which then reacts with $O_2$ to form sulfoxide.

Different isoforms of vanadium-dependent iodoperoxidases were isolated from *S. polyschides*. Such isoforms are found due to the difference of relative molecular weights of the enzyme, different amino acid compositions, specific activity values, reactivation mode, and kinetic properties, which vary due to several glycosylation patterns. Three different isoforms of vanadium-dependent iodoperoxidases have been characterized. They differ from one another with respect to their sources (*S. polyschides, Phyllariopsis brevipes, Pelvetia canaliculata*).

## Vanadium Nitrogenase

### Types of Metallonitrogenases (Nases)

There are three distinct classes of metallonitrogenases available. The $N_2$ fixation process employs one of the metallonitrogenases, and all of them consist of Fe-S cluster-containing subunits. The regulatory genes that encoded these proteins are Nif, Vnf, and Anf. Presence of heterometal–containing protein cofactors makes them to differentiate from each other.

1. *Nif-encoded Nase*: It consists of Mo with Fe as metal cofactor. This type of Nase is also known as Mo-Nase.
2. *Vnf-encoded Nase*: It is also known as V-Nase because of the existence of vanadium and iron in the heterometal cofactor. Even though it is widely distributed, it can be expressed when the Mo nutrient is limited.

3. *Anf-encoded Nase*: During the starvation/absence of Mo and V, this kind of Nase is expressed. Anf-encoded Nase exists with active site of Fe and Fe-S cluster. This Nase is referred as Fe-Nase.

## Vanadium Nitrogenase

Vanadium nitrogenase (V-Nase) is an enzyme containing vanadium (V) as cofactor. It is available in Azotobacter such as *Azotobacter vinelandii* and *Azotobacter chroococcum* and some other species, such as *Rhodopseudomonas palustris* and *Anabaena variabilis*. The nitrogen-fixing bacteria use vanadium nitrogenase pathway as a secondary pathway when molybdenum (Mo) is unavailable to the primary molybdenum nitrogenase pathway.

## Structural Aspects of Vanadium Nitrogenase Metalloprotein

V-Nase is an iron-protein containing vanadium. A ferrodoxin like moiety in the protein acts as electron transport system. This is a specific homodimeric iron vanadium protein having two components. The first component is a single 4Fe-4S cluster (MW = 64 kDa) protein component, and the second component of it consists of VFe protein about 240 kDa (Crans et al. 2004). Two binding sites for MgATP exist in vanadium nitrogenase isolated from Azotobacter species. Further, V-Fe protein with $\alpha_2\beta_2\delta_2$ subunit structure is composed of two P clusters (in between the interface of $\alpha$ and $\beta$ subunits) and two M clusters (exist in the $\alpha$ subunits), where the M clusters act as active site (FeVco) for substrates and the P clusters have double-cubane structure $[\{Fe_4S_3 (C_4S)_2\}_2(M-S)_2(M-C_4S)_2]$.

## Comparison of V-Nase with Mo-Nase

Although, V-Nase is encoded by structural gene homologous to molybdenum nitrogenase (Mo-Nase)-encoding structural genes, but the enzyme is distinct from molybdenum-containing nitrogenase. V-Nase differs from the Mo-Nase in its subunit structure with additional $\delta$-subunits essential for V-Nase function. Thus, it has one $\alpha_2\beta_2\delta_2$ hexameric structure (Eady 2003). Stoichiometrically, metal contents (0.7–2 Vanadium atom, 9–19 Fe atom) vary for the vanadium-containing protein as compared to Mo-Nase in which those values are less. The redox center of the vanadium protein is however very similar to that of the Mo-protein. Chemical structures of the two cofactors (i.e., FeVCo, FeMoCo) are very similar. Distinct biosynthesis pathways are believed to produce those different cofactors in Azotobacter. In reaction with cyanide, V-Nase is found to be less sensitive as compared to Mo-Nase. The former enzyme suffers electron flux inhibition due to the presence of cyanide. Nevertheless, MgATP hydrolysis is not inhibited due to the presence of cyanide as it happens for the Mo-containing protein.

Role of vanadium or molybdenum in nitrogenase whether to bind substrate or to influence the continuation of reaction is yet to be elucidated. Replacement of molybdenum by vanadium shows that the enzyme is still capable of catalyzing reactions however with a lesser degree.

## Function of V-Nase

The enzyme involves in biological nitrogen fixation by reduction of nitrogen into ammonia. V-Nase has intrinsic hydrogenase activity, which results in production of hydrogen. The catalytic reactions of vanadium nitrogenase (Eady 2003) in *Azotobacter chroococcum* are given below.

$$N_2 + 12e^- + 12H^+ + 40MgATP$$
$$\rightarrow 2NH_3 + 3H_2 + 40MgADP + 40P_i$$

During catalysis, two protein components (iron protein with 4Fe-4S cluster and VFe protein) form a complex. An electron is passed from iron protein with 4Fe-4S cluster to VFe protein, which results in dissociation of complex along with the hydrolysis of MgATP. This causes reduction of the oxidized iron protein with 4Fe-4S cluster and charging with MgATP for a further electron transfer cycle. Other than nitrogen, V-Nase also can reduce acetylene and protons. Unlike molybdenum-containing enzyme, the V-Nase enzyme releases 4 electron-reduced products such as $N_2H_4$ during $N_2$ reduction and $C_2H_6$ during $C_2H_2$ reduction.

V-Nase involves in $N_2$ fixation process at ambient temperature and at 0.8 atm $N_2$ pressure where Mo is unavailable. The more active form of dinitrogen fixation can occur at a temperature 5°C below the ambient temperature, as at ambient temperature the V-Nase converts $H^+$ to $H_2$ which results in a less active process. When compared with Mo-Nase, the V-Nase requires more amount of $H_2$ and ATP to produce $NH_3$. V-Nase

**Vanadium Ions and Proteins, Distribution, Metabolism, and Biological Significance, Table 1** Differences between vanabins

| Types of vanabin | Constituents | Location in species | Dissociation constants($K_d$) | Binding ions | Amino acid sequences |
|---|---|---|---|---|---|
| Vanabin-1 | 12-Lysine and 18-cysteine | Cytoplasm and some organelles | $2.1 \times 10^{-5}$ M | Binds with 10 V(IV)ions | 87 amino acid sequences |
| Vanabin-2 | 14-Lysine and 18-cysteine | Cytoplasm and some organelles | $2.4 \times 10^{-5}$ M | 20 V(IV)ions | 91 amino acid sequences |
| Vanabin-3 | 18-Cysteine | Cytoplasm | – | – | – |
| Vanabin-4 | Cysteine | Cytoplasm | – | – | – |
| Vanabin P | – | Blood plasma | $2.8 \times 10^{-5}$ M | 13 V(IV) ions | 20 amino acid sequences |

function is generally carried out under anaerobic conditions. During this process, the V-Nase executes with protection mechanisms to exclude oxygen from the active site and from the redox active cofactor.

### Regulatory Gene of V-Nase

The regulatory gene Vnf-A is required for V-Nase. The binding site of Vnf-A has been identified as 5′-GTAC-N6-GTAC-3′. V-Nase gene has two independently regulated operons *Vnf* HFd and *Vnf* DGK. Subunits α and ß of V-Nase are encoded by *Vnf*DK, while the δ subunit of V-Nase is encoded by *Vnf*G. *Vnf*HFd and *Vnf*DGK operon expressions are independent from each other.

## Vanabins

Vanabin is a vanadium-binding protein isolated from some sea squirts, *Ascidian*. Some of such *Ascidians* like *Ascidia sydneiensis samea* and *Ascidian gammata* can accumulate very high content of vanadium (~350 mM) in them. Another living organism polychaete *Potamilla occelata* also accumulates a very high level of vanadium in a selective manner in their body.

The transport of vanadium from the coelomic fluid (blood plasma) into the vanadium accumulation cells (vanadocytes) takes place through the carrier proteins (vanabins) (Michibata et al. 2002). It is also referred as vanadium-binding proteins or vanadium chromagen. Generally, these types of carrier proteins are present in coelomic fluid of tunicates/sea squirts. The accumulation of vanadium in *Ascididae* family species is excess 350 mM which is about $10^7$-fold higher in concentration than that of vanadium in seawater (Crans et al. 2004). Generally, the *Ascididae* species have signet ring cells (the cell under the microscope looks like a signet ring, with the nucleus pushed over to one side as the stone in the ring and the remaining part of the "ring" resembles the cell membrane or boundary of the cell) with large vacuoles in which vanadium ions are deposited as a fraction of vanadium accumulated in the cells. These kinds of cells are referred as vanadocytes. Vanabins act more as metal carriers, called metallochaperones, than as proteins for metal storage or detoxification. There are different types of vanabins (Michibata et al. 2002) characterized so far, namely, vanabin-1 (12.5 kDa), vanabin-2(15 kDa), vanabin-3(16 kDa), vanabin-4, and vanabin P (recently characterized) (Yoshihara et al. 2005). Vanabin P is a soluble blood protein to carry vanadium through blood and synthesized in many different cells. The differences between the above-mentioned vanabins are listed below (Table 1).

Vanadium accumulation and reduction of vanadium are unusual mechanisms. The biological studies suggest that ferritin plays a wide role in the storage of vanadium in vanadocytes. Ferritin forms a shell-like structure with a hollow interior. It comprises of two types of 24 subunits, along with some of the other subunits (H-subunit).

In the Ascidiidae family species, vanadium is absorbed at +5 oxidation state and stored at +3 oxidation sate in vanadium storage cells (vanadocytes). Vanadium is initially reduced from +5 to +4 oxidation state in cytoplasm and then from +4 to +3 oxidation state in vacuoles (Michibata et al. 2002). NADPH available during pentose phosphate pathway plays a major role for the reduction of vanadium (V) to V (IV) in cytoplasm. It could be tempting to predict that the reduction of vanadium from +5 to +4 oxidation

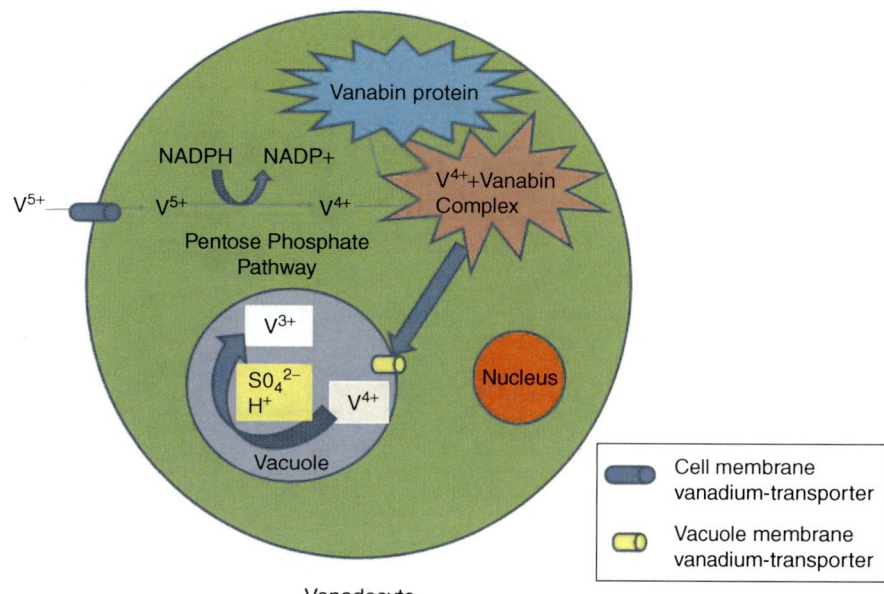

Vanadium Ions and Proteins, Distribution, Metabolism, and Biological Significance, **Fig. 6** Vanadium storage mechanism in Vanadocyte

state and binding to vanabin proteins in cytoplasm may be very much instantaneous. The vacuoles contain high concentrations of protons and sulfate ions which probably take part in the reduction process from +4 to +3 oxidation state (Fig. 6).

In humans, ferritin having subunit His[118] is vanadium-binding site. Generally, the oxygen-carrying proteins in blood (e.g., hemoglobin, hemocyanin, etc.) easily capture the oxygen molecules because of the existence of ferric or cupric ions. Based on existence of a metalloprotein, the color of blood varies from one species to another. Spider and horseshoe crab have blue blood due to the presence of copper-based hemocyanin in their blood. Humans and some other animals have red blood because of the existence of iron-based hemoglobin. Likewise, the vanadium-containing protein, hemovanabin, gives the yellow coloration to the blood (e.g., sea cucumber). However, whether or not vanabin involved in oxygen transport in the species with the vanadium-containing protein remains unknown.

## Vanadium Distribution

As discussed above, vanadium is carried by vanabins to vanadocytes in tunicates. However, in humans, vanadium absorption and distribution depend on the oxidative states of the element. Vanadium possesses an exceptional biochemical characteristic for its anionic and cationic forms, which take part in different biological process (Mukherjee et al. 2004). After absorption, when vanadate anions reach to blood stream, there they convert into vanadyl cations again. A little amount of vanadate can also exist. In the bloodstream, vanadates convert to vanadyl ions by the systemic reducing agents such as glutathione, NADH, ascorbic acid, and L-cysteine. These vanadyl ions bind with serum albumins and a few iron-containing proteins such as transferrin and ferritin. The protein-bound vanadyl ions are rapidly transported to the various tissues (Fig. 7). Upon vanadium supplementation, the vanadyl cations are transported to tissues such as liver, kidney, brain, heart, muscle, and bone at high concentrations. Inside the tissues, vanadyl may be converted back to vanadates by NADPH oxidation system in presence of oxygen. So the endogenous reducing agents and the dissolved oxygen regulate the vanadate cation and anion concentrations in serum, tissues, and cells. Bone is the major reservoir of vanadium. In the intracellular and the extracellular fluid, the vanadium cations and anions typically behave like a transitional metal ion and compete with another metal ion to combine with different biological ligands or compounds such as nucleotides, carbohydrates, and phosphates, to produce its biological effects.

The human serum transferrin binds with vanadium having different oxidation states (+3, +4, +5) at the two metal-binding sites of the protein. It was observed that

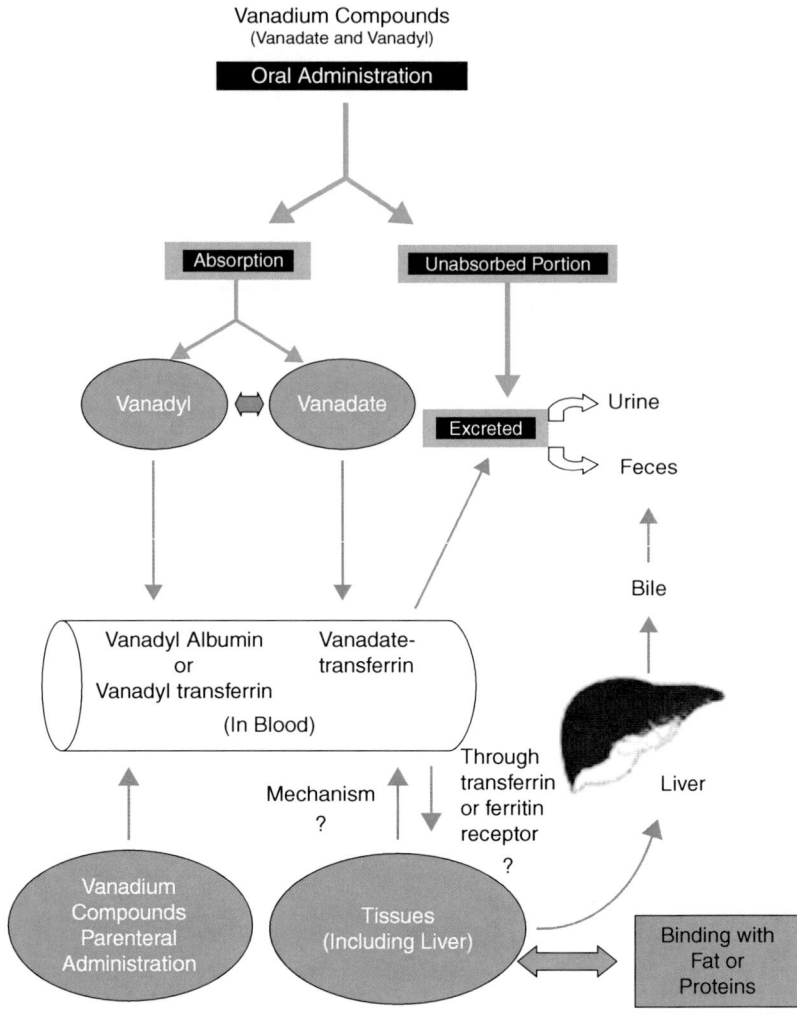

Vanadium Ions and Proteins, Distribution, Metabolism, and Biological Significance, Fig. 7 Major routes of absorption, distribution, and excretion of vanadium ions (Mukherjee B. et al. (2004). Toxicology Letters 150: 135–143, Copyright @ELSEVIER)

vanadium binds with the phenolic group containing amino acid at pH 9.5.

Some studies show that $VO^{2+}$ makes complexes with bovine serum albumin system at the site specific for $Cu^{++}$, which is found at the N-terminal of polypeptide chain. At the carboxylic group of the protein, there are also four to five weaker binding sites for $VO^{2+}$.

Vanadium ion has similarity with phosphate ions ($VO_4^{3-}$ and $PO_4^{3-}$), so the mechanism of accumulation of vanadates to bone and other tissue has resemblance with accumulation of phosphate to those tissues. For these same incorporation sites for phosphate, high concentrations of vanadium in bone and skeletal muscle are reported. However, phosphate is involved only to the surface interaction of $VO^{2+}$ with bone, and it is absorbed on the material surface not incorporated into the apatite lattice. It also interacts with the tropocollagen and the compounds of organic matrix of bone such as chondroitin sulfate A, a mucopolysaccharide present in bone and connective tissue.

### Vanadium Metabolism

As vanadium has different oxidation states, its absorption depends on them. All vanadate ions in dietary components are reduced to vanadyl ions in the acidic pH of stomach. Dietary vanadate anions are well absorbed through the phosphate transport pathway as vanadium has similarity with phosphate. The reduced vanadyl cations remain unabsorbed and well associated with the dietary fibers to excrete unchanged (Mukherjee et al. 2004). Therefore, vanadium anions are three to five times more efficiently absorbed through biological membrane. Studies in rats have shown that vanadium absorption occurs in gastrointestinal tract from 10% to 40%, and the remaining unabsorbed vanadyl ions are excreted with feces by binding with dietary fibers.

Metabolisms of vanadium include different oxidation-reduction reactions of vanadium cations and anions, different complex formation during the transport to different tissues and the accumulation, as well as precipitation reaction to the tissues. The different pathways to reduce vanadium are:

1. Reduction reaction of vanadate ions with glutathione
2. Complexes with L-cysteine and vanadate anions
3. Complexes of L-ascorbic acid with vanadates (+5)
4. Complexes with transferrin and albumin
5. Deposition of vanadium ions in several tissues

### Reduction Reaction of Vanadate Ions with Glutathione

Reduced glutathione has a very important biological role for metabolism of vanadium, as well as detoxification process. It reduces the vanadate anions to vanadyl ion ($VO^{2+}$) within a pH range from 5 to 7.5 and ligand to metal ratio from 10 to 140. The oxidized glutathione (GSSG) has also some action on vanadyl cations. In the pH range 6–7, GSSG is a powerful binder of oxovanadium (+4) than GSH (reduced glutathione). Therefore, both GSH and GSSG take part in stabilizing and transporting of $VO^{2+}$ just after its initial GSH-mediated reduction in the plasma.

### Complexes with L-cysteine and Vanadate Anions

L-cysteine is another powerful reducing agent for vanadates in the body fluid in any pH range. At pH 6–8 and a ligand to metal ratio of 2:1, the reaction may lead to a purple color formation. It is thought that $VO^{2+}$ cation reacts with nitrogen atom of amino group and the deprotonated SH group of amino acid molecules.

### Complexes of L-Ascorbic Acid with Vanadates (+5)

Another natural reducing agent of the biological system is L-ascorbic acid, which reduces vanadates (+5) to oxovanadium (+4) and then forms complexes with the reduced products. L-ascorbic acid gets oxidized itself to dehydroascorbic acid and interacts with $VO^{2+}$, hydrolyses irreversibly with the opening of the lactone ring. It forms a 2:1 ligand to metal complex, where enolized form of 2,3-diketo gluconic acid is bound. A microcrystalline powder can be obtained by directly interacting with sodium metavanadate and ascorbic acid.

### Excretion

Vanadium excretes via bile and urine. Unabsorbed vanadium ions rapidly excrete via feces. Vanadyl strongly associates with dietary fibers and facilitates the fecal excretion of vanadium.

The absorbed vanadium after making complexes with different ligands accumulates in different tissues. The low molecular weight vanadium complexes accumulate in kidneys and finally excrete through urine, whereas high molecular weight protein complexes of vanadium accumulate in liver and finally excrete with feces.

## Biological Significance

In humans, vanadium enters to the body through diet, drinking water, or by inhalation. This trace element in human body with concentration in human blood is about 10–100 μg/mL. Vanadium plays a role in metabolism of carbohydrates and may have functions in cholesterol and blood lipid metabolism.

Many animal studies show that in diabetes, vanadium supplement may have a positive effect in regulating blood glucose levels.

Vanadium mimics the insulin activity. It stimulates glucose uptake into the cells, enhances glucose metabolism, and inhibits catecholamine-induced lipolysis in adipose tissue. It also stimulates glycogen synthesis in liver and inhibits gluconeogenesis.

The studies showed that vanadium prevents heart attack. It inhibits the formation of cholesterol in blood vessels. In animal studies, it shows that vanadium was also involved in building bones and teeth and formation of erythrocytes and thyroid functions.

In animal studies, vanadium deficiency is associated with stunted growth, impaired reproduction, altered red blood cell formation, and iron metabolism.

Vanadium is a relatively controversial dietary supplement for its increasing insulin sensitivity mechanism and for use in bodybuilding.

Vanadyl sulfate is itself important for normal cell function and development. It may help maintaining blood sugar levels already in normal range.

Vanadium appears to interfere with enzymes like different ATPases, protein kinases, ribonucleases, and phosphatases. Several genes seem to be regulated by vanadium ions or by its salts. Some such genes are tumor necrosis factor alpha (TNF-α), interleukin-8 (IL-8), activator protein-1 (AP-1), ras, c-raf.1, mitogen-activated protein kinase (MAPK), p53, etc. All these genes are potential oncogenes, which reflect that vanadium is also emerging as a potent anticarcinogenic agent.

Vanadium plays an important role in ocean environments than on land as its complexes make different proteins and essential enzymes of different sea animals. Ascidiacea (sea squirts) and tunicates such as bluebell tunicate contain vanadium in vanabin, a vanadium chromagen protein, in their blood cells. Vanadium accumulates in high concentration in holothurian (sea cucumber) blood. Amanita muscaria and the related species accumulate vanadium (up to 500 mg/kg in dry weight) to form the fungal fruit bodies amavadin which has the toxin functions or peroxidase enzyme functions. Bromoperoxidases accumulate vanadium as organobromine compounds in algae and act as a vanadium-dependent bromoperoxidase. Organobromine compounds in the sea are synthesized through the action of vanadium-dependent bromoperoxidase.

Some nitrogen-fixing microbes, Azotobacter, also use vanadium-dependent nitrogenase enzymes. Sodium orthovanadate ($Na_3VO_4$) acts as an inhibitor of protein tyrosine phosphatases, alkaline phosphatases, and ATPases and as a phosphate analogue. The $VO_4^{3-}$ ion binds reversibly to the active sites of most protein tyrosine phosphatases. It inhibits endogenous phosphatases present in cell lysate mixture. It is generally used in a concentration range of 1–10 μM. Thus, to prevent the dephosphorylation of protein(s) of interest, $Na_3VO_4$ is added to buffer solutions to be used in protein analysis in molecular biology.

## Cross-References

▶ Lanthanum, Physical and Chemical Properties
▶ Vanadium Ions and Proteins, Distribution, Metabolism, and Biological Significance

## References

Crans DC, Smee JJ, Gaidamauskas E, Yang L (2004) The chemistry and biochemistry of vanadium and the biological activities exerted by vanadium compounds. Chem Rev 104:849–902

Eady RR (2003) Current status of structure function relationships of vanadium nitrogenase. Coord Chem Rev 237:23–30

Lide DR (2004) Vanadium. In: CRC handbook of chemistry and physics. CRC Press, Boca Raton, pp 4–44

Michibata H, Uyama T, Ueki T, Kanamori K (2002) Vanadocytes, cells hold the key to resolving the highly selective accumulation and reduction of vanadium in ascidians. Microsc Res Tech 56:421–434

Mukherjee B, Patra B, Mahapatra S, Banerjee P, Tiwari A, Chatterjee M (2004) Vanadium-an element of atypical biological significance. Toxicol Lett 150:135–143

Winter JM, Moore BS (2009) Exploring the chemistry and biology of vanadium-dependent haloperoxidases. J Biol Chem 284:18577–18581

Yoshihara M, Ueki T, Watanabe T, Yamaguchi N, Kamino K, Michibata H (2005) VanadiumP, a novel vanadium-binding protein in the blood plasma of an ascidian, *Ascidia sydneiensis samea*. Biochim Biophys Acta 1730:206–214

# Vanadium Metal and Compounds, Properties, Interactions, and Applications

Bonex W. Mwakikunga
Council for Scientific and Industrial Research, National Centre for Nano–Structured, Pretoria, South Africa
Department of Physics and Biochemical Sciences, University of Malawi, The Malawi Polytechnic, Chichiri, Blantyre, Malawi

## Synonyms

Erythronium; Panchromium

## Definition

Vanadium, symbol V and atomic number 23, a soft, silvery gray, ductile transition metal is found only in chemically combined form in nature. In 1831, Nils Gabriel Sefström named the element as vanadium after the Scandinavian goddess of beauty and fertility, Vanadis (Freya), attributing to the wide range of colors found in vanadium compounds for its vanadium content. The element occurs naturally in about 65 different minerals and in fossil fuel deposits. Large amounts of vanadium ions are found in a few organisms particularly in the ocean as an active center of enzymes such as the vanadium bromoperoxidase of some ocean algae. Vanadium is probably a micronutrient in mammals, including humans, but its precise role in this regard is yet to be properly elucidated.

### 23: Vanadium          2,8,11,2

**Vanadium Metal and Compounds, Properties, Interactions, and Applications, Fig. 1** Electron shell diagram for a vanadium atom (Commons, Wikimedia)

## General Properties

| | |
|---|---|
| Name, symbol, number | Vanadium, V, 23 |
| Element category | Transition metal |
| Group, period, block | 5, 4, d |
| Standard atomic weight | 50.9415 g·mol$^{-1}$ |
| Electron configuration | [Ar] 3d$^3$ 4s$^2$ |
| Electrons per shell | 2, 8, 11, 2 (Fig. 1) |

## Physical Properties

| | | | | | | |
|---|---|---|---|---|---|---|
| Phase | Solid | | | | | |
| Density (near r.t.) | 6.0 g·cm$^{-3}$ | | | | | |
| Liquid density at m.p. | 5.5 g·cm$^{-3}$ | | | | | |
| Melting point | 2,183°K, 1,910°C, 3,470°F | | | | | |
| Boiling point | 3,680°K, 3,407°C, 6,165°F | | | | | |
| Heat of fusion | 21.5 kJ·mol$^{-1}$ | | | | | |
| Heat of vaporization | 459 kJ·mol$^{-1}$ | | | | | |
| Specific heat capacity | (25°C) 24.89 J·mol$^{-1}$·K$^{-1}$ | | | | | |
| Vapor pressure | P (Pa) | 1 | 10 | 100 | 1 k | 10 k | 100 k |
| | at T (K) | 2,101 | 2,289 | 2,523 | 2,814 | 3,187 | 3,679 |

## Miscellaneous Properties

| | |
|---|---|
| Crystal structure | Body-centered cubic |
| Magnetic ordering | Paramagnetic |
| Electrical resistivity | (20°C) 197 nΩ·m |
| Thermal conductivity | (300°K) 30.7 W·m$^{-1}$·K$^{-1}$ |
| Thermal expansion | (25°C) 8.4 μm·m$^{-1}$·K$^{-1}$ |
| Speed of sound (thin rod) | (20°C) 4,560 m/s |
| Young's modulus | 128 GPa |
| Shear modulus | 47 GPa |
| Bulk modulus | 160 GPa |
| Poisson ratio | 0.37 |
| Mohs hardness | 6.7 |
| Chemical Abstracts Service registry number | 7440-62-2 |

## Atomic Properties

| | |
|---|---|
| Oxidation states | 5, 4, 3, 2, 1, −1 (Amphoteric oxide) |
| Electronegativity | 1.63 (Pauling scale) |
| Ionization energies (more) | 1st: 650.9 kJ·mol$^{-1}$ |
| | 2nd: 1,414 kJ·mol$^{-1}$ |
| | 3rd: 2,830 kJ·mol$^{-1}$ |
| Atomic radius | 134 pm |
| Covalent radius | 153 ± 8 pm |

## Most Stable Isotopes

Main article: Isotopes of vanadium

| Iso | NA | Half-life | DM | DE (MeV) | DP |
|---|---|---|---|---|---|
| $^{48}$V | Syn | 15.9735 d | ε + β$^+$ | 4.0123 | $^{48}$Ti |
| $^{49}$V | Syn | 330 d | ε | 0.6019 | $^{49}$Ti |
| $^{50}$V | 0.25% | 1.5 × 10$^{17}$y | ε | 2.2083 | $^{50}$Ti |
| | | | β$^-$ | 1.0369 | $^{50}$Cr |
| $^{51}$V | 99.75% | $^{51}$V is stable with 28 neutrons | | | |

## About the Metal

Vanadium was originally discovered by Andrés Manuel del Río, a Spanish-born Mexican mineralogist, in 1801. Del Río extracted the element from a sample of Mexican "brown lead" ore, later named vanadinite. Since its salts exhibit a wide variety of colors, he named the element *panchromium* (which means "all colors" in Greek). Later, Del Río renamed the element *erythronium* as most of its salts turned red upon heating.

In 1831, the Swedish chemist Nils Gabriel Sefström rediscovered the element in a new oxide while working with iron ores. In the same year, Friedrich Wöhler confirmed del Río's earlier work. Sefström chose a name beginning with V, which had not been assigned to any element yet. He called the element *vanadium* after Old Norse *Vanadís* (another name for the Norse Vanir goddess Freyja, whose facets include connections to beauty and fertility), because of the many beautifully colored chemical compounds it produces (Commons, Wikimedia).

**Vanadium Metal and Compounds, Properties, Interactions, and Applications, Fig. 2** The *blue-silver-grey* metal (Commons, Wikimedia)

## Characteristics

This soft, ductile, silver-gray metal (Fig. 2) has good resistance to corrosion, and it is stable against alkalis, sulfuric, and hydrochloric acids. It is oxidized in air at about 933 K (660°C, 1,220°F), although an oxide layer forms even at room temperature.

## Isotopes

Naturally occurring vanadium is composed of one stable isotope $^{51}V$ and one radioactive isotope $^{50}V$. The latter has a half-life of $1.5 \times 10^{17}$ years and a natural abundance of 0.25%. $^{51}V$ has a nuclear spin of 7/2 which is useful for NMR spectroscopy. Twenty-four artificial radioisotopes have been characterized, ranging in mass number from 40 to 65. The most stable of these isotopes are $^{49}V$ with a half-life of 330 days and $^{48}V$ with a half-life of 16.0 days. All other vanadium isotopes have half-lives shorter than an hour, most of which are below 10 s. At least four isotopes have metastable excited states. Electron capture is the main decay mode for isotopes lighter than the $^{51}V$. For the heavier ones, the most common mode is beta decay. The electron capture reactions lead to the formation of element 22 (titanium) isotopes, while for beta decay, it leads to element 24 (chromium) isotopes.

## Chemistry and Compounds

The chemistry of vanadium is noteworthy for the accessibility of four adjacent oxidation states. The common oxidation states of vanadium are +2 (lilac), +3 (green), +4 (blue), and +5 (yellow) (Fig. 3). Vanadium(II) compounds are reducing agents, and vanadium(V) compounds are oxidizing agents. Vanadium(IV) compounds often exist as vanadyl derivatives which contain the $VO^{2+}$ center.

Ammonium metavanadate(V) ($NH_4VO_3$) (Fig. 4) can be successively reduced with elemental zinc to obtain the different colors of vanadium in these four oxidation states. Lower oxidation states occur in compounds such as $V(CO)_6$, $[V(CO)_6]^-$ and substituted derivatives.

The vanadium redox battery utilizes these oxidation states; conversion of these oxidation states is illustrated by the reduction of a strongly acidic solution of a vanadium(V) compound with zinc dust. The initial yellow color characteristic of the vanadate ion, $VO_4^{3-}$, is replaced by the blue color of $[VO(H_2O)_5]^{2+}$, followed by the green color of $[V(H_2O)_6]^{3+}$, and then violet, due to $[V(H_2O)_6]^{2+}$.

**Vanadium Metal and Compounds, Properties, Interactions, and Applications, Fig. 3** Changes of color in solution of vanadium in different oxidation states, from *left* +2 (*lilac*), +3 (*green*), +4 (*blue*), and +5 (*yellow*) (Commons, Wikimedia)

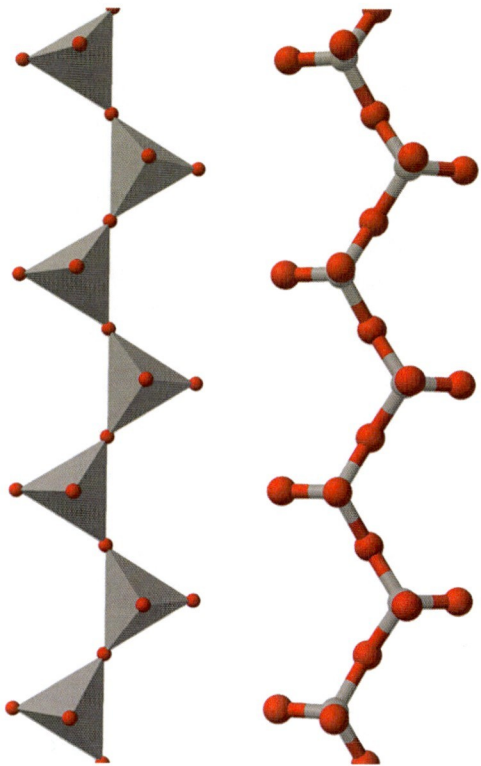

**Vanadium Metal and Compounds, Properties, Interactions, and Applications, Fig. 4** Metavanadate chains – (*left*) the *red spheres* on the apexes of the trihedra pyramids represent oxygen atoms; the vanadium atoms are at the middle of the trihedral pyramids. (*Right*) the *red spheres* represent oxygen, whereas the *gray spheres* represent vanadium atoms (Commons, Wikimedia)

The most commercially important compound is vanadium pentoxide, which is used as a catalyst for the production of sulfuric acid. This compound oxidizes sulfur dioxide ($SO_2$) to the trioxide ($SO_3$). In this redox reaction, sulfur is oxidized from +4 to +6, and vanadium is reduced from +5 to +3:

$$V_2O_5 + 2\ SO_2 \rightarrow V_2O_3 + 2\ SO_3$$

The catalyst is regenerated by oxidation with air:

$$V_2O_3 + O_2 \rightarrow V_2O_5$$

## Oxy and Oxo Compounds

The oxyanion chemistry of vanadium (V) is complex (Fig. 5). The vanadate ion, $VO_4^{3-}$, is present in dilute solutions at high pH. On acidification, $HVO_4^{2-}$ and $H_2VO_4^-$ are formed, analogous to $HPO_4^{2-}$ and $H_2PO_4^-$. The acid dissociation constants for the vanadium and phosphorus series are remarkably similar. In more concentrated solutions, many polyvanadates are formed. Chains, rings, and clusters involving tetrahedral vanadium, analogous to the polyphosphates, are known. In addition, clusters such as the decavanadates $V_{10}O_{28}^{4-}$ and $HV_{10}O_{28}^{3-}$, which predominate at pH 4–6, are formed in which compound is octahedral about vanadium (Commons, Wikimedia).

The correspondence between vanadate and phosphate chemistry can be attributed to the similarity in size and charge of phosphorus(V) and vanadium(V). Orthovanadate $VO_4^{3-}$ is used in protein crystallography to study the biochemistry of phosphate.

Vanadium also forms various peroxo- complexes when treated with hydrogen peroxide. For instance,

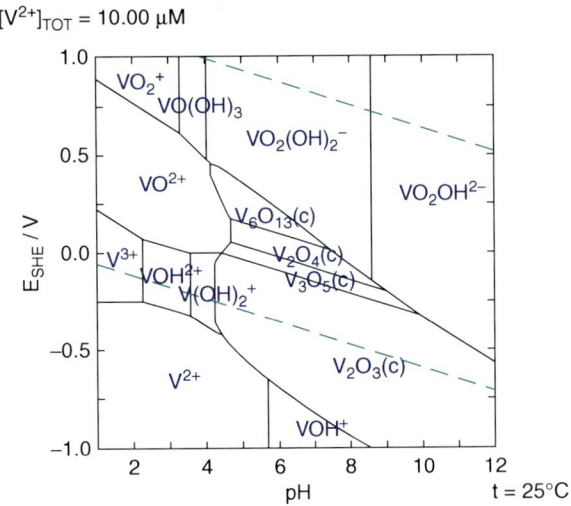

**Vanadium Metal and Compounds, Properties, Interactions, and Applications, Fig. 5** The Pourbaix diagram for vanadium in water (Commons, Wikimedia)

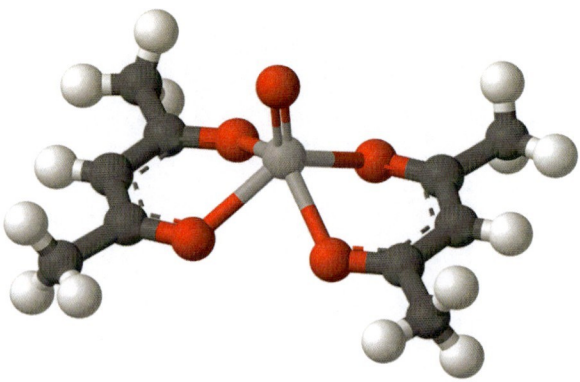

**Vanadium Metal and Compounds, Properties, Interactions, and Applications, Fig. 6** A ball-and-stick model of VO(acac)$_2$ (Commons, Wikimedia)

the yellow oxovanadium(V) ion VO$^{+2}$ in acidic hydrogen peroxide solution forms the brick red peroxovanadium(V) ion, VO(O$_2$)$_2^+$.

## Halide Compounds

Several halides are known for oxidation states +2, +3, and +4. VCl$_4$ is the most important commercially. This liquid is mainly used as a catalyst for polymerization of dienes (Commons, Wikimedia).

## Coordination Compounds

Vanadium's early position in the transition metal series leads to three rather unusual features of the coordination chemistry of vanadium. Firstly, metallic vanadium has the electronic configuration [Ar]4s$^2$3d$^3$, so compounds of vanadium are relatively electron-poor. Consequently, most binary compounds are Lewis acids (electron pair acceptors); examples are all the halides forming octahedral adducts with the formula VX$_n$L$_{6-n}$ (X = halide; L = other ligand). Secondly, the vanadium ion is rather large and can achieve coordination numbers higher than 6, as is the case in [V(CN)$_7$]$^{4-}$. Thirdly, the vanadyl ion, VO$^{2+}$, is featured in many complexes of vanadium(IV) such as vanadyl acetylacetonate (V(=O)(acac)$_2$) (Fig. 6). In this complex, the vanadium is 5-coordinate, square pyramidal, meaning that a sixth ligand, such as pyridine, may be attached, though the association constant of this process is small. Many 5-coordinate vanadyl complexes have a trigonal bypyramidal geometry, such as VOCl$_2$(NMe$_3$)$_2$.

## Organometallic Compounds

Organometallic chemistry of vanadium is well developed, but organometallic compounds are of minor commercial significance. Vanadocene dichloride is a versatile starting reagent and even finds minor applications in organic chemistry. Vanadium carbonyl, V(CO)$_6$, is a rare example of a metal carbonyl containing an unpaired electron but which exists without dimerization. The addition of an electron yields V(CO)$_6^-$ (isoelectronic with Cr(CO)$_6$), which may be further reduced with sodium in liquid ammonia to yield V(CO)$_6^{3-}$ (isoelectronic with Fe(CO)$_5$) (Commons, Wikimedia).

## Biological Role

Vanadium plays a very limited role in biology and is more important in ocean environments than on land.

## Bromoperoxidases in Algae

Organobromine compounds in a number of species of marine algae are generated by the action of a vanadium-dependent bromoperoxidase. This is a haloperoxidase in algae which requires bromide and is an absolutely vanadium-dependent enzyme. Most organobromine compounds in the sea ultimately arise via the action of this vanadium bromoperoxidase.

## Vanabins in Tunicates and Ascidians

In 1911, German chemist Martin Henze discovered vanadium in the blood cells (or coelomic cells) of Ascidiacea (sea squirts) (Fig. 7) where it is essential to ascidians and tunicates (Fig. 8), as vanabins (vanadium chromagen proteins) (Michibata et al. 2002). The concentration of vanadium in their blood is more than 100 times higher than the concentration of vanadium in the seawater around them. Vanadium has been reported in high concentrations in holothurian (sea cucumber) blood. However, other researchers have been unable to reproduce these results. There is no evidence that hemovanadin carries oxygen, in contrast to hemoglobin and hemocyanin, which may also be present in these organisms.

## Nitrogen Fixation

A vanadium nitrogenase is used by some nitrogen-fixing microorganisms, such as *Azotobacter*. In this role, vanadium replaces more common molybdenum or iron and gives the nitrogenase slightly different properties.

## Fungi

Several species of macrofungi, namely, *Amanita muscaria* (Fig. 9), and related species accumulate vanadium (up to 500 mg/kg in dry weight). Vanadium is present in the coordination complex, amavadin, in fungal fruit bodies. However, the biological importance of the accumulation process is unknown. Toxin functions or peroxidase enzyme functions have been suggested.

**Vanadium Metal and Compounds, Properties, Interactions, and Applications, Fig. 7** *Ascidiacea* (sea squirts) contain vanadium as vanabin (Commons, Wikimedia)

**Vanadium Metal and Compounds, Properties, Interactions, and Applications, Fig. 8** Tunicates such as this bluebell tunicate contains vanadium as vanabin (Commons, Wikimedia)

## Mammals and Birds

Rats and chickens are also known to require vanadium in very small amounts, and deficiencies result in reduced growth and impaired reproduction (Soazo and Garcia 2007). Vanadium is a relatively controversial dietary supplement, primarily for increasing insulin sensitivity and body building. Whether it works for the latter purpose has not been proven, and there is some evidence that athletes who take it merely experience a placebo effect. Vanadyl sulfate may improve

**Vanadium Metal and Compounds, Properties, Interactions, and Applications, Fig. 9** *Amanita muscaria* contains amavadin (Commons, Wikimedia)

glucose control in people with type 2 diabetes. In addition, decavanadate and oxovanadates are species that potentially have many biological activities and that have been successfully used as tools in the comprehension of several biochemical processes (Aureliano and Crans 2009).

## Safety

All vanadium compounds should be considered to be toxic (Manfred 2004). Tetravalent $VOSO_4$ has been reported to be over five times more toxic than trivalent $V_2O_3$. The Occupational Safety and Health Administration has set an exposure limit of 0.05 mg/m$^3$ for vanadium pentoxide dust and 0.1 mg/m$^3$ for vanadium pentoxide fumes in workplace air for an 8-h workday, 40-h workweek. The National Institute for Occupational Safety and Health has recommended that 35 mg/m$^3$ of vanadium be considered immediately dangerous to life and health. This is the exposure level of a chemical that is likely to cause permanent health problems or death.

Vanadium compounds are poorly absorbed through the gastrointestinal system. Inhalation exposures to vanadium and vanadium compounds result primarily in adverse effects on the respiratory system. Quantitative data are, however, insufficient to derive a subchronic or chronic inhalation reference dose. Other effects have been reported after oral or inhalation exposures on blood parameters, on liver, on neurological development in rats, and other organs.

There is little evidence that vanadium or vanadium compounds are reproductive toxins or teratogens. Vanadium pentoxide was reported to be carcinogenic in male rats and male and female mice by inhalation in a National Toxicological Program study, although the interpretation of the results has recently been disputed. Vanadium has not been classified to carcinogenicity by the United States Environmental Protection Agency (Duffus 2007).

Vanadium traces in diesel fuels present a corrosion hazard; it is the main fuel component influencing high temperature corrosion. During combustion, it oxidizes and reacts with sodium and sulfur, yielding vanadate compounds with melting points down to 530°C, which attack the passivation layer on steel, rendering it susceptible to corrosion. The solid vanadium compounds also cause abrasion of engine components.

## Vanadium and Diabetes

The insulin-like properties of vanadium salts have attracted much attention firstly, due to its promising role in the management of diabetes in the cases of insulin resistance and insulin deficiency (Chen and Tan 2012) (Fig. 10) and secondly, due to its usefulness in elucidating the molecular mechanism underlying pathology in diabetes. Despite the benefits of insulin therapy, especially in insulin-dependent diabetes mellitus patients, many of those suffering from diabetes have found that insulin is far from being an ideal drug, primarily because of the frequent incidences of insulin resistance. In addition, oral drug therapy aimed at controlling hyperglycemia in non-insulin-dependent diabetes mellitus patients often fails, and most patients require insulin treatment late in the course of their disease (Boden et al. 1996). This progressive deterioration in glucose metabolism is due, in part, to worsening insulin sensitivity that may be ameliorated by the glucose-lowering effect of exogenous insulin therapy. Therefore, agents that could lower the requirements for insulin or augment insulin sensitivity may be useful in treatment of both forms of diabetes mellitus. Another important reason to search for new antidiabetic treatments has to do with the evolving definition of diabetes, which is now perceived not so much as a "high blood sugar" disease but a "small vessel" disease.

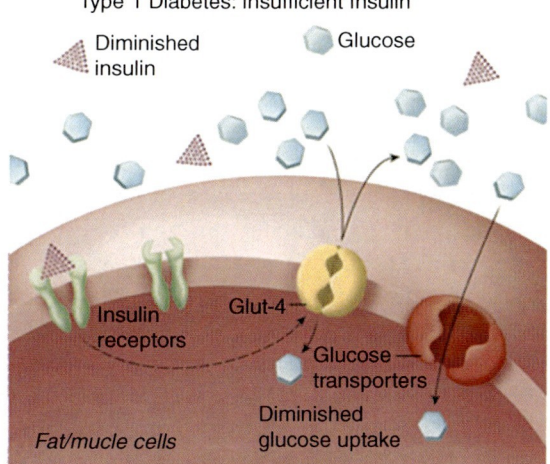

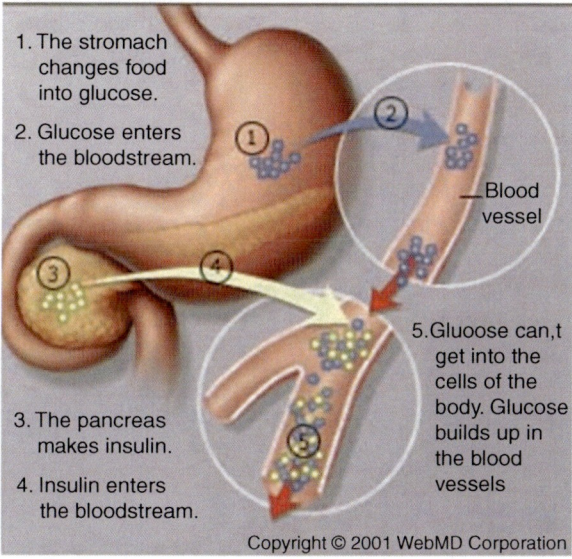

**Vanadium Metal and Compounds, Properties, Interactions, and Applications, Fig. 10** The causes of diabetes type 1 and type 2 illustrated (http://www.medicalook.com/Diabetes/Type_1_Diabetes.html)

Vanadium as a sodium vanadate was first considered therapeutically useful in the management of a variety of infectious and debilitating diseases, including diabetes, in the late nineteenth century. This early 1899 observation revealed that the administration of sodium vanadate to diabetic patients resulted in a decrease in urinary glucose levels. The insulin mimetic action of vanadium was first reported in in vitro experimental model in 1979 by Tolman and colleagues (Thompson and Orvig 2006).

In 1985, Heyliger and colleagues published a similar observation derived from their in vivo experiments. Since those discoveries, vanadium has been considered as an insulin mimicking compound, which may control glucose metabolism by an insulin-dependent and/or insulin independent biochemical pathway(s).

Vanadium belongs to the biologically important group of transition elements. These elements have the ability to exist in a number of different oxidation states and possess the tendency to form complex ions. This latter property enables these elements to form biologically important complexes called coordination compounds with organic carriers, such as proteins. Vanadium has an extremely complex chemistry; it can readily alter its oxidation state and can exist in an anionic or cationic form. The +5 oxidation state (anionic form) is predominant under physiological conditions. Under these conditions, vanadium exists as metavanadate ($VO^-$) or orthovanadate ($H_2VO^-$), which resembles phosphate. The +4 oxidation state manifests in the cationic vanadyl form ($VO^-$), which resembles $Mg^-$. Coordination compounds play several important roles in plants and animals. Functionalities identified include:

- Storage and transportation of oxygen, e.g., iron heme complex in hemoglobin
- Role as electron transfer agents, e.g., iron heme complex in cytochromes
- Role as catalysts, e.g., cobalt in cobalamin
- Role in photosynthesis, e.g., magnesium in chlorophyll

Vanadium is relatively abundant in nature (approximately 0.02%).

Theoretically, vanadium meets several criteria for being considered an essential nutrient (Jan et al. 2012):

- Low molecular weight (atomic weight 50.94)
- Excellent catalytic activity
- Ubiquity in the geosphere and possibly in the biosphere
- Homeostatic regulation by controlled accumulation
- Low toxicity on oral intake (Manfred 2004)

The influence of vanadium on ion transport pathways in biological systems was first demonstrated in 1977. Vanadium was shown to inhibit $Na^+/K^+$ transport adenosine triphosphatase (ATPase) in vitro. Vanadium salts can also inhibit other transport ATPases. However, vanadium deprivation or its physiological essentiality on impaired reproductive performance was yet to be established (Thompson and Orvig 2006).

## Cross-References

▶ Vanadium in Biological Systems
▶ Vanadium Pentoxide Effects on Lungs

## References

Aureliano M, Crans DC (2009) Decavanadate and oxovanadates: oxometalates with many biological activities. J Inorg Biochem 103:536–546

Boden G, Chen X, Ruiz J, van Rossum GD, Turco S (1996) Effects of vanadyl sulfate on carbohydrate and lipid metabolism in patients with non-insulin-dependent diabetes mellitus. Metabolism 45:1130–1135

Chen H, Tan C (2012) Prediction of type-2 diabetes based on several element levels in blood and chemometrics. Biol Trace Elem Res 147(1–3):67–74

Commons, Wikimedia. http://commons.wikimedia.org/wiki/Commons:Reusing_content_outside_Wikimedia

Duffus JH (2007) Carcinogenicity classification of vanadium pentoxide and inorganic vanadium compounds, the NTP study of carcinogenicity of inhaled vanadium pentoxide, and vanadium chemistry. Regul Toxicol Pharmacol 47(1):110–114

Jan K, Irena BB, Izabela G (2012) Biochemical and medical importance of vanadium compounds. Acta Biochim Pol 59(2):195–200

Manfred A (2004) Vanadium – an element both essential and toxic to plants, animals and humans? Anal Real Acad Nac Farm 70:961

Michibata H, Uyama T, Ueki T, Kanamori K (2002) Vanadocytes, cells hold the key to resolving the highly selective accumulation and reduction of vanadium in ascidians. Microsc Res Tech 56(6):421–434

Soazo M, Garcia GB (2007) Vanadium exposure through lactation produces behavioral alterations and CNS myelin deficit in neonatal rats. Neurotoxicol Teratol 29(4):503–510

Thompson KH, Orvig C (2006) Vanadium in diabetes: 100 years from Phase 0 to Phase I. J Inorg Biochem 100:1925–1935

# Vanadium Pentoxide Effects on Lungs

Lisa K. Ryan
New Jersey Medical School, The Public Health Research Institute, University of Medicine and Dentistry of New Jersey, Newark, NJ, USA

## Synonyms

Particulate matter; $PM_{2.5}$; $V_2O_5$; Vanadium

## Vanadium Pentoxide

Vanadium (V) is a transition metal that forms compounds with several valence states: $1^+$, $2^+$, $3^+$, $4^+$, and $5^+$, with the latter three being the most common. The effects of vanadium compounds on human health can be beneficial or harmful depending on the valence state of the vanadium, the solubility, the route of exposure, absorption, and concentration of the compound. Vanadium pentoxide ($V_2O_5$) is one of the most common vanadium compounds. It is the main form of vanadium in particulate matter generated from the burning of oil-fueled power plant emissions (Ghio et al. 2002; Cohen 2004) and has been studied extensively for its detrimental effects on human health, mainly in the lung.

## Chronic Bronchitis, Asthma, Fibrosis, and Airway Remodeling

$V_2O_5$ airway disease associations with bronchitis, fibrosis, and asthma have been reported following both environmental and occupational exposures to this compound (Ghio et al. 2002; Cohen 2004). To understand these disease associations, a rat animal model of inhalation exposure to $V_2O_5$ was developed and demonstrated toxic effects including mucous cell hyperplasia, increased airway smooth muscle mass, and peribronchiolar fibrosis (Bonner 2007). The lung myofibroblasts underwent hyperplasia during fibrogenesis following exposure to $V_2O_5$. Interleukin (IL)-1ß mediated the induction of platelet-derived growth factor receptor-α (PDGF-Rα), which is important in promoting lung myofibroblast hyperplasia during fibrogenesis (Bonner 2007).

The inflammation, fibrosis, and airway remodeling that occur in chronic bronchitis following vanadyl compound exposure have been studied extensively using in vitro studies with human lung fibroblasts. Genomic analysis of human lung fibroblasts exposed to vanadium pentoxide to identify possible mediators involved in the pathogenesis of bronchitis revealed that about 1,400 genes were modulated by 10 $\mu g/cm^2$ $V_2O_5$ (Ingram et al. 2007). Of these, 300 genes had induced expression, while about 1,100 genes had suppressed expression compared with untreated fibroblasts. These represented a wide variety of candidate genes that could mediate $V_2O_5$-induced airway remodeling following exposure. Key genes that were induced were

IL-8, which is involved in neutrophil recruitment in inflammation, PTGS2 (COX-2), which is involved in protection from lung fibrosis, and the genes that code for the peroxide-generating enzymes superoxide dismutase 2 (SOD2) and L-pipecolate oxidase (PIPOX). In addition, interferon-inducible, STAT1-dependent chemokine genes (CXCL9 or MIG and CXCL10), or interferon gamma–induced protein 10 (IP-10) involved in lymphocyte accumulation and dampening of fibroblast proliferation were induced. Surprisingly, transforming growth factor (TGF) β2 and its signaling intermediates SMAD1 and SMURF1 were suppressed along with several major collagen genes (COL1A2, COL1A1, COL3A1), indicating that the fibroblasts are not responsible for generating their own collagen in response to $V_2O_5$ and that other cell types produce collagen-generating mediators. However, suppression of the OXR1 gene by $V_2O_5$ in fibroblasts could further contribute to oxidative damage.

Another study in human fibroblasts revealed that heparin-binding epidermal growth factor–like growth factor (HB-EGF) is the major mitogen for these cells and that $V_2O_5$ induces HB-EGF via extracellular signal-related kinase (ERK)-1/2 and p38 mitogen-activated phosphokinase (MAPK) (Bonner 2007). $H_2O_2$ generation occurred in fibroblasts following $V_2O_5$ stimulation, but it was independent of HB-EGF induction.

However, $V_2O_5$ does not directly stimulate fibroblast proliferation, and collagen deposition is not attributable to fibroblasts (Antao-Menezes et al. 2008). Since lung fibrosis and collagen deposition are induced by $V_2O_5$ in vivo, the growth factors and collagen stimulation are postulated to come from other cells in the lung such as airway and alveolar epithelial cells and alveolar macrophages. The balance of fibrosis versus the resolution of fibrotic lesions is mediated by induction of interferon (IFN)-ß by $V_2O_5$ in fibroblasts (Antao-Menezes et al. 2008). The induction of IFN-α was also observed, but only IFNß was affected by the generation of $H_2O_2$ in fibroblasts by $V_2O_5$. IFNß in turn activates the transcription factor signal transducers and activators of transcription (STAT)-1, which plays a protective role in fibrogenesis by causing growth arrest and increased levels of CXCL10, a potent antifibrotic factor. The IFNß-mediated induction of CXCL10 is dependent on NADPH and xanthine oxidases. Thus, these factors in fibroblasts play a role in limiting fibrogenesis and collagen in the lung in response to $V_2O_5$.

## Effect on Airway Epithelial Cells to Enhance Inflammation

Airway epithelial cells produce mediators that are involved in the inflammatory response. $V_2O_5$ can stimulate the production of these mediators, particularly if airway epithelial cells are grown on an airway-liquid interface. However, many studies utilize conventional submerged epithelial cell cultures with $VOSO_4$ stimulation to investigate the mechanism by which vanadium initiates the cytokines and chemokines utilized in the inflammatory response of the lung. These studies give insight as to how vanadium initiates an inflammatory response and other divergent biological functions, such as insulin-like effects (Chen and Shi 2002).

The effects vary with the form of vanadium utilized to stimulate the cells, the cell line or type used, and the dosage of vanadium used. In many types of cells, vanadium activates a transcription factor for inflammatory cytokine genes in airway epithelial cells nuclear factor (NF)-κB. NF-κB is a heterodimer of two subunits, p50 and p65 (RelA), and is sequestered in the cytoplasm as an inactive complex of proteins. Any one of several inhibitory molecules will keep it in its inactive state: IκBα, IκBß, IκBε, p105, and p100. Many studies report that vanadium causes a cascade of phosphorylation of proteins that culminate with the phosphorylation and degradation of IκB in airway epithelial cells, which lead to the liberation of the active form of NF-κB, allowing it to translocate to the nucleus and bind to the promotor of a variety of genes. These genes include IL-8, a cytokine/chemokine that attracts neutrophils to the site of inflammation, and IL-6. Thus, vanadium increases the production of IL-8 and IL-6 in airway epithelial cells (Chen and Shi 2002).

The mechanism for NF-κB activation by vanadium has been studied. Some studies attribute reactive oxygen species (ROS) to activate the signal transduction pathway to increase NF-κB-mediated transcription. Others have evidence that NF-κB activation occurs independently of IκB degradation. In addition, in some studies, NF-κB activation in cells has been inhibited by vanadium. The final outcome may be due to factors mentioned above, such as cell type and dose and type of vanadium interacting with the cell (Chen and Shi 2002).

Vanadium ($V^{IV}$) is able to activate the epithelial growth factor receptor (EGFR) tyrosine kinase, followed by activation of Ras, then MAPK kinase (MEK), and MAPK (ERK) in human airway epithelial cells (BEAS2B cell line). Activation of NF-κB is EGFR tyrosine kinase and Ras dependent, but is MEK independent, for an EGFR tyrosine kinase inhibitor, PD-98059 failed to inhibit MEK activity and subsequent IκB degradation and NF-κB activation (Bonner 2007).

$VOSO_4$ also induced the increase in endothelin-1 (ET-1), intracellular adhesion molecule-1 (ICAM-1), IL-6, and IL-8 in normal human bronchial epithelial (NHBE) cells, supporting the role of the airway epithelial cell in releasing factors that mediate inflammation (L.K. Ryan, unpublished data).

Vanadium can modulate the initiation of apoptosis in target cells via ROS and direct effects via c-Jun N-terminal kinases (JNK) and via the intrinsic apoptotic pathway in mitochondria. In most instances, vanadium initiates apoptosis. Again, the variable effects can be attributed to cell type, valence of vanadium, and dose of vanadium. In bovine tracheal epithelial cells exposed to 10–20 $\mu g/cm^2$ $V_2O_5$, apoptosis was induced (L.K. Ryan, unpublished data).

## Effect on Human Pulmonary Artery Endothelial Cells and Cardiovascular Health

Systemic exposure to $VOSO_4$ via dietary body-building supplements could lead to the development of pulmonary hypertension over time (Li et al. 2004). Rats fed with vanadium-containing drinking water for 2 months developed pulmonary hypertension (Li et al. 2004). Pulmonary hypertension occurred in a rabbit-isolated perfused lung (IPL) model following of 0.5–50 μM of $VOSO_4$ infusion, with increased pulmonary arterial pressure and vasoconstriction following injection into the pulmonary artery. In pulmonary artery rings, the vasoconstriction also occurred, and following the addition of acetylcholine to relax the artery, $VOSO_4$ shifted the vasorelaxation curve to the right (more acetylcholine was required to relax the pulmonary artery treated with vanadium than with untreated rings). $VOSO_4$ inhibited nitrite/nitrate accumulation (indicative of nitric oxide production) in untreated IPL and in IPL treated with A23187 $Ca^{++}$ ionophore. $VOSO_4$ was added to human pulmonary artery endothelial cells (HPAEC), and endothelial nitric oxide synthase (eNOS) phosphorylation of threonine (Thr) was examined after 10 min. $VOSO_4$ induced an approximately twofold increase in Thr phosphorylation of eNOS compared with untreated control cells. This phosphorylation was partially mediated by protein kinase C (PKC), for calphostin C, a PKC inhibitor, attenuated $VOSO_4$-induced Thr phosphorylation and restored NO production by decreasing Thr phosphorylation of eNOS.

## Effects on Innate Immune Responses

$V_2O_5$ modulates innate immune responses in airway epithelial cells in vitro when incubated for 6 h prior to induction with innate immune stimulators. When pentavalent vanadium compounds were given 6 h prior to washing and stimulating them with either 100 ng/ml IL-1ß or 100 ng/ml lipopolysaccharide (LPS) for 24 h, vanadium suppressed both bovine and human IL-1ß- and lipopolysaccharide (LPS)-induced ß-defensin-2 in a dose-dependent manner (Klein-Patel et al. 2006).

Human ß-defensin-2 (hBD-2) is an antimicrobial peptide that has broad spectrum antibiotic properties (including antiviral and antifungal properties) and also is capable of the recruitment of immune cells involved in adaptive immunity, namely, dendritic cells and memory T cells.

A more soluble tetravalent vanadium compound, $VOSO_4$, also suppressed ß-defensin-2 at similar doses, suggesting that valence was not important in this immune suppressive effect. The mechanism was not dependent on generation of ROS. In human A549 type II alveolar epithelial cells pretreated with $V_2O_5$ and then stimulated with 100 ng/ml IL-1ß, which robustly induces hBD-2, as little as 0.3 $\mu g/cm^2$ significantly suppressed hBD-2 induction by IL-1ß. Significant suppression of hBD-2 also occurred with as little as 145 $ng/cm^2$ $VOSO_4$ in NHBE cells stimulated with 100 ng/ml LPS and as little as 290 $ng/cm^2$ $VOSO_4$ in NHBE cells stimulated with 100 ng/ml IL-1ß (L.K. Ryan, unpublished data). LPS-induced bovine tracheal epithelial cell IL-8 and lactoferrin, as well as IL-1ß-induced human NHBE cell IL-6, were inhibited by preincubation with $VOSO_4$ at these levels. However, in human NHBE cells, both lactoferrin and IL-8 were stimulated by 1.16 $\mu g/cm^2$ $VOSO_4$, and the IL-1ß-induced lactoferrin and IL-8 were not suppressed by $VOSO_4$ treatment (L.K. Ryan, unpublished data). This suggests that at lower concentrations than those tested that cause inflammation (<1 μg/cm$^2$), V$_2$O$_5$ may suppress the innate immune responses of the airway epithelium, leading to greater risk of infection with microorganisms.

In addition to effects on airway epithelium, V$_2$O$_5$ has effects on other lung cells. Acute and subchronic exposure to V$_2$O$_5$ decreased alveolar macrophage phagocytosis and lysosomal enzyme activity and release, reduced proinflammatory cytokine release, reduced bactericidal and tumoricidal factor production in situ and ex vivo, disturbed macrophage Ca$^{++}$ ion balance and interferon-γ-induced major histocompatability Class II antigen expression, altered lung immune cell numbers and types, modified mast cell histamine release, and increased kerotinocyte-derived chemokine (KC or CXCL1) and macrophage inflammatory protein-2 chemokine (MIP-2 or CXCL2) mRNA with subsequent neutrophil recruitment during the inflammatory response (reviewed in Cohen 2004). Many of the inflammatory and immunomodulatory effects of particulate matter containing V$_2$O$_5$ correlate with the vanadium levels in the particles and were reproduced by exposing animals to either VOSO$_4$ or V$_2$O$_5$ at amounts equivalent to that of the vanadium in each type of particulate matter particle under investigation.

Solubility and valence of vanadium compounds became more important when determining *Listeria monocytogenes* burden in the lung following inhalation exposure of rats. Soluble-oxidizing vanadate had the greatest impact on resistance; reducing V$^{III}$ altered resistance to a lesser extent, and V$^{IV}$ and insoluble V$^V$ had no effect on *Listeria* resistance (Cohen et al. 2010) nor did the insoluble vanadium-containing fraction of residual oil fly ash (ROFA) (Cohen et al. 2010). Further investigation with both soluble (NaVO$_3$) and insoluble (V$_2$O$_5$) forms of vanadium pentoxide led to the observation that NaVO$_3$ was the strongest immunomodulator and inhibitor of local bacterial resistance using *Listeria monocytogenes* (Cohen et al. 2010). However, when iron homeostasis was examined in this study, both of the insoluble and soluble vanadium compounds induced dramatic increases in iron levels in the airway and in airway epithelial cells. In addition, there were significant increases in production of cytokines and chemokines that are affected by hypoxia inducible factor (HIF-1α), a factor related to iron status in cells (Cohen et al. 2010). It is conceivable that although V$_2$O$_5$ did not induce resistance in this rat model to *Listeria monocytogenes*, resistance to other bacteria such as *Pseudomonas aeruginosa*, *Klebsiella pneumoniae*, *Staphylococcus aureus*, and *Streptococcus pneumoniae* may be affected.

In vitro addition of soluble sodium vanadate (Na$_3$VO$_4$) to rat alveolar macrophages subsequently induced the binding of the iron response element (IRE)-binding proteins to the IRE on mRNA for transferrin receptors or lactoferrin in alveolar macrophages, due to its ability to compete with the binding of iron to transferrin or lactoferrin in the lining of the lung and to displace iron from these carrier proteins, creating an iron deficiency in the cell (Cohen et al. 2010). This could explain the enhanced iron levels in vivo. Enhanced iron levels in the lung can alter resistance to infection by making Fe$^{3+}$ available for growth of bacteria and enhancement of virulence factors and also can suppress the immune system via modulation of chemokines and cytokines by affecting NF-κB and AP-1, which are transcription factors needed for cytokine induction, or by inducing overproduction of reactive oxygen and reactive nitrogen species, downregulating alveolar macrophage phagocytic activity. Thus, vanadium modulates iron homeostasis, and this in turn is one factor affecting susceptibility to bacterial infection. Further research into the mechanisms of altered iron homeostasis on other types of bacterial infection is needed to determine how vanadium affects immunity to bacteria not so dependent on cellular immunity and the T helper 1 immune response.

Another example of the importance of the physiochemical properties of vanadium influencing the inflammatory or innate immune response is demonstrated by a study where rats intratracheally instilled with 21 or 210 μg/kg NaVO$_3$, 21 or 210 μg/kg VOSO$_4$, or 42 or 420 μg/kg V$_2$O$_5$ had the most immediate, intensive, and prolonged inflammatory responses when the vanadium compound was most soluble in water (Cohen 2004). Neutrophil influx was greatest with vanadyl sulfate, followed by sodium metavanadate and then vanadium pentoxide. Neutrophil influx appeared as early as 4 h with VOSO$_4$ but not until 24 h with V$_2$O$_5$. The VOSO$_4$-induced inflammatory response persisted for 5 days as compared with shorter periods with the other compounds. Neutrophil chemotactic factors MIP-2 and KC were induced as early as 1 h after instillation and persisted through 48 h and were low but detectable at 5 and 10 days after instillation. The major KC source was the alveolar macrophage, which had high levels of KC mRNA detectable by in situ hybridization of bronchoalveolar lavage cells.

## Tumor Promotion

$V_2O_5$ functions as an in vivo tumor promoter in mice susceptible to chemical carcinogenesis (Rondini et al. 2010). Subchronic repeated aspiration exposures of $V_2O_5$-induced pulmonary inflammation and tumors promoted by 3-methylcholanthrene (MCA) in A/J mice and to a lesser extent in BALB/c mice. No tumors were induced by $V_2O_5$ exposure in B6 mice. This susceptibility to inflammation was correlated with the susceptibility of the mice to develop tumors following MCA administration. Tumor development of different mouse strains was positively correlated with elevated levels of KC and MCP-1, higher NF-κB and c-Fos binding activity, and sustained ERK1/2 activation in lung tissue.

When F344/N rats and B6C3F mice inhaled 0.5 and 2.0 mg/m$^3$ $V_2O_5$, A/B neoplasms were increased in all exposed animals except female rats (Rondini et al. 2010). Interstitial fibrosis, chronic inflammation, and alveolar and bronchiolar epithelial hyperplasia were also observed in both rats and mice. However, female rats did not have an increased tumor response, suggesting sex differences were present.

In lung carcinomas of $V_2O_5$-exposed mice, there are a high number of K-*ras* mutations and loss of heterozygosity in the K-*ras* region on chromosome 6. In most of these carcinomas with K-*ras* mutations, phosphorylated MAPK was elevated, suggesting that MAPK is activated during vanadium pentoxide-indued B6C3F1 mouse lung tumorigenesis following K-*ras* mutation and loss of the wild-type K-*ras* allele (Rondini et al. 2010).

However, despite its tumor-promoting properties, $V_2O_5$ in the proper formulation and concentration has the possible beneficial effect of killing tumor cells in vitro. Nanosize particles of $V_2O_5$ have been shown to be cytotoxic for fibrosarcoma (FsaR) and transformed L929 fibroblasts at a concentration of 20 μM, while 100 μM $V_2O_5$ was cytotoxic to V79, SCCVII, and B16F10 cells in addition to the other tumor cells, suggesting that these nanoparticles could be incorporated as a treatment for fibrosarcoma (Ivankovic et al. 2006).

In humans, vanadium-containing air pollutant particles (PM2.5) were associated with DNA damage in lymphocytes via the measurement of 7-hydro-8-oxo-2′-deoxyguanosine (8-oxodG) (Rondini et al. 2010). The association of vanadium exposure measured in personal samples of healthy males and females was positively associated with the formation of 8-oxodG in lymphocytes but independent of the mass concentration of the PM2.5. Epidemiologic studies have found exposure to high concentrations of PM2.5 to be associated with an increase in cancer.

## Systemic Effects

Vanadyl compounds are a component of particulate matter emissions from power plants and in concert with other metals exacerbate existing cardiovascular disease as well as pulmonary immunomodulation. However, individually, the compounds may not have the same systemic toxicity as other metals found in particulate matter. A recent study determined that rats intratracheally instilled with 1 μg/kg of the soluble form of vanadium, tetravalent $VOSO_4$, did not have much of a systemic response to this compound compared with zinc and nickel sulfates (Wallenborn et al. 2009). The level of $VOSO_4$ was higher than estimated human exposure to ambient air particulate matter containing vanadium compounds. Minimal inflammation and minimal increases in the injury markers lactate dehydrogenase, n-acetyl glucoaminidase, and protein in the bronchoalveolar lavage fluid occurred with vanadyl and iron sulfates compared with zinc, nickel or copper sulfates. Metallothionein-1 (MT-1) mRNA in the liver increased significantly upon IT instillation of vanadyl sulfate in the lung. However, no significant increases of MT-1 mRNA occurred in the heart and lung. Zinc transporter 1 (ZnT-1), cardiac cytosolic ferritin, and cardiac mitochondrial ferritin did not change in these organs following instillation of $VOSO_4$.

## Vanadium in Respirable Air Particulate Matter (PM$_{2.5}$)

Vanadium is a known tracer for the combustion of oil and is contained in concentrated fine air particulates (CAPS) from ambient air and in high concentrations in residual oil fly ash (ROFA). Human patients with asthma and chronic obstructive lung disease (COPD) have a greater fraction of deposition and retain a greater proportion of 1 μm fine particles roughly proportionate to the severity of airway obstruction compared with normal, healthy individuals (Morishita et al. 2004). The pulmonary retention of vanadium and other specific anthropogenic trace elements (La, Mn and S) that are present in $PM_{2.5}$ was associated with the exacerbation of ovalbumin-induced allergic airway inflammation in rats exposed to 313 or 676 $\mu g/m^3$ CAPS via inhalation for 5 days, 10 h/day (Morishita et al. 2004). Eosinophils and protein in the BAL fluid, which are markers of allergic inflammation, were elevated only in allergic animals exposed to CAPS. This study also showed that after CAPS exposure, vanadium was deposited in significant amounts in the lungs from rats having allergic inflammation and was barely detected (using inductively coupled plasma-mass spectrometry) with the saline-exposed control rats. This suggests that the physiochemical properties of CAPS were responsible for the exacerbation of allergic disease and that individuals with allergic airway disease have a tendency to retain and localize this potentially toxic element following exposure to urban air, even after a short-term exposure.

When vanadium is in high quantities in $PM_{2.5}$, such as ROFA, vanadium can access into the pulmonary circulation (Cohen 2004). In humans, inhalation of boiler contents high in vanadium has led to increased heart rate variability (Li et al. 2004). In addition, vanadium in all fractions of PM has a strong correlation with the induction of IL-8 with this air pollutant particle in human bronchial airway epithelial cells obtained from brush samples of normal volunteers (Ghio et al. 2002).

## Summary

Vanadium compounds have many health effects, some beneficial and some toxic, depending on the valence, concentration, route, and form of the molecules. Vanadium pentoxide is a pentavalent transition metal compound and a by-product of the burning of fossil fuels. Occupational exposures via inhalation occur in boilermakers, for example. An example of an environmental exposure is inhalation of particulate matter as part of air pollutant emissions from power plants. Vanadium pentoxide is implicated in exacerbating cardiovascular disease as well as inducing bronchitis and pneumonia following exposure. Exposure to pentavalent vanadium compounds suppresses the ability to fight infection. Therefore, there are immunotoxic effects of vanadium compounds in terms of immune suppression to infection and enhancement of asthma. Human epidemiologic and exposure studies and animal studies show the ability of this compound to increase inflammation, decrease innate immune responses at low levels of exposure, increase cardiopulmonary inflammatory mediators, and exacerbate hypertension. In vitro studies to explain mechanisms have been done, such as measurement of beta defensins and cytokines from exposed airway epithelial cells and macrophage phagocytosis and activation. Innate immune function such as macrophage activation and beta defensin gene expression was inhibited, whereas chemokines and proinflammatory cytokines were enhanced following exposure to pentavalent vanadium compounds. Vanadium compounds also have carcinogenic potential in the lung, probably via a nongenotoxic mechanism. Thus, pentavalent vanadium compounds have pulmonary toxicity and immunotoxicity if inhaled and, as part of air pollutant particulate matter, can initiate various serious health effects and exacerbate existing disease. Further research into the mechanisms of these effects is warranted.

## Cross-References

▶ Iron Homeostasis in Health and Disease
▶ Iron Proteins, Transferrins and Iron Transport
▶ Systems
▶ Vanadium in Biological Systems
▶ Vanadium in Live Organisms
▶ Vanadium Ions and Proteins, Distribution, Metabolism, and Biological Significance
▶ Vanadium Metal and Compounds, Properties, Interactions, and Applications
▶ Vanadium Pentoxide Effects on Lungs

## References

Antao-Menezes A, Turpin E, Bost PC, Ryman-Rasmussen JP, Bonner JC (2008) STAT-1 signaling in human lung fibroblasts is induced by vanadium pentoxide through an IFN-ß autocrine loop. J Immunol 180:4200–4207

Bonner JC (2007) Lung fibrotic responses to particle exposure. Toxicol Pathol 35:148–153

Chen F, Shi X (2002) Signaling from toxic metals to NF-kB and beyond: not just a matter of reactive oxygen species. Environ Health Perspect 110(Suppl 5):807–811

Cohen MD (2004) Pulmonary immunotoxicology of select metals: aluminum, arsenic, cadmium, chromium, copper, manganese, nickel, vanadium, and zinc. J Immunotoxicol 1:39–69

Cohen MD, Sisco M, Prophete C, Yoshida K, Chen LC, Zelikoff JT, Smee J, Holder AA, Stonehuerner J, Crans DC, Ghio AJ (2010) Effects of metal compounds with distinct physiochemical properties on iron homeostasis and antibacterial activity in the lungs: chromium and vanadium. Inhal Toxicol 22:169–178

Ghio AJ, Silbajoris R, Carson JL, Samet JM (2002) Biologic effects of oil fly ash. Environ Health Perspect 110(suppl 1):89–94

Ingram JI, Antao-Menezes A, Turpin EA, Wallace DG, Mangum JB, Pluta LJ, Thomas RS, Bonner JC (2007) Genomic analysis of human lung fibroblasts exposed to vanadium pentoxide to identify candidate genes for occupational bronchitis. Respiratory Res 8:34–47

Ivankovic S, Music S, Gotic M, Ljubesic N (2006) Cytotoxicity of nanosize $V_2O_5$ to selected fibroblast and tumor cells. Toxicol In Vitro 20:286–294

Klein-Patel ME, Diamond G, Boniotto M, Saad S, Ryan LK (2006) Inhibition of ß-defensin gene expression in airway epithelial cells by low doses of residual oil fly ash is mediated by vanadium. Toxicol Sci 92:115–125

Li Z, Carter JD, Dailey LA, Huang YCT (2004) Vanadyl sulfate inhibits NO production via threonine phosphorylation. Environ Health Perspect 112:201–206

Morishita M, Keeler GJ, Wagner JG, Marsik FJ, Timm EJ, Dvonch JT, Harkema JR (2004) Pulmonary retention of particulate matter is associated with inflammation in allergic rats exposed to air pollution in urban Detroit. Inhal Toxicol 16:663–674

Rondini EA, Walters DM, Bauer AK (2010) Vanadium pentoxide induces pulmonary inflammation and tumor promotion in a strain-dependent manner. Part Fibre Toxicol 7:9–20

Wallenborn JG, Schladweiler MJ, Richards JH, Kodavanti UP (2009) Differential pulmonary and cardiac effects of pulmonary exposure to a panel of particulate matter-associated metals. Toxicol Appl Pharmacol 241:71–80

## Vascular Remodeling (Change in Vascular Structure)

▶ Magnesium and Vessels

## Vascular Smooth Muscle Cell (VSMC)

▶ Magnesium and Vessels

## Vasculitis

▶ Silicon Exposure and Vasculitis

## Vasoconstriction (Contraction of Vessels)

▶ Magnesium and Vessels

## Vasodilation (Relaxation of Vessels)

▶ Magnesium and Vessels

## Vesicle Fusion

▶ C2 Domain Proteins

## Vesicular Zinc

▶ Zinc-Secreting Neurons, Gluzincergic and Zincergic Neurons

## VGSC

▶ Sodium Channels, Voltage-Gated

## Visualization Method

▶ Labeling, Human Mesenchymal Stromal Cells with Indium-111, SPECT Imaging

## Vitamin B$_{12}$

▶ Cobalt Proteins, Overview

## Voltage-Gated Ion Channels

▶ Potassium Channels and Toxins, Interactions

## VOSO$_4$

▶ Vanadium in Biological systems

## vps15

▶ Phosphatidylinositol 3-Kinases

## vps34

▶ Phosphatidylinositol 3-Kinases

## VTVH, Variable-Temperature Variable-Field

▶ Zinc Aminopeptidases, Aminopeptidase from Vibrio Proteolyticus (Aeromonas proteolytica) as Prototypical Enzyme

# W

## Water Channels

▶ Aquaporins and Transport of Metalloids

## Water Dikinase

▶ Selenophosphate Synthetase

## WD

▶ Zinc and Treatment of Wilson's Disease

## White Muscle Disease

▶ Selenium and Muscle Function

## Wolfram

▶ Tungsten in Biological Systems

## Wolframite

▶ Tungsten Cofactors, Binding Proteins, and Transporters in Biological Systems

## Wound Repair

▶ Zinc and Wound Healing

## WT, Wild Type

▶ Zinc Aminopeptidases, Aminopeptidase from Vibrio Proteolyticus (Aeromonas proteolytica) as Prototypical Enzyme

V.N. Uversky et al. (eds.), *Encyclopedia of Metalloproteins*, DOI 10.1007/978-1-4614-1533-6,

# Xanthine Dehydrogenase (Se-Dependent)

William Self
Molecular Biology & Microbiology, Burnett School of Biomedical Sciences, University of Central Florida, Orlando, FL, USA

Xanthine dehydrogenases (XDH) catalyze the hydroxylation of hypoxanthine to xanthine or hydroxylation of xanthine to uric acid using water as the source of oxygen (Choi et al. 2004). Selenium-dependent xanthine dehydrogenases represent a subclass of the larger xanthine dehydrogenase (or xanthine oxidoreductase) family of enzymes. The selenium-dependent members have been purified and studied from several strictly anaerobic organisms, such as *Clostridium acidiurici, Clostridium purinolyticum, and Eubacterium barkeri* (Alhapel et al. 2006; Dilworth 1983; Schrader et al. 1999; Self and Stadtman 2000; Self et al. 2003; Wagner and Andreesen 1977, 1979). As with other members of the selenium-dependent molybdenum hydroxylases (purine hydroxylase and nicotinic acid hydroxylase), selenium is present as a labile cofactor and not in the form of selenocysteine. Although the role for selenium is poorly understood, the turnover rate of these XDH enzymes is typically 2–3 orders of magnitude faster than its sulfur-dependent counterparts (Choi et al. 2004; Schrader et al. 1999; Self et al. 2003). The role of these enzymes in purinolytic clostridia has been proposed to be the reversible hydroxylation of uric acid to xanthine since these organisms can thrive on uric acid as a carbon and nitrogen source (Durre and Andreesen 1983; Schrader et al. 1999; Self and Stadtman 2000; Wagner and Andreesen 1979).

The xanthine dehydrogenase of *C. purinolyticum* was purified and studied alongside the discovery of the purine hydroxylase (PH) from the same organism (Self and Stadtman 2000). Indeed, the goal of the research initially was to purify the XDH, and the PH was uncovered during anion exchange chromatography as a flavin-containing contaminant with a different substrate specificity. The XDH was found to contain Mo, Fe, and FAD, as is expected in this class of molybdenum hydroxylases, and to catalyze the hydroxylation of hypoxanthine to xanthine or xanthine to uric acid. The presence of approximately four atoms of Fe per holoenzyme suggested that the enzyme contains two separate 2Fe-2S clusters, although sufficient enzyme to confirm this by x-ray absorption spectroscopy was not possible given the low yields. A closely related XDH was also characterized at nearly the same time by Andreesen's group and revealed an interesting twist for this family of enzymes (Schrader et al. 1999). The presence of both tungsten and molybdenum was reported in the purified preparation, in essentially equal amounts, suggesting that either metal could be utilized for formation of the active catalyst. However, both metals were present in the culture medium in the same concentration. Further study is required to determine if the environmental level of each metal alters the final metal content in the protein.

A recent report suggests that a selenium-dependent xanthine dehydrogenase is also present in an aerotolerant opportunistic pathogen, *Enterococcus faecalis* (Srivastava et al. 2011); however, the role for XDH in this organism has yet to be determined. This model system has the potential to elucidate the poorly understood metabolic pathway to insert selenium into the active site of this class of enzymes. Both genetic and biochemical analyses of this pathway are currently under investigation.

## References

Alhapel A, Darley DJ, Wagener N, Eckel E, Elsner N, Pierik AJ (2006) Molecular and functional analysis of nicotinate catabolism in *Eubacterium barkeri*. Proc Natl Acad Sci USA 103:12341–12346

Choi EY, Stockert AL, Leimkuhler S, Hille R (2004) Studies on the mechanism of action of xanthine oxidase. J Inorg Biochem 98:841–848

Dilworth GL (1983) Occurrence of molybdenum in the nicotinic acid hydroxylase from *Clostridium barkeri*. Arch Biochem Biophys 221:565–569

Durre P, Andreesen JR (1983) Purine and glycine metabolism by purinolytic *clostridia*. J Bacteriol 154:192–199

Schrader T, Rienhofer A, Andreesen JR (1999) Selenium-containing xanthine dehydrogenase from *Eubacterium barkeri*. Eur J Biochem 264:862–871

Self WT, Stadtman TC (2000) Selenium-dependent metabolism of purines: a selenium-dependent purine hydroxylase and xanthine dehydrogenase were purified from *Clostridium purinolyticum* and characterized. Proc Natl Acad Sci USA 97:7208–7213

Self WT, Wolfe MD, Stadtman TC (2003) Cofactor determination and spectroscopic characterization of the selenium-dependent purine hydroxylase from *Clostridium purinolyticum*. Biochemistry 42:11382–11390

Srivastava M, Mallard C, Barke T, Hancock LE, Self WT (2011) A selenium-dependent xanthine dehydrogenase triggers biofilm proliferation in *Enterococcus faecalis* through oxidant production. J Bacteriol 193:1643–1652

Wagner R, Andreesen JR (1977) Differentiation between *Clostridium acidiurici* and *Clostridium cylindrosporum* on the basis of specific metal requirements for formate dehydrogenase formation. Arch Microbiol 114:219–224

Wagner R, Andreesen JR (1979) Selenium requirement for active xanthine dehydrogenase from *Clostridium acidiurici* and *Clostridium cylindrosporum*. Arch Microbiol 121:255–260

## YS1: Yellow Stripe 1

▶ Iron Proteins, Plant Iron Transporters

## YSL: Yellow Stripe 1-Like

▶ Iron Proteins, Plant Iron Transporters

## Ytterbium

Takashiro Akitsu
Department of Chemistry, Tokyo University of Science, Shinjuku-ku, Tokyo, Japan

## Definition

A lanthanoid element, the 13th element (yttrium group) of the f-elements block, with the symbol Yb, atomic number 70, and atomic weight 173.04. Electron configuration [Xe] $4f^{14}6s^2$. Ytterbium is composed of stable ($^{168}$Yb, 0.13%; $^{170}$Yb, 3.04%; $^{171}$Yb, 14.28%; $^{172}$Yb, 21.83%; $^{173}$Yb, 16.13%; $^{174}$Yb, 31.83%; $^{176}$Yb, 12.76%) and four synthetic radioactive ($^{166}$Yb; $^{168}$Yb; $^{170}$Yb; $^{171}$Yb) isotopes. Discovered by J. C. G. de Marignac in 1878. Ytterbium exhibits oxidation states III and II; atomic radii 176 pm, covalent radii 188 pm; redox potential (acidic solution) $Yb^{3+}/Yb$ $-2.372$ V; $Yb^{3+}/Yb^{2+}$ $-1.15$ V; electronegativity (Pauling) 1.25. Ground electronic state of $Yb^{3+}$ is $^2F_{7/2}$ with $S = 1/2$, $L = 3$, $J = 7/2$ with $\lambda = -2940$ cm$^{-1}$. Most stable technogenic radionuclide $^{171}$Yb (half-life 1.92 years). The most common compounds: $Yb_2O_3$, $YbO$ and $Yb(OH)_3$. Biologically, ytterbium is of high toxicity to cause human pneumonoultramicroscopicsilicovolcanoconiosis, contained trace amount (less than 1 ppb) in animal tissue (Atkins et al. 2006; Cotton et al. 1999; Huheey et al. 1997; Oki et al. 1998; Rayner-Canham and Overton 2006).

## Cross-References

▶ Lanthanide Ions as Luminescent Probes
▶ Lanthanide Metalloproteins
▶ Lanthanides and Cancer
▶ Lanthanides in Biological Labeling, Imaging, and Therapy
▶ Lanthanides in Nucleic Acid Analysis
▶ Lanthanides, Physical and Chemical Characteristics

## References

Atkins P, Overton T, Rourke J, Weller M, Armstrong F (2006) Shriver and Atkins inorganic chemistry, 4th edn. Oxford University Press, Oxford/New York

Cotton FA, Wilkinson G, Murillo CA, Bochmann M (1999) Advanced inorganic chemistry, 6th edn. Wiley-Interscience, New York

Huheey JE, Keiter EA, Keiter RL (1997) Inorganic chemistry: principles of structure and reactivity, 4th edn. Prentice Hall, New York

Oki M, Osawa T, Tanaka M, Chihara H (1998) Encyclopedic dictionary of chemistry. Tokyo Kagaku Dojin, Tokyo

Rayner-Canham G, Overton T (2006) Descriptive inorganic chemistry, 4th edn. W. H. Freeman, New York

# Yttrium, Biological Effects

Xiao He
CAS Key Laboratory for Biomedical Effects
of Nanomaterials and Nanosafety & CAS Key
Laboratory of Nuclear Analytical Techniques,
Institute of High Energy Physics, Chinese
Academy of Sciences, Beijing, China

## Definition

The biological effects of Yttrium refer to the activity, behavior, and toxicity of Yttrium element or compounds in cells, tissues, organs, and organisms. Yttrium is present in low abundance in soil, water bodies, and organisms. It is not an essential element for organisms and its bioavailability is very low. As a heavy metal, yttrium is not entirely free of toxicity. Due to many chemical similarities, yttrium acts like a lanthanide in respect to its toxicological behavior. Although yttrium possesses potential for therapy of cancer and for diagnosis by radioactive imaging, its biological effects still remain incompletely understand.

The acute toxicity of yttrium ion ($Y^{3+}$), the deposition, retention, metabolism, and clearance have been documented in the previous reports. Hirano et al. (1993) studied time-course and dose-related changes in tissue distribution, subcellular localization, clearance, and acute toxicity of intravenous-injected yttrium chloride ($YCl_3$) in rats. Electron microscopic analyzes revealed that the colloidal yttrium-containing material was taken up by phagocytic cells in the liver and spleen. The elimination half-time of liver Y was 144 days at a dose of 1 mg Y/rat. Acute hepatic injury and transient increase of plasma Ca were found following the injection of $YCl_3$. A significant and tremendous amount of Ca was deposited in the liver (over 10-fold) and spleen (over 100-fold), while Ca concentration was only slightly increased in the lung and kidney (less than 1.5-fold). These results indicate that liver and spleen are primary target organs of iv-injected $YCl_3$.

Another study showed that most of the Y administered was distributed into liver, bone, and spleen (Nakamura et al. 1997). Yttrium disappeared from the blood within 1 day but was retained in the organs for a long time. Changes in Ca concentrations in liver, spleen, and lungs were in accordance with those of Y. Two mechanisms of REE metabolism in the case of intravenous administration are suggested: (1) The REEs may be transported partly contained in serum protein and partly incorporated into phagocytes, and taken mostly into the reticuloendothelial system by the phagocytic mechanism and deposited there. (2) The major route of excretion of REEs may be biliary excretion and the REEs might be excreted gradually into feces. Slight portions of the REEs may be excreted rapidly into urine. Generally, Y resembles to medium REEs in respect to their behavior by intravenous administration, i.e., distribution patterns, Ca-accumulating action, and hepatotoxicity. The categorization seems to coincide with their ionic radii.

Hirano et al. (1990) investigated the time-course and dose-related changes in distribution of Y between lung tissue and bronchoalveolar lavage fluid (BALF) and pulmonary inflammantory responses after yttrium chloride was instilled intratracheally (i.t.) into rats. Pulmonary clearance of Y was very slow with a half-life of 168 days. Transmission electron microscopy and X-ray microanalysis suggested that Y was localized in lysosomes of alveolar and interstitial macrophages and basement membranes. These results clearly explain the long pulmonary half-life of Y. β-Glucuronidase activity and calcium and phosphorous contents in the supernatant of BALF increased significantly. Comparative dose-effect profiles of lactate dehydrogenase activity in BALF supernatant revealed that 1 mol of $YCl_3$ is equivalent to about one-third mol of cadmium compounds and about 3 mol of zinc oxide in the potency for acute pulmonary toxicity.

Marubashi et al. (1998) investigated pulmonary clearance of yttrium and acute lung injury following i.t.-instilled of yttrium chloride in saline- or $YCl_3$-pretreated rats (some rats were pretreated with i.t.-instilled $YCl_3$ solution at a dose of 50 μg Y/rat 30 days before). Y was localized in lysosomes of alveolar and interstitial macrophages and basement membranes. About 67% of the initial dose of Y remained in the lung even 31 days after the intratracheal instillation. LDH, ALP, and β-GLU activities and protein concentration in BALF were addressed in the study, which are considered to be biochemical indices for cell lysis, injury of type II pulmonary epithelial cells, secretion of lysosomal enzymes, and exudation of plasma proteins, respectively. The pretreatment with YCl$_3$ significantly reduced i.t.-YCl$_3$-induced increases in BALF, such as lactate dehydrogenase, β-glucuronidase, and alkaline phosphatase activities and protein concentration, while the pretreatment increased the number of polymorphonuclear leukocyte (PMN) in BALF. The lung Mn-SOD activity in the YCl$_3$-pretreated group was two times higher than that of the saline-pretreated group. The reduction of the increases in those biochemical inflammatory indicators indicated that the Y-loaded rat lungs became more resistant against i.t. challenge of YCl$_3$, which may be due, at least in part, to the increase of manganese-superoxide dismutase (Mn-SOD) activity in the lung tissue. The increased transpulmonary infiltration of PMN was not directly associated with the biochemical inflammatory indicators in YCl$_3$-instilled rat lungs.

Yttrium-doped hydroxyapatites and yttrium tetragonal zirconia polycrystals (Y-TZP)-based materials are introduced for biomedical use in orthopedics for total hip replacement and are successful due to the excellent mechanical properties and biocompatibility of the materials. Y-TZP is also used in all-ceramic fixed partial dentures. The effects of yttrium addition on the microstructure, mechanical properties and biocompatibility of materials were studied in vitro and in vivo, and the toxicity of yttrium to osteoblasts was also studied. Zhang et al. (2010) found $Y^{3+}$ promoted the proliferation of primary osteoblasts at all tested concentrations. However, the effects of $Y^{3+}$ on the differentiation, adipocytic transdifferentiation, and mineralization function of primary osteoblasts depended on the concentration and culture time.

Long-term toxicity of yttrium has also been studied. Schroeder and Mitchener (1971) compared the innate toxic effects of small dose of seven metals in terms of growth and survival. According to the results, oral toxicity in terms of suppression of growth rates were in the order: gallium > yttrium > scandium > indium = chromate > palladium > rhodium. Differences in absorption may account for such effects as were found. There was suggestive evidence that yttrium exerted some carcinogenic activity, but this could not be proven from the available data. The yttrium groups showed no increase in amyloidosis, compared to the control.

Yttrium complex [YR(mtbmp)(thf)] (R = CH$_2$SiMe$_3$; mtbmp = 1,3-dithiapropanediylbis (6-tert-butyl-4-methylphenolato)) belongs to the family of rare-earth metal alkyl complexes that are supported by a linked bis(phenolate). A recent study demonstrated in vitro efficiency of [YR(mtbmp)(thf)] in cells of hematological malignancies and revealed its ability to be a possible agent for polychemotherapy (Lee et al. 2011). Exposure of BJAB cells to [YR(mtbmp)(thf)] led to a death receptor mediated reduction of cell viability and induction of apoptosis. The independence of Bcl-2 expression suggested that the [YR(mtbmp)(thf)]-induced apoptosis was mainly mediated via the extrinsic pathway. The extensive antitumor activity of [YR(mtbmp)(thf)] could be underlined by its capability to overcome multiple drug resistance in leukemic cells (Nalm-6) that were characterized by an overexpression of P-glycoprotein. [YR(mtbmp)(thf)] in combination with the conventional drug vincristine displayed impressive synergistic effects.

In the past decades, with the rapid development of nanotechnology and its broad applications, a wide variety of engineered nanoparticles (NPs) are now used in commodities, pharmaceutics, cosmetics, biomedical products, and industries. For example, rare-earth-ion-doped $Y_2O_3$ nanocrystals are ideal upconverting phosphors (UCPs), which are potentially useful reagents for bioimaging since the use of low energy photons avoids phototoxicity. Because of their unique properties, such as smaller size, larger specific surface area, and greater reactivity, yttrium-contained NPs maybe exhibit drastically different bioeffects from their bulk materials of the same composition. Schubert et al. (2006) found that the exposure to monoclinic $Y_2O_3$ (yttria) NPs sized 12 nm could increase the viability of HT22 cells in a concentration-dependent manner from 2 ng/ml to 20 µg/ml. The results suggested that yttria NPs act as a direct antioxidant to limit the amount of reactive oxygen species required to kill the cells and thereby are neuroprotective. This report has raised concerns about the potential biological effects of yttrium-contained NPs. Andelman et al. (2009) synthesized several $Y_2O_3$ NPs of various morphologies and found their cytotoxicities for human foreskin fibroblasts were different from each other. Cell viability was effectively invariant for the various concentrations of $Y_2O_3$ spherical NPs, with no statistically significant difference between the treated samples and the control. For the rod-like sample, at concentrations of 100 µg/ml and above, cell viability is increased in a statistically significant manner, approximately 1.3–1.5 times the untreated control. For the cells incubated with $Y_2O_3$ platelets, viability decreases in a statistically significant manner with increasing concentrations. However, it is difficult to attribute the cytotoxicity responses to the particle morphology, because differences in the other physicochemical properties are also found between the three types of $Y_2O_3$ NPs. Er-doped $Y_2O_3$ NPs were synthesized and used for bioimaging of M1 cells (macrophages) by Venkatachalam et al. (2009). The cellular uptake of nanoparticles was evidenced from bright field, UC, and NIR fluorescence images of live M1 cells. The exposure caused cell viability losses in a dose-dependent manner and the cytotoxicity could be decreased by surface modification of Er-doped $Y_2O_3$ NPs with PEG-b-PAAc polymer. Autophagy and massive vacuolization were observed in GFP-LC3/HeLa cells treated with $Y_2O_3$ NPs. Further study showed that the $Y_2O_3$-induced vacuolization was size-dependent and independent from autophagic pathway.

## Cross-References

▶ Lanthanide Metalloproteins
▶ Lanthanides, Toxicity
▶ Yttrium, Interactions with Proteins and DNA
▶ Yttrium, Physical and Chemical Properties
▶ Yttrium, Techniques

## References

Andelman T, Gordonov S, Busto G et al (2009) Synthesis and cytotoxicity of $Y_2O_3$ nanoparticles of various morphologies. Nanoscale Res Lett 5:263–273

Hirano S, Kodama N, Shibata K et al (1990) Distribution, localization, and pulmonary effects of yttrium chloride following intratracheal instillation into the rat. Toxicol Appl Pharmacol 104:301–311

Hirano S, Kodama N, Shibata K et al (1993) Metabolism and toxicity of intravenously injected yttrium chloride in rats. Toxicol Appl Pharmacol 121:224–232

Lee S, Peckermann I, Abinet E et al (2011) The rare-earth yttrium complex [YR (mtbmp)(thf)] triggers apoptosis via the extrinsic pathway and overcomes multiple drug resistance in leukemic cells. Med Oncol. doi:10.1007/s12032-010-9787-6

Marubashi K, Hirano S, Suzuki K (1998) Effects of intratracheal pretreatment with yttrium chloride ($YCl_3$) on inflammatory responses of the rat lung following intratracheal instillation of $YCl_3$. Toxicol Lett 99:43–51

Nakamura Y, Tsumura Y, Tonogai Y et al (1997) Differences in behavior among the chlorides of seven rare earth elements administered intravenously to rats. Fundam Appl Toxicol 37:106–116

Schroeder H, Mitchener M (1971) Scandium, chromium (VI), gallium, yttrium, rhodium, palladium, indium in mice: effects on growth and life span. J Nutr 101:1431–1437

Schubert D, Dargusch R, Raitano J et al (2006) Cerium and yttrium oxide nanoparticles are neuroprotective. Biochem Biophys Res Commun 342:86–91

Venkatachalam N, Okumura Y, Soga K et al (2009) Bioimaging of M1 cells using ceramic nanophosphors: synthesis and toxicity assay of $Y_2O_3$ nanoparticles. J Phys Conf Ser 191:012002

Zhang J, Liu C, Li Y et al (2010) Effect of yttrium ion on the proliferation, differentiation and mineralization function of primary mouse osteoblasts in vitro. J Rare Earths 28:466–470

# Yttrium, Interactions with Proteins and DNA

Xiao He
CAS Key Laboratory for Biomedical Effects of Nanomaterials and Nanosafety & CAS Key Laboratory of Nuclear Analytical Techniques, Institute of High Energy Physics, Chinese Academy of Sciences, Beijing, China

## Definition

The interactions of Yttrium with biomolecules refer to the interactions between yttrium element or compounds and biomolecules (e.g., proteins and nucleic acids). Yttrium is not an essential element for organisms; therefore, there is no developed mechanism involving site-specific binding of yttrium in biological system. However, $Y^{3+}$ has a great affinity to many metal-sites of bio-macromolecules, and may thereby disturb the functions of metal-biomolecules complexes. Both experimental data and theoretical simulations demonstrate that yttrium can interact with biomolecules. Understanding the interactions of yttrium with biomolecules is essential for predicting and interpreting the activity and toxicity of yttrium in a living system.

Rosoff and Spencer (1979) found that the number of mol of yttrium bound per mol of α-globulin increased fourfold as the concentration of yttrium was varied from 6.7 to $14.7*10^{-5}$ mol/L. The strong binding of Y to α-globulin appeared to be selective. The binding of $YCl_3$ to α-globulin was at least 20 times greater than that for albumin at the same $Y^{3+}$ concentration. The electrophoresis study shows that most of Y in blood samples bound to α- and β-globulin. Zhang et al. (2008) proposed that $Y^{3+}$ ions could interact preferably with solvent-exposed negatively charged side chains (Asp, Glu) on the protein surface, and thereby modulate the surface charge distribution. The conclusion was also supported by Monte Carlo simulations. The number of bound ions gradually increased with salt concentration.

The strong binding of lanthanides to DNA has been demonstrated before, and they have been shown to chemically precipitate nucleic acid and have been used as a nucleic acid stain. As $Y^{3+}$ and $Ln^{3+}$ are remarkably similar in their ionic behavior, the binding of Y to serum proteins and to nucleic acids has been studied using the methods of equilibrium dialysis and ultrafiltration (Rosoff and Spencer 1979). It was found that $La^{3+}$ the binding to DNA was similar when the concentration of $Y^{3+}$ was $0.8*10^{-5}$ mol/L and that of lanthanum $0.7*10^{-5}$ mol/L. When the concentration of $Y^{3+}$ was $3.2*10^{-5}$ mol/L and that of lanthanum was $3.7*10^{-5}$ mol/L, the lanthanum bound was three times greater than that of yttrium. However, the binding of lanthanum was lower than the binding of yttrium at the higher concentrations.

Bligh et al. synthesized a dimeric $Y^{3+}$ complex of an 18-membered hexaaza macrocycle and found it an efficient catalyst for double-stranded DNA (dsDNA) hydrolysis (Bligh et al. 2001). The dimeric complex could cleave 90–95% of the supercoiled, closed, circular form of plasmid DNA but was slightly less active toward the more conformationally relaxed nicked DNA (60% cleaved). However, monomeric complex of $Y^{3+}$ had no measurable ability to cleave dsDNA. These findings expand the range of new metal complexes which are capable of hydrolyzing double-stranded DNA which may therefore have the potential for therapeutic applications in genetically based disease.

## Cross-References

▶ Lanthanide Metalloproteins
▶ Lanthanides, Toxicity
▶ Yttrium, Biological Effects
▶ Yttrium, Physical and Chemical Properties
▶ Yttrium, Techniques

## References

Bligh A, Choi N, Evagorou E, McPartlin M, White K (2001) Dimeric yttrium(III) and neodymium(III) macrocyclic complexes: potential catalysts for hydrolysis of double-stranded DNA. J Chem Soc, Dalton Trans, 3169–3172

Rosoff B, Spencer H (1979) Binding of rare earths to serum proteins and DNA. Clin Chim Acta 93:311–319

Zhang F, Skoda M, Jacobs R et al (2008) Reentrant condensation of proteins in solution induced by multivalent counterions. Phys Rev Lett 101:148101

# Yttrium, Physical and Chemical Properties

Fathi Habashi
Department of Mining, Metallurgical, and Materials Engineering, Laval University, Quebec City, Canada

Yttrium is a transition metal. It occurs in nature in association with the lanthanides. Only the isotope Y 89 exists. The most important use of yttrium is in making phosphors, such as the red ones used in television set cathode ray tube.

## Physical Properties

| | |
|---|---|
| Atomic number | 39 |
| Atomic weight | 88.91 |
| Relative abundance in Earth's crust, % | $1 \times 10^{-3}$ |
| Density, g/cm$^3$ | |
| at room temperature | 4.472 |
| at the melting point | 4.42 |
| Melting point, °C | 1,526 |
| Boiling point, °C | 3,336 |
| Heat of fusion, kJ·mol$^{-1}$ | 11.42 |
| Heat of vaporization, kJ·mol$^{-1}$ | 365 |
| Molar heat capacity, kJ·mol$^{-1}$·K$^{-1}$ | 26.53 |
| Atomic radius, pm | 180 |
| Crystal structure | Hexagonal |
| Thermal conductivity, W·m$^{-1}$·K$^{-1}$ | 17.2 |
| Electrical resistivity, nΩ·m | 596 |
| Brinell hardness, MPa | 589 |

## Chemical Properties

Yttrium resembles the lanthanides and not scandium. It is stable in air in bulk form, due to the formation of a protective oxide, $Y_2O_3$, film on its surface. When finely divided, however, it ignites in air when heated. Although it has two electrons in the outermost shell yttrium is exclusively trivalent. Like its members of the group it forms only colorless compounds. Yttrium forms a water-insoluble fluoride, hydroxide, and oxalate, but its bromide, chloride, iodide, nitrate, and sulfate are all soluble in water.

Concentrated nitric and hydrofluoric acids do not attack yttrium, but other strong acids do. With halogens, yttrium forms trihalides $YF_3$, $YCl_3$, and $YBr_3$ at temperatures above 200°C. Similarly, carbon, phosphorus, selenium, silicon, and sulfur all form binary compounds with yttrium at elevated temperatures.

## References

Cotton S (2006) Lanthanide and actinide chemistry. Wiley, New York

Habashi F (2008) Researches on rare earths. History and technology. Métallurgie Extractive Québec, Québec City. Distributed by Laval University Bookstore. www.zone.ul.ca

McGill I (1997) Rare earth metals. In: Habashi F (ed) Handbook of extractive metallurgy. Wiley-VCH, Weinheim, pp 1693–1741

# Yttrium, Techniques

Xiao He
CAS Key Laboratory for Biomedical Effects of Nanomaterials and Nanosafety & CAS Key Laboratory of Nuclear Analytical Techniques, Institute of High Energy Physics, Chinese Academy of Sciences, Beijing, China

## Definition

Techniques employing yttrium refer to the applications of yttrium element and compounds in biological analysis, diagnosis, and therapy. Yttrium is of relatively low toxicity, and its metabolism can be manipulated using specific chelators. Its radioisotopes, which emit β, $β^+$, or γ radiation, possess medical potential. The labeling with yttrium is easy and fast to implement. These properties of yttrium appear to lend it to biological and clinical applications.

## $Y_2O_3$: A Digestibility Marker in Nutrition Research

The digestibility of proteins, lipids, energy components, and individual amino acids is normally estimated by the apparent digestibility coefficients. A standard approach in digestibility studies is the incorporation of an unabsorbed maker compound into the diets, normally chromium oxide ($Cr_2O_3$). Recent studies suggested that yttrium oxide is a viable alternative to chromic oxide as an inert digestibility marker. Advantages of yttrium oxide include low marker concentration in the feed; high recovery in diets and feces, with no effects on the nutrients metabolism; and precise digestibility measurements.

## $^{90}$Y-Labeled Microsphere for Radioisotope Therapy

Hepatocellular carcinoma (HCC) is the most common malignant hepatobiliary disease; it is responsible for about one million deaths per year. $^{90}$Y-labeled microsphere is used in the selective internal radiation therapy (SIRT) for HCC (Rossi et al. 2010). $^{90}$Y is a beta radiation emitter and has a half-life of 64 h. $^{90}$Y is transported into the hepatic artery by microspheres with a diameter of 35 μm, a size that allows the radioactive particles to pass through the vascular network of the liver and reach the finest peripheral capillaries. The effect of $^{90}$Y takes about 10 days; then irradiated cells undergo necrosis and within 1 month after surgery, the liver remains as a single scar. SIRT is a new treatment for liver cancer and liver metastases originating from colorectal carcinoma. It represents a new therapeutic option for patients with unresectable hepatocellular carcinoma, and the first clinical studies seem to demonstrate an increase in terms of survival rate when using this technique in combination with systemic chemotherapy. In addition, the use of SIRT tends to reduce the size of the tumor and allows some patients to become eligible for surgical resection. Most patients have not reported any major side effects; in a small percentage, an elevation of liver enzymes, fatigue, and fever have been reported. However, in most cases it is a well-tolerated therapy.

## Radiolabeling of Antibodies and Other Compounds

Targeted delivery of radionuclides for imaging and therapy of cancer (radioimmune diagnosis and radioimmune therapy) has progressed during the past two decades. The development of Y-labeled monoclonal antibodies and other compounds has stimulated interest in imaging and cancer therapy with $^{90}$Y. 1,4,7,10-Tetraazacyclododecane-1,4,7,10-tetraacetic acid (DOTA) and diethylenetriamine pentaacetic acid (DTPA) are the most commonly used chelators for yttrium (Wadas et al. 2010). $^{90}$Y is of much lower energy with respect to $^{131}$I. It provides advantages over $^{131}$I because it delivers, on average, a more energetic tumor-killing $\beta^-$ (935 vs 182 keV for $^{131}$I) and, concomitantly, a longer mean range (3.78 vs 0.36 mm for $^{131}$I). These characteristics improve the ability of a radiolabeled antibody to kill both targeted and neighboring cells, and offer a particular advantage in treating bulky or poorly vascularized disease. In addition, being a $\beta$-emitting radionuclide, $^{90}$Y can be administered on an outpatient basis (Boswell and Brechbiel 2007).

$^{90}$Y-labeled antibodies could be used for radioimmune imaging with SPECT/CT techniques. But the absence of low-energy gamma emission limits accurate dosimetry with SPECT. Most investigators use bremsstrahlung photons to generate SPECT images of $^{90}$Y biodistribution. The decay of $^{90}$Y has a minor branch to the $0^+$ excited state, followed by an internal e+/e- creation which happens in 32 out of one million decays. Consequently, $^{90}$Y PET scan is proposed in order to assess the biodistribution of $^{90}$Y-labeled therapeutic agents (Lhommel et al. 2009). $^{90}$Y-PET better reflects the tumor heterogeneity assessed by FDG PET/CT (a necrotic core surrounded by active tumor margins) than traditional $^{90}$Y-SPECT. Y-labeled antibodies and other compounds could also be used to generate images of gene expression. Schlesinger et al. investigated the application of $^{86}$Y-labeled L-oligonucleotides in mRNA expression imaging (Schlesinger et al. 2008).

Recently, yttrium-based nanomaterials have shown promise in imaging and therapy of cancer. Wu et al. reported a simple apoferritin-based approach to synthesize radioactive yttrium NPs (Wu et al. 2008). Compared to the metal chelation approach, apoferritin-templated yttrium NPs have higher loading capacities for radioactive yttrium and substantially improve chemical stability. Apoferritin with many amino acids at the end of the channel can be easily functionalized with pretargeting conjugates that could bind to specific tumor cells. Therefore, apoferritin-templated yttrium NPs may have high potential for applications in the immunodiagnosis and immunotherapy of cancers.

## Upconverting Luminescent Materials

Upconverting nanoparticles (UCNPs) when excited in the near-infrared (NIR) region display anti-Stokes emission whereby the emitted photon is higher in energy than the excitation energy. The use of the infrared confers benefits to bioimaging because of its deeper penetrating power in biological tissues and the lack of autofluorescence background. Rare earth ion-doped UCNPs emit photons from the outer 4f shells which are underlying 5s and 5p electrons; thus they have very sharp emission peaks (<10 nm) and do not bleach like organic dyes and fluorescent proteins. NaYF4:Yb,Er(Tm) is one of the most efficient upconversion phosphors that can be pumped with an inexpensive and common 980-nm diode laser source and has been actively researched as UCNP. Colloidal Yb/Er and Yb/Tm co-doped NaYF4 UCNPs show seven orders of magnitude stronger upconversion luminescence than CdSe-ZnS-based quantum dots. In order to make UCNPs suitable for bioimaging, surface modification is needed for hydrophobic UCNPs. It was reported that the cit-coated NaYF4:Yb/Er UCNPs could be taken up by HeLa cells and that very bright UCL could be used for detection of cancer cells in biomedical applications (Kim and Kang 2010). PEI-coated NaYF4 UCNPs and core-shell-structured silica/NaYF4 UCNPs also showed good in vitro and in vivo biocompatibilities. Rare earth ion-doped $Y_2O_3$ NPs are attractive for biological imaging applications due to their non-toxicity, resistance to photobleaching, and possibility for upconversion. Zako et al. (2010) proposed a possible application of $Y_2O_3$:YbEr NPs for cancer diagnosis and therapy using NIR-NIR imaging system.

## Cross-References

▶ Lanthanide Ions as Luminescent Probes
▶ Lanthanide Metalloproteins
▶ Lanthanides and Cancer
▶ Lanthanides in Biological Labeling, Imaging, and Therapy
▶ Lanthanides, Toxicity
▶ Yttrium, Biological Effects
▶ Yttrium, Physical and Chemical Properties

## References

Boswell C, Brechbiel M (2007) Development of radioimmunotherapeutic and diagnostic antibodies: an inside-out view. Nucl Med Biol 34:757–778

Kim D, Kang J (2010) Review: upconversion microscopy for biological applications. In: Méndez-Vilas A, Díaz J (eds) Microscopy: science, technology, applications and education. Formatex Research Center, Badajoz

Lhommel R, Goffette P, Van Den Eynde M et al (2009) Yttrium-90 TOF PET scan demonstrates high-resolution biodistribution after liver SIRT. Eur J Nucl Med Mol Imaging 36:1696

Rossi L, Zoratto F, Papa A et al (2010) Current approach in the treatment of hepatocellular carcinoma. World J Gastrointest Oncol 2:348–359

Schlesinger J, Koezle I, Bergmann R et al (2008) An 86Y-labeled mirror-image oligonucleotide: influence of Y-DOTA isomers on the biodistribution in rats. Bioconjug Chem 19:928–939

Wadas T, Wong E, Weisman G et al (2010) Coordinating radiometals of copper, gallium, indium, yttrium, and zirconium for PET and SPECT imaging of disease. Chem Rev 110:2858–2902

Wu H, Wang J, Wang Z et al (2008) Apoferritin-templated yttrium phosphate nanoparticle conjugates for radioimmunotherapy of cancers. J Nanosci Nanotechnol 8:2316–2322

Zako T, Hyodo H, Tsuji K et al (2010) Development of near infrared-fluorescent nanophosphors and applications for cancer diagnosis and therapy. J Nanomater 2010:1–7

# Z

## ZAP1/ZRG10

▶ Zinc Storage and Distribution in S. cerevisiae

## ZEB2 Gene

▶ Zinc and Mowat-Wilson Syndrome

## ZFHX1B Gene

▶ Zinc and Mowat-Wilson Syndrome

## ZiaR

▶ Zinc Sensors in Bacteria

## ZIM17/TIM15

▶ Zinc Storage and Distribution in S. cerevisiae

## Zinc

▶ Zinc in Alzheimer's and Parkinson's Diseases

## Zinc Adamalysins

F. Xavier Gomis-Rüth
Proteolysis Lab, Department of Structural Biology,
Molecular Biology Institute of Barcelona, CSIC
Barcelona Science Park, Barcelona, Spain

### Synonyms

ADAM; ADAMTS; Metzincin; Reprolysin; Secretase; Sheddase; Snake-venom metallopeptidase

### Definition

Adamalysins are zinc metallopeptidases, which belong to the clan of the metzincins. The latter are grouped together with the gluzincin clan – to which thermolysin belongs (▶ Thermolysin) – into the zincin tribe. Adamalysins are mainly found in two major environments: in snake venoms as potential virulence factors acting at post-envenomational events and in vertebrates where they play major roles in processing of extracellular proteins.

### Background

The adamalysins are a family of zinc-dependent metallopeptidases (MPs) named after their structural prototype, adamalysin II, from the snake venom of the Eastern diamondback rattlesnake *Crotalus adamateus* (Bode et al. 1993; Gomis-Rüth et al. 1993). These proteins – the whole family or individual

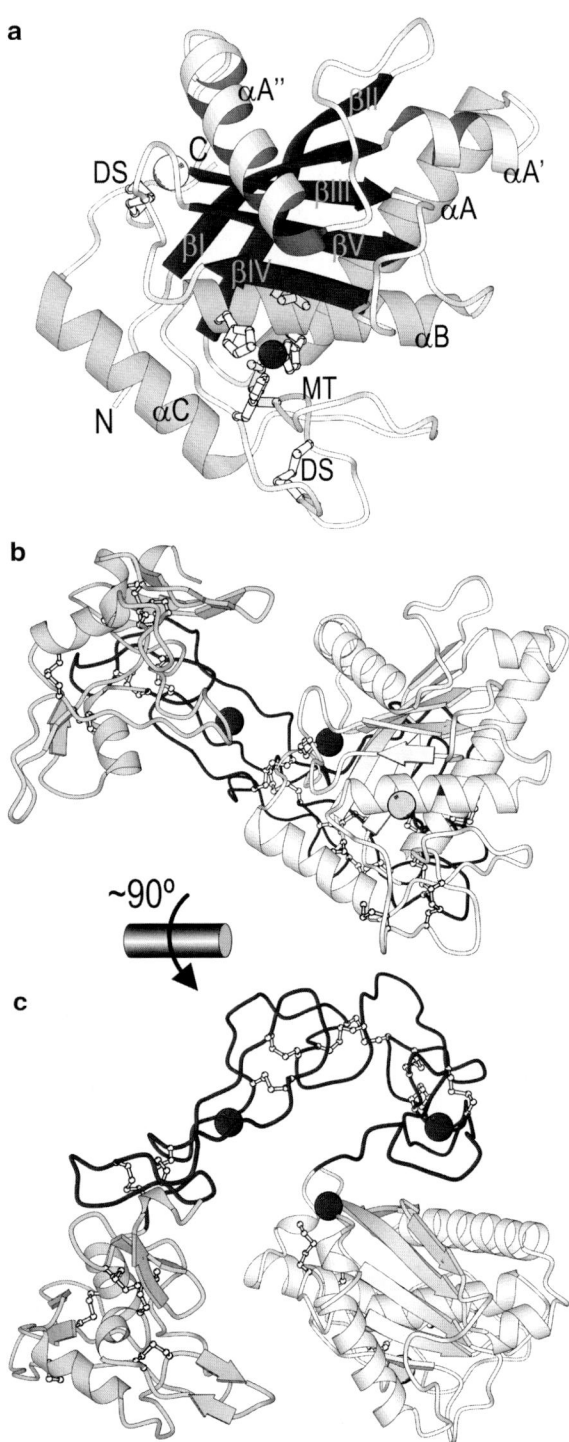

**Zinc Adamalysins, Fig. 1** *Three-dimensional structure of adamalysins.* Ribbon plots – in standard orientation of their MPs (Gomis-Rüth et al. 2012) – of (**a**) adamalysin II catalytic domain (PDB 1IAG; (Gomis-Rüth et al. 1993)) and (**b**) the P-III SVMP catrocollastin/VAP2B (PDB 2DW0; (Takeda et al. 2012)). (**c**) Orthogonal view of (**b**). VAP2B comprises in addition to a catalytic MP domain (*white; right*), a disintegrin domain (*dark gray; center*), and a cysteine-rich domain (*light gray; left*). In (**a**), the zinc-binding histidine residues, the Met-turn methionine (labeled MT), the general base/acid glutamate, and the two disulfide bonds (labeled DS) are shown as *stick* models for their side chains. The catalytic zinc ion is shown in *dark gray*, the structural calcium ion in *white*, and the termini are labeled. Helices are depicted as labeled *light-gray ribbons*, and strands as labeled *dark-gray arrows*. In (**b**) and (**c**), *dark-gray spheres* represent calcium ions, *light-gray spheres* correspond to zinc ions, and the disulfide bonds are shown as *white balls* and *sticks* (Figure prepared with program MOLSCRIPT)

subfamilies – have also been termed ADAMs (from *a d*isintegrin *a*nd *m*etalloprotease), ADAMTSs (from ADAMs with *t*hrombo*s*pondin-like motifs), snake-venom metalloproteinases (SVMPs), reprolysins, or MDCs (*m*etalloprotease-like, *d*isintegrin-like, and *c*ysteine-rich proteins) (Apte 2009; Blobel 2005; Fox and Serrano 2009; Hooper and Lendeckel 2005; Takeda et al. 2012). Adamalysins are extracellular soluble or membrane-anchored multimodular proteins, which can be composed of an N-terminal signal peptide, a prodomain engaged in latency, a catalytic – mostly zinc- and calcium-dependent – MP domain, a disintegrin(–like) domain, and a cysteine-rich domain, which includes a hypervariable region and is subdivided into "wrist" and "hand." In addition, some adamalysins have a snake-venom C-type lectin-like domain, an EGF-like domain, a transmembrane domain, a cytoplasmic domain, a spacer domain, N-glycan-rich modules, thrombospondin-type-1 motifs, immunoglobulin repeats, netrin-like domains, Kunitz-domains, *p*rotease and *lac*unin modules (PLAC), and variable domains. Each of the three adamalysin subfamilies (see below) comprises a different combination of these modules (see respective Fig. 1 in Blobel (2005), Fox and Serrano (2009), and Takeda et al. (2012) and Figs.1 and 2 in Apte (2009)). In a broader sense, adamalysins belong to the metzincin clan of MPs, which also includes astacins, serralysins, matrix metalloproteinases (MMPs), leishmanolysins, snapalysins, pappalysins, archaemetzincins, and fragilysins, and other families that have yet to be structurally characterized (Bode et al. 1993; Gomis-Rüth 2003; Gomis-Rüth 2009). Metzincin catalytic domains are distinguished by an extended zinc-binding motif, HEXXHXXG/NXXH/D (amino-acid one-letter code), which includes three zinc-binding histidines (or two histidines and an aspartate), a catalytic acid/base glutamate required for catalysis, and a hallmark glycine.

## Snake-Venom Metalloproteinases (SVMPs)

This subfamily was the first to be studied, and reports go back to 1881 (Fox and Serrano 2009). Hundreds of SVMPs have been characterized from both hemorrhagic and nonhemorrhagic reptilian poisons. They contribute to the post-envenomation condition through digestion of extracellular matrix (ECM) components surrounding capillaries, such as type IV collagen, nidogen, fibronectin, laminin, and gelatin, as well as digestion of plasma proteins. Further targets include endothelial cells, inflammatory response cells, and platelet-aggregation proteins. Such degradation is exacerbated by disintegrins – peptides inhibiting platelet aggregation – which leads to tissue necrosis. ECM collagen degradation is the main function of MMP family members, and adamalysins share with them a general preference for bulky hydrophobic residues in $P_1'$ of substrates. In addition, SVMPs contribute to the systemic effects of snakebite envenomation, such as hemostatic disturbances leading to pro- or anticoagulatory events and apoptotic and inflammatory effects. According to their domain organization, SVMPs can be divided into three classes: (1) P-I SVMPs only harbor the signal peptide, the prodomain, and the catalytic domain; (2) P-II SVMPs have an additional C-terminal disintegrin(–like) domain, which may be cleaved and give rise to separately acting MP and disintegrin domains; and (3) P-III SVMPs, which also include cysteine-rich domains and have been subdivided into P-IIIa through P-IIId (Fox and Serrano 2009; Takeda et al. 2012).

## Mammalian ADAMs/MDCs

This subfamily was originally described to play a role in fertilization and sperm function in mammalian reproductive tracts. They contain, in addition to the modules found in P-III SVMPs, downstream EGF-like domains, transmembrane domains, and cytoplasmic domains, and so, they are mostly membrane proteins. There are 20 functional ADAMs in humans (see http://degradome.uniovi.es/met.html#M12), which are involved in myogenesis, development and neurogenesis, differentiation of osteoblasts, cell-migration modulation, and muscle fusion (▶ Zinc Homeostasis, Whole Body). In addition, they are also involved in such disorders as asthma, cardiac hypertrophy, obesity-associated adipogenesis and cachexia, rheumatoid arthritis, endotoxic shock, inflammation, and Alzheimer's disease (▶ Zinc Homeostasis, Whole Body). The involvement of the latter is a result of their functions as α-secretases. Most notably, they play a major role in protein ectodomain shedding, and major players are ADAM-10 and ADAM-17. Such proteolytic processing results in the release of membrane-bound proteins, such as cytokines, growth factors, and receptors, thus increasing the levels of their soluble forms in circulation (Blobel 2005).

## Mammalian ADAMTSs

Finally, some soluble family members lacking the transmembrane domain and harboring one or more thrombospondin type 1 repeats, cysteine-rich modules, spacer modules, N-glycan-rich modules, PLAC, Kunitz- or netrin-like domains, N-glycan-rich or immunoglobulin repeats, or CUB domains gave rise to a distinct subfamily of soluble extracellular proteases, the ADAMTSs (Apte 2009). In humans, 19 functional orthologs and seven ADAMTS-like proteins – which lack proteolytic activity – have been described (see http://degradome.uniovi.es/met.html#M12). These enzymes disable cell adhesion by binding to integrins. They are also involved in gonad formation, embryonic development and angiogenesis, and maturation of procollagen and von-Willebrand factor, as well as in ECM processing related to morphogenesis and ovulation (▶ Zinc Homeostasis, Whole Body). This proteolytic potential is also linked to inflammatory processes, cartilage (aggrecan) degradation in arthritic diseases, bleeding disorders, and glioma tumor invasion. Significant roles are attributed to the nonproteolytic domains, propeptide processing, and glycosylation. In contrast, ADATMS-like proteins seem to function in regulatory processes of the ECM (Apte 2009).

## Structure of the Catalytic Domain of Adamalysin

The first catalytic domain structure to be analyzed was that of adamalysin II (Gomis-Rüth et al. 1993), and it has since been considered the prototype for the catalytic domains of the distinct subfamilies. It is a 203-residue compact molecule of oblate ellipsoidal shape, notched at the periphery to render a relatively flat substrate-fixing cleft (Fig. 1a). This cleft separates a large upper subdomain spanning about three quarters of the total length from a smaller lower lobe. The upper subdomain comprises a five-stranded twisted β-sheet at the top. All strands (βI to βV) except the fourth are parallel to each other and to any substrate that is bound in the cleft. The antiparallel strand, βIV, forms the lower edge of this subdomain and creates an upper rim or northern wall of the active-site crevice (see below). This strand binds substrates in an antiparallel manner, mainly on its nonprimed side. The upper subdomain also contains two long α-helices: αA, the *backing helix,* and αB, the *active-site helix,* which provides two zinc-binding histidine residues. These helices are both arranged on the concave side of the β-sheet. Two additional helices, αA′ and αA″ – the *adamalysin helix* – are found preceding αA and βIII, respectively (see Fig. 1a). At the end of αB, the polypeptide chain takes a sharp, downward turn mediated by the glycine of the consensus sequence and disembogues into the lower subdomain. The latter features the third zinc ligand, a histidine that approaches the metal from below. This subdomain contains few regular secondary-structure elements, mainly a *C-terminal helix* αC at the end of the polypeptide chain, which anchors the N-terminus of the mature catalytic domain. Helices αB and αC are connected by an irregular segment that includes a conserved 1,4-β-turn containing a methionine at position three, the *Met-turn.* It is positioned underneath the catalytic zinc, which lies at the bottom of the active-site cleft at half width, forming a hydrophobic pillow, but no direct contact of the methionine side chain with the metal is observed. The zinc, in turn, is coordinated by the Nε2 atoms of the three aforementioned histidines and a catalytic solvent molecule in an approximately tetrahedral manner. The $S_1′$ *specificity pocket* of adamalysin is shaped at the top by a protruding bulge made by the loop connecting strand βIII with βIV and at the bottom by a wall-forming segment consisting of residues intercalated between the Met-turn and the C-terminal helix. The pocket is deep and hydrophobic, reminiscent of some MMPs. Finally, two disulfide bonds cross-link the irregular segment within the lower subdomain and attach αC to the upper subdomain, respectively, and a calcium ion with merely structural function is located on the surface, opposite the active site, and close to the C-terminus (Fig. 1a). A number of catalytic domain structures have since been reported, which include several P-I and P-III SVMPs, as well as ADAM-17, -22, and -33 and ADAMTS-1, -4, and -5 (see Table 2 in Takeda et al. (2012)). These molecules share the essential structural features described for adamalysin II.

## Adamalysin Ancillary Domains

The domains flanking the catalytic domain in adamalysins are engaged in regulation of the hydrolytic activity and in interactions with binding partners and substrates. Latency, for example, is maintained by the prodomain upstream of the catalytic domain, and activation is believed to occur as in MMPs, that is, by cleavage and removal of the prodomain according to a *cysteine switch-like* mechanism (see Gomis-Rüth 2003; Gomis-Rüth 2009). The first structure described containing ancillary domains in addition to the catalytic domain was that of vascular apoptosis-inducing protein-1 (VAP1), and several structures have been reported since (Takeda et al. 2012). Among them is the vascular apoptosis-inducing protein-2 (VAP2B), a P-III SVMP (Fig. 1b, c), which shows that the catalytic, disintegrin(−like), and cysteine-rich domains adopt a C-shaped conformation, with the distal region of the latter domain close to the MP domain. While the fold of each domain is conserved among adamalysins, large variations are found in the relative orientations between the MP and disintegrin-like domains, and so, there is variety in the spacing between the terminal domains. In addition, structural analysis of ADAM22 revealed that the EGF-type domain found in ADAMs was tightly associated with the disintegrin-like and cysteine-rich domains, thus providing a rigid spacer to position the MP and disintegrin-like domains against the membrane. In ADAMTS structures, in contrast, the disintegrin-like module seemed to be an integral part of the peptidase domain engaged in exosite shaping. This is supported by the absence of such a domain in the catalytically inert ADATMS-like proteins (Apte 2009).

## Cross-References

▶ Thermolysin

▶ Zinc Homeostasis, Whole Body

## References

Apte SS (2009) A disintegrin-like and metalloprotease (reprolysin-type) with thrombospondin type 1 motif (ADAMTS) superfamily: functions and mechanisms. J Biol Chem 284(46):31493–31497

Blobel CP (2005) ADAMs: key components in EGFR signalling and development. Nat Rev Mol Cell Biol 6(1):32–43

Bode W, Gomis-Rüth FX, Stöcker W (1993) Astacins, serralysins, snake venom and matrix metalloproteinases exhibit identical zinc-binding environments (HEXXHXXGXXH and Met-turn) and topologies and should be grouped into a common family, the 'metzincins'. FEBS Lett 331:134–140

Fox JW, Serrano SM (2009) Timeline of key events in snake venom metalloproteinase research. J Proteomics 72(2):200–209

Gomis-Rüth FX (2003) Structural aspects of the *metzincin* clan of metalloendopeptidases. Mol Biotech 24:157–202

Gomis-Rüth FX (2009) Catalytic domain architecture of metzincin metalloproteases. J Biol Chem 284:15353–15357

Gomis-Rüth FX, Botelho TO, Bode W (2012) A standard orientation for metallopeptidases. Biochim Biophys Acta 1824:157–163

Gomis-Rüth FX, Kress LF, Bode W (1993) First structure of a snake venom metalloproteinase: a prototype for matrix metalloproteinases/collagenases. EMBO J 12:4151–4157

Hooper NM, Lendeckel UE (2005) The ADAM family of proteases, vol 4. Springer, Dordrecht

Takeda S, Takeya H, Iwanaga S (2012) Snake venom metalloproteinases: Structure, function and relevance to the mammalian ADAM/ADAMTS family proteins. Biochim Biophys Acta 1824:164–176

# Zinc Alcohol Dehydrogenases

Hans Jörnvall and Tomas Bergman
Department of Medical Biochemistry and Biophysics, Karolinska Institutet, Stockholm, Sweden

## Synonyms

Active site zinc; Catalytic site zinc; Catalytic zinc

## Definition

Zinc alcohol dehydrogenases are oxidoreductases belonging to the MDR (medium-chain dehydrogenases/reductases) protein superfamily. They are dimeric or tetrameric and composed of two-domain subunits (catalytic and coenzyme-binding domains, respectively). Each subunit (size about 40 kDa) binds one or two zinc ions. One has a direct functional role and is a major component in the catalytic events; the other, when present, is implicated in the maintenance of structural stability but may also have additional roles.

## Structural and Functional Properties

*Overview.* Alcohol dehydrogenase (ADH) of the medium-chain dehydrogenase/reductase (MDR) type catalyzes reversible oxidation of primary and secondary alcohols to the corresponding aldehydes and ketones, respectively, employing the coenzymes NAD(H) or NADP(H) in the hydride transfer process. ADHs are widespread among animals, plants, fungi, and bacteria and reveal extensive isozyme patterns. In humans and in many organisms, they serve to eliminate toxic alcohols and to generate alcohol, aldehyde, and ketone functional groups in biosynthetic reactions. Vertebrate zinc-containing ADHs have two tetrahedrally coordinated zinc ions per subunit, one catalytic at the active site and one structural at a site influencing structural integrity and subunit interactions. The subunits are organized in a catalytic domain and a coenzyme-binding domain, the latter having a structure typical of the Rossmann fold with two mononucleotide-binding units forming the dinucleotide-binding site. The enzyme active site is located in the cleft between these two domains. Generally, the catalytic zinc interacts with three protein ligands, one His and two Cys residues, while substrate (or water, or in inactive enzyme conformation a Glu) constitutes the fourth zinc ligand. The structural zinc interacts with four protein ligands, frequently Cys residues (cf. ▶ Zinc Structural Site in Alcohol Dehydrogenases).

*Catalysis.* Zinc ADHs reveal an ordered catalytic mechanism where the coenzyme binds before the substrate. Crystallographic data show that binding of the coenzyme ($NAD^+$ or NADH) triggers a large conformational change involving a rotation of the catalytic domain in relation to the coenzyme-binding domain of about 10° (Eklund and Ramaswamy 2008; Plapp 2010). The rotation closes the coenzyme-binding site, trapping the NAD(H) in the protein

**Zinc Alcohol Dehydrogenases, Fig. 1** The horse liver ADH active site with cofactor $NAD^+$ bound and a benzyl alcohol substrate coordinated to the catalytic zinc. *Arrows* indicate three steps of the suggested proton relay (transfer) mechanism (labeled PT1, PT2, and PT3), and the hydride transfer reaction (Reprinted with permission from Agarwal et al. 2000. Copyright 2000, American Chemical Society)

structure. The holoenzyme with bound coenzyme is therefore the closed form and the coenzyme-free apoenzyme the open form. The oxidation of alcohol requires the net removal of two hydrogen atoms in the form of a hydride ion and a proton. The hydride transfer is a direct transfer from the substrate to the nicotinamide ring of $NAD^+$. A proton relay mechanism has been suggested where the proton of the alcohol group is transferred to His51 via a hydrogen-bonded pathway involving His51, Ser48, and the two hydroxyl groups of the coenzyme ribose moiety. There is some debate regarding the order of hydride transfer and proton movement, but a majority of kinetic investigations point at the following consensus mechanism for alcohol oxidation by horse liver ADH (HLADH) (Fig. 1). In the apoenzyme, zinc-bound water is deprotonated via the proton relay system just described to produce a negatively charged zinc hydroxide moiety which attracts the positively charged nicotinamide ring in the coenzyme. The substrate alcohol then enters and displaces the hydroxide, and this replacement of zinc ligand may involve two steps where first the carboxylate group of a Glu residue replaces the hydroxide followed by coordination of the substrate alcohol oxygen to the zinc (Eklund and Ramaswamy 2008; Plapp 2010). The resulting enzyme-$NAD^+$-alcohol complex is now ready for deprotonation via the proton relay system leading to a zinc-bound alkoxide ion that undergoes hydride transfer to $NAD^+$ (Fig. 1). This results in the generation of NADH and a zinc-bound aldehyde or ketone. The catalytic cycle is then completed by the release of the product aldehyde/ketone and NADH. High-resolution crystal structures of HLADH have revealed that the nicotinamide ring of $NAD^+$ becomes puckered and adopts a twisted boat conformation during the catalytic process, interpreted to be important for efficient hydride transfer (Meijers et al. 2007).

*Evolutionary aspects.* Phylogenetic tree constructions from aligned enzyme primary structures correlated with enzyme purifications from many organisms and with corresponding enzyme activity measurements have revealed much of the origin, evolution, and multiplicity for the ADH branch of the MDR superfamily (Danielsson et al. 1994; cf. Persson et al. 2008; Jörnvall et al. 2010). The ancestral origin for the vertebrate ADH is the class III form, with its glutathione-dependent formaldehyde dehydrogenase activity. A gene duplication at about 550 million years ago then started the line leading to the classical liver ADH activity of class I type. Additional gene duplications and subsequent mutational divergence at several stages produced novel forms ("enzymogenesis"), with the more early steps giving the ADH classes (functionally

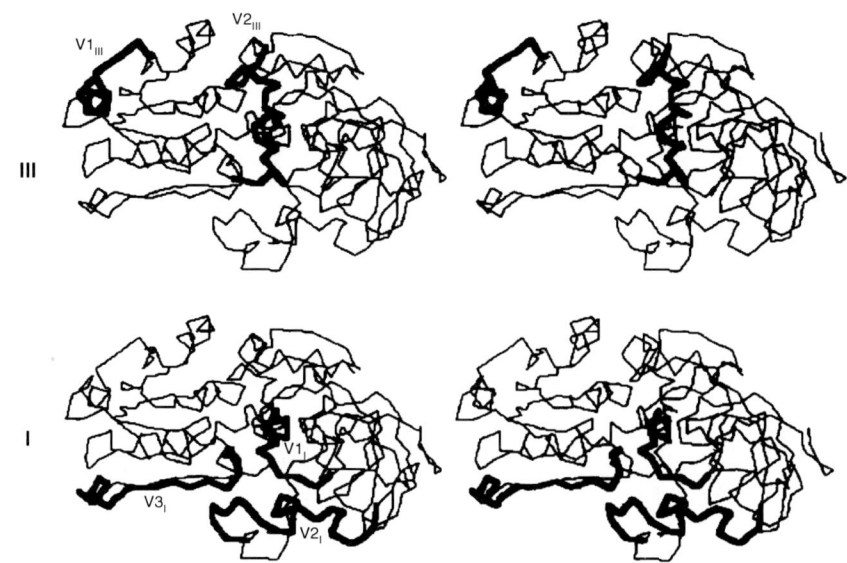

Zinc Alcohol Dehydrogenases, Fig. 2 Differences in patterns of ADH variability for classes III (*top*) and non-III (*bottom*), the latter illustrated for class I. Top two variable regions are superficially positioned and nonfunctional, *bottom* three variable regions are internal in the dimer and functional, corresponding to subunit interactive (*left region*), at the active site (*centre region*), and at the structural zinc-binding loop (*bottom region*) (Modified from Danielsson et al. 1994, with permission)

more separate), later steps the isozymes (functionally distinct but similar in type), and the most recent ones the addition of the various allelic variants ("allozymes"). In humans, five classes are recognized, with considerable differences in expression patterns, the parent class III form of fairly limited but universal expression, the class I form abundant in the liver, and class IV most common in the stomach. Remaining classes appear less distinct, and some are unique to separate vertebrate lines. Isozymes are most frequent in class I, and in humans, those derived from separate expressions of three genes giving the ADH1A, B, and C forms, of which B and C have allelic variants (ADH1B1-3 and ADH1C1-2). The A form has clear fetal expression, while especially, the B forms show extensive differences among human populations, constituting one of the explanations to the separate sensitivities to alcohol consumption among different populations (from a point mutation between B1 and B2 at a residue, position 47, participating in coenzyme binding and therefore influences the speed of the catalytic reaction and hence the alcohol turnover rate). Already just these gene, organ, and function multiplicities illustrate that ADH as enzyme participates in a number of different metabolic reactions, of which some are universal and some quite specific in time, distribution, or species.

In molecular terms, the structural architectures and variability patterns are also well known (Danielsson et al. 1994). The most ancestral class III is the overall most conserved, with largely constant function (formaldehyde metabolism) and evolutionary speed throughout many life forms. Its evolutionary changes that do occur are largely concentrated to surface-positioned residues (Fig. 2) without much functional importance and separate from the active site, zinc-binding sites, and other specific functions, as for basic metabolic enzymes in general. In contrast, the non-III classes evolve faster overall (about three- to fivefold faster than class III), and the structural changes that do occur are concentrated to some "hot spots" that include just functional sites, like the active site, zinc binding (both the "structural" and the "catalytic" zinc sites), and subunit interactional segments (Fig. 2), thus reflecting the enzymogeneses that have occurred for some time. Interestingly, this pattern of separate overall "constant" and "evolving" forms (Fig. 2) is visible with similar evolutionary rate differences also in many other dehydrogenases with class and isozyme multiplicities (Jörnvall et al. 2010), and Zn-ADH is an excellent model of enzyme evolution in general.

In a still wider scheme, Zn-ADHs constitute a part of a still wider evolutionary unit, with at least three major functionally and structurally distantly related lines composed of another superfamily (SDR, for short-chain dehydrogenases/reductases, also including ADH-active forms), the non-zinc MDR enzymes, and the Zn-MDR enzymes. They, too, illustrate several common evolutionary patterns and have ancestrally

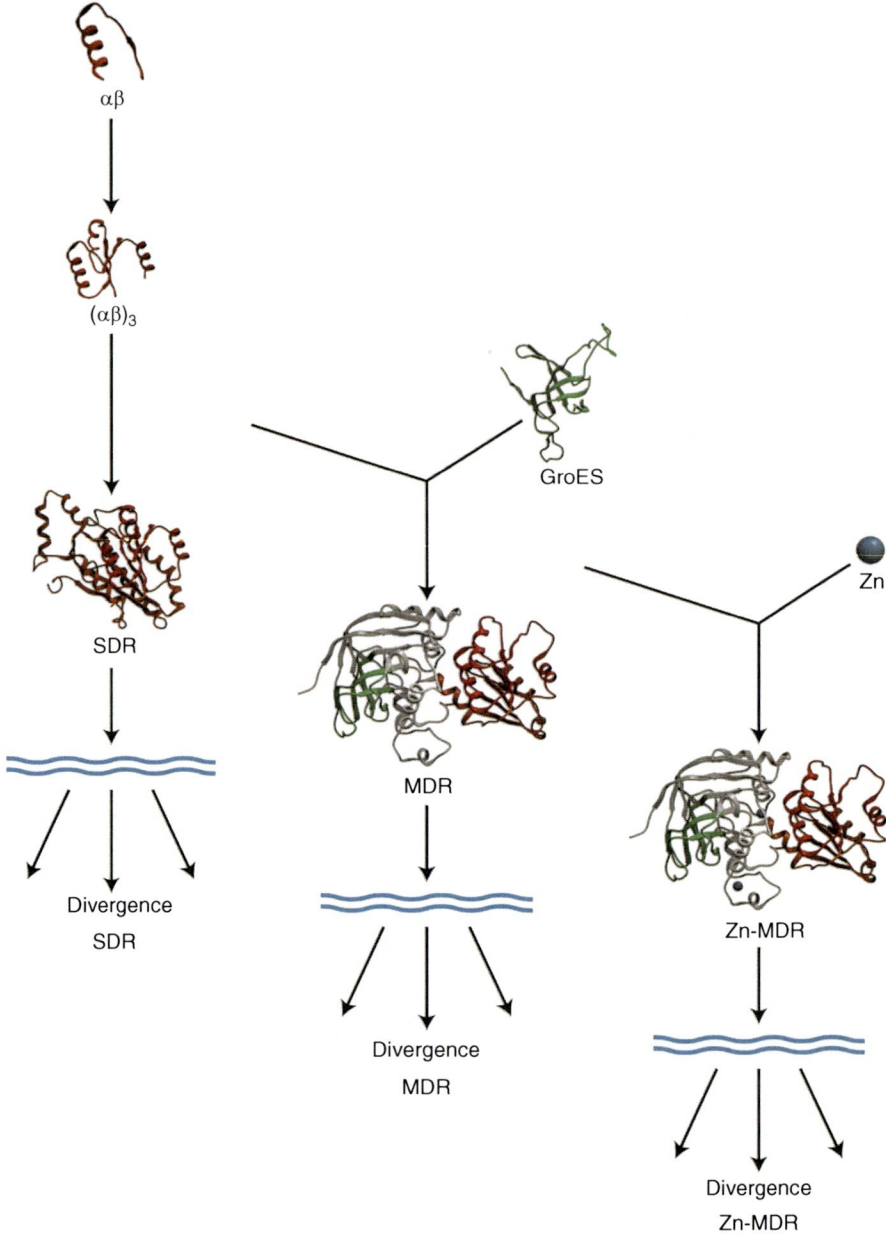

**Zinc Alcohol Dehydrogenases, Fig. 3** Ancestral origins of SDR and MDR forms at separate stages in cellular evolution. SDR has the simplest buildup and widest spread in nature, compatible with an early origin from αβ elements to a Rossmann fold domain of the universal cellular ancestor (wavy *lines*) and subsequent Darwinian evolution in the cells of all three kingdoms of life. MDR proteins are more complex, less widespread, and compatible with a later exit into Darwinian evolution, and still further so for Zn-MDR forms (Reprinted from Jörnvall et al. 2010, with permission from Elsevier)

related building units. At least one line, SDR, appears possible to trace back to an early exit from the last common cellular ancestor (Fig. 3) and is now of widespread occurrence in all types of life forms (one of the three most common protein families), including viruses (Jörnvall et al. 2010). Also the non-zinc MDR line and the Zn-MDR line can be traced far back (Fig. 3) and may reflect later exits from the common cellular ancestor. In fact, these three lines have distributions and occurrences suggesting that they might form separate remnants of the three exits traced to the evolution of all kingdoms of life (Woese 1998), as illustrated in Fig. 3. The origins, stages, and separate functions can be deduced from simple facts, such as present occurrence, abundance, phylogenetic relationships derived from sequence alignments, and average present coenzyme usage (NADPH or $NAD^+$), corresponding to an average catalytic reductive or oxidative direction (Jörnvall et al. 2010). Combined, it appears possible that the Zn-MDR forms, including

Zn-ADH, may represent the recruitment of metal coinciding with the emergence of atmospheric oxygen, or, if then already existing, giving its life form evolutionary advantage to survive at that stage by the fact that zinc was a suitable "choice" of metal since it is redox stable, an important prerequisite for the function of a Lewis acid catalyst in the ADH reaction (see below).

Another interesting feature detected in large-scale sequence comparisons is that MDR forms having a considerable sequence variability, particularly in the coenzyme-binding domain, are more likely to be Zn-MDRs. This phenomenon may indicate a role for zinc and zinc liganding in replacing amino acid conservation with the effect to maintain basic enzyme functions (Jörnvall et al. 2010).

*Structural features of zinc-binding sites and their variability in Zn-ADH.* Catalytic zinc sites in the resting apoenzyme state are generally tetracoordinated employing three protein ligands and a water molecule (Vallee and Auld 1990). The spacing in the primary structure between these amino acid ligands is such that a short and a long spacer motifs are found (cf. ▶ Zinc Structural Site in Alcohol Dehydrogenases). For oxidoreductases like Zn-ADH, the presence of hydrogen-bonding interactions between residues in the short spacer and the relatively large size of the cofactor (NAD$^+$/NADH) may have required an extension of the short spacer up into the range 20–29 residues as is now the case (cf ▶ Zinc Structural Site in Alcohol Dehydrogenases). The structural significances of the short and the long spacer motifs appear to be different, for the short spacer to contribute to a bidentate nucleus for zinc binding, for the long spacer to provide structural flexibility, allowing alignment of residues for interaction with the substrate molecule or the catalytic activity of the zinc ion. The fourth ligand in catalytic zinc sites, the universal water molecule, is activated by ionization, polarization, or poised for displacement (see also below). It has been suggested that the identity of the other three zinc ligands and their spacing pattern has a strong impact on the activation mechanism (Auld and Bergman 2008).

The ADH catalytic zinc-binding site reveals a particular species variability regarding residues and ligand spacing (▶ Zinc Structural Site in Alcohol Dehydrogenases). The distance between the first two protein ligands, the short spacer (Cys46 and His67 in HLADH), varies in the range 20–29 residues, and the distance between the second and the third protein ligand, the long spacer (His67 and Cys174 in HLADH), varies in the range 84–110 residues. The ligands defining the short spacer, Cys and His, reveal a consensus pattern in which the Cys residue is located in a loop structure connecting a β-sheet with an α-helix, and the His residue is always the first residue before a β-sheet structure. The third protein ligand, defining the long spacer of the catalytic zinc site, is frequently a Cys residue, but it varies substantially (▶ Zinc Structural Site in Alcohol Dehydrogenases). In NADP(H)-dependent dehydrogenases, an Asp sometimes replaces the Cys ligand in the long spacer, while this position is occupied by Glu in the polyol dehydrogenase family. The Glu is either the residue adjacent to the His ligand or another Glu residue located 85 residues away. The crystal structures of ADH family members indicate a shift between two types of catalytic zinc liganding correlated with the binding of coenzyme where the fourth ligand is either a water molecule or a Glu residue (corresponding to Glu68 in HLADH). The Glu-coordinated zinc is displaced about 2 Å from the active site, and this type represents an inactive form of ADH, alternatively promotes expulsion of the water/product during the reaction cycle, or promotes formation of the coenzyme-binding site (Auld and Bergman 2008).

*Zinc in enzyme catalysis.* Zinc encompasses distinct physicochemical properties influencing its reactivity and suitability for catalysis (Vallee and Auld 1990). As stated above, the zinc ion (Zn$^{2+}$) is redox stable since it possesses a filled $d$ orbital (d$^{10}$); nevertheless, it participates in enzymatic redox reactions in combination with organic cofactors. The reactivity of Zn$^{2+}$ comes from its potential as a Lewis acid (i.e., an electron-pair acceptor) and its amphoteric properties (i.e., zinc-bound water exists either as a "hydronium" ion or as a hydroxide ion at physiological pH). The fact that the oxidation state of Zn$^{2+}$ does not vary in the highly redox-active environment in cells is particularly important for its function as a Lewis acid catalyst. The protein scaffold supporting zinc sites contributes also to the effect of zinc binding on the entire protein structure. In addition to supplying the specific zinc ligands, the scaffold is a structural basis for interactions between other amino acids and the inner shell ligands. An additional factor influencing the catalytic potential of Zn$^{2+}$ is the rapid ligand exchange which allows fast dissociation of products and high turnover numbers as seen for many Zn-ADHs. Further aspects of the catalytic and structural zinc sites in Zn-ADHs, as well as comparisons of these sites, are given in the adjacent chapter (▶ Zinc Structural Site in Alcohol Dehydrogenases).

In conclusion, Zn-ADH is a structurally, functionally, and evolutionarily well-known enzyme family, constitutes an excellent model of enzyme evolution, illustrates both the catalytic and structural capacity of zinc in protein function and stability, and confers special properties to the MDR-ADH enzyme family, which may have been especially beneficial during the transition from a reductive to an oxidative atmosphere during the evolution of life on earth.

## Cross-References

▶ Zinc, Physical and Chemical Properties
▶ Zinc Structural Site in Alcohol Dehydrogenases
▶ Zinc-Binding Sites in Proteins

## References

Agarwal PK, Webb SP, Hammes-Schiffer S (2000) Computational studies of the mechanism for proton and hydride transfer in liver alcohol dehydrogenase. J Am Chem Soc 122:4803–4812
Auld DS, Bergman T (2008) The role of zinc for alcohol dehydrogenase structure and function. Cell Mol Life Sci 65: 3961–3970
Danielsson O, Atrian S, Luque T et al (1994) Fundamental molecular differences between alcohol dehydrogenase classes. Proc Natl Acad Sci USA 91:4980–4984
Eklund H, Ramaswamy S (2008) Three-dimensional structures of MDR alcohol dehydrogenases. Cell Mol Life Sci 65:3907–3917
Jörnvall H, Hedlund J, Bergman T et al (2010) Superfamilies SDR and MDR: From early ancestry to present forms. Emergence of three lines, a Zn-metalloenzyme, and distinct variabilities. Biochem Biophys Res Commun 396:125–130
Meijers R, Adolph H-W, Dauter Z et al (2007) Structural evidence for a ligand coordination switch in liver alcohol dehydrogenase. Biochemistry 46:5446–5454
Persson B, Hedlund J, Jörnvall H (2008) The MDR superfamily. Cell Mol Life Sci 65:3879–3894
Plapp BV (2010) Conformational changes and catalysis by alcohol dehydrogenase. Arch Biochem Biophys 493:3–12
Vallee BL, Auld DS (1990) Zinc coordination, function, and structure of zinc enzymes and other proteins. Biochemistry 29:5647–5659
Woese CR (1998) The universal ancestor. Proc Natl Acad Sci USA 95:6854–6859

# Zinc Amidohydrolase Superfamily

Frank M. Raushel and Andrew N. Bigley
Department of Chemistry, Texas A&M University, College Station, TX, USA

## Synonyms

Metal-dependent hydrolase

## Definition

A family of structurally and evolutionarily related enzymes that utilize a $(\beta/\alpha)_8$-barrel protein fold as a scaffold for metal-catalyzed hydrolytic, deamination, isomerization, and decarboxylation reactions.

## Amidohydrolase Superfamily

Thousands of proteins have been identified as members of the amidohydrolase superfamily of enzymes (Seibert and Raushel 2005). This family of structurally and evolutionarily related enzymes catalyzes a wide variety of enzymatic transformations in numerous metabolic pathways in every organism examined to date (Gerlt and Raushel 2003). Most of the reactions catalyzed by members of the amidohydrolase superfamily involve the hydrolysis of amide or ester bonds of phosphates and carboxylates (Seibert and Raushel 2005). However, members of this superfamily have also been shown to catalyze isomerization, hydration, and decarboxylation reactions. The defining characteristic of the amidohydrolase superfamily is the core structural fold of the enzyme and location of the active site (Seibert and Raushel 2005). All known amidohydrolases have a distorted $(\beta/\alpha)_8$-barrel structural fold, the most common protein fold observed in nature (Anantharaman et al. 2003).

The structural fold for the $(\beta/\alpha)_8$-barrel is presented in Fig. 1a. At the heart of this structural fold is a central core of eight parallel β-strands that are surrounded by an equal number of α-helices (Gerlt and Raushel 2003). The eight β-strands of the central barrel structure are all oriented in the same direction, and thus the N-terminal ends are found at one end of the barrel and the C-terminal ends at the other. Connecting the β-strands and α-helices are irregular polypeptides that

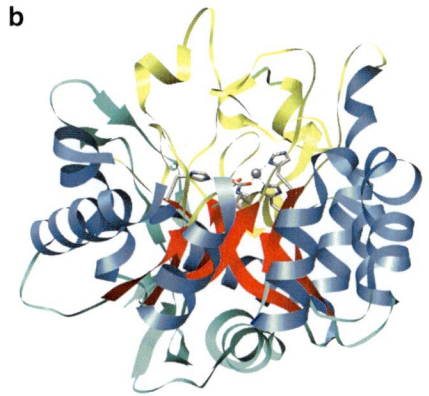

**Zinc Amidohydrolase Superfamily, Fig. 1** Crystal structure of the $(\beta\alpha)_8$-barrel fold of dihydroorotase from *E. coli* (pdb code 1j79). Core structural elements are colored *red* (β-sheets) and *blue* (α-helices). The loops extending from the N-terminal side of the core barrel are colored cyan. Loops extending from the C-terminal side of barrel are colored *yellow*. Side chains of metal-binding ligand are shown as sticks, and the metal ions are shown as spheres. (**a**) Top-down view of the C-terminal end of barrel. (**b**) Side view of barrel structure

form interconnecting loops. In a typical $(\beta/\alpha)_8$-barrel protein, the β-strand/loop/α-helix motif is repeated eight times (Fig. 1b). In all of the $(\beta/\alpha)_8$-barrel proteins examined to date, the active sites are always found at the C-terminal end of the β-barrel. In the amidohydrolase superfamily, the β-barrel is utilized as a scaffold for the binding of one to three divalent metal ions (Seibert and Raushel 2005). The most common metal is $Zn^{2+}$, but other examples have included $Fe^{2+}$, $Ni^{2+}$, or $Mn^{2+}$.

In a typical binuclear metal center, the more buried metal, known as the α-metal, is ligated by two histidines in an HxH motif from β-strand 1 and an aspartate from β-strand 8 (Seibert and Raushel 2005). The more exposed site, known as the β-metal, is ligated by histidines from β-strands 5 and 6. There is typically a carboxylate ligand from the end of β-strand 4 that acts a bridge between the two metals, and hydroxide serves as a second bridging ligand. The substrate-binding site that gives each enzyme its unique specificity is typically made up of residues from the loops connecting the C-terminal end of one β-strand to the following helix of the barrel core. The enzymes of the amidohydrolase superfamily catalyze a variety of reactions involving phosphorus-oxygen, carbon-nitrogen, and carbon-oxygen bond cleavages (Seibert and Raushel 2005). While the basic fold of the enzymes is conserved throughout the family, there are many variations on this theme. Amidohydrolases can differ in the number of metal ions used as well as the precise roles of the metal in the chemical reaction. These differences allow the superfamily to be subdivided according to the structure of the metal center and the type of reaction that is catalyzed. Among the most common chemical reactions are α-β binuclear hydrolysis reactions, α-mononuclear deaminations, α-mononuclear isomerizations, and decarboxylations.

### α-β Hydrolytic Reactions

The amidohydrolases with two divalent cations make up a diverse group of enzymes that hydrolyze ester and amide bonds of phosphates and carboxylates (Seibert and Raushel 2005). As exemplified by the enzyme dihydroorotase (DHO), the binuclear metal center is ligated at the α-site to the HxH motif from β-strand 1 and an aspartate from β-strand 8 (Seibert and Raushel 2005; Porter et al. 2004). The metal at the β-site is coordinated by histidines from β-strands 5 and 6 and has an open coordination site to accommodate substrate binding. The metal center is additionally supported by the presence of a carboxylated lysine from β-strand 4 that bridges the two metal ions. The metals are also bridged by a nucleophilic hydroxide (Fig. 2a). In addition to the direct interaction with the β-metal, substrate binding is facilitated by electrostatic interactions between the substrate carboxylate and an arginine from loop 1, an asparagine from loop 2, and a histidine from loop 8. The reaction catalyzed by DHO is shown in Fig. 3a. The coordination of the substrate to the metal center enables the carbonyl

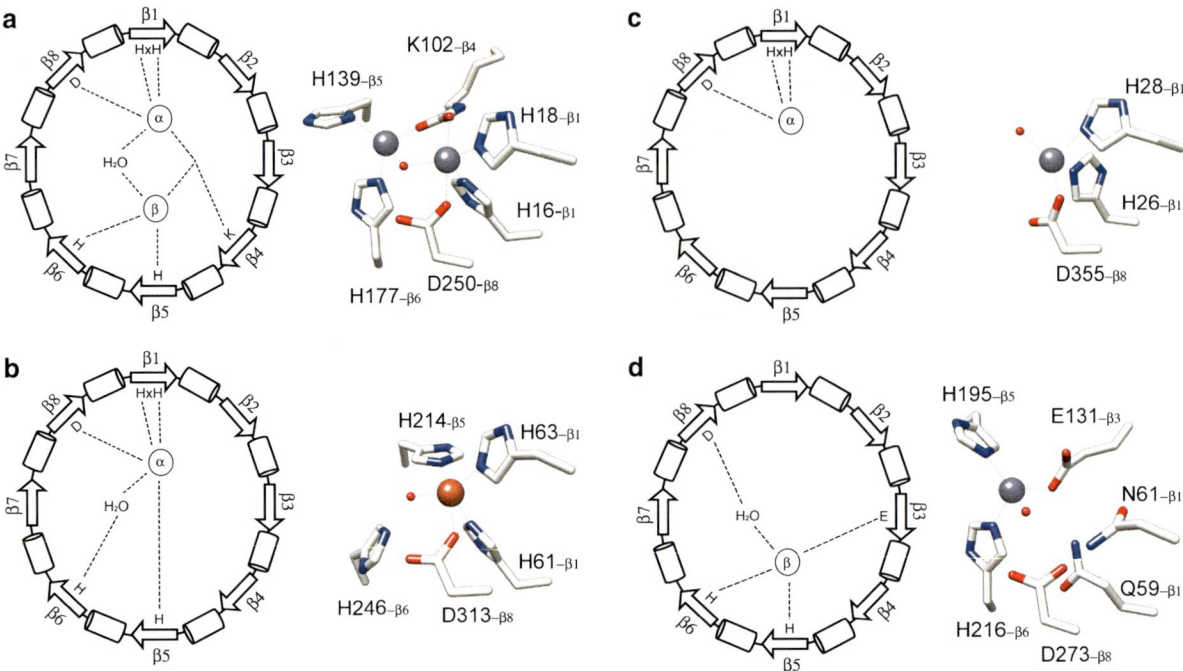

**Zinc Amidohydrolase Superfamily, Fig. 2** Metal coordination patterns for specific examples of amidohydrolases shown as a cartoon wheel diagram and crystal structure for dihydroorotase (**a**), cytosine deaminase (**b**), uronate isomerase (**c**), and N-acetyl-D-glucosamine-6-phosphate deacetylase (**d**)

group to be polarized prior to attack by the nucleophilic hydroxide (Porter et al. 2004). In the proposed reaction mechanism for DHO, the bridging hydroxide attacks the substrate carbonyl group resulting in a tetrahedral transition state which, on collapse, results in the ring-opening cleavage of the carbon-nitrogen bond. The β-metal stabilizes the generation of charge in the transition state. The metal-ligating aspartate residue from the end of β-strand 8 serves as a general base to deprotonate the attacking hydroxide and then donates the proton back to the leaving amine.

### α-Site Deamination Reactions

Many amidohydrolases utilize only a single divalent metal ion for catalysis. The purine and pyrimidine deaminases, including cytosine deaminase (CDA), which converts cytosine to uracil, utilize a mononuclear metal center to catalyze the removal of the amino group from nucleotides (Ireton et al. 2002). Despite having a single metal ion, five of the six ligands present in the binuclear metal hydrolases are in the active site of CDA (Fig. 2b). The HxH motif from β-strand 1 and the aspartate from β-strand 8 are metal-binding ligands. The absence of the β-metal allows for the assignment of other residues to new roles. The histidine from β-stand 5 rotates to ligate the metal at the α-site. The histidine from β-strand 6 does not ligate a metal but serves to hydrogen bond to the nucleophilic water. A bridging ligand from β-strand 4 is absent in CDA. Substrate binding to CDA is facilitated by aromatic stacking with a tryptophan from β-strand 8 and hydrogen bonding with a glutamine from β-strand 3. The histidine from β-stand 6 is thought to deprotonate the nucleophilic water prior to the attack on the substrate (Ireton et al. 2002). A glutamate from β-strand 5 serves as a general acid to protonate the ring nitrogen as the nucleophilic water attacks the substrate to form the tetrahedral transition state. The aspartate from β-strand 8 then deprotonates the oxygen to facilitate the collapse of the transition state and the release of ammonia.

### α-Site Isomerization

In addition to the single metal hydrolysis reactions, the α-site metal can facilitate aldose-ketose isomerizations of some sugars. The enzyme uronate isomerase (URI) converts the six carbon sugars glucuronate and galacturonate to fructuronate and tagaturonate,

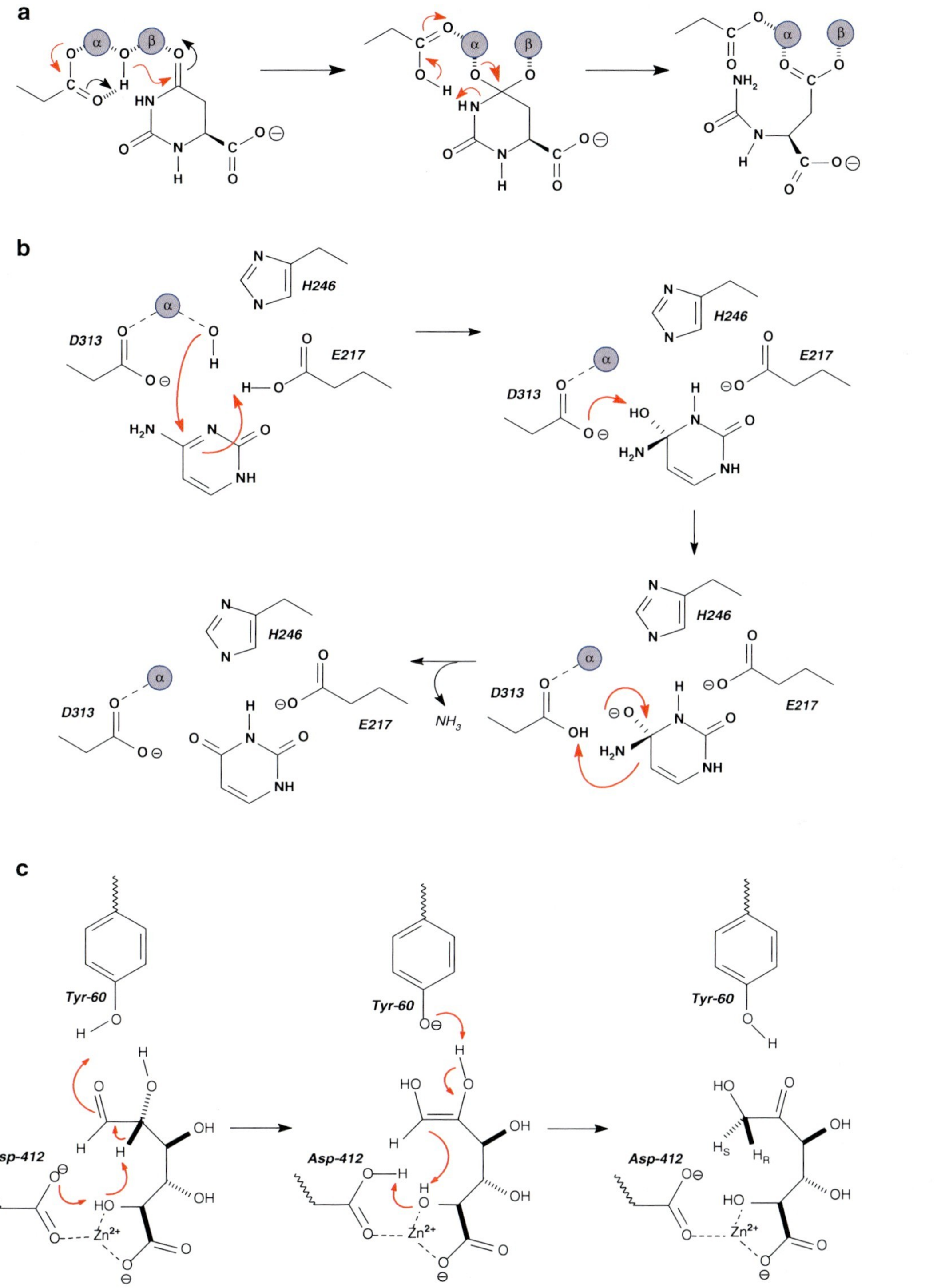

**Zinc Amidohydrolase Superfamily, Fig. 3** Proposed chemical mechanisms for dihydroorotase (**a**), cytosine deaminase (**b**), and uronate isomerase (**c**)

respectively (Nguyen et al. 2009). Unlike CDA, the β-site ligands are not conserved in the structure of URI. The metal is bound via the HxH motif from β-strand 1 and the conserved aspartate from β-strand 8, but there are no equivalent ligands from β-strands 4, 5, or 6 (Fig. 2c). Structural studies have revealed that the unbound enzyme binds water at an open coordination site on the metal, but the substrate-bound structure has no water ligated to the metal center. Substrate binding is facilitated by divalent coordination to the metal center via the C-5 hydroxyl and the C-6 carboxylate (Nguyen et al. 2009). Additionally, the C-6 carboxylate interacts with an arginine from loop 6, and the hydroxyls from C-2 and C-3 hydrogen bond to an arginine from β-strand 8. The hydroxyl from C2 is able to form a hydrogen bond to a histidine from β-strand 1. The catalytic mechanism proposed for URI has been probed by both structural and kinetic studies (Nguyen et al. 2009). The mechanism shown in Fig. 3c maintains the characteristics of the amidohydrolases for the isomerization reaction. The obvious difference in the isomerization is the lack of the nucleophilic water. With URI, it is thought that the C-5 hydroxyl replaces the water, and the coordination to the metal facilitates its deprotonation by the conserved aspartate from β-strand 8. The C-5 hydroxylate can then deprotonate C-2. A tyrosine from β-strand 1 is proposed to act as the general acid to protonate the oxygen at C-1 to form an enol intermediate. The tyrosine then removes the proton from the C-2 hydroxyl, and the C-5 hydroxyl protonates C-1 to complete the conversion from the aldose to the ketose product.

β-Site Hydrolysis

A small number of amidohydrolases catalyze hydrolytic reactions using a β-mononuclear metal center. Among these is the N-acetyl-D-glucosamine-6-phosphate deacetylase (NagA) from *Escherichia coli* (Hall et al. 2007a). The structures of NagA from different species have been solved. In some species, the NagA homolog contains an α-β binuclear center (Hall et al. 2007b). In other species, it is unclear if the metal center is mononuclear or binuclear due to the presence of ligating residues for both sites. In the case of the enzyme from *E. coli*, the mononuclear metal center is located at the β-site with the metal ligated by histidines from β-strands 5 and 6 (Hall et al. 2007a) (Fig. 2d). The bridging carboxylate observed in the binuclear metal sites typically comes from β-strand 4, but in NagA, a glutamate from β-strand 3 occupies this role. The aspartate from β-strand 8 is conserved and interacts with the hydrolytic water. Residues equivalent to the HxH motif from β-strand 1 are present as glutamine and asparagine. Substrate binding to NagA is achieved with a direct ligation of the acetyl group to the metal along with a hydrogen bond between the carboxyl oxygen and a histidine from loop 3. Hydrogen bonding between the C-1 hydroxyl and a histidine from loop 7 positions the acetyl group. There is an additional interaction between the phosphate group and an arginine from loop 6. As is observed with other amidohydrolases, the aspartate from β-strand 8 serves as the base to deprotonate the nucleophilic water preceding the attack on the acetyl carbon. The charge of the tetrahedral transition state is supported by the bidentate ligation to the metal center and the interaction between the acetyl oxygen and the histidine from loop 3. The collapse of the transition state results in the cleavage of the carbon-nitrogen bond with the protonation of the leaving amine by the aspartate from β-strand 8.

## Cross-References

- ▶ Cobalt-containing Enzymes
- ▶ Magnesium in Biological Systems
- ▶ Urease
- ▶ Zinc Aminopeptidases, Aminopeptidase from Vibrio Proteolyticus (Aeromonas proteolytica) as Prototypical Enzyme
- ▶ Zinc-Binding Sites in Proteins

## References

Anantharaman V, Aravind L, Koonin E (2003) Emergence of diverse biochemical activities in evolutionarily conserved structural scaffolds of proteins. Curr Opin Chem Biol 7:12–20

Gerlt J, Raushel F (2003) Evolution of function in (β/α)8-barrel enzymes. Curr Opin Chem Biol 7:252–264

Hall R, Brown S, Fedorov A, Fedorov E, Xu C, Babbitt P, Almo S, Raushel F (2007a) Structural diversity within the mononuclear and binuclear active sites of N-acetyl-D-glucosamine-6-phosphate deacetylase. Biochemistry 46: 7953–7962

Hall R, Xiang D, Xu C, Raushel F (2007b) N-Acetyl-D-glucosamine-6-phosphate deacetylase: substrate activation via a single divalent metal ion. Biochemistry 46:7942–7952

Ireton G, McDermott G, Black M, Stoddard B (2002) The structure of *Escherichia coli* cytosine deaminase. J Mol Biol 315:687–697

Nguyen T, Fedorov A, Williams L, Fedorov E, Li Y, Xu C, Almo S, Raushel F (2009) The mechanism of the reaction catalyzed by uronate isomerase illustrates how an isomerase may have evolved from a hydrolase within the amidohydrolase superfamily. Biochemistry 48:8879–8890

Porter T, Li Y, Raushel F (2004) Mechanism of the dihydroorotase reaction. Biochemistry 43:16285–16292

Seibert C, Raushel F (2005) Structural and catalytic diversity within the amidohydrolase superfamily. Biochemistry 44:6383–6391

# Zinc Aminopeptidases, Aminopeptidase from Vibrio Proteolyticus (Aeromonas proteolytica) as Prototypical Enzyme

Richard C. Holz[1], Anna Starus[1] and Danuta M. Gillner[2]
[1]Department of Chemistry and Biochemistry, Loyola University Chicago, Chicago, IL, USA
[2]Department of Chemistry, Silesian University of Technology, Gliwice, Poland

## Synonyms

AAP, Aminopeptidase from *Vibrio proteolyticus* (*Aeromonas proteolytica*); ACD, N-Acetylcitrulline deacetylase; ALS, Amyotrophic lateral sclerosis; ArgE, ArgE-encoded N-acetyl-L-ornithine deacetylase; Bestatin, N-[(2 S, 3R)-3-amino-2-hydroxy-4-phenylbutyryl]-L-leucin; $CPG_2$, Carboxypeptidase $G_2$ from *Pseudomonas sp.* strain RS-16; CSD, Cambridge Structural Database; DapE, DapE-encoded N-succinyl-L,L-diaminopimelic acid desuccinylase; DFT, Density functional theory; DppA, D-aminopeptidase from *Bacillus subtilis*; EPR, Electron paramagnetic resonance; GCP-II, Glutamate carboxypeptidase II; ITC, Isothermal titration calorimetry; LLL, L-Leucyl-L-leucyl-L-leucine; LPA, L-leucinephosphonic acid; L-*p*NA, L-Leucine-p-nitroanilide; MCD, Magnetic circular dichroism; ORF, Open reading frame; QM/MM, quantum mechanical/molecular mechanical; rAAP, Recombinant leucine aminopeptidase; SAP, Aminopeptidase from *Streptomyces griseus*; TET, Tetrahedral-shaped aminopeptidase; Tris, Tris(hydroxymethyl) aminomethane; UV–vis, Ultraviolet–visible spectroscopy; VTVH, Variable-temperature variable-field; WT, Wild type

## Definition

Aminopeptidases containing zinc(II) ions in their active sites are central to numerous biological processes, and consequently, characterization of their structure and function is a problem of outstanding importance. One of the least explored groups of these enzymes is the aminopeptidases that contain dinuclear metal centers. These enzymes play key roles in carcinogenesis, tissue repair, and protein degradation processes. The determination of detailed reaction mechanisms for these enzymes is required for the design of highly potent, specific inhibitors that can function as potential pharmaceuticals. The aminopeptidase from *Vibrio proteolyticus (Aeromonas proteolytica; AAP)*, one of the best mechanistically characterized hydrolytic enzymes that contains a dinuclear center, serves as a paradigm for these enzymes since they all contain a carboxylate-bridged dinuclear Zn(II) active site with nearly symmetrical coordinating ligands and is, thus, the focus of this chapter.

## Introduction

Metallohydrolases that contain cocatalytic Zn(II) active sites catalyze a set of diverse reactions involving physiologically key biomolecules such as polypeptides, nucleic acids, antibiotics, and phospholipids (Lipscomb and Sträter 1996; Wilcox 1996; Holz et al. 2003; Lowther and Matthews 2002). As a result, they are central to numerous physiological processes including but not limited to carcinogenesis, tissue repair, neurological disorders, bacterial cell wall synthesis, protein maturation, hormone level regulation, cell cycle control, and protein degradation. Some of the metallohydrolases that fall into this group include the leucine aminopeptidases from *Vibrio proteolyticus* (*Aeromonas proteolytica*, AAP) and *Streptomyces griseus* (SAP), the tetrahedral-shaped aminopeptidase (TET) and D-aminopeptidases (DppA). Additional related metallohydrolases include the carboxypeptidase $G_2$ from *Pseudomonas sp.* strain RS-16 ($CPG_2$), the *dapE*-encoded N-succinyl-L,L-diaminopimelic acid desuccinylase (DapE), the *argE*-encoded N-acetyl-L-ornithine deacetylase (ArgE), the N-acetylcitrulline deacetylase (ACD), and glutamate carboxypeptidase II (GCP-II). The importance of understanding the structure and function of metallohydrolases that contain cocatalytic divalent Zn(II) active sites is underscored by their central role in several disease states including stroke, diabetes, cancer, HIV, bacterial infections, and neuropsychiatric disorders associated with the dysregulation of glutamatergic neurotransmission, such as schizophrenia, seizure disorders, and amyotrophic lateral sclerosis (ALS). For these reasons, cocatalytic metallohydrolases have become the subject of intense efforts in inhibitor design.

The X-ray crystal structures for AAP (PDB: 1AMP), SAP (PDB: 1XJO), TET (PDB: 1XFO), DppA (PDB: 1HI9), CPG$_2$ (PDB: 1CG2), DapE (PDB: 3IC1), GCP-II (PDB: 2OOT), and ACD (PDB: 2F7V) have been reported and, interestingly, reveal that each of these enzymes contain nearly identical cocatalytic active sites (Fig. 1) (Holz et al. 2003; Holz 2002). To date, no X-ray crystallographic data has been reported for ArgE; however, all of the amino acids that function as metal ligands in each of the structurally characterized enzymes are fully conserved in ArgE. Each of these cocatalytic metallohydrolases contain an identical set of active site residues [HDDEE (E/D)D] that are capable of binding two Zn(II) ions in vivo. All of the available structures except ACD which is the mononuclear enzyme reveal a nearly identical (μ-aquo)(μ-carboxylato)dizinc(II) core. Despite their ubiquity and the considerable structural information available, there remains a significant gap in knowledge as to how these structural motifs relate to function. Moreover, the mononuclear Zn(II) structures of ACD and DapE underscore the fact that a majority of the metallohydrolases that contain cocatalytic Zn(II) active sites also retain some catalytic activity as mononuclear enzymes but typically exhibit faster rates with dinuclear sites. The fact that some hydrolases utilize a single divalent metal ion while others can function with either one or two divalent metal ions in their active site, and still others require two metal ions to catalyze the same chemical reaction, is not well understood.

This chapter focuses on the recent literature with an emphasis on mechanistic aspects of AAP, which is by far the best characterized enzyme in this group and

**Zinc Aminopeptidases, Aminopeptidase from Vibrio Proteolyticus (Aeromonas proteolytica) as Prototypical Enzyme, Fig. 1** Drawing of the active site of enzymes in this class

serves as a paradigm for these enzymes since they all contain a carboxylate-bridged dinuclear Zn(II) active site with nearly symmetrical coordinating ligands (Holz 2002). In addition, new structural studies will be discussed that provide important mechanistic insights with a focus on protein-inhibitor and protein-substrate interactions. These studies provide an important starting point for the rational design of inhibitors that may function as future pharmaceuticals.

# History

The aminopeptidase from *Aeromonas (Vibrio) proteolytica* (EC 3.4.11.10; AAP) is one of the best mechanistically characterized hydrolytic enzymes that contains a cocatalytic Zn(II) center (Holz 2002). AAP is ideally suited for structure/function studies since it can be obtained in large quantities (>100 mg), can be genetically manipulated, and is a remarkably thermostable 29.5-kDa monomer. Moreover, the native Zn(II) ions can be replaced by several first row transition metal ions in a sequential fashion, providing highly active metal-substituted forms of the enzyme. Therefore, structural properties of each active site divalent metal ion (e.g., coordination geometry and ligand type) regulate the Lewis acidity, which in turn controls the level of hydrolytic activity. Possible catalytic roles for one or both metal ions include (a) to bind and position substrate, (b) to bind and activate a water molecule to yield an active site hydroxide nucleophile, and/or (c) to stabilize the transition state of the hydrolytic reaction. The fact that metal binding is sequential for AAP allows the mechanistic role of each metal ion to be examined independently of one another. Therefore, understanding the reaction mechanism of AAP provides a starting point for the determination of reaction mechanisms for other, less-studied metallohydrolases that contain cocatalytic Zn(II) sites (Holz et al. 2003). As a detailed review of AAP appeared in 2002 (Holz 2002) and a structural review also appeared in 2002 (Lowther and Matthews 2002), data published over the past 9 years only will be reviewed.

## AAP Expression Systems

The breakthrough for further insight into the catalytic mechanism of AAP occurred with the development of an *E. coli* expression system for AAP (Bzymek et al. 2004). The open reading frame (ORF) for AAP encodes a 54-kDa enzyme; however, the extracellular enzyme has a molecular weight of 43 kDa (Fig. 2). This form of AAP is further processed to a mature, thermostable 32-kDa form via the cleavage of both N- and C-terminal propeptides. Overexpression of the full 54-kDa AAP enzyme provided an enzyme that was significantly less active due to a cooperative inhibitory interaction between both propeptides. Overexpression of AAP lacking its C-terminal propeptide provided an enzyme of 43 kDa that exhibited an identical $k_{cat}$ value to WT AAP but an increased $K_m$ value, suggesting competitive inhibition of AAP by the N-terminal propeptide ($K_i \sim 0.13$ nM). Processing of this 43-kDa enzyme by proteinase K removed the N-terminal leader sequence and resulted in a 32-kDa form of AAP that was characterized by kinetic and spectroscopic methods and was shown to be identical to WT AAP. Sonoda et al. (2009) reported another overexpression system for AAP in *E. coli*. In this system, a recombinant vibriolysin was coexpressed with AAP in order to cleave the C-terminal propeptide. The N-terminal propeptide was also absent so it was suggested that AAP may auto process its N-terminus. Bennett and coworkers improved upon all *E. coli* expression systems for AAP by including a polyhistidine tag on the C-terminal of the 32-kDa gene encoding for AAP (Hartley and Bennett 2009). This heterologous expression system rapidly and reproducibly provided 10 mg/L of AAP. Similarly, Rivero and coworkers have designed a recombinant system with an N-terminal (His)$_6$ tag for industrial use in interferon processing (Perez-Sanchez et al. 2011). Finally, Mall et al. (2011) recently reported the design and cloning of the DNA sequence corresponding to the mature and active 32-kDa AAP enzyme. The ease of purification of recombinant forms of AAP has been the primary driver for new detailed mechanistic studies on AAP through site-directed mutagenesis.

**Zinc Aminopeptidases, Aminopeptidase from Vibrio Proteolyticus (Aeromonas proteolytica) as Prototypical Enzyme, Fig. 2** Open reading frame (ORF) for AAP indicating the N- and C-terminal extensions

## AAP Glutamate-151 Mutant Enzymes

Glutamate-151 (E151) had been proposed to act as a general acid or base during the peptide hydrolysis reaction catalyzed by AAP (Fig. 3a), and given its importance in the proposed catalytic mechanism, it was the first AAP mutant reported. E151 was mutated to glutamine (Q), alanine (A), and aspartate (D) (Bzymek and Holz 2004). Upon substitution to D, the Michaelis constant ($K_m$) did not change but the rate of the reaction decreased in the order: E (100% activity), D (0.05%), Q (0.004%), and A (0%). Examination of the pH dependence of the kinetic constants $k_{cat}$ and $K_m$ for E151D-AAP, revealed a change in the p$K_a$ of a group that ionized at pH 4.8 in rAAP to 4.2. The remaining p$K_a$s at 5.2, 7.5, and 9.9 did not change. Proton inventory studies on E151D-AAP indicated that one proton is transferred in the rate-limiting step of the reaction at pH 10.5 for both rAAP and E151D-AAP, but at pH 6.5, two protons and general solvation effects are responsible for the observed effects in the reaction catalyzed by rAAP and E151D-AAP, respectively. Based on these data, E151 was assigned as intrinsically involved in the peptide hydrolysis reaction catalyzed by AAP and plays the role of both an active site general acid and base. The role was corroborated by a quantum mechanical/molecular mechanical (QM/MM) study on AAP, which implicated E151 as an acceptor of a proton from the N-terminus of the incoming peptide and also a proton donor to the newly formed N-terminus (Schürer et al. 2004). The latter step was calculated to be the rate-limiting step. It was also confirmed by density functional theory calculations (DFT) where an accumulated barrier for proton transfer from E151 to the nitrogen of the peptide bond was found to be 18.2 kcal/mol (Chen et al. 2008).

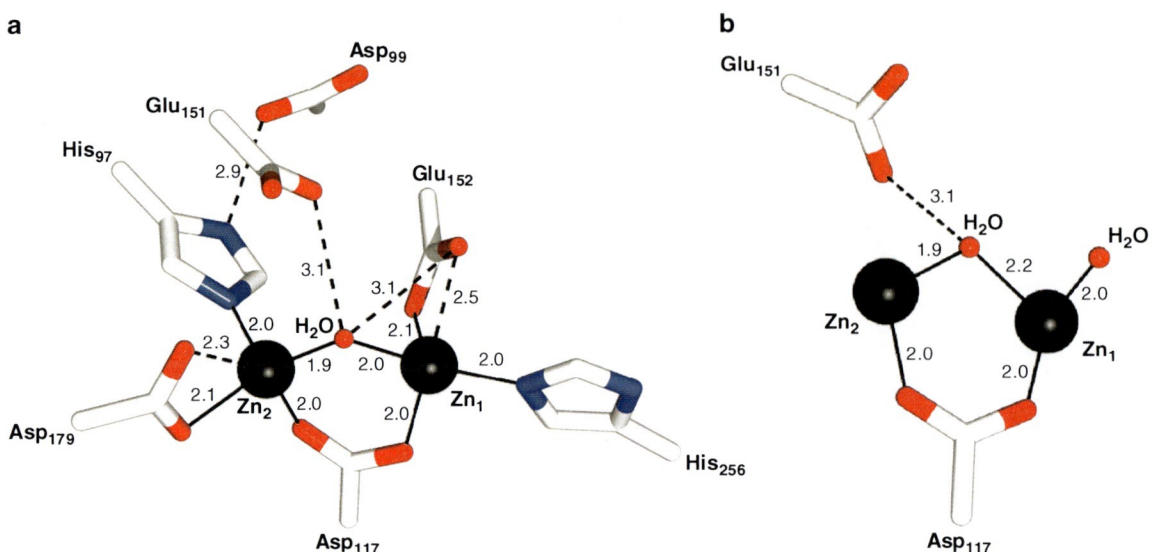

**Zinc Aminopeptidases, Aminopeptidase from Vibrio Proteolyticus (Aeromonas proteolytica) as Prototypical Enzyme, Fig. 3** Active site of AAP at (**a**) 0.95-Å resolution and (**b**) at pH 4.5. All distances are in Å and the terminal ligands have been removed from (**b**) for clarity

The close proximity of residues, such as E151 to the bridging water/hydroxide molecules (2.80 Å and 3.94 Å in carboxypeptidase $G_2$, and 3.30 Å and 3.63 Å in AAP), suggested it may also be involved in stabilizing the dinuclear active site (Holz et al. 2003; Lowther and Matthews 2002; Holz 2002). Therefore, structural perturbations for the E151D-AAP and E151A-AAP enzymes were examined by UV–vis and EPR spectroscopy of the Co(II)-substituted enzymes (Bzymek et al. 2005a). Substitution of E151 to A did not significantly perturb the electronic absorption spectrum of AAP. However UV–vis spectra of mono- and dicobalt(II) E151D-AAP exhibited lower molar absorptivities (23 $M^{-1}$ $cm^{-1}$ and 43 $M^{-1}$ $cm^{-1}$ vs. 56 $M^{-1}$ $cm^{-1}$ and 109 $M^{-1}$ $cm^{-1}$ for E151D-AAP and AAP, respectively) suggesting both Co(II) ions reside in distorted five- or six-coordinate sites. EPR spectra of [Co_(E151D-AAP)], [ZnCo(E151D-AAP)], and [CoCo(E151D-AAP)] were identical, with $g_\perp = 2.35$, $g_\parallel = 2.19$, and $E/D = 0.19$, similar to [CoCo(AAP)]. On the other hand, the EPR spectrum of [Co_(E151A-AAP)] was best simulated assuming the presence of two species with (a) $g_{x,y} = 2.509$, $g_z = 2.19$, $E/D = 0.19$, and $A = 0.0069$ $cm^{-1}$ and (b) $g_{x,y} = 2.565$, $g_z = 2.19$, $E/D = 0.20$, $A = 0.0082$ $cm^{-1}$ indicative of a five- or six-coordinate species. Isothermal titration calorimetry (ITC) experiments revealed a large decrease in metal ion affinities, with $K_d$ values elevated by factors of ~850 and ~24,000 for the first metal binding events of E151D- and E151A-AAP, respectively. These data are consistent with the hypothesis that E151 stabilizes the dinuclear active site in AAP.

It was also shown that E151 is not the only residue that is capable of functioning as a general acid/base during catalysis (Bzymek et al. 2005b). E151 was mutated to a histidine residue which resulted in an active AAP enzyme that exhibited a $k_{cat}$ value of 2.0 $min^{-1}$. This $k_{cat}$ value is over 2,000 times slower than rAAP (4,380 $min^{-1}$). The impetus behind this mutation was the X-ray crystal structure of the D-aminopeptidase from *Bacillus subtilis* (DppA) (PDB: 1HI9), which revealed a histidine residue that resides in an identical position to E151 in AAP. Because the active site ligands for DppA and AAP are identical and AAP is a prototypical cocatalytic Zn (II) protease, the E151H mutant of AAP was used as a model system for DppA. Kinetic data determined as a function of pH revealed a change in an ionization constant in the enzyme-substrate complex from 5.3 in rAAP to 6.4, consistent with E151 in AAP being the active site general acid/base while proton inventory studies at pH 8.50 indicated that one proton is transferred in the rate-limiting step. UV–vis and EPR spectra of Co(II)-loaded E151H-AAP indicated that the first metal ion resides in a hexacoordinate/pentacoordinate equilibrium environment, whereas the second metal ion is six coordinate and ITC data revealed that the two Zn(II) ions bind 330 and 3 times more weakly, respectively, to E151H-AAP compared to rAAP. The X-ray crystal structure of [ZnZn(E151H-AAP)] was also solved to 1.9-Å resolution (PDB: 2ANP) and revealed that alteration of E151 to histidine does not introduce any major conformational changes to the overall protein structure or the dinuclear Zn(II) active site. Therefore, a histidine residue can function as the general acid/base in hydrolysis reactions of peptides, and through analogy of the role of E151 in AAP, H115 in DppA likely shuttles a proton to the leaving group of the substrate.

## X-Ray Crystallographic Studies

The spectroscopically inactive Zn(II) ions in AAP have been routinely substituted with the spectroscopically active Co(II) ions, and these Co(II)-loaded AAP enzymes have been exploited to gain insight via spectroscopic methods to characterize substrate and transition-state analog inhibitors of AAP providing important mechanistic information. However, no structural data existed for a Co(II)-loaded AAP enzyme, which was a significant gap between AAP structures containing Zn(II) ions and spectroscopic data on Co(II)-loaded enzymes. To fill that gap, the X-ray crystal structure of the Co(II)-loaded form of AAP solved at 2.2-Å resolution (Fig. 4) was reported (PDB: 2PRQ). The Co(II)-loaded form of AAP folds in an identical manner to that of the Zn(II)-loaded enzyme so Co(II) binding does not introduce any major conformational changes to the overall protein structure. The amino acid residues that function as ligands in the Zn(II)-loaded structures are also ligands in the Co(II)-loaded enzyme with only the expected minor perturbations in bond lengths. The Co(II)–Co(II) distance is 3.3 Å, and interestingly, tris(hydroxymethyl)aminomethane (Tris) coordinates to the dinuclear Co(II) active site of AAP. One hydroxyl oxygen atom of Tris bridges between the two Co(II) ions, with Co1–O and Co2–O bond distances of 2.2 and 1.9 Å, respectively. Each Co(II) ion resides in a distorted trigonal bipyramidal geometry.

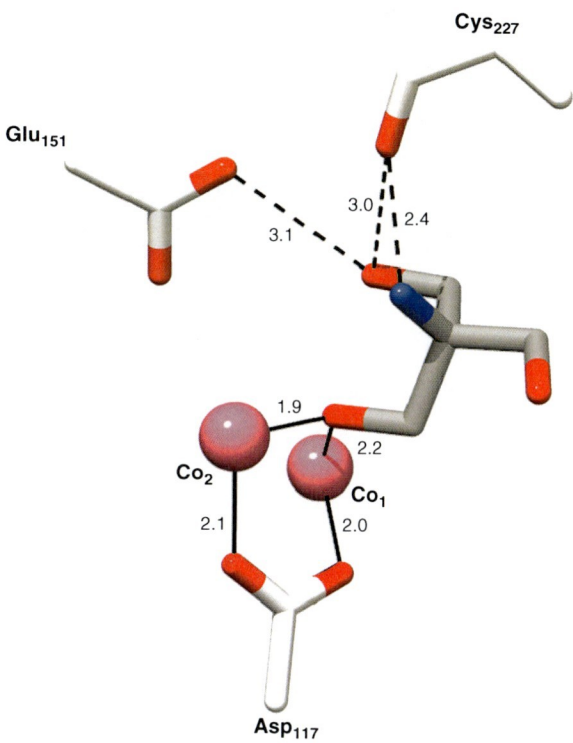

**Zinc Aminopeptidases, Aminopeptidase from Vibrio Proteolyticus (Aeromonas proteolytica) as Prototypical Enzyme, Fig. 4** Active site of the Co(II)-substituted form of AAP bound by a molecule of Tris. All distances are in Å and the terminal ligands have been removed for clarity

An important structural study was reported in which the X-ray crystal structure of AAP was solved at 0.95-Å resolution (PDB: 1RTQ) (Fig. 3a). This structure, which rivals the resolution of small molecules, allowed the precise modeling of atomic positions since the positions of all the atoms were not restrained during refinement providing precise Zn-ligand distances. The protonation state of the bridging oxygen species was proposed by comparing Zn–O distances found in the Cambridge Structural Database (CSD) and those studied by computational methods to the precise Zn–O distances obtained in this structure. These standard values indicate that the average Zn–OH coordination distances are 1.90–2.00 Å while the average Zn–OH$_2$ coordination distances are 2.10 Å to 2.20 Å. The Zn1–O distance in the 0.95-Å resolution AAP structure is 2.01 Å compared to 1.93 Å for Zn2, suggesting the bridging oxygen species is an OH$^-$. Furthermore, a hydrogen-bonding interaction is clearly evident between the bridging oxygen atom and E151 (Bzymek

**Zinc Aminopeptidases, Aminopeptidase from Vibrio Proteolyticus (Aeromonas proteolytica) as Prototypical Enzyme, Fig. 5** Active sites of AAP bound by (**a**) BuBA and (**b**) LPA. All distances are in Å and the terminal ligands have been removed for clarity

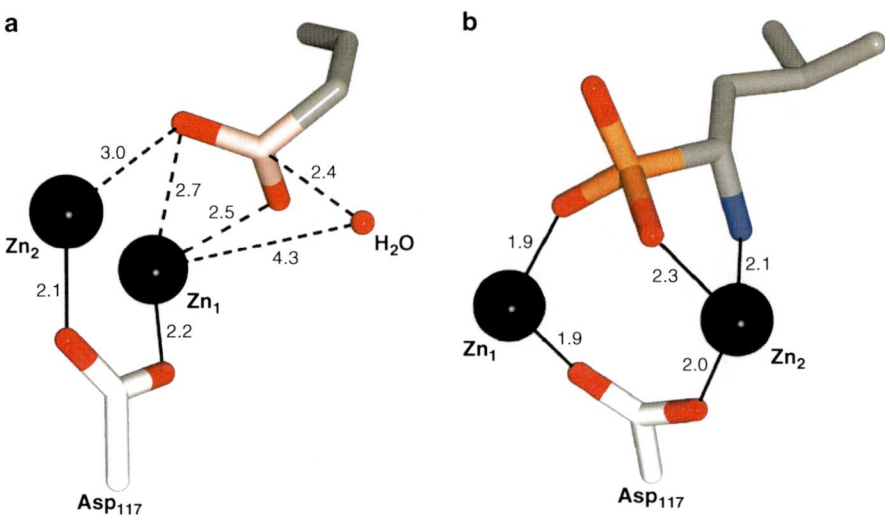

and Holz 2004). Variable-temperature variable-field (VTVH) magnetic circular dichroism (MCD) data recorded on the Co(II) form of AAP indicated that the two Co(II) ions are ferromagnetically coupled with a J of 3.4 cm$^{-1}$ (Larrabee et al. 2009), consistent with EPR data (Bennett and Holz 1997). Comparison of J values for dicobalt(II) centers in proteins and model complexes with bridging water or hydroxide moieties indicated that J is either zero or negative (antiferromagnetic) when a water molecule bridges and positive (ferromagnetic) when a hydroxide bridges between the two Co(II) ions. These data are consistent with the 0.95-Å X-ray structures in that AAP contains a μ-hydroxo-bridging ligand.

To provide additional supporting evidence for an OH$^-$ oxygen bridge in AAP, as suggested by the 0.95-Å structure and MCD data, the 1.24-Å resolution crystal structure of AAP at pH 4.7 (PDB: 2DEA) was solved (Fig. 3b). In this structure, the bridging oxygen is expected to contain two protons. At the level of precision reported, the 0.1–0.2 Å change that should occur in Zn(II)-oxygen distances when the bridging oxygen changes its protonation state from OH$^-$ to OH$_2$ would be observable. In fact, the Zn1–O distance in the pH 4.7 structure is 2.19 Å, an increase of 0.19 Å over its value at pH 7.5; however, the Zn2–O distance remained the same when compared to the Zn(II)-oxygen distances at pH 7.5 (Fig. 3b). Interestingly, Zn1 gained an additional ligand at low pH to become

5 coordinate due to the coordination of a water molecule, which likely contributes to the observed increase in bond length. The coordination number of Zn2 remained unchanged. This low pH structure was suggested to serve as a simple model for the first step in the proposed chemical reaction mechanism of AAP, which proposes the coordination of the carbonyl oxygen of the N-terminal amino acid to Zn1, increasing Zn1's coordination number from 4 to 5, thereby weakening the Zn1–OH(H)–Zn2 bridge. The new water ligand was surmised to be a model for carbonyl binding, resulting in the Zn1–OH(H) bond distance to increase to 2.19 Å, a reasonable bonding distance, but clearly weakened from the pH 7.5 structure.

While these structures provided important insight into these catalytic steps, additional structural data on substrate-analog, transition-state analog, and product-bound forms of AAP have emerged in the past 8 years that have helped to further refine our understanding of the catalytic steps involved in the hydrolysis of peptides. Two important structures reported prior to 2004 provided substrate and transition-state structural information. These two structures were of AAP complexed by the substrate analog inhibitor 1-butaneboronic acid (BuBA) at 1.9-Å resolution (PDB: 1CP6) (Fig. 5a) and the transition-state analog inhibitor L-leucinephosphonic acid (LPA) at 2.1-Å resolution (PDB: 1FT7) (Fig. 5b). Another X-ray crystal structure of a substrate-analog inhibitor, namely, N-[(2 S, 3R)-3-

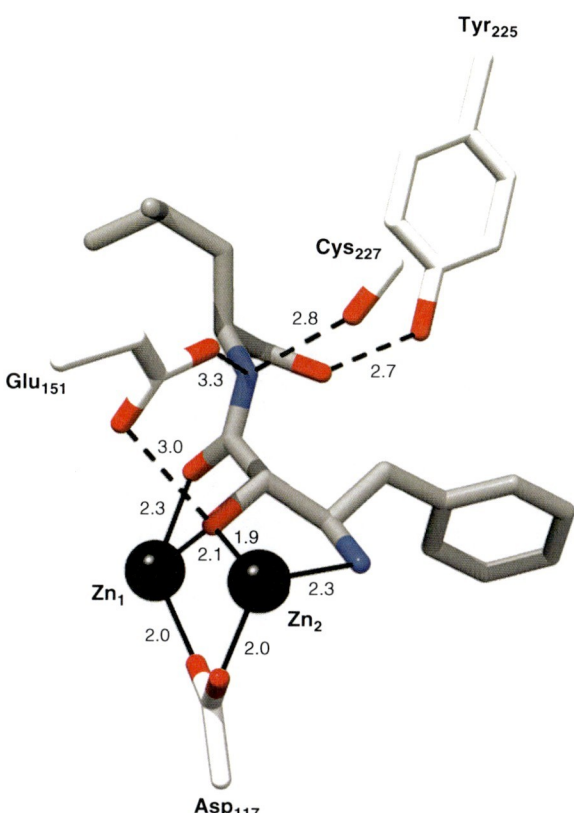

Zinc Aminopeptidases, Aminopeptidase from Vibrio Proteolyticus (Aeromonas proteolytica) as Prototypical Enzyme, Fig. 6 Active site of AAP bound by bestatin. All distances are in Å and the terminal ligands have been removed for clarity

amino-2-hydroxy-4-phenylbutyryl]-L-leucine (bestatin) bound to AAP was reported at 2.0-Å resolution (PDB: 1TXR) (Fig. 6). In this structure, the leucine portion of bestatin extends back toward L155 while the phenylalanine ring extends toward F244, Y251, and F248. The C-terminus of bestatin forms a hydrogen bond with Y225 (2.7 Å). The amino acid residues ligated to the dizinc(II) cluster in AAP are identical to those in the native structure with only minor perturbations in bond length (Fig. 3a). The alkoxide oxygen of bestatin bridges between the two Zn(II) ions in the active site displacing the bridging water molecule observed in the native [ZnZn(AAP)] structure. The M–M distances observed in the AAP-bestatin complex and native AAP are identical (3.5 Å) with alkoxide oxygen atom distances of 2.1 and 1.9 Å from Zn1 and Zn2, respectively. Interestingly, the backbone carbonyl oxygen atom of bestatin is coordinated to Zn1 at a distance of

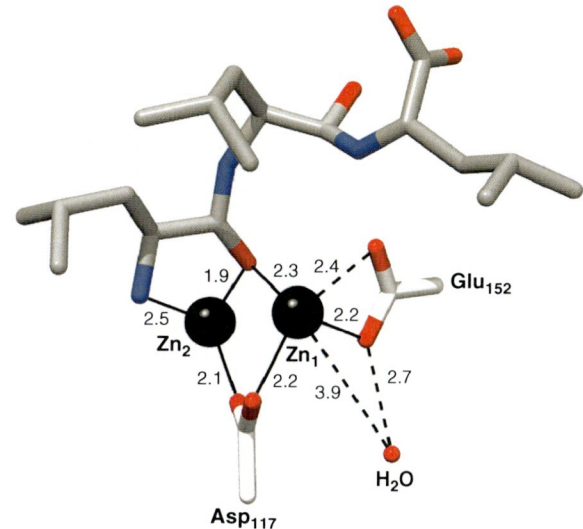

Zinc Aminopeptidases, Aminopeptidase from Vibrio Proteolyticus (Aeromonas proteolytica) as Prototypical Enzyme, Fig. 7 Structure of L-leucyl-L-leucyl-L-leucine (LLL) bound to AAP. All distances are in Å and all of the terminal ligands except E152 have been removed for clarity

2.3 Å. In addition, the NH$_2$ group of bestatin, which mimics the N-terminal amine group of an incoming peptide, binds to Zn2 with a bond distance of 2.3 Å. UV–vis spectra of the catalytic competent [Co_(AAP)], [CoCo(AAP)], and [ZnCo(AAP)] enzymes were also recorded in the presence of bestatin and confirmed that both of the divalent metal ions in AAP are involved in binding bestatin in solution. EPR spectra of [CoCo(AAP)] bound by bestatin recorded in both the perpendicular and parallel modes exhibited no observable signals. These data indicate that the two Co(II) ions in AAP are antiferromagnetically coupled yielding an $S = 0$ ground state, consistent with a single oxygen atom bridges between the two Co(II) ions.

Another very interesting and important X-ray crystal structure of AAP was reported by Kumar et al. (2007) where L-leucyl-L-leucyl-L-leucine (LLL) (PDB: 2IQ6), an actual substrate of AAP, was complexed to the dinuclear active site of AAP (Fig. 7). This is the first AAP structure with a bona fide substrate bound. Based on omit maps and B factors, the electron density observed for this 2.0-Å resolution structure was best described as a mixture of species that included the tripeptide LLL and the products of the hydrolysis reaction LL and L. Interestingly, the N-terminal carbonyl oxygen atom of LLL bridges between the two Zn(II) ions and the carbon atom was

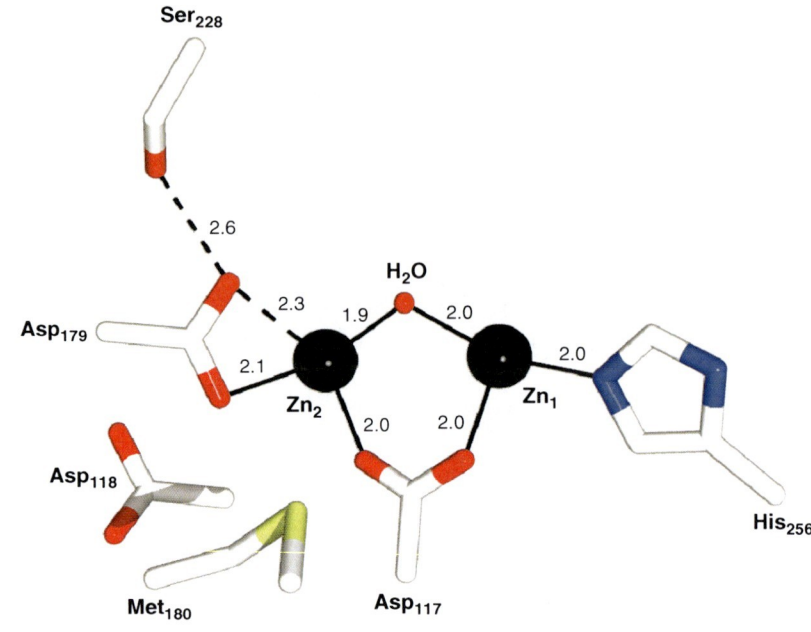

**Zinc Aminopeptidases, Aminopeptidase from Vibrio Proteolyticus (Aeromonas proteolytica) as Prototypical Enzyme, Fig. 8** Active sites of AAP highlighting the positions of S228, M180, and D118. All distances are in Å and all of the terminal ligands have been removed for clarity

suggested to be trigonal planar (sp$^2$), similar to the oxygen atom bridge observed in the AAP-bestatin structure, suggesting that the LLL complex represents pretransition-state species. The P1 amide nitrogen atom is bound to Zn2 similar to that observed in the AAP-LPA structure. No bridging water/hydroxide was observed; however, an active site water molecule was observed hydrogen bound to an oxygen atom of E152 (2.7 Å), which functions as a Zn1 ligand analogous to an active site water molecule observed in the AAP-bestatin structure. The observed active site water molecule in the AAP-LLL structure resides 3.9 Å from Zn1 and is, therefore, not likely a Zn1 ligand. This water/hydroxide is, however, weakly coordinated to a Zn1-binding carboxylate oxygen atom of Glu152 and was proposed to be the nucleophile in the catalytic reaction. The proposal is counter to the previously proposed model where Zn1 functions to deliver the nucleophile but is consistent with the computationally derived mechanism for the class B2 metallo-β-lactamases (Xu et al. 2006). As observed in the AAP-LPA structure, an oxygen atom of E151 is hydrogen bonded (2.8 Å) to the amide group of the P1 leucine moiety. Combination of this structure with spectrokinetic studies including stopped-flow spectroscopy and rapid freeze-quench EPR data of Co(II)-loaded AAP confirmed the formation of a post-Michaelis pretransition-state intermediate with a structure analogous to that of the AAP-LLL and AAP-bestatin complexes.

## Second Sphere Effects

The significant amount of X-ray crystallographic data available for AAP lead to the hypothesis that a combination of several hydrogen-bonding interactions, including second sphere interactions, may act to destroy the local symmetry of the dinuclear active site cluster as well as decrease the Lewis acidity of the Zn2 ion, thus, providing the driving force to break the Zn2–OH(H) bond. Recent PDB surveys show that the second shell (the environment surrounding the metal and its respective ligands) in Zn(II) enzymes is not hydrophobic but rather is more polar than nonpolar. To address the roles of second sphere residues in the catalytic mechanism of AAP, Ataie et al. (2008) substituted the conserved S228 and M180 residues with alanine and the nonconserved residue D118 with an asparagine residue (Fig. 8). M180 is located atop the bridging aspartate ligand (D117) and against one wall of the hydrophobic cavity while D118 is located in back of the active site where it forms hydrogen bonds to the backbone carbonyls of two adjacent loops. The S-atom of Met180 was also shown to be only 3.1 Å from the N-terminal amine group of LPA, suggesting a hydrogen-bonding interaction. On the other hand, S228 forms a hydrogen bond (2.58 Å) with the oxygen atom of D179, which is a Zn2 ligand, while the oxygen atoms of D118 form hydrogen bonds with two loops behind the dinuclear Zn(II)active site of AAP.

The observed catalytic activity of each of these AAP mutant enzymes was diminished compared to WT AAP. The S228A enzyme is 10-fold less active while the M180A enzyme is 100-fold less active than WT AAP. Interestingly, the D118N enzyme exhibits activity levels that are identical to WT AAP. The $K_m$ values for S228A and D118N toward L-leucine-$p$-nitroanilide were not affected compared to WT AAP, but the $K_m$ value obtained for the M180A enzyme was 30-fold higher than WT AAP. The X-ray crystal structure of the S228A mutant enzyme was solved at 1.2-Å resolution (PDB: 3B7I), and additional active site electron density was modeled as a bound isoleucine (Ataie et al. 2008). Alignment of the S228A enzyme with the structure of WT AAP showed very little change in the positions of the metal ions or the protein backbone. The crystal structure did reveal that D179 exhibits monodentate-like rather than equivalent bidentate binding, as observed in WT AAP since the oxygen of D179 moves more than 0.4 Å away from Zn2 in the S228A enzyme. A possible explanation for the observed change in the position of D179 is that without the hydrogen-bonding interaction of S228, D179 is a simple unprotonated carboxylate, which allows negative charge build-up on the Zn2 ion.

The X-ray crystal structure of the M180A AAP mutant enzyme in the absence (1.10-Å resolution; PDB: 3B35) and presence (1.46-Å resolution; PDB: 3B3C) of LPA and the X-ray crystal structure of the D118N enzyme bound by isoleucine (1.17-Å resolution; PDB: 3B3T) have also been reported (Fig. 9) (Ataie et al. 2008). For both M180A and D118N, no major disparities to the locations of the active site ligands are observed; however, in the M180A structure, I255 is rotated ~180° around the β-carbon. Since M180 is strictly conserved, the sulfur atom may be involved in mediating the reactivity of the dinuclear Zn(II) cluster. For D118A only slight changes in ligand coordination are observed even though $K_m$ is markedly perturbed. However, some clear differences are observed such as the binding mode of D179 to Zn2 (which is monodentate-like rather than equivalent bidentate mode as observed in WT AAP) and the 0.1 Å increase in the M–M distance. In the transition-

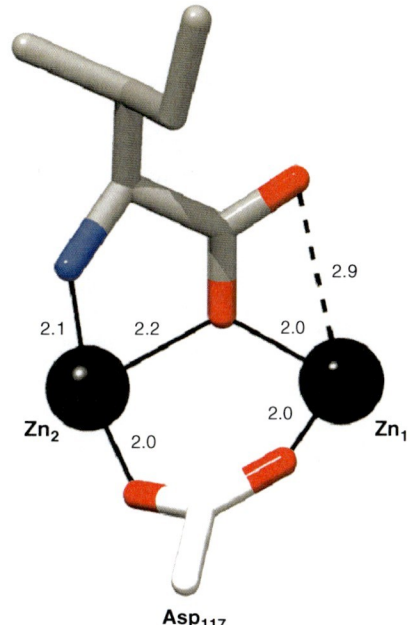

**Zinc Aminopeptidases, Aminopeptidase from Vibrio Proteolyticus (Aeromonas proteolytica) as Prototypical Enzyme, Fig. 9** Structure of the active site of AAP bound by isoleucine. All distances are in Å and all of the terminal ligands have been removed for clarity

state analog inhibitor M180A-AAP complex, the LPA molecule is bound in the same orientation as in the WT enzyme. Surprisingly, in the M180A-LPA structure, D179 is bound in an identical manner to that observed in the WT APP structure and forms a 2.6-Å hydrogen bond with S228; however, LPA binds 50-fold less tightly when M180 is mutated to an alanine.

## Mechanistic Insights

Combination of all of the kinetic, thermodynamic, QM/MM, DFT, spectroscopic, and X-ray crystallographic data reported to date has allowed more than one detailed reaction mechanism to be proposed for the peptide hydrolysis reaction catalyzed by AAP (Fig. 10). Based on kinetic measurements of aliphatic alcohol binding to AAP, the first step in catalysis is likely the recognition of the incoming N-terminal peptide leucine group with the hydrophobic pocket adjacent to the dinuclear active site (Holz 2002). Substrate recognition pockets have been observed or predicted for all metallopeptidases containing similar cocatalytic metallo-active sites. This binding scheme is consistent with the large negative entropy and large positive enthalpy of activation reported for AAP (Holz 2002).

The second step is proposed to be the binding of the carbonyl oxygen atom of the incoming substrate to the dinuclear Zn(II) active site, which can polarize the carbonyl group rendering it susceptible to nucleophilic attack. An issue in this step is if the carbonyl moiety of the substrate binds to Zn1 [species 2a] or bridges between the two Zn(II) ion [species 2b]. Several pieces of data point to Zn1 such as the confirmed binding mode of BuBA and the 1.24-Å resolution X-ray structure of AAP at pH 4.7. Moreover, the carbonyl oxygen atoms of bestatin binds to Zn1 expanding its coordination number to 5. The single piece of data indicating a bridging carbonyl group is the 2.0-Å resolution X-ray structure of AAP bound to LLL which reveals a bridging carbonyl oxygen between the two Zn(II) ions. In the pH 4.5 and bestatin structures, the Zn2–OH(H) bond distance remained the same, suggesting that upon substrate binding, the bridging hydroxide ion shifts electron density to Zn2. For LLL, a bridging water molecule was not observed. Yet another option, based only on DFT calculations involves an initial substrate-binding step where the N-terminal amino group and not the peptide carbonyl, binds first to the dinuclear cluster via Zn2. In this way, Zn2 would play an important role in binding substrate and orienting the peptide bond toward the nucleophile. To date, no experimental evidence has been reported to verify such a substrate-binding mode. Moreover, many pieces of data argue against such as binding step including, but not limited to, the fact that AAP exhibits significant activity in the presence of only one metal ion, metal binding is sequential, and the multitude of X-ray and spectroscopic data that indicate that the carbonyl group initially binds to the dizinc cluster.

If the BuBA, pH 4.5, and bestatin structures serve as models for the initial substrate-binding step in the chemical reaction pathway of AAP, two intermediate species can be proposed (Fig. 10: species 3a or 3b). For intermediate 3a to form, the next step would involve breaking of the Zn2–OH(H) bond, which would be assisted by N-terminal amine binding to Zn2. E151 located near the catalytic active site would assist in the deprotonation of the Zn1–OH(H) moiety to a nucleophilic hydroxo, similar to that of Glu270 in carboxypeptidase A (Christianson and Lipscomb 1989). This intermediate is consistent with the fact that AAP is *ca.* 80% active with only a single Zn(II) ion bound, suggesting that both the substrate and nucleophile must be bound to the same metal ion and also the fact that fluoride binding occurs after substrate binding. At this point, the metal-bound hydroxide can attack the activated carbonyl carbon of the peptide substrate forming a gem-diolate transition-state intermediate 4a that is stabilized by coordination of both oxygen atoms to the dizinc(II) center [species 5], consistent with the [ZnZn(AAP)]-LPA structure.

Alternatively, the 0.95-Å X-ray structure of AAP revealed a weakening in the Zn1–OH(H) bond while the Zn2–OH(H) bond distance remained the same, suggesting that upon substrate binding the bridging hydroxide ion shifts electron density to Zn2. If N-terminal amine binding to Zn2 does not facilitate the loss of a Zn2–OH(H) bond a Zn2 associated hydroxide ion would function as the nucleophile and attack the activated carbonyl carbon of the substrate [species 3b]. This reaction pathway leads to the direct formation of a stable five-membered ring for the tetrahedral intermediate (Fig. 10, species 5) with a structure that would be identical to that of [ZnZn(AAP)]-LPA. The role of Zn1 would then be to assist in stabilizing the bridging hydroxide ion in the native enzyme and in activating the carbonyl carbon of the N-terminal amino acid. Intermediate 3b differs from the intermediate previously proposed in that it reverses the roles proposed for each metal ion in catalysis. If species 2b is the substrate-binding mode, then a nonmetal centered nucleophile would attack the activated carbonyl species of the LLL substrate forming a gem-diolate transition state where both oxygen atoms are bound to the dizinc(II) center [species 5].

In each of these proposed mechanistic routes, the amide nitrogen must be stabilized via a hydrogen bond in order to make it a suitable leaving group. This hydrogen bond would also facilitate the collapse of the transition state. The active site glutamate (E151) likely provides the additional proton to the penultimate amino nitrogen returning it to its ionized state consistent with recent kinetic isotope effect studies (Bzymek and Holz 2004). These data in combination with thermodynamic studies suggest that the product-forming C-N bond-breaking step is the rate-limiting step. The product-bound species 6 is verified by the 1.17-Å

**Zinc Aminopeptidases, Aminopeptidase from Vibrio Proteolyticus (Aeromonas proteolytica) as Prototypical Enzyme, Fig. 10** Proposed catalytic mechanisms for AAP

X-ray structure of isoleucine bound to the active site of AAP. Finally, the cocatalytic Zn(II) site releases the cleaved peptides and adds a water molecule that bridges between the two metal ions regenerating the staring species 1. Thus, both metal ions are required for full enzymatic activity in metallopeptidases containing cocatalytic metallo-active sites, but their individual roles appear to differ markedly.

## Concluding Remarks

While our understanding of the catalytic mechanism of AAP has markedly improved over the past nine years, it has not yet been entirely delineated. For example, several substrate-binding steps have been proposed and the roles of each metal ion during catalysis are still in question. Therefore, more experimental evidence is needed to distinguish between the competing mechanistic proposals for AAP. While AAP is not a specific target for pharmaceuticals, understanding its catalytic mechanism remains an extremely important endeavor as many other metallohydrolases that contain dinuclear active sites and are drug targets, containing identical active sites to AAP. Many of these enzymes are multimeric and/or cannot be obtained in large quantities for mechanistic studies. Therefore, a prototypical hydrolase such as AAP is an exciting and important enzyme for mechanistic studies from both an inorganic/catalytic perspective as well as from a medicinal viewpoint.

**Acknowledgements** This work was supported by the National Institutes of Health (R15 AI085559-01A1, RCH) and the ACS Petroleum Research Fund (ACS PRF 50033-ND4, RCH).

## Cross-References

▶ Cadmium Carbonic Anhydrase
▶ Calcineurin
▶ Thermolysin
▶ Urease
▶ Zinc Amidohydrolase Superfamily
▶ Zinc Beta Lactamase Superfamily
▶ Zinc Carbonic Anhydrases
▶ Zinc Metallocarboxypeptidases
▶ Zinc-Binding Sites in Proteins

## References

Ataie NJ, Hoang QQ, Zahniser MPD, Tu Y, Milne A, Petsko GA et al (2008) Zinc coordination geometry and ligand binding affinity: the structural and kinetic analysis of the second-shell Serine 228 residue and the Methionine 180 residue of the Aminopeptidase from *Vibrio proteolyticus*. Biochemistry 47(29):7673–7683

Bennett B, Holz RC (1997) EPR studies on the mono- and dicobalt(II)-substituted forms of the aminopeptidase from *Aeromonas proteolytica*. Insight into the catalytic mechanism of dinuclear hydrolases. J Am Chem Soc 119:1923–1933

Bzymek KP, Holz RC (2004) The catalytic role of Glutamate-151 in the Leucine Aminopeptidase from *Aeromonas proteolytica*. J Biol Chem 279:31018–31025

Bzymek KP, D'souza VM, Chen G, Campbell H, Mitchell A, Holz RC (2004) Function of the signal peptide and N- and C-Terminal pro-peptide in the leucine aminopeptidase from *Aeromonas proteolytica*. Protein Express Purif 37(2):294–305

Bzymek KP, Swierczek SI, Bennett B, Holz RC (2005a) Spectroscopic characterization of the E151D and E151A altered Aminopeptidases from *Aeromonas proteolytica*. Inorg Chem 44:8574–8580

Bzymek KP, Moulin A, Swierczek SI, Ringe D, Petsko GA, Bennett B et al (2005b) Kinetic, spectroscopic, and X-ray crystallographic characterization of the functional E151H Aminopeptidase from *Aeromonas proteolytica*. Biochemistry 44(36):12030–12040

Chen SL, Marino T, Fang WH, Russo N, Himo F (2008) Peptide hydrolysis by the binuclear zinc enzyme aminopeptidase from *Aeromonas proteolytica*: a density functional theory study. J Phys Chem B 112:2494–2500

Christianson DW, Lipscomb WN (1989) Carboxypeptidase A. Acc Chem Res 22:62–69

Hartley M, Bennett B (2009) Heterologous expression and purification of *Vibrio proteolyticus* (*Aeromonas proteolytica*) aminopeptidase: a rapid protocol. Protein Express Purif 66(1):91–101

Holz RC (2002) The aminopeptidase from *Aeromonas proteolytica*: structure and mechanism of co-catalytic metal centers involved in peptide hydrolysis. Coord Chem Rev 232:5–26

Holz RC, Bzymek K, Swierczek SI (2003) Co-catalytic metallopeptidases as pharmaceutical targets. Curr Opin Chem Biol 7:197–206

Kumar A, Periyannan GR, Narayanan B, Kittell AW, Kim J-J, Bennett B (2007) Experimental evidence for a metallohydrolase mechanism in which the neucleophile is not delivered by a metal ion: EPR spectrokinetic and structural studies of aminopeptidase from *Vibrio proteolyticus*. Biochem J 403:527–536

Larrabee JA, Johnson WR, Volwiler AS (2009) Magnetic circular dichroism study of a Dicobalt(II) complex with mixed 5- and 6-Coordination: a spectroscopic model for Dicobalt(II) Hydrolases. Inorg Chem 48:8822–8829

Lipscomb WN, Sträter N (1996) Recent advances in zinc enzymology. Chem Rev 96:2375–2433

Lowther WT, Matthews BW (2002) Metalloaminopeptidases: common functional themes in disparate structural surroundings. Chem Rev 102:4581–4607

Mall P, Kumar V, Parikh V, Dave D, Tunga B, Tunga R (2011) Production and partial purification of leucine aminopeptidase from *Aeromonas proteolytica* (ATCC 15338). Int J Pharm Bio Sci 2(2):341–351

Perez-Sanchez G, Leal-Guadarrama LI, Trelles I, Perez NO, Medina-Rivero E (2011) High-level production of a recombinant *Vibrio proteolyticus* leucine aminopeptidase and its use for N-terminal methionine excision from interferon alpha-2b. Process Biochem 46:1825–1830

Schürer G, Lanig H, Clark T (2004) *Aeromonas proteolytica* Aminopeptidase: an investigation of the mode of action using a quantum mechanical/molecular mechanical approach. Biochemistry 43(18):5414–5427

Sonoda H, Daimon K, Yamaji H, Sugimura A (2009) Efficient production of active *Vibrio proteolyticus* aminopeptidase in *Escherichia coli* by co-expression with engineered vibriolysin. Appl Micro Biotechnol 84(1):191–198

Wilcox DE (1996) Binuclear metallohydrolases. Chem Rev 96:2435–2458

Xu D, Xie D, Guo H (2006) Catalytic mechanism of class B2 metallo-beta-lactamase. J Biol Chem 281(13):8740–8747

# Zinc and Diabetes (Type 2)

Wolfgang Maret
Metal Metabolism Group, Diabetes and Nutritional Sciences Division, School of Medicine, King's College London, London, UK

## Definitions

In 1936, Sir Harold Percival Himsworth introduced the distinction between type 1 and type 2 for the chronic systemic disorder diabetes mellitus. The two forms became known as IDDM, insulin-dependent diabetes mellitus, and NIDDM, non-insulin-dependent diabetes mellitus, because of the absolute and relative requirements for insulin, respectively. Type 1 is an autoimmune disease, in which the insulin-producing islet cells are destroyed and hence pancreatic insulin secretion ceases. It is also known as juvenile diabetes as it is frequently diagnosed in children and young adults. This entry focuses on type 2 diabetes, which accounts for 90+% of all diabetes cases. In type 2 diabetes, peripheral tissues become insulin-resistant. Initially, the pancreas can compensate for this defect with additional insulin production but this capacity is exhausted eventually and insulin administration becomes necessary. Major health risks are associated with the hyperglycemia that is characteristic for both types of the disease. Consequences are microvascular disease, leading to kidney failure (nephropathy), blindness (retinopathy), and nerve damage (neuropathy), and macrovascular disease, leading to amputations, cardiovascular disease, and stroke. In addition, diabetes is a risk factor for developing Alzheimer's disease and cancer (▶ Zinc in Alzheimer's and Parkinson's Diseases). Significant efforts are being made to understand how the disease develops from predisposing factors, such as obesity, the metabolic syndrome, aging, and a lack of physical activity. Type 2 diabetes mellitus was called "adult-onset" diabetes but is now diagnosed with increasing frequency in younger people, including children. The number of diabetics worldwide is estimated to be 347 million (Danaei et al. 2011). The estimate significantly exceeds estimates made 5 years ago about the extent of the diabetes epidemic. The resulting global economic toll poses largely unresolved challenges to existing health care systems.

## Zinc

The World Health Organization considers zinc deficiency the fifth leading cause of morbidity and mortality in developing countries (▶ Zinc Deficiency). The zinc(II) ion is nutritionally essential as a cofactor of about 3,000 human proteins (▶ Zinc-Binding Proteins, Abundance). It has catalytic functions in zinc enzymes of all enzyme classes and structural functions in the global structure of proteins, in protein domains that interact with other proteins, DNA/RNA, and lipids, and in bridging proteins in homo- and heterooligomerizations. In addition to the functions of zinc in this remarkable number of proteins, zinc(II) ions are discussed as signaling ions in both intercellular and intracellular communication (▶ Zinc and Zinc Ions in Biological Systems). While the number of proteins that are targets of zinc(II) ion signals is presently unknown, regulatory functions of zinc ions are undoubtedly an area of great importance. Proper control of cellular zinc is critical for the balance between health and disease. At least three dozen proteins homeostatically control cellular zinc (▶ Zinc Cellular Homeostasis). They include membrane zinc transporters of the ZnT and Zip families and cytosolic transport proteins, such as metallothioneins (MTs) (▶ Zinc Metallothionein). Together, these proteins keep cytosolic free zinc(II) ions at low picomolar concentrations. One important requirement for this cellular control of zinc is that the zinc homeostatic proteins must be fully functional. Thus, not only any primary nutritional zinc deficiency or overload but also the proper functions of a considerable number of proteins that regulate cellular zinc impinge on diabetes and its consequences, such as poor wound healing (▶ Zinc and Wound Healing).

# Zinc and Diabetes

The interaction of zinc and diabetes at different stages of the disease is quite obvious with regard to the involvement of zinc in virtually all cellular processes and the complexity of diabetes in terms of its effects on carbohydrate, lipid, and protein metabolism (Chausmer 1998; Taylor 2005; Maret 2005). Examples for the interaction include the function of zinc in many proteins relevant to the disease (Table 1). Associations between zinc and diabetes are known for a long time but received relatively little attention because the key functions of zinc in proteins and for life in general were not generally appreciated. It was recognized already in the 1930s that the pancreas of diabetic individuals that had succumbed to the disease has only about half the amount of zinc compared to a healthy pancreas. It was documented repeatedly that diabetics have significant zincuria. However, a zincemia or systemic zinc deficiency in diabetics is not always evident as a reliable measure of cellular zinc status is not available. Blood zinc represents only about 0.1% of total body zinc and is not a sensitive indicator of tissue zinc status. Insights into molecular functions of zinc in specific processes relevant to diabetes also became available only recently. The three areas where molecular studies have significantly strengthened the link between zinc and diabetes are the insulin and zinc transporter biochemistry in beta-cells, insulin signal transduction in peripheral tissues, and the role of zinc in redox metabolism (Chimienti et al. 2011). The new insights in these areas add to an understanding of diabetogenesis and raise expectations for novel therapeutic or preventative interventions.

**Zinc and Diabetes (Type 2), Table 1** Zinc/protein interactions relevant to diabetes

| Protein | Function of zinc |
|---|---|
| Insulin | Structure; insulin crystallization and storage in secretory granules, involvement in the allosteric transition of insulin, possible role in preventing amyloidogenesis of secreted insulin; mutation of BHis-10, the zinc ligand in pro-insulin, affects insulin storage |
| Phosphatases | Regulation; inhibition of the activity of protein tyrosine phosphatase-1B (PTP1B) which controls the insulin receptor |
| Proteinases | Catalysis; carboxypeptidase E: pro-insulin conversion; insulin-degrading enzyme (IDE, insulysin): insulin degradation; insulin-regulated aminopeptidase (IRAP): involvement in glucose transporter (GLUT4) trafficking → zinc metallocarboxypeptidases; zinc aminopeptidases |
| Dehydrogenases | Catalysis and structure; sorbitol dehydrogenase: involvement in the polyol pathway and implicated in diabetic complications → zinc alcohol dehydrogenases |
| Transcription factors | Structure; zinc finger type of proteins, such as peroxisome-proliferator activated receptors (PPARs) |
| Transporters | Transport, sensing, and control of cellular and subcellular zinc concentrations; mutation at position 325 in ZnT8 affects granular zinc storage and secretion; other members of the ZnT and Zip families are also associated with diabetes |
| Metallothioneins | Structure and regulation; linked functions in zinc and redox metabolism, mutation at position 27 in MT1A is associated with type 2 diabetes and coronary heart disease → zinc metallothionein |

## Zinc in Pancreatic Beta-Cell Physiology

In humans, insulin is stored as a zinc complex in the secretory granules of pancreatic beta-cells. Two zinc ions are bound per insulin hexamer. Zinc is involved in the allosteric transition of the insulin hexamer from the R to the T form. It is secreted with insulin from the beta-cells and may have additional functions. One effect, which however is controversial, is on glucagon secretion from alpha-cells. Another is the prevention of amyloidogenesis of proteins co-secreted with zinc. Thus, zinc inhibits the fibrillation of monomeric insulin, which is amyloidogenic. It also inhibits the formation of the dimer of the human islet amyloid polypeptide (IAPP, amylin), which is also amyloidogenic, co-secreted with insulin from the beta-cells, and detected as extracellular amyloid fibers in the pancreas of diabetics.

A specific transporter of the cation diffusion facilitator (CDF, SLC30A) family of zinc transporters, ZnT8, makes zinc available to the insulin-storing secretory granules. It is preferentially expressed in pancreatic islet cells (both alpha and beta) and localizes to the granules. In 2007, several investigators independently reported a strong association of a mutation in ZnT8 with type 2 diabetes in different populations. It was also discovered that the mutant protein is an autoantigen in type 1 diabetes. ZnT8 with an Arg instead of a Trp at position 325 in its cytoplasmic domain increases the risk for diabetes by 53% in carriers that are homozygous for the allele. Significantly, the risk allele is the one prevalent in populations, 55% in Asians, 75% in Europeans, and 95% in Africans. Another mutation in ZnT8 has Gln at position 325. These findings have alerted a large part of the diabetes research community to the role of zinc in diabetes (Rutter 2010). Beta-cell specific, but not alpha-cell specific, knock-out of ZnT8 in mice affects glucose tolerance. The mice have decreased zinc in the beta-cell granules, which become amorphous, and they have altered insulin secretion and mild glucose intolerance. Other zinc transporters of the ZnT and Zip families also participate in beta-cell zinc metabolism. Associations of some of them with diabetes type 2 have been noted.

## Zinc in the Physiology of Cells Targeted by Insulin

A series of experiments that began about 45 years ago demonstrated that zinc has an insulin-sparing effect. Zinc-deficient rats turned out to be much less sensitive to insulin. In the early 1980s, it was shown that zinc stimulates lipogenesis in isolated fat tissue and affects glucose uptake. Remarkably, zinc can replace insulin when some mammalian cells are cultured in serum-free media. Taken together these experiments led to the concept that zinc has an insulin-mimetic effect. The actions of zinc are intracellular. Zinc increases the phosphorylation of the insulin/IGF-1 receptor and hence protein phosphorylation downstream in the insulin signaling pathway. One explanation for these effects is the strong zinc inhibition of protein tyrosine phosphatases, in particular PTP1B, which is a major regulator of the phosphorylation state of the insulin receptor. Zinc inhibits PTP1B with an apparent zinc-binding constant of 15 nM. Whether the insulin-mimetic effect of zinc can be attributed entirely to the inhibition of protein tyrosine phosphatases is a matter of the concentrations of zinc employed in the experiments and the zinc binding constants of the proteins affected. It appears that most if not all the observations relate to zinc-dependent phosphorylation events downstream of the insulin receptor, for example, the adaptor molecule insulin-receptor substrate-1 (IRS-1), the kinases PKB (Akt), PKC, and GSK-3, the phosphatase PTEN, and the transcription factor FOXO1a.

## Zinc and Oxidative Stress

Zinc is frequently described as an antioxidant. Zinc cannot have such a function directly as it does not change its redox state in cells; it is redox-inert and remains always Zn(II). Yet, it has biological effects that resemble that of an antioxidant. Therefore, the term pro-antioxidant is more appropriate (Maret 2008). Zinc functions as a pro-antioxidant only in a certain range of concentrations. Zinc deficiency and zinc overload are pro-oxidant conditions. These opposing effects of zinc demonstrate how important it is for redox signaling to control cellular zinc ion concentrations. The pro-antioxidant effects of zinc are thought to be due to (1) binding to and protecting free sulfhydryls against oxidation, (2) competing with redox-active transition metal ions and suppressing the production of damaging free radicals, and (3) inducing the synthesis of antioxidants, such as the expression of genes for antioxidant enzymes through MTF-1 (metal response element (MRE)-binding transcription factor-1) and Nrf-2 (NF-E2-related factor) dependent gene transcription. A deficiency of zinc compromises these three functions and, therefore, constitutes a pro-oxidant condition. Elevated free zinc ion concentrations inhibit anti-oxidant enzymes and the mitochondrial respiratory chain, thus increasing free radicals, also a pro-oxidant condition. The thresholds of zinc concentrations that determine the switches from pro-antioxidant and cytoprotective to pro-oxidant and cytotoxic functions are not well defined.

Control of the cellular functions of zinc requires zinc buffering. About a third of the cellular zinc buffering capacity relies on sulfhydryl donors as zinc-binding ligands. Environmental agents, such as certain toxins, nutrients, and drugs, react with thiols and make fewer thiols available for zinc buffering. These reactions change the zinc buffering capacity and increase the availability of free zinc ions, which bind to proteins that are not targeted under physiological conditions and affect their functions. This molecular mechanism is important for diabetes where oxidative stress occurs. For example, diabetic hyperglycemia leads to the glycation of proteins. The resulting advanced glycation end products (AGEs) increase carbonyls, generating so-called carbonyl stress. Reactive carbonyls chemically modify sulfhydryl groups that are ligands of zinc and lower the zinc buffering capacity. Oxidative stress is also thought to cause insulin resistance, establishing a causal link to zinc because of the role of zinc in oxidative stress. Oxidative stress decreases the zinc buffering capacity, possibly leading to a loss of zinc from tissues. A resulting cellular zinc deficiency would exacerbate the oxidative stress, since zinc deficiency is a pro-oxidant condition, and initiate a vicious cycle.

Metallothioneins participate in cellular zinc and redox buffering. They are a 60+ amino acid family of human proteins and bind up to seven zinc(II) ions with their twenty cysteines. Their zinc/thiolate coordination environments make MTs redox-active zinc proteins. Oxidation or chemical modification of the cysteine ligands leads to zinc dissociation, while the reduction of the oxidized cysteines or restoration of the thiol functionality generates zinc-binding capacity. This cycling links redox changes and the availability of cellular zinc and positions MT in a signaling pathway where redox signals are converted into zinc signals with a specificity determined by the coordination environments of zinc-binding sites in proteins. Increased amounts of MT protect the pancreas against chemically induced diabetes and protect organs such as the heart and the kidney against injury in diabetes (Li et al. 2007). A polymorphism of MT1A, an Asn27Thr substitution that is thought to change the zinc-binding properties of the protein, is associated with type 2 diabetes and coronary heart disease.

## Prospects for Preventing and Treating Diabetes Through Approaches That Target Zinc Metabolism

From the above discussion, it becomes clear that zinc enhances the insulin action in peripheral tissues, has pro-antioxidant functions in protecting the endocrine pancreas and peripheral tissues, and has a role in insulin storage and secretion in pancreatic beta-cells. In all these cellular actions, metallothioneins are involved. The significance of these molecular actions has been shown for physiology and pathophysiology in both animals and humans. Cellular, genetic, clinical, and epidemiological investigations support the significance of the molecular mechanisms. In rodents, insulin is less effective in zinc deficiency and there is decreased beta-cell granulation, while zinc supplementation has insulin-enhancing effects, lowers blood glucose, and improves beta-cell function. In humans, zinc supplementation alleviates signs of oxidative stress. Clearly, these findings have implications for the consequences of primary zinc deficiency in developing countries, in specific populations such as the elderly, in genetically predisposed individuals, and for those with environmental exposures that lead to perturbations of zinc homeostasis. From knowledge in all three areas, it is becoming evident that it is not only the classical functions of zinc as a catalytic and structural cofactor of proteins but rather the regulatory functions of zinc in metabolism and signaling and the control of cellular zinc that are critically important for health.

Whether zinc deficiency is a cause or a consequence of diabetes remains to be determined. A Cochrane review suggests that there is no evidence for zinc supplementation preventing the development of type 2 diabetes (Beletate et al. 2007). It could mean that additional zinc does not afford protection. However, there were insufficient data to analyze any particular subgroups. The issue remains whether zinc supplementation prevents diabetes in individuals that are zinc-deficient. In contrast to prevention, there is ample evidence for therapeutic effects of zinc. Zinc compounds have antidiabetogenic and insulinomimetic properties (Sakurai and Adachi 2005). Perturbed zinc homeostasis may not be readily restored by giving only zinc. Under conditions of sustained oxidative stress additional zinc may harm rather than benefit because a reduced zinc buffering capacity increases free zinc ion concentrations, allowing zinc to bind to nonphysiological targets. There are reports that diabetic tissues have increased free zinc ion concentrations. Restoration of redox homeostasis may be necessary before zinc supplementation is effective. Interventions have focused too narrowly on zinc itself. An understanding of zinc homeostasis offers novel pharmacological approaches, in particular the targeting of the numerous proteins that are involved in regulating cellular zinc. The zinc/diabetes relationship offers significant and new areas of research to find preventive strategies and treatments for diabetes. Such approaches are direly needed.

## Cross-References

▶ Zinc Alcohol Dehydrogenases
▶ Zinc Aminopeptidases, Aminopeptidase from Vibrio Proteolyticus (Aeromonas proteolytica) as Prototypical Enzyme
▶ Zinc and Wound Healing
▶ Zinc and Zinc Ions in Biological Systems
▶ Zinc Cellular Homeostasis

- Zinc Deficiency
- Zinc in Alzheimer's and Parkinson's Diseases
- Zinc Metallocarboxypeptidases
- Zinc Metallothionein
- Zinc-Binding Proteins, Abundance

## References

Beletate V, El Dip R, Atallah AN (2007) Zinc supplementation for the prevention of type 2 diabetes mellitus. Cochrane Database of Syst Rev 2007, Issue I, article CD005525

Chausmer AB (1998) Zinc, insulin and diabetes. J Am Coll Nutr 17:109–115

Chimienti F, Rutter GA, Wheeler MB et al (2011) Zinc and diabetes. In: Rink L (ed) Zinc in human health. Ios Press, Amsterdam, pp 493–513

Danaei G, Funicane MM, Lu Y et al (2011) National, regional, and global trends in fasting plasma glucose and diabetes prevalence since 1980: systematic analysis of health examination surveys and epidemiological studies with 370 country years and 2.7 million participants. The Lancet 378:31–40

Li X, Cai L, Feng W (2007) Diabetes and metallothionein. Mini-Rev Med Chem 7:761–768

Maret W (2005) Zinc and diabetes. BioMetals 18:293–294

Maret W (2008) Metallothionein redox biology in the cytoprotective and cytotoxic functions of zinc. Exp Gerontol 43:363–369

Rutter GA (2010) Think zinc. New roles for zinc in the control of insulin secretion. Islets 2:1–2

Sakurai H, Adachi Y (2005) The pharmacology of the insulinomimetic effect of zinc complexes. BioMetals 18:319–323

Taylor CG (2005) Zinc, the pancreas, and diabetes: insights from rodent studies and future directions. BioMetals 18:305–312

## Zinc and Immune Defense

- Zinc and Immunity

## Zinc and Immunity

Hajo Haase and Lothar Rink
Institute of Immunology, Medical Faculty, RWTH Aachen University, Aachen, Germany

## Synonyms

Zinc and immune defense; Zinc and the immune system

## Definition

Immunity, i.e., the ability of the body to defend itself against pathogens, is mediated by an intricate system of cellular and soluble factors. Multiple aspects of the immune system depend on zinc, making this trace element a critical nutrient required for protection against infectious diseases.

## Introduction

Immunity is the ability of an organism to defend itself against infection by pathogens, such as bacteria, viruses, fungi, and multicellular parasites. To this end, a complex system of different cells (cellular immunity) and soluble factors (humoral immunity) acts in concert. Together, these components form the immune system. In humans, the immune system consists of two major branches. One is the innate immune system, which forms the first line of defense. The other is the adaptive immune system, which reacts in a delayed manner, but acts specifically toward unique molecular structures of pathogens, so-called antigens. Moreover, the adaptive immune system has the ability to remember these antigens, leading to a faster and far more efficient reaction of the immune system during a subsequent contact with a previously encountered pathogen (Murphy et al. 2008).

Nutritional zinc deficiency is the major reason for an insufficient zinc status and is mainly based on malnutrition. In the developing countries, zinc deficiency is a major cause for infections and parasitic diseases, leading to a significant loss in healthy life years (Rink 2011). Additionally, zinc deficiency may result from increased zinc excretion during inflammatory diseases, such as diabetes or hepatitis, or as a consequence of aging. The dominant feature of zinc deficiency is an increased frequency of infections, indicating that the immune system critically depends on this metal (Haase et al. 2008). Consequently, zinc supplementation can prevent immunodeficiency based on zinc-deficient nutrition and is an effective measure for reducing incidence and severity of infectious diseases in children in the developing world (Bhutta et al. 1999). Virtually all aspects of immunity are affected by zinc, including most cells that belong to the immune system (Table 1).

**Zinc and Immunity, Table 1** Effect of zinc deficiency on cells of the immune system

| Cell type | Function | Effect of zinc deficiency[a] |
|---|---|---|
| NK cells | Killing of tumor cells and virus-infected cells | Loss of specificity |
| | | Impaired killing |
| Monocytes and macrophages | Phagocytosis and killing of bacteria and fungi, presentation of their antigens to T cells | Increased production of proinflammatory cytokines |
| | | Augmented formation in the bone marrow |
| Neutrophil granulocytes | Phagocytosis and killing of bacteria and fungi | Impaired viability, phagocytosis, oxidative burst, and chemotaxis |
| Dendritic cells | Presentation of antigens to T cells | Increased maturation |
| B cells | Production of antibodies | Increased apoptosis of precursor cells |
| | | Reduced formation of antibodies |
| T cells | Killing of tumor cells and virus-infected cells, accessory function to several other cell types | Reduced survival of precursors and impaired function of mature cells |
| | | Shift toward $T_H2$ response |

[a]These effects are based on in vivo observations in humans and experimental animals. In addition, results from in vitro experiments were used in some cases to predict the most likely effects of zinc deficiency when sufficient in vivo observations are not available

## Innate Immunity

Many pathogens are unable to enter the human body because skin and mucosa form a barrier that is impermeable for most infectious agents, unless it is compromised. Insufficient zinc status impairs this defense mechanism. Notably, one hallmark of zinc deficiency is delayed wound healing. Moreover, the inherited zinc deficiency syndrome *Acrodermatitis enteropathica*, which results from a reduced enteric uptake of zinc, based on mutations in the zinc transport protein hZIP4, is characterized by dermatitis (Ackland and Michalczyk 2006). Epithelial barriers of the mucosa are also compromised during zinc deficiency, and cell-cell junctions of mucosal epithelial cell layers have reduced barrier function. Furthermore, antimicrobial peptides are secreted as a form of naturally occurring antibiotics, providing an additional layer of protection for the external barriers. Zinc supplementation has been shown to augment secretion of some of these peptides.

Once the barriers are overcome, pathogens encounter the cells of the innate immune system. Here, specialized cells are in charge of attacking different classes of pathogens.

Cells infected with viruses, and also tumor cells, have to be eliminated in order to preserve the entire organism. Natural killer (NK) cells recognize these cells and kill them by releasing perforin and granzyme, proteins which permeabilize target cell membranes and induce apoptosis, respectively. The activity of NK cells depends on a sufficient availability of zinc, and zinc deficiency is accompanied by reduced NK cell numbers and cytotoxic potential. Moreover, the decision to eliminate target cells is regulated by a competition between activating and inhibiting receptors. Some inhibitory receptors depend on zinc for the detection of the corresponding ligand on target cells. Hence, zinc is required for fine tuning and specific eradication of tumors and virally infected cells, and its deficiency might result in a less specific NK cell action.

Pathogens such as bacteria and fungi are eliminated by neutrophil granulocytes, monocytes, and macrophages. This is mediated through uptake by phagocytosis and killing by releasing reactive oxygen species within the vesicles into which the pathogens had been taken up. This process is called the oxidative burst. Several investigations demonstrated that zinc deficiency leads to reduced phagocytosis by monocytes, which can be corrected by zinc supplementation. The impact of zinc on the oxidative burst is multisided. On the one hand, the NADPH oxidase enzyme complex is inhibited by zinc. NADPH oxidase produces the superoxide radical, which is the compound from which all reactive oxygen species that are involved in the oxidative burst are generated. Conversely, zinc deficiency leads to impaired oxidative burst. Because multiple reports describe an activation of protein kinase C by zinc, which is a potent activator of the NADPH oxidase, zinc may be required for signals leading to the activation of this enzyme.

## Cytokine Secretion

Leukocytes, i.e., the white blood cells, including granulocytes and monocytes, are recruited toward the site of infection by chemotaxis, the directed cell movement along a gradient of chemoattractant peptides, so-called chemokines. The first step in this movement requires the interaction of leukocytes with the endothelial wall of the blood vessels in order to leave the blood and enter the infected tissue. Hereby, the presence of extracellular zinc promotes this interaction, and the release of intracellular zinc is important for signaling events leading to the attachment of monocytes induced by the monocyte chemoattractant protein (MCP)-1. Furthermore, very high doses of zinc (above 0.5 mM) have been reported to induce chemotaxis of neutrophil granulocytes. However, this concentration will not be reached within the interstitial fluid, making a chemokine-like function for zinc ions in vivo extremely unlikely.

The immune response is a coordinated action by many different cells. Therefore, successful clearance of an infection depends on communication between different parts of the immune system. This is mediated by cytokines, soluble proteins which are released by various cells, transmitting signals to or from immune cells by activating receptors on the plasma membrane.

Proinflammatory cytokine production by monocytes, in particular tumor necrosis factor (TNF)-α, interleukin (IL)-1, and IL-6, can be directly induced by incubation with high concentrations of zinc. Zinc is involved in the signal transduction leading to formation of these cytokines, as discussed below. These cytokines can activate other cells, such as T cells, which, in turn, also start producing cytokines. On the other hand, higher zinc concentrations inhibit cytokine production.

In vivo, zinc deficiency leads to elevated production of proinflammatory cytokines in the absence of pathogens. On the other hand, the production of these cytokines is impaired during infection. One example are the elderly, which have elevated basal levels of cytokines such as IL-6 but show reduced production of the same cytokines when their cells are being stimulated ex vivo. The deregulated cytokine production in the elderly can be normalized by zinc supplementation.

In direct comparison, the production of cytokines seems to be more sensitive to zinc deprivation than phagocytosis and oxidative burst. In vitro experiments with the zinc chelator TPEN abrogated the release of TNF-α and IL-1β in response to contact with *Escherichia coli*, whereas phagocytosis and the production of reactive oxygen species remained unaffected.

## Adaptive Immunity

Infections that cannot be controlled by innate immunity trigger the adaptive immune system. It consists of two main cell types (B and T cells), which express antigen-specific receptors. Each one of these cells carries receptors with a different antigen specificity, which has been generated by recombination of the coding DNA sequences during the cells' maturation. A prominent feature of the adaptive immune system is the rapid expansion of those cells that express receptors which correspond to antigens of an infectious agent. Hereby, a high number of specific cells are generated, allowing the targeted elimination of the pathogen. Another notable feature is the existence of memory cells. Some cells persist after the infection has been cleared, allowing a faster and more efficient response during subsequent contacts with the same pathogen. Here, zinc deficiency seems to affect predominantly naïve cells rather than memory cells, indicating that a lack of zinc particularly impairs the immune response toward previously unknown antigens.

The receptors of B cells are secreted in soluble form as antibodies. These antibodies can mark pathogens for phagocytosis by innate immune cells, trigger lysis by the complement protein cascade, neutralize antigens by binding to surface antigens (thereby sterically hindering their function), or induce killing by NK cells. Antibody production is reduced during zinc deficiency, mostly due to reduced B cell numbers, while the production per cell remains unchanged. The formation of antibodies in response to vaccination is also impaired during zinc deficiency. Yet the majority of zinc supplementation studies were unable to demonstrate an improvement of the response to vaccines after zinc supplementation. Successful immunization involves the interaction of B cells with T helper cells, which are discussed below. Current data indicate that zinc supplementation has to start sufficient time prior to vaccination, allowing to correct a T cell deficiency. Only then will these cells provide help to B cells for producing antibodies against the vaccine.

The proliferation of all T cells in response to mitogens, substances that trigger their activation and proliferation, has been shown to depend on zinc, indicating a general requirement for zinc in T cell function. T cells can be further subdivided into two major populations: cytotoxic T lymphocytes (CTL) and T helper ($T_H$) cells. CTL have a function similar to NK cells, i.e., the elimination of cells that are infected with a virus, or tumor cells. Their functions in experimental animal models and zinc-deficient humans are reduced during zinc deficiency. A recent study in primary human T cells and the murine CTL cell line CTLL-2 indicates that zinc signals are involved in signal transduction of the cytokine IL-2, which is crucial for T cell proliferation. Nevertheless, CTL seem to be less affected by zinc deficiency because their numbers are not reduced to the same extent as $T_H$ cells during zinc deficiency.

There are different kinds of $T_H$ cells. So-called $T_H1$ cells support the cellular immune response, e.g., by stimulation of macrophages through the cytokine interferon (IFN)-γ, which triggers the killing of intracellular pathogens. In contrast, $T_H2$ cells augment the humoral immune response, in particular antibody production by B cells. These systems are mutually inhibitory and ensure that only the appropriate reaction toward a specific pathogen is utilized. Zinc deficiency leads to a shift to a $T_H2$-dominated T cell polarization. This is mediated by reduced formation of $T_H1$ cytokines, whereas $T_H2$ cell function seems to be comparatively unaffected. A prominent example for a $T_H1$-mediated immune function is the delayed type hypersensitivity (DTH) reaction. Limited availability of zinc compromises DTH, and studies in different zinc-deficient populations report a restoration of DTH upon zinc supplementation.

In recent years, two additional subpopulations of T helper cells have been described: $T_H17$ and regulatory T cells (Treg). It has already been demonstrated that application of very high amounts of zinc suppresses the $T_H17$ response in vivo. Still, the impact of zinc on these two cell types still needs to be investigated in more detail.

## Development of Immune Cells

Not only the function of immune cells but also their generation and maturation are affected by zinc. Proliferation and differentiation of B and T cell precursors are reduced in zinc-deficient individuals. This is particularly evident in T cells. T cells mature in the thymus, and one hallmark of zinc deficiency is thymic atrophy, indicating the thymus as a highly zinc-dependent organ. Its epithelial cells secrete a peptide hormone, thymulin, which depends on zinc to acquire its biologically active conformation. Thymulin is important for the differentiation of T cells. It has been shown that during zinc deficiency, thymulin is present, but inactive, due to the lack of zinc. It can be activated by addition of zinc in vitro, indicating that the hormone is secreted, but depends on the availability of zinc in order to be functional.

Another important aspect is the survival of lymphocytes. Especially during maturation, these cells are highly sensitive to apoptosis because at this stage, autoreactive cells need to be eliminated in order to prevent autoimmune diseases. This makes precursor cells sensitive to death by zinc deprivation. Moreover, by influencing the sensitivity toward apoptotic stimuli, zinc may have a critical function in setting the threshold for their elimination. The higher it is, the more cells will be generated, but at the cost of reduced specificity. On the other hand, a higher sensitivity to the induction of apoptosis will yield "safer" cells, but at the cost of reduced total numbers. This corresponds to in vivo observations, in which the ratio of newly formed (i.e., naïve) T cells to memory T cells was measured. This ratio decreases during zinc deficiency, indicating that the formation of new T cells is affected to a greater extent than the survival of preexisting cells.

As for B and T cells, the development of NK cells also depends on zinc. The differentiation of hematopoietic precursor cells into mature NK cells was promoted by zinc with respect to cell number and cytotoxic activity. A case report of one zinc-deficient patient described strongly reduced NK cell activity in vivo, which improved upon zinc supplementation.

In contrast to B, T, and NK cells, the number of monocytes increases during zinc deficiency (Fraker and King 2004). In fact, the calcitriol-induced differentiation of promyeloid cells into monocytes involves a reduction of free intracellular zinc levels and is augmented by limited zinc availability. This corresponds well to in vivo observations in mice, where acute zinc deficiency leads to an elevated number of monocytes.

## Molecular Requirements for Zinc in Immunity

The molecular basis for the profound impact of zinc on immunity is still only incompletely understood.

The continuous production of immune cells necessitates a high rate of proliferation. Therefore, the requirement for zinc as a building material in zinc-containing metalloproteins is certainly an important factor. However, recent studies indicate that zinc ions also act as a signaling substance in immune cells, similar to a second messenger, and changes of the intracellular concentration of free zinc regulate a variety of cellular functions, in particular in the immune system (Cousins et al. 2006, Haase and Rink 2009). Some examples for well-investigated molecular functions of zinc in immune cell signaling are given below.

T cells use their T cell receptor (TCR) to recognize their specific antigen when it is presented to them on major histocompatibility complex (MHC) proteins on the surface of antigen-presenting cells. One essential kinase for TCR signaling is the Src kinase Lck. It is bound to the co-receptors CD4 in $T_H$ cells and CD8 in CTL, respectively. The extracelluar domain of the co-receptors attaches to MHC during an interaction between the two cells. Within the cells, this brings Lck in close vicinity to its target sites on the TCR. The protein interface site is a "zinc clasp" structure, based on a bridging zinc ion, bound by two cysteine ligands from each protein. Moreover, a second zinc-bridged protein interface site connects two Lck, thereby facilitating mutual phosphorylation and activation.

Zinc ions are also involved in signal transduction in other immune cells. Receptors which are involved in the recognition of characteristic and conserved chemical structures that are found in pathogens, but not in mammalian cells, the so-called pathogen-associated molecular patterns (PAMP), employ zinc signals. Stimulation of cells with the PAMP lipopolysaccharide, which triggers Toll-like receptor (TLR)-4, leads to a decrease of cellular free zinc in dendritic cells, which is involved in their maturation. In contrast, in monocytes and macrophages, TLR-4 signaling involves elevated free intracellular zinc, which is required for inhibiting protein tyrosine phosphatases, thereby preserving phosphorylation signals that lead to the expression of proinflammatory cytokines. In contrast to the stimulatory effects of these zinc signals, high zinc concentrations can also inhibit cytokine production by monocytes. Several mechanisms for this inhibition have been proposed, including upregulation of the negative regulatory protein A20, or an interaction of zinc with cyclic nucleotide signaling, based on its inhibition of cyclic nucleotide phosphodiesterases.

Several other examples of zinc-regulated signaling pathways exist, and frequently, more than one zinc-dependent event is involved in a particular signaling pathway (Haase and Rink 2009). Consequently, it remains difficult to predict the impact of zinc on immunity based on the knowledge of its molecular functions. On the one hand, many molecular events are still unknown. On the other hand, the complexity of the immune system, with multiple effects of zinc in numerous signaling pathways in different cells, makes it difficult to identify the molecular basis for the outcome of zinc deficiency and supplementation. For example, zinc supplementation has been shown to reduce the duration and severity of common cold. Several mechanistic explanations have been provided for this observation. The effect of CTL and NK cells against cells infected with the rhinovirus certainly depends on zinc because both cell types require this metal for their functions. In addition, a direct antiviral effect of zinc has been postulated, as well as an interference with the expression and function of antiviral cytokines, i.e., type I interferons. Alternatively, it has also been shown that zinc modulates the expression of ICAM-1 on human cells or binding of the rhinovirus to this protein, which is an important step during infection of cells.

## Conclusion

Taken together, current knowledge clearly shows a profound impact of zinc on nearly all aspects of immunity (Prasad 2009). However, much is still to be learned in order to understand the complex interactions of this metal with the processes that are essential for protection against infections and cancer.

## Cross-References

▶ Zinc and Wound Healing
▶ Zinc and Zinc Ions in Biological Systems
▶ Zinc Cellular Homeostasis
▶ Zinc Deficiency
▶ Zinc, Fluorescent Sensors as Molecular Probes
▶ Zinc Homeostasis, Whole Body

## References

Ackland ML, Michalczyk A (2006) Zinc deficiency and its inherited disorders -a review. Genes Nutr 1(1):41–49

Bhutta ZA, Black RE, Brown KH, Gardner JM, Gore S, Hidayat A, Khatun F, Martorell R, Ninh NX, Penny ME, Rosado JL, Roy SK, Ruel M, Sazawal S, Shankar A, Zinc Investigators' Collaborative Group (1999) Prevention of diarrhea and pneumonia by zinc supplementation in children in developing countries: pooled analysis of randomized controlled trials. J Pediatr 135(6):689–697

Cousins RJ, Liuzzi JP, Lichten LA (2006) Mammalian zinc transport, trafficking, and signals. J Biol Chem 281(34):24085–24089

Fraker PJ, King LE (2004) Reprogramming of the immune system during zinc deficiency. Annu Rev Nutr 24:277–298

Haase H, Rink L (2009) Functional significance of zinc-related signaling pathways in immune cells. Annu Rev Nutr 29:133–152

Haase H, Overbeck S, Rink L (2008) Zinc supplementation for the treatment or prevention of disease: current status and future perspectives. Exp Gerontol 43(5):394–408

Murphy KM, Travers P, Walport M (2008) Janeway's immunobiology, 7th edn. Taylor & Francis, London

Prasad AS (2009) Impact of the discovery of human zinc deficiency on health. J Am Coll Nutr 28(3):257–265

Rink L (ed) (2011) Zinc in human health. IOS Press, Amsterdam

# Zinc and Iron, Gamma and Beta Class, Carbonic Anhydrases of Domain Archaea

James G. Ferry
Department of Biochemistry and Molecular Biology, Eberly College of Science, The Pennsylvania State University, University Park, PA, USA

## Synonyms

Anaerobic prokaryotes; Bicarbonate; Carbon dioxide; Enzyme mechanism; Evolution

## Definition

Carbonic anhydrases catalyze the reversible hydration of carbon dioxide to bicarbonate and are abundantly distributed in diverse prokaryotes and eukaryotes. There are five classes ($\alpha,\beta,\gamma,\delta,\zeta$) with no significant sequence or structural identity among them, a remarkable example of convergent evolution. Carbonic anhydrases characterized from the domain *Archaea* are primarily from methane-producing species and of the β-and γ-class. The properties of these enzymes have revealed new paradigms including the role of iron in the active site.

## Background

The five independently evolved classes ($\alpha,\beta,\gamma,\delta,\zeta$) of carbonic anhydrases (CAs) have no significant overall structural or sequence similarity, although, all have an active-site metal and catalyze the reversible hydration of carbon dioxide (1)

$$CO_2 + H_2O = HCO_3^- + H^+ \qquad (1)$$

CAs are found in all three domains of life (*Eukarya*, *Bacteria*, and *Archaea*), underscoring the fundamental importance of this enzyme to life on Earth. It is proposed that the γ-class CA is an ancient enzyme and is thought to have played a prominent role in the evolution of early life. The α-class is restricted mostly to mammals, although a few pathogenic prokaryotes contain isoforms of this class. CAs have a diversity of functions that further underscore the broad importance of this enzyme in nature. The α-class CAs are found in nearly every mammalian tissue for which there are at least 16 isozymes that serve multiple functions, primarily in pH homeostasis and the transport of solutes. The β-class and putative γ-class enzymes are reported to be present in species from all three domains of life. The β-class CAs from aerobic prokaryotes function to maintain internal pH and $CO_2$/bicarbonate balances required for biosynthetic reactions. In anaerobic prokaryotes, β-class CAs are implicated in the transport of $CO_2$ and bicarbonate across the cytoplasmic membrane that regulates internal pH and assists in the transport of solutes into and out of the cell. Plants and phototrophic prokaryotes depend on β-class CAs for the transport and concentration of $CO_2$ required for photosynthesis. Although widely distributed in nature, physiological roles for the γ-class CAs are not as well documented. It is hypothesized that the archetype γ-class CA (Cam) from *Methanosarcina acetivorans*, an anaerobic methane-producing prokaryote from the domain *Archaea*, functions in the transport of $CO_2$ and acetate across the cell membrane. Interestingly, CA activity could not be demonstrated for a γ-class homolog found in plant mitochondria prompting speculation that the homolog possesses an unknown albeit related activity possibly contributing to recycle $CO_2$ in the context of photorespiration.

The vast majority of research has focused on α-class isozymes from mammals and β-class CAs from plants. This entry focuses on prokaryotic β- and γ-class CAs from the domain *Archaea* with emphasis on structure and catalytic mechanisms.

## The γ-Class

The archetype γ-class CA (Cam) from *Methanosarcina thermophila* of the domain *Archaea* was first investigated with the enzyme over produced in *E. coli* and purified in the presence of air which contained zinc in the active site; however, when exchanged with iron in vitro the catalytic efficiency increases threefold over the zinc form of the enzyme. When purified from *E. coli* in an inert atmosphere void of oxygen, ferrous iron is present in the active site that when exposed to air is oxidized to the ferric form with concomitant loss of iron from the active site and loss of activity. Aerobic purification of the iron form produced in *E. coli* results in exchange of iron with contaminating amounts of zinc found in untreated buffers. Thus, when evaluating the metal content of CAs it is imperative that the enzyme is purified in the absence of oxygen if iron is suspected in the active site. Finally, when Cam from *M. thermophila* is overproduced in the close relative *Methanosarcina thermophila*, and purified in an inert atmosphere, the enzyme contains nearly stoichiometric amounts of ferrous iron which establishes iron as the physiologically relevant metal in vivo (MacAuley et al. 2009).

Cam is a homotrimer with an overall left-handed β-helical fold distinct from all other CA structures (Fig. 1) (Iverson et al. 2000). One of three histidine side chains that ligate the active-site metal are contributed by a monomer adjacent to a monomer contributing the remaining two. Although the γ-class is independently evolved from the α-class with no significant sequence or overall structural identity, the three metal-ligating His residues superimpose nearly perfectly onto the His residues ligating the α-class HCAII (Kisker et al. 1996), a remarkable example of convergent evolution. A crystal structure is unavailable for Fe-Cam, although it is proposed to be similar to in vitro cobalt-substituted Cam with a coordination sphere containing three waters in addition to the three His ligands.

Similar to all CAs, kinetic analyses of Cam establish a two-step ping pong mechanism for the reversible hydration of $CO_2$ to $HCO_3^-$ as indicated in (2) and (3) where E represents enzyme residues, M is metal and B is buffer (Zimmerman and Ferry 2006).

$$E - M^{2+} - OH^- + CO_2 = E - M^{2+} - HCO_3^- \quad (2a)$$

$$E - M^{2+} - HCO_3^- + H_2O$$
$$= E - M^{2+} - H_2O + HCO_3^- \quad (2b)$$

$$E - M^{2+} - H_2O = H^+ - E - M^{2+} - OH^- \quad (3a)$$

$$H^+ - E - M^{2+} - OH^- + B$$
$$= E - M^{2+} - OH^- + BH^+ \quad (3b)$$

In the first step, a pair of electrons residing on the metal-bound hydroxide attack the carbon of $CO_2$ (2a) producing metal-bound $HCO_3^-$ that is next displaced by water (2b). In the next step, a proton is extracted from the metal-bound water (3a) and transferred to bulk solvent or buffer (3b). Cam has a $k_{cat} > 10^4 \, s^{-1}$ which is faster than the fastest rate at which protons transfer from zinc-bound water with a $pK_a$ of 7 to bulk solvent. Thus, Cam evolved a mechanism in which Glu84, residing on an acidic loop exposed to solvent, assumes two conformations that act as a shuttle to facilitate proton transfer (Zimmerman and Ferry 2006). Residues Glu106 and Thr199 of the human α-class HCAII, hydrogen bond with the zinc-bound hydroxide by way of the Thr199 Oγ atom. This hydrogen bond network facilitates lowering the $pK_a$ of the metal-bound hydroxide to near neutrality and optimally positions the lone pair of electrons for nucleophilic attack on the carbon of $CO_2$. The backbone amide of Thr199 hydrogen bonds and stabilizes the transition states of $HCO_3^-$. Replacement of active-site residues via site-directed mutagenesis and kinetic analyses of the variant proteins has identified several residues either essential or important for catalysis in γ-class Cam that function analogous to essential residues of the well-characterized α-class HCAII. Residues Gln75 and Asn73 of Cam are proposed to participate in a hydrogen bond network analogous to that of HCAII to orient and increase the nucleophilicity

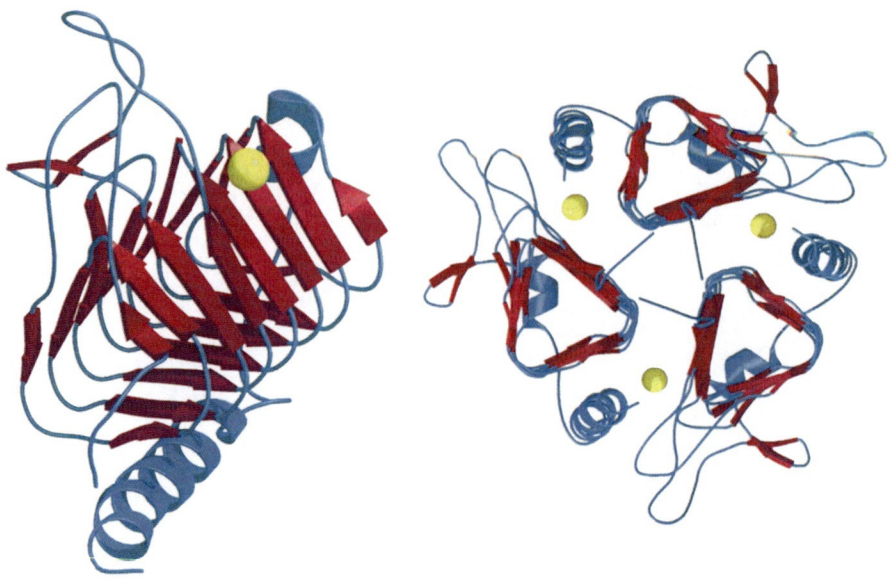

**Zinc and Iron, Gamma and Beta Class, Carbonic Anhydrases of Domain Archaea, Fig. 1** Overall structure of the γ-class CA from *Methanosarcina thermophila*. *Left panel*: Side view of the Cam monomer. β-strands are shown as *curved arrows* in *purple* and α-helices as ribbons in *cyan*. The active site metal is shown with van der Waals surface in *yellow*. *Right panel*: View along the molecular threefold axis (Reproduced by permission Kisker et al. (1996))

of the metal-bound hydroxide (Zimmerman and Ferry 2006). Further, Asn202 is proposed to stabilize the transition state analogous to Thr-199 of HCAII, and Glu62 is essential for $CO_2$ hydration. Based on these results, a catalytic mechanism has been proposed (Zimmerman and Ferry 2006) in which Glu62 extracts a proton from one of two metal-bound waters producing a hydroxyl group which extracts a proton from the adjacent water producing $Zn-OH^-$. The $Zn-OH^-$ attacks the carbon of $CO_2$ producing metal-bound $HCO_3^-$ that displaces a water molecule. The $HCO_3^-$ then undergoes a bidentate transition state wherein the proton either rotates or transfers to the nonmetal-bound oxygen of $HCO_3^-$. Glu62 hydrogen bonds with the hydroxyl of $HCO_3^-$, destabilizing it. An incoming water further destabilizes the $HCO_3^-$ by replacing one of the bound oxygens. A second incoming water displaces $HCO_3^-$ with product removal and regeneration of the water-bound active-site metal.

Cam is the archetype γ-class CA with only two others of this class shown to have CA activity. A search of databases queried with Cam reveals that a majority of the homologs comprise a putative subclass (CamH) in which there is major conservation of all of the residues essential for the archetype Cam except an acidic loop often missing the essential proton shuttle residue Glu84 of Cam (Zimmerman et al. 2010). The CamH homolog from *M. thermophila* overproduced in *E. coli* contains 0.15 iron per monomer and only a trace amount of zinc (Zimmerman et al. 2010). The enzyme losses activity when exposed to air which suggests ferrous iron has a role in the active site. A second CamH subclass CA (CcmM) from *Thermosynechococcus elongatus* (strain BP-1) of the domain *Bacteria* has been characterized (Pena et al. 2010). Either full-length CcmM or a construct truncated after 209 residues (CcmM209) is active. The structure of CcmM209 closely resembles Cam, except that residues 198-207 form a third α-helix stabilized by an essential Cys194-Cys200 disulfide bond. Deleting residues 194-209 (CcmM193) results in an inactive protein, the structure of which shows disordering of the N- and C-termini, and reorganization of the trimeric interface and active site. Under reducing conditions, CcmM209 is similarly partially disordered and inactive underscoring the importance of the disulfide bond. The only other CamH homolog investigated is from *Pyrococcus horikoshii*, an anaerobic species of the domain *Archaea* (Jeyakanthan et al. 2008). The crystal structure of the enzyme overproduced in *E. coli* and purified in the presence of air shows the overall fold similar to Cam with zinc in the active site, although activity was not reported raising the possibility that active-site iron is essential for activity.

## The β-Class

The β-class is further divided into two subclasses, based on the zinc coordination sphere, defined as "canonical" and "noncanonical." In both subclasses the zinc is coordinated by one histidine, two cysteines and a fourth ligand. Water or a substrate analog is the fourth ligand in the canonical subclass whereas an Asp residue, conserved in both subclasses, is the fourth ligand in the noncanonical subclass. The canonical subclass includes enzymes from the common pea plant (*Pisum sativum*), an anaerobic methane-producing thermophilic species from the domain *Archaea* (*Methanothermobacter thermautotrophicus* (*Methanothermobacter thermautotrophicus*, f. *Methanobacterium thermoautotrophicum*) strain ΔH), a human pathogen from the domain *Bacteria* (*Mycobacterium tuberculosis*, homolog Rv1284), and a photosynthetic species from the domain *Bacteria* (*Halothiobacillus neapolitans*). The noncanonical subclass includes CAs from the *Porphyridium purpureum*, *E. coli*, *Mycobacterium tuberculosis* (homolog Rv3588c), and *Haemophilus influenzae*.

The β-class canonical subclass CA (Cab) is from *M. thermautotrophicus*, a methane-producing species. Cab is the first β-class CA characterized from the domain *Archaea* and the first β-class CA from a strictly anaerobic species. The enzyme overproduced in *E. coli* and purified in air contains zinc; however, when purified in an inert atmosphere contains iron and looses activity on short exposure to air (unpublished results). Although Cab has yet to be purified from *M. thermautotrophicus* and characterized, the results are at least consistent with a physiological role for iron in Cab as for Cam from the methane-producing species *M. thermophila* of the domain *Archaea*.

Although the overall fold of Cab is similar to that of other β-class CAs that are tetrameric and octameric, Cab is a dimer in the crystal structure (Strop et al. 2001). Furthermore, although analytical centrifugation indicates Cab is a homotetramer, formation of a tetramer is unlikely in the absence of conformational changes due to steric clashes of opposing helices. The N-terminus of the *P. sativum* and *P. purpureum* structures extend into a long helix that packs against an opposing monomer forming additional dimer interactions. The N-terminus in Cab participates in crystal packing and does not adopt the same conformation as for the *P. sativum* and *P. purpureum* β-class CAs. Cab lacks the extended C-terminus characteristic of β-class CAs and as such is the smallest of the β-class. In the *P. sativum* enzyme, the C-terminus forms a long β strand essential for octamerization. The crystal structure shows that the active site metal of Cab resides at the interface of the two monomers, although the two His and one Cys coordinating the metal originate from the same monomer as opposed to the γ-class CA Cam. The metal coordinating residues superimpose closely to those of the *P. sativum* CA. The crystal structure of the *P. sativum* enzyme shows acetate as the fourth ligand to zinc, although acetate was a component of the crystallization buffer suggesting the possibility of an artifact. Unlike the *P. sativum* CA, acetate is not ligated to the active site metal of Cab; instead, water is the fourth ligand that assumes an orientation similar to acetate (Fig. 2). The coordinating water of Cab hydrogen bonds to an active-site Asp residue conserved in all β-class CAs. This conserved Asp is hydrogen bonded to an Arg conserved in the active sites of β-class CAs. The hydrophobic pocket of Cab proposed to bind $CO_2$ is more open and less hydrophobic than the *P. sativum* CA. Finally, the crystal structure shows the active-site cavity of Cab open to the bulk solvent such that a HEPES buffer molecule is present 8 Å from the catalytic metal (Fig. 2) distinct from the *P. sativum* CA. Although the active site is located at the interface between two monomers consistent with Cab, the *P. sativum* enzyme has a narrower hydrophobic channel leading to the bulk solvent that limits access of molecules larger than water to enter the active site (Kimber and Pai 2000).

The kinetic analysis of Cab is consistent with a two-step metal-hydroxide mechanism proposed for all β-class CAs. The conserved Asp residue in the active site of Cab when replaced with Ala significantly reduces the $k_{cat}/K_m$ value (Smith et al. 2002). When the Arg interacting with Asp is replaced with Ala, an even larger reduction in $k_{cat}/K_m$ results. In this variant, where the interaction between Asp and Arg is disrupted, Asp may swing toward the active site and ligate the catalytic metal in a dead-end inactive conformation. Imidazole rescues the activity of both the variants which suggests a role for the conserved Asp and Arg in the proton transfer step. Thus, an exchange of ligands to Asp during catalysis cannot be ruled out and may be necessary if the $pK_a$ of Asp were decreased when interacting with Arg such that Asp is unable to abstract a proton from the zinc-bound water. Another potential function for Arg may be to bind bicarbonate after release from the catalytic metal, thereby facilitating product removal, as proposed for an Arg residue in the active site of Cam.

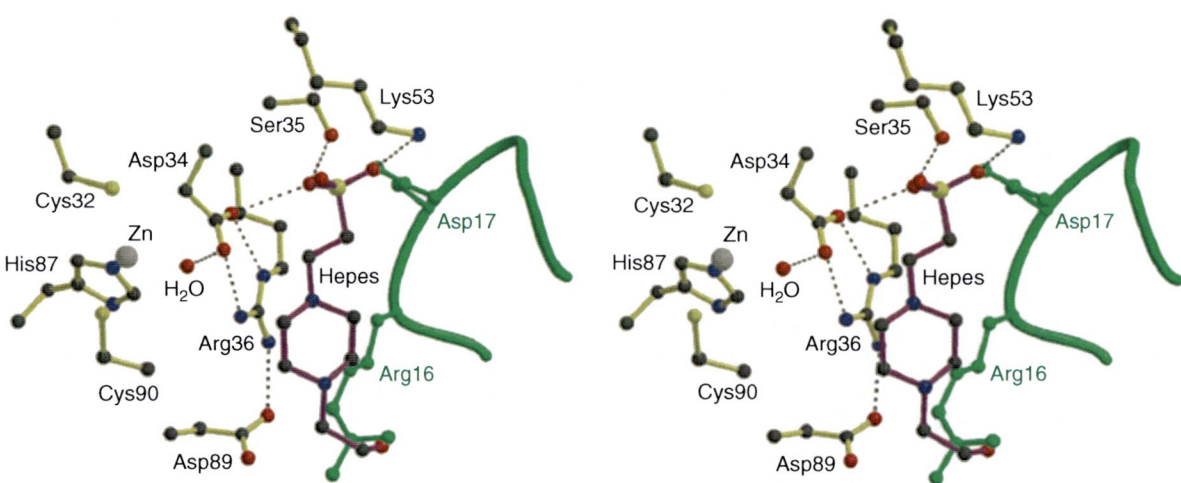

**Zinc and Iron, Gamma and Beta Class, Carbonic Anhydrases of Domain Archaea, Fig. 2** Stereo view of the active-site of the zinc form of Cab from *Methanothermobacter thermoautotrophicus* (Reproduced from Strop et al. (2001))

**Zinc and Iron, Gamma and Beta Class, Carbonic Anhydrases of Domain Archaea, Fig. 3** Zinc coordinating environment of the *Porphyridium purpureum* CA. Hydrogen bonds are depicted by *dotted lines* (Reproduced from Mitsuhashi et al. (2000))

The catalytic mechanism proposed for Cab is significantly different than that proposed for the *P. purpureum* CA which is characterized by very low $CO_2$ hydration activity when assayed below pH 7.0 and a metal coordination sphere distinct from Cab (Mitsuhashi et al. 2000). The side chain of the conserved Asp in the *P. purpureum* CA coordinates zinc (Fig. 3) instead of the water molecule in Cab (Fig. 2). As a consequence, the conserved Asp of the *P. purpureum* enzyme is unable to pair with the conserved Arg that is oriented away from the active site leading to a mechanism distinct from that proposed for Cab (Mitsuhashi et al. 2000). In the mechanism, the zinc-bound Asp extracts a proton from a water molecule hydrogen bonded to the residue yielding a nucleophilic hydroxide. In the next step, the protonated Asp is released from zinc and the nucleophilic hydroxide is transferred to the metal. The zinc-bound hydroxide attacks $CO_2$ generating zinc-bound $HCO_3^-$ and the proton of the protonated Asp is transferred to bulk solvent or buffer. The zinc-bound $HCO_3^-$ is displaced by the deprotonated aspartate releasing $HCO_3^-$. Finally, a water molecule binds to the carboxyl oxygen of the zinc-bound Asp initiating another round of catalysis. The low activity at neutral pH for the *P. purpureum* CA is consistent with this hypothesis since coordination to zinc is essential for the metal to act as a Lewis acid lowering the p$K_a$ of water to near neutrality.

## Conclusions

Although the γ-class is widely distributed among all three domains of life, the biochemical understanding is primarily limited to two examples from the domain *Archaea* and one from the domain *Bacteria* which implores further research on this class of CA. In particular, phylogenetic analyses indicate that the γ-class is dominated by a subclass (CamH) that lacks an acidic loop structure and residues essential for activity of Cam. The finding that iron functions as the physiologically relevant metal in the γ-class Cam and β-class Cab from anaerobic species raises the possibility that iron likely functions in CAs from diverse anaerobes requiring specialized methods for characterization.

## Cross-References

▶ Cadmium Carbonic Anhydrase
▶ Zinc Carbonic Anhydrases
▶ Zinc Sensors in Bacteria
▶ Zinc-Binding Proteins, Abundance

## References

Iverson TM, Alber BE, Kisker C, Ferry JG, Rees DC (2000) A closer look at the active site of γ-carbonic anhydrases: high resolution crystallographic studies of the carbonic anhydrase from *Methanosarcina thermophila*. Biochemistry 39:9222–9231

Jeyakanthan J et al (2008) Observation of a calcium-binding site in the γ-class carbonic anhydrase from *Pyrococcus horikoshii*. Acta Crystallogr D Biol Crystallogr 64(Pt 10):1012–1019

Kimber MS, Pai EF (2000) The active site architecture of *Pisum sativum* β-carbonic anhydrase is a mirror image of that of α-carbonic anhydrases. EMBO J 19(7):1407–1418

Kisker C, Schindelin H, Alber BE, Ferry JG, Rees DC (1996) A left-handed beta-helix revealed by the crystal structure of a carbonic anhydrase from the archaeon *Methanosarcina thermophila*. EMBO J 15:2323–2330

MacAuley SR et al (2009) The archetype γ-class carbonic anhydrase (Cam) contains iron when synthesized in vivo. Biochemistry 48:817–819

Mitsuhashi S et al (2000) X-ray structure of beta-carbonic anhydrase from the red alga, *Porphyridium purpureum*, reveals a novel catalytic site for $CO_2$ hydration. J Biol Chem 275(8):5521–5526

Pena KL, Castel SE, de Araujo C, Espie GS, Kimber MS (2010) Structural basis of the oxidative activation of the carboxysomal γ-carbonic anhydrase. CcmM Proc Natl Acad Sci USA 107:2455–2460

Smith KS, Ingram-Smith C, Ferry JG (2002) Roles of the conserved aspartate and arginine in the catalytic mechanism of an archaeal β-class carbonic anhydrase. J Bacteriol 184:4240–4245

Strop P, Smith KS, Iverson TM, Ferry JG, Rees DC (2001) Crystal structure of the "cab"-type β-class carbonic anhydrase from the archaeon *Methanobacterium thermoautotrophicum*. J Biol Chem 276(13):10299–10305

Zimmerman SA, Ferry JG (2006) Proposal for a hydrogen bond network in the active site of the prototypic γ-class carbonic anhydrase. Biochemistry 45(16):5149–5157

Zimmerman SA, Tomb JF, Ferry JG (2010) Characterization of CamH from *Methanosarcina thermophila*, founding member of a subclass of the γ class of carbonic anhydrases. J Bacteriol 192(5):1353–1360

# Zinc and Mowat-Wilson Syndrome

Livia Garavelli[1], G. Marangi[2], S. Rosato[1] and Marcella Zollino[2]
[1]Struttura Semplice Dipartimentale di Genetica Clinica, Dipartimento di Ostetrico-Ginecologico e Pediatrico, Istituto di Ricovero e Cura a Carattere Scientifico, Arcispedale S. Maria Nuova, Reggio Emilia, Italy
[2]Istituto di Genetica Medica, Università Cattolica Sacro Cuore, Policlinico A. Gemelli, Rome, Italy

## Synonyms

*SMADIP1* gene and *SIP1* gene (for Smad-interacting protein); *ZEB2* gene (for zinc finger E-box protein2); *ZFHX1B* gene (for zinc finger homeobox 1 B)

## Definition

Mowat-Wilson syndrome (MWS; OMIM# 235730) is a recently characterized genetic condition caused by heterozygous deletions or mutations in *ZEB2* gene (GenBank accession number NM-001171653.1).

## Introduction

Consistent clinical manifestations include typical craniofacial anomalies, with sunken, large, deep-set eyes, hypertelorism, epicanthic folds, large, medially flaring and sparse eyebrows, wide nasal bridge, prominent rounded nasal tip prominent columella, open mouth with M-shaped upper lip, pointed triangular chin, linearized mandibular bones, rotated ears with uplifted, fleshy lobes with a central depression, and a high incidence of microcephaly. Additional component manifestations are moderate-to-severe mental retardation, epilepsy, Hirschsprung disease, and multiple congenital anomalies, covering genital anomalies, like hypospadias in males, congenital heart disease, agenesis of the corpus callosum, and eye defects. Since the first delineation by Mowat et al. (1998), over 200 patients with a proven *ZEB2* haploinsufficiency have been reported mainly from Europe, Australia, and from the United States. The basic genomic defect is heterogeneous, including large chromosome 2q21-q23 deletions encompassing *ZEB2* and additional contiguous genes, partial gene deletions, and gene mutations. Loss-of-function mutations of *ZEB2*, as frameshift, nonsense, or splice site mutations, represent the most frequent gene defect. On the contrary, missense mutations occur rarely in association with a typical MWS phenotype, suggesting specific genotype-phenotype correlations.

The causative gene was mapped to chromosome 2q21-q23, on the basis of a cytogenetic deletion within this region detected in just one patient (Mowat et al. 1998) who shared some clinical manifestations with a patient reported earlier with a similar 2q22 deletion. It was hypothesized that mutations of a single gene mapping within the 2q22-q23 region could be the underlying pathogenic mechanism in undeleted patients (Mowat et al. 1998). In 2001, deletions or intragenic mutations of the *ZEB2* gene (*SIP1*) mapping on 2q22-2q23 were found in patients with syndromic HSCR (Wakamatsu et al. 2001; Cacheux et al. 2001). After the definition of the distinct MWS phenotype with or without HSCR caused by mutations in the *ZEB2* gene (Zweier et al. 2002), more than 200 cases have been reported (Mowat et al. 1998, 2003; Cacheux et al. 2001; Wakamatsu et al. 2001; Zweier et al. 2002, 2005; Garavelli et al. 2003, 2007; Dastot-Le Moal et al. 2007).

### The *ZEB2* Gene

The human *ZEB2* gene (for zinc finger E-box protein2) (also known as *ZFHX1B*, for zinc finger homeobox 1 B; *SMADIP1* and *SIP1* for Smad-interacting protein) consists of 10 exons that span 120 kb (Wakamatsu et al. 2001; Dastot-Le Moal et al. 2007) (Fig. 1). The cytogenetic location is 2q22.3, and the genomic coordinates (GRCh37) are 2:145,141,941-145,277,957. The initiation codon lies in exon 2, and the stop codon is located in exon 10. *ZEB2* (*SIP1*) is a member of the *ZFHX1* (*ZEB*) family of two-handed zinc finger/homeodomain proteins, and it was initially characterized as a gene repressor. The *ZEB* family of transcription factors comprises two members, *ZEB1/ dEF1* and *ZEB2/SIP1*.

The *ZEB2* (*SIP1*) protein contains two separated clusters of multiple $C_2H_2$-type zinc finger domains (Postigo and Dean 2000; Verschueren et al. 1999) flanking a homeodomain-like segment, a C-terminal-binding-protein (CtBP) binding site, and a Smad-binding domain (Figs. 1 and 2). The N-terminal cluster (NZF) contains four Zn-fingers (three CCHH fingers and one CCHC finger), while the other cluster (CZF), located in the C-terminal part of the protein, contains three CCHH zinc fingers. Binding with each zinc finger cluster to a tandem of spaced CACCT(G) (Verschueren et al. 1999) or CACANNT(G) sequences in regulatory elements of genes causes *ZEB2* to repress target genes such as *E-cadherin* (a major cell-cell adhesion molecule responsible for strong intercellular interactions and distinct epithelial cell polarity) and *Xenopus Brachyury* (a member of the T-box family of transcription factor and essential in mesoderm formation during early development).

### The Function of the *ZEB2* Family of Zinc Finger Transcription Factors

The *ZFHX1* family (*ZEB* family) of zinc finger transcription factors is essential during normal embryonic development: they induce epithelial to mesenchymal transition (EMT), a process that reorganizes epithelial cells to become migratory mesenchymal cells. This process, in which cells undergo a molecular switch

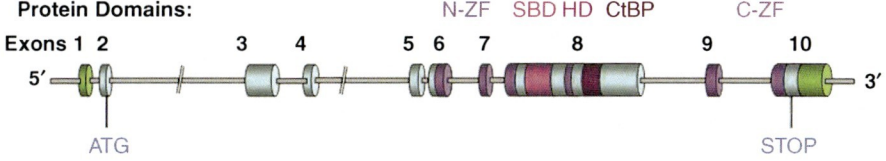

**Zinc and Mowat-Wilson Syndrome, Fig. 1** The *ZEB2* gene: coding sequences (exons 2–10) and functional domains. *N-ZF* N-terminal zinc finger cluster, *SBD* Smad-binding domain, *HD* homeodomain, *CtBP* C-terminal binding protein interacting domain, *C-ZF* C-terminal zinc finger cluster (Adapted from Zweier et al. 2002)

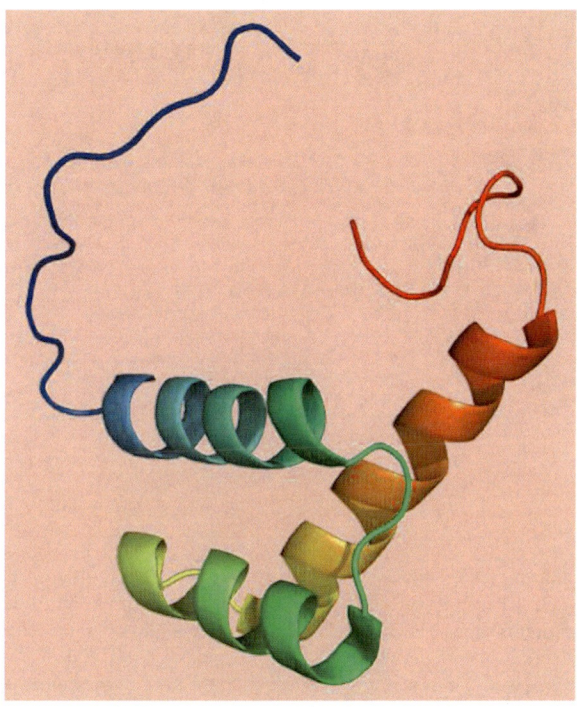

**Zinc and Mowat-Wilson Syndrome, Fig. 2** The ZEB2 protein (from Protein DataBank http://www.pdb.org/pdb/explore/explore.do?structure Id=2da7)

from a polarized, epithelial phenotype to a highly motile, nonpolarized mesenchymal phenotype, is essential for developmental processes such as gastrulation, neural crest formation, heart morphogenesis, and formation of the musculoskeletal system and craniofacial structures. *E-cadherin* is a major target gene of these transcriptional repressors, and this downregulation is considered a hallmark of EMT (Vandewalle et al. 2009).

The reverse process of EMT, known as mesenchymal-epithelial transition or MET, has also been reported. MET occurs during somitogenesis, kidney development, and coelomic-cavity formation. Complete reversal of epithelial to mesenchymal transition (EMT) requires inhibition of both *ZEB* expression and the Rho pathway (Das et al. 2009) (Fig. 3).

The onset of in vitro osteogenic differentiation is associated with the upregulation of the expression of the liver/bone/kidney alkaline phosphatase (LBK-ALP) gene. ZEB2/SIP1 was shown to repress LBK-ALP promoter activity by virtue of its binding to the CACCT/CACCTG sites in the latter promoter.

The Smad-binding domain interacts with TGF-beta family receptor activated Smad proteins (intracellular mediator of transforming growth factor-beta signaling) and acts as transcriptional repressor (Verschueren et al. 1999). ZEB2 contains also a CtBP interaction domain (CID) and seems to be part of a CtBP corepressor core complex that contains ZEB1, the other member of ZEB family, histone deacetylases and histone methyltransferases, chromodomain-containing proteins, coREST, and coREST-related proteins: that is all essential elements for promoter targeting, transcriptional repression, and chromatin remodeling (Shi et al. 2003). In Xenopus, XSIP1 acts as a direct repressor of BMP4, and its efficient repression depends on the interaction of XSIP1 with the corepressor CtBP, thus limiting BMP signaling and subsequent epidermal differentiation. Nevertheless, downregulation of certain epidermal genes by XSIP1 occurs independently of BMP repression (van Grunsven et al. 2007).

Though studies have focused mainly on Zfhx1b repressive function, there is evidence that this factor also activated gene transcription. It was shown that the transcriptional activity of ZEB2 can be modified (attenuated), in a manner that depends on the promoter context, by its posttranslational SUMOylation, mediated by polycomb protein Pc2.

Besides, some classes of microRNAs, specifically miR192, miR141, miR200 family, and miR205, were shown to directly target *ZEB2*, resulting in repression of protein expression.

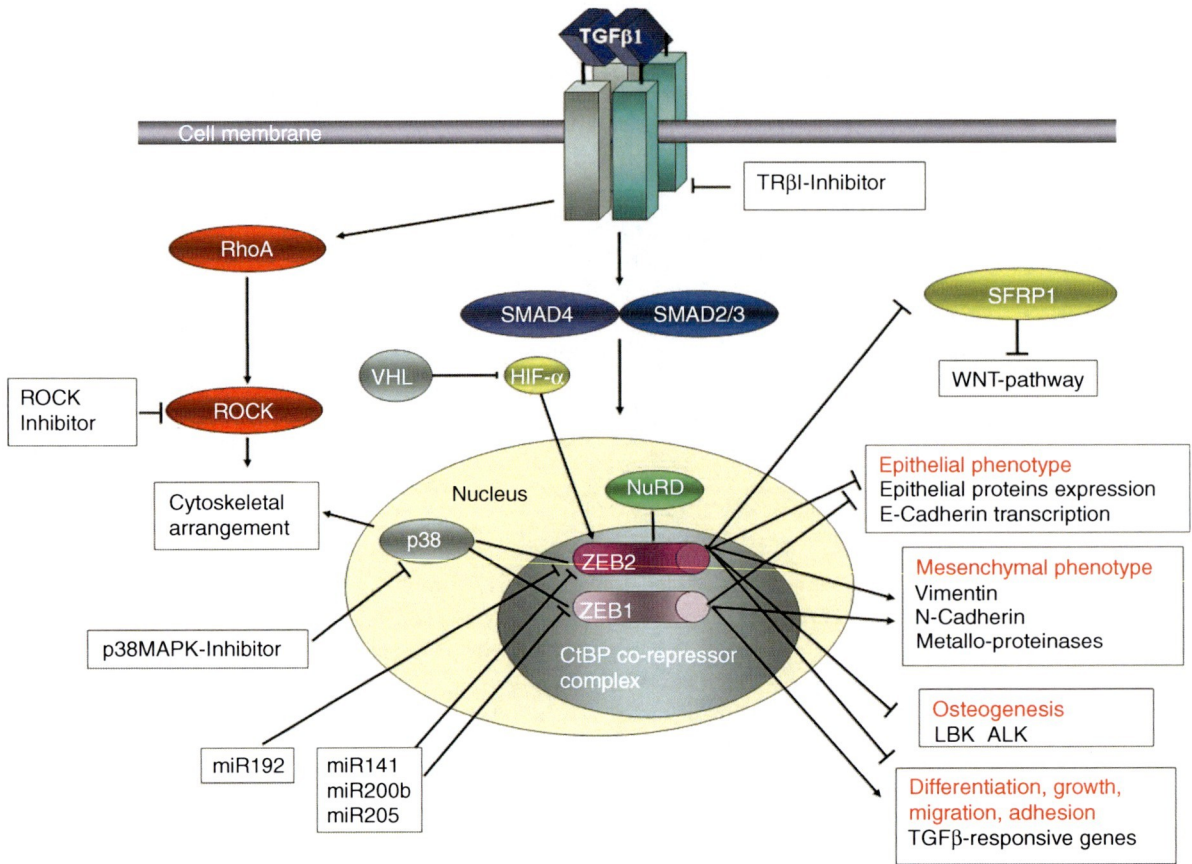

**Zinc and Mowat-Wilson Syndrome, Fig. 3** The ZEB family of transcription factors: *ZEB1/dEF1* and *ZEB2/SIP1*. They are both downstream effectors of TGF-b-mediated EMT but can also be downregulated by TGF-b. For *ZEB2/SIP1*, this occurs by the action of microRNA miR192. In addition, miR141, miR200 family, and miR205 target both *ZEB1/dEF1* and *ZEB2/SIP1*. Complete reversal of epithelial to mesenchymal transition (EMT) requires inhibition of both *ZEB* expression and the Rho pathway. Model for reversal of EMT induced by TGF-β1 (proposed by Das et al. (2009)): to reexpress epithelial proteins such as E-cadherin, a TβRI kinase inhibitor is needed to decrease expression of mesenchymal genes (*ZEB1* and *ZEB2*), while a Rho kinase inhibitor is required to stabilize the epithelial cortical actin (Adapted from Das et al. (2009) and from Vandewalle et al. (2009))

As a matter of fact, biological properties of Zfhx1b are not fully characterized, and some phenotypic manifestations in MWS remain still unclear at a molecular level. Taken together, Zfhx1-type proteins are complex multi-domain proteins that can associate with different partners via different domains, causing selective repression or activation of target genes. The NuRD complex (multiple subunits of the nucleosoma remodeling and histone deacetylation complex) was recently characterized as a novel cofactor of Zfhx1b. Through interaction with its N-terminal domain, NuRD is possibly involved in Zfhx1b-dependent gene repression (Verstappen et al. 2008). It is worth noting that typical MWS patients carry either gene deletions or gene mutations removing the zinc finger domains and the C-terminal segment of the protein. A *ZEB2* mutation disrupting a splice site was recently identified in a patient with a mild form of MWS. The aberrant protein was demonstrated to retain its zinc finger clusters but to have the N-terminal 24 amino acids missing. A defective recruitment of NuRD has been proposed as the most likely underlying mechanism in this mild form of MWS (Verstappen et al. 2008).

From a clinical point of view, it is important to note that *ZEB2* (*SIP1*) has been implicated in neuroectoderm development in Xenopus (van Grunsven et al. 2007), mice (Van de Putte et al. 2007), chick

(Sheng et al. 2003), and humans (Espinosa-Parrilla et al. 2002). In particular, in human embryos, *ZEB2* (referred to as *SMADIP1*) is homogeneously expressed throughout the central nervous system from the mesencephalon to the spinal cord, including diencephalon (clinical correlation with agenesis of corpus callosum), mesencephalon (clinical correlation with epilepsy), and rhombencephalon (clinical correlation with severe intellectual disability). It is also expressed in facial neuroectoderm (correlation with facial dysmorphisms), in the mandible (abnormal mandibular bones in patients), in the middle ear region surrounding the developing ossicles (typical ears conformation), in gut (Hirschsprung disease, severe constipation, pyloric stenosis), in genitalia (hypospadias, cryptorchidism), and in mesonephros and metanephros (kidney and urinary tract abnormalities). Of note, a mouse model involving a conditional mutation in the *Zfhx1b* gene in neural cell precursor cells clearly indicated a crucial role of *Zfhx1b* for proper neural crest cells development (Van de Putte et al. 2007). Based on this evidence, MWS can well be considered a particular form of neurocristopathy. The neural crest is a transient multipotent cell population specific to vertebrates. Formed in the neural ridge at about the fourth week of gestation, these cells detach from the initial position and migrate through the embryo. They differentiate into several structures according to the site of origin along the anterior-posterior axis. Cranial neural crest cells differentiate into connective tissue and skeletal-muscular structures of the head, while vagal neural crest reaches the gastrointestinal tract and the cardiovascular system. Nearly, the totality of MWS-associated clinical signs can be explained by a defect in the induction, migration, and differentiation of neural crest cells.

Specifically, it was demonstrated that the ablation of *Sip1* in the dorsal telencephalon of mice leads to the loss of hippocampus and corpus callosum in adult: this effect is at least in part caused by a strong upregulation of *Sfrp1* (an extracellular antagonist of the Wnt pathway) and an inhibition of *JNK*-dependent noncanonical Wnt signaling pathway, eventually resulting in decreased proliferation of neural progenitors and in apoptosis of postmitotic cells.

Besides, *Zfhx1b* appears to interact with *Sox10* (mutation of the human paralogous that is responsible for Waardenburg syndrome type 2E, with Hirschsprung disease) during all phases of enteric nervous system development (Stanchina et al. 2010), and both proteins are also present in other neural crest derivatives such as melanocytes, cranial ganglia, satellite glial cells, and Schwann cell of the peripheral nervous system. These findings provide further explanation of the mechanisms by which *ZEB2* haploinsufficiency leads to brain anomalies and Hirschsprung disease in Mowat-Wilson syndrome.

Finally, a recent study showed that *Zeb2* plays a key role in embryonic hematopoiesis (Goossens et al. 2011), driving the hypothesis that Mowat-Wilson patients could develop hematological disorders, although no such conditions have been reported in the described patients to date.

Haploinsufficiency of the metalloprotein *ZEB2* appears to cause a severe, multisystemic developmental disruption in humans, featuring the complex MWS phenotype acting at multisystemic levels. Understanding its biological properties and clarifying the molecular pathway of protein-to-protein interaction could allow targeted therapeutic strategies.

## Cross-References

▶ Cadherins
▶ Zinc and Treatment of Wilson's Disease
▶ Zinc Cellular Homeostasis
▶ Zinc Finger Folds and Functions
▶ Zinc Homeostasis, Whole Body
▶ Zinc Matrix Metalloproteinases and TIMPs
▶ Zinc-Binding Proteins, Abundance
▶ Zinc-Binding Sites in Proteins

## References

Cacheux V, Dastot-Le Moal F, Kääriäinen H, Bondurand N, Rintala R, Boissier B, Wilson M, Mowat D, Goossens M (2001) Loss-of-function mutations in SIP1 Smad interacting protein 1 results in a syndromic Hirschsprung disease. Hum Mol Genet 10:1503–1510

Das S, Becker BN, Hoffmann FM, Mertz JE (2009) Complete reversal of epithelial to mesenchymal transition requires inhibition of both ZEB expression and the Rho pathway. BMC Cell Biol 10:94. doi:10.1186/1471-2121-10-94

Dastot-Le Moal F, Wilson M, Mowat D, Collot N, Niel F, Goosens M (2007) ZFHX1B mutations in patients with Mowat-Wilson syndrome. Hum Mutat 28:313–321

Espinosa-Parrilla Y, Amiel J, Augé J, Encha-Razavi F, Munnich A, Lyonnet S, Vekemans M, Attié-Bitach T (2002) Expression of the SMADIP1 gene during early human development. Mech Dev 114(1–2):187–191

Garavelli L, Donadio A, Zanacca C, Della Giustina E, Bertani G, Albertini G, Zollino M, Rauch A, Banchini G, Neri G (2003) Hirschsprung disease, mental retardation, characteristic facial features and mutation in the gene ZFHX1B (SIP1): confirmation of the Mowat-Wilson syndrome. Am J Med Genet 116A:385–388

Garavelli L, Zollino M, Mainardi PC, Gurrieri F, Rivieri F, Soli F, Verri R, Albertini E, Favaron E, Zignani M, Orteschi D, Bianchi P, Faravelli F, Forzano F, Seri M, Wischmeijer A, Turchetti D, Pompilii E, Gnoli M, Cocchi G, Mazzanti L, Bergamaschi R, De Brasi D, Sperandeo MP, Mari F, Uliana V, Mostardini R, Cecconi M, Grasso M, Sassi S, Sebastio G, Renieri A, Silengo M, Bernasconi S, Wakamatsu N, Neri G (2007) Mowat-Wilson syndrome: facial phenotype changing with age: study of 19 Italian patients and review of the literature. Am J Med Genet A 149A(3):417–426

Goossens S, Janzen V, Bartunkova S, Yokomizo T, Drogat B, Crisan M, Haigh K, Seuntjens E, Umans L, Riedt T, Bogaert P, Haenebalcke L, Berx G, Dzierzak E, Huylebroeck D, Haigh JJ (2011) The EMT regulator Zeb2/Sip1 is essential for murine embryonic hematopoietic stem/progenitor cell differentiation and mobilization. Blood 117:5620–5630

Mowat DR, Croaker GDH, Cass DT, Kerr BA, Chaitow J, Adès LC, Chia NL, Wilson M (1998) Hirschsprung disease, microcephaly, mental retardation, and characteristic facial features: delineation of a new syndrome and identification of a locus at chromosome 2q22-q23. J Med Genet 35:617–623

Mowat DR, Wilson MJ, Goossens M (2003) Mowat-Wilson syndrome. J Med Genet 40:305–310

Postigo AA, Dean D (2000) Differential expression and function of members of the zfh-1 family of zinc finger/homeodomain repressors. Proc Natl Acad Sci USA 97:6391–6396

Sheng G, dos Reis M, Stern CD (2003) Churchill, a zinc finger transcriptional activator, regulates the transition between gastrulation and neurulation. Cell 115(5):603–613

Shi Y, Sawada J, Sui G, Affar el B, Whetstine JR, Lan F, Ogawa H, Luke MP, Nakatani Y, Shi Y (2003) Coordinated histone modifications mediated by a CtBP co-repressor complex. Nature 422:735–738

Stanchina L, Van de Putte T, Goossens M, Huylebroeck D, Bondurand N (2010) Genetic interaction between Sox10 and Zfhx1b during enteric nervous system development. Dev Biol 341(2):416–428

Van de Putte T, Francis A, Nelles L, van Grunsven LA, Huylebroeck D (2007) Neural crest-specific removal of Zfhx1b in mouse leads to a wide range of neurocristopathies reminiscent of Mowat–Wilson syndrome. Hum Mol Genet 16:1423–1436

van Grunsven LA, Taelman V, Michiels C, Verstappen G, Souopgui J, Nichane M, Moens E, Opdecamp K, Vanhomwegen J, Kricha S et al (2007) XSip1 neuralizing activity involves the co-repressor CtBP and occurs through BMP dependent and independent mechanisms. Dev Biol 306:34–49

Vandewalle C, Van Roy F, Berx G (2009) The role of the ZEB family of transcription factors in development and disease. Cell Mol Life Sci 66(5):773–787

Verschueren K, Remacle JE, Collart C, Kraft H, Baker BS, Tylzanowski P, Nelles L, Wuytens G, Su MT, Bodmer R et al (1999) SIP1, a novel zinc finger/homeodomain repressor, interacts with Smad proteins and binds to 50-CACCT sequences in candidate target genes. J Biol Chem 274:20489–20498

Verstappen G, van Grunsven LA, Michiel C, Van de Putte T, Souopgui J, Van Damme J, Bellefroid E, Vandekerckhove JI, Huylebroeck D (2008) Atypical Mowat–Wilson patient confirms the importance of the novel association between ZFHX1B/SIP1 and NuRD corepressor complex. Hum Mol Genet 17(8):1175–1183

Wakamatsu N, Yasukazu Y, Kenichiro Y, Takao O, Nomura N, Taniguchi H, Kitoh H, Mutoh N, Yamanaka T, Mushiake K, Kato K, Sonta S, Nagaya M (2001) Mutations in SIP1, encoding Smad interacting protein-1, cause a form of Hirschsprung disease. Nat Genet 27:369–370

Zweier C, Albrecht B, Mitulla B, Behrens R, Beese M, Gillessen-Kaesbach G, Rott HD, Rauch A (2002) "Mowat-Wilson" syndrome with and without Hirschsprung disease is a distinct, recognizable multiple congenital anomalies-mental retardation syndrome caused by mutations in the zinc finger homeobox 1 B gene (ZFHX1B). Am J Med Genet 108(3):177–181

Zweier C, Thiel CT, Dufke A, Crow YJ, Meinecke P, Suri M, Ala-Mello S, Beemer F, Bernasconi S, Bianchi P, Bier A, Devriendt K, Dimitrov B, Firth H, Gallagher RC, Garavelli L, Gillessen-Kaesbach G, Kääriäinen H, Karstens S, Mannhardt A, Mücke J, Kibaek M, Nylandsted Krogh L, Peippo M, Rittinger O, Schulz S, Schelley S, Temple K, Van der Knaap MS, Wheeler P, Yerushalmi B, Zenker M, Lowry RB, Rauch A (2005) Clinical and mutational spectrum of Mowat-Wilson syndrome. Eur J Med Genet 48:97–111

# Zinc and the Decreased Blood Coagulability

▶ Zinc in Hemostasis

# Zinc and the Immune System

▶ Zinc and Immunity

# Zinc and Treatment of Wilson's Disease

Claudia Della Corte, Andrea Pietrobattista and Valerio Nobili
Unit of Liver Research of Bambino Gesù Children's Hospital, IRCCS, Rome, Italy

## Synonyms

Hepatolenticular degeneration; WD; Zn

## Zinc and Treatment of Wilson's Disease

Wilson's disease (WD), also known as "hepatolenticular degeneration," is an autosomal recessive inherited disorder of copper metabolism (Roberts and Schilsky 2008). The disease is caused by a mutation in a gene on chromosome 13 that encodes a copper-transporting P-type adenosine triphosphatase 7B (ATP7B) that mediates the translocation of cations across cellular membrane. ATP7B has two main functions: to translocate copper into the trans-Golgi for the synthesis of ceruloplasmin and to export this metal from the cell (Barte and Lutsenko 2007). Disease-causing mutations lead to impairment in hepatic copper excretion into the bile that leads to progressive copper accumulation in the liver and subsequent deposition in other organs, such as the nervous system, corneas, kidney, bones, and joints (▶ Zinc and Mowat-Wilson's Syndrome). The distribution of the metal in diverse organs over time accounts for the wide range of clinical manifestations (Ala et al. 2007). In fact, although the failure to excrete biliary copper is present from birth, it is widely accepted that symptoms generally do not develop until about 3 years of age. Clinical symptoms develop sequentially based on the pathophysiologic disturbance of copper metabolism. Copper silently accumulates in the liver during childhood. After the liver storage capacity for copper becomes saturated, circulating non-ceruloplasmin-bound "free copper" levels rise and copper is then redistributed systemically, accumulating in various organs. This change in organ distribution of copper over time is paralleled by the clinical presentations of Wilson's disease. Based on data from the combined large WD patient series by Walshe and by Scheinberg and Sternlieb, most of the pediatric WD patients present with liver disease, whereas neuropsychiatric symptoms are more common after age 18 years and are found in only 4–6% of pediatric cases with hepatic onset (O'Connor and Sokol 2007). Wilson's disease is a treatable disorder and its therapy is the more effective the earlier it is instituted. The goal of therapy is to reduce copper accumulation generating initially a negative copper balance (decopperizing phase) and then maintaining an equilibrium of copper level (maintenance therapy) either by increasing its urinary excretion or by reducing its intestinal absorption (Huster 2010). Current treatments are able to stabilize or to revert the symptoms due to toxic copper accumulation in symptomatic patients or to prevent their onset in presymptomatic cases. In fact, it has been demonstrated that appropriately life-long-treated patients with Wilson's disease have an excellent long-term prognosis, with a survival probability not differing from general population age and sex matched (Bruha et al. 2010).

Zinc acts inducing in enterocytes metallothionein that binds copper reducing its absorption into portal circulation. It represents an attractive drug because not only is it more physiological than other used drugs, such as penicillamine or trientine, with a low toxicity, but also it is cheap. The main adverse event of zinc therapy is represented by gastritis due to salt linked to zinc, which usually resolves by changing formulation (from sulfate to acetate) or modifying administration (from breakfast to midmorning or with a small amount of proteins). An immunosuppressant effect has been described with zinc therapy, but in clinical practice no immunological side effects during long-term follow-up have been reported (Brewer et al. 1997). In some cases an isolated increase of serum amylase and lipase levels without clinical and radiological features of pancreatitis has been observed (Yuzbasiyan-Gurkan et al. 1989). Although zinc is currently recommended for the treatment of presymptomatic patients and for maintenance therapy (Roberts and Schilsky 2008), mainly in pediatric population it has been increasingly used as first-line therapy with good outcome in patients with isolated hypertransaminasemia as hepatic presentation (Marcelli et al. 2005; Brewer and Askari 2005).

However, the data on monotherapy with zinc in symptomatic Wilson's disease are controversial. It is to note that, as also reported in a recent systematic review on initial therapy of Wilson's disease (Wiggelinkhuizen et al. 2009), the available data could be biased because zinc is scarcely used in patients with moderate to severe hepatic presentation and so the efficacy of zinc in this group of patients is not well known. In fact, because of its slower onset of action, zinc therapy is generally not used as initial therapy for symptomatic patients. Although some authorities suggest that because children have generally an isolated increase of aminotransferases or a mild hepatopathy, they should not be exposed to the possible side effects of copper chelators. Because these children have not yet accumulated a toxic burden of copper, it is proposed that the treatment with zinc is effective and less toxic. Prospective long-term follow-up of patients with various presentation forms of WD is needed to establish the safest and most effective manner of treating this group of patients (O'Connor and Sokol 2007).

## Cross-References

▶ Biological Copper Transport
▶ Zinc and Mowat-Wilson Syndrome

## References

Ala A, Walker AP, Ashkan K, Dooley JS, Schilsky ML (2007) Wilson's disease. Lancet 369:397–408
Barte MY, Lutsenko S (2007) Hepatic copper-transporting ATPase ATP7B: function and inactivation at the molecular and cellular level. Biometals 20:627–637
Brewer GJ, Askari FK (2005) Wilson's disease: clinical management and therapy. J Hepatol 42:S13–S21
Brewer GJ, Johnson V, Kaplan J (1997) Treatment of Wilson's disease with zinc: XIV. Studies of the effect of zinc on Lymphocyte function. J Lab Clin Med 129:649–652
Bruha R, Marecek Z, Pospisilova L et al (2010) Long-term follow-up of Wilson disease: natural history, treatment, mutations analysis and phenotypic correlation. Liver Int 31:83–91
Huster D (2010) Wilson disease. Best Pract Res Clin Gastroenterol 24:531–539
Marcelli M, Di Ciommo V, Callea F et al (2005) Treatment of Wilson's disease with zinc from the time of diagnosis in pediatric patients: a single-hospital, 10-year follow-up study. J Lab Clin Med 145:139–143
O'Connor JA, Sokol RJ (2007) Copper metabolism and copper storage disorders. In: Suchy FJ, Sokol RJ, Balistreri WF (eds) Liver disease in children. Cambridge University Press, New York, pp 626–659
Roberts EA, Schilsky ML (2008) Diagnosis and treatment of Wilson disease: an update. Hepatology 47:2089–2111
Wiggelinkhuizen M, Tilanus MEC, Bollen CW, Houwen RHJ (2009) Systematic review: clinical efficacy of chelator agents and zinc in the initial treatment of Wilson disease. Aliment Pharmacol Ther 29:947–958
Yuzbasiyan-Gurkan V, Brewer GJ, Abrams GD, Main B, Giacherio D (1989) Treatment of Wilson's disease with zinc. V. Changes in serum levels of lipase, amylase, and alkaline phosphatase in patients with Wilson's disease. J Lab Clin Med 114:520–526

# Zinc and Wound Healing

Magnus S. Ågren
Department of Surgery and Copenhagen Wound Healing Center, Bispebjerg University Hospital, Copenhagen, Denmark

## Synonyms

Wound repair

## Definition

Wound healing is a key survival process to restore to a state of soundness from an interruption in the continuity of external or internal surfaces of the body caused by any type of injury.

## Wound-Healing Process

Injury to tissues starts a repair process characterized by a series of overlapping and chronological phases. The first step is to arrest bleeding. The so formed blood clot composed of fibrin and fibronectin enables an influx of inflammatory cells to the wound site. Invasion by the inflammatory cells is guided by chemoattractive factors liberated during clot formation. The function of inflammatory cells is primarily to kill bacteria. Later, the provisional matrix becomes populated by new blood vessel cells and connective tissue cells (fibroblasts).

Cytokines and growth factors control many of the biological processes. These potent polypeptides are delivered via the blood mostly by the platelets but are also produced by the cells in the wounds. The provisional tissue also provides the substratum for the reestablishment of new epidermis, also named epithelialization, through proliferation and migration of epithelial cells from the wound edges. The epithelial coverage protects from invasion of microorganisms and other foreign bodies as well as reduces water loss. The provisional structural proteins fibrin and fibronectin are gradually replaced by the permanent interstitial collagen molecules that are produced by fibroblasts and render strength to the wound. The collagen molecules are remodeled for years, but the strength of the wound never reaches that of uninjured skin. The reader is further referred to reviews on the wound-healing process (Ågren and Werthén 2007; Singer and Clark 1999).

## Zinc in General

Zinc is a ubiquitous trace element essential for growth. Zinc has multiple biological functions in wound repair known from ancient times (Lansdown et al. 2007). Severe hereditary zinc deficiency (acrodermatitis enteropathica) is caused by impaired zinc uptake in small intestine due to mutation of the ZIP-4 gene, which encodes a zinc membranous transporter protein. Decreased resistance to infections and impaired wound healing are prominent symptoms of acrodermatitis enteropathica.

Zinc is widely distributed in the human environment being found in water, air, and virtually all foodstuffs. The medicinal properties of zinc in the form of calamine were documented more than 3,000 years ago in the Ebers Papyrus and in ancient Ayurvedic manuscripts in early Indian medicine (Prasad 1995). In 1941, Keilin and Mann identified the first metalloenzyme, carbonic anhydrase, with zinc as an essential cofactor, but more recently, zinc has been identified in more than 300 different enzymes.

Of these, the metzincin superfamily of metalloendopeptidases is closely linked to wound healing (Toriseva and Kähäri 2009). The metzincins comprise a large number of zinc-dependent metalloproteinases that are characterized by particular sequence similarities which are highly conserved among its members (Gomis-Rüth 2003). Metzincin substrates include not only extracellular matrix proteins but also numerous cytokines and their receptors and chemokines that promote recruitment of inflammatory cells. Thus, these enzymes are involved in physiological and pathological conditions. There are at least 70 metzincin genes in the human genome, and about 60 of them encode zymogens of active proteinases. The metzincin subgroups matrix metalloproteinases (MMPs), a disintegrin and metalloproteinases (ADAMs), and ADAMs with thrombospondin motifs (ADAMTSs) are all fundamental for different processes during wound healing (Toriseva and Kähäri 2009). MMPs are obligatory for epithelialization, ADAMs govern the release of the proinflammatory cytokine tumor necrosis factor-$\alpha$ and its receptor as well as the shedding of agonists to the epidermal growth factor receptor, and ADAMTSs are pivotal for extracellular processing of procollagen.

Zinc fingers encompass even a larger group of zinc-containing proteins. They were originally discovered by Klug and coworkers in 1986 and are found in the DNA-binding domain of the most abundant family of transcription factors. This overwhelming family comprises more than 700 proteins. Zinc-finger proteins activate transcription of growth factors, cytoprotective proteins, and regulators of adult hematopoietic stem cells. For example, early growth response-1 (Egr-1) transcription factor is an important mediator for tumor growth factor-$\beta$ signaling. Wound healing is delayed in Egr-1-deficient mice and augmented with overexpressed Egr-1 (Wu et al. 2009).

Apart from its importance in protein complexes, the zinc ion is closely involved in not only neurotransmission but also in intracellular signaling much like calcium (Fukada et al. 2011). These signals are mediated and controlled by zinc transporter proteins together with metallothioneins (MTs). MTs complex up to 20% of intracellular zinc and play a central role in zinc metabolism in both health and disease. These ubiquitous, cysteine-rich low molecular weight proteins regulate the intracellular supply of zinc to enzymes, gene regulatory molecules and zinc depots, and protect cells from deleterious effects of exposure to elevated levels of zinc. One MT molecule can bind seven zinc ions.

## Systemic Zinc Therapy

The first positive therapeutic results of zinc supplementation on wound healing were reported 1953 in experimental animals and 1967 in man. After these enthusiastic reports, it is now clear that systemic zinc treatment is beneficial only in the case of zinc deficiency (Wilkinson and Hawke 2000). The critical issue is to determine whether the patient is deficient or not. Diagnosis of zinc deficiency in the human body is complicated by the low concentrations present and by controversy and lack of sensitive indices. Traditionally, serum zinc has been used for diagnosis. In serum, about 60% of zinc is bound to albumin, 30% to $\alpha_2$-macroglobulin, and 10% to various other ligands. Normal serum zinc is considered to be in the range 10–18 µM, with higher levels in men than in women, but variations in published figures may be due to the accuracy of analytical methods used and circadian rhythms of zinc levels. Serum zinc level below 9 µM is biochemically defined as zinc deficiency, but cancers, hyperactivity, stress, trauma, active tuberculosis, skin diseases, chronic wounds, chronic kidney insufficiency, kidney failure, and nephrotic syndrome are predictable causes of low zinc levels. Regardless, it can be assumed that elderly patients are zinc deficient and will benefit from zinc supplementation.

In surgical patients, the degree of serum zinc depression following surgical trauma is important and may be predictive for the outcome of the intervention. Under certain conditions, parenteral zinc administration has been applied to correct the zinc balance and restore healing capability.

Patients with large burns suffer from acute zinc depletion via loss of zinc with the exudate. These patients seem to benefit from zinc supplementation manifested by reduction of infections and improved wound healing. Recent results from animal experiments verify this promising new indication for zinc therapy.

It should be emphasized that systemic zinc administration can lead to copper deficiency due to induction of MTs in enterocytes (intestinal epithelial cells) in the small intestines and colon due their higher affinity for copper than for zinc. On the other hand, one is taking advantage of this fact in the treatment of Wilson's disease characterized by an excess load of copper.

## Topical Zinc Treatment

Zinc is more commonly used topically although it is unclear when zinc was first used in the management of skin wounds (Lansdown et al. 2007). Importantly, as opposed to systemic zinc therapy, newer and compelling evidence suggests that topical zinc is effective *independent* of the nutritional status of the subject. This strongly indicates that apart from being an essential nutrient, zinc exerts a pharmacological action on wound healing. Many of the biochemical and molecular events in wound repair can thus be expedited by addition of supplementary zinc through upregulation of zinc metalloenzymes and MTs. Furthermore, increased expression of zinc-finger transcription factors in RNA-coding of growth factors is consistent with improved wound healing (Wu et al. 2009). Consequently, more recent studies have demonstrated unequivocally that topical zinc therapy reduces dead tissue on wound surface and advances epithelialization in surgical wounds in experimental animals and in humans.

There are many different dressings that deliver zinc in various forms (Lansdown et al. 2007). The mode of delivery is crucial to achieve a beneficial effect. There are two major forms: water-soluble and water-insoluble zinc compounds. Pharmacopoeias list the water-soluble zinc sulfate as a local astringent and antiseptic and the water-soluble zinc chloride as an escharotic and insoluble zinc oxide and calamine as mild antiseptics, astringents, and protective agents, with particular value in treating inflammatory skin conditions and superficial wounds. Experimental wound-healing studies clearly show that the preferable form is zinc oxide which is water-insoluble but in a proteinaceous wound milieu zinc ions are solubilized. When administered locally in the form of zinc oxide, zinc ions are released at a sustained rate. In contrast, when administered as zinc ion solutions, the therapeutic range is achieved with difficulty. At zinc concentrations above physiological, but below pharmacological levels, zinc may inhibit essential cellular and biochemical functions. Zinc oxide, on the other hand, provides a depot that delivers zinc ions over an extended treatment period that is beneficial to wound healing. These zinc levels are below those that cause toxicity but sufficiently high to stimulate some biologic systems during wound healing pharmacologically. Notably, wound repair cells tolerate up to

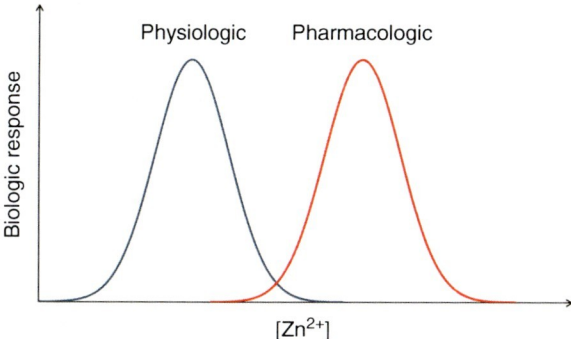

**Zinc and Wound Healing, Fig. 1** Dose-dependent effect of zinc under physiological wound-healing conditions and when added topically eliciting pharmacological effects

1 mM of solubilized zinc oxide which is about 100 times the physiologic serum level (Fig. 1).

There are several possible modes of action to explain promotion of epithelialization with supplementary zinc treatment. Increased nuclear MT in wound marginal epithelial cells and in mitotically active cells of the basal epidermis is a positive indication of the involvement of zinc in DNA polymerases in the burst of mitosis that precedes epithelialization. Zinc accumulates in proliferating as opposed to stationary epithelial cells, and topical zinc oxide increases epithelial proliferation in wounds. Zinc ions mimic the action of growth factors by enhancing intracellular mitogenic signaling pathways, and zinc oxide is capable of upregulating endogenous growth factors, notably insulin-like growth factor-I, which may increase epithelialization. In addition, activation of MMPs with supplementary zinc may contribute to enhanced dissolution of dead tissue that is a barrier to wound healing and epithelial migration. Taken together, topical zinc enhances the rate by which epithelium covers an open wound through a number of different mechanisms.

Zinc is an essential micronutrient also for prokaryotic organisms. Zinc homeostasis in prokaryotes is regulated through a number of specific and nonspecific membrane-bound uptake and efflux pumps. Intracellular free zinc levels are maintained at subtoxic levels through metallothioneins unrelated by evolution to the MT in eukaryotic cells. At superphysiological levels, zinc inhibits growth of several bacterial species found in wounds. Gram-positive bacteria such as *Staphylococcus aureus* are noted to exhibit greater sensitivity to zinc than Gram-negative bacteria exemplified by *Escherichia coli*. There is currently a trend in the increased use of antimicrobial agents like silver and iodine. The effectiveness of antimicrobial therapies depends on host defense mechanisms and virulence factors. Zinc concentrates naturally in tissues showing high cell turnover, including bone marrow and thymus, and is held to be an important regulator for polymorphonuclear leukocytes and macrophages. These cells are paramount for the innate immunity and the killing of colonizing bacteria. Zinc is also capable of inhibiting nitric oxide formation and prevents sulfhydryl groups from oxidation, other possible mechanisms of the anti-inflammatory activity of zinc. Thus, zinc supplements possess mild antimicrobial properties against common wound flora and assist a host's defense system against infections. In line with these mechanisms of zinc, topical zinc oxide inhibits the bacterial growth in wounds to the extent that the need for systemic antibiotics is significantly reduced.

## Putative Physiological and Therapeutic Roles of Zinc in Wound Healing

- Cofactor in enzymes, in particular, the metzincins (MMPs, ADAMs, and ADAMTSs)
- Cofactor in transcription factors (zinc-finger proteins)
- Induces metallothioneins (MTs)
- Antiseptic
- Boosts innate immune response
- Enhances epithelialization

## Cross-References

▶ Copper, Biological Functions
▶ Zinc Adamalysins
▶ Zinc and Immunity
▶ Zinc and Treatment of Wilson's Disease
▶ Zinc and Zinc Ions in Biological Systems
▶ Zinc-Astacins
▶ Zinc Cellular Homeostasis
▶ Zinc Deficiency
▶ Zinc in Superoxide Dismutase
▶ Zinc Matrix Metalloproteinases and TIMPs
▶ Zinc Metallothionein
▶ Zinc, Physical and Chemical Properties

## References

Ågren MS, Werthén M (2007) The extracellular matrix in wound healing: a closer look at therapeutics for chronic wounds. Int J Low Extrem Wounds 6:82–97

Fukada T, Yamasaki S, Nishida K, Murakami M, Hirano T (2011) Zinc homeostasis and signaling in health and diseases: zinc signaling. J Biol Inorg Chem 16:1123–1134

Gomis-Rüth FX (2003) Structural aspects of the metzincin clan of metalloendopeptidases. Mol Biotechnol 24:157–202

Lansdown AB, Mirastschijski U, Stubbs N, Scanlon E, Ågren MS (2007) Zinc in wound healing: theoretical, experimental, and clinical aspects. Wound Repair Regen 15:2–16

Prasad AS (1995) Zinc: an overview. Nutrition 11:93–99

Singer AJ, Clark RA (1999) Cutaneous wound healing. N Engl J Med 341:738–746

Toriseva M, Kähäri VM (2009) Proteinases in cutaneous wound healing. Cell Mol Life Sci 66:203–224

Wilkinson EA, Hawke CI (2000) Oral zinc for arterial and venous leg ulcers. Cochrane Database Syst Rev:CD001273

Wu M, Melichian DS, de la Garza M, Gruner K, Bhattacharyya S, Barr L, Nair A, Shahrara S, Sporn PH, Mustoe TA, Tourtellotte WG, Varga J (2009) Essential roles for early growth response transcription factor Egr-1 in tissue fibrosis and wound healing. Am J Pathol 175:1041–1055

# Zinc and Zinc Ions in Biological Systems

Wolfgang Maret
Metal Metabolism Group, Diabetes and Nutritional Sciences Division, School of Medicine, King's College London, London, UK

## Definitions

When life scientists refer to zinc, they mean the zinc (II) ion as it is the only valence state of zinc relevant for biology. Also, structural biologists refer to a zinc atom in a structure of a biomolecule when in fact it is the zinc(II) ion they are describing. Given this ambiguous terminology and new roles of zinc(II) ions in cellular biology, a need for chemical speciation of zinc in biological systems becomes obvious.

## Introduction: Protein-Bound and Free Zinc (II) Ions

In biology, cellular zinc is redox-inert, remaining always in the Zn(II) valence state. A characteristic feature of zinc metabolism and the control of zinc homeostasis in cells is the tight binding of zinc(II) ions to proteins and the resulting very low concentrations of free zinc(II) ions (▶ Zinc Cellular Homeostasis). Until recently, most of the work in the field of zinc biology focused on protein-bound zinc. However, starting with the discovery of zinc-rich neurons in the brain, a new paradigm emphasizing the functions of free zinc(II) ions evolved. It was found that presynaptic vesicles contain zinc(II) ions that are released into the synapse via vesicular exocytosis. This finding has several implications; functionally, that presynaptically released zinc(II) ions are neuromodulators, and, chemically, that synaptic vesicles contain free zinc(II) ions rather than protein-bound zinc because the vesicular zinc is readily visualized by autometallography or detected with zinc-chelating fluorophores or fluorogenic chelating agents (Pluth et al. 2001; Danscher and Stoltenberg 2005) (▶ Zinc, Fluorescent Sensors as Molecular Probes). A considerable number of other cells/tissues release zinc(II) ions, in particular pancreatic beta-cells, mammary glands, and the prostate. The next step in the chronology of discoveries was the observation that zinc(II) ions are released intracellularly from zinc/thiolate coordination environments by oxidative reactions on the sulfur donors. While originally deemed to be mostly of toxicological significance, evidence has been adduced that this zinc/thiolate chemistry is important under physiological conditions. Physiological oxidants modify the sulfur donors with concomitant zinc dissociation while reducing conditions favor the binding of zinc, thus linking redox and zinc homeostasis and making the availability of zinc(II) ions dependant on the cellular thiol/disulfide redox state. The link also suggests that zinc/thiolate coordination sites are switching points in signal transduction, where redox signals are converted into specific zinc(II) ion signals. The most recent discoveries demonstrate that several stimuli release zinc(II) ions from intracellular vesicular stores. Taken together, these observations led to the concept that zinc(II) ions are intra- and intercellular signaling ions. In parallel developments, significant progress was made in the identification and characterization of the proteins that control cellular zinc homeostasis. At least three dozen proteins participate in this control. In humans, they include ten zinc transporters from the ZnT family, 14 transporters from the Zip family, about a dozen metallothioneins, and zinc-activated transcription factors, such as metal response element (MRE)-binding transcription factor-1 (MTF-1) (▶ Zinc Metallothionein). It became evident that control of cellular zinc homeostasis does not only have a housekeeping function but that it is necessary to control the biological activities of zinc(II) ions in cellular regulation. Zinc(II) ions complement the regulatory functions of the other two redox-inert metal ions magnesium(II) and calcium(II). Thus, metal ions are involved in information transfer over a range of concentrations that spans at least six orders of magnitude. The zinc(II) ion is ideally suited for this task as it binds with high affinity in specific coordination environments.

## Protein-Bound Zinc

Zinc is a constituent of about 3,000 human proteins (▶ Zinc-Binding Proteins, Abundance). In these proteins, zinc has a catalytic, structural, or regulatory function. The coordination environments of zinc in proteins have been tabulated and categorized (Auld 2005) (▶ Zinc-Binding Sites in Proteins). Functional annotations are ongoing and are available for about half of the zinc enzymes and proteins with zinc-binding domains. Zinc enzymes occur in all six enzyme classes, i.e., hydrolases, oxidoreductases, ligases, transferases, lyases, and isomerases. The largest group of zinc enzymes are proteinases. The chemical mechanisms of zinc in zinc enzymes are the activation of a zinc-bound substrate, a zinc-bound water molecule, or a ligand of zinc for the reaction to be catalyzed. Structural zinc has a function in the global architecture of proteins, and, more frequently, in the structure of protein domains. These zinc-binding domains are modules for the interaction of proteins with other biomolecules, such as DNA/RNA, proteins, or lipids. The most prominent group of zinc-binding domains is in proteins with zinc fingers. The name was coined on the basis that these zinc-binding domains bind DNA in a way reminiscent of fingers gripping an object. Zinc-binding domains extend the conformational landscape of proteins significantly. Some of the zinc-binding domains are not purely structural as their function is regulated via their metal content (Maret 2004).

Additional functions of zinc in proteins extend this classification though they remain based on the three elements: catalysis, structure, and regulation. Zinc sites in proteins can be binuclear, trinuclear, or even tetranuclear. In enzymes, the additional zinc(II) ions have been called co-catalytic because they assist catalysis directly or indirectly through constraining the structure of the active site (Auld 2005). Zinc(II) ions also bind between protein subunits, a function that zinc serves exceptionally well because it binds stronger than the other essential metal ions except copper(II) and has the advantage of not being redox-active. In these interface zinc sites, zinc(II) ions have catalytic functions or, more frequently, structural functions in determining quaternary (same subunits) or quinary (different subunits) structure of proteins. Yet other proteins bind zinc transiently. This category includes proteins involved in the regulation of cellular zinc homeostasis and proteins that are regulated by zinc.

## Free Zinc(II) Ions

Speciation of zinc(II) ions is based on the affinities of proteins for zinc. Cellular zinc proteins bind zinc with picomolar dissociation constants (Maret 2004). Hence, concentrations of free zinc(II) ions are commensurately low but not negligible. Estimates are in the range of tens to hundreds of picomolar. Concentrations of free zinc(II) ions depend on the state of the cell and whether or not fluctuations have been induced. The pool of free zinc(II) ions is variously referred to as labile zinc, mobile zinc, or rapidly exchanging zinc. Every term has a different meaning. The terms should not be used interchangeably and terms that lack a chemical definition should be avoided. Structurally, cellular free zinc(II) ions have not been defined – their ligands are unknown. In a biological environment, there are several potential ligands, and therefore, the term does not refer to the aquo complex(es). The term is simply used to distinguish the minute pool of zinc(II) ions that are not bound to proteins from the large pool of protein-bound zinc. This discussion applies to the cytosol of eukaryotic cells. Some cellular vesicular compartments associated with zinc storage and release have significantly higher concentrations of free zinc(II) ions.

### Generating Free Zinc(II) Ions

Aside from the transport of zinc(II) ions across the plasma membrane, three pathways increase the concentrations of zinc(II) ions. One pathway is the release of zinc ions from vesicles into the extracellular space, where zinc(II) ions can be involved in paracrine signaling. The prime example is the release of zinc(II) ions from synaptic vesicles in a subset of glutamatergic neurons (Frederickson et al. 2005). The chemical form of the stored zinc(II) ions is not known, nor is the chemical form of zinc(II) ions in the synapse known. One target of the released zinc(II) ions is the N-methyl-D-aspartate (NMDA) receptor, which zinc inhibits at low nanomolar concentrations. Zinc(II) ions also modulate the activities of other neuroreceptors. Which of these targets of zinc(II) ions are physiologically significant depends on the effective concentrations of synaptic zinc. Release of zinc(II) ions from vesicles also has been observed in a number of exocrine and endocrine glands. Somatotrophic cells in the pituitary gland, pancreatic acinar cells, β-cells of the islets of Langerhans, Paneth cells in the crypts of Lieberkuehn, cells of the tubuloacinar glands of the prostate, epithelial cells of the epididymal ducts, and osteoblasts all secrete zinc(II) ions (Danscher and Stoltenberg 2005). In pancreatic β-cells, zinc(II) ions are used to form and store hexameric insulin. Zinc(II) ions are secreted together with insulin and may have paracrine functions or may be involved in avoiding the formation of amyloid deposits from amyloidogenic proteins such as insulin and amylin (▶ Zinc and Diabetes (Type 2)). In mammary epithelial cells, exocytotic vesicles are loaded with zinc(II) ions and supply the milk with zinc. Mouse oocytes take up significant amounts of zinc at the latest stage of maturation and then arrest after the first meiotic division (Kim et al. 2011). Upon fertilization, the embryos resume progression through the cell cycle and, after well-established calcium(II) ion oscillations, release intracellular zinc(II) ions into the extracellular milieu in what has been named "zinc sparks." Fluctuations of intracellular free zinc(II) ions occur in the mitotic cell cycle with two distinct peaks, one early in G1 and another one early in the S phase (Li and Maret 2009).

The other two pathways involve release of zinc(II) ions inside cells and suggest roles of zinc(II) ions in intracrine signaling. First, zinc(II) ions are released intracellularly from vesicles. Release from a store in the endoplasmic reticulum is mediated by casein kinase 2–induced phosphorylation of the zinc transporter Zip7 (Taylor et al. 2012). In macrophages, release is triggered by stimulation of the immunoglobulin E FcεRI receptor and ERK/IP$_3$ signaling and is preceded by Ca$^{2+}$ release (Yamasaki et al. 2007). This phenomenon has been named a "zinc wave." Lysosomal release in interleukin 2–stimulated T-cells is a required signal for proliferation (Kaltenberg et al. 2010). Second, in addition to the intracellular release from vesicles, oxidative reactions target the sulfur donors in zinc/thiolate coordination environments and release zinc(II) ions from proteins (Maret 2004).

Thus, zinc(II) ion fluctuations have different frequencies and amplitudes: Zinc sparks occur within seconds while zinc waves take minutes to develop.

### Controlling Free Zinc(II) Ions

It is not obvious how zinc(II) ion fluctuations and signaling are possible in the cytosol where zinc is buffered. In a biological system, such as a cell, additional mechanisms contribute to buffering. Zinc (II) ions are bound and transported to cellular compartments or exported from the cell. Such fluxes of ions are referred to as muffling in the field of calcium signaling. Muffling of zinc(II) ions by transport processes increases the buffering capacity of cells well above the buffering provided by ligands that only bind zinc.

Four independent observations define a range of picomolar to low nanomolar concentrations at which zinc(II) ions can serve regulatory functions: (i) the free zinc(II) ion concentrations and their fluctuations, (ii) the affinities of MT for zinc, (iii) the affinities of zinc proteins for zinc, and (iv) the zinc affinities of proteins that are targets of zinc ion fluctuations.

### Targets of Free Zinc(II) Ions: Protein-Bound Zinc

Fluctuations above the steady-state level make it possible for zinc(II) ions to bind to additional proteins and modulate their biological activities. Picomolar to low nanomolar concentrations of zinc(II) ions inhibit some enzymes. Some of these inhibited proteins are not recognized as zinc proteins, mainly because they are isolated with chelating agent to preserve their enzymatic activity. In the case of some protein tyrosine phosphatases, the inhibition is so strong to suggest tonic inhibition (Wilson et al. 2012). A possible activation mechanism involves thionein, the apoprotein of metallothionein, which binds zinc and activates zinc-inhibited enzymes. The phosphorylation of Zip7 and the inhibition of protein tyrosine phosphatases by zinc (II) ions indicate a role of zinc in phosphorylation signaling. In a visionary article written almost 30 years ago, inhibitory control, vesicular sequestration of zinc, and complementarity of zinc and calcium biology in cellular regulation have been given as an answer to the question: "Zinc, what is its role in biology?" (Williams 1984).

## From Molecular to Global Functions of Zinc

Overall, zinc is important for growth and development, proliferation (mitosis and meiosis), differentiation, and controlled cell death. Zinc has critical functions in the immune system and in neurotransmission. Zinc is generally considered to have pro-antioxidant and anti-inflammatory properties. These properties however depend on its concentrations because outside the physiological or pharmacological range, zinc(II) ions have the opposite effect: They become pro-oxidants and are pro-inflammatory. Even with an estimate of the number of zinc proteins in a proteome and a significant functional annotation of zinc proteins and thus considerable insight into the molecular roles of zinc in proteins, it has not been possible to account fully for the global effects of zinc deficiency (Maret 2008, 2010) (▶ Zinc Deficiency). It is not known which molecular processes become limiting under zinc deficiency. The recent attention given to zinc(II) ions as signaling ions offers an alternative view on fundamental functions of zinc. Zinc deficiency could affect primarily cellular signaling functions because zinc proteins will retain their zinc due to their high affinity for zinc while cellular zinc(II) ion stores are expected to be depleted first. The full impact of this concept is emerging with a role of zinc in critical pathways of phosphorylation signaling and in specific aspects of redox homeostasis in eukarya and multicellular organisms.

On a yet higher level, the role of zinc in the origin of life and in the evolution of life is receiving considerable attention (Mulkidjanian et al. 2012). In discussions about the origin of life, the capacities of inorganic zinc(II) ions and zinc sulfide to catalyze reactions that lead to crucial biomolecules, such as intermediates of the citric acid cycle, have been pointed out. The catalytic versatility of zinc is utilized in hundreds of zinc enzymes. In discussions about the evolution of life, it has been noted that zinc-binding makes possible a huge additional landscape of protein conformations. The expanding role of zinc in protein structure is evident in the zinc proteomes of eukarya, suggesting that the acquisition of zinc for such functions was a critical step in the evolution from single cells to multicellular organisms.

## Cross-References

▶ Zinc and Diabetes (Type 2)
▶ Zinc Cellular Homeostasis
▶ Zinc Deficiency
▶ Zinc, Fluorescent Sensors as Molecular Probes
▶ Zinc Metallothionein
▶ Zinc-Binding Proteins, Abundance
▶ Zinc-Binding Sites in Proteins
▶ Zinc-Secreting Neurons, Gluzincergic and Zincergic Neurons

## References

Auld D (2005) Zinc enzymes. In: King RB (ed) Encyclopedia of inorganic chemistry. Wiley, Chichester, pp 5885–5927

Danscher G, Stoltenberg M (2005) Zinc-specific autometallographic in vivo selenium methods: tracing of zinc enriched (ZEN) terminals, ZEN pathways, and pools of zinc ions in a multitude of other ZEN cells. J Histochem 43:141–153

Frederickson CJ, Koh JY, Bush AI (2005) The neurobiology of zinc in health and disease. Nat Rev Neurosci 6:449–462

Kaltenberg J, Plum JL, Ober-Blobaum JL, Honscheid A, Rink L, Haase H (2010) Zinc signals promote IL-2-dependent proliferation of T cells. Eur J Immunol 40:1496–1503

Kim AM, Bernhardt ML, Kong BY, Ahn RW, Vogt S, Woodruff TK, O'Halloran TV (2011) Zinc sparks are triggered by fertilization and facilitate cell cycle resumption in mammalian eggs. ACS Chem Biol 6:716–723

Li Y, Maret W (2009) Transient fluctuations of intracellular zinc ions in cell proliferation. Exp Cell Res 315:2463–2470

Maret W (2004) Zinc and sulfur: a critical biological partnership. Biochemistry 43:3301–3309

Maret W (2008) Zinc proteomics and the annotation of the human zinc proteome. Pure Appl Chem 80:2679–2687

Maret W (2010) Metalloproteomics, metalloproteomes, and the annotation of metalloproteins. Metallomics 2:117–125

Mulkidjanian AY, Bychkov AY, Dibrova DV, Galperin MY, Koonin EV (2012) Origin of first cells at terrestrial, anoxic geothermal fields. Proc Natl Acad Sci USA 109:E821–830

Pluth MD, Tomat E, Lippard SJ (2011) Biochemistry of mobile zinc and nitric oxide revealed by fluorescent sensors. Annu Rev Biochem 80:333–355

Taylor KM, Hiscox S, Nicholson RI, Hogstrand C, Kille P (2012) Protein kinase CK2 triggers cytosolic zinc signaling pathways by phosphorylation of zinc channel ZIP7. Sci Signal 5(210):ra11

Williams RJP (1984) Zinc: what is its role in biology. Endeavour 8:65–70

Wilson M, Hogstrand C, Maret W (2012) Picomolar concentrations of free zinc(II) ions regulate receptor protein-tyrosine phosphatase β activity. J Biol Chem 287:9322–9326

Yamasaki S, Sakata-Sogawa K, Hasegawa A, Suzuki T, Kabu K, Sato E, Kurasaki S, Yamashita M, Tokunaga K, Nishida K, Hirano T (2007) Zinc is a novel intracellular second messenger. J Cell Biol 177:637–645

# Zinc Balance

▶ Zinc Homeostasis, Whole Body

# Zinc Beta Lactamase Superfamily

Carine Bebrone
Centre for Protein Engineering, University of Liège, Sart–Tilman, Liège, Belgium
Institute of Molecular Biotechnology, RWTH–Aachen University, c/o Fraunhofer IME, Aachen, Germany

## Synonyms

Metallo-β-lactamase superfamily; Zinc metallohydrolase family of the β-lactamase fold

## Definition

The metallo-β-lactamase superfamily was first defined in 1997 on the basis of a sequence alignment. The members of this superfamily are characterized by the presence of a common αββα fold and share five conserved motifs: Asp84, His116-Xaa-His118-Xaa-Asp120-His121, His196, Asp221, and His263, which are (with the exception of Asp84) involved in the binding of two metal ions. This superfamily contains now more than 23,900 members divided into 17 biological groups. Besides metallo-β-lactamases which cleave the amide bond of the β-lactam ring of penicillins, cephalosporins, or carbapenems (thus inactivating the antibiotic), the metallo-β-lactamase superfamily includes enzymes which hydrolyze thioester, phosphodiester, and sulfuric ester bonds as well as oxidoreductases.

# Zinc Beta-Lactamases

## What Are Metallo-β-lactamases? Why Are These Enzymes Important?

β-lactamases are enzymes produced by bacteria to counter the action of β-lactam antibiotics such as penicillins, cephalosporins, carbapenems, and monobactams (Fig. 1a). These enzymes are able to cleave the amide bond of the β-lactam ring, thus inactivating the antibiotic (Fig. 1b).

β-lactamases have been divided into four classes, A to D, on the basis of their amino acid sequences. In enzymes of classes A, C, and D, a serine residue performs a nucleophilic attack on the carbonyl carbon atom of the β-lactam ring. The serine β-lactamases and the penicillin-binding proteins (PBPs) involved in the bacterial cell wall biosynthesis and maintenance are structurally related, and the former probably evolved from the latter or from a common ancestor. On the other hand, class B enzymes have no structural similarity with the active-site serine enzymes and require a bivalent metal ion ($Zn^{2+}$ in vivo, although in vitro $Co^{2+}$ or $Cd^{2+}$ forms are also active) for activity. Consequently, they are currently named zinc β-lactamases or metallo-β-lactamases (MBLs). The founding member of this family is the MBL from *Bacillus cereus* (BcII) which was first identified in 1966. Until 1985, it was the only known MBL example and was thus regarded as a biochemical curiosity. Indeed, the second representative of this family (L1 from *Stenotrophomonas maltophilia*) was only discovered 19 years later. However, since, the number of described MBLs has grown rapidly and several pathogens (such as *Bacteroides fragilis*, *Pseudomonas aeruginosa*, *Aeromonas hydrophila*, *Serratia marcescens*, *Elizabethkingia meningoseptica*, *Bacillus anthracis*) are now known to synthesize members of this class. Nowadays, MBLs are widespread and found in Europe, Asia, Australia, and South and North America. The emergence of MBLs represents a real threat to human health since they show a broad spectrum of substrates including last-generation cephalosporins and especially carbapenems (Bush 2010). Indeed, carbapenems (notably imipenem or meropenem) are usually one of the last resort antibiotics for many bacterial infections such as *Escherichia coli* and *Klebsiella pneumoniae*. Moreover, despite the huge number of potential molecules already tested, there is currently no clinically useful MBL inhibitor (Bebrone 2007).

**Zinc Beta Lactamase Superfamily, Fig. 1** (a) Structure of the four main families of β-lactam antibiotics. (b) Hydrolysis of a penicillin antibiotic by a β-lactamase enzyme. (c) Nitrocefin, a chromogenic cephalosporin. Nitrocefin is an atypical substrate due to its highly conjugated bis-nitro-substituted styryl substituent

Many MBL genes are present in environmental species, which thus constitute reservoirs of β-lactam resistance genes. The fact that some of the MBL genes (as VIM-1, IMP-1, and their numerous variants, SPM-1, GIM-1, SIM-1, or the newly identified New Delhi Metallo-β-lactamases NDM-1 or NDM-2) are plasmid-encoded represents an additional cause of concern. Indeed, they can spread from one strain of bacteria to another by horizontal gene transfer. Recently, the large media interest accorded to the so-called NDM-1 superbug has revealed the importance of these enzymes in the antibiotic resistance phenomena to the general public. NDM-1 was first detected in 2008 in a *K. pneumoniae* isolate from a Swedish patient of Indian origin. It was later found in nearly all clinical species of *Enterobacteriaceae*, *Acinetobacter*, and *Pseudomonas* spp. in different countries all over the world (among others in India, Pakistan, the UK, Norway, the Netherlands, Germany, Austria, Singapore, the USA, Canada, Japan, Brazil, France, and Belgium). NDM-1 shares 55% identity and 72% similarity with the MBL from *Erythrobacter litoralis*, a marine isolate, suggesting a possible environmental origin. Furthermore, the presence of NDM-1-producing bacteria in environmental samples in New Delhi (India) has been recently demonstrated. The gene coding for NDM-1 ($bla_{NDM-1}$) is located on remarkably plastic and highly promiscuous plasmids that appear to cross the genus/family barrier with ease. Indeed, $bla_{NDM-1}$ can disseminate among species, including not only enterobacteria but also aeromonads, vibrios, and various nonfermenters. Moreover, plasmids bearing $bla_{NDM-1}$ also contain up to 14 other antibiotic resistance genes (Pillai et al. 2011).

Metallo-β-lactamases are thus now regarded as a therapeutic challenge, and the biochemical and

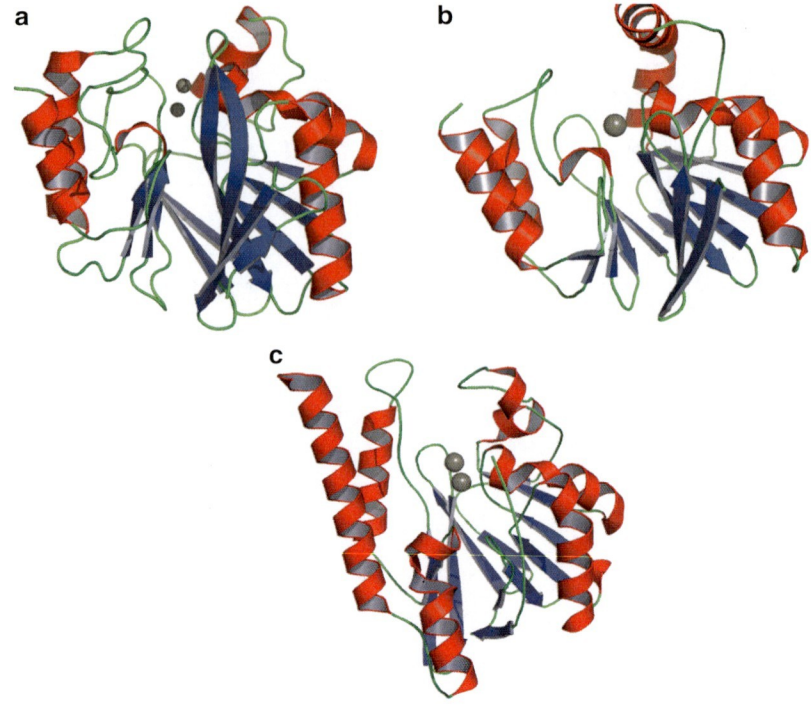

**Zinc Beta Lactamase Superfamily, Fig. 2** Crystallographic structures of MBLs. (**a**) BcII (B1). (**b**) CphA (B2). (**c**) FEZ-1 (B3). Helices, β-sheets, and loops are represented in *red*, *blue*, and *green*, respectively. The zinc ions are shown in *gray*

structural characterization of this increasingly large MBL family presents a great interest since a detailed study of the active site and of the mechanism of action of these enzymes should lead to the rational design of inhibitors that could be coadministered with the antibiotic therapy as well as of antibiotics escaping their hydrolytic action.

## Classification

A classification into three subclasses (B1, B2, and B3) and a standard numbering scheme (the BBL numbering) have been established on the basis of a structural alignment. Functional and mechanistic properties also clearly distinguish the B1, B2, and B3 enzymes from each other. The B1 (including BcII, CcrA from *B. fragilis*, BlaB from *E. meningoseptica*, and the acquired MBLs) and B3 (including L1, FEZ-1 from *Legionella gormanii* and GOB enzymes from *E. meningoseptica*) MBLs are broad-spectrum enzymes and hydrolyze most β-lactam antibiotics. Only the monobactams are neither hydrolyzed nor recognized by these enzymes. On the other hand, the subclass B2 enzymes (CphA from *Aeromonas hydrophila*, ImiS from *A. veronii*, and Sfh-1 from *Serratia fonticola*) only hydrolyze carbapenems efficiently and show a very weak activity, if any, toward penicillins and cephalosporins. Moreover, while the B1 and B3 enzymes exhibit maximum activity as dizinc species, the subclass B2 MBLs are inhibited in a noncompetitive manner upon binding of a second $Zn^{2+}$ ion (for CphA, the dissociation constant $K_{D2}$ is 46 μM at pH 6.5). (Bebrone 2007).

## 3-D Structures and Description of the Zinc-Binding Sites in the Three Subclasses

The crystallographic structure of the monozinc form of BcII was the first to be solved in 1995. The first dizinc structure of the family, that of CcrA, was determined in 1996, followed in 1998 by the structure of the dizinc form of BcII (Fig. 2). Thereafter, the structures of other subclass B1 enzymes (IMP-1, BlaB, VIM-2, VIM-4, SPM-1, and IND-7) and of subclass B3 enzymes (L1, FEZ-1, and BJP-1 from *Bradyrhizobium japonicum*) were determined. The first structure of a subclass B2 enzyme, CphA, was only solved in 2005. The structures of several mutants, metal-substituted forms, or complexes with small molecules were also determined for the three subclasses. Despite the low level of identity between the amino acid sequences (subclass B2 shows only 11% of identity

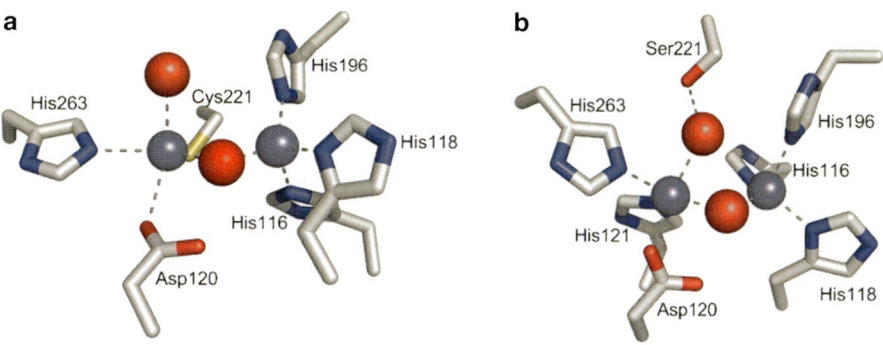

**Zinc Beta Lactamase Superfamily, Fig. 3** Zinc-binding sites. (**a**) BCII (B1). (**b**) L1 (B3). Zinc ions and water molecules are represented as *gray* and *red* spheres, respectively

with subclass B1, while B3 MBLs have only nine conserved residues when compared to the other MBLs), the general tertiary fold of these enzymes is similar and consists of an αββα structure (Bebrone et al. 2011) (Fig. 2).

This structure is composed by two central β-sheets with five solvent-exposed α-helices. The N-terminal and C-terminal parts of the molecule, each of them comprising a β-strand and two α-helices, can be superposed by a 180° rotation around a central axis, suggesting that the complete structure might have arisen from the duplication of a gene. In all known structures, the active site is located at the external edge of the ββ sandwich. All MBLs possess two potential ▶ zinc-binding sites and share a small number of conserved motifs that help to coordinate the zinc ions, notably His/Asn/Gln116-Xaa-His118-Xaa-Asp120 and Gly/Ala195-His196-Ser/Thr197.

Structural analysis of subclass B1 enzymes shows that one zinc ion has a tetrahedral coordination sphere involving His116, His118, His196, and a water molecule or $OH^-$ ion, whereas the other has a trigonal-pyramidal coordination sphere involving Asp120, Cys221, His263, and two water molecules. One water/hydroxide serves as a ligand for both metal ions (Fig. 3a). The two binding sites are named the "HHH" and "DCH" sites, respectively. In the mononuclear structures of B1 enzymes (BcII, VIM-2, VIM-4, and SPM-1; the PDB accession n° are 1BMC, 1KO2, 2WRS, and 2FHX, respectively), the sole metal ion was found to be located in the "HHH" site.

In subclass B3, the "HHH" site is similar to that found in subclass B1. The sole exception is the GOB-type enzymes, which possess a glutamine at position 116 instead of the conserved ligand His116. Recent results suggest that the Q116 residue could play a role in the binding of the zinc ion in a "QHH" site, but the structure of a GOB-type enzyme has yet to be solved to determine precisely the architecture of this new binding site. The second zinc ion is coordinated by Asp120, His121, His263, (this one is thus called "DHH" site in B3) and the nucleophilic water molecule which forms a bridge between the two metal ions. Cys221 is replaced by a serine residue (Ser221) which does not interact with the zinc ion directly but with a second water molecule located in the active site that could serve as proton donor in the catalytic process (Fig. 3b).

The crystallographic structure of the monozinc form of CphA showed that the catalytic zinc ion was located in the "DCH" binding site. In 2009, the crystal structure of the dizinc form of this B2 enzyme was determined. As previously observed in the monozinc form, the first zinc ion binds in the "DCH" site. The tetrahedral coordination sphere is completed by a sulfate ion from the crystallization solution. The second zinc ion occupies a slightly modified "HHH" site, where only the conserved His118 and His196 residues act as metal ligands ("HH" site). In subclass B2 enzymes, the third histidine ligand in B1 and B3 (His116) is not conserved and is replaced by an asparagine residue, which is not involved in the binding of the second zinc ion. The sulfate ion and a water molecule occupy the vacant coordination positions of the second zinc ion, forming a tetrahedral coordination sphere (Fig. 4).

This atypical coordination sphere probably explains the rather high dissociation constant for the second zinc ion (46 μM) compared to those observed in enzymes of subclasses B1 and B3. The value of the dissociation constant for the "HHH" site is 1.8 nM for BcII (a subclass B1 enzyme), and subclass B3 enzymes have high affinity constants for both binding sites ($K_D$ < 6 nM). It was proposed that His118 and His196 residues play a role in the catalytic mechanism

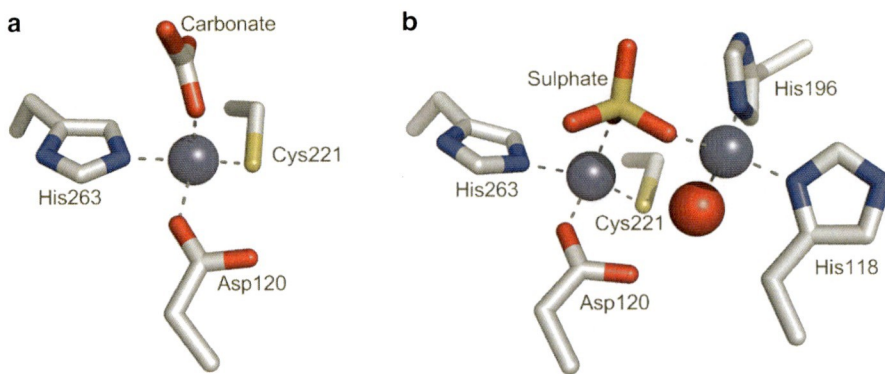

Zinc Beta Lactamase Superfamily, Fig. 4 Zinc-binding sites in B2 MBLs. (a) The active site of the monozinc form of CphA. (b) The active site of the dizinc form of CphA

of CphA. This could explain the inhibition by the binding of the second zinc ion, these residues being prevented from playing their role when they are immobilized as ligands of this second zinc ion in the dizinc-inhibited form of the enzyme.

For more details about the similarities and differences in the structures and the active sites between the three subclasses MBLs or even within a subclass, please refer to Bebrone et al. 2011.

## Protein: Zinc Stoichiometry? Where Is the Zinc Ion in the Mononuclear Forms?

The situation for the subclass B2 enzymes is the clearest. Indeed, these enzymes exist mainly in their active monozinc form (e.g., for CphA, the values of the dissociation constants are 20 nM and 46 µM, respectively, for the binding of the first and the second zinc ion) with the catalytic zinc ion located in the "DCH" binding site.

On the other hand, the number of zinc ions necessary for optimal activity is still controversial in subclass B3 and especially in subclass B1.

While the subclass B3 enzymes were previously known to be dizinc with two high affinity constants for both binding sites, it was recently suggested that the active form of the GOB-18 enzyme contains only one zinc ion, located in the "DHH" site. In contrast, another GOB variant, GOB-1, was newly shown to be produced as a native, fully active dizinc form just as other subclass B3 enzymes, L1 and FEZ-1.

A long debate exists since the beginning of the study of the subclass B1 enzymes to understand what the active form is in vivo. Is it mono- or dizinc or both? Why the presence of two binding sites if only one is needed? If a sole zinc ion is present, which would be the favored site?

Most of the studies have been performed with BcII. An anti-cooperative binding of the two zinc ions has been previously suggested, with a slightly higher activity for the dizinc enzyme. Indeed, fluorescence spectroscopy studies with a chromophoric chelator indicated that BcII had very different dissociation constants for the two metal-binding sites. For the loss of metal ion from the mononuclear enzyme, the dissociation constant, $K_{mono}$, was $6.2 \times 10^{-10}$ M and that for the loss of one metal ion from the dinuclear MBL, $K_{di}$, was $1.5 \times 10^{-6}$ M. $K_{mono}$ decreased significantly, from nanomolar to picomolar values, in the presence of substrate, whereas $K_{di}$ decreased only twofold. This suggested that the monozinc enzyme was responsible for the catalytic activity under physiological conditions, where the concentration of free zinc ions is in the picomolar range. Accordingly, X-ray data indicated a "looser binding" of the metal ion in the "DCH" site, the sole zinc ion being found in the "HHH" site. But the existence of a mononuclear form where the metal ion would be shared between the two binding sites was also supported by several kinetic results, PAC (cadmium) and spectroscopic (cobalt) data. In contrast to the previous proposal of mononuclear B1 MBLs being the sole contributor to the hydrolysis of β-lactams under physiological conditions, recent studies have suggested that the mononuclear form is catalytically incompetent. Indeed, results obtained with cobalt-substituted BcII concluded that only the dicobalt form might be catalytically active. This conclusion was extended to native BcII that would in consequence require two bound zinc ions for activity. On the other hand, it was shown that both the mono- and the dicobalt enzymes are catalytically active with the "DCH" site as the catalytic site. Nevertheless, recent results obtained by mass spectrometry and nuclear magnetic resonance (NMR), together with catalytic activity measurements, indicate cooperative binding of zinc with a $K_{d1}/K_{d2}$ ratio equal to or larger than 5 and, hence, that catalysis is associated with the dizinc enzyme species only. Strong positive cooperativity is also observed for cadmium binding in the presence of the inhibitor thiomandelate.

In agreement with the latest results obtained for BcII, CcrA binds both zinc ions very tightly. Kinetic studies of CcrA with nitrocefin (a chromogenic cephalosporin; Fig. 1c) have shown that only the dizinc species were active and that a previously observed monozinc CcrA enzyme was actually a mixture of the dizinc and the apoenzyme. The activity of the monozinc VIM-4 appears to be very low, binding of a second zinc ion ($K_{d2} = 10$ μM) yielding the fully active enzyme, while the SPM-1 enzyme remains mononuclear even in the presence of 100 μM zinc. Such heterogeneity is surprising for a group of closely related proteins.

As stated above, the activity and even the existence in physiological conditions of monozinc forms of B1 and B3 enzymes remain controversial. As yet, it seems impossible to supply a clear answer to these questions, and it remains possible that minor sequence differences might result in different behaviors. Such binding site heterogeneity, if confirmed, might constitute a significant obstacle to the design of general drug inhibitors of metallo-β-lactamases (Heinz and Adolph 2004; Bebrone et al. 2011).

## Mechanism of Action

The catalytic mechanism of MBLs will be only briefly described here; for an exhaustive explanation, please refer to Badarau and Page 2008; Herzberg and Fitzgerald 2004 and Bebrone et al. 2011. For B1 and B3 enzymes, since mono- and dizinc species are observed, different catalytic mechanisms have been proposed. In the mononuclear enzymes, the zinc-bound water/hydroxide ion is expected to be the attacking nucleophile; this mechanism resembles those of other metalloenzymes such as ▶ carboxypeptidase A and ▶ thermolysin. In this mechanism, the zinc ion acts as a Lewis acid and participates in the ionization of the water molecule. This $OH^-$ ion performs a nucleophilic attack on the carbon of the carboxyl group of the β-lactam. After the hydroxide attack, a tetrahedral intermediate stabilized by its interaction with the zinc ion is formed. Asp120, acting as a general base, deprotonates the –OH group to generate a dianionic tetrahedral second intermediate which is also stabilized by the zinc ion. In a subsequent step, Asp120 gives the proton to the nitrogen of the β-lactam ring and takes part in cycle opening.

In the binuclear enzymes, the water/hydroxide ion that bridges the two zinc ions is expected to be the attacking nucleophile. A mechanism based on the study of the hydrolysis of nitrocefin (Fig. 1c) by the CcrA and L1 enzymes has been proposed for the dizinc form of metallo-β-lactamases. A ring-opened intermediate containing an anionic nitrogen was identified; the protonation of which was rate limiting. It should however be noted that the existence of this intermediate has only been documented with this substrate. When other substrates are used, the cleavage of the tetrahedral intermediate, which occurs concomitantly to the protonation of the nitrogen of the β-lactam ring, appears to be the rate-limiting step.

A different reaction mechanism has been proposed for the monozinc subclass B2 enzymes. The nucleophilic hydroxide is generated by the Asp120 and His118 residues. This nucleophile attacks the β-lactam carbonyl, possibly activated by His196, to generate a tetrahedral intermediate. As observed for B1 and B3 enzymes, the rate-limiting step in the reaction is the C–N bond cleavage.

## The Superfamily

When the crystallographic structure of BcII was determined in 1995, it was the first αββα fold to be observed, and for some time, it was thought that this fold was unique to MBLs. However, subsequent structure determinations of the human glyoxalase II in 1999 (Fig. 5a) and the rubredoxin:oxygen oxidoreductase (ROO) flavoprotein from *Desulfovibrio gigas* in 2000 revealed that these proteins also possess an αββα domain in combination with domains specific to their function (Mannervik 2008; Vicente et al. 2008). By now, besides MBLs (group 1), 8 of the 17 protein groups proposed to contain an MBL αββα fold have been structurally characterized, glyoxalases II (group 2), rubredoxin:oxygen oxidoreductases ROO (group 3), the phosphorylcholine esterase (Pce) (group 9), the methyl parathion hydrolase MPH (group 15), the *N*-acyl homoserine lactone hydrolase (group 12), the alkylsulfatase from *Pseudomonas aeruginosa* SdsA1 (group 13), the tRNA3′-processing endoribonuclease tRNaseZ (group 4), and proteins members of

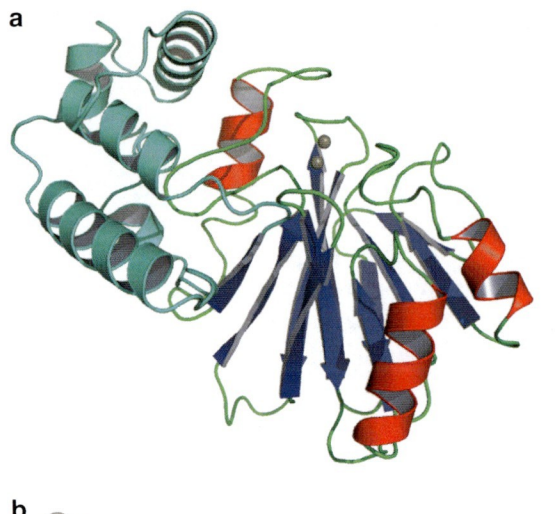

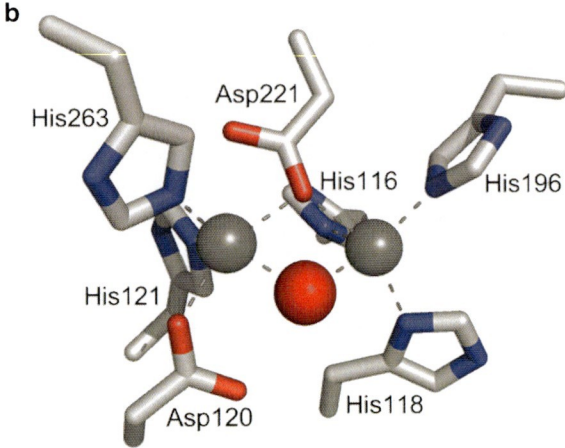

**Zinc Beta Lactamase Superfamily, Fig. 5** (a) Glyoxalase II structure. The MBL-like domain is represented in *red* (helices), *blue* (strands), and *green* (loops and disordered regions). Zinc ions are represented as *gray spheres*. The additional α-helical domain is in *cyan*. (b) Metal-binding sites of glyoxalase II which present the classical structural features of the MBL superfamily. Zinc ions and water molecules are represented as *gray* and *red spheres*, respectively

the β-CASP (CPSF–Artemis–Pso2–Snm1) nucleic acid–processing factors family (groups 6 and 7) (Bebrone 2007). More than 23,900 members of the MBL superfamily span all natural kingdoms and have very different functions related to hydrolysis and redox reactions, DNA repair and uptake, and RNA processing. The largest group, composed of more than 3,200 sequences, is ubiquitous, and the corresponding enzymes are involved in an essential function, tRNA maturation (Vogel et al. 2005). These two features might suggest that a tRNA maturase is the ancestor of the superfamily.

All of these proteins share the five conserved motifs of the superfamily: Asp84, His116-Xaa-His118-Xaa-Asp120-His121, His196, Asp221, and His263, which are (with the exception of Asp84) involved in the binding of two metal ions. The "HHH" and "DHH" sites are found throughout the MBL superfamily (Fig. 5b). For the majority of these enzymes, the first metal ion–binding site is composed of His116, His118, and His 196 ("HHH" site), while the second binding site is composed of Asp120, His121, and His263 ("DHH" site). In all of the above-mentioned structures, the carboxylate group of the conserved Asp221 residue bridges the two active-site metal ions, with the exception of SdsA1 in which Glu221 only binds the zinc into the "DHH" site. In contrast, in the "real" MBLs, a Cys221 residue is found in B1 and B2 β-lactamases and a Ser221 residue in B3. The His121 ligand residue is conserved in B3 enzymes but is replaced by an arginine or a serine residue which is not a zinc ligand in B1 and B2 MBLs.

Interestingly, within the MBL superfamily, the metal requirements vary dramatically among the different enzymes, exhibiting monozinc, dizinc, zinc–iron, di-iron, monomanganese, or dimanganese metal centers. For example, ROO from *D. gigas*, an oxidoreductase enzyme, contains a di-iron center (Vicente et al. 2008). As stated above, all MBLs known so far are zinc-dependent enzymes. Although crystallographic data and metal-bound analysis suggest that the *Streptococcus pneumoniae* teichoic acid phosphorylcholine esterase (Pce) is a zinc-dependent hydrolase, the possibility that, depending on the physiological conditions, the enzyme might contain binuclear zinc–iron or iron–iron centers cannot be discarded. Indeed, the structure of the isolated catalytic module of Pce showed the presence of two bound iron ions. Moreover, the Fe(II)-reconstituted Pce apoenzyme showed an activity about three times higher than the value obtained for the zinc-reconstituted enzyme under the same conditions.

Glyoxalases II, despite exhibiting the same coordination sphere as displayed by other members of the superfamily, are outstanding exceptions regarding the metal ion dependence. Studies on the human enzyme revealed ∼1.5 mol of zinc and 0.7 mol of iron per mol of protein and a 1:1 ratio of zinc to iron was found for the glyoxalase II from *Trypanosoma brucei*. Glyoxalases II from *Arabidopsis thaliana* (Glx2–2 and Glx2–5) and from *Salmonella typhimurium* (GloB) exhibit either a two zinc center, a Fe(III)Zn(II) center, a Fe(III)Fe(II) center, or even a Mn(II)Mn(II) center. It was also shown that *A. thaliana* GLX2–2 exhibits positive cooperativity in metal binding and does not ever appear to exist as a mononuclear metal ion containing enzyme.

The metal ions are expected to be involved in substrate binding to the active site and may activate the nucleophilic agent. Major goals of the future studies will be to identify the factors on which the metal selectivity and the function of these enzymes rely.

## Cross-References

- Thermolysin
- Zinc Carboxypeptidases
- Zinc-Binding Sites in Proteins

## References

Badarau A, Page MI (2008) The mechanisms of catalysis by metallo-beta-lactamases. Bioinorg Chem Appl Article ID 576297

Bebrone C (2007) Metallo-β-lactamases (classification, activity, genetic organization, structure, zinc coordination) and their Superfamily. Biochem Pharmacol 74:1686–1701

Bebrone C, Garau G, Garcia-Saez I, Chantalat L, Carfi A, Dideberg O (2011) X-ray structures and mechanisms of metallo-β-lactamases. In: Frère J-M (ed) β-lactamases, Molecular anatomy and physiology of proteins. Nova Science, Hauppauge

Bush K (2010) Alarming β-lactamase-mediated resistance in multidrug-resistant *Enterobacteriaceae*. Curr Opin Microbiol 13:558–564

Heinz U, Adolph H-W (2004) Metallo – β-lactamases: two binding sites for one catalytic metal ion? Cell Mol Life Sci 61:2827–2839

Herzberg O, Fitzgerald PMD (2004) Metallo-β-lactamases. In: Messerschmidt A, Bode W, Cygler M (eds) Handbook of metalloproteins, vol 3, 3rd edn. Wiley, Chichester, pp 217–234. ISBN ISBN: 0-470-84984-3

Mannervik B (2008) Molecular enzymology of the glyoxalase system. Drug Metabol Drug Interact 23:13–27

Pillai DR, McGeer A, Low DE (2011) *New Delhi* metallo-β-lactamase-1 in *Enterobacteriaceae*: emerging resistance. CMAJ 183:59–64

Vicente JB, Carrondo MA, Teixeira M, Frazão C (2008) Structural studies on flavodiiron proteins. Methods Enzymol 437:3–19

Vogel A, Schilling O, Späth B, Marchfelder A (2005) The tRNase Z family of proteins: physiological functions, substrate specificity and structural properties. Biol Chem 386:1253–1264

## Zinc Carbonic Anhydrases

- Cadmium Carbonic Anhydrase

## Zinc Carboxypeptidases

- Zinc Metallocarboxypeptidases

## Zinc Cellular Homeostasis

Wolfgang Maret
Metal Metabolism Group, Diabetes and Nutritional Sciences Division, School of Medicine, King's College London, London, UK

## Definitions

Walter Bradford Cannon (1871–1945) introduced the concept of homeostasis to describe the constancy of the internal environment in living organisms. It is now well accepted that this concept applies to controlling the nutritionally essential transition metal ions at physiologically viable concentrations. This entry addresses the cellular control of zinc(II) ions. Whole body zinc homeostasis is treated separately ( Zinc Homeostasis, Whole Body).

## Introduction

Zinc(II) ions are nutritionally essential for all forms of life ( Zinc and Zinc Ions in Biological Systems). Living organisms use zinc more widely than any other transition metal. In humans, zinc functions as a structural and catalytic cofactor in about 3,000 proteins, which together constitute the zinc proteome (Andreini et al. 2006) ( Zinc-Binding Proteins, Abundance). Functions in so many proteins require homeostatic control mechanisms and regulated redistribution of zinc in cellular time and space. The significance of cellular zinc homeostasis is that it is not only important for controlling the supply of zinc proteins with zinc but rather that zinc(II) ions also have regulatory functions in metabolism and signal transduction. These functions involve transient binding of zinc(II) ions to proteins and require additional mechanisms for controlling fluctuations of zinc concentrations. Examples include the role of zinc in gene transcription and cellular redox homeostasis. Zinc activates transcription factors, such as metal-response element (MRE)-binding transcription factor-1 (MTF-1). In this way, zinc controls the expression of proteins involved in the antioxidant response. Since zinc is redox-inert in biology, remaining in the zinc(II) valence state, it cannot have an antioxidant function itself. It is a pro-antioxidant. This effect on redox homeostasis critically depends on the concentrations of zinc because zinc is also a pro-oxidant causing oxidative stress in zinc deficiency and zinc overload. These opposing functions of zinc in redox homeostasis demonstrate that controlling cellular zinc homeostasis is generally important for controlling critical regulatory functions of zinc(II) ions in cellular biology. In this regard, zinc resembles calcium rather than the other transition metal ions.

The use of zinc ions in cellular regulation in eukarya differs from the control of zinc in prokarya. This entry focuses on the function of zinc in human cells. Control of cellular zinc in bacteria and plants has been reviewed (Blindauer and Schmid 2010; Hantke 2001).

## The Proteins Involved in Cellular Zinc Homeostasis

The remarkable number of proteins that control the cellular availability and redistribution of zinc(II) ions is an indication of the importance of zinc in cellular biology. These regulatory proteins supply zinc metalloproteins with zinc, scavenge an excess of zinc(II) ions to avoid unspecific reactions, safeguard zinc(II) ions during transit through membranes (zinc transporters) and cellular compartments (zinc metallochaperones), store and release zinc(II) ions (zinc-storage proteins), or sense whether or not zinc concentrations are adequate (zinc-sensor proteins). Zinc transporters and other regulatory proteins are highly integrated into cellular metabolic and signaling pathways. Ten transporters of the ZnT (SLC30A, exporting zinc from the cytosol) family and 14 transporters of the Zip (SLC39A, importing zinc into the cytosol) family control the traffic of zinc through cellular membranes (Fukada and Kambe 2011) (Tables 1 and 2). In addition to controlling zinc efflux and influx through the plasma membrane, members of the two protein families transport zinc intracellularly between cytosol and various compartments. Zinc transporters are extensively regulated transcriptionally, translationally, and at the protein level through heterodimer formation, ubiquitination, phosphorylation, proteolysis, and trafficking in the cell. Three-dimensional structures of Zip proteins are not available. Limited investigations on their transport mechanisms indicate that they

**Zinc Cellular Homeostasis, Table 1** Human ZnT proteins

| ZnT | Properties |
|---|---|
| 1. | Zinc export from cells and into the endoplasmic reticulum (ER) |
| 2. | Transport of zinc into secretory vesicles (e.g., mammary gland) |
| 3. | Transport of zinc into secretory vesicles (e.g., brain) |
| 4. | Transport of zinc into secretory vesicles |
| 5. | Transport through intestinal basolateral membrane/transport into Golgi/trans Golgi network (TGN) |
| 6. | Transport into Golgi/TGN |
| 7. | Transport into Golgi/TGN |
| 8. | Transport of zinc into secretory vesicles (e.g., pancreatic β-cells granules) |
| 9. | Limited information available |
| 10. | Limited information available |

**Zinc Cellular Homeostasis, Table 2** Human Zip proteins

| Zip | Properties |
|---|---|
| 1. | Zinc import through the plasma membrane |
| 2. | Zinc import through the plasma membrane |
| 3. | Zinc import through the plasma membrane |
| 4. | Zinc import through the plasma membrane (intestinal apical membrane) |
| 5. | Zinc import through the plasma membrane |
| 6. | Zinc import through the plasma membrane |
| 7. | Zinc export from the ER into the cytoplasm |
| 8. | Zinc import through the plasma membrane, also transports Mn(II), Cd(II) |
| 9. | Zinc export from the TGN into the cytoplasm |
| 10. | Zinc import through the plasma membrane |
| 11. | Limited information available |
| 12. | Limited information available |
| 13. | Zinc export from the ER into the cytoplasm |
| 14. | Zinc import through the plasma membrane, also transports Cd(II) and non-transferrin-bound iron |

are $Zn^{2+}/HCO_3^-$ symporters. A crystal structure of the *Escherichia coli* Yiip protein, a homologue of the human ZnT proteins, has been determined. Yiip is a $Zn^{2+}/H^+$ antiporter residing in the inner membrane of Gram-negative bacteria and having a transmembrane domain and a cytoplasmic domain. It is a dimer and has several zinc-binding sites. A zinc site between the transmembrane helices in each monomer is involved in zinc transport. The function of a zinc-binding site between the cytoplasmic domain and the transmembrane domain is unknown. Each monomer also has a binuclear zinc site in the cytoplasmic domain. This site is thought to sense cytoplasmic zinc(II) ion concentrations. Zinc binding to this site triggers a conformational change of the protein, allowing transport of zinc through the membrane. Metal-response element (MRE)-binding transcription factor-1 (MTF-1) is another zinc ion sensor. It controls zinc-dependent gene expression of proteins at increased zinc(II) ion concentrations. Sensing is thought to occur through a pair of its six zinc fingers. MTF-1 is an essential transcription factor in development. Its genetic ablation in mice is embryonically lethal. The fourth type of protein involved in cellular zinc homeostasis is metallothionein (MT). In humans, MTs are a family of at least a dozen proteins (Table 3). The gene expression of most but not all of them is under the control of MTF-1, inducing the production of MT and increasing the cellular zinc-binding capacity at elevated cellular zinc(II) ion concentrations (▶ Zinc Metallothionein). Mammalian MTs are relatively small proteins with 60+ amino acids and bind zinc exclusively to the sulfur donors of cysteines in 2 zinc/thiolate clusters, one with 3 zinc(II) ions and 9 cysteine ligands and the other with 4 zinc(II) ions and 11 cysteine ligands. Despite the binding of up to 7 zinc (II) ions and a similar binding in tetrathiolate coordination environments, the binding sites in human MT-2 have different affinities for zinc, ranging from nanomolar to picomolar. With these characteristics, MTs serve as both zinc acceptors and zinc donors in the cell and control zinc availability over a range of zinc(II) ion concentrations commensurate with the affinities of their binding sites for zinc(II) ions. Binding and delivering zinc is a molecular characteristic of the zinc/thiolate clusters and quite different from that of other proteins that bind zinc permanently as a catalytic or structural cofactor. Another remarkable molecular property of MTs is that they are redox proteins. This function also is in contrast to that of other zinc proteins, which are redox-inert if they have nitrogen and oxygen donors in their zinc coordination environments. Since thiolate ligands confer redox activity on the zinc/thiolate clusters of MT, the capacity of MT to serve as a zinc donor increases under more oxidizing conditions while its capacity to bind zinc(II) ions increases under more reducing conditions. In zinc/thiolate coordination environments, the reactivity of the ligand donor instead of the central atom changes the stability of the complex. It is one of biology's solutions to the problem of how to mobilize zinc(II) ions from their high affinity sites in proteins. It links cellular zinc and redox homeostasis by using changes in the cellular redox poise to modulate zinc(II) ion concentrations (zinc potentials, $pZn = -\log[Zn^{2+}]_i$) (Maret 2009).

**Zinc Cellular Homeostasis, Table 3** Human metallothioneins

| MT | Properties |
|---|---|
| 1. | Binds mainly zinc, several gene products, ubiquitous expression |
| 2. | Binds mainly zinc, ubiquitous expression |
| 3. | Binds zinc and copper, expression in the brain |
| 4. | Binds zinc and copper, expression in squamous epithelia |

## Zinc Buffering and Muffling

Quantitative considerations are critical for understanding the control of cellular zinc. Zinc is maintained at specific concentrations to avoid interference with the functions of other essential metal ions. The control prevents coordination of the wrong metal to a zinc metalloprotein and enables zinc-specific reactions. The cell regulates the concentration of total zinc and it buffers zinc, thereby determining the concentration of free zinc(II) ions. The total cellular zinc concentrations are a few hundred micromolar. While they can at least double when cells are cultured at higher extracellular zinc loads, it is not known to which extent they vary under physiological conditions. The affinities of proteins for zinc determine the buffering of zinc. In a given coordination environment, affinities follow the Irving-Williams series, according to which zinc(II) ions bind stronger than the other divalent metal ions of the first transition series, with the exception of copper(II) ions. Zinc affinities of zinc proteins are rather high and in the picomolar range of dissociation constants. Consequently, free zinc(II) ion concentrations are commensurately low as confirmed by direct measurements, which provide estimates of free zinc(II) ion concentrations in the range of tens to hundreds of picomolar (pZn = $-\log[Zn^{2+}]$ = 10–11). The concentrations of free zinc(II) ions are at least six to seven orders of magnitude lower than the total zinc concentrations. Furthermore, they seem to be much more tightly controlled than the total cellular zinc concentrations. Free zinc(II) ions – designated as such to distinguish them from protein-bound zinc – are not devoid of any ligands. The nature of the ligands is not known. Other terms are used for this zinc pool, such as free, readily or easily exchangeable, mobile, or kinetically labile zinc. Each term has a different meaning and the simultaneous use of these terms in the scientific literature is confusing. The important conclusion is that the cellular zinc(II) ion pool is not negligible as it fluctuates under various conditions and is functionally significant.

Zinc buffering is a dynamic process. Both the pZn and the buffering capacity of cells can change. Higher buffer capacity allows the cell to handle more zinc while keeping the pZn constant. But changing the buffering capacity also allows different pZn values. Thus changes of pZn can be effected independently by changing either zinc influx/efflux or the concentrations of buffering ligands. The same cell can have different pZn values when proliferating, differentiating, or undergoing apoptosis. Different pZn values are likely a consequence of an altered zinc-buffering capacity. More data are needed to define to which extent cells differ in their total zinc concentration, buffering capacity, and pZn value. Global or local changes of cellular zinc(II) ion concentrations provide a way of controlling cellular processes, indicating that changes of pZn values are a cause rather than a consequence of an altered state of a cell. The control of zinc(II) ions at the remarkably low concentrations appears to be a way of achieving biological control over the metabolic state or the fate of a cell through zinc targeting highly specific coordination environments and affecting enzyme activities, protein-protein interactions, and gene expression. Cells have the capacity to adapt to different pZn values by regulating zinc transporters and MTs. Many zinc enzymes have binuclear metal-binding sites that bind either two zinc(II) ions or one zinc(II) ion and another metal ion. Investigations of isolated enzymes demonstrate that the occupancy of the second, so-called cocatalytic site with zinc modulates enzymatic activity. The discovery that cellular zinc(II) ion concentrations fluctuate suggests that such a modulation of enzymatic activity is physiologically significant.

If metal affinities were the only factor to control cellular zinc redistribution, zinc(II) ions would move only in one direction determined by the thermodynamic gradient. Cells also restrict the availability of zinc through compartmentalization and kinetic effects. Such control can occur at the protein level when protein conformational changes and associated dynamics of zinc coordination modulate the kinetics of metal association and dissociation (Maret and Li 2009). But it can also occur at the cellular level through the extensive compartmentalization of eukaryotic cells. The transfer of zinc between different compartments contributes to buffering. Such a kinetic component describing the time-dependent transfer of cellular constituents from one compartment to another has been referred to as muffling. Muffling allows cells to tolerate much higher zinc loads than determined by the zinc-buffering capacity at steady state. Thus cellular zinc buffering is a combination of the cytosolic zinc-buffering capacity and the zinc muffling capacity determined by the activities of transporters that remove the surplus of zinc(II) ions from the cytosol to an organelle or to the outside of the cell at pre-steady state and restore the steady state (Colvin et al. 2010). Sequestration of zinc(II) ions in a cellular compartment is one way of storing zinc(II) ions in the cell and then releasing them again. Such a release occurs through stimuli targeting the zinc transporter Zip7 that controls the release of zinc from a store in the ER to the cytoplasm (Hogstrand et al. 2009) or through exocytosis of vesicles that have been loaded with zinc(II) ions by ZnT3 (brain), ZnT8 (pancreatic β-cells), or ZnT2 (mammary gland). Additional cells secrete zinc(II) ions, indicating that cellular zinc homeostasis also has a role in cell-cell communication and in releasing zinc(II) ions for other purposes. This compartmentalization and subcellular distribution is a central aspect of the cellular homeostatic control of zinc and requires the relative large number of zinc transporters. Compartmentalization also resolves a long-standing issue of whether or not cells store zinc, namely, that zinc is stored in vesicles rather than in a protein with a high zinc-binding capacity akin to ferritin in iron-storage. The chemical state of zinc in the vesicles is not known. In some vesicles, it exists at a much higher pZn value and is readily available to chelating agents, indicating different buffering and control of zinc in subcellular compartments. The determination of subcellular pZn values is an area of significant present interest. Muffling solves the issue of how transient increases of free zinc(II) ions are possible in a buffered environment. Zinc(II) ions participate in information transfer as signaling ions within cells and between cells (Haase and Maret 2010).

$$\text{Signal} \rightarrow \text{increase of}\, [Zn^{2+}] \rightarrow \text{targets}$$

For signaling functions, zinc(II) ions must be made available at certain amplitudes and frequencies. Under conditions of stimulation, cellular zinc(II) ion concentrations increase above 1 nM and unless the buffering capacity changes, return within minutes to their resting concentrations. How zinc(II) ions can serve such signaling functions hinges on the fluctuations of zinc(II) ions, how they are generated, controlled, transmitted to targets, and terminated at their targets (▶ Zinc and Zinc Ions in Biological Systems). The amplitudes of these fluctuations must match the affinities of targets for zinc(II) ions. Indeed, picomolar/nanomolar concentrations of zinc(II) ions inhibit the activities of certain enzymes such as protein tyrosine phosphatases. The affinities of zinc proteins for zinc and the affinities of proteins that are targeted by signaling zinc(II) ions define the lower and upper concentrations of the range, in which free zinc(II) ions regulate biological processes.

No proteins other than MTs have been identified in cellular zinc redistribution. In contrast to picomolar free zinc(II) ion concentrations, the micromolar concentrations of MTs in most cells make sufficient zinc available in a deliverable form to meet the demand of many cellular functions for zinc. In this way, MT keeps zinc sufficiently tightly sequestered as a temporary storage for cellular zinc. The zinc-binding characteristics of human MT-2 and its translocation in the cell are in agreement with a role in buffering and muffling zinc, namely, binding zinc(II) ions and delivering them to a cellular compartment. Though bona fide zinc metallochaperones have not been identified, these functions of MT resemble those of metallochaperones, in particular since MTs have the capacity to transfer zinc via protein-protein interactions by an associative mechanism. The high inducibility of MTs by many different cellular signal transduction pathways permits changing the buffering and muffling capacity. Modulating the chelating capacity of MTs through changes of either their total amount or their redox state affords a way of activating zinc-inhibited enzymes and controlling the regulatory functions of zinc(II) ions. Elevated zinc(II) ion concentrations induce MTF-1-controlled gene transcription of thioneins, the apoproteins of MTs, and the zinc transporter ZnT1, resulting in the binding of the surplus of zinc to MTs and export of zinc from the cell. A host of other conditions also induce thioneins, indicating that thioneins are produced for lowering the availability of cellular zinc(II) ions, modulating zinc-dependent processes, and increasing the reducing power of cells.

## The Control of Zinc Homeostasis in Health and Disease

The importance of regulating cellular zinc homeostasis extends well beyond regulating the supply of zinc for metalloproteins. Zinc homeostasis is integrated into cellular metabolism and signal transduction and its control is critically important for cellular functions. Controlling cellular zinc requires several dozen proteins that evolved with specific mechanisms for handling zinc(II) ions at unprecedented low concentrations. An unresolved issue is whether zinc deficiency affects all zinc-dependent cellular processes to the same extent or the cell shunts the available zinc from less important processes to processes that are critical for survival (▶ Zinc Deficiency). The relatively large number of proteins necessary to control cellular zinc homeostasis demonstrates how these proteins and not only zinc itself determine cellular zinc status. Furthermore, following the initial observation that mutations in the zinc transporter Zip4 are responsible for Acrodermatitis enteropathica, a fatal disease of impaired intestinal zinc uptake if the afflicted individuals are not treated with zinc, many mutations and disease associations were found for other zinc transporters (Andrews 2011). If the buffering/muffling capacity is exhausted, zinc(II) ion concentrations increase drastically and zinc(II) ions bind to additional proteins, which do not bind zinc under normal physiological conditions, and affect their functions. Increasing the buffering capacity of a cell, e.g., by introducing agents that bind zinc, decreases zinc availability and generates a state that can be described as a transient cellular zinc deficiency. Compounds in the diet, toxins, and therapeutic drugs are examples. Because of the remarkably low concentrations at which zinc(II) ions are biological effectors, a significant number of compounds perturb zinc metabolism either directly or indirectly via ensuing redox changes. Such perturbations have significant implications for acute and chronic exposures to various chemical agents, and thus for the balance between health and disease.

## Cross-References

▸ Zinc and Zinc Ions in Biological Systems
▸ Zinc Deficiency
▸ Zinc Homeostasis, Whole Body
▸ Zinc Metallothionein
▸ Zinc-Binding Proteins, Abundance

## References

Andreini C, Banci L, Bertini I et al (2006) Counting the zinc-proteins encoded in the human genome. J Proteome Res 5:196–201

Andrews GK (2011) Conclusions from zinc transporter mutations for zinc physiology. In: Rink L (ed) Zinc in human health. IOS Press, Amsterdam, pp 493–513

Blindauer CA, Schmid R (2010) Cytosolic metal handling in plants: determinants for zinc specificity in metal transporters and metallothioneins. Metallomics 2:510–529

Colvin RA, Holmes WR, Fontaine CP et al (2010) Cytosolic zinc buffering and muffling: their role in intracellular zinc homeostasis. Metallomics 2:306–317

Fukada T, Kambe T (2011) Molecular and genetic features of zinc transporters in physiology and pathogenesis. Metallomics 3:662–674

Haase H, Maret W (2010) The regulatory and signaling functions of zinc ions in human cellular physiology. In: Zalups R, Koropatnick J (eds) Cellular and molecular biology of metals. Taylor and Francis/CRC Press, Boca Raton, pp 179–210

Hantke K (2001) Bacterial zinc transporters and regulators. Biometals 14:239–249

Hogstrand C, Kille P, Nicholson RI et al (2009) Zinc transporters and cancer: a potential role for ZIP7 as a hub for tyrosine kinase activation. Trends Mol Med 15:101–111

Maret W (2009) Molecular aspects of human cellular zinc homeostasis: redox control of zinc potentials and zinc signals. Biometals 22:149–157

Maret W, Li Y (2009) Coordination dynamics of zinc in proteins. Chem Rev 109:4682–4707

# Zinc Content

▸ Zinc-Binding Proteins, Abundance

# Zinc Deficiency

Ananda S. Prasad
Department of Oncology, Karmanos Cancer Center, Wayne State University, School of Medicine, Detroit, MI, USA

## Introduction

The essentiality of zinc for the growth of microorganism (*Aspergillus niger*) was recognized in 1869. Fifty years later, it was shown that zinc was essential for the growth of the plants. Later, it was shown to be essential for the growth of the rats, pigs, and chicken. However, until 1960 it was not known if zinc was essential for humans, and its deficiency was unknown.

## Discovery of Human Zinc Deficiency

### Studies in the Middle East

I arrived in Shiraz, Iran, in June 1958 after finishing my formal training as a clinical scientist in medicine under Dr. C.J. Watson at the University of Minnesota Medical School. Dr. Hobart A. Reimann, chief of medicine at the Nemazee Hospital of Pahlevi University in Shiraz, Iran, invited me to join him to set up a medical curriculum for teaching medicine to students and house staff. In Shiraz, I met Dr. James A. Halsted, who was a Fulbright Professor at Pahlevi University, Shiraz, and was primarily involved with Saadi hospital. In the fall of 1958, I was invited by Dr. Halsted to discuss a patient with anemia at the medical center grand rounds at the Saadi Hospital. The case was presented to me by the chief resident, Dr. M. Nadimi, a graduate of the Shiraz Medical School.

The patient was a 21-year-old male, who looked like a 10-year-old boy. In addition to severe growth retardation and anemia, he had hypogonadism, hepatosplenomegaly, rough and dry skin, mental lethargy, and geophagia. The patient ate only bread made from wheat flour, and intake of animal protein was negligible. He consumed nearly 0.5 kg of clay daily. Later, we discovered that the habit of geophagia (clay eating) was fairly common in the villages around Shiraz. Further studies documented the existence of iron-deficiency anemia in our patient, but there was no evidence of blood loss. Inasmuch as 10 additional similar cases were brought to the hospital under my care within a short period of time, hypopituitarism as an explanation for growth retardation and hypogonadism was ruled out. The anemia of the subjects promptly responded to oral administration of iron. The probable factors responsible for anemia in these patients were insufficient availability of iron in the diet, excessive sweating probably causing greater iron loss from the skin than would occur in a temperate climate, and geophagia further decreasing iron absorption.

It was difficult to explain all of the clinical features solely by tissue iron deficiency inasmuch as growth retardation and testicular atrophy are not seen in iron-deficient experimental animals. The possibility that zinc deficiency may have been present was considered. Zinc deficiency was known to produce retardation of growth and testicular atrophy in animals; however, its deficiency in human was not known. Because heavy metals may form insoluble complexes with phosphate, we speculated that some factors responsible for decreased availability of iron in these patients with geophagia may also have decreased the availability of zinc. Phytate (inositol hexaphosphate), which is present in cereal grains, was known to impair the absorption of both iron and zinc. We published the clinical description of the Iranian cases as a syndrome and speculated that zinc deficiency may account for growth retardation and male hypogonadism in human subjects (Prasad et al. 1961).

I left Iran in January 1961 and joined the department of Biochemistry and Medicine at Vanderbilt University under Dr. William J Darby. I moved to US Naval Medical Research Unit No.3 (NAMRU-3), Cairo, Egypt, where I studied zinc metabolism in growth-retarded subjects. My team consisted of Harold Sandstead, M.D., and A. Schulert, Ph.D., both from Vanderbilt University, A. Miale Jr., M.D., from US Navy and Z. Farid, M.D., a local physician, also from NAMRU-3.

In Egypt, subjects similar to the growth-retarded Iranian subjects were encountered. The clinical features were remarkably similar except that the Iranian subjects had more pronounced hepatosplenomegaly, a history of geophagia, and no hookworm infection and the Egyptian subjects had both schistosomiasis and hookworm infestations but no history of geophagia.

We carried out a detailed investigation of the Egyptian cases at NAMRU-3 in Cairo. The dietary history of the Egyptian subjects was similar to that of the Iranians. The consumption of animal protein was negligible. Their diet consisted mainly of bread and beans (Vicia fava). These subjects were shown to have zinc deficiency. The evidences were decreased zinc concentrations in plasma, red cells, and hair, and by studies with zinc-65, we showed that the plasma zinc turnover was greater, the 24-h exchangeable pool was smaller, and the excretion of zinc-65 in stool and urine was less in the dwarfs than in the control subjects (Prasad et al. 1963).

Studies in Egypt also showed that the rate of growth was greater in patients who received supplemental zinc as compared with those receiving iron instead or those receiving only an adequate animal-protein diet (Sandstead et al. 1967). Zinc-supplemented subjects gained 5–6 in. in height in 12 months. Pubic hair appeared in all subjects within 7–12 weeks after zinc supplementation. Genitalia increased to normal size, and secondary sexual characteristics developed within 12–24 weeks in patients who received zinc. In contrast, no such changes were observed in a comparable length of time in the iron-supplemented group or in the group on an animal-protein diet. Thus, the growth retardation and gonadal hypofunction in these subjects were related to zinc deficiency. The anemia was due to iron deficiency and responded to oral iron treatment.

It is now evident that nutritional as well as conditioned deficiency of zinc affecting both sexes may complicate many diseased states in human subjects (Prasad 1993). Zinc deficiency in human populations throughout the world is prevalent, and it may affect nearly two billion subjects. Zinc deficiency should be present in countries where primarily cereal proteins are consumed by the population. One would also expect to see a spectrum of zinc deficiency, ranging from severe cases to marginally deficient examples, in any given population.

In 1973, Barnes and Moynahan (1973) studied a 2-year-old girl with severe acrodermatitis enteropathica who was being treated with diiodohydroxyquinoline and a lactose-deficient synthetic diet.

The clinical response to this therapy was not satisfactory, and the physicians sought to identify contributing factors. The concentration of zinc in the patient's serum was profoundly decreased; therefore, they administered oral zinc sulfate. The skin lesions and gastrointestinal symptoms cleared completely, and the patient was discharged from the hospital. When zinc was inadvertently omitted from the child's regimen, she suffered a relapse; however, she promptly responded to oral zinc again. In their initial reports, the authors attributed zinc deficiency in this patient to the synthetic diet. It soon became clear that zinc might be fundamental to the pathogenesis of this rare inherited disorder and that the clinical improvement reflected improvement in zinc status. This original observation was quickly confirmed in other patients throughout the world. The underlying pathogenesis of the zinc deficiency in these patients is due to malabsorption of zinc caused by a mutation in ZIP4, a zinc transporter.

In 1974, a landmark decision to establish recommended dietary allowances (RDAs) for humans for zinc was made by the Food and Nutrition Board of the National Research Council (NRC) of the US National Academy of Sciences (US NAS).

## Clinical Effects of Zinc Deficiency

### Clinical Spectrum of Human Zinc Deficiency

During the past five decades, a spectrum of clinical deficiency of zinc in human subjects has been recognized. On the one hand, the manifestations of zinc deficiency may be severe; and, on the other end of the spectrum, zinc deficiency may be mild or marginal. A severe deficiency of zinc has been reported to occur in patients with acrodermatitis enteropathica, following TPN (total parenteral nutrition) without zinc, following excessive use of alcohol, and following penicillamine therapy.

### Severe Deficiency of Zinc

#### Acrodermatitis Enteropathica

Acrodermatitis enteropathica (AE) is a lethal, autosomal, recessive trait that usually occurs in infants of Italian, Armenian, or Iranian lineage. This disease is not present at birth but usually develops in the early months of life soon after weaning from breast feeding. The dermatologic manifestations of severe zinc deficiency in patients with AE include bullous pustular dermatitis of the extremities and the oral, anal, and genital areas (around the orifices) combined with paronychia and generalized alopecia. Ophthalmic signs may include blepharitis, conjunctivitis, photophobia, and corneal opacities. Neuropsychiatric signs include irritability, emotional disorders, tremors, and occasional cerebellar ataxia. The patients with AE generally have weight loss and growth retardation, and males exhibit hypogonadism. A high incidence of congenital malformation of fetuses and infants born of pregnant women with AE has been reported.

Patients with AE have an increased susceptibility to infections. In AE, thymic hypoplasia, absence of germinal centers in lymph nodes, and plasmacytosis in the spleen are found consistently. All T-cell-mediated functional abnormalities are completely corrected with zinc supplementation. In general, the clinical course is downhill with failure to thrive and complicated by intercurrent bacterial, fungal, and other opportunistic infections. Gastrointestinal disturbances are usually severe, including chronic diarrhea, malabsorption, steatorrhea, and lactose intolerance. The disease, if unrecognized and untreated, is fatal. Zinc supplementation results in complete recovery.

AE gene has been localized to an $\sim$3.5-cM region on 8q24. The gene encodes a histidine-rich protein, which is now referred to as hZIP-4, which is a member of a large family of transmembrane proteins, some of which are known to serve as zinc-uptake proteins. In patients with AE, mutations in this gene have been documented.

#### Total Parenteral Nutrition (TPN)

Patients on TPN with diarrhea may lose 6–12 mg of zinc/day. This excessive loss of zinc may result in a severe deficiency of zinc. In such cases, not only dermatologic manifestations are seen but also alopecia, neuropsychiatric manifestations, weight loss, and intercurrent infections, particularly involving opportunistic infections, are also observed. Carbohydrate utilization is impaired, and there is a negative nitrogen balance. If zinc deficiency in such cases is not recognized and treated, the condition may become fatal.

#### Alcoholism and Following Penicillamine Therapy

Severe deficiency of zinc has been also observed following excessive alcohol intake and following penicillamine therapy in patients with Wilson's disease.

## Moderate Deficiency of Zinc

A moderate deficiency of zinc has been reported in a variety of conditions. These include nutritional due to dietary factors, malabsorption syndrome, alcoholic liver disease, chronic renal disease, sickle cell disease, and chronically debilitated conditions.

### Nutritional

Growth retardation, hypogonadism in males, poor appetite, mental lethargy, rough skin, and intercurrent infections were the classical clinical features of chronically zinc-deficient subjects from the Middle East as reported by Prasad et al. in the early 1960s (Prasad et al. 1963; Sandstead et al. 1967; Prasad 1993). The basis for zinc deficiency was nutritional inasmuch as zinc was poorly available from their diet due to high content of phytate and phosphate. All the above-mentioned features were corrected by zinc supplementation.

### Gastrointestinal Disorders and Liver Disease

A moderate level of zinc deficiency has been observed in many gastrointestinal disorders. These include malabsorption syndrome, Crohn's disease, regional ileitis, and steatorrhea. A low serum and hepatic zinc and, paradoxically, hyperzincuria were demonstrated in patients with cirrhosis of the liver many years ago.

It is likely that some of the clinical features of cirrhosis of the liver, such as loss of body hair, testicular hypofunction, poor appetite, mental lethargy, difficulty in healing, abnormal cell-mediated immune functions, and night blindness, may indeed by related to the secondary zinc-deficient state in this disease.

### Renal Disease

Patients with chronic renal failure have low concentrations of zinc in plasma, leukocytes, and hair as well as increased plasma ammonia levels and increased activity of plasma ribonuclease (Prasad 1993). Uremic hypogeusia improved following zinc supplementation. Impotence is common in uremic males and is not improved by hemodialysis (HD). Our study showed that zinc deficiency was a reversible cause of sexual dysfunction in uremia.

### Zinc Deficiency in Sickle Cell Disease

Our studies have documented the occurrence of zinc deficiency in adult sickle cell anemia (SCA) patients (Prasad 1993). Growth retardation, hypogonadism in males, hyperammonemia, abnormal dark adaptation, and cell-mediated immune disorder in SCA have been related to a deficiency of zinc. The biochemical evidences of zinc deficiency in SCA included a decreased level of zinc in the plasma, erythrocytes, and hair; hyperzincuria; decreased activities of certain zinc-dependent enzymes such as carbonic anhydrase in the erythrocytes, alkaline phosphatase in the neutrophils, deoxythymidine kinase activity in newly synthesizing skin connective tissue and collagen; and hyperammonemia. Inasmuch as zinc is known to be an inhibitor of ribonuclease (RNase), an increased activity of this enzyme in the plasma of SCA subjects was regarded as an evidence of zinc deficiency. Zinc supplementation to SCA subjects resulted in significant improvement in secondary sexual characteristics, improvement in growth, normalization of plasma ammonia level, and reversal of dark adaptation abnormality. As a result of zinc supplementation, the zinc level in plasma, erythrocytes, and neutrophils increased, and an expected response to supplementation was observed in the activities of the zinc-dependent enzymes. Zinc deficiency in patients with sickle cell anemia was associated with impaired DTH (delayed-type hypersensitivity reactions) and decreased NK (natural killer) cell lytic activity, which were corrected by zinc supplementation.

A 3-month placebo-controlled zinc supplementation trial (25 mg zinc as acetate three times a day) in 36 sickle cell disease patients showed that the zinc-supplemented group had decreased incidence of infections and increased hemoglobin and hematocrit, plasma zinc, and antioxidant power in comparison to the placebo group (Bao et al. 2008). Plasma nitrite and nitrate (NOx), lipid peroxidation products, DNA oxidation products, and soluble vascular cell adhesion molecule-1 decreased in the zinc-supplemented group in comparison to the placebo group. Zinc-supplemented subjects showed significant decreases in lipopolysaccharide-induced tumor necrosis factor-alpha (TNF-$\alpha$) and IL-1$\beta$ mRNAs and TNF-induced nuclear factor of kB-DNA binding in mononuclear cells (MNCs) compared with the placebo group. Zinc supplementation also increased relative levels of IL-2 and IL-2R$\alpha$ mRNAs in PHA-p-stimulated MNCs.

## Mild Deficiency of Zinc

Although the clinical, biochemical, and diagnostic aspects of severe and moderate levels of zinc deficiency in humans are well defined, the recognition of mild levels of zinc deficiency has been difficult. We therefore developed an experimental model of zinc deficiency in order to define mild deficiency of zinc in humans (Prasad 1993). In a group of human volunteers, we induced a mild state of zinc deficiency by dietary means. A semipurified diet, which supplied approximately 3.0–5.0 mg of zinc on a daily basis, was used to produce zinc deficiency.

Decreased serum testosterone level, oligospermia, decreased NK cell activity, decreased IL-2 activity of T helper cells (TH), decreased thymulin activity, hyperammonemia, hypogeusia, decreased dark adaptation, and decreased lean body mass were observed in subjects with mild deficiency of zinc. It is, therefore, clear that even a mild deficiency of zinc in humans affects clinical, biochemical, and immunological functions adversely.

## Major Clinical Effects of Zinc Deficiency in Humans

### Growth Retardation

A meta-analysis of 33 studies of the effect of zinc supplementation on children's growth was reported by Brown et al. (2002). Prospective intervention trials were included if they enrolled a control group and provided suitable data on change in height or weight during the period of observation. The pooled study population included 1,834 children less than 13 years of age, with representation from most regions of the world. The meta-analysis showed that in all children studied, zinc supplementation had a highly statistically significant positive response in height and weight increments respectively.

Nakamura et al. (1993) screened 220 prepubertal subjects with short stature in a hospital clinic for zinc supplementation trial. In zinc-supplemented children, caloric intake ($p < 0.01$); growth velocity ($p < 0.01$); serum zinc, calcium, and phosphorus concentrations; alkaline phosphatase activity ($p < 0.001$); percentage of tubular reabsorption of phosphorus ($p < 0.05$); ratio of maximal tubular reabsorption rate for phosphorus to the glomerular filtration rate ($p < 0.05$); serum osteocalcin level ($p < 0.01$); and plasma insulin-like growth factor-1 (IGF-1) ($p < 0.05$) were significantly increased in comparison to control group, but urinary excretion of growth hormone (GH) was unchanged.

All the above parameters remained unchanged in the controls. Zinc supplementation resulted in a considerable increase in height. Also, these investigators reported that urinary GH excretion and other pituitary hormone levels in the plasma were unchanged with zinc supplementation. However, a significant elevation in serum IGF-1 level after zinc supplementation was observed. Increased plasma levels of IGF-1, despite unchanged GH production, may be explained if zinc has a role in binding GH to GH receptor in hepatic cells. Nakamura et al. (1993) recommended that the children with non-endocrinologic short stature should be treated for at least 6 months with zinc, even if the serum zinc concentration is within normal range and zinc clearance test is not available. These children may have a marginal deficiency of zinc, and they would respond to zinc supplementation with increase in height.

Approximately one-half of the children <5 years of age in developing countries are retarded in statural growth. Nutrient deficiencies underlie the stunting of growth in many of these children. Zinc deficiency has been associated with poor growth, and zinc supplementation of growth-retarded children stimulated growth. Inasmuch as zinc deficiency is associated with cell-mediated immune dysfunctions, and since children with malnutrition show clinical evidences of immunologic deficit correctable by zinc, the high prevalence of infections in malnourished children might also be caused by zinc depletion.

### Zinc and Immunity

Zinc affects multiple aspects of the immune system. Zinc is crucial for normal development and function of cells mediating innate immunity, neutrophils, and natural killer cells. Macrophages are also affected by zinc deficiency. Phagocytosis, intracellular killing, and cytokine production are all affected by zinc deficiency. Zinc deficiency also affects adversely the growth and function of T and B cells. This occurs through dysregulation of basic biological functions at the cellular level. Zinc is needed for DNA synthesis, RNA transcription, cell division, and cell activation. Programmed cell death (apoptosis) is also potentiated in the absence of adequate levels of zinc. Secretion and function of cytokines, the basic messengers of the immune system, are adversely affected by zinc deficiency. The ability of zinc to function as an antioxidant and stabilize membranes suggests that it has a role in prevention of free radical–induced injury during inflammatory processes.

It has been known for many years that zinc deficiency in experimental animals leads to atrophy of thymic and lymphoid tissue. Later studies in young adult zinc-deficient mice showed thymic atrophy, reductions in absolute number of splenocytes, and depressed responses to both T-cell-dependent (TD) and T-cell-independent (TI) antigens.

A decrease in in vivo–generated cytotoxic T killer activity to allogeneic tumor cells in zinc-deficient mice and an impairment in cell-mediated response to non-$H_2$ allogeneic tumor cells in zinc-deficient mice have been reported. Animals maintained on a zinc-deficient diet for as little as 2 weeks developed a severe impairment in their ability to generate a cytotoxic response to the tumor challenge. This was totally reversible by zinc supplementation.

## Studies of Immune Functions in Experimental Human Model

During our studies in the Middle East, we observed that most of the zinc-deficient dwarfs did not live beyond the age of 25 years. The cause of death appeared to be infections. The possibility that zinc deficiency may have played a role in immune dysfunctions in the zinc-deficient dwarfs was considered, but lack of proper facilities prevented us from gathering meaningful data on immune functions in those patients.

We developed an experimental model, which allowed us to study specific effects of mild zinc deficiency in humans on immune functions. When zinc deficiency was very mild (5.0-mg Zn intake during the zinc-restricted period), the plasma zinc concentration remained more or less within the normal range, and it decreased only after 4–5 months of zinc restriction. On the other hand, zinc concentrations in lymphocytes, granulocytes, and platelets decreased within 8–12 weeks, suggesting that the assay of cellular zinc provided a more sensitive criterion for diagnosing mild deficiency of zinc.

We assayed serum thymulin activity in mildly zinc-deficient human subjects. Thymulin is a thymus-specific hormone, and it requires the presence of zinc for its biological activity to be expressed. Thymulin binds to high-affinity receptors on T cells, induces several T-cell markers, and promotes T-cell function, including allogeneic cytotoxicity, suppressor functions, and interleukin-2 (IL-2) production.

As a result of mild deficiency of zinc, the activity of thymulin in serum was significantly decreased and was corrected by both in vivo and in vitro zinc supplementation. The in vitro supplementation studies indicated that the inactive thymulin peptide was present in the serum in zinc-deficient subjects and was activated by addition of zinc. The assay of serum thymulin activity with or without zinc addition in vitro thus may be used as a sensitive criterion for the diagnosis of mild zinc deficiency in humans.

Our studies in the experimental human model showed for the first time that the production of IL-2 and IFN-$\gamma$ was decreased, whereas the production of IL-4, IL-6, and IL-10 was not affected due to zinc deficiency. Our data suggested that cell-mediated immune dysfunctions in human zinc deficiency may be due to an imbalance between TH1 and TH2 cell functions.

## Role of Zinc as an Antioxidant and Anti-inflammatory Agent

Oxidative stress and chronic inflammation are important contributing factors in several chronic diseases, such as atherosclerosis and related vascular diseases, mutagenesis, and cancer, neurodegeneration, immunologic disorders, and the aging process. We administered 45-mg zinc as gluconate daily to 10 volunteers, and 10 subjects received placebo for 8 weeks. The volunteers were healthy, and their ages ranged from 19 to 50 years. In subjects receiving zinc, plasma levels of lipid peroxidation products and DNA adducts were decreased, whereas no change was observed in the placebo group. LPS-stimulated MNC isolated from zinc-supplemented groups showed reduced mRNA for TNF-$\alpha$ and IL-1$\beta$ compared to placebo. Ex vivo, zinc protected MNC from TNF-$\alpha$-induced NF-$\kappa$B (Nuclear Factor-kB [a zinc-dependent transcription factor]) activation. In parallel studies using HL-60, a promyelocytic leukemia cell line, we observed that zinc enhances the upregulation of mRNA- and DNA-specific binding for A-20, a transactivating factor which inhibits the activation of NF-$\kappa$B. Our results suggest that zinc supplementation may lead to downregulation of the inflammatory cytokines through upregulation of the negative feedback loop A-20 to inhibit induced NF-$\kappa$B activation.

Zinc deficiency, cell-mediated immune dysfunction, susceptibility to infections, and increased oxidative stress have been observed in elderly subjects

(>55 years old). We conducted a randomized, double-blind, placebo-controlled trial of zinc supplementation in elderly subjects ages 55–87 years. Fifty healthy elderly subjects were recruited for this study. The supplementation was continued for 12 months. The zinc-supplemented group received zinc gluconate (45 mg elemental zinc) orally daily. Compared with a group of younger adults, at baseline the older subjects had significantly lower plasma zinc and higher plasma oxidative stress markers and endothelial cell adhesion molecules. The incidence of infections and ex vivo generation of TNF-$\alpha$ and plasma oxidative stress markers were significantly lower in the zinc-supplemented group than in the placebo group. Plasma zinc and PHA-induced IL-2 mRNA in isolated PMNC were significantly higher in the zinc-supplemented group than in the placebo group.

In another study, we conducted a randomized, double-blind, placebo trial of zinc supplementation in 40 elderly subjects (ages 56–83 years) and randomly assigned them to two groups. One group received 45 mg elemental zinc daily as gluconate for 6 months, and the other group received placebo. Cell culture studies were also done in order to study the mechanism of zinc action as an atheroprotective agent.

After zinc supplementation, plasma high-sensitivity C-reactive protein (hsCRP), interleukin-6, macrophage chemoattractant protein-1 (McP-1), VCAM-1, secretory phospholipase A2, and MDA + HAE decreased in the elderly subjects in comparison to the placebo group. In cell culture studies, we showed that zinc decreased the generation of TNF-$\alpha$, IL-1$\beta$, VCAM-1, and MDA + HAE and the activation of NF-$\kappa$B and increased A-20 and peroxisome proliferator-activated receptor-$\alpha$ in human monocytic leukemia THP-1 cells and human aortic endothelial cells compared to the zinc-deficient cells. These data suggest that zinc may have a protective effect in atherosclerosis because of its anti-inflammatory and antioxidant functions.

Zinc Deficiency and Cognitive Impairment
Penland et al. (1997) conducted a study in elementary schools in low-income districts of three provinces in China in order to assess the effects of zinc supplementation on growth and neuropsychological functions of children. Three hundred and seventy-two children between the ages of 6 and 9 years were recruited for this study. Supplementations were 20 mg zinc, 20 mg zinc with micronutrients, or micronutrients alone. The micronutrient mixture was based on guidelines of the US NAS/NRC.

Neuropsychological functions were tested by using the cognition psychomotor assessment system revised (CPAS-R) developed by Penland. Zinc-containing treatments improved neuropsychological functions, but micronutrients alone had little effect. Performance after zinc alone and/or zinc with micronutrients was better than after micronutrients alone for continuous performance, perception (matching of complex shapes), visual memory (delayed matching of complex shapes), tracking of a cursor on the computer screen, concept formation (identification of oddity) and key tapping.

It is evident that the Chinese children were deficient also in other nutrients besides zinc inasmuch as the group receiving micronutrients with zinc showed the maximum growth. These results are similar to the report published from Iran, which showed that repletion of latent deficiencies was essential for demonstration of the effects of zinc supplementation on growth.

## Diagnostic Criteria for Zinc Deficiency

Measurement of zinc level in plasma is very useful provided the sample is not hemolyzed and contaminated. In the condition of acute stress or infection, or following a myocardial infarction, zinc from the plasma compartment may redistribute to other tissues, thus making an assessment of zinc status in the body difficult. Intravascular hemolysis would also increase the plasma zinc level inasmuch as the zinc in the red cells is much higher than in the plasma.

Zinc in the red cells and hair also may be used for assessment of body zinc status, however, inasmuch as these tissues turn over slowly, and their zinc levels do not reflect recent changes with respect to body zinc stores. Zinc determination in granulocytes and lymphocytes, however, reflects the body zinc status more accurately and is thus a useful measurement. A quantitative assay of alkaline phosphatase activity in the granulocytes is a useful tool in our experience.

Urinary excretion of zinc is decreased as a result of zinc deficiency. Thus, determination of zinc in 24-h urine may be of additional help in diagnosing zinc deficiency provided cirrhosis of the liver, sickle cell disease, chronic renal disease, and other conditions known to cause hyperzincuria are ruled out.

In experimental human model studies, we observed that a decrease in serum thymulin activity, decreased production of IL-2, decrease in lymphocyte ecto-5′-nucleotidase activity, decrease in intestinal endogenous zinc excretion, and decrease in urinary zinc excretion occurred within 8 weeks of the institution of a zinc-restricted diet (approximately 5-mg zinc daily intake). These changes were observed prior to the changes in plasma zinc concentration and changes in lymphocyte and granulocyte zinc concentration. The decrease in plasma zinc was observed at the end of 20 weeks, and decrease in zinc concentration of lymphocyte and granulocytes was observed at the end of 12 weeks of zinc-restricted diet.

The activities of many zinc-dependent enzymes have been shown to be affected adversely in zinc-deficient tissues. Three enzymes, alkaline phosphatase, carboxypeptidase, and thymidine kinase, appear to be most sensitive to zinc restriction in that their activities are affected adversely within 3–6 days of institution of a zinc-deficient diet to experimental animals. In human studies, the activity of deoxythymidine kinase in proliferating skin collagen and alkaline phosphatase activity in granulocytes was shown to be sensitive to dietary zinc intake. As a practical test, quantitative measurement of alkaline phosphatase activity in granulocytes may be a very useful adjunct to granulocyte zinc level determination in order to assess the body zinc status in man.

We have reported that in zinc-deficient mononuclear cells following PHA (phytohemagglutinin) stimulation, IL-2 in RNA is decreased, but an addition of zinc in vitro in physiological amounts corrects this abnormality. Thus, this test may also be very useful for diagnostic purposes.

## Therapeutic Impact of Zinc

Zinc therapy is very effective as a treatment for acute diarrhea and acute respiratory tract infections in infants and children. The mortality has decreased following zinc therapy significantly, and now millions of children are living. WHO is now implementing zinc treatment for acute diarrhea in infants and children in many developing countries.

Active search for an effective agent for the treatment of common cold has been ongoing for many decades. Use of zinc acetate lozenges (13–14 mg of elemental zinc), one every 3–4 h during waking hours, has been found effective. If started within 24 h of the onset of cold symptoms, the severity and duration of common cold is decreased by 50%.

Wilson's disease is an inherited autosomal disorder of toxic copper accumulation in brain, liver, kidney, eye, and other organs. It is a very serious disorder, and the longevity is sharply decreased. Zinc is very effective in the management and prevention of Wilson's disease (Brewer 1995). This mode of therapy has been approved by FDA in USA.

In a study conducted by Age-Related Eye Disease Study group (AREDS) involving 11 centers in USA and supported by National Eye Institute, NIH used zinc supplementation in a double-masked clinical trial in patients with dry type of age-related macular degeneration (AMD). Three thousand six hundred and forty participants ages 55–85 years were enrolled and followed for an average of 6.3 years. Group taking zinc alone (zinc oxide 80.0 mg/day) reduced the risk of developing advanced AMD by about 21% and vision loss by about 11%. Another interesting observation was that the zinc-supplemented group showed increased longevity. The risk of mortality by zinc treatment was reduced by 27% in AMD participants.

**Acknowledgments** Supported by NIH grant no. 5 R01 A150698-04 and Labcatal Laboratories, Paris, France. I thank Sally Bates for her assistance and clerical help.

## References

Bao B, Prasad AS et al (2008) Zinc supplementation decrease oxidative stress, incidence of infection and generation of inflammatory cytokines in sickle cell disease patients. Transl Res 152:67–80

Barnes PM, Moynahan EJ (1973) Zinc deficiency in acrodermatitis enteropathica. Proc R Soc Med 66:327–329

Brewer GJ (1995) Practical recommendations and new therapies for Wilson's disease. Drugs 2:240–249

Brown KH, Peerson JM et al (2002) Effect of supplemental zinc on the growth and serum zinc concentrations of prepubertal children. Am J Clin Nutr 75:1062–1071

Nakamura T, Nishiyama S et al (1993) Mild to moderate zinc deficiency in short children: efficacy of zinc supplementation on linear growth velocity. J Pediatr 123:65–69

Penland JG, Sandstead HH et al (1997) A preliminary report: effects of zinc and micronutrient repletion on growth and neuropsychological function of urban Chinese children. J Am Coll Nur 16:268–272

Prasad AS (1993) Biochemistry of zinc. Plenum Press, New York

Prasad AS, Halsted JA et al (1961) Syndrome of iron deficiency anemia, hepatosplenomegaly, hypogonadism, dwarfism, and geophagia. Am J Med 31:532–546

Prasad AS, Miale A et al (1963) Zinc metabolism in patients with the syndrome of iron deficiency anemia; hypogonadism and dwarfism. J Lab Clin Med 61:537–549

Sandstead HH, Prasad AS et al (1967) Human zinc deficiency, endocrine manifestations and response to treatment. Am J Clin Nutr 20:422–442

# Zinc Finger Folds and Functions

Jacqueline M. Matthews
School of Molecular Bioscience, The University of Sydney, Sydney, Australia

## Synonyms

Zinc-coordinating domains

## Definition

Zinc fingers (ZnFs) are small (~20–100-residue) domains that coordinate one or more zinc ions, usually through cysteine and histidine sidechains, to stabilize their fold. The term was first used to describe a repeated motif in a zinc-binding region of the Xenopus laevis transcription factor III (Miller et al. 1985) but has since been used to describe many different known and predicted structural classes. Early research focused on the ability of several classes to function in the sequence-specific recognition of double-stranded DNA. However, it is now well established that many ZnFs also bind other ligands such as RNA, lipids, and proteins.

The term "zinc finger" (hereafter ZnF) was proposed for a repeated cysteine/histidine-rich motif observed in Xenopus laevis transcription factor III. Domains of that type are now referred to as "classical" or C2H2 ZnFs, and the term ZnF has expanded to encompass a wide range of small domains and subdomains that use the coordination of Zn(II) ions to stabilize their fold. Note that this feature differentiates them from enzymes that use Zn(II) in catalysis, although many enzymes also contain ZnF structural domains. ZnFs are generally found inside cells, where the reducing environment precludes the ready formation of disulfides bonds; the coordination of a zinc ion allows these small domains to co-opt cysteine residues to form compact structures in the absence of an extensive hydrophobic core.

ZnF-containing proteins are highly represented in eukaryotes, comprising, for example, >2% of the proteins encoded by the human genome. But these proteins are found in all domains of life, including viruses. ZnF proteins have varied roles within cells, but the domains themselves are used to mediate interactions with other biomolecules, including DNA, RNA, proteins, and lipids. Many ZnF proteins contain multiple ZnFs, often from different classes, as well as other types of protein domains allowing the proteins to bring together a range of different molecules.

In the early 2000s, Grishin and colleagues collected the many different extant ZnFs into eight groups based on the structural properties in the vicinity of the zinc-binding site (Table 1) (e.g., Krishna et al. 2003). There are many variations on each of the canonical core folds and atypical examples of most ZnF classes (i.e., with a difference spacing or identity of zinc-coordinating residues). The structures of myriad additional ZnFs have since been determined. Although many newer structures easily fall into the eight original fold groups, others are less easy to classify. Thus, ZnFs are also named for the pattern of cysteine and histidine residues that coordinate the zinc ion (e.g., C4 means a zinc ion coordinated by four cysteine residues).

A large proportion of ZnFs use pairs of short bidentate zinc-coordinating motifs (e.g., $C-X_{2-4}-C$) separated by intervening sequences or loops, the sequences and structures of which vary according to the class of ZnF. These motifs usually take the form of a rubredoxin knuckle (a type of β-turn originally identified in rubredoxin), which is also often referred to as a zinc knuckle.

This article provides examples of many currently published ZnF structures and their functions (where known). This compilation is far from complete, even at the time of writing many additional ZnF structures were available in the Protein Data Bank (PDB), but had not yet been described in journals. Many other predicted ZnFs are found in various databases, for example, >40 types of ZnFs are annotated in UniProtKB. For more information see the following reviews and references therein (Laity et al. 2001; Matthews and Sunde 2002; Gamsjaeger et al. 2007; Matthews et al. 2009).

**Zinc Finger Folds and Functions, Table 1** Fold classes of zinc finger

| Fold or classification class | |
|---|---|
| C2H2 | A β-hairpin containing a zinc knuckle followed by an α-helix, forming a left-handed ββα topology. Zinc-coordinating residues come from the zinc knuckle and the C-terminus of the α-helix |
| Gag knuckle | Two short β-strands connected by a zinc knuckle and followed by a short helix or a loop. Zinc-coordinating residues come from the zinc knuckle the helix/loop |
| Treble clef | A β-hairpin containing a zinc knuckle at the N-terminus and an α-helix at the C-terminus. Zinc-coordinating residues come from the zinc knuckle and the N-terminal loop of the helix. A loop and additional β-hairpin (or one or two helices) between the hairpin and loop vary in length and conformation. Treble-clef ZnFs can ligate one, two[a], or three zinc ions |
| Zinc ribbon | The core structure comprises two β-hairpins, each of which contains a zinc knuckle that contain the zinc-coordinating residues. An additional β-strand often augments the secondary β-hairpin to form a three-stranded antiparallel β-sheet. Structures outside the core fold are very diverse |
| Zn2Cys6 | Two zinc ions are coordinated by six cysteine sidechains – two sidechains each coordinate both zinc ions. The domain also contains two helix-strand motifs (a short α-helix followed by an extended β-strand) that wrap around the metal ions |
| TAZ2-domain-like | Characterized by zinc-coordinating sidechains that are located at the termini of α-helices |
| Zinc-binding loops | Characterized by at least three closely spaced zinc-coordinating sidechains that are not incorporated into regular secondary structural elements. The final zinc-coordinating residue may be distant in sequence from the other three |
| Metallothionein | Cysteine-rich loops that bind multiple zinc or other metal ions |

[a]Two-zinc-coordinating treble-clef ZnFs tend to use pairs of closely spaced zinc-ligands (Zn1–4) separated by longer loops (Loop1–3). The identity and arrangement of the zinc-coordinating residues, loop lengths, and patterns of conserved hydrophobic residues distinguish the different domains. Zinc-coordination can be sequential (Zn1/Zn2 and Zn3/Zn4 sets each coordinate one zinc ion) or alternating (Zn1/Zn3 and Zn2/Zn4 coordination) to form a "cross-brace" topology (Fig. 1)

## Classes of ZnFs

*A20 ZnFs* often have ubiquitin E3 ligase activity, can bind to ubiquitin, and form a C4/treble-clef-like motif with poorly defined β-hairpins.

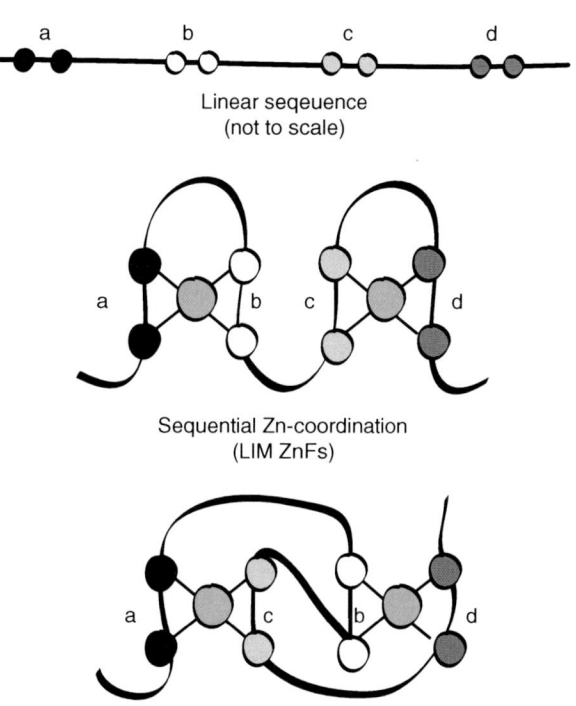

**Zinc Finger Folds and Functions, Fig. 1** Sequential versus cross-brace binding in ZnFs that coordinate two or more zinc ions. Each of (**a**) (*black circles*), (**b**) (*white circles*), (**c**) (*light gray circles*), and (**d**) (*dark gray circles*) represents rubredoxin or zinc knuckles

*B-boxes* bind two zinc ions forming a treble-clef cross-brace structure (Fig. 1) and are thought to bind proteins. They have been separated into two types based on sequence homology. B-box1 domains have the consensus: $C-X_2-C-X_{6-17}-C-X_2-C-X_{4-8}-C-X_{2-3}-C/H-X_{3-4}-H-X_{5-10}-H$ $[C_5(C/H)H_2]$. B-box2 domains have the consensus sequence found in *tri*partite *m*otif (TRIM) proteins that consist of an N-terminal RING finger, followed by one or two B-boxes and a coiled-coil domain: $C-X_{2-4}-H-X_{7-10}-C-X_{1-4}-D/C_{4-7}-C-X_2-C-X_{3-6}-H-X_{2-5}-H$. These domains are similar to canonical RING domains (which were formerly known as A-boxes) and are found in a range of proteins including transcription factors, ribonucleoproteins, and proto-oncoproteins.

*CCCH ZnF* proteins have pivotal roles in differentiation, development, and regulatory processes in eukaryotes. The TZF domain in TIS11 family proteins (which form a significant subset of CCCH-proteins) comprises two C3H ZnFs with the consensus

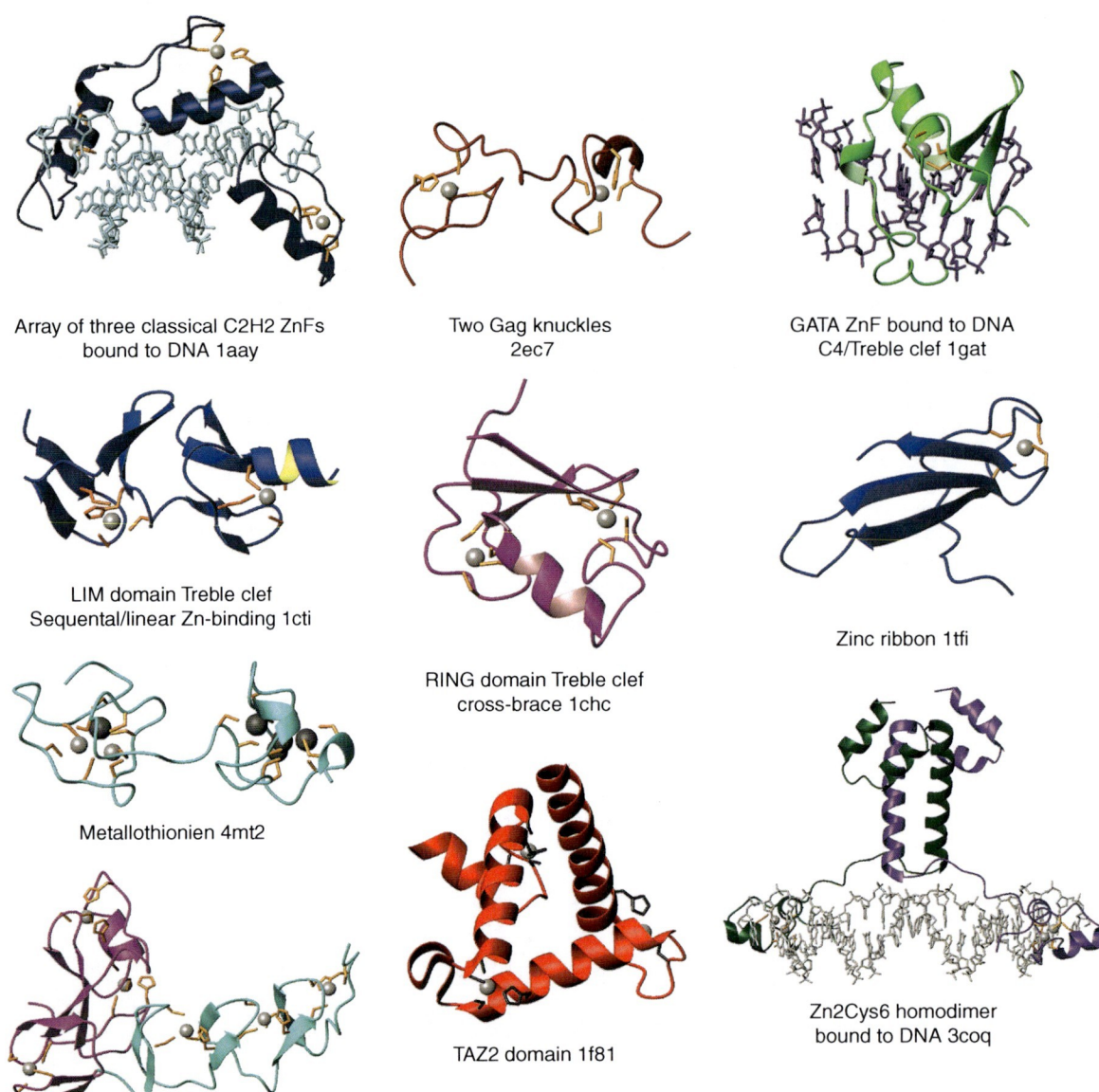

**Zinc Finger Folds and Functions, Fig. 2** Examples of ZnFs. Protein data bank accession codes are given, and figures were generated using the molecular graphics program PyMol

$C-X_8-C-X_5-C-X_3-H$ separated by an 18-residue linker. Each ZnF folds independently to form small, compact domains with little regular secondary structure: a short α-helix follows the first cysteine and a turn of $3_{10}$-helix lies between the second and third cysteine ligands. TIS11 family proteins bind mRNA; an (R/K) YKTEL motif at the N-terminus of each ZnF and the intercalation of aromatic sidechains between bases appear to make major contributions toward binding.

*Classical (C2H2 and variant C2HC) ZnFs* form the majority of ZnFs in the human genome. These ZnFs coordinate zinc via pairs of cysteine and histidine residues forming a simple ββα structure: two β-strands form a β-hairpin which is followed by an α-helix and both elements pack around a zinc ion. Many proteins that contain classical ZnFs are transcription factors that use three or more fingers to target DNA (and potentially RNA, and/or DNA/RNA heteroduplex) sequences in a highly modular fashion (Fig. 2). The ZnFs are often separated by conserved TGEKP linkers that appear to be important for binding to DNA, but the key recognition motifs are the α-helices, which bind in the major groove of DNA and contact ~3 bases per ZnF. Large protein engineering initiatives have developed arrays of modified classical ZnFs that can target many different sequences. The fusion of activation or repression domains allows these arrays to be incorporated into artificial transcription factors, whereas their fusion to endonuclease domains has led to the development of ZnF-nucleases (Sera et al. 2009). Provided a suitable and accessible unique DNA sequence can be identified, artificial transcription factors can be used to regulate the expression of target genes, and ZnF-nucleases can be used to edit genomes (including the targeted disruption of genes to generate knockout cells and animals, correct mutations, and/or incorporate new genes). ZnF-nucleases and artificial transcription factors have been a key development focus of Sangamo (a US-based biotech company). Some examples are in clinical trials for HIV, glioma, and diabetic neuropathies, and ZnF-nucleases as research tools are available commercially.

Some classical fingers can also bind to proteins. For example, the transcriptional regulator FOG-1 (friend of GATA) contains four classical C2H2 ZnFs and five variant C2HC ZnFs (which have the same structure as C2H2 ZnFs); at least five of these ZnFs can independently mediate interactions with either GATA-1 or the centrosomal protein transforming acidic coiled-coil 3. Ikaros family transcription factors use an N-terminal cluster of four ZnFs to mediate binding to DNA and a C-terminal cluster of two ZnFs to mediate homodimerization or heterodimerization with other family members. Many proteins contain a single CCHH or CCHC ZnF that are likely to mediate protein-protein interactions or interact with other biomolecules.

*CHY and CTCHY ZnFs* are two unusual domains found in the N-terminal domain (NTD) of Pirh2 (Fig. 2). They are each named for a conserved sequence motif and pack together to form the NTD. Each contains three zinc-coordination centers. In the CTY ZnF, the first two zinc ions form a cross-brace structure using CHC2/C2H2 motifs which packs up against the third C4 cluster. In the CTCHY ZnF, six short, antiparallel strands form a unique left-handed β-spiral. The zinc ions are each sandwiched between two β-hairpins or, in the case of the final zinc ion, between a β-hairpin and a loop. C2HC motifs are used for coordination, but the spacing of the loops varies. Both CHY and CTCHY are putative protein interaction domains.

*CSL/DPH-type ZnFs* are C4 zinc ribbons that form two β sheets separated by a short α-helix, and the structure is capped by a C-terminal helix. Some proteins that contain these domains are required for diphthamide (DPH) biosynthesis (a posttranslational modification of histidine found on translation elongation factor 2 from archaea and eukaryotes). CSL refers to a conserved sequence motif at the final zinc-coordinating cysteine.

*CW ZnFs* (conserved C4 zinc-coordination motif and tryptophan (W) residues): The first two cysteine residues are separated by two or four amino acids; the third and fourth cysteine residues are separated by an unusually long 6–15 residue loop. The structure forms a two-stranded β-sheet. Both structures also show three short helical elements. The solution structure of ZCWPW1 CW in complex with an H3K4me3 peptide shows that the histone peptide traverses a cleft lined by two of the three conserved tryptophan residues. CW domains are mainly found in vertebrates, higher plants, and other animals. At least some of these domains are H3K4-selective recognition modules, but their preference for the methylation states may vary.

*CXXC ZnFs* bind two zinc ions with C4/C4 motifs in a modified cross-brace-like structure; cysteines 1–3 and cysteine 8 coordinate 1 zinc ion while cysteines 4–7 coordinate the other. These domains have little secondary structure: two alpha helices and one short $3_{10}$ helix linked by two long loops. CXXC ZnFs can recognize CpG-motifs. For example, the crescent-shaped CFP1 CXXC domain fits into the major groove of CpG DNA forming an extended interface characterized by electrostatic and hydrogen bonding interactions. The N- and C-termini from the MLL CXXC ZnF largely make additional contacts with the minor groove of the DNA.

*DNA polymerase-α ZnFs* may be involved in DNA binding. The domain from the C-terminus of human DNA polymerase-α is a C4 ZnF that forms a helix-turn-helix motif in which a kinked N-terminal helix is followed by several turns and a short C-terminal helix. Zinc coordination is mediated by a $C-X_4-C$ motif which causes the kink in the first helix, whereas a $C-X_2-C$ motif spans the N-terminus of the second helix.

*FCS/MYM-ZnFs* are identified in MYM family proteins and have a conserved (KR)FCS sequence. These form an atypical C4/treble clef, in which the two pairs of zinc-ligands are separated by a loop rather than elements of secondary structure. These ZnFs are proposed to bind RNAs. FCS/MYM proteins are often involved in development; orthologues are found in all nucleocytoplasmic large DNA viruses and C-termini of recombinases from some prokaryotic transposons.

*FYVE* (Fab1, YOTB, Vac1, and EEA1) domains are approximately 70 residues in length and are found in eukaryotic proteins involved in membrane trafficking and phosphoinositide metabolism. This domain binds to phosphatidylinositol 3-phosphate [PtdIns(3)P] helping target FYVE proteins to PtdIns(3)P-enriched endocytic membranes. FYVE proteins often contain other phosphoinositide binding, protein interaction, and catalytic domains. These two zinc-coordinating ZnFs form a cross-brace structure using C4/C4 motifs (Fig. 1) that are distinguished by conserved N-terminal WXXD, central RR/KHHCR, and C-terminal RVC motifs that form the PtdIns(3)P binding site.

*GAG-knuckles* are very small ZnFs, usually comprising <20 residues and conforming to the sequence C-$X_2$-C-$X_4$-H-$X_4$-C (Table 1; Fig. 2). These domains are predominantly found in an array of two ZnFs separated by a basic sequence in nucleocapsid proteins of retroviruses (i.e., gag nucleocapsid proteins) where they bind to single-stranded RNA and are required for viral genome packaging and early infection. They are also found in the outer capsid proteins from reoviruses where they appear to modulate binding to other capsid proteins. Examples of this domain are also found in eukaryotic proteins such as RNA polymerase II where they have roles in binding to RNA and or single-stranded DNAs.

*GATA-type ZnFs* bind a single zinc ion using C4 coordination and form the canonical treble-clef fold. GATA proteins, which are named for their ability to recognize GATA-containing DNA sequences (Fig. 2), contain two GATA-type ZnFs separated by a linker; each ZnF is followed by a basic tail that contributes to DNA binding. The N-terminal ZnF contributes to binding at double GATA sites but otherwise binds GATC sequences. GATA-type ZnFs also mediate interactions with other proteins including other ZnFs (e.g., the N-terminal GATA ZnF from GATA-1 can interact with both FOG and LMO2). GATA proteins can both up- and downregulate gene expression in target cell types. The complex roles these proteins play in gene regulation can be largely attributed to the different specificities of its ZnFs for binding to both DNA and proteins. The MED1 GATA-type ZnF has an additional helical region at the C-terminus which appears to contact the major groove of the DNA, explaining the extended DNA consensus sequence observed for MED1.

*HIT domains* bind two zinc ions with C4/C2HC coordination motifs in a cross-brace ZnF architecture. The HIT domain from ZNHIT2 contains two consecutive antiparallel β-sheets followed by a C-terminal short α-helix and $3_{10}$-helix, which pack against the second β-sheet. These domains are predicted to bind proteins.

*HIV-1 integrase H2C2 ZnF* forms a TAZ-like fold in which a single zinc (or cadmium) ion is coordinated at one end of a helix-turn-helix-like fold. This domain forms dimers in solution suggesting that the domain is important for self-association.

*JAZ/dsRNA-binding ZnFs* contain a core C2H2 fold with a kinked C-terminal α-helix and can also contain an additional N-terminal helix and β-strand. JAZ ZnFs are thought to bind dsRNA or RNA/DNA hybrids but have also been proposed as protein interaction domains based on their similarity to protein-binding C2H2 finger from FOG (friend of GATA). These domains are found in multiple copies in JAZ, WIG, and TFA proteins separated by long flexible linkers and are required for nucleolar localization. A ZnF with a similar core structure is the only structured domain in ZNF593, a protein that appears to negatively modulate the DNA-binding activity of the Oct-2 transcription factor, suggesting that these fingers may also bind DNA.

*LIM (Lin-11, Isl1, and Mec3) domains* are ~50–60 residues in length and comprise two treble-clef sequential zinc-binding modules (C2HC/C3Z, where Z is C, H, D, or E) separated by a two-residue linker (Figs. 1 and 2). LIM domains are protein interaction domains. No single-binding mode exists for LIM domain proteins, but limited flexibility between zinc-binding modules within loops and between closely arrayed LIM domains may be important for mediating protein-protein interactions. LIM domain proteins fall into four groups (1–4) depending on sequence similarity and arrangement of domains. LIM proteins from all groups appear to regulate transcription, through either binding to transcription factors and/or binding to DNA through specific DNA-binding domains. Most LIM proteins from groups 2 to 4 are linked to the regulation of the cytoskeleton but can translocate into the nucleus in response to various stimuli apparently providing communication between the cytoskeleton and the nucleus.

*Metallothionein* domains bind zinc ions or other heavy metals such as Cd(II) and/or Cu(I). Mammalian metallothionein proteins comprise two domains: an α-domain with eleven cysteine residues, which coordinates four Zn/Cd ions or six Cu ions, and a β-domain with nine cysteine residues, which coordinates three Zn/Cd ions or six Cu ions. The structure of rat metallothionein bound to four Cd ions in the α-domain and two Zn and one Cd in the a β-domain reveals that each metal ion is coordinated in a tetrahedral manner by four cysteine sidechains (Fig. 2), but many of those sidechains contribute to the coordination of two metal ions such that each cluster of metal ions is held within a network of thiol groups. There are few regular secondary structural elements. It has been proposed that the domains have some flexibility, allowing them to accommodate metals of different sizes. Metallothioneins have been implicated in many biological processes including toxic metal detoxification, protection against oxidative stress, and as a metallochaperones involved in the homeostasis of zinc and copper.

*MYND ZnFs*: Myeloid translocation protein 8/Nervy/ Deformed epidermal autoregulatory factor-1 domains coordinate two zinc ions in a cross-brace topology. The basic structure of the domains is a short β-hairpin with strands coming from Loop1 and Loop2, followed by a kinked α-helix that encompasses Loop3 and the zinc-coordinating motifs that flank this loop; MYND domains are found in many proteins implicated in transcriptional regulation that often play important roles in development. The best characterized specific role of MYND domains is the binding of short proline-rich peptide regions in partner proteins. The structure of the MYND domain from AML/ETO in complex with a proline-rich peptide from SMRT/N-CoR shows the peptide binds in an extended fashion, forming a short β-strand that packs against the β-hairpin in the MYND domain.

*MYT-type ZnFs* (myeloid transcription factor) are C2HC ZnFs that bind DNA. The structure of an MYT domain from MYT1 shows essentially no regular secondary structure.

*Nanos ZnFs*: Nanos binds to the 3′ untranslated region of a messenger RNA and represses its translation and is involved in the development and maintenance of germ cells. The protein has two conserved motifs that are indispensable for its function. These Nanos fingers are similar to GAG-knuckles, coordinating zinc with C2HC motifs, but the N-terminal zinc-binding motif contains 10-residue, rather than a 2–4-residue spacer between the third and fourth zinc-coordinating residues ($C-X_2-C-X_{12}-H-X_{10}-C-X_7-C-X_2-C-X_7-H-X_4-C$), and forms two short helices rather β-structure near the N-terminus. The two ZnFs from zebrafish Nanos appear to be independent folding motifs that pack together to form a bimodal structure with a wide cleft. Conserved, often basic residues, in and around this cleft are implicated in binding to RNA.

*Nuclear hormone/steroid-receptor ZnFs*: Hormones such as vitamin D, 9-*cis* retinoic acid, and thyroid hormone diffuse into the cell and bind to ligand-binding domains on nuclear receptor or steroid receptor proteins. Loaded receptors translocate into the nucleus, where the ZnF domains from two hormone receptor molecules dimerize and bind to DNA to regulate the transcription of target genes in a ligand-responsive manner in diverse physiological processes, including control of embryonic development, cell differentiation, and homeostasis. The ZnFs contain two sequential C4/C4 zinc-coordinating motifs. The first repeat is a typical treble-clef motif, and the second repeat resembles the C-terminal half of a treble clef but lacks a first zinc-knuckle/β-hairpin motif. The first repeat recognizes DNA in a sequence-specific manner whereas the second repeat appears to be involved in dimerization. The ZnF domain may also be involved in recruiting other transcription factors.

*PHD (plant homeodomain) ZnFs* are not homeodomains but ZnFs that bind two zinc ions with C4/HC3 motifs using a cross-brace structure. These domains contain a highly conserved tryptophan in Loop3. The core PHD structure contains a single short β-hairpin comprising strands from Loop1 and Loop2, and the long loops in the PHD (Loop1 and Loop3) can vary considerably in terms of length and sequence. It was possible to modulate the lengths of, and engineer novel protein-binding activity into, Loop3 of a PHD from Mi2β. The main role of PHD domains is protein binding, but some have been shown to bind phosphoinositides (see FYVE domains). Many PHD proteins contain one or two PHD domains and are involved in chromatin regulation. PHD proteins often contain more than one chromatin-binding domain, or require an additional binding partner, to fully recognize histone modifications. They are often part of larger remodeling complexes including NuRD (nucleosome remodeling and deacetylase), NURF (nucleosome remodeling factor), HAT (histone acetyl transferase), and HDAC (histone deacetylase) complexes. Different PHDs recognize different methylated states of the arginine and lysine sidechains of histone three protein (H3) tails. They use conserved features to recognize the constant region of the H3 tail and bind R2 and K4(or K36) sidechains using two roughly parallel binding grooves that either stabilize unmethylated sidechains through networks of salt bridges and/or hydrogen bonds (and prevent the accommodation of methyl groups) or bind methylated sidechains through deep hydrophobic pockets. The binding groove residues are often contributed by the variable long loops of the PHDs. Other PHDs appear to be SUMO E3 ligases, but in many cases, it is not yet clear if a given PHD is involved in H3 recognition, SUMOylation, or another activity, such as phosphoinositide binding.

*PARP ZnFs* have a zinc-ribbon/C4 fold with an unusual spacing of ligands:$C-X_2-C-X_{12}-C-X_9-C$. The structure comprises three antiparallel β-strands plus a helical region. PARP-1 has two of these ZnFs bound that bind to DNA in a bipartite sequence-independent manner. The ZnFs contact the phosphodiester backbone and the hydrophobic faces of exposed nucleotide bases. Although the second ZnF binds to DNA with higher affinity than the first, it is the first ZnF that is absolutely required for function. PARP-1 catalyzes poly(ADP-ribosyl)ation of the acceptor proteins using NAD (+) as a substrate.

*RanBP2 ZnF* is a C4 zinc-ribbon-like fold that lacks a beta-strand compared with most members of the class. Two stacked beta-hairpins are oriented at approximately 80° to each other, sandwiching the zinc ion. zRANBP2 contains two of these ZnFs, which recognize ssRNA in a sequence-specific manner involving intercalation of a flipped tryptophan sidechain between bases.

*RING (Really Interesting New Gene) domains* coordinate two zinc ions using C3H/C4 motifs and form a cross-brace structure (Fig. 1) using the consensus $C-X_2-C-X_{9-39}-C-X_{1-3}-H-X_{2-3}-C-X_2-C-X_{4-48}-C-X_2-C$. The structure comprises a β-hairpin with strands coming from Loop1 and Loop2 and an α-helix from Loop3 (Fig. 2). Often, an additional C-terminal β-strand packs against the Loop1 β-strand to form a short antiparallel β-sheet. RING domain proteins are protein interaction domains, and many form homo- and/or heterodimers where the dimer interfaces comprise the β-sheets, often with contributions from residues that flank the RING domains. RING domain proteins are best known as E3-ubiquitin ligases that target proteins to the proteasome for degradation but can also reversibly ubiquitinate molecules (including themselves) for other functions related to transcription, gene silencing, and DNA repair. RING proteins tend to contain only a single RING domain, although some contain other RING-like domains. The proteins work either as single subunit enzymes in which the RING domain binds a ubiquitin-conjugating E2 enzyme and a separate domain in the same protein recognizes the substrate protein or act as a part of a large multisubunit E3 protein complex. In both cases, RING E3 ligases bind E2 enzymes (using a conserved face) and the substrate in close proximity, without direct contacts between the E3 ligase and ubiquitin, or the E3 RING domain and the active site of the E2 enzyme. It has been suggested that the RING E3 ligases allosterically regulate E2 activity.

*TAZ (transcriptional adaptor ZnF)-like domains*: The two TAZ domains from the CBP/p300 transcriptional regulator family (histone deacetylase proteins that are involved in chromatin remodeling) are protein interaction domains. These domains comprise four amphipathic α-helices arranged to coordinate three zinc ions through HCCC motifs that lie in loop regions at the ends of helices. In CBP-TAZ1 and TAZ2 (Fig. 2), the fourth helix is oriented in opposite directions. Several protein domains that bind TAZ ZnFs are intrinsically unfolded but become folded upon binding such that they wrap around the TAZ domain upon (e.g., the transactivation domains of hypoxia-inducible factor 1 alpha (HIF1α), Cbp/p300-interacting transactivator 2 (CITED), and p 53).

*THAP (thanatos-associated protein) domains* are C2CH/treble-clef ZnFs that conform to the sequence $C-X_{2-4}-C-X_{35-50}-C-X_2-H$ and which bind DNA. THAP proteins are found in proteins involved with transcriptional regulation, cell-cycle control, and chromatin modification.

*U1-C type ZnFs* are a C2H2-type domain in which the two Zn-coordinating histidine residues are separated by a five-residue loop. The domain is also extended at its C-terminus by two additional helices. The first and second helices are broken only by the five-residue zinc-coordination loop, and the final helix is probably loosely associated with the rest of the structure through a set of hydrophobic contacts. These ZnFs are found, sometimes in multiple copies, in several U1 small nuclear ribonucleoprotein C (U1-C) proteins, sometimes as multiple copies of this motif. The proteins bind to the pre-mRNA 5′ splice site at early stages of spliceosome assembly.

*UBP (ubiquitin-binding protein) ZnFs* are found in a small family of ubiquitin C-terminal hydrolases (deubiquitinases) that have isopeptidase-T activity (cleave isopeptide bonds between ubiquitin moieties). In the structure of IsoT/USP5, the UBP domain has C3H zinc coordination and forms a discrete fold within a larger domain. The UBD domain binds to the C-terminus of ubiquitin.

*UBR box/N-recognin ZnFs* are part of E3 ubiquitin ligases known as N-recognases, which provide the mechanistic basis for the N-end rule (the half-life of a protein is related to the identity of its N-terminal residue). The UBR box recognizes the N-terminus of target proteins. The UBR box binds three zinc ions in an unusual cross-brace-like C4/C3H/C2H2 topology. The first two zinc ions form a binuclear cluster, sharing the third cysteine (underlined) in the sequence C-$X_{24}$-C-$X_2$-$\underline{C}$-$X_{21}$-C-X-C-$X_{11}$-C-$X_2$-H. The final zinc ion is coordinated by a C-$X_2$-C motif in the first long loop ($X_{24}$) and an H-X2-H motif in the second long loop ($X_{21}$). The structure contains two antiparallel β-strands, two short helices that lie between the strands, and two long ordered loops at either end of the domain. The substrate peptide binds a preformed pocket in the UBR box.

*$Zn_2Cys_6$ binuclear cluster* (Fig. 2) conforms to the consensus sequence C-$X_2$-C-$X_6$-C-$X_{5-16}$-C-$X_2$-C-$X_{6-8}$-C (see Table 1). These ZnFs are found mainly in transcription factor proteins from fungal species and some amoeboid protozoa that are involved in biosynthetic, metabolic, and catabolic pathways. These proteins tend to bind DNA as homodimers. Each $Zn_2Cys_6$ motif recognizes a CGG half-site, with the relative orientation and spacing of the CGG half-sites contributing to binding specificity; however, some proteins bind their cognate DNA sequences as monomers or heterodimers or recognize non-CGG half-sites. The metal-binding cluster ZnF in Gal4 and related proteins is separated from a helical dimerization domain by a linker region. The CGG-recognition motif is the Zn2Cys6 cluster, which binds the major groove of the DNA, but residues from both the linker and the dimerization domain also contribute to binding.

*ZAD ZnFs* (zinc finger-associated domain ZnFs) are common in insect transcription factors. The ZAD of Grauzone forms an atypical C4/treble-clef fold. A two-stranded β-hairpin is broken by an α-helix, and the domain terminates with an extended α-helix. This domain does not bind to DNA and is thought to mediate dimer formation.

*Zinc-binding loops* are usually found as subdomains in larger proteins such as sorbitol and alcohol dehydrogenases, DNA and RNA polymerases, endonucleases, and single-stranded DNA-binding proteins. In most cases, it is not clear what specific roles are carried out by the zinc-binding loops, but in some cases, they are located at homodimerization interfaces of these larger proteins.

*ZPR1-type ZnFs* consist of a four-stranded antiparallel β-sheet stabilized by a C4 zinc-coordinating site located at one end. This ZnF motif packs against a double-stranded beta helix with a helical hairpin insertion. ZPR1 binds directly to eukaryotic translation elongation factor 1A (eEF1A), assembles into multiprotein complexes with the survival motor neuron (SMN) protein, and accumulates in subnuclear structures such as gems and Cajal bodies. The ZnF is thought to be a protein interaction domain.

*ZZ domains* coordinate two zinc ions using a cross-brace structure with C3/C2H2 motifs. An unusual feature seen in the ZZ domain from human CREB-binding protein involves the final HXH motif; the $N^{\varepsilon 2}$ atom of histidine-40 and the $N^{\delta 1}$ atom of histidine-42 coordinate the zinc ion. The first zinc cluster of the ZZ domain tends to be highly conserved, whereas the second zinc cluster shows variability in the position of the two histidine residues. ZZ domains are thought mediate protein-protein interactions. ZZ proteins have various functions, including chromatin remodeling, cytoskeletal scaffolding, and E3 ubiquitin ligase activity (the ZZ domain is thought to have a scaffolding rather than possessing the ligase activity), and membrane insertion.

## Cross-References

▶ Cadmium and Metallothionein
▶ Metallothioneins and Copper
▶ Zinc and Zinc Ions in Biological Systems

▶ Zinc Finger Folds and Functions
▶ Zinc Metallothionein
▶ Zinc Metallothionein-3 (Neuronal Growth Inhibitory Factor)
▶ Zinc, Physical and Chemical Properties
▶ Zinc Structural Site in Alcohol Dehydrogenases
▶ Zinc-Binding Proteins, Abundance
▶ Zinc-Binding Sites in Proteins

## References

Gamsjaeger R, Liew CK, Loughlin FE et al (2007) Sticky fingers: zinc-fingers as protein-recognition motifs. Trends Biochem Sci 32:63–70

Krishna SS, Majumdar I, Grishin NV (2003) Structural classification of zinc fingers: survey and summary. Nucleic Acids Res 31:532–550

Laity JH, Lee BM, Wright PE (2001) Zinc finger proteins: new insights into structural and functional diversity. Curr Opin Struct Biol 11:39–46

Matthews JM, Sunde M (2002) Zinc fingers–folds for many occasions. IUBMB Life 54:351–355

Matthews JM, Bhati M, Lehtomaki E et al (2009) It takes two to tango: the structure and function of LIM, RING, PHD and MYND domains. Curr Pharm Des 15:3681–3696

Miller J, McLachlan AD, Klug A (1985) Repetitive zinc-binding domains in the protein transcription factor IIIA from Xenopus oocytes. EMBO J 4:1609–1614

Sera T (2009) Zinc-finger-based artificial transcription factors and their applications. Adv Drug Deliv Rev 61:513–526

# Zinc Finger Targeting

▶ Zinc, Metallated DNA-Protein Crosslinks as Finger Conformation and Reactivity Probes

# Zinc Homeostasis, Whole Body

Nancy F. Krebs
Department of Pediatrics, Section of Nutrition, University of Colorado, School of Medicine, Aurora, CO, USA

## Synonyms

Equilibrium; Zinc balance; Zinc physiology; Zn

## Definitions

1. Absorbed zinc: the amount of ingested zinc that is transported from the lumen of the intestine, across the enterocyte, and taken into the circulation and distributed to body tissues over a day (total absorbed zinc (TAZ) in mg/day) or from a single meal or dose (absorbed Zn (AZ) in mg). TAZ is determined by multiplication of diet (or dose) Zn x FAZ.
2. Endogenous zinc losses: zinc from systemic sources that is excreted through gastrointestinal tract, kidneys, menstrual blood losses, semen, or integument (including via sweat).
3. Endogenous fecal zinc (EFZ): zinc that is secreted from systemic sources (e.g., enterocytes or pancreas) into the gastrointestinal tract and then excreted in the feces; some endogenously secreted zinc is reabsorbed in the distal small intestine.
4. Fractional absorbed zinc (FAZ): the fraction or proportion of zinc ingested that is absorbed; FAZ typically measured by use of isotopic (stable or radio) tracers ingested simultaneously with food; assumption is that the isotopic zinc is handled physiologically identically to the zinc in food, once the zinc in food is liberated from the food matrix during the process of digestion.
5. Total dietary zinc: the daily amount of zinc ingested with foods.

## Introduction

Whole-body homeostasis refers to the integrated systems that function to maintain the body in a relatively stable equilibrium with respect to a given nutrient, in this case zinc. Multiple adjustments and regulatory mechanisms make homeostasis possible. This review will consider the factors and processes that impact the uptake of zinc into the body, its distribution once it is absorbed into the circulation and tissues, and the routes of excretion and loss from the body.

## How Does Zinc Enter the Body? What Are the Processes of Uptake?

The primary route of entry is via foods or nutrient supplements taken into the gastrointestinal tract. Some absorption of zinc through the skin and mucous membranes can occur, but in normal circumstances, these provide only minor contributions to total body zinc.

*Sources of Dietary Zinc*: The richest sources of zinc are animal source foods, specifically organs and flesh (liver, meats, poultry, fish, etc.). Milk and dairy products and eggs are medium-rich sources of zinc. Plant foods such as unrefined grains and cereals, seeds and nuts, and legumes are moderately rich in zinc on a mg/g basis, but they generally also contain high amounts of a chemical that binds the zinc: phytate (myoinositol hexaphosphate). This compound is the storage form of phosphorus for plants. As such, it contains up to 6 phosphate groups, each with a negative charge which effectively binds dietary cations; the inositol penta- and hexaphosphate forms particularly impact zinc absorption. Although some animals' gastrointestinal tracts contain phytases, enzymes that can cleave/hydrolyze the phytate molecule, humans do not have these enzymes. The result is that the zinc contained in these plant foods can become bound to the phytate in the lumen of the gastrointestinal tract during the process of digestion. The effects of the zinc-phytate interaction on zinc homeostasis are discussed further below. Refined grains (without enrichment or fortification) and plant foods such as fruits and vegetables are generally quite low in zinc and contribute little to daily intake of this mineral.

Once ingested in the diet, in the process of digestion, zinc is released from food components and is absorbed in the proximal small bowel, primarily the duodenum and proximal jejunum. Although absorption can occur along the entire intestine, if the small bowel is intact, at typical dietary intakes, only a very small percent of ingested zinc will reach the colon for absorption. Zinc is absorbed by both active (saturable transporters) and passive (non-mediated) processes. The protein transporters that enable zinc to be taken into the enterocytes (the cells lining the gastrointestinal tract through which nutrients are absorbed) have been identified. The sum effect of these is the uptake of zinc from the lumen at the apical aspect of the enterocyte. One apical transporter, ZIP4 (SLC39A4), is upregulated in zinc deficiency, and a mutation of this transporter protein is the underlying defect in the classic heritable zinc deficiency condition of acrodermatitis enteropathica (▶ Zinc Deficiency). There are multiple zinc transporters: ones that allow the zinc into the cell (e.g., ZIP4), others that traffic the zinc among organelles inside the cell, and still others that transport it out of the cell into the circulation (▶ Zinc Cellular Homeostasis).

*Zinc Absorption*: Current understanding of the regulation of zinc absorption at the whole-body level – the sum of processes by which dietary zinc gets into the circulation for distribution in the body – is that it is primarily controlled by 2 factors, which explain over 80 % of the variability in zinc absorption. These 2 factors include the *amount of zinc* ingested and the *amount of phytate* concurrently ingested. Notably, the zinc "status" of the host does not appear to alter the amount of zinc that is absorbed. For example, even if a person is zinc deficient, for a given amount of ingested zinc, the amount absorbed will be similar to the amount absorbed by a person that is not zinc deficient. At very low levels of zinc intake, the fraction or percent of zinc absorbed will be quite high, but the fraction will steadily decrease as the amount of ingested zinc increases. However, as the amount of zinc ingested increases, even as the fraction absorbed declines, the total amount that is absorbed is increased. At very high doses of ingested zinc, some leveling off of absorption occurs, and the amount of absorbed zinc does not continue to rise in a linear fashion. At high doses, as in with use of supplements, passive (or non-mediated) absorption plays a greater role and will allow continued uptake. Mathematical models of zinc absorption, based on studies in human subjects, predict that the maximum amount of daily zinc absorbed from diets for adults is approximately 7 mg. The implication of this "saturation response" model of zinc absorption is at least crude protection against excessive zinc absorption. Up and down regulation of absorption can apparently occur quite quickly. A tripling of zinc intake was associated with reduced absorption efficiency within 24 h. This regulation of absorption, combined with the adjustable quantities of zinc secreted and excreted in the intestine, contrasts with iron homeostasis (and that for some other micronutrients), for which absorption is more finely tuned, and excretory pathways are very limited. For diets/meals that are phytate-free, typical fractional absorption is ~0.4 (i.e., absorption of 40 % of zinc ingested). Zinc that is not absorbed from the diet (typically about 60–70 % of what is ingested) is excreted from the gastrointestinal tract in the feces. Reviews of these homeostatic mechanisms are presented in King and Cousins (2013) and Hambidge et al. (2010) below.

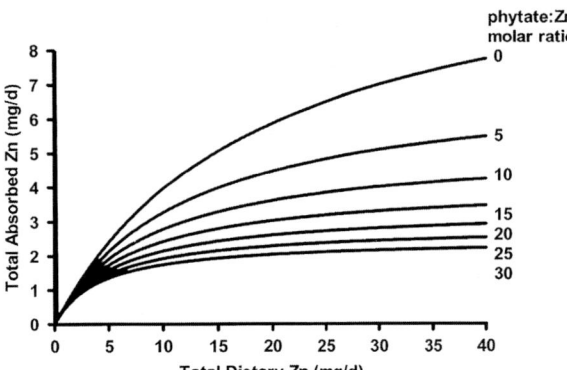

**Zinc Homeostasis, Whole Body, Fig. 1** Curves showing predicted total daily absorbed zinc by the model versus dietary zinc intake, for various phytate to zinc molar ratios

The effect of phytate in the diet is profound and is the major dietary effector of zinc absorption (Fig. 1, Miller et al. 2007). This compound, common in plant-based diets, has been shown in many studies in animal models and humans to significantly interfere with zinc absorption. Mathematical modeling, derived from data from >70 human studies in the literature, predicts a progressively inhibitory effect of dietary phytate. In older literature, the concept of a "threshold effect" was considered likely. More recent data, however, indicate that any phytate in the diet will have an effect, but at low intakes, the impact will be modest. Most commonly, the amount of phytate relative to zinc is quantified in terms of the phytate to zinc molar ratio. This ratio can be calculated as follows: (phytate content of food in mg/660)/(zinc content of food in mg/65.4); 660 and 65.4 represent the molecular and atomic micromolar weights of phytate and zinc, respectively. Ratios <15 will have a measurable but modest effect, whereas ratios >15, which are associated with whole grains, legumes, and seeds, are likely to markedly limit the absorption of dietary zinc.

For example, if 60 % of dietary zinc is obtained from maize, as is common in traditional diets in Central America, the daily amount of zinc consumed might be approximately 7.5 mg/day (4.5 mg/day from maize). Maize is particularly high in phytate, and for the intakes described here, measured daily phytate intake was 3,000 mg/day; the phytate to zinc molar ratio would thus be ~40. Measured fractional absorption was 0.217, compared to FAZ of ~0.32 predicted by the model for a similar zinc intake from a phytate-free diet (Fig. 1). More importantly, this reduction in FAZ (from low to high phytate to zinc molar ratio) translates to a reduction in the amount of absorbed zinc from 2.5 to 1.8 mg/day, approximately a 30 % reduction. Practically, these calculations illustrate how the total amount of daily zinc intake represents only part of the picture in determining the adequacy of that intake; the amount of phytate critically impacts the availability of that zinc for absorption. In many developing countries, where a large percentage of energy intake is from grains or legumes that are high in phytate and the phytate to zinc molar ratio for the diets averages above 20, a total zinc intake that may be "just adequate" becomes "inadequate" when the effect of the phytate on absorption is considered. This effect of phytate is a major factor contributing to designations of bioavailability used for determining recommended dietary zinc intakes; i.e., diets with high phytate levels will be designated low bioavailability and will require a higher total zinc intake to compensate.

Systematic studies of absorption of zinc from oral supplements are quite limited, but data suggest that absorption may be more efficient than when zinc is consumed in food and, as noted above, may be more impacted by passive absorption capacity. This is especially the case when the supplement is in a liquid form. Studies of zinc absorption from zinc lozenges, typically taken for short-term treatment of upper respiratory infections and pharyngitis, are even more limited (▶ Zinc and Immunity). Some investigators have used the increase in plasma or serum zinc after a large oral dose of zinc as an indicator of bioavailability. Although this method may provide useful comparisons among different forms of zinc, the nonphysiologic nature of the study design warrants caution in interpretation.

### Zinc Metabolism

Once zinc enters the circulation via the portal vein, it is distributed to various body pools and tissues. The liver is a major reservoir of zinc, especially of zinc that is used for metabolic processes, including protein synthesis. Other tissues that contain zinc that is easily mobilized for metabolic functions include the bone marrow, pancreas, kidney, brain, blood cells, and some types of muscle tissue (▶ Zinc Cellular Homeostasis). The skeleton contains a very large proportion of total body zinc. The zinc in bone turns over, or circulates, much more slowly and is thought to be less involved in short-term, day-to-day metabolic processes, although at least a small percentage (10–20 %) of the zinc in bone may be more readily mobilized. The zinc that is considered to be readily available for active metabolic processes is estimated to total only approximately 10 % of the total amount of zinc in the body. A portion of the absorbed zinc will be excreted as part of homeostatic regulation. The secretion into the gastrointestinal tract represents (in addition to absorption) the major process by which homeostasis is regulated. This is described in the following section.

**Zinc Excretion**

In contrast to absorption, which does not seem to be tightly regulated according to the person's zinc needs or status, the amount of zinc that is excreted seems to be more regulated and responsive to the individual's chronic zinc absorption and to host zinc status. Hence, if dietary zinc intake is chronically only marginally adequate or frankly inadequate, the body will conserve the amount of zinc that is lost from the intestine. The major source of zinc excretion is the pancreas, via the exocrine pancreatic secretions, and from the intestine via trans-epithelial flux. Biliary secretions also contain some zinc but these are thought to be a minor source. Although transporters have been localized to the pancreas, the actual mechanisms and their regulation are not well understood at this point (▶ Zinc Signal-Secreting Cells).

Some fraction of the zinc that is secreted into the intestine is reabsorbed in the more distal small bowel, while the remainder of the endogenously secreted zinc is excreted in the feces. The amount of zinc that is contained in these pancreatic-biliary secretions and that is excreted daily in the feces can vary by severalfold, depending on the zinc "status" of the individual. Many studies in both animal models and humans have documented a sustained "conservation" of endogenous intestinal zinc losses under conditions of dietary zinc restriction and zinc depletion of the host. In contrast, if a person has a generous daily intake and absorption, more of the zinc in the feces will be from the endogenously secreted zinc (in addition to the unabsorbed dietary zinc).

Our understanding of the quantitative aspects of endogenous intestinal zinc are derived from isotope studies which allow distinguishing zinc in the feces that is derived from endogenous sources versus unabsorbed dietary zinc. Available data suggest that the main control of endogenous intestinal zinc ultimately excreted is the amount that is secreted, although regulation of reabsorption is also possible. Excessive losses of endogenous intestinal zinc have been documented in conditions with fat malabsorption, which suggests impaired reabsorption. Similarly, losses exceed absorbed zinc with acute diarrhea, a phenomenon that is thought to contribute to the etiology of zinc deficiency in young children in developing countries who experience recurrent diarrhea and often have diets only marginally adequate in zinc (▶ Zinc Deficiency). A high phytate intake has also been postulated to interfere with reabsorption of endogenously secreted zinc, but empirical data to support this are very limited.

Other routes of loss of zinc are via the kidney, menstrual blood loss, semen, and the integument (skin), including sloughed cells and sweat. Under most circumstances, these are quantitatively relatively minor losses and do not seem to be altered in response to host "status," except perhaps at the extremes of intake. Specifically, urinary zinc output is low and changes modestly over a wide range of dietary zinc intake. With moderate to severe zinc deficiency, the kidney is able to conserve daily zinc losses in the urine (▶ Zinc Deficiency).

## Homeostasis in Physiologic States and Special Circumstances

*Infancy (Including Prematurity) and Early Childhood:* Zinc requirements are relatively high in conditions of rapid growth, such as in infants and young children. Evidence available to date supports homeostatic mechanisms similar to those described above in adults, although the magnitude may differ, apparently based in part on the absorptive capacity of the immature gastrointestinal tract. Normal breast-fed infants demonstrate very efficient absorption of the zinc available in human milk, and they maintain zinc retention adequately positive to support growth, (especially of lean tissue) by conservation of intestinal endogenous zinc. The zinc in human milk is frequently cited as being highly bioavailable, and indeed human milk contains no phytate. Much of the favorable fractional absorption of zinc from human milk, however, seems to be attributable to the very modest quantities present. Infant formula contains much higher concentrations of zinc, and fractional absorption of that zinc is lower, but the actual amount of zinc absorbed by a formula-fed infant is severalfold higher than that of breast-fed infants. Thus, normal infants' zinc absorption obeys the principles of saturation response. The higher total amount of absorbed zinc in formula-fed infants is accompanied by a higher endogenous fecal zinc, but net retention averages higher for formula-fed compared to breast-fed infants (Hambidge et al. 2006).

Zinc concentrations in human milk decline strikingly from birth through the first 3–4 months postpartum (▸ Zinc Signal-Secreting Cells). This decline is physiologic and not altered by supplementation of the mother or by maternal dietary zinc intake. The literature now convincingly demonstrates that milk zinc concentrations are independent of maternal intake and do not differ among women in many different settings and with different chronic zinc intakes and status. Although the zinc content of human milk in early lactation is generous relative to estimated infant requirements, by approximately 5–6 months postpartum, the zinc concentration is sufficiently low that even 100 % absorption (combined with efficient conservation of endogenous losses) would not be enough to meet the physiologic requirements. Thus, by about 6 months of age, breast-fed infants are reliant on complementary foods to supplement their intake of zinc from milk to meet requirements. In settings where complementary foods are predominantly unfortified plant-based staples, such as grains and legumes, the risk of developing zinc deficiency is relatively high (▸ Zinc Deficiency). This situation is further exacerbated by increased losses of endogenous zinc resulting from diarrhea (Brown et al. 2009; Krebs and Hambidge 2007).

Premature infants are born with relatively low tissue levels of zinc, and their rapid postnatal growth results in a high zinc requirement. Furthermore, immature kidney function and possibly immature gastrointestinal processes result in high urinary and intestinal endogenous losses, respectively. They are thus at high risk for developing zinc deficiency if they do not receive supplemental zinc either from fortified human milk or formulas designed for the preterm infant. The clinical literature contains numerous case reports of severe zinc deficiency in preterm infants, evidence of their susceptibility (▸ Zinc Deficiency).

*Pregnancy and Lactation:* The additional requirement for pregnancy is modest, especially during the first half of pregnancy. With commonly used methods of measuring homeostasis, increases in absorption efficiency and/or reduction in endogenous losses have not been consistently detected. This may reflect the fact that daily increased requirement is small, and the methodologies may not be sensitive enough to detect subtle changes. Early lactation poses a higher demand on maternal homeostasis, by which she must accommodate zinc losses into the milk of 2–3 mg/day for several weeks. From 3 months postpartum onward, zinc losses in milk are 1 mg/day or less. The homeostatic responses to lactation include an increase in zinc intake, increased efficiency in absorption, relatively reduced endogenous intestinal zinc, and possibly mobilization of zinc from bone (King and Cousins 2013; Hess and King 2009).

## Cross-References

▸ Zinc and Immunity
▸ Zinc Cellular Homeostasis
▸ Zinc Deficiency
▸ Zinc Signal-Secreting Cells

## References

Brown KH, Eagle-Stone R, Krebs NF, Peerson JM (2009) Dietary intervention strategies to enhance zinc nutrition: promotion and support of breastfeeding for infants and young children. 2nd IZiNCG technical document. Food Nutr Bull 30:144–171

Hambidge KM, Krebs NF, Westcott JE, Miller LV (2006) Changes in zinc absorption during development. J Pediatr 149(suppl 3):S64–S68

Hambidge KM, Miller LV, Westcott JE, Sheng X, Krebs NF (2010) Zinc bioavailability and homeostasis. Am J Clin Nutr 91(supp):1478S–1483S

Hess SY, King JC (2009) Effects of maternal zinc supplementation on pregnancy and lactation outcomes. 2nd IZiNCG technical document. Food Nutr Bull 30:60–78

King JC, Cousins RJ (2013) Zinc. In: Ross AC, Cabellero B, Cousins RJ, Tucker KL, Ziegler TR (eds) Modern nutrition in health and disease, 11th edn. Lippincott, Williams & Wilkins, Baltimore

Krebs NF, Hambidge KM (2007) Complementary feeding: clinically relevant factors affecting timing and composition. Am J Clin Nutr 85(suppl):639S–645S

Miller LV, Krebs NF, Hambidge KM (2007) A mathematical model of zinc absorption in humans as a function of dietary zinc and phytate. J Nutr 137:135–141

# Zinc in Alzheimer's and Parkinson's Diseases

Scott Ayton, David I. Finkelstein, Robert A. Cherny, Ashley I. Bush and Paul A. Adlard
The Mental Health Research Institute, The University of Melbourne, Parkville, VIC, Australia

## Synonyms

Zinc; Zn

## Definition

The brain has a requirement for zinc in numerous functions. Disorders such as Alzheimer's and Parkinson's diseases could be contributed by zinc dyshomeostasis.

## Introduction

As the role of zinc in neurobiology has been elucidated in recent years ▶ Zn and Zn2+ ions in biological systems, it has become apparent that this "trace metal" may be a potentiating, if not causative, agent in Alzheimer's disease (AD), as well as possibly having a role in the pathogenesis of other neurodegenerative diseases such as Parkinson's disease (PD). This entry will review the role of zinc in both AD and PD.

## Alzheimer's Disease

Alzheimer's disease is the most common neurodegenerative disorder and is the leading cause of dementia in the community, affecting approximately 2% of the population of industrialized countries (~12 million people). The etiology and pathophysiology of AD have been extensively studied, and while there is no definitive understanding of the precise events that precipitate and potentiate the disease, the basic features of AD are well known. The AD brain is characterized by the presence of extracellular deposits of an insoluble fibrillar metalloprotein, ß-amyloid (Aß), and the intracellular accumulation of neurofibrillary tangles, whose primary constituent is abnormally phosphorylated forms of the microtubule-associated protein tau. Other macroscopic manifestations include a thinning of the cortical gray matter, enlargement of the ventricles, and generalized atrophy of the brain. Microscopic changes include a loss of synaptic connections, gliosis, and the presence of abnormal neurites throughout the neuropil. Taken together, this pathology is believed to drive the clinical presentation of AD, which typically involves a progressive decline in cognition and ultimately a loss of all higher order executive functioning.

### Altered Distribution of Zinc in Alzheimer's Disease

Many of the features of AD may be mediated, at least in part, by a failure in metal ion homeostasis. In this regard, a number of studies described below have investigated the zinc content of multiple different tissues in AD, including the CSF, blood, and brain.

#### Blood and CSF Zinc Levels

The zinc levels in AD blood (both serum and plasma) have been variously reported to be either elevated (Rulon et al. 2000), unchanged (Molina et al. 1998; Ozcankaya and Delibas 2002), or decreased (Hullin 1983; Baum et al. 2010; Brewer et al. 2010; Vural et al. 2010) relative to matched healthy controls. The source of this disparity is not known, but may be related to patient demographics. Gonzalez and colleagues (1999), for example, examined serum zinc levels in 51 AD subjects and 40 controls and found that differences in zinc levels were only detectable between groups when stratifying by ApoE4 genotype (ApoE4 positive: n = 34 AD; n = 9 control), with the AD group having higher levels.

While a similar phenomenon may account for the variation observed in reports on CSF zinc levels in AD, these studies have also suffered from often being underpowered and lacking proper controls. Hershey and colleagues (1983), for example, found no difference in CSF zinc levels between 33 cases of dementia of the Alzheimer type and 20 non-age-matched cases without neurological disease (the cohort only included 7 AD and 8 control cases verified at autopsy). A similar finding was observed in a more recent study (Sahu et al. 1988). In contrast, Molina and colleagues (1998) examined AD (n = 26) and control (n = 28) patients, while Kapaki and colleagues (1989) examined a smaller cohort of patients (n = 5 AD and n = 28 non-age-matched controls) which both found decreased zinc levels in the CSF of AD patients.

Thus, there is no clear consensus on whether AD patients have altered zinc levels in the periphery. Unfortunately, a similar lack of consistency is found when reviewing reports on brain zinc levels in AD.

Brain Zinc Levels

Early reports suggested that there was no difference in brain zinc levels between AD and controls, and this was subsequently confirmed in a whole-tissue analysis of frontal cortex. However, when tissues in this latter study were fractionated, it became apparent that there was a significant decrease in zinc levels in the nuclear fraction of AD cases (Wenstrup et al. 1990). A number of other reports have similarly shown AD-related decreases in zinc levels in the neocortex (see Adlard and Bush 2006) and in the superior frontal and parietal gyri, the medial temporal gyrus and thalamus (Panayi et al. 2002), and the hippocampus (Corrigan et al. 1993; Panayi et al. 2002). In contrast, there are numerous reports suggesting that zinc levels are in fact increased in the AD brain, as compared to healthy age-matched controls. Elevated zinc levels have been reported in the amygdala (Thompson et al. 1988; Samudralwar et al. 1995; Deibel et al. 1996; Adlard and Bush 2006), hippocampus (see Adlard and Bush 2006), cerebellum (see Adlard and Bush 2006), olfactory areas (including olfactory bulb, olfactory tract, and olfactory trigone) (Samudralwar et al. 1995), and superior temporal gyrus (Religa et al. 2006).

This disparity may arise from issues noted in the previous section but may also occur from other methodological issues. Fixation of human brain, for example, has been shown to significantly decrease brain zinc levels/detection (Schrag et al. 2010). Thus, bulk tissue analyses may continue to be confounded by a variety of issues that preclude an accurate picture of zinc status in the AD brain from being delineated. This may, at least in part, be remedied by recent studies that have undertaken a more spatially detailed approach to the assessment of cerebral zinc levels in the AD brain.

Plaque and Neuropil Zinc Levels

A microparticle-induced X-ray emission analysis of zinc load within the plaques and surrounding neuropil in the amygdala was undertaken by Lovell et al. (1998) (see Adlard and Bush 2006). This study demonstrated that, as compared to control neuropil, the zinc levels in the AD patients were elevated by twofold to threefold in the neuropil and threefold to fourfold in the plaques themselves. These data are consistent with other histochemical (reviewed in Adlard and Bush 2006), autometallographic tracing (Stoltenberg et al. 2005) and synchrotron-based infrared and X-ray imaging (Miller et al. 2006) analyses. Taken together, these data strongly support the notion that there is both a miscompartmentalization of zinc within the AD brain and a focal accumulation of zinc within plaques.

The cause of this zinc imbalance remains unknown, but is likely to involve a dysregulation in a number of key protein families involved in the regulation of cellular zinc levels ▸ Zinc cellular homeostasis, including the Zrt-, Irt-like protein (ZIP), zinc transporter (ZnT), and metallothionein proteins ▸ Zinc binding sites and proteins. A number of these show brain region-specific alterations (both increases and decreases) in mild cognitive impairment and preclinical, early- and late-stage AD (see Adlard and Bush 2006; Smith et al. 2006; Lyubartseva et al. 2010). The role of these proteins has been extensively reviewed elsewhere and will not be discussed here (Lovell 2009).

The Role of Zinc in AD Pathology

Zinc may be involved in, or modulated by, numerous aspects of the AD cascade. The Aß that comprises the extracellular plaque, for example, is derived from a larger protein, the amyloid precursor protein (APP), which has specific binding sites within its amino-terminal ectodomain for metals such as zinc (reviewed in Adlard and Bush 2006). APP may thus be involved in maintaining zinc homeostasis. Furthermore, the processing of APP to generate varying length Aß species involves the coordinated activities of a number of secretases, which are known to interact with, and be modulated by, metal ions such as zinc (for review, see Adlard and Bush 2006). Once Aß is generated, it too is able to bind zinc (the obligatory zinc binding site has been mapped to region 6–28 of Aß), with potentially three zinc ions bound simultaneously via the histidines located at positions 6, 13, and 14 (Damante et al. 2009). The aggregation of Aß requires zinc binding to the histidine at position 13 (see Adlard and Bush 2006), and furthermore, zinc binding can obscure the proteolytic cleavage site (reviewed in Adlard and Bush 2006) and subsequently inhibit the degradation of Aß by matrix metalloproteases (Crouch et al. 2009). While these are the principle points of intersection between zinc metabolism and Aß pathology, this laboratory has also recently discovered that APP functions as an iron-export ferroxidase whose activity is specifically inhibited by zinc (Duce et al. 2010). Thus, a disturbance in zinc homeostasis may contribute to deficits in ferroxidase activity and subsequent iron accumulation.

In addition to APP and Aß, it is also known that zinc is found within tangle-bearing neurons (see Adlard and Bush 2006) and may be involved in the progression of the tau pathology present in the AD brain. Zinc, for example, coordinates to tau monomers with moderate affinity (via bridging between Cys-291 and Cys-322 as well as two histidine residues) and can induce both the fibrillization and aggregation of the protein (Mo et al. 2009). In addition, zinc can affect the translation of tau and also modulate the activities of glycogen synthase kinase (GSK)-3ß, protein kinase B, extracellular signal–regulated kinase, and c-Jun N-terminal kinase to then also affect the phosphorylation of tau (An et al. 2005; Pei et al. 2006). Together with potential roles for zinc in modifying the conformational state of tau (Boom et al. 2009), these data support the notion that zinc may contribute to the formation of neurofibrillary tangles.

In addition to the effect of zinc on these hallmark AD pathologies, there are numerous other proteins of potential relevance to AD that may also interact with zinc, but these will not be discussed here.

## Zinc as a Potential Therapeutic Target for AD

This section will outline preclinical and clinical studies that provide insight into the role of zinc ion homeostasis in AD. There are numerous studies that more broadly address the role of metal ion dyshomeostasis (primarily copper, zinc, and iron) in the pathogenesis of AD. Principle among these are a series of studies generated from this laboratory investigating the metal chaperone, PBT2. This compound, which is believed to restore metal ion homeostasis in the brain and subsequently then affect multiple cellular cascades crucial to the metabolism of Aß and to cognitive function, has proven to be efficacious in a number of different transgenic mouse models of AD and also in a phase 2a human clinical trial (see Adlard et al. 2011). While such studies are important in validating this therapeutic approach, these are not discussed here as such compounds have not been linked to an effect specifically on zinc.

### Preclinical Studies

Lee and colleagues (2002) (see Adlard and Bush 2006) crossed the Tg2576 mice with ZnT3 KO animals, which have a specific loss of vesicular zinc in glutamatergic synapses ► Zinc release from synapses. The resulting double mutant animals showed a marked reduction in Aß burden, supporting a role for synaptic zinc in the metabolism of Aß. In contrast, a recent study examined the effect of chronic (5 and 11 months) dietary supplementation with zinc in APP transgenic mice (TgCRND8 and Tg2576 mice, respectively) and also found consequent reductions in Aß load (Linkous et al. 2009). While seemingly contradictory to the previous study, it is consistent with work from Stoltenberg and colleagues (Stoltenberg et al. 2007) who demonstrated that APP/PS1 animals maintained on a zinc-deficient diet for 3 months had a significant elevation in plaque burden.

### Clinical Studies

A recent clinical trial utilizing *reaZin* (a dietary supplement containing 150-mg elemental zinc) was completed by Adeona Pharmaceuticals. This study demonstrated that patients in the zinc supplement group (n = 29) had elevations in serum zinc levels, as compared to the placebo group (n = 28). While cognitive scores did not differ between treatment groups, there were a number of trends to improved performance on the ADAS-cog, MMSE, and the Clinical Dementia Rating Scale (sum of boxes subscore) in the zinc treatment group.

### Summary

While these cumulative data contain inconsistencies, this is perhaps indicative of the complexity of the metallome and the difficulties encountered when trying to provide an accurate snapshot of brain biochemistry in an age-related neurodegenerative disease. However, despite the lack of consensus, these studies do highlight the notion that the AD neurodegenerative cascade is sensitive to alterations in zinc levels and suggest that the modulation of zinc may represent a therapeutic target in AD.

## Parkinson's Disease

PD is the second most common neurodegenerative disease after AD and is characterized by a degeneration of neurons in the substantia nigra (SN) and a consequent loss of dopamine. This loss of function in the SN results in the classical PD phenotype of motor deficits (manifesting as rigidity and bradykinesia) that can be ameliorated by the oral administration of the dopamine precursor L-dopa. Although dopamine therapy provides profound symptomatic relief of these motor symptoms, the treatment is not disease modifying and can result in the development of debilitating dyskinesias that often preclude ongoing treatment at more advanced stages of the disease. Other neuroanatomical sites are also affected in PD, as suggested by the varied distribution of Lewy body (LB) inclusions throughout the brain.

In addition to these gross changes, the SN is also characterized by cellular changes, including increased lipid peroxidation, oxidative damage to DNA, and lowered levels of the reduced form of glutathione. These features are all indicative of elevated levels of oxidative stress and likely contribute to the neurodegeneration present in the PD brain. Thus, zinc, as a potential effector or respondent to cellular oxidative stress, warrants investigation for its role in the pathogenesis of PD.

### Zinc Levels in Parkinson's Disease Tissue

The status of zinc levels in PD tissue has been inconsistently reported and remains unclear (Table 1). Early studies by Dexter and colleagues (Dexter et al. 1989, 1991) suggested that there was an elevation in zinc levels in the SN and caudate of PD patients. This initial report, however, was not confirmed in subsequent study that utilized X-ray microanalysis (Hirsch et al. 1991) (a more recent rapid scanning X-ray fluorescence study has been conducted, but is not included here as it only examined two brains). Unpublished findings from this laboratory (Ayton et al. unpublished) are in agreement with Dexter et al., showing an approximate 40% increase in zinc levels in the PD SN (n = 9 control and n = 9 PD; assessed using inductively coupled plasma mass spectrometry (ICPMS)), with no difference in cortex (data not shown).

The level of zinc in PD CSF has been interrogated twice, with contradictory findings. Molina et al. (1998) reported an approximate 40% decrease in CSF zinc levels in PD patients, whereas more recently, Hozumi et al. (2011) reported an almost threefold increase in CSF Zn levels in PD. The earlier report, however, may have been influenced by a disproportionately high variance in the control patients. Findings from this laboratory (Ayton et al. unpublished) show no difference between control (n = 7) and PD (N = 7) CSF (ICPMS analysis; data not shown).

Zinc has also been measured in the serum of PD patients, with three studies reporting no difference between levels found in PD and control patients (Jimenez-Jimenez et al. 1992; Molina et al. 1998; Kocaturk et al. 2000). A more recent publication, however, reported a modest but significant decrease in serum zinc levels in PD (Brewer et al. 2010). Other studies examining zinc levels in urine and hair have found no differences between control and PD subjects.

Given the limited number of investigations and indeed the conflicting findings on zinc levels in PD tissue, further studies are required to determine whether zinc levels are altered in this disease.

### Role of Zinc in Parkinson's Disease Pathology

Despite the unresolved question of whether zinc levels are altered in PD tissues, there have been several investigations into a potential role for zinc in disease pathogenesis. PD, which is characterized by elevated levels of oxidative stress in the brain, is typically modeled in rodents by intoxication with 6-OHDA or MPTP. 6-OHDA is administered by local injection into the striatum; the toxin enters SN neurons that express the dopamine transporter and induces a selective lesion via an elevation in oxidative stress. MPTP is administered peripherally, enters the brain, and is converted to the toxic MPP+ metabolite by glia. MPP+ then enters dopaminergic neurons via the dopamine transporter, where it inhibits complex I of mitochondria, induces oxidative stress, and causes selective death of dopaminergic neurons. In both of these models, there are reports that zinc accumulates following intoxication. In the case of the 6-OHDA lesions, there is an increase in zinc in the SN and lenticular nuclei (Tarohda et al. 2005), while MPTP lesions result in high levels of cytosolic zinc in degenerating neurons (Lee et al. 2009). It has been suggested that this zinc accumulation is a protective response, as zinc protects against 6-OHDA-induced oxidative stress in ex vivo models (Mendez-Alvarez et al. 2002).

There are several mechanisms postulated for the putative protective role of zinc in PD. Zinc is requisite for Cu/Zn superoxide dismutase (SOD) activity ▶ Zinc in superoxide dismutase: Role, and increased activity

**Zinc in Alzheimer's and Parkinson's Diseases, Table 1** Reports of Zinc in Parkinson's disease

| References | Dexter et al. (1989) | Dexter et al. (1991) | Hirsch et al. (1991) | Jimenez-Jimenez et al. (1992) | Molina et al. (1998) | Kocaturk et al. (2000) | Forte et al. (2005) | Gh Popescu et al. (2009) | Brewer et al. (2010) | Hozumi et al. (2011) |
|---|---|---|---|---|---|---|---|---|---|---|
| Technique | ICPMS | ICPMS | X-Ray microanalysis | AAS | AAS | AAS | ICPMS | RS-XRf | AAS | ICPMS |
| No. control | 34 | 34 | 7 | 39 | 37 | 24 | 17 | 1 | 29 | 15 |
| No. PD | 27 | 27 | 5 | 39 | 37 | 30 | 81 | 1 | 30 | 20 |
| Cortex | 94.9 | 94.9 | | | | | | | | |
| Caudate | **135.1** | **135.1** | | | | | | 53.8 | | |
| Putamen | 92.2 | 92.2 | | | | | | 69.3 | | |
| Globos Palidus | 96.4 | 96.4 | | | | | | 80.0 | | |
| SN compactor | **154.1** | **154.1** | 95.0 | | | | | 87.5 | | |
| Cerebellum | 112.0 | 112.0 | | | | | | | | |
| CSF | | | | | **58.8** | | | | | |
| Serum | | | | 102.9 | 106.5 | 97.8 | | | **93.6** | |
| Urine | | | | | | | | | 115.8 | |
| Hair | | | | | | | 108.4 | | | 273.6 |

Numerical values represent zinc in PD tissue as percentage of control. Bold indicates significant difference. *ICPMS* inductively coupled plasma mass spectrometry. *AAS* atomic absorption spectroscopy, *RS-XRF* rapid screen x-ray fluorescence, *CSF* cerebral spinal fluid of this enzyme is likely to be protective against oxidative stress and may ameliorate degeneration. To this end, it has been reported that there is elevated Cu/Zn SOD activity in PD plasma (Kocaturk et al. 2000); however, there is currently no evidence for there being an alteration in Cu/Zn SOD activity in degenerating neurons burdened by redox chemistry. Thus, a role for elevated zinc in providing neuroprotection by oxygen radical quenching has not been established in PD.

Dopamine itself has also been implicated in disease pathogenesis and is capable of redox chemistry. Deamination of dopamine by monoamine oxidase, or by reaction with iron, releases hydrogen peroxide that can cause oxidative stress. Dopamine also causes the aggregation of the Lewy body protein, alpha synuclein, which stabilizes a toxic protofibrillar species. Methamphetamine intoxication has been used as model for dopamine toxicity as it inhibits the sequestration of dopamine into synaptic vesicles and results in a toxic rise in cytosolic dopamine. Pretreatment with zinc has been shown to protect against methamphetamine intoxication of neuronal cultures (Ajjimaporn et al. 2005), possibly by a fortification of antioxidant defenses, the rescue of alpha synuclein overexpression, and/or the delivery of zinc to metallothionein (Ajjimaporn et al. 2008), whose synthesis can be induced by zinc. Metallothionein may be an important player ▶ Zinc Metallothionein, as it binds both copper and zinc with high affinity and acts as a free radical scavenger to then promote the antioxidant defenses of the brain. Thus, metallothionein may mediate the protective effects of zinc in PD models. Metallothionein is also a major cellular sensor of oxidized dopamine. Oxidized dopamine species react readily with Zn–sulfhydryl clusters on metallothionein, resulting in liberation of zinc to the cytosol. Elevated cytosolic zinc can then activate Cu/Zn SOD and, by activating metal transcription factor-1, induce expression of metallothionein and the glutathione (a major cellular antioxidant) producing enzyme gamma-glutamylcysteine synthetase (Haq et al. 2003).

Parkin, an E3 ligase, is another zinc-binding protein that may mediate protection offered by zinc in PD. Mutations in the gene-encoding *Parkin* result in autosomal recessive juvenile PD with symptoms appearing in late teens and early twenties. Interestingly, Parkin binds zinc, and zinc-deficient Parkin causes unfolding of the protein and loss of its ligase activity (Hristova et al. 2009). Reduced bioavailability of zinc for Parkin could, therefore, mimic the familial protein dysfunction and thus participate in PD degeneration. Notably, zinc supplementation improves the lifespan and motor function of Parkin mutant drosophila (Saini and Schaffner 2010).

Chronic zinc elevation, however, may actually induce PD-like degeneration. In a recent study, rats that were treated with zinc for up to 8 weeks had decreased locomotor activity, reflecting reduced dopamine and nigral tyrosine hydroxylase positive cells, together with alterations in various antioxidant systems (Kumar et al. 2010). Any potential PD treatment that targets zinc, therefore, must avoid this potential issue of zinc toxicity while still maintaining/restoring zinc to key proteins and/or fortifying antioxidant defense mechanisms.

## Summary

The scant and disparate reporting of basic zinc levels in PD tissue confounds attempts to pinpoint a role for this metal in the pathogenesis of PD. This notwithstanding, attempts have been made to mechanistically implicate zinc in the pathophysiology of PD, including a role as a cofactor for antioxidant enzymes SOD and metallothionein, as well as the PD-implicated ubiquitin ligase, Parkin. While the restoration of zinc to these ligands has been speculated to confer protection in PD, more research is required to elucidate the many and varied intersections between zinc biology and the cellular cascades that precipitate and potentiate PD.

## Conclusion

In the context of abnormal neurobiology, interest in zinc has most commonly emphasized its potential as a neurotoxin; however, information regarding the role of zinc in normal neuronal function and specifically in neurotransmission is rapidly expanding. Given the diverse biochemical pathways influenced by and dependent upon zinc, it is perhaps unsurprising that sustained alterations to normal zinc trafficking or bioavailability may result in chronic degenerative conditions. In AD in particular, a role for zinc in the neurotoxic cascade has been explicated, in which Aβ, bearing a metal coordinating motif, adopts a toxic oligomeric conformation upon binding zinc, in the process exacerbating an aging-related synaptic zinc deficit. A therapy aimed at interdicting the toxic pathway and restoring neuronal zinc deficits is currently in clinical development. So too in PD, while a holistic hypothesis is yet to emerge, numerous lines of evidence indicate aberrant zinc metabolism contributing to the neurodegenerative cascade in the SN. While therapies aimed at reducing the excess iron known to accumulate in the SN in PD are in development, it may also be worthwhile directing efforts toward remediating zinc dyshomeostasis.

**Acknowledgments** We acknowledge the National Health and Medical Research Council of Australia, the Australian Research Council, and the Joan and Peter Clemenger Trust.

## Cross-References

▶ Zinc-Binding Sites in Proteins
▶ Zinc Cellular Homeostasis
▶ Zinc in Superoxide Dismutase
▶ Zinc Metallothionein
▶ Zinc-Secreting Neurons, Gluzincergic and Zincergic Neurons
▶ Zinc and Zinc Ions in Biological Systems

## References

Adlard PA, Bush AI (2006) Metals and Alzheimer's disease. J Alzheimers Dis 10:145–163

Adlard PA, Bica L, White AR, Nurjono M, Filiz G, Crouch PJ, Donnelly PS, Cappai R, Finkelstein DI, Bush AI (2011) Metal ionophore treatment restores dendritic spine density and synaptic protein levels in a mouse model of Alzheimer's disease. PLoS One 6:e17669

Ajjimaporn A, Swinscoe J, Shavali S, Govitrapong P, Ebadi M (2005) Metallothionein provides zinc-mediated protective effects against methamphetamine toxicity in SK-N-SH cells. Brain Res Bull 67:466–475

Ajjimaporn A, Shavali S, Ebadi M, Govitrapong P (2008) Zinc rescues dopaminergic SK-N-SH cell lines from methamphetamine-induced toxicity. Brain Res Bull 77:361–366

An WL, Bjorkdahl C, Liu R, Cowburn RF, Winblad B, Pei JJ (2005) Mechanism of zinc-induced phosphorylation of p70 S6 kinase and glycogen synthase kinase 3beta in SH-SY5Y neuroblastoma cells. J Neurochem 92:1104–1115

Baum L, Chan IH, Cheung SK, Goggins WB, Mok V, Lam L, Leung V, Hui E, Ng C, Woo J, Chiu HF, Zee BC, Cheng W, Chan MH, Szeto S, Lui V, Tsoh J, Bush AI, Lam CW, Kwok T (2010) Serum zinc is decreased in Alzheimer's disease and serum arsenic correlates positively with cognitive ability. Biometals 23:173–179

Boom A, Authelet M, Dedecker R, Frederick C, Van Heurck R, Daubie V, Leroy K, Pochet R, Brion JP (2009) Bimodal modulation of tau protein phosphorylation and conformation by extracellular $Zn^{2+}$ in human-tau transfected cells. Biochim Biophys Acta 1793:1058–1067

Brewer GJ, Kanzer SH, Zimmerman EA, Molho ES, Celmins DF, Heckman SM, Dick R (2010) Subclinical zinc deficiency in Alzheimer's disease and Parkinson's disease. Am J Alzheimers Dis Other Demen 25:572–575

Corrigan FM, Reynolds GP, Ward NI (1993) Hippocampal tin, aluminum and zinc in Alzheimer's disease. Biometals 6:149–154

Crouch PJ, Tew DJ, Du T, Nguyen DN, Caragounis A, Filiz G, Blake RE, Trounce IA, Soon CP, Laughton K, Perez KA, Li QX, Cherny RA, Masters CL, Barnham KJ, White AR (2009) Restored degradation of the Alzheimer's amyloid-beta peptide by targeting amyloid formation. J Neurochem 108:1198–1207

Damante CA, Osz K, Nagy Z, Pappalardo G, Grasso G, Impellizzeri G, Rizzarelli E, Sovago I (2009) Metal loading capacity of Abeta N-terminus: a combined potentiometric and spectroscopic study of zinc(II) complexes with Abeta (1-16), its short or mutated peptide fragments and its polyethylene glycol-ylated analogue. Inorg Chem 48:10405–10415

Deibel MA, Ehmann WD, Markesbery WR (1996) Copper, iron, and zinc imbalances in severely degenerated brain regions in Alzheimer's disease: possible relation to oxidative stress. J Neurol Sci 143:137–142

Dexter DT, Wells FR, Lees AJ, Agid F, Agid Y, Jenner P, Marsden CD (1989) Increased nigral iron content and alterations in other metal ions occurring in brain in Parkinson's disease. J Neurochem 52:1830–1836

Dexter DT, Carayon A, Javoy-Agid F, Agid Y, Wells FR, Daniel SE, Lees AJ, Jenner P, Marsden CD (1991) Alterations in the levels of iron, ferritin and other trace metals in Parkinson's disease and other neurodegenerative diseases affecting the basal ganglia. Brain 114(Pt 4):1953–1975

Duce JA et al (2010) Iron-export ferroxidase activity of beta-amyloid precursor protein is inhibited by zinc in Alzheimer's disease. Cell 142:857–867

Gonzalez C, Martin T, Cacho J, Brenas MT, Arroyo T, Garcia-Berrocal B, Navajo JA, Gonzalez-Buitrago JM (1999) Serum zinc, copper, insulin and lipids in Alzheimer's disease epsilon 4 apolipoprotein E allele carriers. Eur J Clin Invest 29:637–642

Haq F, Mahoney M, Koropatnick J (2003) Signaling events for metallothionein induction. Mutat Res 533:211–226

Hershey CO, Hershey LA, Varnes A, Vibhakar SD, Lavin P, Strain WH (1983) Cerebrospinal fluid trace element content in dementia: clinical, radiologic, and pathologic correlations. Neurology 33:1350–1353

Hirsch EC, Brandel JP, Galle P, Javoy-Agid F, Agid Y (1991) Iron and aluminum increase in the substantia nigra of patients with Parkinson's disease: an X-ray microanalysis. J Neurochem 56:446–451

Hozumi I, Hasegawa T, Honda A, Ozawa K, Hayashi Y, Hashimoto K, Yamada M, Koumura A, Sakurai T,

Kimura A, Tanaka Y, Satoh M, Inuzuka T (2011) Patterns of levels of biological metals in CSF differ among neurodegenerative diseases. J Neurol Sci 303:95–99

Hristova VA, Beasley SA, Rylett RJ, Shaw GS (2009) Identification of a novel Zn2+ -binding domain in the autosomal recessive juvenile Parkinson-related E3 ligase parkin. J Biol Chem 284:14978–14986

Hullin RP (1983) Serum zinc in psychiatric patients. Prog Clin Biol Res 129:197–206

Jimenez-Jimenez FJ, Fernandez-Calle P, Martinez-Vanaclocha M, Herrero E, Molina JA, Vazquez A, Codoceo R (1992) Serum levels of zinc and copper in patients with Parkinson's disease. J Neurol Sci 112:30–33

Kapaki E, Segditsa J, Papageorgiou C (1989) Zinc, copper and magnesium concentration in serum and CSF of patients with neurological disorders. Acta Neurol Scand 79:373–378

Kocaturk PA, Akbostanci MC, Tan F, Kavas GO (2000) Superoxide dismutase activity and zinc and copper concentrations in Parkinson's disease. Pathophysiology 7:63–67

Kumar A, Ahmad I, Shukla S, Singh BK, Patel DK, Pandey HP, Singh C (2010) Effect of zinc and paraquat co-exposure on neurodegeneration: modulation of oxidative stress and expression of metallothioneins, toxicant responsive and transporter genes in rats. Free Radic Res 44:950–965

Lee JY, Cole TB, Palmiter RD, Suh SW, Koh JY (2002) Contribution by synaptic zinc to the gender-disparate plaque formation in human Swedish mutant APP transgenic mice. Proc Natl Acad Sci USA 99:7705–7710

Lee JY, Son HJ, Choi JH, Cho E, Kim J, Chung SJ, Hwang O, Koh JY (2009) Cytosolic labile zinc accumulation in degenerating dopaminergic neurons of mouse brain after MPTP treatment. Brain Res 1286:208–214

Linkous DH, Adlard PA, Wanschura PB, Conko KM, Flinn JM (2009) The effects of enhanced zinc on spatial memory and plaque formation in transgenic mice. J Alzheimers Dis 18:565–579

Lovell MA (2009) A potential role for alterations of zinc and zinc transport proteins in the progression of Alzheimer's disease. J Alzheimers Dis 16:471–483

Lovell MA, Robertson JD, Teesdale WJ, Campbell JL, Markesbery WR (1998) Copper, iron and zinc in Alzheimer's disease senile plaques. J Neurol Sci 158:47–52

Lyubartseva G, Smith JL, Markesbery WR, Lovell MA (2010) Alterations of zinc transporter proteins ZnT-1, ZnT-4 and ZnT-6 in preclinical Alzheimer's disease brain. Brain Pathol 20:343–350

Mendez-Alvarez E, Soto-Otero R, Hermida-Ameijeiras A, Lopez-Real AM, Labandeira-Garcia JL (2002) Effects of aluminum and zinc on the oxidative stress caused by 6-hydroxydopamine autoxidation: relevance for the pathogenesis of Parkinson's disease. Biochim Biophys Acta 1586:155–168

Miller LM, Wang Q, Telivala TP, Smith RJ, Lanzirotti A, Miklossy J (2006) Synchrotron-based infrared and X-ray imaging shows focalized accumulation of Cu and Zn co-localized with beta-amyloid deposits in Alzheimer's disease. J Struct Biol 155:30–37

Mo ZY, Zhu YZ, Zhu HL, Fan JB, Chen J, Liang Y (2009) Low micromolar zinc accelerates the fibrillization of human tau via bridging of Cys-291 and Cys-322. J Biol Chem 284:34648–34657

Molina JA, Jimenez-Jimenez FJ, Aguilar MV, Meseguer I, Mateos-Vega CJ, Gonzalez-Munoz MJ, de Bustos F, Porta J, Orti-Pareja M, Zurdo M, Barrios E, Martinez-Para MC (1998) Cerebrospinal fluid levels of transition metals in patients with Alzheimer's disease. J Neural Transm 105:479–488

Ozcankaya R, Delibas N (2002) Malondialdehyde, superoxide dismutase, melatonin, iron, copper, and zinc blood concentrations in patients with Alzheimer disease: cross-sectional study. Croat Med J 43:28–32

Panayi AE, Spyrou NM, Iversen BS, White MA, Part P (2002) Determination of cadmium and zinc in Alzheimer's brain tissue using inductively coupled plasma mass spectrometry. J Neurol Sci 195:1–10

Pei JJ, An WL, Zhou XW, Nishimura T, Norberg J, Benedikz E, Gotz J, Winblad B (2006) P70 S6 kinase mediates tau phosphorylation and synthesis. FEBS Lett 580:107–114

Religa D, Strozyk D, Cherny RA, Volitakis I, Haroutunian V, Winblad B, Naslund J, Bush AI (2006) Elevated cortical zinc in Alzheimer disease. Neurology 67:69–75

Rulon LL, Robertson JD, Lovell MA, Deibel MA, Ehmann WD, Markesber WR (2000) Serum zinc levels and Alzheimer's disease. Biol Trace Elem Res 75:79–85

Sahu RN, Pandey RS, Subhash MN, Arya BY, Padmashree TS, Srinivas KN (1988) CSF zinc in Alzheimer's type dementia. Biol Psychiatry 24:480–482

Saini N, Schaffner W (2010) Zinc supplement greatly improves the condition of parkin mutant Drosophila. Biol Chem 391:513–518

Samudralwar DL, Diprete CC, Ni BF, Ehmann WD, Markesbery WR (1995) Elemental imbalances in the olfactory pathway in Alzheimer's disease. J Neurol Sci 130:139–145

Schrag M, Dickson A, Jiffry A, Kirsch D, Vinters HV, Kirsch W (2010) The effect of formalin fixation on the levels of brain transition metals in archived samples. Biometals 23:1123–1127

Smith JL, Xiong S, Markesbery WR, Lovell MA (2006) Altered expression of zinc transporters-4 and -6 in mild cognitive impairment, early and late Alzheimer's disease brain. Neuroscience 140:879–888

Stoltenberg M, Bruhn M, Sondergaard C, Doering P, West MJ, Larsen A, Troncoso JC, Danscher G (2005) Immersion autometallographic tracing of zinc ions in Alzheimer beta-amyloid plaques. Histochem Cell Biol 123:605–611

Stoltenberg M, Bush AI, Bach G, Smidt K, Larsen A, Rungby J, Lund S, Doering P, Danscher G (2007) Amyloid plaques arise from zinc-enriched cortical layers in APP/PS1 transgenic mice and are paradoxically enlarged with dietary zinc deficiency. Neuroscience 150:357–369

Tarohda T, Ishida Y, Kawai K, Yamamoto M, Amano R (2005) Regional distributions of manganese, iron, copper, and zinc in the brains of 6-hydroxydopamine-induced parkinsonian rats. Anal Bioanal Chem 383:224–234

Thompson CM, Markesbery WR, Ehmann WD, Mao YX, Vance DE (1988) Regional brain trace-element studies in Alzheimer's disease. Neurotoxicology 9:1–7

Vural H, Demirin H, Kara Y, Eren I, Delibas N (2010) Alterations of plasma magnesium, copper, zinc, iron and selenium concentrations and some related erythrocyte antioxidant enzyme activities in patients with Alzheimer's disease. J Trace Elem Med Biol 24:169–173

Wenstrup D, Ehmann WD, Markesbery WR (1990) Trace element imbalances in isolated subcellular fractions of Alzheimer's disease brains. Brain Res 533:125–131

# Zinc in Hemostasis

Małgorzata Sobieszczańska[1] and Sławomir Tubek[2]
[1]Department of Pathophysiology, Wroclaw Medical University, Wroclaw, Poland
[2]Institute of Technology, Opole, Poland

## Synonyms

Blood clotting impairments as result of zinc disorders; Effect of nutritional zinc deficiency on hemostasis dysfunction; Participation of zinc in blood coagulation mechanisms; Role of Zinc in Etiopathogenesis of Bleeding; Significance of zinc in hemostasis regulation; Zinc and the decreased blood coagulability

## Definition

Zinc influences hemostasis functioning by affecting the plasma clotting factors, the platelets aggregation intensity, as well as the platelets mutual interactions and their interactions with the endothelial cells. The clinical presentation of nutritional zinc deficiency can be blood coagulation disturbances manifested in various forms of bleeding.

Zinc is a microelement that plays a multifunctional role in the human organism. It is a significant component of about 300 enzymes in which it performs as a catalytic factor, cofactor or as a part of the enzyme's structure (▶ Zinc-Binding Sites in Proteins). Furthermore, zinc is also essential for making the correct quaternary structure of many regulatory proteins as well hormone receptors. Besides, zinc is a part of the DNA structure and has important impact on its metabolism. Three distinct motifs of DNA-binding zinc proteins domains, namely, zinc fingers, zinc clusters, and zinc twists, were recognized (▶ Zinc Finger Folds and Functions). Zinc stabilizes cell membranes and regulates functions of the cells, likewise modifies calcium-dependent cell processes. In addition, both cell growth and cytokinesis depend on cellular zinc content and its transport within the cells (Vallee and Falchuk 1993).

For the first time, the attention to a causative relationship between hemostasis regulation and serum zinc level has been drawn by Gordon et al. who described blood clotting disturbances in adult males, which was proved to have resulted from a nutritional zinc deficiency (Gordon et al. 1982).

The physiological serum zinc level drops within the range of 12.2–21.4 µM. It has been shown that hyperzincemia predisposes to increased coagulability. Contrary to this, hypozincemia appeared to lead to diminished platelet aggregation and prolonged bleeding time. It is commonly known that there are three interrelated phenomena that must be present to maintain blood liquidity inside the blood vessels, i.e., adequate blood flow, correct endothelium state, and proper blood composition. Bearing in mind the diverse biological capabilities of zinc, its influence on the above-mentioned phenomena can be anticipated. The significance of zinc as one of the most essential microelements for human body functioning is very strongly pronounced in relation to the hemostasis because of its important participation in regulation of the equilibrium between pro- and antithrombotic factors originating in the platelets and endothelium (Tubek et al. 2008).

It was shown that zinc strongly modulates hemostasis by impairing platelets aggregation as well as their interactions among themselves and also with endothelial cells. Platelets recruitment at the site of blood-vessel injury is accompanied by a considerable local increase of the zinc concentration. Thrombocytes accumulate zinc ions mainly in cytoplasm and in α-granules, and this store is even up to 30–60 times higher as compared with the normal serum zinc concentration (Gordon et al. 1982).

The platelets aggregation process is initiated by the transfer of $Ca^{2+}$ ions to the thrombocytes platelets from the extracellular space, and this action is directly associated with zinc deficiency and, on the other hand, reversed by zinc supplementation. It was observed that providing a 50 mg dose of zinc per day resulted in the after-meal increase of serum zinc level and subsequent reactivity of platelets (Hughes and Samman 2006).

A crucial enzyme that takes place in the platelet activation acceleration is calcium-dependent protein kinase C (PKC). Activity of this enzyme is depended mainly on the $Ca^{2+}$ concentration in the extracellular space. It is documented that the zinc deficit does not change the PKC concentration; however, through decreasing a platelet $Ca^{2+}$ capture, $Zn^{2+}$ can cause a decrease of the PKC enzyme activity. A putative mechanism of this phenomenon is based on the defective calcium channels, which is due to Zn insufficiency inducing the changes of these channels' conformation by oxidation of the thiol groups. It can be assumed that the zinc binding to the –SH moieties in the calcium channels would prevent its oxidation and subsequent inactivation (Vallee and Falchuk 1993; Hughes and Samman 2006; Tubek et al. 2008).

Moreover, it is known that zinc can also potentiate the intravascular clotting process. It affects hemostasis by modifying the plasma clotting factors activity. Zinc deficiency was found to induce blood coagulation disorders by reducing the activity of factor XII (FXII). It is of significance that Zn reversibly binds itself to specific sites of the clotting factor XII until complete saturation, depending on the extracellular $Zn^{+2}$ ions concentration. A presence of sufficient Zn ions is therefore indispensable for FXII activation after its primary contact with endothelium, which subsequently enables a conversion of prekallikrein to kallikrein (Tubek et al. 2008). Zinc was shown to be helpful in binding the high weight molecular kininogen (HK) and factor XII (FXII) with negatively charged surfaces. However, it was also found that endothelial cells express a specific matrix protein which can also attract the activated factor XII in a zinc-independent manner (Schousboe 2006). Furthermore, zinc is able to enhance a production of the vasoactive peptide bradykinin, engaged in the proper blood circulation (Tubek et al. 2008).

Zinc can alter also the functions of the other clotting factor, i.e., FXI. It is a protein existing in a complex with high weight molecular kininogen (HK) that is activated and transformed to the active form (FXIa) with the help of FXIIa and thrombin, as well using autocatalysis. It is documented that a presence of either HK and $Zn^{2+}$ or prothrombin and $Ca^{2+}$ is necessary for the FXI binding to glycoprotein Ib-IX-V(GPIb-IX-V), a specific surface receptor of the activated platelets that facilitates activation of FXI by thrombin. It was shown in vitro experiments that FXI binds itself to glycocalicin (a component of the GPIb-IX-V receptor) in a zinc-dependent manner, regardless of the presence of HK and prothrombin. The binding of FXI to the above-mentioned receptor in the activated platelets is crucial for the initial consolidation phase of blood clotting and generation of thrombin in the site of platelets aggregation (Baglia et al. 2004; Tubek et al. 2008). Thus, once again, zinc seems to play a significant role in the blood coagulation stimulation. In turn, as zinc is actively involved in this process, its deficit is probable to cause some impairments of hemostasis.

The next important contribution is that zinc considerably reduces the action of the active factor VII (FVIIa) as well of FVIIa/tissue factor(TF) complex, which is not a case for the remaining vitamin K-dependent clotting factors. Replacing $Ca^{2+}$ by $Zn^{2+}$ at the FVIIa active center diminishes its activity through allosteric changes. Thus, similarly as for FXIIa, zinc and calcium exert a contrary effect on the FVIIa functioning, namely, Zn decreases, and Ca enhances the FVIIa activity (Petersen et al. 2000; Tubek et al. 2008).

It was shown that zinc depletion can cause thrombin inactivation, leading to blood clotting disturbances and bleeding. Coagulation may be modulated through histidine-rich glycoprotein (HRG) that competes with thrombin for binding the gamma chain of fibrin. HRG circulates in the blood as the complex with fibrinogen and this complex persists also upon fibrin formation. HRG is able to bind fibrinogen and other plasma proteins (including plasminogen – crucial in fibrinolysis) in a $Zn^{2+}$-dependent manner. Increase of $Zn^{2+}$ amount in conjunction with pH decline can change in a synergistic way the conformation and function of HRG (Vu et al. 2011; Tubek et al. 2008).

HRG seems to be also an important physiological factor that regulates the binding between heparin and antithrombin. Anticoagulation action of heparin through antithrombin III, inhibitor of thrombin, is competitively blocked by HRG. The binding rates of HRG and heparin in vivo increase along with Zn concentration. Interestingly, in vitro, the significant procoagulation effects occur at much higher than physiological zinc concentrations, i.e., over 50 μM. It is known that platelets, both activated and inactive, contain HRG and zinc ions in the amount of even up to 30–60 times more than serum does. Therefore, when platelets are activated, the conditions favor rather the binding between HRG and heparin. This phenomenon can cause the immediate decrease of thrombin inactivation and subsequently intensify the procoagulation action. In contrary, any state of zinc insufficiency would significantly contribute to thrombin inactivation and promote some bleeding (Tubek et al. 2008).

Furthermore, zinc ions have effect on fibrin production. It was shown in vitro that after adding $Ca^{2+}$ ions to a fibrin oligomer solution, a fibrin chains polymerization was intensified, and, what more, this phenomenon was then enhanced several times by insertion of some $Zn^{2+}$ ions. This process took place with the physiological zinc level and consequently rose along with its increase above physiological values. In addition, it was found that the increase of thrombin absorption by fibrin and shortening of blood clotting time are affected by the action of zinc ions in a proportional manner, within the $Zn^{2+}$ concentration range of 0.01–0.1 μM (Tubek et al. 2008).

Zinc has been proved to inhibit fibrinolysis, the process acting against coagulation. Zinc was found at the active center of metallocarboxypeptidase (TAFI/TAFIa), a thrombin-activatable fibrinolysis inhibitor that inhibits production of plasmin and modulates its activity (Tubek et al. 2008) (▶ Zinc Metallocarboxypeptidases).

Moreover, it was found that the binding of $Zn^{2+}$ to fibrinogen decreases its conjunction with decorin. Decorin is a tissue factor influencing the formation of collagen filaments. This process subsequently changes the structure of fibrin, making it more susceptible to tissue-plasminogen activator-dependent proteolysis and finally leading to the clot destruction (Dugan et al. 2006).

On the other hand, zinc seems to play a stimulating role in fibrinolysis that commences with the transformation of plasminogen to plasmin. Plasminogen circulates in the blood in a closed form that prevents an uncontrolled generation of plasmin. The binding of plasminogen to fibrin through histidine-rich glycoprotein (HRC) at the surface of endothelium alters its conformation and makes it more prone to be converted to plasmin. It was shown that zinc can be a kind of regulator in uPA-induced fibrinolysis. The binding between urokinase-type plasminogen activator (uPA) and fibrin depends on the $Zn^{+2}$ concentration in a specific and reversibly manner, and the binding of uPA to fibrin increased rapidly in the 0–50 μM Zn range, reaching then a plateau (Tubek et al. 2008).

It has been recently reported that zinc is also able to efficiently inhibit activated protein C (APC) through inducing a conformational change in the protein with several possible sites for zinc binding suggested. The inhibition mechanism appeared to be noncompetitive both in the absence and presence of $Ca^{2+}$. Thus, zinc seems to reveal the impact on the regulation of anticoagulant activity of APC (Zhu et al. 2010).

Summing up, considering so diversified and efficient role of zinc ions in hemostasis, it should be remembered that blood clotting disturbances with bleeding can be caused by the zinc depletion, and this condition can be normalized using the suitable zinc supplementation.

## Cross-References

▶ Zinc and Immunity
▶ Zinc and Wound Healing
▶ Zinc and Zinc Ions in Biological Systems
▶ Zinc Cellular Homeostasis
▶ Zinc Deficiency
▶ Zinc Homeostasis, Whole Body
▶ Zinc-Binding Sites in Proteins

## References

Baglia FA, Gailani D, Lopez JA, Walsh PN (2004) Identification of a binding site for glycoprotein Ibα in the Apple 3 domain of factor XI. J Biol Chem 279:45470–45476

Dugan TA, Yang VW-C, McQuillan DJ, Höök M (2006) Decorin modulates fibrin assembly and structure. J Biol Chem 281:38208–38216

Gordon PR, Woodruff CW, Anderson HL, O'Dell BL (1982) Effects of acute zinc deprivation on plasma zinc and platelet aggregation in adult males. Am J Clin Nutr 35:849–857

Hughes S, Samman S (2006) The effect of zinc supplementation in humans on plasma lipids, antioxidant status and thrombogenesis. J Am Coll Nutr 25:285–291

Petersen LC, Olsen OH, Nielsen LS, Freskgard P, Persson E (2000) Binding of $Zn^{2+}$ to a $Ca^{2+}$ loop allosterically attenuates the activity of factor VIIa and reduces its affinity for tissue factor. Protein Sci 9:859–866

Schousboe I (2006) Endothelial cells express a matrix protein which binds activated factor XII in a zinc-independent manner. Thromb Haemost 95:312–319

Tubek S, Grzanka P, Tubek I (2008) Role of zinc in hemostasis: a review. Biol Trace Elem Res 121:1–8

Vallee BL, Falchuk KH (1993) The biochemical basis of zinc physiology. Physiol Rev 73:79–87

Vu TT, Stafford AR, Leslie BA, Kim PY, Fredenburgh JC, Weitz JI (2011) Histidine-rich glycoprotein binds fibrin (ogen) with high affinity and competes with thrombin for binding to the gamma'-chain. J Biol Chem 286:30314–30323

Zhu T, Ubhayasekera W, Nickolaus N, Sun W, Tingsborg S, Mowbray SL, Schedin-Weiss S (2010) Zinc ions bind to and inhibit activated protein C. Thromb Haemost 104(3):544–553

# Zinc in Superoxide Dismutase

Melania D'Orazio and Andrea Battistoni
Dipartimento di Biologia, Università di Roma Tor Vergata, Rome, Italy

## Synonyms

Copper, Zinc superoxide dismutase; Cuprein; Cu,Zn-SOD; Cytocuprein; EC 1.15.1.1; Erythrocuprein; Hemocuprein; SOD; SOD-1; Superoxide:superoxide oxidoreductase

## Definition

The copper and the zinc ions present in the active site of Cu,Zn superoxide dismutases are bridged by a common histidine ligand, forming a characteristic metal arrangement which is conserved in bacterial and eukaryotic enzymes of this class and has no counterparts in other proteins. The copper ion constitutes the catalytic center, and zinc-free forms of the enzyme still display a high catalytic rate. In contrast, the removal of zinc is invariantly associated to a consistent loss of stability. However, given the high conservation of the bimetal cluster, the hypothesis that zinc could also play a role in modulating the redox properties of the catalytic copper has been debated for a long time. The identification of an enzyme variant from *M. tuberculosis* which lacks the zinc ion has provided a unique tool to investigate the role of zinc in Cu,Zn superoxide dismutases. Studies on this enzyme have in fact demonstrated that the redox properties of the mycobacterial enzyme are comparable to those typical of the enzymes containing both copper and zinc in the active site. In contrast, this enzyme variant proved to be the less stable between the characterized Cu-containing superoxide dismutases. These observations confirm that the main role of zinc is that of conferring structural stability to the protein.

## The Binuclear Metal Site of Cu,Zn Superoxide Dismutase

Cu,Zn superoxide dismutases (Cu,ZnSODs) are widespread antioxidant enzymes that catalyze the dismutation of the superoxide anion into oxygen and hydrogen peroxide (Fridovich 1995). They are localized in the cytoplasm and in the mitochondrial intermembrane space of all the eukaryotic cells and in the periplasm of several prokaryotes.

The vast majority of known Cu,ZnSODs are homodimeric proteins, although some fully active monomeric enzymes have been identified in prokaryotes. Bacterial and eukaryotic Cu,ZnSODs clearly derive from a common ancestor protein, and despite showing some relevant differences in subunit assembly and active site channel organization, they share a similar structure based on a flattened antiparallel β-barrel fold and an identical arrangement of the binuclear metal center (Bordo et al. 2001). Here, four histidine residues and a water molecule coordinate the oxidized $Cu^{2+}$ ion with a distorted square pyramidal geometry, and three histidine and one aspartate residues coordinate the $Zn^{2+}$ in a distorted tetrahedral geometry. Interestingly, in the oxidized form of the enzyme, the two metal ions are simultaneously coordinated by the imidazole ring of a histidine residue, sometimes referred to as "the histidine bridge" or "the imidazolate bridge" (Fig. 1). During the catalytic process, which involves the cyclic reduction and oxidation of copper upon the successive encounters with the substrate, this histidine loses coordination with copper when it is in the $Cu^+$ state.

While the copper ion plays a central role in the catalytic mechanism, the role of the neighboring zinc ion is less clear as it does not directly participate to the dismutation reaction or to the binding of the substrate. However, since this binuclear metal site, which has no homologs in other characterized proteins, has been strictly conserved during evolution, a possible role of zinc in the modulation of Cu,ZnSOD reactivity has been frequently hypothesized.

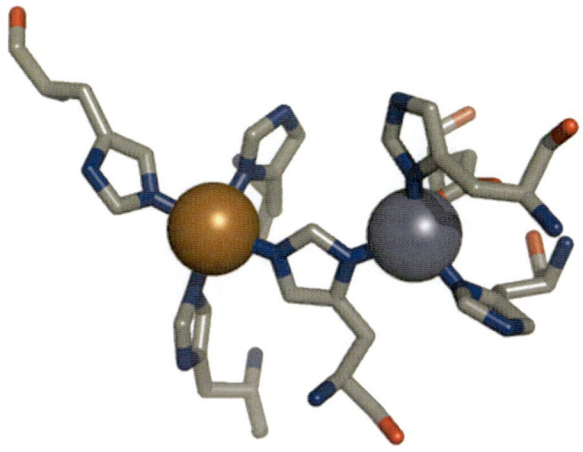

**Zinc in Superoxide Dismutase, Fig. 1** *Schematic drawing of the Cu,ZnSOD binuclear site.* This view puts in evidence the presence of a histidine residue ligating either copper (*gold sphere*) or zinc (*cyan sphere*)

## Studies on the Zinc Role in Cu,ZnSODs

Since its identification as a cofactor of the enzyme in 1970, the presence of zinc in Cu,ZnSODs has been the object of several investigations. Most of these investigations have been carried out on protein derivatives containing accurately controlled metal amounts (Valentine and Pantoliano 1981). These derivatives can be easily prepared by adding stoichiometric amounts of metals to the apoprotein obtained by extensive dialysis of the enzyme at pH 3.8 in presence of a chelating agent (Valentine and Pantoliano 1981; Bonaccorsi et al. 2002). Procedures have been established either to obtain Cu-free and Zn-free forms of the protein or to substitute zinc with other divalent metals.

These studies have clearly established that zinc contributes to the exceptional stability and compactness of the holoenzyme. In fact, it is well known that Cu,ZnSOD is between the most stable globular proteins so far characterized and displays a structural stability comparable to or higher than that typical of many proteins from thermophilic organisms. For example, the bovine enzyme retains its native structure in 8 M urea and 4% SDS, resists to the attack of proteolytic enzymes, and unfolds only at very high temperatures. All these properties are dramatically modified by zinc removal, whereas the binding of zinc to the apoprotein induces a native-like fold and restores a stability that is only marginally lower than that of the fully metalated protein. Moreover, structural studies have shown that the binding of zinc to the apoprotein is sufficient to induce a native-like conformation to the copper ligands, which may facilitate the subsequent copper insertion into the protein (Banci et al. 2003). All together, these investigations demonstrate that zinc has an unequivocal structural role in Cu,ZnSOD. In addition, different metals, including $Co^{2+}$, $Cu^{2+}$, $Cd^{2+}$, and $Ni^{2+}$, can substitute zinc with modest effects on the catalytic activity of the enzyme, which maintains full activity at physiological pH, even in the absence of zinc.

Nonetheless, several lines of evidence have suggested that zinc could subtly modulate the reactivity of the copper ion. For example, unlike the holoenzyme, whose activity is constant over the 5–9 pH range, the activity of the zinc-free enzyme significantly drops above neutral pH. Besides this effect, that can be likely ascribed to pH-dependent conformational changes of the protein scaffold rather than to a direct influence of zinc on copper reactivity, the charge of the zinc ion contributes to the formation of the positively charged electric field around the catalytic copper center, which facilitates the electrostatic guidance of the substrate toward the active site (Bordo et al. 2001). However, despite different studies have shown that changes in the charge distribution around the copper site may significantly affect the efficiency of the dismutation reaction, it is unlikely that this role could justify the evolutionary conservativeness of zinc, as other mutations in the active site channel could compensate for the absence of the charge contribution due to this metal. More interestingly, other studies have pointed to a change in the redox properties of Cu in the absence of zinc.

Unlike the holoprotein, Zn-deficient Cu,ZnSOD is prone to fast reduction by ascorbate in vitro (Trumbull and Beckman 2009). This effect may be explained by large structural changes induced by metal loss in protein structure, which, in the absence of zinc, shows significant disorder in the two large loops that shape the active site channel and normally restrict the access to the catalytic copper only to superoxide. In principle, the abnormal copper reduction by ascorbate or other small molecular weight intracellular reductants makes possible the transfer of electrons from copper to molecular oxygen and the generation of superoxide by the enzyme. It has been hypothesized that such altered reactivity of the reduced Zn-free enzyme could promote formation of the highly toxic peroxynitrite radical in presence of nitric oxide (Trumbull and Beckman 2009). The same Cu,ZnSOD can react with peroxynitrite to catalyze tyrosine nitration, with the zinc-devoid enzyme being able to catalyze this reaction faster than the holoenzyme.

These observations have raised considerable attention in recent years, as mutations in human Cu,ZnSOD are the most common genetic cause of the familial form of amyotrophic lateral sclerosis (ALS). Several mutant Cu,ZnSODs associated to ALS have been shown to possess decreased affinity for the zinc ion (Shaw and Valentine 2007), leading to the suggestion that this feature could contribute to the toxic gain of function causing the disease. In support to this hypothesis, it has been shown that zinc-deficient Cu,ZnSODs induce apoptosis in cultured motor neurons (Trumbull and Beckman 2009). An alternative explanation to justify the involvement of mutant Cu,ZnSODs in the pathology of familiar forms of ALS is that these proteins have a significantly higher tendency to aggregate (Chattopadhyay and Valentine 2009). Although the controversies about the mechanisms relating Cu,ZnSOD to ALS are not resolved, in this context, it is interesting to underline that also Cu,ZnSOD aggregation could be promoted by improper metal binding.

## The Mycobacterial Cu,SOD

The above-mentioned studies indicate that the removal of zinc from Cu,ZnSODs affects either protein stability or copper reactivity, but do not clarify whether zinc is an indispensable requirement for the construction of a fully active Cu-containing superoxide dismutase. A significant improvement in the understanding of the reasons which have favored the evolution of the peculiar binuclear metal site of Cu,ZnSODs has been provided by the identification and molecular characterization of a copper-containing superoxide dismutase from *Mycobacterium tuberculosis* (MtSOD) which lacks the zinc ion. The primary and tertiary structures of this enzyme, which is anchored to the bacterial membrane though lipids attached to the N-terminal cysteine, show evident homology with the other members of the Cu,ZnSOD family. However, it cannot bind

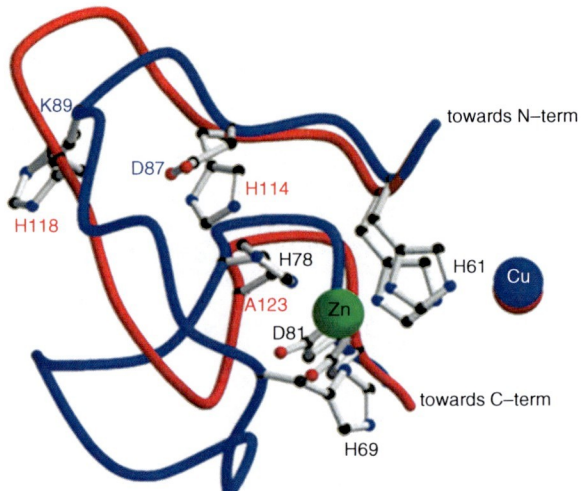

**Zinc in Superoxide Dismutase, Fig. 2** *The active site of Mycobacterium tuberculosis* (red) *and Actinobacillus pleuropneumoniae* (blue) *superoxide dismutases.* Conserved residues (numbering based on the bovine enzyme) are labeled in *black*; residues mutated or unique in MtSOD are labeled in *red*. The zinc subloop of MtSOD is significantly shorter and lacks some of the residues indispensable for zinc binding such as His[69] (missing due to a deletion) and His[78] (mutated to Ala[123]). The positions of two histidine residues (His[114] and His[118]) present in the short zinc loop of MtSOD exclude the possibility of zinc binding (taken from Spagnolo et al. 2004)

zinc due to a large deletion in a loop shaping the zinc-binding site and to other specific mutations affecting the residues involved in the coordination of this metal (Fig. 2). The copper ligands are fully conserved and, similarly to classical Cu,ZnSODs, the copper ion shows a distorted square planar coordination with water as a fifth ligand (Spagnolo et al. 2004). Unlike Zn-deficient Cu,ZnSODs, whose spectral properties are pH dependent, the EPR spectrum of MtSOD is stable over the 6–10 pH range, indicating that zinc is not necessary to obtain a stable copper site. Moreover, MtSOD shows a catalytic rate constant ($k_{cat}/k_m$) above $2 \times 10^9 M^{-1} s^{-1}$, which is a value comparable to that observed for other Cu,ZnSODs of prokaryotic or eukaryotic origin. Although the activity of MtSOD decreases above pH 8.0, this feature cannot be attributed to the absence of zinc as a similar behavior is typical of all bacterial Cu,ZnSODs. To further investigate the possibility that zinc could finely modulate the copper redox properties, the capability of copper in the active site of MtSOD to remain in the oxidized state in presence of ascorbate has been assayed

(D'Orazio et al. 2009). Ascorbate, in fact, can easily reduce Zn-deficient Cu,ZnSOD but not the holoenzyme. In contrast, ascorbate is unable to reduce MtSOD, indicating that its redox properties are remarkably similar to those of fully metalated Cu, ZnSODs (D'Orazio et al. 2009). Overall, these observations indicate that evolution of an active site containing a zinc ion in tight association with copper seems not to be justified by the need of obtaining a better modulation of copper reactivity.

At the same time, MtSOD has proved to be an excellent model also to evaluate the contribution of zinc to the stability of copper-containing superoxide dismutases (D'Orazio et al. 2009). The stability of this enzyme has been studied by differential scanning calorimetry (DSC), one of the most direct and sensitive approaches to obtain quantitative measurements of heat effects on protein stability. DSC has been extensively used by different authors to compare the conformational stability of different Cu,ZnSODs as well as the specific contribution of zinc and copper (Bonaccorsi di Patti et al. 2002). Several thermodynamic parameters can be extrapolated from the analysis of DSC traces, including $\Delta G$, $\Delta H$, and $\Delta S$ of unfolding. However, the most helpful parameter providing indications about the conformational stability of a protein is the transition midpoint ($T_m$) value, which indicates the temperature of maximum heat capacity. The $T_m$ value corresponds to the temperature where 50% of the protein is in its native conformation and the other 50% is denatured. Comparisons between the $T_m$ values of proteins of the same class or of mutated variants or derivatives of a specific one may provide useful information about factors affecting stability. Of course, the $T_m$ values obtained in DSC experiments depend on the experimental conditions and may change as a function of pH, buffer composition, and scan rate. To favor a comparison between different samples, most DSC experiments on Cu,ZnSODs have been carried out in phosphate buffer 100 mM, pH 7.8, and increasing the temperature by 1°C/min. Table 1 reports the available $T_m$ values of several prokaryotic and eukaryotic holo-Cu,ZnSODs.

DSC studies have shown that the presence of the active site metals is one of the most important determinants of the extremely high thermal stability of Cu,ZnSODs. For example, the difference in $T_m$ between the holo and apo forms of bovine Cu,ZnSOD is around 40°C at pH 7.8 (Bonaccorsi di Patti et al. 2002).

**Zinc in Superoxide Dismutase, Table 1** $T_m$ of different Cu-containing SODs

| SOD | $T_m$ (°C) |
| --- | --- |
| Ox | 88.0 |
| Sheep | 87.1 |
| Human | 83.6 |
| Shark | 84.1 |
| Yeast | 73.1 |
| X. laevis A | 71.1 |
| X. laevis B | 76.8 |
| E. coli O157:H7 | 78.0 |
| P. leiognathi | 71.0 |
| E. coli (monomeric) | 65.9 |
| M. tuberculosis | 60.6 |

The reported $T_m$ values are taken from literature data and refer to seven dimeric eukaryotic, two dimeric (E. coli O157:H7 and P. leiognathi), and one monomeric (E. coli) bacterial Cu,ZnSODs and two dimeric CuSOD from M. tuberculosis. Such $T_m$ values are directly comparable as they have been obtained by analyzing protein samples under identical experimental conditions (100 mM phosphate buffer, pH 7.8; scan rate = 60°C/h)

Although both metals contribute to stability, the greatest contribution is provided by zinc. Another important determinant of protein stability is the stable association between subunits. This is exemplified by a comparison of the $T_m$ values of the monomeric Cu,ZnSOD from E. coli with those of the dimeric enzymes from different sources (Table 1). Interestingly enough, the $T_m$ of MtSOD is lower than those obtained for all the other Cu,ZnSODs, including the monomeric E. coli enzyme (Table 1). The remarkable decrease in stability of MtSOD is particularly striking if one considers that the protein, while maintaining the classical β-barrel structure typical of all Cu, ZnSODs, shows a complex rearrangement of the dimer interface which has been interpreted as an attempt to compensate the loss of stability caused by the absence of the zinc ion (Spagnolo et al. 2004). Figure 3 shows that the quaternary structure of MtSOD is characterized by the presence of a long loop that protrudes from the enzyme core in a direction roughly perpendicular to the β-barrel axis. The presence of this peculiar dimerization loop significantly increases the interface area of the MtSOD dimer compared to the other Cu,ZnSODs. In fact, nearly 19% of the total surface area of a subunit is part of the dimer interface, a value which decreases to nearly half in bacterial and eukaryotic Cu,ZnSODs.

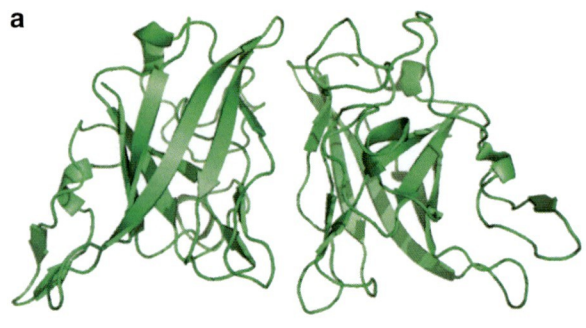

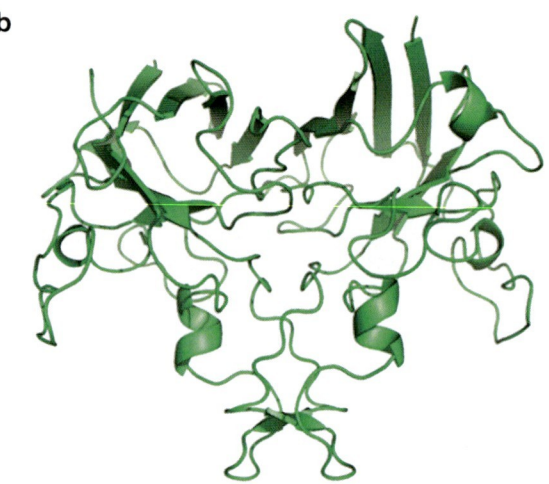

**Zinc in Superoxide Dismutase, Fig. 3** *Dimeric structure of Actinobacillus pleuropneumoniae* (**a**) *and Mycobacterium tuberculosis* (**b**) *superoxide dismutases.* Whereas the structure of *A. pleuropneumoniae* Cu,ZnSOD is characterized by the classical subunit arrangement, the mycobacterial enzyme displays a unique and complex dimeric architecture, with a significant extension of the buried surface area

by the need to avoid the risk of possible side reactions due to copper bound to an unstable protein.

## Cross-References

- Copper, Biological Functions
- Copper-Binding Proteins
- Copper-Zinc Superoxide Dismutase and Lou Gehrig's Disease
- Zinc and Zinc Ions in Biological Systems
- Zinc-Binding Sites in Proteins

## Conclusions

Although zinc removal from the active site of Cu,ZnSODs has a dramatic impact on the redox properties of copper, the hypothesis that the presence of zinc in the active site is necessary to ensure the subtle modulation of protein activity has been challenged by the observation that the CuSOD from *M. tuberculosis* has catalytic properties nearly identical to those of the other proteins of the same class. Most of the unusual properties of Zn-deficient SODs are therefore related to the structural changes following the loss of the zinc ion. In contrast, zinc has an unequivocal role in conferring structural stability. Therefore, it is likely that the early evolution of binuclear metal site in Cu,ZnSODs has been driven

## References

Banci L, Bertini I, Cramaro F, Del Conte R, Viezzoli MS (2003) Solution structure of Apo Cu, Zn superoxide dismutase: role of metal ions in protein folding. Biochemistry 42:9543–9553

Bonaccorsi di Patti MC, Giartosio A, Rotilio G, Battistoni A (2002) Analysis of Cu, ZnSOD conformational stability by differential scanning calorimetry. Methods Enzymol 349:49–61

Bordo D, Pesce A, Bolognesi M, Stroppolo ME, Falconi M, Desideri A (2001) In: Wieghardt K, Huber R, Poulos T, Messerschmidt A (eds) Handbook of metalloproteins. Wiley, London, pp 1284–1300

Chattopadhyay M, Valentine JS (2009) A role for copper in the toxicity of zinc-deficient superoxide dismutase to motor neurons in amyotrophic lateral sclerosis. Antioxid Redox Signal 11:1603–1614

D'Orazio M, Cervoni L, Giartosio A, Rotilio G, Battistoni A (2009) Thermal stability and redox properties of M. tuberculosis Cu-superoxide dismutase. Insights into the role of the zinc ion in Cu, Zn superoxide dismutases. Arch Biochem Biophys 486:119–124

Fridovich I (1995) Superoxide radical and superoxide dismutases. Annu Rev Biochem 64:97–112

Shaw BF, Valentine JS (2007) How do ALS-associated mutations in superoxide dismutase 1 promote aggregation of the protein? Trends Biochem Sci 32:78–85

Spagnolo L, Toro I, D'Orazio M, O'Neill P, Pedersen JZ, Carugo O, Rotilio G, Battistoni A, Djinovic-Carugo K (2004) Unique features of the *sodC*-encoded superoxide dismutase from Mycobacterium tuberculosis, a fully functional copper-containing enzyme lacking zinc in the active site. J Biol Chem 279:33447–33455

Trumbull KA, Beckman JS (2009) A role for copper in the toxicity of zinc-deficient superoxide dismutase to motor neurons in amyotrophic lateral sclerosis. Antioxid Redox Signal 11:1627–1639

Valentine JS, Pantoliano MW (1981) Protein-metal ion interactions in cuprozinc protein (superoxide dismutase), a major intracellular repository for copper and zinc in the eukaryotic cell. In: Spiro TG (ed) Copper proteins. Wiley, New York, pp 292–358

## Zinc in the Nervous System

▶ Zinc-Secreting Neurons, Gluzincergic and Zincergic Neurons

## Zinc Leukotriene A₄ Hydrolase/Aminopeptidase Dual Activity

Jesper Z. Haeggström and Agnes Rinaldo-Matthis
Department of Medical Biochemistry and Biophysics (MBB), Karolinska Institute, Stockholm, Sweden

### Synonyms

Inflammation; Leukotriene B₄; Lipid mediator

### Definition

Leukotriene A₄ hydrolase/aminopeptidase (LTA4h) (EC 3.3.2.6) is a bifunctional zinc metalloenzyme with both an epoxide hydrolase and aminopeptidase activity (Fig. 1). Its primary and most well-recognized function is to catalyze the committed step in the formation of the proinflammatory mediator LTB₄, and the chemistry of this reaction is hydrolysis of the allylic epoxide LTA₄. Both the epoxide hydrolase and the aminopeptidase activity are dependent on a Zn ion bound in the active site.

LTA4h is a monomeric enzyme with a molecular mass of 69 kDa, and its crystal structure has been determined in complex with substrates and inhibitors. It is a widely distributed enzyme that has been detected in almost all mammalian cells.

### Physiological Function

LTA4h was discovered through its epoxide hydrolase activity, i.e., hydrolysis of LTA₄ into LTB₄. The dihydroxy acid LTB₄ is a potent chemoattractant that is secreted by polymorphonuclear leukocytes (PMNs) and mediates inflammation by triggering chemotaxis and adherence of immune cells to the endothelium (Ford-Hutchinson 1990; Samuelsson 1983). LTB₄ also participates in host defense against infections and acts as a key mediator in platelet-activating factor (PAF)-induced lethal shock, vascular inflammation, and arteriosclerosis (Samuelsson et al. 1987).

The precursor of LTA₄ is arachidonic acid, a fatty acid that is metabolized either through the cyclooxygenase pathway (COX) or the 5-lipoxygenase (5-LO) pathway (Fig. 2). The COX pathway generates prostaglandins and thromboxanes, whereas the 5-LO pathway generates leukotrienes from the common intermediate LTA₄ (Samuelsson 1983; Funk 2001). Besides being transformed by LTA4h, LTA₄ can be conjugated with glutathione by LTC₄ synthase to form the spasmogenic LTC₄. The effects of leukotrienes are mediated through G-protein-coupled receptors, and those signaling in response to LTB₄ are denoted BLT1 and BLT2 (Haeggstrom and Funk 2011).

As mentioned earlier, LTA4h combines the LTB₄ epoxide hydrolase activity with an aminopeptidase activity in a common active center. This catalytic architecture was discovered when a consensus Zn-binding motif (HEXXH-$X_{18}$-E) was identified in the primary structure of LTA4h, similar to those present in certain Zn proteases and aminopeptidases. This finding led to the proposal that LTA4h is a Zn-dependent aminopeptidase (Barret et al. 1998). In order to verify the suggested Zn content of LTA4h, atomic absorption spectrometry was performed which revealed the presence of one Zn atom per enzyme molecule. Furthermore, removing the Zn by chelating agents inactivated the enzyme. Subsequent addition of stoichiometric amounts of Zn restored the enzymatic activity, which led to the conclusion that the primary function of the Zn is catalytic rather than structural. The postulated peptidase activity was uncovered by direct activity measurements with p-nitroanilide amino acid derivatives as substrates, and based on its Zn signature and peptide-cleaving activity, the enzyme was subsequently classified as a member of the M1 family of Zn aminopeptidases. The aminopeptidase activity was characterized as an Arg-specific tripeptidase activity stimulated by monovalent anions, e.g., chloride ions (Haeggström 2004).

However, until recently, no physiological substrate for the aminopeptidase activity had been identified. In 2010, Snelgrove and coworkers discovered that extracellular LTA4h was responsible for the efficient hydrolysis and inactivation of the proinflammatory tripeptide Pro-Gly-Pro (PGP) during neutrophil-dependent inflammation in the respiratory tract. PGP is a chemotactic

**Zinc Leukotriene A₄ Hydrolase/Aminopeptidase Dual Activity, Fig. 1** LTA4h is a bifunctional enzyme with an epoxide hydrolase and an aminopeptidase activity

**Zinc Leukotriene A₄ Hydrolase/Aminopeptidase Dual Activity, Fig. 2** Arachidonic acid is the precursor to $LTA_4$, the lipid substrate of LTA4h

mediator for neutrophils generated from the enzymatic breakdown of extracellular matrix (ECM) collagen. It is a biomarker for chronic obstructive pulmonary disease (COPD), and it has been implicated in neutrophil infiltration of the lung (Snelgrove 2011).

Hence, the two enzymatic activities of LTA4h, the epoxide hydrolase and the PGP aminopeptidase activity, have opposing roles in inflammatory reactions. During initiation of inflammation, $LTB_4$ is synthesized and governs neutrophil recruitment to the site of injury. In contrast, LTA4h can participate in the resolution phase of inflammation, by virtue of its capacity to degrade the neutrophil chemoattractant PGP.

## Cellular Location

LTA4h is widely distributed in mammals and the enzyme has been found in practically all cells, organs,

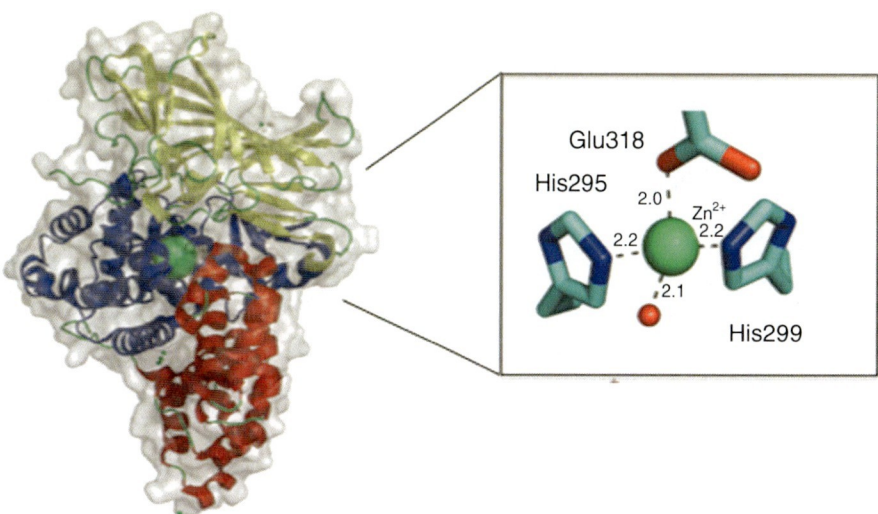

Zinc Leukotriene $A_4$ Hydrolase/Aminopeptidase Dual Activity, Fig. 3 The overall structure of LTA4h has three domains with the active site in between the domains. The active site is situated in a cleft between the domains close to where the Zn ion (*green*) is bound. The inset illustrates the tetrahedral coordination of Zn by His295, His299, Glu318, and a water molecule. The distances are measured in Ångström units

and tissues in man, rat, and guinea pig (Haeggström 2004). $LTB_4$ formation has also been observed in lower vertebrates including birds, fish, and frogs. $LTB_4$ has not been detected in nonvertebrates, although a yeast and a *C. elegans* aminopeptidase have been described to be highly similar in sequence to the human LTA4h. These observations have raised questions about when, in an evolutionary perspective, $LTB_4$ biosynthesis became important (Tholander et al. 2005).

The enzyme LTA4h, as well as the product $LTB_4$, has also been detected in cells devoid of 5-LO; thus, cells unable to provide the substrate $LTA_4$. This broad distribution of LTA4h has been difficult to rationalize from a functional point of view since leukotriene biosynthesis is generally regarded as a process restricted to white blood cells and bone marrow. One explanation for the uneven distribution of 5-LO and LTA4h has been the so-called transcellular metabolism, a phenomenon that occurs in vivo and refers to the transfer of an intermediate or product, such as $LTA_4$, produced in one cell type, that is transferred to another cell type for further enzymatic metabolism (Sala et al. 2010). Another explanation considers the bifunctional property of LTA4h, where the peptide-cleaving activity of LTA4h would account for the uneven distribution of 5-LO and LTA4h.

LTA4h is present in the cytosol of human cells as well as in the extracellular space. The epoxide hydrolase activity is generally assumed to be cytosolic, whereas the aminopeptidase activity is thought to play a major role extracellularly (Snelgrove 2011).

## Structure

In 2001, the crystal structure of LTA4h in complex with the competitive inhibitor bestatin was determined to a resolution of 1.95 Ångströms (Å). The structure revealed a monomeric protein with three domains: an α-/β-saddle like N-terminal part (residues 1–207), a thermolysin-like catalytic domain (208–450), and an α-helical C-terminal domain (461–610) (Fig. 3). The molecular dimensions of the protein are approximately $85 \times 65 \times 50$ Å. The N-terminal domain is structurally similar to bacteriochlorophyll a; the catalytic domain is similar to that of thermolysin, and the C-terminal domain has structural features resembling a so-called armadillo repeat or HEAT region, which implicates that the domain may take part in protein-protein interactions.

The active site is situated in a deep cleft in between the domains. The cleft consists of a hydrophilic part, where the Zn is located, as well as a hydrophobic cavity. The Zn ion is coordinated to His295, His299, and Glu318 (Fig. 3). These three Zn ligands constitute the Zn-binding motif, $HEXXH\text{-}(X)_{18}\text{-}E$, that is conserved among members of the M1 family of the MA clan of metallopeptidases (Barret et al. 1998) to which LTA4h belongs. Examples of other members in this family are the aminopeptidase A, aminopeptidase B, and endoplasmic reticulum aminopeptidases 1 and 2. Common to all members are the conserved stretch of amino acids containing the Zn-binding ligands, a GXMEN peptide-substrate-binding motif

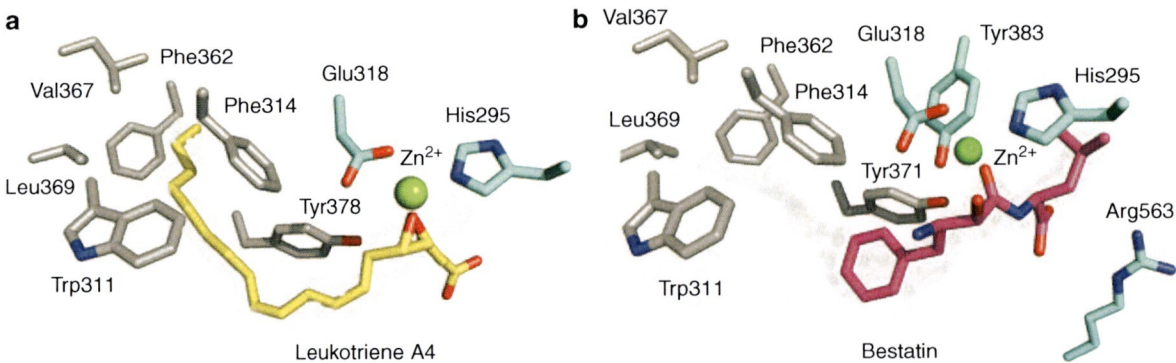

Zinc Leukotriene A₄ Hydrolase/Aminopeptidase Dual Activity, Fig. 4 The active site of LTA4h where the hydrophobic pocket is shown in *gray*, and the hydrophilic cleft is colored in *cyan*. The Zn ion is *green* and the ligand in (**a**) is a *yellow-modeled* $LTA_4$. (**b**) Shows the crystal structure complex of a peptide mimic, bestatin, bound in the active site. The two catalytic activities occupy distinct but overlapping active sites

as well as the catalytic mechanism by which the Zn ion facilitates peptide hydrolysis.

Apart from the Zn ligands, the hydrophilic portion of the active site harbors Glu271, Glu296, and Tyr383, residues that are important for the peptidase activity (Tholander et al. 2008) (Fig. 4). The hydrophobic part of the active site is L-shaped and lined with aromatic and hydrophobic residues such as Leu and Phe. $LTA_4$ modeled into the hydrophobic cavity of LTA4h has its carboxyl group coordinating Arg563, the epoxide coordinating the Zn, and C12, the site for hydroxyl group insertion, close to Asp375 (Haeggstrom and Funk 2011) (Fig. 4).

## Enzymatic Activity

LTA4h is unusually selective for its epoxide substrate $LTA_4$. It only accepts a few other substrates such as the double bond isomers $LTA_3$ and $LTA_5$ although with low efficiency (Haeggström 2004). The intrinsic chemical features of $LTA_4$, which form the molecular basis for recognition by LTA4h, are the carbon chain length (C20), the carboxyl group at C1, the geometry of the conjugated triene (trans, trans, cis), and the position of the epoxide (Fig. 5).

Once $LTA_4$ is bound to the enzyme, the $LTB_4$ formation is thought to proceed through a unique $S_N1$ epoxide hydrolase mechanism (Haeggstrom and Funk 2011). The $Zn^{2+}$ and Glu271 polarize a water and

Zinc Leukotriene A₄ Hydrolase/Aminopeptidase Dual Activity, Fig. 5 The chemical structure of $LTA_4$ with its carboxyl group at C1, an allylic epoxide at C5–C6 and a conjugated double bond system at C7–C12

activate the epoxide by promoting an acid-induced opening of the epoxide, forming an unstable carbocation intermediate (Fig. 6a). The generated charge is delocalized over the conjugated triene system (C6–C12) leaving the planar $sp^2$-hybridized C12 open for nucleophilic attack from either side of the molecule. Subsequently, a water molecule is added at C12 in a stereospecific manner directed by Asp375 to generate the 12(R)-hydroxyl group in $LTB_4$. The epoxide hydrolase reaction of LTA4h is unique in the sense that the hydroxyl group is introduced in a stereospecific manner at a site distant (C12) from the epoxide moiety (C5–C6).

The LTA4h epoxide hydrolase activity is suicide inactivated which leads to covalent attachment of the $LTA_4$ to the protein during catalysis. Tyr378 has been identified as the residue that binds $LTA_4$ during this process (Haeggstrom and Funk 2011).

The LTA4h aminopeptidase activity is less specific towards its substrates as compared with the epoxide hydrolase activity although it prefers to

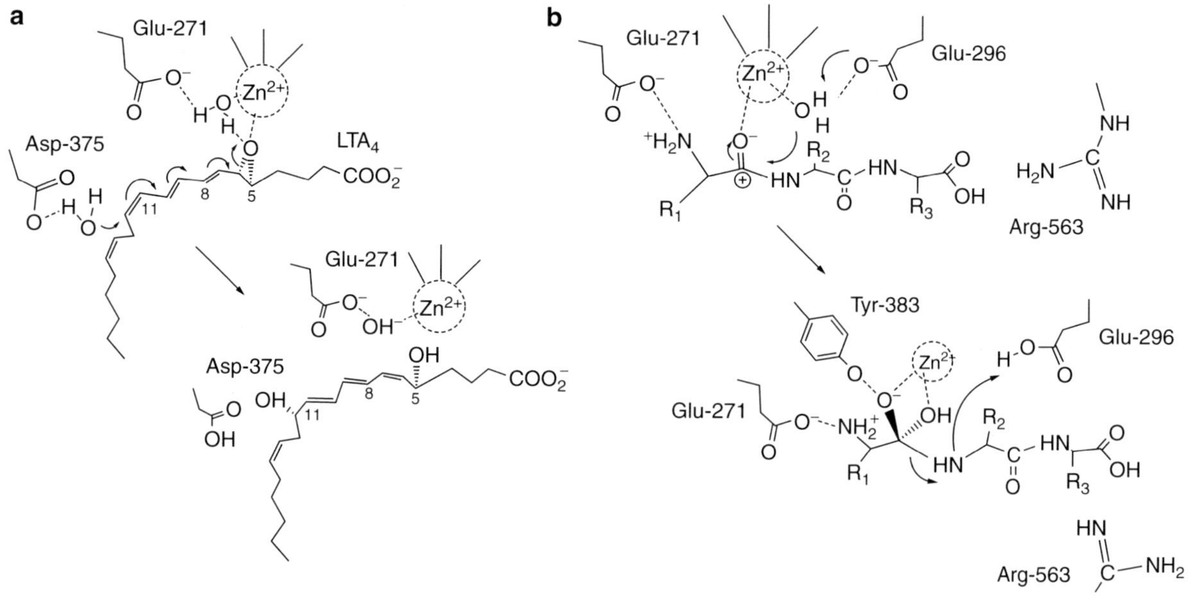

**Zinc Leukotriene A₄ Hydrolase/Aminopeptidase Dual Activity, Fig. 6** The two enzymatic activities of LTA4h where, in (**a**), the epoxide hydrolase activity is shown and, in (**b**), the aminopeptidase catalytic activity is shown

hydrolyze the amino-terminal, positively-charged, residue of di- and tripeptides (Haeggstrom and Funk 2011). As observed in crystal structures of substrate complexes, the tripeptide substrate bound to the hydrophilic part of the active site has its N-terminal amino group coordinated to Glu271, the carbonyl oxygen bound to the $Zn^{2+}$ and the C-terminus coordinated to Arg563. The tripeptidase activity follows a general base mechanism, similar to thermolysin, where Glu296 activates a water molecule for nucleophilic attack on the carbonyl carbon of the scissile peptide bond (Fig. 6b).

The nucleophilic attack is facilitated by the induced partial positive charge at the carbonyl carbon by the coordination to the Zn. The oxyanion intermediate is stabilized by the Zn ion and Tyr383. In the final step, Glu296 acts as an acid, protonating the N-terminus, thus facilitating leaving group departure. The proposed mechanism of LTA4h is based on mutagenic and structural analysis of LTA4h as well as by similarities with thermolysin (Tholander et al. 2008).

The aminopeptidase activity of LTA4h can be conveniently measured using aminoacyl-$p$-nitroanilide (aminoacyl-$p$-NA) derivatives of amino acids by spectrophotometric analysis of $p$-nitroaniline formation at 405 nm (Tholander et al. 2005).

The aminopeptidase and epoxide hydrolase activities are exerted at distinct but overlapping active sites. Certain residues such as Glu271, Arg563, and the Zn ion are needed for both activities, whereas Asp375 is critical only for the epoxide hydrolase reaction, and Glu296 and Tyr383 are necessary for the aminopeptidase reaction (Haeggstrom and Funk 2011).

## LTA4h as a Drug Target

The product of LTA4h, LTB₄, plays a significant role in many inflammatory diseases such as in inflammatory bowel disease, rheumatoid arthritis, psoriasis, asthma, and COPD (Haeggstrom and Funk 2011; Snelgrove 2011). Therefore, LTA4h is considered an attractive target for drug discovery. The recent discovery that LTA4h can degrade the proinflammatory mediator Pro-Gly-Pro has shed new light on how to think when developing drugs against LTA4h. Probably an inhibitor that selectively blocks the epoxide hydrolase activity (LTB₄ synthesis) while sparing the aminopeptidase activity of LTA4h (PGP degradation) would have improved anti-inflammatory properties. The detailed structural information as well as available activity assays of LTA4h will be helpful to guide in the development of this kind of selective inhibitors.

## Cross-References

▶ Angiotensin I-Converting Enzyme
▶ Thermolysin
▶ Zinc Aminopeptidases, Aminopeptidase from Vibrio Proteolyticus (Aeromonas proteolytica) as Prototypical Enzyme
▶ Zinc-Binding Sites in Proteins

## References

Barret AJ, Rawlings ND, Woessner JF (1998) Leukotriene A4 hydrolase. In: Barrett AJ, Rawlings ND, Woessner JF (eds) Handbook of proteolytic enzymes. Academic, San Diego/London, pp 1022–1025

Ford-Hutchinson AW (1990) Leukotriene $B_4$ in inflammation. Crit Rev Immunol 10:1–12

Funk CD (2001) Prostaglandins and leukotrienes: advances in eicosanoid biology. Science 294:1871–1875

Haeggström JZ (2004) Leukotriene $A_4$ hydrolase/aminopeptidase, the gatekeeper of chemotactic leukotriene $B_4$ biosynthesis. J Biol Chem 279:50639–50642

Haeggstrom JZ, Funk CD (2011) Lipoxygenase and leukotriene pathways: biochemistry, biology, and roles in disease. Chem Rev 111:5866–5898

Sala A, Folco G, Murphy RC (2010) Transcellular biosynthesis of eicosanoids. Pharmacol Rep 62:503–510

Samuelsson B (1983) Leukotrienes: mediators of immediate hypersensitivity reactions and inflammation. Science 220:568–575

Samuelsson B, Dahlen SE, Lindgren JA, Rouzer CA, Serhan CN (1987) Leukotrienes and lipoxins: structures, biosynthesis, and biological effects. Science 237:1171–1176

Snelgrove RJ (2011) Leukotriene $A_4$ hydrolase: an anti-inflammatory role for a proinflammatory enzyme. Thorax 66:550–551

Tholander F, Kull F, Ohlson E, Shafqat J, Thunnissen MM, Haeggstrom JZ (2005) Leukotriene A4 hydrolase, insights into the molecular evolution by homology modeling and mutational analysis of enzyme from *Saccharomyces cerevisiae*. J Biol Chem 280:33477–33486

Tholander F, Muroya A, Roques B-P, Fournié-Zaluski M-C, Thunnissen MGM, Haeggström JZ (2008) Structure-based dissection of the active site chemistry of leukotriene A4 hydrolase; implications for M1 aminopeptidases and inhibitor design. Chem Biol 15:1–10

# Zinc Matrix Metalloproteinases and TIMPs

Hideaki Nagase
Kennedy Institute of Rheumatology, Nuffield Department of Orthopaedics, Rheumatology and Musculoskeletal Sciences, University of Oxford, London, United Kingdom

## Synonyms

Collagenase; Enamelysin; Epilysin; Gelatinase; Matrilysin; Matrixin; Metzincin

## Definition

Matrix metalloproteinases (MMPs), also called matrixins, are a family of proteolytic enzymes containing a zinc ion ($Zn^{2+}$) in the active site, and their amino acid sequences are homologous to that of the catalytic metalloproteinase domain of human matrix metalloproteinase 1 (MMP-1). TIMP is an acronym for tissue inhibitor of metalloproteinases which inhibits MMPs by forming a 1:1 molar stoichiometry.

The first MMP was a collagenase found in tadpole tissues undergoing metamorphosis (Gross and Lapiere 1962). The matrixins are now found in both animal and plant kingdoms. The main function of animal MMPs was considered to be the degradation and removal of extracellular matrix (ECM) macromolecules from the tissue, but breakdown of ECM also alters cell-matrix, cell-cell interaction, and the release of growth factors bound to ECM and makes them available to cell receptors (Overall 2002). A number of non-ECM molecules such as cytokines, chemokines, and proteinase inhibitors are also processed. A variety of cell types produce matrixins, but their production is regulated by growth factors, cytokines, physical stress to the cell, hormones, cell-cell, cell-matrix interactions, and oncogenic transformation (Sternlicht and Werb 2001). Under physiological conditions, MMPs are important in embryonic development, organ morphogenesis, tissue remodeling and repair, and angiogenesis and apoptosis. Aberrant MMP activities are associated with diseases such as cancer, arthritis, cardiovascular disease, neurodegenerative disease, tissue ulcers, and fibrosis.

## MMP Family

There are six *Mmp* genes in nematode (*Caenorhabditis elegans*), two genes in fruit fly (*Drosophila melanogaster*), and five genes in plant (*Arabidopsis thaliana*). The human genome has 24 *MMP* genes, of which two genes encode an identical MMP-23 protein. Thus, there are 23 different MMPs in humans. MMPs are classified as the matrixin subfamily of metalloprotease family M10 in the MEROPS database (http://merops.sanger.ac.uk/search.shtml). Vertebrate matrixins are assigned with MMP numbers and trivial names in many cases (Table 1). Some MMP numbers (i.e., MMP-4, -5, -6, and -22) are missing in the list because they are identical to other MMPs.

Common domains in MMPs are a signal peptide, a pro-domain, and a catalytic metalloproteinase domain. In addition, many have a hemopexin domain attached to the C-terminal of the catalytic domain via a linker (or hinge) peptide (Fig. 1). Several of them are glycoproteins. In particular, MMP-8, MMP-9, and MMP-23 in humans are highly glycosylated, with approximately 20–45% of their molecular mass being carbohydrate. All pro-domains, except that of MMP-23, have a so-called "cysteine switch" motif with amino acid sequence PRCGXPD whose cysteine residue interacts with the catalytic $Zn^{2+}$ as its fourth ligand. This Cys-$Zn^{2+}$ coordination keeps proMMPs inactive. The catalytic domain has a zinc-binding motif HEXGHXXGXXH, where the three histidines coordinate $Zn^{2+}$. It also contains a conserved methionine, forming a "Met-turn" eight residues after the zinc-binding motif, which forms a base to support the structure around the catalytic zinc.

The zinc-binding motif and the Met-turn are also conserved in members of the ADAM (*a d*isintegrin *a*nd *m*etalloproteinase) family, the ADAMTS (ADAM with *t*hrombo*s*pondin motifs) family, the astacin family, the bacterial serralysin family, a protozoan proteinase leishmanolysin, and a pregnancy-associated plasma protein A (PAPP-A, pappalysin), and they are collectively called "metzincins" (Gomis-Rüth 2009). While the primary structures of these metalloproteinase domains have little homology among the families, the overall protein folds are similar.

On the basis of substrate specificity, sequence similarity, and domain organization, vertebrate matrixins are subgrouped into collagenases, gelatinases, stromelysins, matrilysins, membrane-type (MT)-MMPs, and others. Figure 1 shows variations in MMP domain arrangement, and Table 2 describes the composition of each MMP member.

### Collagenases

This subgroup consists of MMP-1 (collagenase 1), MMP-8 (collagenase 2), MMP-13 (collagenase 3), and MMP-18 (collagenase 4) in *Xenopus*. The common feature of these enzymes is their ability to cleave interstitial collagens I, II, and III into characteristic ¾ and ¼ fragments. They also can digest other ECM molecules and soluble non-ECM proteins. MMP-2 (gelatinase A) and MMP-14 (MT1-MMP) have collagenolytic activity, but they are grouped into other subgroups due to their domain compositions.

### Gelatinases

MMP-2 (gelatinase A) and MMP-9 (gelatinase B) are in this subgroup. The term "gelatinases" was coined for them as they have potent gelatinolytic activity, which depends on their three fibronectin type II repeats inserted in the catalytic domain (Fig. 1). They also digest a number of ECM molecules. MMP-2 digests collagen I, II, and III in a similar manner to the collagenases, but collagenase activity of MMP-9 is controversial. Both enzymes participate in activation and inactivation of cytokines and chemokines.

### Stromelysins

MMP-3 (stromelysin 1), MMP-10 (stromelysin 2), and MMP-11 (stromelysin 3) are in this subgroup. MMP-3 and MMP-10 have similar enzymatic properties, including digestion of a number of ECM molecules, and activate several proMMPs. MMP-11 has a divergent amino acid sequence and restricted substrate specificity. Human MMP-11 has little activity toward ECM molecules, but inactivates serpins. Alternative splicing and promoter usage generates an intracellular form of MMP-11.

**Zinc Matrix Metalloproteinases and TIMPs, Table 1** Matrix metalloproteinases (MMPs) and their substrates

| MMP | Enzyme | Chromosomal location (Human) | Representative substrates |
|---|---|---|---|
| MMP-1 | Interstitial collagenase (collagenase-1) | 11q22–q23 | Collagens I, II, III, VIII, and X; gelatin; aggrecan; versican; proteoglycan link protein; casein; $\alpha_1$-proteinase inhibitor; $\alpha_2$-M; pregnancy zone protein; ovostatin; nidogen; MBP; proTNF; L-selectin; proMMP-2; proMMP-9; IGFBP-2, -3; MCP-1, -3, -4; CDF-1; IL-1$\beta$ |
| MMP-2 | Gelatinase A | 16q13 | Collagens I, IV, V, VII, X, XI, and XIV, gelatin, elastin, fibronectin, aggrecan, versican, proteoglycan link protein, MPB, proTNF, $\alpha_1$-proteinase inhibitor, proMMP-9, proMMP-13, poly(ADP-ribose)polymerase, IGFBP-3, -5, MCP-3, SDF-1, IL-1$\beta$ |
| MMP-3 | Stromelysin-1 | 11q23 | Collagens III, IV, IX, and X; gelatin; aggrecan; versican; perlecan; nidogen; proteoglycan link protein; fibronectin; laminin; elastin; casein; fibrinogen; antithrombin-III; $\alpha_2$M; ovostatin; $\alpha_1$-proteinase inhibitor; MBP; proTNF; proMMP-1; proMMP-7; proMMP-8; proMMP-9; proMMP-13; IGFBP-3; MCP-1, -2, -3, -4; CDF-1 |
| MMP-7 | Matrilysin-1 (PUMP-1) | 11q21–q22 | Collagens IV and X, gelatin, aggrecan, proteoglycan link protein, fibronectin, laminin, entactin, elastin, casein, transferrin, MBP, $\alpha_1$-proteinase inhibitor, proTNF, proMMP-1, proMMP-2, proMMP-9, E-cadherin, syndecan 1 |
| MMP-8 | Neutrophil collagenase (collagenase-2) | 11q21–q22 | Collagens I, II, III, V, VII, VIII, and X, gelatin, aggrecan, $\alpha_1$-proteinase inhibitor, $\alpha_2$-antiplasmin, fibronectin |
| MMP-9 | Gelatinase B | 20q11.2–q13.1 | Collagens IV, V, VII, X, and XIV, gelatin, elastin, aggrecan, versican, proteoglycan link protein, fibronectin, nidogen, $\alpha_1$-proteinase inhibitor, MBP, proTNF, MCP-1, CDF-1, IL-8 |
| MMP-10 | Stromelysin-2 | 11q22.3–q23 | Collagens III, IV, and V, gelatin, casein, aggrecan, elastin, proteoglycan link protein, fibronectin, proMMP-1, proMMP-8 |
| MMP-11 | Stromelysin-3 | 22q11.2 | $\alpha_1$-proteinase inhibitor |
| MMP-12 | Macrophage metalloelastase | 11q22.2–q22.3 | Collagen IV, gelatin, elastin, $\alpha_1$-proteinase inhibitor, fibronectin, vitronectin, laminin, proTNF, MBP |
| MMP-13 | Collagenase-3 | 11q22.3 | Collagens I, II, III, and IV, gelatin, plasminogen activator inhibitor 2, aggrecan, perlecan, tenascin, MCP-3, CDF-1 |
| MMP-14 | MT1-MMP | 14q11–q12 | Collagens I, II, and III, gelatin, casein, elastin, fibronectin, laminin B chain, vitronectin, aggrecan, dermatan sulfate proteoglycan, MMP-2, MMP-13, proTNF, MCP-3, CDF-1, CD44, tissue transglutaminase |
| MMP-15 | MT2-MMP | 15q13–q21 | proMMP-2, gelatin, fibronectin, tenascin, nidogen, laminin |
| MMP-16 | MT3-MMP | 8q21 | proMMP-2 |
| MMP-17 | MT4-MMP | 12q24.3 | Gelatin, fibrinogen/fibrin |
| MMP-18 | Xenopus collagenase | Not found in humans | Collagen I |
| MMP-19 | | 12q14 | Collagen IV, gelatin, laminin, nidogen, tenascin, fibronectin, aggrecan, COMP |
| MMP-20 | Enamelysin | 11q22.3 | Amelogenin |
| MMP-21 | XMMP (Xenopus) | | Gelatin, $\alpha_1$-proteinase inhibitor |
| MMP-23 | CA-MMP | 1p36.3 | gelatin |
| MMP-24 | MT5-MMP | 20q11.2 | proMMP-2, proMMP-9, gelatin |
| MMP-25 | MT6-MMP | 16p13.3 | Collagen IV, gelatin, fibronectin, fibrin |
| MMP-26 | Matrilysin-2, endometase | 11p15 | Collagen IV, fibronectin, fibrinogen, gelatin, $\alpha_1$-proteinase inhibitor, proMMP-9, estrogen receptor $\beta$ |
| MMP-27 | | 11q24 | |
| MMP-28 | Epilysin | 17q21.1 | Casein |

$\alpha_2$-M $\alpha$2-macroglobulin, *CDF-1* stromal cell-derived factor-1, *COMP* cartilage oligomeric matrix protein, *IGFBP* insulin-like growth factor-binding protein, *MBP* myelin basic protein, *MCP* monocyte chemoattractant protein, *TNF* tumor necrosis factor

# Zinc Matrix Metalloproteinases and TIMPs

**Zinc Matrix Metalloproteinases and TIMPs, Fig. 1** Domain structures of the mammalian MMP family. See Table 2 for the domain arrangement for each MMP. *SP* signal peptide, *Pro* pro-domain, *FN* fibronectin type II domain, *L1* linker 1, *Hpx* hemopexin domain; *L2* linker 2, *Mb* plasma membrane, *TM* transmembrane domain, *Cy* cytoplasmic tail, *CysR* cysteine-rich domain, *Ig* immunoglobulin domain, *GPI* glycosylphosphatidylinositol anchor (Courtesy of Alan Lyons)

**Zinc Matrix Metalloproteinases and TIMPs, Table 2** Matrix metalloproteinases and their domain composition

| Enzyme | MMP | SP | Pro | CS | RX[R/K]R | Cat | FN2 | Lk 1 | Hpx | Lk 2 | TM | GPI | Cyt | CysR-Ig |
|---|---|---|---|---|---|---|---|---|---|---|---|---|---|---|
| *Collagenases* | | | | | | | | | | | | | | |
| Insterstitial collagenase; Collagenase 1 | MMP-1 | + | + | + | − | + | − | + | + | | | | | |
| Neutrophil collagenase; Collagenase 2 | MMP-8 | + | + | + | − | + | − | + | + | | | | | |
| Collagenase 3 | MMP-13 | + | + | + | − | + | − | + | + | | | | | |
| Collagenase 4 (Xenopus) | MMP-18 | + | + | + | − | + | − | + | + | | | | | |
| *Gelatinases* | | | | | | | | | | | | | | |
| Gelatinase A | MMP-2 | + | + | + | − | + | + | + | + | | | | | |
| Gelatinase B | MMP-9 | + | + | + | − | + | + | + | + | | | | | |
| *Stromelysins* | | | | | | | | | | | | | | |
| Stromelysin 1 | MMP-3 | + | + | + | − | + | − | + | + | | | | | |
| Stromelysin 2 | MMP-10 | + | + | + | − | + | − | + | + | | | | | |
| *Matrilysins* | | | | | | | | | | | | | | |
| Matrilysin 1 | MMP-7 | + | + | + | − | + | − | − | − | | | | | |
| Matrilysin 2 | MMP-26 | + | + | + | − | + | − | − | − | | | | | |
| Stromelysin 3 | MMP-11 | (+) | (+) | + | + | + | − | + | + | | | | | |
| *Membrane-type MMPs* | | | | | | | | | | | | | | |
| *A) Transmembrane type* | | | | | | | | | | | | | | |
| MT1-MMP | MMP-14 | + | + | + | + | + | − | + | + | + | + | − | + | |
| MT2-MMP | MMP-15 | + | + | + | + | + | − | + | + | + | + | − | + | |
| MT3-MMP | MMP-16 | + | + | + | + | + | − | + | + | + | + | − | + | |
| MT5-MMP | MMP-24 | + | + | + | + | + | − | + | + | + | + | − | + | |
| *B) GPI –anchored* | | | | | | | | | | | | | | |
| MT4-MMP | MMP-17 | + | + | + | + | + | − | + | + | + | − | + | − | |
| MT6-MMP | MMP-25 | + | + | + | + | + | − | + | + | + | − | + | − | |
| *Others* | | | | | | | | | | | | | | |
| Macrophage elastase | MMP-12 | + | + | + | − | + | − | + | + | | | | | |
| – | MMP-19 | + | + | + | − | + | − | + | + | | | | | |
| Enamelysin | MMP-20 | + | + | + | − | + | − | + | + | | | | | |
| – | MMP-21 | + | + | + | + | + | − | + | + | | | | | |
| CA-MMP | MMP-23 | + | + | − | + | + | − | − | − | | | | | + |
| – | MMP-27 | + | + | + | − | + | − | + | + | | | | | |
| Epilysin | MMP-28 | + | + | + | + | + | − | + | + | | | | | |

Subgroups of MMPs are listed with their trivial names. *SP* signal peptide, *Pro* pro-domain, *CS* cysteine switch motif, *RX[R/K]R* proprotein convertase recognition sequence, *FN2* fibronectin type II motif, *LK* linker, *TM* transmembrane domain, *GPI* glycosylphosphatidylinositol anchoring sequence, *Cyt* cytoplasmic domain, *CyR-Ig* cysteine-rich and immunoglobulin domains

## Matrilysins

MMP-7 (matrilysin 1) and MMP-26 (matrilysin 2) are in this subgroup. They lack a hemopexin domain. MMP-7 is synthesized by epithelial cells and is secreted apically. Besides cleaving ECM components, it processes cell surface molecules such as pro-α-defensin, Fas-ligand, pro-tumor necrosis factor α (TNFα), and E-cadherin. MMP-26 is expressed in normal cells such as those of the endometrium and in some carcinomas. It digests several ECM molecules, but is largely stored intracellularly.

## Membrane-type MMPs (MT-MMPs)

There are six MT-MMPs in humans. They are numerically assigned from MT1-MMP to MT6-MMP, and four are type I transmembrane proteins (MMP-14, -15, -16, and -24), and two are glycosylphosphatidylinositol-anchored proteins (MMP-17 and -25). They all have a furin recognition sequence RX[R/K]R at the C-terminus of the pro-domain and can therefore be activated intracellularly, and active enzymes are expressed on the cell surface. All MT-MMPs, except MMP-17 (MT4-MMP), can activate proMMP-2. MMP-14 (MT1-MMP) digests interstitial collagens effectively. Postnatal skeletal abnormalities in MMP-14-null mice are attributed to the lack of collagenolytic activity.

## Other MMPs

Seven MMPs are not grouped in the above categories, even though MMP-12, MMP-20, and MMP-27 have similar structures and chromosome location as the stromelysins (see Table 1).

MMP-12 (metalloelastase) is expressed primarily in macrophages. It digests elastin and a number of ECM molecules. It is essential in macrophage migration. The hemopexin domain has antimicrobial activity against Gram-negative and Gram-positive bacteria.

MMP-19 digests many ECM molecules, including the components of basement membranes. It was also called RASI (rheumatoid arthritis synovial inflammation) as it was found in the activated lymphocytes and plasma from patients with rheumatoid arthritis as an autoantigen in patients with rheumatoid arthritis and systemic lupus erythematosus. It is, however, widely expressed in many organs including proliferating keratinocytes in healing wounds.

MMP-20 (enamelysin) is expressed in newly formed tooth enamel and digests amelogenin.

MMP-21 was originally found in *Xenopus*, and later in mice and humans. It is expressed in various fetal and adult tissues and in basal and squamous cell carcinomas. It has a similar domain organization as most MMPs with a furin activation sequence in the propeptide, but it lacks the linker region. It digests gelatin and inactivates α1-proteinase inhibitor.

MMP-23 is mainly expressed in reproductive tissues. It is a unique member of the matrixins. Two identical copies of the MMP-23 gene are located on human chromosome at the 1p36 locus, in a head to head arrangement. It lacks the cysteine switch sequence in the propeptides and has a unique C-terminal cysteine-rich immunoglobulin-like domain instead of a hemopexin domain. It is proposed to be a type II membrane protein, having a transmembrane domain at the N-terminal of the propeptide, but the enzyme is released from the cell as the membrane anchored propeptide is cleaved intracellularly by a proprotein convertase. It digests gelatin, but physiological substrates of this enzyme are not known.

MMP-27 was first found in chicken embryo fibroblasts. Chicken MMP-27 digests gelatin and casein and causes autolysis of the enzyme, but little information is available on the activity of mammalian enzymes.

MMP-28 (epilysin) is expressed in many tissues such as lung, placenta, heart, gastrointestinal tract, and testis. The enzyme expressed in basal keratinocytes in skin is considered to function in wound repair. It is also elevated in cartilage from patients with osteoarthritis and rheumatoid arthritis.

## Three-Dimensional (3D) Structures

The crystal structure of the proMMP-2-TIMP-2 complex (Morgunova et al. 2002) and each representative domain which constitutes proMMPs are presented in Fig. 2. The propeptide domain of ~80 amino acids plays a key role in maintaining the MMPs in their zymogen forms (proMMPs) by blocking the active site. It consists of three α helices and connecting loops (Fig. 2a, b). The cysteine switch motif interacts with the substrate-binding cleft of the catalytic domain by forming β structure-like hydrogen bonds in a manner similar to a substrate, but in the opposite direction. The SH group of the cysteine coordinates with the catalytic $Zn^{2+}$ and prevents a water molecule essential for catalysis from binding to the $Zn^{2+}$.

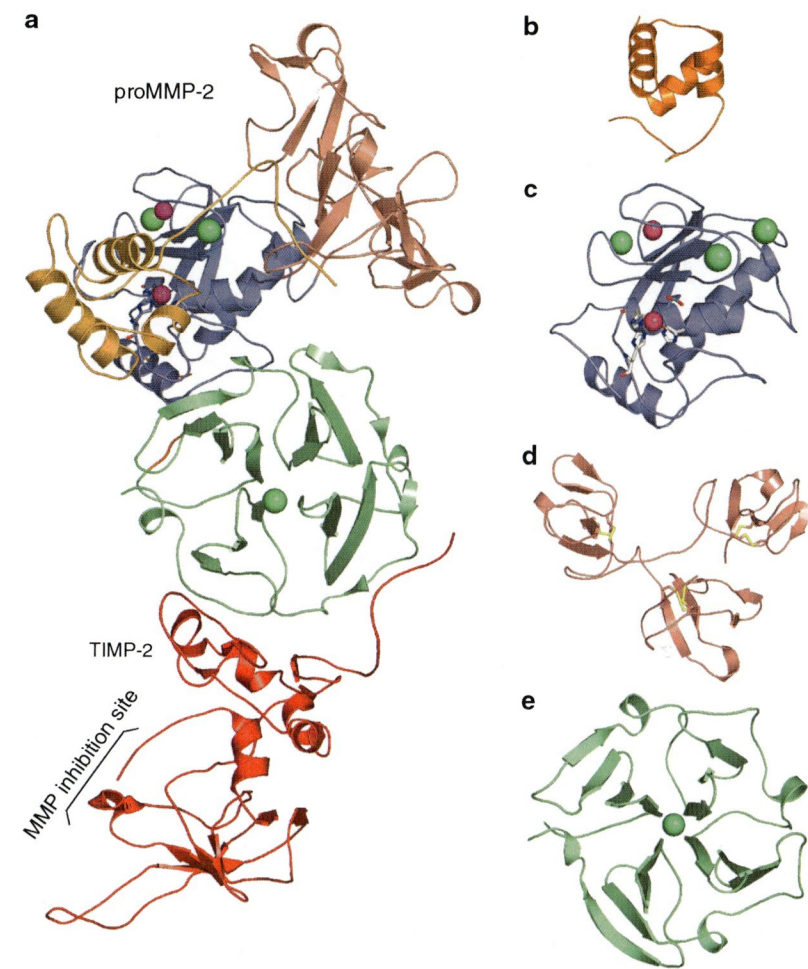

**Zinc Matrix Metalloproteinases and TIMPs, Fig. 2** Three-dimensional structures of MMPs. (**a**) The proMMP-2-TIMP-2 complex. (**b**) Pro-domain of proMMP-2. (**c**) The catalytic domain of MMP-1. (**d**) The three fibronectin domains of MMP-2. (**e**) The hemopexin domain of MMP-1. The pro-domain is in orange; the catalytic domain in *blue*; fibronectin domains in *pink*; hemopexin domain in *green*; TIMP-2 in *red*; $Zn^{2+}$ in *read sphere*; $Ca^{2+}$ in *green sphere* (Courtesy of Robert Visse)

Upon activation of proMMP, the cysteine switch dissociates from the active site.

The catalytic domain has ~170 amino acids containing the zinc-binding motif (HEXGHXXGXXH) and a conserved "Met-turn" structure. The domain consists of three α-helices, a five-stranded β sheet, and connecting loops (Fig. 2a, c). It contains two zinc ions; one is essential for catalysis, and the other together with 2–3 calcium ions stabilizes the structure. The overall polypeptide folds of catalytic domains of MMPs are essentially identical and homologous to those of other metzincin metalloproteinases. In the orientation presented in Fig. 3, a peptide substrate binds to the active site of the catalytic domain from left (N-terminal of the substrate) to right (C-terminal of the substrate), and the binding is dictated by the structure of the substrate-binding site, including a site called the "S1' pocket" located to the right of the catalytic $Zn^{2+}$. This pocket, which accommodates the P1' residue (the residue located on the C-terminal to the cleaved peptide bond), is hydrophobic in nature, but variable in depth and shape among MMPs. It is therefore one of the determining factors for substrate specificity of MMPs. For protein substrate recognition, particularly for ECM molecules, substrate-binding subsites may extend not only in the catalytic domain but also to sites in the non-catalytic domains, called exosites.

The hemopexin domain consists of ~190 amino acids and forms a four-bladed β-propeller structure arranged almost symmetrically around a central axis, and each blade is comprised of a four-stranded antiparallel β sheet (Fig. 2a, e). The domain is stabilized by a calcium ion and a disulfide bond between the 1st blades and 4th blade. The hemopexin domains are essential for collagenases to degrade native triple

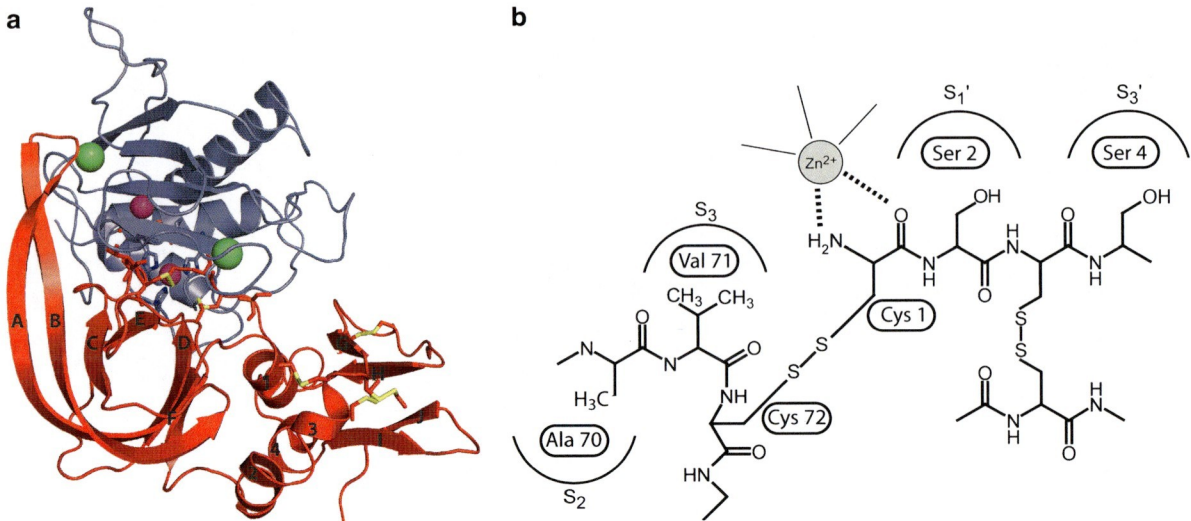

**Zinc Matrix Metalloproteinases and TIMPs, Fig. 3** The model of interaction between the catalytic domain of MMP-14 and TIMP-2. (**a**) The 3D structure of the MMP-14-TIMP-2 complex. MMP-14 is shown in *blue*; TIMP-2 in *red*; $Zn^{2+}$ in *read sphere*; $Ca^{2+}$ in *green sphere*. Strands and helices in TIMP-2 are labeled as A-J and 1-4, respectively. (**b**) The model of interaction between TIMP-2 and the catalytic domain of MMP-14. The N-terminal α-amino and carbonyl groups of $Cys^1$ bidentately chelate the catalytic $Zn^{2+}$ and $Ser^2$, and $Ser^4$ interacts with S1' and S3' pockets, and $Ala^{70}$ and $Val^{71}$ interact with S2 and S3 pockets, respectively (Courtesy of Robert Visse)

helical collagens I, II, and III and for MMP-14 to dimerize on the cell surface. The 3rd and 4th blades of proMMP-2 interact with the C-terminal domain of TIMP-2 (Fig. 3a). This proMMP-2-TIMP-2 complex formation is essential for proMMP-2 activation by dimeric MMP-14 on the cell surface. The hemopexin domains of proMMP-2 also bind to TIMP-3 and TIMP-4 and that of proMMP-9 to TIMP-1 and TIMP-3 in a similar manner. Thus, TIMPs bound to proMMP-2 or -9 retain their activity to inhibit MMPs.

The catalytic domain and the hemopexin domain are connected by a "linker" peptide of variable length which does not have specific structure. Linkers are considered to be flexible, but they are rich in proline. Thus, they may have some structural constraint suited for specific functions. Correct movement and rearrangement between the catalytic and hemopexin domains are considered to be required for collagenases to express collagenolytic activity.

Three repeats of the fibronectin type II motif are found in MMP-2 and MMP-9. A single type II motif consists of two double-stranded antiparallel β-sheets and two large loops (Fig. 2a, d). Each β-sheet has one disulfide bond. Fibronectin motifs enhance their activities on gelatin, collagen IV, and laminin substrates.

## Mechanism of Peptide Bond Hydrolysis

The active site of MMPs consists of the catalytic $Zn^{2+}$ bound to three imidazoles of histidines and the glutamate in the HEXGHXXGXXH motif and one water molecule bound to the $Zn^{2+}$. A proposed mechanism of peptide bond hydrolysis is that the binding of a peptide substrate to the active site and the interaction of the carbonyl group of the P1 residue (the residue located on the N-terminal to the cleaved peptide bond) with the catalytic $Zn^{2+}$ displace the water molecule from the catalytic $Zn^{2+}$ toward the carboxyl group of the glutamate. This polarizes the water molecules and promotes its nucleophilic attack on the carbonyl carbon of the scissile bond, resulting in formation of a hemiketal intermediate and the subsequent hydrolysis of the peptide bond. The glutamate functions as a general acid/base during catalysis. Mutation of the glutamate to aspartate reduces the enzymatic activity to less than 1%, and mutation to alanine reduces to 0.01% activity of the wild-type.

## Activation of ProMMPs

MMPs are synthesized as pre-proMMPs and co-translationally processed to inactive zymogens (proMMPs). Among 23 human MMPs, 10 of them (6 MT-MMPs, MMP-11, -21, -23, and -28) have furin-like proprotein convertase susceptible sequence at the C-terminal end of the pro-domain (see Table 2), and they are activated intracellularly. Secreted proMMPs are activated extracellularly by proteinases or by chemical and physical means.

### Activation by Proteinases

ProMMPs secreted from cells can be activated by tissue or plasma proteinases or opportunistic bacterial proteinases. One of the key initiators of proMMP activation is the urokinase-type plasminogen activator/plasmin system on the cell surface which generates pericellularly active MMPs. In many cases, these proteinases attack a proteinase susceptible "bait" region located in the middle of the pro-domain and remove a part of the pro-domain rather than removing the entire propeptide. The complete removal including the cysteine switch is carried out in trans by the action of the MMP intermediate or by other active MMPs. This mechanism is referred to as "stepwise activation," and it has been demonstrated for many of the proMMPs. ProMMP-2 is most effectively activated on the cell surface by MMP-14 in a stepwise manner. This activation process is assisted by binding of TIMP-2 to the cell surface-expressed MMP-14. The C-terminal domain of TIMP-2 in this complex is exposed and bind to the hemopexin domain of proMMP-2 (see Fig. 3a). This ternary complex formation presents the propeptide of proMMP-2 to an adjacent active MMP-14 and subsequent activation of proMMP-2. ProMMP-13 is also activated by MMP-14, but this does not require TIMP-2.

### Activation by Chemical and Physical Means

One property unique to matrixins is that secreted proMMPs can be activated in vitro by SH-reactive agents (e.g., 4-aminophenylmercuric acetate, iodoacetamide, oxidized glutathione), chaotropic agents (e.g., urea, $SCN^-$, $I^-$), sodium dodecylsulfate, reactive oxygen, low pH, and heat treatment. This is due in part to the disruption of the $Cys-Zn^{2+}$ interaction formed between the propeptide and catalytic domains and subsequent removal of the propeptide by autoproteolysis. Oxidation by reactive oxygen species activates some proMMPs in the tissue, but it may also destroy the enzyme.

### Activity and Substrate Specificity

MMPs degrade ECM and non-ECM proteins (Table 1). The potential substrates of more recently discovered MMPs (higher numbers) have not been well characterized. In general, the substrate needs to be six amino acids or longer, and the enzymes have a strong preference for hydrophobic residues at the P1' site such as Leu, Ile, Met, Phe, or Tyr, but the P1 site is not specified. MMPs exhibit optimal proteolytic activities at around pH 7.0–8.0 for most substrates, but in some cases they exhibit a broader pH range. An exception is human MMP-3 which has optimal activity at around pH 5.5–6.0 with a shoulder of activity at pH 7.5–8.0. Non-catalytic domains greatly influence substrate specificity for some MMPs when substrates are extracellular macromolecules, e.g., fibronectin domain repeats of MMP-2 and MMP-9 for the activity on type IV collagen, gelatin, and elastin, and the hemopexin domain of collagenases is essential for the collagenolytic activity.

MMPs are inhibited by chelating agents such as EDTA, 1-,10-phenanthroline, cysteine, and dithiothreitol. Based on substrate specificity, numerous synthetic inhibitors of MMPs, many aimed for therapeutic purposes, have been designed. They commonly contain a zinc-chelating moiety such as a hydroxamic acid, a carboxyl, a thiol, or a phosphorus group. Some selective inhibitors for specific MMPs have been generated, but many broadly inhibit MMPs and other metalloproteinases. Phosphoramidon is not inhibitory. TIMPs (see below), α2-macroglobulins, and ovomacroglobulins (ovostatins) are the known natural protein inhibitors.

### Biological Activities of MMPs

The degradation of ECM molecules by MMPs is essential for tissue remodeling and repair, but it also expresses new biological functions of these molecules. MMPs can also release growth factors bound to ECM and from other carrier proteins or inactivate inflammatory cytokines and chemokines, which influences cellular behavior. For example, the cleavage of type I collagen is associated with keratinocyte migration during wound healing, activation of osteoclasts during bone remodeling, and induce apoptosis of amnion epithelial cells before the onset of labor. Cleavage of laminin 5 and CD44 by MMPs promotes cell migration. Shedding of syndecan 1/chemokine complex by MMP-7 in response to injury promotes inflammation and tissue repair. Inactivation of monocyte chemoattractant proteins or interleukin 1β by MMPs regulates inflammatory process. The release of transforming growth factor β and vascular endothelial growth factor (VEGF) from the matrix by MMPs modulates ECM synthesis and angiogenesis, respectively, which are key events in development, morphogenesis, reproduction, tissue remodeling, and repair. These events occur in extracellular milieus, and many take place pericellularly. While MT-MMPs are membrane anchored, secreted MMPs are recruited to the cell surface by interacting with cell surface receptors or pericellular matrix molecules (e.g., integrins, CD44, tetraspanins (CD63, CD151), EMMPRIN (CD147), cell membrane cholesterol sulfate, cell membrane Ku, neutrophil surface, glycosaminoglycans) (Murphy and Nagase 2011). These interactions are specific and introduce intricate regulatory mechanism that generates a gradient or directionality of MMP activities. Low density lipoprotein-related protein-mediated endocytosis of MMP-2, -9, and -13 downregulated their activities.

Although MMPs function extracellularly in most cases, MMPs-2, -3, -9, -13, and -14 are found in nucleus of a number of cells including heart myocytes, brain neurons, endothelial cells, fibroblasts, and hepatocytes (Hadler-Olsen et al. 2011). The mechanisms by which secreted MMPs enter into the cell are poorly understood, but they are associated with cellular apoptosis resulted from the cleavage of poly(ADP-ribose) polymerase. MMP-3 participates in transcriptional activation of the connective tissue growth factor (CCN2) gene expression in chondrocytes.

## TIMPs

TIMPs are structurally related proteins which inhibit MMPs by forming a 1:1 molar stoichiometric complex. The human genome has four TIMP genes, and there are two genes in *C. elegans* and one in *Drosophila*. The properties of four human TIMPs are described in Table 3 (Brew and Nagase 2010). The four isoforms (TIMP-1 to -4) consist of 184–194 amino acids, subdivided into N-terminal and C-terminal domains, each having three disulfide bonds. The N-terminal domain is the inhibitory domain and it binds to the active site of the MMPs. All four TIMPs inhibit MMPs, but their affinities vary depending on the enzyme-inhibitor pairs. The C-terminal domain of TIMP-2, TIMP-3, and TIMP-4 binds to the hemopexin domains of progelatinases, and these complexes retain MMP inhibitory activity since their N-terminal inhibitory domains are free. Among the four MMPs, TIMP-3 has the broadest inhibition spectrum as it inhibits several members of ADAMs and ADAMTSs. It also differs from the others in being tightly bound to the ECM. TIMP-1 inhibits ADAM10.

### Inhibition Mechanism

The inhibitory domain of TIMPs has an OB-fold structure with a reactive ridge formed by the N-terminal segment $Cys^1$-Ser-Cys-Ser-$Pro^5$ and the $Ala^{70}$-Val-$Cys^{72}$ (residues are in TIMP-2), where $Cys^1$ and $Cys^{72}$ are disulfide linked. This ridge slots into the active site cleft of the MMPs (Fig. 3), and the first five TIMP-2 residues bind to that active site of MMP similar to P1, P1′, P2′, P3′, and P4′ peptide substrate residues forming interchain hydrogen bonds, whereas the $Ala^{70}$-$Val^{71}$ segment interacts with subsites S2 and S3 in an arrangement nearly opposite from that of a substrate. The nitrogen of the N-terminal $NH_2$-group and the carbonyl oxygen of $Cys^1$ bidentately chelate the $Zn^{2+}$ in the active site of the enzyme. The $Ser^2$ side chain of TIMP-2 extends toward the S1′ pocket of the MMP with its Oγ pointing to the carboxylate of the catalytic Glu (Fernandez-Catalan et al. 1998). TIMP-1 and TIMP-3 interact with a target metalloproteinase in a similar manner. MMPs do not form covalent bonds with TIMPs, nor do they cleave TIMPs, but form tight complexes with inhibition constants in the subnanolar range.

### Biological and Pathological Functions of TIMPs

Many biological and pathological functions have been demonstrated from studies with mice deficient in a specific TIMP gene (Table 4; see references in Brew and Nagase (2010)).

TIMP-1$^{-/-}$ mice exhibit alteration in processes in reproduction and steroid genesis, in remodeling of heart, liver and lung tissues, resistance to corneal and pulmonary infection by *Pseudomonas Aeruginosa*, decrease in adipose tissue weight, and impaired learning and memory. TIMP-1 contributes the regulation of feeding and energy balance.

**Zinc Matrix Metalloproteinases and TIMPs, Table 3** Properties of the four human TIMPs

| Property | TIMP-1 | TIMP-2 | TIMP-3 | TIMP-4 |
|---|---|---|---|---|
| Glycosylation | Yes | No | Partial | No |
| pI | 8.47 | 6.48 | 9.14 | 7.21 |
| No. residues[a] | 184 | 194 | 188 | 194 |
| $Mr$[b] | 20,709 | 21,755 | 21,690 | 22,329 |
| MMP inhibition | Weak for MMP-14, -16, -19, -24 | All | All | Most |
| Other MP inhibition | ADAM-10 | ADAM-12 | ADAM10, 12, 17, 28, and 33; ADAMTS -4, -5; ADAMTS -2 (weak) | ADAM17 and 28 |
| ProMMP interactions | proMMP-9 | proMMP-2 | proMMP-9, proMMP-2 | proMMP-2 |
| Other partners | CD63, LRP-1 (MMP-9 complex) | $\alpha_3\beta_1$ integrin LRP-1 | EFEMP1 VEGFR2 Angiotensin II receptor LRP-1 | |
| Apoptotic effects | Negative | Negative | Positive | |
| Angiogenesis | Negative | Positive | Negative | Negative |
| Chromosomal location: Human | X11p11.23–11.4 | 17q23–25 | 22q12.1–q13.2 | 3p25 |
| Genetic diseases | | | Sorsby's fundus dystrophy | |

[a]Mature protein
[b]Excluding any glycans

TIMP-2$^{-/-}$ mice exhibit deficiency in neurological disorder, especially those related synaptic plasticity, decreased cerebellar neurite outgrowth, and delayed neuronal differentiation. These effects are considered to be due to an abnormal turnover of neuronal ECM. Altered myotube formation and weakened muscle are also observed.

TIMP-3$^{-/-}$ mice studies indicate that TIMP-3 is a key regulator of metalloproteinases involved in ECM turnover/degradation, inflammation, innate immunity, and cancer progression. Unbalanced matrix degradation is observed in many organs and tissues such as heart, lung, liver, articular joints, and during mammary gland involution. When challenged with disease models linked to inflammation and ECM modeling, the inflammatory process and tissue destruction are accelerated due to a large increase in metalloproteinase activities. For example, liver regeneration after partial hepatectomy in TIMP-3$^{-/-}$ mice was impaired because of abnormal inflammation that increases the release of TNFα, suggesting that ADAM17 is one of the major target of TIMP-3 in this system.

## TIMPs Are Multifunctional Proteins: Biological Activities Independent of Metalloproteinase Inhibition

TIMPs have biological activities that are not associated with their inhibition of metalloproteinases. These include effect on cell growth, differentiation, cell migration, apoptosis, and angiogenesis.

TIMP-1 and TIMP-2 have erythroid-potentiating activity and growth-promoting activity on many cell types. Both TIMP-1 and TIMP-2 increase the levels of Ras-GTP, but TIMP-1 activates the tyrosine kinase/mitogen-activated protein kinase pathway, while TIMP-2 activates protein kinase A which is directly involved in Ras/phosphoinositide 3 kinase complex, but the receptors involved in these processes are not identified.

On the other hand, TIMP-1 suppresses the growth of human breast epithelial cells by inducing cell cycle arrest at G$_1$ phase through downregulation of cyclin D$_1$ via upregulation of cyclin-dependent kinase inhibitor p27$^{KIP1}$. TIMP-1 also exhibits anti-apoptotic effect which is mediated through the tetraspanin, CD63 by forming a TIMP-1-CD63-integrin β1 complex, which constitutively turns on survival signals. The combination of cell cycle arrest and anti-apoptotic activities of TIMP-1 leads to cellular transformation. TIMP-1 positive Burkitt's lymphoma cells are resistant to Fas-dependent, and Fas-independent apoptosis and TIMP-1 promote plasmablastic differentiation of Burkitt's cells.

TIMP-2 suppresses vascular endothelial growth factor (VEGF)- or fibroblast growth factor 2 (FGF-2)-stimulated cell proliferation. This effect is mediated by

**Zinc Matrix Metalloproteinases and TIMPs, Table 4** Mouse phenotypes resulting from individual TIMP gene deletion

| Genotype | Phenotypes |
|---|---|
| TIMP-1$^{-/-}$ | Decreased serum testosterone |
| | Reduced serum progesterone during corpus luteum development |
| | Enhanced estrogen-induced uterine edema |
| | Altered reproductive cycle and uterine morphology |
| | Accelerated endometrial gland formation during early postnatal uterine development |
| | Increased resistance to infection by *Pseudomonas aeruginosa* due to increased inflammatory and complement-dependent immune response |
| | Impaired learning and memory |
| | Reduced airway fibrosis after tracheal transplantation |
| | Enhanced acute lung injury after bleomycin exposure |
| | Increased HGF activity in regenerating livers |
| | Decreased adipose tissue development in dietary-induced obesity |
| | Altered left ventricular (LV) geometry and cardiac function |
| | Exacerbated LV remodeling after myocardial infarction |
| | Hyperphagia and obesity in female mice |
| TIMP-2$^{-/-}$ | Impairment of proMMP-2 activation by MMP-14 |
| | Motor dysfunction |
| | Delayed neuronal differentiation |
| | Defect in pre-attentional sensorimotor gating |
| | Weakened muscle and reduced fast-twitch muscle mass |
| | Altered myotube formation |
| TIMP-3$^{-/-}$ | Spontaneous air space enlargement and impaired lung function in aged mice |
| | Accelerated apoptosis during mammary gland involution |
| | Impaired bronchiole branching morphogenesis |
| | Chronic hepatic inflammation and failure of liver regeneration |
| | Spontaneous dilated cardiomyopathy |
| | LV dilation and dilated cardiomyopathy following aortic banding |
| | Accelerated cardiac dilation and matrix disruption following a myocardial infarction |
| | Resistance against oxygen-glucose deprivation-induced neuronal cell death and increased Fas ligand shedding |
| | Abnormal vascularization with dilated capillaries and increased angiogenesis due to an unbalanced VEGF signaling |
| | Enhanced tumor growth and angiogenesis |
| | Increased susceptibility to LPS-induced mortality |

*(continued)*

**Zinc Matrix Metalloproteinases and TIMPs, Table 4** (continued)

| Genotype | Phenotypes |
|---|---|
| | Increased inflammatory response to intra-articular antigen-induced arthritis |
| | Increased pulmonary compliance following LPS challenge |
| | Spontaneous development of arthritis in aged mice |
| TIMP-3$^{+/-}$ | Accelerated type II diabetes with insulin receptor heterozygosity |

$\alpha_3\beta_1$ integrin, which induces de novo synthesis of p27$^{KIP1}$, resulting in inhibition of cyclin-dependent kinases. TIMP-2 expressed in postmitotic neurons interacts with $\alpha_3\beta_1$ integrin and promotes neuronal cell differentiation and neurite outgrowth as a result of cell cycle arrest. Overexpression of TIMP-2 prevents the apoptosis of murine B16F10 melanoma cells, macrophages, and macrophage-derived foam cells, but TIMP-2 induces apoptosis of activated T cells. TIMP-3 induces apoptosis in a number of cell types, including rat vascular smooth muscle cells. It is thought to be due to the inhibition of death receptor shedding.

A number of MMPs are involved in endothelial cell migration and capillary formation, and inhibition of MMPs by TIMPs can prevent angiogenesis. However, TIMP-2 and TIMP-3 can inhibit angiogenesis without inhibiting MMPs. The engagement of TIMP-2 to $\alpha_3\beta_1$ integrin reduces Src tyrosine kinase levels, which results in enhancement of RECK (the reversion-inducing-cysteine-rich protein with Kazal motif) expression, which results in loss of endothelial cell migration. Antiangiogenic activity is localized in the C-terminal domain of TIMP-2. This region may interact with $\alpha_3\beta_1$ integrin. TIMP-3 inhibits angiogenesis by direct binding to VEGF receptor 2 and blocks the action of VEGF.

## Human Genetic Disorders and DNA Polymorphisms

Unregulated MMP activities are implicated in the progression of a variety of diseases. In many cases, their activities are controlled at the cellular level by various stimulatory factors. In addition, a few inherited human disorders associated with the modification of MMP and TIMP genes have been described.

Mutations in human MMP-2 resulting in the absence of active MMP-2 are linked to a rare autosomal recessive genetic disorder, multicentric osteolysis, that causes destruction of the affected bones and joints. A mutation in the coding region for MMP-13 is associated with spondyloepimetaphyseal dysplasia, a dwarfism resulted from abnormal development of the vertebrae and extremities. A genetic disorder, amelogenin imperfecta, is caused by defective enamel formation. This is due to a mutation of the substrate amelogenin at MMP-20 cleavage sites. Mutations in the coding region of the *TIMP3* gene leading to the presence of an extra Cys in the C-terminal region of the expressed protein cause the rare autosomal dominant Sorsby's fundus dystrophy, a form of blindness characterized by death of the retinal pigment epithelium.

Various DNA polymorphisms in MMP genes are associated with a number of diseases. In many cases, polymorphic sites are found in the promoter regions of MMP genes: Polymorphisms in the *MMP1* gene are associated with coronary heart disease, rheumatoid arthritis, periodontitis, and cancers (e.g., breast, cervix, ovary, colon, kidney, lung, and skin); in the *MMP2* gene with the risk of developing lung cancer; in the *MMP3* gene with coronary heart disease, coronary aneurysms, and cancers (colon and breast); in the *MMP9* gene with coronary heart disease, aneurysms, nephropathy, and preterm prenatal rupture of the fetal membrane; and in the *MMP12* gene with coronary atherosclerosis.

## Cross-References

▸ Calcium and Extracellular Matrix
▸ Zinc Adamalysins
▸ Zinc Meprins
▸ Zinc-Astacins
▸ Zinc-Binding Sites in Proteins

## References

Brew K, Nagase H (2010) The tissue inhibitors of metalloproteinases (TIMPs): an ancient family with structural and functional diversity. Biochim Biophys Acta 1803:55–71

Fernandez-Catalan C, Bode W, Huber R, Turk D, Calvete JJ, Lichte A, Tschesche H, Maskos K (1998) Crystal structure of the complex formed by the membrane type 1-matrix metalloproteinase with the tissue inhibitor of metalloproteinases-2, the soluble progelatinase a receptor. EMBO J 17:5238–5248

Gomis-Rüth FX (2009) Catalytic domain architecture of metzincin metalloproteases. J Biol Chem 284:15353–15357

Gross J, Lapiere CM (1962) Collagenolytic activity in amphibian tissues: a tissue culture assay. Proc Natl Acad Sci USA 48:1014–1022

Hadler-Olsen E, Fadnes B, Sylte I, Uhlin-Hansen L, Winberg JO (2011) Regulation of matrix metalloproteinase activity in health and disease. FEBS J 278:28–45

Morgunova E, Tuuttila A, Bergmann U, Tryggvason K (2002) Structural insight into the complex formation of latent matrix metalloproteinase 2 with tissue inhibitor of metalloproteinase 2. Proc Natl Acad Sci USA 99:7414–7419

Murphy G, Nagase H (2011) Localizing matrix metalloproteinase activities in the pericellular environment. FEBS J 278:2–15

Overall CM (2002) Molecular determinants of metalloproteinase substrate specificity: matrix metalloproteinase substrate binding domains, modules, and exosites. Mol Biotechnol 22:51–86

Sternlicht MD, Werb Z (2001) How matrix metalloproteinases regulate cell behavior (Review). Ann Rev Cell Dev Biol 17:463–516

# Zinc Meprins

Judith S. Bond
Department of Biochemistry and Molecular Biology, Pennsylvania State University College of Medicine, Hershey, PA, USA

## Synonyms

Endopeptidase 24.18; Endopeptidase-2; Meprin A complex peptidase; PABA-peptide hydrolase

## Definition

Meprins are multidomain glycosylated metalloproteinases that belong to the "astacin" evolutionary family and the "metzincin superfamily." Meprin A is composed of $\alpha\alpha$ or $\alpha\beta$ subunits; the homo- or heterodimers are disulfide-linked. Heteromeric meprin A dimers are membrane-bound and tend to form tetramers ($\alpha_2\beta_2$). Homomeric meprin A dimers are secreted enzymes and tend to form high molecular mass complexes. Meprins are expressed in epithelial cells of the kidney, intestine, and skin, as well as leucocytes and cancer cells. The enzymes play important roles at sites of infection and inflammation.

## Background

Meprins are rather unique proteases that have properties that distinguish them from all other metalloproteinases. They are multidomain oligomeric proteases that are highly glycosylated and exist as secreted and cell surface isoforms. They are members of the "metzincin superfamily" and the "astacin" evolutionary family of metalloproteinases. The protease domains of meprin subunits (α and β subunits) have approximately 35% amino acid sequence identity with the crayfish enzyme astacin, the prototype of the evolutionary family. However, meprins, with molecular sizes of 340,000 Da to 6 million Da, are much more complex than the monomeric astacin which has a molecular size of approximately 20,000 Da. There have been several reviews on the structure and function of meprins, the most recent one by Sterchi, Stocker, and Bond (2008). A 2005 review focused on the roles of meprins in disease (Bond et al. 2005). In addition, there are two chapters in the Handbook of Proteolytic Enzymes, 3rd Edition, on Meprins A and Meprin B that were just revised (Bertenshaw and Bond 2011). This entry will give a brief overview of the structure, function, and involvement of meprins in health and disease.

## Discovery of Meprins

Meprins were discovered by three groups working independently, led by John Kenny (UK), Judith Bond (USA) and Robert Beynon (UK) working together, and Erwin Sterchi (Switzerland). John Kenny described an endopeptidase in mouse kidney in 1974 that he called NEP-2 (to distinguish it from NEP-1, now called neprilysin) and characterized this "endopeptidase-2" from rat kidney in 1987. The first purification and characterization of the meprins was by Beynon, Shannon, and Bond (1981) who found that the mouse kidney (from BALB/c mice) had particularly high proteolytic activity compared to other tissues using azocasein as a substrate. In 1982, the Swiss group discovered a "non-pancreatic hydrolase" in human intestine that they called PABA-peptide (*N*-benzoyl-L-tyrosyl-*p*-aminobenzoic acid-peptide) hydrolase; the name was based on the substrate used to identify the enzyme. It was not until the early 1990s when meprins were cloned and sequenced that it became clear that the mouse and human metalloproteinases were products of the same gene and, along with bone morphogenetic protein-1, were members of a new evolutionary family that was named the "astacin family of metalloproteinases" (Dumermuth et al. 1991).

In the 1980s, information accumulated on the localization of meprins in the plasma membrane of epithelial cells (particularly mouse kidney proximal tubule cells and human intestinal cells) and on substrate specificity. It also was discovered that some strains of inbred mice (such as C3H/HeJ mice) had a latent metalloproteinase activity using azocasein as substrate (originally thought to be mice "deficient" in meprins). The mice that were used were from Jackson Labs and had been bred for differences in the major histocompatibility genes (H-2 genes) for transplantation studies. It was the realization that different inbred mouse strains expressed different levels of azocaseinase activity that led to the discovery that a gene (the *Mep-1* gene) on mouse chromosome 17, near the H-2 complex, was responsible for the level of expression of meprin activity in mouse kidney. It is now known that the *Mep-1a* gene codes for the meprin α subunit, and it was the keen eye of a mouse geneticist, Chella David from the Mayo Clinic, who recognized the connection between the H-2 complex and the pattern of enzyme expression in mice.

All the early work on meprins (in the 1980s) was with the membrane-bound isoforms, and we know now that all membrane-bound forms contain the β subunit. The mouse strains that have "high meprin activity" in the kidney (e.g., Balb/c, C57BL/6 mice) express both the α and β subunit; those that have "low meprin activity" (e.g., C3H/He, CBA mice) express only the β subunit. The reason for the lack of expression of the α subunit in adult mice is unknown, but it is known that this is developmentally regulated because both the α and β subunits are expressed in all mouse strains in the embryonic kidney and until puberty. Another difference between the meprin α and β subunits in mouse kidney is that the α subunit is fully active at the plasma membrane while the β subunit is predominantly latent. The β subunit can be fully activated by trypsin-like enzymes and under some circumstances, can be activated and shed from the plasma membrane. The latent activity of meprin β at the kidney plasma membrane is species specific in that it appears to be active on the membrane in the rat kidney and in the human intestine.

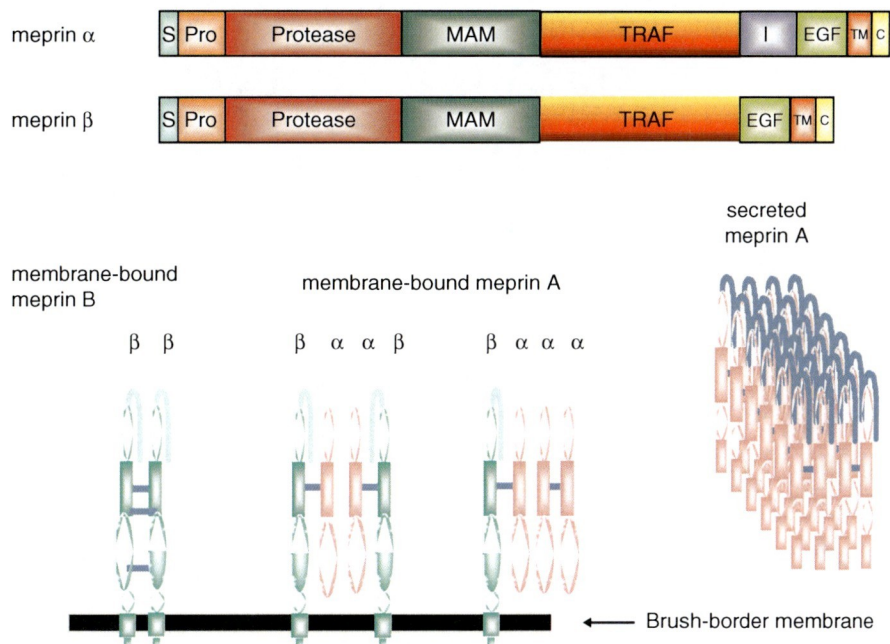

**Zinc Meprins, Fig. 1** *The domain structure of the meprin α and β subunits and their associated meprin A and B isoforms.* Meprin metalloproteases are composed of meprin α and meprin β subunits. Both subunits contain a signal sequence (S), prosequence, protease domain, meprin A5 protein tyrosine phosphatase μ (MAM) domain, tumor necrosis factor receptor-associated factor (TRAF) domain, EGF-like domain, transmembrane (TM) domain, and a COOH-terminal (C) domain. The meprin α subunit contains an additional inserted (I) domain between EGF and TRAF domains. The I domain is necessary and sufficient for meprin α to lose its C-terminal end during maturation, resulting in the three isoforms of meprin: membrane-bound meprin B (dimer of meprin β subunits), membrane-bound meprin A (tetramer of α and β subunits, found in $α_2β_2$ or $α_3β_1$ ratios), or secreted meprin A (multimer of α-α dimers) (Taken from Keiffer (2011))

## Structural Aspects of Meprins

### Domain Structure

The domain structure of meprins is shown in Fig. 1.

Meprins are composed of two evolutionary related subunits, α and β. Meprin α and β subunits are 42% identical in amino acid sequence and have similar domain structures. Both subunits are multidomain proteases containing a signal sequence, prosequence, protease domain, meprin A5 protein tyrosine phosphatase μ (MAM) domain, tumor necrosis factor receptor-associated factor (TRAF) domain, EGF-like domain, transmembrane domain, and a C-terminal domain.

The signal sequence directs the protein into the rough endoplasmic reticulum (ER) during biosynthesis. The prosequence must be removed by proteolysis for the enzyme to be active against peptides and protein substrates. The proteases that activate meprin subunits in vivo are unknown; however, for the α subunit in mouse kidney, immunohistochemistry studies indicate that activation occurs after the subunit passes through the Golgi apparatus and near or at the plasma membrane. For the intestine, it is likely that trypsin in the lumen of the gut activates the subunits. Mutational analyses have shown that astacins are processed to a precise residue in the mature form that results in maximal activity and stability.

The protease domain of meprins contains about 200 amino acids and is 30–40% identical to crayfish astacin in amino acid sequence. There are at present no crystal structures of meprins. However, homology models using the crayfish astacin indicate that the protease domain contains a five-stranded β-sheet, three α-helices, and a characteristic Met-turn. There are two disulfide bonds within the protease domain. This domain contains the signature sequence HEXXHXXGFXHEXXRXDRD, found in all metzincins. Atomic absorption analyses revealed the presence of 1.1 mol zinc and 2.75 mol calcium per mol of meprin α (subunit size of 85,000 Da). Based on the

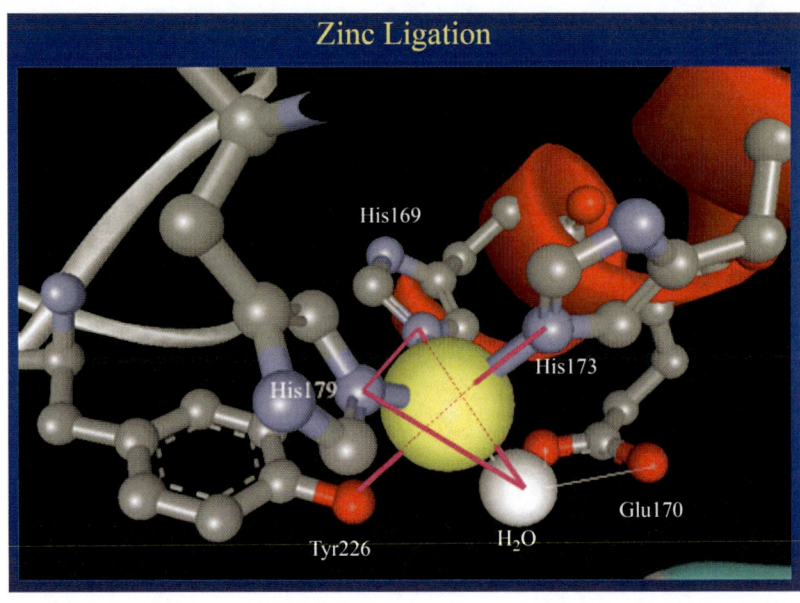

**Zinc Meprins, Fig. 2** *Zinc binding by pentacoordination.* Diagram of the residues in the active site of the mouse meprin α subunit responsible for zinc (*yellow*) binding. The homology model, based on the crayfish astacin crystal structure, shows the zinc ligands consisting of three His residues, a Tyr residue, and a catalytic water molecule coordinated by a Glu residue

crystal structure of the crayfish astacin, zinc is bound by pentacoordination of five ligands: three His residues within the signature sequence and a Tyr residue located approximately 60 residues COOH-terminal to the signature sequence (Fig. 2). In addition, a Glu residue is proposed to polarize a zinc-bound water molecule for nucleophilic attack of the scissile peptide bond during catalysis. Mutational analyses with mouse meprin α indicated that the Tyr residue involved in the pentacoordination is not essential for zinc binding; however, it is essential for full enzymatic activity and stability of the enzyme.

The MAM (*m*eprin, *A*5 protein, receptor protein tyrosine phosphatase *μ*) domain has been suggested to be involved in protein-protein interactions and signal transduction. The MAM domain is present in the extracellular region of several functionally diverse proteins, such as protein phosphatases, ALK (anaplastic lymphoma kinase), and EGFL6 (epidermal growth factor-like protein). Meprins are the only membrane proteins that have an extracellular TRAF (*t*umor *n*ecrosis *r*eceptor-*a*ssociated *f*actor) domain. There are several intracellular proteins containing TRAF domains, and they are involved in protein-protein interactions and transmitting signals from the cell surface to the nucleus in processes such as cell proliferation and cell death. The TRAF domains in meprins are thought to be involved in oligomerization of meprin subunit. Crosslinking studies of meprin B (consisting of b-b subunits) indicated there are three intersubunit disulfide bonds, two between MAM domains and one between TRAF domains. Mutation of any one of the Cys residues involved in these disulfide bridges resulted in no S-S bridges between the subunits, and the meprin β subunit was expressed as a monomer rather than a dimer. Monomers are enzymatically active, after removal of the prosequence; however, they are much less stable than dimers. The function of the EGF-like domains in meprins is not clear. They are not required for correct folding of the subunits, stability, or activity. This is in contrast to the MAM and TRAF domains that are required for correct folding, secretion, and activity of the meprin subunits as well as oligomer formation. The meprin α and β subunit C-terminal tails differ in that the former contains only 6 amino acids and the latter 26–28 amino acids. The β subunit C-terminal tail contains a potential phosphorylation site and potential interaction sequences for proteins. Immunoprecipitation studies using the meprin β C-terminal tail have identified several interacting proteins including the γ subunit of the epithelial sodium channel (ENaC) and the cytoskeletal proteins villin and actin (Ongeri et al. 2011).

The meprin subunits are highly glycosylated. Approximately 20–25% of the mass of the subunits is N-linked carbohydrate, either high mannose or complex-type. There are ten potential Asn glycosylation sites on the mouse meprin α subunit, and nine of them are glycosylated. When these Asn residues were singularly mutated to Gln residues to prevent glycosylation, the subunits were secreted as the WT subunits, but several mutants had decreased enzymatic activity and stability. Multiple Asn mutations did prevent secretion of the subunits. Additional studies indicated that glycans are at or near the noncovalent interface of meprin subunits. Glycans in the protease domain are particularly important for intersubunit disulfide bond formation, noncovalent associations, and enzymatic activity. The data indicate that the carbohydrates are critical for folding, activity, and stability of meprin subunits and oligomers (Ishmael et al. 2006).

Both meprin α and β are synthesized with EGF, transmembrane (TM), and C-terminal domains. However, the meprin α subunit contains an inserted domain (I domain) that contains 56 amino acids $NH_2$-terminal to the EGF domain. This domain, plus a few amino acids in the COOH-terminus of the TRAF domain, is proteolytically cleaved during biosynthesis, which removes the subunit from membrane interactions. Thus, meprin isoforms that contain only α subunits are secreted from cells. The meprin β subunit retains its TM domain through biosynthesis and remains membrane-bound at the cell surface. Mutational studies have demonstrated that if the I domain is removed from the α subunit, the subunit is retained in the endoplasmic reticulum, demonstrating that release from the membrane is necessary for the α subunit to move through the secretory pathway. Retention is due to sequences in the transmembrane and C-terminal domains. In addition, if the I domain is inserted into the β subunit after the TRAF domain, this subunit is proteolytically released from the membrane and secreted from the cell. These studies demonstrate the critical importance of the I domain in determining whether the subunits are secreted or membrane-bound.

## Oligomeric Structure

The basic unit of meprin enzymes is a disulfide-bridged dimer. The dimers may be hetero- or homomeric, and they may form noncovalent associations with other dimers (see Fig. 1). The meprin isoforms that result are heteromeric meprin A (consisting of $α_2β_2$ or $α_3β_1$), meprin B ($β_2$), and homomeric meprin A ($α_2$). Because heteromeric

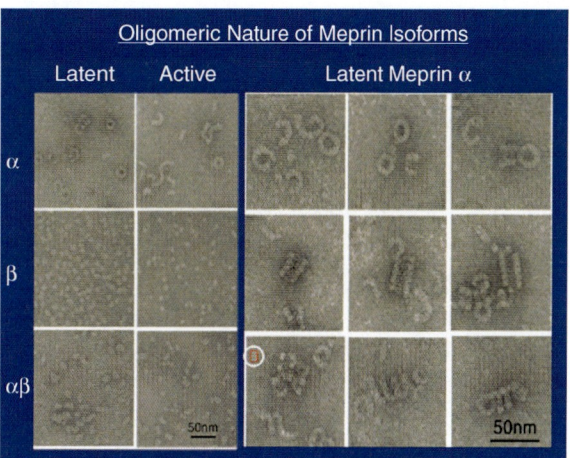

**Zinc Meprins, Fig. 3** *Electron micrographs of rat meprins.* Negatively stained samples of purified preparations of meprin A and B. The oligomerization status depends on several factors including protein concentration, activation status, salt concentration. The number of interacting subunits can vary from 2 to 100 subunits. The *left panel* shows latent and active forms of homomeric meprin A (α), meprin B (β), and heteromeric meprin A (αβ). *The right panel* (latent meprin A) shows multiple structures of latent homo-oligomeric rat meprin A. The top group shows closed ring structures and crescents containing primarily 10–12 subunits of meprin α. The *bottom panels* show tube and spiral structures that form with latent homomeric meprin A with large numbers of subunits (molecular sizes up to six billion daltons). The *bottom panel* also shows an inserted picture of a proteasome core particle (~150 × 115 Å) in the white circle, for size comparison (Taken from Bertenshaw et al. 2003)

meprin A and meprin B contain the β subunit, they are membrane-bound isoforms. Homomeric meprin A is secreted from cells. It is clear that meprin dimers differ markedly in their ability to oligomerize. The dimers of homomeric meprin A are homophilic and tend to form higher order multimers, i.e., they form noncovalent interactions with other dimers. While meprin B forms only dimers, heteromeric meprin A tends to form tetramers. Homomeric meprin A displays a great variety of oligomeric forms, crescent-shaped, and ring structures as well as tubes and spiral-shaped structures composed of as many as 100 subunits (Fig. 3). The homophilic nature of the meprin α dimers results in some of the largest known secreted proteases.

The oligomeric structures of meprins have been determined by electrophoresis, light scattering, size exclusion chromatography, electron microscopy, and crosslinking studies (Fig. 3).

Meprin B has only been found as a disulfide-linked dimer of two β subunits. The intersubunit S-S bridges have been mapped by crosslinking studies; there are two disulfide bonds that link the MAM domains and one that links TRAF domains of the subunits.

Heteromeric meprin A tends to form tetramers containing αβ disulfide-linked dimers. Recombinant heteromeric meprin A was produced by transfecting 293 HEK cells with both meprin α and β cDNA, and only αβ dimers and αββα tetramers were observed. In studies where heteromeric meprin A was isolated from kidney membranes, $α_3β_1$ tetramers have also been observed. We suggest that when both the α and β subunits are expressed by cells, the heterodimer is preferred form. However, when the α subunit is more abundantly expressed, homodimers of the α dimer will associate with the heteromeric αβ dimer and result in the $α_3β_1$ tetramer.

Homomeric meprin A structures are dependent on salt and protein concentration and whether the subunits have been activated with trypsin or trypsin-like enzymes. The latent oligomers are much larger than activated forms. The self-association of the meprin α dimers concentrates the protease in the extracellular environment and allows for autocompartmentalization. This might be very important for the stability and action of these proteases particularly at sights of infection and harsh environments of the intestine.

## Functional Aspects of Meprins

### Peptide Bond and Substrate Specificity

Meprins are capable of hydrolyzing a wide variety of substrates, from peptides to proteins. They prefer peptides containing over six amino acids, and the α and β subunits have different peptide bond specificities. In studies of mouse meprins (homomeric meprin A and B), using peptide libraries (dodecapeptides), it was found that meprin B has a clear preference for acidic residues at the P1 and P1′ sites of the substrate. The homomeric meprin A, by contrast, preferred to cleave peptide bonds containing small or hydrophobic amino acids in the P1′ site and selects for a proline residue in the P2′ site. These differences imply different functions for meprin α and β, and indeed there are clear differences when biologically active peptides are examined. For example, meprin B cleaves gastrin and osteopontin, whereas homomeric meprin A does not; homomeric meprin A cleaves bradykinin and substance P, whereas meprin B does not. Meprin B also cleaves cell surface proteins, such as E-cadherin and ENaC (epithelial sodium channel) proteins, and thereby can effect cell-cell interactions and ion transport. Among the cytokines, meprin A inactivates MCP-1, RANTES, and MIP-1, whereas meprin B activates IL-1β and IL-18. Because both meprin A and B have been found to have roles in inflammation, the balance of the meprin subunits could have important implications in cytokine profiles and the progression of inflammatory responses.

There are overlapping specificities of the meprin subunits as well. Some of the best substrates are cleaved by both meprins A and B. For example, gastrin-releasing peptide, cholecystokinin, and interleukin-6 (IL-6) are hydrolyzed by all meprin isoforms, as are extracellular proteins such as collagen IV, nidogen, laminin, and fibronectin. The cytoskeletal proteins villin and actin are also good substrates for meprin A and B. While these intracellular proteins would ordinarily not be encountered by meprins whose active sites are outside the cell, when there is tissue stress or injury (as in ischemia/reperfusion), the meprins can be found in the cytosol and could cause damage by cleaving these proteins (Ongeri et al. 2011).

Recent proteomic analyses indicate that human meprins A and B as well the crayfish astacin select for acidic residues in the P1′ site. While this preference was previously established for mouse meprin B, it was not observed for the mouse homomeric meprin A isoform. The preference of the human meprin α subunit for acidic residues may well be specie specific. It is unusual for metalloendopeptidases to cleave peptide bonds with acidic amino acid residues. However, aggrecanases and stromelysin (MMP-3) cleave at acidic residues in the P1 site. Astacin family proteases are unique in cleaving bonds containing acidic residues at the P1′ site.

### Inhibitors

Meprins are inhibited, as other metalloproteinases, by zinc-chelating reagents (EDTA, *o*-phenanthroline). Meprins are not inhibited by the tissue inhibitors of metalloproteinases (TIMP 1–4), phosphoramidon or captopril, potent inhibitors of several other metalloproteinases such as matrix metalloproteinases, neprilysin, and angiotensin-converting enzyme, respectively. Perhaps the best inhibitor known to date for meprins is actinonin, a hydroxamate produced by actinomyces. Actinonin has an apparent Ki of 150 nM for mouse meprin A and 20 nM for human meprin A; the Ki for mouse meprin B is 300 nM. Fetuin-A and cystatin C have been reported to be endogenous inhibitors of human meprin metalloproteinases, and mannan-binding lectin (MBL) has been reported as an endogenous inhibitor of mouse meprins. The evidence that these proposed endogenous inhibitors regulate meprins in mammalian tissues and fluids is not well established as yet. The predominant forms of regulation appear to be targeting meprin proteases to specific locations (epithelial membranes, lumen of the kidney and gastrointestinal tract, inflammation sites), activation of the proteases by removal of the prosequence, and transcriptional regulation. There is also preliminary evidence that meprin subunits are posttranscriptionally regulated by RNA binding proteins.

## The Role of Meprins In Vivo

To explore the role of meprins in vivo, the meprin subunits have been disrupted in mice. Targeted disruption of the meprin α and β genes has resulted in meprin α, meprin β, and double αβ knockout (KO) mice (e.g., Banerjee et al. 2009). The mice do not display an obvious phenotype, but react to a number of stresses differently from wild-type (WT) mice. Thus, the KO mice generally grow at similar rates to their wild-type counterparts, they can reproduce, and no structural abnormalities have been noted. However, when the KO mice are subjected to a stress such as kidney ischemia/reperfusion (a model for acute renal failure) or dextran sulfate in the drinking water (which results in an experimental model of inflammatory bowel disease) or are exposed to endotoxin in the bladder (a model for urinary tract infections), there are marked differences in the reaction of the KO mice compared to WT mice. For example, meprin β KO mice have less severe damage to the kidney after ischemia/reperfusion (I/R) than WT mice. The meprin β mice KO mice have no meprin protein at the brush border membrane, because this is the meprin subunit that anchors the meprins to the membrane. Several studies indicate that during the stress of I/R, membrane-bound meprins are redistributed into cytosol and other subcellular compartments and cause significant kidney damage by hydrolyzing cytoskeletal proteins such as villin and actin (Ongeri et al. 2011). From the studies of experimental inflammatory bowel disease (a model for chronic inflammation), it has been found that meprin α KO mice develop more severe inflammation than WT mice. In this instance, the lack of meprin α results in a decrease in mobility of leukocytes to the site of inflammation and a decrease in the degradation and inactivation of cytokines at the site of inflammation. From studies with an experimental model of urinary tract disease, the lack of meprin α decreased the severity of the response to endotoxin, indicating an impairment of the immune response in this acute form of inflammation (Yura et al. 2009). Thus, the in vivo studies using different disease/pathological models indicate that meprins are important players in the response to injury or infection and have a role in the acute and chronic inflammatory response.

## Genes and Relation of Meprins to Disease

The structural genes for meprin subunits are the following: for meprin α, on mouse chromosome 17 and human chromosome 6, near the major histocompatibility complex and for meprin β, on chromosome 18 for both mouse and man. There must be some coordination for the expression of the meprin α and β genes because they are co-expressed in many instances. However, they are also regulated independently. For example, in the adult kidney of certain inbred strains of mice, meprin β but not α is expressed, and in human skin, the subunits are expressed in different cells in the epidermis. Several cancer cells (e.g., intestinal, breast, bone) express one or the other meprin subunits. Also, in recombinant mice, disruption of one meprin subunit gene does not affect the expression of the other subunit.

Polymorphisms have been identified in the human meprin α gene that is associated with inflammatory bowel disease (Banerjee et al. 2009). A single-nucleotide polymorphism in the 3′-untraslated region showed a particularly strong association with ulcerative colitis, and meprin α mRNA was decreased in the inflamed mucosa of IBD patients. These studies identified the meprin α gene as a susceptibility gene for IBD in humans. The results are consistent with those from studies of meprin α KO mice that display more severe inflammation than WT mice when subjected to an experimental model of IBD.

Polymorphisms have also been found in the human meprin β gene that is associated with diabetic nephropathy. In studies of Pima Indians, a population that has a high incidence of type 2 diabetes, diabetic patients that had a series of single-nucleotide polymorphisms in the meprin β gene had a higher incidence of kidney disease than their siblings. These studies led to the suggestion that the identified polymorphisms affect the transcription and trafficking of the meprin β protein and ultimately result in enhanced kidney fibrosis.

Meprin genes are expressed in several tumors and cancer cell lines and appear to be associated with metastases and invasiveness. For example, meprins are overexpressed in colorectal tumors, hepatomas, and breast cancer cells. In colorectal cancer, meprin α is overexpressed and appears to promote tumor progression by its pro-migratory and pro-angiogenic activities. In breast cancer and osteosarcoma cell lines, unique forms of meprin β mRNA are upregulated and are responsible for increased expression of the meprin β protein.

Meprins also play a role in urinary tract infections. Human urine normally has very low to non-detectable levels of meprin protein. However, in patients with urinary tract infections, the levels are markedly elevated. The increased meprins in the urine are likely due to leukocytes called to the site of infection to combat the infection. Studies with meprin α KO mice have demonstrated that these mice are hyporesponsive to an endotoxin challenge compared to WT mice in that when *E. coli* toxin was administered to the lumen of the bladder, there was less leukocyte infiltration and bladder permeability. These and other studies indicate that meprins have an important role in leukocyte movement to sites of infection and inflammation.

Recent studies in vitro and in vivo have demonstrated that amyloid precursor protein (APP) expressed in brain tissue is a substrate for meprin B (Jefferson et al. 2011). Meprin B generates nontoxic APP fragments, and this is particularly important because APP processing is critical in the development of Alzheimer's disease.

## Cross-References

▶ Zinc Adamalysins
▶ Zinc Matrix Metalloproteinases and TIMPs
▶ Zinc-Astacins
▶ Zinc-Binding Sites in Proteins

## References

Banerjee S, Oneda B, Yap LM, Jewell DP, Matters GL, Fitzpatrick LR, Seibold F, Sterchi EE, Ahmad T, Lottaz D, Bond JS (2009) MEP1A for meprin A metalloprotease subunit is a susceptibility gene for inflammatory bowel disease. Mucosal Immunol 2:220–231

Bertenshaw GP, Bond JS (2011) Meprin A; Meprin B. In: Rawlings N, Salvessen G (eds) Handbook of proteolytic enzymes. Elsevier, Oxford

Bertenshaw GP, Norcum MT, Bond JS (2003) Structure of homo- and hetero-oligomeric meprin metalloproteases: dimers, tetramers, and high molecular mass multimers. J Biol Chem 278:2522–2532

Beynon RJ, Shannon JD, Bond JS (1981) Purification and characterization of a metalloproteinase from mouse Kindney. Biochem J 199:591–598

Bond JS, Matters GL, Banerjee S, Dusheck RE (2005) Meprin metalloprotease expression and regulation in kidney, intestine, urinary tract infections and cancer. FEBS Lett 579:3317–3322

Dumermuth E, Sterchi EE, Jiang W, Wolz RL, Bond JS, Flannery AV, Beynon RJ (1991) The astacin family of metalloendopeptidases. J Biol Chem 266:21381–21385

Ishmael SS, Ishmael FT, Jones AD, Bond JS (2006) Protease domain glycans affect oligomerization, disulfide bond formation, and stability of the meprin A metalloprotease homooligomer. J Biol Chem 281:37404–37415

Jefferson T, Causevic M, Auf dem Keller U, Schilling O, Isbert S, Geyer R, Maier W, Tschickardt S, Jumpertz T, Weggen S, Bond JS, Overall CM, Pietrzik CU, Becker-Pauly C (2011) Metalloprotease meprin beta generates nontoxic N-terminal amyloid precursor protein fragments in vivo. J Biol Chem 286:27741–27750

Keiffer TR (2011) Meprin metalloproteases cleave and inactivate interleukin-6. Penn State University Thesis

Ongeri ME, Anyanwu O, Reeves WB, Bond JS (2011) Villin and actin in the mouse kidney brush border membrane bind to and are degraded by meprins, an interaction that contributes to injury in ischemia reperfusion. Am J Physiol Renal Physiol 301:F871–F882

Sterchi EE, Stöcker W, Bond JS (2008) Meprins, membrane-bound and secreted astacin metalloproteases. Mol Aspects Med 29:309–328

Yura RE, Bradley SG, Ramesh G, Reeves WB, Bond JS (2009) Meprin A metalloproteases enhance renal damage and bladder inflammation after LPS challenge. Am J Physiol Renal Physiol 296:F135–F144

# Zinc Metallocarboxypeptidases

Joan L. Arolas and F. Xavier Gomis-Rüth
Proteolysis Lab, Department of Structural Biology,
Molecular Biology Institute of Barcelona,
CSIC Barcelona Science Park, Barcelona, Spain

## Synonyms

Cowrins; Funnelins; Zinc carboxypeptidases

## Definition

Matellocarboxypeptidases are mostly zinc-dependent peptidolytic enzymes that cleave C-terminal residues from protein and peptide substrates. As such, they are involved in a plethora of biological functions including digestion of intake proteins, regulation of fibrinolysis, prohormone and neuropeptide processing, and inflammation.

## Background

Endopeptidases cleave their substrates in the middle, while exopeptidases cleave them at the end. The latter are called aminopeptidases if they cut the substrate at the N terminus and carboxypeptidases if they cut at the C terminus. In addition, peptidases are classified into aspartic, cysteine, glutamic, asparagine, serine, and threonine, depending on the protein residues that are essential for catalysis (Barrett et al. 2004). A further class comprises the metallopeptidases, in which divalent metal ions participate in catalysis. Metallocarboxypeptidases (MCPs) are thus metal-dependent exopeptidases that catalyze the hydrolysis of peptide bonds at the C terminus of protein or peptide substrates (see Chapters ▸ Sarcoplasmic Calcium-Binding Protein Family: SCP, Calerythrin, Aequorin, and Calexcitin and ▸ Sodium as Primary and Secondary Coupling Ion in Bacterial Energetics in Barrett et al. 2004 and Auld 1997). These peptidases have one or two metal ions, mostly zinc, inside the active site, and they cleave peptide bonds by a general base/acid mechanism. A metal-bound solvent molecule attacks the scissile peptide bond with the assistance of a strictly conserved glutamate residue (Auld 1997; Gomis-Rüth 2008; Matthews 1988). MCPs are found in a wide variety of organisms, ranging from bacteria to mammals, and they participate in a variety of biological processes ▸ Zinc Homeostasis: Whole Body, such as protein digestion, prohormone and neuropeptide processing, blood coagulation ▸ Zinc in Hemostasis: Role and fibrinolysis, and inflammation (Arolas et al. 2007). MCPs are strictly regulated, and deregulation may have drastic consequences, such as microbial infection, neurological failure, thrombotic and cardiovascular disease ▸ Zinc in Hemostasis: Role, inflammation and autoimmune disorders, and cancer. In this context, there is a need to understand the function and three-dimensional structure of these enzymes, which may contribute to the development of therapeutic strategies against their potential negative effect. MCPs are divided into funnelins and cowrins on the basis of structural resemblance to a sphere with a funnel-like opening at the top or to a conch shell or cowry, respectively (Gomis-Rüth 2008). The following sections focus on the biochemical properties of the MCPs that have been described to date.

## Diversity and Function of Funnelins

Funnelins are classified as A/B or N/E MCPs, also known as pancreatic and regulatory MCPs, respectively (Aviles et al. 1993). They are included within subfamilies M14A and M14B of the MEROPS database (http://merops.sanger.ac.uk/) (Rawlings et al. 2010). A/B MCPs are normally synthesized with a signal peptide and secreted as inactive zymogens with an N-terminal proregion that participates in the folding and latency of the peptidase (Vendrell et al. 2000). A-like enzymes cleave hydrophobic C-terminal residues, with A1-type forms preferring aliphatic and small aromatic residues, and A2-type enzymes cleaving bulky aromatic side chains. B-like enzymes cleave only basic C-terminal residues (lysine and arginine). In mammals, there are nine A/B MCPs, namely, CPA1, CPA2, CPA3, CPA4, CPA5, CPA6, CPB, CPU, and CPO (Table 1). CPA1 was one of the first peptidases to

**Zinc Metallocarboxypeptidases, Table 1** Metallocarboxypeptidases described in the present study

| Name | Substrate specificity | Localization | Structure (PDB code) |
| --- | --- | --- | --- |
| *Funnelins A/B type* | | | |
| CPA1 | Hydrophobic | Exocrine pancreas, duodenum | 1PYT[a] |
| CPA2 | Hydrophobic | Exocrine pancreas, duodenum | 1AYE[a] |
| CPB1 | Basic | Exocrine pancreas, duodenum | 1KWM[a] |
| CPA3 | Hydrophobic | Mast cells | – |
| CPA4 | Hydrophobic | Broad | 2BOA[a] |
| CPA5 | Hydrophobic | Brain, pituitary, and testis | – |
| CPA6 | Hydrophobic | Extracellular matrix | – |
| CPO | Acidic | Intestine | – |
| CPU | Basic | Liver, blood | 3D66[a] |
| CPT | Hydrophobic and basic | (*Thermoactinomyces vulgaris*) | 1OBR |
| CPAHa | Hydrophobic | (*Helicoverpa armigera*) | 1JQG |
| CPBHz | Basic | (*Helicoverpa zea*) | 2C1C |
| *Funnelins N/E type* | | | |
| CPE | Basic | Neuroendocrine | – |
| CPN | Basic | Liver, blood | 2NSM |
| CPM | Basic | Broad | 1UWY |
| CPD | Basic | Broad | 1QMU |
| CPZ | Basic | Placenta and embryonic tissues | – |
| CPX1 | (Inactive) | mRNA in placenta | – |
| CPX2 | (Inactive) | mRNA in heart and other tissues | – |
| ACLP | (Inactive) | Heart and other tissues | – |
| *Cowrins* | | | |
| ACE2 | Hydrophobic and basic | Broad | 1R42 |
| TcMCP-1 | Basic | (*Trypanosoma cruzi*) | 3DWC |
| TcMCP-2 | Hydrophobic | (*Trypanosoma cruzi*) | – |
| TaqCP | Broad | (*Thermus aquaticus*) | – |
| PfuCP | Broad | (*Pyrococcus furiosus*) | 1K9X |
| TthCP | Broad | (*Thermus thermophilus*) | 3HOA |
| BsuCP | Broad | (*Bacillus subtilis*) | 3HQ2 |

[a]Zymogens

be studied for both its catalytic mechanism and its three-dimensional structure. CPA1, CPA2, and CPB are produced by the exocrine pancreas and secreted to the digestive tract, where they are activated by trypsin to participate in the degradation of dietary proteins (Arolas et al. 2007). A-type zymogens can be secreted as monomers or in binary/ternary complexes with chymotrypsinogen C and/or proproteinase E. CPA3 is located in the secretory granules of mast cells and presumably functions following the action of chymase, that is, it is probably involved in anaphylactic and inflammatory processes. CPA4, originally named CPA3, is associated with hormone-regulated tissues, suggesting that it may have a role in cell growth and differentiation. Interestingly, it has been related with prostate cancer aggressiveness. CPA5 could be involved in the intracellular processing of pituitary peptides like proopiomelanocortin-derived peptides. CPA6 has been implicated in Duane syndrome and is thought to function in the extracellular processing of neuropeptides. CPO is highly expressed in intestinal epithelial cells and has preference for acidic residues. CPU, also known as CPB2, CPR, plasma CPB, and thrombin-activatable fibrinolysis inhibitor (TAFI), which is produced by the liver and secreted to the bloodstream, has been extensively investigated due to its important role in clot fibrinolysis and inflammation ▶ Zinc In Hemostasis: Role (Arolas et al. 2007).

Other peptidases included in subfamily M14A are those from the bacterium *Thermoactinomyces vulgaris* (CPT) and the insects *Helicoverpa armigera* (CPAHa) and *Helicoverpa zea* (CPBHz) (Rawlings et al. 2010).

Unlike A-/B-type MCPs, N/E forms are not synthesized as zymogens and rely on their substrate specificity and subcellular compartmentalization to regulate their activity (Reznik and Fricker 2001). These regulatory MCPs contain a C-terminal domain which is believed to contribute to protein folding. In mammals, there are eight N/E MCPs, five of which (CPE, CPN, CPM, CPD, and CPZ) show stringent carboxypeptidase activity toward basic residues, and three (CPX1, CPX2, and aortic carboxypeptidase-like protein; ACLP) are catalytically inert but may act as binding proteins (Table 1). CPE, also known as CPH and enkephalin convertase, contributes to the biosynthesis of peptide hormones and neurotransmitters following the action of prohormone convertases (Arolas et al. 2007). The catalytic step occurs within the secretory pathway of neuroendocrine cells. CPN and CPM may also participate in the processing of peptide hormones, but in contrast to CPE, they operate outside the cell. CPN is produced by the liver and secreted to the bloodstream, where it may inactivate or alter the specificity of vasoactive peptides such as kinins and anaphylatoxins. CPM is membrane bound, although soluble forms of this protein have been detected in various fluids. CPD has a broad tissue distribution and most likely functions in the processing of proteins that are initially cleaved by furin or related endopeptidases of the *trans*-Golgi network. The *Drosophila* homolog was found to be encoded by the *silver* gene, which upon mutation, alters the color of the cuticle and the wing shape. CPZ appears to be located in the extracellular matrix and contains an N-terminal domain that is homologous to Wnt-binding proteins, which are important signaling molecules during development (Reznik and Fricker 2001).

Other members of the family M14, classified as M14C and M14D, include gamma-D-glutamyl-(L)-meso-diaminopimelate peptidase I from *Bacillus sphaericus* and cytosolic/Nna1-like carboxypeptidases (Rawlings et al. 2010). The first seems to be related with sporulation. The latter is an emerging group of multidomain peptidases that would play a role in the modification of tubulin and protein turnover. However, their specific proteolytic activity remains to be determined.

## Three-Dimensional Structure of Funnelins

A/B MCPs contain a 90–95-residue proregion that folds into a globular-independent unit connected to the 300–310-residue catalytic domain by a short segment. A number of X-ray crystal structures are available for these peptidases, in both their zymogenic and active forms. In addition, the ~80-residue globular part of the proregion has been the subject of structural studies in solution by NMR spectroscopy (Vendrell et al. 2000). Despite the low sequence identity observed in some cases, the proregions share a similar structural topology. This consists of an open sandwich with a twisted four-stranded antiparallel β-sheet and two nearly parallel α-helices on top (Fig. 1) (Gomis-Rüth 2008). An extended C-terminal α-helix connects the globular domain to the catalytic moiety. The end of this connecting segment is the target of limited proteolysis, which activates the enzyme. The prodomain/catalytic domain interaction involves a large interface but few direct contacts, which accounts for the capacity of A-like and TAFI zymogens to cleave small substrates or even peptides. The prodomain blocks access to the funnel-like opening of the mature enzyme, which is reminiscent of the shape obtained when a cone is extracted from a sphere (Gomis-Rüth 2008). The active site cleft is rather shallow, which explains why these peptidases act on a large variety of well-folded proteins. The funnelin catalytic domains share a common globular fold, formed by a central mixed parallel/antiparallel eight-stranded β-sheet, with a twist of 130° between the first and the last strand, over which eight α-helices pack on both sides (Fig. 1). The active site is located in a cavity formed by the internal β-sheet, where the catalytic zinc ion resides pentacoordinated by atoms Nδ1 of His69 and His196, Oε1 and Oε2 atoms of Glu72, and a water molecule, giving rise to a distorted tetrahedral + 1 coordination sphere. The active site and catalytic mechanism of funnelins has been studied in depth by using bovine CPA1 as a model protein. The conserved residues that are involved in zinc binding and participate directly in catalysis are His69, Glu72, Arg127, Asn144, Arg145, His196, Tyr248, and Glu270 (HXXE + R + NR + H + Y + E), which are distributed in five subsites that bind the zinc and accommodate the substrate and the reaction intermediate. On the other hand, the difference in substrate

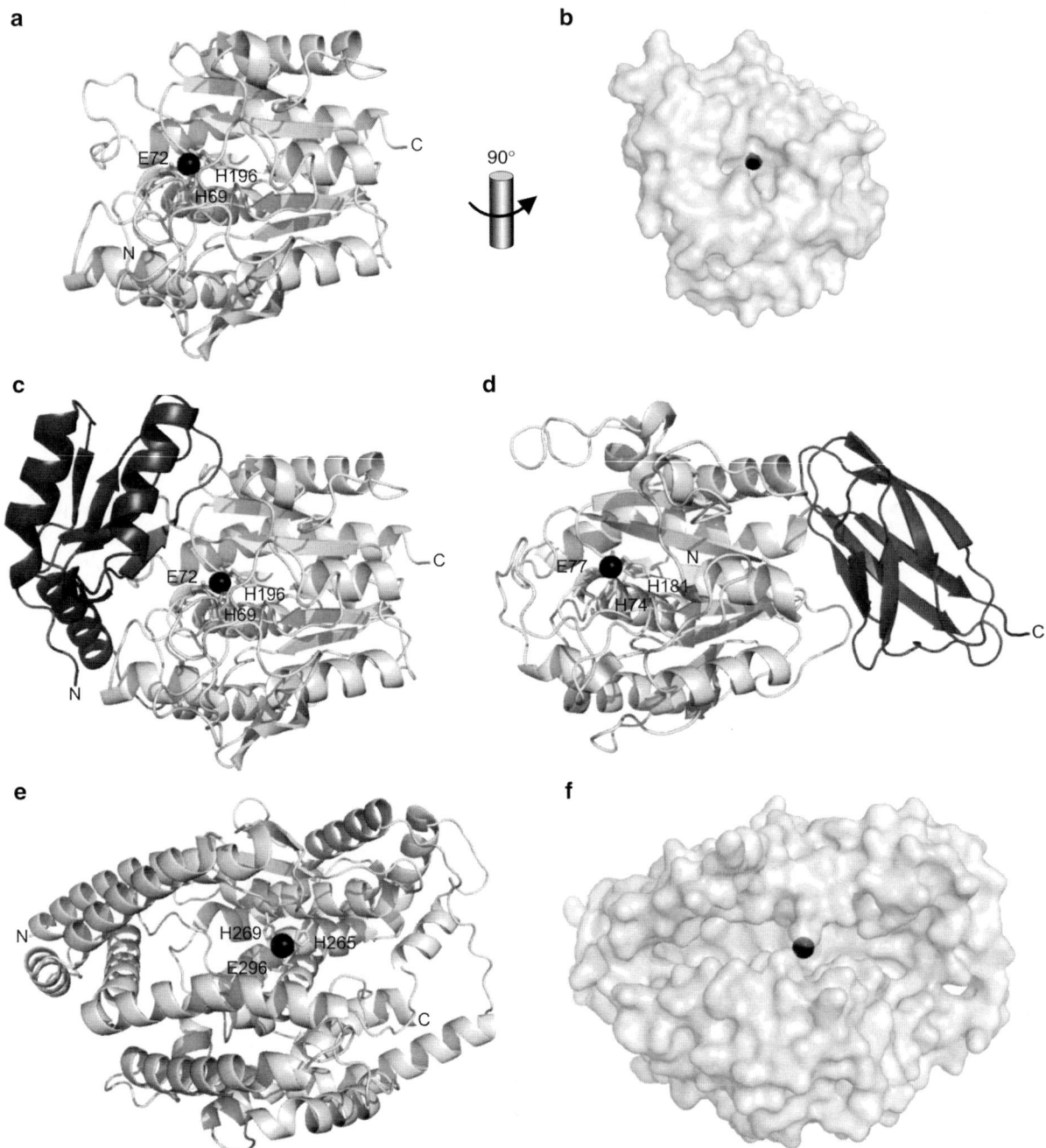

**Zinc Metallocarboxypeptidases, Fig. 1** *Three-dimensional structure of funnelins and cowrins.* Ribbon plots in standard orientation of (**a**) an A-/B-type funnelin (human CPA1; PDB code 2V77; metal ion as a *black sphere*), (**c**) an A-/B-type funnelin zymogen (bovine pro-CPA1; PDB code 1PYT; prodomain in dark gray, catalytic domain in light gray, metal ion as a *black sphere*), (**d**) an N-/E-type funnelin (duck CPD domain II; PDB code 1QMU; catalytic domain in light gray, transthyretin-like domain in dark gray, metal ion as a *black sphere*), and (**e**) a cowrin (TcMCP-1; PDB code 3DWC; metal ion as a *black sphere*). The three protein ligands of the catalytic ion are shown as sticks and labeled, as are the N and C termini of each peptidase. (**b**) and (**f**) display molecular surface representations showing the funnel-like and cowrin-like apertures of the active site cleft of human CPA1 and TcMCP-1, respectively. Figure prepared with program PyMOL specificity of A-like and B-like enzymes is mainly defined by the residue at position 255, which is an isoleucine or an aspartate, respectively. The catalytic domain of these peptidases is resistant to extreme pH values and proteolysis, with the exception of TAFI, which is highly unstable and shows a short half-life in the bloodstream, probably as a result of functional requirements (Arolas et al. 2007).

N/E MCPs are less well characterized in terms of their three-dimensional structure and catalytic mechanism than A/B MCPs. The X-ray crystal structures of the catalytic domain of human CPN and CPM, duck CPD domain II, and *Drosophila* CPD short variant 1B show a similar overall fold to that of A/B MCPs (Fig. 1) (Gomis-Rüth 2008). The only major differences are found in the length of the loops between the β-strands and α-helices, which account for a more restrictive substrate-binding pocket and provide the structural basis to understand the inability of protein inhibitors of A/B MCPs to affect the N/E counterparts. Some of the residues involved in substrate binding in A/B MCPs are different in N/E MCPs. While B-like enzymes contain an aspartate at position 255, the N/E members contain either a glutamine (CPE, CPN, CPM, and CPD) or a serine (CPZ) at the corresponding position. In addition, the presence of a glutamine at the position equivalent to Ser207 in CPB1 appears to contribute to the specificity of N/E MCPs for C-terminal basic residues. The inactive members of N/E MCPs lack one or more catalytic residues and show important differences in critical substrate-binding residues. The aforementioned structures reveal a conserved ~80-residue domain downstream of the catalytic domain that folds into a seven-stranded flat β-sandwich made up by two layers of three mixed and four antiparallel strands held together by a hydrophobic core (Fig. 1) (Gomis-Rüth 2008). This domain has topological similarity with the plasma protein prealbumin, also known as transthyretin, but its biological role remains to be established. It is still a challenge to determine the structure of full-length CPN, CPD, and CPZ, which have particularly interesting features. Thus, CPN is a dimer of heterodimers, with each heterodimer containing one catalytic subunit and one noncatalytic regulatory subunit, which is highly glycosylated. CPD is a three-repeat carboxypeptidase: the first two repeats show complementary pH optima, and the third is catalytically inactive. The function of the latter is unknown, although in duck, it has been reported to act as a receptor for the preS envelope protein of hepatitis B virus. Finally, CPZ displays a cysteine-rich N-terminal domain that has sequence identity with proteins that bind Wnt/wingless, such as the frizzled receptors.

## Inhibition of Funnelins

A/B MCPs are kept inactive by the presence of a proregion that acts as an autologous inhibitor, shielding the $S_2$, $S_3$, and $S_4$ subsites of the catalytic cleft. In addition, the proteolytic activity of this type of MCP is regulated by heterologous protein inhibitors, which have been identified from diverse organisms: *Solanaceae* plants (potato and tomato), intestinal parasites (*Ascaris*), hematophagous animals (leech and tick), and mammals (mouse and human) (Arolas et al. 2007). Latexin, alias endogenous carboxypeptidase inhibitor (ECI), with about 220 residues is the largest reported inhibitor. It is broadly distributed in human tissues, which suggests it does more than just inhibit digestive MCPs. The X-ray crystal structure of latexin in complex with human CPA4 revealed two topologically equivalent domains reminiscent of cystatins. Latexin binds to the target enzyme through the interface of the two domains in a way that mimics the autologous inhibition of the proregion, that is, with a large contact area and few specific interactions (Gomis-Rüth 2008). The first protein inhibitors to be described were those from potato (PCI) and tomato, which may protect these plants against insects. The inhibitor from the roundworm *Ascaris* (ACI) also seems to protect against digestive MCPs, whereas those from leeches and ticks (LCI and TCI) may prevent blood coagulation through inhibition of TAFI. All these exogenous inhibitors are small proteins (39–75 residues), whose structure is mainly maintained by the presence of multiple disulfide bonds (three in PCI, four in LCI, five in ACI, and six in TCI). Their inhibition mechanism is different from that of latexin, as inferred from the crystal structures of their complexes with different A/B MCPs: they insert their C-terminal tails inside the peptidase active site in a substrate-like manner. The C-terminal residues are cut off by the enzyme, and the newly generated C-terminal residues coordinate the catalytic zinc ion and create several specific interactions with residues of subsites $S_1$' and $S_1$ (Gomis-Rüth 2008; Vendrell et al. 2000). The enzyme/inhibitor complexes are further stabilized by secondary contacts from other segments of the inhibitor. Notably, TCI, which consists of two domains connected by a short flexible link, interacts in a double-headed manner: while the C-terminal domain contacts the active site, the N-terminal domain is anchored to the peptidase surface, thus defining an exosite located around subsite $S_4$.

No protein inhibitors of N/E MCPs have been reported to date, which is consistent with their physiological function. This type of MCP is only inhibited by small molecules such as guanidinoethyl-mercaptosuccinic acid (GEMSA) (Arolas et al. 2007).

## Diversity and Function of Cowrins

Cowrins are found within families M2, M3, and M32 of the MEROPS database, and they are therefore less related to each other than funnelins (Gomis-Rüth 2008; Rawlings et al. 2010). The three most symbolic mammalian members, angiotensin I converting enzyme (ACE), neurolysin, and thimet oligopeptidase, are peptidyl dipeptidases or oligopeptidases. However, a human homolog of ACE, ACE2, shows strict carboxypeptidase activity. ACE2 has been implicated in cardiac dysfunction, hypertension, heart failure and ventricular remodeling, diabetes, obesity, and most remarkably, as a receptor for the coronavirus that causes severe acute respiratory syndrome (Guy et al. 2005). ACE2 cleaves a wide variety of biologically active peptides with the consensus sequence Pro-X-Pro-hydrophobic/basic, but it does not cleave the archetypal ACE substrates. It is also insensitive to the potent ACE inhibitors lisinopril and captopril, but it is dependent on pH and monovalent anions. Much effort has been devoted to the search for selective inhibitors of ACE2; these include a series of peptidic and nonpeptidic compounds, which were identified through screening of constrained peptide libraries. ACE2 is localized predominantly in the plasma membrane, but like ACE, it undergoes ectodomain shedding, which yields a soluble form. Interestingly, preliminary studies have reported that the soluble form of ACE2 prevents coronavirus infection. $Ace2^{-/-}$ mice suffer from severe cardiac dysfunction, which is rescued by ablation of the ACE gene, thus suggesting that ACE2 regulates the effects of ACE in the renin-angiotensin system. In addition, it seems that, in the absence of ACE2, mice are more susceptible to angiotensin II–induced hypertension.

ACE2, which belongs to family M2, was the first mammalian MCP to be shown to contain the characteristic motif of the zincin tribe of metalloproteases ▶ Thermolysin, HEXXH, instead of the typical funnelin MCP motif, $HXXE(X)_{123-132}H$ (Gomis-Rüth 2008). The cowrins embrace more distant relatives from unicellular organisms, which are included in family M32 and putatively function in protein degradation (Table 1) (Rawlings et al. 2010). In this family, there are two MCPs from *Trypanosoma cruzi*, the etiological agent of Chagas disease (TcMCP-1 alias LmaCP1, and TcMCP-2). These peptidases are localized in the cytosol and show different expression patterns: TcMCP-1 is present at all life stages of the parasite, while TcMCP-2 is mainly restricted to the stages that occur in the vector insect. The specificities of these MCPs are complementary: TcMCP-1 acts preferentially on basic C-terminal residues, while TcMCP-2 prefers aromatic and aliphatic residues. Accordingly, they match the specificity of A-type and B-type MCPs, respectively. Cowrin MCPs have also been discovered in the extremophilic bacteria *Thermus aquaticus* (TaqCP), *Pyrococcus furiosus* (PfuCP), *Thermus thermophilus* (TthCP), and *Bacillus subtilis* (BsuCP) (Lee et al. 2009). TaqCP and PfuCP are stable at high temperatures and exhibit broad substrate specificity, although the former does not cleave proline at the C terminus of peptides. There is growing biotechnological interest in such hyperthermophilic enzymes since they are normally more stable in severe environmental conditions such as high temperatures, denaturing agents, and organic solvents. PfuCP shows a temperature optimum of 90–100°C and is thus suitable for use in proteomics (e.g., C-terminal sequencing) and other biotechnological applications.

## Three-Dimensional Structure of Cowrins

ACE2 is an 805-residue membrane-bound glycoprotein with an N-terminal catalytic domain followed by a C-terminal domain that is similar to collectrin, an enzymatically inactive glycoprotein of unknown function (Guy et al. 2005). Despite low sequence identity, the catalytic domain of ACE2 is structurally homologous to that of its protozoan and bacterial counterparts. As mentioned above, the catalytic domains of funnelins are more closely related to each other than those of cowrins. Viewed in standard orientation for structural representations of metalloproteases, cowrin catalytic domains are prolate spheroids that resemble an elongated cowry with a long active site cleft mimicking the aperture (Fig. 1) (Gomis-Rüth 2008). The cleft lies between two halves of similar size, that is, an upper and a lower moiety. Interestingly, the structures reported evince varying degrees of aperture of the cleft, suggesting that it opens and closes like a clamshell. The catalytic domains of cowrins share a core of 17 $\alpha$-helices and 3 $\beta$-strands, with additional secondary structure elements and segments that account for differences in peptidase length (500–700 residues). The structural core is defined by the active site helix, which traverses the molecule almost horizontally, as well as the glutamate helix and the tyrosine helix, both of which cross the molecule front to back.

The active site cleft completely traverses the catalytic domain from left to right and features a deep narrow canyon that has been implicated in the cleavage of oligopeptides or partially unfolded substrates. This cleft is different from the cul-de-sac hallmark displayed by funnelins, being more reminiscent of that of endopeptidases (Gomis-Rüth 2008). The active site includes one metal ion that is a zinc in all structures except TcMCP-1 and PfuCP, which are most likely cobalt dependent. The compatibility of cobalt with peptide bond hydrolysis is supported by biochemical studies on TaqCP and endopeptidases like thermolysin and astacin. The catalytic metal ion is coordinated by the N$\epsilon$2 atoms of the two histidine residues imbedded in the HEXXH motif ▸ Thermolysin, which is contained in the active site helix, by a glutamate within the glutamate helix and by a single solvent molecule. This coordination displays distorted tetrahedral geometry. Cobalt-containing peptidases, however, may differ in the coordination sphere and contain additional solvent molecules. Other residues implicated in substrate binding and catalysis include a serine or a glycine, a conserved histidine, and two tyrosine residues or a tyrosine and an arginine from the tyrosine helix, thus giving rise to a consensus signature, HEXXH + EXXS/G + H + Y/R + Y. Several studies have revealed that some cowrins form dimers in solution (Lee et al. 2009). This finding is consistent with the quaternary structure of PfuCP and TcMCP-1, which are homodimers. These proteins dimerize through a vertical twofold axis, giving rise to an overall shape that resembles a butterfly.

## Cross-References

▸ Thermolysin
▸ Zinc Homeostasis, Whole Body
▸ Zinc in Hemostasis

## References

Arolas JL, Vendrell J, Aviles FX, Fricker LD (2007) Metallocarboxypeptidases: emerging drug targets in biomedicine. Curr Pharm Des 13(4):349–366

Auld DS (1997) Zinc catalysis in metalloproteases. Struct Bond 89:29–50

Aviles FX, Vendrell J, Guasch A, Coll M, Huber R (1993) Advances in metallo-procarboxypeptidases. Emerging details on the inhibition mechanism and on the activation process. Eur J Biochem 211(3):381–389

Barrett AJ, Rawlings ND, Woessner JF (2004) Handbook of proteolytic enzymes. Academic, London

Gomis-Rüth FX (2008) Structure and mechanism of metallocarboxypeptidases. Crit Rev Biochem Mol Biol 43(5):319–345

Guy JL, Lambert DW, Warner FJ, Hooper NM, Turner AJ (2005) Membrane-associated zinc peptidase families: comparing ACE and ACE2. Biochim Biophys Acta 1751(1):2–8

Lee MM, Isaza CE, White JD, Chen RP, Liang GF, He HT, Chan SI, Chan MK (2009) Insight into the substrate length restriction of M32 carboxypeptidases: characterization of two distinct subfamilies. Proteins 77(3):647–657

Matthews BW (1988) Structural basis of the action of thermolysin and related zinc peptidases. Acc Chem Res 21:333–340

Rawlings ND, Barrett AJ, Bateman A (2010) MEROPS: the peptidase database. Nucleic Acids Res 38(Database issue): D227–D233

Reznik SE, Fricker LD (2001) Carboxypeptidases from A to Z: implications in embryonic development and Wnt binding. Cell Mol Life Sci 58(12–13):1790–1804

Vendrell J, Querol E, Aviles FX (2000) Metallocarboxypeptidases and their protein inhibitors. Structure, function and biomedical properties. Biochim Biophys Acta 1477(1–2):284–298

# Zinc Metallohydrolase Family of the β-Lactamase Fold

▶ Zinc Beta Lactamase Superfamily

# Zinc Metallothionein

Wolfgang Maret
Metal Metabolism Group, Diabetes and Nutritional Sciences Division, School of Medicine, King's College London, London, UK

## Definitions

In the middle of the last century, at a time when new trace elements were discovered in biology, Vallee and his group were interested in the biological chemistry of cadmium. They isolated a cadmium(II)- and zinc(II)-binding protein from equine renal cortex (Margoshes and Vallee 1957) (See entry ▶ Metallothioneins as Cadmium-Binding Proteins). The protein was later named metallothionein (MT), with the prefix *metallo* referring to the fact that it contained both cadmium (2.9% per gram of dry protein) and zinc (0.6%) and the suffix *thionein* alluding to the relatively high sulfur content of the protein (4.1%) (Kaegi and Vallee 1960). When isolated from equine or human liver, MT contains mostly zinc and virtually no cadmium. The sulfur content turned out to be due to cysteine, which accounts for about one third of its amino acids. The protein is not encoded by a single gene. In humans, it is a family of at least a dozen proteins with 60–68 amino acids (Li and Maret 2008).

The word "zinc metallothionein" is a pleonasm. Yet, it is used more frequently than "zinc thionein" because the protein is known as metallothionein, whereas the metal-free protein is referred to as thionein (T). It is critically important to recall that the name was given to a mammalian protein because it was subsequently adopted for other metal-binding proteins from various species, including proteins without sequence homology to mammalian MTs. *Thus, metallothionein became a generic name for structurally unrelated proteins.* Proteins that have been named metallothioneins occur in many phyla, including plants and bacteria, and are structurally extremely diverse (Blindauer and Leszczyszyn 2010). Almost any article on metallothionein starts with an attempt to provide a definition. *However, there is no single definition that holds for all proteins that have been named metallothioneins or metallothionein-like proteins* (Table 1). Without an agreement on what defines a metallothionein, it seems pointless to address a function that is common to all of them. In fact, the indiscriminate use of the name metallothionein in connection with implied structure/function relationships to mammalian MT is confusing and should be avoided. For this reason, this article focuses on those proteins for which the name was originally coined. A PubMed search for "metallothionein" in October 2011 retrieves 9,503 references in the biomedical literature alone and illustrates the scope of research on these proteins.

Human MTs have been investigated in detail and therefore will be used to present the state of knowledge of mammalian MTs. How far the analogy applies within the animal kingdom is not clear because evolutionary trees are unrooted and a common ancestor has not been identified.

## Multiple Gene Products, Metal Content, and Structures of MTs

Humans have at least a dozen MT genes. The proteins are classified as MT-1, -2, -3, and -4. The additional proteins are grouped under MT-1 and are given letters after the number one. This classification is based on a difference at position 11, which in human MT-2 is an aspartate and imparts a different charge to the protein. The human MT-1 proteins differ in about 3–13 amino acids out of a total of 61. MT-3, a 68-amino acid protein, was discovered by employing a biological assay (Uchida et al. 1991) (See entry ▶ Zinc Metallothionein-3 (Neuronal Growth Inhibitory Factor)). The investigators noted a lack of an antiproliferative activity in brain tissue from Alzheimer disease victims. This activity was used as an assay to isolate MT-3, and it adduced the reason for naming MT-3 growth inhibitory factor (GIF). The molecular basis for the activity remains unknown. Isolated MT-3 from different mammalian species contains four copper ions and three to four zinc ions. Copper is bound in the copper(I) valence state in the N-terminal β-domain

**Zinc Metallothionein, Table 1** The validity of general assumptions about metallothioneins

| Statements | Comments |
|---|---|
| Metallothioneins | |
| Are metal-binding proteins | Metal binding is a general property of many proteins with cysteine-containing ligand signatures. Redox functions of MTs are linked to the cysteines when they are not bound to metals |
| Have zinc/thiolate clusters | Zinc/thiolate clusters are not a unique property of MT because the $Zn_3S_9$ cluster is present in some lysine methyl transferases. Some MTs have a mononuclear zinc finger-type of zinc-binding site |
| Utilize only cysteine ligands for metal binding | Some MTs employ histidine |
| Are sulfur(cysteine)-rich proteins | The percentage of cysteines in some MTs is significantly lower than in mammalian MTs and does not differ from the cysteine content of other cysteine-rich proteins |
| Are low-molecular-weight proteins | Some proteins classified as MTs, such as MT from *Tetrahymena thermophila*, have as many as 162 amino acids |

and shielded from solvent in such a way that it is not redox-active. The primary structure of MT-4 is also significantly different from the other MTs (Quaife et al. 1994). When isolated from mouse tongue, MT-4 contains zinc and copper in the ratio 2.6:1. While MT-1 and MT-2 are present in all cells, MT-3 and MT-4 have a more restricted tissue distribution. In the fetal or neonatal liver, MT proteins also contain copper. The involvement of MTs in copper physiology needs further scrutiny (See entries ▸ Metallothioneins and Copper, ▸ Copper, Biological Functions). The cadmium content of renal MT appears to be related to the accumulation of cadmium in this organ. It may affect the primary function of the protein in the metabolism of zinc and possibly copper, both essential nutrients. Without toxic exposures to cadmium, there is very little cadmium in human MTs, albeit cadmium accumulates in MT with age. Therefore, certain tissues of older individuals contain increased amounts of this toxic and nonessential metal ion (See entry ▸ Cadmium Exposure, Cellular and Molecular Adaptations). Similar considerations apply to exposure to toxic metal ions, such as mercury and lead (See entries ▸ Cadmium Exposure, Cellular and Molecular Adaptations, ▸ Metallothioneins and Lead, ▸ Metallothioneins and Mercury).

Metal ions determine the tertiary structure of MT. The metal-free protein, thionein, has the structural characteristics of an intrinsically disordered protein. The cysteines in the sequence of thionein bind zinc ions with a complex, nonsequential pattern that differs from the sequential pattern typically observed in most other metalloproteins. A titration end point and a defined 3D structure are obtained with seven zinc (II) ions per protein, three in the N-terminal β-domain and four in the C-terminal α-domain. Four thiolates coordinate each zinc ion in a tetrahedral geometry. Seven zinc ions employ exclusively the sulfur donors of the 20 cysteines. However, some of the cysteines form ligand bridges in so-called zinc/thiolate clusters because there are only 20 (21 in the case of MT-1B) cysteines and not 28 that would be required if the seven zinc ions were to bind in *isolated* tetrathiolate sites. The peptide chain wraps around the two zinc/thiolate clusters, clockwise in the N-terminal domain and anti-clockwise in the C-terminal domain when looking at the molecule from the N-terminus. In this way, the metals ions are buried in the interior of the two domains. The $Zn_3S_9$ cluster in the N-terminal β-domain has a chair conformation with three sulfur bridges and six terminal ligands. Each zinc ion has two bridges and two terminal ligands. The $Zn_4S_{11}$ cluster in the C-terminal α-domain has five sulfur bridges, and there are two different coordination environments. Two zincs have three bridges and only one terminal ligand, whereas the remaining two have two bridges and two terminal ligands.

The crystal structure was determined for rat liver MT-2 that had been induced with cadmium. The protein is dumbbell-shaped with two domains and contains five cadmium ions and two zinc ions. Both zinc ions and one cadmium ion are located in the β-domain, and the remaining four cadmium ions are in the α-domain of the protein. Solution structures were determined for rabbit, human, and rat liver MT-2 with seven cadmium ions, and later with seven zinc ions, demonstrating essentially the same overall structure. Cadmium was used because the $^{113}$Cd isotope allows cadmium NMR spectra to be recorded, which turned out to be a critical step in solving the protein's 3D structure. The derivatives with seven cadmium ions used for structure determination do not reflect the metal composition of the

**Zinc Metallothionein, Table 2** The validity of general assumptions about mammalian metallothioneins

| Statements | Comments |
| --- | --- |
| Mammalian metallothioneins | |
| Bind seven divalent metal ions | MTs can bind fewer or more than seven metal ions |
| Have sulfur only in the -2 valence state in cysteine | MTs are redox proteins. Disulfide bonds or other forms of cysteine sulfur oxidation are possible |
| Have no aromatic amino acids | Aromatic amino acids are rare but are present in some MTs |
| Are cytosolic proteins | MTs occur extracellularly and translocate from the cytosol to different cellular organelles |

protein under normal physiological conditions or even under conditions of environmental exposure to cadmium. In fact, zinc and copper are the native metal ions. 3D structures have been determined for derivatives containing seven copper ions, though it is unknown whether MT ever contains only copper and no zinc. The structures of the two individual domains of mouse MT-1, which usually contain mostly zinc, form stable folds with three and four bound copper(I) ions, respectively. The structures of the copper-containing peptides differ significantly from those of the zinc-containing peptides as copper(I) prefers a trigonal geometry, while zinc(II) prefers a tetrahedral geometry. However, mammalian MT-1/-2 bind additional copper (I) ions. When 12 Cu(I) ions are bound, they are organized in $Cu_6S_9$ and $Cu_6S_{11}$ clusters. Two intermediate $Cu_4S_6$ clusters form prior to full metal loading. Both diagonal and trigonal coordination geometries have been suggested for these species. An updated review of the structural biochemistry of mammalian MTs is available (Vasak and Meloni 2011).

When isolated from natural sources, MTs are often heterogeneous in terms of their metal content and sulfur redox state. It is not clear whether this is an inherent property of the protein or an artifact of the isolation. To prepare homogeneous material, all the metal ions are removed from the protein, all cysteine sulfurs reduced, and the protein is then reconstituted with a particular metal ion. Many isolated metalloproteins are promiscuous in terms of binding different metal ions. The fact that MTs have defined 3D structures with different metal ions does not answer questions about which and how many metal ions are bound to MTs in cells (Table 2).

## Zinc Load and Sulfur Redox State of Isolated MTs

When examining the primary structures of mammalian MTs, the conservation of the 20 cysteines is the most obvious characteristic. The mammalian proteins are rich in cysteines. Their cysteine content is 30% versus 2.6% on average for a cytosolic protein.

*Zinc Binding.* The sulfur coordination environments of the seven zinc ions differ, though each zinc ion is formally in a tetrathiolate site. Each metal ion is in a unique environment as is evident from seven distinct resonances in $^{113}$Cd-NMR spectra. It was thought that all seven zinc ions are bound with high affinity. However, human MT-2 has four sites with high affinity (log $K = 11.8$), two sites with intermediate affinity (log $K \approx 10$), and one site with even lower affinity (log $K = 7.7$). This finding leads to a description of human MT-2 with at least four forms: $Zn_7T$, $Zn_6T$, $Zn_5T$, and $Zn_4T$. The molecular structures of the species with less than seven zinc ions are not known. The affinity for cadmium (II) ions is on average about four orders of magnitude higher, while the affinity for copper (I) ions is believed to be even higher.

*Redox Reactions.* The other important property of the cysteine thiols in MT is their reactivity with electrophiles and concomitant dissociation of zinc. Zinc is redox-inert in biology, remaining always $Zn^{2+}$. Therefore, zinc proteins are generally not considered to be redox-active. However, the thiolate ligands of zinc in MT confer redox activity on the zinc clusters, making zinc binding to MT dependent on the redox environment. Hence, in contrast to other redox-active metalloproteins, the ligands and not the central metal ions make MT a redox protein. The redox potential of the thiols in MT is $< -366$ mV and allows biological oxidants, such as disulfides, to react with MT. The metal load of MT determines this reactivity. Four zinc ions quench the reactivity with a disulfide significantly, but even a large excess of zinc ions does not abolish the reactivity completely. A host of other thiol-reactive electrophiles releases zinc from MT. Covalent modifications and noncovalent interactions with other biomolecules modulate both zinc binding and thiol reactivity of MTs. In addition, MTs aggregate; form intermolecular disulfides; bind phosphate, chloride, sodium, or other ions; and interact with ligands, such as glutathione or ATP, and other proteins. *The two properties of the cysteines in MTs, metal binding and redox reactions, are linked* (Maret 2011).

## Zinc Load and Sulfur Redox State of MTs in Cells and Tissues

Chemical modification assays that differentiate free thiols from metal-bound thiolates have shown that the protein has free thiol(ate)s and does not bind seven metal ions in cells and tissues. T (no metal bound) or MT (seven metals bound) do not exist under steady-state conditions in the cell, although both species can be prepared in the laboratory. Therefore, the name metallothionein becomes ambiguous when applied to the protein in cells and tissues and used with the assumption that the protein has the same structure as the one for which the solution and crystal structures were determined. Less than 10% of the thiols can be oxidized, corresponding to the formation of about one disulfide bridge in thionin, the name given to the oxidized form of thionein. The amount of thionin increases under conditions of oxidative stress. The formation of specific disulfides among the 20 cysteines could not be confirmed, however. The redox state of the thiols in MT correlates with the glutathione/glutathione disulfide (GSH/GSSG) redox potential in growth-arrested, proliferating, differentiated, and apoptotic cells. In addition, when the amount of oxidized MT increases, a more oxidized GSH/GSSG redox ratio and higher free zinc ion concentrations are observed, indicating that the redox state is a major factor in determining the availability of cellular zinc ions. According to this metabolic link, oxidative conditions make more zinc available while reductive conditions make less zinc available. In summary, investigations of the isolated protein and speciation of the protein in cells and tissues demonstrate that MT differs from other zinc proteins in terms of its variable zinc content and its *dynamic protein structure* that depends on the biological environment. The dependence of MT structure on zinc availability and the redox state could explain the heterogeneity of isolated MTs. Additional investigations are necessary to determine the metal content of the individual MT gene products in different tissues.

## Functions of MTs

MT is a prime example for the difficulties in relating protein structure to protein function. Yet functions are regularly discussed without specific reference to a molecular function or even without knowledge about the metal content of the protein. Approaches within single disciplines failed to establish a function of MT. Thus, chemical characterizations usually neglect the biological environment in which MT functions while biological approaches rarely address the chemical properties of variable zinc affinities, low redox potentials, and the relationship between the two. Speculations about the biological functions of MT linger since its discovery. They include roles in zinc and copper homeostasis (See entry ▶ Zinc Cellular Homeostasis), the protection against toxic metal ions and oxidative stress, and roles in the proliferation, differentiation, and apoptosis of cells, including a role in protecting the mammalian brain against neurodegeneration after injury or in disease. Main issues in addressing the function(s) of MT are how the global biological activities of MTs in the life and death of cells relate to their chemical reactivities and whether multiple general functions are based on a single molecular mechanism of action.

The finding that MT-1/-2-null mice are viable and reproduce is consistent with a role of MT in zinc metabolism because the mice are more sensitive to zinc deficiency (See entry ▶ Zinc Deficiency) and zinc overload. Initially, the phenotype of the mice was tested only with regard to the sensitivity to cadmium. Gene knockout experiments are of limited value when phenotypes do not change drastically. In the case of the MT-1/-2-null mice, redundancy is apparent only and reflects the lack of applying tests for altered phenotypes relating to zinc metabolism. MT-3-null mice are more prone to kainate-induced seizures than wild-type mice. A genetic knockout of MT-4 or a genetic knockout of all MTs has not been reported.

A relationship of MTs to zinc metabolism will be discussed as it offers one major mechanism of action. Though MT has been related to zinc metabolism ever since it was shown to contain zinc, a specific role is emerging only in the last few years as an understanding emerges of how cellular zinc is controlled. Such a role could not have been discussed earlier when the specific characteristics of MT, such as its redox activity and variable zinc affinities, and quantitative aspects of cellular zinc homeostasis were not known. Functions need to consider the biological environment of these proteins and the extensive regulation of their gene expression.

## Regulation of Gene Expression

The different human MT proteins are extensively regulated. The types of MT genes expressed differ in tissues, and expression levels vary over 100-fold. Various forms of chemical and physical stress, the acute phase response, and inflammation induce MTs. Interferons, cytokines, growth factors, ligands of nuclear hormone receptors, and adenylate cyclase, phospholipase C, and Jak/Stat signaling pathways are all involved in the expression of MTs. The specificity and extent of regulation of the MT family alone would seem to support a central role in metabolism rather than a role in detoxification of a metal ion such as cadmium. The availability of essential metal ions is tightly regulated in the cell. Therefore, it is not a foregone conclusion that any newly synthesized T is immediately converted into MT. In fact, the primary function seems to be to produce T for zinc binding and abrogating zinc-dependent processes rather than to produce MT for supplying and activating zinc-dependent processes. With the extremely limited availability of free zinc ions in the cell, mechanisms must exist to provide zinc for binding to T. In fact, zinc itself is involved in the control of MT-1/-2 transcription, not only in transcription factors, such as Sp1 and nuclear receptors, but also in the metal sensor MTF-1 (metal response element (MRE)-binding transcription factor-1). MTF-1 is not a transcription factor for expression of MT-3, though. MTF-1 is itself a zinc protein and binds DNA with its six zinc fingers. Its genetic ablation is embryonically lethal. There is ample evidence that zinc controls signaling pathways and is part of the following signaling pathway:

$$\text{"signal} \rightarrow \text{reactive species MT} \rightarrow Zn^{2+} \rightarrow MTF_1 \rightarrow \text{gene transcription"}$$

## MTs at the Intersection of Zinc and Redox Metabolism

Functions of MTs in zinc metabolism become apparent when their chemical properties are interpreted in the context of how cells control zinc. The differential zinc affinities allow MTs to redistribute zinc rather than to serve as a thermodynamic sink for zinc. Cytosolic proteins bind zinc with high (picomolar) affinity. Free zinc ion concentrations in a variety of cultured mammalian cells are also in the range of picomolar. They are controlled in this range in order not to interfere with other essential metal ions and to make the functions of zinc as a regulatory ion possible (See entry ▶ Zinc and Zinc Ions in Biological Systems). The zinc affinities of proteins and the estimated free zinc ion concentrations in cells match the free zinc ion concentrations calculated to be in equilibrium with zinc bound to MT. This relationship strongly supports a role of MT in cellular zinc redistribution. MT binds zinc tightly but provides a micromolar pool of zinc that can become available upon demand. In cultured cells, the cellular concentrations of MT are too low for serving as the only cellular zinc buffer, even under conditions of its induction. This conundrum is solved when considering MT in the transport of cytosolic zinc in addition to the transport of zinc through cellular membranes by specific transporters. Controlling ion concentrations kinetically complements thermodynamic buffering and is referred to as muffling. Binding zinc, moving zinc in the cell, and delivering zinc to cellular compartments where zinc transporters are located are all biological properties of MT, which can be extracellular, cytosolic, endosomal, nuclear, and mitochondrial. The factors for transport of MT have not been identified aside from the energy requirements for translocation from the cytoplasm to the nucleus and the interaction of the small GTPase-protein Rab3a with MT-3 in a pathway of zinc reuptake into synaptic vesicles. Cellular uptake of other MTs involves scavenger receptors, such as the megalin receptor on the plasma membrane.

The relation of MTs to both zinc and redox metabolism has been formulated as a redox cycle, in which the oxidoreduction of MTs is biochemically coupled to other redox pairs with concomitant association and dissociation of zinc ions. Many publications lack a specific reference to MTs' involvement in either zinc metabolism or redox metabolism. When the reactivity of the thiolates in MT is implicated in antioxidant functions, zinc dissociation and modulation of zinc-dependent processes are necessary consequences because zinc ions are potent effectors of proteins at picomolar to nanomolar concentrations.

The functions of MT at the intersection of zinc and redox metabolism are fundamentally important when interpreting the effects of the myriad of chemical (e.g., toxins, drugs, and nutrients), physical/mechanical (e.g., UV light, temperature), and biological (e.g., inflammation) factors on MT gene expression. All these interactions affect the redox and metal buffering capacities of cells. The affinity of MTs for metal ions and the reactivity of its cysteines provide the molecular basis for the protection afforded when increased MT concentrations scavenge metals, free radicals, or other reactive substances.

The recent advances in understanding cellular zinc homeostasis and its redox link would seem to reenforce the conclusion reached by the late Bert L. Vallee 20 years ago: "It seems reasonable to conclude that metallothionein is the keystone that has been endowed with all the necessary attributes to safeguard and regulate zinc homeostasis" (Riordan and Vallee 1991).

## Diseases

MTs have been implicated in numerous diseases (Zatta 2008). Almost always the disease association remains a phenomenon. Findings are largely limited to observing altered mRNA levels and are usually not explored in terms of what they could mean for the perturbation of cellular zinc and redox homeostasis. MT is involved in numerous diseases with oxidative stress, which releases its metal ions, generating potent cellular effectors. This link between MT and redox stress is particularly relevant for aging because biomarkers detect an oxidative stress and zinc deficiency in the elderly. Altered MT levels in many diseases indicate that measurements of MT mRNA and MT protein have considerable potential for the diagnosis or prognosis of such diseases. Evidence has been provided that manipulation of MT levels can be exploited in preventative or therapeutic interventions. The presence of polymorphic forms of MTs increases the complexity of the already relatively large family of a dozen human MTs. An allelic isoform of MT-1 is associated with the risk for diabetes type 2 and heart disease. It is expected that additional polymorphic forms of MT affect zinc metabolism and many of the chemical and biological activities that have been and are being studied.

## Methods

An entire volume of Methods in Enzymology is dedicated to metallothionein. It is an invaluable source for the methodologies employed in MT research (Riordan and Vallee 1991).

## Applications

MTs provide ample opportunities for applied research. Their metal-binding capacity can be employed in phytoremediation and in the design of metal-sensors and biomarkers of environmental exposure to metal ions. Their changes in pathophysiology make them important biomarkers for prognosis or diagnosis of diseases. One area that has received particular attention is their sulfur reactivity with some compounds used in cancer chemotherapy. Increased levels of MT bind these compounds, quench their reactivity, and decrease their efficacy in treatment. Similar considerations apply for the reaction of MT with metal drugs in treatment, such as bismuth drugs for ulcers, gold drugs for rheumatoid arthritis, and platinum drugs for cancer. This interference is contrasted by protective effects of MT against a large number of toxicants and types of injuries, including the amelioration of toxic effects of inorganic cancer drugs in some organs.

Space restrictions allow only key references. All the material discussed here can be found in these references, in particular the October 2011 issue of the Journal of Biological Inorganic Chemistry, edited by Milan Vasak.

## Cross-References

▶ Cadmium and Metallothionein
▶ Cadmium Exposure, Cellular and Molecular Adaptations
▶ Copper, Biological Functions
▶ Metallothioneins and Copper
▶ Metallothioneins and Lead
▶ Metallothioneins and Mercury
▶ Zinc and Zinc Ions in Biological Systems
▶ Zinc Cellular Homeostasis
▶ Zinc Deficiency
▶ Zinc Metallothionein-3 (Neuronal Growth Inhibitory Factor)

## References

Blindauer CA, Leszczyszyn OI (2010) Metallothioneins: unparalleled diversity in structure and functions for metal homeostasis and more. Nat Prod Rev 27:720–741

Kaegi JHR, Vallee BL (1960) Metallothionein: a cadmium- and zinc-containing protein from equine renal cortex. J Biol Chem 235:3460–3465

Li Y, Maret W (2008) Human metallothionein metallomics. JAAS 23:1055–1062

Maret W (2011) Redox biochemistry of mammalian metallothioneins. J Biol Inorg Chem 16:1079–1086

Margoshes M, Vallee BL (1957) A cadmium protein from equine kidney cortex. J Am Chem Soc 79:4813

Quaife CJ, Findley SD, Erickson JC et al (1994) Induction of a new metallothionein isoform (MT-IV) occurs during differentiation of stratified *Squamous epithelia*. Biochemistry 33:7250–7259

Riordan JF, Vallee BL (eds) (1991) Metallobiochemistry Part B Metallothionein and related molecules, vol 205, Methods in Enzymology. Academic, San Diego

Uchida Y, Takio K, Titani K et al (1991) The growth inhibitory factor that is deficient in the Alzheimer's disease brain is a 68 amino acid metallothionein-like protein. Neuron 7:337–347

Vasak M, Meloni G (2011) Chemistry and biology of mammalian metallothioneins. J Biol Inorg Chem 16:1067–1078

Zatta P (ed) (2008) Metallothioneins in biochemistry and pathology. World Scientific Publishing, Singapore

# Zinc Metallothionein-3 (Neuronal Growth Inhibitory Factor)

Milan Vašák[1] and Gabriele Meloni[2]
[1]Department of Inorganic Chemistry, University of Zürich, Zürich, Switzerland
[2]Division of Chemistry and Chemical Engineering and Howard Hughes Medical Institute, California Institute of Technology, Pasadena, CA, USA

## Synonyms

Growth inhibitory factor (GIF)

## Definition

Mammalian metallothionein-3 (MT-3), a small intra- and extracellularly occurring cysteine-rich metal-binding protein (7–8 kDa), belongs to the vertebrate family 1 of the metallothionein superfamily which is composed of four major isoforms designated MT-1 through MT-4. As isolated, MT-3 contains four Cu(I) and between three and four Zn(II) ions. The metal binding occurs via thiolate ligands of the cysteine residues forming two sulfur-based metal-thiolate clusters. The MT-3 expression is predominantly confined to the brain and is localized in both neurons and astrocytes. MT-3, but not MT-1/MT-2, inhibits survival and neurite formation of cortical neurons in vitro. The neuronal growth inhibitory activity of MT-3 led to its other designation as the growth inhibitory factor (GIF).

## Occurrence and Biological Properties of Mammalian Metallothionein-3

The present chapter focuses on mammalian metallothionein-3 (MT-3), a small intra- and extracellularly occurring cysteine-rich metal-binding protein whose expression is predominantly confined to the brain (Uchida et al. 1991). However, in the brain besides MT-3, also the MT-1/MT-2 isoforms exist. Since the results of many studies on MT-3 have been related to those obtained with MT-1/MT-2, the latter isoforms are also discussed when required. In contrast to noninducible MT-3, the biosynthesis of MT-1/MT-2 is inducible by a wide range of stimuli including metals, drugs, and inflammatory mediators (Vašák and Meloni 2009). Based on the classification, the brain MTs belong to the vertebrate family 1 which is characterized by the consensus sequence K-X(1,2)-C-C-X-C-C-P-X(2)-C, where X stands for other amino acid than cysteine (Binz and Kägi 1999). These proteins are composed of a single polypeptide chain of 61–68 amino acids. They are characterized by the presence of a conserved array of 20 cysteine residues and the absence of aromatic amino acids and histidine. The conserved Cys residues show distinct characteristic clustering into Cys-X-Cys, Cys-Cys and Cys-X-X-Cys motifs (Table 1). Whereas MT-1/MT-2 contain 61–62 amino acids, MT-3 contains 68 amino acids and exhibits 70% sequence identity with MT-1/MT-2. Compared with the amino acid sequences of MT-1/MT-2, the MT-3 sequence shows two inserts, a Thr in the position 5 followed by a conserved $C_6PCP_9$ motif in the N-terminal region and an acidic hexapeptide in the C-terminal region.

A number of molecular biological studies aimed at the understanding of the tissue-specific expression of MT-3 have been carried out. In case of MT-1/MT-2, the binding of the metal-regulatory transcription factor 1 (MTF-1) to proximal metal responsive elements (MREs), located at the promoter region of the *MT-1/MT-2* genes, is essential for their basal expression and induction by zinc (Miles et al. 2000). However,

**Zinc Metallothionein-3 (Neuronal Growth Inhibitory Factor), Table 1** Alignment of amino acid sequences of four MT-3 isoforms compared to human MT-2 sequence using ClustalW. The conserved cysteine residues are highlighted in *yellow*, and the peptide sequences characteristic of MT-3 are indicated by *red boxes*. Generated with SeaView4 (Gouy et al. 2010)

```
                    1         10        20        30        40        50        60      68
MT-3 Homo sapiens   MDPETCPCPS GGSCTCADSC KCEGCKCTSC KKSCCSCCPA ECEKCAKDCV CKGGEAAEAE AEKCSCCQ
MT-3 Mus musculus   MDPETCPCPT GGSCTCSDKC KCKGCKCTNC KKSCCSCCPA GCEKCAKDCV CKGEEGAKAE AEKCSCCQ
MT-3 Equus caballus MDPETCPCPT GGSCTCSGEC KCEGCKCTSC KKSCCSCCPA ECEKCAKDCV CKGGEGAEAE AEKCSCCQ
MT-3 Sus scrofa     MDPETCPCPT GGSCTCAGSC KCEGCKCTSC KKSCCSCCPA ECEKCAKDCV CKGGEGAEAE EEKCSCCQ
MT-2 Homo sapiens   MDPN-CSCAA GDSCTCAGSC KCKECKCTSC KKSCCSCCPV GCAKCAQGCI CKGA------ SDKCSCCA
                              β-domain                                α-domain
```

although the upstream region of the *MT-3* gene also contains multiple MREs, they are presumably nonfunctional as no in vitro binding of MTF-1 occurs. Although in several studies different regulatory sequences upstream from the transcription site have been suggested to be responsible for the tissue-specific expression of the *MT-3* gene, the so far obtained data are inconclusive. As the regulation of tissue-specific *MT-3* gene expression in humans does not appear to involve a repressor, other mechanisms such as chromatin organization and epigenetic modifications may account for the presence or absence of *MT-3* transcription (Vašák and Meloni 2009).

The most distinctive biological property of MT-3 is its extracellular growth inhibitory activity in neuronal primary cultures, leading to its other designation as the neuronal growth inhibitory factor (GIF). Thus MT-3, but not MT-1/MT-2, antagonizes the ability of Alzheimer's disease (AD) brain extract to stimulate survival and neuritic sprouting of cultured neurons. This bioactivity has been found for the native MT-3 and recombinant $Zn_7$MT-3 and linked to specific structural features brought about by the unique $T_5CPCP_9$ motif (Table 1). Native MT-3 has been isolated from human and bovine brains as a monomeric and air-stable protein containing four Cu(I) and between three and four Zn(II) ions, i.e., $Cu(I)_4Zn_{3-4}$MT-3. However, comparative biological studies on well-defined metalloforms of MT-3 are currently lacking. Other studies showed that only the $Zn_7$MT-3 isoform protects neuronal cells from the toxic effect of amyloid-beta (Aβ) peptide, a major component of the extracellular amyloid plaques in AD brains. These two distinct findings led to the hypothesis that MT-3 may be involved in pathogenic processes leading to AD. Evidence that MT-3 displays biological properties not observed for MT-1/MT-2 was clearly documented by in vivo studies, in which mice overexpressing MT-3 in most organs died as a result of pancreatic atrophy, whereas expression of similar amounts of MT-1 had no effect. Although the reason for MT-3 toxicity is unknown, these results provide biological evidence that the MT isoforms have different functional properties. Other in vivo studies in support of this conclusion showed that in a mouse model of brain injury exogenously administered MT-3, in contrast to human MT-2, does not affect inflammation, oxidative stress, and apoptosis (West et al. 2008; Vašák and Meloni 2009).

## Expression Pattern of Metallothionein-3 and Its Suggested Biological Roles

Numerous studies investigated the localization of MT-1/MT-2 and MT-3 in the central nervous system (CNS). Whereas MT-1/MT-2 are present throughout the brain and spinal cord with the astrocytes being the main cell type expressing both isoforms, MT-3 is localized in both neurons and astrocytes where it appears to play an important role in the homeostasis of copper and zinc. All three MT isoforms have been reported to be secreted, suggesting that they may play different biological roles in the intra- and extracellular spaces (West et al. 2008). MT-3 is present in several brain regions at different concentrations, being particularly concentrated in presynaptic terminals of zinc-enriched neurons (ZEN) of cerebral cortex and hippocampus. ZEN belong to a subset of glutamatergic neurons and as such contain presynaptic zinc vesicles in which 10–15% of the total Zn(II) in the brain is present. The colocalization of MT-3 and zinc vesicles in ZEN led to the suggestion that the protein may contribute to the utilization of Zn(II) as a neuromodulator. The fact that Zn(II) can act not only as a major regulator of neuronal physiology but also as an important intracellular signaling agent similar to calcium has emerged rather recently (Fukada et al. 2011). A support for the role of MT-3 as a neuromodulator came from several studies. Thus, the studies of $Zn_7MT$-3 and zinc transporter 3 (ZnT3) knockout mice revealed that ZnT3, a transporter that concentrates Zn(II) in presynaptic vesicles of ZEN, and $Zn_7MT$-3 function in the same pathway. In other studies, the direct interaction of $Zn_7MT$-3 with Rab3A, a small GTPase involved in the regulation of synaptic vesicle cycle, has been shown (Vašák and Meloni 2009). Besides the discussed studies toward the intracellular role(s) of MT-3, its extracellular role(s) have also been investigated. During synaptic signaling, i.e., upon zinc vesicle fusion with the membrane and the cargo release, the free Zn(II) concentration in the synaptic cleft can reach up to 300 µM. In this regard, a possible reversible switch between the $Zn_7$- and $Zn_8MT$-3, via the specific and reversible binding of one extra Zn(II) to $Zn_7MT$-3 ($K_{app}$ ~ 100 µM), may play a role as a zinc buffer or in an interaction with binding partner(s). In more recent studies, a number of other proteins interacting with $Zn_7MT$-3 have been identified by immunochemical and mass spectroscopic techniques. These proteins together with those previously identified have been classified into three functional groups: transport and signaling, chaperoning and scaffolding, and glycolytic metabolism. Clearly, studies directed toward elucidating the physiological role of these interactions are required. In addition, low expression levels of MT-3 have also been reported in pancreas, kidney, stomach, heart, salivary glands, organs of the reproductive system, and maternal deciduum, but the biological function of MT-3 in these tissues remains elusive (Vašák and Meloni 2009).

## Structure of Mammalian Metallothionein-3

The studies toward the structure of both bioactive metalloforms of MT-3, i.e., recombinant $Zn_7MT$-3 and native $Cu(I)_4Zn_{3-4}MT$-3, have been carried out by using various spectroscopic techniques. In both instances, the metal binding to the full length protein and its individual β- and α-domain (Table 1) have been investigated. The characterization of recombinant $Zn_7MT$-3 and $Cd_7MT$-3 revealed a monomeric protein of dumbbell-like shape determined by the presence of

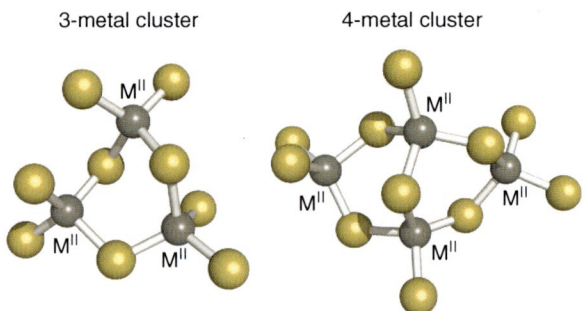

**Zinc Metallothionein-3 (Neuronal Growth Inhibitory Factor), Fig. 1** Schematic structure of the two metal-thiolate clusters with divalent metal ions present in mammalian $M^{II}_7MTs$. The metal ions ($M^{II}$) are shown as *gray spheres*, and the sulfur atoms (S) of cysteine ligands as *yellow spheres*. The cyclohexane-type cluster is located in the N-terminal β-domain and the adamantane-type cluster in the C-terminal α-domain of the protein

two mutually interacting protein domains, with each domain encompassing a metal-thiolate cluster. These data established the presence of a 3-metal cluster in the N-terminal β-domain (residues 1–31) and a 4-metal cluster in the C-terminal α-domain (residues 32–68) of $M^{II}_7MT$-3, a metal partitioning found also in $M^{II}_7MT$-1/MT-2 (Fig. 1). In both clusters, the metal ions are tetrahedrally coordinated by both terminal and µ-bridging thiolate ligands.

From the detailed $^{113}Cd$ NMR studies of human $^{113}Cd_7MT$-3, evidence for unprecedented dynamic processes within the $Cd_3Cys_9$ and $Cd_4Cys_{11}$ cluster have been obtained. Thus, besides the presence of a fast exchange between conformational cluster substates occurring in both clusters, additional very slow exchange processes between configurational cluster substates take place in the $Cd_3Cys_9$ cluster of the β-domain. The changes in conformational substates may be visualized as minor dynamic fluctuations of the metal coordination environment, and those of the configurational substates as major structural alterations brought about by temporarily breaking and reforming of the metal-thiolate bonds. The existence of interchanging configurational cluster substates of comparable stability has also been demonstrated for inorganic adamantane-like metal-thiolate clusters with the general formula $[M_4(SPh)_{10}]^{2-}$ (M = Cd(II), Zn(II), Co(II), and Fe(II)). The NMR studies toward the determination of the three-dimensional (3D) structure of mouse and human $Cd_7MT$-3 have also been undertaken. However, because of dynamic processes in the

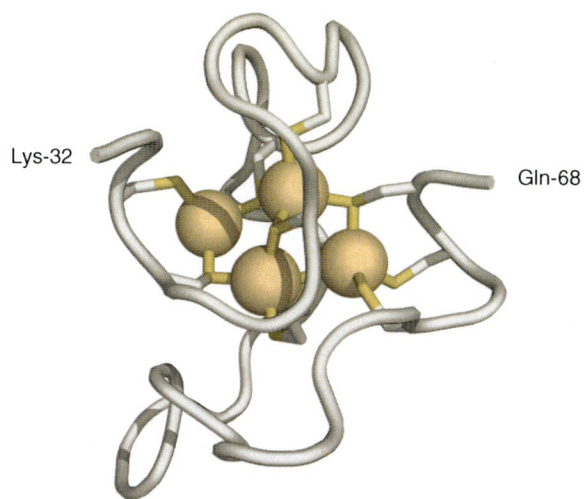

**Zinc Metallothionein-3 (Neuronal Growth Inhibitory Factor), Fig. 2** The NMR solution structure of the α-domain of human $^{113}$Cd$_7$MT-3. The Cd(II) ions are shown as *light-orange spheres* connected to the protein backbone by cysteine thiolate ligands. The model was generated with PyMOL using the Protein Data Bank coordinates 2F5H

β-domain of Cd$_7$MT-3, only the 3D structure of the C-terminal α-domain (residues 32–68), containing an adamantane-like Cd$_4$Cys$_{11}$ cluster, could be determined by NMR. The structure of this domain reveals a peptide fold and cluster organization very similar to that found in mammalian Cd$_7$MT-1/MT-2, with the exception of an extended flexible loop encompassing the acidic hexapeptide insert (Fig. 2). In the absence of a complete set of NMR data for the β-domain, the organization of the 3-metal cluster in Cd$_7$MT-3, resembling that present in mammalian M$^{II}_7$MTs (Vašák and Romero-Isart 2005), was obtained through modeling (Fig. 1). To account for slow dynamic events centered at the 3-metal cluster of MT-3, a partial unfolding of the β-domain, whose kinetics could be determined by the *cis/trans* interconversion of Cys-Pro amide bonds in the C$_6$PCP$_9$ motif, has been suggested. A support for this model came from the X-ray absorption fine structure (EXAFS) studies of Zn$_7$MT-3, and its single Zn$_3$β-domain as well as from molecular dynamics (MD) simulations. The former technique allows the precise determination of metal-metal and metal-ligand distances. In the Zn$_3$β-domain, the determined Zn···Zn distance of 3.28 Å is inconsistent with the cyclohexane-like Zn$_3$Cys$_9$ cluster structure present in the β-domain of mammalian MT-2, where a Zn···Zn distance of about 3.8 Å exists (Fig. 1). The different structure of the 3-metal cluster in MT-3 may reflect the presence of one cluster configuration under the condition of EXAFS measurements (at 77 K). The MD simulation of the partial unfolding supported the proposed role for *cis/trans* interconversion of Cys-Pro amide bonds in the folding/unfolding process of the β-domain of Cd$_7$MT-3. Furthermore, as a consequence of the higher calculated energy, due to structural changes induced by the C$_6$PCP$_9$ motif, the β-domain in MT-3 appears to be looser and less constrained when compared with those of MT-1/MT-2. In other MD simulations, a correlation between the number of backbone amide hydrogen bonds to metal coordinated cysteines, NH–S$^\gamma$ hydrogen bonds, in mammalian Cd$_7$MTs, has been proposed to represent a controlling factor regulating the metal-thiolate cluster dynamics (Vašák and Meloni 2011).

In more biological studies, the structural features critical for the neuroinhibitory activity of Zn$_7$MT-3 have been sought. Thus, the mutation of conserved proline residues in the C$_6$PCP$_9$ motif of MT-3 to the corresponding polypeptide sequence present in MT-2 (Table 1) abolished the neuroinhibitory activity and cluster dynamics. Besides this structural motif, the importance of other amino acids in the β-domain along with domain-domain interactions, mediated partly through the acidic hexapeptide insert in the α-domain of MT-3, has also been shown. Although the mechanisms underlying these biological effects remain to be elucidated, the results obtained so far suggest that the structure of the β-domain of Zn$_7$MT-3 is subjected to a fine tuning. Taken together, the structural and mutational studies led to the conclusion that both the specific structural features and the structure dynamics are necessary prerequisite for the extracellular biological activity of Zn$_7$MT-3.

The studies toward the structure of native Cu(I)$_4$Zn$_{3-4}$MT-3 have also been carried out. The EXAFS studies of Cu(I)$_4$Zn$_{3-4}$MT-3 revealed the presence of two homometallic clusters, a Cu(I)$_4$-thiolate cluster and a Zn$_{3-4}$-thiolate cluster. In these clusters, the Zn(II) ions are tetrahedrally coordinated by four thiolate ligands, and the Cu(I) ions are diagonally and/or trigonally coordinated by two or three thiolate ligands. By using domain-specific antibodies, the localization of the Cu(I)$_4$-thiolate cluster has been shown in the β-domain of the protein. A striking feature distinguishing Cu(I)$_4$-thiolate cluster in MT-3 from most biologically occurring Cu(I)-thiolate clusters, including those present in the transcription factors Ace1 and Mac1, is its remarkable stability against molecular oxygen. Detailed information on the interaction of Cu(I) with MT-3 came from the studies of the stepwise Cu(I) binding into the metal-free synthetic domains and the full-length protein. These studies provided also a plausible explanation for the air stability of the Cu(I)$_4$-thiolate cluster. Thus, the Cu(I) binding to the β-domain (residues 1–32) resulted in the successive formation of two cluster forms in which all nine cysteine ligands of this domain were involved. In this case, the Cu(I)$_4$S$_9$ cluster was formed first in a cooperative manner. The binding of two additional Cu(I) led to the expansion of the Cu(I)$_4$S$_9$ core yielding a Cu(I)$_6$S$_9$ cluster. The major differences in the respective spectroscopic features between the Cu(I)$_4$ and Cu(I)$_6$ cluster forms were observed in the low-temperature Cu(I) luminescence emission spectra at 77 K. Thus, while the Cu(I)$_6$ cluster exhibited only a single emissive band at 600 nm, in the case of the Cu(I)$_4$ clusters, two emissive bands at 420 and 610 nm were discerned. The presence of two emissive bands in the Cu(I)$_4$ cluster has been correlated with short intranuclear Cu···Cu distances (<2.8 Å), allowing metal-metal interactions due to a d$^{10}$-d$^{10}$ orbital overlap. Accordingly, the high-energy emissive band has been assigned to $^3$CC (cluster centered) origin and that at low energy to triplet charge-transfer. That the metal-metal interactions in the Cu(I)$_4$ cluster are contributing to its stability in air is supported by the susceptibility of the larger Cu$_6$S$_9$ cluster to oxidation (Vašák and Meloni 2009).

## Reactivity of Mammalian Metallothionein-3

The MT-3 structure, like those of other mammalian MTs, shows a highly unusual reactivity. Despite a high thermodynamic stability of metal-thiolate clusters, and as a consequence of dynamic properties of its structure, they are kinetically very labile, i.e., the thiolate ligands undergo metallation and demetallation rapidly. The experimentally observed kinetic lability allows (1) a rapid metal transfer and (2) metal substitution reactions which follow the order of increasing metal affinity of inorganic thiolates (Cu(I) >> Cd(II) > Zn (II)). Although the thiol groups in the protein are masked through their interaction with metal ions, they retain a substantial degree of nucleophilicity seen with the metal-free protein. This property is reflected by the extremely high reactivity of the coordinated cysteines with electrophiles such as alkylating and oxidizing agents and with radical species such as hydroxyl (OH$^{\bullet}$), superoxide (O$_2^{\bullet-}$), and nitric oxide ($^{\bullet}$NO) (Vašák and Romero-Isart 2005). The free radical attack occurs at the metal-bound thiolates and results in the thiolate oxidation and/or modification with a subsequent metal release. Zn$_7$MT-3 was found to be significantly more reactive than Zn$_7$MT-1/MT-2 toward S-nitrosothiols, while the reactivity of all three isoforms toward H$_2$O$_2$ was comparable. S-nitrosylation occurs via S-transnitrosation, a mechanism that differs from direct attack of thiolate sulfur by NO. The increased reactivity of Zn$_7$MT-3 with free NO and S-nitrosothiols has led to the proposal that Zn$_7$MT-3 may specifically convert NO signals to zinc signals (Vašák and Meloni 2009). The generation of the reactive oxygen species (ROS) through the redox cycling of free or abnormally bound Cu(II) in a Fenton-type reaction is harmful in a number of neurodegenerative disorders. In view of the extracellular occurrence of MT-3, the reaction of Zn$_7$MT-3 with free Cu(II) has been investigated. Zn$_7$MT-3 was able to scavenge free Cu(II) ions through their reduction to Cu(I) and binding to the protein, forming the Cu$_4$,Zn$_4$MT-3 species together with two disulfides bonds. The air-stable Cu$_4$-thiolate cluster, localized in the N-terminal β-domain of Cu$_4$Zn$_4$MT-3 and presumably containing the disulfide bonds, is formed in a cooperative manner. Since in this process the copper-catalyzed production of ROS is quenched, an important protective role of Zn$_7$MT-3 from free Cu(II) toxicity in the brain has been suggested (Vašák and Meloni 2009).

## Roles of Metallothionein-3 in Metal-Linked Neurodegenerative Disorders

In the normal brain, a high concentration of essential transition metal ions such as Zn, Cu, and Fe is present. Dysregulated metal homeostasis, abnormal metal-protein interactions, and the associated oxidative stress, protein misfolding, and aggregation are critical common pathological hallmarks of the progression of several metal-linked neurodegenerative disorders. Insoluble deposits of individual amyloidogenic proteins/peptides and diffusible oligomers are observed in the brains of patients afflicted by a number of neurodegenerative pathologies. These include the amyloid-β peptides (Aβ) in extracellular amyloid plaques in Alzheimer's disease (AD), prion protein (PrP) deposits typical of Creutzfeldt–Jakob and other prion disorders, α-synuclein (α-Syn) in intracellular Lewy body in Parkinson (PD), superoxide dismutase (SOD-1) aggregates in amyotrophic lateral sclerosis (ALS), and Huntington inclusions in cases of Huntington disease (Barnham et al. 2004). In these neurodegenerative diseases, the expression of MT-3 has been found downregulated or altered (Hozumi et al. 2004), and changes in normal homeostasis of essential transition metals such as zinc and copper have been implicated as possible etiological factors. In contrast to redox-inert zinc, aberrantly bound redox-active copper to amyloidogenic proteins can react with molecular oxygen ($O_2$) resulting in the production of ROS. As mentioned above, only the $Zn_7$MT-3 isoform protects neuronal cells from the toxic effect of $A\beta_{1-40}$ peptide. Studies aimed at the understanding of the protective effect of human $Zn_7$MT-3 from the $A\beta_{1-40}$ toxicity demonstrated that the protein can efficiently remove copper via a metal swap reaction not only from soluble $A\beta_{1-40}$–Cu(II) oligomers but also from insoluble aggregates. In this process, Cu(II) is reduced by protein thiolates, forming the stable Cu(I)$_4$Zn$_4$MT-3 species described above and the nonredox active $A\beta_{1-40}$–Zn(II). This metal swap quenches the ROS production mediated by Cu(II) bound to $A\beta_{1-40}$ and occurs in vitro and in human neuroblastoma cell culture whereby the toxic effect of $A\beta_{1-40}$–Cu(II) is abolished. A similar protective effect of $Zn_7$MT-3 has also been demonstrated for other neurodegenerative pathologies including PD and transmissible spongiform encephalopathies (prion diseases). Thus, in vitro studies have shown that $Zn_7$MT-3, through Cu(II) removal from α-Syn and PrP and the formation of stable Cu(I)$_4$Zn$_4$MT-3 species, efficiently prevents the deleterious redox activity of these protein-Cu(II) complexes, their misfolding/oligomerization, and oxidative modification of critical amino acid residues (Vašák and Meloni 2011). In view of widely different Cu(II) binding motifs in Aβ, α-Syn, and PrP, a general protective role of $Zn_7$MT-3 from Cu(II) toxicity in the brain can be envisaged.

## Cross-References

▶ Zinc Metallothionein

## References

Barnham KJ, Masters CL, Bush AI (2004) Neurodegenerative diseases and oxidative stress. Nat Rev Drug Discov 3: 205–214

Binz PA, Kägi JHR (1999) Metallothionein: molecular evolution and classification. In: Klaassen C (ed) Metallothioenein IV. Birkhäuser, Basel, 7–13

Fukada T, Yamasaki S, Nishida K et al (2011) Zinc homeostasis and signaling in health and diseases: zinc signaling. J Biol Inorg Chem. 16:1123–1134

Gouy M et al (2010) SeaView version 4: A multiplatform graphical user interface for sequence alignment and phylogenetic tree building. Mol Biol Evol 27:221–224

Hozumi I, Asanuma M, Yamada M et al (2004) Metallothioneins and neurodegenerative diseases. J Health Sci 50:323–331

Miles AT, Hawksworth GM, Beattie JH et al (2000) Induction, regulation, degradation, and biological significance of mammalian metallothioneins. Crit Rev Biochem Mol Biol 35: 35–70

Uchida Y, Takio K, Titani K et al (1991) The growth inhibitory factor that is deficient in the Alzheimer's disease brain is a 68 amino acid metallothionein-like protein. Neuron 7:337–347

Vašák M, Meloni G (2009) Metallothionein-3, zinc and copper in the central nervous system. In: Sigel A, Sigel H, Sigel RKO (eds) Metallothioneins and related chelators. Royal Society of Chemistry, Cambridge, 319–351

Vašák M, Meloni G (2011) Chemistry and biology of mammalian metallothioneins. J Biol Inorg Chem. 16:1067–1078

Vašák M, Romero-Isart N (2005) Metallothioneins. In: King RB (ed) Encyclopedia of inorganic chemistry, 2nd edn. Wiley, New York, 3208–3221

West AK, Hidalgo J, Eddins D et al (2008) Metallothionein in the central nervous system: roles in protection regeneration and cognition. Neurotoxicology 29:489–503

# Zinc Physiology

▶ Zinc Homeostasis, Whole Body

# Zinc Regulation in Fish Biology

Christer Hogstrand
Metal Metabolism Group, Diabetes and Nutritional Sciences Division, School of Medicine, King's College London, London, UK

## Synonyms

Slc30a = ZnT; Slc39a = Zip

## Definitions

Paralogs: Homologous DNA sequences generated by genome duplication.

Orthologs: Genes in two species that originate from the same gene in the most recent common ancestor.

96-h LC50: Concentration of a substance that kills 50% of a test population within 96 h.

Water hardness: A measure of the amount of calcium and magnesium ions in water according to the formula Hardness (mg $L^{-1}$ as $CaCO_3$) = 2.5 × $[Ca^{2+}]$ + 4.1 × $[Mg^{2+}]$.

Slc30: Solute carrier family 30 (zinc transporter) family.

Slc39: Solute carrier family 39 (zinc transporter) family.

Sp1: Sp1 transcription factor.

## Introduction

An estimated 10% of all proteins in eukaryotic cells bind zinc, and there are approximately 3,000 zinc proteins in humans. ▶ Zinc Binding Proteins: Abundance DNA sequencing data suggest that this is also a reasonable approximation for the numbers of proteins present in different fish species. Almost all of the zinc in cells is bound to proteins, peptides, and amino acids, but there is a small fluctuating pool of free cytosolic $Zn^{2+}$ which is involved in cell signaling pathways.

Transport of Zinc across the plasma membrane and between cellular compartments is mediated by two large families of proteins, called Znt (Slc30) and Zip (Slc39), which together have at least 21 members in fish. With few exceptions, Znt and Zip proteins in fish map directly onto their mammalian orthologs. Other proteins central to zinc regulation in mammals are also present in fish.

An important difference between fish and land-living vertebrates is that fish take up metals across the gill in addition to the intestine. The gill transports zinc and regulates its uptake in the animal. This presents a useful in vivo model for studies of transepithelial zinc transport and its control.

Compared to other transition metals, dietary zinc has low toxicity to vertebrates including fish. In contrast, fish and other water-breathing animals are moderately sensitive to waterborne zinc with acutely (96 h exposure) lethal concentrations being higher than for metals such as silver, cadmium, and copper, but lower than those for manganese and nickel.

## Zinc-Regulatory Proteins

### Zinc Transporters

Transport of zinc across the plasma membrane and between cellular compartments is mediated by two families of proteins called Slc30 (Znt) and Slc39 (Zip). There are at least 8 Znt paralogs and 13 Zip paralogs in fish and in almost all cases these map directly onto their mammalian orthologs (Hogstrand 2012). Thus, additional paralogs of the genes coding for these proteins created through genome duplications in fish must have been suppressed through evolution. Members of the Zip family of zinc transporters move zinc into the cytosol, either from the exterior or from organelles. Znt proteins transport zinc away from the cytosol and either into organelles or out of the cell. However, at least in mammals, one of the Znt proteins, namely, a splice variant of Znt5, can function as a cellular zinc importer (Valentine et al. 2007). Of all Znt proteins only Znt1 has been shown to operate as a zinc export system. All of the Slc30 paralogs, except Znt5, have six transmembrane domains, with N- and C- termini being located on the cytoplasmic side. Znt5 is much larger and has up to 12 transmembrane domains.

Most Zip proteins, with the exceptions for Zip7, -9, and -13, are functional in the plasma membrane and mediate tissue-specific zinc uptake. Zip7 releases zinc from the ER, and Zip9 and -13 from the trans-Golgi network. Zip8 may transport zinc into some cells and out of lysosomes in other tissues. Zip proteins have eight transmembrane spanning domains with both N- and C-termini located away from the cytosol. They all have a long cytoplasmic loop (between transmembrane domains III and IV), which typically has several histidine residues located in clusters. These may function as temporary binding sites for zinc as it transverses the protein. Members of a subfamily of Zip proteins, called the LIV-1 subfamily, have a long extracellular N-terminal stretch which is also rich in histidine residues and in some members contains a putative proteolytic cleavage site, which may be of importance for posttranslational processing. Several Zip Zinc transporters have been functionally characterized in fish and all three were shown to mediate Zinc import when ectopically expressed in cells or *Xenopus* oocytes. *Zip3* is one of the most abundantly expressed zinc importers in gills and intestine of zebrafish and the $Zn^{2+}$ ion is the likely transported species for Zip3. Furthermore, Zip3-mediated zinc transport is stimulated by a slightly acidic medium (pH 5.5–6.5) and inhibited by $HCO_3^-$.

Although there is considerable redundancy in terms of Zinc transporter function, many of the described zinc transporters have distinct biological functions relating to their specific distribution among tissues and organelles (Hogstrand 2012). Thus, the distribution of zinc within the fish and between cellular compartments is regulated by differential expression and activation of the 21 zinc transporters.

In addition to the Znt and Zip protein families, calcium transporters are generally permeable to zinc. For fish, the competition of $Zn^{2+}$ with $Ca^{2+}$ for the epithelial calcium channel (Ecac/Trpv6), located on the apical membrane of the gill, is of particular importance because this contributes to the well-known protective effect of hardness against $Zn^{2+}$ toxicity (Hogstrand 2012). Ecac (Trp6) belongs to the Transient Receptor Potential (TRP) family of proteins. It is primarily expressed in the gill where it is responsible for apical calcium entry (Qiu and Hogstrand 2004). While Ecac (Trpv6) is probably a zinc uptake pathway when zinc concentrations in the water are elevated, it is not known whether or not Ecac contributes to nutritional zinc uptake.

## Metallothionein

Metallothioneins (MT) are important zinc binding proteins in fish cells. Vertebrate MT is a homologous family of proteins with only about 60 amino acids of which 19 are cysteinyl residues (Fig. 1a). ▸ Zinc Metallothionein All these cysteines are involved in binding zinc or other metals with thiolate bonds and the protein can contain up to seven zinc atoms. While mammals have four major isoforms of MT, fish have only one or two. A model of their individual relationships by multiple sequence alignment suggests that all fish and mammalian MT sequences may have evolved from a common ancestral gene and that MT-IV is the most divergent form of vertebrate MT (Fig. 1b). Based on their general tissue distribution and metal inducibility, fish MT are functionally equivalent to MT-I and MT-II isoforms in mammals. MT can protect cells against metal insult and function as a redox switch to release $Zn^{2+}$ ions for zinc signaling events (Maret 1995). The fraction of total tissue zinc bound to MT varies enormously among fish tissues and with zinc content, and may be as little as 6% in the gill of an unexposed rainbow trout or as much as 74% in the liver of female squirrelfish and soldierfish of the *Holocentridae* family (Hogstrand 2012). Small molecules, such as glutathione, cysteine, and histidine, are probably important zinc ligands in cells but the quantitative roles of these in zinc binding in fish tissues is not well researched.

## Zinc Signaling

Zinc regulates transcription of genes through metal-responsive transcription factor-1 (Mtf1), which is conserved through evolution and is present in most animals including fish. This protein has about 590 amino acids including six zinc-finger domains, two of which bind zinc with lower affinity than the others (Li et al. 2006). If there is a rise in the cytosolic $Zn^{2+}$ concentration, the low-affinity zinc binding sites will be filled enabling Mtf1 to associate with its cognate DNA motif (5'-TGCRCNC-3') known as a Metal Response Element (MRE) and thereby regulate transcription. Other mechanisms, such as phosphorylation, may modulate Mtf1 activity. MTF1 was first discovered as the transcription factor responsible for induction of the mouse MT gene in response to metal exposure. Several searches have been carried out to identify Mtf1 target genes and at least one such hunt was performed on fish (Hogstrand et al. 2008). Using microarray technology with a multifactorial experimental design, including RNAi knockdown of *mtf1* and zinc treatments in zebrafish ZF4 cells, it was shown that regulation of over 1,000 genes was Mtf1 dependent. However, sequence analysis of these genes showed that only 43 of them contained MRE in configurations and location compatible with Mtf1 responsiveness and it was concluded that remaining genes were likely induced as downstream ripples of the signaling cascade. Of the 43 putative Mtf1 targets, 19 were genes involved in development. There was also a staggering over representation of transcription factors, which made up almost half (48%) of the identified genes. The results strongly suggest that zinc signaling via Mtf1 is of particular importance during embryonic development.

In terms of cellular zinc homeostasis, the zinc exporter, Znt1, is positively regulated at the transcriptional level by Mtf1. However, Mtf1 can also be a transcriptional repressor. Expression of the zinc importer Zip10 is downregulated in gills of zebrafish treated with high levels of zinc in the water. The negative zinc regulation of expression of *zip10* is mediated by Mtf1, which functions as a repressor for the gill transcript of *zip10* (Hogstrand et al. 2008; Zheng et al. 2008). A cluster of three MREs is responsible for Mtf1 transcriptional repression in *zip10* and these are straddling the transcription initiation site. These MREs overlap with Sp1 sites and binding of Mtf1 to these MREs may therefore block assembly of the transcription initiation complex. Interestingly, the *zip10* transcript expressed in kidney is regulated by an alternative promoter, which is positively regulated by zinc and Mtf1.

Zinc activation of Mtf1 is of importance in the defense against free radical stress in fish as well as in humans. Mt is an important antioxidant and it reduces free radicals while metal thiolate bonds are oxidized and zinc released (Maret 1994). This results in a transient increase in the labile $Zn^{2+}$ concentration of the cell, activation of Mtf1, and consequential expression of several key antioxidant genes, including glutathione peroxidase (*gpx*), glucose-6-phosphate dehydrogenase (*g6pd*), glutathione-S-transferase (*gst*), and *mt* itself (Hogstrand 2012).

Zinc can also regulate cellular signaling pathways without direct involvement of gene transcription. One of the most persuasive examples of non-genomic zinc signaling comes from a study on zebrafish, in which it was shown that the epithelial-to-mesenchyme transition (EMT) in the gastrula

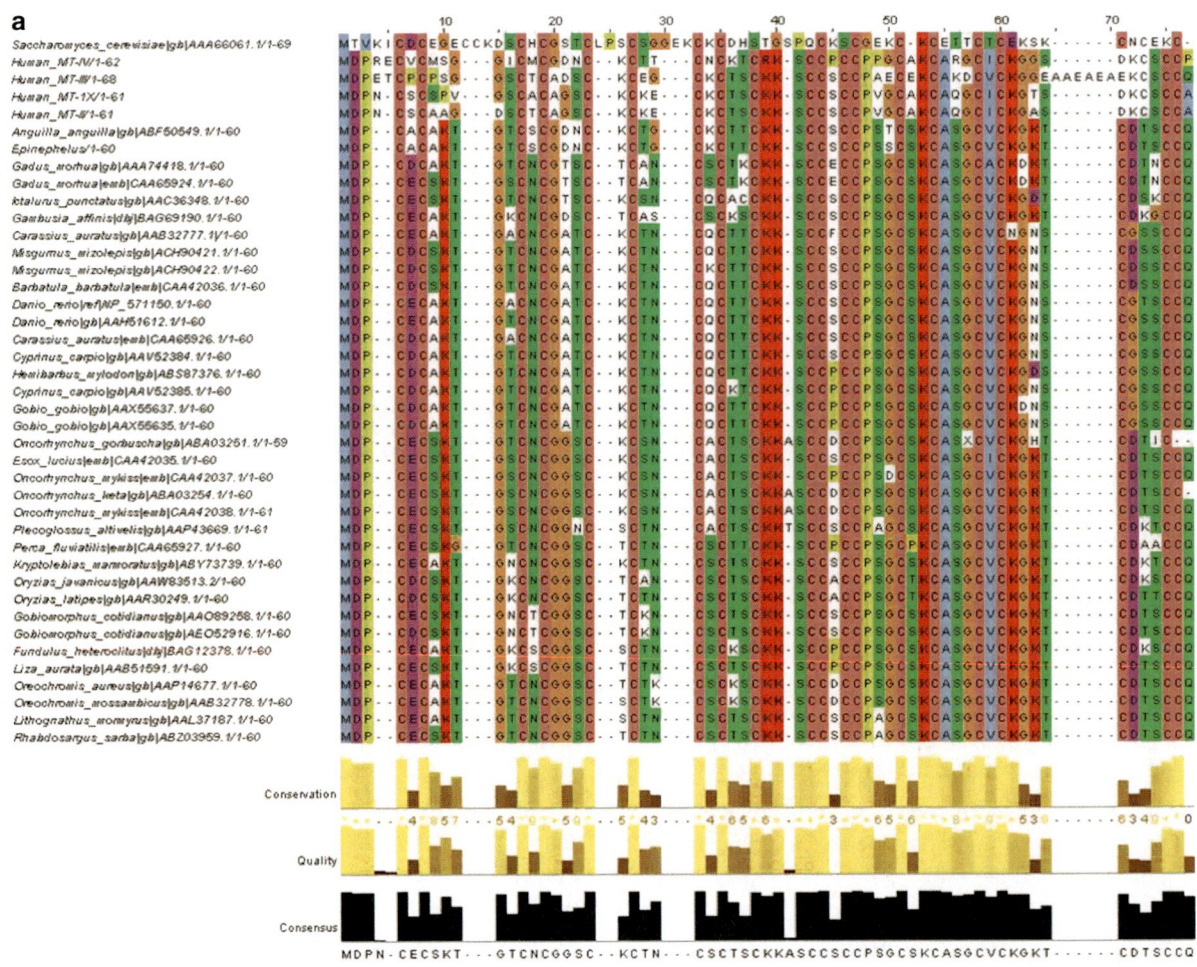

**Zinc Regulation in Fish Biology, Fig. 1** (continued)

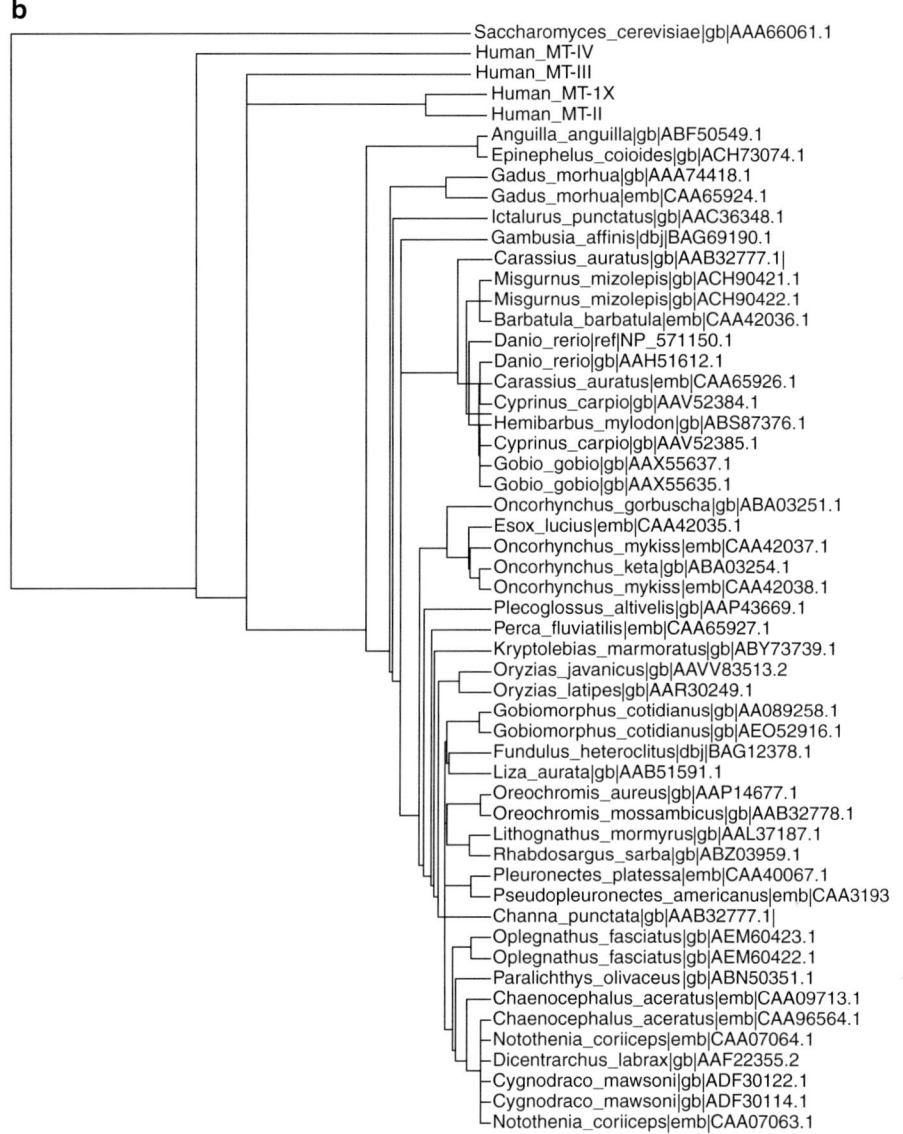

**Zinc Regulation in Fish Biology, Fig. 1** Alignment of metallothionein protein sequences from fish, human and yeast (*Saccharomyces cerevisiae*) using MUSCLE as hosted by the European Bioinformatics Institute (EBI). (**a**) Sequence alignment is annotated with species, sequence accession number, sequence length, degree of conservation between species, indication of the quality of the alignment, and the consensus sequence along all species. Note the high conservation of cysteinyl residues among vertebrate metallothioneins. (**b**) Phylogenetic tree of the metallothionein sequences shown in (**a**). *Saccharomyces cerevisiae* metallothionein was included as an out-group. Note that all fish sequences cluster separately from human MT-1X, MT-II, and MT-III, indicating a common ancestral gene for these sequences and fish metallothioneins. Human MT-IV is indicated to be divergent from all other forms in human and fish

organizer is dependent on the presence of the zinc importer, Zip6 (aka Liv-1/Slc39a6) (Yamashita et al. 2004). During EMT, cells downregulate expression of the cell-cell adhesion protein, E-cadherin (Cdh1), allowing them to migrate into new positions. In the zebrafish embryo this process leads to inward migration and formation of the mesenchyme germ layer. Incidentally, EMT is pathologically activated during progression and metastasis of cancers; human *ZIP6* was originally discovered because it is one of the most upregulated genes in estrogen-dependent breast cancers.

## Zinc Requirements

While fish can supplement their zinc supply by uptake directly from the water across the gills, dietary zinc is still essential. The Board of Agriculture, NAS, has set nutritional requirements of zinc for fish. The nutritional requirement of zinc for channel catfish was set to 20 mg kg$^{-1}$ feed while rainbow trout, common carp, and tilapia were considered to require 30 mg kg$^{-1}$. However, partially because of its perceived additional benefits and concerns about bioavailability (e.g., high levels of hydroxyapatite and/or phytate) and batch variability, fish feeds in the EU are allowed to contain a maximum of 200 mg kg$^{-1}$ feedstuff. This ration provides approximately 10 mg zinc kg$^{-1}$ day$^{-1}$, which is almost 100 above the recommended daily intake for an adult human in the UK. Zinc deficiency in fish causes anorexia, poor growth, bone deformations, reduced survival, cataracts, and exaggerated startling response (Hogstrand 2012).

## Zinc Toxicity

One of the most defining differences in toxicological effects between fish and terrestrial vertebrates is due to the presence of gills. Because of the low metal buffering capacity of many natural waters (compared with, for example, gut content), fish and other water-breathing organisms are prone to toxicity from zinc and other waterborne chemicals. Zinc is therefore regarded as an important environmental pollutant. Acute toxicity values (96-h LC50) that formed the basis for the US Water Quality Criteria for zinc in freshwater ranged from 66 to 40,900 µg L$^{-1}$ with the lowest value recorded in a test with rainbow trout in moderately soft water (water hardness = 92 mg L$^{-1}$ as CaCO$_3$). Acute zinc toxicity in freshwater fish is primarily caused by reduced entry of Ca$^{2+}$ across the gill, resulting is severe hypocalcemia (Hogstrand 2012). Zn$^{2+}$ inhibits Ca$^{2+}$ uptake across the apical membrane (facing the water) of the gill epithelial cells through competitive inhibition of the apical epithelial calcium channel (Ecac/Trpv6) and mixed inhibition of the high-affinity Ca$^{2+}$-ATPase (Pmca), which extrudes Ca$^{2+}$ across the basolateral membrane (facing the interior of the animal) of gill epithelial cells. Pmca is extremely sensitive to Zn$^{2+}$ with inhibition in rainbow trout occurring at a cytosolic free Zn$^{2+}$ activity of 100 pM (6.5 ng L$^{-1}$). This concentration is within the range of measured free Zn$^{2+}$ ion concentrations in mammalian cells raising the intriguing possibility that zinc is an endogenous regulator of Pmca activities in cells.

There is little experimental evidence regarding the potential importance of blocked calcium uptake as a major cause of chronic toxicity beyond that occurring during the initial weeks of exposure. The most sensitive documented endpoints observed in chronic toxicity studies of different freshwater species are diverse: survival, growth, reproduction, and hatching (Hogstrand 2012). The lowest NOEC value included the European Risk Assessment for zinc was for mottled sculpin (*Cottusbairdi*), which when exposed to zinc for 30 days in soft water (water hardness = 46 mg L$^{-1}$ as CaCO$_3$) showed a NOEC for reduced survival of 16 µg L$^{-1}$ (Bodar et al. 2005).

Avoidance may be the most sensitive behavioral response to waterborne zinc exposure (Hogstrand 2012). Rainbow trout avoid waterborne zinc concentrations orders of magnitude below those causing lethality. Other fish species also show avoidance reaction to modest concentrations of zinc with lowest observed effect concentrations (LOEC) for lake whitefish (Coregonusclupeaformis), Atlantic salmon (Salmosalar), and vimba bream (Vimbavimba) being 10, 53, and 220 µg L$^{-1}$, respectively.

Surprisingly little is known about zinc toxicity to seawater fish. The range of available 96-h LC50 values for zinc in seawater span from 190 µg L$^{-1}$ in cabezon (*Scorpaenichthysmarmoratus*) to 83,000 µg L$^{-1}$ recorded in a test with mummichog (*Fundulus heteroclitus*).

## Zinc Uptake

Zinc can be absorbed by intestine as well as the gill of fish (Hogstrand 2012). During most conditions uptake of zinc occurs primarily over the gut and the gill is an auxiliary organ for zinc uptake. However, especially in freshwater fish uptake across the gill can contribute significantly (>50%) to total zinc absorption if the zinc concentration in the water is high or that in the diet low.

Uptake of zinc across the gill and intestinal uptake epithelia is achieved through the cooperation of apical zinc import proteins and a basolateral extrusion system

(Hogstrand 2012). While the latter function is likely served by ZnT1 in either epithelium, the setup of importers differs slightly. Zip3, Zip10, and Znt5 are highly expressed in both gill and intestine. In contrast, Zip4 is highly expressed in the gut, but it has little or no expression in the gill. The epithelial calcium channel, Ecac (Trpv6), is highly expressed in the gill and significantly contributes to branchial zinc uptake when the zinc concentration in the surrounding water is high. In fish, Ecac does not appear to be expressed in the intestine.

In zebrafish znt1, znt5, zip3, and zip10 are all regulated at the mRNA level in response to zinc depletion or supplementation. Expression of znt1 increases in zinc supplemented fish and decreases during zinc depletion. Zinc depletion results in upregulation of zip3, zip10, and znt5, which is consistent with these transporters operating as zinc importers at the apical membrane of the gill. Zinc supplementation causes decreased expression of znt5 and zip10, which serves to limit zinc uptake during zinc excess to prevent overload and toxicity. The expression of the basolateral zinc efflux protein Znt1 is increased in gill when availability of zinc is in excess, presumably to protect the epithelial cells from zinc toxicity.

Under most conditions, the intestinal zinc absorption may be more important than that across the gills in terms of whole body zinc uptake. However, the latter has received considerable attention in fish partially because of its significance for zinc toxicity and also because of the ease by which gill uptake can be manipulated experimentally. Compared with the gill, the fish intestine has low affinity and high capacity for zinc uptake. Like in mammals, intestinal uptake of zinc is highest in the small intestine. Uptake of zinc in the gut starts with the diffusion into the unstirred layer followed by binding to the mucus of the intestinal epithelium. The metal is then transported into the epithelial cells, primarily by means of zinc transporters.

Zinc uptake is inhibited by several other cations as well as by high phosphorous in the form of hydroxyapatite and phytate contents of the diet. Amino acid chelates of zinc are often used as feed additives for farmed animals, including fish, in attempt to improve the efficiency of zinc absorption. However, it cannot be categorically concluded that the use of organic forms of zinc result in consistent improvement of absorption and performance indicators above that obtained with inorganic forms of the metal. Zinc bound to L-histidine was taken up by trout intestine at least as efficiently as unbound zinc, but through histidine-facilitated pathways. Chelation of zinc by L-cysteine appeared to increase zinc uptake in trout and sea bream, compared with that for unbound zinc. Interestingly, the presence of histidine and cysteine strongly influenced the distribution of the newly accumulated zinc in the body. Histidine promoted accumulation of zinc in the intestinal tissue whereas cysteine caused zinc to specifically accumulate in the blood.

Zinc uptake in fish is stimulated by the hormones cortisol and vitamin D (1,25-dihydroxycholecalciferol) and reduced by Stanniocalcin. However, it is not known if these hormones are important physiological regulators of zinc homeostasis.

## Distribution of Zinc in Fish

The total zinc content of a fish is about 10–40 mg kg$^{-1}$ wet mass, depending on species (Hogstrand 2012). The largest amounts of zinc in the body are found in muscle, bone, and skin, which combined make up 60% of the body's zinc content. The highest concentrations may be present in the eye, where up to 30 mg g$^{-1}$ dry weight has been recorded. Most parts of the eye are very high in zinc and in some cell types this may be related to the coordination of melanin to zinc. This is not a peculiarity for fish, but a common theme among vertebrates although the exact function(s) of zinc in the eye remains uncertain.

There are no specialized storage organs for zinc, with the exception for the liver, which in female fish stores up zinc ahead of redistribution to the ovaries for incorporation into the developing eggs. This takes extreme proportions in females of the squirrelfish family (*Holecentridae*), which accumulates copious amounts of zinc in the liver and later delivers this to the eggs (Hogstrand 2012). The highest zinc concentration measured in a squirrelfish liver is 4.6 mg g$^{-1}$ wet weight, which corresponds to about 1.8% of the dry mass. To avoid toxicity from such an enormous tissue concentration of zinc, female squirrelfish have also extremely high levels of Mt in the liver. Males, on the other hand, have liver zinc concentrations similar to those of other fish species as do immature females. The reason for this extraordinary behavior remains unsure, but might relate to the unusually large eyes of the embryos in this family.

Typical zinc concentrations in gill, intestine, and ovary among different species (from both zinc contaminated and pristine environments) are 14–130 µg g$^{-1}$ wet mass (Hogstrand 2012). In most species and most conditions, the kidney contains 20–100 µg Zinc g$^{-1}$ wet weight but in yellowfin tuna (*Thunnus albacares*) the dry mass zinc content of the kidney has been reported to reach 23,500 µg g$^{-1}$. Zinc concentrations in edible muscle of different fish are 4–40 µg g$^{-1}$ wet mass.

From the points of uptake, the intestine and the gill, zinc is distributed by the circulatory system and taken up by different tissues. Average concentrations of zinc in plasma of several fish species have been determined to range from 6.3 to 23 mg L$^{-1}$, which is one order of magnitude higher than those found in humans (0.65–1.3 mg L$^{-1}$). The zinc concentration in red blood cells ranges from being about the same as that in plasma to about three times the plasma level.

There is no known specific plasma protein, which distributes zinc among tissues. Instead, as mentioned earlier, distribution of zinc in the body is managed by a comprehensive set of zinc transporters and their expression patterns and activities dictate zinc uptake in different tissues. In rainbow trout plasma, 0.2% has been estimated to be unbound and this corresponds to about 22 µg L$^{-1}$. Most of the plasma zinc is bound to albumin, and the rest to alpha 2-macroglobulin. Although it may not be a specific plasma protein for zinc, the egg yolk protein, vitellogenin (Vtg), does bind zinc and transports it to the developing oocytes.

## Zinc Excretion

Elimination of zinc from gills, liver, and kidney is fast, but whole body excretion is slow with a biological half-life in excess of 200 days. Relatively little is known about excretion routes of zinc in fish, but based on limited data it is possible that the gill is a possible major excretory route. Zinc is also secreted into the intestine, but that the relative contribution of the biliary route might be small. Likewise, excretion of zinc with the urine may represent <1% of total zinc losses.

## Cross-References

▶ Zinc and Diabetes (Type 2)
▶ Zinc and Immunity
▶ Zinc and Wound Healing
▶ Zinc and Zinc Ions in Biological Systems
▶ Zinc Cellular Homeostasis
▶ Zinc Deficiency
▶ Zinc Metallothionein
▶ Zinc-Binding Proteins, Abundance
▶ Zinc-Binding Sites in Proteins
▶ Zinc-Secreting Neurons, Gluzincergic and Zincergic Neurons

## References

Bodar CW, Pronk ME et al (2005) The European Union risk assessment on zinc and zinc compounds: the process and the facts. Integr Environ Assess Manag 1(4):301–319

Hogstrand C (2012) Zinc. In: Wood CM, Farrell AP, Brauner CJ (eds) Homeostasis and toxicology of essential metals. Academic, London, pp 136–200

Hogstrand C, Zheng D et al (2008) Zinc-controlled gene expression by metal-regulatory transcription factor 1 (MTF1) in a model vertebrate, the zebrafish. Biochem Soc Trans 36:1252–1257

Li Y, Kimura T et al (2006) The zinc-sensing mechanism of mouse MTF-1 involves linker peptides between the zinc fingers. Mol Cell Biol 26(15):5580–5587

Maret W (1994) Oxidative metal release from metallothionein via zinc-thiol/disulfide interchange. Proc Natl Acad Sci USA 91(1):237–241

Maret W (1995) Metallothionein/disulfide interactions, oxidative stress, and the mobilization of cellular zinc. Neurochem Int 27(1):111–117

Qiu AD, Hogstrand C (2004) Functional characterisation and genomic analysis of an epithelial calcium channel (ECaC) from pufferfish, Fugu rubripes. Gene 342(1):113–123

Valentine RA, Jackson KA et al (2007) ZnT5 variant B is a bidirectional zinc transporter and mediates zinc uptake in human intestinal Caco-2 cells. J Biol Chem 282(19):14389–14393

Yamashita S, Miyagi C et al (2004) Zinc transporter LIVI controls epithelial-mesenchymal transition in zebrafish gastrula organizer. Nature 429(6989):298–302

Zheng D, Feeney GP et al (2008) Regulation of ZIP and ZnT zinc transporters in zebrafish gill: zinc repression of ZIP10 transcription by an intronic MRE cluster. Physiol Genomics 34((2):205–214

## Zinc Release (If Different Title Is Chosen)

▶ Zinc-Secreting Neurons, Gluzincergic and Zincergic Neurons

# Zinc Sensors in Bacteria

Nigel J. Robinson and Ehmke Pohl
Biophysical Sciences Institute, Department of Chemistry, School of Biological and Biomedical Sciences, Durham University, Durham, UK

## Synonyms

AdcR; CzrA; SmtB; ZiaR; Zinc-dependent activators; Zinc-dependent corepressors; Zinc-dependent derepressors; ZntR; Zur

## Definition

Zinc sensors detect when critical thresholds in the amounts of zinc within cells have been surpassed. The sensors detect either (1) zinc excess to trigger removal, use, or storage of surplus atoms or (2) zinc deficiency to trigger acquisition of more atoms or to reduce demand.

## Overview of Zinc Sensors in Bacteria

Bacteria possess multiple types (families) of DNA-binding, zinc-sensing regulators of gene transcription (Ma et al. 2009). Other family members detect other metals. More than one type of zinc sensor is commonly present in a cell. Membrane-associated two-component sensors which transduce the metal signal to a separate DNA-binding transcriptional regulator are known for some metals but are either absent or yet to be described for zinc.

Bacterial zinc sensors bind to the operator-promoter regions of genes encoding zinc export proteins (often $P_1$-type ATPases such as ZiaA and ZntA), zinc import proteins (often ABC-type ATPases such as ZnuABC), zinc-sequestering metallothioneins (SmtA and BmtA), or enzymes whose exploitation alters cellular demand for zinc (Ma et al. 2009). The latter include non-zinc-requiring alternatives to zinc enzymes, the alternates being expressed when zinc is limiting. For example, a search of the *Bacillus subtilis* genome for DNA-binding sites for the zinc sensor Zur identified the promoter of a gene encoding a ribosomal protein YtiA, which is expressed under zinc-deficient conditions to replace the zinc-requiring ribosomal protein L31 (RpmE). Since cells contain large numbers of ribosomes, this represents a substantial economy in the overall cellular budget for zinc. Similarly, the zinc-dependent enzyme of folate biosynthesis FolE (MtrA) is replaced by the YciA enzyme when low zinc is sensed (Lee and Helmann 2007).

Bacterial zinc sensors fall into three categories: derepressors, corepressors, and activators, with the three founding sensors being SmtB, Zur, and ZntR, respectively (Fig. 1). SmtB and ZntR control the expression of genes whose products are required when cytosolic zinc is high, while conversely, Zur controls the expression of genes whose products are required when cytosolic zinc is low.

## Zinc-Dependent Derepression

### SmtB

SmtB was the first zinc-sensing transcription regulator to be cloned and sequenced in 1993, immediately followed in the same year by the mammalian zinc-sensor MTF1, and the yeast zinc-sensor Zap1 was cloned and described 4 years later (as cited in Waldron et al. 2009). The original *SmtB* gene is divergently transcribed from a gene-encoding bacterial metallothionein SmtA which binds and sequesters surplus zinc. Cyanobacterial mutants missing the *SmtA* gene have increased sensitivity to elevated concentrations of zinc. The *Smt* divergon was discovered in the cyanobacterium *Synechococcus PCC 7942* (related strains known as *Synechococcus elongatus*, *Synechococcus PCC 6301*, and *Anacystis nidulans* R2). Genes encoding zinc sensors, and sensors for other metals, related to SmtB have since been described in a huge diversity of other bacteria (Osman and Cavet 2010). The species distribution of homologues of the *SmtA* gene is more restricted. These genes are most prevalent within other cyanobacteria and some pseudomonad species. The small size and low sequence complexity of metallothioneins, due to their high content of cysteine residues, could mean that related genes are widespread but easily overlooked in data base searches (Waldron and Robinson 2009).

The shared operator-promoter region between the *SmtA* and *SmtB* genes contains multiple binding sites for the SmtB protein. DNA binding by SmtB was first

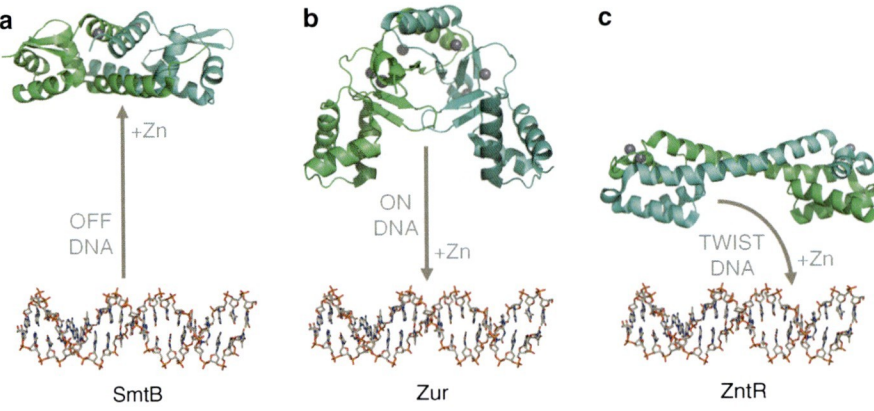

**Zinc Sensors in Bacteria, Fig. 1** Representation of three types of sensors (**a**) zinc-dependent transcriptional derepressor SmtB (1R22), (**b**) zinc-dependent transcriptional corepressor Zur(3MWH), and (**c**) zinc-dependent transcriptional activator ZntR (1Q08)

observed in gel retardation assays. Subsequently, such interactions have been biochemically characterized in detail by using equilibrium assay systems that exploit fluorescence anisotropy to follow the decline in the rate of rotation of fluorescently tagged DNA upon protein binding. Addition of the zinc chelator 1–10 ortho-phenanthroline to DNA-binding reactions enhances SmtB-DNA complex formation while treatment with zinc has the converse effect, impairing SmtB-DNA complex formation. This was the opposite of what had previously been observed for large numbers of eukaryotic zinc-dependent DNA-binding proteins in which occupancy of a zinc finger was necessary for DNA binding. Formation of a zinc-SmtB complex inhibits formation of a DNA-SmtB complex. Mutants missing SmtB have elevated constitutive gene expression from the *Smt* operator-promoter region. In normal cells containing SmtB, expression of SmtA from the *Smt* operator-promoter region is only elevated when cells are treated with surplus zinc. Thus, SmtB binds zinc to alleviate repression of expression of the zinc-sequestering metallothionein which in turn binds and buffers the excess zinc atoms. An SmtB-like protein, ZiaR, from the cyanobacterium *Synechocystis PCC 6803* controls expression of a zinc-exporting $P_1$-type ATPases ZiaA, while SmtB homologues in some other cyanobacteria control expression of zinc-sequestering metallothioneins as well as zinc-exporting $P_1$-type ATPases (Ma et al. 2009; Waldron and Robinson 2009; Osman and Cavet 2010).

The crystal structures of apo-SmtB and zinc-SmtB reveal a winged helical homodimer (Fig. 2). DNA binding is achieved through a helix-turn-helix structural motif (Osman and Cavet 2010). Two pairs of zinc-binding sites have been identified through a combination of site-directed mutagenesis, spectroscopic methods, and structural biology. One pair of sites is located between the fifth (and carboxyl-terminal) pair of α-helices (α5) which are aligned (one helix from each subunit) in an antiparallel manner. The second pair of sites exploits two ligands from helix α3, which is proximal to the DNA-binding region, and two ligands from the amino-terminal region of the second subunit (designated α3N). Site-directed mutagenesis revealed that only the α5 sites of SmtB are obligatory for zinc-mediated gene derepression in vivo. In other related proteins, the α3N metal-binding sites are active in inhibiting DNA binding. It is easy to envisage how metal binding in the vicinity of the helix-turn-helix motif might disrupt DNA binding. Structural studies of the homologous protein CzrA have uncovered the more elaborate allosteric mechanism which is required to transduce the zinc-binding signal from remote α5 metal-binding sites to the DNA-binding region (Ma et al. 2009).

## CzrA

In common with SmtB, CzrA from *Staphylococcus aureus* binds and senses zinc via sites located on carboxyl-terminal α5 helices, distal from the DNA-binding helix-turn-helix regions of the homodimer. Somehow, the two regions must be allosterically coupled such that zinc binding impairs DNA binding and hence alleviates repression. One of the four liganding amino acids, histidine$_{97}$, which contributes to the α5 zinc sites of CzrA, plays a special role in creating this connection. Zinc forms a bond with the $N^{\delta 1}$ atom of histidine$_{97}$ imidazole which positions the ring such that $N^{\epsilon 2}$ forms a hydrogen bond with oxygen from the main chain carbonyl of histidine$_{67}$ derived

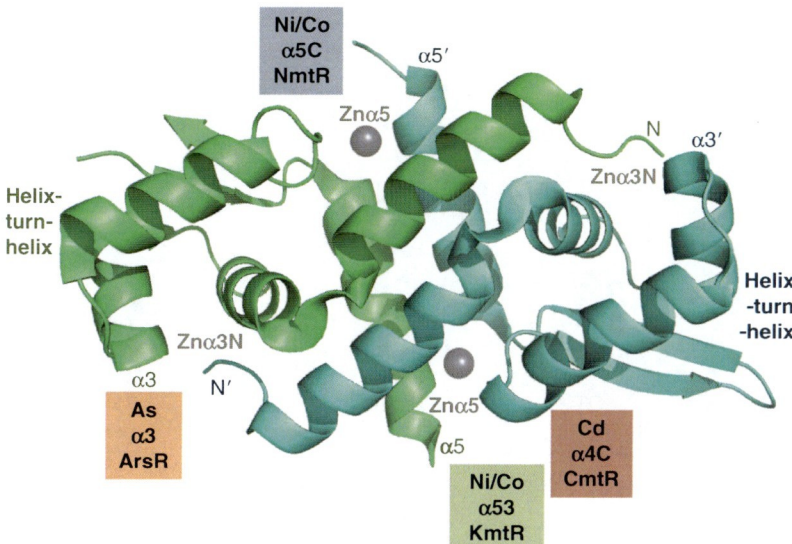

**Zinc Sensors in Bacteria, Fig. 2** Two sets of zinc-binding sites in SmtB, with ligands derived from antiparallel α5 helices (Znα5) or α3 helices plus additional residues from the amino-terminal region (Znα3N) with helix α3 contributing toward the DNA-binding helix-turn-helix region (1R22). Only metals at α5 were visible in the crystal structure. The diversity of locations of sensory metal sites in proteins related to SmtB are also indicated. In the nickel and cobalt sensors NmtR and KmtR, the ligands are derived from α5 helices plus a carboxyl-terminal extension; in cadmium-sensing CmtR, they are from α4 helices plus the carboxyl-terminal region; in oxyanion-sensing ArsR, they are from α3 helices

from the opposing subunit of the homodimer (Fig. 3). An analogous hydrogen bond forms in SmtB albeit exploiting the carbonyl of arginine$_{87}$. A network is established which propagates to the DNA-binding region via a second hydrogen bond from histidine$_{67}$ to leucine$_{63}$ (leucine$_{83}$ in SmtB) (Ma et al. 2009). This model for allosteric coupling is rather static. It envisages the binding of zinc driving the conformation of the protein from a form that is ideally suited to DNA binding to a second form which is less well suited to DNA binding. However, solution structural studies have painted a more dynamic picture in which the metal-bound form of the protein samples a spectrum of conformers which are distinct from those sampled by apo-CzrA, with some analogy to observations made with the related cadmium sensor CmtR. The range of conformers sampled by apo-CzrA encompasses forms which are best suited to DNA binding.

## Zinc-Dependent Corepression

### Zur

Zur enables genes to be expressed in response to zinc starvation. The first representatives of the Zur family to

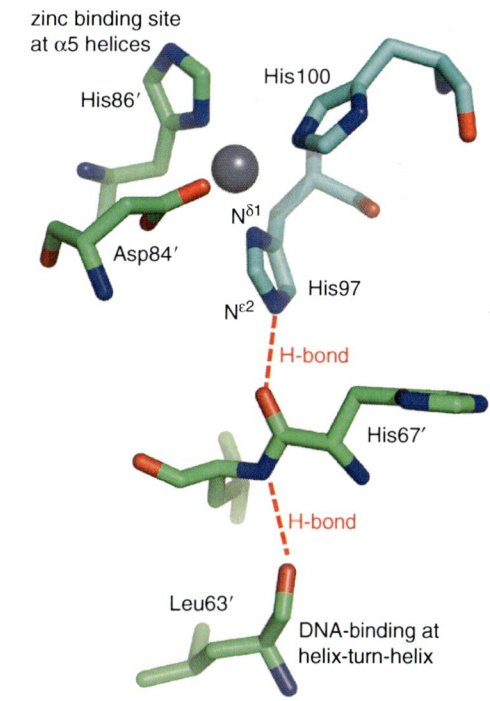

**Zinc Sensors in Bacteria, Fig. 3** A hydrogen-bond network connects zinc-coordinated histidine$_{97}$ from the α5 helix of CzrA to the DNA-binding helix-turn-helix region (1R1U)

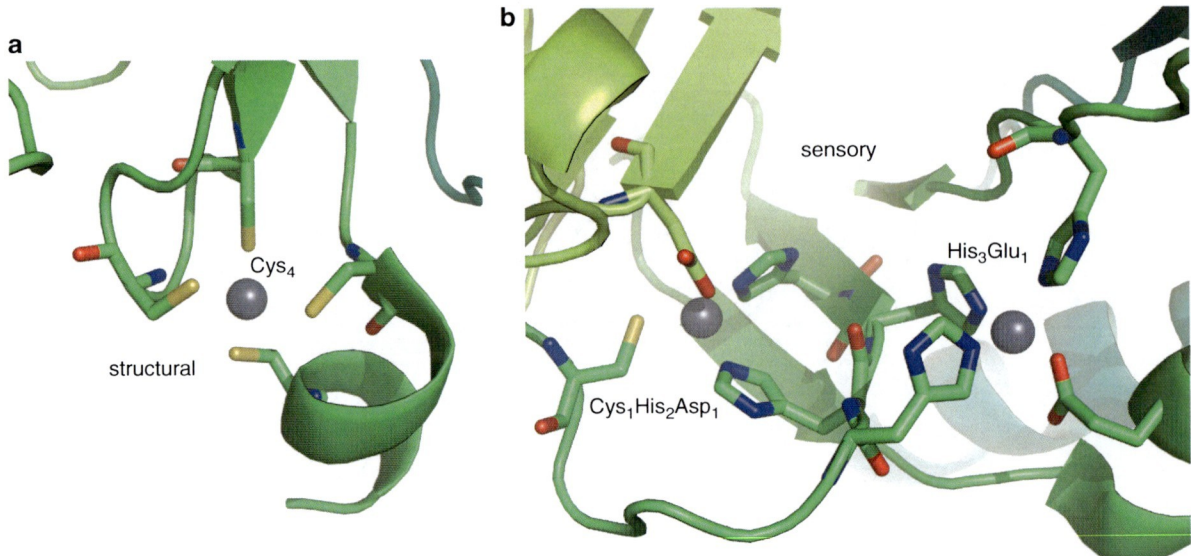

**Zinc Sensors in Bacteria, Fig. 4** The distinct coordination environments of the structural zinc site (**a**) and the exchangeable sensory zinc sites (**b**) of Zur (3MWH)

be discovered were the iron-sensing Fur proteins (Lee and Helmann 2007). Representatives of the Fur family that sense zinc were described in 1998 simultaneously from *E. coli* and from *B. subtilis*. Fur and Zur proteins are generally transcriptional repressors, but some target genes have been found to lose expression in mutants missing the metal sensor suggesting a role as an activator. In some cases, at least for Fur proteins, this mode of regulation is indirect and mediated by the antisense RNA RyhB, with the antisense RNA in turn being repressed by Fur. However, there are exceptions where Fur directly activates gene transcription upon binding its cognate metal: For example, ferritin and superoxide dismutase (*sodB*) expression in *Helicobacter pylori* are both transcriptionally activated by metal-bound Fur.

Zur has an amino terminal winged helical DNA-binding domain connected to a carboxyl-terminal dimerization domain (Fig. 1b). Both *E. coli* Fur and Zur proteins contain structural zinc sites in addition to their sensory metal-binding sites. DNA binding by Zur requires occupancy of both sites (Lee and Helmann 2007). The crystal structure of *M. tuberculosis* Zur (Lucarelli et al. 2008), a protein originally designated Fur$_B$, visualized the structural zinc atom coordinated to four cysteine residues in a tetrahedral geometry, and similar structural sites have since been identified in homologous proteins from other bacteria (Fig. 4a). The structural zinc ties the carboxyl-terminus to the dimerization domain. Unlike the structural zinc ion, the sensory zinc atoms (Fig. 4b) are in exchange with solution in the folded form of the protein. Estimates of the zinc affinities of the sensory sites of *E. coli* Zur made using TPEN to create zinc buffers are extremely tight, sub-femtomolar, which has been used to draw inferences about the disposition of zinc within bacterial cells, considered in the final section of this entry (Outten and O'Halloran 2001).

### AdcR

The AdcR protein from *Streptococcus pneumoniae* controls expression of genes encoding a high-affinity ABC zinc uptake system during host infection, but the precise connection with the infection process is unclear. This bacterial zinc sensor is a member of the MarR family of repressors. In common with Zur, DNA binding by AdcR is promoted by zinc. Loss of AdcR leads to constitutive expression of the zinc import system even in elevated zinc, and hence growth of such mutants is impaired in zinc-supplemented medium (Reyes-Caballero et al. 2010).

## Zinc-Dependent Transcriptional Activation

### ZntR

Transcriptional activation by MerR family proteins, of which ZntR is a representative, involves an unusual

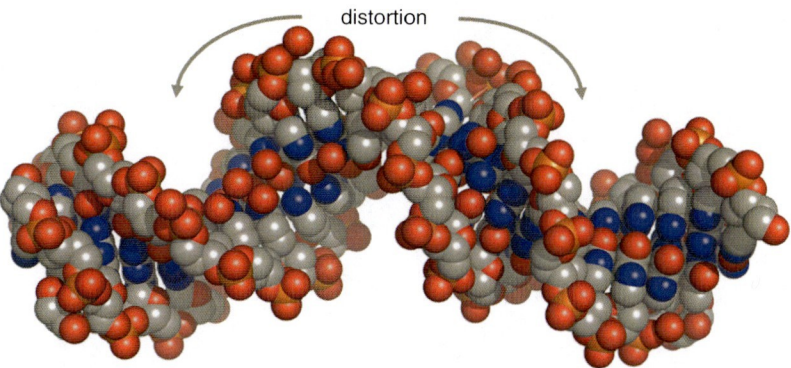

**Zinc Sensors in Bacteria, Fig. 5** ZntR alters the conformational of DNA upon binding zinc to realign suboptimally spaced promoter elements in a manner analogous to other members of the MerR family of transcriptional activators. The diagram shows the distortion of DNA caused by the activated form of a related, albeit not metal-sensing, protein BmrR (1R8E)

mechanism of DNA distortion (Fig. 5). The promoter regions of bacterial genes commonly contain two nucleotide sequences that are simultaneously recognized by RNA polymerase; the sequences are positioned around 10 and 35 nucleotides upstream of the transcription start site. Promoters recognized by this family of activators have a slightly longer than normal gap between these two sequences, usually about 19 or 20 nucleotides rather than 17 (Ma et al. 2009). Because the double helix has ten nucleotides per turn, an extra spacing of two nucleotides will offset the $-10$ and $-35$ elements by approximately $72°$ ($2 \times 360 \div 10$). In consequence, RNA polymerase cannot readily bind to both elements at once. ZntR and other MerR-like proteins bind to hyphenated inverted repeat sequences located within these abnormal promoter regions. They bind both in the absence and in the presence of the sensed metal. However, the metal-bound forms of the sensors in effect underwind the promoter region such that the $-10$ and $-35$ elements become realigned to positions that are more optimal for RNA polymerase recognition (Fig. 5). Thus, mutants missing ZntR lose activation of the target genes in surplus zinc.

ZntR, in common with all structurally characterized MerR proteins, has an amino-terminal winged helical domain which includes a long dimerization helix at its carboxyl end (Fig. 1c). Although SmtB, ZntR, and Zur do not share significant sequence similarity, a winged-helix DNA-binding domain is a common feature of all three. Two symmetrical sets of metal-binding sites are positioned at either end of the dimerization helix in ZntR. In the zinc-sensor ZntR, these sites can each coordinate a pair of zinc ions (Fig. 6). These dinuclear sites involve a histidine and three cysteine residues from one subunit plus, crucially, a further cysteine from the second protomer which provides bridging thiolate ligands to coordinate both of the zinc atoms (Changela et al. 2003; Ma et al. 2009). In all MerR-like proteins characterized to date, regardless of whether they sense zinc or some other metal ion, the metal-sensing sites are located in this same position at the two ends of these long dimerization helices (Ma et al. 2009). This is distinct from SmtB-like proteins where, as discussed in the next section of this entry, the sensory sites have evolved at a diversity of locations within different members of the family. Here, it is proposed that this reflects the greater constraint placed upon a metal site which must correctly organize the sensor to distort the DNA to a form which is ideal for activating transcription (as in ZntR) versus one which must merely reorganize the sensor to anything other than the set of conformers best suited to DNA binding (as in SmtB).

## Metal Selectivity

Each bacterial zinc sensor is a member of a family in which other representatives sense metals other than zinc: SmtB (NmtR-, KmtR-, CzrA-, CmtR-, CadC-, ArsR-sensing metals or metalloids such as cobalt, nickel, cadmium, lead, or oxyanions of arsenic); Zur (Fur-, Mur-, Nur-sensing metals such as iron, manganese, and nickel); ZntR (MerR-, CueR-, PbrR-, CoaR-sensing metals such as mercury, copper, lead, and

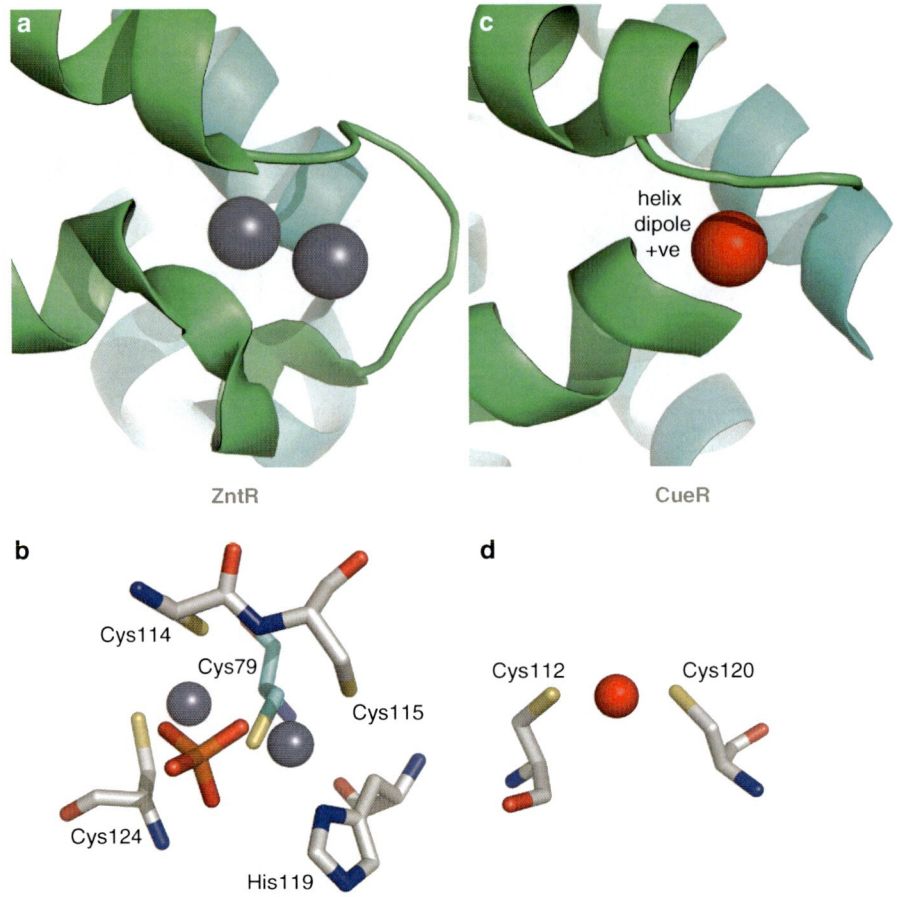

**Zinc Sensors in Bacteria, Fig. 6** Two zinc atoms bind to the sensory site of ZntR (1QO8) (**a**), each via tetrahedral coordination with cysteine$_{79}$ providing bridging sulfur ligands to both atoms (**b**). In the related copper sensor CueR, a helix dipole proximal to the copper site is thought to repel divalent metal ions (1QO5) (**c**) and in this manner is unlike ZntR (1QO8). Copper is bound in a diagonal site in CueR (**d**)

cobalt) (Waldron and Robinson 2009; Ma et al. 2009). At present, we cannot confidently predict which metal is sensed by a homologue through bioinformatics, and many of the assignments in databases are unsound. Typically, additional experimental data are required although specificity for zinc can sometimes be inferred from gene context: For example, if the gene for the deduced metal sensor is divergently transcribed from, or co-transcribed with, a gene encoding a known zinc enzyme or protein of zinc homeostasis.

In addition to the α3N and α5 metal-sensing sites of SmtB-like sensors described previously, other locations for sensory sites in this family are helix α5 plus extra ligands from a carboxyl-terminal region (α5C and α53), helix α4 plus carboxyl-terminal ligands (α4C), and helix α3 (without additional ligands from the amino-terminal region) (Osman and Cavet 2010) (Fig. 2). Based on the homologues which have so far been characterized in vivo, there is a good correlation between the location and characteristics of the sensory site and the metal sensed. In contrast, there is imperfect correlation between the sensory sites and the phylogenetic relationships of the proteins generated from overall sequence similarities (Osman and Cavet 2010). This implies that the sensory sites have either evolved multiple times, a case of convergent evolution, or that concerted evolution within each organism has obscured the evolutionary paths. The nature of the sensory metal-binding sites, location and ligand set, offers the best "guess" as to which metals are sensed.

The mechanisms by which each sensor discerns the correct, sensed metal from all other ions have been subdivided into three categories (Waldron et al. 2009; Waldron and Robinson 2009):

## Affinity

The properties of the first or second coordination sphere may favor binding of the correct, sensed metal. For example, a helix dipole is oriented toward the cuprous-copper site in CueR, but not toward the analogous zinc-sensing site in ZntR. This introduces positive charge into the CueR metal coordination sphere which disfavors binding of divalent metals (Fig. 6). Many metal sensors have been shown to bind metals with an order of affinity which correlates with the Irving Williams series (magnesium and calcium binding weakly, then manganese and ferrous iron, then cobalt and nickel, and finally zinc and cupric ions binding most tightly, with cuprous ions also binding tightly to sensors with thiolate ligands). For those sensors which detect the weaker-binding metals, absolute metal affinity alone cannot explain selectivity.

### Allostery

In some cases, a metal which is not sensed binds tightly but using a coordination geometry, or using a partial complement of ligands, which is ineffective at triggering the allosteric metal-sensing mechanism. For example, SmtB binds zinc in a tetrahedral geometry which readily impairs DNA binding and alleviates repression. NmtR also binds zinc in a tetrahedral geometry and more tightly than cobalt, but cobalt is the more effective metal in triggering the allosteric mechanism because cobalt prefers to bind in an octahedral geometry, and DNA binding by NmtR is more effectively impaired when six ligands (including two ligands from an extra carboxyl-terminal extension relative to SmtB) are recruited to the octahedral metal coordination site of NmtR.

### Access

A metal which can trigger the allosteric mechanism of a sensor in vitro may not be detected in vivo, because the sensor never gains access to a sufficiently high intracellular concentration of the ion to allow binding. This requires that other metal homeostasis proteins maintain the ion at some lower threshold concentration.

## Inferences Made from Zinc Affinities of Sensors

Metal sensors have been called the arbiters of metal sufficiency. This is because when the sensors become occupied by their cognate metal, they trigger the expression of genes whose products act to prevent metal concentrations from becoming any higher or lower, depending upon the metal threshold which has been exceeded (Waldron and Robinson 2009). What are these metal thresholds?

The zinc affinities of the *E. coli* zinc sensors Zur and ZntR have been estimated to be in the femtomolar range (Outten and O'Halloran 2001). One atom per *E. coli* cell volume equates to nanomolar. This implies that these zinc sensors inhibit production of zinc import proteins (zinc-Zur represses expression of ZnuABC zinc import) and activate expression of a zinc exporter (zinc-ZntR activates expression of the ZntA $P_1$-type ATPase) at a concentration of zinc a million times less than one free atom per cell ($10^{-9} \div 10^{-15}$). This has been used to infer that all zinc ions within the *E. coli* cytosol are tightly bound and buffered (Outten and O'Halloran 2001). This provides one explanation for how sensors for weaker-binding metals can be prevented from gaining access to tighter-binding metals based upon the relative metal affinities of the set of metal sensors. However, a living cell is not at thermodynamic equilibrium. Thus, how closely the metal affinity of each metal sensor approximates to the buffered cell concentration of the "sensed" metal remains to be established. There is a need for future studies that consider factors such as: interactions between metal sensors and delivery proteins such as metallochaperones, metal importers, or other metal-containing proteins; the abundance of each metal sensor in the cell; metal-dependent change in stability of the apo- and the metallated forms of the sensors (Pruteanu and Baker 2009); and the contributions of small molecules that could promote swift kinetics of associative metal transfer within cells. A more complete understanding of the actions of bacterial zinc sensors will ultimately require that their properties are considered in the context of those of the full set of metal sensors, plus many other components of zinc homeostasis, in the same cell.

## Cross-References

▶ Bacterial Mercury Resistance Proteins
▶ NikR, Nickel-Dependent Transcription Factor

## References

Changela A, Chen K, Holschen J, Outten CE, O'Halloran TV, Mondragon A (2003) Molecular basis of metal-ion selectivity and zeptomolar sensitivity by CueR. Science 301:1383–1387

Lee J-W, Helmann JD (2007) Functional specialisation within the Fur Family of metalloregulators. Biometals 20:485–499

Lucarelli D, Vasil ML, Meyer-Klaucke W, Pohl E (2008) The metal-dependent regulators FurA and FurB from *Mycobacterium tuberculosis*. Int J Mol Sci 9: 1548–1560

Ma Z, Jacobsen FE, Giedroc DP (2009) Coordination chemistry of bacterial metal transport and sensing. Chem Rev 109:4644–4681

Osman D, Cavet JS (2010) Bacterial metal-sensing proteins exemplified by ArsR-SmtB family repressors. Nat Prod Rep 27:668–680

Outten CE, O'Halloran TV (2001) Femtomolar sensitivity of metalloregulatory proteins controlling zinc homeostasis Science 292:2488–2492

Pruteanu M, Baker TA (2009) Proteolysis in the SOS response and metal homeostasis in Escherichia coli. Res Microbiol 160:677–683

Reyes-Caballero H, Guerra AJ, Jacobsen FE, Kazmeirczak KM, Cowart D, Koppolu UMK, Scott RA, Winkler ME, Giedroc DP (2010) The metalloregulatory zinc site in Streptococcus pneumoniae AdcR, a zinc-activated MarR family repressor. J Mol Biol 403:197–216

Waldron KJ, Rutherford JC, Ford D, Robinson NJ (2009) Metalloproteins and metal sensing. Nature 460:823–830

Waldron KJ, Robinson NJ (2009) How do bacterial cells ensure that metalloproteins get the correct metals? Nat Rev Microbiol 7:25–35

# Zinc Signaling

▶ Zinc-Secreting Neurons, Gluzincergic and Zincergic Neurons

# Zinc Signal-Secreting Cells

Christopher J. Frederickson
NeuroBioTex, Inc, Galveston Island, TX, USA

## Synonyms

Cells that exocytose zinc – Cells that exocytose $Zn^{2+}$

## Definition

Research on zinc's roles in biology, medicine, and health is dominated by studies on the roles of zinc-containing and zinc-modulated proteins, of which there are literally thousands in vertebrate tissues. As much as 3 % of the human genome is estimated to be devoted to proteins that contain or require zinc. These zinc-proteins are in virtually every organelle of every cell type in every vertebrate, and they are the subject of many entries in this encyclopedia. This entry is about the "*other*" pool of zinc, the "free" or "weakly bound" or "rapidly exchangeable" zinc that is located almost exclusively in the secretory granules (also called "vesicles") of cells that secrete zinc as an ionic, intercellular signal: $Zn^{2+}$. This "secretory" zinc was actually studied fairly intensively in the early 1900s, and was the subject of several Nobel prize nominations in that era. Secretory zinc was prominent again in the 1940s when the new histochemical zinc staining serendipitously revealed the "free" zinc in secretory cells (and fortunately did not show the protein-bound zinc, which is ubiquitous). However, it was not until the 1970s–1980s that the discovery of the cortical brain cells that secrete zinc began to generate substantial interest in cells that secrete zinc signals.

## Early History

Papyrus scrolls recovered from ancient Egypt described preparation of salves comprised of zinc ores and animal fat for the treatment of superficial skin wounds. This use of zinc has continued, more or less uninterrupted, through to the present time. Zinc is a major ingredient of myriad powders and lotions for diaper rash and scratches and itches, as a trip down the relevant isle in the drug store will quickly show. Indeed, "Calamine Lotion®," used in America since the mid-1800s for itches, abrasions, stings, and scratches, is named from the eighteenth century French word for zinc ore, "calamine" which at that time was mined near the town of La Calamine, in Belgium. Thus, although this was not a discovery of a natural biological role of zinc, the antimicrobial potency of zinc and zinc ions has been implicitly recognized for several thousand years.

Perhaps the largest volume of early published research on secreted zinc was done at the Pasteur Institute in the early 1900s by Dr. Delezenne. Delezenne studied two fluids, snake venom and pancreatic juice, which comes from the exocrine, acinar portion of the pancreas, and found both fluids highly enriched in zinc. The pancreatic juice that flows into the gut contains around 100 μM of zinc. Over the first 2 weeks after an i.v. $^{65}$Zn infusion, about 10 % of the tracer is secreted in pancreatic juice, with a total zinc excretion of 8–14 mg in those same 2 weeks. (Failure to reabsorb this zinc challenges zinc status.) Snake venom, in comparison, appears to have the highest concentration of zinc in any fluid or tissue ever tested in nature, with concentrations as high as 50 mM of zinc recorded for some species. Delezenne was nominated for the Nobel Prize repeatedly for these discoveries, but never received the prize. Intriguingly, it is still not known how much of the zinc in pancreatic juice or snake venom is "free" and how much is bound to which proteins. (Some zinc-containing enzymes in venom and pancreatic juice are known, however.)

Following Delezenne's work, the next surge in research on secreted zinc came from the serendipitous discoveries of Okamoto and Maske in Germany and McNary and McLardy, in Boston, USA, in the late 1940s and 1950s. All of those investigators used an intravital stain for zinc, diphenythiocarbazone (dithizone), which is dark green in color and strongly lipophilic. Intravital injections of dithizone solutions produce an immediate distribution of the dye throughout the circulatory system and the organism. Zinc: dithizonate is bright red, so whenever "free" or "rapidly exchangeable" zinc was present, the bright zinc:dithizonate chromogen formed in situ (Fig. 1). Using this method, Okamoto reported on bright red zinc staining in the secretory granules of the endocrine pancreas (Islets of Langerhans) the salivary glands, and the Paneth cells of the intestine. In addition, Maske reported that there was also a bright red band of staining in the hippocampal formation of the brain (Fig. 1).

In Boston a few years later, McNary showed the red zinc:dithizonate staining of granulocytes from the blood, and Turner McLardy undertook a series of studies of the zinc:dithizonate staining in Maske's bright red band in the brain. McLardy showed convincingly that the zinc was almost certainly located in the giant secretory terminals ("boutons" or "presynaptic" terminals) of a band of "mossy" axons in the hippocampus. Most importantly, McLardy also showed that after prolonged seizure activity in the hippocampus, the zinc staining was dramatically reduced. In his singular prescience, McLardy correctly guessed that the zinc was localized in the presynaptic (secretory) vesicles of

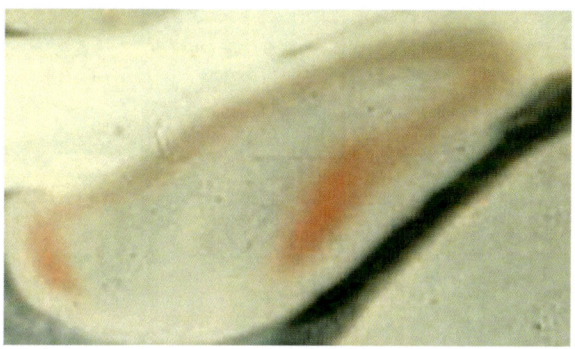

**Zinc Signal-Secreting Cells, Fig. 1** The *red* band is zinc: dithizonate staining in the mossy axons of the hippocampus of the cat brain (about 30×)

the mossy nerve endings and further speculated that the zinc was part of the mix of neurosecretory synaptic signals released from those mossy axons.

McLardy's speculation was proven correct at the Institute of Neurobiology at Arhus University in Denmark where first Theodore Blackstadt, then Gorm Danscher, led for 30 years a team of neuroscientists who perfected the silver-staining of secretory zinc (Timm-Danscher method) and then proceeded to map the entire cerebrocortical zinc-containing neuronal network or the vertebrate brain (including man's). The synaptic release of the silver-labeled zinc from the presynaptic terminals was also preliminarily affirmed by the Aarhus group. In 1984, the first papers confirming McLardy's 1962 opinion and demonstrating directly that zinc was released from the mossy nerve terminals on an impulse-by-impulse basis put zinc firmly in the list of synaptic neurotransmitters and neuromodulators of the mammalian cortex. Since then just under 1,000 scientific papers on "synaptic zinc" (Medline) and just under 1,000,000 popular discussions of "synaptic zinc" on the world-wide web (Google) have appeared.

## Nineteen Types of Zinc-Secreting Cells

As of 2012, 19 types of cells that secrete significant amounts of zinc are known. Some, like the submandibular salivary gland cells of rodents, are no doubt homologs of others, such as the venom gland of the venomous snake (Fig. 2). Others, like the amebocyte-hemocytes antimicrobial cells of mollusks, are phylogenetic precursors of such antimicrobial cells in man, such as the granulocytes in vertebrate blood. There are some cells

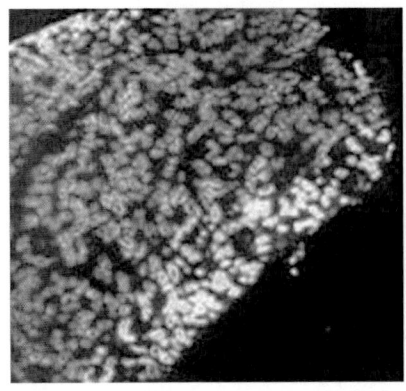
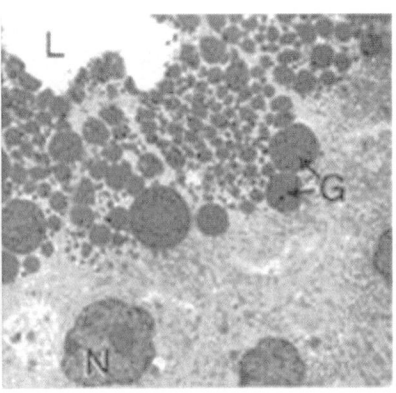

**Zinc Signal-Secreting Cells, Fig. 2** The submandibular salivary gland of a male mouse stained for zinc with TSQ (*left*) and an electron microscopic detail of two salivary cells with their granules (G) stained for zinc by the Timm-Danscher silver method (*right*). The whole gland is magnified about 30×. *N* nucleus, *L* lumen

for which the presence of "free" zinc has not been confirmed (tears gland, sweat gland, and acinar pancreatic gland), but in virtually all, it is known that a substantial portion of the secreted zinc is "free" or "rapidly exchangeable" zinc. The fact that the zinc is "free" is, after all, the reason that all of these cells were originally found with the dithizone, fluorescence, or silver-sulfide zinc staining methods. All of the zinc stains are (necessarily) selective for the tiny fraction of all zinc that is "free" zinc (also known as "histochemically reactive" zinc); none of them will stain the vast preponderance of zinc that is tightly bound to proteins. (If protein-bound zinc reacted with the common zinc stains, the stains would be useless, because all organelles of all cells and tissues would stain – rather like a stain for carbon or hydrogen. Instead, in healthy, normal tissue, only secretory granules (and their precursors in the Golgi-ER) stain for zinc. Compromised tissues, (which includes virtually all cells and tissues grown in vitro) on the other hand, show numerous "injury" and "distress" zinc-stained organelles and compartments, but that staining is an artifact and is not found in intact, healthy tissue in vivo).

As Table 1 shows, *the zinc secretion of* 11 of the 19 different cells has been suggested to have a primary or secondary antimicrobial function. The remainder participate in sperm-oocyte interactions, nutrient (zinc) supply to breast milk, and, quite prominently, the cognitive and mnemonic functions of the cerebral cortex of the brain.

## Zinc-Secreting Endocrine Cells

Perhaps the best characterized zinc-secreting cell of all is the *beta* cell of the islets of Langerhans in the pancreas. These are the insulin-secreting cells, and the insulin is stored in these secretory granules as a hexamer coordinated around zinc atoms. However, in addition to the zinc atoms coordinated to the insulin, there is about a 10-fold molar excess of additional zinc which is so weakly bound that no measurement of the Kd was possible in one attempt to measure it. It is this 10-fold molar excess of the "free" zinc that stains vividly with all of the zinc stains, and has also been imaged as an extracellular "puff" of free zinc when islet cells have been induced to degranulate in vitro, in the presence of zinc fluorescent probes. The ZnT8 transporter keeps the zinc concentration in the granules high.

Though the exocytosis of the zinc signal is unambiguous, the physiological/biological function of that signal is still not clear. One observation is that the zinc functions as an amacrine signal, affecting both neighboring beta cells and neighboring amacrine cells in the exocrine portion of the pancreas. Another noteworthy observation is that serum zinc concentrations are typically depressed in pancreatic cancer, suggesting the possibility that the pancreas has a role in maintaining serum zinc homeostasis. Another speculation is that the excess zinc functions to briefly elevate serum free zinc concentration in the microenvironment of the secreted zinc:insulin complexes so that the zinc:insulin complexes will not dissociate "explosively" upon exocytosis into the blood. Regardless of the function of the secreted zinc, it is interesting to note that the zinc would be relatively easy to image and quantify in vivo in man with near-infrared, NIR, tomography, and an i.v. NIR zinc probe, thus allowing the secretion of the organ to be monitored in real time, noninvasively.

The pituitary gland is the other major zinc-secreting endocrine gland, and, unfortunately, little is known

**Zinc Signal-Secreting Cells, Table 1** Cell types secreting zinc

| Cell type | Secreting cell | Secreting into/onto | Zinc signal function | Possible anti-microbial role | "free" $Zn^{2+}$ |
|---|---|---|---|---|---|
| *Exocrine* | | | | | |
| Digestion | Pancreatic Acinar cells | stomach | | * | |
| | Salivary | Oral cavity | | * | * |
| | | | | * | |
| Reproduction | Prostate gland | Semen | Timed release of spermatozoa | * | * |
| | Fertilized ovum | Fallopian tube | | * | * |
| | Mammary gland | milk | Infant nutrition | | |
| Other | tears | | | * | |
| | sweat | | | * | |
| *Endocrine* | | | | | |
| | Pancreatic beta cell | Circulation Local interstices | Amacrine effects of beta cells | | * |
| | Pituitary cells (3 types) | | | | * |
| *Amacrine* | | | | | |
| | Gluzinergic cortical neurons | synapses | Short- and long-term modulation of glutamatergic synapse | | * |
| | Retinal photoreceptors | synapses | Modulation of retinal funtion | | * |
| *Immune* | | | | | |
| | Granulocytes (3 types) | Interstices/pathogen | antimicrobial | * | * |
| | Mast cells | Interstices/pathogen | antimicrobial | * | * |
| | Paneth cells | Intestinal crypts | | * | * |
| | Haemocyte/ amoebocyte | Oysters, snails | Antimicrobial | * | * |

about the possible functions of those zinc secretions. The problem with all endocrine zinc secretions is, of course, that the secreted $Zn^{2+}$ would be rapidly scavenged by serum albumin and other serum zinc buffers, so the $Zn^{2+}$ signal could only act on receptors encountered close in distance and time to the actual degranulation/secretion.

## Zinc as an Endogenous Antimicrobial

Marine biologists, aquatic environmentalists, and marine toxicologist have long appreciated that the $Zn^{2+}$ ion in aqueous media is a potent, lethal toxin. However, the toxicity of the free $Zn^{2+}$ ion has only recently become widely appreciated by molecular, cell, and animal biologists. In fact, as many have shown, only ~100 nM of free $Zn^{2+}$ ion in a cell culture medium (or open water) is toxic to virtually all cells, and 1 μM of $Zn^{2+}$ will kill even vertebrate fish within 24 h. (Fortunately, all of the commercial tissue media (and all media with blood serum) have a rich abundance of zinc-buffering molecules (e.g., albumin), so that the addition of 10 or even 100 μM of zinc, accidentally or on purpose, does not kill the cells, because the free $Zn^{2+}$ is simply buffered to ~10 nM, the physiologically correct concentration. This is why experimenters typically have to add several hundred micromolar or more of $Zn^{2+}$ to saturate the buffers before – abruptly – they obtain effects upon their cells or tissues in culture media.)

Given the cytotoxicity of the zinc ion, it follows that $Zn^{2+}$ secreted into the blood or into another lumen of the body would be toxic to cells directly exposed to even low micromolar concentrations of the ion. In turn, this implies that a cell that can secrete a high concentration "puff" of $Zn^{2+}$ onto or into the milieu surrounding a pathogen can potentially kill or injure the pathogen with that zinc. This would be especially true if the pathogen membranes were rendered porous by defensins co-secreted along with the zinc. This is because the cytolethal load of free zinc in intracellular cytosol is in the mid-pM range. (Naturally, the buffers in blood and tissue will terminate cell-secreted $Zn^{2+}$ "puffs" over short times and distances.)

As mentioned above, this antimicrobial potency of the zinc ion has been implicitly understood since ancient Egyptian times and has continued into the present in the common use of zinc-laced ointments in salves, and, more recently in treatments that shorten the reign of the rhinovirus in the oropharyngeal cavities (e.g., Coldeze) and of zinc in vaginal inserts and lubricants introduced to reduce STD transmission (*). As shown by the work of the Population Council, NY, NY (Singer et al. 2012). Similarly, it is the cytolethal action of $Zn^{2+}$ that has led to the development of zinc-based antifouling (i.e., antimicrobial) bottom paints that are now replacing the more expensive classical copper-based paints for boat hulls.

Intriguingly, the antimicrobial defense cells, called "hemocytes," or "amebocytes" of snails and oysters sequester large amounts of free zinc in their secretory granules. One assumes that they secrete the sequestered zinc against pathogens as part of the antimicrobial defense. In fact, one pioneering study showed that oysters living free in natural waters with higher zinc concentrations were able to defend more effectively against pathogens than the lower-zinc dwelling oysters, presumably because of the more fully laden zinc defense granules in their hemocytes (Fig. 2) (Fisher 2004).

In man and other vertebrate subjects, it is practically a truism that zinc deficiency impairs immune function and that zinc supplementation remediates against this. How much of that effect is due to the "unloading" and "reloading" of the secretory granules of immune cells with secretable free zinc is not clear, but it certainly seems one plausible mechanism for the strong effects of dietary zinc.

$Zn^{2+}$ ions are almost certainly co-secreted by one subset of the leukocytic immune cells, the granulocytes, all three types of which have been stained for granular secretory zinc by dithizone, silver-sulfide, or both

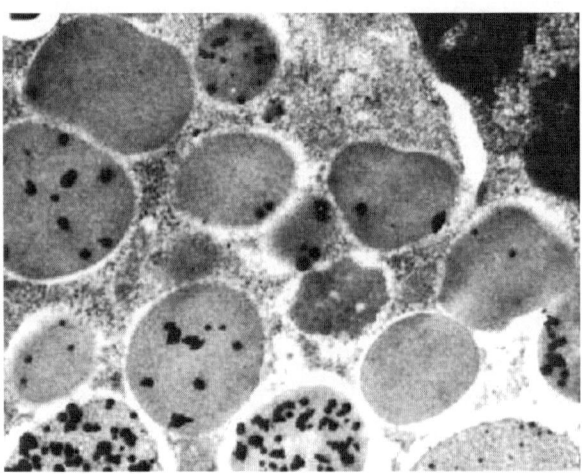

**Zinc Signal-Secreting Cells, Fig. 3** Mast cell secretory granules stained for zinc by silver method. Note the variability in staining. About 50,000×

(Smith et al. 1969). So far as it is known, the actual exocytosis of granulocytic zinc has never been directly observed or recorded, but the co-localization of the zinc with the other secretory products in the granules makes such a result all but certain.

The mast cell, which is nearly ubiquitous in vertebrate tissues, and the Paneth cells (which nestle in the crypts of Lieberkuhn that line intestinal walls) are also generally accepted as specialized parts of the immune or antimicrobial cellular defense system. Both of those types also sequester dramatic amounts of free zinc in their secretory granules (Figs. 3, 4). As with the leukocytic granulocytes, there is no doubt that the mast and Paneth cells degranulate in response to microbial pathogen signals, but, again, published examples showing the presumed "puff" of free zinc that would accompany the degranulation are not known. The secretion of the zinc out of the Paneth cells into the lumen of the Crypts of Lieberkuhn shown in Fig. 4 is an example of what one would expect to see in response to a pathogen signal.

Another very recent antimicrobial action of "secreted" zinc worth mentioning is the intracellular release of zinc (presumably from metallothionein) within human macrophages which has been observed as a presumed antimicrobial response to pathogens endocytosed by the macrophages (Botella et al. 2011). Though intracellular, this "release" of zinc as an antimicrobial is parallel to the extracellular releases from other antimicrobial cells.

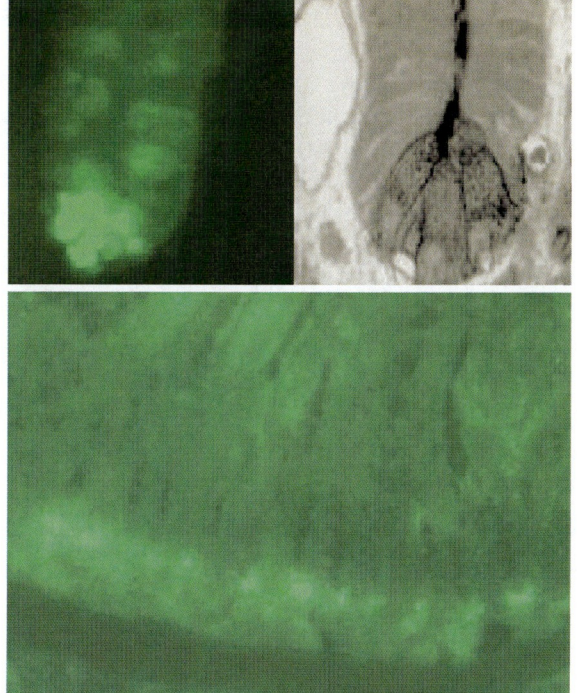

**Zinc Signal-Secreting Cells, Fig. 4** The *bottom* panel shows a transverse section through the intestinal wall with a row of 10–12 brightly-fluorescent, zinc-filled cells at the bottoms of the crypts. *Upper left* shows a single crypt with 3–4 cells filled with zinc-rich fluorescing granules, and the *upper right* is a similar view showing 3 Paneth cells at the bottom of a crypt. The *black, silver zinc* staining shows zinc recently exocytosed into the lumen of that crypt as well (From Frederickson and Danscher unpublished)

Whether the zinc that is secreted into semen, into tears, saliva, and into sweat has any significant antimicrobial function is unknown. In fact, except for the zinc in semen (which is buffered by the 10-fold molar excess of the citrate ion at about 20 μM "free" $Zn^{2+}$), it is not known what the free zinc concentration actually is in tears, sweat, or saliva. Obviously, if the zinc in those fluids is tightly bound to proteins, then it would have no antimicrobial function.

Submandibular salivary glands in mice have free zinc staining in the secretory granules, and the zinc is exocytosed during secretion (Fig. 2), but interestingly, only male mice have the free zinc, the females do not.

Finally, the pZn (i.e., free zinc concentration, analogous to pH) of zinc-rich snake venom is also unknown. The venom gland is a specialized salivary gland, but given the sexual dimorphism of zinc in mouse salivary glands, extrapolation is problematic. It may be noted, however, that the enormously high zinc concentration of some venoms (50 mM!) would be cytotoxic and cytolethal if injected into tissue with even 0.02 % of that zinc ion solution (10 μM; pZn = 5). Once the zinc buffering of the local tissue was exceeded, local necrosis from such a zinc ion injection would be pronounced.

## Zinc Secretion in Reproduction

Second only to snake venom glands in zinc-secreting capacity is the vertebrate prostate gland which, in man, secretes ∼ 10 mM of zinc into the prostatic fluid. The zinc is stored in secretory granules along with other secretory products, including PSA (prostate specific antigen), citrate, and acid phosphatase, and all are exocytosed into the prostatic ducts constitutively. In the fluid of the tubule, the zinc is about 10 mM; the citrate, 100 mM; and the PSA, 2–3 mM. Upon ejaculation, the fluid is expelled from the tubules and ducts into the urethra, mixed with sperm from the testes and the "clot" proteins from the seminal vesicles, and the mixture is expelled to land on the vaginal wall.

The three ingredients – zinc, PSA, and the clot proteins (semenogelins) – act in concert to produce a "timed-release" of the spermatozoa (Fig. 5) (Jonsson 2005). Thus, after ejaculation, (1) spermatozoa remain "stuck" to the vaginal walls for 10–15 min and then (2) the "clot" liquefies, releasing the sperm cells to swim and (potentially) find ova. The role of the citrate is to keep the zinc from precipitating. (The solubility of $Zn^{2+}$ in aqueous media at normal pH is only about 20 μM).

After dilution into the other fluids, the solutes from prostatic fluid have final concentrations of about 2 mM (zinc), 20 mM (citrate), and 0.4 mM (PSA). The sperm cells are bound to the clot proteins by receptor-ligand binding, and the "sticky" clot stays on the vaginal tissues during and after intercourse. The role of the PSA, a protease, is to cleave the semenogelin "clot," causing liquefaction and also denaturing the receptors so that they no longer bind to the ligands expressed on sperm surfaces. The PSA, however, is normally inhibited by the zinc bound to it, so the cleaving cannot begin until the zinc (which is buffered in solution by the citrate, a weak binder, to ∼ 20 μM) is removed

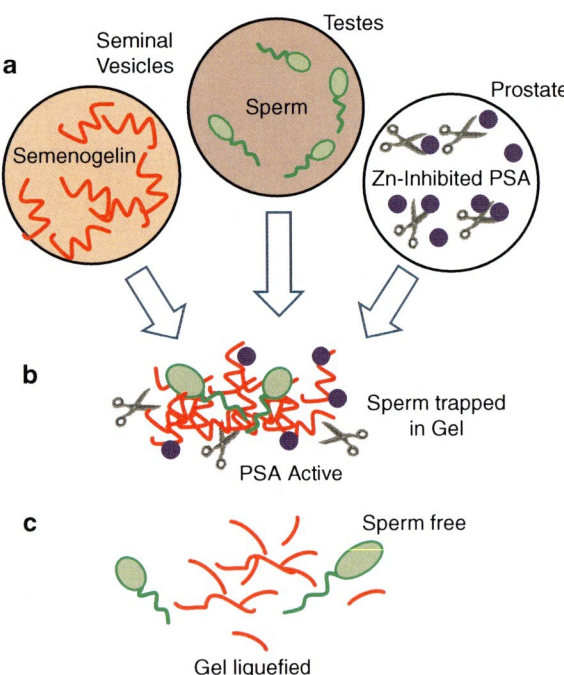

**Zinc Signal-Secreting Cells, Fig. 5** The zinc-controlled timed-release of spermatozoa from the seminal "clot" is illustrated. In (**b**), the clot is stuck to the vaginal wall holding the spermatozoa while intercourse ends and the tissues relax. In (**c**), the liquefaction has been completed by the disinhibited PSA, and the zinc is bound to the semenogelin fragments

from the solution and, finally, from the PSA enzyme itself. It is the many zinc-binding sites on the semenogelins which account for zinc removal. Intriguingly, when the semenogelins are cleaved, the fragments retain their high-affinity zinc-binding sites, even though the receptors to which the spermatozoa are bound are allosterically "broken" by the cleaving.

Once the zinc is cleared from the PSA, the semenogelins are cleaved, the clot is liquefied, and the sperm cells are released (Fig. 5). It is often asserted that the zinc in semen acts as an antimicrobial to reduce pathogens in the male urethra and then vaginal and fallopian tubes. Given that the free zinc concentration is buffered to about 20 μM by the citrate (a strongly cytotoxic concentration), this could certainly be a nontrivial, auxiliary role of the zinc in semen. Likely, this is why others have introduced the use of zinc salts in sexual lubricants against STDs (as mentioned above).

As a "finale" to the zinc signaling that controls the timed-release of the spermatozoa from the clot, it turns out there is another $Zn^{2+}$ signal event that occurs upon fertilization of an oocyte by a sperm cell. Thus, as O'Halloran's group has beautifully illustrated: Oocytes accumulate dramatic quantities of zinc from their extracellular milieu, finally ceasing further development, and pausing to await fertilization once they reach a critical concentration (Kim et al. 2011). Then, within tens of minutes after the sperm cell penetrates and fertilization occurs, the fertilized egg "puffs" out a great cloud of $Zn^{2+}$ that surrounds the entire egg (Fig. 6). The exact function of this "cloud" is not certain, but one theory comes from the well-demonstrated inhibition of vertebrate sperm swimming by high $Zn^{2+}$ in the medium. Perhaps the "cloud" serves to prevent multiple sperm attachments to the egg (?).

This zinc-sperm "dance" has phylogenetically ancient harbingers, although the more ancient role of zinc appears to be initiating, rather than inhibiting, sperm motility. In two echinoderms, the starfish and sea urchin, zinc has been found to provide an "on" switch to initiate swimming in the initially immobile ejaculated spermatozoa. The initiation of sperm motility is triggered by adding as little as 100 nM of $Zn^{2+}$ to clean sea water. Given that natural sea water is typically below 1 nM in $Zn^{2+}$, this implies that a zinc signal is needed. Because the sperm becomes motile in the immediate vicinity of the eggs, it is presumed that the egg jelly contains sufficient zinc to provide the $\sim 100$-fold increase in $Zn^{2+}$ that initiates swimming.

## Zinc-Secreting Neurons

Though not the first zinc-secreting cells discovered, the zinc-secreting neuron is undoubtedly the most thoroughly studied and best understood example. A separate entry is dedicated to these neurons, so only a few aspects are mentioned here.

A major factor drawing attention to these neurons from the very beginning of the research in the 1960s has been the fact that the neurons that use zinc as a synaptic signal molecule (signal ion !) are found nearly exclusively in the most highly evolved, complex parts of the brain, the cerebral cortex of the vertebrate. This is all the more striking because the main "messenger" (i.e., the primary neurotransmitter) in the zinc-containing neurons is glutamate, which is the so-called utility transmitter of the nervous system. Glutamate is used throughout phylogeny and throughout the nervous system. But those neurons that

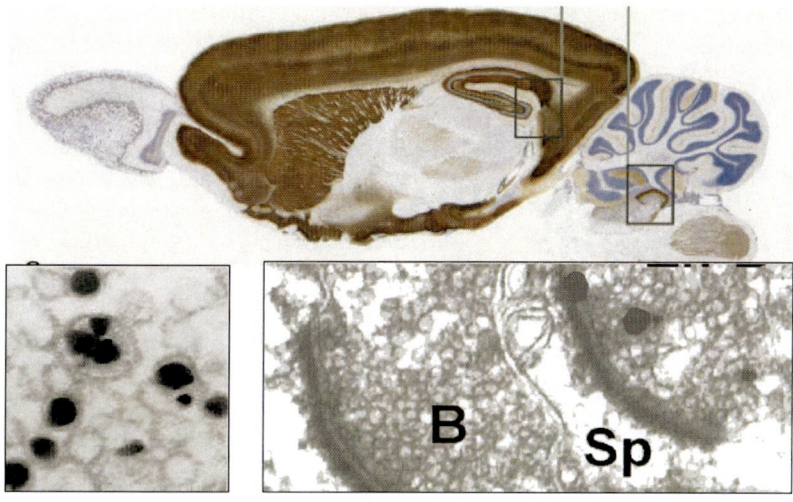

**Zinc Signal-Secreting Cells, Fig. 6** Four sequential images of a recently fertilized vertebrate egg are shown. In the second one, an intracellular calcium "spark" is revealed by an intracellular fluorescent probe. Immediately thereafter, in the third frame, the egg "puffs" a fairly massive cloud of free zinc ion out into the surrounding fluid, where an extracellular zinc fluorescent probe shows that signal (Courtesy TV O'Halloran)

**Zinc Signal-Secreting Cells, Fig. 7** The *top* image is a sagittal view of a mouse brain in which the zinc-containing terminals are *brown*-to-*black* because of silver staining. Note that the "roof" of the forebrain (cortex) is all stained, whereas other regions have no brown or black whatsoever. The *Right lower* panel shows boutons (B) around a neuronal spine (Sp), with the zinc-silver grains marking the *right* bouton. The *left* panel shows that the silver grains are inside individual vesicles (where the zinc is sequestered) (Courtesy G. Danscher)

sequester and secrete glutamate *and* zinc (so-called gluzinergic neurons) are almost 100 % restricted to the cerebral cortex. A rat brain image shown (Fig. 7) illustrates that the zinc-containing synaptic terminals are all either in cortical structures or (in the striatum) the output terminals of fibers that originate in the cortex. In other words, the gluzinergic "voice" is exclusively that of the cortical assemblies that are the signature structures of *Homo sapiens*.

Not surprisingly, the fact that the gluzinergic neurons are in the brain regions most expanded in man has led to intensive searching to see if the zinc is directly involved in the rapid "one trial" episodic memory characteristic and indispensable to us, *Homo sapiens*.

In fact, this appears to be the case, and the synaptically released zinc appears to be directly involved in the modification of cortical synapses in the service of the cortical type of learning (Motta and Dingledine 2011; Pan et al. 2011).

## Cross-References

► Zinc and Zinc Ions in Biological Systems
► Zinc, Fluorescent Sensors as Molecular Probes
► Zinc, Visibility
► Zinc-Secreting Neurons, Gluzincergic and Zincergic Neurons

## References

Botella H, Peyron P et al (2011) Mycobacterial P1-type ATPases mediate resistance to zinc poisoning in human macrophages. Cell Host & Microbe 10:248–259

Fisher WS (2004) Antimicrobial activity of copper and zinc accumulated in eastern oyster amebocytes. J Shellfisheries Res 23:321–332

Jonsson M, Linse S, Frohm B, Lundwall A, Malm J (2005) Semenogelins I and II bind zinc and regulate the activity of prostate-specific antigen. Biochem J 387:447–453

Kim AM, Bernhardt ML, Kong BY, Ahn RW, Vogt S, Woodruff TK, O'Halloran TV (2011) Zinc sparks are triggered by fertilization and facilitate cell cycle resumption in mammalian eggs. ACS Chem Biol 6:716–723

Meldrum BS (1965) The actions of snake venoms on nerve and muscle. The pharmacology ff phospholipase A and of polypeptide toxins. Pharmacol Rev 17:393–445

Motta DD, Dingledine R (2011) Unraveling the role of zinc in memory. Proc Natl Acad Sci 108:3103–3104

Pan E, Zhang XA, Huang Z, Krezel A, Zhao M, Tinberg CE, Lippard SJ, McNamara JO (2011) Vesicular zinc promotes presynaptic and inhibits postsynaptic long-term potentiation of mossy fiber-CA3 synapse. Neuron 71:1116–1126

Singer R, Mawson P, et al. (2012) An intravaginal ring that releases the NNRTI MIV-150 reduces SHIV transmission in Macaques. Sci Transl Med (Epub) 4:150–153

Smith GL, Jenkins RA, Gough JF (1969) A fluorescent method for the detection and localization of zinc in human granulocytes. Histochem Cytochem 17:749–750

# Zinc Storage and Distribution in *S. cerevisiae*

David J. Eide
Department of Nutritional Sciences, University of Wisconsin-Madison, Madison, WI, USA

## Synonyms

ADH1/ADC1; SOD1/CRS4; ZAP1/ZRG10; ZIM17/TIM15; ZRC1/OSR1

## Definitions

- CDF, cation diffusion facilitator. Metal transporters that transport their substrates from the cytoplasm into intracellular vesicles or out of the cell. Also known as SLC30A family transporters.
- SLC30A, solute carrier family 30A. Metal transporters that transport their substrates from the cytoplasm into intracellular vesicles or out of the cell. Also known as CDF family transporters.
- SLC39A, solute carrier family 39A. Metal transporters that transport their substrates into the cytoplasm from within intracellular vesicles or from outside of the cell. Also known as ZIP family transporters.
- ZIP, ZRT/IRT-like protein. Metal transporters that transport their substrates into the cytoplasm from within intracellular vesicles or from outside of the cell. Also known as SLC39A family transporters.

Zinc is an essential cofactor required for the structure or catalytic function of many proteins. The trafficking and utilization of this essential nutrient require precise control of its uptake, storage, and intracellular distribution to meet cellular needs while avoiding toxicity due to zinc overload. Arguably, these processes are best understood in the yeast *Saccharomyces cerevisiae*. The proteins involved and their functions described here are summarized in Table 1.

## Zinc Transport Across the Plasma Membrane in Yeast

The primary uptake system for zinc in *S. cerevisiae* is the Zrt1 transporter that is required for growth in low zinc conditions (Eide 2006). Zrt1 is a member of the ZIP family of metal ion transporters, many of which are involved in zinc uptake in other organisms (Gaither and Eide 2001; Kambe et al. 2006). In mammals, ZIP family members are designated as SLC39A-type proteins. Zrt1 abundance is upregulated by zinc deficiency at the transcriptional level by the Zap1 zinc-responsive activator protein (Eide 2009). Expression of *ZRT1* mRNA in zinc-limited cells is induced approximately 30-fold over zinc-replete expression levels, and protein levels are altered to a similar degree. This transporter has a remarkably high affinity for zinc with an estimated apparent $K_m$ of 10 nM for free $Zn^{2+}$ ions. Because of Zrt1's activity, yeast can accumulate zinc as much as 2,000-fold over the available extracellular levels. Zrt1 is also regulated posttranslationally. In high zinc, Zrt1 is ubiquitinated, endocytosed from the cell surface, and degraded in the vacuole to decrease zinc uptake and prevent zinc overload. Zrt1 is also capable of cadmium transport, so zinc-limited cells are especially sensitive to exogenous cadmium.

**Zinc Storage and Distribution in S. cerevisiae, Table 1** Proteins involved in zinc homeostasis in *Saccharomyces cerevisiae*

| Protein | Location | Transporter family[a] | Function in zinc homeostasis |
|---|---|---|---|
| Zap1 | Nuclear | NA | Zinc-responsive transcription factor, regulates *ZRT1, ZRT2, FET4, ZRG17, ZRC1, ZRT3* |
| Zrt1 | Plasma membrane | ZIP | High-affinity zinc uptake transporter, regulated by Zap1 |
| Zrt2 | Plasma membrane | ZIP | Low-affinity zinc uptake transporter, regulated by Zap1 |
| Fet4 | Plasma membrane | FET4 | Low-affinity zinc, iron, and copper uptake transporter, regulated by Zap1 |
| Pho84 | Plasma membrane | MFS | High-affinity phosphate uptake transporter, possible low-affinity zinc transporter |
| Msc2 | Endoplasmic reticulum | CDF | Subunit of the zinc transporter complex supplying the endoplasmic reticulum |
| Zrg17 | Endoplasmic reticulum | CDF | Possible regulatory subunit of the ER zinc transporter complex |
| Zrc1 | Vacuole | CDF | Zinc transport into the vacuole, also plays minor role in supplying the secretory pathway |
| Cot1 | Vacuole | CDF | Zinc transport into the vacuole, also plays minor role in supplying the secretory pathway |
| Yke4 | Endoplasmic reticulum | ZIP | Possible bidirectional zinc transporter mediating both ER zinc influx and efflux |
| Zrt3 | Vacuole | ZIP | Transport of stored zinc out of the vacuole into cytosol for use |

[a]*ZIP* Zrt-Irt-like protein, *MFS* major facilitator superfamily, *CDF* cation diffusion facilitator family, *NA* not applicable

A second ZIP protein, Zrt2, has a lower affinity for $Zn^{2+}$ (apparent $K_m = \sim100$ nM) and plays a role in zinc uptake under less severe zinc-limiting conditions (Eide 2006). Zrt2 is induced under conditions of mild zinc limitation but then repressed under more severe zinc-limiting conditions. Both activation and repression of the *ZRT2* gene are mediated by Zap1 (Eide 2009). This intriguing pattern of regulation involves the locations of three Zap1-binding sites within the *ZRT2* promoter. Two binding sites are located upstream of the TATA box and serve to activate gene expression. Zap1 has high affinity for these sites, so they are occupied by Zap1 under moderate zinc-limiting conditions. A third binding site is located downstream of TATA near the start site of transcription, and Zap1 binding at this site blocks *ZRT2* transcription. This repressive site is a low-affinity site, so Zap1 only binds there when zinc deficiency is severe. Thus, the locations and sequences of these binding sites determine the pattern of regulation by Zap1.

Mutational inactivation of Zrt1 and Zrt2 revealed the existence of additional zinc uptake systems with still lower affinity that function in zinc-replete cells (Eide 2006). One such system is the Fet4 transporter, which is involved in the low-affinity uptake of iron and copper as well as zinc. Fet4 is not a ZIP protein, and surprisingly, related proteins are only found in ascomycete fungi such as *S. cerevisiae*, *Schizosaccharomyces pombe*, *Candida albicans*, and *Aspergillus fumigatus*. The *FET4* gene is also a target of Zap1 transcriptional control (Eide 2009). A fourth system of zinc uptake in yeast is likely to be the Pho84 high-affinity phosphate transporter. Pho84 is a member of the major facilitator superfamily of transporters. Mutation of the *PHO84* gene confers zinc tolerance and reduces the accumulation of zinc consistent with Pho84 transporting zinc in addition to phosphate. While likely to have a low affinity for zinc, Pho84 is highly induced by phosphate limitation, so it may play a more important role in zinc homeostasis under low-phosphate growth conditions. No zinc efflux transporters in the plasma membrane of *S. cerevisiae* have been discovered, and biochemical studies suggest that none exist.

## The Level and Distribution of Zinc in the Cytosol

After entering the cell, $Zn^{2+}$ is quickly bound by zinc-requiring proteins or is rapidly transported into intracellular compartments. The net result of these processes is that the level of free or labile zinc in the cytosol is very low. The minimum number of zinc atoms per cell required for optimal yeast growth, referred to as the "zinc quota," is $5 \times 10^6$. Below that value, cell growth is reduced due to incomplete metalation of zinc-dependent proteins. The vast majority of the total zinc in a cell is bound by proteins. Based on research using other types of eukaryotic cells, the free or labile zinc pool in yeast is predicted to be 1 nM or less (Colvin et al. 2010; Vinkenborg et al. 2010). One nanomolar zinc in a yeast cell corresponds to less than 0.001% of the zinc quota. This comparison highlights the precise control of free zinc within cells of all types and the delicate balance that exists between zinc transport (i.e., uptake and organellar sequestration) and the binding of zinc by newly synthesized metalloproteins.

The population of zinc metalloproteins in a cell has been referred to as the "zinc proteome." While the zinc proteome of yeast has not yet been described in detail, there is sufficient information available to estimate where much of this zinc is bound and to what level. In *S. cerevisiae*, major zinc-binding proteins in the cytosol include Cu/Zn superoxide dismutase (Sod1) and the predominant isozyme of alcohol dehydrogenase (Adh1). Sod1 accumulates to about $5 \times 10^5$ molecules per cell, while Adh1 is around $7.5 \times 10^5$ molecules per cell and binds two atoms of zinc per monomer. Also, six atoms of zinc are bound to each ribosome accounting for a total of approximately $6 \times 10^5$ atoms of ribosomal zinc per cell. These three examples alone account for almost 50% of the zinc quota of the yeast cell. In addition to these high-affinity sites, low-affinity zinc-binding sites are abundant in the cell. These include sites on proteins, lipids, and DNA, as well as small molecular weight compounds, such as organic anions (e.g., citrate), amino acids (e.g., histidine), and glutathione.

## The Transport of Zinc into and out of Intracellular Organelles

### The Secretory Pathway

Many zinc-dependent enzymes are secreted or are resident within the secretory pathway. Furthermore, the lumen of the ER is the site for the initial folding and posttranslational modification of proteins within secretory pathway, and many of these processes are zinc dependent. For example, glycosylphosphatidylinositol phosphoethanolamine transferases (GPI-PETs) in the endoplasmic reticulum act in the biosynthesis of GPI anchors that attach many proteins to the cell surface. GPI-PETs have a zinc-binding motif like that found in alkaline phosphatases and similarly require zinc for their function. The yeast DnaJ homolog Scj1 is a protein co-chaperone in the ER that contains two zinc fingers and requires zinc for its function. Disrupting Scj1 function causes defects in protein folding and less effective degradation of aberrant proteins by ER quality control mechanisms. Similarly, the ER chaperones, calnexin and calreticulin, require zinc for their function.

There is little information regarding where in the secretory pathway zinc is delivered to resident and secreted apoproteins. The zinc requirement of luminal ER proteins such as DnaJ-like chaperones and GPI-PETs indicates that some metalation occurs in the ER. Zinc transporters responsible for delivery of zinc to the secretory pathway have been recently identified, and these proteins belong to the CDF family of metal ion transporters (Gaither and Eide 2001; Kambe et al. 2006). CDF family members in mammals are designated as SLC30A-type proteins. In yeast, the heteromeric Msc2/Zrg17 CDF complex mediates zinc transport into the ER (Eide 2006). Mutational inactivation of either or both subunits of this complex results in pleiotropic defects in ER function, including induction of the unfolded protein response (UPR) and defects in ER-associated protein degradation. Cell wall biogenesis, a process highly dependent on secretory pathway function, is also disrupted in *msc2* and *zrg17* mutants.

The Msc2/Zrg17 transporter complex transports zinc into the ER under low zinc conditions. Consistent with this role, *ZRG17* is a Zap1 target gene, and its expression is high in zinc-limited cells and reduced in replete cells (Wu et al. 2011). It was proposed that Zrg17 is the regulatory subunit of the complex, and its level alters Msc2's transporter function. When higher zinc levels are provided in the medium, the ER dysfunction phenotypes of *msc2* and *zrg17* mutants are completely suppressed. This indicates that other transport mechanisms are present to supply zinc to the ER. The vacuolar CDF transporters Zrc1 and Cot1 (see below) were also shown to contribute to ER zinc. These proteins may be active in the early secretory pathway soon after their synthesis and before they transit to the vacuole. Surprisingly, in a mutant disrupted for Msc2, Zrg17, Zrc1, and Cot1, increased exogenous zinc could still suppress the defects in ER function. Thus, additional pathways of ER zinc transport exist in yeast.

Supply of zinc to the Golgi may occur by vesicular trafficking of ER zinc to that compartment. Alternatively, zinc transporters may be active in the Golgi as well. In yeast, not all of the Msc2/Zrg17 complex localizes to the ER suggesting that some fraction may be present in the Golgi. Zinc may also be exported from the Golgi. The mammalian Zip7/SLC39A7 protein localizes to the Golgi and appears to mediate transport of secretory pathway zinc back to the cytoplasm. This may be a means to recover unused zinc in the secretory pathway prior to its loss by secretion. It is not clear if *S. cerevisiae* uses a similar strategy, but a yeast ZIP transporter related to Zip7, called Yke4, could fulfill this function. Yke4 has been localized to the ER membrane.

## The Yeast Vacuole

The vacuole of yeast is the major site of zinc sequestration and detoxification (Eide 2006). Wild-type *S. cerevisiae* can tolerate exogenous zinc concentrations as high as 5 mM. When the transport systems for vacuolar zinc sequestration are mutated (see below), the maximum level of tolerable zinc drops by 100-fold to ~50 μM. Chronic exposure to millimolar levels of zinc is rarely encountered by yeast growing in their normal environment. Thus, vacuolar sequestration is likely to be more important to the cell for growth in other situations. One such condition is "zinc shock." Zinc shock occurs when cells are grown under zinc-limiting conditions and then resupplied with even low levels of zinc. Zinc-deficient cells are poised to accumulate substantial amounts of zinc due to the upregulation of the plasma membrane transporter Zrt1. When resupplied with zinc, these cells accumulate large amounts of the metal ion before transcriptional and posttranslational mechanisms that regulate Zrt1 can shut off additional uptake. Mutant cells unable to sequester zinc in the vacuole during zinc shock are sensitive to zinc concentrations as low as 0.1 μM. Given that microbial cells growing in the wild live an existence of sporadic nutrient availability, the primary function of the vacuole in zinc detoxification is likely to be during zinc shock caused by fluctuations in zinc availability.

Zinc sequestered within the vacuole also serves as a storage pool of zinc that can be mobilized under deficient conditions for use by the cell. Zinc can accumulate in the vacuole to as high as 100 mM ($7 \times 10^8$ atoms of vacuolar zinc per cell) (Simm et al. 2007). This is sufficient to supply the zinc quota of over 100 daughter cells in the absence of any exogenous supply.

Zinc uptake into the yeast vacuole is mediated by two members of the CDF family, Zrc1 and Cot1. Zrc1 is a $Zn^{2+}/H^+$ antiporter that allows zinc accumulation in the vacuole to be driven by the proton concentration gradient provided by the vacuolar $H^+$-ATPase. The Cot1 protein may act in a similar fashion, although its biochemical properties have not been analyzed. Zrc1 is also capable of cadmium transport into the vacuole, and Cot1 can transport cobalt. Thus, mutations in these genes also disrupt the resistance to these other metals.

Zinc release to the cytosol under conditions of zinc deficiency is mediated by Zrt3, a member of the ZIP family of transporters. Zrt3 expression is upregulated under low zinc conditions to facilitate this mobilization of vacuolar zinc. Other mechanisms of zinc efflux from the vacuole exist, but the genes and proteins responsible have not yet been identified.

Zinc stored within vacuoles may not accumulate simply as free $Zn^{2+}$ ions. X-ray absorption spectroscopy analysis of vacuolar zinc in plants indicates that $Zn^{2+}$ may instead accumulate bound to organic anions. For example, in *Arabidopsis halleri*, a plant species capable of hyperaccumulating large amounts of zinc, the metal was found complexed primarily with phosphate and/or organic acids such as malate and citrate depending on the growth conditions and the plant tissues examined. Similar results were obtained with *Thlaspi caerulescens*, another zinc hyperaccumulating plant species. Binding of vacuolar zinc by such compounds may provide a means of accumulating large amounts of zinc within the vacuolar compartment for storage and detoxification. Similarly, the glutamate levels in synaptic vesicles are proportional to their zinc accumulation, suggesting that the anion influences the zinc capacity of the vesicle. It is unclear at this time if vacuolar zinc in yeast is similarly bound to anions, but it seems a likely possibility given that they accumulate to high levels in the yeast vacuole.

## Mitochondria

Zinc is required in the matrix of the mitochondria for function of proteins within that compartment. In yeast, these include alcohol dehydrogenase 3 (Adh3) which converts acetaldehyde to ethanol as a means of shuttling NADH across the inner membrane. The Leu4 protein, alpha-isopropylmalate synthase, mediates the first step of leucine synthesis. Leu4 is found in the matrix and is a zinc metalloenzyme. Zim17 is a zinc–finger matrix protein required for protein import. These examples indicate that mechanisms must exist to get zinc into the mitochondria. Zinc must first traverse the outer membrane, most likely through the porin channels. Then, the zinc must cross the inner membrane via transporter proteins. The identity of these transporters is not yet known. However, mitochondrial zinc has recently been shown to be homeostatically regulated (Atkinson et al. 2010; Simm et al. 2007). A labile, low molecular weight pool of mitochondrial zinc has been discovered that appears to be critical for mitochondrial function. The $Zn^{2+}$ in this pool is ligand bound and cationic in nature. The identity of the ligand is not yet known nor is it known if this pool's major role is zinc detoxification or providing zinc for binding by metalloproteins.

## Cross-References

▶ Cadmium Absorption
▶ Cobalt Transporters
▶ Zinc Cellular Homeostasis
▶ Zinc Deficiency

## References

Atkinson A, Khalimonchuk O, Smith P, Sabic H, Eide D, Winge DR (2010) Mzm1 influences a labile pool of mitochondrial zinc important for respiratory function. J Biol Chem 285:19450–19459

Colvin RA, Holmes WR, Fontaine CP, Maret W (2010) Cytosolic zinc buffering and muffling: their role in intracellular zinc homeostasis. Metallomics 2:306–317

Eide DJ (2006) Zinc transporters and the cellular trafficking of zinc. Biochim Biophys Acta 1763:711–722

Eide DJ (2009) Homeostatic and adaptive responses to zinc deficiency in *Saccharomyces cerevisiae*. J Biol Chem 284:18565–18569

Gaither LA, Eide DJ (2001) Eukaryotic zinc transporters and their regulation. Biometals 14:251–270

Kambe T, Suzuki T, Nagao M, Yamaguchi-Iwai Y (2006) Sequence similarity and functional relationship among eukaryotic ZIP and CDF transporters. Genomics Proteomics Bioinformatics 4:1–9

Simm C, Lahner B, Salt D, LeFurgey A, Ingram P, Yandell B, Eide DJ (2007) *Saccharomyces cerevisiae* vacuole in zinc storage and intracellular zinc distribution. Eukaryot Cell 6:1166–1177

Vinkenborg JL, Koay MS, Merkx M (2010) Fluorescent imaging of transition metal homeostasis using genetically encoded sensors. Curr Opin Chem Biol 14:231–237

Wu YH, Frey AG, Eide DJ (2011) Transcriptional regulation of the Zrg17 zinc transporter of the yeast secretory pathway. Biochem J 435:259–266

# Zinc Structural Site in Alcohol Dehydrogenases

Tomas Bergman and Hans Jörnvall
Department of Medical Biochemistry and Biophysics, Karolinska Institutet, Stockholm, Sweden

## Synonyms

Non-catalytic zinc; Second zinc ion; Structural zinc site; Structural zinc

## Definition

Zinc alcohol dehydrogenases are enzymes of the oxidoreductase class belonging to the MDR (medium-chain dehydrogenases/reductases) protein superfamily. They are dimeric or tetrameric and composed of two-domain subunits (catalytic and coenzyme-binding domains, respectively). Each subunit (size about 40 kDa) binds one or two zinc ions: One is required for the catalytic activity, and the other, when present, has been assigned a structural role in stabilizing the tertiary fold and in promotion of subunit interactions. While catalytic zinc sites in MDR enzymes are characterized by three protein ligands (from widely separated positions), with the substrate (or a water molecule, or when in inactive conformation, a Glu) as the fourth ligand, structural zinc sites are composed of four protein ligands, frequently Cys residues and covering a local segment of the protein chain.

## Properties and Functional Aspects

### Overview

Alcohol dehydrogenase (ADH) of the medium-chain dehydrogenase/reductase (MDR) type (Persson et al. 2008) catalyzes reversible oxidation of primary and secondary alcohols to the corresponding aldehydes and ketones, respectively, employing the coenzymes NAD(H) or NADP(H) in the hydride transfer process (Eklund and Ramaswamy 2008). ADHs are widespread among animals, plants, fungi, and bacteria and reveal extensive isozyme patterns. In humans and in

**Zinc Structural Site in Alcohol Dehydrogenases, Table 1** Protein ligands and number of intervening residues for selected zinc sites of MDR-enzymes

| Enzyme type | Structural zinc | Catalytic zinc |
| --- | --- | --- |
| Vertebrate ADH[a] classes I-IV | Cys 2 Cys 2 Cys 7 Cys | Cys 20-21 His 106-110 Cys |
| Bacterial ADH[b] | Cys 2 Cys 2 Cys 7-8 Cys | Cys 20-22 His 86-101 Cys/Asp |
| Archaeal ADH[c] | Glu/Asp 2 Cys 2 Cys 7 Cys | Cys 23-29 His 84-88 Cys/Asp |
| NADP(H)-dependent ADH[d] | Cys 2 Cys 2 Cys 7 Cys | Cys 20-21 His 90-105 Cys/Asp |
| Polyol dehydrogenases[e] | Cys 2 Cys 2 Cys 7 Cys | Cys 24-26 His 0 Glu |

Protein data bank entries:
[a]Human (1HSO,1UST,3HUD,1HSZ,1HDY,1HTB,1HT0,1U3W,1TEH,1M6H,2FZE,1AGN), Horse (8ADH,2JHG), Mouse (1E3E), Cod (1CDO)
[b]Pseudomonas (1LLU,1KOL), Bacillus (1RJW), Thermotoga (1VJ0)
[c]Sulfolobus (1R37,1JVB,2H6E), Aeropyrum (1H2B)
[d]Amphibian (1P0F), Arabidopsis (2CF5), Populus (1YQD), Saccharomyces (1Q1N,1PS0), Entamoeba (1Y9A,1OUI), Clostridium (1JQB,1KEV), Thermoanaerobacter (1YKF)
[e]Human sorbitol dehydrogenase (1PL8), Bemisia (1E3J), Sulfolobus (2CD9)

many organisms, they serve to eliminate toxic alcohols and to generate alcohol, aldehyde, and ketone functional groups in biosynthetic reactions. Zinc-containing ADHs frequently have two tetrahedrally coordinated zinc ions per subunit, one catalytic at the active site and one non-catalytic at a site influencing structural integrity and subunit interactions.

This entry summarizes general properties of ADH structural zinc sites. In addition, it reviews an experimental approach where zinc binding of peptide variants of the horse liver ADH (HLADH) structural zinc site have been studied (Bergman et al. 2008). The results reveal a different zinc coordination in the free, native sequence peptide versus that of the same peptide locked in the folding pattern of the protein. This difference indicates an energetically strained conformation of the protein structural zinc site. Such a strain is a characteristic of an entatic state (Vallee and Williams 1968), which in turn implies the possibility of a functional nature of also this ADH zinc site (Bergman et al. 2008).

## Structural Zinc Sites

Four protein ligands and no zinc-bound water are characteristics of structural zinc sites, and the most common ligand is Cys. However, any combination of four Cys, His, Glu, and Asp residues has the potential to coordinate zinc at these sites (Auld and Bergman 2008). Folding patterns of more than ten dozen structural zinc sites have been reported with representatives in all enzyme classes. While the purpose of such sites initially was thought to affect local protein conformation only, it now appears that some of these sites may also influence enzyme catalysis through the actions of amino acid residues localized within the spacer arms of the zinc ligands.

## The Structural Zinc Site of ADH

The first structural zinc site to be recognized in any protein was that of mammalian ADH. Crystallographic data show that the HLADH structural zinc site is composed of four Cys residues closely spaced at positions 97, 100, 103, and 111 in a separate loop of the protein (Eklund and Ramaswamy 2008). The general design of this site is highly conserved among ADH family members, and the typical spacing pattern between the four zinc ligands is 2:2:7 residues, respectively (Table 1).

In HLADH, the first Cys ligand is positioned after a β-sheet and before an α-helix, the second immediately before an α-helix, and the third within an α-helix, while the fourth comes from a largely irregular polypeptide segment. Although Cys is the dominating ligand, some variation has been found, e.g., in the archaeal type of ADHs, where the first ligand in the structural zinc site is Glu or Asp (Table 1).

Early experiments showed that the zinc ligands are susceptible to oxidation which could lead to disulfide bridge formation and, hence, loss of zinc. This type of oxidoreductive interplay between thiols/disulfides is now known to be a general phenomenon associated with zinc transport in metallothionein and metalloproteins in general (Maret and Vallee 1998; Maret 2004). Results of early studies removing the non-catalytic zinc by dialysis established its structural importance. Further indications in this direction are that enzymes of related type but not containing this zinc atom, e.g., sorbitol dehydrogenase, differ in quaternary structure and that recombinant enzyme variants of ADH, with one each of the four zinc ligands exchanged from Cys to Ala, are labile (Jeloková et al. 1994).

Since the ligand-altered enzymes are unstable and easily denature, they are not suitable for evaluation of zinc binding. Instead, peptides synthesized to correspond to the zinc-binding HLADH segment, with and without amino acid replacements/deletions, can be examined (Bergman et al. 2008). The peptide replica (23 residues) and analogs were found to mimic the metal-binding stoichiometry of HLADH, and for all peptide variants, the zinc-binding constants were determined in titrations using the metallochromic chelator 4-(2-pyridylazo)resorcinol (PAR) to extract zinc from metal-saturated holopeptides.

PAR-titration and X-ray absorption fine structure (XAFS) analysis revealed differences in zinc ligation between the peptide analogs and between the peptide replica and the corresponding HLADH segment. These differences suggest that the structural zinc site in the conformation observed in HLADH represents a state of higher energy relative to that of the peptide replica zinc site which is free from all interactions with a protein scaffold. This situation resembles an entatic state, a concept introduced for energized protein environments (Vallee and Williams 1968; Williams 1995) and defined as an intrinsic property of an atom, group, area, or region of a protein that through its binding pattern attains a geometry or electronic configuration suitable for function and is energetically poised for catalytic action already before substrate is added.

## Zinc Binding to ADH Structural Zinc Site

Results from analysis of the HLADH structural zinc site using the peptide approach (see above) reveal that the number of Cys residues, the presence of His, and the spacing between Cys/His residues have a clear impact on zinc-to-peptide stoichiometry, zinc coordination, and zinc affinity. Zinc-binding constants were found to vary in an ordered fashion from tight to low binding (from $10^{10}$ to $10^6$ M$^{-1}$) depending on the number and position of ligating residues.

Peptide analogs to the HLADH structural zinc site representing every combination of four, three, and two Cys residues all generated significant zinc binding. However, XAFS analysis of the peptide/zinc complexes showed that the single His residue, corresponding to His105 in the HLADH structure, is a zinc ligand in the peptides. Apparently, the presence of His results in a three Cys/one His coordination in the peptide replica (zinc-binding constant $7.0 \times 10^9$ M$^{-1}$), rather than the four-Cys coordination found in the protein. Moreover, when His is replaced by Ala in an analog of the peptide replica, zinc is still bound but with higher affinity ($1.3 \times 10^{10}$ M$^{-1}$), and the XAFS analysis reveals a distinct four-Cys coordination (Bergman et al. 2008).

The situation with a three Cys/one His coordination in the peptide replica leaves one nonbinding Cys which most likely is the Cys corresponding to Cys103 in the HLADH structure. When this Cys is replaced by Ala, the resulting peptide analog reveals a zinc affinity very similar ($6.9 \times 10^9$ M$^{-1}$) to that found for the peptide replica ($7.0 \times 10^9$ M$^{-1}$, above), supporting a nonbinding nature of the equivalent to Cys103 in the protein. Furthermore, XAFS data for the peptide replica and this Cys-to-Ala analog are close to identical. This is consistent with the finding that replacement of the single His residue (corresponding to His105 in HLADH) with Ala in the peptide replica results in increased zinc affinity to the four Cys residues as observed in the XAFS analysis (see above). This indicates a condition of lower energy and more stable zinc coordination when His is not present in the peptide.

The fact that a His, together with three Cys residues, coordinates zinc in the free peptide while it is substituted by a Cys (most likely Cys103) in the protein site provides novel insight into the nature of this and similar structural zinc sites. The situation in the protein is likely due to forces acting upon Cys103 and/or His105 that result from structural interactions in the three-dimensional protein space. Examination of ADH family member sequences and available three-dimensional structures show that the side chain of Cys103 points toward the interior of the protein and ligates with zinc while the residue at position 105 (nonconserved, Asn > Ser > His) has its side chain directed toward the exterior in contact with the surrounding solvent and the nitrogen or oxygen atom (the potential ligand) is positioned 11.5–13 Å from the zinc, precluding zinc coordination. A possible explanation is that an α-helix comprising 14 residues extends from amino acid 324 to 337 and places the ε-amino group of Lys323 (conserved) within hydrogen-bonding distance to the backbone amide carbonyl oxygen of residues

**Zinc Structural Site in Alcohol Dehydrogenases, Fig. 1** The structural zinc site of horse liver alcohol dehydrogenase (PDB# 1BTO). The helix ending with Ala337 positions the ε-amino group of Lys323 within hydrogen-bonding distance of the backbone carbonyls of amino acid residues Cys103, His105, and Gly108 (Reproduced with kind permission from Springer Science + Business Media: Figure 6 in Bergman et al. 2008)

103, 105, and 108, which likely stabilizes the conformation of the zinc site in the enzyme (Fig. 1). Although the residues at positions 105 and 108 are not conserved, the interaction between their amide carbonyl oxygen and the ε-amino group of Lys323 is preserved throughout the available structures. This interaction places the side chain of residue 105 in contact with the exterior solvent far away from the structural zinc. The fact that the peptide data show that the His residue corresponding to His105 in HLADH is a zinc ligand together with three Cys residues thus indicates the presence of structural strain and increased energy in the protein structural zinc site (Bergman et al. 2008).

## Evolutionary Aspects

The zinc-containing MDRs are of later origin than the zinc-free forms (cf. ▶ Zinc Alcohol Dehydrogenases). The structural zinc resides in a region absent or structurally different in related enzymes lacking this zinc such as sorbitol dehydrogenase. It is usually found in addition to the catalytic zinc, but exceptions exist where the protein contains just the structural zinc (Jörnvall et al. 2010). Domain variability and zinc-ligating patterns detected in large-scale sequence comparisons of MDR enzymes support a structural role for the non-catalytic zinc with effects on inter-domain interactions, coenzyme binding, and substrate pocket formation. Hence, the structural zinc, in addition to the catalytic zinc which is of earlier origin, appears to compensate for disturbances from domain variability and structural alterations. Since MDR forms containing only the structural zinc do exist, it seems that either zinc type may emerge or be lost during molecular evolution and that both types to various degrees contribute to structural stability and activity. The differences observed between structural and catalytic zinc ions may therefore be of relative importance only, and involve several states, all supporting enzyme activity and specificity (Jörnvall et al. 2010). Interestingly, although catalytic zinc sites of MDR enzymes basically are built from three protein ligands and a water molecule (Table 1), the crystal structures of ADH family members indicate the existence of a switch between two types of catalytic zinc liganding where the water is replaced by a Glu residue as the fourth ligand. Since the Glu-coordinated zinc is displaced about 2 Å from the active site, this type has been suggested to represent an inactive form of ADH. However, when the catalytic zinc is coordinated to four protein ligands with no zinc-bound water, the site appears to go into a "standby" mode charged with energy and catalytic potential already before the substrate has entered. In other words, this mode represents an entatic state, and the zinc-binding pattern is that of a structural zinc site (four protein ligands and no zinc-bound water, Table 1). When the substrate enters, the catalytic zinc site goes into action, and the proton and hydride transfer processes start (cf. ▶ Zinc Alcohol Dehydrogenases).

## Functionality and Potential Reactivity in Structural Zinc Sites

At present, one can only speculate about specific reactivity and function connected to ADH structural zinc sites, and some possible scenarios are suggested. Residue 105 can potentially, in combination with other surface located residues, participate in protein-protein interactions and formation of protein complexes. A more direct involvement of the structural zinc site in protein function would be modification of the Cys sulfur atoms in nucleophilic substitution reactions. From analysis of zinc-finger structures, it is known that hydrogen bonds between neighboring amides and thiolate ligands electronically stabilize the thiolates. A single hydrogen bond reduces the reactivity toward an electrophile by up to two orders of magnitude. This suggests that the zinc thiolates in the

ADH structural site should be unreactive towards electrophiles since there are two to four amide N-H bonds within 3.1–3.6 Å of each of the zinc-bound thiolates (Auld and Bergman 2008). However, in contrast to the deactivating effect of neighboring amide N-H on zinc thiolates, the presence of basic side chains within 6.5 Å of Cys residues is known to increase the nucleophilicity of thiolate groups, possibly due to enhanced deprotonation and formation of the electron dense thiolate anion. This requires that a Cys ligand in the protein is released from the zinc. In HLADH, Cys97 is a candidate since this residue is close to the exterior solvent and 4.43 Å from the ε-amino group of Lys113. Furthermore, the equivalent Cys residue in a structural zinc site in *E. coli* threonine dehydrogenase has been implicated in the formation of an air-dependent disulfide bond without loss of enzymatic activity (cf. Auld and Bergman 2008).

In other words, the molecular architecture of the HLADH structural zinc site seems to promote some flexibility in the zinc-liganding pattern making the distinction to catalytic zinc sites less sharp. Notably, a shift in zinc coordination from three to four protein ligands, the latter with characteristics of an entatic state, has been observed at the ADH catalytic site (see above). It is thus possible that the potentially zinc-binding residues lining the HLADH structural site, particularly His105, create the tension and higher energy necessary to make the coordination to the regular zinc ligands flexible, thus explaining the existence of an entatic state rendering one of the participating Cys thiolates potentially active in nucleophilic substitution reactions, enhanced by the high frequency of basic residues in this segment of the protein.

Support for a similarly increased flexibility and metalloactivation of a Cys thiolate exists in the literature. In many zinc-containing proteins that have structural-like zinc sites with multiple Cys ligands, the site promotes alkylation of a zinc-bound thiolate. One example of nucleophilic attack by a zinc-liganding Cys residue has been reported for the *E. coli* Ada DNA-repair protein (Myers et al. 1993). In this case, the methyl group of methylphosphotriesters in damaged DNA molecules is transferred to a Cys residue, converting Ada to a sequence-specific DNA-binding transcription factor which induces genes conferring resistance to methylating agents. The reactive Cys residue is part of a tetracoordinated zinc site with four conserved Cys ligands where the zinc has been suggested to play an activating role through coordination of the nucleophilic Cys thiolate which lowers the pKa and increases the reactivity of this group. Furthermore, enhancement of thiol nucleophilicity at neutral pH by direct zinc metalloactivation has been suggested to reflect a novel catalytic function of zinc metalloproteins (McCall et al. 2000).

In conclusion, structural zinc sites are defined by four protein ligands and no zinc-bound water. They are found in proteins with diverse functions and play a role both for structural integrity and for additional functions. The peptide data for the HLADH structural zinc site show that the zinc ligands are different in the free peptide (three Cys and one His) and in the protein (four Cys), suggesting structural strain and higher energy in the protein zinc site characteristic of an entatic state which indicates potential reactivity and function also in this ADH region.

## Cross-References

▶ Zinc Alcohol Dehydrogenases
▶ Zinc-Binding Sites in Proteins

## References

Auld DS, Bergman T (2008) The role of zinc for alcohol dehydrogenase structure and function. Cell Mol Life Sci 65:3961–3970

Bergman T, Zhang K, Palmberg C, Jörnvall H, Auld DS (2008) Zinc binding to peptide analogs of the structural zinc site in alcohol dehydrogenase: Implications for an entatic state. Cell Mol Life Sci 65:4019–4027

Eklund H, Ramaswamy S (2008) Three-dimensional structures of MDR alcohol dehydrogenases. Cell Mol Life Sci 65:3907–3917

Jeloková J, Karlsson C, Estonius M, Jörnvall H, Höög J-O (1994) Features of structural zinc in mammalian alcohol dehydrogenase. Site-directed mutagenesis of the zinc ligands. Eur J Biochem 225:1015–1019

Jörnvall H, Hedlund J, Bergman T, Oppermann U, Persson B (2010) Superfamilies SDR and MDR: From early ancestry to present forms. Emergence of three lines, a Zn-metalloenzyme, and distinct variabilities. Biochem Biophys Res Commun 396:125–130

Maret W (2004) Zinc and sulfur: a critical biological partnership. Biochemistry 43:3301–3309

Maret W, Vallee BL (1998) Thiolate ligands in metallothionein confer redox activity on zinc clusters. Proc Natl Acad Sci USA 95:3478–3482

McCall KA, Huang C-c, Fierke CA (2000) Function and mechanism of zinc metalloenzymes. J Nutr 130:1437S–1446S

Myers LC, Terranova MP, Ferentz AE, Wagner G, Verdine GL (1993) Repair of DNA methylphosphotriesters through a metalloactivated cysteine nucleophile. Science 261: 1164–1167

Persson B, Hedlund J, Jörnvall H (2008) The MDR superfamily. Cell Mol Life Sci 65:3879–3894

Vallee BL, Williams RJP (1968) Metalloenzymes: the entatic nature of their active sites. Proc Natl Acad Sci USA 59:498–505

Williams RJP (1995) Energised (entatic) states of groups and of secondary structures in proteins and metalloproteins. Eur J Biochem 234:363–381

# Zinc, Fluorescent Sensors as Molecular Probes

Kalyan K. Sadhu[1], Shin Mizukami[1,2] and Kazuya Kikuchi[1,2]
[1]Division of Advanced Science and Biotechnology, Graduate School of Engineering, Osaka University, Suita, Osaka, Japan
[2]Immunology Frontier Research Center, Osaka University, Suita, Osaka, Japan

## Synonyms

Fluorescent probes for $Zn^{2+}$

## Definition

Fluorescent probes for intracellular second messengers have been proved to be useful for bioimaging studies (Demchenko 2009). These fluorescence-based probes have been studied to turn up the biomolecules within living samples (Kikuchi 2010). Direct visualization of the cellular processing in biology helps to obtain information about molecular functions. The development of fluorescence-based chemosensors for $Zn^{2+}$ ion was found to be useful for the detection of $Zn^{2+}$ in vitro and further live cell or tissue slice imaging studies. These probes clarified various functions and behaviors of $Zn^{2+}$ in living systems.

## Introduction

In this present scenario, it is very important to clarify the biological significance of intracellular metal ions directly within living cells. During the last two decades, significant advancements have occurred in the case of fluorescence-based sensors for detecting metal ions. This visualization has helped to shed light on the biological roles of these cations. For these purposes, small tailor-made molecular chemical probes are very useful to convert the input biological signal to chemical outputs. An ideal fluorescent chemosensor consists of a receptor part linked to a fluorophoric part which translates the recognition event at the receptor region into the fluorescence signal alteration. Therefore, a typical fluorescent chemosensor must have a receptor with a strong affinity to the relevant target. To obtain the maximum output on the basis of good binding-selectivity, the effect of environmental interference such as photobleaching of the fluorescence signal should be minimal. The concentration of sensor along with the local environment including pH, polarity, and temperature around the sensor molecule plays important roles in determination of the selectivity issue. The commonly known fluorophore–spacer–receptor scaffold may be found to be inefficient in terms of imaging the target of interest in a biological environment. A thorough understanding of the available concepts can help to enlighten and to ameliorate the design of chemosensors. The output signal from a fluorescent chemosensor is usually measured through the change in fluorescence intensity, fluorescence lifetime, or a shift of fluorescence wavelength.

## Designing Principle

Receptor designing for the specific metal ion involves knowledge about ionic, covalent, and supramolecular interactions. Metal chelation effect is very important for cation receptor design. For communication between receptor and fluorophore, charge transfer (CT), photoinduced electron transfer (PET), resonance energy transfer (RET), and monomer-excimer conversion are the main mechanisms to direct the design of target-oriented fluorescence sensors. However, the fluorescence measurement by an increase of the fluorescence intensity without a large shift of either excitation or emission wavelength can be influenced by several factors. To reduce the influence of such factors, ratiometric measurement is utilized. In this technique, simultaneous recording of the fluorescence intensities at two wavelengths and calculation of their ratio provides greater precision than measurement at a single wavelength, and is found suitable for cellular imaging studies. Ratiometric measurement is possible in the cases where the probe exhibits a large shift in its emission or excitation spectrum after its interaction with the target molecule.

## Importance of Zinc Ion Sensing

After iron, zinc is the second most abundant transition metal ion in biological systems (McRae et al. 2009; Nolan and Lippard 2009). For more than 300 DNA-binding proteins and enzymes, zinc plays the role of an essential structural element. It plays a significant role in neurotransmission, signal transduction, gene expression, brain function, pathology, immune function, and mammalian reproduction. A number of disorders, including Alzheimer's disease, diabetes, cancer, epilepsy, ischemic stroke, and infantile diarrhea, are related to the disruption of zinc homeostasis. The lack of tools for assessing alterations of intracellular zinc concentrations with high spatial and temporal fidelity created a problem in understanding the molecular mechanisms of its physiology and pathology prior to the last two decades. The total concentration of zinc ions in human blood serum is $\sim 12\ \mu M$. The amount of free or rapidly exchangeable zinc is only at a picomolar level since a major portion of zinc ions are coordinated within proteins and cell organelles. However, free zinc pools have been observed in some tissues such as the brain, intestine, pancreas, and retina (Potyrailo and Mirsky 2009). A highly specific and sensitive fluorescent sensor is required to measure these concentrations with picomolar sensitivity. The $d^{10}$ electron configuration of $Zn^{2+}$ makes it spectroscopically silent, with the exception of the case of atomic absorption spectroscopy (AAS). Thus, some fluorescent chemosensors for the detection of $Zn^{2+}$ have been studied intensively. A recent tutorial review (McRae et al. 2009) describes the zinc chemosensors based on receptor types and elucidates the design and application of fluorescent chemosensors for zinc ion in a lucid way to new comers in this field. Scientists have focused on zinc metalloneurochemistry aspect with small-molecule zinc fluorescent chemosensors (Nolan and Lippard 2009).

The chemosensors, applicable for sensing $Zn^{2+}$ in biological samples, have to satisfy several major criteria. First of all, the sensor should detect $Zn^{2+}$ selectively among all other constituents in the biological environment, including higher concentrations of $Na^+$, $K^+$, $Ca^{2+}$, and $Mg^{2+}$. Secondly, the sensor should provide $Zn^{2+}$ detection with spatial and temporal resolution. The affinity of the probe toward $Zn^{2+}$ is measured by its dissociation constant ($K_d$), keeping in mind that it should approach the median concentration of $Zn^{2+}$ in the sample for monitoring its flux. Other desirable characteristics of the sensors include fast and reversible $Zn^{2+}$ coordination, excitation and emission wavelengths in the visible/near-IR regions for single photon excitation, a bright (quantum yield ($\Phi$)$\times$extinction coefficient ($\varepsilon$)) signal, water-solubility, nontoxicity, and photostability. The strong understanding of ingestion and distribution of sensors within cells controls the intracellular or extracellular application of the same sensor.

The ligand designing for specifically $Zn^{2+}$ metal ion depends on zinc-chelator denticity, sterics, and chelate ring size, resulting in $Zn^{2+}$-responsive sensors that can detect $Zn^{2+}$ from a very low to high level of concentrations such as $10^{-10}$–$10^{-3}$ M. The well-known employed mechanisms for transduction of fluorescence signaling in the design of $Zn^{2+}$ chemosensors are PET and intermolecular charge transfer (ICT). For a PET-based chemosensor, a fluorophore is generally connected through a spacer to a receptor containing a comparatively high-energy nonbonding electron pair which can transfer an electron to the excited state of the fluorophore, resulting in weak fluorescence. In the presence of the metal ion, the electron pair is engaged in coordination of the cation and the redox potential of the receptor is raised sufficiently so that the HOMO of the receptor becomes lower in energy than that of the fluorophore. Thus, the effective PET process from the receptor to the fluorophore is no longer in action and strong fluorescence is observed. Most of the chemosensors based on the PET mechanism for sensing $Zn^{2+}$ result in fluorescence enhancement signal. However, ratiometric measurements, that is, the changes in the ratio of the intensities of the absorption or the emission at two wavelengths, have found to be more favorable for measuring sensitivity of the sensor molecule. The ICT mechanism has been mostly applied in the excogitation of ratiometric fluorescent chemosensors. Direct covalent linkage between a fluorophore with a receptor without any spacer can form a $\pi$-electron conjugation system with electron-rich and electron-poor terminals of the probe. In this situation, ICT from the electron donor to electron

**Zinc, Fluorescent Sensors as Molecular Probes, Fig. 1** The structures of Zinquin, ZnAF-2, FluoZin-3, FuraZin, and RhodZin-3

acceptor would be enhanced upon excitation by light. The interaction between metal ions with receptor reduces the electron-donating character of the receptor and a blueshift of the emission spectrum is expected. On the other hand, if a cation receptor acts as an electron receptor, the interaction between the receptor and the cation would further strengthen the push–pull effects. This reflects in a redshifted emission from the fluorophoric part. This strategy has been successfully applied to quinoline-based molecular probes for $Zn^{2+}$. The coordination of $Zn^{2+}$ by quinoline derivatives can induce a redshifted ratiometric fluorescence signal.

## Examples

It is worth mentioning that the combining effect of PET and ICT mechanisms in the design of chemosensors was found to be valuable, since a wavelength-shifted fluorescence intensity enhancement will exaggerate the recognition outcome to a greater magnitude. Successful examples include the evolution of zinc sensors based on the merging of dipicolylamine, 8-substituted quinoline, iminodiacetic acid derivatives as receptors with various fluorophores. About two decades ago, Zinquin (Fig. 1) was successfully used to show apoptosis as a function of $Zn^{2+}$ concentration in biology. But the major success in the $Zn^{2+}$ sensor was observed within only the last 10 years; during this period, rapid improvements have been made in the development of $Zn^{2+}$-specific sensor molecules. Among several approaches, major parts of fluorescent sensor molecules were reported for the selective detection of chelatable $Zn^{2+}$ in biological environment. Some of these $Zn^{2+}$ sensor molecules like ZnAF-2, FluoZin-3, FuraZin, and RhodZin-3 (Fig. 1) were used to lead new biological functions related to $Zn^{2+}$ in living systems (Qian et al. 2009). These fluorescent sensor molecules, which allow visualization of $Zn^{2+}$ in living cells by fluorescence microscopy, are useful tools for studying biological systems.

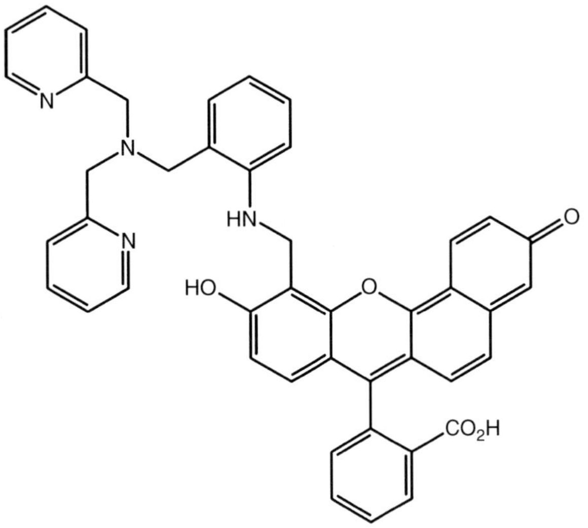

**Zinc, Fluorescent Sensors as Molecular Probes, Fig. 2** The structure of ZNP1

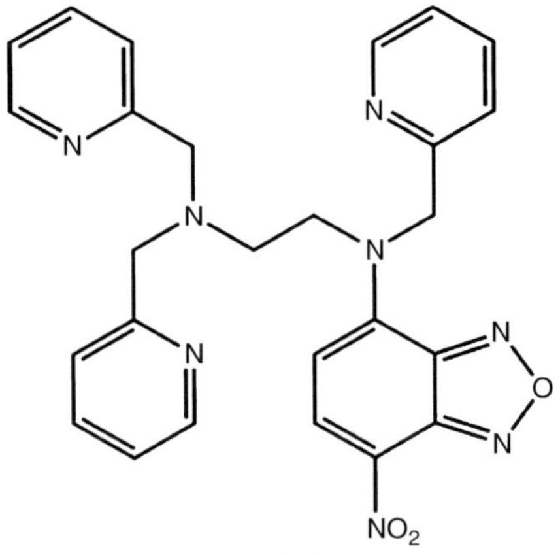

**Zinc, Fluorescent Sensors as Molecular Probes, Fig. 3** The structure of NBD-TPEA

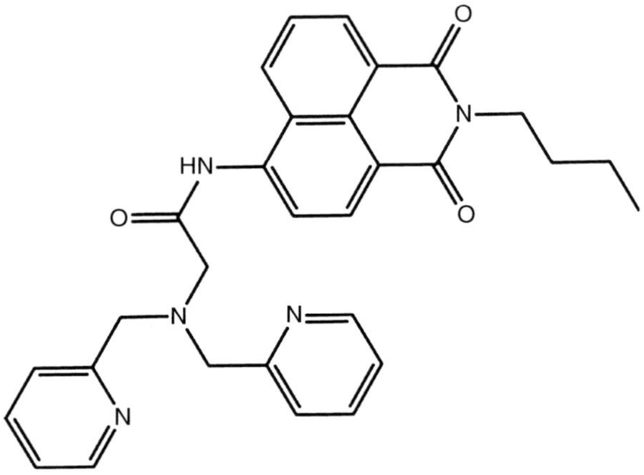

**Zinc, Fluorescent Sensors as Molecular Probes, Fig. 4** The structure of ZTRS

A single-excitation, dual-emission ratiometric zinc probe (ZNP1, Fig. 2) was reported, where the probe uses the mechanism of controllable $Zn^{2+}$-induced switching between the fluorescein and naphthofluorescein tautomeric forms of the dye (Nolan and Lippard 2009). The probe exhibited two fluorescence maxima at about 530 and 600 nm when it was excited at a single wavelength of 499 nm. In the presence of nanomolar concentrations of zinc ions, there was a shift of the long-wavelength band and the emission was observed at 624 nm, with a dramatic increase of relative intensity. The ratio between two different emission intensities ($F_{624}/F_{528}$) varies about 18 times in the absence and presence of $Zn^{2+}$. The best feature of the probe was visible excitation and emission profiles with excellent selectivity toward intracellular $Zn^{2+}$ over competing $Ca^{2+}$ and $Mg^{2+}$ ions in living mammalian cells with a dissociation constant ($K_d$) of <1 nM for $Zn^{2+}$. For ratiometric zinc measurements, iminocoumarin as a fluorophore and (ethylamino)-dipicolylamine as a $Zn^{2+}$ chelator have also been suggested.

In a recent report, a well-known ICT-based fluorophore (4-amino-7-nitro-2,1,3-benzoxadiazole) of smaller aromatic plane was used to construct visible light–excited fluorescent $Zn^{2+}$ sensor (NBD-TPEA, Fig. 3) (Que et al. 2008). This sensor demonstrates a visible ICT absorption band, a large Stokes shift, and biocompatibility. Low emission of the free probe was observed within a pH range 7.1–10.1 with an emissive band at 550 nm. The NBD-TPEA displays distinct 15 times amplification of fluorescence intensity in the presence of $Zn^{2+}$ selectively with a shift in emission maxima from 550 to 534 nm. This change is due to the synergic $Zn^{2+}$ coordination by the outer bis(pyridin-2-ylmethyl) amine (BPA) receptor and involvement of the nitrogen at the 4-position of the fluorophore. A preferential accumulation of the NBD-TPEA staining was exhibited at lysosome and Golgi with dual excitability at either 458 or 488 nm when the successful imaging on living cells was performed with intracellular $Zn^{2+}$. The intact in vivo imaging application on zebra fish embryo or larva stained due to $Zn^{2+}$ fluorescence with

**Zinc, Fluorescent Sensors as Molecular Probes, Table 1** Properties of selected fluorescent $Zn^{2+}$ sensors

| Probe | $K_d$ for $Zn^{2+}$ | $\lambda_{ex}$, nm, Free (Probe + $Zn^{2+}$) | $\lambda_{em}$, nm, Free (Probe + $Zn^{2+}$) | $\Phi$ Free (Probe + $Zn^{2+}$) |
|---|---|---|---|---|
| Zinquin | 0.2 nM | 370 (370) | 490 (490) | N/A[a] |
| ZnAF-2 | 2.7 nM | 490 (492) | 514 (514) | 0.023 (0.36) |
| FluoZin-3 | 15 nM | 488 (488) | 515 (515) | N/A[a] |
| FuraZin | 3.4 µM | 378 (330) | 510 (510) | N/A[a] |
| RhodZin-3 | 65 nM | 545 (545) | 575 (575) | N/A[a] |
| ZNP1 | 0.55 nM | 503, 539 (547) | 528, 604 (545, 624) | 0.02 (0.05) |
| NBD-TPEA | 2 nM | 469 (469) | 550 (534) | 0.003 (0.046) |
| ZTRS | 5.7 nM | 371 (348) | 483 (514) | 0.016 (0.36) |

[a]Data not available

NBD-TPEA revealed two zygomorphic luminescent areas around its ventricle which suggests the possible $Zn^{2+}$ storage area for the zebra fish development.

A recent work combined with novel amide-containing receptor for $Zn^{2+}$ (ZTRS, Fig. 4) and a naphthalimide fluorophore has been published (McRae et al. 2009). The interaction between the probe and $Zn^{2+}$ has been confirmed by fluorescence measurement, absorption detection, NMR, and IR studies. The studies indicated that ZTRS binds with $Zn^{2+}$ through the imidic acid tautomeric form of the amide/di-2-picolyamine receptor in aqueous solution, while most other heavy and transition metal ions were bound to the sensor in an amide tautomeric form. This different binding mode of ZTRS reflects in redshift emission of the fluorophoric part from 483 to 514 nm with excellent enhanced fluorescence selectivity (22-fold) for $Zn^{2+}$ over most competitive metal ions. The same ZTRS sensor can differentiate the binding to $Zn^{2+}$ and $Cd^{2+}$ by green and blue fluorescence respectively. The ratiometric detection of $Zn^{2+}$, via a $Cd^{2+}$ displacement approach, results with a large emission wavelength shift from 446 to 514 nm. This cell permeable sensor can be applied to monitor trace amounts of zinc ions during the development of a living organism and has been tested to detect zinc ions during the development of living zebra fish embryos.

A brief list of properties for selected small-molecule probes for $Zn^{2+}$ ion has been mentioned in Table 1. It provides the information of the emission spectral shift along with the efficiency of the probes in terms of enhancement of fluorescence quantum yield in presence of $Zn^{2+}$ and the dissociation constant values of the $Zn^{2+}$ complexes of the probes.

In addition to small molecular probes, some peptide-based zinc sensors [e.g., FS01DMB with Ac-B(DMB)ACDICGKNFSQSDELTTHIRTHT-NH$_2$ sequence, FS02DNS with Ac-B(DNS) ACDIHGKNFSQSDELTTHIRTHT-NH$_2$] have also been developed (Tomat and Lippard 2010). These include the alleged "zinc-finger" segments of the DNA-binding proteins in the structure. The response in the presence of $Zn^{2+}$ is the reflection of the conformational change in peptide upon $Zn^{2+}$ binding. This conformational change results in the fluorescence property of the dye due to the change of environment of the dye attached to peptide or two dyes come close together to make effective FRET. The successful study reveals that the binding of zinc in the active site of carbonic anhydrase enzyme is explored by ratiometric fluorescent sensor. Zinc ion–dependent FRET between the two dyes is responsible for the generation of the wavelength-ratiometric fluorescence response. This type of picomolar concentration–sensitive sensor molecules is well adapted for zinc imaging in a physiologically condition.

## Future Direction

In spite of the development of several $Zn^{2+}$ sensors including imaging of mobile $Zn^{2+}$ (Xu et al. 2010), there are nevertheless a lot of possibilities to improve the receptors for zinc, keeping in mind criteria like immediate chelation, eminent selectivity, and extravagant biologically suitable fluorescence imaging capacity. For this purpose, the design of the small molecular probe for the zinc receptor is the most significant consequence. To satisfy the requirement, receptor designing should be more diverse through the modification of the coordination approach with proper alterations in the binding site geometry of probe molecules.

## Cross-References

▶ Zinc and Zinc Ions in Biological Systems
▶ Zinc, Physical and Chemical Properties

## References

Demchenko AP (2009) Introduction to fluorescence sensing. Springer Science+Business Media B.V, The Netherlands

Kikuchi K (2010) Design, synthesis and biological application of fluorescent sensor molecules for cellular imaging. Adv Biochem Eng/Biotechnol 119:63–78

McRae R, Bagchi P, Sumalekshmy S, Fahrni CJ (2009) In situ imaging of metals in cells and tissues. Chem Rev 109:4780–4827

Nolan EM, Lippard SJ (2009) Small-molecule fluorescent sensors for investigating zinc metalloneurochemistry. Acc Chem Res 42:193–203

Potyrailo RA, Mirsky VM (eds) (2009) Combinatorial methods for chemical and biological sensors. Springer Science+Business Media, LLC, New York

Qian F, Zhang C, Zhang Y, He W, Gao X, Hu P, Guo Z (2009) Visible light excitable $Zn^{2+}$ fluorescent sensor derived from an intramolecular charge transfer fluorophore and its in vitro and in vivo application. J Am Chem Soc 131:1460–1468

Que EL, Domaille DW, Chang CJ (2008) Metals in neurobiology: probing their chemistry and biology with molecular imaging. Chem Rev 108:1517–1549

Tomat E, Lippard SJ (2010) Imaging of mobile zinc in biology. Curr Opin Chem Biol 14:225–230

Xu Z, Yoon J, Spring DR (2010) Fluorescent chemosensors for $Zn^{2+}$. Chem Soc Rev 39:1996–2006

# Zinc, Metallated DNA-Protein Crosslinks as Finger Conformation and Reactivity Probes

Sarah R. Spell, Samantha D. Tsotsoros and Nicholas P. Farrell
Department of Chemistry, Virginia Commonwealth University, Richmond, VA, USA

## Synonyms

Anticancer metallodrugs; Antiviral metallodrugs; Cobalt Schiff base; Metallated DNA; Polynuclear Pt complexes; Pt-DNA adducts; Ternary crosslinks; Transcription factor hijacking; Zinc finger targeting

## Definition

This entry describes the interactions of zinc finger peptides with metallated DNA and small molecule platinum-metal probes. The interactions have ▶ biological consequences from sequence-specific inhibition of zinc finger function to hijacking of transcription factors by binding to platinated DNA. Small molecule interactions have chemical consequences because the formal analogy between alkylation and metallation suggests the possibility of electrophilic attack of the cysteine and histidine residues with eventual zinc ion ejection and loss of tertiary structure and consequently biological function. These approaches create new opportunities to investigate the chemical properties of zinc fingers and may possibly produce new leads for disease intervention where zinc finger function is implicated.

## Introduction

In biology, zinc serves either a structural or catalytic purpose. When zinc acts in a catalytic manner, it is bound to a water molecule that is converted to a hydroxyl ligand. Structural zinc is characterized by coordination to four amino acids, generally histidine and cysteine. These types of zinc sites include zinc fingers (ZF), which compose 2–3% of the human genome (Klug 2010). The most common role of zinc fingers in biology is in the binding of DNA to transcription factors and they are also involved in RNA packaging, transcriptional activation, regulation of apoptosis, and protein folding and assembly.

Zinc fingers are defined by the number and type of amino acids coordinated to the central zinc atom (Fig. 1).

Analysis of $Cys_2His_2$ zinc finger proteins has allowed for elucidation of DNA recognition mechanisms and their role in transcriptional regulation. The $Cys_3His$ motif is involved in RNA and DNA recognition, cell-cycle control, and cellular signaling pathways. The third common zinc $Cys_4$ finger motif is found in human DNA repair enzymes and many nuclear localized systems.

---

Sarah R. Spell and Samantha D. Tsotsoros contributed equally.

**Zinc, Metallated DNA-Protein Crosslinks as Finger Conformation and Reactivity Probes, Fig. 1** The three main coordination environments of zinc fingers: $C_2H_2$, $C_3H$, and $C_4$

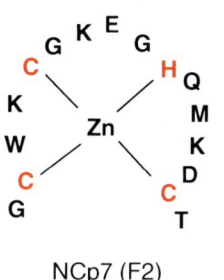

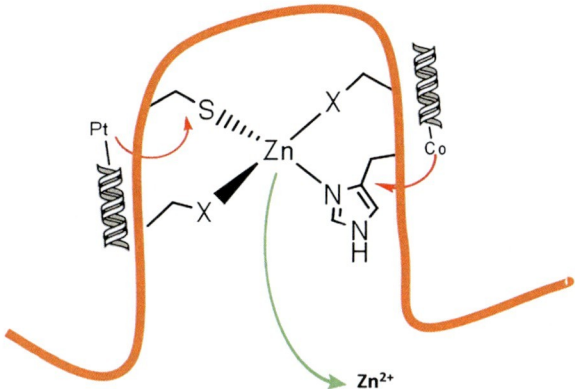

**Zinc, Metallated DNA-Protein Crosslinks as Finger Conformation and Reactivity Probes, Fig. 2** General schematic of electrophilic agent attack on a zinc finger. Platinum agents attack at cysteine residues, while cobalt agents attack at histidine residues. X = (Cys)(His) or $Cys_2$

molecule metal-based electrophiles and formation of DNA-protein crosslinks as probes for zinc finger conformation and reactivity will be discussed. These approaches create new opportunities to investigate the chemical properties of zinc fingers and may possibly produce new leads for disease intervention where zinc finger function is implicated.

## Targeting of Zinc Fingers by Cobalt Schiff Base Complexes

Cobalt(III) Schiff base (Co-sb) complexes may disrupt the structure of a zinc finger peptide by axial ligation of the Co(III) ion to the nitrogen of the imidazole ring of a histidine residue (Louie and Meade 1998). The reaction of Co(III)-sb with histidine is a dissociative ligand exchange process, with loss of the labile axial $NH_3$ ligands by $H_2O$ facilitating histidine binding. The reaction of Co(III)-sb at the zinc finger site irreversibly inhibits protein activity as a result of zinc ejection due to the binding of cobalt to histidine (Harney et al. 2009). Zinc ejection causes loss of tertiary structure and therefore loss of function.

Coordination to zinc enhances the nucleophilicity of the Zn-Cys bond, allowing for attack by small electrophiles. The chemical reactivity is dependent on the nature of the zinc coordination sphere ($Cys_2His_2$; $Cys_3His$, $Cys_4$) (Quintal et al. 2011). The interaction of zinc finger proteins with DNA and RNA can be inhibited upon chemical modification of the zinc binding ligands. There is a formal analogy between alkylation and metallation, where in the latter case chemical modification refers to electrophilic attack of the cysteine and histidine residues by Pt and Co agents, respectively (Fig. 2).

In both cases, however, eventual loss of zinc is seen and the proteins lose tertiary structure and therefore loss of biological function. Tethering of the electrophilic agent to a sequence-specific DNA may allow for targeting and differentiation between different zinc fingers and enhancement of selectivity. In this review, the approaches to the use of small

Initial studies showed that cobalt complexes inhibit binding of the human transcription factor Sp1, to its consensus sequence. DNA-coupled conjugates of the cobalt complexes selectively inhibited Sp1 in the presence of several other transcription factors. The Snail family of $Cys_2His_2$ zinc finger transcription factors, Slug, Snail, and Sip1, are emerging as anticancer drug targets as they are implicated in tumor metastasis through the regulation of epithelial-to-mesenchymal (EMT) transitions. Slug, Snail, and Sip1 bind to the Ebox consensus sequence CAGGTG in the promoter region of target genes with high specificity to mediate transcriptional repression. To improve specificity for the Snail family of transcription factors, an oligonucleotide containing the DNA consensus sequence was conjugated to the Co(III) complex yielding Co(III)-Ebox (Fig. 3) (Harney et al. 2009).

No significant changes in the structure of DNA occurred upon conjugation. Through addition of the Ebox DNA sequence to the Co(III) metal complex, peptide-binding specificity was improved 150-fold over Co(III)-sb. Co(III)-Ebox was found to effectively inhibit the Slug, Snail, and Sip1 zinc finger transcription factors from binding to their DNA targets with eventual loss of $Zn^{2+}$ (Harney et al. 2009). Studies demonstrate that neither the oligo nor the Co(III) Schiff base complex alone is sufficient for transcription factor inactivation at concentrations where the conjugated complex mediates inhibition.

Subsequently, a Co(III) Schiff base DNA conjugate has been designed to target the Gli $C_2H_2$ transcription factors in the Hedgehog (Hh) pathway, which has been implicated in the formation and development of certain cancers such as medulloblastomas and basal cell carcinomas (Hurtado et al. 2012). In order to test the specificity and efficacy of Co(III)-DNA complex, the authors tested the Ci consensus sequence alone, Co(III)-sb, Co(III)-Ci, and Co(III)-Mut (one base pair mutation). The complete structure of Co(III)-Ci was found to be the most effective at inhibiting Gli family ZF transcription factors. The mutation of one base pair in Co(III)-CiMut inhibited its specific and potent activity. The activity of Co(III)-Ci was specific, as no other zinc finger transcription factor tested was inhibited by the addition of the complex (Hurtado et al. 2012). These results show that subtle changes in the oligomer attached to Co(III) can modulate the specificity and activity of these Co(III)-DNA conjugates both in vitro and in vivo.

The studies performed using Co(III)-DNA complexes demonstrate the development of a versatile class of specific and potent complexes that may be used to study zinc finger proteins and may prove valuable as an experimental tool and as anticancer therapeutics.

## Pt Complexes as DNA/Protein Crosslinkers

The structure and nature of DNA adducts of platinum complexes may be manipulated to facilitate ternary DNA-Pt-protein crosslinks. Table 1 summarizes the compounds studied. One of the first examples is the use of *trans*-diamminedichloroplatinum(II) (TDDP). In order to map the section of HIV1 RNA that was recognized by the nucleocapsid NC protein, TDDP was used as a crosslinking agent. After digestion of a crosslinked solution of HIV1 RNA and NC protein, the crosslink was shown to form between positions 315–324 of the RNA. This suggests the possible recognition site for the NC protein on HIV1 RNA (Darlix et al. 1990).

Various mononuclear and dinuclear platinum complexes have been analyzed for their crosslinking abilities to zinc finger containing proteins such as the DNA repair proteins, UvrA and UvrB, and Sp1. The repair of the DNA adducts of the anticancer agent *cis*-diamminedichloroplatinum(II) (cisplatin, CDDP) was examined with the bacterial UvrABC complex (Lambert et al. 1995). A sterically hindered analog, *exo*-[N-2-methylamino-2,2,1-bicyclohepane] dichloroplatinum(II) (Pt-BCH) was compared for its protein crosslinking abilities. UvrAB proteins recognized the mono- and diadducts of Pt-BCH with a higher affinity than those of cisplatin. Analysis of the crosslinks showed the involvement of UvrB in the ternary nucleoprotein complexes. In general, UvrB has a greater affinity for the adducts formed by Pt-BCH over those formed by cisplatin (Lambert et al. 1995).

Dinuclear platinum complexes bind to DNA in a manner that is unique from cisplatin and other mononuclear complexes, the main adducts being long-range (Pt,Pt) interstrand crosslinks (Kloster et al. 2004). Dinuclear complexes may be formally bi, tri, or tetrafunctional (depending on number of substitution-labile chlorides present – see Table 1). Even in the presence of a CDDP-like *cis*-[$PtCl_2$(amine)$_2$] coordination sphere the (Pt,Pt)-interstrand crosslinks form preferentially (Fig. 4).

Thus, upon bifunctional binding to DNA, the third $Cl^-$ site can then bind to a protein, preferentially at Cys, His, or Met residues (Kloster et al. 2004). Platinum complexes are known to bind in the major groove of DNA and the ability of the dinuclear complexes to crosslink UvrA and UvrB suggests that the proteins contact the major groove of the DNA helix within 4–4.5Å (Van Houten et al. 1993). The flexibility of the dinuclear platinum adducts may be important in their crosslinking to Sp1 due to the DNA bending induced upon Sp1 binding (Kloster et al. 2004).

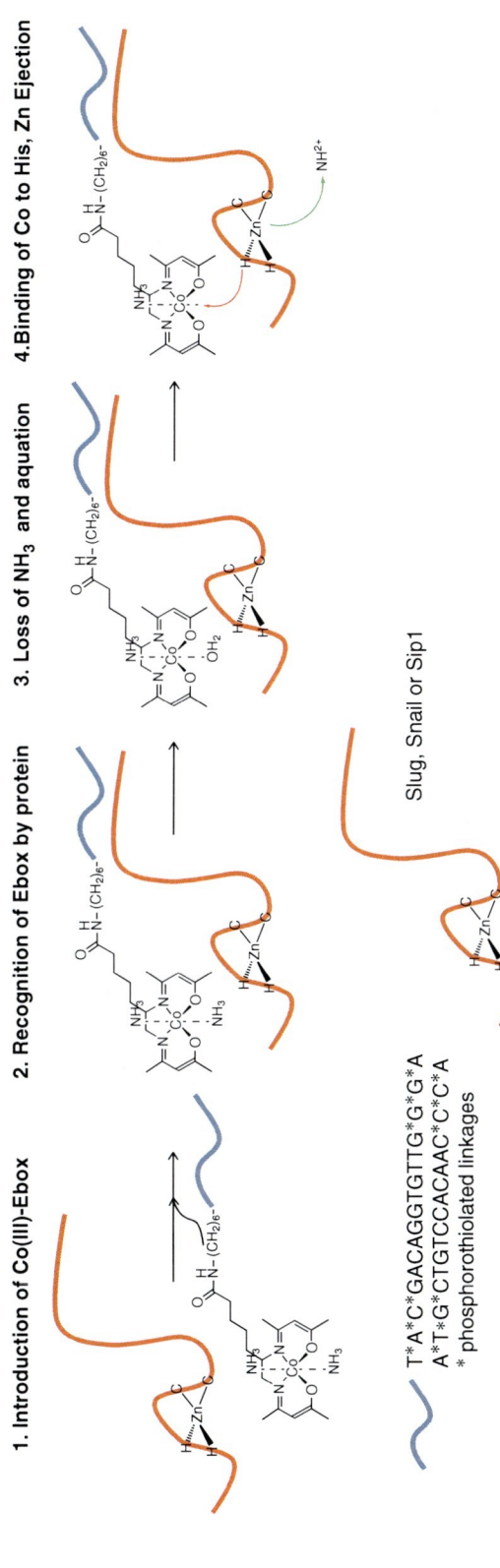

**Zinc, Metallated DNA-Protein Crosslinks as Finger Conformation and Reactivity Probes, Fig. 3** Scheme for zinc finger-specific inactivation using Co(III)-sb-DNA conjugates. Co(III)-Ebox is recognized by the Snail family of transcription factors. Loss of axial ligands and subsequent binding of histidine results in eventual ejection of $Zn^{2+}$ and transcription factor inactivation

**Zinc, Metallated DNA-Protein Crosslinks as Finger Conformation and Reactivity Probes, Table 1** Various metal compounds form crosslinks with zinc finger proteins and DNA

| Compound | Structure | DNA studied | Zn finger studied | Crosslink | References |
|---|---|---|---|---|---|
| CDDP | | Plasmid pSP65 | UvrAB | DNA monoadduct crosslink formation of UvrB | Lambert et al. 1995 |
| TDDP | | $^{32}$P HIV-1 RNA (311-415) | HIVI NC | Crosslink formation between RNA (between positions 315-324) and HIVI NC | Darlix et al. 1990 |
| Pt-BCH | | Plasmid pSP65 | UvrA UvrB | DNA monoadduct crosslink formation of UvrB | Lambert et al. 1995 |
| (Pt-Pt)-2,2/c,c | | 49 bp duplex | UvrA UvrB Sp1 | Crosslink formation between platinated DNA and UvrA and UvrB | Van Houten et al. 1993 |
| (Pt-Pt)-1,1/t,t | | 49 bp duplex | UvrA UvrB Sp1 | Crosslink formation between platinated DNA and UvrA and UvrB | Van Houten et al. 1993; Kloster et al. 2004 |
| (Pt-Pt)-1,2/t,c | | Plasmids pSP73KB and pUC19 | Sp1 | Crosslink formation between platinated DNA and Sp1 | Kloster et al. 2004 |
| (Pt-Pt)-1,2/c,c | | Plasmids pSP73KB and pUC19 | Sp1 | Crosslink formation between platinated DNA and Sp1 | Kloster et al. 2004 |
| (Pt-Ru) | | 49 bp duplex | UvrA UvrB | Crosslink formation between platinated DNA and UvrA and UvrB | Van Houten et al. 1993 |

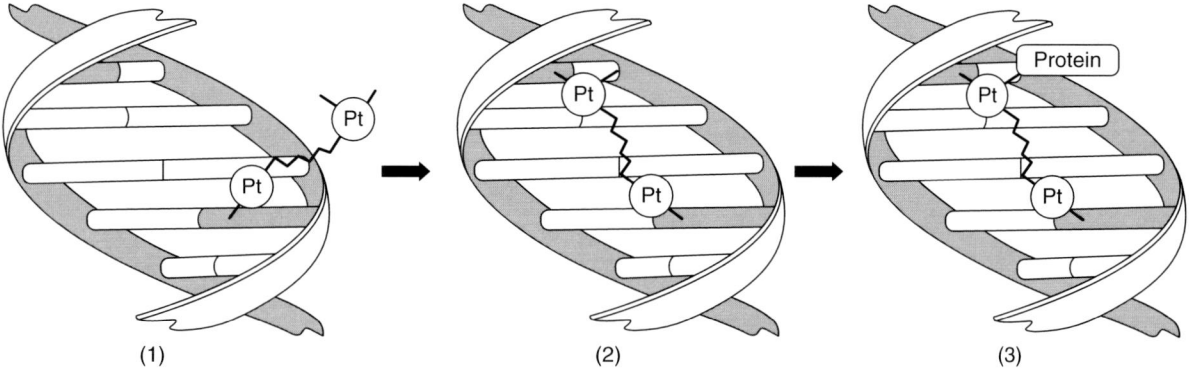

**Zinc, Metallated DNA-Protein Crosslinks as Finger Conformation and Reactivity Probes, Fig. 4** The binding of dinuclear platinum complexes to DNA. (1) Initial monofunctional binding, (2) long-range bifunctional interstrand crosslink formation with second platinum unit and (3) binding of protein at the third active site

The results of these platinum crosslinking studies suggest that the formation of metal-mediated DNA-protein complexes may play a role in the cytotoxic properties of these compounds due to the irreversible crosslinking, and therefore sequestering, of repair proteins or transcription factors to DNA. The differences between mononuclear and dinuclear platinum complexes may also explain the different cytotoxic properties of the two classes of anticancer drugs. The differences in DNA adduct structure between mononuclear and dinuclear complexes may also be used to probe protein recognition of structurally different DNA crosslinks (Kloster et al. 2004).

## Interaction of Platinum Molecules with Zinc Fingers

The demonstration of zinc finger crosslink formation by platinated DNA is fundamentally an example of coordination compounds acting as electrophiles toward the peptide cysteine residues. Small molecule platinum-metal agents have therefore been explored for their ability to target the HIV nucleocapsid retroviral protein NCp7. Figure 5 shows the structures of the metal complexes studied as well as that of NCp7 nucleocapsid protein.

Again, *functionality* can be studied by use of structurally distinct compounds such as CDDP, *trans*-[PtCl(9-EtGua)(pyr)$_2$]$^+$, [PtCl(terpy)]$^+$, and [PtCl(dien)]$^+$ (Quintal et al. 2012). Further, the reactivity of the central metal ion may be altered by use of the general structure [MCl(chelate)] (M = Pt(II), Pd(II), or Au(III) and chelate = diethylenetriamine, dien or 2,2′;6′,2″-terpyridine, terpy). Interestingly, both Pt(II) and Pd(II)-dien species showed evidence of adduct formation on ZF2 by replacement of the M-Cl bond with zinc-bound thiolate. Eventual loss of the dien ligand, as well as zinc ejection, was observed. Due to the strong thiol affinity of gold drugs both Au(III) compounds reacted extremely fast producing only "gold fingers." For all terpy compounds metal exchange occurred producing the product [Zn(terpy)]$^{2+}$. These results show that by modifying both the metal and ligand of the reacting compound, the chemical reactivity of zinc finger proteins can by altered (Quintal et al. 2011).

A major challenge for all small molecule electrophiles is selectivity. To enhance the selectivity of these drugs, it is important to understand the details of the zinc finger – DNA interaction. The recognition of DNA or RNA by the HIV nucleocapsid protein NCp7 is dominated primarily by π-stacking between the purine nucleic acid bases and the planar aromatic amino acid residues, especially between guanine and tryptophan37 (Quintal et al. 2011). Fluorescence quenching studies have shown that metallation of nucleobase compounds significantly enhances the π–π stacking interactions with L-tryptophan. Metallation modifies frontier orbital properties, lowering the π-acceptor LUMO of the metallated nucleobase, thus improving the overlap toward the π-donor HOMO of the tryptophan. Study of the biologically relevant C-terminal peptide (ZF2) of the HIV-NCp7 zinc knuckle showed that [Pt(dien)(9EtGua)]$^{2+}$

**Zinc, Metallated DNA-Protein Crosslinks as Finger Conformation and Reactivity Probes, Fig. 5** Structures of HIV NCp7 nucleocapsid protein (ZF; C-terminal finger, ZF2, shown in *dashed box*) and the platinum-metal complexes studied for molecular recognition and electrophilic attack

(9EtGua = 9-ethylguanine) and [Pt(dien)(5′GMP)] (5′GMP = guanosine 5′-monophosphate) quench tryptophan fluorescence with $K_a$ association constants of $7.5 \times 10^3$ and $12.4 \times 10^3$ M$^{-1}$, respectively. Though there is little difference between their binding with the simple amino acid tryptophan, the extra phosphate group on the 5′GMP causes a significant increase in binding with ZF2 (Anzellotti et al. 2006). ESI-MS of both [Pt(dien)9EtGua]$^{2+}$ and [Pt(dien) (5′GMP)] with ZF2 showed formation of a 1:1 adduct between the peptide and Pt-complex. It is most likely that this adduct formation is facilitated by the π–π stacking interactions between the tryptophan and the platinated nucleobase. Incubation of *cis*-[Pt(NH$_3$)$_2$(Guo)$_2$]$^{2+}$ (Guo = guanosine) with ZF2 showed significantly less formation of the 1:1 adduct, possibly due to steric hinderance from the mutually *cis*-oriented purines (Anzellotti et al. 2006).

The monofunctional *trans*-[PtCl(9-EtGua)(pyr)$_2$]$^+$ (pyr = pyridine) has high reactivity toward sulfur over nitrogen, and it has also been shown to cause Zn ejection in model chelates (see below). In the presence of a 1:1 stoichiometric ratio of *trans*-[PtCl(9-EtGua) (pyr)$_2$]$^+$, the ESI-MS of ZF2 showed numerous peaks consistent with multiple adduct formation. The peak corresponding to the adduct ([ZF2(Pt[pyr]2)2]-Zn) shows two [Pt(pyr)]$_2$ units bound to the protein with resultant loss of Zn. CD experiments revealed changes in the three-dimensional structure of the protein, consistent with the loss of tertiary structure due to Zn ejection. Therefore, the proposed mechanism for attack of Pt-nucleobase compounds on zinc fingers consists of two steps: (1) noncovalent recognition through π–π stacking of tryptophan with platinated nucleobase and (2) covalent interaction with Pt-S bond formation followed by zinc ejection (Anzellotti et al. 2006).

## Platinated DNA Affects Zinc Finger Conformation

To expand on the studies of platinum-nucleobase complex – zinc finger interactions, the interaction of Pt(dien)6-mer (6-mer = d′(5′-TACGCC-3′)) with the C-terminal finger of the HIV NCp7 zinc finger, ZF2, was studied by NMR Spectroscopy. In this case the Pt(dien) moiety was bound to the single guanine residue (Fig. 6a). The solution structures of ZF2, the 6-mer/ZF2, and Pt(dien)6mer/ZF2 adducts were calculated from NOESY-derived distance constraints. For the 6-mer/ZF2 structure, stacking with tryptophan was observed as well as showed that there was an interaction between Trp37 and the ribose protons

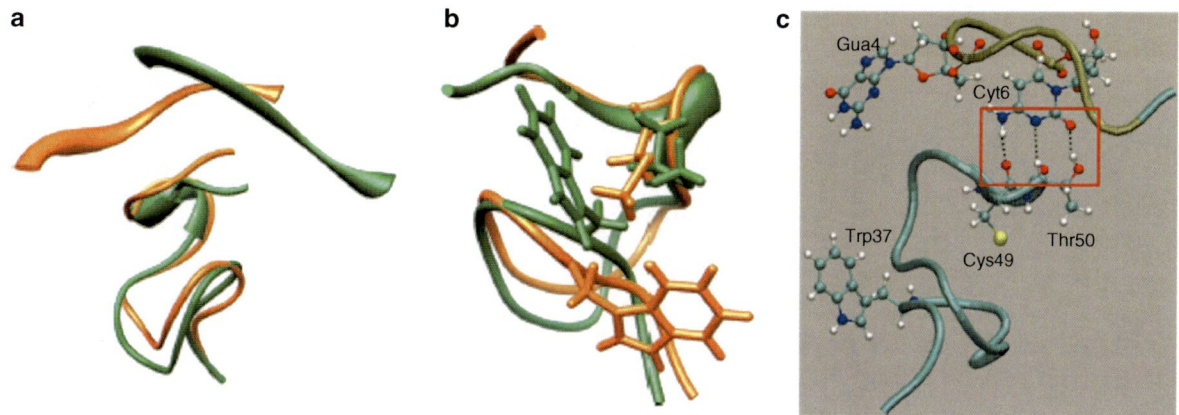

**Zinc, Metallated DNA-Protein Crosslinks as Finger Conformation and Reactivity Probes, Fig. 6** (a) Superposition of the minimized structures of the 6-mer/ZF2 adduct (*green*) and the Pt(dien)(6-mer)/ZF2 adduct (*orange*). (b) Change in conformation of Trp37 upon platination of Gua4. (c) Stabilizing interactions of Cyt6 with the protein residues, Cys49 and Thr50, in the Pt(dien)(6-mer)/ZF2 adduct (hydrogen bonding highlighted in *red box*)

of the carbohydrate moiety of Gua4, confirming the carbohydrate-aromatic ring interaction as a key recognition site between NCp7 protein and oligonucleotides. The NMR spectrum of ZF2 in the presence of Pt(dien)6-mer showed weakening of the Trp37-Gua4 contact attributed by the steric effects caused by platination of Gua4. The aromatic ring of the tryptophan residue changes orientation (Fig. 6b), causing the DNA to be in a completely different position than when platinum is absent.

Molecular dynamics calculations of the DNA-protein interactions showed that the π-stacking between Trp37 and Gua4 as well as the hydrogen bonding between the pentose and phosphate oxygen(s) of Cyt5 and Gua4 both help stabilize the interactions between the 6-mer and ZF2. Once platinum is bound, the 6-mer is less flexible and stays in one stable conformation on the surface of the zinc finger. The Gua4 − Trp37 interaction is disrupted, resulting in hydrogen bonding interactions between Cyt6 and Cys49 and Thr50. Additionally, the backbone CH group is close to N3 of Cyt6. With the addition of this third hydrogen bond, the bonding mode imitates the three intermolecular Cyt−Gua base pair hydrogen bonds (Fig. 6c).

In conclusion, these results show for the first time structural characterization of platinated single-stranded DNA interacting with a zinc finger protein, resulting in conformational change of the peptide. These results further demonstrate the feasibility of using DNA-tethered coordination compounds to target specific zinc finger proteins (Quintal et al. 2012).

## Small Molecule Models for Platinated DNA-Zinc Finger Interactions

To provide a small molecule model for the reactions of platinum complexes with ZFs, the chelate N,N′-bis (2-mercaptoethyl)-1,4-diazacycloheptanezinc(II), [Zn(bme-dach)]$_2$, was studied. Structures for model metal/DNA/protein crosslinks and zinc ejection were deduced from ESI-MS spectra of this [Zn(bme-dach)]$_2$ with *trans*-[PtCl(9-EtGua)(pyr)$_2$]$^+$, a model for a monofunctional adduct of platinum on DNA. The schematic of the reaction is shown in Fig. 7.

The intermediate species [Zn(bme-dach)-Pt (9-EtGua)(pyr)$_2$]$^+$ was observed and after 20 h was identified as the major species present. This is considered to be the first heterodinuclear Pt,Zn monothiolate bridged species to be identified using biologically relevant species. MS experiments showed the final product to be [Pt(bme-dach)]$^+$, giving proof of Zn ejection followed by platinum replacement (Liu et al. 2005). Similar experiments were performed for [Pt(dien)Cl]Cl and [Pt(terpy)Cl]Cl. Results showed formation of mono- and dithiolate bridged intermediates as well as metal exchange and multimetallic aggregate species (Quintal et al. 2011). These results represent suitable models for the molecular structure of

**Zinc, Metallated DNA-Protein Crosslinks as Finger Conformation and Reactivity Probes, Fig. 7** Reaction pathway of [Zn(bme-dach)]$_2$ with trans-[PtCl(9-EtGua)(pyr)$_2$]$^+$ with formation of monothiolate bridged and metal exchanged species (Almaraz et al. 2008)

ternary DNA-protein complexes involving zinc finger proteins discussed above (Liu et al. 2005).

**Acknowledgments** This contribution was supported by NSF CHE-1012269 and NIH RO1-CA78754.

## Cross-References

▶ Platinum- and Ruthenium-Based Anticancer Compounds, Inhibition of Glutathione Transferase P1-1
▶ Platinum Anticancer Drugs
▶ Platinum Complexes and Methionine Motif in Copper Transport Proteins, Interaction
▶ Platinum-Containing Anticancer Drugs and Proteins, Interaction
▶ Terpyridine Platinum(II) Complexes as Cysteine Protease Inhibitors
▶ Zinc and Zinc Ions in Biological Systems
▶ Zinc Finger Folds and Functions
▶ Zinc-Binding Sites in Proteins

## References

Almaraz E, de Paula Q, Liu Q, Reibenspies J, Darensbourg M, Farrell NJ (2008) Thiolate bridging and metal exchange in adducts of a zinc finger model and Pt(II) complexes: biomimetic studies of protein/Pt/DNA interactions. Am Chem Soc 130:6272–6280

Anzellotti A, Liu Q, Bloemink M, Scarsdale J, Farrell N (2006) Targeting retroviral Zn finger-DNA interactions: a small-molecule approach using the electrophilic nature of trans-platinum-nucleobase compounds. Chem Biol 13:539–548

Darlix J-L, Gabus C, Nugeyre M-T, Clave F, Barre-Sinoussi F (1990) Cis elements and trans-acting factors involved in the RNA dimerization of the human immunodeficiency virus HIV-1. J Mol Biol 216:689–699

Harney A, Lee J, Manus L, Wang P, Ballweg D, LaBonne C, Meade T (2009) Targeted inhibition of Snail family zinc finger transcription factors by oligonucleotide-Co(III) Schiff base conjugate. Proc Natl Acad Sci USA 106:13667–13672

Hurtado R, Harney A, Heffern M, Holbrook R, Holmgren R, Meade T (2012) Specific inhibition of the transcription factor Ci by a Cobalt(III) Schiff base–DNA conjugate. Mol Pharm 9:325–333

Kloster M, Kostrhunova H, Zaludova R, Malina J, Kasparkova J, Brabec V, Farrell N (2004) Trifunctional dinuclear platinum complexes as DNA-protein cross-linking agents. Biochemistry 43:7776–7786

Klug A (2010) The discovery of zinc fingers and their applications in gene regulation and genome manipulation. Annu Rev Biochem 79:213–231

Lambert B, Jestin J-L, Brehin P, Oleykowski C, Yeung A, Mailliet P, Pretot C, Le Pecq J-B, Jacquemin-Sablon A, Chottard J-C (1995) Binding of the Escherichia coli UvrAB proteins to the DNA mono- and diadducts of cis-[N-2-amino-N-2-methylamino-2,2,1- bicycloheptane]dichloroplatinum (II) and cisplatin. J Biol Chem 270:21251–21257

Liu Q, Golden M, Darensbourg M, Farrell N (2005) Thiolate-bridged heterodinuclear platinum-zinc chelates as models for ternary platinum-DNA-protein complexes and zinc ejection from zinc fingers. Evidence from studies using ESI-mass spectrometry. Chem Commun 34:4360–4362

Louie AY, Meade TJ (1998) Isolation of a myoglobin molten globule by selective cobalt(III)-induced unfolding. Proc Natl Acad Sci USA 95:6663–6668

Quintal S, dePaula Q, Farrell N (2011) Zinc finger proteins as templates for metal ion exchange and ligand reactivity. Chemical and biological consequences. Metallomics 3:121–139

Quintal S, Viegas A, Erhardt S, Cabrita E, Farrell N (2012) Platinated DNA affects zinc finger conformation. Interaction of a platinated single-stranded oligonucleotide and the C-terminal zinc finger of nucleocapsid protein HIVNCp7. Biochemistry 51:1752–1761

Van Houten B, Illenye S, Qu Y, Farrell N (1993) Homodinuclear (Pt,Pt) and heterodinuclear (Ru,Pt) metal compounds as DNA-protein cross-linking agents: potential suicide DNA lesions. Biochemistry 32:11794–11801

# Zinc, Physical and Chemical Properties

Fathi Habashi
Department of Mining, Metallurgical, and Materials Engineering, Laval University, Quebec City, Canada

Zinc is a less typical metal, i.e., when it loses its two outermost electrons, it will not have the 8-electon octet typical of the electronic structure of inert gases but it will have an 18-electron shell. This renders the metal less reactive than the typical metals. Zinc occurs in nature mainly in the form of the mineral sphalerite, ZnS.

## Physical Properties

| | |
|---|---|
| Atomic number | 30 |
| Atomic weight | 65.39 |
| Relative abundance in Earth's crust, % | $8 \times 10^{-3}$ |
| Crystal structure | |
| α-Zn | Face-centered cubic |
| β-Zn | Tetragonal |
| Transformation temperature, °C | |
| α-Zn ⇌ β-Zn | 13.2 |
| Heat of transformation, J/gram atom | 1,966 |
| Density of β-Zn, g/cm$^3$ | |
| At 20°C | 7.286 |
| At 100°C | 7.32 |
| At 230°C | 7.49 |
| Density of α-Zn at 25°C g/cm$^3$ | 5.765 |

(*continued*)

| | |
|---|---|
| Density of liquid zinc, g/cm$^3$ | |
| At 240°C | 6.993 |
| At 400°C | 6.879 |
| At 800°C | 6.611 |
| At 1,100°C | 6.484 |
| Melting point, °C | 232.6 |
| Heat of melting, J/mol | 7,029 |
| Boiling point, °C | 906 |
| Heat of vaporization, kJ/mol | 295.8 |
| Vapor pressure at 827°C, mbar | $9.8 \times 10^{-6}$ |
| At 1,527°C | 7.5 |
| At 1,827°C | 83.9 |
| At 2,127°C | 512 |
| Thermal conductivity of β-Zn at 0°C, W cm$^{-1}$ K$^{-1}$ | 0.63 |
| Electrical resistivity, Ω m | |
| α-Zn at 0°C | $5 \times 10^{-6}$ |
| β-Zn at 25°C | $11.15 \times 10^{-6}$ |
| Superconductivity critical temperature, K | 3.70 |

The low boiling point of 906°C creates problems in pyrometallurgical zinc production since the metal will be in the vapor state during reduction of its oxide. That is why hydrometallurgical route followed by electrowinning has now nearly displaced the thermal route.

## Chemical Properties

Zinc has an oxidation state of 2+ in all its compounds. It forms complexes with ammonia, cyanide, and halide ions. It dissolves in mineral acids with evolution of hydrogen, but in nitric acid with evolution of NO$_x$. Zinc is resistant to air because of the protective coating formed. It is also resistant to halogens but is rapidly corroded by HCl gas. Because of its high surface area, zinc dust is much more reactive and can even be pyrophoric, e.g., reacting vigorously at elevated temperatures with the elements oxygen, chlorine, and sulfur. Zinc is used as a reducing agent in chemical processes, mainly in the form of powder or granules.

Zinc chloride and sulfate are soluble in water, while the oxide, sulfide, carbonate, phosphate, silicate, and organic complexes are insoluble or sparingly soluble. Zinc sulfide undergoes aqueous oxidation at 130°C and under oxygen pressure in a pressure reactor to form elemental sulfur and zinc ion:

$$ZnS + 2H^+ + \tfrac{1}{2}O_2 \rightarrow Zn^{2+} + S + H_2O$$

**Zinc, Physical and Chemical Properties, Table 1** Part of the electrochemical series showing metals that can be displaced by zinc

| Group | Metals |
|---|---|
| Less reactive metals | Zinc |
|  | Gallium |
|  | Iron |
|  | Cadmium |
|  | Indium |
|  | Thallium |
|  | Cobalt |
|  | Nickel |
|  | Tin |
|  | Lead |
|  | Hydrogen |
| Least reactive or noble metals | Copper |
|  | Mercury |
|  | Silver |
|  | Platinum-group |
|  | Gold |

This is the basis of the new technology for treating directly zinc sulfide concentrate.

Zinc has a high electrode potential. Besides displacing hydrogen from dilute acids, it can displace all metals below it in the electrochemical series (Table 1). A zinc coating on steel ensures lasting protection against the corrosive effects of the weather. The coating is made by immersing the steel in molten zinc; the process is called galvanization. This is one of the major applications of zinc.

Zinc can also displace gold from a cyanide solution – a process that is used on industrial scale:

$$2[Au(CN)_2]^- + Zn \rightarrow 2Au + [Zn(CN)_4]^{2-}$$

Zinc-copper alloys known as brass were known since Roman times. Zinc oxide in powders and ointments has bactericidal properties. Soluble zinc is toxic in large amounts, but the human body requires small quantities (10–15 mg/day) for metabolism.

## References

Graf GG et al (1997) Zinc. In: Habashi F (ed) Handbook of extractive metallurgy. Wiley-VCH, Weinheim, pp 641–688

Habashi F (2003) Metals from Ores. An introduction to extractive metallurgy, Métallurgie Extractive Québec, Quebec City. Distributed by Laval University Bookstore « Zone », www.zone.ul.ca

# Zinc, Visibility

David S. Auld
Harvard Medical School, Boston, MA, USA

## Definitions

Zinc retains a filled d-shell in all stable derivatives. The maximum oxidation state is +2, arising from the loss of two s electrons. Zinc does not undergo oxidation or reduction, thus providing a stable metal ion species in a biological medium whose redox potential is in constant flux. However, the filled d-shell has hindered mechanistic studies of its function since it has no useful chromophoric properties like copper and iron. X-ray absorption fine structure, XAFS, provides an experimental means to directly visualize the zinc in proteins in solution.

## Background

Zinc, a ubiquitous element, is known to be indispensable to growth and development and transmission of the genetic message. It does this through a remarkable mosaic of zinc-binding motifs that orchestrate all aspects of metabolism (► Zinc-Binding Sites in Proteins). Since zinc has a filled d-shell, it does not have oxidation/reduction properties like that of Cu and Fe, whose oxido-reductive properties are essential to their function. This same property has hindered direct examination of the role of zinc in biological systems since a metal with a full d-shell has no useful chromophoric properties, like cobalt, copper, or iron, to reveal its presence.

However, X-ray absorption fine structure, XAFS, experiments do not need a metal with an unfilled d-shell in order to examine the spectral properties of the metal in different environments (Stern and Heald 1983). It is a quantitative experimental technique that can probe the local environment around metal ions in the solution or solid state. This is demonstrated by our XAFS studies on the solution form of zinc carboxypeptidase A, ZnCPD. Our initial studies focused on the effect of $H^+$ ion, substrate and inhibitor binding on the coordination properties of the catalytic zinc site (Auld 1997; Zhang and Auld 1993, 1995).

## XAFS Studies of Zinc Enzyme in Solution

Examination of the XAFS spectrum of the zinc enzyme over the pH range 7–11 reveals a near-edge feature that titrates with pH (Fig. 1). The intensities of the double peak feature located at 9,659 and 9,664 eV at pH 7.2 invert at pH 10.6. Analysis of the data yields two distributions of atoms in the first coordination shell of ZnCPD at all pH values and at both 150 and 297 K (Zhang and Auld 1993). Direct comparison of the first and higher coordination shells of ZnCPD reveals structural differences between pH 7.2 and 10.6 that are principally reflected in the average inner coordination distance decreasing from 2.024 to 2.002 Å. The structural analyses of ZnCPD at intermediate pH values show that this distance decreases as a function of the increase in pH (Zhang and Auld 1993). The plot of the normalized spectral differences between the peaks located at 9,659 and 9,664 eV of the absorption edge as a function of pH conforms to a theoretical pH-titration curve with a $pK_a$ of 9.49 at −4°C (Fig. 2). This value corresponds to that obtained by extrapolation of the temperature-dependent value of $pK_{EH}$ obtained by kinetic analysis of the CPD-A-catalyzed hydrolysis of tripeptides (Auld and Vallee 1971) (Fig. 3). The 0.022 ± 0.006 Å average distance change for the four atoms is equivalent to a 0.09 ± 0.03 Å distance change for a single ligand, provided that the other three ligands remain at the same distance. The pH-dependent shorter metal-ligand distance can be attributed to ionization of the water coordinated to the catalytic zinc ion (Zhang and Auld 1993). Since the kinetics indicate that this ionization leads to weaker substrate binding, it is reasonable to assign $pK_{EH}$ in the kinetic profiles to the ionization of the metal-bound water (EH $\rightleftharpoons$ E, Fig. 4). The Zn-O distance is expected to be shorter for zinc-bound hydroxide and its displacement by substrates and inhibitors should be more difficult.

## XAFS Studies of Binary and Ternary Complexes

Examination of the effect of pH on the XAFS spectra of binary and ternary product complexes of ZnCPD provides further evidence for the role of unionized water in catalysis (Zhang and Auld 1995). Inhibition studies show that L-Phe, the amino acid product of the most rapidly turned over peptide substrates of CPD-A,

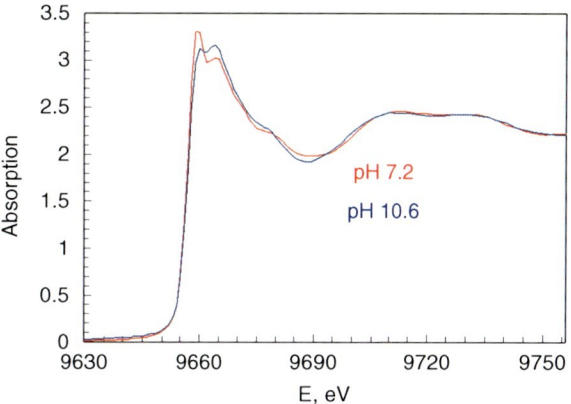

**Zinc, Visibility, Fig. 1** Zn K-edge XAFS spectra of carboxypeptidase A at pH 7.2 (*red line*) and 10.6 (*blue line*) at −4°C, the freezing point of the buffered solution. The double peak feature located at 9.659 and 9.664 KeV of the absorption edge can be used to follow pH-dependent changes in the zinc coordination sphere

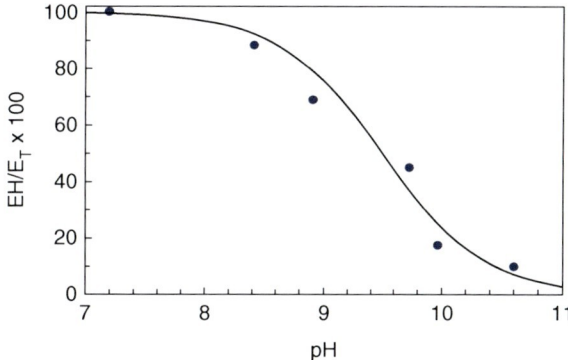

**Zinc, Visibility, Fig. 2** Plot of the normalized spectral difference between the peaks located at 9.659 and 9.664 KeV of the absorption edge (Fig. 1), compared with the theoretical titration curve with a $pK_a$ equal to 9.49. Experimental details are given elsewhere (Zhang and Auld 1993)

shifts $pK_{EH}$ from 9 to 7.5 at 25°C (Auld et al. 1986). Increasing the pH from 7 to 10 for the ZnCPD•L-Phe complex results in the same type of progressive spectral changes in the near-edge XAFS spectrum as are seen for ZnCPD, but the changes are complete more than 1 pH value below that observed for ZnCPD (Fig. 5). The XAFS results show that the average interatomic distance, R, for the zinc ligands of the ZnCPD•L-Phe complex decreases by 0.023 Å with an increase in pH, essentially identical to that obtained for the EH and E forms of the native enzyme (Zhang and Auld 1995).

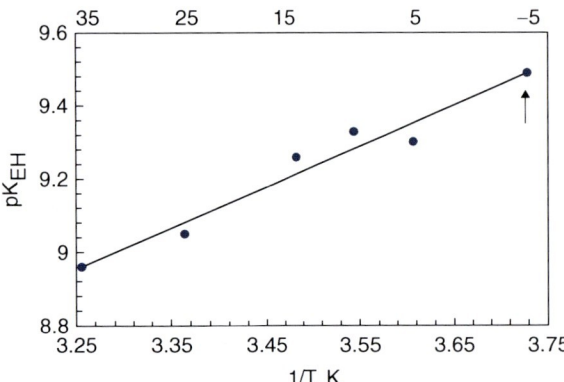

**Zinc, Visibility, Fig. 3** The temperature dependence of $pK_{EH}$ as determined for the carboxypeptidase A catalyzed hydrolysis of tripeptides (Auld and Vallee 1971). The value of $pK_{EH}$ from the XAFS titration curve in Fig. 2 is indicated by the *arrow*. At the top are temperature values in degrees Centigrade

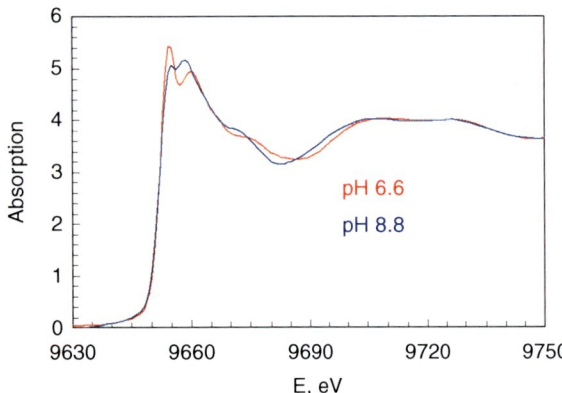

**Zinc, Visibility, Fig. 4** The active site ionizable groups critical to carboxypeptidase A catalysis

**Zinc, Visibility, Fig. 5** Zn K-edge X-ray absorption near-edge spectra for the binary complex of ZnCPD and L-Phe at pH 6.6 at 25°C (*red line*) and pH 8.8 (*blue line*). Experimental details are given in Zhang and Auld (1995)

The results give a structural basis for the kinetic studies which show that E binds the protonated form of L-Phe more tightly than EH, thus in effect decreasing the value of $pK_{EH}$ (Auld et al. 1986). The protonated amino group of L-Phe likely forms a salt bridge with the carboxylate of Glu-270 thus breaking up the hydrogen bond between the carboxylate and the metal-bound water (Fig. 4). While no X-ray crystallographic results have been reported for the L-Phe complex, they have for the D-Phe complex (Christianson et al. 1989). In this case, the α-amino group of D-Phe hydrogen bonds to the carboxylate of Glu-270 which no longer interacts with the metal-bound water. Similar interactions for L-Phe might, therefore, be anticipated. Since the zinc-bound water is now no longer H-bonded to the ionized carboxylate of Glu 270 it is free to ionize at a lower pH value. The apparent $pK_a$ of the zinc-bound water in this inhibitor complex is about 1 unit higher than that observed in carbonic anhydrase II (Kiefer et al. 1995). The higher $pK_a$ value for ZnCPD probably reflects the replacement of a His ligand in carbonic anhydrase by a Glu ligand in carboxypeptidase. The presence of the Glu ligand should reduce the charge on the zinc and, thus, make it more difficult for the zinc-bound water to ionize.

Addition of azide to the ZnCPD·L-Phe complex at pH 7 markedly changes the zinc coordination sphere from 4 N/O atoms at 2.024 Å and 1.3 N/O atoms at 2.54 Å to 3.9 N/O atoms at 1.995 Å (Zhang and Auld 1995). Examination of the higher coordination shell between 2.8 and 4.0 Å reveals that marked changes occur when azide is bound, providing further support for this assignment. This is the region where the second and third nitrogens of the zinc-bound azide would be expected to contribute to the XAFS spectrum. Both single and multiple scattering paths of these nitrogens can lead to a focusing effect that can account for the observed spectral changes. The decrease of about 0.03 Å in R for the ZnCPD·L-Phe·Azide complex is probably due to the ligand exchange from a neutral water to an anion. The XAFS spectra of the ternary complex are pH-independent from pH 7 to 9 (Fig. 6) in agreement with the ionization of the water being the source of the spectral changes in the free enzyme and its binary L-Phe complex. Since azide has displaced the water from the zinc coordination sphere no pH-dependent distance change is expected. The ZnCPD·L-Phe·azide complex is probably bound in a manner analogous to that expected for a post-transition state in a bi-product complex for peptide hydrolysis, that is, the carboxylate anion of the peptide bound to the Zn and the protonated form of L-Phe hydrogen bonded to the catalytic Glu-270 carboxylate

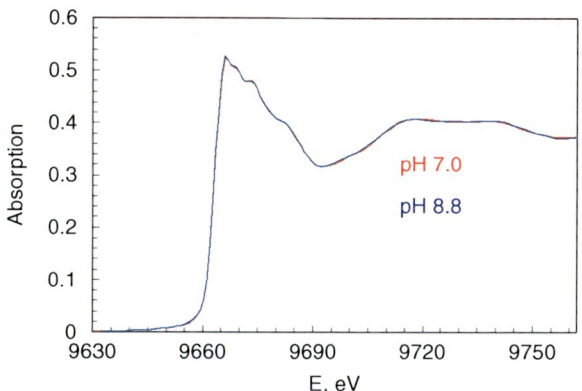

**Zinc, Visibility, Fig. 6** Zn K-edge XAFS spectra for the ternary complex of ZnCPD, L-Phe, and azide at pH 7.0 (*red line*) and 8.8 (*blue line*). Experimental details are given in Zhang and Auld (1995)

(Auld 1997). The XAFS results on the "spectroscopically silent" ZnCPD well complement the results of nuclear magnetic resonance and electron absorption studies on cobalt-substituted carboxypeptidase A (Auld 2004). These studies revealed the identity of the group responsible for the alkaline $pK_a$ in activity profiles and determined the structure of intermediates during catalysis. The results indicate that the zinc-bound water plays a central role in catalysis.

Dynamic information on zinc sites is now also possible through the integration of a low-temperature stop-flow instrument with the synchrotron beamline and a X-ray fluorescence detector system (Zhang et al. 2004). The capability of operating at sub-zero temperatures will lengthen the lifetime of reaction intermediates. Since XAFS can be performed on the metal environment in solution and solid states it has also been able to monitor the zinc coordination sites during oogenesis of *Xenopus laevis* (Auld et al. 1996).

## Cross-References

▶ Zinc, Fluorescent Sensors as Molecular Probes
▶ Zinc Metallocarboxypeptidases
▶ Zinc-Binding Sites in Proteins

## References

Auld DS (1997) Zinc catalysis in metalloproteases. Struct Bond 89:29–50
Auld DS (2004) Catalytic mechanisms for metallopeptidases. In: Barrett AJ, Rawlings ND, Woessner JF (eds) Handbook of proteolytic enzymes. Academic, London, pp 268–289
Auld DS, Vallee BL (1971) Kinetics of carboxypeptidase A, pH and temperature dependence of tripeptide hydrolysis. Biochemistry 10:2892–2897
Auld DS, Larsen K, Vallee BL (1986) Active site residues of carboxypeptidase A. In: Bertini I, Luchinat C, Maret W, Zeppezauer M (eds) Zinc enzymes. Birkhauser, Boston, pp 133–154
Auld DS, Falchuk KH, Zhang K, Montorzi M, Vallee BL (1996) X-ray absorption fine structure as a monitor of zinc coordination sites during oogenesis of *Xenopus laevis*. Proc Natl Acad Sci USA 93:3227–3231
Christianson DW, Mangani S, Shoham G, Lipscomb WN (1989) Binding of D-phenylalanine and D-tyrosine to carboxypeptidase A. J Biol Chem 264:12849–12853
Kiefer LL, Paterno SA, Fierke CA (1995) Hydrogen bond network in the metal binding site of carbonic anhydrase enhances zinc affinity and catalytic efficiency. J Am Chem Soc 117:6831–6837
Stern EA, Heald SM (1983) Basic principles and applications of XAFS. In: Koch EE (ed) Handbook on synchrotron radiation. North-Holland, Amsterdam/New York, pp 995–1014
Zhang K, Auld DS (1993) XAFS studies of carboxypeptidase A: detection of a structural alteration in the zinc coordination sphere coupled to the catalytically important alkaline pKa. Biochemistry 32:13844–13851
Zhang K, Auld DS (1995) Structure of binary and ternary complexes of zinc and cobalt carboxypeptidase A as determined by X-ray absorption fine structure. Biochemistry 34:16306–16312
Zhang K, Liu R, Irving T, Auld DS (2004) A versatile rapid-mixing and flow device for X-ray absorption spectroscopy. J Synchrotron Radiat 11:204–208

# Zinc-Astacins

Walter Stöcker
Johannes Gutenberg University Mainz, Institute of Zoology, Cell and Matrix Biology, Mainz, Germany

## Synonyms

Astacin; *Astacus* protease; Crayfish small-molecule proteinase; Low molecular weight protease

## Definition

Astacin is the prototype of a family of extracellular zinc peptidases termed the "astacin family" and of the metzincin superfamily, which combines the matrix metalloproteinases (MMPs), the bacterial serralysins, the adamalysins/reprolysins/ADAMs (a disintegrin and metalloprotease), the pappalysins, and other zinc proteases. Astacins are found in animals and in microorganisms. All are secreted to the extracellular space or stay membrane bound at the cell surface. Their key feature is a catalytic domain of about 200 amino acid residues in a typical fold including a five-stranded β-sheet and two prominent α-helices. The zinc-binding moiety comprises three imidazole side chains from histidines, a glutamic acid-bound water molecule, and a tyrosine, which is part of a unique methionine containing tight β-turn, the Met-turn, beneath the active site.

## The Astacin Family of Zinc Proteases

The digestive protease astacin was first purified in the late 1960s from the European noble crayfish, *Astacus astacus* L., and originally termed "low molecular protease" or "*Astacus* protease." In the 1980s, amino acid sequencing and metal analysis revealed the *Astacus* protease as the orphan prototype of a new family of zinc peptidases. The second family member to be discovered was bone morphogenetic protein 1 (BMP1), a protease co-purified with TGFβ-like growth factors also termed bone morphogenetic proteins due to their capability to induce ectopic bone formation in mice. In the following years, a variety of other related proteases were reported, most of them multi-domain proteins, all containing a proteolytic domain similar to the *Astacus* protease. Therefore, this group of proteolytic enzymes was designated "astacin family."

For a more detailed view, the reader is referred to the MEROPS protease database (http://merops.sanger.ac.uk), which compiles astacin proteases in the M12A family. (For reviews, see also Stöcker et al. 1993; Bond and Beynon 1995; Stöcker et al. 1995; Sterchi et al. 2008; Gomis-Rüth 2009; Muir and Greenspan 2011; Gomis-Rüth et al. 2012).

## Distribution and Physiological Role of Astacins

The astacin gene of the crayfish *Astacus astacus* spans 2,616 bp and comprises five exons, four introns, and a TATA-box in the promoter region (EMBL genomic database: X95684). *Astacus* astacin is translated as a pre-pro-protein comprising 251 residues including a signal sequence of 15 and a prodomain of 34 residues, respectively (EMBL cDNA database: AJ242595; UniProt: P07584).

Mature, proteolytically active crayfish astacin is a single-domain protein. Most other members of the astacin family have a multi-modular structure including one astacin-like catalytic domain, which exhibits at least about 30% sequence identity with the crayfish enzyme.

There are six mammalian astacin genes (http://degradome.uniovi.es/met.html). Three of these form the tolloid subgroup *bmp1*, *tll1*, and *tll2*, coding for the procollagen C-proteinase, also termed BMP1 (i.e., bone morphogenetic protein 1) and the mammalian tolloid-like proteins (Tll1, Tll2). Two, *mep1a* and *mep1b*, encode the meprin subunits α and β and one, named *astl*, encodes ovastacin.

BMP1-, Tll1-, and Tll2-deficient mice suffer from impaired development of connective tissue and heart and skeletal muscle because these enzymes are crucial for correct processing of fibrillar procollagens during extracellular matrix construction. They also cleave precursors of many other matrix proteins such as proteoglycans, laminins, and anchoring fibrils. Furthermore, tolloids are important for proteolytic trimming of growth factors and their antagonists, involved in embryonic body axis establishment (Muir and Greenspan 2011). This developmentally important group of enzymes is present in all animal phyla. *Drosophila* tolloid – the prototype if this subgroup – was named after a mutation in the fruit fly, which is lethal in the embryo due to a severe defect in dorso-ventral patterning.

Meprin-α and meprin-β have so far been detected only in vertebrates. They are translated as membrane-bound multi-domain proteins. The meprin subunits have different fates due to different proteolytic processing. While meprin-β homodimers and α/β heterodimers stay mostly cell-surface bound, meprin-α is proteolytically cleaved within the so-called inserted domain on the secretory pathway and forms large soluble multimers of mega Dalton size. Meprins cleave gastrin, cholecystokinin, substance P, cytokines, and chemokines important for tissue differentiation and signaling (Sterchi et al. 2008). Other meprin substrates are components of the basal lamina, procollagens, and cell-adhesion proteins. In vivo proteomics of physiologically relevant target substrates unraveled the vascular endothelial growth factor, amyloid precursor protein, procollagens I and III, interleukin-1β, interleukin-18, prokallikrein-7, and fibroblast growth factor 19.

The so-called hatching enzymes comprise an important subgroup of astacins observed in both invertebrates and vertebrates. They are present in a variety of different types, especially in water living organisms, whose early embryogenesis takes place outside of the maternal body. Their function is the degradation of egg envelopes during the hatching process, which releases the first larval stages or young adults into the ambient water. Examples are crayfish embryonic astacin, fish alveolin, and low and high choreolytic enzymes (LCE, HCE), frog UVS.2, and bird CAM1. In mammals, there is only a single astacin related to hatching enzymes. This protein has been named ovastacin, since it is expressed predominantly in the developing oocyte and in the early embryo. Recent reports suggest ovastacin to be involved in a proteolytic network responsible for preventing polyspermy upon fertilization by hardening of the egg envelope through cleavage of zona pellucida protein 2 (Burkart et al. 2012). Proteolytic networks regulating egg development and fertilization have also been detected in other species, e.g., in the fruit fly *Drosophila* (LaFlamme et al. 2012).

Other vertebrates and invertebrates generally contain considerably more astacin genes than mammals. For example, there are about 18, 25, and 40 in fishes, insects, and nematodes, respectively. This may partly be owed to the parasitic nature of some of these organisms, which utilize proteases to degrade the connective tissue of their hosts.

## Domain Structure of Astacins

The smallest astacins are found in bacteria and consist of a blank protease domain of approximately 200 amino acid residues. Eukaryotic astacins additionally have signal peptides and propeptides amino-terminally of the protease domain, rendering the proenzymes secreted and inactive, respectively. A variety of different domains (or modules) have been identified in astacins carboxy-terminally (i.e., downstream) of the protease domain (Fig. 1).

Typical domains are EGF-like modules (epidermal growth factor-like) and/or one or more copies of CUB modules (found in complement proteins C1r and C1s, sea urchin uEGF, and BMP1). In the tolloid-like astacins, CUB and EGF domains seem to be important for selective substrate recognition, additionally triggered by calcium ions. In other developmental astacins such as sea urchin proteins SPAN and BP10, unique serine/threonine-rich regions may serve as potential targets of *O*-glycosylation. Several *Caenorhabditis elegans* astacins contain thrombospondin type 1 repeats (TSP1), which are also found in the ADAMTS peptidases (soluble ADAMs containing thrombospondin repeats), yet another family of metzincins. Some cnidarian and nematode astacins contain ShK toxin domains, which were originally identified in metridin, a potassium channel-blocking toxin from the sea anemone *Stichodactyla helianthus*. Other important domains are TRAF (tumor necrosis factor receptor-associated factor) and MAM (meprin, A5 receptor protein, tyrosine phosphatase-μ), both found in meprins and MAM alone in HMP2 from hydra, LAST-MAM from the horseshoe crab and so-called myosinases from squid. Meprins are composed of prodomains, astacin domains, MAM domains (meprin, A5 protein, and receptor protein tyrosine phosphatase-μ), TRAF domains (tumor necrosis factor receptor-associated factor), and EGF-like, transmembrane, and cytosolic domains. Meprin-α and meprin-β differ in the presence (α) or absence (β) of an additional inserted domain (I), which is cleaved on the secretory transport from the endoplasmic reticulum via the Golgi to the cell membrane. Consequently, meprin-α, but not meprin-β, is constitutively secreted. Furthermore, regions of generally low compositional complexity and similarity to other protein modules have been discovered and termed LH domains. Such regions have been observed in prokaryote astacins, in *C. elegans* astacins and in the sea urchin astacins SPAN/BP10. Moreover, mouse and human ovastacins contain a distinct C-terminal domain of approximately 150 residues with little similarity to other reported proteins (Stöcker et al. 1995; Gomis-Rüth et al. 2012).

## Structure of the Catalytic Domain of Astacins

Several X-ray crystal structures of mature astacin catalytic domains have been solved to atomic

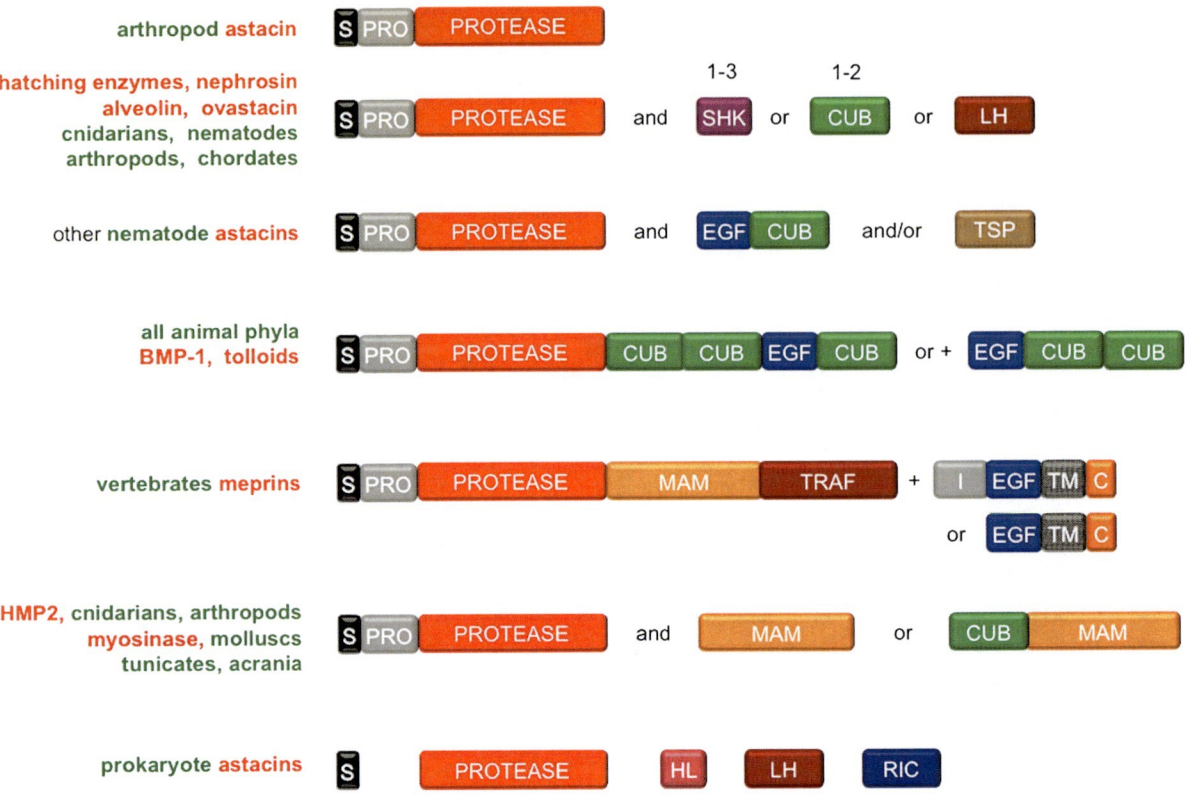

**Zinc-Astacins, Fig. 1** *Most significant protein domains observed in astacin proteases.* S signal peptide, PRO prodomain, PROTEASE astacin-like catalytic domain, PFAM accession number: PF01400, TM trans membrane, C cytoplasmic, I inserted, EGF epidermal growth factor-like, PF00008; CUB complement C1r/1s – sea urchin Uegf – BMP1, PF00431, TSP thrombospondin type 1 repeat, PF00090, ShK *Stichodactyla helianthus* potassium (K) channel toxin, PF01549, MAM meprin, A5 receptor protein, tyrosine phosphatase μ, PF00629, TRAF tumor necrosis factor receptor-associated factor, PF00917, HL H-type-Lectin, PF09458, LH low homology, RIC ricin B-type lectin, PF00652

resolution, including crayfish astacin, human BMP1, human Tll1, and zebrafish hatching enzyme ZHE1. Astacin catalytic domains are kidney shaped and divided into two sub-modules of almost equal size by a deep cleft. The upper amino-terminal part (standard view, "astacin" terminology) is formed by a five-stranded ß-sheet comprising four parallel strands (I, II, III, V), one antiparallel strand (IV), and two long α-helices (helices A and B) (Fig. 2).

The lower sub-domain is folded less regularly except of a short $3^{10}$-helix (helix C) and the longer helix D. The latter is attached to the amino-terminal sub-domain via the disulfide bond C42-C198 (Figs. 2 and 3). The second disulfide bond C64-C84 fixes β-strand IV to the loop between strand V and helix B. The sequential arrangement of consensus motifs and conserved cysteines is depicted in Fig. 3. The alignment with astacins of unknown structure suggests these two disulfide bonds to be conserved throughout the astacin family. In addition, selected members may show additional SS-bridges. ZHE1 (DRE_HCE in Fig. 3, a zebrafish hatching enzyme) shows a cross-link between two additional cysteine residues close to the amino terminus of the catalytic domain (C5-C10 according to astacin numbering; see Protein Data Bank (PDB) access code 3LQB). On the other hand, BMP1 and TLL1 contain an additional pair of cysteines in the edge strand IV. This causes rearrangement of the disulfides by formation of the new links C64-C65 and C62-C85 according to astacin numbering; (PDB 3EDH; Fig. 3). This cysteine-rich loop is unique for the tolloid subfamily (Fig. 3) and should influence enzyme substrate interactions significantly.

The catalytic zinc is located in the basement of the active site cleft in the center of the molecule (Figs. 2 and 4). The metal is coordinated by three histidine

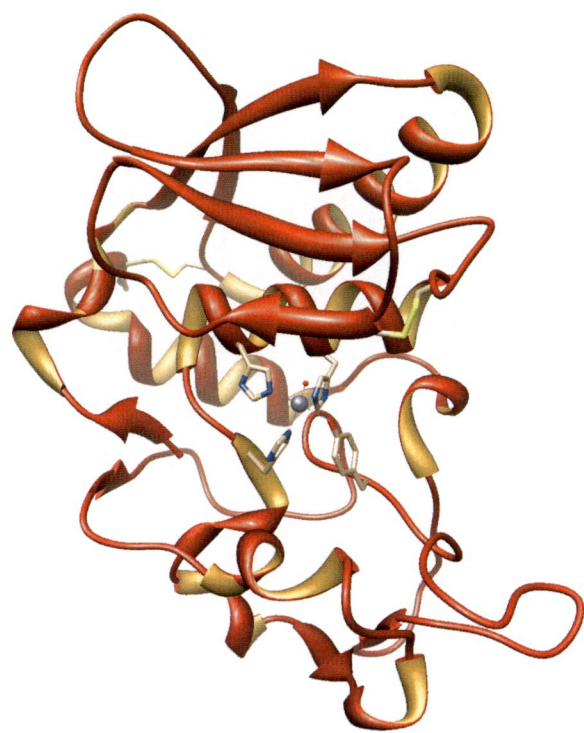

**Zinc-Astacins, Fig. 2** *Ribbon representation of mature crayfish astacin* based on the X-ray crystal structure (PDB accession code: 1AST; modeled with CHIMERA, http://www.cgl.ucsf.edu/chimera/). Side chains shown include the three zinc-binding histidines, Tyr149 of the Met-turn, and two conserved disulfide bonds between Cys42-Cys198 and Cys64-Cys85. The zinc-bound water is symbolized by a *red dot*

residues, a tyrosine residue and a water molecule, which is hydrogen bonded by E93 and Y149, resulting in a coordination sphere with H96Nε and the phenolic oxygen of Y149 at the vertices of a trigonal bipyramid and H92Nε and H102Nε and the water molecule Sol300 forming the trigonal plane in the center (Fig. 4). H92 and H96 are separated by a single turn of helix B (Fig. 4) at distances of 2.0 Å and 2.2 Å removed from the metal, respectively. The third zinc ligand, H102 (2.0 Å), follows six residues downstream of the second, which is facilitated by G99, which breaks helix B and enables the chain to bend sharply (Figs. 2 and 4). The fourth and the fifth zinc ligands are provided by the hydroxyl oxygen of Y149 and a water molecule at distances of 2.5 Å and 2.0 Å removed from the metal, respectively (Fig. 4).

The imidazole zinc ligands and the catalytically important glutamic acid are combined in the distinguishing consensus motif HEXXHXXGXXH (single letter code, X is any amino acid residue), seen in all astacins and almost any other metzincins (Fig. 3). At a distinct distance downstream of the third histidine, the polypeptide chain again returns to the zinc-binding site, thereby generating another key consensus motif, the so-called Met-turn with the underlying consensus sequence SXMHY within the astacin family (Figs. 3 and 4). This 1,4-β-turn provides the fifth zinc ligand, Y149, and the eponymous methionine residue. The tyrosine ligand is only present in the astacins and serralysins, not in other metzincins. Four strands of the β-pleated sheet and long helices A and B of the amino-terminal sub-domain of the metzincin superfamily exhibit topological similarity to the archetypical zinc-proteinase thermolysin. This includes the central helix B, whose counterpart in thermolysin carries two of the zinc-binding histidines in the more general consensus motif HEXXH, shared by the majority of metalloproteases of the zincin group.

## Metal Derivatives of Astacin

Astacin preparations purified from the natural source have single zinc per catalytic domain as determined by atomic absorption spectrometry; further metals or cofactors were not detectable. Other astacins may have additional zinc ions in their catalytic domain and calcium ions in their C-terminal EGF-like domains.

Zinc-free apo-astacin can be produced by dialysis against neutral buffers containing metal chelating agents like 1,10-phenanthroline. Full catalytic activity (100%) is restored by stoichiometric amounts of zinc (II)-ions. Also copper(II)-ions (37%) and cobalt(II)-ions (140%), but not mercury(II), nickel(II), or calcium(II) (among others), reactivate the apo-enzyme. X-ray crystal structures of native zinc(II)-, copper(II)-, cobalt(II)-, mercury(II)-, nickel(II)-, and apo-astacin were solved to resolutions of 1.8 Å (PDB 1AST), 1.90 Å (PDB 1IAA), 1.79 Å (PDB 1IAB), 2.10 Å (PDB 1IAC), 1.83 Å (PDB 1IAD), and 2.30 Å (PDB 1IAE), respectively. They exhibit an identical overall framework, but the metal coordination in their active sites differ. The active copper astacin and cobalt astacin contain trigonal bipyramidally coordinated metal as it is seen in the zinc enzyme, whereas mercury astacin has tetrahedrally coordinated metal and the nickel derivative is octahedrally coordinated.

| | Astacin | Asp | | | | Zinc | | Met | S1' | |
|---|---|---|---|---|---|---|---|---|---|---|
| AAS_ASTA | FEG**D**IKL...ARVG | **A**AILGDEY---LWS... | **C**... | SG**C**W... | **C**... | **HELM**HAIGFY**H**EHTRMDRD... | SI**M**HY... | DP-YDK... | D... | **C** |
| HSA_MEPα | FQG**D**ILL...QKSR | **N**GLRDPNT---RWT... | **C**... | DG**C**W... | **C**... | **HEIL**HALGFY**H**EQSRTDRD... | SL**M**HY... | IG---QR... | D... | **C** |
| HSA_MEPβ | FEG**D**IRL...AQIR | **N**SIIGEKY---RWP... | **C**... | SG**C**W... | **C**... | **HEFL**HALGFY**H**EQSRSDRD... | SV**M**HY... | IG---QR... | D... | **C** |
| HVU_HMP2 | FGG**D**ILL...RQKR | **A**ALSNSSI---LWL... | **C**... | SG**C**W... | **C**... | **HEIG**HAMGFY**H**EQNRPDRD... | SY**M**QY... | IG---QR... | D... | **C** |
| LPO_ASTM | FEG**D**IAG...YADK | **A**IVDHTL---LWP... | **C**... | DG**C**W... | **C**... | **HELG**HAVGFW**H**EQNRADRD... | SV**M**LY... | IGPVWK... | D... | **C** |
| CEL_NAS13 | FEG**D**IAN...GVQR | **N**AVRQTYL---KWE... | **C**... | DG**C**Y... | **C**... | **HELM**HAVGFF**H**EQSRADRD... | SV**M**HY... | IG---QR... | D... | **C** |
| TPA_MYO1 | FQG**D**IEM...PFSR | **S**VVGDLNK---RWP... | **C**... | TG**C**R... | **C**... | **HEVL**HTLGFY**H**EQSRPDRD... | SL**M**HY... | IG---QR... | D... | **C** |
| ATE_TLL | FVG**D**IAL...RTKR | **A**ATARPER---LWD... | **C**... | **C**G**CC**... | **C**... | **HELG**HVVGFW**H**EHTRPDRD... | SI**M**HY... | IG---QR... | D... | **C** |
| HSA_BMP1 | FLG**D**IAL...RSRR | **A**ATSRPER---VWP... | **C**... | **C**G**CC**... | **C**... | **HEVL**HVVGFW**H**EHSTRPDRD... | SI**M**HY... | IG---QR... | D... | **C** |
| SPU_SPAN | FQG**D**MML...KKRK | **A**TIYESQR---WS... | **C**... | NG**C**W... | **C**... | **HEIG**HAIGFH**H**EQSRPDRD... | SI**M**HY... | LG---QR... | D... | **C** |
| OLA_HCE1 | LEG**D**LVA...T-NR | **N**AMK**C**WSSS**C**FWK... | **C**... | TG**C**Y... | **C**... | **HELN**HALGFQ**H**EQTRSDRD... | SI**M**HY... | IG---QR... | D... | **C** |
| DRE_HCE | FEG**D**VVL...K-NR | **N**ALI**C**EDKS**C**FWK... | **C**... | DG**C**Y... | **C**... | **HELN**HALGFY**H**EQSRSDRD... | SL**M**HY... | IG---QR... | D... | **C** |
| HSA_OVAST | IEG**D**IIR...SPFR | **L**LSAASNK---WP... | **C**... | YG**C**F... | **C**... | **HELM**HVLGFW**H**EHTRADRD... | SV**M**HY... | IG---QR... | D... | **C** |
| | -21 | 1 | 10 | 42 | 64 | 85 92 96 102 | 149 | | | 198 |

**Zinc-Astacins, Fig. 3** *Astacin consensus sequences.* The *left column* lists selected astacins: AAS_ASTA (UniProt Accession P07584), crayfish astacin; HSA_MEPα (Q16819) human meprin-α; HSA_MEPβ (Q16820) human meprin-β; HVU_HMP2 (Q9XZG0) from hydra; LPO_ASTM (B4F320) from horse shoe crab; CEL_NAS13 (Q20191) from nematode; TPA_MYOI (Q8IU46) from squid; ATE_TLL (Q75UQ6) from spider; HSA_BMP1 (P13497) human BMP1; SPU_SPAN (P98068); OLA_HCE1 (P31580) from medaka fish; DRE_HCE (=ZHE1) (Q75NR9) from zebrafish; HSA_OVAST (Q6HA08) human ovastacin. Residues (in single letter code) are numbered in the *bottom line* according to their position in mature crayfish astacin, starting with A1 of the mature active protease. The preceding propeptide is numbered with negative sign starting at G-1. *Dots* indicate omitted residues. *Asp* aspartate switch region in the propeptide, with the zinc ligand D-21; *light blue*: mature amino terminus and conserved E103, forming a water-mediated salt bridge; *yellow*: cysteine residues; *white*: zinc binding residues; *green*: methionine of the Met-turn; *pink*: residues forming $S_1'$ subsite (nomenclature according to Schechter and Berger)

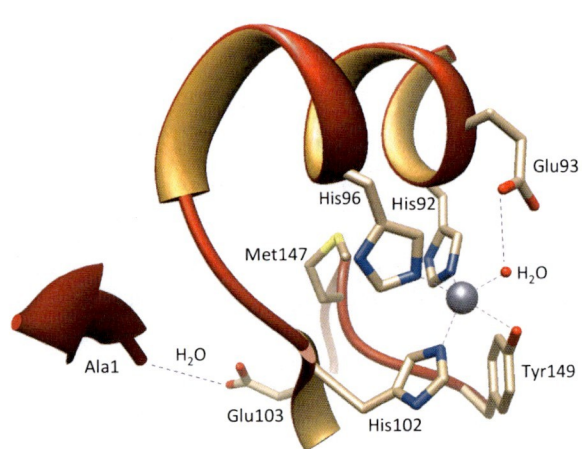

**Zinc-Astacins, Fig. 4** *Zinc-binding region of astacin.* Shown are the zinc ligands, the catalytically indispensable E93, the methionine residue of the Met-turn backing the zinc site, and E103 engaged in a water-mediated salt bridge to the amino-terminal alanine. Modeling of PDB 1AST was performed with CHIMERA (http://www.cgl.ucsf.edu/chimera/)

UV-VIS and electron paramagnetic resonance spectra of copper and cobalt astacins show similar structures as the corresponding metal derivatives of carboxypeptidase or thermolysin. There is one significant difference, however, regarding a strong phenol charge-transfer band in copper astacins at 445 nm ($\varepsilon = 1,900$ M$^{-1}$ cm$^{-1}$) and an additional band at 325 nm ($\varepsilon = 1,600$ M$^{-1}$ cm$^{-1}$), due to the additional tyrosine metal ligand of astacin. Interestingly, the intense charge-transfer band at 445 nm is abolished upon inhibitor binding, which spectroscopically visualizes the "tyrosine switch" during catalysis (see below). Interestingly, a copper(II)-acetate complex of the heteroscorpionate ligand (2-hydroxyl-3-*t*-butyl-methylphenyl)bis(3,5-dimethylpyrazolyl)methane, synthesized as a model of the trigonal bipyramidal zinc coordination seen in astacin and serralysin, reproduces these spectroscopic features.

## Amino-Terminal Water Cavity in Mature Astacins

A unique structural feature of mature astacin is a water-filled cavity, which harbors the two amino-terminal residues A1 and A2 (Figs. 2 and 4). Ala1 is linked via a water-mediated salt bridge to the carboxylate of E103, neighboring zinc-ligand H102. Although reminiscent to the activation mechanism of trypsin-like proteases, this salt bridge seems to be not essential to gain catalytic activity, since site-directed mutagenesis of E103 in astacin and the corresponding residue in mouse meprin-α did not abolish activity of the protease, but rather rendered the enzyme extremely unstable. Hence, the conserved salt bridge seems to provide structural stability rather than catalytic function. Most remarkable is the location of the amino terminus in the hatching enzymes, which like ZHE1 contain an additional disulfide bond in the N-terminal segment (see above). In these enzymes, conserved residues of the helix D seem to be involved in structural stabilization.

## Structure of Pro-astacin and Zymogen Activation

Eukaryotic astacins are synthesized as inactive zymogens (pro-enzymes), which require proteolytic removal of an N-terminal segment to gain full catalytic activity. Although the pro-segments of astacins vary considerably in length (from 34 to 393 residues), they have in common a conserved aspartate residue embedded in a short consensus motif FXGDU (single letter code, U = unipolar residue) (Fig. 3). In the known structure of pro-astacin from crayfish (Fig. 5), the side chain carboxylate of this aspartate bound to the zinc ion, thereby blocking access of substrates to the active site. The pro-segment runs across in opposite orientation compared to a substrate, thus preventing self-cleavage. This "aspartate switch" of astacin proteases resembles the "cysteine switch" of MMPs and ADAMs. However, removal of the pro-segment induces a large conformational rearrangement of a so-called activation domain between residues D129 and G138, which undergoes a flap-like upward movement in order to close the gap left open by the removed propeptide.

The activation of pro-astacin is brought about by a two-step proteolysis. In the first step, catalyzed by a trypsin-like protease, intermediate species are generated starting at V-2. Subsequent trimming to the mature amino terminus A1 is done by astacin itself (Fig. 3). In many other pro-astacins like meprin proteases, the residue in position-1 is a basic residue, and activation can directly be achieved by trypsin-like enzymes such as kallikreins, plasmin, or trypsin. Others like the BMP1/tolloid-like astacins contain furin activation cleavage sites and can be activated constitutively on the secretory pathway (see Fig. 3).

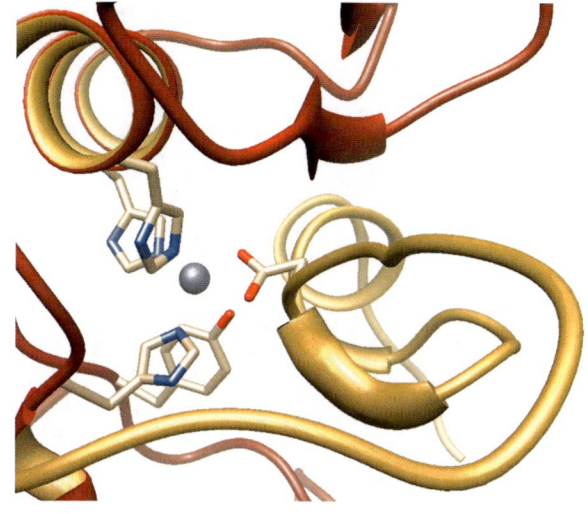

**Zinc-Astacins, Fig. 5** *Ribbon representation of the zinc-binding region of inactive pro-astacin, the zymogen form of crayfish astacin based on the X-ray crystal structure (PDB accession code: 3LQ0; modeled with CHIMERA, http://www.cgl.ucsf.edu/chimera/). The propeptide (yellow) fills the active site cleft and provides D-21 as an additional zinc ligand, which prevents activity*

## Catalytic Function and Substrate Specificity

The pH optimum of astacins ranges between pH 6 and pH 8, depending on the substrate. Several chromogenic and fluorescent substrates have been designed and utilized for characterizing astacin proteases. In addition, native and denatured physiological substrates have been employed including proteins of the extracellular matrix (Stöcker and Zwilling 1995).

It has been suggested that the catalytic mechanism of astacin zinc peptidases relies on the polarization of the metal-bound water between the zinc(II)-ion, acting as a Lewis-acid, and the glutamic acid residue (E93), acting as the general base (Gomis-Rüth et al. 2012). However, the catalytic water is also bound to Y149 (Figs. 2, 4, 6). To clarify the roles of E93 and Y149, both residues were mutated. Replacement of Y149 by phenylalanine still retained weak activity, whereas the E93A mutant was completely inactive (review: Gomis-Rüth et al. 2012). Hence, E93 of astacin most likely acts as general base and as a proton shuttle to the leaving amino group during catalysis in analogy to E143 of thermolysin or E270 of carboxypeptidase A. Y149 on the other hand seems to stabilize the transition state similar to H231 in thermolysin and Y248 of carboxypeptidase A.

**Zinc-Astacins, Fig. 6** *Complex of astacin with a transition state analog inhibitor.* Astacin complexed with the phosphinic pseudopeptide BOC-PKRΨ(PO$_2$CH$_2$)AP-OCH$_3$ (BOC = benzoyl-oxycarbonyl) ($K_i$ = 14 μM). Crystals were obtained by soaking native crystals in a 4.0-mM solution of the inhibitor. Shown are the inhibitor (*in gray*), the zinc-binding imidazoles, Y149 (hydrogen bonding the lower phosphinyl oxygen), and the catalytic base E93. The metal-chelating phosphinyl group adopts the geometry of a tetrahedrally coordinated carbon during peptide bond cleavage (PDB accession code: 3QJI; modeled with CHIMERA, http://www.cgl.ucsf.edu/chimera/)

The specific interactions of astacin with the transition state analog inhibitor BOC-PKRΨ(PO$_2$CH$_2$)AP-OCH$_3$ have allowed to get insight into the catalytic mechanism of astacin-like enzymes (Fig. 6). In this inhibitor, the scissile peptide bond is replaced by a phosphinic pseudopeptide bond Ψ(PO$_2$-CH$_2$), which is resistant to enzymatic hydrolysis. However, the phosphinyl group of the inhibitor PO$_2$ mimics the tetrahedrally coordinated carbon of the high-energy transition state during peptide bond cleavage.

Astacins bind their substrates in extended conformation by antiparallel attachment to the upper-rim β-strand IV through inter-main-chain interactions (Figs. 2 and 6). The inhibitor side chains amino-terminally of the pseudopeptide are hydrogen bonded via backbone carbonyl and amide groups and the proline ring slots into a niche of the β-edge strand IV (Fig. 6). The two phosphinyl oxygens form a *gem*-diol chelating complex with the zinc ion. Remarkably, upon binding of the transition state analog inhibitor, the tyrosine side chain moves away from the zinc into a new position, about 5.0 Å removed from the metal, and becomes hydrogen bonded with the PO$_2$ group, which mimics a water-attacked peptide bond (Fig. 6). This "tyrosine switch" is a unique feature of the astacin-like proteinases and also of the serralysins.

The side chain carboxy-terminally of the cleavage site is an alanine methyl group in BOC-PKRΨ(PO$_2$CH$_2$)AP-OCH$_3$ (Fig. 6). In zinc endopeptidases, this is the most important side chain since it points directly into the active site cleft. In the nomenclature of Schechter and Berger, this is the P$_1'$ position of the substrate, which is harbored in the corresponding S$_1'$ subsite of the enzyme. Most interestingly, 90% of all astacins deposited in the MEROPS database share a similar S$_1'$ region. The corresponding consensus motif is underlaid pink in Fig. 3. The motif contains a conserved arginine shaping the bottom of a deep S$_1'$-pocket in the known structures of ZHE1, TLL1, and BMP1. Crayfish astacin is an exception in this respect, with a rather shallow pocket shaped by P176. This explains why most family members, including ZHE1, BMP1, TLL1, meprin-α and meprin-β, and horseshoe-crab LPO_MAM, prefer aspartate residues in P$_1'$, while astacin prefers small aliphatic residues. A comprehensive proteomic approach has shown that many astacins indeed prefer aspartate in the P$_1'$ position of their substrates. This preference is even more pronounced in meprin-β and tolloids (e.g., BMP1), whose S$_2'$ subsite additionally contributes to the binding of acidic substrate side chains (Gomis-Rüth et al. 2012).

## Protein Inhibitors

Astacins are not inhibited by the tissue inhibitors of metalloproteases (TIMPs), presumably due to the limited accessibility of their active site cleft, which is much narrower than in MMPs and ADAMs. The most effective general antagonist of astacins is the plasma protein α$_2$-macroglobulin. However, α$_2$-macroglobulin will only entrap endopeptidases of limited size, like crayfish astacin or the tolloid protease BMP1. Large oligomeric astacins, such as the meprins, are resistant to inhibition by this regulator of vascular and interstitial proteolysis. Interestingly, another vertebrate protein named fetuin A has recently been discovered as a specific inhibitor of astacins. Fetuins are large plasma proteins with manifold functions. They regulate calcification and they contain cystatin-like domains, which are inhibitors of cysteine cathepsins. Both fetuin A and cystatins have been shown to act as cross-class protease inhibitors, being active against cysteine proteases, serine proteases, as well as metalloproteases (Gomis-Rüth et al. 2012).

## Conclusive Remarks

Astacin proteases have emerged recently as a family of extracellular metalloproteases occurring throughout the animal kingdom and also in microorganisms. They are important new members of the complex protease web on the cell surface and in the extracellular space. Astacins are involved in embryonic development, tissue differentiation, extracellular matrix assembly, and transcellular signaling. Aside of transcriptional and translational control, their regulation relies on posttranslational zymogen activation and antagonizing protein inhibitors. Their unique cleavage specificity for acidic residues in $P_1'$ and $P_2'$ links these proteases to Alzheimer's disease, connective tissue disorders, and fibrosis and might be a clue for the design of specific small-molecule inhibitors.

## Cross-References

▸ Cadherins
▸ Calcium in Biological Systems
▸ Thermolysin
▸ Zinc Cellular Homeostasis
▸ Zinc Homeostasis, Whole Body
▸ Zinc in Hemostasis
▸ Zinc Matrix Metalloproteinases and TIMPs
▸ Zinc Meprins
▸ Zinc Metallocarboxypeptidases
▸ Zinc-Binding Proteins, Abundance
▸ Zinc-Binding Sites in Proteins

## References

Bond J, Beynon R (1995) The astacin family of metalloendopeptidases. Protein Sci 4:1247–1261
Burkart AD, Xiong B, Baibakov B et al (2012) Ovastacin, a cortical granule protease, cleaves ZP2 in the zona pellucida to prevent polyspermy. J Cell Biol 197:37–44
Gomis-Rüth FX (2009) Catalytic domain architecture of metzincin metalloproteases. J Biol Chem 284:15353–15357
Gomis-Rüth FX, Trillo-Muyo S, Stöcker W (2012) Functional and structural insights into astacin metallopeptidases. Biol Chem, published online 24/04/2012, DOI: 10.1515/bc-2012-0149
LaFlamme BA, Ravi Ram K, Wolfner MF (2012) The drosophila melanogaster seminal fluid protease "seminase" regulates proteolytic and post-mating reproductive processes. PLoS Genet 8:e1002435
Muir A, Greenspan DS (2011) Metalloproteinases in drosophila to humans that are central players in developmental processes. J Biol Chem 286:41905–41911
Sterchi EE, Stöcker W, Bond JS (2008) Meprins, membrane-bound and secreted astacin metalloproteinases. Mol Aspects Med 29:309–328
Stöcker W, Zwilling R (1995) Astacin. Meth Enzymol 248:305–325
Stöcker W, Gomis-Rüth FX, Bode W et al (1993) Implications of the three-dimensional structure of astacin for the structure and function of the astacin family of zinc-endopeptidases. Eur J Biochem 214:215–231
Stöcker W, Grams F, Baumann U et al (1995) The metzincins–topological and sequential relations between the astacins, adamalysins, serralysins, and matrixins (collagenases) define a superfamily of zinc-peptidases. Protein Sci 4:823–840

# Zinc-Binding Proteins, Abundance

Claudia Andreini[1,2] and Ivano Bertini[1,2]
[1]Magnetic Resonance Center (CERM) – University of Florence, Sesto Fiorentino, Italy
[2]Department of Chemistry, University of Florence, Sesto Fiorentino, Italy

## Synonyms

Bioinformatics; *In silico* prediction; Metalloproteomes; Zinc content

## Definition

The essential physiological role of zinc mainly depends on its widespread association with many different proteins, whose number can be estimated by bioinformatics approaches. The zinc proteome represents about 9% of the entire proteome in eukaryotes, but it ranges from 5% to 6% in prokaryotes, therefore indicating a substantial increase of the number of zinc proteins in higher organisms.

Zinc is indispensable to all forms of life. The first demonstration of the essential importance of zinc for the growth and development of living organisms dates back to 1869, when the requirement for zinc by the filamentous fungus *Aspergillus niger* was shown (▶ Zinc Deficiency). However, a precise biological function for zinc was first discovered only in 1940, when it was shown to be necessary for the catalytic activity of carbonic anhydrase. Since then, it has become increasingly clear that the requirement for zinc by living organisms mostly results from its association with a multiplicity of proteins responsible for many different physiological functions, including, for example, carbohydrate, lipid, and protein syntheses and degradation; cell differentiation; DNA metabolism and repair; vision; and neurotransmission (Vallee and Falchuk 1993; Maret and Sandstead 2006). After nearly 150 years since the study on *Aspergillus niger*, it is now well established that zinc is a key component of a large number of proteins involved in a wide range of biological processes: The Protein Data Bank (PDB), for example, at the end of 2010 contained over 6,000 zinc protein structures, approximately corresponding to 10% of the entire PDB archive. In eukaryotic organisms, zinc is the second most abundant trace metal, second only to iron. The whole body zinc content of a 70-kg adult human is 2–3 g, corresponding to approximately 30–40 mg/kg (or ppm). It is found in all tissues, with about 85% of the total in muscle and bone. In blood plasma, zinc is found at a concentration of 12–16 μM, and is largely bound to albumin (60%), macroglobulin (30%), and transferrin (10%) (Rink and Gabriel 2000).

Zinc in proteins invariably occurs as Zn(II), and generally plays one of the two major roles, i.e., catalytic (when it is directly involved in the reaction mechanism of an enzyme) and structural (when it stabilizes the tertiary or the quaternary structure of a protein) (▶ Zinc-Binding Sites in Proteins). Zinc is the second most abundant metal found in enzymes, and zinc enzymes are present in all the six major classes of enzymes (oxidoreductases, transferases, hydrolases, lyases, isomerases, and ligases) (Auld 2001; Andreini et al. 2008). The main reason for the widespread use of zinc as a catalytic cofactor in proteins lies in its distinctive chemical properties, which combine Lewis acid strength, lack of redox reactivity, and fast ligand exchange. Specifically, the analysis of Metal-MACiE, a database collecting functional information on the catalytic metal ions found in enzymes, showed that zinc is the most used metal to induce formation of nucleophilic species for attack of the enzyme substrate, and that in some cases (such as in carboxypeptidase A and thermolysin), it is involved not only in generating the attacking nucleophile but also in enhancing the electrophilicity of the substrate via polarization of C-O or P-O bonds (Andreini et al. 2008). The lack of redox activity is also an important factor in the use of zinc as a structural component of proteins. Other zinc properties relevant to this function are its filled *d* shell and its borderline hard-soft acid-base character, which underlie the capability of zinc to be bound specifically in a variety of (generally tetrahedral) protein sites (Auld 2001; Andreini et al. 2008). Within the wide range of protein domains whose structures are stabilized by zinc, a central place is occupied by the so-called zinc fingers, which are small, often modular protein domains that can fold stably only upon zinc binding. Originally discovered as DNA-binding motifs, zinc fingers are now known to mediate also protein-RNA and protein-protein interactions (▶ Zinc Finger Folds and Functions).

Since the discovery of the catalytic function of zinc in carbonic anhydrase, the importance of zinc at the molecular level has been established, and increasingly appreciated, by the collection of a wealth of structural and functional information accumulated over the years for hundreds of individual zinc proteins (Berg and Shi 1996). In the last 10–15 years, the rapid progress of high-throughput experimental technologies providing omics-type data (e.g., genomics, proteomics) offered a new dimension to the study of zinc proteins and to biological research in general, opening the possibility of obtaining a virtually complete list of the cellular molecular components of interest in a single experiment. In particular, whole-genome sequencing projects are revealing the full set of genes of an ever-increasing number of organisms, at a rate that far exceeds the capability to characterize experimentally the products of these genes. On the other hand, the identification, localization, and quantification of metalloproteins at the proteome level are still challenging tasks, despite the significant improvements in the development of metalloproteomics techniques. In this framework, the use of computational approaches to predict whether a protein binds zinc from its amino acid sequence (as derived from the translation of nucleotide sequences) is of special value, because the application of such predictive tools to all the proteins encoded by the genome of an organism results in the estimate of the zinc proteome of that organism (Andreini et al. 2009).

A robust bioinformatics approach has been recently used to estimate the number of zinc proteins encoded in the genomes of 57 representative organisms from the three domains of life, including 40 bacteria, 12 archaea, and 5 eukaryotes (Andreini et al. 2006b). It was shown that zinc proteins are indeed widespread across all living organisms, and that within each superkingdom, a good correlation exists between the number of zinc proteins and the total number of proteins encoded in the genome (Fig. 1).

On average, prokaryotic organisms have a lower fraction of zinc proteins (i.e., 6.0% ± 0.2% of the entire proteome in archaea and 4.9% ± 0.1% in bacteria) than eukaryotic organisms, whose zinc proteome represents 8.8% ± 0.4% of the entire proteome (Andreini et al. 2006b). For humans, this percentage amounts to about 10%, which corresponds to the remarkable number of approximately 2,800 zinc proteins (Andreini et al. 2006a). Among the representative organisms taken into account, the organism with the lowest share of zinc proteins (3.9%) is the marine bacterium *Rhodopirellula baltica*. In archaea, the relatively smallest zinc proteomes are found in the hyperthermophilic crenarchaeon *Pyrobaculum aerophilum* and in the anaerobic methanogen *Methanosarcina acetivorans* (5.0% in both cases). At the other end of the range, the highest fraction of zinc proteins in prokaryotic proteomes occurs in the bacteria *Mesoplasma florum* (an obligate parasite with a very small proteome, consisting of only 682 proteins) and *Thermoanaerobacter tengcongensis* (a thermophilic anaerobe), which both have 7.9% of their proteome consisting of zinc proteins, and in the archaeon *Pyrococcus abyssi*, a deep-sea hyperthermophilic anaerobe, which has 8.2%. In eukaryotes, the share of the zinc proteome ranges between 8.0% (in the worm *Caenorhabditis elegans* and the small flowering plant *Arabidopsis thaliana*) and 10.2% (in the fruit fly *Drosophila melanogaster*) (Andreini et al. 2006b).

Approximately two thirds of the prokaryotic zinc proteins have homologues in eukaryotes, while the remaining third comprises zinc proteins that are present only in prokaryotes. On the other hand, three quarters of the eukaryotic zinc proteins have no prokaryotic homologues, suggesting that they are evolutionarily more recent. The reason for this observation becomes clear by examining the functional diversification of the eukaryotic and prokaryotic zinc proteomes (Fig. 2).

The largely prevalent use of zinc proteins in prokaryotes is for performing enzymatic catalysis, whereas the majority of zinc proteins in eukaryotes are almost equally divided into enzymes and proteins regulating DNA transcription (Fig. 2a). Most zinc-dependent enzymes have homologues in both eukaryotes and prokaryotes (Fig. 2b), indicating that zinc has been recruited as a cofactor in the catalytic sites of enzymes prior to the branching of the three domains of life. Conversely, zinc-binding transcription factors, most commonly containing zinc finger domains (see above), appear to occur almost exclusively in eukaryotes. Therefore, whereas zinc enzymes appear to derive from an ancestral zinc proteome, zinc-binding transcription factors appear to have evolved in response to the increased needs of higher organisms in the context of gene regulation, including that involved in complex processes like cell compartmentalization and, for multicellular organisms, cell differentiation (Andreini et al. 2006b). The extensive use of zinc in the regulation of cellular processes makes highly proliferating cell systems like the immune system, the skin, and the reproductive system the most sensitive indicators of zinc deficiency in humans, as cell proliferation has an absolute requirement for zinc (Rink and Gabriel 2000). Prokaryotic zinc proteins without homologues in eukaryotes, on the other hand, are most commonly found in only a few bacterial classes, and are likely to result from environmental adaptation to specific niches. In this respect, it was observed that the share of zinc proteins in the proteome progressively increases going from psychrophilic (4.5% ± 0.2%) to mesophilic (5.3% ± 1.0%) to thermophilic (6.0% ± 1.0%) to hyperthermophilic organisms (7.0% ± 1.1%), suggesting that organisms living at higher temperatures make a wider use of zinc to enhance the structural stability of proteins (Andreini et al. 2006b).

Not differently from other predictive computational approaches, the methods applied to estimate zinc proteomes and derive the data outlined above ultimately depend on the knowledge available, in that they can detect only zinc proteins containing either a known zinc-binding domain or a known zinc-binding sequence motif (Andreini et al. 2009). In particular, it is now emerging that zinc homeostasis is a complex process, which is needed to avoid dysfunction determined by depletion or overload of this essential metal. While the physiological effects of

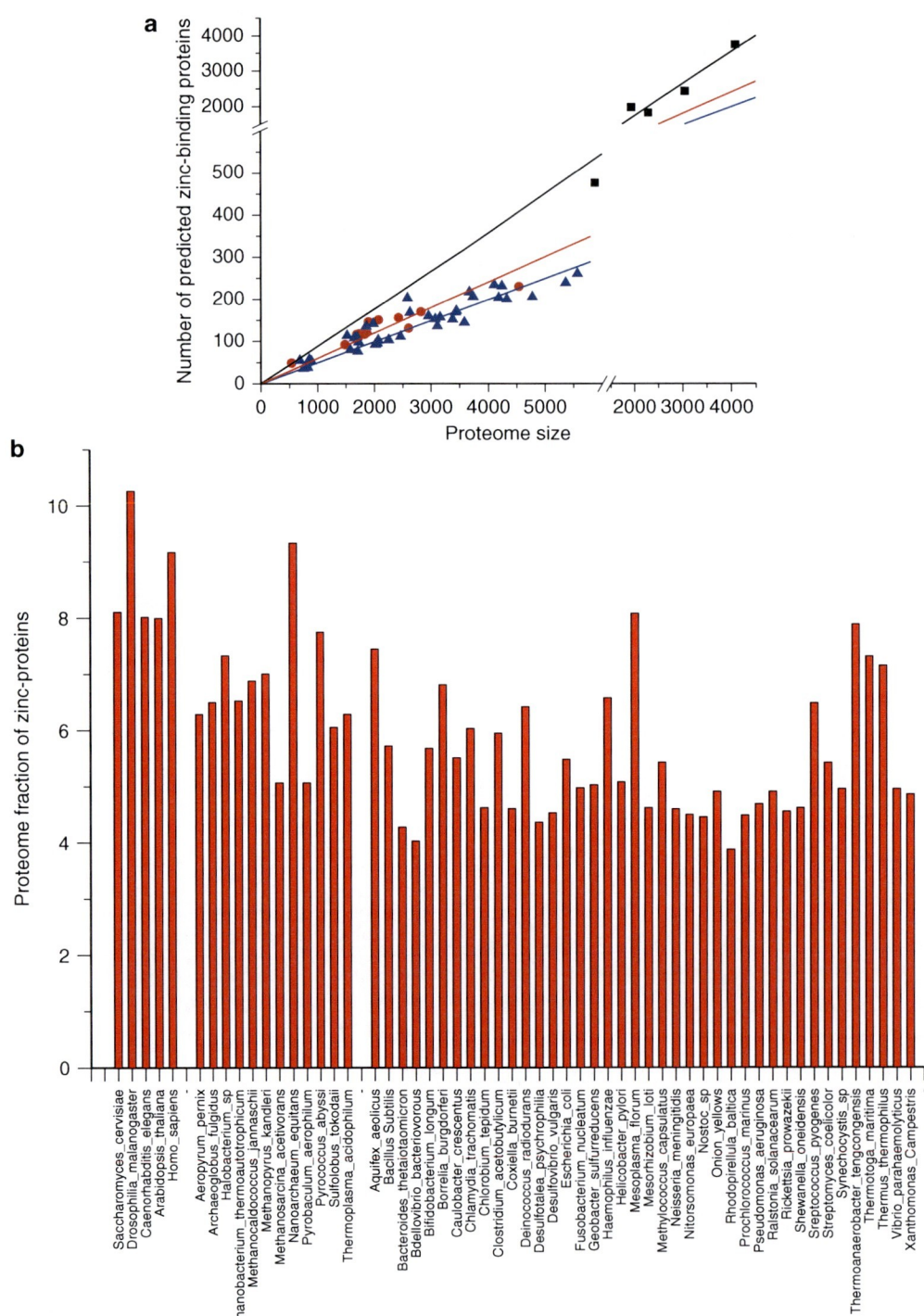

**Zinc-Binding Proteins, Abundance, Fig. 1** (a) Number of putative zinc proteins encoded in genomes (*squares*, eukaryotes; *circles*, archaea; *triangles*, bacteria) as a function of the proteome size. Best-fit lines for each of the three superkingdoms are also shown. (b) Percentage of zinc proteins identified in each organism (Reprinted with permission from "Andreini C, Banci L, Bertini I, Rosato A (2006) Zinc through the three domains of life. J Proteome Res 5(11):3173–3178." Copyright 2006 American Chemical Society)

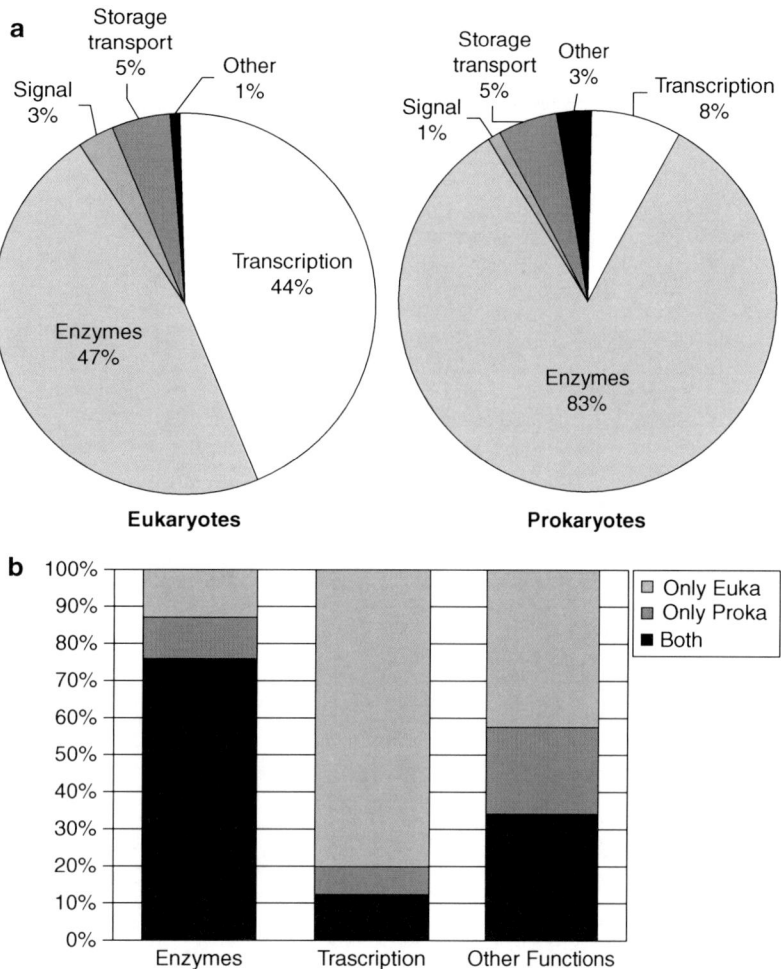

**Zinc-Binding Proteins, Abundance, Fig. 2** (a) Distribution of the functions of zinc proteins in eukaryotes and prokaryotes. (b) Fraction of zinc proteins (separated into enzymes, transcription factors, and proteins with other functions) that have homologues in both prokaryotes and eukaryotes or are specific to one of the two groups (Reprinted with permission from "Andreini C, Bertini I, Rosato A (2009) Metalloproteomes: a bioinformatic approach. Acc Chem Res 42(10):1471–1479." Copyright 2009 American Chemical Society)

zinc deficiency are well studied, the hazards associated with zinc excess are still far from being understood in detail: Until 10–15 years ago, the consensus was that zinc is essentially nontoxic, and the only established effect of exceedingly large zinc intakes was that of causing copper deficiency by interfering with the mechanisms of copper absorption (Vallee and Falchuk 1993). In recent years, zinc toxicity has been mostly investigated in neuronal systems, and disruption of zinc homeostasis is being increasingly associated with a wide range of neurological disorders (Roberts 2011). At the molecular level, zinc homeostasis is likely to involve many proteins for zinc sensing, transport, buffering, and storage. Only a few of these proteins are currently known, also because most of them are expected to bind zinc only transiently, making their recognition as zinc proteins difficult. It is therefore likely that the size of zinc proteomes is larger than what is currently realized, and that future estimates may yield higher numbers as the molecular mechanisms of cellular zinc homeostasis are progressively elucidated. Still, even at the present level of accuracy, the possibility to detail the zinc metalloproteome of an organism is of major importance for systems-wide approaches aimed at elucidating clinical findings at the molecular level, such as in the study of the hepatotoxicity mechanisms in Wilson's disease (Roberts 2011).

## Cross-References

- ▶ Zinc and Zinc Ions in Biological Systems
- ▶ Zinc Cellular Homeostasis
- ▶ Zinc Deficiency
- ▶ Zinc-Binding Sites in Proteins

## References

Andreini C, Banci L, Bertini I, Rosato A (2006a) Counting the zinc-proteins encoded in the human genome. J Proteome Res 5(1):196–201

Andreini C, Banci L, Bertini I, Rosato A (2006b) Zinc through the three domains of life. J Proteome Res 5(11):3173–3178

Andreini C, Bertini I, Cavallaro G, Holliday GL, Thornton JM (2008) Metal ions in biological catalysis: from enzyme databases to general principles. J Biol Inorg Chem 13(8):1205–1218

Andreini C, Bertini I, Rosato A (2009) Metalloproteomes: a bioinformatic approach. Acc Chem Res 42(10):1471–1479

Auld DS (2001) Zinc coordination sphere in biochemical zinc sites. Biometals 14(3–4):271–313

Berg JM, Shi Y (1996) The galvanization of biology: a growing appreciation for the roles of zinc. Science 271(5252):1081–1085

Maret W, Sandstead HH (2006) Zinc requirements and the risks and benefits of zinc supplementation. J Trace Elem Med Biol 20(1):3–18

Rink L, Gabriel P (2000) Zinc and the immune system. Proc Nutr Soc 59(4):541–552

Roberts EA (2011) Zinc toxicity: from "no, never" to "hardly ever". Gastroenterology 140(4):1132–1135

Vallee BL, Falchuk KH (1993) The biochemical basis of zinc physiology. Physiol Rev 73(1):79–118

# Zinc-Binding Sites in Proteins

David S. Auld
Harvard Medical School, Boston, MA, USA

## Definitions

The inner shell coordination properties of zinc enzymes have led to the identification of four types of zinc-binding sites: *Catalytic*, *Cocatalytic*, *Structural*, and *Protein Interface*. Zinc has also been found to form clusters in a number of proteins involved in DNA/RNA binding and in distribution of zinc to other proteins.

## Background

Zinc, copper, and iron are the three most prevalent metals in biological systems. The proof of the nutritional importance of zinc spanned nearly 100 years from the first studies in the fungus *Aspergillus niger* by Raulin in 1869 to that in humans by Prasad in 1963

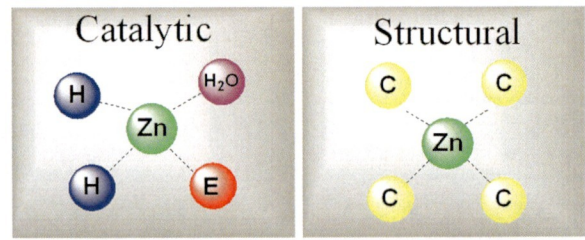
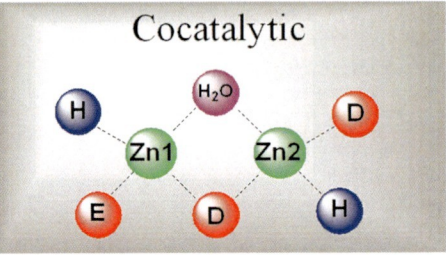

**Zinc-Binding Sites in Proteins, Fig. 1** Zinc-binding sites in metalloproteases. Schematic shown for the *Catalytic* Zinc site for human carboxypeptidase A1 (pdb 2V77), the *Structural* Zinc site for the human alcohol dehydrogenase ADH1B ($\beta_1\beta_1$) (pdb 1HSZ) and the *Cocatalytic* Zinc site for the aminopeptidase *Streptococcus pneumoniae* PepA (pdb 3KL9). Amino acid ligands are denoted by the single amino acid code, *H* for histidine, *C* for cysteine, *E* for glutamic acid, and *D* for aspartic acid

(▶ Zinc Deficiency). The rapid changes in technology over the last several decades have led to establishing its importance on the molecular level. Protein-bound zinc is involved in a wide variety of metabolic processes including carbohydrate, lipid, nucleic acid and protein synthesis, and degradation (Auld 2005). Zinc is the only metal to have representatives in all six of the International Union of Biochemistry, IUB, classes of enzymes. The function of these enzymes is, therefore, related to the chemistry of zinc, which is quite versatile. It has a remarkably adaptable coordination sphere that allows it to accommodate a broad range of coordination numbers and geometries.

Our first attempts at classifying the manner in which zinc binds to enzymes were made on the basis of the known zinc ligands and coordination properties of about one dozen zinc enzymes during the period 1990–1993 (Vallee and Auld 1990, 1993) (Fig. 1). This led to the recognition of three types of zinc-binding sites: *Catalytic*, *Cocatalytic*, and *Structural*. A fourth type of zinc-binding site, *Protein Interface*, was first identified based on the observations that zinc can have an impact on the quaternary structure of a protein (Auld 2001). This type of zinc-binding site is composed of amino acid ligands that reside in the

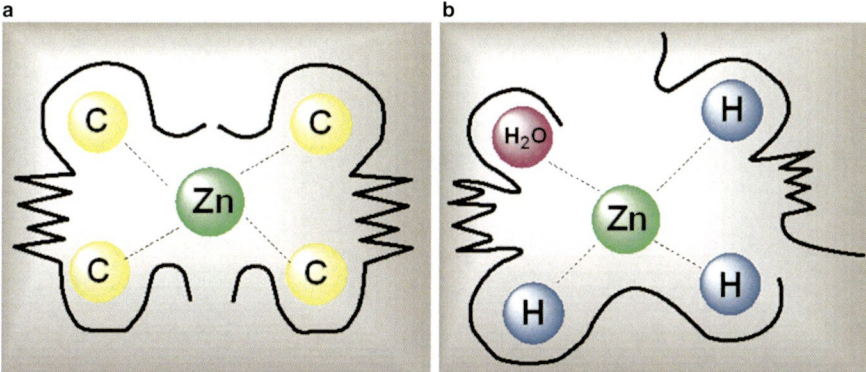

**Zinc-Binding Sites in Proteins, Fig. 2** *Protein Interface Sites* A structural like Zn site occurs in the nitric oxide synthase family (**a**) A zinc ion is tetrahedrally coordinated to a pair of symmetry-related Cys residues (Cys96 and Cys101) near the bottom of dimer interface of the human endothelial nitric oxide synthase, eNOS or NOS-3 (pdb 3NOS). (**b**) A catalytic like Zn site occurs in the *Methanosarcina thermophila* ($\gamma$-class carbonic anhydrase (pdb 1QRM)). The zinc is coordinated between two monomers with His81 and His122 contributed by one monomer, and His117 contributed from an adjacent monomer. Two water molecules complete a trigonal bipyramidal coordination geometry

binding surface between two protein subunits or interacting proteins (Fig. 2). The resulting site still retains the functional properties of the first three identified sites.

The knowledge we have about these zinc sites has grown enormously and the assignments have held up very well. The zinc proteome represents about 9% of the entire proteome in eukaryotes (▶ Zinc-Binding Proteins, Abundance). Three-dimensional structures existed for 450 zinc-binding proteins in 2005 (Auld 2005). Based on the rapidly increasing information on protein structure this number is estimated to be about 1,500 today. There were only five examples of catalytic zinc protease (enzyme Class III hydrolases) sites in 1990, while today there are three-dimensional structures known for 129 catalytic, 19 structural, and 39 cocatalytic zinc protease sites (Auld 2012). Similarly there were only two known examples of the catalytic and structural zinc sites in alcohol dehydrogenase (enzyme Class I dehydrogenases) in 1990. A more recent review found over three dozen structures of zinc sites for this class of zinc enzymes (Auld and Bergman 2008) (▶ Zinc Alcohol Dehydrogenases, ▶ Zinc Structural Site in Alcohol Dehydrogenases).

The most frequent amino acid ligands are His, Glu, Asp, and Cys (Auld 2005). Catalytic zinc sites are generally composed of three protein ligands, two of which come from a short amino acid spacer, and a bound water molecule (Fig. 1). His (usually the NE2 nitrogen) may be chosen because of its capacity to disperse charge through H-bonding of its non-ligating nitrogen. While in principle the third ligand could be supplied from either the C- or N-terminal side of the first two ligands, the C-terminus is preferred by a ratio of about 5:1. The coordination number for such sites is usually 4 or 5 and the geometry in the inhibitor/substrate-free state is frequently distorted tetrahedral or trigonal-bipyramidal. Water is always a ligand to the catalytic zinc. The zinc-bound water can be involved in catalysis as a nucleophile through its direct ionization, or it can be activated through polarization by a neighboring amino acid residue acting as a general acid/base (Fig. 3). The zinc can function as a Lewis acid by a substrate ligand displacement of the water or expansion of the coordination shell (Auld 2005). The chemical nature of the three protein ligands, their spacing and scaffolding interactions with neighboring amino acids in conjunction with the vicinal properties of the active center created by protein folding are all critical for the various mechanisms through which zinc can be involved in catalysis as well as its binding strength to the protein.

Structural zinc sites contain four protein ligands and no metal-bound water (Fig. 1). While Cys is the most frequent ligand of these sites, any combination of four Cys, His, Glu, and Asp residues in principle can form this type of zinc site. Twelve combinations of the 22 permutations of these four ligands had been observed in 2005 (Auld 2005). The role of the structural zinc site is to maintain the localized structure of

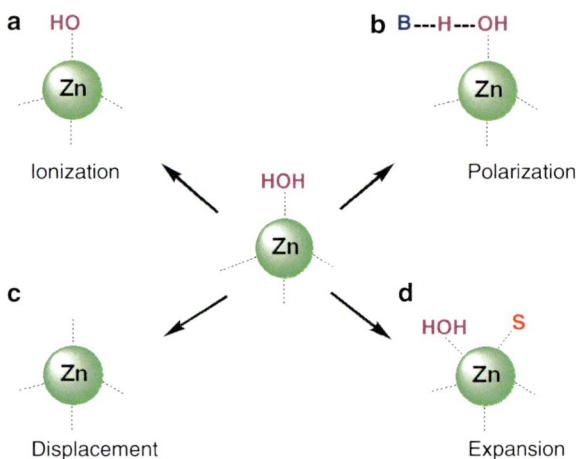
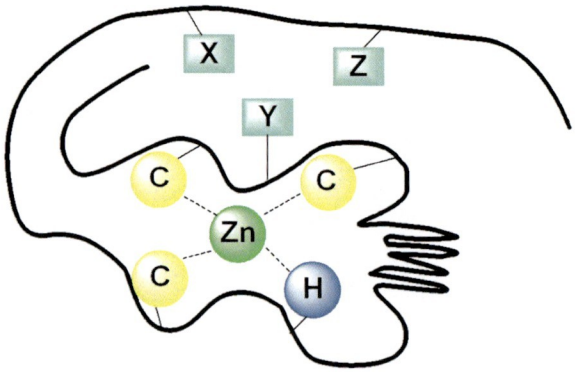

**Zinc-Binding Sites in Proteins, Fig. 3** The role of the zinc and its bound water in catalysis. The water can ionize either (**a**) without or (**b**) with the help of an adjacent base, B, supplied by an amino acid residue. The zinc-bound water can also (**c**) be displaced by substrate, S or (**d**) the coordination sphere be expanded upon interaction with substrate

**Zinc-Binding Sites in Proteins, Fig. 4** Schematic of a structural zinc-binding site that provides active-site residues from amino acids located within the metal-coordination spacers. Other active-site residues are often provided by the peptide sequence N-terminal to the structural zinc site. The symbol Y can represent more than one amino acid

the protein which could in turn influence protein folding or function by supplying residues involved in catalysis that arise from within the spacer arms (Fig. 4). This is similar in what happens in DNA/RNA-binding zinc proteins (▶ Zinc Finger Folds and Functions) where amino acids that interact with the nucleotides often reside in the spacers separating zinc ligands.

Cocatalytic zinc sites are found in enzymes containing two or more zinc and/or other transition metals in close proximity to each other that operate in concert as a catalytic unit (Auld 2005; Vallee and Auld 1993) (Fig. 1). The distance between the metals is determined by the type of amino acid (Asp, Glu, His, or a carboxylated Lys) that bridges the two metals. Sometimes a water molecule forms a bridge between the metal atoms in a cocatalytic zinc site. Asp and His are the most frequent ligands in this type of site. The ligands to these sites often come from nearly the whole length of the protein. In these types of zinc sites, the metals are important not only to catalytic function but to protein folding.

In protein interface zinc sites the zinc can function as it does in all the zinc-binding sites mentioned above. In this capacity, it may also influence the quaternary structure of proteins. Zinc binding to ligands supplied from protein molecules at their interface contact surface has been observed in increasing numbers in the last few years (Auld 2001, 2005; Maret 2004). These interactions can lead to dimers or trimers of the same protein or link two different proteins through the intermolecular ligands (Fig. 2). The amino acid residues His, Glu, and Asp primarily supply the ligands to these sites but Cys-containing sites are also found (Auld 2005). The ligands are generally contributed by some form of secondary structure with β-sheets predominating. The most common types of these sites are formed from either each protein molecule providing two zinc ligands or one protein moiety providing three ligands and the second providing one. However, zinc has been observed to cross-link three or four polypeptide chains with a coordination number of 5 or 6 as well as cross-link two polypeptide chains with coordination numbers of 2, 3, or 4 (Maret 2004).

Once the zinc ligands and spacers are known families of zinc enzymes can be formed (Vallee and Auld 1990). In 1990, a potential zinc-binding site signature existed within astacin, HExxHxxGxxH, that was also present in about a dozen known protease sequences (Stocker et al. 1990). By 1992 the use of this putative zinc signature led to the identification of 33 proteases that defined four major groups of homologous proteins (Auld 1992). Today there are 3,000 sequences of these four types of zinc proteases (Rawlings et al. 2010) (http://merops.sanger.ac.uk/index.shtml). Remarkably the ligands and spacers of this zinc site are completely conserved in these proteins.

## Some General Characteristics of Zinc-Binding Sites in Enzymes

The secondary support and scaffolding in combination with the direct ligands of the zinc allows fine tuning of the role of the zinc and its neighboring amino acids in the protein function. A number of general observations pertain to the structural properties of zinc-binding sites found in enzymes and other zinc-binding proteins. (1) Nearly all the zinc sites contain at least one secondary structural element. (2) β-sheets supply ligands to a wide variety of zinc-binding sites. (3) Ligands frequently come from the first, last, or 1 or 2 amino acids before or after a β-sheet or α-helix. (4) Small loop regions (about six amino acids) between two types of secondary structure can also supply the ligands. This may allow more freedom in the coordination properties of the zinc site. (5) Short spacers of one generally use a β-sheet to supply the ligands while a spacer of three uses an α-helix. (6) An α-helix is used to supply the short spacer ligands most frequently in hydrolytic enzymes (in particular, the metalloproteases). (7) The non-ligating imidazole NH of the His ligands is often H-bonded to carboxyl groups of Asp or Glu or carbonyl groups from amides. (8) Hydrophobic amino acids likely are involved in stabilizing the zinc-binding site through orientation of the ligands.

## Outer Shell Interactions of Zinc-Binding Sites in Enzymes

Catalytic zinc-binding sites are often composed of His and Glu/Asp residues that coordinate the zinc by imidazole and carboxyl groups. The zinc dissociation constant for these protein sites is in the nM to pM range. Since the log of the zinc-binding constants for imidazole and acetate is about 2.3 and 1.5, respectively, the theoretical value for a zinc dissociation constant for two imidazoles and one acetate is approximately in the l M range (Auld 2009). The much lower zinc dissociation constants for protein-bound zinc reflects the proximity and restricted mobility of the ligands in the protein. The first kind of stabilization comes from the secondary support structure that supplies the ligands to the zinc. In catalytic zinc sites, an α-helix or a β-sheet frequently supplies the ligands in the short spacer (Auld 2005). There is strong correlation

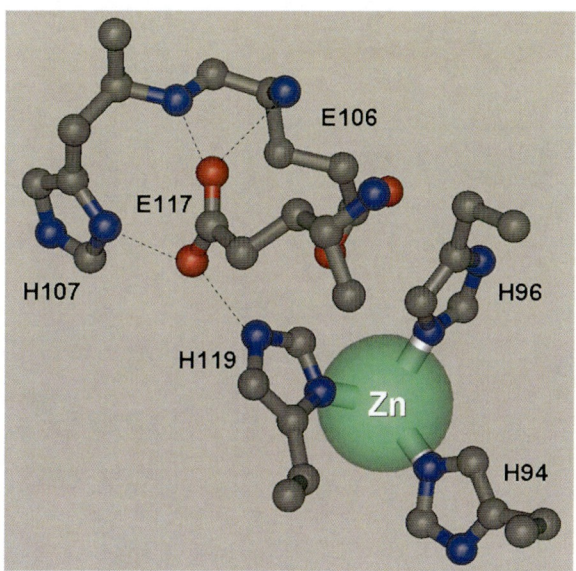

**Zinc-Binding Sites in Proteins, Fig. 5** The human carbonic anhydrase II (pdb 1CA2) carboxyl oxygens of Glu117 hydrogen bond the imidazole NH of His107, the non-zinc-binding NH of His119 and amide NHs of His106 and Glu106 (Figure prepared using the program Cn3D of the National Center for Biotechnology Information (Auld 2009))

between the short spacer length and the type of secondary support structure (see above). The second level of restricted mobility comes from outer shell interactions. Outer shell coordination can influence the stability of the zinc site and its function as exemplified by the zinc sites in carbonic anhydrase, promatrix metalloproteases, and alcohol dehydrogenase (Auld 2009). Thus, the carboxylate of the highly conserved Glu117 in carbonic anhydrase H-bonds to the non-bonding NH of the His119 ligand of the catalytic Zn ion (Fig. 5). Elimination of this H-bond results in a 300-fold increase in the off rate constant for zinc but has a negligible effect on the $pK_a$ of the zinc-bound water (Kiefer et al. 1995). Both of these results can be explained by inspection of the secondary interactions that Glu117 has with the zinc ligand His119 and neighboring amino acids (Auld 2009). There are several potential H-bonds in the vicinity of the carboxylate of Glu117 in addition to the one with His119 (Fig. 5). Thus, the carboxylate OE2 of Glu117 is 2.57 and 2.93 Å from the NE2 of His119 and ND1 of His107, respectively, while the OE1 of Glu117 is 2.83 and 2.87 Å from the amide nitrogens of His107 and Glu106, respectively. This complex pattern of H-bonds would be expected to disperse the negative

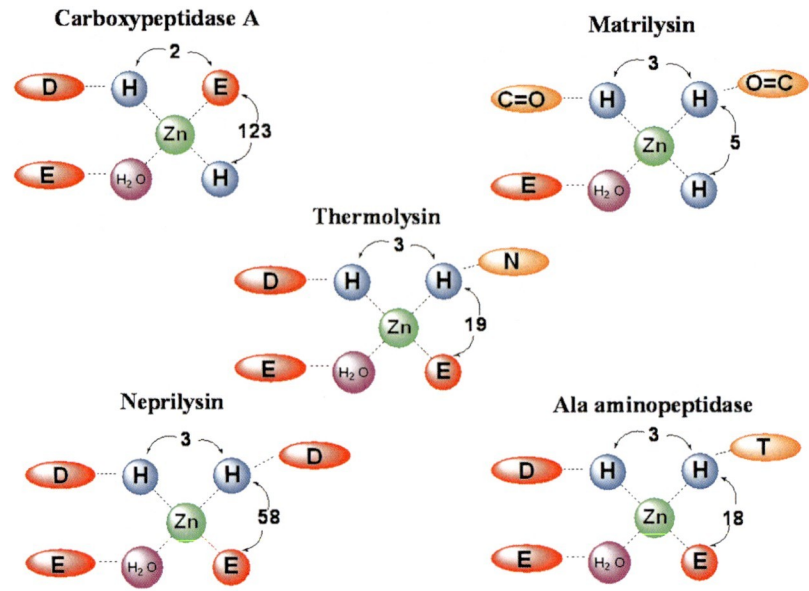

**Zinc-Binding Sites in Proteins, Fig. 6** Scaffolding in catalytic zinc sites of the metalloproteases, carboxypeptidase A (pdb 3CPA), matrilysin (pdb 1MMP), thermolysin (pdb 1LND), neprilysin (pdb 1DMT), and Ala aminopeptidase (pdb 2DQ6)

charge of the Glu117 carboxylate onto several residues instead of concentrating it on the His119 residue. Thus, removal of this H-bonding interaction should not greatly change the charge on the His119 ligand or in turn affect the $pK_a$ of the zinc-bound water in agreement with the obtained results. The intricate pattern of H-bonds may also explain why the disruption of the Glu117 interaction with His119 has such a large effect on the off rate constant for zinc since several H-bonds are involved in stabilizing the interaction between these two amino acids.

A large number of the catalytic zinc sites in metalloproteases have either two His and a Glu or an Asp or three His ligands (Auld 2012). It is, therefore, generally assumed that these sites behave the same with respect to the stability and catalytic function of the metal. However, the scaffolding differs even in relatively closely related family groups (Fig. 6). Carboxypeptidase, thermolysin, neprilysin, and Ala aminopeptidase all have inner shell coordinations of one glutamate and two histidyl ligands. Matrilysin representing the superfamily of matrix metalloproteinases, snake venoms (ADAMS), astacin and serriatia all have an HExxH metal-binding sequence containing the catalytic group Glu that occurs in thermolysin, neprilysin, and Ala aminopeptidase. Nevertheless, they all differ in the type of H-bonding interactions that occur with the His ligands of the short spacer. Thus, H-bonds come from two Asp carboxylates in neprilysin, one Asp carboxylate and a Thr hydroxyl in Ala aminopeptidase, an Asp carboxylate and an Asn terminal carboxyamide in thermolysin, two polypeptide backbone amide oxygens in matrilysin, and a single Asp carboxylate in the case of carboxypeptidase A (Fig. 6). These secondary H-bonds are highly conserved. Thus, the H-bonding Asp residues in 980 neprilysin sequences are conserved in all the sequences (where Asp is 66 amino acids removed from His ligand $L_1$) and 955 sequences (where Asp is two amino acids removed from His ligand $L_2$) in this family of M13 metalloproteases, while both the H-bonding Asp residues (29 amino acids removed from His ligand $L_1$) and Thr (21 amino acids removed from His ligand $L_2$) are conserved in all 288 sequences in the Ala aminopeptidase M1 family of metalloproteases (Auld 2012). Such changes in charge and strength of H-bonds can lead to profound differences in the stability of the zinc site, $pK_a$ of the zinc-bound water and the Lewis acid properties of the zinc ion.

## Other Zinc-Binding Sites in Proteins

Structural studies in recent years have revealed a number of zinc-binding sites in proteins of diverse function (Auld 2005). The great majority of these zinc sites are formed from four protein side-chain ligands to yield a tetrahedral binding site. Their function may, therefore, be related to the local and/or overall structure since the sites span from about 20 to 200 amino acids. As in the case of structural zinc sites in enzymes the vast majority of the sites contain at least one short spacer. The residues are generally supplied from β-sheets or reside within one or two residues of such a secondary structure. The brevity of this entry does not allow a detailed description of all of these binding sites. Some of the major classifications and their structures are described in detail in other entries in this encyclopedia. For those involved in interactions with DNA and RNA see ▶ Zinc Finger Folds and Functions. Zinc exists in complex cluster structures that are described in ▶ Zinc Metallothionein, ▶ Zinc Metallothionein-3 (Neuronal Growth Inhibitory Factor). For zinc proteins involved in zinc sensing and transport see ▶ Zinc Sensors in Bacteria and ▶ Zinc Storage and Distribution in S. cerevisiae.

## Cross-References

- ▶ Angiotensin I-Converting Enzyme
- ▶ Zinc Adamalysins
- ▶ Zinc Alcohol Dehydrogenases
- ▶ Zinc Amidohydrolase Superfamily
- ▶ Zinc Aminopeptidases, Aminopeptidase from Vibrio Proteolyticus (Aeromonas proteolytica) as Prototypical Enzyme
- ▶ Zinc and Iron, Gamma and Beta Class, Carbonic Anhydrases of Domain Archaea
- ▶ Zinc Beta Lactamase Superfamily
- ▶ Zinc Deficiency
- ▶ Zinc Finger Folds and Functions
- ▶ Zinc Leukotriene $A_4$ Hydrolase/Aminopeptidase Dual Activity
- ▶ Zinc Matrix Metalloproteinases and TIMPs
- ▶ Zinc Meprins
- ▶ Zinc Metallocarboxypeptidases
- ▶ Zinc Metallothionein
- ▶ Zinc Metallothionein-3 (Neuronal Growth Inhibitory Factor)
- ▶ Zinc Structural Site in Alcohol Dehydrogenases
- ▶ Zinc-Astacins
- ▶ Zinc-Binding Proteins, Abundance

## References

Auld DS (1992) Astacin family of zinc proteases. Faraday Discuss 93:117–120

Auld DS (2001) Zinc coordination sphere in biochemical zinc sites. BioMetals 14:271–313

Auld DS (2005) Zinc enzymes. In: King RB (ed) Encyclopedia of inorganic chemistry. Wiley, Chichester, pp 5885–5927

Auld DS (2009) The ins and outs of biological zinc sites. Biometals 22:141–148

Auld DS (2012) Catalytic mechanisms for metallopeptidases. In: Rawlings ND, Salvesen GS (eds) Handbook of proteolytic enzymes. Elsevier, Oxford 370–396

Auld DS, Bergman T (2008) Medium- and short-chain dehydrogenase/reductase gene and protein families: the role of zinc for alcohol dehydrogenase structure and function. Cell Mol Life Sci 65:3961–3970

Kiefer LL, Paterno SA, Fierke CA (1995) Hydrogen bond network in the metal binding site of carbonic anhydrase enhances zinc affinity and catalytic efficiency. J Am Chem Soc 117:6831–6837

Maret W (2004) Protein Interface zinc sites: the role of zinc in the supramolecular assembly of proteins and in transient protein-protein interactions. In: Messerschmidt A, Bode W, Cygler M (eds) The handbook of metalloproteins. Wiley, Chichester, pp 432–441

Rawlings ND, Barrett AJ, Bateman A (2010) MEROPS: the peptidase database. Nucleic Acids Res 38:D227–D233

Stocker W, Ng M, Auld DS (1990) Fluorescent oligopeptide substrates for kinetic characterization of the specificity of astacus protease. Biochemistry 29:10418–10425

Vallee BL, Auld DS (1990) Zinc coordination, function, and structure of zinc enzymes and other proteins. Biochemistry 29:5647–5659

Vallee BL, Auld DS (1993) New perspective on zinc biochemistry: cocatalytic sites in multi-zinc enzymes. Biochemistry 32:6493–6500

# Zinc-Coordinating Domains

▶ Zinc Finger Folds and Functions

# Zinc-Dependent Activators

▶ Zinc Sensors in Bacteria

# Zinc-Dependent Corepressors

▶ Zinc Sensors in Bacteria

# Zinc-Dependent Derepressors

▶ Zinc Sensors in Bacteria

# Zinc-Finger-Based Artificial Transcription Factors

Takashi Sera
Department of Applied Chemistry and Biotechnology,
Graduate School of Natural Science and Technology,
Okayama University, Okayama, Japan

## Synonyms

Zinc-finger-based engineered transcription factor; Zinc-finger-based synthetic transcription factor

## Definition

Zinc-finger-based artificial transcription factors (ATFs) have been developed to modulate target gene expression in eukaryotes. The ATFs principally comprise three functional domains: (1) an artificial zinc-finger protein (ZFP) as a DNA-binding domain; (2) a transcriptional regulatory domain; and (3) a nuclear localization domain. An artificial ZFP is required for ATFs to specifically bind to the promoter of a target gene and is the most important and critical component among those of ATFs.

## Introduction

A variety of biological processes including development, differentiation, and disease are regulated through gene expression. Gene expression is mainly modulated by endogenous transcription factors. This suggests that "artificial" transcription factors (ATFs) designed to bind to a promoter region of a gene of interest could modulate the target gene expression independently of organisms, thereby giving new insights into molecular biology and presenting novel gene therapy protocols. Transcription factors, in principle, comprise (1) a DNA-binding domain, (2) a transcriptional regulatory domain or effector domain, and (3) a nuclear localization signal (NLS); a DNA-binding domain is required to bind to the promoter of a target gene, an effector domain to up- or downregulate the target gene, and an NLS to deliver an ATF into nuclei (because eukaryotic transcription occurs in nuclei) (Fig. 1). Therefore, if an artificial

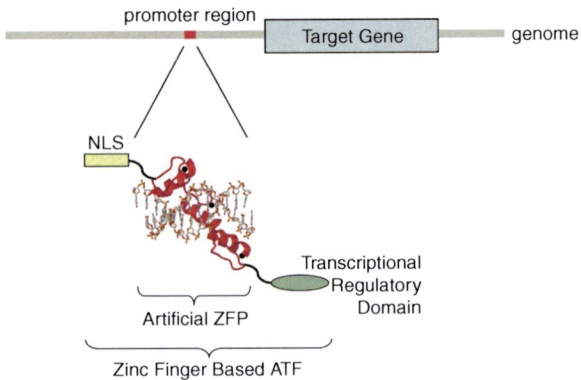

**Zinc-Finger-Based Artificial Transcription Factors, Fig. 1** Basic structure of zinc-finger-based ATF. The ATF is composed of three domains, an artificial ZFP as a DNA-binding domain, a transcriptional regulatory domain, and a NLS. The ATF enters a nucleus via the NLS, binds to a promoter region of a target gene, and then up- or downregulates the target gene expression in a manner dependent on the transcriptional regulatory domain

DNA-binding protein (or domain) that recognizes a target DNA specifically can be designed and constructed, a desired artificial transcription factor can be created.

Until now, many artificial DNA-binding proteins or small molecules such as pyrrole–imidazole polyamides have been reported. Among them, the $Cys_2His_2$-type zinc-finger protein (ZFP) is the most promising candidate for the DNA binder ▶ Zinc Finger Folds and Their Functions. The zinc-finger DNA-binding motif comprising multiple repeats of approximately 30 amino acids $Xaa_2$-Cys-$Xaa_{2-4}$-Cys-$Xaa_{12}$-His-$Xaa_{3,4}$-His-$X_{2-6}$ was first discovered in the transcription factor TFIIIA of *Xenopus laevis* in 1985. To date, many of the motifs have been identified in many eukaryotes (e.g., >700 ZFPs for humans), and it is now one of the most common DNA-binding motifs. The unique features are (1) ZFPs bind to their DNA targets as a monomer, and (2) ZFPs contain several finger domains, generally from three to more than nine domains. The first feature is very important because almost DNA regions in genomes are non-palindromic, which means that motifs that bind to DNA as homodimers, such as the helix-turn-helix motif, are not adequate for genomic DNA recognition. ZFPs have such no restriction. The second feature is also important for genome recognition. To target a specific genomic site, recognition of a long DNA sequence is necessary. For example, it is theoretically necessary to recognize >16-bp DNA for specific recognition of one DNA region in human cells because the genome size is $3 \times 10^9$ bp. Because ZFPs are known to contain many finger domains and recognize long DNA sequences, the consecutive linking of the motif seems to perturb the whole structure relatively little compared with other DNA-binding motifs, and thereby potentially recognize >16-bp targets with high affinities.

## Short History

The first study of gene regulation by engineered transcription factors in living cells (yeast) was reported in 1992 by Young's group. They mutagenized the $Cys_2His_2$ zinc-finger domains of the yeast transcription factor ADR1 to alter the binding specificities and demonstrated that one of the ADR1 mutants recognized their target DNA and transiently activated a reporter gene harboring its target DNA in yeast.

In 1994, gene regulation using a ZFP designed to recognize a 9-bp genomic sequence in mammalian cells was reported by Klug's group. In their study, a three-finger ZFP alone (but not an ATF!) was used for gene regulation. The ZFP was designed to bind to a unique 9-bp region of a BCR–ABL fusion oncogene, and indeed it repressed $p190^{BCR-ABL}$ expression in murine cells by blocking RNA polymerase movement. This study demonstrated for the first time that ZFP could modulate endogenous gene expression in living cells. Concurrently, they also demonstrated for the first time that an ATF, in which the engineered ZFP was fused to a VP16 transcriptional activation domain, could activate CAT-reporter gene expression transiently in living cells. After these reports, many groups reported engineered three-finger ATFs and evaluated their functions in transient reporter assays.

However, ATFs are required to specifically recognize >16-bp genomic sequences in human cells due to the genome size (i.e., $3 \times 10^9$ bp) in order to modulate their target gene expression. In 1997, Barbas's group reported the first six-finger ATF applicable to endogenous gene regulation in human cells. They generated a six-finger ZFP by assembling modular blocks recognizing 5'-GNN-3' triplets and demonstrated that the six-finger ATF modulated expression of a reporter gene under the control of an artificial promoter harboring six tandem copies of its 18-bp binding site. In 1998, the same group succeeded in modulation of a reporter gene under the control of a native erbB-2/HER-2 promoter by using a six-finger ATF.

Then, in 2000, Barbas's group reported that their six-finger ATFs specifically up- and downregulated the endogenous erbB-2 gene in living cells (Beerli et al. 2000). At almost the same time, Juliano's group reported repression of the endogenous MDR1 multidrug resistance gene by a five-finger ATF. After these studies, various ATFs were generated to modulate endogenous gene expression both in vitro and in vivo.

## Components of ATFs

ATFs, in principle, comprise (1) an artificial ZFP, (2) a transcriptional regulatory domain or effector domain, and (3) a NLS.

An artificial ZFP is required for ATFs to specifically bind to the promoter of a target gene and is the most important and critical component among those of ATFs (Mackay and Segal 2010). The component determines the overall specificities of gene regulation by ATFs. Currently, many good reviews on the design and construction of artificial ZFPs are available (e.g., please see Jamieson et al. 2003). The simplest method for construction of ZFPs via modular assembly is described in the next section.

A transcriptional regulatory domain is necessary to up- or downregulate the target gene. Transcriptional regulatory domains used for ATFs are listed in Table 1. Among them, the most frequently used activation domain is the herpes simplex virus VP-16 activation domain. The most popular repression domain is a Krüppel-associated box (KRAB) domain of KOX1.

An NLS is also necessary to deliver an ATF into nuclei because eukaryotic transcription occurs in nuclei. The short peptide PKKKRKV from the simian virus 40 large T antigen has been used most extensively as the NLS, and it works well.

To monitor ATF expression, several epitope tags such as FLAG, c-myc, and HA are incorporated at N-or C-termini (or both) of ATFs.

Optionally, a switch domain is connected to ATF to regulate the timing of gene modulation by addition of a small molecule to the medium. To date, fusion of ligand-binding domain(s) of hormone receptors such as an estrogen receptor, thyroid hormone receptor α, or a progesterone receptor to ATF was reported. Barbas's group also generated single-chain ligand-binding

**Zinc-Finger-Based Artificial Transcription Factors, Table 1** Transcriptional regulatory domains used for ATFs

| | Name | Comments |
|---|---|---|
| Activation domain | VP16 | C-terminal transcriptional activation (approximately 75 amino residues) of herpes simplex virus VP-16 transactivator, very effective and most popular |
| | VP64 | Tetrameric repeat of DALDDFDLDML, more active than VP16 |
| | p65 | C-terminus of the human NK-κB transcription factor p65 subunit (residues 288–548) |
| | S3H | Chimeric activation domain comprising the p65 subunit (residues 281–551) and human heat shock factor 1 (residues 406–529) |
| Repression domain | KRAB | Krüppel-associated box domain of KOX1 (residues 1–75), most popular and powerful |
| | SID | Residues 1–36 of the Mad mSIN3 interaction domain |
| | ERD | Residues 473–530 of the est2 repressor factor |
| | vErbA | Ligand-binding domain of the thyroid hormone receptor v-erbA |

domains derived from an estrogen-receptor homodimer and an ecdysone-receptor/retinoid X receptor heterodimer and observed ligand-dependent gene regulation in mammalian cells. Pollock's group reported a gene switch using a small-molecule heterodimerizer. In their system, a ZFP is fused to FK506-binding protein (FKBP), and a transcriptional regulatory domain is fused to the rapamycin-binding domain of FKBP rapamycin-associated protein. The heterodimerizer assembles the ZFP and the transcriptional regulatory domain noncovalently, and the resulting ATF can modulate target gene expression in a heterodimerizer-dependent manner.

Further, a cell-penetrating peptide (CPP; reviewed in Zorko and Langel 2005) or protein transduction domain is conjugated to ATF, rendering ATF molecules able to permeate cells (Tachikawa et al. 2004).

## Design and Characterization of ZFPs

After choosing an endogenous target gene for modulation, the genomic region to be targeted has to be determined. This step is probably the most important or critical one to obtain good ATFs. As Wolffe's group showed, the accessibility of genomic DNA in chromatin to ATFs seems to be very important (Liu et al. 2001). Before their study, transient reporter assays were often used to select good ATFs. In the reporter assays, a reporter gene such as luciferase was placed under the control of the promoter of the target gene. However, as shown in the study done by Wolffe's group, certain ATFs that showed good performance in transient reporter assays did not modulate the endogenous gene expression effectively, indicating that the chromatin structures of promoters in reporter plasmids are different from those in actual chromosomes. Therefore, they first identified DNase I hypersensitive sites and then constructed ATFs targeting these sites.

The same group also reported that their ATFs could activate the expression of silenced genes, in which the target sites were not hypersensitive to DNase I digestion, suggesting that certain ATFs may be able to sneak in and bind to their target sites crowded in chromosomes. Therefore, the most practical way to obtain effective ATFs is to screen a panel of ATFs that individually target different sequences in the promoter region of the gene of interest in a high-throughput manner, as they reported (e.g., please see Yokoi et al. 2007).

After determining the sequence and length [i.e., 3 N- or (3 N + 1)-bp, where N is the number of finger domains used] of a target site for an ATF, the site is divided into N pieces of 3-bp or overlapping 4-bp DNA segments, where the last base of each 4-bp target is the first base of the next 4-bp target. (Note: Currently, many papers and reviews still report that one zinc-finger domain recognizes 3 or 4 bp. The binding mode may depend on the finger framework used for ZFP design.) Then, an individual finger domain recognizing each 3- or 4-bp segment is selected from building modules, such as Barbas's modules for 5′-GNN-3′, 5′-ANN-3′, 5′-CNN-3′, and a partial 5′-TNN-3′, Kim's modules, or Young's modules for 5′-GNN-3′ and a partial 5′-TNN-3′. Barbas's modules can be also designed by using the zinc-finger tools website. An alternative method to construct each finger domain is to choose four recognition amino acids of positions −1, 2, 3, and 6 of the α-helix region of the zinc-finger domain from recognition code tables. Klug's and Pabo's groups individually constructed degenerate recognition code tables based on finger domains selected by their phage display, which enable generation of selected finger domains. Sera's group constructed a nondegenerate recognition code table, based on known and potential DNA base–amino acid interactions. Because the table is nondegenerate, it is easy to make all finger domains for specific 4-bp targets, enabling generation of artificial ZFPs in a high-throughput manner. The artificial ZFPs can recognize DNA sequences containing more than three guanines (at any positions) in the first 9-bp of 10-bp targets (Sera 2009). It was named "3 G rule."

DNA fragments encoding selected zinc-finger domains are assembled by PCR to generate DNA fragments encoding artificial ZFPs. In many cases, the canonical peptide linker of TGEKP works well to link finger domains. However, certain linkers are known to enhance the binding affinities of ZFPs. For the detailed information on engineered linkers, please see Sera 2009 for examples.

Selection of ZFPs from a library is a powerful alternative approach to generation of ZFPs. In addition to phage display selection (Rebar and Pabo 1993) and ribosome display, other selection systems have been applied or newly developed to generate ZFPs. A yeast one-hybrid system has been used to select ZFPs. A bacterial two-hybrid system was developed by Pabo's group. In their system, two expression plasmids were used for fusion of ZFP–Gal11P and fusion of Gal4 with the α-subunit of RNA polymerase. The specific interaction between Gal11P and Gal4 recruits RNA polymerase to activate a selection marker gene in vivo. By using this system, Young's group selected ZFPs successfully. Chandrasegaran's group reported a bacterial one-hybrid selection system, where the α-subunit of RNA polymerase was directly fused to a ZFP. Furthermore, Choo's group exploited cell-free selection of ZFPs by using emulsion-based in vitro compartmentalization.

After cloning DNA encoding a ZFP into an *E. coli* expression plasmid such as pET (Novagen), the ZFP is expressed and purified. DNA-binding properties of purified ZFPs are examined in several ways. The most popular method to determine the apparent dissociation constant ($K_d$) is an electrophoretic mobility shift assay (EMSA) or gel shift assay. The assay also enables examination of the specificity by using mutant DNA probes. Apparent $K_d$s are also determined by using surface plasmon resonance or a BIAcore system. This assay gives us the kinetic information on the DNA binding at the same time.

The DNA-binding specificities can also be examined by enzyme-linked immunosorbent assay (ELISA). A 96-well format enables evaluation of relative affinities against multi-mutant probes. The use of systematic evolution of ligands by exponential enrichment (SELEX) or cyclic amplification and selection of targets (CAST) presents all possible binding sequences of a ZFP by the combination of EMSA with a permutated DNA probe library and PCR. The use of DNA microarrays to determine the DNA-binding specificities was also reported. A bacterial one-hybrid system, in which ZFP is fused to the α-subunit of RNA polymerase, was also reported.

## Evaluation of ATFs

Finally, the resulting ATF is cloned into a mammalian expression plasmid or viral vector. The efficiency of the ATF is investigated by transient reporter assays or more ideally by analyzing expression levels of endogenous target genes. The mRNA levels of target genes are analyzed by Northern blotting or quantitated by real-time PCR. The protein levels of target genes are analyzed by flow cytometry, Western blotting, or ELISA. The transfection or transduction efficiencies of ATF-expression plasmids or viral vectors and expression levels of ATFs should be normalized when several ATFs are compared. DNA microarray analysis is a very powerful method to evaluate specific modulation of target genes by ATFs. Further, binding of ATFs to their target sites in cells is examined by in vivo DNase I footprinting or chromatin immunoprecipitation (ChIP). Now ChIP assays are used more frequently.

## Applications

ATFs have successfully modulated expression of a variety of genes, including ErbB, MRD1 multidrug resistance gene, erythropoietin, vascular endothelial growth factor A, PPARγ, IGF2, Bax, Oct-4, checkpoint kinase 2, cholecystokinin 2 receptor, parathyroid hormone receptor 1, γ-globin, mammary serine protease inhibitor, utrophin, and pigment epithelium–derived factor. In in vitro experiments, ATFs demonstrated highly specific gene regulation. DNA microarray analysis revealed that some ATFs modulated expression of only a single target gene without affecting other unrelated gene expression (Sera 2009).

ATFs up- or downregulated endogenous target genes not only in vitro but also in vivo. The first in vivo application was reported using a mouse ear angiogenesis model (Rebar et al. 2002). In the study, a six-finger ATF was designed and constructed to target the mouse vascular endothelial growth factor A (VEGF-A) gene. As shown in their photographs of ATF-treated ears, injection of the ATF-expressing adenoviral vector into the external ear of CD-1 mice induced neovascularization. Immunohistochemical counting of vessels revealed the 2.5-fold enhancement of vessel formation by the ATF. They also demonstrated acceleration of wound healing by the six-finger ATF in a mouse model. Furthermore, they evaluated the in vivo efficacy of their ATF in rats and rabbits. The ATF is in a phase II trial now.

Another potential method for the delivery is the direct introduction of ATF proteins into target cells or tissues. To achieve this, Sera's group conjugated CPPs to ATF proteins (Tachikawa et al. 2004). It was demonstrated that CPP–ATFs exogenously added to the medium entered the nuclei and effectively up- or downregulated the endogenous human VEGF-A gene expression depending on transcriptional regulatory domains used. These cell-permeable ATFs were called designed regulatory proteins (DRPs). DRPs provide a convenient way to modulate the direction, time, location, magnitude, and duration of the desired change in gene expression.

## Future Developments

Many laboratories worldwide are working to refine the technology further. The following issues/technologies are expected to be improved or developed in the future:
1. ZFP design: further improvement of the DNA-binding specificities and expansion of the targetable DNA sequences
2. Delivery of the genes or proteins to specific cells or tissues
3. Tissue-specific expression of ATFs: development of tissue (or cancer)-specific promoters
4. Regulation of the timing of gene regulation: development of gene-switch domains

## Summary

ATFs are potentially a powerful molecular tool to modulate endogenous target gene expression in living cells and organisms. To date, many DNA-binding molecules have been developed as the DNA-binding domains for ATFs. Among them, ATFs comprising $Cys_2His_2$-type zinc-finger proteins (ZFPs) as the DNA-binding domain have been extensively explored. The zinc-finger-based ATFs specifically recognize targeting sites in chromosomes and effectively up- and downregulate expression of their target genes not only in vitro but also in vivo. Zinc-finger-based ATF technology is growing and becoming more and more important as an effective gene therapy protocol as well as a molecular tool giving new insights into molecular biology. Certain ATFs are or will soon be in clinical trials.

## Cross-References

▶ Zinc Finger Folds and Functions

## References

Beerli RR, Dreier B, Barbas CF III (2000) Positive and negative regulation of endogenous genes by designed transcription factors. Proc Natl Acad Sci U S A 97:1495–1500

Jamieson AC, Miller JC, Pabo CO (2003) Drug discovery with engineered zinc-finger proteins. Nat Rev Drug Discov 2:361–368

Liu P-P, Rebar EJ, Zhang L, Liu LQ, Jamieson AC, Liang Y, Qi H, Li P-X, Chen B, Mendel MC, Zhong X, Lee X-L, Eisenberg SP, Spratt SK, Case CC, Wolffe AP (2001) Regulation of an endogenous locus using a panel of designed zinc finger proteins targeted to accessible chromatin regions. J Biol Chem 276:11323–11334

Mackay JP, Segal DJ (eds) (2010) Engineered zinc finger proteins. Humana Press, New York

Rebar EJ, Pabo CO (1993) Zinc finger phage: affinity selection of fingers with new DNA-binding specificities. Science 263:671–673

Rebar EJ, Huang Y, Hickey R, Nath AK, Meoli D, Nath S, Chen B, Xu L, Liang Y, Jamieson AC, Zhang L, Spratt SK, Case CC, Wolffe A, Giordano FJ (2002) Induction of angiogenesis in a mouse model using engineered transcription factors. Nat Med 8:1427–1432

Sera T (2009) Zinc-finger-based artificial transcription factors and their applications. Adv Drug Deliv Rev 61:513–526

Tachikawa K, Schroder O, Frey G, Briggs SP, Sera T (2004) Regulation of the endogenous VEGF-A gene by exogenous designed regulatory proteins. Proc Natl Acad Sci U S A 101:15225–15230

Yokoi K, Zhang HS, Kachi S, Balaggan KS, Yu Q, Guschin D, Kunis M, Surosky R, Africa LM, Bainbridge JW, Spratt SK, Gregory PD, Ali RR, Campochiaro PA (2007) Gene transfer of an engineered zinc finger protein enhances the anti-angiogenic defense system. Mol Ther 15:1917–1923

Zorko M, Langel U (2005) Cell-penetrating peptides: mechanisms and kinetics of cargo delivery. Adv Drug Deliv Rev 57:529–545

# Zinc-Finger-Based Engineered Transcription Factor

▶ Zinc-Finger-Based Artificial Transcription Factors

# Zinc-Finger-Based Synthetic Transcription Factor

▶ Zinc-Finger-Based Artificial Transcription Factors

# Zinc-Secreting Neurons, Gluzincergic and Zincergic Neurons

Yang V. Li[1] and Christopher J. Frederickson[2]
[1]Department of Biomedical Sciences, Heritage College of Osteopathic Medicine, Ohio University, Athens, OH, USA
[2]NeuroBioTex, Inc, Galveston, TX, USA

## Synonyms

Synaptic transmission; Synaptic zinc; Vesicular zinc; Zinc in the nervous system; Zinc release (if different title is chosen); Zinc signaling

## Definition

Zinc ions ($Zn^{2+}$) are secreted by many types of cells, including pancreatic, prostatic, GI tract cells, and various kinds of leucocytes. The most intricate and complex zinc-secreting system is the network of zinc-secreting neurons that dominates the newly evolved, cortical regions of the brain. These neurons, which secrete the transmitter glutamate along with zinc ions from their presynaptic terminals, are called "gluzincergic," or "zincergic." Zinc that is found in the synaptic vesicles is referred to as synaptic zinc. Virtually all of the connections from one cortical region to another (cortical associational pathways) are gluzincergic. Even more intriguing, the hippocampal and the amygdalar structures of the brain, key structures for processing emotional meaning and for conscious remembering, are interconnected with each other and with the entire the cerebral cortex by a vast gluzincergic network. Virtually none of the "old brain" of subcortex and spinal cord nor any of the nervous systems of sub-vertebrates have gluzincergic neurons.

The gluzincergic neurons have a few millimolar of relatively weakly bound zinc in their synaptic vesicles. That is nearly one billion-fold more weakly bound zinc than that is found outside of the vesicles in the cytoplasm ($\sim$5 picomolar), and about one million-fold more than is found in the brain's extracellular spaces ($\sim$1 nM). Synaptic zinc-containing neurons must concentrate zinc in the secretory vesicles that are destined for release. This job is performed by the zinc transporter ZnT-3, which is expressed on the membranes of the vesicles.

Zinc and glutamate are co-released by exocytosis of the zinc- and glutamate-containing vesicles. The zinc release is nonlinear with frequency of firing: disproportionately more zinc is released during very fast firing compared to slow firing. As with other synaptic neurotransmitters and neuromodulators, the rapid removal of released zinc from the synaptic cleft is critical. Zinc is removed from the cleft by four mechanisms: (a) diffusion (coupled by binding to ligands in the extracellular fluid), (b) reuptake into presynaptic terminal by transporters, (c) translocation into postsynaptic neurons, or (d) (speculatively) uptake into glia.

Not surprisingly, synaptically released zinc ions are potent modulators of glutamate receptors. In particular, NMDA receptors, which are pivotal in triggering long-term changes in synaptic efficacy, have several modulatory zinc receptor sites. Beyond the glutamatergic synapse, there are dozens of membrane-spanning proteins that are sensitive to zinc diffusing from the gluzincergic clefts, including

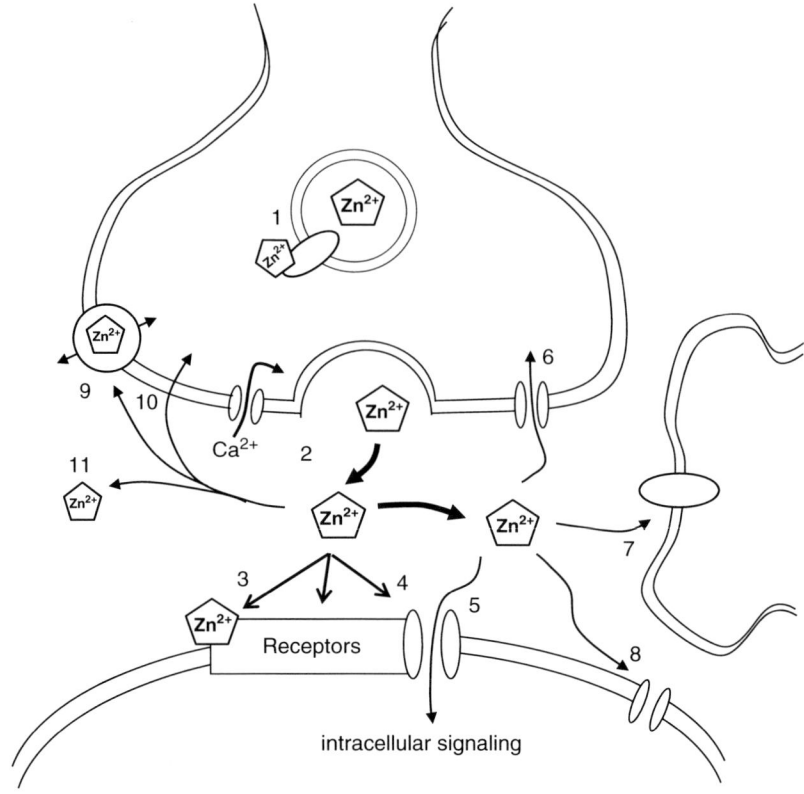

**Zinc-Secreting Neurons, Gluzincergic and Zincergic Neurons, Fig. 1** *Action and cycle of zinc at gluzincergic synapse.* Zinc transporter ZnT-3 pumps zinc into vesicle (*1*). Calcium- and impulse-dependent zinc release (*2*). The released zinc can modulate channels and receptors in postsynaptic and presynaptic membranes (*3,4*). If neuron was given stronger stimulation, a substantial amount of released zinc can translocate into postsynaptic neuron (or presynaptic terminals) (*4, 5, 6*), inhibit the uptake of glutamate into glia (*7*), or "spill over" a few tens of µM to interact with a distant receptor (*8*). Zinc removal from the cleft by diffusion down to concentration gradients (*9*), zinc transporters (*10*), and diffusion and endogenous chelation (*11*)

receptors, channels, pumps, and proteins. Among these, the GABA and glycine receptor has received especially active attention and has proved to be sensitive to zinc. The synaptically released zinc is almost certainly involved in the cellular mechanisms of neuronal plasticity, learning, and memory. Zinc also permeates into the postsynaptic neurons through ligand- and voltage-gated channels, and these "transcellular" signals have many important roles.

## Introduction

Gluzincergic neurons are those glutamatergic neurons that sequester weakly bound zinc ions in the vesicles of their presynaptic glutamatergic terminals. Historically, it was the fact that the vesicular zinc ions are only weakly bound to other molecules that led to the discovery in 1955 of gluzincergic neurons. For unlike 95% of the zinc in the brain, the 5% that is in gluzincergic vesicles is so "loosely" or "weakly" bound to other molecules that it is highly reactive to stains for zinc. In the half century since the pioneering work of McLardy, Haug, Danscher, and others, our knowledge of synaptic zinc-containing neurons has expanded dramatically (Haug et al. 1971; Frederickson et al. 2005).

The original descriptions of the zincergic brain systems from the 1970s have been reaffirmed by more recent methods, and the entire system has been mapped and described in several species, including our own. In overview, one sees that this gluzincergic neuronal system comprises a vast cortical–limbic associational network that dominates limbic and cerebrocortical functions. This entry describes the distribution, the release, and the action of synaptic zinc.

The molecular biology of zinc sequestering and pumping into secretory vesicles as they are emerging from the Golgi apparatus is also coming into focus. Proteins and genes specific for sequestering brain zinc and for pumping zinc into vesicles have been identified, sequenced, and cloned. At the distal end of the neuron (the synaptic terminal), the details of zinc storage in the vesicle, of exocytotic release, and of reuptake are gradually being revealed as well. Across on the postsynaptic side of the synapse, zinc-specific recognition sites that modulate the responsiveness of both excitatory and inhibitory amino acid receptors have been identified and characterized. Also elucidated are the processes by which the synaptically released zinc ions move through transmembrane channels in the postsynaptic spines and dendrites (and back into the presynaptic terminals) under specific conditions (Fig. 1).

## Distribution of the Gluzincergic Neuronal System

### Type of Synapses

Zinc storage in and release from vesicles that contain glutamate along with the zinc is well established. Indeed, gluzincergic neurons are among the most numerous in the mammalian cerebral cortex. There are also scattered reports of zinc in the vesicles of non-glutamatergic neurons, but the implications of these reports remain little explored (Frederickson et al. 2000; Sindreu et al. 2003).

### Distributed Regions

The neuronal cell bodies from which the zinc-containing axonal terminals arise are located almost exclusively in the forebrain, in the cerebral cortex (including allocortex), and in the amygdalar nuclei. Efferent zinc-containing fibers from these regions are in turn directed almost exclusively to (1) other cortical areas and the amygdala (2) striatum or (3) "limbic" targets such as septum, nucleus of the diagonal band, and medial hypothalamus. The amygdalar complex and the hippocampal formation (including the subiculum) are also major nodes of convergence within the zinc-containing neuronal network. All amygdalar nuclei receive synaptic zinc-containing input from cortex, and most of the nuclei also send zinc-containing efferents to both local and remote targets. Amygdalofugal zinc-containing systems project broadly to the bed nucleus of stria terminalis, pyriform cortex, striatum, and periamygdalar cortices. Like the amygdalar complex, the hippocampal formation is another intriguing case. It appears to have four sets of exclusively zinc-containing neurons (dentate granule neurons, CA3 pyramidal neurons, CA1 pyramidal neurons, and prosubicular neurons) arranged in a serial circuit that project to a large number of targets, including septum, subiculum, and several hippocampal fields. Overall, the subiculum is one of the most densely innervated by gluzincergic inputs of all brain regions. Intriguingly, however, the output neurons of the subiculum are all non-zinc-containing glutamatergic neurons. Thus the "final output" of the hippocampal formation is zinc-free. Another allocortical region with a prominent population of exclusively zinc-containing neurons is the pyriform cortex. In fact, all of the pyriform output pyramids are zinc containing (Frederickson et al. 2000).

The dorsal cochlear nucleus has the only laminar gluzincergic fiber system in the brain stem. The "associational" granule neurons of the deep nucleus project a distinct band of gluzincergic parallel fibers onto the dendrites of the cochlear pyramidal cells. Interestingly, these zinc-containing granule neurons are embryological sisters of the cerebellar granule neurons (Frederickson et al. 2000). Both innervate their respective principal cells with parallel fibers and both are glutamatergic, yet one group is gluzincergic (the cochlear) and the other is zinc-free. Functional comparisons of these two pathways could be a very powerful research tactic for elucidating zinc's role. Beyond the cochlear nucleus, there are also some sparse, scattered gluzincergic terminals in parts of the thalamus and brain stem.

### Type of Fibers

One organizational feature of the synaptic zinc-containing neuroarchitecture of the telencephalon is that synaptic zinc-containing fiber systems tend to be corticofugal associational, rather that direct, long-pathway system (Frederickson et al. 2000). In particular, the major sensory and motor pathways that run to and from thalamic, subthalamic, brain stem, or spinal structures are virtually all zinc-free. These are so-called type I fibers with large cell bodies, and long axon are generally precocious and genetically "hardwired." Instead, synaptic zinc-containing neurons are those with short axon type II fibers that are generally later born and more plastic in both number and connections. The synaptic zinc-containing systems are generally corticocortical, corticolimbic, or limbic-cortical.

Several examples of this pattern may be noted. In primary sensory cortex, for example, zinc-containing boutons are conspicuously absent in layer IV, where the primary thalamocortical afferents terminate (Dyck et al. 1993; Frederickson et al. 2000). In contrast, the corticocortical fibers that arise from the small cortical pyramidal neurons are synaptic zinc containing. The laminar segregation of zinc-containing and non-zinc-containing fibers is especially interesting in view of the fact that both sets of afferent fibers are glutamatergic. On the corticofugal, efferent side, it is generally true that axons destined for thalamic, brain stem, or spinal targets are never zinc containing, whereas those destined for cortical or "limbic" targets (amygdalar, septal, or hypothalamic) are in part zinc containing. The same pattern of segregation can be seen in the hippocampal formation, where the neurons giving rise to purely intrinsic fibers as well as projecting to other cortical and limbic targets are all synaptic zinc-containing, but the neurons that have extensive basal forebrain and brain stem projections (subicular neurons) are entirely of the non-synaptic zinc-containing variety. The pyriform cortex illustrates the same principle in yet another way: the principal pyramidal neurons of the pyriform cortex project exclusively to telencephalic targets (mostly cerebrocortical), and all of those pyramidal neurons are gluzincergic. The implication of this organizational pattern is that vesicular zinc is somehow involved in the plasticity of synaptic connections of the type II neurons.

## Release of Synaptic Zinc

### Vesicle

Synaptic zinc-containing neurons must transport zinc into the cell and then concentrate the zinc further within the secretory vesicles that are destined for the presynaptic terminals at the ends of the axons. The zinc transporter ZnT-3, which is expressed exclusively in the brain, serves to pump zinc into the vesicles (Palmiter et al. 1996). Knockout mice lacking the ZnT-3 gene have no histochemically reactive zinc in their vesicles. Like the ZnT-3 knockouts, the AP-3-deficient mutants (unable to insert the ZnT-3 protein in the vesicles) also fail to sequester zinc in the pre-synapses of their synaptic zinc-containing neurons. The amount of zinc in vesicles has never been directly measured, but estimates are in the range of 1–10 mM. The availability of zinc in the cytosol for transport into vesicles by ZnT-3 is governed by (1) passive channel-mediated flux into or out of the cell, (2) active uptake into the cell, (3) export out of the cell, and (4) intracellular zinc-binding ligands. Passive flux through the neuronal membrane occurs when the extracellular level of free zinc is high, as happens when the synaptic zinc-containing pre-synapses "dump" their vesicular zinc during synaptic transmission. Other zinc transporters (both ZnT and ZIP families) along with $Na^+$–$Zn^{2+}$ exchanger can move zinc in or out of cells, across the plasma membrane (Frederickson et al. 2005). These pumps are important not only in gluzincergic neurons but also in neurons that receive gluzincergic input (i.e., gluzinoceptive neurons).

ZnT-3 pumps zinc into vesicles, but the mechanisms and structures involved in the cellular trafficking of zinc are not completely understood. One storage "depot" for zinc in the cell is the metallothioneins (MTs), from which zinc can be released rapidly by nitrosylation or oxidation of the thiol ligands. Thioneins are small proteins (3,000 Da) that contain several cysteine residues that allow them to bind metals, including zinc. They function physiologically by accepting zinc from other zinc-binding ligands, including proteins. Thionein can bind seven zinc atoms through 20 cysteine residues in zinc clusters. Oxidation or nitrosylation of cysteine residues in the zinc cluster results in the release of zinc, so these proteins can function as zinc donors to other zinc-binding proteins, moving zinc from likely sites of carrier-mediated uptake in the cell body to distant sites of zinc sequestration in synaptic vesicles of axon terminals or pre-synapse.

One fascinating question concerns the timing of zinc transport into vesicles. For though the ZnT-3 transporter is clearly embedded in the membranes of vesicles at their point of origin in the Golgi apparatus, the stainable or "free" zinc cannot be seen in the cell bodies at all, but only begins to be detectable in the vesicles as they travel down the distal axons and, of course, once they reach the ends of the axons to rest in terminals. Whether this means that the zinc is pumped into the vesicles in transit or instead that the zinc changes from being bound tightly to weakly in transit is unknown.

### Release

Besides calcium and magnesium, which co-localize in vesicles, zinc is the first metal ion ever proposed to be a synaptically released neuromodulator. As is the case with the release of glutamate, the release of zinc from the presynaptic vesicles of gluzincergic neurons is depolarization and calcium dependent (Howell et al. 1984). The amount of zinc release is nonlinear with respect to the frequency of firing. High-frequency stimulation such as 100 Hz produces disproportionately high quantity of zinc release. The implication of this is that zinc could have a "unique" modulatory effect during very fast firing that would be absent or minimal during slower firing. At the same time, individual action potentials produce a quantal release of zinc at synaptic zinc-rich synapses, which is highly correlated with glutamate release and depends on vesicular exocytosis. Zinc is co-released with glutamate and regulates the glutamatergic receptors. The resting or "tonic" level of ionic zinc in the extracellular fluids of the brain is about 1 nM. Physiologically, this 1 nM is adequate to give tonic partial activation of zinc receptors, such as the inhibitory site on the NMDA receptor.

Different methods have been used to show synaptic zinc release: before-and-after imaging of zinc in the presynaptic terminals, analytical detection of zinc released into perfusates, and directly observed calcium-dependent zinc release into perfusates of electrically or chemically stimulated brain tissue in vivo and in vitro. Most recently, the use of novel tools, including specific zinc-sensitive probes and genetically modified mice, has allowed the physiology and biophysics of zinc release to be thoroughly elucidated. Zinc release is absent in the brain tissues from ZnT-3 KO mice, and the knockout mice have disrupted learning and memory and are also especially seizure prone.

If a single presynaptic terminal [e.g., 1 $\mu m^3$ ($10^{-15}$ L) in volume] had 1 mmol/L average zinc content in the terminal, it would release $\approx 1 \times 10^{-18}$ moles or $\approx 10^5$ zinc ions. (This assumes a massive, complete "dump" of all the zinc of the bouton.) Within only a 15-$\mu$m radius from the bouton, the ions will be diluted by 1:$15^3$ ($\approx$1:3,000), that is, from 1 mM down to $\approx$300 nM. Because only $\approx$10% of that volume is extracellular fluid, the extracellular concentration could reach $\approx$3 $\mu$M. Many fluorescent imaging studies of synaptic zinc release have confirmed that such low-to-mid $\mu$M concentrations of free zinc are, in fact, released during the brief "puffs" of synaptic zinc release. These concentrations are locally and transiently 1,000-fold above the postsynaptic zinc receptor affinities. Empirically, one sees that these zinc concentrations quickly return to the single nM levels. Presumably, the binding of the zinc to extracellular ligands such as albumin and glutathione (both present at $\approx$1 $\mu$M in the cerebrospinal fluid and both with $K_d$ values in the nanomolar range) as well as uptake into cells is $\approx$3 $\mu$M zinc to the low nanomolar range.

## Fate of Synaptically Released Zinc

As is the case for all synaptically released neurotransmitters and modulators, the timely removal of synaptic zinc from the synaptic cleft is critical to maintain normal neuronal function. Zinc is removed from the cleft by probably three mechanisms (diffusion, reuptake into presynaptic terminals, and translocation into postsynaptic neurons) and a possible fourth uptake into glia. Zinc can diffuse away from the cleft in extracellular medium. Zinc-binding ligands in the extracellular medium (albumin, histidine, glutathione, and so on) will avidly bind zinc, reducing the true concentration of free zinc ion back to the resting $\sim$1 nM concentration.

The cytoplasmic levels of zinc homeostasis are regulated by zinc influx and efflux transporters. Synaptically released zinc, after being released, may undergo reuptake to presynaptic synapses. The presence of the high-affinity zinc-specific transporter ZnT-3 helps to maintain a low concentration of cytoplasmic zinc by loading free zinc into synaptic vesicles. Active sequestration and vesicularization of zinc consequently create a concentration gradient across the membrane, with cytosolic concentrations of free zinc in the picomolar range. If synaptically released zinc at the cleft reaches as high as 1–10 $\mu$M, the concentration difference between the cytosol of the presynaptic boutons and the extracellular space could be as high as 1,000-fold (nM to $\mu$M). Zinc uptake following neurotransmission could therefore be like that described for other ions: a passive process in which zinc moves from an area of high concentration in the synaptic cleft to an area of low concentration within the presynaptic cytosol, a process expected to be mediated by as yet unidentified ion pores or channels.

Unlike "classical" neurotransmitters, synaptically released zinc is routinely translocated from the cleft into postsynaptic neurons. The translocation is heavily through zinc-permeable gated channels, which include the NMDAR, voltage-gated calcium channels, and other divalent ion channels. Under favorable conditions, zinc will translocate from inside vesicles of a presynaptic neuron to inside a postsynaptic neuron. Because both glutamate and depolarization open the zinc-permeable channels, maximum zinc translocation would be expected during intense neuronal activity such as during high-frequency stimulation. Co-released glutamate can promote zinc entry into the neuron by opening zinc-permeable channels directly or indirectly through depolarization of the membrane. Thus, zinc translocation into postsynaptic neurons would thus not take place unless both presynaptic release and postsynaptic channel opening occurred simultaneously. The translocation of synaptically released zinc raises the distinct possibility that zinc may interact with many cytosolic macromolecules in the postsynaptic (or presynaptic) cell. This makes the zinc ion a novel, orthograde transcellular messenger.

## Action of Synaptic Zinc

### Interaction with Membrane Proteins or Receptors
Virtually all zinc-containing neurons are glutamatergic though only some glutamatergic neurons are zinc containing. This relationship points toward a role of zinc in modulation of glutamate receptors. Concerning the role or roles of zinc in the cleft, it is quite clear that zinc ions are powerful modulators of both ionotropic and metabotropic receptors for glutamate. In particular, NMDA receptors (NMDARs), which are pivotal in triggering long-term changes in synaptic efficacy, have several modulatory zinc binding sites that enable them to "sense" extracellular zinc over a wide range of concentrations (from nanomolar to micromolar concentrations). In the sub-micromolar range, zinc mediates noncompetitive (allosteric) inhibition of NMDARs, whereas at higher concentrations it acts as an NMDAR pore blocker (although some zinc may also permeate). Beyond the glutamatergic synapse, there are dozens of membrane-spanning proteins sensitive to zinc, including receptors, channels, transporters, and other proteins (Frederickson et al. 2005). Among these, the GABA and glycine receptor has received especially active attention and has proved to be sensitive at least to high concentrations of zinc. Zinc has been proposed to affect aminergic, purinergic, and cholinergic receptors. It has also recently been shown that the proton receptors are sensitive to zinc. The recently described zinc-sensing receptor, which is a membrane-spanning protein that is sensitive to zinc under physiological conditions, also merits further attention. Several types of voltage-gated channels (calcium channels, ATP-sensitive potassium channels) and transporters (glutamate-uptake transporters, cocaine-sensitive site of dopamine-reuptake transporters) have also been shown to be sensitive to exogenous zinc. Matrix metalloproteinases (MMPs) are zinc-dependent endopeptidases that may be affected by synaptic zinc. Ultimately, which of the zinc-sensitive proteins in the brain will actually be modulated by zinc will depend on whether zinc ions could ever reach the proteins in question. In this regard, it is useful to consider concentrations (see above). The outcome of these interactions is dependent on not only the amount of synaptic zinc being released but also the expression of a particular set of membrane proteins in synaptic membranes.

### Physiological Function
Long-term potentiation (LTP) is a lasting enhancement of synaptic potency that many investigators use as a model for the synaptic changes underlying natural learning and memory. As of the summer of 2012, there were over 50 publications in the MEDLINE database on zinc and LTP. Most papers reported that zinc released from or applied to gluzincergic synapses does indeed modify LTP (Li et al. 2003; Pan et al. 2011). Thus, there is generic support for the notion that synaptic zinc is involved in the cerebrocortical mechanisms of memory. The molecular mechanism by which zinc modulates LTP is not simple, however. This is because there are many different types of LTP, involving different molecular machinery and signal pathways, and each expressed differentially at different types of synapses. Zinc affects each of these distinct pathways of LTP differently. Thus, one can confidently say that "zinc is involved in the synaptic events mediating memory," but one must also add, "in myriad ways."

### Neurotoxicity and a Therapeutic Target
Synaptically released zinc is directly implicated in the phenomenon of excitotoxicity, in which brain tissue

deprived of oxygen, glucose, or both begins to release massive amounts of excitatory amino acid transmitter and (from the gluzincergic synapses) massive amounts of zinc (Li 2012). In this death scenario, neurons are tonically depolarized and cell death, both immediate and delayed, ensues. There is abundant evidence that during ischemia (stroke), seizures, or mechanical head injury, a "flood" of zinc is released from boutons that (1) depletes the boutons of zinc and (2) allows toxic excesses of zinc to enter postsynaptic neurons, causing injury or death. The depletion of zinc from the synaptic terminals after excitotoxic insults as well as the appearance of abundant, anomalous free zinc in the injured and dying cells has been verified for ischemia, seizures, and traumatic head injury.

In addition to the translocated zinc, there appears in injured neurons a "flood" of free zinc that is liberated off zinc storage sites (MT in particular) during the injury cascade. Nitrosylation and oxidation of thiol MT ligands is a key step producing these intracellular free-zinc "floods."

Both the influxing zinc ions and the intracellular zinc ions mobilized of MT can be rendered visible for research with fluorescent zinc probes. More importantly, they can be effectively "removed" by zinc chelators. Such therapeutic chelation has been shown to reduce neuron death by as much as 80% in rat studies of ischemia, and there are clinical trials in progress now showing good benefits of zinc chelation for patients suffering stroke (Li 2012).

Synaptically released zinc ions are also implicated in the pathophysiology of Alzheimer's disease. One of the pathological hallmarks of Alzheimer's disease is the marked accumulation of amyloid-β (Aβ) protein, in the form of senile plaques and cerebrovascular amyloid deposits. There is considerable evidence that free zinc in the extracellular fluid can interact with Aβ protein, causing the latter to precipitate into plaques and perivascular angiopathy. The major points of evidence are that (1) plaques form preferentially in brain regions densely innervated by gluzincergic fibers; (2) micromolar amounts of zinc precipitate Aβ, potentially causing the formation of Aβ -rich dense-core plaques in the brains of patients with Alzheimer's disease; (3) zinc chelation can solubilize plaque material from brain homogenate; (4) plaques in brain tissue obtained at autopsy from patients with Alzheimer's disease are enriched with zinc; and (5) both the tissue distribution of zinc and the histochemical staining for zinc have been reported to be disturbed in Alzheimer's disease. Attenuating amyloid pathology and congophilic angiopathy crossing ZnT-3−/− mice with Tg2576 β-amyloid precursor protein (APP) transgenic mice further support that presynaptic zinc release causes amyloid formation. The therapeutic inventions based on the metals' hypothesis are currently being tested in clinical trials. This remains a promising candidate treatment of Alzheimer's disease (Frederickson et al. 2005).

## Cross-References

▶ Calcium in Nervous System
▶ Zinc Cellular Homeostasis
▶ Zinc in Alzheimer's and Parkinson's Diseases
▶ Zinc Signal-Secreting Cells

## References

Dyck R, Beaulieu C, Cynader M (1993) Histochemical localization of synaptic zinc in the developing cat visual cortex. J Comp Neurol 329:53–67

Frederickson CJ, Suh SW, Silva D, Thompson RB (2000) Importance of zinc in the central nervous system: the zinc-containing neuron. J Nutr 130:1471S–1483S

Frederickson CJ, Koh JY, Bush AI (2005) The neurobiology of zinc in health and disease. Nat Rev Neurosci 6:449–462

Haug FM, Blackstad TW, Simonsen AH, Zimmer J (1971) Timm's sulfide silver reaction for zinc during experimental anterograde degeneration of hippocampal mossy fibers. J Comp Neurol 142:23–31

Howell GA, Welch MG, Frederickson CJ (1984) Stimulation-induced uptake and release of zinc in hippocampal slices. Nature 308:736–738

Li YV (2012) Zinc overload in stroke. In: Li YV, Zhang JH (eds) Metal ion in stroke, 1st edn. Springer Science + Business Media, New York, pp 167–189

Li YV, Hough CJ, Sarvey JM (2003) Do we need zinc to think? Sci STKE 2003:pe19

Palmiter RD, Cole TB, Quaife CJ, Findley SD (1996) ZnT-3, a putative transporter of zinc into synaptic vesicles. Proc Natl Acad Sci USA 93:14934–14939

Pan E, Zhang XA, Huang Z, Krezel A, Zhao M, Tinberg CE, Lippard SJ, McNamara JO (2011) Vesicular zinc promotes presynaptic and inhibits postsynaptic long-term potentiation of mossy fiber-CA3 synapse. Neuron 71:1116–1126

Sindreu CB, Varoqui H, Erickson JD, Perez-Clausell J (2003) Boutons containing vesicular zinc define a subpopulation of synapses with low AMPAR content in rat hippocampus. Cereb Cortex 13:823–829

# Zirconium, Physical and Chemical Properties

Fathi Habashi
Department of Mining, Metallurgical, and Materials Engineering, Laval University, Quebec City, Canada

## Physical Properties

Zirconium is a refractory metal with a high melting point. Its chief mineral is zircon, $ZrSiO_4$, which always contains about 1% hafnium – another refractory metal with a high melting point. Both zirconium and hafnium are transition metals, i.e., they are less reactive than the typical metals but more reactive than the less typical metals. The transition metals are characterized by having the outermost electron shell containing two electrons and the next inner shell an increasing number of electrons. Although zirconium [and hafnium] has two electrons in the outermost shell and it would have been expected that to have valency of 2, yet its valency is 4.

| | |
|---|---|
| Atomic number | 40 |
| Atomic weight | 91.224 |
| Relative abundance in Earth's crust, % | $2.2 \times 10^{-2}$ |
| Atomic radius, nm | 15.90 |
| Ionic radius ($Zr^{4+}$), nm | 7.5 |
| Standard potential $M/MO_2$, V | 1.53 |
| Melting point, °C | $1,852 \pm 2$ |
| Boiling point, °C | 3,850 |
| Crystal structure | |
| ω-Zr | Hexagonal open |
| α-Zr | Hexagonal dense |
| | a = 32.3 mm |
| | c = 51.5 nm |
| β-Zr | Body-centered cubic |
| | a = 36.1 nm |
| Transformation temperatures, °C | |
| ω → α | −73 |
| α → β | $862 \pm 5$ |
| Heat of fusion, J/mol | $2.30 \times 10^4$ |
| Heat of evaporation, J/mol | $5.96 \times 10^5$ |
| Electrical resistivity, Ω cm | $3.89 \times 10^{-5}$ |
| Thermal conductivity, $W\ m^{-1}\ K^{-1}$ | |
| At 25°C | 21.1 |
| At 100°C | 20.4 |
| At 300°C | 18.7 |
| Specific heat, $J\ g^{-1}\ kg^{-1}$ | |
| At 25°C | 0.285 |
| At 865°C | 0.335 |
| Thermal expansion coefficient, $K^{-1}$ | |
| α, bulk, at 25°C | $5.89 \times 10^{-6}$ |
| α-Zr, parallel to the C axis | $6.4 \times 10^{-6}$ |
| α-Zr perpendicular to C axis | $5.6 \times 10^{-6}$ |
| β-Zr | $9.7 \times 10^{-6}$ |
| Density, $g/cm^3$ | |
| α-Zr | 6.50 |
| β-Zr | 6.05 |
| Effective cross-section for thermal neutrons, $m^2$ | $1.9 \times 10^{-29}$ (0.19 barns) |

Although zirconium is a high-melting metal, its mechanical properties are similar to those of much lower melting metals: Its elastic modulus is quite low, and its strength diminishes rapidly with increasing temperature. The mechanical properties of zirconium are strongly dependent on purity, especially the oxygen and nitrogen contents, the amount of cold work, and the crystallographic texture.

## Chemical Properties

The ionic radii of hafnium and zirconium are almost identical because of the lanthanide contraction (Fig. 1). Both elements exhibit a valence of four. Therefore, the chemistry of zirconium is similar to that of hafnium.

Both metals are used in nuclear reactors: zirconium as a container for uranium fuel element because of its low-neutron cross-section, and hafnium as control rods because of its high-neutron cross-section.

Zirconium is a reactive metal that, in air or aqueous solution, immediately develops a surface oxide film. This stable, adherent film is the basis for zirconium's corrosion resistance. The corrosion resistance in hydrochloric acid is excellent at temperatures up to 130°C and concentrations up to 37%. However, even small amounts of ferric or cupric ions will lead to severe pitting and stress cracking.

The corrosion resistance in nitric acid is excellent at all concentrations up to 90% and temperatures up to 200°C. In concentrations above 65%, stress corrosion cracking may occur if high tensile stresses are present. The corrosion resistance in phosphoric acid is

*(continued)*

**Zirconium, Physical and Chemical Properties, Fig. 1** The similarity of atomic radii of Zr-Hf, Nb-Ta, Mo-W, etc., due to the lanthanide contraction

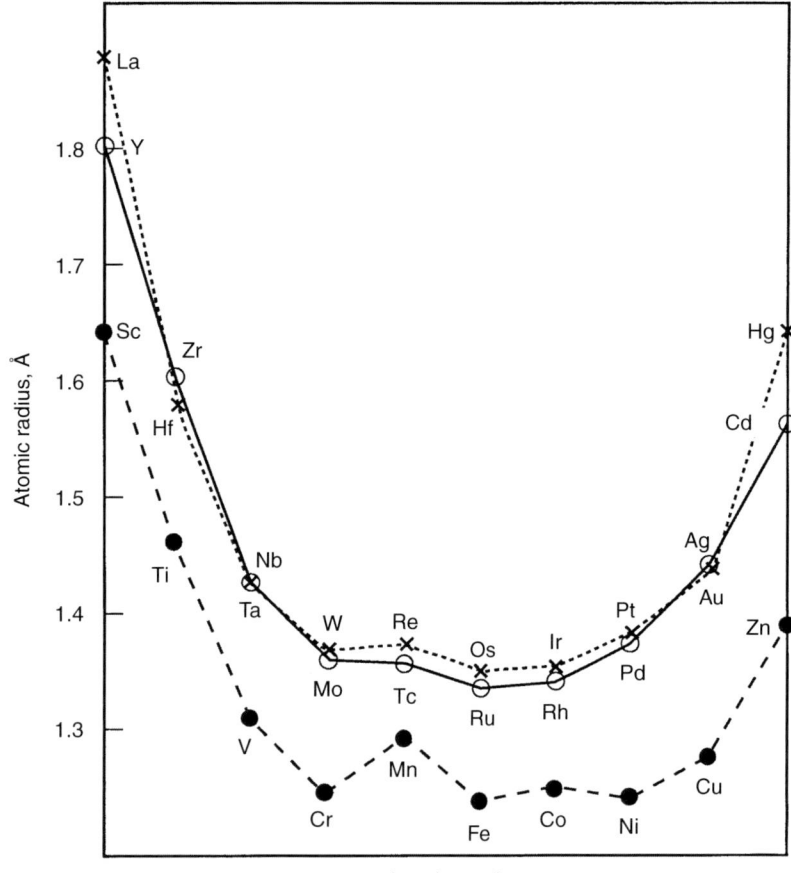

excellent in all concentrations up to 65°C, and, at concentrations below 40%, up to 185°C. Zirconium is rapidly attacked by hydrofluoric acid, even at concentrations below 0.1%.

The corrosion resistance in sodium hydroxide and potassium hydroxide is excellent at all concentrations, up to the boiling temperature. Zirconium is resistant to molten sodium hydroxide up to 1,000°C. Zirconium is very resistant to most organic compounds at all concentrations and temperatures. But, when air or moisture is not available to reform the surface oxide film, zirconium is attacked by anhydrous chlorinated organics at elevated temperatures and is etched by bromine or iodine dissolved in anhydrous organics. Stress corrosion cracking may also occur. Zirconium reacts with most gases at relatively low temperatures.

## References

Habashi F (2003) Metals from ores. An introduction to extractive metallurgy. Métallurgie extractive Québec, Quebec City, Canada. Distributed by Laval University Bookstore, www.zone.ul.ca

Nielsen RH (1997) Zirconium and zirconium compounds. In: Habashi F (ed) Handbook of extractive metallurgy. Wiley-VCH, Weinheim, pp 1431–1458

# Zn

▶ Zinc and Treatment of Wilson's Disease
▶ Zinc in Alzheimer's and Parkinson's Diseases

## ZntR

▶ Zinc Sensors in Bacteria

## ZRC1/OSR1

▶ Zinc Storage and Distribution in S. cerevisiae

## Zur

▶ Zinc Sensors in Bacteria